SATURN

SATURN

Edited by

Tom Gehrels
Mildred Shapley Matthews

With 78 collaborating authors

THE UNIVERSITY OF ARIZONA PRESS
TUCSON, ARIZONA

LC

SPACE SCIENCE SERIES
Tom Gehrels, General Editor

PLANETS, STARS AND NEBULAE, STUDIED WITH
  PHOTOPOLARIMETRY, T. Gehrels, ed., 1974, 1133 pp.
JUPITER, T. Gehrels, ed., 1976, 1254 pp.
PLANETARY SATELLITES, J. A. Burns, ed., 1977, 598 pp.
PROTOSTARS AND PLANETS, T. Gehrels, ed., 1978, 756 pp.
ASTEROIDS, T. Gehrels, ed., 1979, 1181 pp.
COMETS, L. L. Wilkening, ed., 1982, 766 pp.
THE SATELLITES OF JUPITER, D. Morrison, ed., 1982, 972 pp.
VENUS, D. M. Hunten et al., eds., 1983, 1143 pp.
SATURN, T. Gehrels & M. Matthews, eds., 1984, 968 pp.
PLANETARY RINGS, R. Greenberg & A. Brahic, eds., 1984, 784 pp.

About the cover: A wide-angle view of Saturn from the outskirts of the ring system. A sparse distribution of ring particles and two larger moonlets are prominent in the foreground. Hypothetical out-of-plane braided or wave structure can be seen in ring segments in the middle distance. Backlighting of the atmosphere, polar aurorae, and reflected light of the rings are visible on the globe of Saturn itself. Painting by William K. Hartmann.

THE UNIVERSITY OF ARIZONA PRESS

This book was set in 10/12 Linotron 202 Times Roman
Manufactured in the U.S.A.

Library of Congress Cataloging in Publication Data

Main entry under title:

Saturn.

    (Space science series)
    Bibliography: p.
    Includes index.
    1. Saturn (Planet)—Addresses, essays, lectures.
I. Gehrels, Tom, 1925-   II. Matthews, Mildred Shapley. III. Series.

QB671.S23 1984   523.4'6   84-2517

ISBN 0-8165-0829-1

10-9-85

To
Marshall Townsend

CONTENTS

Part VII—ORIGIN

COLOR SECTION, BIBLIOGRAPHY, GLOSSARY, ACKNOWLEDGMENTS, AND INDEX

COLLABORATING AUTHORS

S. K. Atreya, *239*
R. F. Beebe, *195*
K. W. Behannon, *416*
J. J. Caldwell, *88*
J. F. Carbary, *416*
D. L. Chenette, *354*
J. E. P. Connerney, *354*
B. J. Conrath, *195*
G. Consolmagno, *811*
D. P. Cruikshank, *640*
J. N. Cuzzi, *463*
L. Davis, Jr., *354*
M. D. Desch, *378*
T. M. Donahue, *239*
L. W. Esposito, *463*
D. R. Evans, *378*
A. Eviatar, *416*
F. M. Flasar, *671*
L. A. Frank, *318*
D. Gautier, *88*
F. Genova, *378*
C. K. Goertz, *546*
R. Greenberg, *593*
E. Grün, *546*
D. A. Gurnett, *318, 760*
R. E. Hartle, *760*
J. R. Hill, *546*
J. B. Holberg, *463*
W. B. Hubbard, *47*
G. E. Hunt, *195*
D. M. Hunten, *671*
A. P. Ingersoll, *195*
W.-H. Ip, *546*
T. V. Johnson, *609*
M. L. Kaiser, *378*
B. N. Khare, *788*
W. S. Kurth, *378*
L. J. Lanzerotti, *318*
H. P. Larson, *88*

A. Lazarus, *318*
L. A. Lebofsky, *640*
A. Lecacheux, *378*
R. P. Lepping, *416*
J. S. Lewis, *788*
E. A. Marouf, *463*
J. C. McConnell, *239*
D. A. Mendis, *546*
D. Morrison, *609*
A. F. Nagy, *239*
F. M. Neubauer, *760*
G. S. Orton, *150*
T. C. Owen, *3*
B. M. Pedersen, *378*
J. B. Pollack, *811*
C. C. Porco, *463*
R. G. Prinn, *88*
C. Sagan, *788*
R. E. Samuelson, *671*
F. L. Scarf, *318*
A. W. Schardt, *416*
J. D. Scudder, *760*
E. M. Shoemaker, *609*
G. L. Siscoe, *416*
E. C. Sittler, Jr., *318*
B. A. Smith, *609*
L. Soderblom, *609*
D. J. Stevenson, *47, 671*
E. C. Stone, *3*
D. F. Strobel, *671*
V. G. Tejfel, *150*
P. Thomas, *609*
M. G. Tomasko, *150, 671*
G. L. Tyler, *463*
J. A. Van Allen, *281*
A. Van Helden, *23*
J. Veverka, *609, 640*
J. H. Waite, Jr., *239*
R. A. West, *150*

PREFACE

The Saturn system is the most complex in the solar system and this book is to summarize it all: the planet, rings, satellites, the magnetospheres, and the interaction with the interplanetary medium. The effective date of the material is approximately November 1983.

The book is dedicated to Marshall Townsend who was until his retirement the Director of the University of Arizona Press. He came in 1965 to a small struggling Press and in his eighteen years of service developed it into a leader among University Presses in this country. Likewise his vision made this Space Science Series possible. The idea to develop a new type of advanced textbook, as described in the Introduction of "Protostars and Planets," appealed to Marshall from the beginning and he helped us execute it.

We are grateful for the essential support given for the production of the book by the Planetary Program Office at NASA Headquarters, by the Voyager Program Office at the Jet Propulsion Laboratory, and by the Pioneer Program Office at the Ames Research Center.

Tom Gehrels
Mildred Shapley Matthews

PART I
Introduction

THE SATURN SYSTEM

E. C. STONE
California Institute of Technology

and

T. C. OWEN
State University of New York, Stony Brook

Saturn is a giant planet surrounded by numerous rings, many satellites, and a large magnetosphere. Although the Saturn system bears a general resemblance to the Jovian system, it has many unique attributes which provide new insight into the formation and evolution of planetary systems. This introductory chapter provides an overview of the results of recent studies of the Saturn system which are described in detail in the following chapters.

Saturn was the most remote planet known to man when it became a unique object of scientific study with the discovery of its rings by Galileo in 1610 (see the chapter by Van Helden). By the 1970s, when the first space probes were launched toward the outer solar system, Saturn was already known to be a giant gaseous planet encircled by three main rings containing centimeter-sized ice-covered particles and surrounded by at least ten satellites. Titan, the largest of these, was known to be a planet-sized body with a substantial atmosphere containing methane, and five others were known to be intermediate-sized icy satellites. It was not yet known, however, whether or not there was a planetary magnetic field with the associated plasmas, trapped particles, and radio emissions.

With the Pioneer and Voyager flybys of Saturn in 1979, 1980, and 1981, the last of the planets visible to the naked eye was visited by spacecraft. As described in the various chapters in this book, these flybys, in conjunction with improved groundbased observations and increasingly sophisticated theoretical analyses, have greatly increased our knowledge and understanding of the Saturn system. In the following sections we briefly summarize those chapters, indicating what is known about the planet, the magnetosphere, the rings, the satellites, and the origin of the Saturn system. Extensive discussions and references will be found in the relevant chapters.

I. THE PLANET

The spacecraft encounters with Saturn altered previous ideas about the significance of apparent similarities between Saturn and Jupiter. These two giants share many characteristics, yet there are basic differences between them that have become more apparent from the detailed scrutiny provided by the spacecraft. Saturn's special qualities seem to be associated with its smaller mass and greater distance from the Sun.

A. The Interior

We encounter these differences as soon as we begin studying models for Saturn's internal structure (see the chapter by Hubbard and Stevenson). Both Saturn and Jupiter are composed predominantly of hydrogen by mass, unlike Uranus and Neptune where compounds of carbon, nitrogen, and oxygen dominate. Yet the heavy-element core of Saturn is proportionately larger than that of Jupiter, and the hydrogen-helium envelope is thinner. On Saturn, the transition from molecular to metallic hydrogen occurs deeper in the envelope than on Jupiter.

The ratio of Saturn's thermal emission and that which it absorbs from the Sun is 1.79 ± 0.10. Although Jupiter has a similar radiation excess, the energy sources probably differ in part. On Saturn, the precipitation of helium in metallic hydrogen is possible, and this process will lead to the liberation of thermal energy through viscous dissipation. This situation arises because Saturn's smaller mass has allowed this planet to cool more rapidly than Jupiter, which is still radiating primordial heat. The review in the chapter by Hubbard and Stevenson suggests that there is also a significant contribution from primordial heat to Saturn's thermal flux.

The rotational period for the conducting region of the core of 10 hr 39 min 24 ± 7 s is derived from observations of Saturn kilometric radiation. The composition and structure of the core are still poorly defined. Models with rock and rock plus ice with various ratios of He/H in the envelope or in a fluid outer core can all satisfy available constraints.

B. The Atmosphere

Composition.   The composition of Saturn's atmosphere (see the chapter by Prinn, Larson, Caldwell, and Gautier) offers immediate support to the idea of helium differentiation in the planet's interior. The helium mass fraction is only 0.13 ± 0.04 for Saturn compared with the 0.20 ± 0.04 for Jupiter, which agrees with the solar value. This depletion of helium in the observable envelope of Saturn is consistent with predictions based on the observed thermal flux, assuming the latter is augmented by raining out of helium in the interior. Other elements in Saturn's atmosphere appear to be enriched compared with solar values. The situation is clearest for carbon, since methane, the dominant carbon-containing gas, does not condense. This enrichment in turn suggests a heterogeneous model for Saturn's formation, in which the accretion of a core of ~ 10 $M_\odot$ is followed by very rapid contraction of the surrounding envelope of nebular gases.

All of the gases found on Saturn are present on Jupiter, but the reverse is not yet true. In particular, the observability of GeH_4 and CO on Jupiter but not on Saturn despite the detection of PH_3 on both planets may imply significant differences in the interiors of these two planets. On the other hand, PH_3 was detected near 10 μm, where there is high signal-to-noise ratio in the infrared spectra, whereas GeH_4 and CO must be observed at 5 μm, where the lower temperature of Saturn makes it much more difficult to observe than on Jupiter. Photochemical models can successfully predict the abundances of nonequilibrium species at higher altitudes, although the vertical distribution of PH_3 remains a problem. A value of $D/H = 2.6^{+3.7}_{-2.0} \times 10^{-5}$ has been derived from Voyager observations of CH_3D. This is similar to the value found on Jupiter, suggesting that this is likely the value of D/H in the primordial solar nebula. This number can thus be used to constrain models for "bigbang" nucleosyntheses.

Cloud Properties. The clouds of Saturn (see the chapter by Tomasko, West, Orton, and Tejfel) are less colorful and form fewer discrete features than on Jupiter. Nevertheless, the Voyager cameras were able to record a variety of cloud systems including a wave-like feature at 46°N, several ovals with white, brown, and reddish colors, and tilted filaments. It is commonly assumed that frozen ammonia is the main constituent of the visible clouds, but an unambiguous spectral identification to support this assumption is still lacking. As for Jupiter, compounds containing sulfur or phosphorus as well as carbon-nitrogen compounds have been invoked to explain the observed coloration. Again there is a lack of spectral signatures. There is an ultraviolet-absorbing layer of small aerosols ($r \sim 0.1$ mm) at high altitudes (pressures $\sim$ 20 mbar) which is particularly prominent over Saturn's poles. These aerosols may have the same composition as a similar layer found on Jupiter, where charged particle precipitation was invoked to explain the association with the poles.

Structure and Dynamics. The thermal structure of Saturn's atmosphere (see the chapter by Ingersoll, Beebe, Conrath, and Hunt) has been studied from a variety of Earth-based and spacecraft observations. It is now possible to define the mean vertical structure with some authority, and to discuss dynamical structures with horizontal scales from 60 to 60,000 km. Motions in Saturn's atmosphere are dominated by the strong, axisymmetric, predominantly eastward zonal flow. Saturn's east-moving equatorial current is twice as broad and four times as fast as Jupiter's. These differences in the currents in the observable cloud layers may be caused by differences in the deep circulation of the atmosphere. In such a model the zonal flow patterns of the cloud motions on both planets may actually extend to great depths, the rotation rate being constant on coaxial cylinders. The dimensions of these cylinders are established by the depth of the molecular envelope in each case.

On the other hand, models for shallow circulation have been proposed that can also account for the observations. Studies of the time-dependent behavior of vortices may prove to be an important diagnostic tool in distinguishing among the various models used to explain the dynamics and in constraining assumptions about flow in the deep interior. At present there are two schools of thought about these vortices: one suggesting they are solitary waves or solitons, the other that they are strongly interactive features called modons. The average eddy velocities on Saturn are less than half as large as those on Jupiter. The planets are further distinguished by the presence of a strong north-south thermal asymmetry in Saturn's upper troposphere, indicating the effect of a seasonal difference in insolation owing to Saturn's higher obliquity ($26°7$ vs. $3°$ for Jupiter).

There also appears to be atmospheric lightning as evidenced by Saturn Electrostatic Discharges (SED), which may emanate from the equatorial atmosphere which has a rotation period of 10 hr 10 min (see the chapter by Kaiser, Desch, Kurth, Lecacheux, Genova, Pedersen, and Evans).

C. Upper Atmosphere and Ionosphere

The atmosphere of Saturn appears to be well-mixed up to a level of about 10^{-9} bars (1 nbar), the homopause. The thermosphere and exosphere encompass the region above 10 nbar (see the chapter by Atreya, Waite, Donahue, Nagy, and McConnell). Voyager studies yield an exospheric temperature of 600 to 800 K, some 400 K lower than the comparable temperature on Jupiter. The average temperature of 140 K in the stratosphere and mesosphere is similar to the Jovian value. The vertical transport of species in the upper atmosphere is governed by the eddy mixing coefficient, K. At the homopause on Saturn, $K \sim 10^8$ cm^2 s^{-1}. This is similar to the value derived for Titan, but two orders of magnitude larger than the Jovian value for reasons that are not yet clear.

The Pioneer and Voyager measurements of the peak electron density in Saturn's ionosphere yield values near 10^4 cm^{-3}, ten times smaller than the predictions of a variety of theoretical models. A likely explanation of this discrepancy involves a new loss mechanism for H$^+$ by collisions with vibrationally excited H_2. The primary transfer of energy to the thermosphere appears to be Joule heating caused by the departure of the magnetospheric plasma from corotation with the planet. Unlike Jupiter, the energy from charged particle impact in Saturn's auroral zone does not make a significant contribution to the heating of the thermosphere/exosphere region. Saturn's upper atmosphere thus shares the characteristic of the other regions of the planet in appearing similar but being distinctly different from that of Jupiter.

II. THE MAGNETOSPHERE

The Pioneer 11 encounter in 1979 and the Voyager encounters in 1980 and 1981 revealed a magnetosphere similar to that of the Earth or Jupiter, but with physically significant differences ranging from the unusual symmetry of the Saturnian magnetic field and the sources and sinks of plasma and energetic particles to the generation of kilometric radio waves.

A. Magnetic Field

Saturn has an internal planetary magnetic field (see the chapter by Connerney, Davis, and Chenette) which is confined and distorted by the dynamic pressure of the impinging solar wind, resulting in an extended magnetotail. Saturn's internal dipole moment produces a 0.21 Gauss magnetic field at 1 R_S (60,330 km), similar to that at the Earth's surface. However, the Saturnian magnetic field is unusual in that it is highly axisymmetric with no measureable tilt ($<1°$) between the dipole and rotational axes. There is, however, hemispherical asymmetry due both to a northward offset of 0.04 R_S of the dipole and to higher-order zonal harmonics. Models of the field yield estimates of surface fields of $\sim$ 0.83 Gauss and $\sim$ 0.69 Gauss for the north and south polar regions.

Although the *in situ* observations indicate an axisymmetric magnetic field, observations of periodicities in the burst activity of Saturn's kilometric radiation (SKR) and in the formation of spokes in the B Ring suggest that there may be longitudinal asymmetries closer to the planet. Both phenomena occur preferentially at the same longitude of $\sim$ 115° SLS (Saturn longitude system) with the same periodicity, but at different local times. A local anomaly in the field could not be too localized, however, since the SKR originates in the auroral zone at high latitudes and the spokes occur on mid-latitude magnetic field lines.

In addition to the internal magnetic moment, there is an equatorial ring current of $\sim 10^7$ A which contributes to the observed magnetic field. The ring current is modeled by an eastward flowing current confined to a 5-R_S-thick ring at distances between $\sim$ 8 and $\sim$ 15 R_S. The total kinetic energy in the ring current is 5×10^{-4} of the magnetic energy in the field beyond Saturn's surface, a ratio similar to that for Jupiter and Earth.

Beyond $\sim$ 15 R_S the subsolar magnetic field is noticeably modified by the presence of the magnetopause current system. For typical solar wind conditions, the nose of the magnetopause is expected to be at $\sim$ 24 R_S. As the dynamic pressure p of the solar wind varies, the location of the magnetopause varies as $p^{1/6}$, similar to the response of the Earth's magnetopause and much stiffer than the $p^{1/3}$ response of the plasma-dominated Jovian magnetodisk.

In the antisolar direction, the magnetotail current system becomes increasingly important beyond $\sim$ 10 R_S, and by 25 R_S the magnetotail diameter is typically 80 R_S and the tail lobe field is $\sim$ 3 nT. The corresponding region of open field lines in the polar region extends down to an auroral zone at latitudes between $\sim$ 75° and $\sim$ 80°.

B. Plasma and Plasma Waves

Several general characteristics of magnetospheric plasma at Saturn resemble those of Jupiter rather than Earth (see the chapter by Scarf, Frank, Gurnett, Lanzerotti, Lazarus, and Sittler). For example, at Saturn the icy satellites and Titan are significant internal sources of plasma which essentially corotates with Saturn, forming an equatorially confined plasma sheet rather than a plasmasphere. However, the rate at which the Saturn sources load mass onto the corotating magnetic field is $\leq 10^{-2}$ of the mass-loading rate provided by Io, so that a plasma-dominated Jovian-like magnetodisk does not form.

The plasma sheet at Saturn consists of at least three distinct regions: an inner plasma torus ($\leq$ 7 R_S), an extended plasma sheet ($7 \leq R \leq 15\ R_S$), and a hot outer magnetosphere ($\geq$ 15 R_S). There is a systematic increase in electron temperature with radius, ranging from $\leq$ 1 eV at 4 R_S in the inner plasma torus and increasing to $\geq$ 500 eV in the hot outer magnetosphere. There is a corresponding increase in the thickness of the disk, with scale heights ranging from 0.2 R_S at the inner edge of the inner plasma torus to $\sim$ 3 R_S in the outer magnetosphere.

Although the exact nature of the sources and sinks of plasma is unresolved, several possibilities have been suggested for the different regions. In the inner plasma torus the abundance of 0^+, and possibly 0^{2+} and 0^{3+}, reaches a maximum between Dione and Tethys, consistent with the icy satellite surfaces being a plasma source. There is an indication that the high-energy tail of the oxygen plasma may extend above 30 keV.

The inner torus region also contains significant plasma-wave activity, the strongest being chorus emission at $\leq$ 1 kHz. However the amplitudes of the waves are insufficient to cause strong pitch angle diffusion or major precipitation of keV electrons. The prevalent waves are electrostatic electron cyclotron harmonics with frequencies greater than several kilohertz. These emissions occur within the inner torus where the plasma density is high enough so that the plasma frequency is greater than the electron gyrofrequency. These waves may be driven by suprathermal plasma electrons.

The inner plasma torus cannot be the sole source of the extended plasma sheet, since the total plasma content in the torus is smaller than that of the plasma sheet. The inner edge of the extended plasma sheet coincides with that of the cloud of neutral hydrogen which has escaped from Saturn's or Titan's atmosphere. The cloud is $\sim$ 16 R_S thick and extends out to $\sim$ 25 R_S. Ionization of the cloud provides a source of 2×10^{26} H^+ s^{-1} in the region beyond 8 R_S.

The extended plasma sheet also contains heavy ions, although it is not clear whether they are nitrogen or oxygen. Titan would be a likely source for nitrogen, but oxygen would have to originate from the icy satellites, the rings, or Saturn's ionosphere. At least some ionospheric contribution is required by the observation of energetic molecules of H_2 and H_3 in the outer magnetosphere.

In the outer magnetosphere there is considerable variability in plasma density and temperature on short time scales, possibly resulting from the onset of contrifugal instability at the outer edge of the extended plasma sheet and the subsequent outward radial flow of blobs of colder plasma sheet ions and electrons. Titan also sheds a plume of cold nitrogen ions which contributes to the complexity of the outer region. Ionization of the neutral hydrogen cloud can provide the warm hydrogen ions in this region, but the source of the hot electrons (up to 800 eV) is not known. There are also significant fluxes of energetic ions ($>$40 keV) in the outer region with energy densities approaching that of the magnetic field.

The flow of the plasma is essentially corotational, with little evidence for convective or radial flows. Beyond $\sim$ 8 R_S, the plasma velocity falls 10% to 20% below full corotation, consistent with a modest, but nonnegligible, estimated mass-loading rate of $\sim$ 7 kg s^{-1} from Titan's atmosphere. Thus, corotation in the outer magnetosphere may be only marginally maintained by the torque supplied by Saturn's ionosphere.

C. Energetic Particles

The nature of the trapped radiation is determined by the sources, sinks, and motions of the energetic particles (see the chapters by Van Allen and by Schardt, Behannon, Lepping, Carbary, Eviatar, and Siscoe). The principal source for electrons and protons with energies of 30 keV to 2 MeV is in the outer magnetosphere, possibly thermalized solar wind or solar energetic particles. The fluxes of these particles are highly variable in the hot outer magnetosphere ($\geq$ 15 R_S). Inside this radius, the particles diffuse inward across the region of the extended plasma sheet, undergoing few losses until reaching the satellites in the inner magnetosphere ($\leq$ 8 R_S).

In the inner region (see the chapter by Van Allen) the particles undergo significant losses as they diffuse inward and are swept up by the succession of satellites, resulting in phase space densities at 4 R_S that are $\leq 10^{-3}$ of those beyond 8 R_S. Electrons with $\sim$ 1 MeV are less attenuated, however, because their drift velocities relative to Saturn nearly match the orbital velocities of the satellites. As a result, they are less likely to collide with a satellite in the time it takes to diffuse across its orbit. Thus, the system of satellites behaves as a bandpass filter. The particles that do diffuse past the inner satellites without loss are totally absorbed upon reaching the outer edge of the A Ring.

The average rate of loss of particles to satellite sweeping can be used to estimate the radial diffusion rate, as can the rate at which particles fill in the void created by individual satellites. Estimates of diffusion coefficients range from 10^{-8} to $10^{-10} R_S^2 s^{-1}$ in the inner magnetosphere. Although the nature of the diffusion process is not known, the derived radial dependence of the diffusion coefficient ($\sim L^3$) in the 4 to 8 R_S region is more consistent with diffusion driven by ionospheric winds than with centrifugally driven diffusion as occurs in the Io plasma torus.

Particle absorption effects can also be used for other studies, such as determination of the tilt and offset of the magnetic field (see the chapter by Connerney et al.) and the search for previously unknown satellites and rings. Absorption signatures provided the first evidence for several such objects (see the chapter by Van Allen).

Although inwardly diffusing particles are totally absorbed by the A Ring, the rings are also sources of trapped particles. Significant fluxes of trapped high-energy (≥ 16 MeV) protons result from cosmic ray albedo neutron decay (Crand). The energetic albedo neutrons are produced in the interaction of cosmic ray nuclei ($E \geq 20$ GeV) with the icy ring material; they decay during flight, becoming trapped protons with typical energies of ~ 40 to ~ 100 MeV.

There is also evidence for other sources of particles in the magnetosphere. In the inner region, there are indications of sources of low-energy ions (~ 30 keV) and electrons (~ 500 keV), although the nature of such sources is unknown. There are also indications that 200 keV electrons are accelerated in the magnetotail as at Earth and Jupiter, although the observations are again too limited to determine the acceleration process. The magnetotail may also be involved in the acceleration of MeV ions (H, H_2, H_3, He, C, O) observed in the outer magnetosphere. The molecular ions are likely of ionospheric origin, while the He, C, and O ions are of solar wind origin.

D. Saturn's Kilometric Radiation (SKR)

Kilometric radiation is the principal radio emission from Saturn, occurring over a frequency range from 3 kHz to 1.2 MHz, with peak intensity at 175 kHz (see the chapter by Kaiser et al.). The strongest source is in the northern auroral region, with maximum intensity occurring when a particular range of longitudes (100° to 130° SLS) is near local noon. There is evidently an anomaly in the magnetic field at this longitude which allows access of solar wind plasma electrons deep into the polar cusp where radiation is generated near the local electron gyrofrequency. There is a similar, but weaker, source region in the southern auroral zone. Radiation observed in the northern hemisphere is right-hand polarized, probably circular, while that in the southern hemisphere is left-handed.

The localization of the source region in both longitude and local time results in the occurrence of SKR episodes with an average periodicity of 10 hr 39 min 24 $\pm$ 7 s, which is presumed to be the period of rotation of Saturn's magnetic field and deep interior. This SKR period is the basis for a new Saturn longitude system (SLS).

The energy source for SKR is the impinging solar wind, and changes in solar wind pressure or speed produce marked changes in SKR power. For example, a pressure increase by a factor of 150 results in a 10 times increase in SKR. The typical SKR power is 200 MW, corresponding to $\sim 5 \times 10^{-5}$ of the average solar wind power incident on the magnetosphere. Peak SKR power ranges up to 50 GW.

There was a period of 2 to 3 days immediately following the Voyager 2 encounter, however, when the SKR was undetectable ($\leq 10^{-4}$ of nominal), possibly due to the absence of solar wind flux resulting from the immersion of the Saturnian magnetosphere in the extended Jovian magnetotail. The magnetosphere was observed to be greatly inflated during this time and Jovian-like continuum radiation was detected. These observations of solar wind control of SKR are consistent with a model involving transfer of solar wind particles deep into the magnetosphere via the cusp region.

E. Titan Magnetosphere Interaction

The immersion of Titan in Saturn's corotating magnetosphere provided the opportunity for a unique study by Voyager 1 of the interaction of a plasma wind and a planetary atmosphere (see the chapter by Neubauer, Gurnett, Scudder, and Hartle). The incident magnetospheric plasma velocity of ~ 120 km s^{-1} was transalfvénic ($M_A \sim 1.9$) and subsonic ($M_s \sim 0.57$), a condition under which no bow shock occurs and which had not been previously observed.

In the resulting smooth flow around Titan, the ambient Saturnian magnetic field is loaded with H$^+$, N$^+$, and N$_2^+$ or H$_2$CN$^+$ ions from Titan's exosphere. The mass loading from the heavy ions slows the regions of the magnetic field lines closest to Titan down to < 10 km s^{-1}, causing them to drape behind Titan in a comet-like tail. The induced magnetotail has a neutral sheet separating northern and southern lobes which are surrounded by a magnetopause plasma of heavy ions.

Mass loading is most effective on the sunlit side of Titan, resulting in an asymmetric plasma flow and magnetotail. The plasma interaction with the colder nightside atmosphere may be somewhat similar to that of the solar wind with Venus, while that with the hotter day side may resemble more the interaction of the solar wind with a comet.

III. THE RINGS

Since they were first observed by Galileo in 1610, the nature of Saturn's rings has been a continuing challenge to observation and theory (see the chapter by Van Helden). The structure of the rings is determined by their origin and by dynamical processes which depend upon the sizes and collisional properties of the ring particles, on the gravitational effects of the satellites and of Saturn's oblateness, and on electromagnetic processes (see the chapters by Esposito, Cuzzi, Holberg, Marouf, and Porco, and by Mendis, Hill, Ip, Goertz, and Grün). Determining the origin of the rings depends on understanding these dynamical processes and physical properties.

A. General Properties

The classical ring system consists of three broad rings (A, B, and C) occupying the region between 1.23 R_S and 2.67 R_S in Saturn's equatorial plane (see Table I). With the exception of the E Ring, the others are too diffuse or too narrow to be observed from Earth. The D Ring fills much of the region between the C Ring and the top of Saturn's atmosphere, while the E, F, and G Rings lie beyond the main rings. Both the F and G Rings are relatively narrow, while the E Ring occupies an extended region about the orbit of Enceladus.

The optical albedo of the A Ring and B Ring is ~ 0.6 and the microwave albedo is nearly unity. The C Ring and the Cassini Division have a somewhat lower optical albedo of ~ 0.2. The color of individual particles in these regions may be similar, although the thinner C Ring and Cassini Division appear to be less reddish. The particles in the main rings have icy surfaces, consistent with an assumed bulk composition of water ice. They range in size from millimeters up to ~ 5 m in radius, with particle numbers decreasing approximately as a^{-3} with increasing radius a. It is estimated that the mass of the rings is $\sim 6 \times 10^{-8}$ M_S, about that of an icy satellite such as Mimas. However, this mass is somewhat uncertain, since the bulk of it is in the B Ring for which the relevant particle-size distributions have not yet been directly determined.

The observed segregation of the main rings into large, distinct regions is a result not only of the primordial distribution of the matter, but also of the dynamical processes which have maintained the segregation. As described below, satellite resonances form an effective barrier to the outward diffusive flow of particles. There is, however, no known physical mechanism which prevents the inward flow of particles into adjacent ring regions.

B. Specific Properties

A Ring. This outermost of the classical rings is relatively unstructured with a typical normal optical depth of ~ 0.5. Although most of the opacity is due to particles with radii $a > 1$ cm, of which there are typically ~ 20 m^{-2}, about one quarter is due to a larger number of millimeter-sized particles.

TABLE I
Dimensions of the Rings of Saturn

Feature Distance from Saturn Center	$(R_S)^a$
D Ring inner edge	1.11
C Ring inner edge	1.23
Maxwell Gap	1.45
B Ring inner edge	1.53
B Ring outer edge	1.95
Huygens Gap	1.95
Cassini Division	1.99
A Ring inner edge	2.02
Encke Gap	2.21
Keeler Gap	2.26
A Ring outer edge	2.27
F Ring center	2.33
G Ring center	2.8
E Ring inner edge	3
E Ring outer edge	8

[a] $1 R_S = 60,330$ km. Distances are given for the center of gaps and divisions.

The outer edge of the A Ring occurs at a radial location where the orbital period of the ring particles is 6/7 that of Janus. This orbital resonance provides a mechanism for exchanging angular momentum between the ring particles and Janus, effectively forming a barrier to their outward motion. It is not understood, however, why this exchange has not forced Janus outward, since its angular momentum should noticeably increase in $< 10^7$ yr.

There are many other locations in the A Ring where there are weaker orbital resonances, most of them with S15 and S16, the small shepherd satellites of the F Ring. Perturbations in the density of ring particles are generated at the locations of these weaker resonances, launching outward-moving spiral density waves. The wavelength of the waves depends on the surface mass density, which is typically ~ 50 g cm^{-2}.

There are also spiral bending waves in the A Ring generated by Mimas which has a slightly inclined orbit and therefore perturbs the motion of the ring particles perpendicular to the ring plane. The vertical amplitude of these waves (~ 0.7 km) could be a major factor in the apparent thickness of the rings when viewed edge on. The rate of damping of these waves indicates that the dispersion velocity of the particles is ~ 4 mm s^{-1}, leading to a dynamical ring thickness of ~ 35 m which is consistent with the upper limit of 200 m determined at sharp boundaries in the ring.

There are two narrow gaps in the outer portion of the A Ring which are not due to resonances with external satellites, but are likely due to undiscovered moonlets within the gaps. The Encke Gap has wavy edges and contains a kinky ringlet, further indications of several imbedded moonlets.

B Ring. This ring, which is separated from the A Ring by the Cassini Division, is the largest and brightest of the three classical rings, with optical depths ranging from 0.7 to > 2. Almost half of the opacity is due to subcentimeter-sized particles, indicating their greater relative abundance than in the A Ring. The surface mass density of the B Ring is not well known, but is estimated to average $\sim$ 100 g cm^{-2}, twice that of the A Ring.

The outer boundary of the B Ring, which occurs at the Mimas 2:1 resonance, has a double-lobed pattern that precesses with Mimas' angular velocity as expected theoretically. Thus, the outward diffusion of particles in the B Ring is inhibited by the resonant transfer of angular momentum to Mimas, which subsequently transfers it to Tethys through another 2:1 resonance (see the chapter by Greenberg).

Although there are few other significant resonances within the B Ring, it displays much more structure than the A Ring. The structure is essentially chaotic on all scales < 15 km. There is currently no physical model for the structure, although diffusional instability may have some role. Since there are no gaps in the B Ring, the structure is not that of numerous discrete ringlets. Plasma instabilities and meteoroidal erosion may contribute to the largest scale structure (see the chapter by Mendis et al.).

Spokes are another B Ring phenomenon for which a complete physical model is lacking, although their tendency to occur with the rotational period of the magnetic field and at the same longitude as Saturn kilometric radiation suggests that electromagnetic processes are involved. A number of proposed models are discussed in the chapter by Mendis et al. The spokes are cloud-like distributions of micron-sized particles that appear sporadically in the region between $\sim$ 1.75 R_S and $\sim$ 1.9 R_S and are most often apparent in the morning sector of the rings. Some of the spokes appear to form radially over thousands of kilometers in < 5 min, with subsequent Keplerian motion producing wedge-shaped patterns.

C Ring and Cassini Division. There are a number of similarities in these two regions. Both have typical optical depths of $\sim$ 0.1, with several clear gaps and eccentric opaque ringlets, and both have relatively fewer particles with radii a < 1 cm. The Cassini Division contains five broad, diffuse rings separated by narrow gaps. The wider Huygens Gap occurs at the inner edge of the division and contains an opaque eccentric ringlet that precesses at a rate governed by Saturn's oblateness. The maintenance of such a gap and narrow ringlet would seem to require imbedded moonlets, although none have been found. Similar opaque ringlets are located in the Maxwell Gap in the C Ring and in a gap at the Titan apsidal resonance at 1.29 R_S. There are several other dense ringlets in the C Ring at the boundaries of gaps occurring at resonances with Mimas. No mechanism is known at present for producing such ringlets.

Other Rings. The E Ring is broad and diffuse, composed predominantly of micron-sized particles. It occupies the region between 3 R_S and 8 R_S and is several thousand kilometers thick. The optical thickness of the ring is only $\sim 10^{-7}$, and it can be observed from Earth only when viewed edge on. The maximum density occurs near Enceladus' orbit, suggesting that this satellite is the source of the E Ring. Such small particles have lifetimes of $< 10^4$ yr before being destroyed by charged particle sputtering and must be resupplied, possibly by meteoroidal impacts (see the chapter by Mendis et al.).

The F Ring also contains numerous micron-sized particles which are distributed in multiple narrow strands over a region several hundred kilometers wide. The core of the F Ring does include centimeter-sized particles, and there is evidence for larger objects or clumps. The ring particles are shepherded between S15 and S16, receiving angular momentum from the inner shepherd and transferring it to the outer shepherd. However, the processes responsible for the multiple strands and for the occasional kinkiness of the F Ring are not yet understood.

Much less is known about the D Ring and G Ring. Both have very small optical depths and cannot be seen from Earth. The broad D Ring has relatively few micron-sized particles, while the optical depth of the narrow G Ring is dominated by such small particles. Neither the structure in the D Ring nor the narrowness of the G Ring are understood.

IV. SATELLITES

Saturn has 17 satellites with orbits that are currently well known (see Table II), as well as several others that have been less well determined (see the chapters by Morrison, Johnson, Shoemaker, Soderblom, Thomas, Veverka, and Smith, and by Cruikshank, Veverka, and Lebofsky). The group of seventeen appears to contain four distinct classes of objects. Only Titan is a major satellite, comparable in size to Ganymede, but with the distinctive attribute of a dense atmosphere. Six others, Mimas, Enceladus, Tethys, Dione, Rhea, and Iapetus, represent a new class of intermediate-sized icy satellites, with radii ranging from 197 to 765 km. Except for Phoebe, the remainder comprise another new class of objects that probably are icy fragments of larger bodies. Phoebe is the only one thought to be a captured object. As described in the chapter by Greenberg, orbital resonances have been important in determining the structure of the satellite system.

TABLE II
Satellites of Saturn

Satellite	a (R_S)	R (km)
S17 Atlas	2.276	20 × ? × 10
S16 1980S27	2.310	70 × 50 × 37
S15 1980S26	2.349	55 × 45 × 33
S10 Janus	2.51	110 × 95 × 80
S11 Epimetheus	2.51	70 × 58 × 50
S1 Mimas	3.08	197 ± 3
S2 Enceladus	3.95	251 ± 5
S3 Tethys	4.88	530 ± 10
S13 Telesto	4.88	15 × 10 × 8
S14 Calypso	4.88	12 × 11 × 11
S4 Dione	6.26	560 ± 5
S12 1980S6	6.26	17 × 16 × 15
S5 Rhea	8.73	765 ± 5
S6 Titan	20.3	2575 ± 2
S7 Hyperion	24.6	205 × 130 × 110
S8 Iapetus	59	730 ± 10
S9 Phoebe	215	110 ± 10

A. Intermediate-Sized Icy Satellites

As a class, Mimas, Enceladus, Tethys, Dione, Rhea, and Iapetus share a number of common characteristics. All have average densities between ~ 1.2 and ~ 1.4 g cm^{-3}, indicating an ice/rock mixture which is likely 60% to 70% H_2O ice, consistent with the spectral identification of H_2O ice on their surfaces and with a high albedo of ~ 0.4 to 1 for all but the dark side of Iapetus. Generally, there appears to be more frost on their surfaces than on Ganymede and Callisto and variations in albedo are thought to result partly from differences in the admixture of neutrally colored dust. Unlike the Galilean satellite system, there is no indication of major changes in bulk composition with radial distance. There is, however, an increase in size with distance from Saturn.

The five inner intermediate-sized satellites show varying evidence for endogenic activity such as surface flooding, troughs, ridges, and grooves, most having occurred in the first few hundred million years of the satellites' geologic history. It has been suggested that liquid magma resulted from a low-density water-ammonia eutectic with a low melting point (173 K), since it would be more readily melted by the limited amount of radioactive heating in such small, icy bodies and would naturally rise through fractures in the denser ice.

Endogenic activity has been even more vigorous on Enceladus, with the most recent activity occurring in the last 10^9 yr, possibly even during the last 10^8 yr. The high albedo of this satellite ($p_V \sim 1.0$) and its association with the E Ring also suggest internal activity which produces fine particles that populate the ring and keep the surface bright. Five distinct terrains have been identified, with the youngest plains regions containing grooves which may be extensional fractures similar to those on Ganymede. The degree of relaxation of impact craters also differs markedly in different regions, indicating large differences in viscosity due to differences in composition and heat flow. The only known source of heat sufficient to cause the level of activity on Enceladus is tidal dissipation, although the current rate of dissipation is inadequate to cause melting (see the chapter by Greenberg) and may be only marginally adequate to maintain a liquid interior that might have been melted during an earlier epoch.

The dark side of Iapetus may also result from endogenic processes which bring a dark magma (albedo < 0.05) to the surface of the icy satellite. However, the symmetrical placement of the dark material with respect to Iapetus' leading hemisphere seems to require either that the distribution of endogenic material is controlled by impacts or that the dark material is of external origin. Currently the origin of the dark material remains undetermined.

The surfaces of the intermediate-sized satellites also indicate the importance of exogenic processes in the Saturnian system. All six satellite surfaces show extensive impact cratering, with at least two crater populations produced by different groups of impacting objects: population 1, containing a relatively large abundance of craters with diameters $D > 20$ km, and population 2, containing an abundance of craters $D < 20$ km and few larger.

Population 1 craters are thought to have been created during the tail-off of a postaccretional bombardment, possibly by bodies of external origin. Subsequently, population 2 craters may have been created by secondary objects in orbit about Saturn, possibly the debris from earlier collisions. There is an abundance of population 1 craters on Rhea, Dione, Tethys, and the bright side of Iapetus. The younger plains regions on Dione and Tethys contain population 2 craters, as does most of the surface of Mimas. Enceladus shows only population 2 craters, indicating that its entire surface has been modified since the postaccretional bombardment.

The abundance of population 1 craters on Iapetus has significant implications for the history of the inner satellites if the impacting bodies were of external origin and if their size distribution extended to larger diameters. In such a case, gravitational focusing would have resulted in several impacts that were large enough to disrupt the inner satellites which would then have subsequently reaccreted. Although the small satellites and the rings may be the remains of such disruptive collisions, the implied flux of impacting objects is much larger than is currently estimated for comets, the only known external source. If, instead, the impacting objects were lower-velocity objects associated with Saturn, then disruptive collisions might not have been prevalent. In this case, population 1 and 2 craters may result from different temporal behavior of large and small objects in orbit about Saturn.

B. Small Satellites

With the exception of Phoebe, the small satellites are irregularly shaped objects which are likely fragments of larger bodies. Hyperion is the largest of the small satellites with a major dimension only slightly smaller than the diameter of Mimas. Although Hyperion is not heavily cratered, the presence of population 1 craters indicates a relatively old surface consistent with fracture of the parent body near the end of postaccretional bombardment. Although Hyperion's density is unknown, there is dirty water ice on its surface, suggesting an icy bulk composition like that of the larger satellites. The uniformly reddish color of Hyperion's surface is similar to, but brighter than, the dark side of Iapetus, possibly resulting from similar carbon-bearing materials. Hyperion's lack of hemispherical asymmetry such as is observed on Iapetus is now understandable with the recognition that the pendulum-like motion associated with its elongated shape results in chaotic rotational motion. Thus, it is possible that some of the dark material swept up by Iapetus has also coated Hyperion, if the exogenic origin of Iapetus' albedo asymmetry is indeed correct.

The other irregularly shaped satellites are smaller and are closer to Saturn. All have relatively bright surfaces with albedos greater than 0.4 and colors like those of the intermediate-sized icy satellites, suggesting a similar surface composition of dirty water ice. The presence of impact craters indicates the surfaces are at least several billion years old. Both Epimetheus and Janus show heavy cratering which is characteristic of the end of the postaccretional bombardment.

These smaller satellites are in dynamically interesting orbits (see the chapter by Greenberg). For example Calypso and Talesto are located in the two stable regions associated with the Lagrangian points in Tethys' orbit, S12 resides in a similar region in Dione's orbit, Janus and Epimetheus are coorbitals which periodically exchange energy so as to avoid collision, and S15 and S16 are shepherd satellites, dynamically constraining the narrow F Ring between them. As described above, several of the smaller satellites have important dynamical effects on the A Ring.

Phoebe's retrograde inclined orbit is also of interest because it indicates that Phoebe is likely a captured object and may be fundamentally different from the other Saturnian satellites. Although small, Phoebe is approximately spherical, and its surface is dark and somewhat patchy with a reddish color suggestive of that of a C-type asteroid and unlike that of the dark side of Iapetus. However, the lack of knowledge of Phoebe's bulk composition and uncertainties in its surface properties preclude a determination of the nature and origin of this intriguing object.

C. Titan

The Voyager encounters with Titan have effectively added a new world to the solar system (see the chapters by Hunten, Tomasko, Flasar, Samuelson, Strobel, and Stevenson, and by Sagan, Khare, and Lewis). This satellite possesses a predominantly nitrogen atmosphere denser than Earth's, in which a fascinating variety of chemical and physical phenomena are occurring. The surface pressure on Titan is 1.5 bar, the temperature 94 ± 2 K. While methane was first detected in Titan's atmosphere by G. P. Kuiper in 1944, the exact abundance of this gas is still poorly known. There is apparently not enough methane in Titan's lower atmosphere to permit the formation of a global methane ocean, but lakes and seas of this hydrocarbon cannot be ruled out. A global ocean of ethane is more likely, however. Argon 36 and 38 may also be present, at a level of a few percent, since the uncertainty in the mean molecular weight derived from the radio occultation and infrared experiments would permit a significant amount of some very volatile, cosmically abundant species that is heavier than nitrogen and spectroscopically undetectable. Argon satisfies all of these constraints and would be expected if the main source of Titan's atmosphere is the decomposition of clathrate hydrates.

Currently, twelve other species besides molecular nitrogen and methane have been identified in Titan's spectrum. With the possible exception of carbon monoxide (which may be primordial), all of these compounds are produced by chemical reactions in Titan's atmosphere. Solar ultraviolet and the bombardment of electrons from Saturn's magnetosphere furnish the necessary energy. These trace constituents include hydrogen cyanide, an important compound in simulations of prebiological organic chemistry on the early Earth.

The atmospheric chemistry produces increasingly complex molecules that comprise the ubiquitous aerosol totally hiding Titan's surface from view. The uppermost particles in this aerosol layer have mean sizes of a few tenths of a micron. Their composition remains uncertain; both theoretical calculations and laboratory simulations point toward a mixture of organic polymers. The surface of Titan must therefore be covered with a layer of aerosols deposited from the atmosphere as well as solid and liquid hydrocarbons. If ethane is the dominant end product, a global ethane ocean with a depth of 1 km could be present. The concomitant destruction of methane implies a source of this gas on or in the satellite. Seas of liquid methane or an ocean of ethane with dissolved methane may buffer the system. A steady-state concentration of H_2, another product of methane photochemistry, is present in Titan's atmosphere, while photochemically-produced hydrogen continually escapes the satellite's weak gravitational field, contributing to a torus around Saturn.

The absence of features in the aerosol layers has prevented any mapping of atmospheric winds. The temperature gradient in the atmosphere implies a cyclostrophic circulation and the north-south asymmetry in the aerosol brightness is another clue that motions are in fact occurring in the atmosphere. At the surface, a modest greenhouse effect maintains a nearly-uniform temperature $\sim 10°$ above the solar equilibrium value.

The mean density of Titan indicates the presence of a rocky core surrounded by ice. This ice could include clathrate hydrates and/or ammonium monohydrate. The size of the rocky core indicates an enrichment of silicates compared to predictions from cosmic abundances. Models for the formation of Titan include the possibility of a hot, postaccretional phase in which a massive atmosphere containing ammonia could exist. If this phase lasted sufficiently long, it could provide another pathway toward the formation of the present N_2 atmosphere by photodissociation of the NH_3.

V. ORIGIN AND EVOLUTION

Many authors have suggested that both Jupiter and Saturn formed as a result of the condensation of giant, gaseous protoplanets from the solar nebula. In recent years, this hypothesis has been challenged and support has grown for a heterogeneous accretion model in which all four of the giant planets formed by the same process: accretion of a core dominated by rock and/or ice followed by hydrodynamic collapse of an envelope of nebular gases (see the chapter by Pollack and Consolmagno). The model predicts that all four planets should have cores of approximately the same size ($10-20$ $M_\odot$), indicating the mass at which instability is induced in the nebula. The smaller masses of Uranus and Neptune are then the result of a smaller "captured" nebular envelope, perhaps because the collapse phase took place at a time when much of the nebular gas had been dissipated. Furthermore, the enrichment of heavy elements in the atmospheres of all four planets—manifested most clearly by the value of CH_4/H_2—is also consistent with this model.

There are, however, many differences between Jupiter and Saturn. Chief among these are the larger relative core size of Saturn, the probable dominance of helium precipitation as a source of internal energy, and the ubiquity of ice in the system of icy satellites and the magnificent icy rings. All the satellites except Phoebe presumably formed with the planet. Phoebe's retrograde orbit and great distance from Saturn both suggest that it is a captured object. Capture would be easiest at a very early stage in the system's evolution, when a greatly extended "atmosphere" was present to slow down an object passing through it.

The nature of the internal source of heat as well as the relative importance of ice in the Saturn system seem closely related to the smaller mass of the planet. Whatever process led to its formation, Saturn evidently originated from a region of the solar nebula with less total mass available to the forming planet than was available at Jupiter. Model calculations suggest that at the end of the hydrodynamical collapse (either of the condensing protoplanet or the nebular envelope onto the accreted core), the radius of Saturn was several times the present value and a nebular disk existed in the equatorial plane. The regular satellites and perhaps the rings developed from this disk, some 10^8 yr after the planet began to form. From this point, further contraction first led to an increase in luminosity, then to a decrease as the planet cooled down. Homogeneous contraction and thermal cooling produced excess internal energy that adequately explains the infrared flux observed from Jupiter, but the value predicted for Saturn is only marginally compatible with the observations. Hence the preference for the helium precipitation theory, which is further substantiated by the observed depletion of helium in the planet's atmosphere.

Formation of the rings and satellites in the nebular disk left in the equatorial plane will be determined by local pressure and temperature. Unlike Jupiter's satellites, Saturn's moons show no evidence of a radial gradient in mean density, thereby providing a useful constraint on models of this disk. Given the predicted luminosity of Saturn, it is possible to construct models that allow the progressive condensation of ices as temperatures fall in the nebular disk in response to the cooling of the planet. An example of such a model predicts water ice at Mimas and Enceladus, ammonium monohydrate at Tethys and Dione, with a mixed clathrate of methane, argon, nitrogen (and possibly CO) becoming likely at the distances of Titan and Iapetus. This condensation sequence would not affect the observed mean densities (rocky material would be available throughout), but does provide the wherewithal for Titan's atmosphere. The latter is certainly not a captured remnant of the primordial nebula since it contains less than one percent neon. The smaller sizes of Mimas and Enceladus may result from later accretion owing to the delay in cooling of the inner region of the nebula.

The observed modification of the surface of Enceladus remains an enigma. The existing orbital resonance between Enceladus and Dione will not produce enough tidal heating to cause melting, even if an ammonia eutectic composition is invoked. Either the forced eccentricity of the orbit was greater in the recent past ($<10^9$ yr), or some other energy source must be identified.

The formation of the rings is viewed as a special case of the same condensation and accretion processes that produced the regular satellites. They would have begun to form as soon as the temperature in the disk at the position of the rings fell below about 240 K, some 5×10^6 yr after the termination of the collapse phase. Since the ring material was inside the Roche limit, the gradient of Saturn's gravitational field was larger than the mutual gravity between any two ring particles, preventing the formation of large objects. Additional studies of the rings are required to set an upper limit on the sizes of objects that did form within this limit. Possible imbedded satellites remain an attractive explanation for some of the observed gaps in the rings, but no such satellites have yet been observed.

VI. CONCLUSION

The chapters in this book describe in detail the current state of knowledge of the Saturn system. Not only has there been a great increase in our understanding of this unique system, but enough is now known to be able to pose basic questions which previously would have been based on little more than speculation. Further analysis of available data and new theoretical considerations will address some of these questions, as will new groundbased and Space Telescope observations. However, the answers to many of the new questions will have to await the eventual return to the Saturn system by orbiting spacecraft and atmosphere probes.

SATURN THROUGH THE TELESCOPE: A BRIEF HISTORICAL SURVEY

ALBERT VAN HELDEN
Rice University

This chapter is a survey of telescopic studies of Saturn, its ring system, and its satellites, from 1610 to about 1900. It covers early observations of the Saturn system and the changing beliefs concerning the constitution of the rings. It shows that what an observer sees in the heavens depends not only on the quality of his instruments, but also to a surprising degree on theories and expectations. Thus, as early as 1616 Galileo was drawing what could easily be interpreted as a ring, but the ring solution to the problem of Saturn's appearances was not found until 1655/56 (by Huygens) when the rings were invisible. Because he believed the rings to be solid, William Herschel did not recognize the crepe ring and saw no trace of it in the ansae, but when the notion of solid rings fell out of favor half a century later his successors easily saw and recognized the crepe ring. They found it hard to believe that Herschel with his large reflectors did not recognize something so manifest to them, and so they thought that the crepe ring itself had become more noticeable. Likewise, Laplace's theory of many solid rings led many observers to see permanent divisions, while Maxwell's particle theory led their successors to see more subtle, less permanent, maxima and minima of particle densities.

The history of astronomy can be divided into three epochs: naked-eye astronomy, from the dawn of civilization to 1609; telescopic astronomy, from 1609 to about 1960; and space craft astronomy (with continued groundbased astronomy), since about 1960. Whereas the origins of the first epoch are lost in preliterate culture, the beginning of the second, the subject of this survey, can be dated rather precisely. On 2 October 1608 the States-General of the Netherlands discussed the application of Hans Lipperhey of Middelburg for a patent on a device for seeing faraway things as though they were nearby (Van Helden 1977, pp. 20–21,36). The news of this gadget spread very quickly, and by August of the following year Thomas Harriot (1571–1628) in London was

[23]

observing the Moon through a six-powered spyglass (North 1974, p. 136); early in December of that year Galileo Galilei (1564–1642) in Padua was doing the same with a 20-powered instrument (Drake 1978, pp. 137–146). Early in January 1610 Galileo discovered four satellites of Jupiter, and his *Sidereus Nuncius*, which appeared in March, was the first publication of the new science of telescopic astronomy (Drake 1978, pp. 146–154,157–158). As the other planets came into convenient positions, Galileo extended his investigations to them as well.

I. THE RINGS

A. Early Observations

Galileo directed his telescope to Saturn for the first time in the middle of July 1610, and to his surprise he saw the planet not round and single, but triple. In order to protect himself from false claims of priority by others, he hid his discovery in an anagram, *smaismrmilmepoetaleumibunenugttauiras*, and he asked others to come forward with discoveries of their own (Carlos 1880, p. 88; Galileo 1890–1909, vol. X pp. 409–410). In November of that same year he gave the solution to the riddle (Galileo 1890–1909, vol. X, p. 474):

> The letters, then, combined in their true sense, say thus: *Altissimum planetam tergeminum observavi.* [I have observed the highest planet to be triple-bodied.] This is to say that to my very great amazement Saturn was seen by me to be not a single star, but three together, which almost touch each other. They are completely immobile [with respect to each other] and are situated in this manner,

> the one in the middle rather larger than the lateral ones. These latter are situated one to the East and the other to the West in the same straight line to a hair."

These lateral bodies were very different from Jupiter's moons; they were large and motionless with respect to the central body. The ring system was approaching its edgewise aspect, and Galileo was observing with a 20-powered instrument with an aperture of perhaps 2 cm and very imperfect optics (North 1974, p. 149). Not knowing what he was supposed to see, he saw this triple-bodied image.

So unchanging was Saturn's appearance, that Galileo lost interest in the planet and observed it only occasionally. In the autumn of 1612, however, when he had not observed Saturn in several months, he found to his surprise that the planet was "solitary without the assistance of the supporting stars, and, in sum, perfectly round and clearly defined like Jupiter" (Galileo 1890–1909, vol. V, pp. 237–238). Although he could not offer an explanation of this disappearance of the lateral bodies, Galileo did predict their return and went on to give a complicated sequence of disappearances and reappearances for the next several years (Galileo 1890–1909, vol. V, pp. 237–238; Drake 1978, p. 198). The lateral bodies did reappear in the summer of 1613, but they did not disappear and reappear again as Galileo had predicted.

Neither Galileo nor any of his contemporaries observed Saturn frequently enough to notice gradual changes in its appearance. In September 1616 he was, therefore, again surprised to find that the lateral bodies had grown so that the entire formation now appeared quite different. He wrote (Galileo 1890–1909, vol. XII, p. 276; Van Helden 1974a, pp. 109–110) that the

> two companions are no longer two small perfectly round globes . . . but are at present much larger and no longer round . . . that is, two half eclipses with two little dark triangles in the middle of the figures and contiguous to the middle globe of Saturn, which is seen, as always, perfectly round.

The best sketch (drawn that same year) that can be associated with this description is found elsewhere in Galileo's manuscripts and can easily be interpreted by the modern reader as representing Saturn and its ring system. At this stage Galileo's telescopes, poor as they were, could obviously give an observer the correct information; henceforth, the problem was the interpretation of this information. The appendages seen in this form were often referred to as "handles" (L. *ansae*).

Although many observers examined Saturn from time to time through their telescopes, the planet's appearances remained a curiosity more than a scientific problem for some time. Thus, although we can point to a fair number of observations between 1610 and 1642, we cannot find any development or progress. If the disappearance of the lateral bodies in 1612 had surprised astronomers, the solitary appearance of 1626 was hardly noticed and did not lead to an understanding of the periodicity of the phenomena. Not until the solitary appearance of 1642 (the year of Galileo's death) did astronomers begin to take an interest in the sequence of appearances. Saturn then became a worthy subject of research.

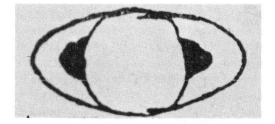

Fig. 1. Saturn as sketched by Galileo in 1616 (Galileo 1890–1909, vol. xii, p. 276, note 1).

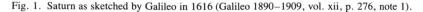

B. Theories

Between the solitary appearances of 1642 and 1656 a systematic body of observations was accumulated by several astronomers, notably Pierre Gassendi (1592–1655) (Gassendi 1658) and Johannes Hevelius (1611–1687) (Hevelius 1647, pp. 41–44;1656). Although this information was inconsistent and often contradictory (Figs. 2–7), it became the basis for several theories.

Gassendi died in 1655, leaving many observations of Saturn but no idea as to the causes of its varying appearances (Gassendi 1658) *ibid.*; Van Helden 1974*b*, pp. 111–112,118). Hevelius in Gdansk published the correct periodicity of the phenomena in 1656 and tried to explain the appearances by postulating an ellipsoidal central body to which two crescents were attached (Fig. 8). The entire formation rotated on its minor axis with respect to the Earth (Hevelius 1656; Van Helden 1974*b*, pp. 156–157). A similar hypothesis was published the following year by Giovanni Battista Odierna (1597–1660) in Sicily. Odierna suggested that Saturn had an egg-shaped body with four dark spots on it (Odierna 1657).

A more interesting theory was formulated by Christopher Wren (1632–1723) in 1658. In an unpublished tract Wren argued that an elliptical corona was attached to Saturn, and that planet and corona rotated or librated about the major axis of the corona, as the planet traveled through the zodiac (see Fig. 9). The corona was so thin that it was invisible when only its edge was presented to the Earth (Van Helden 1968). Over the next decade several other astronomers arrived at much the same theory independently (Van Helden 1974*b*, pp. 163,166,169).

The geometrical problem of explaining the appearances was related to the physical problem of how such shapes could exist in the heavens. Wren believed that the corona was a fluid structure caused by vaporous emanations from the planet's equatorial or "torrid" zone. The mathematician Gilles Personne de Roberval (1602–1675) likewise invoked exhalations: for every appearance there was an appropriately shaped cloud about Saturn (Huygens 1888–1950, vol. XV, pp. 288–290).

Christiaan Huygens (1629–1695) began his study of Saturn early in 1655 with his first good telescope, a 12-foot astronomical telescope with a simple ocular and a magnification of 50. In March of that year he discovered a satellite of the planet (see Sec. II). His first recorded observation of the planet, as it was approaching its solitary appearance again (Fig. 10), showed Huygens that it was not the length of the ansae that diminished, but rather their width (Huygens 1888–1950, vol. I, p. 322; vol. XV, pp. 228–230, 238–239). In the winter of 1655/56, when the rings were invisible, Huygens arrived at his theory by the following reasoning process:

1. As a follower of Descartes (1596–1650), he believed that each planet was the center of its own vortex of matter.

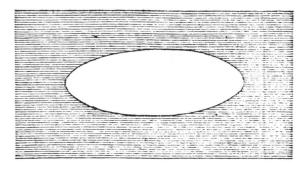

Fig. 2. Observation of Pierre Gassendi, 1634 (Gassendi 1658, p. 183).

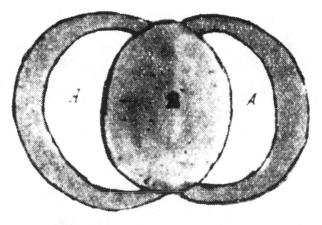

Fig. 3. Observation of Francesco Fontana, 1638 (Fontana 1638).

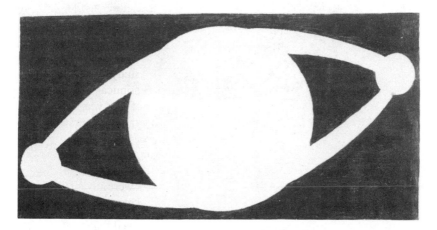

Fig. 4. Observation of Francesco Fontana, 1645 (Fontana 1646, p. 141).

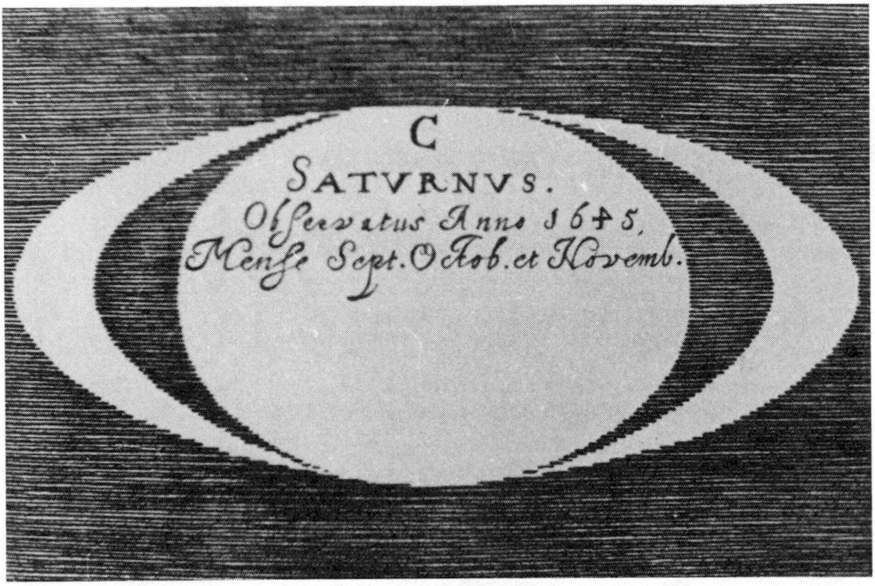

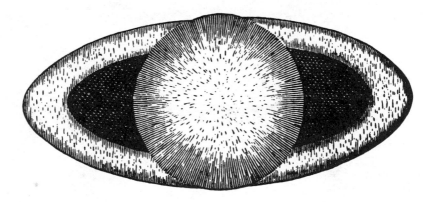

Fig. 5. Observation of Johannes Hevelius, 1645 (Hevelius 1647, p. 42).

Fig. 6. Observation of Eustachio Divini, 1649 (Govi 1887).

Fig. 7. Observation of Giovanni Battista Odierna, 1657 (Odierna 1657, title page).

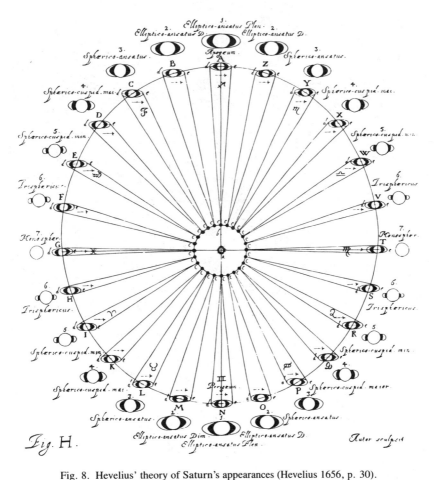

Fig. 8. Hevelius' theory of Saturn's appearances (Hevelius 1656, p. 30).

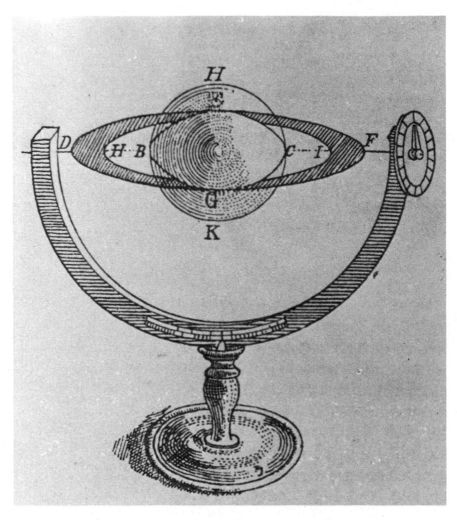

Fig. 9. Wren's theory (Huygens 1888–1950, vol. III, p. 424).

Fig. 10. Christiaan Huygens' first recorded observation of Saturn, March 1655 (Huygens 1888–1950, vol. I, p. 322).

2. In the vortex of the Earth, the Moon has a period of a month while the Earth turns on the same axis in one day. Saturn's newly found moon had a period of 16 days. Therefore, by analogy, Saturn must turn on the same axis in a much shorter period (say, half a day), and all the matter in between Saturn and its moon must circle the planet in its vortex in intermediate times.

3. The axis in question was perpendicular to the ansae, but no changes could be seen in the ansae in 16 days or less. Therefore there had to be rotational symmetry about this axis, and a ring was the obvious shape to satisfy this requirement (Huygens 1888–1950, vol. XV, pp. 294–297).

Huygens announced his ring theory in the form of an anagram in March 1656 in *De Saturni Luna Observatio Nova*, the tract in which he finally published his discovery of the satellite (Huygens 1888–1950, vol. XV, pp. 172–177). Although he privately communicated the solution of the anagram to some colleagues, his ring theory did not become generally known until the appearance of *Systema Saturnium* in 1659.

While a number of scientists accepted the ring theory quickly, there was also a great deal of opposition centered on two aspects of Huygens' theory, the thickness of the ring and its inclination. Huygens argued that the ring was a solid structure of perceptible thickness, whose edge was invisible either because it absorbed all light or because it was so smooth that it reflected the Sun's light from only one point (Huygens 1888–1950, vol. XV, pp. 318–320). On this issue he rapidly became isolated. Those who accepted the ring explanation believed the ring was so thin that an edgewise presentation made it invisible. Huygens insisted on a thick ring until his death. In his posthumously published *Cosmotheoros* (1696) he assigned it a thickness of at least 600 German miles, more than 2500 English miles (Huygens 1888–1950, vol. XXI, p. 786).

By analogy with the Earth, Huygens gave the axis of the ring an inclination of 23°5 with Saturn's orbital plane. This value was too small and it led to errors in predicting the width of the ansae at their greatest inclination. Not until Huygens made an adjustment to the inclination in 1668 were other astronomers entirely happy with his model of the ring (Huygens 1888–1950, vol. XV, pp. 483–484). The prediction given in *Systema Saturnium* for the next edgewise appearance in 1671 turned out to be reasonably accurate, and when the ansae disappeared only to reappear quickly and disappear again, Huygens' ring theory could explain this phenomenon elegantly (Cassini 1671*a,b*). By 1671 Saturn's ring had become a scientific fact.

C. Constitution of the Ring

Huygens' thick ring was a solid structure, stable because of its rapid rotation about the planet (Huygens 1888–1950, vol. XV, pp. 298–300). Most other scientists rejected the thick ring for two reasons: it required an *ad hoc* hypothesis to explain its periodic invisibility, and the physics of the configuration did not feel right. (Note that Newtonian dynamics would not appear until 1687.) A very thin structure would explain the periodic invisibility much more elegantly, and Wren had already made his corona so thin that it could be considered "a mere surface" (Van Helden 1968, pp. 221–222). As astronomers transferred this thinness to Huygens' ring, the question of the stability of a very thin ring arose.

Two solutions presented themselves: to consider the ring a cloud of vapor or milky liquid (Van Helden 1973, pp. 251–253); or to suppose that it was made up of a large number of very small satellites. The latter notion was first suggested by the French poet Jean Chapelain (1595–1674) in 1660 (Huygens 1888–1950, vol. III, p. 35), and it grew in popularity during the rest of the century.

Huygens' thick ring did not fit well with new discoveries. In 1664 Giuseppe Campani (1635–1715), the Roman telescope maker, published an engraving (Fig. 11), showing the outer half of the ring less bright than the inner half (Campani 1664, pp. 8, 18; Huygens 1888–1950, vol. V, p. 118). It was with Campani telescopes that Giovanni Domenico Cassini (1625–1712) made many discoveries, including four satellites of Saturn (see Sec. II below). In 1675 Cassini discovered that the brighter and dimmer portions of the ring are separated by a dark band, which he interpreted as a gap (Cassini 1676,1677). This meant that there were two rings, and the idea that they could be solid spinning disks was not very appealing. Thus, by the end of the century, the consensus was that Saturn's rings consisted of a large number of very small satellites revolving around the planet. In the words of Christopher Wren (Gregory 1702),

> . . . the Ring of Saturn is a number of small moons like a swarm of Bees and which for their swift motion look like an united body like a burnt stick quickly circumrotated.

By 1700 the pace of celestial discoveries had fallen off. The refractor, uncorrected for chromatic and spherical aberrations, with lengths approaching 200 feet and aperture ratios of f/200 and more, had reached its limit of usefulness. Further increases in magnification and light-gathering power, leading to new discoveries in the heavens, did not come until the introduction of the reflecting telescope in the second half of the eighteenth century.

William Herschel (1738–1822) began his study of Saturn in 1774. Since Newton had been silent on the construction of Saturn's ring, Herschel took his cue from Huygens and believed that the ring was solid and of perceptible thickness. In his opinion it had a bevelled edge that did not reflect much light. He clearly saw the dark band on the ring discovered by Cassini, but believed it to be a feature of the ring's northern surface only (W. Herschel 1790a, pp. 3–5). When after the edgewise appearance of 1789 he found the same zone on the southern face, he was finally convinced that it was a true division

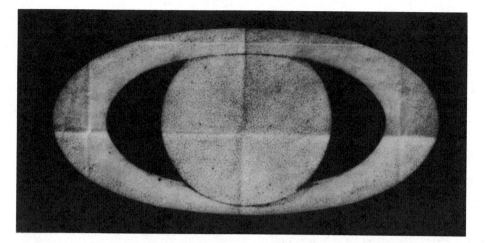

Fig. 11. Giuseppe Campani's observation, 1664 (Huygens 1888–1950, vol. V, p. 118).

(W. Herschel 1792, pp. 1–6), but he would not allow other divisions. The telescope maker James Short (1710–1768) claimed to have seen many divisions in the brighter inner ring, and Herschel himself had seen faint dark lines on four separate occasions; however he considered such other divisions unproven (W. Herschel 1792, pp. 7–11).

In spite of Herschel's great authority, multiple divisions in the ring system were accepted in the first half of the nineteenth century, because of the work of Pierre Simon de Laplace (1749–1827). In a memoir written in 1787 and published two years later, Laplace investigated the stability of Saturn's rings and concluded that Cassini's Division could by no means be the only division. The ring system had to consist of a large number of thin, concentric, solid rings, irregular in width and density, so that their centers of gravity did not coincide with their geometric centers. These centers of gravity could then be considered as so many satellites moving about Saturn with their appropriate speeds (Laplace 1878–1912, vol. XI, pp. 275–292). Laplace repeated this mathematical argument in his very influential *Exposition du Système du Monde* (Laplace 1878–1912, vol. VI, pp. 290–292), and *Mécanique Céleste* (Laplace 1878–1912, vol. II, pp. 166–177).

It would seem a small step from Laplace's many thin hoops to a system of rings consisting of particles or a fluid. Laplace himself had found the equilibrium conditions by supposing that the ring system was covered by an infinitely thin film of fluid that remained undisturbed. He had also put forward his tremendously influential nebular hypothesis on the origin of the solar system, in which Saturn's ring represented the unique case of rings of nebular material solidifying without disintegrating. This could only be the case if a fluid ring was stable during the cooling phase (Laplace 1878–1912, vol. VI, pp. 498–

509; Jaki 1976). Yet he insisted on solid rings, and the consensus among astronomers in the first half of the nineteenth century was that, however many rings made up the system, they were solid. J. H. Schröter (1745–1816) and his assistant K. L. Harding (1765–1834) near Bremen, Germany, even went so far as to claim, in 1808, that Saturn was surrounded by a single, solid ring that did not revolve; they had observed a stationary bright spot on one of the ansae (Schröter 1808, pp. v–vi, 65–66; Möller 1819, p. 6).

As can be seen in the case of Huygens' Cartesian convictions, theory influences observations. Laplace's theory of many narrow solid rings directed the attention of observers to possible divisions in the ring system. In 1825 Henry Kater (1777–1835) saw three divisions in the A Ring, the middle one much wider than the other two. One of his companions saw no fewer than six divisions (Kater 1830,1831). Neither John Herschel (1792–1871) with his large reflectors, nor Wilhelm Struve (1793–1864), director of the Dorpat Observatory with its 9.5-inch Fraunhofer refractor, could verify Kater's observation (Kater 1831; Struve 1826). However, in 1837 Johann Franz Encke (1791–1865), director of the Berlin Observatory, observed with a 9-inch Fraunhofer refractor a dark band on the ring that matched the division claimed by Kater (Encke 1838). The existence of the gap, now known as Encke's Division, was quickly verified by others (De Vico 1838, p. 10), although it was not always visible even with the best telescopes.

The number of markings on both the A and B Rings reported by astronomers from Rome to Boston to St. Petersburg now increased rapidly, reaching a high point in the second quarter of the century. In 1856 Warren De la Rue (1815–1889), a very fine observer and later a pioneer of astronomical photography, circulated an engraving of Saturn (Fig. 12) showing a large number of divisions in the ring system, as he had observed them in the spring of that year (De la Rue 1856). The ring system appeared here as though it had been designed by Laplace himself, and it bears a striking resemblance to the recently obtained close-ups of Saturn.

In 1856, however, Laplace's theory had already fallen out of favor. Several observers had noticed that the inside edge of the B Ring was not sharply defined. In 1837 Encke's assistant Johann Galle (1812–1910) had commented on this and had shown a dark band across the planet's body immediately adjacent to the interior edge of the B Ring (Encke 1838). In November 1850 William C. Bond (1789–1865) and his son George P. Bond (1825–1865) at the Harvard College Observatory saw the same dark band with the 15-inch Merz refractor. Charles W. Tuttle (1829–1881), their assistant, suggested that this appearance might be caused by a dusky ring inside the B Ring (W. C. Bond 1851,1857, p. 48). Two weeks later the sharp-sighted William R. Dawes (1799–1868) in Wateringbury, England, independently came to the same conclusion using a 6.75-inch Merz refractor (Dawes 1850). The dusky or crepe ring had, in fact, been seen but not recognized as early as 1664 by Giuseppe Campani (see Fig. 11) and 1665 by Robert Hooke (1635–1703)

Fig. 12. Observation of Warren De la Rue, 1856 (Chambers 1867, p. 122).

(Hooke 1666). Otto W. Struve (1819–1905) first suggested the system in which the rings are designated by the letters A, B, and C (Lassell 1852).

In drawings of Saturn published by W. Herschel, his successors could now see the crepe ring as a dark belt on the planet, but even with his largest telescopes Herschel had apparently never noticed any trace of this dark ring in the ansae themselves. Moreover, measurements of the widths of the rings made at different times did not agree with each other. All this suggested that changes in the ring system had made the crepe ring more noticeable since Herschel's days. Such changes strongly contradicted Laplace's theory of many solid rings, but they did fit his nebular hypothesis. As early as 1848, Edouard Roche (1820–1883) had reexamined the question of the constitution of the rings and had concluded that a fluid satellite had approached Saturn so closely that it had been torn apart by tidal forces, thus forming the ring system (Roche 1849). Roche's ideas which directly contradicted the nebular hypothesis, did not become influential until later.

By about 1850, then, solid rings were beginning to fall out of favor for various reasons, and the question of the constitution of the rings was again open. In 1850 G. P. Bond reexamined Laplace's theory and concluded that a system of narrow solid rings could not be stable; the rings had to be fluid. The many different divisions seen by various observers were, in all likelihood, temporary phenomena (G. P. Bond 1851). His Harvard colleague, the mathematician Benjamin Peirce (1809–1880), confirmed Bond's conclusions in 1851, after his own investigation (Peirce 1851,1856,1866). When in 1852

several observers noticed that the limbs of the planet could be seen through
the C Ring, any belief in solid rings became very difficult to defend (Lassell
1852; Jacob 1853).

In 1855 the University of Cambridge made the stability of Saturn's rings
the subject of the Adams Prize Essay. The prize was awarded in 1857 to James
Clerk Maxwell (1831–1879). In his essay Maxwell disposed of both solid and
fluid rings: the ring system could be stable only if it consisted of an indefinite
number of particles (Maxwell 1859). Maxwell's treatment was definitive for
several generations, and from 1859 onward his results ruled the astronomical
community. They were verified observationally in 1895 when James E. Keeler
(1857–1900) at the Allegheny Observatory, obtained two spectrograms of Saturn and its rings, showing that the outside of the ring system moves more
slowly than the inside (Keeler 1895a,b). Within weeks Keeler's results were
confirmed by William W. Campbell (1862–1938) at Lick Observatory (Campbell 1895), and H. A. Deslandres (1853–1948) at Meudon (Deslandres 1895).

With the particle view of the rings firmly established after 1859, an explanation was needed for the gaps like Cassini's Division and the many other
dark bands like Encke's Division. In 1866 Daniel Kirkwood (1814–1895)
pointed out that a particle with one-third the period of Enceladus would revolve about Saturn exactly at the distance of Cassini's Division. The major
application of this resonance theory, the mechanism of which remained in
doubt, was to account for the gaps in the asteroid belt between Mars and Jupiter. For Kirkwood, both the asteroid belt and Saturn's ring system were evidence for the meteoric origin of the solar system in his modified version of the
nebular hypothesis (Kirkwood 1866). In 1872 he associated both Cassini's Division and Encke's Division with resonances of all four interior satellites,
Mimas, Enceladus, Tethys, and Dione. Although apparently Kirkwood supplied a rationale for other possible permanent gaps, he in fact only allowed the
permanence of Cassini's Division. According to him, it was probable but not
certain that Encke's Division was never entirely closed, and extremely doubt-
ful that other gaps were permanent (Kirkwood 1872).

In his influential book *Saturn and its System* (Proctor 1865, 1st ed.;
1882, 2nd ed.), Richard A. Proctor (1837–1888) presented a fully elaborated
explanation of the nature of the rings. According to the particle theory, the
C Ring was relatively dark because the particle density was lower there than
in the brighter parts of the ring system. Likewise, the temporary dark bands
were annuli of low particle densities, not true gaps. Their temporary nature
was explained by invoking wave phenomena due to collisions of the particles,
or perturbations by satellites and even other planets. This explanation, which
allowed only one true gap, Cassini's Division, but many temporary dark minima which might sometimes be considered gaps, became the reigning consensus (Proctor 1865; 1882, Ch. 4).

Observers now began to describe features of the ring system differently.
Where before they had reported gaps, now they saw darker and brighter, or

rarer and denser circles. Although all considered Cassini's Division a true gap, there was now room for some particles in it. Encke's Division remained controversial. In 1888, using the new 36-inch Lick refractor, Keeler saw it as a broad feature bounded on the outside by a very thin but sharply defined dark line (Keeler 1888,1889). Using the same instrument, in 1889 E. E. Barnard (1857–1923) saw it as "a rather feeble dark line *on* the ring"; in 1894 he could not see it at all; and in 1895 he saw it as a faint broad band almost as wide as Cassini's Division (Barnard 1895; Osterbrock and Cruikshank 1983). Various other lighter and darker bands continued to be seen in both the A and B Rings.

By the second decade of the twentieth century the particle nature of the ring system with its varying density had led to expectations of the translucency of even the bright A and B Rings. This was supported by Barnard's photographs taken with the Mount Wilson 60-inch reflector in 1911, in which the planet was visible through the A Ring (Alexander 1962, p. 339). Rare occultations of stars by Saturn's rings in 1917 and 1920 proved this translucency dramatically. In 1920, even as the ring system was approaching its edgewise aspect, a seventh-magnitude star whose light passed very obliquely through the B Ring remained clearly visible except for one momentary flicker (Alexander 1962, pp. 338–351).

Several trends in twentieth-century astronomy have affected the visual study of Saturn and its rings. With the rise of astrophysics, interest in the planets lagged, and planetary observations at major observatories became less frequent and systematic. Even with the best instruments, observers who examine the planets only occasionally will not see much subtle detail. Moreover, as it became clear that no canals exist on Mars, the legitimacy of the visual search for delicate planetary details was called into question (Hoyt 1976). The objective evidence of a photograph, even if it showed little detail, was to be preferred over the subjective impressions of a single observer shown in a drawing. One important exception to these two trends was the planetary work done by Bernard Lyot (1897–1952) at the Pic-du-Midi Observatory. His drawing of Saturn and its rings, as seen in 1942 with the 24-inch refractor is shown in Fig. 13.

II. THE SATELLITES

Until dry-plate photography revolutionized the search for new bodies in the solar system, at the end of the nineteenth century, satellites of Saturn were discovered only when the rings were at or near their edgewise aspect. At those times the brightness of the rings does not interfere with the visibility of objects near the planet, and the satellites' back and forth motions along the line of the ansae make them easier to recognize. Moreover, as the rings approach their edgewise aspect, observers often interrupt other projects in order to witness the disappearance of the rings and to make observations best done at this time, e.g., measuring the diameter of the planet.

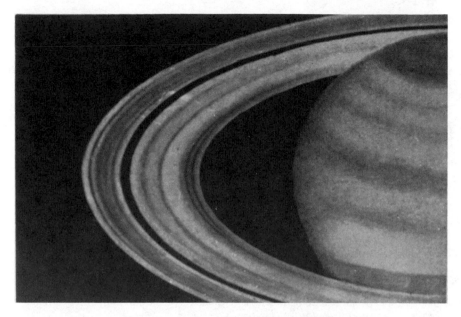

Fig. 13. Observation of Bernard Lyot, 1943 (Dollfus 1961, plate 41).

Huygens' 12-foot refractor with a magnification of 50, with which he discovered Titan in 1655, was his first research telescope, and it was by no means exceptional in power and quality. The satellite had, in fact, been perceived earlier by Hevelius in Poland and Wren in England, but in the absence of any clues as to its nature they had taken it to be a fixed star (Hevelius 1659; Huygens 1888–1950, vol. II, p. 306). Huygens, a neophyte at this business, directed his telescope to Saturn on 25 March 1655 and saw a little star 3 min from the planet on the line of the very narrow ansae. He immediately guessed that it was a satellite and confirmed his guess within a few days (Huygens 1888–1950, vol. XV, pp. 173–176).

Cassini was doubly lucky in 1671. When he observed Saturn late in October, Iapetus was not only on the extension of the narrow ansae within a few minutes of the planet, but also to the west of the planet where it is brightest. When Cassini tried to observe it near its greatest eastern elongation in 1672, he could not see it, even though he knew where it was. To explain this mysterious behavior, he postulated that this satellite has a lighter and a darker hemisphere and that it always keeps the same face turned to Saturn, just as our Moon always presents the same face to us. His discovery of Rhea, late that same year, and Tethys and Dione just before the edgewise appearance of 1684, were made possible by the lack of interfering brightness of the very narrow ansae (Cassini 1673*a,b*,1686*a,b*). Rhea, Tethys, and Dione were extremely difficult to see with the long refractors of that period. Between Cassini and Herschel, a century later, only a handful of observers were able to see them (e.g., Pound 1718). Huygens never saw Tethys and Dione (Huygens 1888–1950, vol. XXI, pp. 193–194,302,778–779), and there is no evidence that John Flamsteed (1646–1720), the first Astronomer Royal, ever saw any satellite of Saturn except Titan.

In 1787 W. Herschel suspected that he had observed a new satellite of Saturn with his 20-foot, 18.5-inch aperture reflector. He did not, however, pursue the matter until near the edgewise appearance of 1789, when he had also finished his 40-foot reflector with its 48-inch aperture. With this new instrument he verified the existence of the satellite now known as Enceladus, on 28 August. Two weeks later he found another satellite with the more convenient smaller instrument and confirmed its existence with the larger (W. Herschel 1790a). Enceladus and Mimas, nearer to Saturn than Tethys and Dione, were extremely difficult to see even with the largest telescopes. J. Herschel could not see Mimas with his 20-foot reflector even with the better seeing conditions at the Cape of Good Hope (Evans et al. 1969, p. 301). Not until the middle of the nineteenth century did observation of Mimas become at all routine with the best instruments.

Up to this point the identification of satellites had presented few problems. Although Galileo named the four Jovian satellites collectively the Medicean Stars after his patrons, he did not name them individually. Instead, he numbered them beginning with the satellite closest to Jupiter (Drake 1960, p. 1). This system became standard until the second half of the nineteenth century. As early as 1614, however, Simon Marius (1570–1624) had suggested several alternatives, one of which he outlined as follows (Prickard 1916, p. 380):

> Jupiter is much blamed by the poets on account of his irregular loves. Three maidens are especially mentioned as having been clandestinely courted by Jupiter with success. Io, daughter of the River Inachus, Callisto of Lycaon, Europa of Agenor. Then there was Ganymede, the handsome son of King Tros, whom Jupiter, having taken the form of an eagle, transported to heaven on his back, as poets fabulously tell . . . . I think, therefore, that I shall not have done amiss if the First is called by me Io, the Second Europa, the Third, on account of its majesty of light, Ganymede, the Fourth Callisto. . . .
>
> This fancy, and the particular names given, were suggested to me by Kepler, Imperial Astronomer, when we met at Ratisbon fair in October 1613. So if, as a jest, and in memory of our friendship then begun, I hail him as joint father of these four stars, again I shall not be doing wrong.

In the case of Saturn, from 1684 onward the five satellites were numbered starting with the one closest to the planet, following the example of Jupiter. But the two satellites discovered by W. Herschel were both interior to the five known before, and confusion now set in. Should one number the satellites in order of discovery or in order of their distance from the planet? W. Herschel himself mixed the two approaches by keeping the existing numbers of the five previously known and assigning the newly discovered interior satellites the numbers 6 and 7, in order of their discoveries, so that starting from the planet the satellites were numbered 7, 6, 1, 2, 3, 4, 5 (W. Herschel 1790a, Table III). When, in 1847, J. Herschel published the results of his observations at the Cape of Good Hope, made the previous decade, he suggested that an end be put to the confusion in the numbering system of Saturn's satellites by assigning them individual names. Referring to the example of Marius, he proposed that Saturn's satellites be named after deities associated with the god Saturn (J. Herschel 1847, p. 415):

> As Saturn devoured his children, his family could not be assembled around him, so that the choice lay among his brothers and sisters, the Titans and Titanesses. The name of Iapetus seemed indicated by the obscurity and remoteness of the exterior satellite, Titan by the superior size of the Huyghenian, while the three female appellations [Rhea, Dione, and Tethys] class together the three intermediate Cassinian satellites. The minute interior ones seemed appropriately characterized by a return to male appellations [Enceladus and Mimas] chosen from a younger and inferior (though still superhuman) brood.

His proposal was favorably received by the astronomical community. William Lassell (1799–1880) began using the nomenclature immediately (Lassell 1848a), and when G. Bond at Harvard and Lassell in Liverpool independently discovered a new satellite in September 1848 (as the rings were approaching their edgewise aspect), Lassell proposed the name Hyperion (Lassell 1848b) and Bond agreed (W. C. Bond 1848; Everett 1849). J. Herschel now included the new satellite under that name in his nomenclature, which he systematically advocated in his influential *Outlines of Astronomy* starting in 1849 (e.g. J. Herschel 1859, p. 367).

An entirely new era of solar-system discovery was inaugurated in 1891, when for the first time an asteroid was discovered by means of photography; in 1892 a new comet was discovered photographically (Clerke 1893, pp. 347,447). In 1896 the 24-inch Bruce astrograph began its career of photographing the southern skies at Harvard's southern station in Arequipa, Peru. On plates taken in August 1898, W. H. Pickering discovered, in April 1899, an outer satellite of Saturn, and named it Phoebe (E. C. Pickering 1899a,b). Confirmation of this object did not come until 1904, when it was found on other plates taken with the Bruce telescope, and when it was finally detected visually by Barnard with the 40-inch Yerkes refractor (E. C. Pickering 1904; W. H. Pickering 1905a). It turned out that Phoebe's motion around Saturn is retrograde, the first instance of retrograde motion found in the solar system. Throughout the nineteenth century the direct motions of all planets and satellites had been cited as strong evidence for the nebular hypothesis; Phoebe's retrograde motion now cast doubt on that theory.

The new method was, however, not perfect. In 1905 W. H. Pickering announced the discovery of yet another satellite of Saturn. He assigned it a period of 21 days (about the same as Hyperion) and named it Themis (W. H. Pickering 1905b; Alexander 1962, p. 268). Although this satellite was included in the literature for some time (e.g., Newcomb 1911), confirmation of its existence remained lacking, and it was eventually disowned by the astronomical community.

III. THE PLANET

A. Shape

During the early days of telescopic observations, the planet itself was often rendered in anomalous shapes due to the poor optics of the instruments as well as the confusing brightness of the appendages. Francesco Fontana (d. 1656) sometimes showed the central body prolate (Fig. 3), while Hevelius showed it oblate (Figs. 5,8). Hevelius developed a theory in which the central body was, in fact, egg-shaped. When the ansae were invisible, however, this body appeared perfectly round. Not until after the micrometer was introduced into the telescope was the flattening of Jupiter at its poles detected by Jean Picard (1620–1682), in 1673 (Le Monnier 1741, p. 28), and not until a century later did W. Herschel diagnose the same condition in Saturn (W. Herschel 1790a, pp. 17–18). The delay was due in part to the smallness and dimness of Saturn's disk compared to Jupiter's, and in part to the ring which usually obscures much of the outline of the planet; only at or near the edgewise position of the ring can the planet's oblateness be securely discerned.

W. Herschel first suspected that Saturn's polar diameter was less than its equatorial diameter in 1776. However, he had to wait until 1789, when the ring was again approaching its edgewise aspect, before he could be certain. In September 1789, when the rings were nearly invisible, Herschel measured Saturn's polar diameter to be 20″.61 and its equatorial diameter 22″.81 (W. Herschel 1790a, p. 17). His measurements were confirmed by Friedrich Wilhelm Bessel (1784–1846) at Königsberg in a series of measurements from 1830 to 1833 (Bessel 1835) and Robert Main (1808–1878) at Greenwich in 1848 (Main 1853).

Bessel, Main, and others did not, however, confirm another contention of Herschel's. In 1805 he published a paper in which he argued that "[t]he flattening of the polar regions [of Saturn] is not in that gradual manner as with Jupiter, it seems not to begin till at a high latitude, and there to be more sudden than it is towards the poles of Jupiter" (W. Herschel 1805, p. 274). In this so-called "square-shouldered" appearance, shown in Fig. 14—in fact an optical illusion caused by the belts and zones—the polar regions and the equatorial regions were flattened and the longest diameters were to be measured

Fig. 14. "Square shouldered" appearance seen by William Herschel, 1805 (W. Herschel 1805, p. 280).

diagonally from 45 north or south latitude (W. Herschel 1805, p. 280). The measurements made by Bessel and Main proved this to be illusory.

B. Surface Markings and Rotation

In a brilliant series of observations, using Campani telescopes in the 1660s, Cassini found surface markings on both Mars and Jupiter and determined their rotation periods (Cassini 1665a,b,1666a,b). But Saturn was a much more difficult subject. Not until the mid 1670s did he manage to find any surface details on this planet, an equatorial belt (Cassini 1676,1677). And although he might agree with Huygens that Saturn must turn on its axis in a period comparable to Jupiter's, he could not make an actual determination.

W. Herschel finally managed to determine a rotation period for the ring as well as the planet. He had seen luminous points on the ring, and after eliminating satellites, the remaining points supported a period of revolution of 10 h 32 m (W. Herschel 1790b, pp. 481–487,494). Since these luminous points are illusory, however, we may speculate that Herschel unconsciously manipulated his data to yield a rotation period predicted by Kepler's Third Law. He was on firmer footing when it came to the rotation of the planet itself. Late in 1793 and early in 1794 he kept careful track of small irregularities on what he called the "quintuple belt," and from their returns deduced a rotation period of 10 h 16 m (W. Herschel 1794). In both cases Herschel made extremely difficult observations at the limit of discrimination of his very powerful telescopes. The first determination was, in the long run, rejected by the astronomical community. It was not until the last quarter of the nineteenth century that delicate spots on the globe came within the reach of the best telescopes. The first rough confirmation of Herschel's value came in 1876/7, when Asaph Hall (1829–1907) determined a rotation period of 10 h 14 m 23.8 s from a bright spot near the planet's equator (Hall 1878). In the 1890s several observers were able to make determinations of the rotation periods of a number of spots, further confirming these values (Williams 1894,1895; Flammarion 1901). Lingering doubts about the validity of these observations were dispelled in 1895 by Keeler's calculation of the tangential velocity of the planet's limb, ~ 10 km s^{-1}, from the Doppler shift of the spectrum of Saturn's globe (Keeler 1895a, p. 425).

Acknowledgments. The author wishes to thank D. E. Osterbrock and D. Morrison for their helpful comments. Figures 1, 5, 9, and 13 were photographed by A. M. Nelson and T. Clark. L. Linton kindly supplied Fig. 9 from his personal collection.

PART II
Saturn

INTERIOR STRUCTURE OF SATURN

W. B. HUBBARD
University of Arizona

and

D. J. STEVENSON
California Institute of Technology

We summarize the principal observational data that constrain interior models of Saturn, and explain why they are relevant. We discuss the behavior of hydrogen, Saturn's major constituent, at pressures on the order of 0.1 to 10 Mbar and temperatures on the order of 10^4 K. Possible behavior and distributions of minor constituents are also considered, along with processes for their transport. We interpret Saturn's external gravitational and magnetic fields in terms of interior structure, and discuss the relationship between atmospheric zonal flows and the deep interior. The constraint imposed by tidal evolution considerations is evaluated. Calculations for the thermal evolution of Saturn are presented, both with and without consideration of possible gravitational unmixing. Possible scenarios for Saturn's mode of origin and their implications for presently observed atmospheric abundances are discussed.

From the point of view of interior structure, Saturn occupies an intermediate position among the four giant planets. Like Jupiter, Saturn appears to be predominantly composed of hydrogen by mass, but it also contains a mass fraction of denser substances substantially larger than the corresponding fraction in Jupiter. This trend is continued in the lower-mass giant planets Uranus and Neptune, in which hydrogen only comprises a small fraction of the total mass and the predominant constituents are probably water and methane (Fig. 1).

[47]

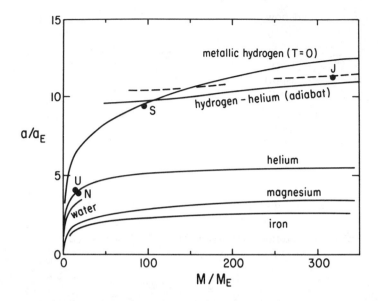

Fig. 1. Plot of equatorial radius a (in units of Earth radius, a_E) vs. mass M (in units of Earth mass M_E) for zero-temperature metallic spheres of hydrogen, helium, magnesium, and iron, and a curve for H_2O computed using an empirical equation of state. Observed values for the giant planets (J, S, U, N) are shown. For reference, the figure also shows a radius-mass curve for spheres of solar composition (hydrogen-helium) having an adiabatic temperature distribution starting at 140 K at 1 bar. Dashed curves show corrections to the hydrogen-helium curve for rotation periods of 10 hr 39 min (near Saturn point) and 9 hr 55 min (near Jupiter point). The figure shows that Saturn is primarily composed of hydrogen.

Nevertheless, Saturn's structure is analogous in many ways to Jupiter. A substantial interior energy source argues for relatively high internal temperatures, and suggests that convective heat transport may be important in the interior. The appropriate temperature regime appears to require that most likely components of Saturn's outer layers be in the liquid state, but at the same time, temperatures may be low enough for immiscibility of major components to play a role in the planet's secular energetics. Comparison of the thermal evolution of Jupiter and Saturn suggests that Saturn's evolution may be qualitatively different from Jupiter's, less star-like and more planet-like.

The goal of interior modeling of Saturn is to synthesize a large number of disparate observational data points on heat flow, gravity field, magnetic field, atmospheric abundances, and satellite characteristics into a coherent picture of the planet's present and past interior state. We are still some distance from achieving this, for there are still differences of opinion on substantial issues. Yet, consensus does exist on a number of points. We will discuss the present state of synthesis of our knowledge of Saturn's interior, indicating those aspects which are considered to be relatively certain and those which are still controversial, and how further observational measurements or calculations may resolve existing uncertainties.

We first review the observational boundary conditions on Saturn interior models (Sec. I). In Secs. II and III we discuss relevant high pressure thermodynamics and transport processes within Saturn. Section IV is a review of static models, and in Secs. V and VI we discuss use of the gravitational and magnetic fields, respectively, to constrain models. In Sec. VII we discuss tidal dissipation in Saturn's interior, and in Sec. VIII we consider evolutionary models of the planet.

I. OBSERVATIONAL BOUNDARY CONDITIONS

A. Atmospheric Abundances

1. Major Expected Constituents. We will consider various models of the interior and evolution of Saturn. To constrain such models one would like to have the planetary mass fraction of each important chemical constituent in Saturn, relative to the corresponding mass fraction in the primordial solar nebula. Fig. 2 shows the most important (according to mass) constituents in the primordial nebula and schematically indicates how they may be redistributed in primordial Saturn. We want to determine how these proportions have been mapped onto the present structure of Saturn.

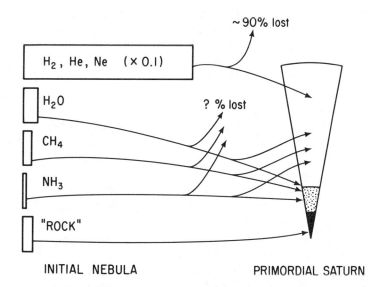

Fig. 2. Bars on left show major constituents of the initial solar composition nebula from which Saturn presumably formed. Models indicate that most of the initial hydrogen and helium was not incorporated in the planet. Some smaller fraction of the H_2O, CH_4, and NH_3 may also have been lost, not necessarily in equal proportions for all three.

Obviously, the mass fraction detectable in the atmosphere will differ in general from the bulk mass fraction in the planet. Here we will define Saturn's atmosphere to be that part of the planet for which the pressure equation of state is well approximated by the ideal gas law. For the expected temperature regime in the atmosphere, a few of the constituents should maintain a constant mixing ratio at least up to the tropopause, due to their noncondensibility and the mixing effects of tropospheric convection. These constituents are H_2, He, and CH_4.

Determination of the most fundamental abundance ratio in Saturn's atmosphere, the He/H_2 number ratio (this molecular number ratio will be simply denoted by He/H_2, and a similar convention will be followed for other molecules), has been accurately carried out recently on the basis of data from the Voyager infrared spectrometer. This result gives He/H_2 as 0.06 (Hanel et al. 1981a), compared with a typical solar value of $0.13 \pm .02$ (see also Gautier and Owen [1983a] and chapter by Prinn et al. in this book). Since there are no known processes for fractionating He with respect to H_2 in Saturn's atmosphere, a discrepancy between the observed value and the presumed primordial value may indicate that processes leading to atmospheric helium depletion are occurring in the nonideal bulk of the planet. For comparison, the Jupiter value for the same ratio, obtained with the same instrument (Gautier et al. 1981), is 0.13 ± 0.04. The best prior measurement of the Saturnian He/H_2 ratio was derived from Pioneer 11 data and gave a value of 0.11 ± 0.04 (Orton and Ingersoll 1980). This result is approximately consistent with the subsequent Voyager measurement, though the uncertainty in the Pioneer measurement is somewhat too large to reveal a difference between Jupiter and Saturn.

Results for CH_4/H_2 are enigmatic. Recent measurements of CH_4/H_2 give values of $(4 \pm 2) \times 10^{-3}$ (Buriez and de Bergh 1981) and $(2.0 + 1.0, -0.8) \times 10^{-3}$ (Encrenaz and Combes 1982), compared with a solar CH_4/H_2 ratio of 0.9×10^{-3} (Lambert 1978). We conclude that there is good evidence for an enhancement over the primordial ratio, but not by a large factor. The enhancement ratio is on the order of 2, and is thus similar to the CH_4/H_2 enhancement observed in Jupiter's atmosphere (Gautier et al. 1982). Because of the relatively low temperatures required to condense methane in the primordial solar nebula, it seems plausible that CH_4 should be present in primordial Saturn in solar proportions relative to H_2. Atmospheric enrichment of methane, if it is present, is probably not the result of differentiation processes at high pressures (see Sec. II); extra methane was probably added by subsequent accretion of planetesimals that formed farther out in the solar system.

NH_3 is not readily observable in Saturn's upper atmosphere, for it condenses at higher pressures than in the warmer Jovian atmosphere. However, microwave brightness temperatures are consistent with an approximately solar ratio of NH_3/H_2 and an adiabatic temperature profile in Saturn's troposphere (Gulkis and Poynter 1972; Marten et al. 1980). This constraint is so crude, however, that at best it only rules out departures from the solar ratio of more than about an order of magnitude.

No results for H_2O/H_2 are currently available. As is predicted by atmospheric models (Lewis 1969), H_2O is essentially unobservable in Saturn's presently accessible atmosphere and thus offers no direct constraint on interior models.

As Fig. 2 indicates, we have now accounted for the principal expected constituents in a collapsing proto-Saturn. The main constraint on interior models presented by these abundance measurements is that we must account for the observed helium depletion and methane enhancement either by differential processes at the time of planetary accumulation, or by processes of differentiation within a planet with an overall solar value for He/H_2 and CH_4/H_2.

2. Minor Constituents. Beside the elements mentioned above, the only additional element detected in Saturn's atmosphere is phosphorus in the form of PH_3 (Larson et al. 1980). The observed PH_3/H_2 ratio of $\sim 10^{-6}$ is on the order of the solar value or perhaps a factor of 2 larger (Larson et al. 1980). From the point of view of interior models, the abundance of this molecule in the observable atmosphere may need to be interpreted in a manner similar to the methane abundance.

Other cosmically abundant elements include silicon and iron. Silicon has been sought in Saturn in the form of SiH_4, but was not detected; Larson et al. (1980) conclude from this that its mixing ratio in the accessible part of Saturn's atmosphere is ~ 5 orders of magnitude below the solar value. This result is not surprising from the point of view of interior modeling, as essentially all the silicon and iron in Saturn is expected to be bound up in a dense iron and magnesium-silicate inner core.

3. Isotopes. Both deuterium ($^2H = D$) and ^{13}C have been identified in Saturn's atmosphere. Neither isotope at present offers a strong constraint on Saturnian interior structure. Hubbard and MacFarlane (1980a) have shown that if H_2O was condensed at the time of Saturn's formation and if the deuterium concentration was able to equilibrate between the gas and the coexisting vapor, then one would expect at least an order of magnitude enhancement of D/H (in the form of HDO/H_2O) relative to the primordial D/H value of $\sim 2 \times 10^{-5}$ (Black 1973). If temperatures were low enough for NH_3 to condense, the equilibrium concentration of deuterium in the condensate would become even higher, and it would become higher still if CH_4 could condense (which is unlikely). If the deuterium thus captured in the condensed core of proto-Saturn is then able to re-equilibrate with the hydrogen-helium envelope, one would expect an enhancement of the atmospheric values of CH_3D/CH_4 and HD/H_2, with the enhancement factor depending on the mass ratio of the core to the envelope. However, for plausible values of this ratio, discussed in Sec. IV.B, the predicted enhancement is a factor of < 2, and is therefore not expected to be detectable at the present level of precision: the observed values for D/H in Saturn are $\sim 2 \times 10^{-5}$ (Fink and Larson 1978) and $(5.5 \pm 2.9) \times 10^{-5}$ (Macy and Smith 1978).

The $^{13}C/^{12}C$ ratio in Saturn's atmosphere seems approximately equal to the values in Earth and in Jupiter (Combes et al. 1977). Interior modeling gives no reason to expect otherwise.

B. Atmospheric Structure and Dynamics

For interior modeling, the outer thermal boundary condition is a specified entropy. This is appropriate because the interior is expected to be close to adiabatic (see Sec. III). It is convenient to characterize the entropy by a specified temperature at 1 bar. The chosen pressure reference level is suitable because it is deep enough that adiabaticity probably applies, yet shallow enough to be accessible by remote sensing. Voyager infrared data (Hanel et al. 1982) constrain the temperature at 0.5 bar to be 110 ± 5 K and indicate that adiabaticity is likely at deeper levels. The slight extrapolation on an adiabat yields 135 ± 5 K at 1 bar. The radio occultation data (Tyler et al. 1982a) are consistent with the infrared data but extend to a slightly deeper level, indicating 143 ± 6 K at 1.2 bar (equivalent to 135 K at 1 bar). All these data show that latitudinal temperature gradients are small at the 1 bar level. Interior models should therefore conform to $T = 135 \pm 10$ K at $P = 1$ bar. (A doubling of the error bar from 5 K to 10 K is warranted because the extraction of temperature profiles from the data requires a modeling effort and assumptions about composition.)

To relate this observation to an interior adiabat, we must allow for the latent heat of constituents which condense at $P \gtrsim 1$ bar, especially water. If the mass fraction of water in Saturn's envelope is x, then the effective boundary condition at 1 bar (defined as the dry adiabatic extension of an interior adiabat) is $(135 - 200\ x)$ K. A water-rich envelope implies a colder interior, other factors being equal. Since x could be as large as 0.1, this latent heat correction is potentially significant. There may also be small corrections associated with disequilibrium of ortho and para populations of H_2 (Massie and Hunten 1982; Gierasch 1983). Further discussion of the thermal structure can be found in the chapter by Ingersoll et al.

Saturn, like Jupiter, possesses a strong, stable zonal wind structure. The prograde equatorial jet has maximum velocity of ~ 500 m s^{-1} (relative to the SKR-defined rotation of 10 hr 39.4 min) and extends to $\pm 35°$ in latitude. Weaker jets (~ 100 m s^{-1}) are found at $45-50°$ and $60-65°$ in both southern and northern hemispheres. These flows are observed by tracking cloud features that form at a few tenths of a bar pressure. There is no consensus or observational information concerning the depth to which the zonal flows extend. Busse (1976) proposed that the atmospheric structure of Jupiter may be the surficial manifestation of deep-seated thermal convection, constrained to a columnar structure by the Coriolis effect (Taylor-Proudman theorem). A similar argument would apply to Saturn. Ingersoll has quantified the depth of the zonal winds (Smith et al. 1982; Ingersoll and Pollard 1982) by showing that the computed height of constant pressure surfaces near Saturn's cloud tops decreases by two scale heights from equator to pole. Since the equator-to-pole temperature gradient is small at this level, it follows that the heat capacity of the mass involved in the zonal motions is large compared to the mass of the atmosphere at and above the cloud tops. This is consistent with the hypothesis that the zonal winds extend into the interior on cylindrical surfaces. It is also consistent with the thin-layer meteorology provided there exists a large reservoir of latent heat in the atmosphere (e.g., the water clouds at $P \sim 10$ bar); this has been proposed by Allison and Stone (1983).

The zonal wind structure is relevant to interior modeling in two respects. First, differential rotation extending into the deep interior modifies the gravitational moments (J_2, J_4) and therefore must be included in analyzing the relation between internal density structure and external gravity field; this is considered further in Sec. V. Second, the latitudinal extent and structure of the zonal winds, if they extend downward into the interior on cylinders, are related to the magnetic field and electrical conductivity within the planet. A wind with velocity 100 m s^{-1} and characteristic radial length scale $\sim 10^3$ km can be hydromagnetically affected even if the conductivity of the fluid is ten orders of magnitude less than that of a good metal. The latitudinal extent of the equatorial jets on both Jupiter and Saturn may be determined by hydromagnetic effects (Kirk and Stevenson 1983).

C. Heat Flow

The previous evolutionary history of Saturn, and models for the present state of its interior, are strongly constrained by the observed value for the intrinsic luminosity of the planet. Two parameters are important for this purpose. First, it is necessary to determine the total thermal infrared luminosity of Saturn,

$$L_t + L_i = 4\pi a_S^2 \sigma T_e^4 \tag{1}$$

where a_S is Saturn's mean radius defined in terms of the surface area and T_e is effective temperature. L_t is the infrared luminosity resulting from the conversion of part of the incident solar flux into thermal radiation, and L_i is the intrinsic energy flow from deep in the planet's interior. L_t can be readily determined by measuring Saturn's luminosity at visual wavelengths due to reflected sunlight L_r, and then subtracting this quantity from the total solar energy incident on Saturn L_s:

$$L_s = L_r + L_t. \tag{2}$$

If $L_i = 0$ then the effective temperature reduces to T_s, defined by Eq. (1) in terms of L_t alone. The ratio $(T_e/T_s)^4$ is just the ratio $(L_t + L_i)/L_t$. Now, the structure of Saturn's atmosphere in the vicinity of the photosphere (at pressures on the order of 1 bar) is determined by T_e, i.e., by the total energy flow through the atmosphere; in turn, the interior temperature distribution may be strongly affected by the temperature distribution around 1 bar pressure. Thus, we need to know not only the intrinsic energy flow but also the total energy flow in order to model the planet's thermal state.

Once the crucial parameters for constraining Saturn's interior thermal structure become known, we can address the fundamental questions of the origin of the intrinsic heat flow, and of whether continuous chemical differentiation of Saturn's interior is implied by the measurements. As we shall discuss, the earliest measurements of Saturn heat flow parameters implied that chemically homogeneous models of Saturn with simple cooling were unrealistic, for they typically evolved through the presently observed heat flow state in a short time compared to the age of the solar system. The existence of additional heat sources such as gravitational differentiation was therefore strongly implied. However, as in the case of Jupiter, the more recent determinations of Saturn's intrinsic heat flow have tended to be scaled down with respect to the earlier measurements, and the discrepancy between observations and the homogeneous cooling models has been reduced. Rieke (1975) obtained $T_e = 95.6 \pm 5$ K with $T_s = 76 \pm 4$ K, yielding $(L_t + L_i)/L_t = 2.50$, while more recent measurements from the Voyager 1 spacecraft (Hanel et al. 1983) find $T_e = 95.0 \pm 0.4$ K, consistent with Rieke's value, but obtain $T_s = 82.3 \pm 0.9$ K. Thus, although the total infrared luminosity of Saturn has not been substantially revised, the component due to intrinsic luminosity is now found to be substantially lower, such that $(L_t + L_i)/L_t = 1.78 \pm 0.09$. In Sec. VIII we will discuss the implications of this downward revision of Saturn's intrinsic luminosity for interior thermal models.

D. Gravity Field

The external gravitational potential of Saturn, which satisfies Laplace's equation, is expressed in the standard form:

$$V_e = (GM/r)\left[1 - \sum_{\ell=1}^{\infty} J_{2\ell}(a/r)^{2\ell} P_{2\ell}(\cos \theta)\right] \qquad (3)$$

where G is the gravitational constant, M is the total mass, r is distance from the center of mass, the J_2 are the dimensionless zonal harmonic coefficients, a is an arbitrary normalizing radius taken to be $\approx a_S$, $P_{2\ell}$ are Legendre polynomials, and θ is the colatitude measured from the rotation axis. For a liquid planet in hydrostatic equilibrium (generalized to include the case of permanent differential rotation on cylinders; see Sec. V.B), the external potential is independent of longitude and invariant with respect to exchange of the north and south hemispheres. Thus, Eq. (3) is the most general expansion of such a potential. In the following, we will use $a = 60,000$ km to normalize Eq. (3). This is to be distinguished from the equatorial radius of Saturn at 1 bar pressure; the latter has been taken to be 60,330 km by the Voyager project.

The most recent determination of Saturn's zonal harmonics is by Null et al. (1981), who used spacecraft Doppler data from the 1979 Pioneer 11 Saturn flyby along with Earth-based observations of Saturn satellites. They found $J_2 = 0.016479 \pm 0.000018$ and $J_4 = -0.000937 \pm 0.000038$, for an assumed value of $J_6 = 0.000084$. As we shall discuss, these results provide fundamental constraints on acceptable interior models, because the zonal harmonics are proportional to various multipole moments of the interior mass distribution, or equivalently, in the pressure-density equation of state in the planet's deep interior.

E. Magnetic Field

The existence of an intrinsic magnetic field on Saturn is significant for study of the interior in two respects. First, the revolution of the magnetosphere and its trapped particles provides a precise reference for the rotation period of Saturn, needed to interpret the gravity field via potential theory. According to Desch and Kaiser (1981a), this period is 10 hr 39 min 24 $\pm$ 7 s. Since the magnetic field is presumably tied to the rotation of the conducting core of Saturn, this period should be representative of the bulk of the planet's mass.

Second, the Saturnian magnetic field must be excited by dynamo action in the conducting core. The strength and geometry of the field thus provide constraints on the dynamo parameters. In particular, an interior model for the dynamo must be able to explain the remarkable symmetry of the field (with tilt angle $< 1°$) and its weakness (by about an order of magnitude) compared with the Jovian field (Smith et al. 1980b).

II. HIGH PRESSURE THERMODYNAMICS

Thermodynamic variables are best determined from a free energy functional since this ensures thermodynamic consistency and enables one to choose the phase(s) with lowest Gibbs energy. From a practical standpoint, however, interior modeling depends primarily on pressure P, Grüneisen parameter γ, and specific heat C_V, each of which are functions of temperature T, density ρ and compositional variables $\{x_i\}$. The pressure function $P(\rho, T, x_i)$ enters directly into the equation of hydrostatic equilibrium; $\gamma \equiv (\partial \ln T / \partial \ln \rho)_{ad}$ is used to determine thermal gradients in convective regions, and C_V is needed to evaluate total thermal content and thermal evolution time scales. Although we emphasize the practical parameters (P, ρ, C_V), it is important not to lose sight of the fundamental role of the free energy, especially in determining the state of matter (liquid vs. solid, homogeneous vs. phase separated).

Consider an element of matter of cosmic composition that undergoes adiabatic compression from a starting state corresponding to that of the deep atmosphere of Saturn ($T \sim 135$ K at $P = 1$ bar). For a convective planet, this compression would represent approximately the locus of thermodynamic states in the interior. The initial state would consist primarily of an almost ideal $H_2 - He$ gas. The small amount of condensed material (e.g., water ice) would vaporize upon adiabatic compression. For the first few decades of pressure increase, the fluid element is close to an ideal gas and highly compressible. Since $\gamma \sim 0.4$, the temperature rises to $> 10^3$ K and no first-order phase transition of the major constituents is encountered. At kilobar pressures, nonideal effects become increasingly important as the molecule orbitals begin to overlap. Temperatures are high enough to prevent freezing and to excite vibrational modes within molecules. At $P \sim 10^5$ bar the thermal pressure becomes less important than the interaction pressure, and substantial reduction of the entropy of molecular rotation may occur despite the high temperature. As P approaches 10^6 bar, T approaches 10^4 K, and the combined effects of pressure and temperature may lead to a significant occupation of conduction levels and dissociation of H_2 molecules. Eventually a metallic state occurs, possibly though not necessarily associated with the complete dissociation of H_2 molecules. At much higher pressure ($P > 10^8$ bar; not encountered in Saturn) the fluid element may be well approximated by a dense Coulomb plasma, i.e., nuclei immersed in a degenerate electron gas.

The most important issues for Saturn concern the thermodynamics for 0.1 Mbar $\lesssim P \lesssim 10$ Mbar, a regime which encompasses $\gtrsim 70\%$ of Saturn's mass and is unfortunately poorly understood. We will describe asymptotic states (low P, high P) and then consider interpolation schemes and bounds on their uncertainties, with emphasis on physical principles rather than on detailed parametrizations. Since earlier work has been covered well in other publications (see *Jupiter* [Gehrels 1976] in this series of books; Zharkov and Trubitsyn 1978; Stevenson 1982b), particular attention is given to recent developments including aspects of the thermodynamics not yet incorporated in interior models, which must be kept in mind when assessing the accuracy of interior models.

A. Hydrogen

Consider an assemblage of protons and electrons. One could imagine many different states of this system: bound molecules, bound atoms, an alkali metal, an antiferromagnetic insulator, a dense molecular state in which conduction could occur by band overlap, or even a polymerized state in which n-mers of hydrogen atoms form. The thermodynamic ground state is well established only in the limits of zero pressure and infinite pressure. At zero pressure, it is clearly the well-studied molecular state. At infinite pressure, no state can be envisaged that has significantly lower energy than the simplest imaginable monatomic (alkali metal) state: protons immersed in a uniform electron gas. We consider each of these limiting cases and discuss how one might interpolate or make the transition between them.

1. Molecular Hydrogen. ·The full, many-body problem of an assemblage of H_2 molecules is often replaced by that of spherical entities which interact only in pairs. The interaction energy is then evaluated by performing lattice sums, for a solid, or by Monte Carlo simulation (Slattery and Hubbard 1976) or liquid perturbation theory (Ross 1974; Stevenson and Salpeter 1976) in the more relevant fluid state. Despite recent shock-wave data (Nellis et al. 1983) there are still uncertainties in the H_2 pair potential, corresponding to $\pm 15\%$ uncertainty in the density at $P=1$ Mbar. The models reported by Hubbard and Horedt (1983) (see Sec. IV) employ a pair potential consistent with the shock-wave data (Ross 1974). *Ab initio* theoretical potentials (Ree and Bender 1979,1980) tend to be harder (more repulsive) than the effective potentials inferred for shock-wave data, and are consequently of limited usefulness. There is no simple parameterized form for the equation of state, either theoretical or empirical. The parameter $(\partial \ln P/\partial \ln \rho)_{ad}$ increases steadily with pressure from ~ 1.4 (the ideal gas limit) to ~ 3 (at $P \sim 1$ Mbar) and is ~ 2 over a substantial pressure range. The polytropic equation of state $P = K\rho^2$, with $K \simeq 2 \times 10^{12}$ in cgs units, is a convenient crude approximation.

Low-temperature data obtained statically, mostly in a diamond cell, provide additional constraints on the pair potential and essential insights into the rotational and vibrational degrees of freedom. The $P-\rho$ relationships found (Shimizu et al. 1981; van Straaten et al. 1982) are compatible with shock-wave data. The rotational degrees of freedom are frozen out at room temperature at $P > 100$ kbar, and the H–H bond softens at $P > 300$ kbar (Sharma et al. 1980; Wijngarden et al. 1982). This is important for γ and C_V. Using a simplified high-temperature free energy (cf. Landau and Lifshitz 1969), it is simple to show that

$$C_V \simeq \left(C_0 + \frac{3}{2} \right) k,$$

$$\gamma \simeq \left(C_0 \gamma_0 + \gamma_R + \frac{1}{2} \gamma_V \right) / \left(C_0 + \frac{3}{2} \right),$$

$$\gamma_{R,V} \equiv \text{d} \ln \theta_{R,V}/\text{d} \ln \rho, \tag{4}$$

where C_0 is the translational contribution to the specific heat in units of Boltzmann's constant per molecule, γ_0 is the translational contribution to the Grüneisen parameter, and θ_R, θ_V are characteristic rotational and vibrational temperatures. At low density, $\theta_R \to 85$ K and $\theta_V \to 6100$ K, but at high density ($\rho \sim 1$ g cm^{-3}) one expects $\theta_R \to \theta_V \sim 2000$ K because all modes are strongly coupled. It follows that γ_R could be as large as 2.5 and γ_V as low as -1 in the 0.1 to ~ 3 Mbar pressure region. For likely values of $\gamma_0 \sim 0.6$ and

$C_0 \sim 2.5$, it is probable that $\gamma \sim 0.7-0.8$ and $C_V \sim 4\ k$. However, we argued below that dissociation or semiconduction may substantially modify these estimates. Note also that the imprecision (as in "$\gamma_0 \sim 0.6$," rather than "$\gamma_0 = 0.63$" or something) is deliberate.

The melting curve of H_2 at $P < 0.1$ Mbar is well constrained by diamond anvil experiments and associated theoretical calculations (Mao and Bell 1979; Young and Ross 1981). Extrapolation based on the potential of Eq. (4) and liquid perturbation theory indicates that the melting temperature $T_m < 1000$ K at 1 Mbar, comfortably lower than the actual temperature within Saturn.

2. Metallic Hydrogen.   The high-pressure limit of protons immersed in an almost uniform, degenerate electron gas has no experimental data base but has received extensive theoretical attention. Unlike the molecular limit, the free energy can be formulated from first principles in a relatively simple form, at least for a rigid lattice (see Ross and McMahan 1976).

Thermal corrections for the fluid phase can be obtained from Monte Carlo simulations (DeWitt and Hubbard 1976) or fluid perturbation theory (Stevenson 1975). The former is more accurate for linear screening, whereas the latter is intrinsically less accurate but has the advantage of including higher order terms; for most purposes, the differences are too small to have a serious impact on interior models. Simulations (Slattery et al. 1982) of the unscreened limit (referred to as the one-component plasma) indicate that the melting temperature of metallic hydrogen is bounded above by 1500 $\rho^{1/3}$ K, so the fluid state is clearly most relevant.

The equation of state for $T = 6000$ K can be approximated as

$$P = 9.95\ \rho^{5/3}\ (1 - 0.909\ \rho^{-1/3} + 0.614\ \rho^{-2/3} - 0.021\ \rho^{-1}) \qquad (5)$$

(Stevenson 1975) where P is in Mbar. This should be accurate within $1-2\%$. The Grüneisen parameter γ is remarkably insensitive to T, and lies in the range 0.60 to 0.65. The similarity of this result to the Fermi gas value of $2/3$ is fortuitous. The specific heat C_V is also rather constant and within 10% of $2k$/proton. Pressures at other temperatures can be deduced from Eq. (5) using the thermodynamic identity $(\partial P/\partial T)_V = \gamma N C_V$, where N is the number density of protons.

3. Molecular-Metallic Transition.   Many estimates have been made of the pressure at which hydrogen transforms from diatomic (molecular) to monatomic (metallic). The most recent estimates are ~ 3 Mbar (see, e.g., Nellis et al. 1983) but values as high as 5 Mbar and as low as 1 Mbar have been suggested. The problem with all these estimates is that they assume the electronic natures of the two phases. It is not known, for example, to what extent a mixture of atoms and molecules is possible, nor to what extent a band overlap (i.e., divalent) molecular metal is possible. Band structure calculations

(Friedli and Ashcroft 1977) indicate that such a state may occur at 1 Mbar, a lower pressure than for the transition to the monatomic (monovalent) metal. Density functional calculations (Chakravarty et al. 1981) provide additional support for the possibility of two transitions. At the high temperatures encountered in the giant planets, the situation is further complicated by the likelihood that these transitions are continuous (i.e., supercritical) rather than first-order.

A number of arguments for or against the first-order character of $H_2 \rightarrow$ 2 H at $T \sim 10^4$ K, $P \sim 3$ Mbar have been made. The most recent attempt to resolve this issue is by Robnik and Kundt (1983). As with previous analyses, the conclusions for or against first-order character are sensitive to the assumed form of the interactions. The following novel argument is not rigorous but suggestive. The emplacement of H_2 molecules in a monatomic, metallic environment is a less severe perturbation of the electron gas than the emplacement of helium atoms. (A hydrogen molecule is just like a helium atom, except that the nuclear charge has been split. Thus the perturbation of the electron gas is reduced. To be precise, the Fourier transform of the nuclear potential is $8\pi q^{-2} \sin(\mathbf{q} \cdot \mathbf{d})/\mathbf{q} \cdot \mathbf{d}$, where $\mathbf{q}$ is the wavevector and $\mathbf{d}$ is the H–H bond length. This is weaker than $8\pi/q^2$, the α-particle potential.) If one treats H_2 as a weakened He, then it is straightforward to show, from an extension of the theory for H–He discussed further in the next section, that the critical temperature for the first-order character of $H_2 \rightarrow 2H$ is much lower than that for immiscibility in H–He. This would suggest that the pure hydrogen transition is continuous for the conditions encountered in the giant planets.

If the transition is continuous, then constraints can be placed on the nature of the equation of state in the interpolation region (e.g., $0.6 \lesssim \rho \lesssim 1.5$ g cm^{-3}). Details will be presented elsewhere (Stevenson, in preparation) but the point is that the actual free energy must be lower than any trial free energy such as that obtained by assuming the hydrogen to be either entirely molecular and insulating or entirely monatomic metal. Consequently, the pressure must decrease below that predicted for the pure molecular phase as one enters the interpolation region, and must be higher than that predicted for the pure monatomic metal as one leaves the interpolation region on the high-pressure side. In other words, the parameter $(\partial \ln P / \partial \ln \rho)_{ad}$ is anomalously low in the interpolation region. The specific heat and Grüneisen parameter may be even more anomalous, because of the entropy arising from a mixed state of molecules, atoms and electrons. The specific heat may be larger than usual and γ may be smaller, possibly even negative. None of these potential effects has yet been incorporated in giant planet models.

B. Hydrogen-Helium Mixtures

In molecular H_2–He fluids, it is straightforward in principle to evaluate the free energy and all thermodynamic properties by an extension of the techniques discussed in Sec. II.A.1 above. Three interparticle potentials (hydrogen-hydrogen, hydrogen-helium, helium-helium) are needed. The uncertainties in the free energy obtained from Monte Carlo simulations (Slattery and Hubbard 1976) or liquid perturbation theory (Stevenson and Salpeter 1976) are greater than for pure hydrogen, because the hydrogen-helium interaction is unconstrained by high-pressure (shock-wave) experiments and must be chosen from theoretical estimates. However, the shapes of the three potentials are expected to be very similar, and the helium mole fraction is quite low in a cosmic mixture ($\sim$ 12%), so the form of the equation of state is very similar to that for H_2. The values of γ and C_V are also rather similar at high density, except for a slight reduction due to the additional degrees of freedom (rotational and vibrational) available to H_2 but not to He.

Helium is known to have limited solubility in liquid molecular hydrogen at low pressure and temperature (Streett 1974), and theoretical calculations (Stevenson and Salpeter 1977a) suggest the persistence of limited miscibility at high pressure. However, the critical temperature above which H_2 and He can mix in all proportions is a few thousand degrees at most, and almost certainly less than the actual temperature within Jupiter or Saturn.

The behavior of hydrogen-helium mixtures at $P \gtrsim 1$ Mbar is less well understood, but helium is probably less soluble in metallic hydrogen than in molecular hydrogen. This possibility and its implications were first noticed by Smoluchowski (1967). Pure helium does not undergo pressure metallization until $\sim$ 100 Mbar (Young et al. 1981) and a helium atom immersed in metallic hydrogen is likely to retain localized states (i.e., states lower in energy than the bottom of the hydrogen conduction band) until about this pressure. For Thomas-Fermi screening, Lam and Varshni (1983) find that the ionization energy of a helium atom in a plasma goes to zero at $r_s \sim 0.8$ ($P \sim 80$ Mbar). Closed shell atoms are highly insoluble in metals of high electron density, and helium is probably not an exception.

Stevenson (1979) considered a pseudopotential model in which the helium atom is only weakly perturbed from its free state but interacts strongly and repulsively with the conduction electron sea of metallic hydrogen, essentially because these conduction electrons must orthogonalize their wave functions to the helium core states. In the low electron density limit, the immersion energy required to insert a helium atom in an electron gas of density n_e is given by the Born approximation:

$$\Delta E = n_e <0|V_{ps}|0> \tag{6}$$

where V_{ps} is the helium-electron pseudopotential and $<0|$ denotes a low-energy plane wave state. This energy is experimentally measurable as the potential barrier of liquid helium to electrons (in which case $\Delta E \simeq 1.3$ eV; see Jortner et al. 1965). At higher electron densities ΔE is overestimated by Eq. (6) and actually begins to decrease at $r_s \sim 1$, from a peak value of ~ 5 eV. This behavior is shown in Fig. 3. The solubility of helium in an electron gas is

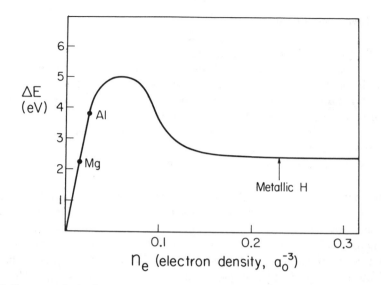

Fig. 3. Energy required to immerse a helium atom in an electron gas, as a function of the electron gas density. The slope as $n_e \rightarrow 0$ is experimentally determined and the asymptote as $n_e \rightarrow \infty$ is theoretically well established. The intermediate maximum is suggested by pseudopotential calculation (see text).

given approximately by $\exp(-\Delta E/kT)$ where ΔE is strictly the Gibbs energy of immersion (i.e., allowance is made for volume changes as well as internal energy changes).

In the high-pressure limit in which a helium atom enters as an α-particle and two free electrons, immiscibility still persists and the critical temperature has been estimated at $\sim 10^4$ K (Stevenson 1975; Hansen and Vieillefosse 1976; Straus et al. 1977; Pollock and Alder 1977; Firey and Ashcroft 1977). The critical temperature is extremely sensitive to the electron distribution, however, and 3-dimensional Thomas-Fermi calculations by MacFarlane and Hubbard (1983) for a solid, ordered H–He alloy even admit the possibility of no immiscibility. Although the relevance of this calculation to Saturn is unclear, it serves as a caveat: it is not yet possible to precisely quantify the solubility of helium in metallic hydrogen. One interesting feature does emerge from Fig. 3; the least solubility of helium occurs for an electron density less than that encountered in pure metallic hydrogen, but comparable to that encountered for an intermediate, partially metallized state. The insolubility within Saturn may be localized to a rather narrow range of pressure and radius within which the hydrogen is undergoing transition. Under these circumstances, a phase separation into a metallic helium-poor and an insulating (or semiconducting) helium-rich phase would occur. The possible evolutionary implications for Saturn are discussed in Sec. VIII.

C. Other Constituents

1. Solubility in Hydrogen: Thermodynamics.   In a cosmic mixture, constituents other than hydrogen and helium have too low an abundance to markedly change the thermodynamics. However, Saturn is clearly not cosmic (see Sec. IV) and water, especially, may be as abundant by mass as helium. Since these constituents could either be present as a separate core or layer or be mixed with hydrogen, it is necessary to consider their behavior both as pure components and as impurities.

Water is a complicated substance. Shock-wave data (Mitchell and Nellis 1982; Lyzenga et al. 1982) can be interpreted in terms of dissociation of water, perhaps to a state similar to an ionic melt, $H_3O^+ + OH^-$ (Hamann and Linton 1966; Stevenson and Fishbein 1981; Ree 1982). However, the equation of state inferred from shock wave data is not consistent with low-temperature diamond-anvil data at $P < 500$ kbar (Liu 1982), indicating the difficulty of constructing theoretical models when the electronic state is poorly characterized. Existing giant planet models use the shock-wave data (see Hubbard and MacFarlane 1980b), an appropriate choice since actual temperatures are close to shock-wave temperatures. At much higher pressures, water is expected to metallize and the equation of state should eventually approach the Thomas-Fermi-Dirac or Quantum Statistical limit (Zharkov and Trubitsyn 1978). However, interpolations between shock-wave data and this high-pressure limit have much greater uncertainties than the corresponding uncertainties for hydrogen. The mixing properties of water with hydrogen are also poorly known. Existing systematics and theory (Shmulovich et al. 1980; Stevenson and Fishbein 1981) indicate that water has probably the lowest solubility in hydrogen of any plausible constituent. Even so, the adiabat within Saturn is likely to be supercritical (i.e., the only phase separation of water is the expected water cloud formation at $P \sim 20$ bar). Phase separation of water from hydrogen is more likely in Uranus and Neptune, where the temperatures are lower and the water abundance is higher.

Shock compression data for methane (Nellis et al. 1981) can be interpreted as indicating dissociation into elemental carbon (diamond or metal) and molecular hydrogen at $P > 100$ kbar (Ree 1979; Ross 1981). These data can also be interpreted to indicate the formation of a more compact carbon-hydrogen mixture. In any case, the important point is that phase separation of carbon or methane from hydrogen is unlikely within Saturn because of the small carbon abundance. If the energy favoring separation is ΔE, then the solubility of carbon in hydrogen would be $\exp(-\Delta E/kT)$. For this to be less than the observed (atmospheric) mole fraction of $\sim 2 \times 10^{-3}$, ΔE would have to exceed ~ 4 eV at $P \sim 0.3$ Mbar ($kT \sim 0.6$ eV). This is highly unlikely; estimates based on the extrapolation of shock-wave data suggest $\Delta E \sim 2$ eV at most (ΔE could also be ~ 0).

Ammonia has some characteristics in common with water. It tends to dissociate (NH_4^+ ion formation), as indicated by high electrical conductivity under shock compression (Mitchell and Nellis 1982). In any phase separation, ammonia is likely to follow water. Its low relative cosmic abundance ($\sim$ 1/6 that of water) makes it less important in modeling, but since it can be more readily measured in the atmosphere it could be an important diagnostic of internal processes and mixing ratios.

The remaining constituents, loosely characterized as rock, are primarily magnesium, silicon, iron and whatever oxygen and sulfur these elements incorporate and retain at high pressure. Neither the composition nor the equation of state for a given composition are well known for the rock component. However, this is probably not a serious problem for Saturn models because the rock component is likely to be a small fraction of the total mass. The mixing properties of rock in the ices or hydrogen are also poorly known, although there are theoretical arguments to suggest that iron, at least, is soluble in metallic hydrogen (Stevenson 1977).

Rock layers (or cores) and ice layers are usually modeled by assuming volume additivity. The density of the mixture is then assumed to be given by

$$\rho(P)^{-1} = \sum_i x_i \rho_i(P)^{-1} \tag{7}$$

where x_i is the mass fraction of constituent i which has equation of state $\rho_i(P)$. The errors associated with this procedure could be substantial ($\sim$ 10%), even at Mbar pressures, but no better procedure exists.

2. Liquid or Solid Core?   The question of whether Saturn's core is solid or liquid is relevant to at least one observable phenomenon. Dermott (1979) has shown that a possible mechanism to produce a value of the effective tidal Q on the order of $\sim 10^5$ in a Jovian planet is dissipation of elastic strain energy in a solid core. Since there appears to be no chance of significant tidal dissipation in the present liquid envelope of Saturn if the envelope is exactly homogeneous, we must consider the possible existence of a solid core with the required elastic properties.

A rigorous calculation of the stable phase for a given core layer would require knowledge of the composition of the layer and the Helmholtz free energies $F(T,V,N_i)$ for both the solid and liquid phases. At present we cannot calculate these thermodynamic quantities accurately enough to hope to derive solidus and liquidus curves. In any case, the information sought is much cruder: Are plausible temperatures in Saturn's core above or below the range of plausible melting temperatures of likely core constituents?

Figure 4 shows estimates of melting temperatures as a function of pressure for pure atomic species, obtained using the Thomas-Fermi model of a compressed solid and the Lindemann melting criterion (Hubbard 1981). Although this model is crude, the results should be qualitatively correct at pressures approaching 50 Mbar. Also shown is a temperature profile for Saturn from a recent model calculated by Hubbard et al. (1980). The temperature in

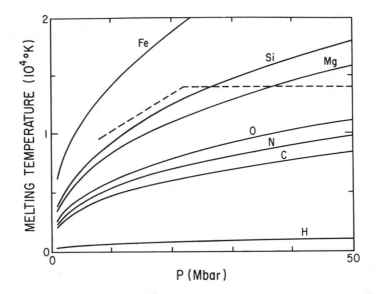

Fig. 4. Melting temperatures of pure elements, estimated using the Thomas-Fermi and Lindemann approximations. Dashed line shows the approximate course of Saturn interior temperatures.

this model is assumed to follow an adiabat in a solar composition mixture of hydrogen and helium from 140 K temperature at 1 bar pressure down to the surface of an outer core composed primarily of H_2O, CH_4, and NH_3. The temperature is assumed constant from this point to the center of the planet, although it must actually continue to rise with a smaller slope. The melting temperature of a mixture of atomic species will depend upon the molecular state of the mixture, but we will assume that it is in any case unlikely to be higher than the highest melting temperature of the individual atomic constituents. Thus, Fig. 4 shows that any outer core composed of H_2O, etc., is almost certain to be molten, and that it is reasonable to assume that such a core would differentiate from the much denser rock-iron component. A pure iron inner core is predicted to be solid in this model, while magnesium silicates are borderline. Continuation of this model down to a pressure of only 1.4 Mbar (well below its range of validity) would predict that the surface of the Earth's core should solidify before the bottom of the mantle, contrary to observation. However, the Earth's outer core is liquid because of the eutectic properties of the core-forming alloy.

III. TRANSPORT WITHIN SATURN

A. Convection

The existence of an intrinsic heat flow comprising 2000 erg cm^2 s^{-1} at the surface of Saturn has important implications for the interior temperature profile. We first conclude that internal temperatures, at least in the hydrogen-rich zones, must substantially exceed melting temperatures, so that essentially all the hydrogen is present as a strongly coupled supercritical liquid. Arguments concerning this point are given by Hubbard et al. (1974), and are based upon two lines of reasoning: (a) the ordinary thermal conductivity of dense hydrogen is so low that the initial accretional heat should be largely preserved in the planet up to the present, in the form of a temperature profile produced by adiabatic compression, unless the planet can efficiently transport the heat by convection; (b) convection is demanded by the low conductivity, low viscosity, and large amount of stored heat in Saturn, thus leading in any case to a nearly adiabatic temperature profile in much of the interior. A discussion of this point in the context of Jupiter given by Stevenson and Salpeter (1976) is largely applicable to Saturn as well. Note that the argument hinges on the values of the relevant thermal conductivities (due to both molecular and radiative transport), which are well known in the metallic hydrogen phase but not in the molecular hydrogen phase. However, a higher conductivity is not conceivable for molecular hydrogen (Stevenson and Salpeter 1977a). Assuming negligible viscosity, the condition for convective instability is

$$| (dT/dr)_{\text{actual}} | > | (dT/dr)_{\text{ad}} | \qquad (8)$$

where dT/dr is the radial temperature gradient. Thus the maximum heat flux which could be carried by conduction without exciting convection is

$$Q_{\text{max}} = K | (dT/dr)_{\text{ad}} | \qquad (9)$$

where K is the thermal conductivity and $(dT/dr)_{\text{ad}}$ is the temperature gradient for an adiabatic temperature distribution. Assuming an adiabatic temperature profile in Saturn, a typical value of $| (dT/dr)_{\text{ad}} |$ would be $\sim 1.7 \times 10^{-6}$ K cm^{-1}. Thus, to assure the existence of convection in extensive regions of the planet, we need $K < 10^9$ erg K^{-1} cm^{-1} s^{-1}. This condition appears to be satisfied in metallic hydrogen (Stevenson and Ashcroft 1974) and in impure molecular hydrogen (Stevenson and Salpeter 1977a).

B. Relationship of Atmospheric Zonal Structure to Interior

If Saturn's interior is largely liquid and convective, how closely does the temperature profile follow the adiabatic distribution? The deviation is conveniently measured in terms of the quantity

$$\epsilon = \left[|(dT/dr)_{\text{actual}}| - |(dT/dr)_{\text{ad}}| \right] \Big/ |(dT/dr)_{\text{ad}}|. \qquad (10)$$

ϵ has been estimated using mixing-length theory, which provides a dimensional relationship relating ϵ to the heat flux via a poorly known parameter, the mixing length. Taking the effects of planetary rotation into account but neglecting differential rotation, Ingersoll and Pollard (1982) have shown that the mixing length is on the order of 60 km, the mean convective velocity is on the order of 2 cm s^{-1}, and $\epsilon \sim 10^{-5}$ to 10^{-6}. With such a small value of ϵ, any differential rotation must take place on coaxial cylinders (see Sec. V below), but the velocity and mixing length imply a very large eddy viscosity, sufficient to quickly obliterate deep differential rotation. Ingersoll and Pollard therefore propose that the lifetime of a convective element is in fact limited by the shear time associated with observed mean zonal flow in Saturn's atmosphere. With this assumption, the convection velocity is reduced to 0.1–1 cm s^{-1} and ϵ increases to $\sim 10^{-2}$. This value of ϵ still implies that deviations from an adiabatic gradient in Saturn's interior are only $\sim 1\%$. If we then assume that the near-adiabatic region in Saturn extends from the tropopause at least to the core boundary, typical deep interior temperatures on the order of 10,000 K are implied.

Thus the zonal flow pattern seen in Saturn's atmosphere may extend deep within the planet, with the rotation rate being constant on coaxial cylinders. The convective circulation pattern takes the form of columnar eddies with the axes of the columns parallel to the rotation axis (Busse 1976). According to this picture, energy is fed *into* the large-scale zonal flows from such convection.

C. Convective Redistribution of Minor Constituents

The superadiabaticity parameter ϵ defined above has a compositional analog χ, defined as

$$\chi = -(H_T/\alpha\rho) \left(\frac{\partial\rho}{\partial x}\right)_{ad} \left(\frac{dx}{dr}\right), \tag{11}$$

where $H_T = C_P/\alpha g$ is the (adiabatic) temperature scaleheight, α is the coefficient of thermal expansion, and x is the mass fraction of minor constituents. The sign convention is chosen so that $\chi > 0$ denotes a stabilizing gradient, and the definition is constructed so that $\epsilon = \chi$ represents a neutrally stable situation. A generalization of mixing-length theory (Stevenson and Salpeter 1977*b*) shows that the heat flux now scales as $\epsilon (\epsilon - \chi)^{1/2}$, rather than the usual $\epsilon^{3/2}$, so that ϵ must increase for a given heat flow if χ is positive. More importantly, the ratio of work done against gravity in redistributing the heavy minor constituents upwards is of order $(\ell/H_T)F$, where F is the heat flux and ℓ is the mixing length. For $\ell \sim H_T$, the available work can be large and many Earth masses of minor constituents can be redistributed during the planet's thermal evolution (Stevenson 1982*a*).

This formulation breaks down if the planet is layered. Suppose, for example, that the primordial state of Saturn included a fluid ice layer overlaid by a less dense hydrogen-helium fluid mantle. Assuming that the ice is fully soluble in the overlying mantle, the upward mixing is limited by the diffusivity of the ice, in accordance with the well-studied phenomenon of double-diffusive convection (Turner 1973; Stevenson 1982a). The physical picture is that in a time τ, heat diffuses upwards from the interface by a distance $(K'\tau)^{1/2}$ (K' is the thermometric conductivity, equal to $K/\rho C_P$), whereas ice diffuses upwards by a distance $(D\tau)^{1/2}$ (D is the diffusion coefficient). This diffusive boundary layer then becomes unstable (typically for $\tau \sim$ minutes), carrying away thermal buoyancy and compositional (negative) buoyancy in the ratio $K'^{1/2}/D^{1/2}$. The whole process is then repeated. In consequence, the available work for redistributing the minor constituents is reduced by a factor $(D/K')^{1/2}$ $\lesssim 10^{-1}$ relative to the mixing-length prediction (or, equivalently, the upper bound predicted by the Second Law of Thermodynamics).

The implication of this inefficiency of double-diffusive convection is that a core of rock or ice could persist throughout the age of the solar system, despite being soluble in an overlying hydrogen-helium mantle. If the envelope of Saturn is water-rich then this is probably an indication of icy planetesimals accreted during or after the collapse of a gaseous envelope, rather than the upward mixing of a primordial, icy nucleus. It is very likely that Saturn possesses both these ice reservoirs.

IV. STATIC MODELS OF SATURN

A. Earlier Work

Many investigators have contributed to the development of static models of Saturn's interior, but the modern era of Saturn interior models essentially began with the classical paper by DeMarcus (1958), who showed that the mass and radius of Saturn (as well as of Jupiter) were roughly consistent with a zero-temperature object having approximately solar composition overall. DeMarcus' model was among the first to incorporate modern concepts about the behavior of dense hydrogen, and he showed that his proposed interior density distribution could be used to closely predict the planet's response to its large rotational perturbing potential, as expressed in the external gravitational harmonics and surface figure. His model has a substantial dense core overlain by an extensive hydrogen-rich envelope comprising most of the mass, and is mostly in the solid phase except for a thin gaseous atmosphere.

The major revision to Saturn models since DeMarcus' work has come from the realization first made by Peebles (1964) that internal temperatures must be quite high, on the order of 10^4 K, and that substantial thermal effects must be taken into account in the equations of state. As discussed in the previous section, this is a natural consequence of the existence of an internal energy source. Recent models of Saturn (e.g. Podolak and Cameron 1974; Podolak 1978; Slattery 1977; Zharkov et al. 1974a,b; Zharkov and Trubitsyn 1978; and Grossman et al. 1980) incorporate numerous experimental and theoretical advances in the equations of state of dense hydrogen-rich liquids at the appropriate temperatures and pressures. These models differ in detail, but have many common features. First, all are calculated for the liquid phase of dense hydrogen, and all assume that the temperature distribution follows an adiabat throughout the hydrogen region, beginning at a point close to the planetary photosphere. We note, however, that the temperature distribution may not follow the same adiabat across the molecular hydrogen-metallic hydrogen interface if the latter involves a first-order phase transition (see Stevenson and Salpeter 1976). This complication, which involves several poorly known factors, has not yet been included in a detailed modeling study.

Second, all models postulate the existence of a dense, massive central core with a mass of 10–20 Earth masses (or ∼ 10–20% of Saturn's total mass). Such a core seems to be required in all models, to obtain a satisfactory fit to Saturn's gravitational quadrupole moment. A chemically homogeneous mixture of hydrogen and other elements can be contrived to yield a coreless model which fits Saturn's mean density, but the quadrupole moment is then far too large. Since a rocky or rocky-icy core in a Saturn of solar composition could have at most a mass of ∼ 2–3% of Saturn's mass, all models agree that Saturn is not precisely of solar composition, but has an enhancement on the order of ≥ 10 in at least some elements other than hydrogen and helium.

Third, all the models agree that, as in Jupiter, some of the hydrogen in Saturn attains sufficient pressures to be converted to the metallic state. The pressure or range of pressures over which this transition occurs is still very uncertain, so there is corresponding uncertainty about the mass of metallic hydrogen in Saturn. In a typical model, the transition region occurs at ∼ 50% of the planetary radius, compared with ∼ 80% of the planetary radius in Jupiter.

B. Current Models

The principal uncertainty in current static models of Saturn concerns the relative composition of the core and envelope. Figure 5 shows three conceivable distributions of components in the interior. Figure 5(a) is the simplest case: only the most refractory, rocky components are differentiated into a core. The rest of the components are uniformly distributed throughout the planet's mantle, although the transition from H_2 to metallic hydrogen (H^+) takes place at about the location of the dashed line. If the latter is a first-order transition, however, then abundances of the other chemical constituents must change across it and case (a) is inadmissible. Figure 5(b) shows a more complicated model in which liquid metallic hydrogen undergoes a first-order phase transition into a hydrogen-rich liquid phase and a helium-rich liquid phase. The helium-enriched layers accumulate above the rocky core. Finally, in Fig. 5(c) it is assumed that the core consists not only of rocky components but also of materials that would condense as ice under nebular conditions (H_2O, NH_3, and at sufficiently low temperature CH_4). It is assumed that not all the latter components can redissolve into the hydrogenic envelope, for if they could their mass would rival that of a solar complement of helium, sig-

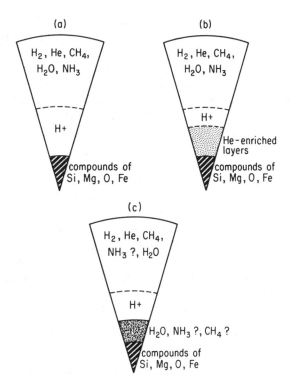

Fig. 5. Three conjectural models for the present distribution of chemical constituents in Saturn's interior.

nificantly enhancing the density of the envelope. Obviously, many other models can be generated by incorporating various combinations of elements from these three models and others.

One of the major goals of static modeling is to provide guidance about which class of interior models most closely fits all available constraints. To be acceptable, a model of one Saturn mass must have the correct equatorial and polar radius at 1 bar pressure, must reproduce, within the error bars, the observed values of J_2, J_4 (and perhaps eventually J_6), and must have a composition and thermal structure in its outer layers consistent with observational constraints. There has recently been substantial improvement in all these constraints, and some earlier Saturn models may now be inadmissible.

The modeling technique just described might be called a direct approach. One computes a planetary model and then accepts or rejects it according to its agreement with observational constraints. Alternatively, one may use an inverse approach by solving for particular density distributions that are in exact agreement with the constraints. This approach makes it easier to assess the degree of uniqueness of a given class of solutions.

Table I presents a recent interior model of Saturn (Hubbard and Horedt 1983) which exactly satisfies the gravitational field and rotation period constraints given in Sec. I. The model also includes corrections for deep differential rotation on cylindrical surfaces, as discussed in Sec. V.B. The helium mass fraction in the hydrogen-rich envelope is assumed to be constant and equal to 0.11, consistent with the Voyager result given in Sec. I. Certain other parameters in this model have been adjusted to provide an exact fit to observational constraints. These parameters include the total mass of a dense core, and two parameters that adjust the form of the pressure-density relation in the hydrogen-rich envelope at pressures > 13 kbar. Thus the pressure-density relation is basically empirical, obtained by requiring a fit to the gravitational constraints. The $P(\rho)$ relation can be independently tested by comparing it with experimental data on compressed hydrogen, and with theoretical relations for metallic hydrogen. Figure 6 shows the inferred $P(\rho)$ relation (solid curve) for the model of Table I, together with some points (crosses) computed from the theory of liquid metallic hydrogen-helium mixtures (for $P \geq 1$ Mbar) and from Monte Carlo runs using intermolecular potentials derived from shock data (for $P < 200$ kbar). These points are in good agreement with the inferred $P(\rho)$ relation. The other curve and points shown in Fig. 6 are discussed below.

Other characteristics of the model of Table I are as follows. Of a total planetary mass 95 M_E (M_E = Earth mass), the dense core comprises 19 M_E. This core is subdivided into two regions, an inner core with a rock-like $P(\rho)$ relation and an outer core with an ice-like $P(\rho)$ relation (Hubbard et al. 1980). The mass ratio of the former to the latter region is constrained to be $\frac{1}{3}$, as would be approximately true if the core contained all abundant constituents in a solar composition mixture, such as Fe, Si, Mg, O, C, and N. Now if helium is depleted in the hydrogen-rich envelope of Saturn, it must be enriched elsewhere in the planet. The total mass of H in the envelope is 68 M_E. The solar composition complement of He to this amount would be $\sim$ 17 M_E, but the model only includes 8 M_E in the envelope, which implies that $\sim$ 9 M_E of helium must be present in the core. The outer core, with total mass $\sim$ 14 M_E, therefore includes substantial amounts of He as well as substances such as H_2O, CH_4, and NH_3. All these constituents have roughly similar $P(\rho)$ relations, so the assumption is not inconsistent.

How unique is the model in Table I? In order to investigate this problem, Hubbard and Horedt (1983) calculated a model similar to this one in all respects except that the envelope helium mass fraction was required to be 0.26, substantially greater than the Voyager result for the atmospheric helium abundance and only just within the error bars of the Voyager result for the Jovian atmosphere. This abundance was imposed by requiring the inferred $P(\rho)$ relation to match the computed hydrogen-helium adiabat at pressures < 13 kbar and > 5 Mbar. The relation which yields an exact fit to the gravitational moments, mass, and 1-bar surface is shown in Fig. 6 (dashed curve). Compari-

TABLE I
Saturn Interior Model Optimized to Fit Gravitational Field Constraints[a]

β[b]	P (Mbar)	$\rho(\text{g cm}^{-3})$
0.0	47.8	19.7
0.05	43.4	19.0
0.10	31.6	16.8
0.1275	23.5	15.0
boundary between inner "rock" core and outer "ice" + He core		
0.1275	23.5	7.69
0.15	20.3	7.24
0.20	14.2	6.28
0.25	9.03	5.29
0.2693	7.26	4.88
boundary between "ice" + He outer core and H-rich envelope		
0.2693	7.26	1.73
0.30	6.39	1.65
0.35	5.24	1.54
0.40	4.29	1.43
0.45	3.49	1.33
approximate location of boundary between metallic-hydrogen inner envelope and molecular-hydrogen outer envelope (the transition may be a gradual one)		
0.50	2.80	1.22
0.55	2.19	1.11
0.60	1.66	0.99
0.65	1.21	0.87
0.70	0.84	0.74
0.75	0.54	0.60
0.80	0.31	0.45
0.85	0.16	0.32
0.90	0.062	0.20
0.95	0.012	0.094
1.00	0.000001	0.00018

[a] Model envelope is composed of 89% hydrogen and 11% helium by mass and is chemically uniform and isentropic. Temperatures in the metallic-hydrogen region can be approximately computed from the relation $T = 7000 \text{ K } \rho^{0.64}$, where ρ is in g cm^{-3}.
[b] The parameter β is the average radius of a level surface, in units of the average radius of the one-bar pressure level.

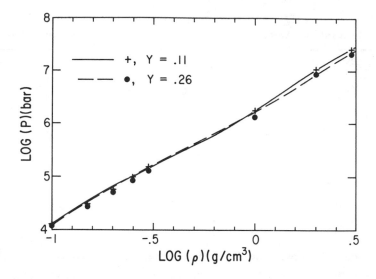

Fig. 6. Two barotropic relationships for Saturn's envelope which yield an exact fit to its external gravity field. Also shown are points calculated from high-pressure theory (right) and from experimental and theoretical data on intermolecular potentials (left). Solid curve represents the inferred $P(\rho)$ relation for the model of Table I. (+) represent some points computed from the theory of liquid metallic hydrogen-helium mixtures (for $P \geq 1$ Mbar) and from Monte Carlo runs using intermolecular potentials derived from shock data (for $P < 200$ kbar). Dashed curve represents the high-helium model of Hubbard and Horedt (1983) (see text) and the (●) are points computed from theory similarly as for the (+).

son points obtained from theory and experiment, as before, are also shown (dots). Although the inferred relation is not in quite as good agreement with the comparison points as is the low-helium model, we cannot argue that this model is unacceptable. The largest disagreement with the comparison points occurs at density 1 g cm^{-3}, where the inferred pressure is $\sim 25\%$ greater than the pressure calculated from theory. At this relatively low density, the theoretical point could well be in error by this amount. Perhaps more importantly, when the gravitational harmonics (J_4 in particular) are permitted to vary across their observational limits, the inferred $P(\rho)$ for Saturn performs excursions comparable in magnitude to the differences between the solid and dashed curves, and between these curves and their comparison points.

In this high-helium model, the total core mass is reduced slightly to 17 M_E. The total mass of hydrogen in the planet is thus ~ 58 M_E, or ~ 10 M_E less than in the helium-depleted model of Table I.

In matching the mean density and gravitational harmonics of Saturn, we are of course sensitive only to the density distribution in the planet and not to the chemical composition. Thus, an enhancement or depletion of helium could be compensated by a depletion or enhancement of some other dense component. The point made by Fig. 6 is that, given the current uncertainties in the gravitational moments of Saturn and in the pressure-density relation for hydrogen-helium mixtures at a pressure around 1 Mbar, it is not yet possible to conclude, solely from static gravitational modeling, whether Saturn's envelope is enhanced or depleted in dense components. This is in contrast to the case for Jupiter, where the gravitational field coefficients are known with much greater precision and the poorly understood region of the equation of state corresponds to a much smaller volume of the planet. For Jupiter, there is some evidence for a density profile in the envelope which either corresponds to solar composition or is slightly enhanced in density relative to it (Hubbard and Horedt 1983). Further progress on Saturn can be made when more precise values of the gravitational harmonics become available.

V. SATURN'S GRAVITY FIELD

A. Effects of Equation of State on Higher Harmonics

Saturn has the most rotational distortion of any planet. This distortion is conveniently represented in terms of a dimensionless parameter

$$q = \omega^2 a^3 / GM \qquad (12)$$

where ω is the angular rotation rate of the planet's deep interior and a is the reference radius defined in Sec. I.D, equal to 60,000 km. For a planet in strict hydrostatic equilibrium, rotating as a solid body at a rate ω, we then have (Zharkov et al. 1972; Zharkov and Trubitsyn 1978),

$$J_2 = \Lambda_{2,0}q + \Lambda_{2,1}q^2 + \ldots, \qquad (13)$$

$$-J_4 = \Lambda_{4,0}q^2 + \Lambda_{4,1}q^3 + \ldots \text{ etc.,} \qquad (14)$$

where the Λ's are dimensionless coefficients that depend on the relative distribution of mass within the planet. The more centrally condensed the mass distribution is, the smaller the coefficients (at least the initial ones). Only even zonal harmonics appear in the external gravity field, and these alternate in sign. Furthermore, the higher-order coefficients $\Lambda_{2n,1}$, $\Lambda_{2n,2}$, etc., generally form a rapidly decreasing sequence. In effect, the Λ's are calculated when a model planet is calculated. The parameter q is clearly crucial to this process, and so determining the fundamental rotation rate of the planet is very important. As discussed in Sec. I.E, the rotation rate of Saturn's magnetic field seems appropriate for this purpose, and leads to $q = 0.1527$. For comparison, Jupiter's $q = 0.0888$, and all other planets have much smaller values.

Ignoring higher-order terms in Eq. (13), one can estimate $\Lambda_{2,0} \simeq J_2/q = 0.108$ for Saturn. As we mentioned above, this parameter becomes smaller with increasing central concentration in the planetary mass distribution. For

Jupiter, the parameter is 0.166, while for a uniform-density sphere it increases to its maximum theoretical value of 0.5. We thus conclude that Saturn is much more centrally condensed than Jupiter; the simplest explanation of this difference is to assume that both planets have dense central cores of roughly the same mass, but that the remaining mass, composed mostly of hydrogen, comprises a much more tenuous envelope in Saturn than in Jupiter.

The coefficient J_4 is also significant for constraining interior models. The outermost layers of mass density are more heavily weighted in their contribution to this coefficient than to the lower-degree coefficient J_2, so that J_4 primarily serves as a constraint on the form of the pressure-density relation in the outer envelope. However, J_4 is strongly correlated with the value of J_2, which means that it must be known quite accurately to be useful as an independent constraint. With a current error bar of $\sim 4\%$, the uncertainty in J_4 is $\sim 1/3$ of the maximum possible excursion in J_4 consistent with a fixed value of J_2.

B. Effects of Differential Rotation

As discussed in Sec. III, recent models of convective circulation in Saturn's interior imply that the strong differential zonal motions observed in Saturn's cloud layer may correspond to rotation on cylinders. Large amounts of mass in the interior may therefore rotate with periods shorter than the magnetic field rotation period, and this effect must be taken into account to accurately match models to the observed gravitational coefficients.

For a planet with an adiabatic interior temperature distribution, the pressure is a function of the density only, and the planet is said to be a barotrope. It is straightforward to show that for a barotrope, and for time-constant inviscid fluid motions, all fluid motions must be constant on cylindrical surfaces centered on the rotation axis. The total force acting on a fluid element is then the sum of the gravitational force exerted by the mass distribution and the centrifugal acceleration due to the rotation of the element's cylinder. This total force can be obtained by taking the gradient of a potential determined from the mass distribution and from the rotation law for the cylinders. Since the rotation law for the cylinders is known from the atmospheric motions, it is only necessary to find a self-consistent interior density distribution, or equivalently an appropriate barotropic $P(\rho)$ relation that yields the observed gravitational moments and is consistent with the rotation law.

The first quantitative work on the effects of differential rotation on the gravitational field of a planet in hydrostatic equilibrium was carried out by Trubitsyn et al. (1976) and by Vasil'ev et al. (1978). A subsequent study by Hubbard (1982) was based on detailed differential rotation measurements carried out by the Voyager spacecraft. In the latter work Hubbard found that inclusion of differential rotation increases all the gravitational coefficients in absolute value above their values computed using the magnetospheric rotation period. It also causes the planet to contract at the poles by a few km and to expand at the equator by several tens of km. The effect on the lower-degree gravitational moments is small, despite the fact that the range in rotation periods is $\sim 10\%$. Differential rotation on cylinders causes Saturn's J_2 to increase by $\sim 0.5\%$, $|J_4|$ to increase by $\sim 2.5\%$, and J_6 to increase by $\sim 10\%$. These corrections turn out to be essentially independent of the interior pressure-density relation, so that the observed values of the gravitational coefficients could be reduced by these percentages for the purpose of comparison with a model rotating as a solid body with the magnetospheric period. Whether such corrections are appropriate depends on whether Saturn is rotating on cylinders as assumed, a question that requires further study by fluid dynamicists.

VI. SATURN'S MAGNETIC FIELD

The observed Saturnian field (described in Sec. I.E) provides important constraints on internal dynamics and electronic state. Four issues are addressed here.

1. How is the field generated?
2. Where is it generated?
3. Why is the dipole tilt so small?
4. What time variability and relation to observable (atmospheric) dynamics can be expected?

The time history of the magnetic field $\mathbf{H}$ is governed by the induction equation (often called the dynamo equation)

$$\frac{\partial \mathbf{H}}{\partial t} = \lambda \, \nabla^2 \mathbf{H} + \nabla \times (\mathbf{v} \times \mathbf{H}), \tag{15}$$

where t is time, λ is the magnetic diffusivity ($\equiv c^2/4\pi\sigma$ if c is the velocity of light and σ is the electrical conductivity in electrostatic units), and $\mathbf{v}$ is the fluid motion relative to the rigidly rotating frame defined by the external magnetic field. If the induction term $\nabla \times (\mathbf{v} \times \mathbf{H})$ is small, then the field undergoes diffusive decay on a time scale $\sim L^2/\pi^2\lambda$, where L is a characteristic length scale of the field. For even the most optimistic estimates of the conductivity (Stevenson and Salpeter 1977a) this time scale is only $\sim 10^8$ yr, over an order of magnitude less than the age of the solar system. It is therefore unlikely that the present field is a diffusive remnant of a primordial field. It follows that, to maintain the field, the induction term must offset the degenerative (Ohmic) effect of the diffusion term. Like all planets with substantial magnetic fields, Saturn requires a hydromagnetic dynamo.

Several dimensionless numbers characterize a planetary dynamo (see Stevenson 1983a for a review) but the most important is suggested by dimensional analysis of Eq. (15) and is $R_m = vL/\lambda$, the magnetic Reynolds number. If $R_m > 10$, then dynamo generation of the field is possible. For plausible convective velocities v $\sim$ 0.1–1 cm s^{-1}, suggested by mixing-length theory or modifications to this theory (Sec. III.B), dynamo generation is possible for $\lambda \lesssim 10^7$ cm^2 s^{-1}. Liquid metallic hydrogen has $\lambda \sim 4 \times 10^2$ cm^2 s^{-1} (Stevenson and Salpeter 1977a) and semiconducting molecular hydrogen may have $\lambda \sim 10^4 - 10^6$ cm^2 s^{-1} at $P \sim 1$ Mbar. Since diffusion is needed for a dynamo (see Moffatt 1978, p. 112) the metallic region may not be the best region for dynamo generation because R_m is too large. However, turbulence may reduce the length scale L, so the importance of metallic hydrogen in the dynamo process remains an open question. The possible role of the molecular region was first suggested by Hide (1967) and subsequently supported by Smoluchowski (1975). If the dynamo region encompasses the metallic core only, then it extends to between 0.45 a_S and 0.55 a_S. If a substantial part of the molecular region is sufficiently conductive, the dynamo region may extend to 0.6 a_S or even 0.65 a_S.

In principle, the strength and multipolarity (but not the tilt) are related to the radial extent of the dynamo region. A number of theoretical arguments (see Stevenson 1983a) suggest that a characteristic field strength in this region is $(8\pi\rho\lambda\omega)^{1/2}$ Gauss, where ω is the planetary rotation and ρ is the fluid density. Since all the parameters entering into this estimate are essentially the same for Jupiter and Saturn, the observed ratio of their dipoles (Jupiter: Saturn 10:1) might be a measure of $\eta_J^3 : \eta_S^3$, where the radial extent of the dynamo region is $\eta_J a_J$ for Jupiter and $\eta_S a_S$ for Saturn. Even for $\eta_J = 0.9$, probably an overestimate, this would imply $\eta_S = 0.33$, suggesting a very deep-seated source. However, there are many possible pitfalls in this argument; perhaps the biggest is our complete ignorance of the time variability of the giant planet magnet dipoles. Another way of assessing source depth relies on the empirical observation that higher-order multipoles may contribute a field amplitude comparable to the dipole at the surface of the source region (Elphic and Russell 1978). This argument has modest success when applied to Jupiter and Earth. It cannot be applied with confidence to Saturn because the existing field inversions disagree on the relative magnitudes of the multipoles. The most recent analysis (Connerney et al. 1982b) allows for large axisymmetric quadrupole and octupole components. A third method of assessing source depth (Hide 1978) is potentially the most powerful, and is based on a theorem which states that the total (unsigned) flux out of a conducting region does not change over a time that is small compared with the magnetic diffusion time. This theorem cannot be applied yet, since it requires observing field changes on a convective time scale (years or decades).

The small dipole tilt of Saturn's magnetic field is its most intriguing feature. Since axisymmetric and nonaxisymmetric dipole components both decay as r^{-3} outside the conducting region, the magnitude of the tilt has nothing to do with the depth of the conducting core. It is also not directly related to the value of R_m. While it is true that a large R_m allows the possibility of a nearly axisymmetric dynamo (Braginskii 1965; Todoeschuck et al. 1981), there is no plausible way that the relevant R_m could be much larger for Saturn than for Jupiter, whose dipole has a large tilt $\sim 10°$. The small dipole tilt is also probably not a chance alignment since the *a priori* probability of aligning the dipole and rotation axis to within an angle δ (in degrees) is $\sim 10^{-2} \delta^2$.

A possible explanation for the small tilt is the electromagnetic skin effect. Suppose that Saturn's dynamo region, like Jupiter's and Earth's, generates a field with a tilt $\sim 10°$, but that only Saturn has a spin-axisymmetric differentially rotating, conducting layer between the dynamo region and the planetary exterior. This layer would preferentially filter the nonspin axisymmetric components of the field and leave the spin axisymmetric components unaltered.

This layer could be present for Saturn because of phase separation of helium from hydrogen (Stevenson 1980,1982c). In the region of helium raindrop formation, a stable conducting layer forms which is expected to undergo differential rotation because of an equator-to-pole temperature difference. From the point of view of a fluid element in the differentially rotating region, the nonspin axisymmetric components of the field look like a time-varying field and undergo attenuation as in the conventional electromagnetic skin effect.

The consequences can be quantified approximately (Stevenson 1982c) by considering a local Cartesian coordinate representation of the field $\mathbf{H} = (f,g,h) \, e^{ipx}$, where x is azimuthal, y is meridional, and z is vertical. The dynamo equation for a purely zonal flow $\mathbf{v} = z\omega(y)\mathbf{x}$ becomes

$$\lambda \, \frac{\partial^2 f_p}{\partial z^2} + \omega h_p = 0, \tag{16}$$

$$\lambda \, \frac{\partial^2}{\partial z^2} \begin{Bmatrix} g_p \\ h_p \end{Bmatrix} - ip\omega z \begin{Bmatrix} g_p \\ h_p \end{Bmatrix} = 0. \tag{17}$$

The azimuthal field f can be amplified, but the $p \neq 0$ contributions to the poloidal field are decaying Bessel functions of order 1/3 and undergo exponential attenuation in the radial direction by an amount dependent on $\omega p L^3/\lambda$, where L is the thickness of the shear zone. For plausible parameter choices, the attenuation is one or two orders of magnitude, more than enough to explain the dipole tilt of $\lesssim 1°$. This model predicts that higher order azimuthal harmonics ($m \geq 2$, where m is the azimuthal order in a spherical harmonic expansion) undergo even greater attenuation than the $m = 1$ terms responsible for dipole tilt.

Although the helium differentiation is an attractive (albeit indirect) explanation for axisymmetrization, it is not unique. Any region that is conducting, stable (not having the vertical motions needed for a dynamo), and undergoing spin-axisymmetric differential rotation can achieve the same effect. The weakly conducting molecular region (above the level at which phase separation might occur) might have important hydromagnetic effects. The zonal winds observed in the atmosphere may extend to deep levels (see Sec. I.B) and provide an axisymmetrization. In this case, the difference in dipole tilts between Jupiter and Saturn could arise because the zonal winds on Saturn are stronger and more nearly axisymmetric. Clearly, observations over a period of years to decades are needed to establish whether atmospheric dynamics and magnetic field are correlated.

VII. TIDAL DISSIPATION IN SATURN

The tidal bulge raised on Saturn by satellites involves a fluid flow that extends into the deep interior. The dissipative nature of this flow can provide additional constraints on the internal properties. A measure of the tidal dissipation is the quality factor Q, defined as

$$Q = \frac{2\pi \text{ (peak energy stored)}}{\text{(energy released per cycle)}}. \tag{18}$$

The lower bound on Q is $\sim 5 \times 10^4$ (Goldreich and Soter 1966), based on the requirement that the orbital evolution of the satellites not be excessive. An upper bound is more difficult to establish, but the evidence for tidal heating of Enceladus indicates a Q close to the lower bound (Yoder 1981). If the commensurability relation of the orbits of Enceladus and Dione was formed by tidal evolution, then $Q < 10^6$.

Although this Q is much greater than the value for Earth (~ 10), implying much lower specific dissipation, it is not easy to find a sufficiently dissipative process in an essentially fluid planet. One possibility is to invoke processes near or within the rocky core, which may be solid or partially solid. The same inelastic processes responsible for dissipation of solid Earth tides would, if applicable to the innermost region of Saturn, provide a $Q \sim 10^5$– 10^6 (Dermott 1979). Turbulent skin friction at the surface of this core (a modified application of a suggestion by Goldreich and Soter 1966) might also be important. These possibilities cannot be quantified with any confidence because of our ignorance concerning conditions near the center of Saturn.

Dissipative processes involving large-scale fluid motions are unpromising. Turbulent eddy viscosity arising from convection (Hubbard 1974) can only provide a $Q \sim 10^{10}$, because the convective time scales are much longer than the tidal period (Goldreich and Nicholson 1977). Excitation and subsequent dissipation of inertial gravity waves in the atmosphere (Houben and Gierasch, personal communication) is also inadequate. Excitation of large-scale inertial oscillations is conceivable but has not been quantified.

The most recently suggested mechanism for Q (Stevenson 1983b) involves the consequences of helium raindrop formation. If helium raindrops are forming in a metallic hydrogen zone, because of the limited solubility of helium their radii grow and diminish by diffusion as the oscillatory tidal pressure perturbation is applied. The same phenomenon occurs as a sound wave passes through a water fog (Epstein and Carhat 1953). The associated time-variable diffusive effects cause irreversible entropy production. (An equivalent way of quantifying this effect is to evaluate the time delay between application of a pressure pulse and the reestablishment of thermodynamic equilibrium between droplet interior and the surrounding fluid.) The resulting dissipation E is dominated by the effects of finite solute diffusivity D and is given by

$$\dot{E} = \int n\rho D \left(\frac{\partial \mu}{\partial x}\right)_{T,P} |\nabla x|^2 \, dV, \tag{19}$$

where n is the droplet number density and μ is the chemical potential of the solute of concentration x. If the tidal frequency Ω is high enough so that each droplet can be considered independent, yet low enough that the droplet size s is small compared to $(D/\Omega)^{1/2}$ the characteristic diffusion length, then E is independent of frequency and $Q \propto \Omega$. However, if the frequency is very low then diffusion can "saturate" the interdroplet medium, $|\nabla x| \rightarrow 0$ and $Q \propto \Omega^{-1}$. The optimal frequency (i.e., minimal Q) is given by the criterion that the diffusive effects of all particles within one diffusion length should affect the interdroplet medium concentration by an amount comparable to the oscillatory component of concentration at the surface of a single droplet. Under these circumstances, E is still near its maximum value and the effects of saturation are only beginning to be important. Since $|\nabla x| \sim \Delta x(s/r)$ where Δx is the oscillatory contribution to x and $r > s$ is the distance from the droplet center, the minimal Q occurs when

$$\int_{0}^{(D/\Omega)^{1/2}} 4\pi r^2 \, dr \, (s/r) \sim 1. \tag{20}$$

The "resonant" frequency implied by this equation is $\Omega_0 \sim 4\pi Dn\bar{s}$, where $\bar{s}$ is the average droplet radius. If the region containing droplets is a substantial fraction of the planetary mass, then $Q(\Omega_0) \sim 10^2 - 10^3$, independent of the raindrop characteristics and population dynamics. The big uncertainty is in Ω_0, since neither n nor $\bar{s}$ are well constrained. Several arguments based on rainfall velocities and thermal evolution suggest $\bar{s} \sim 10^{-2} - 10^{-3}$ cm, with small regions of the two-phase medium being near resonance and other regions being away from resonance. An interesting consequence of this mechanism is that Q may be noisy (i.e., the value at any frequency may fluctuate on a convective time scale). Since the tidal heating of satellites depends on the planetary Q, this may have implications for their thermal evolution.

VIII. EVOLUTIONARY MODELS OF SATURN

A. Homogeneous Evolution

The simplest evolutionary models of Saturn's interior assume that the planet evolves essentially as a low-mass sub-main-sequence dwarf. Such a model is completely convective because of its high interior opacity, and the interior temperature distribution follows an isentrope whose initial condition is defined by the pressure-temperature location of the planetary photosphere. The model evolves step by step from one isentrope to an adjacent lower one, with the time step defined by the luminosity of the object and the energy evolved in the course of the transition. The model is assumed to be homogeneous, meaning that although it may have layers of different chemical composition the mass and location of these layers do not change with time. The intrinsic luminosity of the planet is then given by

$$L_i = - \int dm \left(\frac{dE}{dt} + \frac{P}{\rho^2} \frac{d\rho}{dt} \right) \tag{21}$$

where E is the internal energy per gram, P is the pressure, ρ is the mass density, and the integral is carried out over the planet's mass. In this class of model the total luminosity $L_i + L_t$ is related to the interior temperature distribution because the latter is defined by an isentrope whose starting point is established by the energy flow through the atmosphere. The specific relationship can be obtained by computing a grid of model atmospheres with different values of T_e and surface gravity g. According to Hubbard (1977), the grid of atmospheres calculated by Graboske et al. (1975) can be related to temperatures in the deep interior by

$$T = 66.8 \ g^{-1/6} \ T_e^{1.243} \rho^\gamma \tag{22}$$

where T is the temperature within the planet in K, as a function of the mass density ρ, and all quantities are expressed in cgs units. Here γ is the Grüneisen parameter, which can be taken as ~ 0.64 to good approximation for either Saturn or Jupiter. Assuming an adiabatic temperature distribution everywhere, Eq. (22) is fairly accurate for the metallic core, but somewhat underestimates temperatures (by $\sim 20\%$ at 0.1 Mbar) in the molecular envelope. Thus the result for the cooling time which we give below will be slightly underestimated as well. Using standard thermodynamic identities and substituting Eq. (22) in Eq. (21), the differential equation governing the planet's evolution becomes (Hubbard 1977)

$$4\pi a_S^2 \sigma (T_e^4 - T_s^4)$$

$$= - [1.243 \ (d \ln T_e/dt) - \frac{1}{6} (d \ln g/dt)] \int dm \ C_V T \tag{23}$$

where C_V is the heat capacity at constant volume per gram, and $T(T_e,\rho)$ is given by Eq. (22). To obtain an estimate of the time required for the planet to cool from an arbitrarily high initial formation temperature to its present value of T_e, we will assume that Saturn cools at constant radius, which is not strictly true but valid to within 5–10% during the last few 10^9 yr of planetary evolution. With this approximation, we may take a_S to be constant and equal to its present value, while the term in d ln g/dt can be neglected. Eq. (23) is then integrated to give the age of the planet as a function of T_e, defined such that $t = 0$ when $T_e = \infty$. The result is expressed in the form

$$t = (\alpha/2.757) \; T_e - 2.757 \; [1 +$$

$$0.41 \; (T_s/T_e)^4 + 0.26 \; (T_s/T_e)^4 + \ldots], \tag{24}$$

where α is the planet's thermal time constant, given by an integral over the internal distribution of C_V and ρ. The heat capacity per gram for a hydrogen-helium mixture is taken to be $1.66 \; k/m_H$, where k is Boltzmann's constant and m_H is the mass of a hydrogen atom. This result is accurate to within $\sim 10\%$ for both the metallic hydrogen and molecular hydrogen phases (the latter region is much more important in Saturn than in Jupiter). The constant α was evaluated numerically using an interior model of Saturn, with the dense core included in the calculation but assumed to be isothermal, with the result, in cgs units

$$\alpha = 7.32 \times 10^{22}. \tag{25}$$

From an analogous calculation, Hubbard (1977) found $\alpha = 2.79 \times 10^{23}$ in cgs units for Jupiter.

Eq. (24) then yields $t = (3.9 \pm 0.1) \times 10^9$ yr for the cooling age of Saturn, using the values of T_e and T_s obtained by Hanel et al. (1983). The uncertainty in t reflects the uncertainty in the observational parameters, not the uncertainty of the theoretical calculation, which is considerably greater.

Previous investigations of the homogeneous evolution of Saturn models (Pollack et al. 1977; Grossman et al. 1980) have found values for t substantially smaller than the above estimate, typically around (2 to 2.5) $\times$ 10^9 yr. These earlier calculations include a considerably more elaborate treatment of the planetary evolution, taking into account the effects of variable radius and interior density distribution. Part of the discrepancy is attributable to the latter's use of an older and smaller value of T_s and a slightly larger value of T_e, but some of the discrepancy may be due to other causes. Using $T_s = 75$ K, the analytic theory indicates that the time required to cool from $T_e = 123$ K to $T_e = 97$ K is 1.8×10^9 yr, while Grossman et al. obtain a corresponding time interval of $\sim 1.0 \times 10^9$ yr. In this interval, they find that the radius changes by $\lesssim 5\%$, so the assumptions of the analytic theory should be valid. This suggests that there are substantial differences, perhaps approaching a factor of 2, in such critical thermodynamic variables as the heat capacity.

Our conclusion from the analysis of homogeneous cooling models of Saturn is that such models appear to evolve somewhat too fast to adequately explain the observed heat flow, but this shortcoming may not be as large as some previous results suggest, and could conceivably be absorbed into the uncertainty in our knowledge of the finite-temperature thermodynamics of Saturn's large molecular hydrogen envelope. Although energy release mechanisms other than homogeneous cooling may now be occurring in the interior of Saturn, we may not need to postulate such mechanisms to explain the heat flow.

B. Heterogeneous Evolution

The thermal energy output of Saturn can be greatly modified by redistribution of constituents. Increased energy output can arise if a heavy constituent such as helium becomes insoluble and rains out in the gravitational field, creating thermal energy by small-scale viscous dissipation. Decreased energy output can arise if a stable, primordial layering is partially homogenized by convective dredging of soluble constituents. The latter is discussed in Sec. III.C and by Stevenson (1982a). The emphasis in this section is on the former result.

Consider a Saturn that formed hot, with the hydrogen and helium uniformly mixed. As the planet cools through geologic time, losing thermal energy in the way described in Sec. VIII.A, a time will be reached when helium becomes supersaturated in metallic hydrogen. The hydrogen-helium phase diagram is not known precisely enough to predict when or even whether this occurred, but the inferred atmospheric depletion of helium (Sec. I.A) supports this possibility. The onset of insolubility would occur first in the uppermost "metallic" (not necessarily monatomic) region at $P = 1-2$ Mbar, in accord with the arguments presented in Sec. II.A. This probably occurs at temperatures for which the molecular-metallic hydrogen transition is continuous.

Droplets of helium-rich fluid form, growing by diffusion until they are large enough to sink at Stokes velocities that exceed local convective velocities. For velocity ~ 1 cm s^{-1}, this implies droplets ~ 0.1 cm in radius. For a likely phase diagram (Fig. 7), the droplets probably dissolve at some deeper level in the planet. The hydrogen-helium envelope of Saturn develops a 3-layer structure: a helium-depleted molecular envelope; a stable metallic or semi-metallic intermediate zone; and a helium-enriched metallic core region. The outer part of the envelope, including the atmosphere, can continuously supply more helium to the intermediate rain-forming region because of convective circulation. The energy released by this process can be found by calculating the change in total (gravitational plus internal) energy as the differentiation proceeds.

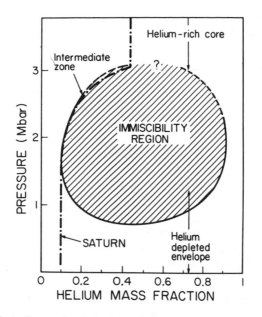

Fig. 7. Likely phase diagram for the hydrogen-helium system if conditions within Saturn are appropriate for phase separation. A helium concentration profile is superimposed. This situation corresponds to Fig. 5(b).

Although detailed evolutions have not been calculated, energy differences between homogeneous and differentiated static models indicate that the energy released per unit mass of displaced helium is $\sim 0.4 \, \bar{g}H$, where $\bar{g}$ is the average gravitational acceleration in the region, H is the distance between the centers of mass of the helium-depleted and helium-enriched zones, and the droplets are assumed to be twice as dense as the coexisting fluid. In an $n = 1$ polytropic model, $\bar{g}$ is similar to the surface g and is ~ 1000 cm s^{-2} for Saturn. However, the presence of a central high density core increases this value significantly to ~ 1500 cm s^{-2}. The value of H is $\sim 2 \times 10^9$ cm and the specific energy release ϵ_{gr} is consequently $(1.5 \pm 0.7) \times 10^{12}$ erg (g-He)$^{-1}$. If we suppose that the outer half of the planetary mass has undergone differentiation from an initial helium fraction of 0.2 by mass to a present fraction of x, and assume that this has occurred over a period τ, then the average resulting surface heat flow $\bar{F}_{gr}$ from gravitational energy release alone is

$$\bar{F}_{gr} = 0.5 \, (0.2 - x) \, \epsilon_{gr} M / 4\pi a_S^2 \tau. \qquad (26)$$

The solubility of helium is dictated by a formula of the form $x \simeq \exp(-T_1/T)$ with $T \simeq 10^4$ and $T_1 \simeq 2 \times 10^4$ K. The change in x during time τ is therefore related to the change in thermal energy. It is important to realize that $\bar{F}_{gr}$ cannot be added to the heat flow predicted by homogeneous cooling models, because differentiation changes the cooling rate. In fact, the deeper regions of the planet can even heat up during differentiation, because the intermediate rain-forming region is stably stratified and may be superadiabatic (Stevenson and Salpeter 1977b). The conservative assumption adopted here is that only the cooling required in the outer envelope to reduce x is a contribution to the total average internal heat flux F_i. We then find that

$$\bar{F}_i = \bar{F}_{gr} + \bar{F}_{th},$$

$$\bar{F}_{th} \simeq \left(\frac{1}{\ln 5} + \frac{1}{\ln x} \right) \epsilon_{th} M/4\pi a_S^2 \tau, \qquad (27)$$

where $\epsilon_{th} \simeq 4 \times 10^{11}$ erg g^{-1}. The total intrinsic heat flux given by Eqs. (26) and (27) is an average over the time τ. To obtain the present intrinsic heat flux from these values, a multiplicative correction factor is needed to allow for the higher heat flow at earlier times. This correction factor is approximately $1-3 (0.2 - x)$.

The present-day internal heat flux is shown in Fig. 8 for differentiating models, as a function of present atmospheric helium content and time since

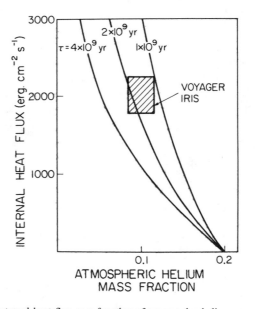

Fig. 8. Present-day internal heat flux as a function of present-day helium mass fraction in the envelope of Saturn, for three theoretical models labeled by τ. This time scale is the time elapsed since the onset of helium differentiation. The Voyager IRIS results for heat flow and atmospheric composition are shown to be consistent with $\tau \sim 2 \times 10^9$ yr.

onset of differentiation. The observational values are indicated, and the differentiating model is clearly consistent provided τ is $\sim 2 \times 10^9$ yr. A precise theoretical determination of τ is not possible because of uncertainties in the thermodynamics and phase diagram for the deep interior, but this value is plausible in the sense that it is not so large as to be inconsistent with the lack of substantial differentiation inferred for Jupiter, nor so small that one would have to argue that we are living at a special time.

There are a number of potential difficulties with this approximate analysis. First, the helium droplets may not be twice as dense as the coexisting metallic hydrogen fluid, because they may contain substantial amounts of molecular hydrogen. In a helium-rich fluid, the creation of occupied conduction electron states may be energetically prohibitive (see Fig. 3); this would reduce the energy output. Second, there may be other constituents (especially water) which partition during the droplet formation and modify the differentiation process. For example, water, methane and ammonia might partition upwards (decreasing the energy output because work must be done to mix them upwards) or even partition downwards (increasing the energy output). Third, the phase diagram could be much more complicated than that shown in Fig. 7 (Stevenson and Salpeter 1977b). More accurate abundances in the atmospheres of both Jupiter and Saturn will be needed to understand the heterogeneous aspects of evolution.

C. Implications for Origin

The above summaries of both homogeneous and heterogeneous evolution assume that Saturn started out much hotter than at present. Although this is not demanded by the observations, it is reasonable because there is no known way to efficiently eliminate all the heat generated during formation, and only $\sim 10\%$ of this heat needs to be retained to give Saturn an initial internal temperature twice its present value. The most interesting clues to the origin of Saturn are compositional. The central core of rock and/or ice is on the order of ten Earth masses, similar to Jupiter's and about equal to that required to nucleate a hydrodynamic collapse of the primordial solar nebula (Mizuno 1980; Stevenson 1982a).

As discussed in Sec. III.C, convective redistribution of these constituents is likely to be inefficient. However, the envelopes of Saturn and Jupiter are probably enriched in some constituents (especially water) relative to cosmic abundance. This suggests accretion of icy planetesimals after the formation event.

The following scenario (Stevenson 1982a) is consistent with these inferences.

1. A rock/ice planet is formed in the presence of the solar nebula. (This must occur quickly, in $< 10^6$ yr. No dynamical model is currently available to explain this step.)

2. When this planet reaches ~ 10 Earth masses, the greatly extended but gravitationally bound atmosphere becomes unstable and collapse begins.
3. A proto-Saturn is formed, and angular momentum conservation implies formation of a gaseous disk from which the satellite system forms.
4. Proto-Saturn has an initial heat content two or three times the present heat content, and therefore has a much higher luminosity. It is also near the rotational limit, but is despun to its present rotation by "viscous" (actually turbulent) and hydromagnetic effects.
5. In the subsequent few 10^8 yr, there is continued accretion of icy planetesimals scattered into the Saturn zone from beyond Neptune. This material, together with a small amount of convective dredging and mixing from the central core, causes the outer envelope to be enriched in heavier constituents (notably ices) relative to cosmic abundance.
6. After $\sim 2 \times 10^9$ yr, differentiation of helium from hydrogen may be initiated, causing a further slight redistribution of minor constituents.

Although this scenario is consistent, it may not be unique. It is offered as a focus for debate and analysis.

IX. CONCLUSIONS

Despite the complexity of Saturn's interior structure, certain common features emerge from all proposed models. We shall summarize these features before passing on to more controversial aspects of interior studies.

DeMarcus' original conclusion that Saturn is predominantly of hydrogen but not of solar composition has been supported by all subsequent investigations. The external gravity field, along with the general properties of the hydrogen equation of state, demands the presence of a massive, dense core. The precise mass and composition of the core remain uncertain, but typical values for its mass are ~ 15 M_E. Thus there is an enhancement in Saturn of the ratio of heavy materials to hydrogen by ~ 1 order of magnitude relative to solar composition.

It is generally agreed that the most plausible class of interior models involve hot ($T \sim 10^4$ K) liquid hydrogen mantles, with the temperature gradient close to adiabatic through much of the mantle but possibly not throughout. Observed zonal motions in the atmosphere may extend deep within the interior, and such flows may produce a small but detectable effect in the external gravity field.

Under Saturnian interior conditions, the major transition of hydrogen from the molecular to the metallic phase may not be first order. If this is the case, certain thermodynamic parameters such as the heat capacity and Grüneisen parameters may exhibit anomalous behavior. Such effects are difficult to quantify, but would have a major effect on thermal models.

The most uncertain aspects of Saturn interior studies involve quantitative comparison between predictions of interior models and observations of the atmospheric chemical composition and heat flow. A priori, it is not hard to appeal to high-pressure phase transitions and immiscibility behavior to account for departures of atmospheric chemical ratios from solar values. The best-studied process of this type involves the system of metallic hydrogen and helium, but even here, quantitative prediction of phase boundaries requires much more accurate statistical-mechanical calculations than does the calculation of lower-order quantities such as pressure and heat capacity. Thus, we cannot say with certainty that hydrogen-helium immiscibility is occurring in Saturn, although a number of calculations indicate approximate correspondence between expected critical temperatures for liquid-liquid phase separation and Saturn interior temperatures.

It is not difficult to account for the Saturn heat flow in terms of interior models, and the recent downward revision of this quantity makes the task easier. Typical models involving phase separation of hydrogen from helium require the differentiation process to have occurred over a reasonable period of $\sim 2 \times 10^9$ yr. Separation of other constituents possibly nearly as abundant as helium in Saturn could also play a role in the energetics. Finally, even homogeneous models that account for the heat flow by a simple cooling mechanism now appear to be marginally viable.

Acknowledgments. Some of the work discussed in this review was supported by various grants and contracts from the National Aeronautics and Space Administration.

COMPOSITION AND CHEMISTRY OF SATURN'S ATMOSPHERE

RONALD G. PRINN
Massachusetts Institute of Technology

HAROLD P. LARSON
University of Arizona

JOHN J. CALDWELL
State University of New York, Stony Brook

and

DANIEL GAUTIER
Observatoire de Meudon

We provide a comprehensive discussion and review of the chemistry and com-position of Saturn as determined by Earth-based, Earth-orbital, and Voyager 1 and 2 spectroscopic observations. The observations imply that there are impor-tant differences between the actual composition of Saturn's atmosphere and that predicted for a homogeneous solar-composition planet. In particular, the H_2, He, CH_4, NH_3, and PH_3 volume mixing ratios differ by factors on the order of 1.06, 0.6, 2–5, 1–3, and 1–4, respectively, from the expected solar composi-tion ratios. These results imply that during its formation Saturn accreted a sig-nificant amount of ice and rock in addition to its solar-composition endowments of these heavy materials. The depletion of He in the visible atmosphere suggests that this element has preferentially differentiated toward the center of the planet. The D to H ratio inferred from CH_3D observations is $\sim 2 \times 10^{-5}$ which is similar to that on Jupiter and has important cosmological implications. Vol-ume mixing ratios for C_2H_6 and C_2H_2 are $\sim 4 \times 10^{-6}$ and 8×10^{-8}, respec-tively, at pressures less than 20 mbar but decrease rapidly at higher pressures.

[88]

This is consistent with the theoretically expected photochemical sources for these gases. The mixing ratios of NH_3 and PH_3 decrease rapidly with altitude in the stratosphere undoubtedly due to condensation and photochemical destruction, respectively. The visible clouds are probably predominantly ammonia crystals but the evidence is not firm. The faint colors are likely due to various allotropes of phosphorus produced from PH_3 photodissociation; however, the evidence is again largely circumstantial. Both equilibrium and nonequilibrium chemical processes are important on Saturn: equilibrium processes in the deep warm atmospheric regions and nonequilibrium processes (photochemistry and vertical mixing in particular) at higher cooler levels. The vertical eddy mixing coefficient $K \simeq 10^5 \exp[z/2H]$ cm^2 s^{-1} where z is altitude above the Saturnian tropopause and H is the density scale height. The observability of GeH_4 on Jupiter but not on Saturn poses important questions concerning the comparative state of the interiors of these two planets. Continued search for CO, H_2S, H_2O, N_2, and C_2H_4 on Saturn is worthwhile because their detection would lead to further understanding of the chemistry of the atmosphere. The role played by seasonal effects in atmospheric chemistry also needs further investigation.

In a remarkable two-year time span, Saturn was visited by three space-craft: Pioneer 11 on 1 September 1979; Voyager 1 on 12 November 1980; and Voyager 2 on 26 August 1981. The infrared and ultraviolet spectrometers on board the two Voyagers have, in particular, enabled the analysis of the composition of Saturn's atmosphere with spatial and spectral coverage and spatial resolution not previously possible from Earth. These spacecraft observations together with some notable prior and contemporary groundbased and Earth-orbital observations have now, in sum, supplied us with at least a first-order picture of Saturn's atmospheric composition. This new compositional knowledge has in turn provided renewed interest in studies of the atmospheric chemistry of this giant planet.

Well before this spacecraft epoch it was realized that Saturn and Jupiter were structurally and compositionally brethren planets, probably the two most similar planets in our solar system. While this realization has not been altered by the more recently obtained information, it is also very clear now that these two giant planets are not identical. There are quantitative differences in the fractional elemental abundances of He, C, N, and P in their atmospheres; in the relative distributions and colorations of their clouds, in their respective atmospheric temperature profiles; and in minor-species concentrations in their visible atmospheres (e.g., the species C_2H_2, C_2H_6, GeH_4, PH_3). Thus while discussions of the composition and chemistry of Saturn will bear many analogies to their Jovian counterparts, there will be important differences and these differences can in principle provide excellent tests of our understanding of the chemistry and evolution of both of these planets.

This chapter is devoted to the composition and chemistry of Saturn. It begins with a brief review of the chemically and spectroscopically relevant aspects of atmospheric structure. This is followed by an extensive, detailed discussion and review of the gaseous composition of Saturn determined by Earth-based, Earth-orbital, and Voyager 1 and 2 spectroscopic observations.

The determination of the H to He ratio is given particular attention due to its importance to discussions of the origin, internal structure and energy balance of the planet (see chapters by Hubbard and Stevenson, Ingersoll et al., and Pollack and Consolmagno in this book) and to its significant cosmological implications (Gautier and Owen 1983a). Observational constraints on the composition of Saturn's clouds are then addressed. With these various compositional observations as a suitable preface, chemical models of the atmosphere are reviewed and further developed. We finally conclude this chapter with a hopefully seminal review of what we perceive to be the outstanding problems and questions, both observational and theoretical, that remain to be addressed in future investigations.

I. ATMOSPHERIC STRUCTURE

There is now little doubt that the energy flux which Saturn emits as infrared radiation to space is greater by a factor of ~ 2 than the energy flux which it absorbs as solar radiation. More specifically, Earth-based observations had indicated quite early that the latter factor was ~ 2.5 (Rieke 1975), Pioneer 11 measurements were later interpreted to yield a factor of ~ 2.8 (Orton and Ingersoll 1980), and most recently, Voyager 1 data indicate a factor of ~ 1.8 (Hanel et al. 1983). The nature of the implied internal energy source on Saturn is discussed in the chapter by Hubbard and Stevenson. For our purposes in this chapter, the requirement that the internal heat be transported upward to the visible layers necessitates that the atmosphere be convective between the 1 and 10^4 bar pressure levels; the work of Stone (1973) further implies that the vertical temperature gradient be close to adiabatic in this region.

In the atmospheric region above ~ 10 bar the convective heat flux from the deep atmosphere is supplemented by the deposition of solar radiation and above ~ 1 bar infrared radiation begins to challenge convection as a mode of upward heat transport. The atmospheric temperature profile above ~ 1 bar can be measured remotely by observations of the Saturnian infrared emission spectrum and also by radio occultation techniques. The inversions of Earth-based infrared observations to obtain Saturnian temperature profiles by various investigators have been reviewed by Tokunaga (1978). Analyses of radio occultation and infrared observations by Pioneer 11 to yield temperature profiles were carried out by Kliore et al. (1980b) and Orton and Ingersoll (1980), respectively. Finally, preliminary temperature profiles derived from Voyager 1 and 2 radio occultation and infrared spectroscopic data have been presented by Tyler et al. (1981,1982a) and Hanel et al. (1981a,1982), respectively. The latter Voyager 1 and 2 vertical and horizontal temperature profiles are summarized in Fig. 1. In general, the tropopause or temperature minimum tends to lie at ~ 100 mbar pressure level with the atmospheric vertical temperature gradient approaching the adiabatic value of ~ 0.9 K km^{-1} at pressures $\gtrsim 500$

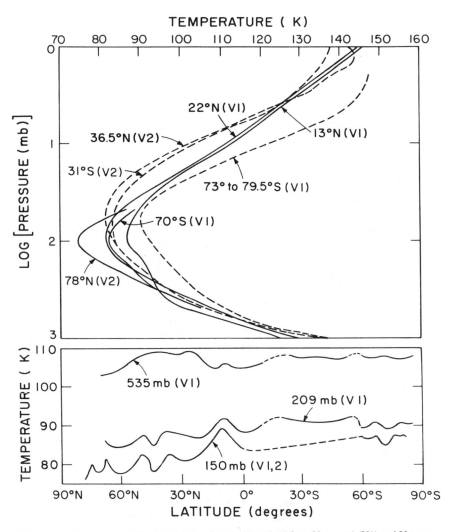

Fig. 1. Spatial variations of temperature on Saturn determined from Voyager 1 (V1) and Voyager 2 (V2) observations. Vertical profiles in upper graph are from Hanel et al. (1981a,1982) [solid lines] and Tyler et al. (1981,1982a) [dashed lines]. Horizontal profiles in lower graphs are from Hanel et al. (1981a,1982).

mbar. Temperatures at pressures of 10 to 300 mbar were generally a few to 10 K less at northern latitudes (which had just entered the ~ 7-yr-long Saturnian spring at the time of the Voyagers) than at the same southern hemispheric latitudes. The same trend is observed at the ~ 500 mbar level but with an anomalously warm region superimposed at northern mid-latitudes.

In our discussions of the composition and chemistry in this chapter, these various predicted and observed temperatures are relevant to the interpretation of spectroscopic data, to the determination of cloud condensation levels, and to the prediction of chemical reaction rates. A more detailed discussion of the observed temperature profiles and their important implications for the meteorology of the planet may be found in the chapter by Ingersoll et al.

II. ATMOSPHERIC COMPOSITION FROM SPECTROSCOPY

Spectroscopic observations of Jupiter, Saturn, Uranus, and Neptune have and still do provide the only means of unambiguously analysing the composition of these four giant planets. Fortunately, our traditional spectroscopic arsenal has been reinforced over the last 15 yr by the development and implementation for astronomical applications of very high resolution Fourier Transform Spectrometers, by airborne (e.g., Kuiper Airborne Observatory) and Earth-orbiting (e.g., IUE) observatories, and by the Voyager instruments. In this section we discuss the application of spectroscopic techniques to the problem of definition of the composition of Saturn's atmosphere. We first discuss Earth-based and Earth-orbital investigations and then follow with Voyager investigations. In each case infrared, visible (Earth-based only), and ultraviolet spectroscopy of Saturn are discussed with due attention to the apparent agreement or disagreement among the conclusions derived from the various spectral regions, observational platforms, and methods of interpretation. As noted earlier, the H to He ratio and cloud composition are both selected for separate attention.

A. Earth-based Infrared Spectroscopy

In general the history of spectroscopic observations of Saturn closely follows that for Jupiter, although progress has been slower because of Saturn's lower flux levels at all wavelengths compared to Jupiter's. In the infrared spectral region, defined here as $1 < \lambda < 100$ μm, this is a serious problem because beyond ~ 2.5 μm Saturn's signal must be detected in the presence of much higher thermal background flux levels that are emitted by our atmosphere, the telescope, and the spectrometers themselves. Thus, the acquisition of high-quality, high-resolution spectra of Saturn has depended not only upon the development of more sensitive detectors and very efficient spectrometers, but also upon the construction of telescopes optimized for infrared observations and located on high-altitude sites where atmospheric interference is minimized. This is clearly revealed in the summary of published spectroscopic observations of Saturn in Table I. In the near infrared ($\lambda < 2.5$ μm) a low-resolution spectrum of Saturn was first produced in 1947, and by the early 1970s much higher resolution data had been recorded at several conventional optical telescopes. Not until 1974 was Saturn's spectrum reported beyond 2.5 μm, and since ~ 1978 infrared data have increased dramatically. Most of these observations were made from special high-altitude facilities

TABLE I
Infrared Spectroscopic Studies of Saturn's Atmosphere

Spectral Region (cm^{-1})	Spectral Resolution (cm^{-1})	Instrument/ Telescope[a]	Spectral Analysis	Reference
4000–13000	$\simeq 80$	Prism		Kuiper 1947
6300–9100	$\simeq 80$	Grating/SAI 1.2, 2.6 m	CH_4, NH_3	Moroz 1966
4000–9000	1.7	FTS/OHP 1.9 m		Connes et al. 1969
2400–8300	8.0	FTS/UAO 1.5 m		Johnson 1970
9057–9133	0.36	Grating/Mc-Donald 2.1 m	CH_4	Bergstralh 1973
4000–12000	0.26	FTS/OHP 1.9 m	CH_4, $^{13}CH_4$, H_2	de Bergh et al. 1973; de Bergh et al. 1974; Combes et al. 1975
9057–9133	0.46	Grating/Mc-Donald 2.7 m	CH_4	Trafton 1973
740–1330	$\simeq 15$	CVF/UCSD-Minn 1.5 m	NH_3, PH_3, C_2H_6	Gillett and Forrest 1974
910–1000	$\simeq 4$	Grating/Lick 3 m	PH_3	Bregman et al. 1975
900–1000	2.5	FTS/ESO 1.5 m	C_2H_4	Encrenaz et al. 1975
780–860	1.0	FTS/SAO 1.5 m	C_2H_6	Tokunaga et al. 1975
6000–11800	0.2	FTS/Palomar 5 m	NH_3, H_2S, H_2, CH_4	Lecacheux et al. 1976; Owen et al. 1977; Combes et al. 1977; de Bergh et al. 1977; Buriez and de Bergh 1981
4000–6000	0.5	FTS/Hawaii 2.2 m	H_2	Martin et al. 1976
400–590	22	FTS/SAO 1.5 m	Thermal structure	Tokunaga et al. 1977
100–450	5.0	FTS/KAO 0.9 m	Internal luminosity	Erickson et al. 1978
1700–6500	2.0	FTS/UAO 1.5 m	CH_3D, PH_3, GeH_4, SiH_4	Fink and Larson 1978; Larson et al. 1980

TABLE I (continued)
Infrared Spectroscopic Studies of Saturn's Atmosphere

Spectral Region (cm^{-1})	Spectral Resolution (cm^{-1})	Instrument/ Telescope[a]	Spectral Analysis	Reference
4000–15400	150	CVF/Hawaii 2.2 m	Spectral albedos	Clark and Mc-Cord 1979
4000–12500	3.6	FTS/KPNO 4 m	Spectral albedos, CH_4, NH_3, H_2S, COS, C_2H_6, C_2H_2, CH_3OH, CH_3NH_2	Fink and Larson 1979
2800–6200	0.5	FTS/KAO 0.9 m	PH_3, C_2H_6, CH_4, NH_3 (gas and solid), HCN	Larson et al. 1980; Bjoraker et al. 1981
1700–2300	2.0	FTS/KAO 0.9 m	H_2O	Larson et al. 1980
930–1000	2.3	Grating/Shane 3 m	PH_3, NH_3, C_2H_4	Tokunaga et al. 1980c
738–781	0.1	FPS/IRTF 3 m	C_2H_2	Tokunaga et al. 1980a
180–2500	4.3	FTS/Voyager	H_2, He, NH_3, PH_3, CH_4, C_2H_2, C_2H_6, C_3H_4, C_3H_8	Hanel et al. 1981a; Courtin 1982a
1250–2000	$\simeq 25$	CVF/KAO 0.9 m	C_2H_6, NH_3 (gas and solid)	Witteborn et al. 1981
1800–2200	1.2	FTS/IRTF 3 m	NH_3 (gas and solid)	Fink et al. 1983

[a] *Abbreviations*: CVF—Circular Variable Filter; ESO—European Southern Observatory; FPS—Fabry Perot Spectrometer; FTS—Fourier Transform Spectrometer; IRTF—Infrared Telescope Facility; KAO—Kuiper Airborne Observatory; KPNO—Kitt Peak National Observatory; OHP—Observatoire d'Haute Provence; SAI—Sternberg Astronomical Institute; SAO—Smithsonian Astrophysical Observatory; UAO—University of Arizona Observatories.

such as the Kuiper Airborne Observatory, from the Voyager spacecraft, and from infrared telescopes at high altitude groundbased sites (Mt. Lemmon in Arizona and Mauna Kea in Hawaii). In general the Voyager infrared spectra provided an unprecedented combination of continuous spectral coverage with moderate spectral resolution and high spatial resolution. Earth-based observations, on the other hand, often have higher spectral resolution and more intensive coverage of specific spectral regions, but with only modest degrees of spatial resolution.

In Table I we list the molecules that were analyzed in each infrared spectrum of Saturn. The early work in the near infrared concentrated on the abundant constituents CH_4 and NH_3, while more recent studies have involved a much broader range of trace constituents, either as detections or as upper limits. Note the dominant use of Fourier Transform Spectrometers (FTS) for observations of Saturn, beginning with Connes et al. (1969). Their first high-resolution spectrum of Saturn received little attention because of the even more striking results for Venus and Mars, but adaptations of the technique were subsequently responsible for exploring the entire 1 to 100 μm region of Saturn's infrared spectrum. In the following section we present the overall characteristics of Saturn's infrared spectrum, followed by a review of the compositional information derived from Earth-based observations.

1. Overview of Saturn's Infrared Spectrum.   At very low spectral resolution Saturn's infrared spectrum resembles Jupiter's because of the gross spectral similarities of mixtures of H_2 and CH_4, the spectroscopically most active constituents in both atmospheres (see Ridgway et al. [1976] for a review of Jupiter's infrared spectrum). These gases produce a series of discrete planetary transmission windows, much as H_2O and CO_2 define regions of transparency in our atmosphere at infrared wavelengths. Figure 2 illustrates both of these characteristics. The composite spectrum of Saturn was assembled from six independent, overlapping data sets, each converted to a common spectral resolution ($\simeq 25$ cm^{-1}) and photometrically scaled using lunar and solar-type stellar calibration data. These data refer to the center of Saturn's disk with no contribution from its rings. Since the individual data sets were obtained with various fields of view, these differences were removed to first order by assuming a uniform disk brightness.

Figure 2 also includes a transmission spectrum of the Earth's atmosphere, itself a series of windows. Note the awkward placement of some of the planetary windows, like the one at 2.7 μm, with respect to regions of transmission in our atmosphere. The high-resolution data in Fig. 3 illustrate more dramatically what this means for compositional analyses of planetary atmospheres. This comparison reinforces the previous comment that some infrared observations of the outer planets can only be conducted at special high-altitude sites.

Saturn's spectrum in Fig. 2 is produced by two very different mechanisms of spectral line formation, reflected solar flux and planetary thermal emission, each offering important opportunities for studying the composition and structure of Saturn's atmosphere at various levels. To illustrate this point, curves denoting unit albedo and blackbody emission at various relevant temperatures are shown in this figure. Between 1 and 3.3 μm Saturn's five prominent windows involve reflected sunlight. The shapes of the three high-albedo windows between 1 and 1.6 μm were derived from the groundbased observations of Fink and Larson (1979), while those between 2 and 3.3 μm come

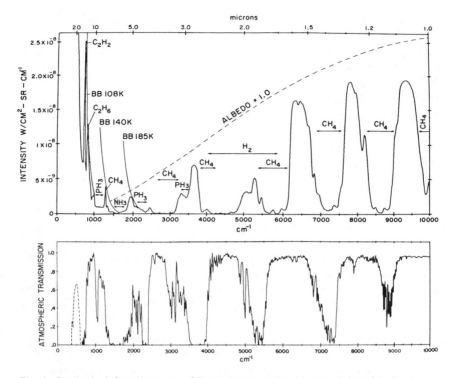

Fig. 2. Composite infrared spectrum of Saturn (upper graph) and transmission of the Earth's atmosphere (lower graph). See text for references.

from the airborne data of Larson et al. (1980). The relative intensities in each data set were scaled to geometric albedos reported by Clark and McCord (1979) for the visible and near-infrared spectral regions. These latter three windows lie between strong H_2 and CH_4 absorptions, with solar flux apparently penetrating to a cloud deck at the $\sim P \simeq 0.5$ atm, $T \simeq 110$ K level. The compositional information derived from the near-infrared windows ($\lambda \lesssim 2.5$ μm) concerns mostly the H_2 and CH_4 abundances. In spite of resolving powers up to $\sim 5 \times 10^4$, no new trace constituents have been detected in this spectral region. This somewhat disappointing return on major efforts to acquire such data is due to several factors. The spectrum of planetary CH_4 itself is complex and cannot be properly simulated in laboratory experiments, so there is always uncertainty over assigning observed features on Saturn to CH_4 or to new constituents. Also, most of the candidate trace species do not have strong vibration-rotation bands at wavelengths below 2.5 μm. On the positive side, the planetary window centered around 3 μm has been more productive by providing detections of PH_3 and C_2H_6 and by setting limits to the role of gaseous and solid NH_3 down to the level of the radiative-convective boundary.

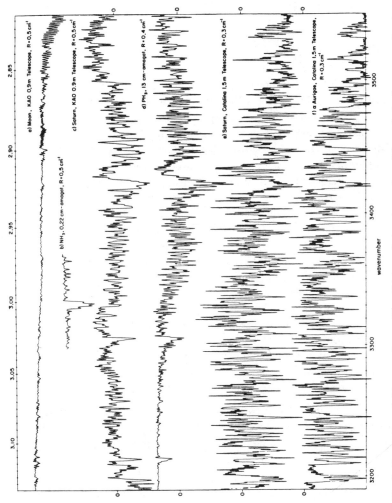

Fig. 3. Comparison of airborne and groundbased high-resolution observations of Saturn in the 3-μm spectral region. No compositional analyses can be made using the groundbased data because of interference from the terrestrial H_2O spectrum. At aircraft altitude, however, several constituents of Saturn's atmosphere stand out clearly, including PH_3 (2.85–3.05 μm) and C_2H_6 (3.05–3.10 μm). Upper limits determined in this spectral region include gaseous and solid NH_3 and HCN. The potential of this spectral region for compositional information cannot be exploited on Jupiter because of its stronger NH_3 bands (figure reproduced from Larson et al. 1980).

Beyond $\sim 6\,\mu$m Saturn's spectrum involves planetary thermal radiation. Most spectral features beyond 6 μm appear in emission rather than absorption because of the upper atmospheric temperature inversion noted in Fig. 1. The data in Fig. 2 between 6 and 8 μm are from airborne observations by Witteborn et al. (1981), and the section from 8 to 50 μm is from Voyager (Hanel et al. 1981a). The prominent spectral features are assigned to H_2, CH_4, PH_3, and NH_3. In addition, gaseous hydrocarbons produced by photodissociation of CH_4 at stratospheric levels show prominent emission features even at low spectral resolution. The thermal structure (see Sec. I) and H/He ratio (see Sec. II.F) in Saturn's atmosphere were both first established from spectral analyses in this wavelength region.

Between ~ 4 and $7\,\mu$m there is an important transition region that by analogy with Jupiter might present a similar opportunity on Saturn to probe spectroscopically to greater atmospheric depths than at any other wavelength. An essential condition is that there be breaks in Saturn's cloud cover to permit transfer of thermal radiation from lower depths. Observationally, the 5-μm region of the thermal infrared is optimal for searching for these "hot spots" because of the coincidentally high transmission of H_2, CH_4, and NH_3 between ~ 4.7 and 5.4 μm. The brightness temperature of Saturn at 5 μm is, in fact, quite high (190 K according to Rieke [1975]; up to 210 K estimated by Hanel et al. [1981a]), but unlike Jupiter the flux is apparently not associated with visual markings on the planetary disk. The shape of Saturn's 5-μm window in Fig. 2 was derived from observations reported in Fink and Larson (1978) and in Larson et al. (1980) (see Fig. 4). Since these data extend into the near infrared as well, the albedos of Clark and McCord (1979) were used to provide absolute calibration of the 5-μm flux. Using CH_3D as a spectroscopic probe of Saturn's 5-μm window, Fink and Larson (1978) found that the mean level of spectral line formation was at a temperature $T \simeq 175$ K and a pressure $P \simeq 1.8$ atm, thus confirming expectations that Saturn's 5-μm flux is thermal emission from atmospheric levels below the cloud tops that cannot be reached at other wavelengths by remote spectroscopic techniques.

2. Recent Compositional Analyses.   Current efforts to interpret Saturn's spectrum are frequently limited by two problems endemic to studies of the outer planets: lack of appropriate laboratory comparison data; and lack of an atmospheric model that accounts for the observed presence and abundance of constituents in different wavelength regions. Some insights gained from addressing similar problems in interpreting Jupiter's spectrum can be applied to Saturn, but, increasingly, the differences rather than the similarities between these two planets seem the most interesting to understand. Thus, in order to preserve as open a perspective as possible, the following discussion of results is presented more in the spirit of a progress report than a critical review.

Hydrogen.   Molecular hydrogen, the most abundant constituent in Saturn's atmosphere, is spectroscopically inactive under normal laboratory con-

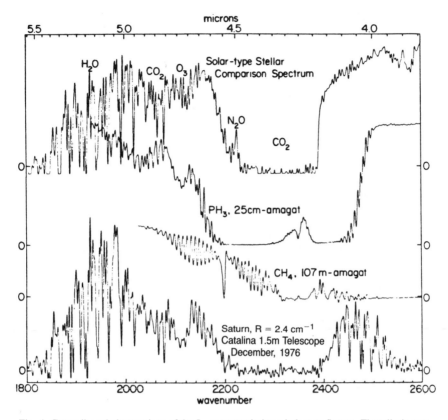

Fig. 4. Groundbased observations of the 5-μm transmission window on Saturn. The telluric spectrum is very complex in this spectral region, but two planetary constituents have been detected in these data. Analysis of the CH_3D band at 4.5 μm gave a rotational temperature of $\sim$ 175 K, indicating that Saturn's 5-μm flux is from deep atmospheric levels. Other prominent features in Saturn's 5-μm spectrum were assigned to PH_3, whose complex spectrum now impedes the search for other trace constituents in this wavelength region (figure reproduced from Larson et al. 1980).

ditions of abundance and pressure. However, a variety of spectral features appear as a consequence of the long absorbing paths in the atmospheres of the outer planets including pressure-induced rotational spectra in the 10–100 μm region, pressure-induced vibration-rotation spectra in the 1–2.5 μm region, and quadrupole lines in the 0.6–1.1 μm region. These features are in principle highly diagnostic of many of Saturn's atmospheric properties such as temperature, density, collision partners, and thermal structure. It is especially important to relate the observed strengths of planetary H_2 bands to column abundances above various atmospheric levels in order to express elemental abundances as mixing ratios relative to H for comparison with solar and other values. The measured H_2 abundances for Saturn are listed in Table II. We

TABLE II

H₂, CH₄, and C/H Measurements on Saturn from Infrared Spectra

H₂ Feature	$\lambda(\mu m)$	Abundance (km-amagat)	CH₄ Feature	$\lambda(\mu m)$	Abundance (m-amagat)	C/H ($\times 10^{-3}$)	Type of Model	Reference
			$3\nu_3$ band	1.1	51–86		Reflecting	Bergstralh 1973
			$3\nu_3$ band	1.1	42±11		Reflecting	de Bergh et al. 1973
CH₄ line widths	1.1	85–150	$3\nu_3$ band	1.1	55–68	0.28	Reflecting	Trafton 1973
2-0 induced dipole	1.1	63±23					Reflecting	de Bergh et al. 1974
4-0 quadrupole lines	0.63	17±3	$3\nu_3$ band	1.1	60–190	3.5	Scattering	Trafton and Macy 1975
2-0 induced dipole	1.1	63^{+13}_{-8}	$3\nu_3$ band	1.1	59^{+15}_{-7}	$0.47^{+0.20}_{-0.13}$	Reflecting	Lecacheux et al. 1976
3-0,4-0 quadrupole lines	0.63,0.8	18–27	$3\nu_3$ band	1.1	100	1.05–1.4	Scattering	Macy 1976
1-0 induced dipole	2.1	25^{+10}_{-9}					Reflecting	Martin et al. 1976

TABLE II (continued)
H_2, CH_4, and C/H Measurements on Saturn from Infrared Spectra

H_2 Feature	$\lambda(\mu m)$	Abundance (km-amagat)	CH_4 Feature	$\lambda(\mu m)$	Abundance (m-amagat)	C/H $(\times 10^{-3})$	Type of Model	Reference
2-0 quadrupole line	1.24	25^{+9}_{-6}					Reflecting	de Bergh et al. 1977
3-0,4-0 quadrupole lines	0.53,0.8	70 ± 20					Reflecting	Trafton 1977
			weak infrared bands	1.3,1.6	160	1.15^{a}	Reflecting	Fink and Larson 1979
2-0 induced dipole	1.1	14 ± 5	$3v_3$ band plus interlopers	1.1	50 ± 20	2 ± 1	Scattering	Buriez and de Bergh 1981
Rotational lines	17,28		H/He $\simeq 12.7^{+7.6}_{-3.6}$				Thermal	Conrath et al. 1983
			v_4 band	7.7	$1.0^{+0.75}_{-0.3}\,^{b}$		Thermal	Courtin 1982a

[a] H_2 abundance from Trafton (1977).
[b] Above the haze layer only.

include in this table the Voyager observations of the H_2 pressure-induced rotational lines S(0) at 17 μm and S(1) at 28 μm which will be discussed later (see Sec. II.D and Fig. 10).

In general, the observed abundances summarized in the tables in this chapter should not be used without reference to the original papers for details concerning the spectral analyses. These measurements are model-dependent and different values may be reported for the same observed spectrum when different models are used. We indicate in Table II the general type of model (reflecting, scattering, thermal) used for determining each entry, but the authors employ so many variations that there is no succinct way to summarize the results. Most of the entries in this table refer to a simple reflecting-layer model even though it has long been considered inadequate. Two trends characterize recent work; more realistic scattering models are being developed with the help of improved observational and laboratory spectra, and more sophisticated computer techniques are being used (see Buriez and de Bergh 1981). Alternatively, Encrenaz and Combes (1982) have developed new criteria for selecting spectral data for abundance analyses to render ratio measurements less sensitive to *a priori* knowledge of aerosol distribution and vertical structure in a planetary atmosphere.

Methane. This molecule is responsible for the positions and shapes of most of Saturn's atmospheric transmission windows in Fig. 2. It is important as a source of opacity in establishing Saturn's thermal balance, and as an important reactant in atmospheric chemical processes (discussed in Sec. III). The abundance measurements from infrared spectra are also summarized in Table II. All entries cluster around 1 μm, the only wavelength region where Saturn's CH_4 band centers are not heavily saturated. As is the case for CH_4 bands on Jupiter (see discussion in Ridgway et al. 1976) most attention has been given to Saturn's $3\nu_3$ CH_4 band. This band is attractive for quantitative analyses because of its relatively simple structure, known quantum number assignments, and low degree of saturation. The spectra in Fig. 5 illustrate the high quality of astronomical observations of this band. This section of Saturn's spectrum in the 1.1-μm region displays strong, narrow absorptions in the $3\nu_3$ band of CH_4 and broad, continuum absorption by the 2-0 induced dipole spectrum of H_2. These overlapping H_2 and CH_4 bands sample different levels of Saturn's atmosphere because their absorption coefficients depend differently upon pressure. Thus, the abundances of H_2 and CH_4 on Saturn, and especially its C/H ratio, cannot be measured directly from observations without supporting laboratory calibrations and a realistic atmospheric structure model. As noted in the discussion below, these requirements generate most of the continuing research and debate over measurements of abundances in planetary atmospheres by remote methods.

As an example of one complication, it has long been recognized in spectra of the outer planets that the weaker CH_4 bands at visible wavelengths (see

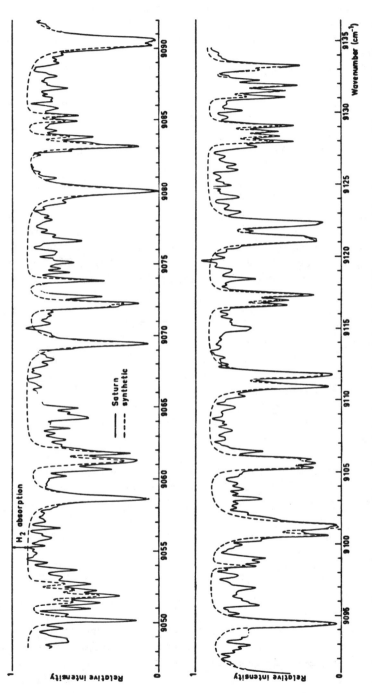

Fig. 5. A portion of Saturn's spectrum at 0.2 cm^{-1} resolution in the region of the $3\nu_3$ band. The strong, narrow features are CH$_4$ to which has been fit a synthetic spectrum using a scattering model. The 2-0 pressure-induced dipole spectrum of H$_2$ is very broad, appearing in this figure only as a gap between the model fit and the observed continuum (figure from Buriez and de Bergh 1981).

Sec. II.B) yield higher abundances than the $3\nu_3$ band. Thus the 150 m-amagat CH_4 determination at 0.6 μm by Lutz et al. (1976) exceeds by a factor of 2.5 the average value of 60 m-amagat from the $3\nu_3$ band at 1.1 μm. This discrepancy can be attributed to still unresolved interpretive problems associated with the $3\nu_3$ band, including saturation, line profile dependence upon pressure, poorly defined continuum, and overlying planetary and telluric lines. Fink and Larson (1979) investigated this possibility by searching for other, weaker CH_4 bands beyond 1 μm that would share some of the presumably more desirable characteristics of the weak visible bands. They found two weak CH_4 bands at 1.27 and 1.56 μm exhibiting pressure-independent curves of growth. Their analysis of these bands on Saturn gave a CH_4 abundance of 160 m-amagat, remarkably similar to that ($\simeq$ 150 m-amagat) from the visible bands.

In addition to problems in choosing appropriate bands for spectral analyses, the degree to which scattering processes affect the measurements has had a long history of debate. Opinions on the need to include scattering mechanisms in spectral analyses range from: unnecessary (Trafton and Macy 1975); important at the base of the line-forming region (Macy 1976); essential up to the inversion level (Buriez and de Bergh 1981); unnecessary when the pairs of lines used in abundance ratio calculations are chosen judiciously (Encrenaz and Combes 1982). It is clear that the CH_4 abundance on Saturn still represents an active area of research, and available measurements should not be used blindly. This is particularly true for calculating Saturn's C/H value. It is not correct to construct this ratio from arbitrary combinations of the entries in Table II. Rather, the ratio should be expressed in terms of measurements made at the same wavelength and for similar mechanisms of line formation. The published values of Saturn's C/H ratio are included in this table. It may be significant that all entries are approximately solar or higher, but it should also be noted that none of these ratios may satisfy the criteria listed above. Given all the problems surrounding the individual H_2 and CH_4 measurements, a "best bet" C/H ratio on Saturn is not actually in Table II, but would use the H_2 quadrupole lines of Trafton (1977) and the weak visible CH_4 lines of Lutz et al. (1976). This ratio is 1.07×10^{-3}. Using similar, but more stringent criteria, Encrenaz and Combes (1982) propose a C/H value of 1.05×10^{-3}. Thus, it appears that the C/H ratio may be converging to the same value for both Jupiter and Saturn, and is enhanced by at least a factor of 2.5 over the solar value (Cameron 1982), consistent with predicted enhancements (Podolak and Danielson 1977).

Ammonia. In principle, the ammonia molecule is another important spectroscopic probe of some of the bulk properties of Saturn's atmosphere. Its abundance should be controlled at least in part by its saturated vapor pressure in Saturn's troposphere and stratosphere, and it should therefore serve as an indicator of the depth of spectral line formation at different wavelengths. Am-

TABLE III
Ammonia Abundances on Saturn
from Infrared Spectra

Wavelength Region (μm)	Abundance (cm-amagat)	Reference
3.0	< 0.5	Martin 1975
1.6	< 15	Owen et al. 1977
1.6	< 10	Fink and Larson 1979
3.0	< 0.1	Larson et al. 1980
10.3	< 0.05	Tokunaga et al. 1980c
50	—	Hanel et al. 1981a; Courtin 1982a
6.3	4	Witteborn et al. 1981
5.3	200	Fink et al. 1983

monia is also a probable candidate for hazes and clouds on Saturn, and its opacity makes it important to Saturn's thermal balance. The infrared and visible NH_3 abundances for Saturn summarized in Table III (and Table VII in Sec. II.B) confirm at least one of these expectations: they range over a factor of 10^4 between the largest detection and the most stringent upper limit, thus defining a very model-dependent interpretive problem.

These disparate results can be understood qualitatively in terms of the degree to which NH_3 condenses or is photochemically destroyed at different levels of Saturn's atmosphere. For example, observations at 10 μm probe Saturn's cold stratosphere where the NH_3 abundance may be greatly undersaturated because of ultraviolet photolysis. This expectation is consistent with the fact that the most stringent upper limit in Table III to NH_3 on Saturn is from 10 μm data. In the upper troposphere photodissociation may become less effective and the NH_3 abundance will be more closely controlled by its saturated vapor pressure. Spectral lines formed in this region will yield widely different abundances if the line-forming regions have different effective temperatures for observations at different wavelengths. Thus, the upper limits at 1.6 and 3.0 μm and the low abundance detected at 6.3 μm (see Table III) may characterize different levels within this region of Saturn's atmosphere. Finally, at deep tropospheric levels Saturn's NH_3 abundance should approach its bulk planetary value. The abundances measured at 0.645 μm (see Table VII in Sec. II.B) and 5.3 μm (Table III) and that inferred from inversion of microwave data (Marten et al. 1980) are high enough to permit comparison of Saturn's N/H ratio to the solar value. In a more quantitative sense, however, there is not yet available for Saturn an atmospheric model that uses thermochemical, photochemical, and scattering processes to predict the observed abundances of NH_3 or, for that matter, any other trace constituent.

TABLE IV
PH_3 Abundances on Saturn from Infrared Spectra

Wavelength Region (μm)	Abundance (cm-amagat)	Mixing Ratio ($P/H_2 \times 10^{-6}$)	Reference
10	2	1.0	Gillet and Forrest 1974
10	0.5	0.2	Bregman et al. 1975
3	16 ± 7	0.9	Larson et al. 1980
5	34 ± 15	1.7	Larson et al. 1980
10	3.2–7.1	0.8–1.6	Tokunaga et al. 1980c,1981b
10	—	1	Hanel et al. 1981a
10	—	2	Courtin 1982a

Phosphine. Less than a decade ago, spectroscopically detectable amounts of PH_3 in the atmospheres of the outer planets were considered impossible according to the thermochemical equilibrium models discussed in Sec. III. Now, however, independent observations support surprisingly high PH_3 abundances on both Jupiter and Saturn. An overview of these results is provided by Larson (1980). The PH_3 abundances for Saturn are summarized in Table IV, and examples of PH_3 spectra are included in Figs. 3 and 4. All PH_3 abundances are consistent with a mixing ratio equal to or larger than the solar value. Although PH_3 is common to the atmospheres of Jupiter and Saturn, its presence may require rather different mechanisms. On Jupiter the high PH_3 abundance was attributed to rapid vertical convection that cycled PH_3 upward from deep atmospheric levels on a time scale short enough to minimize its chemical conversion to P_4O_6. This mechanism was supported by other detections (e.g., GeH_4 and perhaps CO) that required a similar explanation. On Saturn, however, neither GeH_4 nor CO have been detected. As discussed in Sec. III, the absence of these latter compounds may signal important differences between the deep atmospheres of Saturn and Jupiter.

Gaseous Hydrocarbons. The simple gaseous hydrocarbons C_2H_2 and C_2H_6 have now been detected on Jupiter, Saturn, Titan, and Neptune. They are seen in emission at wavelengths $\lambda \geq 6$ μm, strongly suggesting a stratospheric origin above a thermal inversion. Their production mechanism appears to involve photodissociation of CH_4 by solar ultraviolet light. The various studies of gaseous hydrocarbons on Saturn are listed in Table V. The mixing ratios are poorly determined in part because of inadequate knowledge of the thermal structure of the hydrocarbon production zone, but as discussed in Sec. II.D.3 there is reasonable agreement between these observations and photochemical model predictions. These latter models also predict a low abundance of ethylene (C_2H_4); however, this species has not yet been detected

TABLE V
Gaseous Hydrocarbons on Saturn from Infrared Spectra

Molecule	Wavelength Region (μm)	Mixing Ratio	Reference
C_2H_2	13.7	—	Tokunaga et al. 1980a
C_2H_2	13.7	2×10^{-8}	Hanel et al. 1981a
C_2H_2	13.7	$(1.2 \pm 0.2) \times 10^{-7}$ northern hemis. $(0.5 \pm 0.1) \times 10^{-7}$ southern hemis.	Courtin 1982a
C_2H_6	12.2	—	Gillett and Forrest 1974
C_2H_6	12.2	—	Tokunaga et al. 1975
C_2H_6	3.07	5×10^{-7}	Bjoraker et al. 1981
C_2H_6	12.2	5×10^{-6}	Hanel et al. 1981a
C_2H_6	12.2	$(6.0 \pm 1.1) \times 10^{-6}$ northern hemis. $(3.1 \pm 0.6) \times 10^{-6}$ southern hemis.	Courtin 1982a
C_2H_6	6.8	$10^{-4} - 10^{-5}$	Witteborn et al. 1981

on Saturn. The features tentatively assigned to C_2H_4 on Saturn by Encrenaz et al. (1975) were most likely due to PH_3 instead.

Only one study in Table V involved detection of a methane-derived hydrocarbon in absorption (Bjoraker et al. 1981). This work permitted a much less model-dependent study of the vertical distribution of C_2H_6 from the stratospheric production region down to the cloud tops. Photochemical models with constant mixing ratios or constant number densities were less consistent with the observations than was a model incorporating height-dependent mixing ratios and other parameters including the eddy diffusion coefficient.

Isotopic Species. Two isotopes have been detected from Earth at infrared wavelengths in Saturn's atmosphere: $^{13}CH_4$ and CH_3D. These represent useful opportunities for comparing abundance ratios of planetary isotopes with cosmological theories. The $^{12}C/^{13}C$ ratio on Saturn is 89^{+25}_{-18} (Combes et al. 1975,1977). This ratio is very close to values derived from meteorites, lunar samples, Mars, and terrestrial samples. However, it is quite different from the $^{12}C/^{13}C$ ratio of 160^{+40}_{-55} recently derived for Jupiter from Voyager spectra of the ν_4 band of CH_4 (Courtin et al. 1983 a,c).

The analysis of CH_3D by Fink and Larson (1978) yielded two results: a D/H ratio on Saturn of 2×10^{-5}; and a rotational temperature $T \simeq 175 \pm 30$ K for Saturn's 5-μm window. This D/H ratio is similar to the Jovian value, but substantially lower than the terrestrial one. This result is consistent with Voy-

TABLE VI
Upper Limits to Trace Constituents on Saturn from Infrared Spectra

Molecule	Wavelength Region (μm)	Upper Limit (cm-amagat)	Reference
H_2O	5	15 (prec. μm H_2O)	Larson et al. 1980
HCN	3	0.1	Larson et al. 1980
	13.5	2.5×10^{-2}	Tokunaga et al. 1981a
H_2S	1.6	15	Martin 1975
	1.6	10	Fink and Larson 1979
	1.6	1	Owen et al. 1977
COS	1.6	10	Fink and Larson 1979
SiH_4	5	2.5×10^{-2}	Larson et al. 1980
GeH_4	5	2×10^{-3}	Larson et al. 1980
CH_3OH	1.6	10	Fink and Larson 1979
CH_3NH_2	1.6	2	Fink and Larson 1979

ager data discussed in Sec. II.D and supports the idea that the D/H value for these giant planets is closer to the primordial ratio with the higher terrestrial and meteoritic values due to enrichment processes. The observed CH_3D rotational temperature $T \simeq 175$ K is consistent with the high brightness temperatures of Saturn at 5 μm ($T \simeq 190$ K: Rieke 1975; T ≤ 210 K: Hanel et al. 1981a).

Upper Limits. A number of upper limits to trace constituents in Saturn's atmosphere have accumulated in the course of analyzing Earth-based infrared spectra. These are collected in Table VI. Some of these species have been detected on Jupiter (e.g., H_2O, HCN, and GeH_4), and their absence on Saturn cannot always be explained by temperature differences between the two atmospheres. The limit to GeH_4, for example, suggests that vertical transport may not be as vigorous in Saturn's deep atmosphere as on Jupiter. The limits to H_2S place limits on the importance of colored sulfur compounds as cloud chromophores. It appears that on both Jupiter and Saturn colored phosphorus compounds may be more abundant at tropopause levels than colored sulfur compounds.

B. Earth-based Visible Spectroscopy

In the visible part of Saturn's spectrum, which for convenience here is defined as the region from 0.3 to 0.8 μm, three species have been observed: H_2, NH_3 and CH_4. A selection of abundance estimates for these species is provided in Table VII. Of the tabulated species H_2 is overwhelmingly the most abundant, but in a sense it is also the least interesting. From observations of

TABLE VII
Species Abundances on Saturn from Visible Spectra

Species	Wavelength (μm)	Abundance (km-amagat)	Reference
H_2	0.63	17 ± 3	Trafton and Macy 1975
	0.63, 0.8	$18 - 27$	Macy 1976
	0.63, 0.8	70 ± 20	Trafton 1977
	0.63, 0.8	77 ± 20	Encrenaz and Owen 1973
	0.63, 0.8	57	Caldwell 1977a
CH_4	$0.4 - 0.6$	0.15	Lutz et al. 1976
NH_3	0.645	0.002 ± 0.0005	Encrenaz et al. 1974; Woodman et al. 1977
	0.645	$0.0004 - 0.005$	Smith et al. 1980

its 4-0 and 3-0 quadrupole lines at 0.63 and 0.8 μm, it is possible to derive H_2 abundances, but such values are actually determinations of the depths of the reflecting visible clouds on Saturn. The abundance results are highly dependent on the scattering properties of the cloud particles in visible light, and in the context of compositional analysis are mainly useful in determining the ratio of other species to hydrogen.

One example of an analysis of an H_2 visible quadrupole line is that of Encrenaz and Owen (1973), who derived an abundance of 77 ± 20 km-amagat, based on a reflecting layer model of their observations. Caldwell (1977a) found that the same data were consistent with a model having 57 km-amagat of H_2 above a thick cloud, but with an isotropically scattering haze permeating the bottom 19 km-amagat of H_2 above the cloud. Recent laboratory studies of the quadrupole line strength (Brault and Smith 1980) suggest that these quoted abundances should be reduced by $\sim 30\%$.

Observations of NH_3 in visible spectra are also somewhat difficult to interpret. In Sec. II.G evidence is cited to the effect that there is significantly less gaseous NH_3 in the visible atmosphere of Saturn than in that of Jupiter, with the missing gaseous NH_3 undoubtedly being in a condensed phase. The observability of gaseous NH_3 on Saturn is further complicated by both spatial and temporal variations (see a review by Trafton 1981). For example, Woodman et al. (1977) clearly detected the visible 645 nm band of gaseous NH_3, whereas Owen et al. (1977) did not detect the infrared 1560 nm band, although the longer wavelength band is intrinsically stronger. This must be due to some combination of temporal variability and the wavelength dependence of the cross sections of particulate scatterers and continuum absorbers. The effects of the cloud particle scattering are especially large because almost all of the small amount of remotely detectable gas phase NH_3 exists in the same altitude range as the clouds.

In any case, it is clear that condensation in the very cold Saturnian visible atmosphere dominates the observability of NH_3 in the visual and near infrared spectrum. To make progress on the cosmogonically significant question of the departure, if any, of the Saturnian N/H ratio from the corresponding solar value, it is necessary to make observations of the lower atmosphere where NH_3 is not saturated. Because of the absence of prospects for a Saturnian entry probe within the foreseeable future, such observations are now possible only with microwave radiometry.

Trafton (1981) has estimated that the NH_3/H_2 ratio for Saturn's deep atmosphere is $\gtrsim 3 \times 10^{-4}$. Trafton revised the conclusion of Ohring and Lacser (1976), which came from an inversion of the microwave spectrum but which depended on obsolete compositional information, to make his estimate. Klein et al. (1978) determined a mixing ratio of $NH_3/H_2 \sim 5 \times 10^{-4}$ also from 1 to 100 cm wavelength observations. These authors were the first to consider the emission of the rings in analyzing such data for atmospheric abundance purposes, and so their results should be superior. The apparent conclusion from the work of Klein et al. (1978) is that N/H is enhanced in the atmosphere of Saturn by a factor ~ 3 with respect to the solar value (see Table VIII). However, if there exist microwave absorbers in addition to gaseous NH_3 on Saturn, then this inferred NH_3 enhancement could be too high. Water droplets are a possible additional absorber. Klein et al. (1978) found that a cloud of H_2O droplets at the level where they condense (~ 273 K) is, however, quantitatively incapable of producing agreement between observations at 21 cm and a model containing approximately solar NH_3; the model is too bright. If the droplets were allowed to be mixed upward from the 273 K level by convection in an attempt to increase microwave opacity at higher altitudes, they would freeze (see Fig. 14 in Sec. III.A) and become much less efficient microwave absorbers (see Fig. 3.7 of Klein et al. 1978). Although inclusion of *ad hoc* opacities can certainly improve agreement between models with solar NH_3 and observation, no completely convincing demonstration of this alternative has yet been made.

Quantitative analysis of visible and infrared CH_4 observations must, like NH_3, also take account of scattering by cloud particles (Lutz et al. 1982). Although some weak CH_4 bands near 725 nm show anomalous variations with respect to stronger CH_4 bands in Saturn's infrared spectrum, according to Lutz et al. (1982) there are no outright contradictions analogous to the 645– 1560 nm case for NH_3; i.e., there are no cases of observable weak CH_4 bands and undetectable strong ones. This is presumably due to the fact that, unlike NH_3, CH_4 does not condense on Saturn. There are therefore significant amounts of the gas above the haze and cloud layers so that while the scattering properties of the cloud particles influence the CH_4 spectral features, they do not overwhelm them.

TABLE VIII
Ratios Relative to H_2 of the 30 Most Abundant Elements in a Solar-Composition Planetary Atmosphere[a]

Element	Cameron (1982)	Cameron (1973)[b]	Element	Cameron (1982)	Cameron (1973)[b]
(H_2)	(1)	(1)	Cr	9.55×10^{-7}	7.99×10^{-7}
He	1.35×10^{-1}	1.39×10^{-1}	Mn	6.99×10^{-7}	5.85×10^{-7}
O	1.38×10^{-3}	1.35×10^{-3}	P	4.89×10^{-7}	6.04×10^{-7}
C	8.35×10^{-4}	7.42×10^{-4}	Cl	3.56×10^{-7}	3.58×10^{-7}
Ne	1.95×10^{-4}	2.16×10^{-4}	K	2.63×10^{-7}	2.64×10^{-7}
N	1.74×10^{-4}	2.35×10^{-4}	Ti	1.80×10^{-7}	1.75×10^{-7}
Mg	7.97×10^{-5}	6.67×10^{-5}	Co	1.65×10^{-7}	1.39×10^{-7}
Si	7.52×10^{-5}	6.29×10^{-5}	Zn	9.47×10^{-8}	7.82×10^{-8}
Fe	6.77×10^{-5}	5.22×10^{-5}	F	5.86×10^{-8}	1.54×10^{-7}
S	3.76×10^{-5}	3.14×10^{-5}	Cu	4.06×10^{-8}	3.40×10^{-8}
Ar	7.97×10^{-6}	7.37×10^{-6}	V	1.91×10^{-8}	1.65×10^{-8}
Al	6.39×10^{-6}	5.35×10^{-6}	Ge	8.80×10^{-9}	7.23×10^{-9}
Ca	4.70×10^{-6}	4.54×10^{-6}	Se	5.04×10^{-9}	4.23×10^{-9}
Na	4.51×10^{-6}	3.77×10^{-6}	Li	4.51×10^{-9}	3.11×10^{-9}
Ni	3.59×10^{-6}	3.02×10^{-6}	Kr	3.11×10^{-9}	2.94×10^{-9}

[a] Values used in this chapter are based on Cameron (1982). Values used by Barshay and Lewis (1978) and Fegley and Lewis (1979) are based on Cameron (1973).
[b] Boron is the 26th most abundant element in this 1973 compilation. Its abundance relative to H_2 has now been revised downwards by a factor of 33 and it does not appear in the 1982 compilation.

If we combine the results of the study of the weak visible CH_4 bands by Lutz et al. (1976) with the visible H_2 3-0 and 4-0 quadrupole line analysis by Trafton (1977), we obtain an estimate for the CH_4/H_2 ratio of $\left(2.1\,^{+0.9}_{-0.4}\right) \times 10^{-3}$ (error bars do not include uncertainty in CH_4 abundance). This result may be compared with an analogous combination of the H_2 2-0 induced dipole and CH_4 $3\nu_3$ absorptions in the near infrared by Buriez and de Bergh (1981). These workers had access to laboratory studies of a wide range of the relevant spectral lines thereby enabling them to draw more accurate conclusions than possible in the past. Their result is $CH_4/H_2 = (4 \pm 2) \times 10^{-3}$, where much of the quoted error is due to uncertainties in the effects of scattering. Nevertheless, the lower error bar still gives a C/H ratio which exceeds the corresponding solar C/H ratio of 4.2×10^{-4} by a factor of > 2. This and other infrared studies of CH_4 on Saturn, with usually similar results, are listed in Table II. The determination of the CH_4 abundance from Voyager observations is discussed in Sec. II.D.

To summarize the situation for NH_3 and CH_4 on Saturn, available visible, infrared, and microwave spectroscopic evidence suggests that both of these substances are enhanced by factors > 2 with respect to solar elemental ratios of C and N to H. The evidence for this enhancement is stronger in the case of CH_4 than that of NH_3. Chemical and cosmogonical implications of enhanced C/H and N/H ratios are discussed in Sec. III.

The relative abundance of deuterium to hydrogen is also of great interest to studies of the origin and evolution of Saturn and to cosmology (see Sec. II.D). Two published measurements of the D/H ratio using visible quadrupole lines (Trauger et al. 1977; Smith and Macy 1977) give a mean value of 6×10^{-5}. More recently, W. H. Smith (personal communication) has suggested that a value of $\sim 9 \times 10^{-5}$ is preferable, mostly because of a revision in the intrinsic strength of the 4-0 S(1) H_2 quadrupole line on which the ratio is based. These D/H values may be compared to the D/H ratio of $\sim 2 \times 10^{-5}$ derived from infrared observations of CH_3D and CH_4 discussed in the previous section. Both estimates are less than the terrestrial oceanic D/H ratio of 1.4×10^{-4}.

C. Earth-Orbital Ultraviolet Spectroscopy

The region from ~ 160 to 300 nm contains electronic transitions of many important molecules, and it is therefore potentially of great interest for compositional inferences through absorption spectroscopy. This part of the ultraviolet spectrum of Saturn has been observed in some detail by a series of Earth-orbiting satellites including OAO-2, TD1A, and IUE. This spectral region is the subject of this section.

The long wavelength boundary of the 160–300 nm region is determined by the onset of terrestrial atmospheric transmission, permitting groundbased observations at wavelengths > 300 nm. The short wavelength boundary is loosely set by the physics of Saturn. Below 160 nm, the reflected sunlight component of Saturn's spectrum is very weak, both because of the weakness of the solar flux there and because of the preponderant absorption by CH_4. The observed emission from Saturn below 160 nm is dominated instead by auroral emission processes involving H_2 and H excited by Saturnian magnetospheric particles. Not coincidentally, that region is well covered by the Voyager ultraviolet spectrometer (UVS) instrument.

The region between 160 and 300 nm is the one in which Rayleigh scattering from H_2 becomes the dominant continuum opacity source. At a wavelength $\lambda = 300$ nm, the Rayleigh scattering coefficient of H_2 is ~ 0.035 (km-amagat)$^{-1}$ while at $\lambda = 160$ nm it is ~ 0.6 (km-amagat)$^{-1}$. The scattering coefficient varies by a factor of seventeen due to its strong wavelength dependence (λ^{-4} plus higher order terms).

At longer wavelengths, there is another continuum opacity source that has not yet been identified directly. This source of opacity causes the reflectivity to decrease from visual wavelengths down to ~ 350 nm, at which point the decrease stops and the reflectivity increases again toward shorter wavelengths (Caldwell 1977*b*). The short wavelength increase is due to Rayleigh scattering from the layer of H_2 that lies above the absorber causing the longer wavelength decrease in reflectivity.

The reflectivity minimum near 300 nm corresponds to a geometric albedo of ~ 0.2. A conservatively scattering atmosphere would have a geometric albedo of ~ 0.7. The absorber that causes this low reflectivity is probably distributed in altitude (although it cannot persist to the top) and it probably scatters as well as absorbs incident sunlight. Nevertheless, an approximate limit on its location can be estimated by assuming that the source of long wavelength opacity forms a sharp, nonreflecting boundary in the atmosphere of Saturn. In this simple approximation, the observed geometric albedo of 0.2 at 300 nm can be attributed to a Rayleigh scattering optical depth of 0.5, corresponding to an H_2 column abundance of ~ 14 km-amagat. More realistic modeling of this absorber, including its scattering and its vertical distribution, would tend to lower this number somewhat, thereby raising the altitude on Saturn, for which ultraviolet photons provide remotely interpretable information, to above the 14 km-amagat (~ 150 mbar) level. It should therefore be kept in mind in the following discussion that quantitative results apply only to approximately the uppermost ~ 10 km-amagat layer of the atmosphere of Saturn, and that at wavelengths below 200 nm the results are relevant to even thinner layers as the Rayleigh scattering cross section increases. A positive aspect of ultraviolet remote sensing is that the results may be applied with some confidence to a well-defined region of the atmosphere, without confusion from lower altitudes. Furthermore, Rayleigh scattering is a tractable problem which does not introduce multiple free parameters into models, in contrast to other types of scattering. The negative aspect of remote sensing by ultraviolet spectroscopy is that there is no information about lower altitudes in the atmosphere.

Above 200 nm wavelength, most of the observed gases on Saturn (with the exception of NH_3) do not have well-defined absorption features. Rather, if they are active spectroscopically above 200 nm, they generally have very broad features, typically several tens of nanometers wide. For example, PH_3 has recently been determined (D. Judge, personal communication, 1981) to have a broad absorption maximum that peaks near 190 nm and decreases monotonically to near zero at 230 nm. Winkelstein et al. (1983) have recently used this feature to choose between two contradictory estimates of the vertical distribution of PH_3 in Saturn's atmosphere derived from infrared observations.

The first of these estimates is by Courtin (1982*a*), who reduced Voyager infrared observations near 10 μm (see Sec. II.D). He derived a mixing ratio $PH_3/H_2 = 2.0 \times 10^{-6}$, which persists from low altitudes up to a pressure level of 4 mbar (0.4 km-amagat H_2) with essentially no PH_3 above this level (see Fig. 6). However, when this amount of PH_3 is incorporated into an ultraviolet reflectivity model it fails to match the observations, in this case from IUE.

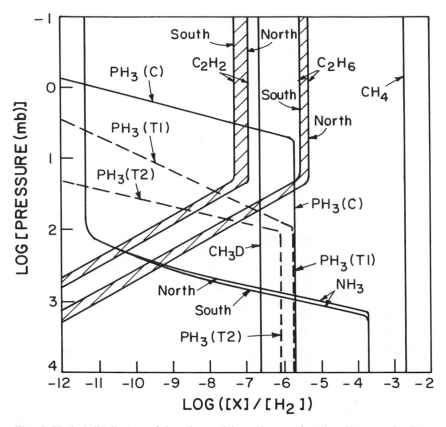

Fig. 6. Vertical distributions of the volume mixing ratios as a function of pressure for CH_4, CH_3D, C_2H_2, C_2H_6, PH_3, and NH_3. The C_2H_2 and C_2H_6 values are retrieved for both northern and southern hemispheres from a selection of 512 spectra in the north (see Fig. 9) and 244 spectra in the south. The PH_3, CH_4, and CH_3D mixing ratios are derived from the north hemisphere spectra only. NH_3 distributions are calculated following the saturation law using thermal profiles retrieved from northern and southern hemisphere spectra (adapted from Courtin 1982a). Also shown are two PH_3 profiles derived by Tokunaga et al. (1980c,1981b) denoted here as T1 and T2 to distinguish them from the Courtin (1982a) profile denoted as C.

The disagreement is considerable as shown in Fig. 7. It is not possible to put that much PH_3, with an absorption cross section of $\sim 10^{-18}$ cm^2 at 206 nm, that high in the atmosphere of Saturn, without causing too much absorption.

The second estimate by Tokunaga et al. (1980c,1981b) resulted in two suggested profiles also shown in Fig. 6. These profiles were derived using a thermal structure model together with groundbased observations of PH_3 at wavelengths near 10 μm that were very similar to the Voyager infrared data. However, these profiles, while similar in the troposphere, are very different in the stratosphere from the results of Courtin (1982a). The Tokunaga et al.

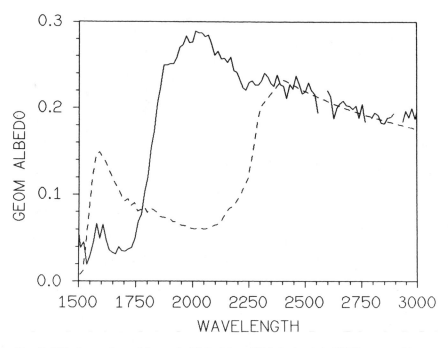

Fig. 7. IUE observations of Saturn (solid line) from Winkelstein et al. (1983) compared to an atmospheric model (broken line) containing the PH$_3$ vertical distribution of Courtin (1982a) shown in Fig. 6 (figure reproduced from Winkelstein et al. 1983).

(1980c, 1981b) profiles have in particular much less PH$_3$ above the 100-mbar level than does the Voyager infrared profile of Courtin (1982a), and they have as much or more PH$_3$ at lower levels. Both of the profiles by Tokunaga et al. (1980c, 1981b) are moreover consistent with the ultraviolet results because the PH$_3$ in the stratosphere for these profiles is sufficiently depleted that Rayleigh scattering from H$_2$ accounts for the observed reflectivity at 210 nm.

What is the reason for the differences between these two estimates? The infrared observations on which both estimates are based must be simulated by integrating the thermal emission of the atmosphere over a physical distance that is more than an order of magnitude greater than the effective altitude range of the ultraviolet observations. In this larger region, the vertical temperature gradient changes sign (see Fig. 1) so the integrated emission as seen from space includes segments where specific PH$_3$ transitions are seen in absorption and other segments where the same transitions appear as emission features. It can be difficult to infer uniquely the molecular distribution from the integrated line-of-sight infrared emission in this circumstance. Perhaps the differences between the candidate PH$_3$ distributions evident in Fig. 6 result from this lack of a unique solution to the infrared inversion problem. Ultraviolet observations have no such ambiguity and since the estimates by Tokunaga et al. (1980c,1981b) are consistent with the IUE results these estimates are preferred although not demonstrably optimal. In particular, latitudinal and seasonal changes in PH$_3$ photodissociation rates and stratospheric vertical transport rates could lead to significant variability of the PH$_3$ vertical profile; this would make a comparison of the Courtin (1982a) and Tokunaga et al. (1980c,1981b) results that refer to different times and different parts of the planet much more difficult to interpret. Strong seasonal changes in tempera-ture are observed in the upper troposphere (e.g., Conrath and Pirraglia 1983) and predicted in the stratosphere (Carlson et al. 1980).

The data in Figs. 7 and 8 were obtained by IUE in 1979 and 1980. Ear-lier observations from 1969 and 1970 with the OAO-2 were reported by Cald-well (1977b, Fig. 6 therein). The earlier data covered a similar wavelength range (200–350 nm) but the results were qualitatively different from those of the IUE. The OAO-2 data showed an albedo maximum at 250 nm, below which it quickly dropped by ∼ 15%. It then maintained a level plateau to ∼ 210 nm and dropped sharply at shorter wavelengths.

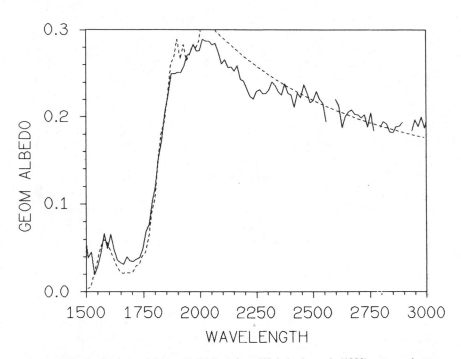

Fig. 8. IUE observations of Saturn (solid line) from Winkelstein et al. (1983) compared to an atmospheric model (broken line) containing the C$_2$H$_2$ and C$_2$H$_6$ vertical distributions of Cour-tin (1982a) shown in Fig. 6, as well as a trace of H$_2$O. Auroral emission from H$_2$ at 1600 Å has not been removed from the observations (figure reproduced from Winkelstein et al. 1983).

It has previously been impossible to evaluate the significance of this difference, because one data set (OAO-2) included light from the rings which had to be removed by a model-dependent process, and the other data set was independent of ring effects. In the earlier work, the ring reflectivity was simply assumed to be flat, following the expected reflectivity of H_2O ice between 200 and 300 nm (Pipes et al. 1974). However, there was no way to confirm this model independently for Saturn's rings in the ultraviolet until recently. Now, preliminary reduction of as yet unpublished spatially resolved IUE spectra of Saturn's rings indicates that the ring reflectivity is indeed very flat between 200 and 300 nm. This in turn suggests that Saturn's albedo near 200 nm is intrinsically variable. Such variability could be due to a changing vertical distribution of a minor constituent in the atmosphere. The effect could be seasonal since the two epochs of observation differ by ~ 10 yr which is $\sim 1/3$ of Saturn's year.

Visual inspection suggests that the onset of the short wavelength absorption in the OAO-2 data occurs at a wavelength (250 nm) that is too long to be associated with PH_3. These data have been modeled approximately with a hypothesized altitude-independent mixing ratio of $[H_2S]/[H_2] = 1.4 \times 10^{-8}$ (Caldwell 1977b). Unfortunately, no attempt has yet been made to model the ultraviolet characteristics of an atmosphere with depletion of H_2S toward higher altitudes. If observable amounts of H_2S are indeed present on Saturn, such high-altitude depletion would certainly be expected due to photodissociation.

Turning now to other species, Caldwell (1977b) has shown that there is no evidence for the characteristic predissociation bands of NH_3 near 210 nm in high-resolution data from the TD1A satellite. The IUE data shown in Figs. 7 and 8 also provide no indication of an NH_3 influence. This is consistent with the vertical NH_3 distribution shown in Fig. 6. Indeed, there are in fact no identifiable molecular absorptions in the IUE data longward of 200 nm. The IUE data do, however, extend below 200 nm where there is strong evidence for high-altitude absorption.

Moos and Clarke (1979) were the first to explore the region below 200 nm, identifying the prominent absorption bands of C_2H_2 (acetylene) between 170 and 190 nm. Clarke et al. (1982) later showed that C_2H_2 absorption is stronger on Saturn than on Jupiter. They attempted in particular to model the Saturnian albedo with a homogeneous mixture of C_2H_2, but found it difficult to match the observed depths of the absorption peaks with any value of the mixing ratio. They suggested that the absorption coefficients they were using might not be accurate and since the comparison laboratory observations were made at room temperature, this is certainly a plausible source of error.

Winkelstein et al. (1983) have also modeled the region below 200 nm. They began with the vertical distributions of C_2H_2 and C_2H_6 determined by the Voyager infrared data discussed in Sec. II.D, but excluded PH_3 as discussed above. With the C_2H_2 providing absorption between ~ 170 and 200 nm and the C_2H_6 absorbing below 160 nm, the Voyager-derived abundances and vertical distributions of these gases shown in Fig. 6 are consistent with the IUE data. However, an additional major absorber is required at wavelengths below 180 nm to bring the model albedos down to the level of the IUE albedo observations; this conclusion is in basic agreement with the work of Clarke et al. (1982).

In an effort to identify the above additional absorber, Winkelstein et al. (1983) tried various hydrocarbons, including C_4H_2 and C_3H_4, and also frosts. For various reasons, each were unsatisfactory alone to make the model agree with the observations (note that mixing ratios of $\sim 10^{-9}$ for gases like C_4H_2 and C_3H_4 are not precluded by IUE data). These workers also tested to see whether H_2O vapor could be the unknown absorber. In particular, they found that a column abundance of 6×10^{-3} cm-amagat of H_2O at the top of the atmosphere, together with C_2H_2 and C_2H_6, produced the relatively good fit to IUE data shown in Fig. 8. This amount of H_2O does not violate spectral constraints at longer wavelengths; however, it is very unlikely that any H_2O from the interior is mixed upward through Saturn's very cold tropopause ($T \sim 80$ K). If H_2O does contribute to the ultraviolet spectrum of Saturn, an external oxygen source is strongly suggested (which is why the H_2O was placed at the top of the above model). Perhaps icy particles from the rings and the inner icy satellites are the required source. Further, an external source of oxygen has been suggested to explain the infrared detection of CO_2 on Titan (see chapter by Hunten et al.).

Finally, we note that the observations of Saturn in the ultraviolet accomplished to date do not represent the limit of what can be learned from this spectral region. No spatial resolution across the disk has been achieved; the spectral resolution has been low (generally ~ 1 nm) and the noise level has been typically high. In all these areas we can expect major improvement when the Space Telescope is launched later in this decade. Many questions concerning the nature and distribution of upper atmospheric species may then be answered with much greater certainty than is now possible.

D. Infrared Spectroscopy from the Voyagers

One of the objectives of the Voyager infrared investigation was to determine the atmospheric composition of the giant planets and of Titan (Hanel et al. 1977). The Voyager infrared interferometer spectrometer (IRIS) has been described by Hanel et al. (1980). The Michelson interferometer was designed to cover the spectral range $200-4000$ cm^{-1} with a spectral resolution of 4.3 cm^{-1}. In practice, the useful range is limited by the fast decrease of the planetary thermal emission at the higher wavenumbers up to ~ 2500 cm^{-1} for Jupiter and to 1500 cm^{-1} for Saturn, although for Saturn some information can be obtained at 2000 cm^{-1} from an average of several thousand spectra.

The Voyager IRIS measurements provide significant advantages over previous infrared groundbased and airborne data. First, the wide spectral range of IRIS allows a simultaneous retrieval of temperature structure and composition, thereby eliminating uncertainties due to temporal or spatial variations or due to the use of a temperature model. Second, the absolute calibration of the spectra is much more accurate than for groundbased measurements (Hanel et al. 1980). Third, because of the high spatial resolution, a large number of spectra can be obtained at various locations on the disk of the planet, thus allowing the reliable detection of spatial variations. This third advantage, although fully exploited during the analysis of the Jupiter IRIS data (Conrath et al. 1981a; Flasar et al. 1981a; Gautier et al. 1981,1982; Marten et al. 1981; Kunde et al. 1982) is, however, less exploitable in the case of Saturn. This limitation is caused by the necessity for averaging many Saturn spectra in order to improve the signal-to-noise ratio (since the thermal emission of Saturn is weaker than that of Jupiter). Below 600 cm^{-1}, however, the signal-to-noise ratio is sufficiently large to permit analysis from a small number of spectra. Significant spatial resolution is thus available for the derived thermal structure (Hanel et al. 1981a,1982; Pirraglia et al. 1981; Conrath and Pirraglia 1983) and the H to He ratio (Hanel et al. 1981a; Conrath et al. 1983).

For the determination of the abundance of very minor constituents, the highest possible signal-to-noise ratio (and thus the addition of as many spectra as possible) is desirable. However, Courtin et al. (1982b) conclude, for reasons detailed below, that the data from the northern and southern hemispheres of Saturn must be analysed separately. An average of 512 individual spectra in the northern hemisphere, recorded between 200 and 1450 cm^{-1} is shown in Fig. 9. In the 200–700 cm^{-1} range, the main opacity is due to the pressure-

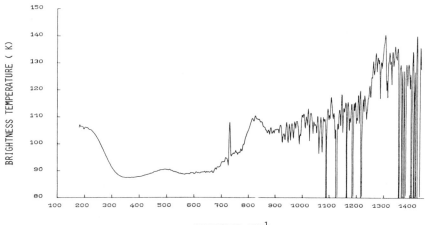

Fig. 9. Average of 512 individual Voyager 1 IRIS spectra selected from the northern hemisphere of Saturn between 10° and 30°N (from Courtin 1982a).

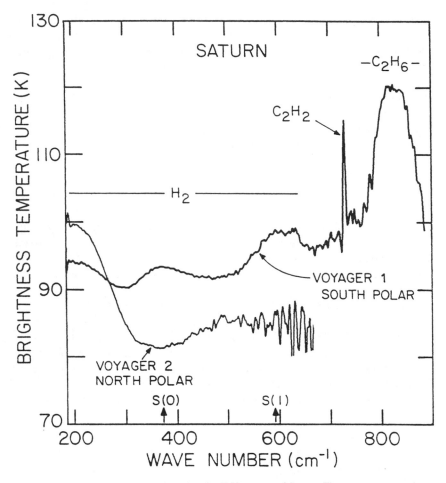

Fig. 10. Comparison of north and south polar IRIS spectra of Saturn. The upper spectrum is an average of 33 individual Voyager 2 spectra acquired at about 78°N. The lower spectrum is an average of 30 individual Voyager 1 spectra grouped at 70°S and 88°S. The mean emission angle for the north polar spectrum is 59° while that for the south polar spectrum is 76° (from Hanel et al. 1982).

induced absorption arising in the collisions between H_2 and H_2 molecules, and H_2 and He molecules (Birnbaum 1978; Birnbaum and Cohen 1976). In particular, the S(0) and S(1) lines centered at 354 and 588 cm^{-1}, respectively, appear in absorption on spectra located at mid-latitudes in both northern and southern hemispheres. They also appear in emission for high emission angles in the southern hemisphere, evident in Fig. 10, and in this case they provide information on the lower stratospheric temperature.

The pure rotational lines of NH_3 corresponding to $J = 9$ and $J = 10$ appear as weak absorption features at ~ 197 and 217 cm^{-1}. The ν_5 band of C_2H_2 centered at 729 cm^{-1} and the ν_9 band of C_2H_6 at 821 cm^{-1} are clearly seen in emission (from the stratosphere). Above 900 cm^{-1}, the absorption is dominated by the ν_2 and ν_4 bands of PH_3 centered at 992 and 1118 cm^{-1}, respectively, while the core of the ν_4 band of CH_4 appears in emission around 1305 cm^{-1}.

The strategy used to infer the gas distributions from the IRIS data (Courtin 1982a; Courtin et al. 1983b,c) first involves the determination of the H to He ratio from the $200-600$ cm^{-1} range (see Sec. II.F). Then, using the derived value of the H_2/He ratio, the tropospheric thermal profile is retrieved by means of inversion techniques (Gautier et al. 1977b; Conrath and Gautier 1980) from the $200-600$ cm^{-1} range. Similarly, the stratospheric thermal profile is retrieved from the core of the ν_4 band of CH_4 at 1305 cm^{-1}, for a given value of the CH_4 mixing ratio. After the complete temperature profile is thus established, the gas abundances are determined from a best-fit approach by comparing the observed spectrum with the spectra calculated using the retrieved temperature profile and various assumed gas distributions. The strong asymmetry between the temperature profiles in the northern and southern hemispheres (Hanel et al. 1981a,1982; Pirraglia et al. 1981) necessitates separate consideration of the spectra in the two hemispheres. Courtin (1982a) selected, in particular, 512 spectra between 10°N and 30°N and 244 spectra between 20°S and 70°S. This represents only a fraction of the total number of available spectra, so the results obtained by Courtin (1982a) should be regarded as preliminary and subject to future revisions. With these general comments as a suitable preface, we will now discuss the derived results for minor constituents under a few convenient headings.

1. Distributions of C_2H_2, C_2H_6 and PH_3. The vertical profiles of the PH_3, C_2H_2 and C_2H_6 mixing ratios derived by Courtin (1982a) are shown in Fig. 6. The sharp decreases in the mixing ratios of C_2H_2 and C_2H_6 below the 20 to 50 mbar pressure levels shown in Fig. 6, provide the best fits to the observed profiles in the wings of the ν_5 band of C_2H_2 and the ν_9 band of C_2H_6. These decreases are also consistent with the photodissociation of methane producing these hydrocarbons at high altitudes as discussed in Sec. III.

The best agreement in the Q branches of the observed bands of PH_3 is obtained with a distribution strongly depleted above the 3 to 9 mbar pressure level. As already noted and discussed in Sec. II.C, this profile differs markedly from the PH_3 profile of Tokunaga et al. (1980c,1981b) in the stratosphere. The tropospheric values of the PH_3 mixing ratio obtained by Courtin (1982a) and Tokunaga et al. (1980c,1981b) are, however, in reasonable agreement and both suggest a P/H ratio significantly higher than the value corresponding to solar composition (Cameron 1982). This result suggests that the P/H value in the deep interior is also larger than the solar-composition value.

An important difference is evident between the abundances of C_2H_2 and C_2H_6 in the northern and southern hemispheres. Error analyses show that the observed difference cannot be attributed to a thermal structure effect in spite of the uncertainties in the retrieved stratospheric temperatures. The production of hydrocarbons seems thus to be different in the stratospheres of the northern and southern hemispheres; perhaps this is a result of seasonal effects. Further analyses of the IRIS data are in progress. Of particular concern are the latitudinal variations of C_2H_2 and C_2H_6, the detection of possible additional minor components, and the study of the upper tropospheric clouds.

2. Detections of NH₃, C₃H₄ and C₃H₈. The weakness of the spectral features of NH_3 at 200 cm^{-1} undoubtedly results from the strong depletion of NH_3 by condensation in the Saturnian upper troposphere as a consequence of low temperatures. A reasonably good fit to the NH_3 lines is obtained in the northern hemisphere by using the relevant retrieved thermal profile together with an NH_3 vertical distribution calculated according to the saturation law. It is not necessary to introduce an additional continuum opacity in the $\sim 50\ \mu$m wavelength region as was the case for cloudy regions of Jupiter (Marten et al. 1981). On the other hand, the situation is somewhat obscure for the southern hemisphere where the relatively low signal-to-noise ratio of the average spectrum prevents firm conclusions. Analyses based on a larger number of spectra are in progress.

From presently available results by Courtin (1982*a*) we conclude that the NH_3 vertical distribution in the upper troposphere of Saturn seems to follow the saturation law, at least in the northern hemisphere (a result also consistent with the ultraviolet observations discussed in Sec. II.C). Also there is no evidence of any infrared absorption due to clouds or aerosols above the 400 mbar pressure level. In particular, temperature profiles retrieved from IRIS data allow a fit to the IRIS spectra without inclusion of particulate absorbers, even in the spectral range not used to retrieve the temperature profiles. Moreover, synthetic spectra calculated using the Voyager 2 radio occultation ingress profile at 36°5 N (Tyler et al. 1982*a*) agree well with the IRIS spectra recorded in the same region using only gas-phase absorbers; this good agreement is not obtained using the egress profile at 31°S, but agreement here would be made even worse if nongaseous absorbers were included (Conrath et al. 1983).

At pressures $> 300-400$ mbar, Conrath and Pirraglia (1983) note that the temperature gradients derived from high spatial resolution IRIS data do not correlate with the zonal jets as expected. They suggest that this could result from erroneously ignoring the presence of particulates below the 400 mbar (100 K) level and thus causing incorrect temperature retrievals for these deeper levels. Examination of their Fig. 1 (Conrath and Pirraglia 1983) implies that this is a distinct possibility at latitudes except those around $25-45°$N where a cloud-free zone appears to extend down to at least the 730 mbar (120 K) level. As noted above, it is at 36°5N that good agreement is obtained with IRIS data using the ingress occultation temperatures without inclusion of any particulate absorbers. In summary, the IRIS data are compatible with the presence of particulates up to the 400 mbar level except in the 25–45°N region; however, the evidence is indirect and no correlation between these high-level particulates and Voyager images is observed.

As far as other weak absorption features are concerned, Hanel et al. (1981a) tentatively identify the presence of methylacetylene (C_3H_4) and propane (C_3H_8) from the weak features at 633 and 748 cm^{-1}, respectively. Additional calculations are in progress to test this possibility.

3. *Determination of CH_4 and of CH_3D.* The mixing ratios of CH_4 and CH_3D can be determined in principle from a best fit of the Voyager IRIS spectrum in the 1200–1250 cm^{-1} and 1400–1450 cm^{-1} ranges as was previously done for Jupiter (Gautier et al. 1982; Kunde et al. 1982). This determination, however, is more difficult for Saturn because of the lower signal-to-noise ratio. The preliminary results of Courtin (1982) obtained by degrading the spectral resolution of IRIS lead to a CH_4/H_2 ratio of $\left(2.0^{+1.5}_{-0.6}\right) \times 10^{-3}$, if we use the most recent determination by Brown (1982) of the intensity of the ν_4 band of CH_4 (namely, 125.3 ± 2 cm^{-2} atm^{-1} at 300 K). Some improvement in the accuracy of the determination of the CH_4 abundance is expected from selecting a larger number of Voyager spectra but the present results already suggest that the visible atmosphere of Saturn is substantially enriched in carbon compared to solar abundance (the solar abundance value as given by Cameron (1982) would correspond on Saturn to $CH_4/H_2 = 0.84 \times 10^{-3}$). The existence of this enrichment was suspected prior to the Voyager encounter (see Table II) but there were large uncertainties and dispersions in the pertinent groundbased measurements. The present result can also be compared to the determination of the Jovian C/H ratio from Voyager (Gautier et al. 1981). These authors derived $CH_4/H_2 = (1.95 \pm 0.22) \times 10^{-3}$, using the value of 140 ± 14 cm^{-2} atm^{-1} at 300 K (Chedin et al. 1978) for the intensity of the ν_4 band of CH_4. Using the aforementioned more precise determination of this intensity (Brown 1982), the CH_4/H_2 ratio on Jupiter is thus $(2.18 \pm 0.16) \times 10^{-3}$. These observed enrichments in carbon in both Jupiter and Saturn favor nonhomogeneous formation models for these giant planets involving accretion of solid planetesimals from the primordial nebula (see e.g., Hubbard and MacFarlane 1980a; Gautier et al. 1981).

The preliminary results of Courtin et al. (1983b) on the Saturn CH_3D/H_2 ratio, obtained by analyzing the observed Voyager spectra in the 1100–1200 cm^{-1} range, indicate a value of $\left(2.6^{+1.8}_{-1.2}\right) \times 10^{-7}$. The resulting D/H ratio in Saturn is thus $D/H = \left(2.6^{+1.8}_{-1.2}\right) \times 10^{-7}/\left[\left(2.0^{+1.5}_{-0.6}\right) \times 10^{-3} \times 4f\right] = \left(2.4^{+2.4}_{-1.3}\right) \times 10^{-5}$ assuming the fractionation factor $f = 1.37$ as derived by Beer and Taylor (1973, 1978) for Jupiter. This result is close to the determination of the D/H ratio by Fink and Larson (1978) from groundbased measurements of 2×10^{-5}, and also to the D/H value on Jupiter derived by Kunde et al. (1982)

from Voyager of $\left(3.2^{+0.9}_{-1.25}\right) \times 10^{-5}$ after correcting for the revised value of the CH_4/H_2 ratio mentioned above.

The cosmological implications of the determination of the D/H ratio in the giant planets are considerable, in particular for the big-bang theory as discussed by Kunde et al. (1982) and Gautier and Owen (1983a). The observed D/H values on Jupiter and Saturn provide a value for the deuterium abundance in the primitive solar nebula. Through models of the chemical evolution of galaxies, it is then possible to obtain an estimate of the deuterium abundance at the time the universe was created, and thus, through the big-bang theory, a value for the ratio of baryons to photons in the universe. The baryon density so derived is insufficient to close the universe. However, current estimates of the helium abundance at the time of creation do not lead to the same baryon density as that derived from the deuterium abundance unless a destruction of deuterium much stronger than commonly envisaged has occurred between the times of the creation of the universe and the formation of the solar system. This difficulty could question the validity of the standard model of the big-bang theory (Gautier and Owen 1983a).

E. Ultraviolet Spectroscopy from the Voyagers

Voyagers 1 and 2 carried two instruments relevant to ultraviolet wavelengths: the UVS experiment (Broadfoot et al. 1981; Sandel et al. 1982) and the photopolarimeter (Lane et al. 1982b). The UVS experiment detected emissions from H, He, and H_2, and CH_4 concentrations were inferred from absorption of 1200 Å light from the ultraviolet star δ-Scorpii occulted by Saturn at the time of Voyager 2 encounter. The photopolarimeter had a channel at 2640 Å that is useful in probing particulate concentrations and properties in the upper atmosphere. We will not discuss these experiments here (although the UVS in particular is relevant to any discussion of Saturnian composition and chemistry) since chapters by Atreya et al. and Tomasko et al. will address these Voyager observations and their relevance to the upper atmosphere and ionosphere and to the Saturnian clouds.

F. The Hydrogen-to-Helium Ratio

The determination of the H/He ratio in the atmosphere of Saturn is crucial for understanding the internal structure of the planet (see chapter by Hubbard and Stevenson) as well as for its possible cosmological implications (discussed briefly in Sec. II.D). Basically, two hypotheses can be envisaged. In the first, hydrogen and helium are uniformly mixed throughout the whole interior of the planet and in this case a measurement of the helium abundance in the outer atmosphere provides a determination of H/He in the primitive solar nebula. Alternatively, the present visible atmospheric helium abundance on Saturn may differ from the planet's bulk composition because helium has differentiated from hydrogen in the interior during the planet's evolution. In particular, over a range of temperature and pressure relevant to Saturn's interior, helium and hydrogen could become immiscible and droplets of helium thus migrate towards the center of the planet. This would lead to an enrichment of the deep interior and to a depletion of the outer atmosphere in helium. In such a case, minor components could also be redistributed within the interior (Stevenson and Salpeter 1977a,b).

Since the results of the Voyager measurements are consistent with a uniform mixing of hydrogen and helium within Jupiter (Gautier et al. 1981), a comparison of the Saturn Voyager He/H measurement with that of Jupiter permits a discrimination between the above two possible hypotheses. The determination of the Saturn helium abundance is, however, somewhat more complicated than Jupiter's for reasons explained below.

As for Jupiter, two methods for determining He/H are available. The first method involves comparing IRIS spectra to synthetic spectra calculated using thermal profiles derived from radio-occultation measurements and various assumed values of the H/He ratio. The true value of the ratio is that which provides the best fit to the observed spectrum (Gautier et al. 1981) in the range where the absorption is due to hydrogen and helium only. For Saturn this range is theoretically from 200 to ~ 700 cm^{-1} but it is usually limited to 200 to ~ 600 cm^{-1} by the experimental noise (cf. Fig. 10). This is a broader interval than on Jupiter where the spectral range 200–280 cm^{-1} cannot be used because of the presence of significant gaseous NH$_3$ opacity in this region relative to Saturn. The calculated spectra are highly sensitive to the H/He ratio since the retrieved temperature and pressure are directly proportional to the mean molecular weight and thus to the hydrogen mole fraction q. Moreover, a change in q results in changes in both the temperature profile and the atmospheric transmission which act in the same sense. This method is thus very sensitive for the determination of the H/He ratio but it requires a high accuracy in the radio occultation temperature profiles that are obviously limited to only two points on the planet (Voyager spacecraft ingress and egress).

The second method is far more subtle and deserves some explanation for the case of Saturn. Initially proposed by Gautier and Grossman (1972), this method takes advantage of the different spectral dependences of the H$_2$–H$_2$ and H$_2$–He absorption coefficients in the far infrared range (see Fig. 11). Within the broad S(0) and S(1) lines of the H$_2$ spectrum, several wavelength ranges exist wherein the observed radiances come from essentially the same atmospheric temperature levels and thus for which the atmospheric opacities must be the same. The corresponding calculated opacities depend upon the hydrogen mole fraction q and they are equal only for the true value of q. These opacities are determined by an inversion method described in Gautier et al. (1981) which provides a set of thermal profiles for various values of q. The best fit to the observed spectrum yields the best estimate of the thermal profile and the best estimate of q.

Since a large number of spectra were obtained by the IRIS Voyager ex-

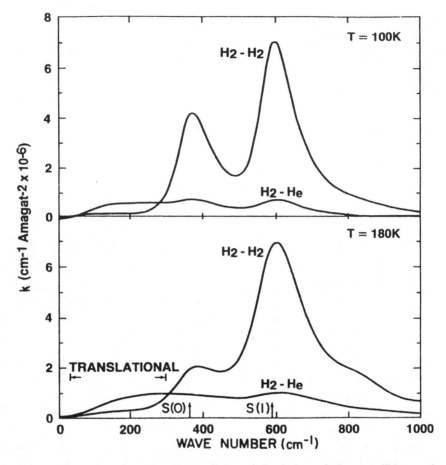

Fig. 11. Pressure-induced absorption coefficients k for hydrogen-hydrogen collisions and
hydrogen-helium collisions at Saturn-like temperatures (top) and Jovian-like temperatures
(bottom) as given by Gautier et al. (1981).

periment at many locations on the disk of Saturn and with high spatial resolution, the second method described above permits us to infer in principle a
large number of independent estimates of q, which after averaging considerably reduces the effects of random errors. This was the main advantage of the
second method for He/H compared to the first (radio occultation) method in
the case of Jupiter. Systematic errors may however still exist with the second
method, mainly due to the presence of aerosols or to uncertainties in absorption coefficients.

The fact that the whole 200–600 cm^{-1} range is available to infer q by the
second method is an advantage for this method on Saturn compared to Jupiter
since from Fig. 11 the H$_2$–He absorption coefficients predominate over the

H_2–H_2 coefficients below 300 cm^{-1} which obviously improves the sensitivity of this method for the determination of q. However, some disadvantages also exist compared to the Jupiter case. First, as evident in Fig. 11 the strengths of the S(0) and S(1) lines at their formation temperatures on Saturn exhibit a much smaller contrast than at Jovian temperatures. This behavior is amplified by the small temperature lapse rate at the relevant levels of Saturn's atmosphere. As a result, the spectrum is rather flat in the range of the cores of the S(0) and S(1) lines (see Figs. 9 and 10) making it more difficult to retrieve the H_2/He ratio independently from the thermal profile. In other words, the retrieval of q is more dependent for Saturn than for Jupiter on the initial guess used to start the inversion process. This is a serious limitation on the accuracy of the method for Saturn. A second disadvantage is the absence of emission at 5 μm and the weakness of the NH_3 absorption at 50 μm which precludes an easy discrimination of cloud areas from clear areas as was possible for Jupiter (Marten et al. 1981; Kunde et al. 1982). The presence of any infrared opacity due to aerosols which was not taken into account in the inversion process, would lead to an erroneous value of the helium abundance. Fortunately, as mentioned in Sec. II.D there is no obvious evidence of solid particle absorption in the IRIS spectra of Saturn so this disadvantage may be minor. A third disadvantage is that the theoretical calculations of the H_2–H_2 and H_2–He absorption coefficients assume that the hydrogen ortho-para ratio is determined by equilibrium at the local temperature. In the conditions of Saturn's atmosphere, as analyzed by Massie and Hunten (1982), a significant departure from equilibrium could occur in the upper troposphere. Introducing this departure into the opacity formulation would lead us to derive a smaller helium abundance from Voyager data than that derived with the "equilibrium" formulation. As a consequence, the helium abundance could then be very small or even negligible. However, a comparison of the shape of IRIS spectra at the Voyager 2 ingress occultation point located in the northern hemisphere, with spectra calculated using the radio-occultation temperature profile obtained by Tyler et al. (1982a), seems to exclude a significant departure from the equilibrium state for the ortho-para ratio in Saturn's upper troposphere (Conrath et al. 1983).

Taking into account the above considerations, the analysis of Voyager data by Conrath et al. (1983) leads to the following conclusions:

1. The relative abundance of hydrogen by volume in Saturn's outer atmosphere derived from the radio-occultation method is (including all random and systematic errors) $q = 0.932 \pm 0.025$ where q is specifically defined as the ratio of the number density of H_2 to the number density of H_2 plus He. The corresponding helium mass fraction y is related to q by $y = (1 - q)/(1 - 0.59) = 0.13 \pm 0.04$ (for Saturn). This result assumes an uncertainty of ± 2 K on the accuracy of the Voyager 2 radio-occultation ingress profile at a latitude 36°5 N (Tyler et al. 1982a). The value obtained from the inversion method, using the radio-occultation profile as an initial

guess, is very close to 0.135. All these results have been obtained from data located in the northern hemisphere (the interpretation of southern hemisphere data is not yet complete). The present determinations are compatible with the values $q = 0.9 \pm 0.03$, $y = 0.18 \pm 0.05$ derived by Orton and Ingersoll (1980) by combining Pioneer 11 broadband radiometer measurements and radio-occultation data, particularly since the latter results did not include any possible systematic errors.

2. The present derived value of $y = 0.135 \pm 0.045$ for Saturn is significantly less than both the estimated helium abundance at the birth of the universe ($y \geq 0.21$ to 0.22) and the helium abundance on the Sun for which $y \geq 0.22$ seems reasonable (for a detailed discussion see Gautier et al. [1981] and Conrath et al. [1983]). There is a slight overlap of the Saturn value with the Jovian value (Gautier et al. 1981) since a Jovian y value as low as 0.17 is possible once we take into account the final error bars on the radio-occultation profile given by Lindal et al. (1981). However, values of y in Jupiter < 0.18 to 0.19 would imply a depletion of helium in the outer atmosphere compared to its solar abundance. This would result in a predicted outward flux of internal energy from Jupiter greater than that determined from Voyager data (Hanel et al. 1981b). The present measured luminosity of Jupiter is in fact in very good agreement with the predictions of homogeneous evolutionary models of Jupiter. However, these models do not agree with the observed luminosity of Saturn which requires an additional source of internal energy. This additional source could be provided by the migration of helium droplets towards the center of the planet (Stevenson 1980). Helium differentiation would also provide an explanation for the low He abundance measured in Saturn's visible atmosphere and perhaps for the observed characteristics of its magnetic field (Stevenson 1980).

G. Cloud Composition

Saturn's internal energy source and its infrared opacity ensure that there is a strong temperature gradient and convection between the interior and a pressure level ~ 1 bar. Given the presence of NH_3 (and presumably of H_2S and H_2O as well) one therefore expects to find a series of condensation clouds, the bottoms of which occur where these volatile substances reach their respective saturation vapor pressures over solid or liquid phases. Cloud particles are also mixed upward from the level at which they form to altitudes determined by the particle sizes and by the vigor of the convective motions. In this section we will address observations relevant to cloud composition; a detailed discussion of the structure and physics of the clouds is provided in the chapter by Tomasko et al.

Theoretical models based on thermochemical equilibrium discussed in Sec. III suggest that the outermost cloud layer on Saturn consists of solid NH_3 particles with a base for these clouds at a pressure ~ 1.4 bar in a solar-composition atmosphere. If, as discussed in Sec. II.E, these clouds extend up to the ~ 0.4 bar pressure level, then they will be much thicker than the corresponding NH_3 clouds on Jupiter. However, the evidence supporting this simple concept of a thick NH_3 cloud on Saturn is somewhat equivocal.

There are a number of observations which provide indirect evidence for the presence of NH_3 clouds. Both ultraviolet observations near 210 nm (Caldwell 1977b; Combes et al. 1981) and infrared observations from 40 to 50 μm (Hanel et al. 1981a, Fig. 1 therein) indicate that the gas phase NH_3 abundance is much reduced on Saturn with respect to Jupiter. The most straightforward explanation for this difference is that most of the NH_3 in Saturn's visible atmosphere is in a condensed phase. Witteborn et al. (1981) have observed absorption features on Saturn at 6.3 and 7.1 μm, which could be compatible with either gaseous or solid NH_3 (see Fig. 12). Fink et al. (1983) have interpreted absorptions seen in high-resolution spectra of Saturn at ~ 5.3 μm as due to gaseous NH_3 together with a small amount of solid NH_3. Gillett and Forrest (1974) observed that Saturn has a broad absorption feature at 9.4 μm and suggested that solid NH_3 might be the cause. Caldwell (1977a) was able to model this feature quantitatively as solid NH_3 absorption, but pointed out that various published absorption coefficients were very different from each other. This led to the suggestion that the conditions of formation of solid NH_3 could influence its observable spectral properties, and that one must be careful to use formation conditions in the laboratory which correspond closely to the planetary conditions that are being investigated.

An alternative explanation of the 9 to 11 μm spectral region has been given by Tokunaga et al. (1980c,1981b) who observed Saturn with a spectral resolution that was an order of magnitude higher than that of Gillett and Forrest (1974). The later investigators found clear evidence for seven spectroscopic features of the ν_2 fundamental vibration-rotation band of PH_3 within the spectral range of their observations (10.0 to 10.7 μm). They were able to achieve a good fit to the low-resolution data of Gillett and Forrest (1974) from 9 to 11 μm using only gas phase PH_3 absorption. Gillett and Forrest also recognized this possibility qualitatively. The broad feature at 9.4 μm is due entirely to overlapping P- and R-branch lines of the ν_2 and ν_4 fundamental bands of PH_3 at 10.1 and 8.9 μm, respectively, in this interpretation.

The gaseous PH_3 model of Tokunaga et al. (1980c,1981b) removes the absolute requirement for absorption by solid NH_3 particles on Saturn at 9.4 μm, but the two models are not mutually exclusive. The PH_3/H_2 mixing ratio for Saturn derived by Tokunaga et al. (1981b), of $0.8 - 1.6 \times 10^{-6}$, exceeds the value for that ratio found by Tokunaga et al. (1979) for Jupiter by a factor of 4 and also exceeds the value corresponding to the solar elemental mixing ratio by a factor of $1.2 - 2.4$. A compromise model, with the two sources of opacity each contributing partially to Saturn's spectral character near 10 μm, could therefore reduce the amount of PH_3 required by Tokunaga et al. (1980c, 1981b) and bring Saturn's PH_3/H_2 ratio closer to that of Jupiter.

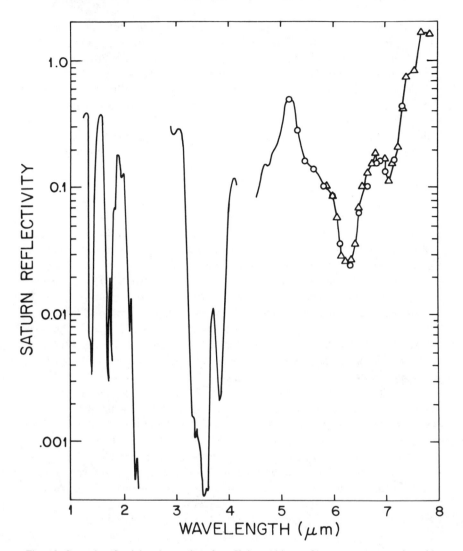

Fig. 12. Saturn's reflectivity observations from Kuiper Airborne Observatory communicated by
Witteborn. This figure has a modification in the ordinate scale compared to a similar figure
published by Witteborn et al. (1981). The reflectivity scale used here is equivalent to geometric
albedo so that thermal emission by CH$_4$ near 8 μm leads to a reflectivity which exceeds unity.

A serious objection to the NH$_3$ cloud model was first raised by Martin et
al. (1976). They pointed out that Saturn is bright at 3.0 μm (Fig. 12) (see
Larson et al. 1980) in apparent conflict with various laboratory spectra of
solid NH$_3$ that show a strong absorption band at this wavelength. Mindful of
this dilemma and of the caution recommended by Caldwell (1977a) on possi-

ble intrinsic variance of solid NH_3 spectra, Slobodkin et al. (1978) made laboratory observations of the 3-μm spectrum of solid NH_3 that had a thermal history similar to that expected for cloud particles on Saturn. In particular, they found that for a sample formed at 150 K, the strong 3-μm feature was shifted to 3.29 μm, and that subsequent cooling to 100 K did not produce further wavelength changes. This cooling sequence was designed to mimic the formation of NH_3 cloud particles at the cloud base followed by convective upwelling of these particles to the higher, cooler altitudes where they are visible. According to Slobodkin et al. (1978) the 3-μm absorption feature of solid NH_3 is not visible on Saturn because it is shifted and hidden by very strong absorption by the ν_3 fundamental band of gaseous CH_4 centered at 3.3 μm.

Later, Sill et al. (1981) showed that this wavelength shift was also found in the ammonia hemi-hydrate, $2\,NH_3 \cdot H_2O$, formed by heating an NH_3/H_2O condensate to temperatures > 146 K and then recooling. According to the theoretical models discussed in Sec. III, condensation of ice crystals occurs well below the visible clouds but it is conceivable that these ice crystals can be convected up to the region where NH_3 condensation is occurring. This would enable ammonia hydrate formation at temperatures > 146 K and thus ammonia hemi-hydrate crystals could contribute conceivably to the reflectivity of Saturn at 3 μm. If this is the case, then NH_4HS crystals which are also predicted in the above theoretical models, should be investigated for their possible contribution to the 3-μm albedo of Saturn.

In summary, there is some circumstantial evidence for the presence of solid NH_3 clouds in Saturn's atmosphere but to date there has been no unambiguous spectral feature observed that proves this conclusion. Further work is needed on the reflection spectrum of solid NH_3 subjected to relevant temperature histories; some H_2O contamination in the laboratory results of Slobodkin et al. (1978) cannot be ruled out (Slobodkin, personal communication). The idea that H_2O and NH_4HS particles can be mixed up to visible levels to complicate spectral analyses (e.g., through ammonia hydrate formation) must be explored. The degree to which laboratory frosts are valid analogues to planetary clouds needs careful critical analysis; if they are not good analogues, then laboratory measurements of particle shapes and bulk optical properties will be needed as inputs to appropriate multiple-scattering models.

III. CHEMICAL MODELS

The atmosphere of Saturn, like that of Jupiter, is believed to be convective to great depths and to possess an adiabatic lapse rate for temperatures below the tropopause. At sufficiently high temperatures, well beneath the visible portion of the atmosphere, the kinetic energy per molecule begins to approach the typical activation energies of thermochemical reactions and thermochemical equilibrium becomes a reasonable assumption. The observations discussed in the earlier sections suggest also that the elemental ratios of

H : C : N : P on Saturn are close to (but not exactly equal to) the ratios expected for a solar composition planet. In this section we will explore the implications of an adiabatic model of thermochemical equilibrium, near-solar composition, for the deep Saturnian atmosphere. At the same time, for the upper layers of the atmosphere we recognize that thermochemical equilibrium may prove to be a hazardous assumption since several potentially disequilibrating mechanisms including rapid vertical transport, photodissociation, and energetic ion, electron, and neutral impact dissociation can be identified in these upper layers. Indeed, the observed presences on Saturn of PH_3, C_2H_2 and C_2H_6 at concentrations far exceeding their thermochemical equilibrium values, together with the observation of colored particulates which are not expected in equilibrium, are both clear manifestations of disequilibrating processes on this planet. We will therefore also explore in this section the implications of specific disequilibrium models for the visible portion of Saturn's atmosphere.

A. Thermochemical Equilibrium Models

Models for the origin of Saturn in the primitive solar nebula fall into two general categories: a one-step process involving simultaneous collapse of both nebula gases and grains due to gravitational instability forming the planet directly, or a two-step process involving accumulation of a massive solid preplanetary core followed by hydrodynamic collapse of nebula gases onto this core. Both models presumably lead to a planet that is sufficiently massive that it survives the subsequent dissipation of the nebula and prevents significant escape of even the lightest element H over geologic time. Since the one-step model can also subsequently capture solid material, both models lead to the prediction that the bulk composition of Saturn is similar to that of the nebula with varying degrees of enhancements of the condensible elements. The magnitude of the possible enhancement can be gauged from the work of Stevenson (1982a) who predicts that Saturn began as a rock-plus-ice core of 15–25 Earth mass compared to its total final mass of 95 Earth mass. The possible composition of such an ice-and-rock core can be gauged from the nebula chemistry studies of Lewis (1972b), Prinn and Lewis (1973), Lewis and Prinn (1980), and Prinn and Fegley (1981); significant enhancements of both ice-forming (O, C, N, etc.) and rock-forming (Si, Mg, Fe, S, O, etc.) elements are plausible.

The elemental composition of the nebula has been largely inferred from data on the composition of the Sun for the volatile elements and of type I carbonaceous chondrites for the less volatile ones (e.g., Cameron 1973). Recent estimates of elemental abundances in this so-called solar-composition mix obtained by Cameron (1982) are given in Tabie VIII in Sec. II.B. Models of the giant planets which assume solar composition or near-solar composition provide a useful basis for any discussion of the chemistry of the giant planets.

While detailed computations of thermochemical equilibrium assuming solar-composition and an adiabatic lapse rate are available for Jupiter (Barshay and Lewis 1978; Fegley and Lewis 1979), similar calculations for Saturn have not yet been published. The results from the above Jovian studies are qualitatively but not quantitatively applicable to Saturn. First, the pressure at a given temperature beneath the tropopause on Saturn is ~ 2.2 times greater than that on Jupiter. Second (see Sec. II), the H_2 and CH_4 volume mixing ratios on Saturn are, respectively, ~ 1.06 and 2.5 times the solar-composition mixing ratios (the NH_3 and PH_3 mixing ratios may also be, respectively, $\lesssim 3$ and $\lesssim 4$ times the corresponding solar-composition ratios). Third, better estimates are now available for some of the thermodynamic constants used in the above Jovian calculations. Finally, the latest estimates of solar abundances (see Table VIII) differ from those used in the Jovian calculations (also given in Table VIII). We will therefore begin with a discussion of how the above Jovian theoretical results should be scaled for application to Saturn.

For purposes of discussion we will assume that the CH_4/H_2 ratio on Saturn is 2.5 times the value of 8.3×10^{-4} in a solar-composition atmosphere. The studies of nebula chemistry cited earlier suggest that enhancement of carbon on Saturn should be accompanied by similar enhancements in many of the rock- and ice-forming elements. We will assume here that the ratios relative to H_2 of all elements other than the noble gases and carbon are either all solar or all 2.5 times solar. The available observations of NH_3 and PH_3 imply this is a reasonable assumption for nitrogen and phosphorus but no relevant observations are available for the other elements.

For carbon-containing gases on Saturn, CH_4 is the dominant predicted species at all temperatures $T < 2000$ K (as it is on Jupiter). On Jupiter, the predicted next-most abundant carbonaceous gases are C_2H_6 when temperatures $T < 900$ K, and CO when $T > 900$ K. The relevant equilibria are

$$2CH_4 \leftrightarrows C_2H_6 + H_2 \tag{1}$$

$$CH_4 + H_2O \leftrightarrows CO + 3\,H_2 \tag{2}$$

and imply that the Jovian C_2H_6 mixing ratio at a given temperature computed by Barshay and Lewis (1978) should be multiplied by a factor of $(2.5 \times 1.13)^2/1.06 = 7.5$ for use on Saturn. For CO, the factor is $1.02 \times 2.5 \times 1.13/(1.06^3 \times 2.2^2) = 0.5$ for solar H_2O and $0.5 \times 2.5 = 1.2$ for 2.5 times solar H_2O. The simple unsaturated hydrocarbons C_2H_4 and C_2H_2 are both very minor and the equilibria

$$2CH_4 \rightleftarrows C_2H_4 + 2H_2 \tag{3}$$

$$2CH_4 \rightleftarrows C_2H_2 + 3H_2 \tag{4}$$

imply that the C_2H_4 and C_2H_2 mixing ratios on Saturn should be respectively $(2.5 \times 1.13)^2/(1.06^2 \times 2.2) = 3.2$ and $3.2/(1.06 \times 2.2) = 1.4$ times the values computed by Barshay and Lewis (1978).

For nitrogen-containing gases on Saturn, NH_3 is predicted to dominate at all temperatures $T < 2000$ K. The next-most abundant nitrogenous gas is N_2. The relevant equilibrium is

$$2NH_3 \rightleftarrows N_2 + 3H_2 \tag{5}$$

implying that the Barshay and Lewis (1978) N_2 mixing ratios at a given temperature should be multiplied by $0.74^2/(1.06^3 \times 2.2^2) = 0.095$ and $0.095 \times 2.5^2 = 0.59$ for solar and 2.5 times solar NH_3 abundances respectively. A minor nitrogenous gas is CH_3NH_2 and the equilibrium

$$CH_4 + NH_3 \rightleftarrows CH_3NH_2 + H_2 \tag{6}$$

dictates that the Jovian CH_3NH_2 mixing ratios should be multiplied by $2.5 \times 1.13/1.06 = 2.7$ (solar NH_3) or $2.7 \times 2.5 = 6.7$ (2.5 times solar NH_3).

On Jupiter, PH_3 gas, P_4O_6 gas, and $NH_4H_2PO_4$ solid are the predicted dominant phosphorus-containing compounds in the temperature ranges 2000–1000, 900–400, and 380–300 K, respectively. For $400 < T < 900$ K on Saturn the relevant equilibrium for determining the PH_3 abundance is

$$P_4O_6 + 12H_2 \rightleftarrows 4PH_3 + 6H_2O \tag{7}$$

so that the Jovian PH_3 mixing ratios of Barshay and Lewis (1978) should be multiplied by $(0.81 \times 1.06^{12} \times 2.2^3)^{1/4}/(1.02)^{3/2} = 2$ and $2/(2.5)^{5/4} = 0.63$ respectively for solar and 2.5 times solar P and H_2O mixing ratios.

According to Barshay and Lewis (1978) the dominant oxygen and sulfur-containing Jovian species for all $T < 2000$ K are, respectively, H_2O and H_2S; the same result is expected for Saturn. For $1500 \gtrsim T \gtrsim 300$ K the most abundant gaseous halogen-containing compound expected on both Jupiter and Saturn is HF. At $T \lesssim 300$ K, HF gas is quantitatively removed as NH_4F crystals on both planets.

Fegley and Lewis (1979) predict for Jupiter that the dominant selenium compound at $T \gtrsim 500$ K is H_2Se gas followed closely by GeSe gas. They also predict that the dominant germanium compound at $T \gtrsim 650$ K is GeS gas followed by GeH_4 and GeSe gases. For Saturn, the equilibrium

$$GeS + 3H_2 \rightleftarrows GeH_4 + H_2S \tag{8}$$

implies that the ratio of GeH_4 to GeS at a particular temperature predicted for Jupiter should be multiplied by $1.06^3 \times 2.2^2/1.2 = 4.8$ and $4.8/2.5 = 1.9$ for solar sulfur and 2.5 times solar sulfur, respectively. For selenium the two equilibria

$$H_2Se + GeS \rightleftarrows GeSe + H_2S \tag{9}$$

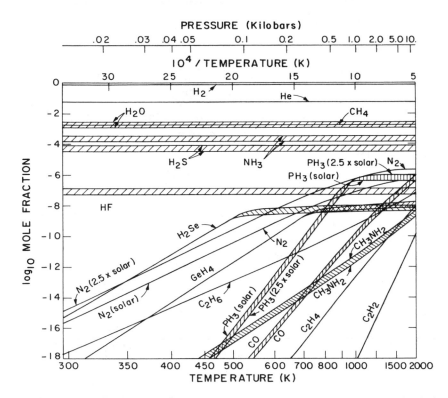

Fig. 13. Predicted compounds and their volume mixing ratios (mole fractions) in thermochemical equilibrium given as a function of temperature and pressure in Saturn's lower atmosphere (after Fegley and Prinn 1984).

$$H_2Se + GeH_4 \rightleftarrows GeSe + 3H_2 \qquad (10)$$

both indicate that H_2Se will be more stable relative to GeSe on Saturn than on Jupiter. Note that while GeH_4 has been observed on Jupiter, it has not yet been observed on Saturn where it is predicted to have an even greater abundance.

Calculations based on the above considerations and including the latest chemical thermodynamic data have recently been completed by Fegley and Prinn (1984). The various predicted equilibrium concentrations for gases in the deep Saturnian atmosphere are summarized in Fig. 13. It is immediately apparent that, if chemical equilibrium applies, the only potentially observable gases by a Saturnian entry probe penetrating to e.g. the 300 K level are H_2, He, H_2O, CH_4, NH_3, Ne, H_2S, Ar, and HF. The fact that we observe PH_3, C_2H_2, and C_2H_6 at even cooler temperatures in the visible part of the atmosphere therefore provides direct evidence of disequilibrium chemistry.

Calculations of the expected cloud composition and vertical structure on Jupiter assuming thermochemical equilibrium and a solar-composition atmo-

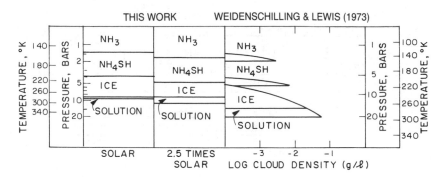

Fig. 14. Predicted positions of cloud bases using Voyager observations of atmospheric temperatures and assuming solar and twice solar NH_3 on Saturn. Also shown are predicted cloud positions and concentrations using an earlier (arbitrary) temperature profile by Weidenschilling and
Lewis (1973).

sphere have been carried out by Weidenschilling and Lewis (1973); their results are shown in Fig. 14. However, their assumed mixing ratios for H_2O,
CH_4, and H_2S were $\sim 14\%$ less than the Cameron (1982) ratios. They also
assumed a temperature $T \sim 108$ K at the pressure $P = 1$ bar level whereas the
subsequently observed temperature at this level was ~ 136 K; i.e., assuming
a dry adiabatic lapse rate we have

$$\ell nP \simeq 3.1 \, \ell n(T/136). \tag{11}$$

Using the Clausius-Clapeyron equation (see also Fig. 2 in Weidenschilling
and Lewis [1973]), we deduce that the pressures and temperatures at the
NH_3, NH_4SH, and H_2O (ice or water) cloud bases in a solar or near-solar
composition atmosphere obey the approximate formula

$$\ell nP \simeq 21 \, \ell n(T/T_0) \tag{12}$$

where $T_0 = 150, 158, 198, 208, 252$ and 262 K for solar NH_3; 2.5 times solar
NH_3; solar NH_3 and H_2S; 2.5 times solar NH_3 and H_2S; solar H_2O and 2.5
times solar H_2O, respectively. Equating (11) and (12), we thus conclude that
for solar composition the cloud bases on Saturn are at: 153 K, 1.4 bar (NH_3);
211 K, 3.9 bar (NH_4HS); and 280 K, 9.4 bar (H_2O). For mixing ratios 2.5
times solar the bases instead lie at: 162 K, 1.7 bar (NH_3); 224 K, 4.7 bar
(NH_4HS); and 294 K, 10.9 bar (H_2O). These levels are indicated in Fig. 14
where appropriate; we note that by using the updated Saturnian temperature
profiles the aqueous NH_3 clouds are much thinner.

As on Jupiter, the cloud materials predicted in equilibrium models
(namely, ammonia crystals, ammonium hydrosulfide crystals, ice crystals,
and aqueous NH_3 droplets) are all colorless. The (albeit faint) colorations as-

sociated with the belts and spots on Saturn (like C_2H_6, C_2H_2, and PH_3) must therefore be a product of various nonequilibrium reactions which we discuss next.

B. Nonequilibrium Phenomena

Some general considerations concerning nonequilibrium phenomena in planetary atmospheres are provided by Prinn and Owen (1976). In general, nonequilibrium conditions may be induced in an air parcel through the occurrence of irreversible chemical reactions or through the exchange of material in the air parcel with its surroundings. The degree to which nonequilibrium tendencies can prevail on Saturn is exemplified by the observation that PH_3, C_2H_6, and C_2H_2 have mixing ratios in the visible atmosphere, respectively, some 24, 12, and 48 orders of magnitude larger than their mixing ratios predicted at the 300 K level in thermochemical equilibrium (see Fig. 13).

Prediction of concentrations of species under nonequilibrium conditions requires solution of an inherently 3-dimensional continuity equation which necessitates extensive knowledge of both vertical and horizontal atmospheric motions. On all planets other than Earth we possess insufficient knowledge of these motions to justify in general the use of a 3- or even 2-dimensional approach. One-dimensional models in which the globally-averaged vertical transport is parameterized as a Fickian diffusion process are clearly the model of last resort but fortunately have enabled at least first-order analyses of most of the available observations of Saturn. In such models the continuity equation, for example, for a species with mixing ratio f_i in an atmospheric region where its chemical production is zero and its chemical destruction time τ is constant is

$$\frac{\partial [i]}{\partial t} = - \frac{[i]}{\tau} + \frac{\partial}{\partial z} \left(K[M] \frac{\partial f_i}{\partial z} \right). \tag{13}$$

Here the square brackets denote molecular concentrations, M is any molecule, and K is the so-called eddy diffusion coefficient. In a steady-state $\partial [i]/\partial t = 0$ and chemical destruction is therefore exactly balanced by convergence of the vertical flux of i. If f_i is finite at large altitudes and prescribed at altitude $z = 0$ the solution to Eq. (13) is

$$f_i(z) = f_i(0) \exp [-z/h]$$

$$h = \left[- \frac{1}{2H} + \left(\frac{1}{4H^2} + \frac{1}{K\tau} \right)^{1/2} \right]^{-1} \tag{14}$$

where h is the scale height of f_i and H the density scale height. If $\tau \ll 4H^2/K$ then $h \simeq \sqrt{K\tau} \ll H$ and f_i decreases very rapidly with height while if $\tau \gg 4H^2/K$ then $h \simeq K\tau/H \gg H$ and $f_i \simeq f_i(0)$. In other words, the vertical distribution of a particular species depends sensitively on both the strength of vertical transport and on its chemical destruction rate.

In this chapter we will restrict our discussion to the chemistry of neutral species below the turbopause. The latter level which lies at a pressure of $\sim 4 \times 10^{-9}$ bar on Saturn (Atreya 1982) denotes the altitude above which molecular diffusion (measured by the molecular diffusion coefficient D) exceeds bulk atmospheric motion (measured by K) for vertical transport of species. Various *a priori* estimates of K on Saturn are possible. In the deep atmosphere free convection is expected to be a dominant mode for vertical transport. If we ignore Coriolis effects, the appropriate formula for K is

$$K \simeq H\left(\frac{\phi}{\rho\gamma}\right)^{1/3} \tag{15}$$

($\simeq 10^8 - 10^9$ cm^2 s^{-1}) where γ is the ratio of the heat capacity at constant volume to the gas constant, ρ is the atmospheric density, and ϕ is the upward heat flux carried by the free convection (which is approximately 0.8 times the solar constant at Saturn). By analogy with the discussion of Jupiter by Stone (1976), we might expect in the region where solar energy is deposited on Saturn (the cloud-containing region between ~ 11 bar and ~ 400 mbar) that free convection dominates vertical transport at low latitudes while baroclinic eddies with K given roughly by

$$K \simeq 0.23 \frac{fH^2}{1 + \text{Ri}} \tag{16}$$

($\gtrsim 10^7$ cm^2 s^{-1}) dominate at higher latitudes. In Eq. (16) we have used values for the Saturnian Coriolis parameter f and Saturnian atmospheric Richardson number Ri as given by Stone (1973). Finally, above the tropopause (located at ~ 100 mbar) we might hypothesize that transient vertically-propagating internal waves dominate vertical transport (cf. Lindzen 1971 for Earth). For these waves

$$K \simeq K(0)\left[\frac{\rho(0)}{\rho(z)}\right]^{1/2} \simeq 10^5 \exp\left[\frac{z}{2H}\right] \tag{17}$$

in cm^2 s^{-1}, where $z = 0$ is the tropopause. This formula provides a reasonable fit to estimates of K at the 30 mbar level (which we derive later [see Eq. 30]) and also at the turbopause (see Atreya 1982).

Our discussion of disequilibrium chemistry on Saturn is subdivided here into three convenient headings: rapid vertical transport; photochemistry; lightning discharges and other processes. As discussed by Prinn and Owen (1976), for breaking chemical bonds there is on Jupiter $\sim 10^3$ times more energy available as solar ultraviolet radiation than as electric currents and acoustic waves in lightning or as ion fluxes from the radiation belts. A similar statement can be made for Saturn; the principal disequilibrating processes appear to be the breaking of chemical bonds by photodissociation and rapid vertical mixing of metastable species from the deep atmosphere.

1. Rapid Vertical Transport.   As first noted by Prinn and Lewis (1975), the discovery of PH_3 on Jupiter provided direct evidence that the oxidation of this gas to P_4O_6 in the deep atmosphere must occur on a time scale well in excess of typical vertical mixing time scales. The observation of PH_3 on Saturn discussed earlier in this chapter leads us to a similar conclusion for Saturn. In particular, referring to Fig. 13 we see that PH_3 in thermochemical equilibrium attains an abundance similar to its observed abundance only at pressures and temperatures exceeding 1000 K and 1300 bar. At lower temperatures and pressures net oxidation of PH_3 begins. The overall reaction is

$$4PH_3 + 6H_2O \rightarrow P_4O_6 + 12H_2 \tag{18}$$

and a possible mechanism for this reaction involves thermal decomposition of PH_3 and H_2O to yield PH and OH followed by the elementary reactions:

$$PH + OH \rightarrow PO + H_2 \tag{19}$$

$$PO + OH \rightarrow PO_2 + H \tag{20}$$

$$PO + PO_2 + M \rightarrow P_2O_3 + M \tag{21}$$

$$2P_2O_3 + M \rightarrow P_4O_6 + M. \tag{22}$$

An approximate method for treating disequilibration by rapid vertical mixing has been devised by Prinn and Barshay (1977) and we can use their method here. For purposes of discussion we will assume that Reaction (19) is the slow or rate-determining step with a rate constant k at $T > 1000$ K of 10^{-10} cm^3 s^{-1}; a more definitive treatment will require laboratory studies of the kinetics of Reaction (18) that are presently unavailable. The chemical lifetime of PH_3 is therefore

$$t_{chem} = \frac{[PH_3]}{k[PH][OH]} \tag{23}$$

resulting in $\simeq (4$ or $11) \times 10^6$ s for $T = 1250$ K where we have used the chemical equilibrium values for the concentration of OH and the ratio $[PH_3]/[PH]$ at 1250 K and solar or 2.5 times solar H_2O. This t_{chem} value can be compared to the time constant for vertical mixing defined by

$$t_{conv} \simeq \frac{H^2}{K} \tag{24}$$

giving $\simeq 6 \times 10^6$ s (1250 K) where we have assumed a K value of 3×10^8 cm^2 s^{-1} based on Eq. (15). Since $t_{chem} \simeq t_{conv}$ at 1250 K, this temperature level in the atmosphere is then the quench level for the oxidation of PH$_3$ to P$_4$O$_6$. At higher temperatures chemical equilibrium prevails. At lower temperatures the mixing ratio of PH$_3$ decreases at a rate depending on the chemical scale height

$$h_{chem} = - \frac{[PH_3] t_{chem}^{-1}}{\dfrac{d}{dz} \left([PH_3] t_{chem}^{-1} \right)} \tag{25}$$

giving $\simeq 10$ km for $T = 1250$ K. Since $h_{chem}^2 \ll H^2$ then the solution to the PH$_3$ continuity equation implies that the mixing ratio of PH$_3$ at $T < 1250$ K remains essentially the same as that at 1250 K; in effect oxidation of PH$_3$ in upward-moving air parcels at 1250 K is quenched by moving a distance much less than the mixing length H. For this reason, we expect to observe PH$_3$ in Saturn's visible atmosphere with an abundance appropriate to that at 1250 K, i.e., with a ratio relative to H$_2$ of 4.9×10^{-7} to 1.2×10^{-6} corresponding to Saturnian P/H ratios of 1 to 2.5 times solar.

At present PH$_3$ is the only nonequilibrium gas observed on Saturn whose presence is interpreted as due to rapid vertical mixing from the deep atmosphere. On Jupiter, CO and GeH$_4$ are observed in addition to PH$_3$ and their presence, like that of PH$_3$, is believed to be due to rapid mixing (see Prinn and Owen 1976; Prinn and Barshay 1977; Barshay and Lewis 1978). In addition, Prinn and Olaguer (1981) have shown that the most abundant nonequilibrium species expected in Jupiter's visible atmosphere as a result of the rapid mixing process is in fact N$_2$; while this gas is difficult to detect by remote sensing, it is potentially detectable by the neutral mass spectrometer on the upcoming Galileo entry probe.

On Saturn, nonequilibrium amounts of N$_2$, CO, and GeH$_4$ are all expected to be present in the visible atmosphere. For N$_2$ the work of Prinn and Olaguer (1981) for Jupiter can be roughly scaled for Saturn taking into account the equilibrium N$_2$ concentrations in Fig. 13. Mixing ratios for N$_2$ of 6×10^{-8} to 6×10^{-6} are predicted for Saturn depending on the precise values assumed for the N/H ratio and the eddy diffusion coefficient K, and on the presence or absence of effective catalysis of the overall N$_2$ destruction reaction

$$N_2 + 3H_2 \rightarrow 2NH_3 \tag{26}$$

by metallic iron particles. For CO the Jovian results of Prinn and Barshay (1977) when similarly scaled to Saturn imply a CO mixing ratio $\sim (5-12) \times 10^{-10}$ (for solar to 2.5 times solar H$_2$O) with at least a factor of three uncertainty. For GeH$_4$ no quantitative studies have been carried out due to the total lack of kinetic data for its appropriate destruction reactions. Its observed mixing ratio on Jupiter is 6×10^{-10} (Fink et al. 1978). This GeH_4 mixing ratio is expected on Jupiter in equilibrium at the 800 K level according to Fegley and Lewis (1979). If we interpret this 800 K level as the quench level on both Jupiter and Saturn, then the Saturnian equilibrium GeH_4 abundances in Fig. 13 imply a GeH_4 mixing ratio in Saturn's visible atmosphere of $\sim 4 \times 10^{-9}$. The observed upper limit for its mixing ratio (see Sec. II) is, however, only $\sim 10^{-10}$ which appears to imply that vertical mixing rates in the deep Saturnian atmosphere may be an order of magnitude or so less than those on Jupiter. Note that this would not affect our conclusions concerning PH_3 on Saturn given earlier; even if we decrease K from 3×10^8 cm^2 s^{-1} to 3×10^5 cm^2 s^{-1} the PH_3 oxidation quench level only moves up from the 1250 K level to the 1100 K level and PH_3 from Fig. 13 is still the dominant P compound at the latter level.

2. Photochemistry.   The photochemistry of Saturn bears many resemblances to that of Jupiter but there are important differences which derive principally from the lower temperatures on Saturn. We will specifically discuss here the neutral photochemistry of CH_4, NH_3 and PH_3 and briefly address the possible photochemistry of H_2S. The photochemistry of H_2 and of the ionosphere are discussed in the chapter by Atreya et al. Our discussion of Saturnian photochemistry will emphasize the use of predictive models based on reaction kinetics and those that include transport. The alternative approach using laboratory simulations can also play a role but there are problems in this approach which are discussed fully by Prinn and Owen (1976).

Methane photochemistry.   The treatment of CH_4 photochemistry on Jupiter by Strobel (1975) is also qualitatively applicable to Saturn. Methane is dissociated by ultraviolet wavelengths < 1600 Å with 70% attributable to the solar H Lyman-α radiation alone. The latter radiation yields $^1CH_2$ 92% of the time and CH 8% of the time. Subsequent reactions lead to the net production of higher hydrocarbons including C_2H_2 and C_2H_6 in particular; as discussed in Sec. II, both of these species have been observed on Saturn and their presence there is directly attributable to CH_4 photodissociation. The chemical cycle for CH_4 is summarized in Fig. 15 and calculations by Strobel (1978) for Saturn assuming two different hypothesized expressions for the vertical K profile are shown in Fig. 16. Strobel (1978) predicts mixing ratios at the 18 mbar level (where $[M] = 10^{18}$ molecule cm^{-3}) for C_2H_6, C_2H_4, and C_2H_2 of 2×10^{-6}, 6×10^{-12}, and 6×10^{-9}, respectively. Since these predicted mixing ratios have at least a factor of 3 uncertainty (associated with the uncertainty in K), the agreement between these predictions and the C_2H_6 and C_2H_2 observations discussed in Sec. II should be considered good. These predicted mixing ratios may also be compared to those predicted in thermochemical equilibrium given in Fig. 13. In particular, the photochemical mixing ratios are not

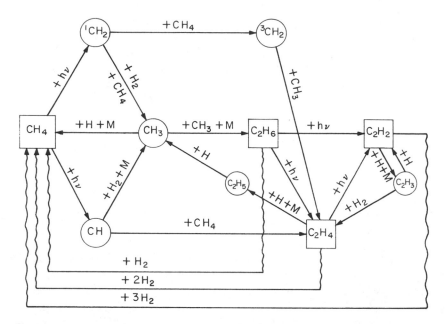

Fig. 15. A summary of CH_4 photochemistry on Saturn. The recycling of CH_4 from C_2H_6, C_2H_4, and C_2H_2 is accomplished in part in the deep hot lower atmosphere. Vertical transport between upper and lower atmosphere is designated by vertical wavy lines. Reactive radicals are enclosed in circles and more stable molecules enclosed in squares (after Prinn and Owen 1976).

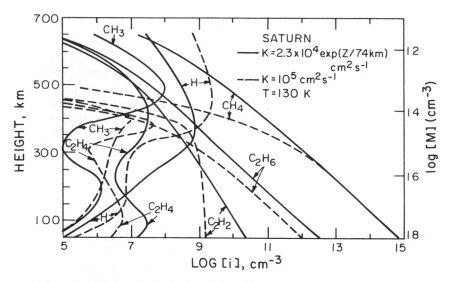

Fig. 16. Predicted hydrocarbon molecular densities [i] on Saturn for two assumed profiles of the vertical eddy diffusion coefficient K are given as functions of both height (with an arbitrary origin) and total molecular number density [M] (after Strobel 1978).

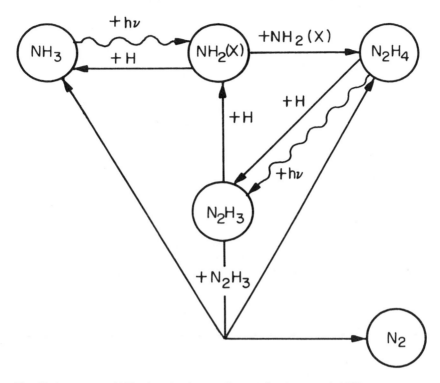

Fig. 17. A summary of NH_3 photochemistry on Saturn (after Atreya et al. 1978).

equalled until $T > 2000$ K for C_2H_6 and C_2H_2 and until $T \sim 1100$ K for C_2H_4. Since pyrolysis of these various hydrocarbons is expected at $T > 1000$ K, it is apparent that downward-moving air parcels containing the photochemically predicted mixing ratios of C_2H_6 and C_2H_2 (and to a lesser extent C_2H_4) will have their C_2H_6 and C_2H_2 pyrolysed to recycle the CH_4 from which they were derived. This recycling is therefore included in the Fig. 15 schematic.

Ammonia photochemistry. The photochemistry of NH_3 on Saturn has been studied by Atreya et al. (1980). The important reactions are outlined in Fig. 17 while the predicted concentrations of the various photochemical products as a function of altitude are summarized in Fig. 18. As expected, the much lower temperatures at a given pressure on Saturn than on Jupiter should severely restrict the amount of NH_3 gas in Saturn's upper atmosphere; it is largely precipitated out as NH_3 crystals beginning at the ~ 1.6 bar level as seen in Fig. 14. However, Atreya et al. (1980) argue that the photons which dissociate NH_3 at wavelengths < 2200 Å will merely penetrate to somewhat higher pressures (and temperatures) on Saturn than on Jupiter; a significant amount of NH_3 photodissociation can therefore still proceed. This argument

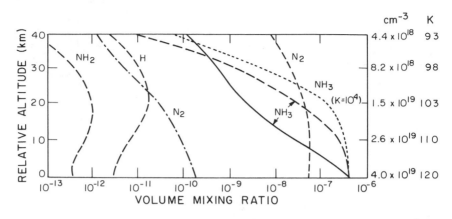

Fig. 18. Predicted mixing ratios of NH_3 photochemical byproducts on Saturn are given as functions of: altitude z above the cloud tops (see left-hand ordinate); and total molecular number density and temperature (see two right-hand ordinates). Dashed lines denote calculations assuming NH_3 and N_2H_4 do not condense above $z = 0$ and that K obeys Eq. (17) with K at $z = 0$ being 2×10^3 cm^2 s^{-1}. Dotted line as for dashed line but with $K = 10^4$ cm^2 s^{-1} throughout. Solid and dashed-dotted lines denote calculations assuming NH_3 and N_2H_4 condense above $z = 0$ yielding saturation mixing ratios for both (figure after Atreya et al. 1980).

is however complicated by the fact that both PH_3 and colored cloud particles will compete with NH_3 for absorption of the wavelengths < 2200 Å. Atreya et al. (1980) also suggest that NH_3 may be supersaturated in Saturn's upper atmosphere. Both these assumptions provide the distinct potential for overestimating the rate of NH_3 photolysis and Fig. 18 should be interpreted with this in mind. The principal predicted product of NH_3 photolysis is N_2 with maximum mixing ratios between 1.8×10^{-10} and 6×10^{-8} with the lower of these values applying if there is no supersaturation. Referring to Fig. 13, these photochemical N_2 mixing ratios equal the thermochemical equilibrium ratios in the $500-700$ K region. Since pyrolysis of N_2 to recycle NH_3 is unlikely at these temperatures, photochemically produced N_2 should slowly accumulate (in contrast to C_2H_2 and C_2H_6 discussed above). In any case, rapid vertical mixing discussed in Sec. III.B.1 yields predicted N_2 mixing ratios in the visible atmosphere of 6×10^{-8} to 6×10^{-6}, significantly larger than the photochemically predicted ratios.

Condensed hydrazine (N_2H_4) is an alternative to N_2 as the dominant product of NH_3 photolysis provided particle nucleation is efficient in the Saturnian stratosphere. On Jupiter, hydrazine particles may provide an important source of opacity at near-ultraviolet wavelengths (see Prinn and Owen 1976). However, the lower stratospheric NH_3 abundances (and thus the lower potential N_2H_4 particle densities) on Saturn argue against hydrazine aerosols being important ultraviolet absorbers on this planet.

Phosphine photochemistry.   Phosphine is dissociated by wavelengths $<$ 2350 Å and since only wavelengths $>$ 1600 Å (which are not absorbed by CH_4) penetrate to levels on Saturn where PH_3 is abundant, the principal immediate products are PH_2 and H. The photochemistry of PH_3 was first studied on Jupiter by Prinn and Lewis (1975). Their results are also applicable to Saturn but with certain modifications.

First, the H produced from PH_3 photodissociation and the NH_2 produced from NH_3 photodissociation can both increase the PH_3 destruction rate through the reactions (Strobel 1977)

$$H + PH_3 \rightarrow PH_2 + H_2, \qquad (27)$$

$$NH_2 + PH_3 \rightarrow PH_2 + NH_3. \qquad (28)$$

For Saturn, Reaction (27) essentially doubles the PH_3 dissociation rate from that predicted by photolysis alone leading to a PH_3 lifetime ~ 58 days in the upper atmosphere (Strobel 1978). Reaction (28) is not as important on Saturn as it is on Jupiter due to the lower NH_3 mixing ratios in Saturn's stratosphere; however, if NH_3 on Saturn is highly supersaturated as Atreya et al. (1980) suggest, then Reaction (28) could play a role in addition to Reaction (27). Note that on both Saturn and Jupiter PH_3 is not depleted in the upper atmosphere by condensation whereas NH_3 is.

Second, the mechanism of P_2 and P_4 formation suggested by Prinn and Lewis (1975) which was based on the experiments of Norrish and Oldershaw (1961) needs modification. In particular, Ferris and Benson (1981) have discovered that P_2H_4 is an important intermediate in the production of P_4 from PH_3 photolysis. They suggest that subsequent destruction of P_2H_4 (by a presently unknown mechanism) is the principal source of P_2 rather than the reaction

$$PH + PH \rightarrow P_2 + H_2. \qquad (29)$$

Third, Ruiz and Rowland (1978) have suggested that C_2H_2 may scavenge some of the radicals produced by PH_3 photolysis thus inhibiting P_4 formation presumably in favor of formation of organo-phosphorus compounds. However, since C_2H_2 production rates on both Jupiter and Saturn are much less than PH_3 photodissociation rates, this must represent a relatively minor path. Finally, Howland et al. (1979) have suggested that recycling of P_4 to PH_3 may proceed through reactions in the H_2O-NH_3 solution clouds rather than deeper in the atmosphere; we have no way to quantitatively test this suggestion at present.

Our present understanding of the phosphorus cycle on Saturn is outlined in Fig. 19. The principal product of PH_3 photolysis is elemental phosphorus presumably in its red triclinic crystalline form denoted in this figure as $P_4(s)$, although other forms are also likely (Noy et al. 1981). Elemental phosphorus

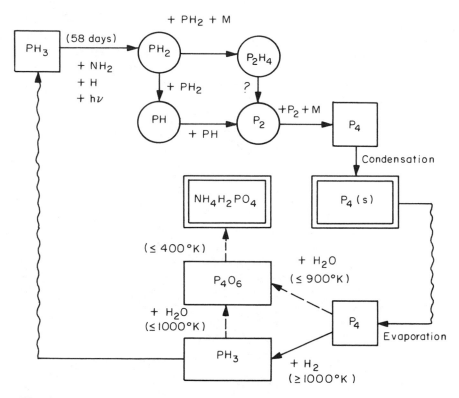

Fig. 19. A summary of PH_3 photochemistry and thermochemistry on Saturn with reactive radicals enclosed in circles, more stable molecules enclosed in squares, and condensates enclosed in double squares. Vertical transport is denoted by wavy lines. Reactions slow with respect to vertical transport are denoted by dashed lines.

particles are a prime candidate for explaining the faint red or orange colorations on Saturn. Recycling of elemental phosphorus through mixing down to deep levels where it is pyrolysed yielding PH_3 is a likely way of closing the phosphorus cycle. As noted earlier, conversion of PH_3 in upward-moving air parcels into P_4O_6 and $NH_4H_2PO_4$ is apparently too slow, relative to vertical mixing for oxidized phosphorus compounds, to play a significant role on Saturn.

Equation (14) may be used to model the vertical distribution of photolysed PH_3 and, when combined with the observations of this distribution, may be used to estimate the vertical eddy diffusion coefficient K in Saturn's stratosphere. In particular, the preferred distribution from Tokunaga et al. (1980c) (marked T1 in Fig. 6) has $h \simeq 10$ km at the 30 mbar level. Also from our discussion above, $\tau = 58$ days, and H at the 30 mbar level is 30 km. Thus rearranging Eq. (14) we have

$$K = \frac{h}{\tau}\left[\frac{1}{h} + \frac{1}{H}\right]^{-1} \tag{30}$$

giving $1.5 \times 10^5 \text{ cm}^2 \text{ s}^{-1}$ which agrees with our suggested general expression for K given by Eq. (17).

Hydrogen Sulfide. While hydrogen sulfide is potentially remotely observable through breaks in the NH_3 clouds on Jupiter, it is very unlikely that it will ever be detected by remote sensing on Saturn. The mixing ratio of H_2S should decrease rapidly above the NH_4SH cloud base which is situated at ~ 4.3 bar level according to Fig. 14. Using the formula derived by Prinn and Owen (1976), its mixing ratio scale height $h \simeq 3$ km just above the NH_4HS cloud base and its mixing ratio will therefore decrease by ~ 10 orders of magnitude between the NH_4HS cloud base and the NH_3 cloud base at 1.6 bar. If the colored impurities in the upper ammonia clouds on Saturn do not totally absorb the radiation between 2300 and 2700 Å, then some fraction of this radiation will be able to penetrate to depths where H_2S abundances are appreciable. As reviewed by Prinn and Owen (1976), H_2S photodissociation is expected to lead to particulates of elemental sulfur, and hydrogen and ammonium polysulfides. In addition, irradiation of NH_4HS particles themselves can also lead to various particulate polysulfides. Such irradiation could occur through upward transport of NH_4HS particles into the visible atmosphere where the required 2300–2700 Å photons have high intensities. Some of these particulate polysulfide compounds are colored yellow, orange, or brown and upward mixing of the smallest particles may occur and thus provide some coloration to the upper visible clouds. Detection of H_2S on Saturn, and thus any validation of this hypothesis will undoubtedly have to await an entry probe mission to this planet.

3. Lightning and Other Processes. As noted earlier, lightning and energetic particles dumped into Saturn's atmosphere from its radiation belts are both minor factors in the bulk disequilibrium chemistry of Saturn relative to ultraviolet radiation. They do, however, have potential for influencing the chemistry in Saturnian auroral zones and thunderstorms, and for the production of interesting chemical compounds including biologically interesting compounds. Production of such compounds in experiments simulating lightning discharges and particle dumping (and also ultraviolet photolysis) also lead to the controversial question of whether there is any biological activity on Saturn. Since all these matters have already been addressed in detail by Prinn and Owen (1976) for Jupiter, we will not attempt an extensive and largely repetitive review here except for further relevant work since then which we briefly review below.

The discovery of CO on Jupiter mentioned earlier led to suggestions for a source for this species other than rapid vertical mixing. In particular, extraplanetary sources of oxygen in Jupiter's upper atmosphere were proposed in the form of micrometeoroids by Prather et al. (1978) and in the form of dumping from the Io torus by Strobel and Yung (1979). The extent to which extraplanetary sources are important in Saturn's upper atmosphere is at present unclear but the presence of the icy rings suggests that the matter should be investigated further.

Borucki et al. (1982) have investigated the rate of lightning discharges on Jupiter using optical and radio frequency detections of lightning from Voyager 1. They conclude that the fraction of the total atmospheric convective energy flux converted into lightning is $(2.7-5) \times 10^{-5}$. In a study of the chemistry of Jovian lightning discharges in which the above fraction was assumed to be 10^{-5}, Lewis (1980) concluded that production of CO, N_2, HCN, and simple hydrocarbons in these discharges was unimportant when compared to the photochemical and deep-atmospheric sources for these species. For example, his predicted HCN mixing ratio from lightning was only $\sim 3 \times 10^{-17}$ (this number should be increased by a factor of $2.7-5$ to take into account the Borucki et al. [1982] results). It is therefore clear that lightning is not playing an important role in producing biologically interesting compounds on Jupiter. It is difficult to imagine why similar conclusions about the role of lightning should not hold also for Saturn.

IV. CONCLUSION

Though our understanding of the composition and chemistry of Saturn probably cannot improve as dramatically in the near future as in the past few years, we can look forward to advances in several areas of research: the future use of the Space Telescope; the Galileo entry probe into Jupiter; continued improvements in spatial and spectral resolution and coverage from ground-based and airborne observatories; acquisition of further laboratory data on the spectra and kinetics of relevant species; and development of atmospheric line-formation and chemical models which provide an optimal fit to all available data. Each of these areas provides potential for important advances.

We can usefully review and pinpoint here some particular topics requiring future resolution. We have noted the peculiar observability of NH_3 at visible but not at infrared wavelengths; what is this telling us about the clouds and line-formation process on Saturn? Are the differences between the D/H ratios derived from CH_3D and HD real and if so, what does this tell us about D-H exchange reactions? Can we identify CH_4, NH_3, PH_3, and H_2 spectral lines with sufficiently similar formation processes to better define the C/H, N/H, and P/H ratios? We have emphasized the fact that PH_3 provides an ideal tracer for vertical transport in Saturn's stratosphere; which of the two very different estimates of the stratospheric distribution of PH_3 is correct? Are the visible clouds composed of NH_3 ice and if so, why is Saturn so bright at 3.0 μm? Germane (GeH_4) is observed on Jupiter but not on Saturn despite the fact that its concentration in the deep atmosphere of Saturn is predicted to be greater than that on Jupiter. What is this telling us about vertical transport and chemistry in the deep atmosphere? We do not know whether CO, H_2S, H_2O, or C_2H_4 are present in the atmosphere (CO, C_2H_4, and H_2O have all been observed on Jupiter). Are the pale colors seen in Saturn's clouds merely due to various allotropes of phosphorus? How does the chemistry of Saturn change from season to season? Are the Voyager data useful here and can further Earth-based observations help?

We could continue this list much further but the point to be made is simply that despite recent progress, Saturn continues, as it has in the past, to provide us with plenty of challenging scientific problems to test our wits. We hope that this review will prove germinal in the future resolution of at least a few of these problems.

Acknowledgments. This research was supported in part by the Atmospheric Chemistry Program of the National Science Foundation through a grant to the Massachusetts Institute of Technology. We thank R. A. West for his constructive comments on the original manuscript.

CLOUDS AND AEROSOLS IN SATURN'S ATMOSPHERE

M. G. TOMASKO
University of Arizona

R. A. WEST
University of Colorado

G. S. ORTON
Jet Propulsion Laboratory

and

V. G. TEJFEL
Astrophysical Institute, Alma-Ata

The physical properties and spatial distribution of aerosol and cloud particles in Saturn's atmosphere are important in many physical processes, but are still only incompletely known. We begin by outlining the connections between the observable radiation scattered or emitted by Saturn and the distribution and optical properties of cloud and aerosol particles. Next we review the observations available for constraining the properties of the aerosols. These include: continuum photometry from the ultraviolet to the near infrared at various spatial resolutions and scattering geometries; photometry in gaseous absorption features of various strengths at different locations on the disk in the near-infrared; maps of the linear polarization of reflected sunlight at phase angles from 10 to 150° in red and blue light, as well as at many wavelengths at small phase from the Earth; maps of the planetary thermal emission as functions of emission angle and wavelength; and infrared spectroscopy of the planet at high spectral resolution. We discuss the implications of the observational data for the vertical and horizontal distributions, single-scattering optical properties

and, so far as possible, the composition of the various aerosol and cloud parti-
cles. A layer of small aerosols (radii ~ 0.1 μm) which absorb strongly in the
ultraviolet is found at very high altitudes (pressures < 20 mbar) in the polar
regions, as well as at lower altitudes (pressures 30–70 mbar) at equatorial and
temperate latitudes. These particles are thought to be produced in photochemi-
cal reactions, and are probably similar to the aerosols seen at high altitudes on
Jupiter (and possibly Titan). At deeper levels (pressures >100 mbar), a diffuse
cloud is seen which may be partly composed of ammonia crystals. Within about
20° of the equator, where the zonal winds are especially strong, the cloud has
an optical depth (in red light) as great as 4 between the radiative-convective
boundary near 400 mbar and its top at about a pressure scale height above. At
latitudes >30°, the red optical depth above 400 mbar is only ~ 0.1. Over much
of the planet, the cloud seems most dense between 400 and ~ 700 mbar, reach-
ing optical depth ~ 5 by this level. Some data hint at a relatively clear region at
greater pressures until still deeper clouds are reached. We identify several
areas in which new planetary observations, laboratory measurements, or the-
oretical analyses could significantly advance our understanding of Saturn's
clouds and aerosols.

Like those of its neighbor Jupiter and its moon Titan, the atmosphere of
Saturn contains a variety of particulate material. The aerosols and clouds on
Saturn play important roles in many physical processes. For example, the
Bond albedo of the planet is largely determined by the single-scattering properties of the aerosol and cloud particles and their distribution with height and
latitude. The ability of the stratospheric aerosols to absorb sunlight and emit
thermal radiation can be important in determining the temperature of the atmosphere at pressures < 100 mb. At deeper levels, the cloud particles can
play significant roles in the atmospheric heat budget through their thermal
opacity, latent heat, absorption or reflection of sunlight, and equilibration of
the ortho and para states of H_2.

The variations in local heating or cooling rates with latitude provided by
the aerosols at high altitudes, or by the deeper clouds, act as forcing mechanisms for atmospheric dynamics. They combine with the planet's rotation and
the escape of internal heat to produce Saturn's dramatic jet-stream winds in
ways that are still incompletely understood. In fact, the winds themselves are
tracked primarily by following the motions of the few distinct cloud features
seen on the disk. The vertical location of these cloud features is often poorly
known.

The importance of scattering to the determination of molecular mixing
ratios cannot be overstated. Regrettably, mixing ratios are often inferred by
calculating the apparent column abundance above a reflecting surface in a reflecting layer model atmosphere. Since scattering affects strong and weak
lines differently, the inferred mixing ratios will generally be incorrect unless
particular care has been taken in choosing the lines (Encrenaz and Combes
1982). Both the center-to-limb behavior of the absorption (West et al. 1982)
and the relative strengths of strong versus weak lines (De Bergh et al. 1977),
prove the obsolescence of the reflecting layer model and demand a more so-

phisticated approach throughout the entire spectral region where scattered sunlight is the dominant source of radiation. Scattering is probably important in the 5 μm region as well.

Finally, the aerosols and clouds play a role in the chemistry of the atmosphere. Considerable effort is being expended to understand the photochemistry of CH_4 in the atmospheres of the outer planets which leads to higher hydrocarbons in the observed abundances as well as solid aerosols with the observed properties. Work is also needed on the physical chemistry of compounds that could form Saturn's condensation clouds and their coloring agents.

Depending on its exact mixing ratio, a cloud of solid ammonia crystals is expected to form with its base at a pressure level of ~ 1.4 bars, well into Saturn's troposphere. Convection could carry these particles at least as high as the upper boundary of the convective region near 400 mbar (Appleby 1980), forming an extended hazy region. An extended haze is indicated by many observations of Saturn, but it has proven difficult to obtain definite evidence that the haze consists of NH_3 crystals. Even if NH_3 crystals are a major constituent of this haze, they cannot be the only constituent because in their pure form they do not have the colors observed for Saturn. The identity and source of the coloring agent is a longstanding open question. Further, absorptions predicted for some types of NH_3 crystals have not been observed, leading some authors to question whether the visible haze contains NH_3 crystals at all, or is primarily due to some other condensable constituent.

At the level of Saturn's tropopause and above, solid particulates are found which are not believed to be atmospheric condensation products. These aerosols absorb strongly in the ultraviolet but are almost invisible at red wavelengths. At high latitudes, the absorption in the ultraviolet is considerably stronger than in equatorial or temperate regions. Are the ultraviolet absorbers the same in the two locations? How are the aerosols in Saturn's atmosphere related to the absorbers found in Jupiter's polar regions? Are they similar to the aerosols which are so predominant on Titan?

Obviously, we can raise many more questions concerning Saturn's clouds and aerosols than we can answer at the present time. Nevertheless, with many new results from the recent space missions in hand it may be useful to summarize the available observations which bear on the nature of aerosols, and to attempt to outline the present information concerning the optical properties, distribution, and composition of this important component of Saturn's atmosphere.

Since there is yet to be a Saturn entry probe mission, our current knowledge of the aerosol and cloud particles comes entirely from interpretation of remote measurements of scattered solar radiation and thermal radiation emitted from Saturn's atmosphere. Accordingly, we begin in Sec. I with a brief overview of the relations between particle properties and the scattered and emitted radiation field. The longer Sec. II contains a review of Saturn observations at wavelengths from the ultraviolet to the thermal infrared which bear on the aerosol and cloud properties. In Sec. III we discuss the single-scattering properties of the particles deduced from observations, and comment on their possible composition. In Sec. IV we review the implications of the observations for the vertical and horizontal distribution of the clouds and aerosols. We summarize current information concerning the clouds and aerosols and outline several areas requiring further work in Sec. V.

I. INFLUENCE OF PARTICLES ON SCATTERED AND EMITTED RADIATION

Before describing the observations and their interpretation, a discussion of the relations between particle properties and the observable radiation field is in order. An important parameter in radiative transfer is τ, the normal optical depth. The equation

$$d\mathbf{I}(\tau, \mu, \mu_0, \varphi - \varphi_0)$$

$$= \frac{d\tau}{\mu} \cdot \left[-\mathbf{I}(\tau, \mu, \mu_0, \varphi - \varphi_0) + \mathbf{J}(\tau, \mu, \mu_0, \varphi - \varphi_0) \right] \quad (1)$$

defines τ, where $\mathbf{I}$ is the four-element Stokes vector (erg cm^{-2} s^{-1} sterad^{-1} Hz^{-1}) traveling in a direction θ with respect to the surface normal (the vertical direction). With sunlight impinging on the atmosphere at angle θ_0 with respect to the surface normal, $\mu = \cos(\theta)$, $\mu_0 = \cos(\theta_0)$, φ = azimuthal angle of the scattered radiation projected onto the planetary surface, and φ_0 = azimuthal angle of the incident radiation projected onto the planetary surface.

In Eq. (1) the first term in brackets accounts for energy loss by absorption and scattering out of the beam, and the second describes addition of energy into the beam from scattering. For wavelengths $\lesssim 5$ μm, solar radiation is much greater than the thermal radiation emitted by Saturn, and

$$\mathbf{J}(\tau, \mu, \mu_0, \varphi - \varphi_0)$$

$$\equiv \int_{-1}^{1} \int_{0}^{2\pi} \mathbf{P}(\mu, \mu', \varphi - \varphi') \mathbf{I}(\tau, \mu', \mu_0, \varphi - \varphi_0) d\mu' d\varphi'$$

$$+ \frac{\widetilde{\omega}}{4} e^{-\tau/\mu_0} \mathbf{P}(\mu, \mu_0, \varphi - \varphi_0) \mathbf{F}_0. \quad (2)$$

In Eq. (2), the albedo for single scattering, $\widetilde{\omega} = \tau_{scat}/(\tau_{scat} + \tau_{abs})$, τ_{scat} = scattering optical depth, τ_{abs} = absorption optical depth, $\pi\mathbf{F}_0$ = incident solar flux, and $\mathbf{P}$ = the 4×4 phase matrix which describes the single scattering properties of the particle. (See Hansen and Travis [1974] for a detailed description of the treatment of scattering in planetary atmospheres.)

Note that $\tau_{\text{scat}} = \sigma_{\text{scat}} \int n dz$, and similarly for τ_{abs}, where n is the number of particles per unit volume and the integral extends over the altitude range covered by the cloud. Often σ_{scat} and σ_{abs} are given as dimensionless efficiency factors, Q_{scat} and Q_{abs} times the projected geometrical cross section of the particle (πr^2 for spheres). For spheres, Mie theory can be used to evaluate the efficiency factors and the phase matrix for particles of arbitrary complex refractive index and for any size (usually given as a dimensionless size parameter $x = 2\pi r/\lambda$, where λ is the wavelength and r is the radius of the particle). Thus, if observations at a range of scattering angles are used to constrain the angular dependence of the phase matrix, comparison with phase matrices computed from Mie theory can yield constraints on the size and refractive index of the particles *if the particles are spherical*. For nonspherical particles, comparison of the phase matrix with laboratory measurements or other types of calculations (such as geometrical ray tracing for large particles) is necessary before the physical properties of nonspherical particles (size, refractive index) can be deduced.

The strength and center-to-limb behavior of molecular absorption contains information on the vertical distribution of aerosols. As a rule of thumb, the depth to which one sees in the atmosphere is about $\tau/\mu = 1$, and depends on wavelength. Thus, measuring the center-to-limb variation of reflected intensity gives information on the vertical structure of the atmosphere as μ ranges between 0.0 at the limb and 1.0 at the disk center. For a purely absorbing gas above a reflecting layer, Eq. (1) has the solution

$$I = I_0 \exp[-\tau \cdot (1/\mu + 1/\mu_0)]. \tag{3}$$

The amount of absorption grows exponentially with $1/\mu + 1/\mu_0$ from the disk center to the limb. For a diffuse-scattering atmosphere, where particles are uniformly mixed with gas, the absorption decreases from the center of the disk to the limb because of the decrease in the ratio of multiply scattered light to singly scattered light and the corresponding decrease in effective optical path as the limb is approached. For Saturn, the observed center-to-limb behavior is intermediate between the reflecting layer and diffuse scattering cases.

Scattering by molecules of H_2 and He is important at short wavelengths. The total pressure and Rayleigh optical depth can be related by the equation of hydrostatic equilibrium. With

$$P = Mg = AL_0 \, \bar{m}g/N_a \text{ and } \tau = N\sigma = L_0 A\sigma \tag{4}$$

we have

$$\tau = P N_a \sigma/(\bar{m}g) \tag{5}$$

where P = pressure (dyne cm^{-2}), A = column abundance (cm-amagat H_2 + He), L_0 = Loschmidt's number (2.68719×10^{19} cm^{-3}), σ = extinction cross section, $\bar{m}$ = mean molecular weight = 2.12 for a 94%/6% H_2-He mixture, g = surface gravity, N_a = Avogadro's number (6.02486 × 10²³ gm-mol⁻¹), and M = surface density (g cm⁻²) of gas above the level with pressure P. The extinction cross section for scattering by H_2 is given by Dalgarno and Williams (1962). The effective gravity g is the sum of the gravity field due to the mass distribution in the planet and the centrifugal term $\Omega^2 \mathbf{R}$, where $\mathbf{R}$ is the position vector from the axis of rotation. The gravity varies from ~ 900 to 1180 cm s⁻² from equator to pole (see chapter by Hubbard and Stevenson). The choice of g = 950 cm s⁻² yields

$$\tau_R \approx 0.0244(1 + 0.016 \lambda^{-2})P\lambda^{-4} \tag{6}$$

where λ is in μm and P in bar. In Eq. (6) we have neglected the term of order λ^{-8} given by Dalgarno and Williams. At the one-bar level, the column abundance of H_2 + He is 111 km-amagat, and the Rayleigh optical depth at 0.264 μm (the wavelength of the Voyager photopolarimeter measurements) is $\tau_R = 6.2$.

Photometric observations are usually reported as reflectivity (I/F) as a function of position (spatially resolved) or geometric albedo p for the integrated disk at full phase (phase angle $\alpha = 0.0$) where

$$p = \frac{1}{\sigma_s F} \int_{\text{disk}} I(\mu, \mu_0, \varphi - \varphi_0) d\Omega. \tag{7}$$

Here σ_s is Saturn's geometric area projected on the plane of the sky and the incident solar flux is πF. At phase angles $> \theta$, integrated disk brightness is given by $p \, \Phi(\alpha)$ where the phase law $\Phi(\alpha)$ is unity at $\alpha = 0$. The geometric albedo and phase law can be calculated for a horizontally homogeneous scattering model from Horak's (1950) cubature formulas. The total amount of solar energy absorbed by the planet is fundamental in the heat balance equation, and is given by

$$E_{\text{abs}} = \pi F \sigma_s \int A(\lambda) d\lambda \tag{8}$$

where $A(\lambda)$ is spherical albedo, $A(\lambda) = p(\lambda) q(\lambda)$, and $q(\lambda)$ is the phase integral

$$q(\lambda) = 2 \int_0^\pi \Phi(\alpha) \sin \alpha \, d\alpha. \tag{9}$$

At wavelengths $\gtrsim 5$ μm the thermal emission for Saturn greatly exceeds the incident energy received from the Sun. Accordingly, the second term on the right-hand side of Eq. (2), which refers to the incident solar flux, can be ignored and replaced by the Planck function, $B[T(P),\lambda]$, at the temperature of the pressure level at which optical depth τ is reached. If the aerosols are sufficiently absorbing at the thermal wavelength considered, their ability to scatter light will be small and the first term on the right-hand side of Eq. (2) can be neglected as well. In this case the intensity emerging from the top of the atmosphere at angle θ to the vertical is given by

$$I(P = 0, \lambda, \theta) = \int_{P' = 0}^{\infty} B[T(P'), \lambda]\exp[-\tau(P', \lambda, \theta)]\mathrm{d}\tau(P', \lambda, \theta). \quad (10)$$

Here the explicit dependence of optical depth τ on θ is retained because the optical depth of the gas at thermal wavelengths is generally nonlinear in gas abundance and hence slant path.

If cloud opacity is ignored, and the composition and hence opacity of the gaseous atmosphere is known, observations of the emerging intensity at a variety of wavelengths which reach $\tau = 1$ at different pressure levels can be inverted for information regarding the $T(P)$ structure of the atmosphere (see, e.g. the chapter on atmospheric structure by Ingersoll et al.). On the other hand, the optical depth τ in Eq. (3) includes the optical depth of the clouds as well as the gaseous atmosphere, and if the $T(P)$ structure is known from other information (such as infrared data at other wavelengths or a radio occultation), the emerging thermal intensity can be used to constrain the pressure level and optical depth of cloud particles at the wavelength of the observation.

II. OBSERVATIONAL CONSTRAINTS

A. Ultraviolet Observations

The first measurements of Saturn and its rings in the ultraviolet below the terrestrial cutoff near 0.32 μm were reported by Bless et al. (1968) from rocket filter photometry in the spectral region 0.245 to 0.295 μm. Subsequently the unresolved Saturn-plus-rings system was observed from OAO-2 (Wallace et al. 1972; Caldwell 1975,1977b), and the TD1A satellite (Caldwell 1977b). Spectra have been obtained with the International Ultraviolet Explorer (IUE) (Winkelstein et al. 1983). Spatially resolved measurements are available from instruments on Pioneer 11 (Judge et al. 1980) and the Voyager 1 and 2 spacecraft (Lane et al. 1982b; Sandel et al. 1982b).

The ultraviolet measurements encompass a broad range of phenomena, including auroral processes and physics of the exosphere in the extreme ultraviolet ($\lambda < 1216$ Å), photochemistry in the stratosphere (primarily for $\lambda < 2000$ Å), and scattering and absorption of sunlight by gas and aerosols in the stratosphere and upper troposphere for $\lambda > 2000$ Å. The extreme ultraviolet observations are discussed in the chapter by Atreya et al. and will not be mentioned further, except to note that auroral processes may contribute to the formation of ultraviolet-absorbing aerosols in polar regions. The spectrum between 1500 and 2000 Å obtained with IUE (Winkelstein et al. 1983) is valuable in understanding the stratospheric composition (mainly the abun-

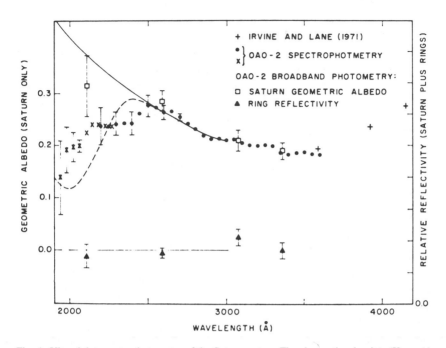

Fig. 1. Ultraviolet spectrophotometry of the Saturn system. The observational points (X, ●, +) are described by the right-hand ordinate, which is the ratio of the brightness of the planet plus rings divided by that of the Sun. Regular OAO-2 spectrophotometry (●), replotted from Wallace et al. (1972), and long integration time spectrophotometry (X), are shown with error bars only because of background uncertainty. The ring reflectivity (▲) error bars are only due to the uncertainty in extrapolating the total brightness to zero ring inclination. The Saturn broadband filter photometry points (□) were used to normalize the geometric albedo scale on the left. This scale is valid only shortward of 3500 Å. The solid curve is a model with 7 km-amagat H_2 above a Lambert reflecting layer with reflectivity 0.17. The dashed curve includes the effect of absorption by 0.01 cm-amagat of H_2S (from Caldwell 1977b).

dance of the trace components C_2H_2, C_2H_6, PH_3, and NH_3) and is discussed in more detail in the chapter by Prinn et al. We wish to focus on the region λ > 2000 Å where the aerosol properties dominate the reflectivity and center-to-limb variations, both in the continuum and in molecular absorption bands.

Saturn's geometric albedo in the spectral region from 0.222 μm to 0.35 μm has been discussed by Caldwell (1977b) who summarized observations from the OAO-2 and TD1A satellites; the results are shown in Fig. 1. The geometric albedo has a relative minimum near 0.35 μm. The broadband OAO-2 photometry measurement at 0.211 μm is discordant with the OAO-2 spectrophotometry (0.002 μm spectral resolution) which decreases below ~ 0.260 μm; the discrepancy is 30% at 0.211 μm.

Saturn's geometric albedo is similar to Jupiter's, and is much lower than that for a semi-infinite conservatively scattering molecular atmosphere. Axel

(1972) showed that, for Jupiter, a stratospheric haze of small particles with a strong wavelength dependence in extinction cross section could account for the spectral variation in geometric albedo over a broad range in wavelengths from the ultraviolet to the near infrared. More recently Macy (1977) and Podolak and Danielson (1977) derived models of the altitude and optical depth of a "Danielson dust" layer. The dust becomes monotonically more strongly absorbing at shorter wavelengths, but the wavelength dependence of the dust (λ^{-1}) is not as steep as the λ^{-4} dependence of molecular Rayleigh scattering from the gas above the dust, and in which the dust is mixed. By choosing the correct optical depth, altitude, and wavelength dependence for the dust, one can reproduce the wavelength dependence of the geometric albedo for $\lambda > 0.25$ μm. Both Macy (1977) and Podolak and Danielson (1977) were able to fit the ultraviolet geometric albedo of Caldwell's *broadband* measurements. As noted above, the OAO-2 spectrophotometry measurements diverge from the broadband filter measurements (and therefore also the models) below 0.25 μm.

More spatially resolved measurements of Saturn's reflectivity are needed. Caldwell (1975) estimated the effect of the rings in his spatially unresolved measurements by fitting a straight line to the brightness as a function of sin (B) where B is the mean of Saturnicentric declinations of the Earth and Sun. Without spatially resolving the disk and rings, it is impossible to estimate the contribution of latitudinal and temporal variations to the net change in brightness during the 3.5 yr period when sin (B) changed from 0.2 to 0.45. Latitudinal and seasonal variations are important in the ultraviolet.

The Voyager photopolarimeter experiment obtained spatially resolved measurements of Saturn in 1981 in broadband $(\Delta\lambda \sim 0.03$ μm) filters centered at effective wavelengths 0.264 μm and 0.750 μm. Preliminary results for Saturn were reported by Lane et al. (1982*b*) and West et al. (1983*a*). A comprehensive analysis of the data was presented by West et al. (1983*c*). Included in the data are central meridian scans in the northern hemisphere, which show latitudinal variations in the reflectivity and limb-to-terminator scans for the northern hemisphere at spatial resolution $\sim$ 8000 km at 10° phase angle, and at higher spatial resolution and higher phase angles ($\leq$ 68°) for the Equatorial Zone. In the ultraviolet the polar region is very dark and the North Equatorial Belt near 20° latitude is bright; the reflectivity at short wavelengths is anticorrelated to that at long wavelengths. The ultraviolet photometry and polarimetry data support the idea that small particles (Danielson dust) in the stratosphere are responsible for the low ultraviolet albedo, but the vertical distribution and total optical depth, discussed in more detail below, differ from the distributions proposed by Macy (1977) and Podolak and Danielson (1977). A tropospheric absorber is also important in the near ultraviolet.

Longward of 0.32 μm, groundbased observations are possible and generally are made with spatial resolution. The absolute groundbased reflectivities shortward of 0.37 μm are fairly uncertain, with accuracies no better

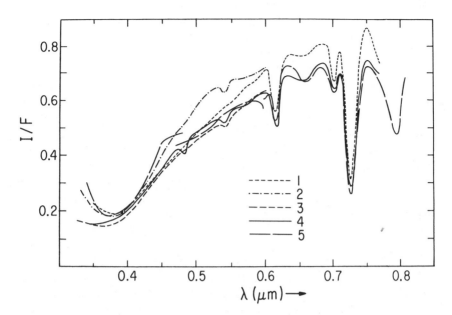

Fig. 2. Spectral reflectivity (*I/F*) of the equatorial region of Saturn according to the absolute
  spectrophotometric observations: (1) — in 1971 by Bugaenko; (2) — in 1979 by Gajsin; (3) —
  in 1981 by Cochran; (4) — in 1982 by Aksenov, Vdovichenko, and Solodovnik; (5) — in 1979
  by Berstralh et al.

than 30% to 50%. Nevertheless, near the center of the disk the groundbased
measurements confirm the decrease in reflectivity to a minimum of 0.15 near
0.37–0.38 μm, before increasing toward still longer wavelengths to values of
0.60–0.75 at 0.60 μm (see Fig. 2). Some versions of the curve of spectral
reflectivity have a hint of a break at 0.52 μm (McCord et al. 1971; Bugaenko
1972; Krugov 1972; Tejfel and Kharitonova 1974; Gajsin 1978,1979; Bergstralh et al. 1981; Cochran 1982). This peculiarity is also present in the spectral reflectivity of Saturn's rings (Kharitonova and Tejfel 1973).

The strong anticorrelation between the brightness of features at 0.264
μm and the visible washes out between the near ultraviolet and the blue region. In the ultraviolet, the equatorial regions appear as a dark belt at latitudes
≲ 10 or 15°, while the temperate belts are the brightest parts of the planet. At
0.35 μm the temperate belts are still at least 30% brighter than the equatorial
regions, but by 0.43 μm, the reflectivities are the same (Tejfel 1974*a,b*).
Some temporary or seasonal changes in the meridional brightness distribution
on Saturn were observed at the shortest groundbased wavelengths, in comparing the photometric measurements in 1966 (Martin 1968) and 1980 (Price and
Franz 1980; Steklov et al. 1983).

The polar darkening observed by Voyager 2 at 0.264 μm (Lane et al.
1982*b*) continues into the near ultraviolet. The same effect is observed on

Jupiter, but the boundary of the dark polar regions there lies at latitudes of
$\sim 45-50°$ instead of $60-65°$ as on Saturn. The limb brightening observed in
temperate and equatorial regions at 0.264 μm continues into the near ultraviolet before being replaced by limb darkening at $\lambda > 0.38$ μm.

Thus we have clear evidence that material with strong ultraviolet absorption is present high enough in Saturn's atmosphere to be visible as dark features on the planetary disk at short wavelengths. The increase in reflectivity
into the ultraviolet in the presence of the limb brightening is due to increased
Rayleigh scattering by gas above the tropospheric cloud (Avramchuk and
Krugov 1972; Avramchuk 1973; Tejfel et al. 1973; Tejfel 1974a,b; West et al.
1983c), as discussed in more detail below. In view of the strong brightness
contrasts with latitude, especially toward the poles, the altitude and probably
also the thickness of the haze are variable with latitude.

B. Visible Observations

The geometric albedo p of Saturn's disk (without the rings) is essentially
unchanged from year to year in the visible spectral region (see Fig. 3). The
estimates of p from photoelectric photometry in 1963–1965 (Irvine and Lane
1971) and in 1980 (Tejfel and Kharitonova 1981) coincide within the probable
error limits ($\sim \pm 3-5\%$). The peak value of p is observed at ~ 0.6 μm where
$p = 0.48-0.50$; the lowest value, $p \sim 0.18$, corresponds to the ultraviolet
reflectivity minimum at $0.36-0.38$ μm (except for the centers of the strong
infrared methane absorption bands).

The reflectivity of individual cloud bands from $0.40-0.70$ μm depends
strongly on wavelength and latitude. The yellow equatorial zone is the most
colored band on Saturn, and moderate latitudinal color differences are ob-

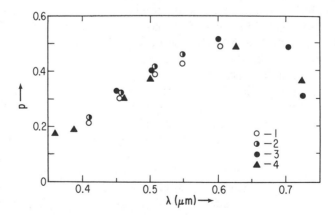

Fig. 3. Geometric albedo of Saturn's globe. (1) February 15–16, 1980; (2) February 18–19,
    1980; (3) March 24–25, 1980 (Tejfel and Kharitonova 1981); (4) 1963–1965 (Irvine and Lane
    1971).

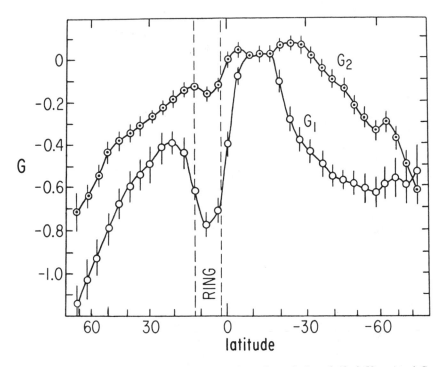

Fig. 4. Variations of the relative spectrophotometric gradients G_1 ($\lambda = 0.40-0.50$ μm) and G_2 ($\lambda = 0.50-0.70$ μm) along the central meridian of Saturn in 1979 (Tejfel and Kharitonova 1982).

served over the entire disk. These differences may be expressed quantitatively by the values of the relative spectrophotometric gradient

$$G = d\ell n(I/I_0)/d(1/\lambda) \qquad (11)$$

which should be determined separately for the spectral regions $0.40-0.50$ μm and $0.50-0.70$ μm. Here I_0 is the intensity of the region chosen as the standard for comparison (near the equator, in Fig. 4). The shortwave gradients G_1 decrease more abruptly with increasing latitude (Tejfel and Kharitonova 1982).

The meridional variations of these gradients G_1 and G_2 (see Fig. 4) are due to various effects: changes in the cloud and haze aerosol properties and altitudes, the Rayleigh scattering in the atmosphere above the aerosol layers and geometry effects (the spectral dependence of the limb-darkening coefficient which decreases towards the short wavelength).

The distribution of the intensity across Saturn's disk may be well described for each latitude by Minnaert's formula

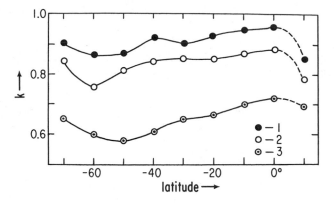

Fig. 5. Latitudinal variations of the limb darkening exponent on Saturn in 1971 at three wave-lengths (Grigorjeva and Tejfel 1979). (1) 0.620 μm; (2) 0.556 μm; (3) 0.434 μm.

$$I/F = R_0 \mu_0^k \mu^{k-1} \qquad (12)$$

at $\mu \geq 0.4$, and the limb darkening exponent k is a very convenient parameter to use together with the normal albedo R_0 (which is equal to I/F when $\mu = \mu_0 = 1$) for comparing the diffuse reflection properties of different bands on Saturn, as well as for comparison of the cloud characteristics of the major planets.

The latitudinal variations of k on Saturn (see Fig. 5) are significantly smaller than its changes with wavelength (Lumme and Reitsema 1978; Grigorjeva and Tejfel 1979). At all latitudes the limb-darkening exponents are less than unity even in red light, i.e. the diffuse reflection from the cloud cover is not Lambertian. It is especially important to take into account the decrease of k toward the violet and ultraviolet when considering the simultaneous effects of the cloud reflectivity and Rayleigh scattering in the upper atmosphere on the observed optical properties of Saturn's belts or the integrated disk. Binder and McCarthy (1973) have found that the limb-darkening exponents for three latitudinal bands on Saturn are smaller than on Jupiter. The relation between k and R_0 is also very different for the two planets (Tejfel 1975).

In contrast to groundbased observations, limited to phase angles $\lesssim 6°\!.4$ for Saturn, measurements from space probes can be made to quite high phase angles depending on the trajectory. The imaging photopolarimeter (IPP) experiment on the Pioneer Saturn mission obtained maps of the brightness and linear polarization of reflected sunlight in red and blue passbands at phase angles from ~ 10 to $150°$. The vidicon imaging system on the Voyager 1 and 2 missions and the photopolarimeter experiment on Voyager 2 also returned measurements of the brightness and polarization of sunlight reflected over a wide range of phase angles.

Observations of the brightness and polarization of the multiply scattered light at a range of phase angles α still contain information on the single-scattering intensity or polarization properties of the clouds or aerosols at the corresponding scattering angles $(180° - \alpha)$, although the structure of the single-scattering functions is appreciably diluted by the relatively neutral higher-order scattering. Nevertheless, sufficient structure remains in the photometry from the Pioneer Saturn mission to permit comparison with multiple-scattering models to yield a single-scattering phase function for Saturn's cloud particles. This function has a significant forward-scattering peak and is relatively flat at scattering angles > 70° (Tomasko et al. 1980b; Tomasko and Doose 1984; West et al. 1983c). In contrast, the phase function derived for Jupiter from similar data (Tomasko et al. 1978) has significant back scattering as well as stronger forward scattering than derived for Saturn.

The photometry from the Pioneer Saturn mission was in two bands centered at ~ 0.44 and 0.64 μm that are relatively free of gaseous absorption, and therefore gives little direct information on the vertical distribution of the cloud particles. However, the polarization produced by the gas molecules in a single scattering at 90° phase is 100% and the Rayleigh scattering optical depth is strongly dependent on wavelength, while the polarization produced by large cloud particles tends to be much smaller and less dependent on wavelength. Thus, the polarimetry in red and blue light near 90° phase provides useful leverage for separating the effects of clouds and gas.

The Pioneer polarimetry in blue light appears to contain a strong component due to Rayleigh scattering by the gas. The polarization at the center of the disk is small at small phase, increases to a maximum value near 90° phase, and returns to small values at large phase. The direction of maximum electric vector is always perpendicular to the scattering plane (positive polarization). The degree of polarization increases rapidly toward the limb and terminator (with increasing slant path) about as expected for Rayleigh scattering from clean gas above a neutral cloud deck.

In red light, the polarization at the center of the disk is relatively small at all phase angles, and is negative (maximum component of the electric vector in the scattering plane) near 90° phase, indicating that the cloud or aerosol particles are much more important relative to the gas at this wavelength. Nevertheless, the polarization even in the red increases very rapidly with large slant paths—much faster than would be expected for the amount of gas expected above the clouds, based on the blue polarimetry. This effect is seen as evidence for the presence of a thin layer of highly polarizing aerosols in the gas above the clouds. A sample of the Pioneer polarimetry is shown in Figs. 6 and 7, from Tomasko and Doose (1984). The constraints on the single-scattering properties of Saturn's cloud particles derived by these authors, and a comparison of these functions with laboratory measurements of the single-scattering properties of various candidate cloud crystals, are discussed below.

While polarimetry of Saturn from the Earth is confined to the small

PIONEER 11
PHASE = 50°

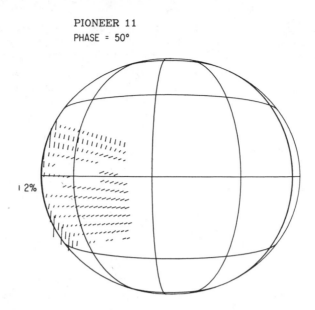

RED POLARIZATION

Fig. 6. The polarization of Saturn in red light (0.64 μm) measured by Pioneer at phase angle 50°, represented by lines with length proportional to the degree of polarization and orientation indicating the direction of maximum electric vector vibration. Vertical lines indicate vibration perpendicular to the scattering plane (positive polarization), and horizontal lines indicate vibration parallel to the scattering plane (negative polarization). Near the bright limb (toward the left) the polarization is large and positive. Away from the limb the direction rotates toward negative polarization except near 10°N. A small band near the equator is blocked by the rings (from Tomasko and Doose 1984).

range of phase angles ≲ 6°4, such measurements have been made over a wide range of wavelengths (Dlugach et al. 1983; Bugaenko and Galkin 1972; Bugaenko and Morozhenko 1981a,b; Santer and Dollfus 1981). Several investigations concerning the size and possible nonspherical nature of the aerosols in the upper atmosphere of Saturn have been discussed by these authors, as outlined below.

The brightness of Saturn in the methane bands of various strengths in the visible and near infrared can be used to probe the vertical distribution of Saturn's cloud particles at pressure levels from ~ 50 mbar to 1 bar. Studies of the variations of the strengths of these features over the disk—which are due both to probing to different effective levels at the changing geometries of illumination and observation (center-to-limb variations) and to real physical changes with latitude—can give additional information on cloud structure. Temporal changes probably also exist (Kharitonova 1976; Trafton 1977; Trafton et al.

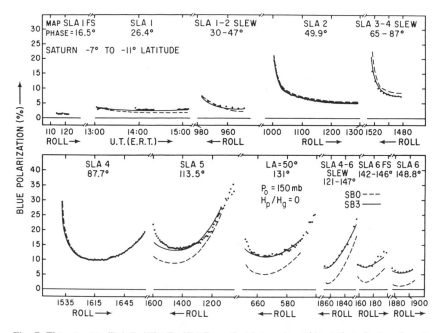

Fig. 7. The center-to-limb variation in the linear polarization of the 7°S to 11°S latitude region on Saturn in blue light at phase angles from 16°.5 to 148°.8 as measured by Pioneer. The abscissa indicates relative position across the disk (in terms of spacecraft rolls) with the bright limb toward the left and the terminator toward the right in each map. The dots are observed values, and the curves are models for clouds located beneath 150 mbar of Rayleigh scattering gas. The model curves both use the correct photometric properties, but produce different amounts of polarization in a single scattering. The dashed curve is for a cloud that is unpolarizing at all scattering angles. The polarization observed at phase angle $\gtrsim 90°$ requires some positive polarization to be introduced by the cloud particles, as for the model shown by the solid curve, whose polarizing properties are given by the curve labeled "150 mb" in Fig. 12 (Tomasko and Doose 1984).

1979) but the data are insufficient to determine accurately their amplitude and regularity, particularly on seasonal scales.

The longitudinal variations were studied using temporally limited observational data for the equatorial belt and low latitudes (Kharitonova 1976; Cochran and Cochran 1981). These variations are small and perhaps not significant. Using dispersion analysis, Kharitonova (1976; personal communication, 1982) found the observed longitudinal variations of the methane absorption to exceed the expected noise level of the measurements by no more than 1%.

The center-to-limb variations of the intensities in the methane bands are more difficult to study for high latitudes than for relatively low latitudes (equatorial and temperate belts) because of the increasingly limited range of incidence and emergence angles available toward the poles in groundbased observations.

After the pioneering work of Hess (1953), investigations of the distribution of molecular absorption on Saturn's disk were resumed only in the 60s (Younkin and Munch 1964; Avramchuk 1967,1968; Owen 1969; Tejfel 1967). These first studies detected a clear distinction between the equatorial zone of Saturn and temperate latitudes: the methane absorption near the equator was noticeably smaller than on other regions of the disk.

More detailed study of the methane absorption bands and the morphology of their variations on Saturn's disk in the 70s (Tejfel and Kharitonova 1970,1972a,b,1974; Tejfel et al. 1971a,b,1973; Bugaenko et al. 1972; Ibragimov 1974; Tejfel 1974a,1977a,b; Zasova 1974; Vdovichenko et al. 1979; Kuratov 1979) has allowed a general picture of the distribution of molecular absorption on Saturn to be constructed.

The smallest amount of absorption is observed in the equatorial region of Saturn for the weak and moderate methane bands (at 0.543, 0.619 and 0.725 μm) as well as for the strong near-infrared bands (at 0.886 and 0.899 μm). The absorption sharply increases at temperate latitudes ($> 30-40°$) and remains constant or decreases slightly toward the poles ($> 60°$ latitude) as shown in Figs. 8 and 9. The center-to-limb variations are weak. There is no strong increase of absorption from the central meridian to the limb along fixed latitudes according to the secant law. This shows directly that the formation of the methane absorptions occurs not only in the atmosphere above the cloud cover of Saturn but also within the clouds. The recent data of West et al. (1982) indicate that the absorption in the weaker 0.619 μm methane band decreases slightly towards the limbs in the equatorial and temperate regions, while a noticeable increase of absorption toward the limbs is observed for the stronger methane bands at 0.725 and 0.8996 μm. The same effect of decreasing intensity towards the limb was detected for a weak isolated methane line at 0.61968 μm (W. H. Smith et al. 1981).

Latitudinal variations of the methane absorption are also not the same in weak and strong bands, leading to difficulties in interpretation because it is not easy to distinguish between several possible reasons for these variations, especially at high latitudes.

There have been a few investigations of the variations of very weak absorption over Saturn's disk in the quadrupole lines of H_2 and in the NH_3 band at 0.645 μm (Bugaenko et al. 1972; Woodman et al. 1977; Aksenov et al. 1978). The absorptions in these bands tend to decrease towards the limbs, but detailed interpretations are made difficult by the relatively low accuracy of the available data. Long-term variations of the H_2 line intensities are probably very small or absent, as shown by Trafton (1979) from high-resolution spectroscopy during 1971 to 1975. Some more significant longitudinal variations were noted by Cochran and Cochran (1981) for the NH_3 absorption band at 0.645 μm, and the real variations in these data amount to $\sim$ 17% according to

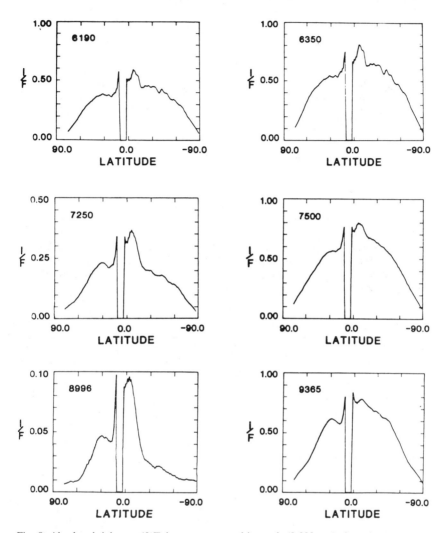

Fig. 8. Absolute brightness (I/F) in narrow spectral intervals (0.003 μm) along the central meridian of Saturn in three methane bands (0.619, 0.725, and 0.8996 μm) and three nearby continuum regions. A correction for blurring due to seeing was made, and the contribution due to light from the rings was subtracted. The rapid decrease in brightness at latitudes $> 20°$ in the strong methane band at 0.8996 μm indicates the increase in effective cloud-top pressure at these latitudes (from West et al. 1982).

the dispersion analysis carried out by Kharitonova (personal communication, 1982). Some observations have been made of the equivalent widths of the very weak blue and green methane bands on Saturn (Lutz et al. 1976, 1982) but no information is available yet on their variations over Saturn's disk.

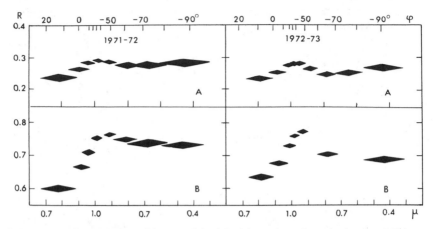

Fig. 9. Latitudinal variations of the central depth R of the methane absorption bands at 0.619 μm (A) and 0.725 μm (B) on the central meridian of Saturn in 1971–72 and 1972–73 (Tejfel and Kharitonova 1974; Tejfel 1977a). The scale along central meridian is linear with respect to μ, the cosine of the local zenith angle to the observer.

C. Infrared Observations

In principle, observations at longer wavelengths in the infrared are capable of providing substantial quantitative information about aerosols and clouds in the atmosphere of Saturn. The large variation of gaseous opacity for well-mixed constituents, like methane, can be used to gain very useful information on particle size distributions. At wavelengths where there is much more thermal emission than reflected sunlight emerging from the atmosphere, additional information about the vertical distribution of particles can be obtained using correlations with a known temperature structure.

However, only recently has any information become available for this spectral region for Saturn, and the data are much scarcer than for shorter wavelengths. This is especially true for observations capable of resolving various regions across the disk of the planet. Fink and Larson (1979) presented a spectrum of the geometric albedo for the 0.8 to 2.5 μm (4000–12500 cm^{-1}) region with resolution 40 cm^{-1}, in addition to higher resolution spectra without an absolute calibration reference. Higher resolution observations in the 2 to 6 μm region have been presented by Larson et al. (1980), Bjoraker et al. (1981), and Fink et al. (1983). While these high-resolution spectra are useful in searching for gaseous constituents and estimating their abundances relative to known absorbers, their lack of absolute calibration makes them much less helpful for quantitative analysis of atmospheric particulates.

Calibrated spectra at lower resolution have been obtained by Russell and Soifer (1977) in the 5–8 μm region, and by Witteborn et al. (1981) in both the 5–8 μm and the 1–3 μm regions. The 1–6 μm spectrum of Saturn is dominated by the signature of strong CH_4 absorption bands with atmospheric spectral windows centered roughly around 1.05, 1.38, 1.9, 2.5, 2.7–3.2, and 4.7–5.4 μm. These windows are dominated by reflected solar radiation up through the 5-μm window, in which both thermal emission and reflected solar radiation make substantial contributions. Spatially resolved observations of Saturn in the 0.6–2.5 μm region were obtained by Apt and Singer (1982), and calibrated on an absolute scale by reference to the observations of Bergstralh et al. (1981) in the overlapping spectral region. These authors also provided the only quantitative modeling of the infrared reflectivity of cloud properties, although this involved only the 0.6–1.3 μm region.

Further spatially resolved observations of the 5 μm region are available from the Voyager 1 and 2 infrared interferometer spectrometer (IRIS) experiment (see Hanel et al. 1981a) in which anomalously strong 5 μm emission appears in discrete areas over the disk, similar to the Jovian hot spots at the same wavelength but greatly reduced in intensity. For these and other regions, the analysis of relative contributions of reflected solar and planetary thermal radiation in terms of a consistent physical cloud model are still in an early stage (see e.g., Bjoraker et al. 1982).

Other regions at longer wavelengths in the thermal spectrum are possibly also sensitive to the presence of particles in the atmosphere. A region near 9 μm might be sufficiently distant from lines of gaseous absorbers to be sensitive to the presence of atmospheric particles, analogously to the atmosphere of Jupiter (Orton et al. 1982). In fact, Caldwell's (1977a) initial analysis of the Gillett and Forrest (1974) spectrum of Saturn invoked a model with ammonia ice particles forming an opaque cloud at the top of the convective region, overlain by an optically thinner haze of the same particles in order to match the 9 μm region of the spectrum. However, the influence of gaseous absorption from CH_3D and PH_3 lines was not included in the quantitative analysis, which accounted only for CH_4 and NH_3 absorption near 9 μm. Farther in the infrared, at and beyond 40 μm, again analogously to Jupiter, the H_2 collision-induced absorption may become low enough that the outgoing thermal radiation would be sensitive to the presence of particles of sufficient size.

No Earth-based spatially resolved observations of Saturn have been published for the 9 μm region. Observations in one or both of these spectral regions have been made, however, by the Voyager IRIS experiment and at 45 μm by the Pioneer Saturn infrared radiometer (IRR) experiment. In the analysis of thermal radiation it is difficult to separate the spatially variable influence of cloud properties from those of gaseous absorbers and kinetic temperatures. This is especially true when dealing with absorbers or scatterers that are spectrally continuous. The wide spectral coverage of the IRIS observations of Saturn should provide the best source of information for separating the effects of cloud and gaseous opacity in the infrared. However, the analysis of IRIS spectra of Saturn has not yet focused on the determination of atmospheric aerosol properties.

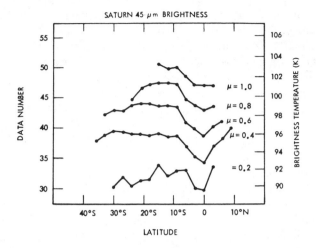

Fig. 10. Brightness of Saturn at 45 μm wavelength. Raw data numbers are shown at left, brightness temperature scale at right. Five values of emission angle are shown (from Orton and Ingersoll 1980).

Some preliminary work has been done in this regard with the Pioneer IRR data. There is an obvious depression in both channels (at roughly 20 and 45 μm) of the Pioneer Saturn IRR experiment which correlates extremely well with the visible morphology of the planet. The region lies within $\sim$ 10° in latitude on either side of the equator, as shown by the 45 μm data in Fig. 10, and its visible appearance is somewhat brighter than the regions immediately adjacent. A similar appearance is indicated by the Voyager IRIS measurements. This is superficially similar to Jovian zones that are probably areas of substantial cloud concentration, which appear bright at visible wavelengths. The straightforward inference is that the thermal infrared equatorial depression is largely due to cloud scattering and absorption rather than to a drop in the true kinetic temperatures in this region. This view is substantiated by near infrared measurements (Apt and Singer 1982) indicating roughly a 20% decrease in the cloud top pressure from mid-latitudes in the southern hemisphere to the equatorial region.

III. DEDUCED CLOUD PROPERTIES

A. Single-Scattering Properties

It would be useful to convert the host of observational material into information regarding the single-scattering optical properties of the cloud and aerosol particles. Such information could be used both to calculate the penetration and absorption of solar and thermal radiation in various parts of Saturn's atmosphere with accuracy, and to constrain the physical properties of size, shape, refractive index, and eventually composition of Saturn's aerosols and clouds.

Several attempts to constrain the optical properties of Saturn's particulates from groundbased measurements have been made. However, because several effects such as variation with depth of the albedo and phase function of the particles, as well as the detailed shape of the particle phase function, combine to produce the observed limb darkening at low phase, it is not possible to determine unambiguously the single-scattering properties of Saturn's aerosols using only center-to-limb variation measured from the Earth. The only relevant quantities which can be measured reliably are the phase behavior and polarization in the phase angle range $0° \leq \alpha \leq 6°4$, and multiple scattering models must be constructed to invert the observations to derive the single-scattering phase matrix in the very restricted angular domain mentioned above.

Linear polarization measurements have been reported by Lyot (1929), Hall and Riley (1969,1974), Bugaenko et al. (1971,1975), Bugaenko and Galkin (1973), Kemp et al. (1978), Sigua (1978), Bugaenko and Morozhenko (1981a,b), Santer and Dollfus (1981), and Dlugach et al. (1983). Circular polarization measurements were discussed by Swedlund et al. (1972) and Smith and Wolstencroft (1983) (see also Kawata 1978).

While the groundbased polarization measurements are quite interesting, they have not yet added to our understanding of the size or composition of the aerosols. Some authors have derived particle size and refractive index from comparison with spherical particles using Mie theory, but these results depend on the particles being spherical or nearly so, which seems highly unlikely. The observations may be helpful at a future time when laboratory measurements are available for NH_3 ice crystals and Danielson dust candidate aerosols at the observed scattering angles and wavelengths.

From the behavior of the polarization across the disk we have learned that the polar regions differ from the rest of the planet. At yellow and red wavelengths the linear polarization is much stronger in the polar region than anywhere else. The sign of linear polarization is positive (the polarization vector is perpendicular to the scattering plane at phase angles $\alpha > 3°$, and radial near $\alpha = 0°$). An abundance of small ($r < 0.1\ \mu$m) absorbing haze aerosols above the clouds would be consistent with the polarization and with observations of polar darkening in the ultraviolet.

If scattering in the atmosphere occurs in a homogeneous medium, the limb polarization at full phase ($\alpha = 0°$) will be radial for a Rayleigh-type phase matrix, and tangential for a phase matrix which is negatively polarizing over a broad range of scattering angles centered around 90°. This phenomenon occurs because second-order scattering dominates the polarization at $\alpha = 0°$, and photons scattered tangential to the limb have a higher probability of scattering in the backward direction than do photons scattered in the radial direction. The direction of polarization is indeed mostly radial at short wave-

lengths (Hall and Riley 1969), suggesting that Rayleigh scattering is important at those wavelengths.

The groundbased polarimetry is slightly peculiar. At short wavelengths
($\lambda < 0.55$ μm) the polarization near the equatorial limb is inclined to the radial direction at $\sim 20°$ within latitudes $\pm 30°$ (Hall and Riley 1969; Bugaenko
and Galkin 1973; Bugaenko and Morozhenko 1981a,b; Dlugach et al. 1983).
At $\lambda > 0.55$ μm the polarization is tangential, indicating negative polarization near 90° phase angle. The 20° difference from radial at short wavelengths
has been interpreted to indicate oriented nonspherical particles, but the orientation mechanism proposed (alignment by winds) is not likely, since winds do
not align particles. Furthermore, it has not been demonstrated that the observations are inconsistent with detailed radiative transfer calculations including
vertical inhomogeneity.

While it is difficult to constrain the single-scattering properties of Saturn's aerosols from the groundbased observations, the spacecraft observations
should permit this to be done straightforwardly. The photometry and polarimetry of Saturn from the Pioneer spacecraft have been presented and discussed by Tomasko et al. (1980), and Tomasko and Doose (1984). These authors used the observations of the multiply-scattered light at phase angles
from 10 to 150° to constrain the shape of the single-scattering phase function
at the corresponding scattering angles (170° to 30°). The phase function was
parameterized by a double Henyey-Greenstein function, $P(\theta) = f P(g_1,\theta) + (1-f) P(g_2,\theta)$, where $P(g,\theta) = (1-g^2)/(1+g^2-2g \cos \theta)^{3/2}$. Here the parameter g_1 controls the shape of the forward-scattering peak, g_2 (taken to be
negative) controls the shape of the back-scattering peak, and f controls the
relative amount of each. Multiple-scattering calculations based on the layer-
doubling and adding method were done for various trial values of g_1, g_2, f,
and the single-scattering albedo for comparison with the observations at a
range of phase angles. The shape of the phase functions derived are shown in
Fig. 11.

This determination of the particle phase function leads directly to the
value of the phase integral q. The zone phase functions give $q = 1.43$ and
1.41 in blue and red, respectively, while in the belt the corresponding values
are 1.45 and 1.41. Hanel et al. (1983) adopted 1.42 as an estimate of the average phase integral over the globe and over wavelength to derive a Bond albedo
of 0.342 from their measured bolometric geometric albedo of 0.242. Together
with their measured effective temperature of $T_E = 95.0 \pm 0.4$ K for Saturn,
Hanel et al. (1983) concluded that Saturn radiates 1.78 ± 0.09 times as much
energy as it absorbs from the Sun.

Measurements of the degree and orientation of the linear polarization in
the multiply-scattered sunlight can be used to constrain other elements in the
4×4 single-scattering phase matrix that relates the Stokes vector (I,Q,U,V)
of the radiation incident on an individual scattering particle to the Stokes vector (I',Q',U',V') of the light scattered through an angle θ. When measure-

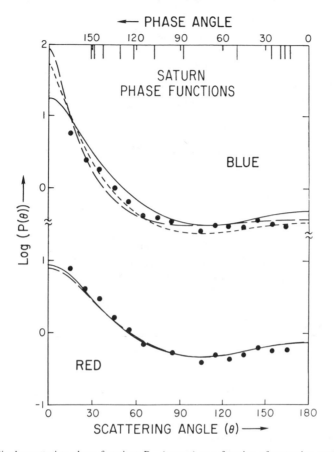

Fig. 11. Single-scattering phase functions P_{11} (curves) as a function of scattering angle in red (0.64 μm) and blue (0.44 μm) light from the Pioneer Saturn data in a relatively bright zone, 7°S to 11°S (solid curve) and in a darker belt, 15°S to 17°S (dashed curve). Because the photometry data used to determine these phase functions are at continuum wavelengths, the derived curves are relatively insensitive to the cloud-top pressure; all curves were derived with the cloud top at 150 mbar except for the blue belt model (long dashes), where the cloud top was at 250 mbar. The planetary observations were obtained at the phase angles marked at the top of the plot and are not necessarily well described by the smooth functions at scattering angles < 30°. Also shown are measurements (dots) of the single-scattering phase functions of ammonia crystals grown at 150 K. The red and blue curves are all displaced one decade for clarity (from Tomasko and Doose 1984).

ments of the degree of circular polarization are not available, it has been shown (Hansen 1971) that the phase matrix can be reduced to a 3 × 3 form by dropping the fourth row and column with little loss of accuracy. Further, if the particles obey fairly general symmetry laws (e.g. that there are equal numbers of randomly oriented particles of arbitrary shape and their mirror images) then the only nonzero elements of the single-scattering phase matrix, in addition to the diagonal elements P_{11}, P_{22}, and P_{33} are the elements P_{12} and P_{21} which are equal. For spheres the elements P_{11} and P_{22} are equal, and P_{33} is similar in magnitude. Tomasko and Doose assumed that these three elements could be approximated as equal for a first analysis of the Pioneer Saturn data. Thus, in addition to the phase function P_{11} (described as the double Henyey-Greenstein function shown in Fig. 11) the only other element to be obtained is P_{12}. The ratio $-P_{12}(\theta)/P_{11}(\theta)$ is the degree of polarization introduced in a single-scattering event for unpolarized incident light, and this function was parameterized by straight-line segments between the phase angles of the Pioneer data.

Tomasko and Doose found a relation between the amount of single-scattering polarization required of the cloud particles and the pressure level at which the cloud top was reached. If the cloud is placed too deep, the cloud particles must be very negatively polarizing at 90° scattering angle to compensate for the positive polarization produced by Rayleigh scattering in the gas above the cloud. As indicated in Fig. 12, it was generally possible to choose the pressure level of the cloud so that the cloud particles could have fairly neutral single-scattering polarizing properties. The family of solutions relating the single-scattering polarizing properties of the cloud particles and their vertical location can be narrowed by using measurements in molecular absorption bands to provide independent constraints on the cloud-top location, as discussed below.

The rapid increase in polarization from the center of the disk toward the limb and terminator in blue light can be reproduced simply by the Rayleigh scattering of the gas above the clouds. However, in red light the observed increase is much too great to be accounted for in this way. A thin layer of highly positively polarizing aerosols (optical depth of 0.02 at $\lambda = 0.64$ μm) seems to be required in the gas above the cloud deck (which is slightly negatively polarizing at 90° scattering angle). Thus the Pioneer polarimetry, like the ultraviolet photometry and the groundbased polarimetry, indicates the presence of an optically thin (at red wavelengths) layer of aerosols in Saturn's upper atmosphere.

While the single-scattering phase function of the cloud particles has been effectively determined in the visible by the Pioneer measurements, it can be expected to change significantly toward the infrared as the wavelength becomes comparable to the particle size. The general problem of realistically determining the scattering properties of irregular particles, NH_3 crystals in particular, is severe in such a case. Several levels of approximation for the single-scattering properties have been used in radiative transfer calculation involving multiple scattering. One of the most common approaches is to assume that the properties associated with spherical particles, as derived from Mie theory, are sufficient for a first-order estimate. Such an approach was adopted for the Jovian atmosphere by Marten et al. (1981) who also assumed that the scattering function was isotropic. Another possible approach was

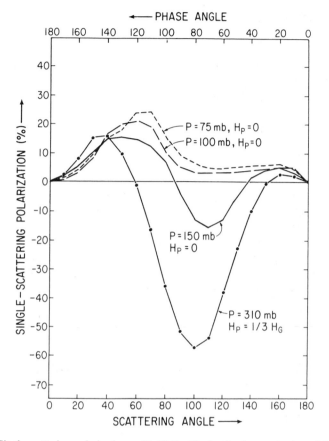

Fig. 12. Single-scattering polarization, $-P_{11}(\theta)/P_{12}(\theta)$, for the Saturn cloud particles in blue light for the reflectivity bright zone at 7°S to 11°S, derived from the Pioneer Saturn data. The curves indicate the family of solutions possible for different choices for the pressure at which optical depth = 0.4 of the cloud is reached and for the ratio H_p/H_g of particle to gas scale heights (from Tomasko and Doose 1984).

adopted by Orton et al. (1982) who used the semi-empirical irregular particle theory of Pollack and Cuzzi (1980) as the basis for scattering properties. The empirical parameters describing the irregular particle perturbation of Mie theory were fit to the laboratory results of Holmes et al. (1980) and Holmes (1981) for NH_3 ice particles at visible wavelengths, and subsequently used for extrapolating to longer wavelengths. Orton et al. found that there were minimal differences between calculation of the outgoing thermal radiance using Mie theory and using the semi-empirical theory when the single-scattering albedo was low (e.g. < 0.80).

However, these approximations are no substitute for laboratory measurements or accurate calculations. Laboratory measurements of the scattering properties of NH_3 ice are underway in the visible (Holmes et al. 1980; Holmes 1981; Tomasko and Stahl 1982), and are needed in the infrared. The existing measurements indicate that NH_3 crystals tend to assume a tetrahedral shape. Once the crystal shape is known, rigorous scattering calculations become possible, at least in principle. One approach to such calculation involves the point-dipole array method to approximate the continuous surface of a dielectric (see e.g., Purcell and Pennypacker 1973). However, larger particles would necessarily imply 3-dimensional arrays of substantial proportions in order to reduce phase shift errors.

B. Possible Composition

One of the long-standing unsolved problems of the Jupiter and Saturn atmospheres is the composition of the aerosols which absorb visible and ultraviolet light, and the chemical and physical processes responsible for their creation and destruction. Since pure NH_3 ice, which is thought to be the main constituent of the upper cloud, is nonabsorbing in the visible and near ultraviolet down to 0.220 μm (Pipes et al. 1974), visible chromophores and ultraviolet absorbers must be products of disequilibrating processes, i.e. processes not in thermochemical equilibrium with the ambient atmosphere. The chapter by Prinn et al. includes a review of the chemistry and possible compositions of the absorbing particles.

The photometry and polarimetry observations place important constraints on the nature of the aerosols. At least three kinds of absorbing aerosol populations are suggested (West et al. 1983c). The first is a tropospheric chromophore which absorbs at visible and ultraviolet wavelengths. The relatively flat limb darkening at blue wavelengths implies that these absorbers are not concentrated above bright lower clouds but are mixed in the clouds themselves (Tomasko et al. 1980b). This component accounts for the contrast between belts and zones for $\lambda > 0.38$ μm and for the general decrease in planetary albedo from the red to the blue and near ultraviolet (West et al. 1983c). The source of the tropospheric absorber is probably within or below the NH_3 cloud, and the aerosols may serve as condensation nuclei for NH_3 ice crystals.

Two stratospheric aerosols, separated in latitude and altitude, are also present. The polar regions (above $\sim$ 60° latitude) have an abundant ultraviolet absorber which extends well above 20 mbar (West et al. 1983a). Lane et al. (1982b) have suggested charged particle bombardment from the magnetosphere (Scattergood et al. 1975) as a possible contribution to the formation of the polar aerosols, and Broadfoot et al. (1981) reported observations of polar aurorae. A detailed mechanism for aerosol production from charged particle precipitation has not been formulated, but presumably CH_4 destruction by protons initiates the process. Some doubt has been expressed concerning the energetics. The best supporting evidence for the process has been reported for Jupiter by Caldwell et al. (1983) who observed temporally varying polar infrared emission in the 7 μm CH_4 band.

In mid-latitudes the stratospheric haze is concentrated in the 30 to 70 mbar region (West et al. 1983c) and the optical depth in the Equatorial Zone is > 0.4 at 0.264 μm, and decreases by at least a factor of 10 to 0.670 μm (D. W. Smith et al. 1981), supporting the small particle Danielson dust idea. If the particles scatter as spheres with a real refractive index $n_r = 1.5$, their mean radius is near 0.1 μm and the imaginary refractive index must be quite large ($n_i \sim 0.4$) at 0.264 μm. West et al. (1983c) estimate a total column density near 10^9 cm^{-2} above the tropopause in the Equatorial Zone. The derived optical depth is below upper limits expected for photochemical processes in Jupiter's stratosphere (Prinn and Owen 1976; Gladstone 1982).

It seems likely that most of the high aerosols which have strong ultraviolet absorption are due to photolysis or energetic particle impact on methane (Scattergood and Owen 1977). It has been suggested that some portion of the aerosols may result from the ablation or disintegration of micrometeoroid particles within the upper atmosphere of Saturn (Tejfel 1974a,b), but detailed estimates of the properties and amount of aerosols that could result from this process remain to be calculated.

Interpretation of the chromophores' composition would benefit from a comprehensive laboratory program to measure the spectral reflectivity and scattering properties (including polarization) of candidate materials. For example, $P_4(s)$ has been suggested (see the chapter by Prinn et al.) from photodissociation of PH_3 which is observed spectroscopically. While P_4 may be a good candidate for the stratospheric haze above 100 mbar, its role in the troposphere is probably very minor. As Fig. 2 shows, the strongest gradient in Saturn's spectral reflectivity occurs in the wavelength range 0.4 to 0.6 μm, while the laboratory spectrum of red P_4 frost is flat in that region (J. Gradie, personal communication). The strongest gradient in the laboratory spectra occurs between 0.6 and 0.7 μm where Saturn's spectral reflectivity is relatively flat. Of course, before P_4 can be dismissed as a candidate for the tropospheric chromophore, we must measure the wavelength dependence of the scattering from small particles rather than from a frost surface, and for a variety of P_4 allotropes subject to formation conditions relevant to the atmospheric conditions. Measurements of the complex refractive indices are insufficient because, as Huffman and Bohren (1980) show, Mie theory fails to account for the spectral properties of nonspherical particles.

Deeper in the atmosphere, NH_3 is known to exist as a gas, and several analyses of infrared measurements to date have concentrated on NH_3 as a candidate for Saturn's clouds. NH_3 is expected to condense out of the gas phase above the 1.4 bar level into particles of NH_3 ice, provided that H_2O is sufficiently depleted at this altitude that hydrates of ammonia do not form instead (Weidenschilling and Lewis 1973). This is expected to be the case, since H_2O is not detectable in spectral regions sensitive to the deep atmosphere where NH_3 is detectable in substantial quantities (Fink et al. 1983). Thus, clouds of NH_3 are expected to form in the atmosphere in regions of upwelling convection, with the cloud-bottom level determined by the local temperature and mixing ratio of the gas in the deeper atmosphere. The cloud thickness and cloud-top location depend on the local temperature lapse rate above the condensation level, the rate of particle size growth and precipitation, and the efficiency of convection for moving the particles up to higher altitudes.

Figure 11 shows a comparison of the single-scattering phase function derived from the Pioneer data with phase functions measured for scattering by NH_3 ice crystals grown at $\sim$ 150 K in a laboratory cloud chamber (Tomasko and Stahl 1982). The agreement is remarkable, although it hardly represents proof of the composition of Saturn's clouds. Several characteristics of these curves such as the strength of the forward peak and relatively flat behavior for scattering angles $\gtrsim 70°$ are found for several types of irregularly shaped particles. Still, it is interesting that the curves differ significantly from those derived for Jupiter (Tomasko et al. 1978) and for those of hexagonal plates or prisms computed by geometrical optics (Cai and Liou 1982) or measured in the laboratory (Stahl et al. 1983).

More stringent constraints on composition should be possible using other elements in the single-scattering phase matrix, such as the single-scattering polarization $(-P_{12}/P_{11})$ shown in Fig. 12. The laboratory measurements of NH_3 crystals to date show little single-scattering polarization ($<$ $\pm 20\%$), but do not have the detailed shape shown in Fig. 12. However, considerable work remains to be done both in laboratory measurements (data are needed at temperatures below 150 K) and in analysis of the Pioneer measurements, in which the polarizing properties of the thin, upper stratospheric haze must be separated from those of the lower tropospheric cloud. Models with essentially unpolarizing cloud particles (similar to some of the NH_3 laboratory measurements rather than to the shapes shown in Fig. 12) are capable of matching the Pioneer polarimetry at phase angles of $\leq 90°$, and the data obtained at larger phase angles are known to sample higher regions of the atmosphere where the observed polarization may be caused primarily by photochemical aerosols rather than NH_3 crystals.

On the other hand, ammonia ice has yet to be identified conclusively in the spectrum of Saturn. Slobodkin et al. (1978) reported an identification of NH_3 ice, on the basis of a correlation between features in the observed spectrum of Saturn and features in their laboratory frost spectra which are highly dependent on the availability of certain particular temperature and deposition rate conditions. Unfortunately, these features are not present in the NH_3 ice absorption spectra (see e.g. Sill et al. 1980), which cover a wide spectral range and from which the relevant indices of refraction for the ice are derived. Given the lack of consensus on the reasons for the incompatible laboratory observations and their applicability to the relevant astronomical observations near 3 μm (Witteborn et al. 1981), some degree of caution seems reasonable regarding this putative identification.

Fink et al. (1983) report that the addition of NH_3 ice absorption would provide the additional atmospheric absorption, which appears to be unavailable from gaseous absorbers, to fit their spectral observations near 5.3 μm. On the other hand, stronger absorption features are present near 28, 9.5, and 3.0 μm. Caldwell (1977a) continued the suggestion by Gillett and Forrest (1974) that a broad absorption feature near 9.5 μm in their thermal spectrum of the center of Saturn's disk was due to NH_3 ice absorption. Caldwell suggested a model with a haze of NH_3 ice particles, small enough that purely absorbing properties can be assumed. These particles were distributed vertically with a scale height equal to the gas scale height from 110 K down to 113 K. Below this level, corresponding to the top of the convective region, he placed an optically opaque cloud, presumed to be a very dense cloud of NH_3 ice particles.

However, Witteborn et al. (1981) do not consider the spectrum of NH_3 ice, even under the conditions suggested by Slobodkin et al. (1978), to be compatible with their own observations in the 3 μm region. Similarly, no features are obvious in the IRIS spectrum covering the relevant spectral region at 28 μm or at longer wavelengths where hydrogen absorption is much reduced (see e.g. Hanel et al. 1981a). A more quantitative investigation of this region, however, has not yet been made.

There are several possible reasons why NH_3 ice clouds, while present in the atmosphere of Saturn, might not be positively identifiable. In fact, the Jovian atmosphere presents a similar problem. For Jupiter, several lines of circumstantial evidence point to the existence of NH_3 ice clouds near the 600 mbar level, at least in zones (visibly bright, thermally cool areas), although there has been no positive identification of spectral features belonging to NH_3 ice anywhere on the planet, in spite of high-quality observations in spectral regions where strong ice features exist, such as at 28 and 9.5 μm. Marten et al. (1981) have interpreted this as an indication of the large size of *spherical* cloud particles. Large particles have the property of "smearing" resonant absorption features in the ice, as shown, for example, by Fig. 3 of Taylor (1973) in his spectral plots of scattering parameters for NH_3 ice, derived on the basis of Mie theory. Marten et al. interpret the available IRIS spectra for Jupiter as implying sizes of ammonia ice particles near 100 μm, although Orton et al. (1982) demonstrated that this value is a strong function of the assumed thickness of the cloud.

Another possibility has not been explored fully for a wide range of particle sizes. Huffman and Bohren (1980) demonstrated that, in the Rayleigh limit, irregular particles suppress the resonant absorption features associated with spherical particles of equal volume. The tetrahedral habit of NH_3 crystals departs from sphericity enough to imply that spectral features could easily be suppressed in particles much smaller than the wavelength. Whether this is the case for particles larger with respect to the wavelength remains an open question.

Finally, it is possible that NH_3 ice features are missing because the particles are either coated or mixed with impurities which mask spectral features sufficiently to prevent identification. This consideration, while appealing in its realistic assessment of probable atmospheric conditions, must somehow be made consistent with the extremely high visible and near infrared single-scattering albedos associated with the bright regions in the Jovian atmosphere that are commonly interpreted as NH_3 clouds.

Again as in the case of the Jovian atmosphere, many of the clues for the presence of NH_3 ice clouds in the atmosphere of Saturn are indirect, involving the expectation of solid NH_3 somewhere in the physical chemistry of the atmosphere. The best evidence to date is inferred from the physical morphology of the planet at thermal infrared wavelengths, from comparison between the results of the Voyager radio subsystem (RSS) occultation experiment and infrared results from the IRIS experiment and the earlier Pioneer IRR experiment, and from models of the vertical cloud structure in absorption bands formed in reflected sunlight. All these data indicate the presence of significant cloud opacity in Saturn's upper troposphere, where NH_3 is expected to condense.

IV. DISTRIBUTION OF THE PARTICULATES

A. Vertical Distribution

Modern notions about the aerosol distribution in Saturn's atmosphere are based on the interpretation of the observational data in terms of various more or less simplified models. It is now evident that the simplest models, consisting of a reflecting cloud layer beneath an absorbing gas layer or a vertically homogeneous distribution of scatterers, can only describe satisfactorily very limited portions of the observations. The growth of observational information has required a movement to more complicated models of both the vertical distribution of the aerosols and the spectral dependence of the optical parameters of the clouds.

Apart from data in the molecular absorption bands, where the strength of the absorption limits the level probed, there is a general tendency for observations at the shortest wavelengths to be sensitive to the uppermost part of the atmosphere; longer wavelengths penetrate to progressively deeper levels. The shortest wavelength data that have been analyzed in some detail are those obtained by Voyager 2 at 0.264 μm. West et al. (1983a,c) prefer a model with optical depth > 0.4 at this wavelength at pressure levels of 30–70 mbar (corresponding to abundances of clean overlying H_2 of 3–7 km-amagat) at low latitudes, which rises to pressure levels of $\leq$ 20 mbar at latitudes $> 60°$. Analyses of groundbased data in the near ultraviolet and visible indicate greater estimates of the amount of clean H_2 above the stratospheric aerosols. The abundance of H_2 above the aerosols at equatorial and temperate latitudes has been estimated respectively as 8 $\pm$ 12 and 12 $\pm$ 3 km-amagat (Tejfel 1975), 13.5 $\pm$ 1.0 and 19 $\pm$ 2 km-am (Tejfel 1974a,b; Tejfel and Kharitonova 1974), 15 and 31 km-amagat (Price and Franz 1980), 19 and 27 km-amagat (Macy 1977). These estimates are greater than those made by Voyager partly because of the decrease in absorption caused by the stratospheric aerosol particles toward longer wavelengths, but a real decrease in scattering cross section with increasing wavelength through the visible occurs as well. Such a decrease in cross section would be expected for small particles, and particle radii on the order of 0.1 μm have been estimated for the aerosols in the pressure region 30–70 mbar (Tomasko and Doose 1984; West et al. 1983c).

The pressures derived for the effective top boundary of the tropospheric aerosol layer at low latitudes from groundbased photometry vary between 80 mbar and 300 mbar in various studies. This is somewhat higher than the upper boundary of the convective region in Saturn's atmosphere near 400 mbar, but convective overshooting might be capable of lifting some of these particles as much as a scale height above this level. Some of these aerosols undoubtedly extend to pressures as low as 100 mbar near Saturn's temperature minimum, where the high static stability strongly inhibits further vertical motions.

In order to determine the vertical distribution of the cloud at deeper levels, additional information is required which may be obtained from analysis of the molecular absorption bands over different regions of Saturn. Reviews of the molecular absorption investigations concerning Saturn and their interpretation have been published by Hunt (1975), Trafton (1978), and Tejfel (1980). All these studies reach the general conclusion that both the simple reflecting layer model and the homogeneous scattering layer model fail to reproduce the observed variations of the molecular absorptions over Saturn's disk.

Better agreement is obtained for reflecting-scattering models which consist of a semi-infinite homogeneous diffuse cloud layer beneath an absorbing pure gas layer. The formation of the molecular absorption occurs not only in the upper gas layer but also by multiple scattering within the clouds, with weak and moderately strong molecular bands being formed mostly within the cloudy region of the atmosphere. Tejfel (1977b) used such a model to derive estimates of the optical thickness τ of clear gas over the clouds and the specific abundance of methane W_{CH_4} in a mean free scattering path in the clouds at different latitudes. The variations in altitude of the effective upper boundary of the aerosol (cloud) layer derived from observations of the CH_4 absorption bands at 0.619 and 0.725 μm are very similar to the results of interpretation of the groundbased data in the near ultraviolet, but are greater than the corresponding values derived from Voyager.

For some latitudes (such as 15°S to 17°S) the inclusion of the still stronger methane band near 0.89 μm results in some inconsistencies when analyzed in terms of such a 2-parameter model. In these cases, the estimates of methane abundance above the diffuse cloud obtained from center-to-limb variations of the residual intensity in the centers of strong methane bands are significantly less than for moderate and weak bands. This has been noted by Tejfel (1976) and confirmed by the analysis of CCD observations of West et al. (1982) carried out in terms of a 2-layer model (Tejfel and Kharitonova 1983, in preparation; West 1983). The correlation of W_{CH_4} and other parameters with the strength of the absorption band might be explained by the presence of a semi-transparent or optically thin aerosol layer (as required by the ultraviolet data) or by vertical inhomogeneity in the lower continuous cloud-haze medium.

Several authors have attempted to minimize such discrepancies by using a 3-layer model: semi-infinite cloud layer, optically thin aerosol layer and clear atmosphere above the clouds. Models of this kind were proposed by Macy (1976,1977), Caldwell (1977a), and Buriez and de Bergh (1981), although in special cases the simplest models were also used by these authors for the interpretation of spectral measurements. Trafton and Macy (1975), and Ibragimov and Avramchuk (1971) have used a model of a semi-infinite homogeneous scattering atmosphere. Buriez and de Bergh (1981) and Caldwell (1977a) have analyzed 3-layer models, as have Franz and Price (1979) who obtained the best agreement of their photometry of Saturn in the ultraviolet with a model which included a clear gas layer, a conservative haze layer, and a thick cloud with Lambertian diffuse reflectivity.

A model proposed by Macy (1977) is among the most complicated models used to date. It consists of four layers:

1. Upper atmosphere of clear gas,
2. Layer of aerosols with strong ultraviolet absorption (probably PH_3 or H_2S) and with particles of radii < 0.2 μm, volume density $\sim 10^3$ cm^{-3}, which lies between the pressure levels 200–300 mbar and 400 mbar,
3. Ammonia haze, and
4. Dense cloud layer with upper boundary at the pressure level ~ 1100 mbar.

This model was able to reproduce many observations of Saturn such as the equivalent widths of the H_2 quadrupole lines and the $3\nu_3$ methane manifolds. However, the shape of the continuum limb-darkening at short wavelengths in Macy's model is steeper than observed, due to the concentration of the blue absorbers in the upper portions of the model (Tomasko et al. 1980b).

In attempting to reconcile the observed strengths and center-to-limb variations of the near infrared methane bands, it is important to remember that the strength of these bands varies by more than a factor of 20 from the weak 0.619 μm band to the strong band near 0.89 μm. Thus, it may well be that the aerosol-to-gas mixing ratio is different at the very different altitudes probed by these bands. Tomasko and Doose (1984) suggest that the intensities near the center of the disk and near the limb in the CH_4 bands at 0.619, 0.725, and 0.890 μm may be used to find the pressure boundaries and gas abundance in a mean free scattering path in each of the three different regions of the atmosphere probed by these bands, beginning at the top of the atmosphere with the strongest band. Figure 13 shows a family of cloud structures derived

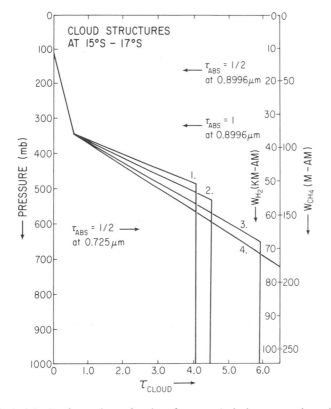

Fig. 13. Optical depths of aerosols as a function of pressure (or hydrogen or methane abundance) derived from the data of West et al. (1982) for the latitude interval from 15°S to 17°S. The structure above 350 mbar has been fixed to fit the band/continuum ratio for the strong 0.8996 μm methane band (see the curve labeled 8996/9365 in Fig. 14). The family of structures labeled (1) through (4) all also fit the band/continuum ratio for the moderately strong band at 0.725 μm (see the curve labeled 7250/7500 in Fig. 14). The depths of penetration of radiation in the 8996 and 7250 μm bands are indicated roughly by the arrows. The blue polarimetry near 90° phase is sensitive to the hydrogen abundance above the level at which cloud optical depth is 0.4. The vertical lines in structures (1)–(3) indicate a clear space down to the 4 bar level. The cloud in structure (4) continues at the indicated slope to large optical depth (from Tomasko et al. 1984).

in this way for the latitude region from 15°S to 17°S from the methane band observations of West et al. (1982). The pressure at the bottom of a clear gas layer (~ 100 mbar) and the specific abundance in the cloud layer below are selected to fit the strong methane band data at 0.89 μm. The family of solutions (labeled 1–4) below 350 mbar all fit the weaker 0.725 μm methane band data as well. Figure 14 shows the fit to the band/continuum ratios in all three bands as a function of position across Saturn's disk for this family of cloud

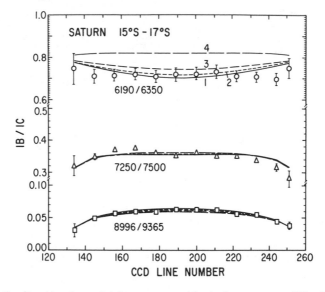

Fig. 14. Ratio of band/continuum brightness computed for the four structures of Fig. 13 for three
methane bands as marked, compared to the observations of West et al. (1982). Cloud structures
similar to curves (1), (2) or (3) in Fig. 13 provide one way to reconcile the observations in
methane bands spanning a large range of strengths.

structures. Structure 4, which has an infinitely thick cloud below 350 mbar ·
and fits the two strong bands, does not give enough absorption in the weak
0.619 μm band. The structures labeled 1, 2 or 3 in Fig. 13, which have a clear
space between $\sim$ 500 or 600 mbar and a deeper diffusely scattering cloud
(placed at 4 bars to simulate a possible NH_4HS cloud) offer one way of recon-
sidering these observations in bands of widely differing strengths.

Figure 15 shows the cloud structure derived in this way from the data of
West et al. (1982) at three latitude intervals, compared to the temperature-
pressure structure obtained from the Voyager 2 radio occultation experiment
(Tyler et al. 1982a). Note that the cloud "top" in red light occurs at about the
100 mbar level (near the temperature minimum) at all three locations. The
models at 27°S–34°S and 15°S–17°S are both considerably more diffuse
above the radiative convective boundary near 400 mbar than below this level.
The observations at 7°S–11°S have more uncertainty due to their proximity to
the rings, and a single specific abundance of $\sim$ 10 km-amagat of H_2 per mean
free scattering path is permitted in an infinitely thick cloud extending beneath
the 100 mbar level at this latitude. At both other locations, the data suggest
that the clouds are most dense between 350 and $\sim$ 600 mbar, with a relative
clearing at greater pressures until the next cloud layer is reached at still deeper
levels.

If nitrogen is present in solar proportions, thermodynamic equilibrium

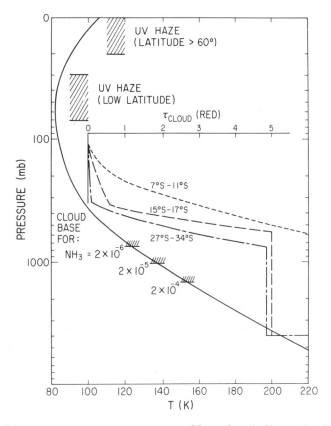

Fig. 15. Solid curve: temperature-pressure structure of Saturn from the Voyager 2 radio occultation experiment (Tyler et al. 1982*a*). Dashed curves: cloud optical depths derived for three
latitudes on Saturn from the methane band data of West et al. (1982) (from Tomasko et al.
1984). The models for two of the latitudes contain a clear space between ~ 600–700 mbar and
4 bar. The local mixing ratio of NH_3 is shown for saturated conditions at an NH_3 cloud base at
each of three possible locations. The vertical location of the haze seen in the ultraviolet is also
indicated schematically for high and low latitudes.

calculations predict an NH_3 partial pressure of ~ 2×10^{-4} times the total
pressure in the deep atmosphere (Weidenschilling and Lewis 1973), and close
to the saturated vapor pressure at total pressures $\lesssim 1.4$ bars. An ammonia
cloud base at 600 or 700 mbar rather than 1.4 bars could result from precipitation or other dynamical processes acting to reduce the ammonia partial pressure somewhat below the saturated values between 600 mbar and 1.4 bars,
and need not imply a depletion of NH_3 by a factor between 100 and 1000
times the value expected for solar proportions in the deep atmosphere.

We should stress that other effects, such as increasing particle absorption
with depth in the atmosphere, remain to be investigated and could also be

important in reconciling the observed behavior of strong and weak absorption bands. Thus, the possible presence of a deep clear space in Saturn's atmosphere requires considerable further work to be established with certainty. Nevertheless, the above results indicate our current ability to extract the vertical distributions of the clouds from available optical observations. Increasing the number of layers in such models too far clearly leads to an excessive number of free parameters to be determined, but the actual vertical distribution of aerosols in the atmosphere is known to be complex. Models with three or four layers, each with different particle-to-gas mixing ratios, beneath the ultraviolet haze near 50 mbar, are probably necessary to reproduce the observed equivalent widths and center-to-limb variations of the several absorptions with widely different strengths for which data are currently available.

B. Variations with Latitude

It has been known for some time (Owen 1969; Macy 1977; Tejfel 1977a,b) that unit aerosol optical depth is at a higher level in the Equatorial Zone than in mid-latitudes. The variation with latitude is now better understood, thanks to high spatial resolution measurements of the polarization and ultraviolet photometry available from spacecraft, and the study of both hemispheres made possible when the ring inclination became favorable between 1979 and 1980 (West et al. 1982; Apt and Singer 1982). A hemispheric asymmetry is clearly present in the strong methane band images of West et al. (1982), and in the Voyager ultraviolet (0.35 μm) narrow-angle camera images. The sense of the asymmetry is such that the aerosols in the southern hemisphere tend to be somewhat deeper than those at the same latitude in the north (the south is more reflective in ultraviolet and more absorbing in methane than the north). Since the high-altitude haze in mid-latitudes has a very small optical depth in the red and near infrared, the qualitative correlations between the ultraviolet images and the methane images suggests that latitudinal variations in the tropospheric cloud are primarily responsible for the difference at latitudes $\lesssim 60°$. Above 60° latitude the haze may be thick enough to influence the methane band reflectivity in the near infrared.

A crude estimate of the variation of cloud altitude with latitude can be gained by examining the behavior of the photometry and polarization along the central meridian. The estimates are rough because the center-to-limb variations which constrain the vertical structure have been analyzed at only a few discrete latitudes. Approximations, described below, have been made to extrapolate the results over a continuum of latitudes covering the observed range.

The Pioneer 11 polarization data in blue light have been used to constrain a 2-parameter vertical structure model where P_1 is the pressure level, at which $\tau_{cloud} = 0.4$, and H_p/H_g is the ratio of particle to gas scale heights (Tomasko and Doose 1984). With H_p/H_g assumed to be independent of latitude, P_1 can be derived as a function of latitude to fit the observed polarization along the central meridian. Similarly, the ratio of the 0.8996 μm CH$_4$ band divided by the 0.9365 μm continuum from the images of West et al. (1982) gives a value to $\tau_{\text{cloud}} = 1$, assuming that the methane abundance per unit cloud optical depth is equal to the abundance above the cloud top in a model where the top of the atmosphere is free of aerosols and the semi-infinite layer on the bottom has $H_p = H_g$. To convert methane abundance to total pressure, a value of [CH$_4$]/[H$_2$] $= 2.5 \times 10^{-3}$ will be assumed.

The reflectivity at 0.264 μm along the central meridian has been modeled by West et al. (1983a) as a conservatively scattering gas layer of optical depth τ_R above a semi-infinite scattering-absorbing layer with effective single-scattering albedo $\widetilde{\omega}$. The pressure level of absorption optical depth (τ_{abs}) $= 1$ is given by P_1(mbar) $= 162[\tau_R + \widetilde{\omega}/(1-\widetilde{\omega})]$. This type of model is not preferred by West et al. (1983c), but serves to convey the character of the latitudinal variations.

The variation of P_1 with latitude is shown in Fig. 16, for $\widetilde{\omega} = 0.5$ at 0.264 μm. The altitudes deduced from the three independent methods are in remarkably good agreement, considering the rough assumptions which have been made to derive the results. The following picture of cloud altitudes emerges. The Equatorial Zone and the latitude band from 45°N to 50°N (near the high-speed zonal jet; see the chapter by Ingersoll et al.) are regions of high clouds. The North Equatorial B (NEB) and South Equatorial B (SEB) at ±20° latitude are relatively deep. Unit cloud optical depth in southern mid-latitudes is $\sim$ 100 mbar deeper than at the same latitude in the north. A high-altitude ultraviolet absorber is abundant in polar regions.

The observation that unit cloud optical depth occurs at higher altitudes in northern mid-latitudes than in southern mid-latitudes may reflect hemispheric differences in lapse rate due to the seasonal insolation asymmetry, which is amplified by ring shading (Brinkman and McGregor 1979). Although the altitude of the NH$_3$ ice cloud base expected near 1.4 bar is controlled by phase equilibria, the cloud-top height depends on the dynamics. Specifically, convection (which transports aerosol upward) decays above the altitude where the lapse rate becomes subadiabatic (the tropopause). The curves in Fig. 16 are a measure of the tropopause altitude and its variation with latitude.

At deeper levels, another pattern can be derived from the available infrared observations. If the changes in thermal radiance morphology observed by the IRR are entirely due to cloud influences, then some constraints can be placed on the cloud properties by examining the extent of the drop in radiance and its dependence on outgoing emission angle. Pursuing this idea quantitatively, Orton (1983) used a temperature profile derived from a relatively clear atmospheric region sensed by the IRR near 15°S latitude to model cloud properties required to create the observed radiance depression near the equator. It was found that a cloud model could be created to match the radiances at the nadir, but the variation of radiance with emission angle cosine could not be matched without a cloud particle scale height greater than the gas scale

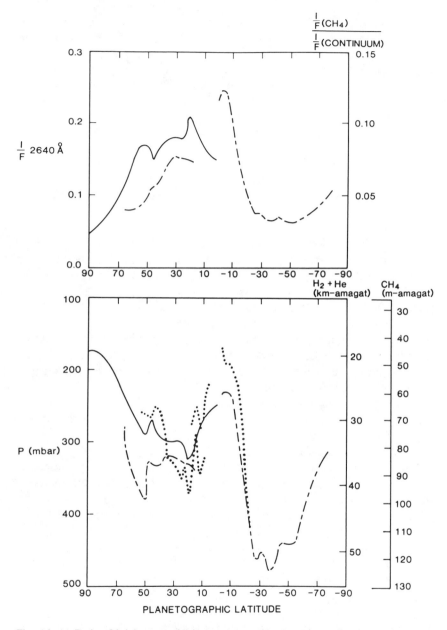

Fig. 16. (a) Ratio of brightness at 0.264 μm and methane band to continuum brightness as a function of latitude. (b) Rough estimates of cloud-top pressure as a function of latitude derived from the 0.264 μm intensities (solid curve), methane band to continuum ratio (dot-dashed curve), and the Pioneer polarimetry (dotted curve) using various simple models as described in the text.

height, which was considered improbable. A further difficulty may be implied by the IRIS observations of the northern hemisphere, a region not sampled by the IRR, which indicate that the 15°S latitude area is cooler than an area at a similar northern latitude, raising the possibility that the clear region used to derive a temperature profile could be influenced substantially by cloud obscuration.

An alternative approach is similar, but uses the RSS occultation results for the neutral atmosphere temperature structure (Tyler et al. 1982); these data should be extremely insensitive to the presence of particles in the atmosphere. In fact, the similarity of Voyager 2 results for ingress at 31°5N and egress at 31°S latitude, especially in the region for pressures > 100 mbar, is intriguing in view of the substantial difference in IRIS results for the same two regions (B. Conrath, personal communication). Further, the extremely good correspondence between RSS and IRIS results for the northern region is in marked contrast to the disagreement with the southern region.

While no work has yet been done with the IRIS spectra to study the cause of the disparity with the RSS results, Orton (1983) has made a preliminary study of the differences which are replicated to some extent in the more limited IRR data, obtained some 24 months before the RSS data acquisition. Assuming that the RSS temperature profile in the troposphere was valid at all latitudes in the region covered by the IRR, Orton derived cloud models which provided a good match to the 45 μm channel IRR data for the region near 15°S previously thought to be clear, and for the cool equatorial region (Fig. 17). For the equatorial region the derived models show a very thick cloud extending from the saturation level near 1.4 bar up to the temperature minimum in the vicinity of 100 mbar, and a vertical distribution of the particles characterized by a scale height nearly equal to the gas scale height. In contrast, the region around 15°S required a cloud with a lower number density and a smaller vertical extent, modeled equally well by a particle scale height ~ 0.15 times the gas scale height with particles extending up to the 100 mbar level, or by a particle scale height equal to the gas scale height with particles extending up to the 400 mbar level (approximately the upper boundary of the convective region). The picture this presents is of a slowly upwelling atmosphere at 15°S, with NH_3 particles condensing out in local equilibrium with the available partial pressure of the condensed gas; a rapid and vigorous convection process near the equator which transports particles as high as possible; and a relatively clear atmosphere in parts of the northern hemisphere, implied by the close agreement between the IRIS and RSS results there.

Unfortunately, this approach, while possibly quite valid, is not without problems. There is a disagreement between RSS temperature structure and the 20 μm channel IRR results in the context of these cloud models, which is replicated by the IRIS results. The infrared results imply that the temperatures near the 200 mbar level and higher should be substantially warmer than indicated by the RSS results. If this is not the case, then it indicates that the

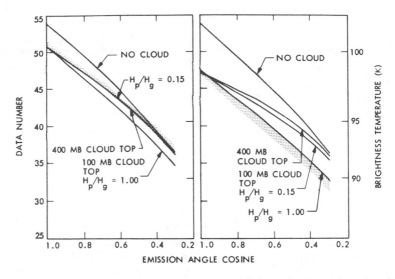

Fig. 17. Brightness temperature of Saturn at 45 μm wavelength as a function of the cosine of the emission angle for 15°S latitude (*left*) and near the equator (*right*). Observations by the Pioneer Saturn IRR are shown as the shaded region, and are compared to various clear and cloudy models as discussed in the text (from Orton 1983).

gaseous absorption, in this case H_2, is rather inaccurate, which is considered quite likely. Alternatively, there could be another source of atmospheric opacity influencing the 20 μm spectral region. The possibility of aerosols in the thermally inverted stratosphere of Saturn is thus raised, although it may be difficult to make particles large enough to affect 20 μm radiation and small enough to remain buoyant at pressures of $\leq$ 100 mbar. Furthermore, if such particles are NH_3 ice, they must be able to withstand destructive photo-chemical processes.

Conrath and Pirraglia (1983) note that temperatures at the 150 mbar level measured by IRIS show a hemispheric asymmetry, which they attribute to the seasonal insolation asymmetry and thermal phase lag. Fine-scale latitudinal gradients are superimposed on the overall thermal profile. The fine-scale thermal structure is correlated with the zonal jet structure and ultraviolet reflectivity. Conrath and Pirraglia showed that vertical motion associated with the jet system may influence the altitude profile of aerosols and the thermal gradient consistent with the observations.

Other evidence for seasonal variations is slowly accumulating. Trafton (1977) suggested a seasonal effect in CH_4 absorption strength in observations taken over a period of a few years. Reflectivity variations in the short wavelength continuum appear to be present in the photographic record (Johnson 1979; Suggs 1982). Systematic spatially resolved photometric observations in the continuum and in molecular absorption bands during the next several decades would be very useful.

The zonal wind field is only slightly asymmetric with respect to the equator in mid-latitudes (see chapter by Ingersoll et al.). The relationship between cloud altitudes inferred from photometry and polarimetry to wind fields is as yet unclear. The wind field measurements are obtained from discrete clouds embedded in a thick, diffuse aerosol layer. It is conceivable that the altitudes of clouds derived by photometric methods are different than the altitudes of clouds used to define the zonal velocity profile. If the latter small clouds are at the same altitude in both hemispheres, the clouds in the south should be more visible, especially near the limb, than the clouds in the north because there is less high diffuse cloud material obscuring those features in the south.

V. SUMMARY AND CONCLUSIONS

Despite the substantial uncertainties, some firm information regarding the aerosol and cloud particles on Saturn has been accumulated.

1. A layer of aerosols which absorb strongly in the ultraviolet exists at pressure levels of 30–70 mbar at low latitudes and at pressures of ≤ 20 mbar at latitudes $> 60°$. At low latitudes, the optical depth of these aerosols is > 0.4 at 0.264 μm. At 0.64 μm these aerosols are seen to produce a rapid increase in polarization toward the planetary limb and terminator near 90° phase. The small optical depth required in the red (20 times less than in the ultraviolet) together with the strong polarization they produce implies that the aerosols are fairly small, with radii estimated at 0.1 μm. They are probably the result of photochemical processes (and possibly energetic particle bombardment at high latitudes) and are probably closely related to the aerosols with similar properties seen in the corresponding part of Jupiter's atmosphere.

2. At longer wavelengths, the effective top boundary of the aerosol region appears to be at deeper levels of the atmosphere, reflecting the decreasing cross section of the small particles above. The top of the cloud as sensed by the strong 0.89 μm methane band is near 100 mbar at both equatorial and temperate latitudes (see Fig. 15). However, the specific abundance of H_2 in a mean free scattering path increases from ~ 10 km-amagat at latitudes $< 10°$ to > 40 km-amagat at latitudes $> 20°$ in the region from 100 to 300 mbar (Tomasko et al. 1984). Cloud optical depth 0.4 (in the visible) occurs near 150 mbar at equatorial latitudes and at perhaps twice this value at temperate latitudes. The particles in this region of the atmosphere absorb in the near ultraviolet and blue, and absorption at these wavelengths does not cease at these levels but persists to deeper levels. At equatorial latitudes, the particles in this region of the atmosphere may be largely crystals carried up from below. The top of this region even in equatorial

latitudes is $\lesssim 1$ scale height above the commonly estimated location of the radiative-convective boundary; convective overshooting and vertical motions associated with the organized zonal flow may be responsible for lifting the crystals to these levels. Recent estimates for the tops of Jupiter's presumably NH_3 clouds are also a factor of 2 to 3 lower in pressure than estimates of Jupiter's radiative-convective boundary (K. Baines, personal communication; Smith and Tomasko 1983). Further, the particles at cloud optical depth ~ 0.4 which occur at 150 mbar at equatorial latitudes and 300 mbar at higher latitudes have similar polarizing properties (somewhat negatively polarizing near 90° scattering angle in red light), implying that the higher particles at equatorial latitudes are similar to the ones found near or below the radiative-convective boundary at somewhat higher latitudes.

3. At pressures of 300 (at equatorial latitudes) or 400 to 500 mbar (temperate latitudes) the cloud optical depth reaches 2. The photometry as a function of phase has been used to give the single-scattering phase function of the particles in this region of the atmosphere in red and blue light, and the corresponding values of the planetary phase integral. The phase functions in red and blue are similar and show modest forward peaks and a rather flat behavior at scattering angles $\gtrsim 70°$. The phase function resembles those measured for NH_3 crystals larger than several microns. The clouds at several locations on the planet (especially the equatorial latitudes, but also at south tropical and temperate latitudes) produce noticeable thermal opacity at 45 μm wavelength by this depth.

4. The cloud structure at deeper levels is less clear. The base of an NH_3 cloud is expected at a pressure of ~ 1.4 bar, and at many latitudes the weak molecular absorption bands, such as the H_2 quadrupole lines and the weak CH_4 bands, indicate that the cloud continues to those levels, attaining optical depths on the order of $10-20$ (corresponding to isotropic optical depths of $5-10$) at red and near infrared wavelengths. On the other hand, the weak CH_4 bands at some latitudes seem to require a clearer region at pressures > 500 mbar, with a total optical depth of the cloud to this level (in the visible) of ~ 5.

Despite these results, many questions remain. For example, correlations between cloud altitudes and other observable meteorological quantities can only be qualitative unless information on the physics of cloud formation is available. One of the most important parameters is the latent heat release in a parcel saturated with NH_3 or some other condensable. Barcilon and Gierasch (1970) and Gierasch (1976) suggested that H_2O condensation could be important in driving the Jovian meteorology. The situation for Saturn is somewhat different. Although the latent heat of vaporization of NH_3 is about half that for H_2O, latent heat from NH_3 condensation may be relatively more important for Saturn than for Jupiter, because Saturn's lower temperatures permit

NH_3 to condense over a larger portion of the atmosphere, and hence a much larger mass of NH_3 is able to condense in the upper cloud on Saturn.

The cloud microphysics has not been adequately studied. Many of the arguments developed by Rossow (1978) for Jupiter's clouds should apply to Saturn if basic differences in pressure, temperature, and scale height are taken into account. Rossow concludes that the condensate clouds in Jupiter's atmosphere are cooling clouds resembling massive terrestrial water and ice clouds. Questions concerning the transport of particles and vapor, laminar versus turbulent flow, time scales for condensation and sedimentation and nucleation processes are largely unanswerable until entry probe data is available.

Despite the substantial body of available observational data, a variety of new remote observations are needed. For example, the variations over the disk of the strength of weak molecular absorptions (like H_2 quadrupole lines) is not well known. Also, many types of existing observations need to be extended to a much longer time base to separate and understand the significant expected seasonal and other temporal variations.

Even within the bounds of existing data, several lines of new analysis as well as additional laboratory data are needed. Until improved laboratory measurements of the scattering properties of actual (nonspherical) candidate cloud and aerosol particles are made, it will remain difficult to relate aerosol optical properties derived from planetary observations to constraints on physical properties such as size and composition. In the visible and near infrared, the interpretations of the data of West et al. (1982) and Apt and Singer (1982) presently assume no temperature dependence for the strength of the methane bands used in those studies. The derived vertical structure is obviously sensitive to that assumption; laboratory spectra at low temperature are required before full confidence can be placed in the models. Further work remains in combining the analysis of ultraviolet and visible photometry and polarimetry, along with the brightness and center-to-limb variations in the near infrared absorption bands, in terms of a single cloud model.

Further work should also be pursued both observationally and analytically for the near infrared region between 1.2 and 5 μm. The high resolution spectroscopy of Larson et al. (1980) and Fink et al. (1983) cover the available spectral windows, but their results are not presented in terms of absolute radiance or geometric albedo. The Saturn spectra presented by Fink and Larson (1979) and by Apt and Singer (1982) could be more useful from the viewpoint of quantitative analysis of cloud properties, because they provide the reflectivity calibration which is extremely important in radiative transfer modeling. Further work in this region, similar to work in the visible and photographic infrared, can easily take advantage of the large variation of CH_4 gaseous absorption in a region where its temperature dependence is more easily modeled in order to provide information on the vertical distribution of particles. Even more information can be gained from emission angle dependence (center-to-limb behavior) and, using spacecraft measurements, from phase angle dependence over a large range of angles.

Several lines of inquiry are indicated by the very preliminary work done in the thermal infrared. The correspondence of the RSS and IRIS results should be pursued further. At the very least, it will eliminate the possible uncertainty involved in the time separation between the Pioneer IRR and the Voyager RSS data sets which were used in Orton's study. Studies similar to those of Marten et al. (1981) and Orton et al. (1982) should be performed for the available IRIS spectra of Saturn to establish the criteria for both the far infrared and 9.5 μm for minimum particle size (or minimum irregularity) provided by the absence of readily identifiable NH_3 ice features. Work must continue on the scattering properties of irregular particles on the order of the wavelength or larger in size, to determine the extent to which the minimum size derived by Marten et al. is valid. Further work can be accomplished in the 5 μm region, both in and out of the warm spot areas (Hanel et al. 1981a). Though this is complicated by the mixing together of reflected solar and planetary thermal radiation, these components can be separated by comparing spectra from the day and night sides (Bjoraker et al. 1982). In any case, a competent physical model should be able to account for both components of the outgoing radiation. Analysis of observations over a wide range of wavelengths is essential for even a rudimentary understanding of the particle size distribution for the major aerosol and cloud layers, and it is possible that extensions to still longer wavelengths will increase the vertical range over which remote observations are sensitive.

Acknowledgments. The authors gratefully acknowledge financial support for the work presented in this chapter from several different grants from the National Science Foundation and the National Aeronautics and Space Administration as listed at the back of the book. We also are grateful to L. R. Doose for access to model calculations in advance of publication and to K. Baines for helpful discussions. One of us (M.G.T.) wishes to acknowledge especially the important support received from the Pioneer Project Office. This program has made possible an appreciable fraction of the work described in this chapter concerning the structure and nature of clouds on Saturn.

STRUCTURE AND DYNAMICS OF SATURN'S ATMOSPHERE

ANDREW P. INGERSOLL
California Institute of Technology

RETA F. BEEBE
New Mexico State University

BARNEY J. CONRATH
Goddard Space Flight Center

and

GARRY E. HUNT
Imperial College, London

The atmosphere of Saturn exhibits dynamical structures (jets, bands, spots, eddies) with horizontal scales ranging from one scale height (60 km) up to the planetary radius (60,000 km). Although the kinds of structures are the same as those on Jupiter, there are quantitative differences. First, Saturn's equatorial jet speed is four times greater and the jet width is two times wider than on Jupiter. Eastward jets at higher latitudes are also speedier and wider, but westward jets are slower compared with Jupiter. There are fewer spots in all size ranges, and no spot comparable in size to Jupiter's Great Red Spot. The rms eddy wind speed is lower, as is the conversion of eddy kinetic energy to zonal mean kinetic energy (the measured rate of conversion is positive but not significantly so). Saturn's convective regions, where eddy lifetimes are a few days or less, are more isolated and cover less area than on Jupiter. The basic causes of these differences have not been quantitatively identified. Basic differences between the two planets include Saturn's lower heat source, lower gravity, and consequently greater depth of atmosphere and deeper nonconducting fluid layers.

Observed temperatures imply that the zonal winds decrease from cloud top upwards, but change more gradually below the cloud tops. One possibility is that zonal winds are constant on concentric cylinders that extend throughout the planet. However the observations, which are limited to cloud-top altitudes and above, do not preclude other configurations. It is remotely possible that Saturn's internal rate of rotation is different from that of its radio emissions. Dynamical models differ in their assumption about the interior and its dynamical interaction with the atmosphere. The time-dependent behavior of the large-scale structures offers a promising way of testing the models and choosing among the various assumptions.

This chapter treats the large-scale structure and dynamics of Saturn's atmosphere, as revealed in the visible markings, the wind patterns, and the horizontal variations of temperature. Examples of this structure include the cloud bands, the zonal jets, the long-lived oval spots, and the active convective regions, all of which are large compared to the atmospheric scale height (of order 60 km) but small compared to the planetary radius ($\sim$ 60,000 km). The equator-to-pole temperature contrast and interhemispheric seasonal contrasts are examples of planetary-scale structure. All of these give evidence of a complicated and interesting dynamics.

By far the bulk of our knowledge of large-scale structure comes from Voyager. Under the best conditions, one or two bands are faintly visible from Earth; spots appear occasionally, but usually the planet appears zonally symmetric (Alexander 1962). During the past 100 yr or more, only ten spots yielded reliable estimates of zonal wind speed (Alexander 1962; Dollfus 1963; Reese 1971). Magnetospheric emissions were not observed from Earth, so the rate of rotation of Saturn's interior was unknown until the Voyager encounters (Desch and Kaiser 1981*a*). Pioneer 11 gave only faint evidence of banding (Burke et al. 1980; Ingersoll et al. 1980), and even Voyager found Saturn to be much less photogenic than Jupiter (B. A. Smith et al. 1981,1982; Hanel et al. 1981*a*,1982). Not only are intensities lower in both the visible and infrared parts of the spectrum, but contrasts are also lower than on Jupiter, particularly in visible light (Smith et al. 1981).

Generally the questions raised by the large-scale structure observed in Saturn's atmosphere are the same as those for Jupiter: Why are the bands and spots so stable and long-lived? Why is the structure so axisymmetric? What processes control the number of bands and magnitude of the velocities? How does the dynamics interact with the chemistry to produce the visible patterns? As with Jupiter, the structure at depth and dynamical interaction with the fluid interior are major unknowns.

The large-scale thermal structure is treated in Sec. I. The wind measurements are discussed in Sec. II. Feature morphologies (spots, waves, eddies) are discussed in Sec. III. Dynamical models are reviewed in Sec. IV. The observations discussed in this chapter refer to pressure levels from $\sim$ 1 mbar to several bars. The upper atmosphere (pressures $<$ 1 mbar) is discussed in the chapter by Atreya et al. The internal structure of Saturn and the internal heat source are treated in the chapter by Hubbard et al. Gaseous abundances and cloud composition are treated in the chapter by Prinn et al. Cloud physical properties are treated in the chapter by Tomasko et al. The subject of this chapter is therefore dynamic meteorology with emphasis on the upper troposphere, to which most of the observations refer.

I. THERMAL STRUCTURE

A considerable body of information now exists on the atmospheric thermal structure of Saturn, derived from both groundbased observations and spacecraft data. Most of this information is confined to the upper atmosphere where ultraviolet spectroscopy can provide data, and to the region between $\sim$ 1 mbar and 1000 mbar which is accessible to both infrared measurements and the radio occultation technique. The upper region is discussed in the chapter by Atreya et al., while the 1 to 1000 mbar region will be considered in this section. Little direct information is available for the deep interior. However, knowledge of the structure of the observable portions of the atmosphere, besides furthering the understanding of the radiative properties, cloud physics and dynamics of these regions, may also serve to establish constraints on models of the deeper levels.

A. Groundbased Observations

Early attempts to deduce the atmospheric thermal structure of Saturn made use of both radiative-convective equilibrium models constrained by groundbased observations (Wallace 1975; Caldwell 1977a; Tokunaga and Cess 1977), and direct inversions of groundbased infrared data. In the equilibrium models, pressure induced hydrogen transitions were assumed to be the primary source of infrared opacity, with solar energy absorbed by near-infrared bands of methane and in some cases an additional ultraviolet absorber (Axel 1972; Podolak and Danielson 1977). With estimates of the effective emission temperature of the planet used to specify an internal heat flux, equilibrium profiles were then obtained.

In the second approach to the problem, infrared measurements are incorporated directly into the radiative transfer equation which is then inverted to retrieve the atmospheric temperature as a function of pressure level. The earliest attempt to apply this approach to Saturn appears to have been that of Ohring (1975) who obtained an estimate of the lower stratospheric structure by inversion of groundbased data in the ν_4 methane band. Gautier et al. (1977a) carried out an extensive study of the then available groundbased data using inversion techniques and obtained a family of possible profiles each of which corresponded to a different set of assumptions concerning the data and constraints. These results were further constrained by Caldwell et al. (1978) using drift scans along the equator of Saturn at 17.8 and 22.7 μm.

B. Spacecraft Observations

The Pioneer 11 Saturn encounter in 1979 provided significant new information on the thermal structure from two separate experiments. Radio occultation measurements provided an estimate of the temperature profile between $\sim$ 2 and 180 mbar pressure levels near $-10°$ latitude (Kliore et al. 1980b). In addition the Pioneer infrared radiometer obtained spatially resolved measurements in two broad spectral bands centered near 20 and 45 μm (Ingersoll et al. 1980). These measurements spanned a sufficient range in emission angles to provide vertical thermal structure information between $\sim$ 70 and 500 mbar. Orton and Ingersoll (1980) retrieved profiles by inversion in the latitude interval $-30°$ to $+10°$. Minimum temperatures were found centered on the equator between $-10°$ and $+10°$ latitude with essentially no horizontal temperature gradients between $-30°$ and $-10°$.

The Saturn encounters of Voyager 1 in November 1980 and Voyager 2 in August 1981 have provided extensive new information on the thermal structure of the planet. The Voyager 1 ingress radio occultation provided a temperature profile at high southern latitudes (Tyler et al. 1981) while profiles were obtained at latitudes of $+36°\!.5$ and $-31°$ from the Voyager 2 ingress and egress measurements, respectively (Tyler et al. 1982a). In addition, the Voyager infrared spectrometer (IRIS) provided spatially and spectrally resolved measurements of thermal emission which yield information on the temperature structure (Hanel et al. 1981a,1982). Data between frequencies of 200 and 700 cm^{-1} within the S(0) and S(1) pressure induced hydrogen rotation lines and a portion of the hydrogen translational continuum provide information between approximately the 80 and 700 mbar levels, while measurements within the ν_4 CH$_4$ band centered near 1304 cm^{-1} provide information within a relatively thick region centered near 1 mbar. Using techniques discussed by Conrath and Gautier (1980), these data have been inverted to obtain temperature profiles; the resulting thermal structure data base extends over a substantial fraction of the planet.

C. Mean Vertical Structure

We shall first consider the planet-wide average vertical thermal structure. The IRIS derived profile shown in Fig. 1 represents an area weighted global mean of profiles obtained from spatially resolved spectral measurements. The data used for constructing this mean profile came primarily from a mapping sequence in which the planet was repeatedly scanned from north to south during one complete planetary rotation. This basic uniform data set was augmented by additional data from other sequences acquired at high latitudes. The vertical resolution of this profile in the region between 100 and 700 mbar corresponds to approximately one half of a pressure scale height. While the formal error due to instrument noise propagation in this average profile (which made use of over 350 individual IRIS spectra) is only $\sim$ 0.1 K, the largest source of uncertainty at the higher pressure levels is the effect of un-

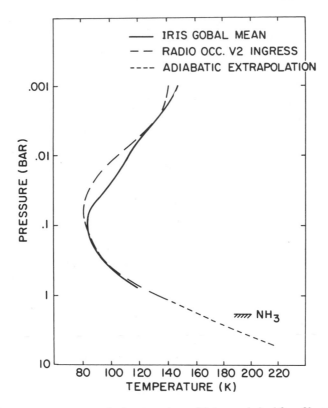

Fig. 1. Temperature vs. pressure in the atmosphere of Saturn as derived from Voyager infrared (IRIS) and radio occultation observations. The IRIS profile is an area-weighted global mean of profiles obtained from spatially resolved spectral measurements. The RSS profile is the Voyager 2 ingress profile at 36°5N latitude. In deriving these profiles, a hydrogen mole fraction (H_2 to H_2 + He) of 0.94 has been assumed. The base of the NH_3 cloud is computed assuming solar composition.

known contributions to the opacity due to hazes and clouds. In the strato-sphere (here defined as that region above the temperature minimum), the IRIS data contain little information on the temperature profile in the region be-tween ~ 10 and 80 mbar so the solution obtained at these levels is essentially an interpolation. However, temperatures above this region up to ~1 mbar are well constrained by measurements in the ν_4 methane band.

Also shown in Fig. 1 is the Voyager 2 ingress profile at +36°5 latitude. Although this profile pertains to only one specific location on the planet, it is in good agreement with the IRIS global average profile between 100 and 700 mbar. In the lower stratosphere, the apparent disagreement between the IRIS and occultation profiles probably reflects the lack of information in the in-frared spectra on this portion of the atmosphere as discussed above. In deriving both the radio occultation and IRIS profiles, a hydrogen mole fraction of 0.94 has been assumed (see chapter by Prinn et al.).

The Voyager 2 ingress profile as given by Tyler et al. (1982a) extends to 1200 mbar. At deeper levels in Fig. 1 we have extrapolated the temperatures along an adiabat appropriate for an atmosphere with helium and hydrogen mole fractions of 0.06 and 0.94, respectively. The ratio of ortho hydrogen to para hydrogen has been assumed to be the equilibrium value at the local temperature. However, the latter assumption may not be valid because of the very long equilibrium times required for ortho-para conversion in the absence of catalytic processes. Hydrogen transported vertically in the atmosphere from warmer or cooler levels may retain its initial ortho-para ratio rather than equilibrating at the local temperature. Massie and Hunten (1982) suggest that hydrogen may be transported upward while retaining "normal" 3:1 ortho-para ratio characteristic of the higher temperature at deep levels. When levels are reached where aerosols are present, partial equilibration may occur due to catalytic processes on the surface of the particulates. This will result in an ortho-para ratio intermediate between normal and equilibrium and a change in the adiabatic lapse rate through modification of the specific heat. In addition the lapse rate may be further modified due to the release of energy of ortho-para conversion (Gierasch 1983). The latter effect will tend to reduce the lapse rate below that predicted by a simple adiabat. Thus, extrapolation of the thermal structure below the directly observable regions is uncertain at the present time.

The condensation level for NH_3 is shown in Fig. 1 for reference. This level was determined under the assumption that the mole fraction for NH_3 in the well-mixed layers is consistent with a solar elemental abundance. The vertical distributions of hazes and clouds are discussed in the chapter by Tomasko et al.

D. Horizontal Temperature Structure: Seasons

In addition to the mean vertical thermal structure, information on horizontal gradients has been obtained by the Pioneer 11 infrared radiometer and especially the Voyager infrared spectrometers. Unfortunately, these data give no direct information on the horizontal structure below the cloud tops; nevertheless, knowledge of the structure of the upper troposphere and lower stratosphere aids in our understanding of the radiative and dynamical processes associated with this portion of the atmosphere and may ultimately contribute to our understanding of the deeper layers.

Data acquired by IRIS on both Voyager 1 and 2 near closest approach to Saturn have been used by Conrath and Pirraglia (1983) in an effort to establish the detailed latitude dependence of the upper tropospheric thermal structure. Approximately 200 individual spectra were inverted to obtain temperature profiles. The data selected have a spatial resolution corresponding to ~ 4° of latitude or better. The latitude dependence of temperatures for three levels within the upper troposphere are shown in Fig. 2. Because of the limited vertical resolution of the retrieved profiles, these temperatures should be regarded as averages over layers approximately one half scale height thick, centered at the indicated pressure levels. The error in these retrieved temperatures due to instrumental noise propagation is $\pm$ 0.5 K.

The temperature at 730 mbar does not show a major equator-to-pole gradient in either hemisphere; however, the retrieved temperatures at mid-latitudes in the north are $\sim$ 4 K warmer than at either low or high latitudes. One possible explanation is that additional opacity sources not included in the retrievals are influencing the spectral measurements near 200 cm^{-1} (50 μm) which are used to provide information on this portion of the atmosphere. Hazes and clouds are, of course, likely candidates for such sources. While the behavior of the retrieved profiles suggests the absence of cloud decks at pressures $\lesssim$ 750 mbar which are completely opaque near 200 cm^{-1}, partially transparent hazes or broken cloud layers could introduce perturbations of a few degrees in the inferred temperatures. If such an effect is in fact occurring, the 730 mbar structure would indicate that the far infrared particulate opacity must be a minimum at mid northern latitudes. It is interesting to note that the opacities inferred in the ultraviolet, blue, and near infrared, as discussed in the chapter by Tomasko et al., are also relatively low in this region. However, these results also indicate even less opacity at midlatitudes in the southern hemisphere while the 730 mbar temperatures show no obvious structure in this region.

At 150 mbar both large-scale hemispheric temperature differences and relatively small-scale latitudinal structure are observed. In order to emphasize both the differences and similarities between hemispheres, a curve representing a visual fit to the northern hemisphere data has been reflected into the southern hemisphere in Fig. 2. Note that poleward of about 25° latitude, the southern hemisphere is on the average 6 to 8 K warmer than the north. This is qualitatively consistent with a lagged response to the seasonally varying solar forcing which results from the 27° obliquity of Saturn. Cess and Caldwell (1979) have theoretically studied the seasonal variation of the stratospheric thermal structure, and their results appear to be essentially consistent with Voyager observations (Hanel et al. 1981a). The response of the upper troposphere has been examined by Hanel et al. (1982) and by Conrath and Pirraglia (1983) in terms of the parameter $\Omega_S \tau$, where Ω_S is the seasonal frequency and τ is the radiative relaxation time. To a first approximation the phase lag of the thermal response to the solar forcing is given by $\tan^{-1} \Omega_S \tau$, and the amplitude of the response is proportional to $(1 + \Omega_S^2 \tau^2)^{-1/2}$. Estimates of τ based on the calculations of Gierasch and Goody (1969) for hydrogen atmospheres with a scaling appropriate to Saturn indicate a lag of approximately one sixth of a Saturn year or 5 terrestrial years. The Sun crossed from the southern hemisphere into the northern hemisphere approximately 6 months prior to the Voyager 1 encounter and 15 months before that of Voyager 2; therefore, it is an-

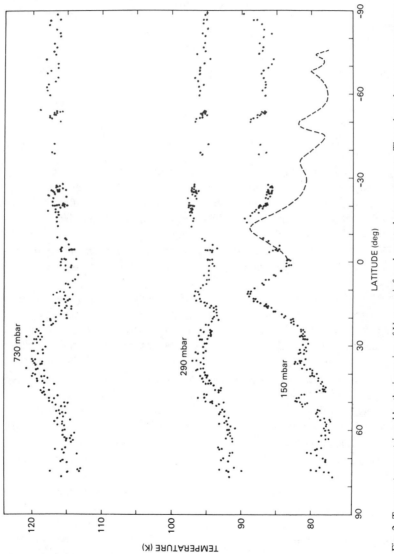

Fig. 2. Temperatures retrieved by the inversion of Voyager infrared spectral measurements. The values shown represent mean values for atmospheric layers approximately one-half scale height thick. The broken curve is a fit to the northern hemisphere 150-mbar temperatures which has been folded over at the equator to permit a comparison with southern hemisphere results (from Conrath and Pirraglia 1983).

ticipated that the northern hemisphere should be cooler by a fractional amount $\sim [1 - (\cos 54°)^{1/4}] (1 + \Omega_S^2 \tau^2)^{-1/2} \approx 6\%$. At levels deeper than ~ 300 mbar the value of $\Omega_S \tau$ becomes sufficiently large that the seasonal response should be greatly reduced, consistent with the relatively small differences observed between hemispheres at 290 mbar.

E. Horizontal Temperature Structure: Jets

The relationship of the smaller scale structure in the 150 mbar temperatures to data from other experiments is of considerable interest. Even though there are gaps in the high spatial resolution data in the southern hemisphere, comparison with the northern hemisphere temperatures strongly suggests that the smaller scale structure is approximately symmetric with respect to the equator. As pointed out in the discussion of the 730 mbar temperatures, opacity effects not included in the analysis in the spectral regions near 350 and 600 cm^{-1} could lead to the retrieval of spurious 150 mbar temperatures. Very large variations in opacities at these frequencies would be required since the lapse rate in this portion of the atmosphere is relatively small (Fig. 1). It appears more likely that the observed structure is primarily due to actual variations in temperature along constant pressure surfaces. If this is in fact the case, then under the reasonable assumptions of hydrostatic equilibrium and geostrophic balance, the latitudinal temperature gradients should be related to the eastward wind component $\bar{u}$ through the thermal wind equation (Holton 1979)

$$\frac{\partial \bar{u}}{\partial z} = - \frac{R}{fa} \frac{\partial T}{\partial \lambda} \tag{1}$$

where z is height in units of pressure scale height, R is the gas constant, f is the Coriolis parameter, a is the planetary radius, T is temperature and λ is latitude. To test this hypothesis, Pirraglia et al. (1981) and Conrath and Pirraglia (1983) have estimated thermal wind shears, $\partial \bar{u}/\partial z$, from the 150 mbar temperatures of Fig. 2 with the results shown in Fig. 3. Comparison of these results with the cloud top zonal winds obtained from Voyager images (Fig. 4), which are discussed in detail below (Sec. II.D), suggests a correlation with the symmetric jet system in both hemispheres. The sense of the correlation between $\bar{u}$ and $\partial \bar{u}/\partial z$ is such that the peak-to-peak amplitude of the jet system decreases with height, although very slowly (in one scale height the fractional change of $\bar{u}$ is of order 0.2). An exception to this is the region at 10° to 25° north latitude where a portion of the eastward equatorial jet increases in speed with height.

These considerations suggest that the smaller scale thermal structure in the upper troposphere must result either directly or indirectly from dynamical processes. One possibility is that we are observing the thermal response of a statically stable, dissipative upper troposphere forced at its lower boundary by

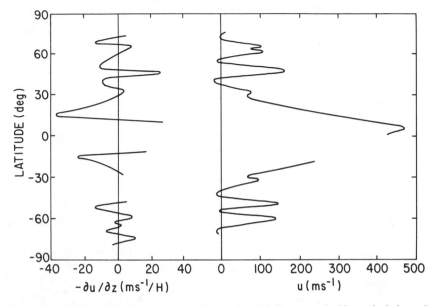

Fig. 3. Vertical thermal wind shear at the 150-mbar level (*left*) compared with zonal wind speed at the cloud tops (*right*). Thermal wind shear was calculated from the temperatures shown in Fig. 2, and the zonal winds were obtained from observations of cloud motion (B. A. Smith et al. 1982). Note that the negative of the shear has been plotted for this comparison (from Conrath and Pirraglia 1983).

an imposed zonal jet system. In such a situation the deceleration (of order $\nu u/L^2$) on the jets due to horizontal eddy viscosity would be balanced to lowest order by the Coriolis acceleration (of order fv) acting on a meridional flow. Mass continuity requires an accompanying vertical motion (of order vH/L), with the upward leg producing adiabatic cooling (of order $w\Delta\vartheta_v/H$) on the equatorward side of an eastward jet and the downward leg producing adiabatic heating on the poleward side of the jet. This heating and cooling is balanced by radiative sources and sinks (of order $\Delta\vartheta_h/\tau$) that tend to return the upper tropospheric temperatures to their normal, unforced values. Here u, v, w are the eastward, northward and upward velocities, respectively; ν is the horizontal eddy viscosity; L and H are the horizontal and vertical length scales, respectively; $\Delta\vartheta_h$ and $\Delta\vartheta_v$ are the horizontal and vertical potential temperature differences, respectively; and τ is the radiative time constant.

Assuming $\Delta\vartheta_h/\Delta\vartheta_v \gtrsim u/fL \approx 0.1$, this model requires $\nu \gtrsim L^2/\tau \approx 3 \times 10^4$ m^2 s^{-1}, for $L \approx 3000$ km and $\tau \approx 3 \times 10^8$ s. This estimate is probably uncertain by two orders of magnitude. Yet the above minimum ν is well below the estimate $\nu \approx v_e L_e \approx 5 \times 10^5$ m^2 s^{-1} obtained from direct observation of eddy velocities and length scales, $v_e \approx 5$ m s^{-1} and $L_e \approx 100$ km, respec-

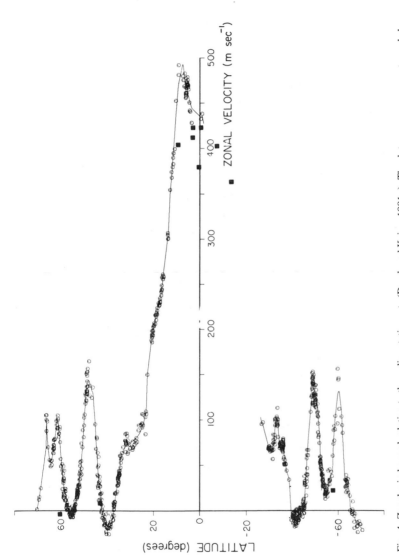

Fig. 4. Zonal wind speed relative to the radio rotation rate (Desch and Kaiser 1981*a*). The dots are measurements made by tracking individual cloud features in pairs of Voyager 2 images separated by 10 to 11 hr at resolutions from 50 to 200 km. This is a reworked and augmented version of the data shown in B. A. Smith et al. (1982). The smooth curve is a least-squares fit to these data. The portion from the equator to 25°S latitude was obscured by rings at the time of the Voyager 2 encounter. The solid squares are groundbased observations of spots larger than ~ 5000 km, and represent > 100 yr of observation as summarized by Dollfus (1963) and Reese (1971).

tively. In other words, the radiative adjustment rate $1/\tau$ is so weak that even a weak eddy viscosity (compared to $v_e L_e$) can have a large effect on upper tropospheric temperatures.

It is also possible that a part or all of the observed upper tropospheric structure results from variations in the amount of infrared radiation upwelling from below or in the amount of *in situ* absorption of sunlight. Such variations could be controlled by latitudinal variations in aerosol abundances, which are related to the jet system through accompanying vertical motion. More aerosols (thicker clouds) at lower levels would imply less upwelling radiation and lower temperatures in the upper troposphere (Orton et al. 1981). However, more aerosols in the upper troposphere might imply more absorption of sunlight and higher temperatures. Thus aerosols are both an indicator of vertical motion and a potential cause of vertical motion through their effects on the radiation. Furthermore, the radiative effects of aerosols can be of either sign, leading either to a net heating or to a net cooling.

The 150 mbar temperature structure appears to be correlated with smaller scale structure in the 2640 Å absorption vs. latitude curve inferred from the Voyager 2 photopolarimeter (see chapter by Tomasko et al.). The sense of the correlation is such that higher temperatures are associated with regions of lower opacity. This is consistent with the view that the thermal structure results directly from the meridional circulation associated with the jet system, and the haze or clouds producing the 2640 Å absorption is denser in regions of upwelling and cooling. However, the optical depth inferred from Pioneer 11 photopolarimeter measurements in the blue (chapter by Tomasko et al.) appears at least in some regions to be correlated in the opposite sense, with areas of lower optical depth corresponding to lower temperatures. This is more nearly consistent with the direct solar absorption hypothesis.

Regardless of the details of the processes producing the small-scale upper-tropospheric thermal structure, the picture which emerges is one in which the observed gradients are related in some way to the underlying jet system. Superposed on this structure is a seasonally varying modulation which may result in minor modifications to the zonal wind in this part of the atmosphere, such as the increase in wind speed with height at low latitudes in the cooler hemisphere, but otherwise has little influence on the deep-seated dynamical system.

II. WIND OBSERVATIONS

At present the only wind data come from tracking cloud features in series of images of the same area at different times. With the exception of several dozen spots seen from Earth, all of the data come from Voyager. Features ranging in size down to about 100 km have been tracked, usually in pairs of images taken on successive rotations of the planet. This method yields estimates of horizontal velocity to an accuracy of about 10 m s^{-1}, depending on the sharpness and steadiness of the features.

Here we review first the Earth-based data, then the Voyager data, including the radio observations from which the internal rate of rotation is inferred, the zonal wind profile, the eddies and eddy transports, and finally the problems with the data and whether they can be overcome by further analysis.

A. Observations from Earth

Viewed from the inner solar system a feature 10^4 km in size on Saturn subtends about 1.4 seconds of arc, in which close to the limit of Earth-based telescopic resolution. Features of this size or greater occur only rarely. Those that last for 10 d or longer may be used to infer a local rate of rotation, from which the zonal wind may be inferred to an accuracy of about 10^4 km/10^6 s $\approx$ 10 m s^{-1}. Features that are visible from Earth do not drift in latitude, so meridional motion cannot be inferred.

Dollfus (1963) and Reese (1971) have reviewed the very best observations. They included only the well-marked spots seen on photographs and in detailed visual drawings. The spots must have adequate contrast, reasonably long lifetimes, constancy in size and shape, and must be distinguishable from neighboring spots on separate nights of observation. The latter requirement is especially important. Multiple outbreaks of spots at the same latitude occur as often as isolated spots. Misidentification usually leads to larger errors than uncertainties in timing or location, and may have occurred in many of the reported observations.

Alexander (1962) describes $\sim$ 30 spot sightings from which rates of rotation were obtained. Reese (1971) and Dollfus (1963) consider $\sim$ 10 of these to be reliable. For comparison with the Voyager data, the rates are expressed as zonal (eastward) wind speed relative to a coordinate system rotating uniformly at the "radio" rotation period of 10 hr 39.4 min determined by Voyager (Desch and Kaiser 1981a). In this coordinate system a zonal wind speed of 450 m s^{-1} at the equator corresponds to a period of 10 hr 11.4 min. The zonal wind profile, $\bar{u}$ vs. latitude, is shown in Fig. 4. Earth-based data are the solid squares.

The Earth-based data are significant because they show how little the general features of the zonal wind profile have changed during 100 yr or more, a time interval 100 times greater than that between the Voyager 1 and Voyager 2 encounters (12 November 1980 and 26 August 1981, respectively). The Earth-based data show the equatorial jet with velocities approaching 450 m s^{-1} extending to latitudes of $\pm$ 10° and the regions of low velocities at latitudes beyond $\pm$ 40°. The magnitudes of the winds are in good agreement with the Voyager data. The latitudes of the Earth-based points are uncertain by $\sim$ 5° (equivalent to 5000 km at normal incidence). Thus, there is no significant difference between the latitudes of the low-velocity regions in the Voyager data and the latitudes at which low velocities were measured from Earth.

B. The Radio Rotation Rate

Earth-based observations do not define a reference frame in which to measure winds. Any period of rotation from 10 hr 10 min to 10 hr 40 min seems equally possible. However, if most of the planet were in solid body rotation, with the differential rotation confined to a thin cloudy zone (thickness ~ 250 km; see chapter by Prinn et al.), then large shears would have to exist from top to bottom across this thin "weather layer." Both the sign and magnitude of these shears as functions of latitude would depend on the value of the internal period of rotation.

At a distance of 3 AU, the planetary radio astronomy instrument aboard Voyager 1 began detecting kilometric radiation from Saturn (Kaiser et al. 1980; Warwick et al. 1981,1982a; also see chapter by Kaiser et al.). In many ways Saturn kilometric radiation (SKR) resembles Jupiter's decametric radiation, from which that planet's internal rate of rotation was first derived. SKR is very strongly modulated (tenfold or more) at a 10 hr 39.4 min period that is presumed to be that of the planet's magnetic field (Desch and Kaiser 1981a). Spoke activity in Saturn's B Ring is also modulated at this period (Porco 1983). Direct measurements of Saturn's magnetic field, however, show no evidence of a magnetic anomaly which could be associated with this modulation (E. J. Smith et al. 1980b; Acuña et al. 1980; Connerney et al. 1982b).

The symmetry of the field leads to speculation that asymmetries in the rings cause the SKR modulation. If this were so, the planet's internal rate of rotation would still be unknown. However the source of SKR is at high northern and southern latitudes (Kaiser and Desch 1982). Field lines originating at these latitudes cross the equator well outside the outer edge of the A Ring, where the orbital period of particles is several days. The situation is somewhat enigmatic, but there appear to be no serious models for which the SKR period is not that of the electrical conducting regions deep within the planet (see chapters by Connerney et al. and by Hubbard et al.).

A different possibility, suggested by the fact that the radio period is significantly longer than the latitudinally averaged period for atmospheric features (~ 10 hr 30 min), is that the asymmetries of the field outside the atmosphere are somehow slipping relative to asymmetries within the planet. This possibility would require a nonlinear mechanism whereby the frequency is changed as field lines pass from the planetary interior to the SKR source regions. Such a down-shifting in frequency is fundamentally different from a simple phase lag, and no mechanism has been suggested for it. The simplest interpretation is again that the SKR period is that of the electrically conducting interior. This interpretation has been adopted by the Voyager imaging team (B. A. Smith et al. 1981,1982), and will be adopted here.

C. Analysis of Voyager Images

Wind measurements were made with the atmospheric feature tracking system (AMOS) of JPL's Image Processing Laboratory (Yagi et al. 1978).

The approach was similar to that used in analyzing the Jupiter data (Beebe et al. 1980; Ingersoll et al. 1981). The goal was to determine the mean zonal wind $\bar{u}$ and its derivatives $d\bar{u}/dy$ and $d^2\bar{u}/dy^2$, to search for significant mean meridional velocities $\bar{v}$, to estimate the root mean squared (rms) fluctuations δu and δv (departures from $\bar{u}$ and $\bar{v}$, respectively), and to search for a significant mean correlation $\overline{u'v'}$ between fluctuations of u and fluctuations of v. Here y is the northward coordinate, and the overbar denotes an average with respect to longitude. Averaged quantities are therefore functions of latitude. For each individual measured velocity, u' and v' are the residuals from $\bar{u}$ and $\bar{v}$, respectively, at the measurement latitude.

The altitude is that of the visible clouds, meaning that it is rather uncertain and quite variable. As shown in Fig. 1, the base of the ammonia cloud is estimated to lie near the 1-bar level, but the clouds and haze at other levels may also contribute to the visible appearance (see chapters by Prinn et al. and by Tomasko et al.). Saturn is different from Jupiter in that there are fewer features at all scales suitable for tracking. Contrast is lower. Features that last for two or more rotations are rare, and the regions where eddies are active cover a much smaller fraction of the planet.

Twenty-six Voyager 2 image pairs of Saturn were analyzed. These yielded about 1000 features whose displacements could be measured. For Jupiter, 25 image pairs yielded over 10,000 feature displacements. The data set described here is an extension of that described in papers following the Saturn encounters (B. A. Smith et al. 1981,1982). For the northern hemisphere the resolution was usually about 50 km per picture element (pixel) pair. For the southern hemisphere the resolution was usually about 150 to 200 km. The principal results, some of which have not been published elsewhere, are shown in Figs. 4 and 5 and Table I. Images of the type used in the analysis are shown in Figs. 6–10 (Sec. III). Additional images have been published by B. A. Smith et al. (1981,1982).

As with Jupiter (Ingersoll et al. 1981), there are two kinds of errors: those associated with feature identification and those associated with image navigation. Image navigation is the process whereby camera pointing is determined to an accuracy of one or two pixels by locating the bright limb of the planet in the narrow-angle image or in the simultaneously shuttered wide-angle image. Feature identification errors are greater for Saturn than for Jupiter because of lower feature contrast, shorter feature lifetimes, and other reasons mentioned above. Image navigation errors are at least as large for Saturn, and have a greater impact on the results than for Jupiter.

Neither error is particularly significant in the determination of $\bar{u}$ vs. latitude (Fig. 4) because zonal velocities are generally much larger than the errors. The errors are significant, however, in the determination of $\bar{v}$, δu, δv, and $\overline{u'v'}$ (Fig. 5), since meridional velocities and eddy velocities are small. Navigation errors have a greater impact for Saturn than for Jupiter because the mean zonal velocities $\bar{u}$ are larger and the eddy velocities are smaller. Thus a

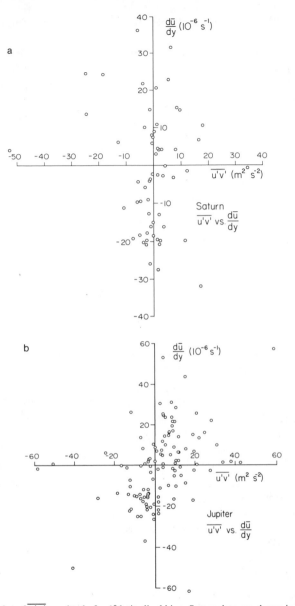

Fig. 5. Scatter plot of $\overline{u'v'}$ vs. $d\bar{u}/dy$ for 1° latitudinal bins. Saturn data are shown in (a) using 61 bins for which there were 5 or more individual measured features. Jupiter data are shown in (b) using 106 bins for which there were 10 or more features. Here u and v are the zonal (eastward) and meridional (northward) velocity components, respectively, $\bar{u}$ is the longitudinal mean of u, and u' and v' are departures from the longitudinal means. Only features for which u' and v' were less than $2\delta u$ and $2\delta v$ were used in these plots, where δu and δv are the global rms averages of u' and v', respectively. For Saturn the lack of correlation ($r = 0.01$) indicates that the global mean of $\overline{u'v'} \cdot d\bar{u}/dy$ is not significantly different from zero. For Jupiter the correlation is significant at the 1% level ($r = 0.40$). The measured rate of conversion $\{K'\bar{K}\}$ of eddy kinetic energy to zonal mean kinetic energy is therefore significant for Jupiter but not for Saturn. The data set is a reworked and augmented version of the set used in B. A. Smith et al. (1982) and Ingersoll et al. (1981).

TABLE I
Peak Values of $d^2\bar{u}/dy^2$ Compared to β[a]

Latitude	$d^2\bar{u}/dy^2$	β
56°	6.1	3.1
40°	7.8	4.2
29°	~13	4.8
17°	~5	5.2
−30°	~9	4.7
−41°	8.6	4.1
−55°	~12	3.1

[a] Units are $10^{-12} \text{ m}^{-1}\text{s}^{-1}$

small error in camera pointing causes part of $\bar{u}$ to appear as a spurious meridional velocity v or eddy velocity (u', v'), especially near the limb of the planet where curvature is important.

The Saturn results presented here have been filtered in two ways: First, 5 of the original 26 image pairs were rejected because meridional velocities were of one sign over large areas (several belts and zones) of the planet. Second, 10% of the velocity vectors in the remaining images were rejected because either u' or v' was larger than $2\delta u$ or $2\delta v$, respectively. Some valid measurements of flow around the larger, more energetic eddies were rejected by the latter filtering procedure. However, the same procedures do not change our results for Jupiter. That is, the principal conclusions of Beebe et al. (1980) and Ingersoll et al. (1981) are the same whether 1% or 10% of the vectors are rejected.

D. Mean Zonal Velocity

The solid line in Fig. 4 shows the zonal velocity profile $\bar{u}$ as a set of quadratic polynomials. Each polynomial is a fit to the data between a minimum of $\bar{u}$ and the adjacent maximum. The profile is consistent with Fig. 4 of B. A. Smith et al. (1982), which shows individual measurements, and with Fig. 5 of Ingersoll and Pollard (1982), which shows a free-hand rendering of the B. A. Smith et al. (1982) data. The dots of Fig. 4 are the individual measurements. The scatter is low, and the quadratics generally provide a good fit to the data.

Interesting features include the strength of the equatorial jet, the symmetry about the equator, and the preponderance of eastward flow in the coordiserved in a band between 6° and 10°N latitude. The wind speed at the equator is $\sim 430 \text{ m s}^{-1}$. It is not known whether the peak wind pattern repeats in the south, because the latitudes from 0° to 15°S were obscured by rings at the time of the Voyager encounters.

The small jog in the profile at 30° latitude is interesting for two reasons.

First, it appears in both northern and southern hemispheres. Second, it resembles a feature in the Jupiter profile at 13° latitude in both hemispheres. In general the latitudinal wavelength of the profile variation is twice as large for Saturn as for Jupiter. For Saturn the three westward jets occur at latitudes around 40°, 55°, and 70°. For Jupiter there are more jets, but the first three occur at 18°, 32°, and 39° (Ingersoll et al. 1981).

The midlatitude eastward jets on Saturn are faster than those on Jupiter. The peak wind on Jupiter is 140 m s^{-1}, which occurs at 23°N. This jet is similar to the 150 m s^{-1} jet at 46°N latitude on Saturn. The latter, however, is dwarfed by Saturn's equatorial jet, which is relatively less significant on Jupiter.

The westward velocities on Saturn are about one-half as large as those on Jupiter. They are about one-tenth as large as the eastward jet velocities. The preponderance of eastward flow on Saturn (more so than on Jupiter) is one of the surprises of the Voyager encounters. To make the Saturn profile look like a rescaled version of the Jupiter profile, one would choose a rotation period of about 10 hr 30 min (vs. the 10 hr 39.4 min period derived from SKR modulations). One is tempted to doubt the relevance of SKR as an indicator of internal rotation, but as already discussed no alternate theories have been put forward.

Table I compares the curvature $d^2\bar{u}/dy^2$ of the zonal velocity profile with $\beta = 2\Omega\cos\lambda/a$, the planetary vorticity gradient at the latitudes of the westward jet maxima. Here Ω is the planetary rotation rate, λ is latitude, and a is the planetary radius. For Jupiter, $d^2\bar{u}/dy^2$ was found to be about two times β at the westward jet maxima (Ingersoll et al. 1981). For Saturn, the second derivative was estimated by fitting second- and fourth-order polynomials to the data in latitude bands centered on the westward jets and extending about halfway to the adjacent eastward jet maxima. The difference between second- and fourth-order fits gives an idea of uncertainty. As on Jupiter, $d^2\bar{u}/dy^2$ exceeds β at the westward jet latitudes. The barotropic stability criterion is violated on Saturn as it is on Jupiter. However, the relevance of this criterion as a condition for stability depends on the temperature profile below the levels of observation and on the depth of the zonal flow. As discussed in Sec. IV, a different stability criterion applies when the zonal flow extends through the planet on coaxial cylinders (Ingersoll and Pollard 1982). The above results concerning the zonal velocity profile and its curvature are confirmed in a recent study of Saturn data by Sromovsky et al. (1983).

The mean meridional velocity $\bar{v}$ is seriously affected by navigation error. For Saturn we found that the absolute value of $\bar{v}$ is less than 4 m s^{-1} at all latitudes, in agreement with other studies for both Jupiter and Saturn (Ingersoll et al. 1981; Limaye et al. 1982; Sromovsky et al. 1983). This is not a very stringent upper bound, since it only excludes circulation times from zones to belts ($\sim$ 10,000 km) that are less than one month. This is 10 to 100 times less than the expected time based on the radiative time constant.

Saturn's rms eddy velocities δu and δv are about 8 m s^{-1} and 3 m s^{-1}, respectively (vs. 15 m s^{-1} and 6 m s^{-1} for Jupiter), according to the present analysis. A similar result follows from the work of Sromovsky et al. (1983). Sromovsky et al. do not consider the covariance $\overline{u'v'}$ of eddy velocities on Saturn.

E. Eddy Momentum Transport

One of the most interesting results of the Jupiter wind analysis concerns the eddy momentum transport $\overline{u'v'}$, which is toward the latitudes where mean zonal momentum $\bar{u}$ is greatest. Thus $\overline{u'v'}$ and $d\bar{u}/dy$ tend to vary together (in phase) as functions of latitude. If pairs of values are computed for latitudinal bins 1° wide, the correlation coefficient between the two variables is about 0.4 (Beebe et al. 1980; Ingersoll et al. 1981), and the global integral $\{K'\bar{K}\}$ of $\overline{u'v'} \cdot d\bar{u}/dy$ is positive. The associated conversion of eddy kinetic energy to zonal mean kinetic energy is more than 10% of the power radiated to space, when the conversion is averaged over the cloudy layer. Normalized in this way, Jupiter's kinetic energy conversions are 100 times more efficient than the Earth's.

Other terms in the energy and momentum cycles were not measured, for several reasons: First, the mean meridional velocity $\bar{v}$ is undetectably small relative to observational uncertainties. Second, temperature differences could not be measured except at scales much larger than the eddies. And third, all quantities could only be measured at cloud-top levels and above; interactions with deeper levels are unknown.

There is currently a debate concerning $\overline{u'v'}$ for Jupiter. Limaye et al. (1982) and Sromovsky et al. (1982) repeated the earlier Jupiter analysis using a data set derived from images at four times poorer resolution than those used by Beebe et al. (1980) and Ingersoll et al. (1981). Initially using the same "target of opportunity" approach to feature selection as Beebe and Ingersoll had done, they obtained results similar to theirs. Then using a principle of "one feature per gridpoint" on a uniform latitude-longitude grid, they obtained no significant correlation between $\overline{u'v'}$ and $d\bar{u}/dy$. Sromovsky et al. point out that uneven sampling of a perfectly symmetrical eddy could lead to a spurious $\overline{u'v'}$. Aliasing due to unresolved eddies could also lead to spurious results. The issue can be settled when every significant eddy of whatever size is adequately resolved and uniformly sampled. The Voyager data set does not allow such a sampling, but studies of isolated regions at high resolution (25 km per pixel pair) are in progress.

For Saturn, the eddy momentum transport $\overline{u'v'}$ is again of interest. A typical absolute value of $d\bar{u}/dy$ is 2×10^{-5} s^{-1}, which is only slightly larger than the typical absolute value for Jupiter. We would therefore expect the global mean of $\overline{u'v'} \cdot d\bar{u}/dy$ to be of order $r\delta u\delta v |d\bar{u}/dy| \approx 7 \times 10^{-5}$ m^2 s^{-3}, where $r \approx 0.15$ is a dimensionless constant chosen to fit the Jupiter data. In fact, the global mean $\{K'\bar{K}\}$ is $\sim 2.0 \times 10^{-5}$ m^2 s^{-3}, which is 10 times smaller than the Jupiter value and 3.5 times smaller than the expected Saturn value based on the magnitudes of δu and δv. Figure 5a shows a scatter plot of $\overline{u'v'}$ vs. $d\bar{u}/dy$ for Saturn. Each point represents the average over a 1° bin of latitude with at least 5 measured features per bin. In all, 882 individual features contributed to the plot. The correlation coefficient of these points is 0.01, and the mean of the distribution $\{K'\bar{K}\}$ is not significantly different from zero. Figure 5b shows a similar scatter plot for Jupiter but with 10 or more features per 1° bin. The correlation coefficient is 0.40, which is significant at the 1% level.

F. Conclusions and Discussion

The differences between Jupiter and Saturn are perhaps the most interesting result of this analysis. The equatorial zonal jet speed is about 500 m s^{-1} (vs 125 m s^{-1} for Jupiter). The latitude at which the equatorial jet falls to zero is about 37° (vs. 16° for Jupiter). The rms eddy velocities δu and δv are 8 m s^{-1} and 3 m s^{-1}, respectively (vs. 15 m s^{-1} and 6 m s^{-1}, respectively, for Jupiter). The global mean of $\overline{u'v'} \cdot d\bar{u}/dy$ is $\{K'\bar{K}\} \equiv 2.0 \times 10^{-5}$ m^2 s^{-3} (vs. 2.0×10^{-4} m^2 s^{-3} for Jupiter), although the value for Saturn is not significantly different from zero. The energy conversion time, defined as the ratio of zonal mean kinetic energy $\frac{1}{2}\,\bar{u}^2$ to the energy conversion rate $\{K'\bar{K}\}$ is about 10 yr for Saturn (vs. 75 day for Jupiter).

The small value of $\{K'\bar{K}\}$ for Saturn is not too surprising when allowance is made for the greater mass/area of Saturn's cloud layer and the smaller value of Saturn's radiated power/area. First, cloud base occurs at higher pressures (assuming solar composition for each planet), because Saturn's temperatures are lower. Second, at the same pressure level the overlying mass/area is greater than on Jupiter because of Saturn's lower gravity. If the value of $\{K'\bar{K}\}$ measured at cloud-top level were characteristic of the entire cloud layer, the power/area P would be $\{K'\bar{K}\}M$, where M is mass/area of the layer. For Jupiter, M is of order 10^4 kg m^{-2} (Ingersoll et al. 1981), and for Saturn M is of order 4×10^4 kg m^{-2}. Thus P for Jupiter is about 2 W m^{-2}, which is about 15% of the total radiated power/area. Similarly P for Saturn is about 0.8 W m^{-2}, which is about 18% of Saturn's radiated power/area. The difference between 18% and 15% is not significant.

In other words, one should not expect $\{K'\bar{K}\}$ to be as large for Saturn as for Jupiter. The fact that the measured value for Saturn is not significantly different from zero then follows from the smaller value of $\{K'\bar{K}\}$ and the larger errors of measurement. This argument assumes, of course, that the energy conversion $\{K'\bar{K}\}$ occurs over the entire cloud layer on both planets, and that the power/area associated with this conversion should be the same fraction of the total radiated power/area on both planets. The fact that the eddy velocities δu and δv are smaller on Saturn compared to Jupiter is at least consistent with this view.

The greater width of Saturn's equatorial jet and the greater jet speed could both be related to the greater depth of Saturn's atmosphere, which follows from the lower gravity. The width of Saturn's jet is slightly more than double that of Jupiter, as is the depth of the cloud zone and the depth of the nonconducting outer envelope (see chapters by Prinn et al. and Hubbard et al.). This fact is consistent with the hypothesis that jet width is proportional to depth. Velocity should then scale as (width)^2, if $d^2\bar{u}/dy^2$ is the same fraction of $\beta = 2\Omega\cos\lambda/a$ on the two planets. The latter follows from the theory of barotropic instability and other basic ideas of meteorology (see Sec. IV of this chapter).

III. SATURN CLOUD MORPHOLOGIES

On a large scale, the atmosphere of Saturn is similar to that of Jupiter, with a generally banded structure. However, telescopic observations found only a small number of nonaxisymmetric features (Alexander 1962; Reese 1971), so that in comparison with Jupiter, considerably less was known of their morphologies and associated motions. The detail now revealed from the Voyager imaging observations (B. A. Smith et al. 1981,1982) has shown a wide range of cloud systems: long-lived oval spots, tilted features in east-west shear zones, rapidly varying features in "convective" regions, and a wave-like feature (the ribbon) reminiscent of certain laboratory phenomena. In this section we discuss the types of cloud morphologies seen in the Saturn atmosphere and their relationship with the characteristics and stability of the atmospheric winds at cloud-top level.

A. Observability of Features

The observability of small-scale features seems to be related to latitude through the zonal velocity profile. As shown in Fig. 4, more features suitable for tracking occur at the latitudes of the zonal velocity extrema than in the regions of strong shear. This observation may indicate that the lifetime of small-scale eddies on Saturn is controlled by the shear. The shear tends to pull the eddies apart in a time of order $|d\bar{u}/dy|^{-1}$, which for Saturn in regions of strong shear is about 12 hr. The fact that Saturn's eddies are weaker means that this dispersive effect of the shear is relatively more important than on Jupiter. The prevalence on Saturn of tilted filaments (northeast-southwest in regions of clockwise shear, northwest-southeast in regions of counterclockwise shear) is consistent with this view (see Fig. 3 of B. A. Smith et al. 1982).

The above interpretation treats the origin of the eddies as a phenomenon separate from the shearing action of the zonal flow. For example, the eddies might be convective in origin, drawing their energy from buoyancy associated with vertical heat transport in the cloud zone. The numerous transient eddies in Jupiter's atmosphere seem to fit this model. They appear suddenly as intense, bright, puffy features; they expand rapidly over the next few Jovian rotations, and are eventually sheared apart by the zonal flow. On Saturn the origin of the eddies is more difficult to observe directly.

An alternate hypothesis associates the origin of the eddies with the shear flow itself. There are two possibilities, but neither is entirely consistent with the observations. If the shear flow were barotropically unstable where its curvature $d^2\bar{u}/dy^2$ exceeds β, more eddies would appear at the latitudes of the zonal wind minima, as observed. The appearance of eddies at the zonal wind maxima, however, would be unexplained. Or if the shear flow were baroclinically unstable where its vertical shear $|d\bar{u}/dz|$ is greatest, more eddies would appear at the maxima of $|\bar{u}|$. The appearance of eddies where $|\bar{u}|$ is close to zero would then be unexplained. In fact, the weak westward jets of Saturn exhibit the most eddy activity, and the strong eastward jets exhibit enhanced activity to a lesser extent (Fig. 4).

B. Isolated Spots

Consider the northern hemisphere first, starting at the equator. In the region extending from the equator to about 10°N wispy cloud features were observed by B. A. Smith et al. (1981,1982). The brightest parts of these features were located at 7°N, where the velocities were found to reach a maximum of 500 m s^{-1}. As with Jupiter (Ingersoll et al. 1981), there would seem to be a minimum in the velocity at the equator.

Several symmetric ovals of various colors (white, brown, red) have been observed at different latitudes. Around 30° north and south latitudes the zonal velocity profile has an inflection or possibly a weak pair of extrema before continuing on its decline to zero just below 40° latitude. Near 27°N there is a long-lived bright oval feature, the UV Spot, so named because it is especially prominent in ultraviolet images. The feature was observed during both the Voyager encounters, although the region in which it was situated altered noticeably (Fig. 6).

Three brown spots are situated at 42°N: Brown Spot 1 (Fig. 7) is 5000 by 3000 km, and B. A. Smith et al. (1982) found that its anticyclonic circumferential flow reaches 30 m s^{-1} and its mean vorticity is 5×10^{-5} s^{-1}, which is approximately a fifth of the local planetary vorticity. These flow characteristics are similar in magnitude to those for the Jovian white ovals (Mitchell et al. 1981).

A complex interaction between two white spots moving westward at 15 and 20 m s^{-1}, respectively, and Brown Spot 1 moving at 5 m s^{-1} in an eastward direction were observed by Smith et al. (1982). In the beginning the two white spots were at approximately the same latitude and about 10,000 and 15,000 km to the east of the brown spot. All the spots were seen to exhibit an anticyclonic circulation. At the beginning of the sequence, the fartherest white spot (WS 1) appeared to be slightly south of the second feature (WS 2), and if they had followed the mean flow, they would have collided as WS 2 overtook WS 1 from the west. However, instead of colliding, WS 1 moved farther south

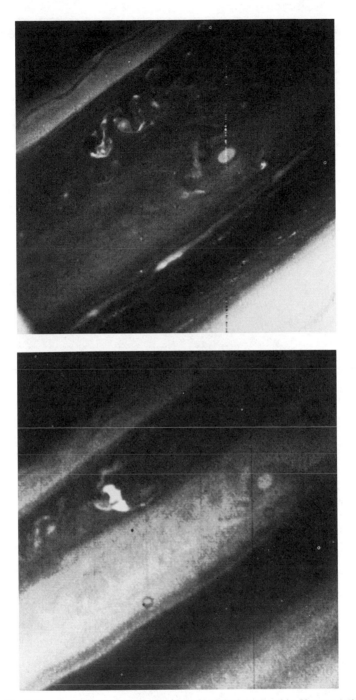

Fig. 6. The UV spot at 27°N latitude observed by Voyager 2 (*top*) and by Voyager 1 (*bottom*), showing changes that occurred during the 9.5-month interval between encounters. North is to the upper left. A convective feature is visible at 39°N in both images. The UV spot moves eastward at 75 m s^{-1}. The convective features move westward at speeds up to 10 m s^{-1}. Individual vortices appear to detach from the main convective feature and move eastward relative to it (figure after B. A. Smith et al. 1982).

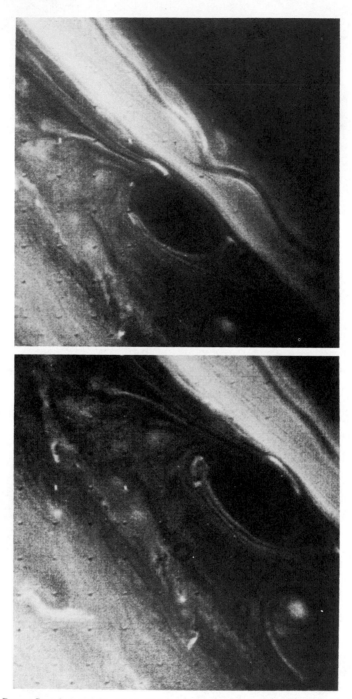

Fig. 7. Brown Spot 1 on two successive Saturn rotations, showing anticyclonic (clockwise in the northern hemisphere) rotation around its periphery. North is to upper right. The spot is centered at 42°5N latitude and it moves to the east at 5 m s^{-1}. The major diameter of the spot is 5000 km. A green filter was used in these narrow-angle images taken ~ 2 d before Voyager 2's encounter with Saturn.

and went around WS 2. This orbiting in a clockwise direction about a common center is consistent with the idea that each spot is a clockwise vortex moving under the influence of the other.

Four days later, WS 2 passed WS 1. While some merging did take place, there appeared to be a connected circulation between the two spots. The dark band between them stretched out considerably during the next few days as WS 2 continued to move in an eastward direction relative to WS 1. After six days the band stretched more than 5000 km and during the next two rotations of Saturn the dark band became much narrower, while the brown spot approached to within a few thousand kilometers of the two white spots. The band between these spots appeared to be on the verge of disappearing a day later.

The major feature observed in the southern hemisphere is a large red spot, Anne's Spot, situated at 55°S, measuring 5000 by 3000 km. Its color resembles that of Jupiter's Great Red Spot. The feature was first observed by Smith et al. (1981) in August 1980 and it was tracked throughout both Voyager encounters until September 1981. It resides in a region where there is an eastward flow of about 20 m s^{-1}.

C. Convective Regions

On Jupiter, actively changing (convective) cloud systems are located in the cyclonic shear zones and are especially prominent to the west of the long-lived anticyclonic ovals. Saturn's atmosphere also seems to contain convective cloud systems, but with a very different visible appearance. The convective clouds are primarily located in the region of the westward jet at 39°N. From measurements of their changing areas during the brief observational sequence, Hunt et al. (1982a) estimated the divergence of these systems to be 5 × 10^{-5} s^{-1}. If we assume that the vertical motion creating these changes in the observed cloud structures is associated with the upper scale height of the atmosphere, Hunt et al. (1982a) indicate that this divergence is consistent with a vertical motion of 1 m s^{-1}. This is about an order of magnitude greater than the estimated vertical motions in the Jovian plumes (Hunt et al. 1982b). These divergence estimates should be viewed with caution, however. A fundamental assumption of cloud tracking analysis is that clouds are a passive tracer of winds. That assumption is clearly violated in regions of active cloud formation such as those from which these measurements were made.

At the time of the Voyager 2 encounter, the region of the westward jet near 40°N had changed its visible appearance, and then displayed a wealth of different cloud systems. Amongst the individual cloud systems observed, one, which at first resembled a knot, was followed for a period of several weeks (Fig. 8). In the original observation, the feature was ~ 3000 km in diameter. Within 64 hr, the local shear coiled the cloud system cyclonically into the shape of a "6" before the cloud ultimately detached itself from the jet to the north. This cloud appeared to have a lifetime of about two weeks.

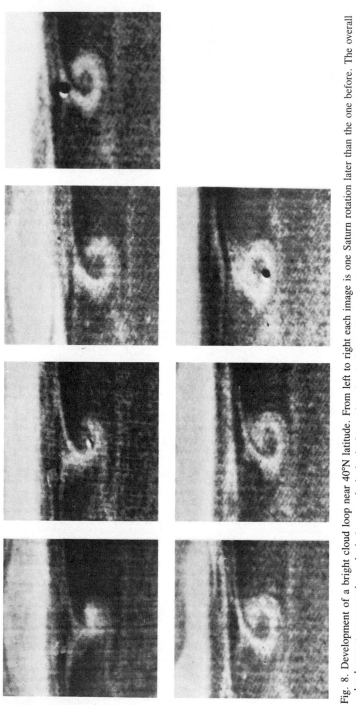

Fig. 8. Development of a bright cloud loop near 40°N latitude. From left to right each image is one Saturn rotation later than the one before. The overall development appears to be cyclonic (counterclockwise in the northern hemisphere), although anticyclonic motion is seen in the northern part of the loop. The diameter of the feature is ~ 3000 km.

B. A. Smith et al. (1981,1982), have found numerous unstable cloud systems in this region where the wind speeds are close to zero in the radio rotation frame. In one of these, a westward-moving dark spot appeared to be shedding a series of cloud vortices that extended to the east (Fig. 9). This time-dependent behavior resembles a vortex street which has been observed in the laboratory (Batchelor 1967; Tritton 1977), and in the Earth's atmosphere in the wake of islands (Chopra and Hubert 1965; Berlin 1981). The presence of a wake-type feature on Saturn is surprising. In their analysis, Godfrey et al. (1983) have compared the characteristics of the Saturn vortex street with those of the terrestrial and laboratory systems. While the Saturnian systems are larger in spatial scale, the derived Von Karman parameter (width to spacing ratio) is similar in magnitude to that measured in wakes. Alternatively, the vortices may be a localized form of convection in a deep, rotating atmosphere.

If the phenomenon is indeed a wake, the obstacle must be a convective system moving at a different (higher westwards) velocity relative to the unperturbed flow at the latitude of the vortex street. The presence of the vortex street must exert a considerable drag on the obstacle so that it must have its own source of westward momentum. However, as the observed zonal flow has its maximum westward velocity at this latitude, this excess westward momentum can only come from above or below. Since the flow above does not appear to be disturbed, this westward momentum probably originated from below. There the flow is presumably to the west at the speed of the obstacle (15 m s^{-1} relative to the radio system, 5 m s^{-1} relative to the line of vortices trailing to the east).

D. The Ribbon

The ribbon-like feature at 46°N may be unique to Saturn (Fig. 10). This dark, wavy line moves with the peak eastward wind of about 150 m s^{-1} at this latitude. Each crest or trough is about 5000 km long (one-half the wavelength of the ribbon) in an east-west direction. To the north of the ribbon, cyclonic vortices nestle in the troughs and their filaments spiral towards the center in a counterclockwise direction. To the south, under the crests, are anticyclonic vortices which spiral inward in a clockwise direction. The pattern resembles those of the classic "dishpan" experiments (e.g., Hide and Mason 1970; Read and Hide 1983), in which a baroclinic wave arises in response to a meridional temperature gradient. These ribbons may be associated with strong meridional heat transport in the atmosphere, for they seem to correspond with the location of large temperature gradients (Fig. 2). The division in the ribbon, which appears as a dark line, is actually several hundred kilometers across. It may therefore mark the division between cells in the circulation system.

In general, Saturn's atmosphere appears to be less active than Jupiter's. Spots and eddies are relatively less abundant. Tilted filaments, which may be pulled-apart remains of the eddies, are relatively more abundant. Compared

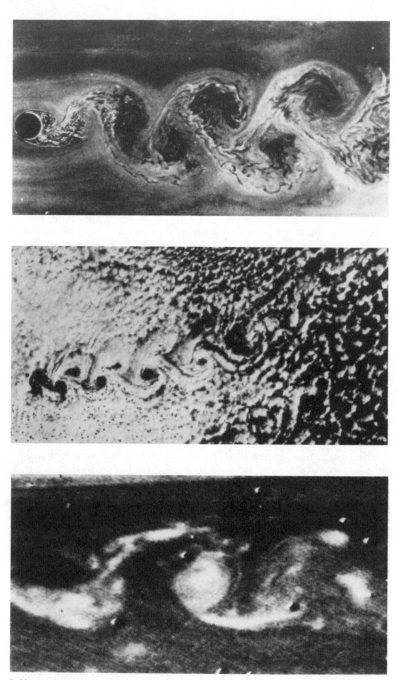

Fig. 9. Vortex streets (*top*) in the laboratory (Scorer 1978), and (*middle*) in the Earth's atmosphere, compared with (*bottom*) a convective feature near 40°N latitude in Saturn's atmosphere.

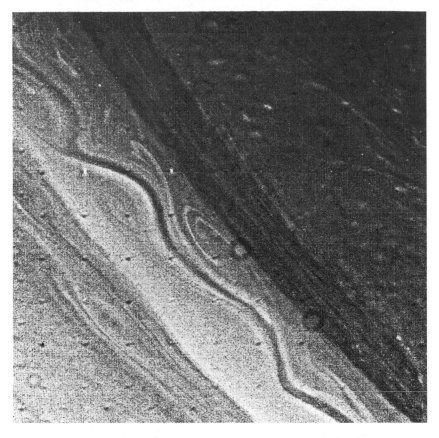

Fig. 10. Details of the ribbon at 46°N latitude. North is to the upper right. The wavelength of the ribbon is ~ 10,000 km, and the troughs and crests move eastward with the flow at 150 m s^{-1} (from B. A. Smith et al. 1982).

with Jupiter's atmosphere, that of Saturn seems to be dominated by the strong, axisymmetric zonal flow.

IV. MODELS OF THE ATMOSPHERIC CIRCULATION

P. H. Stone (1976), summarizing his review of meteorological models for the book *Jupiter*, stated, "Theoretical calculations so far shed little light on the nature of Jupiter's circulations . . . . Analytical models of (the inertial instability and forced convection) regimes and of finite amplitude rotating convection in the deep atmosphere are needed to further the deductive approach to Jovian meteorology. Measurements of the latitudinal variation of the internal heating are particularly important for developing realistic models . . . .

To evaluate different speculations for the causes of the banded structure and nonsymmetric features, it would be valuable to have measurements of meridional velocities, measurements of the lapse rates, and static stability in both the cloud layers and the layers immediately below, and to have high-resolution visual observations, capable of resolving small-scale convection."

Now, after eight years, following the four Voyager encounters with Jupiter and Saturn, it is time for another assessment. The basic problem is still to understand the observed latitudinal banding, the zonal jets, the long-lived ovals, and the distributions of temperature, cloud properties and radiative heat flux. Some of the observations listed by Stone could not be made by Voyager or any remote sensing instrument. Thus many unanswered questions remain. However, other key observations have proven to be extremely useful.

In a sense, there is no purely deductive approach to atmospheric dynamics, particularly the dynamics of Jupiter and Saturn. For reasons of economy and simplicity, theoretical models begin with a set of restrictive assumptions, e.g., about the wind field and temperature structure below the level of observations, about small-scale processes (often parameterized as an eddy viscosity and eddy thermal conductivity), and about the latitudinal distribution of internal heat impinging on the atmosphere from below. Constraining these assumptions is just as fundamental to our understanding as reproducing features of the observed circulations. The models provide specific questions, a framework for analyzing the data. The observations provide a challenge, namely to find out which assumptions of the models do not apply to Jupiter or Saturn.

A. Depth of the Zonal Flow

The depth of the flow is a fundamental unknown of all models. The interior of the planet, i.e., the metallic core, is assumed to rotate at the radio rotation rate. For Jupiter, the period is 9 hr 55 min 29.71 s (Carr and Desch 1976) and was derived from Earth-based observations spanning more than 20 yr. For Saturn, the period is 10 hr 39.4 min (Desch and Kaiser 1981a) and was derived from several months of Voyager observations. The observed radio emission arises from electrons trapped in the planet's magnetic field, and is modulated as a result of asymmetries in the field, e.g., a tilt of the field axis relative to the planetary rotation axis. Winds in the atmosphere are measured relative to this uniform rotation, and the fundamental question is: At what level, if any, does the wind profile fall to zero?

For Saturn, the answer seems to be that the level of no motion is deep, where pressures are 10^3 to 10^4 bar or greater (B. A. Smith et al. 1982). This conclusion is derived from the zonal wind profile observed at cloud top (Fig. 4), upper bounds on the equator-to-pole temperature gradient observed at cloud top (Fig. 2), and the thermal wind equation of meteorology (e.g., Holton 1979). That equation is derived by assuming the zonal winds are geostrophically and hydrostatically balanced. The conclusion is illustrated in Fig. 11, which shows the height of a typical constant pressure surface in the visible

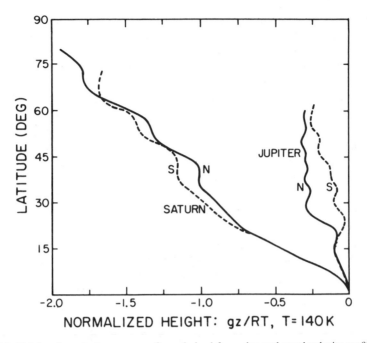

Fig. 11. Height of constant pressure surfaces derived from observed zonal velocity profiles of Jupiter and Saturn (after B. A. Smith et al. 1982). Height is computed by means of the geostrophic relation, and is shown here in units of the pressure scale height RT/g at cloud-top level ($T = 140$ K). The computed heights are measured relative to constant pressure surfaces of a hypothetical atmosphere rotating uniformly at the radio rotation rate. The total equator-to-pole drop in height is 0.35 scale heights for Jupiter and 2.0 scale heights for Saturn. The difference reflects the predominance of strong eastward winds on Saturn. Smith et al. argue that these strong eastward winds cannot fall to zero in a thin layer immediately below the clouds; pressure at the level of no motion must be at least 1000 times the pressure at cloud base.

layers. This height is measured relative to the equipotential surfaces $\Phi =$ constant, where Φ is the sum of gravitational and centrifugal potentials of the oblate planet rotating at angular velocity Ω. For Saturn this decrease in height is extremely large, approaching two scale heights. For Jupiter the decrease is about one-third of a scale height. A scale height is here defined as the pressure e-folding height at a temperature of 140 K, which is typical of the cloud tops for both planets. The curves are computed from the geostrophic relation. The Saturn height decrease is larger because the observed winds are stronger and more uniformly eastward than Jupiter's winds.

At levels where the fluid rotates uniformly at angular velocity Ω, pressure surfaces and equipotentials must coincide. The thickness of the layer between two constant pressure surfaces, one in the visible layers and the other at the level of no motion, must therefore decrease from equator to pole by the amount shown in Fig. 11, about 2 scale heights. Therefore if the layer were 5 scale heights thick at Saturn's equator, its thickness would decrease 40% from equator to pole. If the layer were 50 scale heights thick, its thickness would decrease by only 4%. B. A. Smith et al. (1982) argue that the former possibility is inconsistent with the observed upper bound on the equator-to-pole temperature difference. The argument hinges on the fact that the thickness of a gaseous layer between constant pressure surfaces in a hydrostatic fluid is proportional to the absolute temperature T of the layer. Observed horizontal variations of T at 730 mbar, the deepest levels sounded by the Voyager infrared instrument, are no larger than several percent (Fig. 2). If the atmosphere below were adiabatic, as expected for a planet with an internal heat source, fractional variations of temperature would be small at all levels. According to this argument, for Saturn the level of no motion is at least 2000 km below the visible clouds, where pressures are 10^3 to 10^4 bar, or 1000 times the pressure at cloud base.

The above argument, which was first advanced by the Voyager imaging team (B. A. Smith et al. 1982), has stimulated some debate. Allison and Stone (1983) point out that a thin-layer configuration would be consistent with the data if the internal rotation period were 6 to 8 min shorter than the SKR period, a point already noted by B. A. Smith et al. (1982, p. 507) and discussed in Sec. II.B. Allison and Stone offer no alternative explanation of the SKR periodicity, however. They also note that the depth of the zonal flow need be only a few hundred bars if the fractional temperature change with latitude were larger than the 4% upper bound chosen by B. A. Smith et al. (1982). Finally, they point out that the fractional temperature change could be hidden entirely from the Voyager instruments if it were concentrated at a deep, stably stratified interface that slopes upward from the ~100 bar level at the equator, reaching the visible clouds only at the poles. Such a large stable stratification at depth could occur if Saturn's molecular envelope had ten times more water than a solar composition mixture (Stevenson 1982b). The postulated slope of the interface could arise if these large water abundances were present at the equator but not at the poles.

Gierasch (1983) argues that variations in the relative abundance of orthohydrogen (parallel proton spins) to parahydrogen (antiparallel proton spins) could account for large temperature differences at depth. He computes the temperature difference between two vertical adiabatic profiles, with abundance ratios frozen at the equilibrium values corresponding to the low stratospheric temperature and to the high temperature of the deep layers, respectively. The frozen-in ratios are assumed to persist to pressures of several hundred bars and temperatures above 1000 K. The low- and high-temperature profiles are assumed to apply on opposite sides of each zonal jet. With these assumptions the pressure differences at cloud-top levels are sufficient to balance the observed winds, with no motion at depths greater than several hundred bars.

Like Smith et al. and Allison and Stone, Gierasch offers a kinematic model. The dynamical processes that might produce the assumed distributions are not specified. He shows that ortho/para abundance variations could account for the multiple-jet structure, i.e., the small-scale oscillations in the curves of Fig. 11 (wavelength ~ 13,000 km). However, these abundance variations fail by a factor of 30 to account for the total equator-to-pole height variation shown for Saturn in Fig. 11.

B. Upwelling and Downwelling

The thermal wind equation was traditionally used for Jupiter to compute the belt-zone temperature gradient, assuming a level of no motion near cloud base (Hess and Panofsky 1951; Ingersoll and Cuzzi 1969; Stone 1976). The fact that the small excursions to the right in Fig. 11 (pressure surfaces displaced upward) usually occur at the latitudes of the light-colored zones, was interpreted to mean that the zones are warmer than the dark-colored belts. Such a temperature difference would have to occur in the layer between the visible cloud tops and the level of no motion.

Voyager observations (Hanel et al. 1979; Flasar et al. 1981a; Conrath et al. 1981a; West et al. 1981), Pioneer observations (Tomasko et al. 1978; Orton et al. 1981), and supporting groundbased observations (Terrile and Beebe 1979; West and Tomasko 1980) provide evidence that the clouds of Jupiter are higher and thicker in the zones. This interpretation is consistent with the zones being warmer, which could then account for the upwelling and cloud formation. The traditional view is therefore that all features of the banded structure are driven by higher temperatures and upwelling in the zones, with divergence and meridional motion at cloud top and subsequent augmentation of the zonal jet motion.

There are also reasons for doubting the traditional view. First, the analogy with Saturn suggests that the zonal flow could extend far below the cloud zone, involving at least 1000 times more mass than the mass within the clouds. Second, the long-term changes of the belt-zone structure (cloud-top temperatures, cloud colors and scattering properties) often take place without a corresponding change in the zonal velocity profile (Smith and Hunt 1976). Both these facts are consistent with the idea that the zonal flow is deep, and is driven at least partly by processes occurring well below the clouds (Ingersoll et al. 1981). Furthermore, for Saturn the correlation between light and dark markings and the zonal jet structures is uncertain at best (B. A. Smith et al. 1982).

Theories of zonally symmetric circulations have been developed for the Earth's atmosphere and for other planets including Jupiter and Saturn. Williams and Robinson (1973) treated the belts and zones as large-scale convection cells in an unstably stratified atmosphere heated from below. This model was criticized by Stone (1976) who pointed out that the assumed eddy diffusivity (necessary to achieve the large scale) is impossibly large, and the assumed east-west orientation is opposite to that of free convection in a rotating sphere. For these reasons, all other zonally symmetric models propose a stably stratified atmosphere or one with substantial latitudinal temperature gradients.

The heating mechanisms and other parameters of these models are often somewhat *ad hoc*. Gierasch et al. (1973) proposed that increased (decreased) radiative cooling associated with a depressed (elevated) cloud layer could drive a meridional circulation that is self-sustaining. Barcilon and Gierasch (1970) and Gierasch (1976) proposed that convergence below the zones could increase the rate of latent heat release as in the Earth's intertropical convergence zone. Stone (1972) showed that a combination of large equator-to-pole temperature gradients and very weak stable stratification could lead to symmetric baroclinic instabilities. Hathaway et al. (1979) show that a large equator-to-pole temperature gradient and weak unstable stratification can lead to symmetric convective instabilities similar to Stone's baroclinic instabilities. Like the modes of free convection in Williams and Robinson's model, these tend to have horizontal scales comparable to the vertical scale, which is likely to be small. Since certain key free parameters of these models are still unknown, it is difficult to assess properly their relevance to Jupiter and Saturn.

Attempts to observe zonally symmetric upwelling and mean meridional motion have met with little success. Voyager imaging studies of Jupiter (Beebe et al. 1980; Ingersoll et al. 1981) reveal considerable north-south motion (rms velocities of order 6 m s^{-1}) associated with mesoscale (radii of order 1000 km) eddies, but little evidence of mean north-south motion (average meridional velocities in 1-degree latitude bins are no larger than the measurement uncertainty of 4 m s^{-1}). This upper bound is not very significant, because it only rules out the shortest meridional turnover times ($<$ 1 month). For Jupiter and Saturn, whose radiative time constants are $\geq$ 1 yr (Gierasch and Goody 1969), much longer turnover times are expected.

The problem is not just in the Voyager images, although uncertainties of camera pointing and orientation do exist (Ingersoll et al. 1981). Another problem is in Jupiter and Saturn themselves. Sharp, well-defined cloud features simply do not last long enough to give precise displacement over long time intervals. One of the surprises revealed by Voyager was the vigorous eddy activity and the transient nature of small-scale features. The traditional view of large-scale, quiescent upwelling and slow meridional motion does not seem to apply.

The only possible evidence of upwelling comes from the temperature structure above the clouds (Conrath et al. 1981*a*). Here, in the upper troposphere the temperatures of the zones are low. In a stably stratified atmosphere, such a distribution could arise when fluid with lower potential temperatures moves upward. Similarly, in the belts downwelling might be carrying fluid with higher potential temperature downward. Other interpretations that do not involve vertical motion also exist, as discussed in Sec. I. For instance, a high-altitude haze could mimic a low kinetic temperature in the zones, or the higher clouds in the zones could cause a reduction of long wave emitted flux and decrease of kinetic temperature in the gas above (Orton et al. 1981). Nevertheless, vertical motion in the belts and zones may account for most of the cloud properties and upper tropospheric temperatures of those features, but much of the zonal velocity structure seems to arise at deeper levels below the clouds.

C. Deep Convection

Theories of convection in rapidly rotating fluid spheres are of several types, each with its strengths and weaknesses. Busse's (1970,1976) approach is analytic; he solves a set of linear problems by treating the amplitude of the motion as a small parameter. To first order the flow is a periodic array of convection cells, with columns aligned parallel to the axis of rotation (Fig. 12a). Busse shows that these cells can drive a weak second-order mean flow that consists of differentially rotating coaxial cylinders (Fig. 12b). He associates the pattern of columns and cylinders with the giant planets' belts and zones. Gilman's (1977,1979) approach is numerical. He solves the nonlinear equations on a discrete grid. He also finds periodic convection cells and mean zonal flows, and these have finite amplitudes. However, he cannot treat the extreme cases of low viscosity, high velocity, and high rotation rate, for which the convection cell width is less than the grid size.

In both theories, the convection cells draw their energy from the internal heat source, the gradual cooling of the planet. Rotation causes the alignment parallel to the axis. The mean flow draws energy from the convection cells by

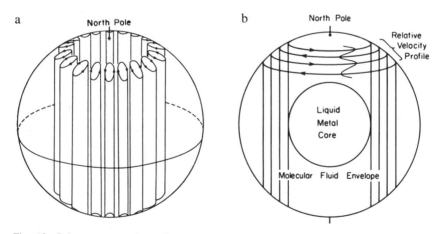

Fig. 12. Columnar convection cells (a) and cyclindrical zonal flow (b). As shown by Busse (1976), the columnar mode is the preferred form of convective instability in a uniformly rotating, viscous, conducting fluid. The cyclindrical mode is the most general form of steady zonal motion in an inviscid adiabatic fluid. The interaction of these two modes is analogous to the behavior of transverse convective disturbances in a sheared horizontal layer, according to Ingersoll and Pollard (1982).

tilting them, causing a radial transfer $\overline{u'v'}$ of eastward momentum against the mean momentum gradient $d\bar{u}/dr$. Here $\bar{u}$ is the mean eastward velocity, r is cyclindrical radius, u' and v' are departures from the longitudinal means of the eastward and radial velocities, respectively, and the overbar denotes a longitudinal mean. The fact that $\overline{u'v'}$ tends to be of the same sign as $d\bar{u}/dr$ means that the eddies are adding to the kinetic energy of the mean flow. Examples of this nonlinear counter-gradient momentum transfer abound in meteorology and oceanography, and in other theories of the Jovian and Saturnian atmospheres (Starr 1968; Rhines 1975; Williams 1978,1979).

A worrisome fact about both theories is that they do not give multijet zonal wind profiles of the sort observed on Jupiter and Saturn. One explanation is that the theories describe only the onset of convection, where the columns are essentially confined to a neighborhood of a certain distance from the axis. The case of fully developed convection is beyond the domain of both theories. Then, presumably, the convection columns would fill the container at all distances from the axis, and multijet profiles would occur. This explanation is not entirely fair to Gilman's numerical model, which describes a nonlinear unsteady flow, although it is still in the laminar viscous regime. Another deficiency of the models, which is being rectified, is the Boussinesq approximation, a restriction whereby density is nearly constant throughout the sphere. Thus it is premature to conclude from these models that deep convection has little to do with Jupiter's and Saturn's multijet profiles.

Ingersoll and Pollard (1982) adopt a more phenomenological view. They assume that the zonal flow exists and ask under what conditions it would extend into the interior. From the works of H. Poincaré, G. I. Taylor, I. Proudman, and others, it was known that any steady flow of an inviscid, barotropic fluid in a rotating, axisymmetric container must have cylindrical structure (Fig. 12b). A compressible fluid is barotropic when surfaces of constant density, pressure, temperature, specific entropy, etc. coincide. An adiabatic fluid, whose specific entropy is constant, is therefore also barotropic. Convection tends to maintain the fluid close to an adiabatic state but also tends to increase the dissipation and apparent eddy viscosity. The question is therefore whether Jupiter's and Saturn's convecting interiors are sufficiently close to being adiabatic and inviscid that the zonal flow would extend into the interior.

The limitation on the viscosity arises from the second law of thermodynamics. Each shell of fluid in which temperature decreases by a factor of two, for example, can only dissipate one-half the power per unit area flowing through it (this is the maximum efficiency of a heat engine operating at an intake temperature of twice the outflow temperature). The dissipation (power per unit mass) depends on the product of the eddy viscosity ν_e and the mean square vorticity $|d\bar{u}/dr|^2$. Since the internal heat flux is known, an upper bound on the eddy viscosity is obtained from the hypothesis that the vorticity of the interior is comparable to that of the visible atmosphere.

A separate estimate of the eddy viscosity is obtained by assuming that it is equal to the eddy thermal diffusivity. The latter is proportional to the internal heat flux divided by the potential temperature gradient $|d\vartheta/dr|$. Equating this estimate of ν_e to the upper bound obtained from the second law, Ingersoll and Pollard derive the condition Ri $= N^2 |d\bar{u}/dr|^{-2} \gtrsim 1$ for the zonal flow to exist in the interior. Here N^2 is the square of the Brunt-Vaisala frequency and is equal to $agT|d\vartheta/dr|/\vartheta$, where a is the thermal coefficient of expansion and g is the gravitational acceleration. Ri is the Richardson number, and the condition Ri ~ 1 is precisely that for which transverse convection cells do occur in the presence of a mean shear, according to theoretical studies by Kuo (1963) and Lipps (1971). Therefore, Ingersoll and Pollard argue that the necessary condition on the viscosity for the existence of deep zonal flow could be met in Jupiter's and Saturn's interiors.

The condition on departures from adiabaticity must also be met. Here Ingersoll and Pollard show that Ri ~ 1 implies that potential temperature variations are large enough to modify the cylindrical structure but not to destroy it; that is, the zonal velocity must extend into the interior if this condition is satisfied, but $\bar{u}(r)$ could vary significantly from its value at the surface in the northern hemisphere to its value at the surface in the southern hemisphere. These arguments demonstrate that the zonal flow of the atmosphere could exist in the interior, but they do not demonstrate that it must exist. Such a demonstration would probably require a numerical model of fully developed convection in a rotating system.

D. Eddy-Mean-Flow Interactions and Earth Analogy

Williams (1975,1978,1979) has emphasized the importance of the eddy stress term $\overline{u'v'}$ as a means of feeding energy into the zonal jets. His numerical studies are for thin layers (β-planes) on a rotating planet. The eddies are introduced in a variety of ways. In his barotropic model, vorticity is introduced in a small-scale checkerboard pattern and is removed by the eddy viscosity (Fig. 13). The mean flow is initially small, but is amplified by the tendency of the $\overline{u'v'}$ term to have the same sign as $d\bar{u}/dy$, where y is the northward horizontal coordinate and v is the corresponding velocity. In his baroclinic model the eddies arise spontaneously, as instabilities that draw their energy from the mean equator-to-pole temperature gradient. The eddy-mean flow interactions are initially as in the barotropic model, but then tend to oscillate as energy is exchanged back and forth from eddies to mean flow.

Voyager also observed a positive correlation between $\overline{u'v'}$ and $d\bar{u}/dy$ on Jupiter (Beebe et al. 1980; Ingersoll et al. 1981). For a layer occupying a twofold range of temperature, which includes most of the clouds, the implied rate of energy transfer is more than 10% of the emitted heat flux. If this transferred energy were dissipated in the layer, the thermodynamic efficiency would be close to the theoretical maximum. On Earth the eddy transfer term is only $\sim 0.1\%$ of the emitted heat flux, and the total dissipation of all forms

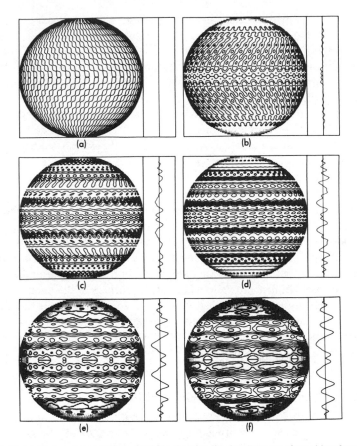

Fig. 13. Numerical model of Williams (1978) showing the development of zonal jets from eddies
in a thin atmosphere on a rotating planet. In this simulation the flow is assumed to be barotropic
(no horizontal density contrasts). The initial stream-function pattern (a) reflects the pattern of
mechanical forcing. Later stream-function patterns (b)–(f) show the emerging zonal structure.
The longitudinal mean zonal velocity is shown to the right of each pattern.

of kinetic energy is only ~ 1% of this flux. Thus the thermomechanical cycle
of Jupiter's atmosphere appears to be different from that of the Earth.

The circulation in Williams' (1979) model is driven entirely by the Sun.
A solid lower boundary completely decouples the atmosphere from the interior. The model operates, in fact, as a general circulation model of the Earth.
Analogy with the Earth's ocean is cited as the basis for the solid boundary
assumption. In the oceans the upper layers are effectively decoupled from the
lower layers by the stable stratification of the main thermocline. However, the
oceanic heat source is at the top, and heat is generally mixed downward by
wind-driven turbulence. Neither of these conditions applies to Jupiter or Sat-

urn. Significant heat is introduced at the bottom of the atmosphere, so the stratification (departure from adiabaticity) is likely to be neutral, at least up to the base of the clouds (e.g., Stevenson and Salpeter 1977a). As we have seen earlier Saturn's zonal flow overlaps into this adiabatic region, so decoupling of the cloud zone and the interior is unlikely.

Another weakness of the Earth analogy was pointed out by Ingersoll (1976) and Ingersoll and Porco (1978). Since the mass and therefore the heat capacity of the interior is a million times greater than that of the cloud zone, the interior is much more able to respond to the uneven rate of cooling of the surface layers. Unmeasurably small potential temperature gradients must have arisen in the interior to deflect heat upward and poleward so as to balance the net radiative emission required at each latitude. Such a process leaves the visible clouds with no net heating and no tendency to transport heat poleward. These ideas are consistent with the lack of observable equator-to-pole temperature gradients on Jupiter, and on Saturn below the level of seasonal heating (Fig. 2).

E. Long-Lived Ovals and the Zonal Velocity Profile

Equally dissimilar models have arisen to explain the numerous oval features that persist for years, decades or centuries in these planets' atmospheres. Most of the ovals are anticyclonic (Mitchell et al. 1981), but some are cyclonic (Hatzes et al. 1981). Saturn's largest ovals (B. A. Smith et al. 1981, 1982) are smaller than Jupiter's, the largest being the three White Ovals and the Great Red Spot. These vortices sit in anticyclonic shear zones, and tend to roll like ball bearings between oppositely moving conveyor belts. Unlike ball bearings, however, they have several times more vorticity than that of the ambient regions (Mitchell et al. 1981; Hatzes et al. 1981). Pairs of vortices have been observed to orbit around each other (Reese 1970; Smith and Hunt 1976; B. A. Smith et al. 1982), to merge (Reese and Smith 1968; Smith et al. 1979a,b) and perhaps to pass through each other without either orbiting or merging (Maxworthy and Redekopp 1976; Maxworthy et al. 1978).

Theories of isolated flow features fall into two distinct classes (Flierl et al. 1980). The first are solitary waves or solitons. These structures are the finite amplitude extension of weakly dispersive linear waves. As with non-dispersive linear waves, they propagate indefinitely as a single crest or trough, passing through each other with no net interaction. The second class are called modons, after Stern's (1975) oceanographic example. Modons are inherently nonlinear, i.e., they do not exist at arbitrarily small amplitudes. Mathematically their solutions are nonanalytic, e.g., there is a discontinuity of vorticity on some closed surface. Unlike solitons, they tend to interact strongly with each other (Flierl et al. 1980; McWilliams and Zabusky 1982).

Maxworthy and Redekopp (1976) and Maxworthy et al. (1978) proposed that Jupiter's isolated vortices and related features are solitary Rossby waves on a shear flow. The solutions are strictly valid only for infinitely long, infi-

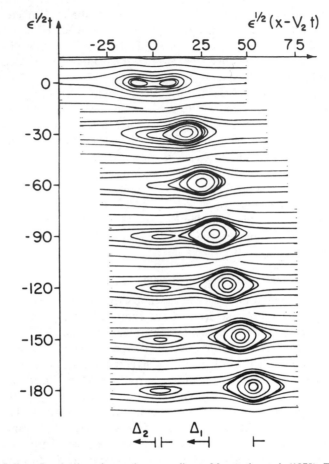

Fig. 14. Solitary Rossby wave interaction, according to Maxworthy et al. (1978). The contours represent the instantaneous streamlines obtained at equal time intervals (indicated along the ordinate). From bottom to top the larger feature, moving westward, overtakes the smaller feature, which is initially stationary in this reference frame. After the interaction the pattern repeats with left and right directions interchanged and time increasing from top to bottom. Thus the larger feature continues moving westward, having suffered only a net displacement Δ_1, during the encounter. The smaller feature remains stationary, having suffered a displacement Δ_2.

nitely weak perturbations, but a reasonable choice of parameters leads to very realistic flow patterns and flow behavior (Fig. 14). The equations are inviscid and adiabatic, and the solutions are either steady or steadily propagating. The mechanism by which the flows are maintained and the stability of the flows are not considered. The flow patterns, however, are not chosen arbitrarily, but emerge as part of the solution.

Ingersoll (1973) and Ingersoll and Cuong (1981) proposed that Jupiter's and Saturn's isolated vortices are modons. The solutions are obtained numerically, and are introduced as initial conditions in a time-dependent calculation. In this way they can study the form of the steady solutions and their stability, as well as more complicated time-dependent behavior (Fig. 15). The equations are again inviscid and adiabatic, so the mechanism by which the flows are maintained cannot be investigated. On the other hand, the vortices can grow by merging. Ingersoll and Cuong suggest that the larger vortices maintain themselves against dissipation by absorbing the smaller convectively-driven eddies.

In both the soliton and the modon theories, the ambient shear flow $\bar{u}(y)$ plays an important role. There are critical layers; i.e., the vortices are stationary with respect to the shear flow at intermediate latitudes between the extrema of the jets. This fact means that these vortices are fundamentally different from the less enduring terrestrial analogs. The shear flows themselves are also rather special. Both theories require that the shear flow must violate the barotropic stability criterion, in other words, that the second derivative $\bar{u}_{yy}$ must exceed $\beta = 2\Omega\cos\lambda/a$ at some latitudes. For horizontal flow on the surface of a sphere, violation of this criterion means that the basic shear flow is unstable. In fact, for Jupiter and Saturn $\bar{u}_{yy}$ is generally $\sim 2\beta$ at the latitudes of the westward jets (Ingersoll et al. 1981; B. A. Smith et al. 1981,1982; Table I in Sec. II.C). Violation of this criterion appears paradoxical, since unstable flows rarely exist in uniform steady configurations.

Resolution of the paradox could take several forms. First, the flow may not be confined to the surface layers. Ingersoll and Pollard (1982) show that a rotating cylinder shear flow (Fig. 12b) obeys a different barotropic stability criterion. Projected onto the planetary surface, a stable velocity profile will have $\bar{u}_{yy} \gtrsim -3\beta$, according to this criterion. Observational data (Ingersoll et al. 1981; B. A. Smith et al. 1982; Ingersoll and Pollard 1982) suggest that this criterion is more nearly satisfied than the one ($\bar{u}_{yy} \lesssim \beta$) for flow in the surface layers.

Second, the fact that these shear flows coexist with vigorous eddies may be important. The numerical experiments of Rhines (1975) and Williams (1978,1979) are relevant to this point. They consider flow confined to the surface layers (β-planes), and find that the resulting mean flow that equilibrates with the eddies tends to have $\bar{u}_{yy} \lesssim \beta$. (Rhines expresses this as $k \lesssim k_\beta$, where k is the horizontal wavenumber, k_β is $(\beta/2u)^{1/2}$, and u is the rms wind speed. Williams shows graphs from which $\bar{u}_{yy}$ may be computed.) Nevertheless, it is possible that a more vigorous eddy input could lead to greater curvature of the velocity profile, e.g., peak values of $\bar{u}_{yy}$ equal to 2β.

A third possibility, somewhat similar to the first, is that the great depth of the fluid interior stabilizes the surface flows, even when the surface flows are confined to a thin layer (Gierasch et al. 1979; Conrath et al. 1981b; Ingersoll and Cuong 1981). These studies show, at least, that the growth rates of un-

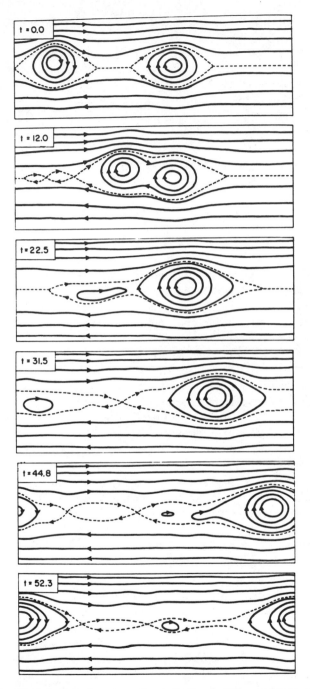

Fig. 15. Interaction of two vortices, according to the "modon" theory of Ingersoll and Cuong (1981). Because the solution is periodic in the east-west direction, the two vortices that appear to be moving apart at $t = 22.5$ and $t = 31.5$ then collide, causing a smaller vortex to form at $t = 44.8$. Eventually the large vortex sweeps up all the smaller vortices and the flow is steady. This behavior contrasts with the noninteracting solitary waves of Maxworthy et al. (1978) shown in Fig. 14.

stable waves in the surface layers are severely reduced by interaction with a deep adiabatic interior.

Regardless of the cause, the multijet shear flow with $\bar{u}_{yy}$ varying between about -3β and $+2\beta$ seems to be present on both Jupiter and Saturn. This jet structure may be the special environment that allows long-lived vortices to exist. The available theories of Jovian and Saturnian vortices use this fact about the structure.

Ingersoll and Cuong derive the governing equations by assuming quasi-geostrophic, hydrostatic flow on a β-plane, with the vortices and nonzonal flow confined to a less dense upper layer on top of an infinitely deep lower layer. The lower-layer velocity profile $\bar{u}(y)$ plays the role of the basic baro-tropic flow introduced by Maxworthy and Redekopp. The equations used by Maxworthy and Redekopp can be derived from those used by Ingersoll and Cuong. The crucial parameter is $k = L/L_D = Lf/(NH)$, where $L = (\bar{u}/\bar{u}_{yy})^{1/2}$ is the maximum profile radius of curvature, $f = 2\Omega\sin$ (latitude) is the Co-riolis parameter, N is the Brunt-Vaisala frequency of the upper layer, and H is its thickness. For this simple two-layer model, N^2 is given by $g\Delta\rho/(\rho H)$, where ρ is the density of the lower layer and $\Delta\rho$ is the density difference. L_D is called the radius of deformation. Only the upper-layer velocity is allowed to vary with time and longitude. Because of its great depth, the lower layer does not respond readily to effects from above. The nonzonal components of ve-locity in the lower layer are therefore assumed to be small. The latter assump-tion is not inconsistent with a theorem by Flierl et al. (1983) that the sum of the angular momenta of the two layers must be zero.

The basic state is one in which the upper- and lower-layer flows $\bar{u}(y)$ are the same. For $k < 1$ (large N, strong stratification) this upper-layer flow is unstable for profiles of the Jovian type. This unstable property seems to be necessary for soliton solutions to exist in the upper layer. Instability exists over a finite band of perturbation wavelengths. Maxworthy and Redekopp base their solution on the infinitely long-wavelength end of this band, ignoring the rest. The solution they consider is a stationary neutral wave; its finite-amplitude interaction with itself exhibits soliton behavior. They neglect solu-tions which grow exponentially in time.

For $k > 1$ (small N, weak stratification) the basic upper-layer flow is stable. There are no stationary neutral waves. The only allowed linear distur-bances moving with the flow are spatially decaying exponentials. These are patched together in a nonlinear closed-streamline region in the model of In-gersoll and Cuong. A discontinuity of vorticity gradient exists at the bound-ary of the region. The resulting vortices are isolated and compact, and survive both small and large perturbations. As with most vortices on Jupiter and Sat-urn, when one overtakes another they tend to merge (Fig. 15).

The instability of the flow for $k < 1$ tends to invalidate the soliton solu-tions, according to Ingersoll and Cuong. These solutions do not exist in a time-dependent numerical model. For $k < 1$, the perturbed upper-layer flow quickly evolves into a very unrealistic-looking pattern (Fig. 2 of Ingersoll and Cuong), whereas for $k > 1$ the upper-layer solutions persist in spite of large amplitude perturbations.

This difference in stability arises because the basic shear flow observed on Jupiter and Saturn violates the barotropic stability criterion for thin layers. For small k (large stratification) the thin upper layer is nearly decoupled from the deep lower layer. The thin-layer stability criterion applies, and the flow is unstable. For large k (small stratification) the upper-layer flow is strongly coupled to the lower layer. But for some reason, the lower-layer flow is not unstable. As discussed above, its stability might be governed by deep cylinder dynamics, for which the observed profiles are stable.

In summary, the special properties (longevity, compactness, stationarity with respect to opposing zonal currents) of Jovian and Saturnian vortices may originate from the great depth of the fluid interiors. The existence of these vortices may prove to be a special diagnostic that reveals how the atmosphere and interior interact. Our hope for understanding the dynamics of these atmospheres lies in finding a sufficient number of these diagnostics that only the correct model will reproduce the observed behavior. Perhaps now, after Voyager, that hope may be realized.

Acknowledgments.   We thank M. S. Matthews for expert editorial assistance. This research was made possible largely by the successful Voyager missions to Jupiter and Saturn. The support of the National Aeronautics and Space Administration (API, RFB, BJC) and the UK Science and Engineering Research Council (GEH) are gratefully acknowledged.

THEORY, MEASUREMENTS, AND MODELS OF THE UPPER ATMOSPHERE AND IONOSPHERE OF SATURN

S. K. ATREYA
The University of Michigan

J. H. WAITE, JR.
Marshall Space Flight Center

T. M. DONAHUE and A. F. NAGY
The University of Michigan

J. C. McCONNELL
York University

The structure and composition of the thermosphere, exosphere and ionosphere of Saturn have been determined from observations at optical and radio wavelengths, principally by instruments aboard Voyager spacecraft. Interpretation of these observations yields an average neutral temperature of 140 K in the stratosphere and mesosphere, a thermospheric temperature gradient of ~ 1.25 K km^{-1}, and an exospheric temperature between 600 K and 800 K for the equatorial region. The amount of power deposited in auroral bands is ~ 2 × 10^{11} W, which is insufficient for the thermospheric heating observed in nonauroral regions. Joule heating and energy deposition from inertia-gravity waves are important candidates as sources of the relatively high exospheric temperature on Saturn. A methane mixing ratio of 1.4 × 10^{-4} measured 965 km above the 1-bar level is indicative of the depletion of this species in the upper atmosphere due to photolysis and diffusion. The strength of vertical mixing is deduced from an analysis of the atomic hydrogen and helium Lyman-α airglow, and from a study of the photochemistry of methane. The value of the eddy mixing coefficient at the homopause is found to be on the order of 10^8 cm^2 s^{-1}; the corresponding pressure is a few nanobars. A model of the ionosphere is developed and com-

pared with the results of radio occultation measurements. There is a discrepancy in the altitude and the magnitude of the peak electron concentration. Of several possibilities discussed, the loss of topside protons, reacting with vibrationally excited molecular hydrogen, along with vertical ion drifts, are the most plausible explanations of the discrepancy. It should be emphasized, however, that a meaningful comparison between models and the measured ionospheric profiles will be possible only after data for the lower ionosphere (below ~ 2000 km) have been analyzed. Although there are many apparent similarities between the aeronomy of Saturn and Jupiter, there are distinct differences in terms of the quantitative behavior of the dynamics, energy budget and the plasma processes.

Groundbased spectroscopic measurements in the visible, infrared and microwave during the past two decades and prior to the Voyager observations provided some information on the bulk composition of the Saturn atmosphere. A summary of these measurements is presented in Table I. The Voyager infrared observations (Hanel et al. 1981a) have yielded extensive data on

TABLE I
Pre-Voyager Composition Measurements of the Saturn Atmosphere

Species	Spectral Region	References
H_2	S(0) & S(1) quadrupole lines of the (4,0) & (3,0) rotational-vibrational system	Münch and Spinrad (1963); Giver and Spinrad (1966); Owen (1969); Encrenaz and Owen (1973).
CH_4	$3\nu_3$ band in the 1.1 μm region	Trafton (1973); Trafton and Macy (1975); Lecacheux et al. (1976); Combes et al. (1977).
$^{13}CH_4$	1.1 μm	Combes et al. (1977).
C_2H_6	ν_9 at 12.2 μm	Gillett and Forrest (1974); Tokunaga et al. (1975).
PH_3	10–11 μm 5 μm	Encrenaz et al. (1975). Fink and Larson (1977).
CH_3D	5 μm	Fink and Larson (1977).
HD	$P_4(1)$ at 0.7467 μm $R_5(0)$	Trauger et al. (1977). Smith and Macy (1977).
NH_3	0.6450 μm 1.56 μm Radio	Woodman et al. (1977). Owen et al. (1977). Gulkis et al. (1969); Gulkis and Poynter (1972).
H_2S	1.59 μm 0.21–0.25 μm	Owen et al. (1977): upper limit. Caldwell (1977b): upper limit.
H	0.1216 μm	Weiser et al. (1977); Barker et al. (1980); Clarke et al. (1981).

the spatial and temporal variations of the above-mentioned species and many more. The first *in situ* measurements of the temperature structure of the upper atmosphere and the distribution of neutral species were provided by the Voyager ultraviolet spectrometer which monitored the sunlight scattered from Saturn's atmosphere, and also sunlight and starlight absorbed by atmospheric species. Complementary information on the structure and composition of the upper atmosphere was provided by the Voyager infrared and radio science investigations. Section I is devoted to the discussion of the neutral upper atmosphere.

The physics and chemistry of the neutral atmosphere and the ionosphere of Saturn are strongly coupled. Although the presence of an extensive ionosphere on Saturn was predicted by several theoretical models, measurements were possible only from the Pioneer and Voyager spacecraft by radio occultation. The first confirmation and measurement of a magnetic field on Saturn was made by the Pioneer vector helium and fluxgate magnetometers (E. J. Smith et al. 1980; Acuna and Ness 1980). Section II deals with the ionospheric measurements, modeling, and the coupling with the neutral atmosphere. Some discussions in this chapter are frequently complementary to those in the chapters on the lower atmosphere (see e.g. chapters by Prinn et al. and Ingersoll et al.).

I. UPPER ATMOSPHERE

The atmosphere of Saturn above the ammonia cloud tops may be conveniently divided into three regions: (1) troposphere with pressures $\gtrsim 1$ mbar; (2) middle atmosphere or the region of "information gap" with pressures between 1 mbar and 10 nbar; and (3) thermosphere and exosphere with pressures $\lesssim 10$ nbar. The atmosphere is mixed to the homopause having pressures ≈ 1 nbar (Sec. I.B). However, photochemical processes cause departure from a mixed atmospheric distribution of certain species, such as NH_3, CH_4 and PH_3. Discussion of the neutral atmosphere in this chapter is limited to processes occurring primarily in the thermosphere and exosphere, although photochemistry in the mesosphere is included when it is important for understanding the aeronomy of the upper atmosphere. We discuss, in the following sections, techniques for the determination of the vertical profiles of temperature and density, and atmospheric vertical mixing.

A. Temperature and Density Distributions in the Upper Atmosphere

There have been no stellar occultations suitable for groundbased studies of the upper atmosphere of Saturn, thus there are no groundbased data to provide vertical profiles of the temperature or density in the upper atmosphere of Saturn. Results obtained from the inversion of the ionospheric radio occultation data (Kliore et al. 1980*a*; Tyler et al. 1981,1982*a*) are dependent on assumptions regarding ion drifts, ionospheric composition and ion-electron energy loss processes. Therefore, the most direct means for determining neutral upper atmospheric characteristics was provided by the Voyager ultraviolet spectrometer when it monitored the absorption of sunlight or starlight passing through the atmosphere. Supplementary information was obtained from the interpretation of the reflected sunlight measured by the ultraviolet spectrometer during and before the encounter. Measurement techniques and results are discussed below.

1. Stellar and Solar Occultation. The technique of ultraviolet stellar and solar occultations has been used successfully to determine height profiles of ozone, oxygen, chlorine, and molecular hydrogen in the terrestrial atmosphere (Hays and Roble 1973*a,b*; Atreya 1981). In this technique, the atmosphere acts as an absorption cell providing a long pathlength to the ultraviolet radiation from a suitable bright source such as the Sun or a star. By monitoring the tangent rays before and after they pass through the "cell," it is possible to determine the distribution (hence scale height) of the absorbing gases.

If I_0 represents unattenuated flux at wavelength λ and I_z is the flux at some tangent altitude z after absorption by a given molecular species, then

$$I_z = I_0 \exp(-\tau) \tag{1}$$

where $\tau = N \sigma_a$, N is the line-of-sight column abundance of the absorbing gas, and σ_a is its absorption cross section at wavelength λ. I_z and I_0 are monitored by the spectrometer; σ_a is measured in the laboratory. The unknown quantity N is thus

$$N = \frac{1}{\sigma_a} \ln (I_0/I_z). \tag{2}$$

One can invert N by Abel inversion, or numerical inversion techniques (Atreya 1981) to yield local number density n of the absorbing gas. Using a set of different wavelengths, one can determine n at different heights in the atmosphere, if the absorption cross sections are different at the different wavelengths selected. The scale height, hence the temperature, can then be determined by using the hydrostatic law,

$$n_2 = n_1 \exp (-\Delta z/H). \tag{3}$$

Δz can be determined by knowing the rate of descent of the minimum tangent altitude in the atmosphere, which in turn is obtained from the spacecraft/satellite velocity data.

The above technique works best in the region of continuum absorption by a single species. In the outer planets, solar occultation experiments thus provide useful information on the H_2 density from the analysis of continuous absorption in H_2 below $\sim$ 845 Å. In the case of stellar occultations, the flux below 911 Å is negligible due to the interplanetary/interstellar absorption by hydrogen. Thus, the upper atmospheric H_2 density and temperature determinations require unfolding of absorptions in the Lyman and Werner bands of H_2, as discussed later. Band absorptions in the hydrocarbons pose another challenge, that of isolating absorptions by individual species; the details will be discussed below.

Using the Sun and stars as light sources, it is possible to determine the structure of the upper atmosphere of Jupiter (Broadfoot et al. 1979; Sandel et al. 1979; Atreya et al. 1979a; Atreya et al. 1981; Festou et al. 1981). Similar techniques were used on both Voyagers 1 and 2 to determine density and temperature profiles of Saturn's upper atmosphere (Broadfoot et al. 1981; Sandel et al. 1982b). At the very short wavelengths used, absorption is much more important than refraction, which is responsible for the effects studied in the visible and radio ranges (Hunten and Veverka 1976). The most successful of these experiments was exit occultation of the star δ Scorpii (Dzuba), in which the passage of tangent rays through the atmosphere of Saturn was unaffected by absorption due to the rings of Saturn.

The ultraviolet flux of star δ Sco (type B0) is much greater than that of α Leo which was used in the Voyager/Jupiter occultation experiment (Festou et al. 1981; Atreya et al. 1981). As a consequence, better statistics resulted in the Voyager/Saturn data. We show in Fig. 1 the unattenuated spectrum of δ Sco (Festou and Atreya 1982) recorded by the Voyager 2 ultraviolet spectrometer on the day of the occultation, 25 August 1981. The data have been corrected for the instrumental scattering and fixed pattern noise. The effective wavelength resolution between 500 Å and 1700 Å is 25 Å; the height resolution is 3.2 km. The exit occultation occurred very near the equator, at 3°8N. The sharp Lyman cutoff below 911 Å in Fig. 1 is an indication of the strong interstellar hydrogen absorption. A decrease in the instrument sensitivity at $\sim$ 1000 Å is reflected by the reduced signal beyond 1060 Å in Fig. 1.

The principal absorbers of the ultraviolet radiation in the upper atmosphere are H_2, CH_4, C_2H_2, and C_2H_6. Absorption of the signal below 1200 Å is due mainly to molecular hydrogen, and above 1200 Å to the hydrocarbons in the upper atmosphere of Saturn. The data in the vicinity of 1216 Å are not usable due to the strong interplanetary/interstellar Lyman-α absorption. Shown in Figs. 2 and 3 are exit occultation data for H_2 and the hydrocarbons, respectively. The lower abscissa in Figs. 2 and 3 represents altitudes above the 1-bar atmospheric pressure level. Δz is the altitude above the 1-bar level where total extinction in the H_2 absorption channels (here, 939–1023 Å) was recorded by the ultraviolet spectrometer. The spacecraft trajectory information appropriate to the geometry of the δ Sco occultation yields a value for the radius of Saturn at the 1-bar level of 60,246 $\pm$ 10 km (Festou and Atreya 1982). Radii of the tangent rays measured from the planetary center are indicated on the upper abscissa of Figs. 2 and 3.

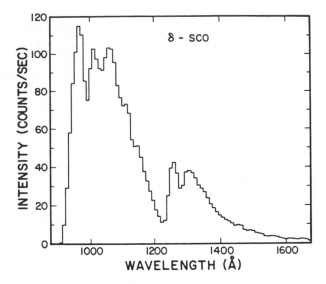

Fig. 1. Unattenuated spectrum of star δ Sco as seen by the Voyager ultraviolet spectrometer. The effective spectral resolution is ∼ 25 Å. Background has been removed and the various instrumental corrections applied (after Festou and Atreya 1982).

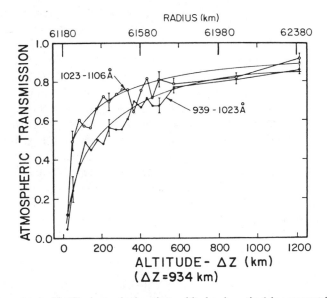

Fig. 2. Absorption by H_2. The lower abscissa shows altitudes above the 1-bar pressure level while the upper abscissa gives corresponding planetocentric radii of the tangent ray points. The height resolution in this figure is 3.2 km, except for points above (Δz + 500) km which have been obtained by averaging 40 consecutive spectra recorded over a 128-km height range. The continuous lines are the absorption curves which are used to derive the temperature and density profiles of H_2 (after Festou and Atreya 1982).

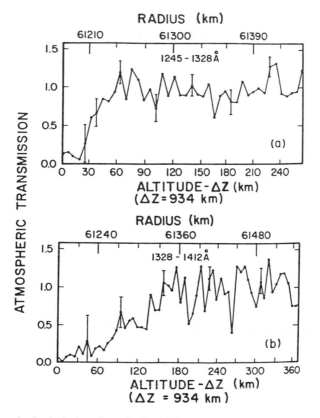

Fig. 3. Absorption by the hydrocarbons. In plot (a) the absorption is attributed to CH₄; in (b), the absorption is due to an unidentified species, alone or in mixture with the hydrocarbons. Altitude scale is the same as in Fig. 2 (after Festou and Atreya 1982).

In the 912–1200 Å range, molecular hydrogen absorbs in the Lyman and Werner bands that connect the ground state $X^1\Sigma_g^+$ with the excited state $B^1\Sigma_g^+$ and $C^1\Pi_u$, respectively. Assuming a Voigt profile (natural and Doppler broadening) for each vibrational-rotational line in the Lyman and Werner band systems, one calculates atmospheric transmission as a function of frequency; the theoretical calculations done by Festou et al. (1981) are shown in Fig. 4. Using simulated transmissions and the lightcurves shown in Fig. 2, Festou and Atreya (1982) have shown that the best fit to the data requires a temperature of 800^{+150}_{-120} K at and beyond 1540 km; the H_2 density at this altitude is found to be $5^{+3.6}_{-1.8} \times 10^9$ cm^{-3}. A temperature gradient of $1.25^{+0.05}_{-0.07}$ K km^{-1} is obtained from approximately 950 to 1540 km.

Absorption shown in the lightcurve for 1245–1328 Å (Fig. 3a) can be explained satisfactorily by methane alone (Festou and Atreya 1982). This assumption is further substantiated by the fact that on examining absorption characteristics of individual 9 Å wide channels within the wavelength range covered in Fig. 3a, one finds that a given optical depth always occurs at the same altitude. Since absorption cross sections of both C_2H_2 and C_2H_4 have large variation in this wavelength range, they cannot account for the observed transmission characteristics. C_2H_6 is not acceptable either since at shorter wavelengths it produces more than the observed absorption around 985 km. Inversion of the absorption characteristics shown in Fig. 3a facilitates the determination of the CH_4 density in the short altitude interval ranging from approximately 934 km to 994 km. The CH_4 density determination, however, is most accurate in the middle of this range where the optical depth is 0.5. Festou and Atreya find $[CH_4] = 1.9^{+0.9}_{-0.3} \times 10^8$ cm^{-3} at an altitude of 966 km where $[H_2] = 1.2 \times 10^{12}$ cm^{-3}. At the same H_2 density, Smith et al. (1983) find the CH_4 density to be within a factor of 2 of Festou and Atreya's value.

The lightcurve, 1328–1412 Å (Fig. 3b) poses an unresolved dilemma.

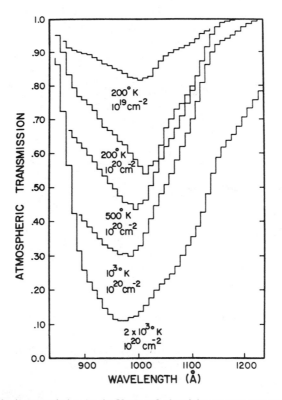

Fig. 4. Atmospheric transmission (as the Voyager 2 ultraviolet spectrometer would measure it) for isothermal lines of sight characterized by the indicated temperature and column densities (after Festou et al. 1981).

The onset of absorption at higher levels than in Fig. 3a and absorption to deeper and deeper levels (when one examines individual channels in 1328–1412 Å range) are not representative of the usual Jovian hydrocarbons, CH_4, C_2H_2, C_2H_6, C_2H_4 or any combination thereof. Although one cannot readily reject high-altitude haze (West et al. 1983a) as the possible absorber, it is difficult to reconcile the fact that it is not required below 1328 Å, but is needed above this wavelength to account for the observations. So far, the absorber responsible for attenuation of the flux shown in Fig. 3b has not been identified.

The analysis of the H_2 continuum absorption around 600 Å in the Voyager 2 solar occultation data at 29°.5N yielded a temperature of 420 ± 30 K down to ∼ 1600 km (Smith et al. 1983). The H Lyman-α analysis gives a thermospheric/exospheric temperature of 600^{+250}_{-150} K (G. R. Smith, personal communication, 1982). The temperatures in the homopause region deduced by Smith et al. from the stellar occultation data are in statistical agreement with those found by Festou and Atreya (1982). Unlike Festou and Atreya (1982), the stellar occultation analysis of Smith et al. above the homopause is dependent on the exospheric temperature they deduce from their solar occultation data. Note that the analysis of Smith et al. (1983) includes the newly published Rydberg series bands of H_2 (Shemansky and Ajello 1983). This, however, has no significant effect on the δ Sco stellar occultation results of Festou and Atreya (1982) for the following reasons:

1. As shown by Smith et al. (1983), inclusion of the additional Rydberg bands affects the H_2 absorption characteristics below 900 Å, and that too with a high H_2 column abundance of 10^{19} cm^{-2}. The δ Sco analysis of Festou and Atreya (1982) is for wavelengths > 939 Å, and the line-of-sight H_2 column abundance does not exceed 4×10^{18} cm^{-2} in the region where the exospheric temperature is reached;

2. As shown by Shemansky and Ajello (1983), inclusion of the Rydberg bands does not change the shape of the H_2 band structure for $H_2 > 10^{16}$ cm^{-2}. An important caveat in the δ Sco stellar occultation analysis of Smith et al. (1983) is that they use a single wavelength range, 939–1041 Å. Festou and Atreya (1982) have demonstrated that a single wavelength range yields a combination of line-of-sight H_2 column abundance and the exospheric temperature.

Indeed all temperatures between 400 K and 1000 K give a satisfactory fit to the lightcurves in the exospheric region. To define the temperature uniquely, one must analyze at least an additional wavelength range, as was done by Festou and Atreya (1982). The single most disconcerting uncertainty in the solar occultation analysis arises from the nonuniformity of the extreme-ultraviolet distribution on the solar disk (Atreya et al. 1979a); large areas of the Sun have been found to be 2 to 3 times brighter (and variable) than the average extreme-ultraviolet intensity of the Sun. The Sun subtends ∼ 150 km in Saturn's atmo-

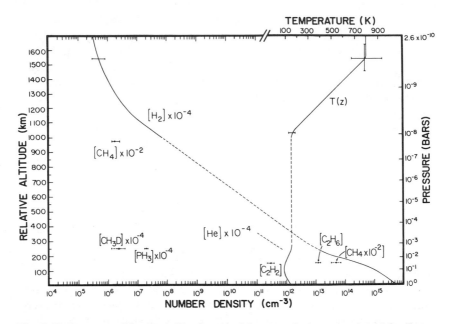

Fig. 5. Temperature and density profiles above the 1-bar atmospheric pressure level (after Festou and Atreya 1982). Atmospheric pressure corresponding to altitudes (left ordinate) is shown on the right ordinate. The interpolated regions of density and temperature at 250–950 km, are shown by broken lines. The helium density profile shown here is simply an illustration of its 6% volumetric mixing ratio in the homosphere. C_2H_2, C_2H_6, CH_4, CH_3D and PH_3 mixing ratios in the northern hemisphere are indicated by broken horizontal bars in the deep atmosphere, and taken from Courtin et al. (1982a).

sphere which is of the same order as the scale height for the 400 K exospheric temperature—further reducing the accuracy of the solar occultation analysis.

Until the discrepancy between exospheric temperatures derived from the solar occultation (Smith et al. 1983) and the stellar occultation (Festou and Atreya 1982) is satisfactorily resolved, it is reasonable to assume that there is a possibility of different exospheric temperatures at the 30° (solar occultation data), and the 4° (stellar occultation data) latitudes.

2. *Temperature and Density Profiles.*   The distribution of temperature and densities in the upper atmosphere of Saturn derived by Festou and Atreya (1982) is shown in Fig. 5. The information above $\sim$ 900 km is from the δ Sco stellar occultation experiment, while that in the lower stratosphere and troposphere is from Voyager infrared and radioscience experiments. There are no measurements in the middle atmosphere shown by a broken line interpolation in the temperature and H_2 density (10 nbar $\leq P \leq$ 1 mbar). The temperature at either end of this information gap is approximately 140 K (Festou and Atreya 1982; Hanel et al. 1981a; Tyler et al. 1981). From the hydrostatic equation, the average temperature of the information gap, 700 km, is 142 K. By

way of comparison, the information gap region on Jupiter is 275 km high with an average temperature of $\sim$ 175 K. In terms of scale heights, the information gaps on the two planets are nearly equivalent: 11 to 12 scale heights high.

Studies of electron and ion energy loss processes indicate virtual equilibrium between plasma and neutral temperatures (Henry and McElroy 1969; Nagy et al. 1976; Waite 1981) at the heights of electron peaks. It is therefore instructive to compare the thermospheric neutral temperatures on Saturn with the plasma temperatures determined from the ionospheric scale height deduced by the Voyager radio occultation experiment: 565 K at 2800 km (ingress, 36°5N), and 617 K at 2500 km (egress, 31°S). The average low-to-midlatitude electron/ion temperature is then on the order of 600 K assuming that H^+ is the major topside ion. The error bars on the plasma scale height have not yet been published. An important caveat in the deduction of the plasma temperatures from the scale height data is that the determination of the average scale height is complicated by the presence of considerable structure in the ionosphere above the peak. Furthermore, the identity of the topside ion has not been determined. Note, however, that any ion other than H^+ would result in a greater temperature than the above-mentioned values.

Since the thermospheric energy budget is directly coupled to ionospheric processes, the discussion of the mechanisms for upper atmospheric heating is deferred to Sec. II.E, which follows the discussion of ionospheric calculations.

In addition to the upper atmospheric H_2 and CH_4 densities obtained from the δ Sco data, Fig. 5 also shows the homospheric densities of CH_4, C_2H_2, C_2H_6, CH_3D and PH_3 for the northern hemisphere (Courtin et al. 1982a), and the helium abundance (Hanel et al. 1981a) from the Voyager infrared measurements. The homospheric mixing ratios of the species are summarized in Table II. As evident in this table, there is a large latitudinal variation in the

TABLE II
Voyager Infrared Measurements of Mixing Ratios

Species	Mixing Ratio by Number ($[X_i]/[H_2]$)	
	Northern Hemisphere [a]	Southern Hemisphere
CH_4	$1.85(+1.2, -0.5) \times 10^{-3}$	
C_2H_2	$1.2(+0.3, -0.3) \times 10^{-7}$	$0.5(\pm0.1) \times 10^{-7}$
C_2H_6	$5.0(+1.1, -1.1) \times 10^{-6}$	$3.1(\pm0.7) \times 10^{-6}$
CH_3D	$2.3(+1.5, -1.0) \times 10^{-7}$	
PH_3	$2.0(+0.3, -0.3) \times 10^{-6}$	
He	0.06	0.06

[a] The mixing ratios of C_2H_2 and C_2H_6 decrease sharply below the 20–50 mbar pressure level. The CH_4 mixing ratio may be as large as 5 times the solar, and there is a possible detection of Allene, $CH_2 = C = CH_2$ with an abundance of 0.3 cm atm (D. Gautier, personal communication, 1983).

mixing ratios of C_2H_2 and C_2H_6; abundances of the other hydrocarbons and PH_3 have not yet been determined for the southern hemisphere. The volume mixing ratio of helium at Saturn of 6% is approximately half this value at Jupiter. The depletion of helium in the upper atmosphere of Saturn is indicative of its probable condensation at the top of the metallic hydrogen zone, and the subsequent rain out of helium droplets toward the core of the planet (Ingersoll 1981).

B. Eddy Diffusion

In aeronomical problems, it is important to know the magnitude of vertical mixing or transport in a planetary atmosphere to determine the altitude distribution of minor species. Vertical mixing is generally expressed in terms of an all-encompassing parameter, the eddy diffusion coefficient K. At the homopause, the eddy diffusion coefficient is equal to the molecular diffusion coefficient. Beyond the homopause, diffusion processes are controlled by the molecular weights of individual species. Three methods have been used to determine the value of the eddy diffusion coefficient on Saturn: from the planetary Lyman-α albedo, from distribution of methane, and from He 584 Å airglow. We discuss below these techniques, their merits and deficiencies, and the results.

1. Lyman-α. Nonauroral Lyman-α emission from the outer planets is principally the result of resonance scattering of the solar Lyman-α photons by hydrogen atoms that lie above the methane homopause. Resonance scattering of interplanetary Lyman-α may contribute to the observed planetary emission for certain geometries. Hydrogen atoms below the homopause do not contribute appreciably to the Lyman-α albedo since methane is a strong absorber of the Lyman-α photons. Hunten (1969) adapted the terrestrial $O-O_2$ diffusion problem (Colegrove et al. 1965) to the atmosphere of Jupiter assuming a cold isothermal exosphere. Wallace and Hunten (1973) considered the problem of atomic hydrogen radiative transfer and chemistry by including previously neglected effects such as H production from CH_4 and some important scattering approximations. These calculations did not consider the possibility of high exospheric temperature and a gradient in the thermospheric temperature, since no clues to the Jovian upper atmospheric temperature were yet available. After the Voyager flyby of Jupiter, Waite (1981) considered these effects and also coupled the homosphere to the ionosphere where hydrogen atoms are produced either by photodissociation or dissociative photoionization of H_2. Once produced, the hydrogen atoms flow down to the deeper denser atmosphere where they are lost by 3-body recombination reactions.

The calculations of Waite (1981) for column abundance of hydrogen atoms as a function of the homopause eddy diffusion coefficient are shown in Fig. 6. These calculations are appropriate for the nonauroral region of Saturn with solar extreme ultraviolet alone responsible for the H production. Voy-

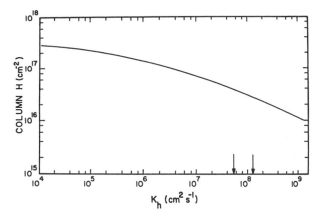

Fig. 6. Column abundance of atomic hydrogen above the unit optical depth level in methane as a function of the homopause value of the eddy mixing coefficient. These calculations assume the appropriate atmospheric temperature structure and the solar fluxes. The arrows represent the situations with and without the contributions of the interplanetary/interstellar Lyman-α backscattering to the observed planetary Lyman-α airglow. The corresponding values of K_h are 1.4 $\times$ 10^8 cm^2 s^{-1} and 8 $\times$ 10^7 cm^2 s^{-1}, respectively (adapted from Waite 1981).

ager's ultraviolet spectrometer measured between 3 and 3.3 kR of Lyman-α emission from the equatorial to midlatitude region (Broadfoot et al. 1981; Sandel et al. 1982b) implying atomic hydrogen column abundance of 4 $\times$ 10^{16} cm^{-2} above the level of the unit optical depth (τ) in methane. This value is uncertain to within a factor of 2 because of uncertainties in the solar Lyman-α flux. The observed planetary Lyman-α, however, is expected to include approximately 0.5 kR of the interplanetary/interstellar Lyman-α (Atreya 1982; Sandel et al. 1982a) so that only $\sim$ 2.5 kR could be attributed to the actual Saturnian Lyman-α. This latter value imples an H column abundance of approximately 2.7 $\times$ 10^{16} cm^{-2} above the τ = 1 level in methane. From Fig. 6, we find that the homopause eddy diffusion coefficients K_h corresponding to the above-mentioned hydrogen abundances are 8 $\times$ 10^7 cm^2 s^{-1} and 1.4 $\times$ 10^8 cm^2 s^{-1}, respectively. Sandel et al. (1982b) arrived at a similar value for K_h from their preliminary analysis of Saturn's Lyman-α albedo.

Sandel et al. (1982a) have also been able to reproduce the observed Lyman-α intensity with a value of 5 $\times$ 10^7 cm^2 s^{-1} for K_h and a simplified model for the hydrocarbon chemistry. They find that for a range of reasonable values of the solar Lyman-α flux at 1 AU of 6 $\times$ 10^{11} photons cm^{-2} s^{-1} (H. E. Hinteregger, personal communication, 1979) to 4 $\times$ 10^{11} photons cm^{-2} s^{-1} (Mount and Rottman 1981), the Saturn Lyman-α intensities obtained lie in the range 2.9 to 2.1 kR which are in reasonable agreement with the observations.

The major uncertainties associated with the above analysis, are: sources of H production other than the solar extreme ultraviolet, and the solar flux at

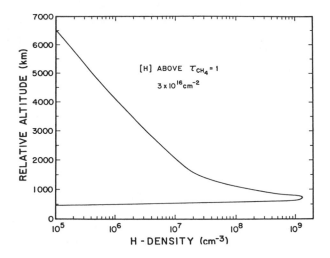

Fig. 7. Calculated atomic hydrogen density as a function of altitude for nonauroral region on
Saturn (after Waite et al. 1983a).

Lyman α. Despite these complicating factors, we are confident that the above
procedure of relating K_h to the Lyman-α emission rate is a reasonably good
approximation. For example, calculations show (Fig. 7) that the solar extreme
ultraviolet alone would produce approximately 3×10^{16} cm^{-2} hydrogen
atoms above the $\tau = 1$ level in methane (Waite et al. 1983a) which is in good
agreement with the above-mentioned value implied by the observed equatorial Lyman-α on Saturn. Thus, additional sources of H atoms are not required to explain the observations unlike on Jupiter at the time of the Voyager
encounter.

    Furthermore, the globally diluted downward flux of the H atoms pro-
duced in the narrow auroral region at Saturn (78–81° latitude, Sandel et al.
[1982b]) is much smaller than that produced on the extreme-ultraviolet dissociation of H_2. For example, calculations by Waite et al. (1983a) yield an H
atom flux of 5×10^8 cm^{-2} s^{-1} due to the extreme ultraviolet, while the flux
resulting from the deposition of 1 to 10 keV electrons of 0.67 erg cm^{-2} s^{-1}
energy is 3×10^{10} cm^{-2} s^{-1}. Although thermospheric winds on Saturn would
tend to distribute globally the H atoms produced in the aurora, their contribution to the H flux due to the extreme-ultraviolet dissociation of H_2 would be
negligible.

    The Saturn Lyman-α emission has been monitored since 1977 and it ap-
pears to be more or less correlated with the variation in the solar Lyman-α
flux or the solar extreme ultraviolet (Atreya et al. 1982a). One can, therefore,
in principle exploit these data to determine the temporal variation in the upper
atmospheric mixing. The pre-Voyager observations of the Lyman-α emission,

however, were carried out from Earth orbit, thus introducing ambiguities in the actual planetary Lyman-α emission due to the absorption of the Lyman-α photons in the interplanetary medium, and the possible contribution from the auroral Lyman-α.

2. Methane. The most direct method of determining the homopause level in an H_2 atmosphere is by monitoring the distribution of a heavier gas, such as methane whose density drops rapidly in the vicinity of the homopause. By comparing the measured CH_4 profiles with profiles calculated by varying K, one can determine the eddy diffusion coefficient. Again, one assumes that the CH_4 distribution is not affected by charged particle precipitation. The δ Sco ultraviolet stellar occultation experiment on Voyager 2 determined CH_4 density in the upper atmosphere (Festou and Atreya 1982). Assuming a reasonable variation of the eddy diffusion coefficient with the atmospheric number density M, such as $K \propto M^{-1/2}$ (which was found to be valid for Jupiter [Atreya et al. 1981]), and by varying K, Atreya (1982) calculated numerous theoretical models for the distribution of methane with photochemical and transport processes included. Shown in Fig. 8 are the H_2 densities at the altitude of unit vertical optical depth in methane at Lyman-α ($\tau_{CH_4}^{\uparrow}$ = 1) vs. eddy diffusion coefficient at the homopause K_h. The $\tau_{CH_4}^{\uparrow}$ = 1 level is found to be at an altitude of 800 km where the H_2 density is 1.6×10^{13} cm^{-3} (Festou and Atreya 1982). This implies K_h of 1.7×10^8 cm^2 s^{-1}, indicated by an arrow in Fig. 8. The altitude of the $\tau_{CH_4}^{\uparrow}$ = 1 level, however, is uncertain so

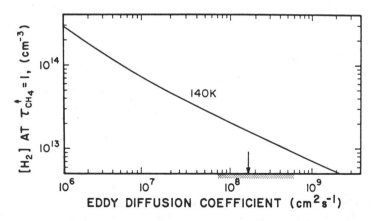

Fig. 8. H_2 density at the altitude of unit vertical optical depth in methane at Lyman-α vs. eddy diffusion coefficient at the homopause, K_h. The arrow represents the value of K_h corresponding to the central value of the H_2 density at the $\tau_{CH_4}^{\uparrow}$ = 1 level determined from the δ Sco occultation data. The shaded area on the X axis corresponds to statistical uncertainty in the determination of the $\tau_{CH_4}^{\uparrow}$ = 1 altitude level in the data (after Atreya 1982).

that the range of H_2 density at this level is between 9×10^{12} cm^{-3} and 2.5 $\times 10^{13}$ cm^{-3}; this would imply (from Fig. 8):

$$6 \times 10^8 \leq K_h \leq 7 \times 10^7 \text{ cm}^2 \text{ s}^{-1}. \tag{4}$$

This range for K_h is shown by the hatched area on the abscissa of Fig. 8. The homopause for CH_4 is determined to be at an altitude of 1110 km where the H_2 density is 1.2×10^{11} cm^{-3} and an atmospheric temperature of 250 K (Atreya 1982). Note that the $\tau_{CH_4}^{\uparrow} = 1$ level lies somewhat below the homopause, as was the case also for Jupiter (Atreya et al. 1981) due to the photolysis of methane.

3. He 584 Å. In a model developed for Jupiter, McConnell et al. (1981) have shown that analysis of the He 584 Å airglow data can yield information on the relationship between the eddy diffusion coefficient K_h at the homopause and the temperature structure. As K_h is increased, more He is transported into the upper atmosphere; the result is more scattering of the solar 584 Å line. As the temperature is increased, the relative amounts of He above the base of the scattering layer decreases by the change in the molecular diffusion coefficient with temperature. The results for Saturn, taken from Sandel et al. (1982a), are shown in Fig. 9. These results use the Voyager infrared interferometer spectrometer (IRIS) mixing ratio of 0.06 for helium (Hanel et al. 1981a).

Broadfoot et al. (1981) presented the first He 584 Å airglow data for the Voyager 1 encounter with Saturn. Since then the effects of the Jovian radiation belts on the calibration of the instrument have been better characterized (Holberg et al. 1982a); currently the best estimates of the He 584 Å intensity at the center of the disk are 3.1 ± 0.4 and 4.2 ± 0.5 R for Voyager 1 and Voyager 2 encounters, respectively. Since the solar flux at 584 Å does not appear to have changed (Sandel et al. 1982a), the implication of the change in the intensity is that either the temperature of the scattering region has changed by a factor of 2 or else, more likely, K_h has increased by a factor of 2. With the CH_4 homopause around 1110 km (Sec. I.B.2), the temperature in the scattering region is close to 250 K (Fig. 5) implying $7 \times 10^7 < K_h < 10^8$ cm^2 s^{-1} for the range of eddy coefficient (Fig. 9). The He homopause could, however, be located at a different altitude than the CH_4 homopause implying a somewhat different range of K_h.

Sandel et al. (1982a) have analyzed the δ Sco occultation data for the hydrocarbons in a different manner than Atreya (1982) to arrive at K_h; they have further attempted to relate it to the He 584 Å data discussed in Sec. I.B.4. The δ Sco occultation data used in this analysis refer to the 1160–1263 Å range (Fig. 10). The method adopts a lower atmosphere CH_4 mixing ratio of 1.6×10^{-3}. For an isothermal atmosphere, the distribution of CH_4 in the vicinity of the homopause is closely approximated by the formulae of Wallace and Hunten (1973). Referred to the homopause as the reference altitude, the

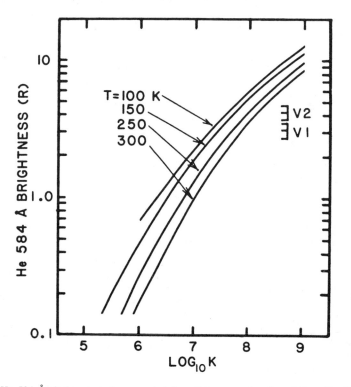

Fig. 9. He 584 Å airglow emission expected from Saturn as a function of the eddy diffusion coefficient at the homopause and the temperature in the scattering region. The brightnesses measured by Voyagers 1 and 2 are shown with their uncertainties on the right (after Sandel et al. 1982a).

shape of the CH_4 mixing ratio is invariant versus K in a density scale height frame. However, the attenuation offered to a solar beam is very sensitive to K (cf. Fig. 10). Fitting the data to the isothermal model the variation of K required to fit the data versus temperature can be obtained (Sandel et al. 1982a). This is illustrated in Fig. 11 which also shows the variation of K with T for the Voyager 2 He 584 Å airglow data. As can be seen, the CH_4 and He data are complementary, from which a unique value of K and T may be obtained. The results that Sandel et al. (1982a) obtained are given below.

$$T = 125^{+40}_{-25} \text{ K} \qquad (5)$$

$$4 \times 10^7 < K_h < 1.2 \times 10^8 \text{ cm}^2 \text{ s}^{-1}. \qquad (6)$$

The range of values for K_h obtained in this manner is in reasonable agreement with those obtained by Atreya (1982) (see Sec. I.B.2), in spite of the differences in analysis. From their recent solar and stellar occultation data, Smith et

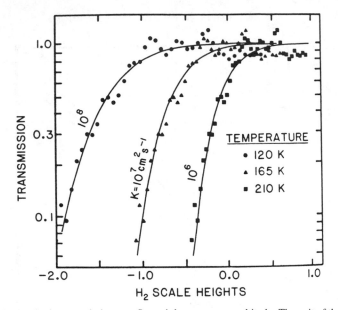

Fig. 10. Atmospheric transmission near Saturn's homopause vs. altitude. The unit of the abscissa is the H_2 scale height. The data points, which have been 1-2-1 smoothed, show the transmission measured in the 1160 to 1263 Å range plotted at scales corresponding to three temperatures. The solid curves are the transmission expected for CH_4 density distributions corresponding to the indicated eddy diffusion coefficients K. The three pairs of curves and data points represent pairs of T and K that are compatible with the measured CH_4 density distributions (after Sandel et al. 1982*b*).

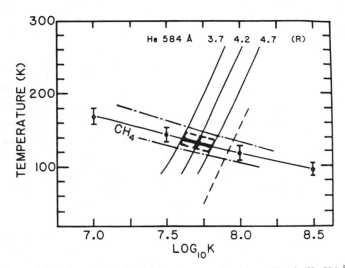

Fig. 11. Plots of the pairs of T and K satisfying the constraints imposed by the He 584 Å airglow brightness and measurements of the CH_4 density distribution. The two curves intersect at the unique pair that satisfies both sets of data and hence represents the conditions that hold near Saturn's homopause. The stippled area shows the uncertainty in T and K corresponding to the uncertainty in the two observations (after Sandel et al. 1982*b*).

al. (1983) derive $K_h = 5 \times 10^6$ cm^2 s^{-1}. This value is unacceptable since Smith et al. did not carry out a complete photochemical and transport modeling including coupling to the ionosphere, as done by Atreya et al. (1981) and Atreya (1982).

4. Comparison of K_h with Other Planets.   Table III presents the best current values for the homopause values of eddy diffusion coefficients on Saturn, Jupiter, Titan, Earth, Mars and Venus. These values are pertinent to solar maximum conditions.

It is evident from Table III that during the Voyager encounters, the eddy coefficient on Saturn was greater than that on Jupiter or Earth at corresponding atmospheric densities. Such a difference could be the result of a different thermal structure in the middle atmosphere (D. M. Hunten, personal communication, 1982), or a more vigorous tropospheric dynamics on Saturn, perhaps driven by helium condensation (Atreya 1982). In any event, the Saturn measurements refer to a single point at one altitude and as such may not be representative for extrapolation to the lower atmosphere. There does not appear to be an appreciable difference between the homopause values of K_h on Saturn from the equatorial to the midlatitudes, as is evident from the results obtained from the δ Sco (equatorial) and the He 584 Å and Lyman-α (low to midlatitudes) analyses.

II. IONOSPHERE

Measurements of the ionospheric structure of Saturn were made by the Pioneer Saturn and Voyagers 1 and 2 spacecraft using the technique of radio occultation discussed in Sec. II.A. The measurements are reviewed in Sec. II.B. Theoretical models of the Saturn ionosphere are essentially similar to those of Jupiter. A review of the models is presented in Sec. II.C, while attempts to interpret these measurements with models are discussed in Sec. II.D. Finally, mechanisms of the thermospheric heating are discussed in Sec. II.E.

A. Ionospheric Measurement Technique

The structures of the ionospheres and tropospheres of the major planets have been successfully measured by the technique of radio occultation employed on both the Pioneer and Voyager spacecraft. Information on the gaseous envelope is obtained from measurements of Doppler frequency shift, group delay, intensity and polarization of the radio signal when the spacecraft swings behind the planetary body and undergoes occultation as viewed from the Earth (Eshleman 1973; Fjeldbo 1973; Hunten and Veverka 1976). The Pioneer measurements were carried out using a single frequency (2.293 GHz or S-band at 13 cm) while Voyager employed dual frequency (S-band, and X-band at 3.5 cm) radio links. The Voyager dual-frequency technique is particu-

TABLE III
Eddy Diffusion Coefficient

	K_h (cm^2 s^{-1})	Density[a] at Homopause (cm^{-3})	Altitude[b] of Homopause (km)	Atmospheric Pressure at Homopause (bar)	References
Saturn	$1.7(+4.3, -1.0) \times 10^8$ $8.0(+4.0, -4.0) \times 10^7$	1.2×10^{11}	1110	4×10^{-9}	Atreya (1982). Sandel et al. (1982a).
Jupiter	$1.4(+0.8, -0.7) \times 10^6$	1.4×10^{13}	440	10^{-6}	Atreya et al. (1981); McConnell et al. (1982a).
Titan	$1.0(+2.0, -0.7) \times 10^8$	2.7×10^{10}	3500	6×10^{-10}	E. J. Smith et al. (1982).
Earth	10^6	10^{13}	100	3×10^{-7}	Hunten (1975).
Venus	10^7	7.5×10^{11}	130–135	2×10^{-8}	Von Zahn et al. (1980).
Mars	$(1.3-4.4) \times 10^8$	$\sim 10^{10}$	135	2×10^{-10}	Nier and McElroy (1977).

[a] Density: H_2 for Jupiter and Saturn; atmospheric for others. Densities at the homopause correspond to the central values of K_h.
[b] Altitude: For Jupiter and Saturn, the altitudes are above the 1-bar atmospheric pressure level in the equatorial region; some previous publications had the cloud tops or the 10^{19} cm^{-3} level as the reference. For Titan, Earth, Venus and Mars, the altitudes are above the surface,

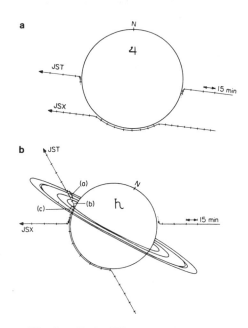

Fig. 12. *Upper diagram*: View from Earth of Voyager occultation at Saturn. The spacecraft radio images follow the indicated paths for the Jupiter-Saturn-Titan (JST) and Jupiter-Saturn-Uranus (JSX) trajectories. Note that there is a combination of near central and more grazing occultations. For JST at Saturn, region (a) provides a clear occultation of the rings and (b) a clear atmospheric occultation, while (c) is a combined ring and atmospheric occultation (after Eshleman et al. 1977). *Lower diagram*: Side view of Voyager occultations at Saturn. The trajectories are plotted in a rotating plane that instantaneously contains the Earth, the spacecraft, and the center of the planet. The pre-encounter values of latitudes of occultation immersions and emersions are shown, and regions (a), (b), and (c) of the upper view are also illustrated here (after Eshleman et al. 1977).

larly important for the Jovian and Saturnian ionospheres where multi-mode propagation of the beam is caused by sharp ionospheric layers (Eshleman et al. 1977). Furthermore, the signal-to-noise ratio for Voyager exceeds the Pioneer S-band values by 10 dB at S-band frequencies, and 23 dB at X-band frequencies (Eshleman et al. 1977). The occultation geometries at Saturn for the nominal Voyager Jupiter-Saturn-Titan (JST) and Jupiter-Saturn-Uranus (JSX) trajectories as viewed from the Earth are shown in Fig. 12. The latitudes of observations shown in the lower diagram of Fig. 12 are pre-encounter values, the actual latitudes, listed in Table II, are somewhat different.

B. Ionospheric Characteristics

The ionosphere of Saturn was probed on six occasions between 1979 and 1981. With the exception of the Voyager 1 exit data which have not been fully analyzed, major characteristics of all other measurements are listed in Table

TABLE IV
Ionosphere Observations of Saturn

Observation Date	Technique	Latitude	Solar Zenith Angle	Peak Electron Concentration (cm^{-3})	Altitude[a] of the peak above 1-bar level (km)	Plasma Scale Heights (km)
1979 Sept. 1	Pioneer Saturn[b] S-band					
	Ingress	11°6S	89°2 Terminator	1.1×10^4	1900	?[b]
	Egress	9°7S	90°9 Terminator	$\sim 1 \times 10^4$	2900	
1980 Nov. 12	Voyager 1 S- and X-bands					
	Ingress	73°S	89° Late afternoon	2.3×10^4	2500	560 km
1981 Aug. 26	Voyager 2 S- and X-bands					
	Ingress	36°N	87° Late afternoon	6.4×10^3	2850	1000 km, topside 260 km, lower
	Egress	31°S	93° Pre-dawn	1.7×10^4	2150	1100 km, topside

[a] 1-bar level is ~ 75 km below the level at which the atmospheric density of 10^{19} cm^{-3} is reached, and 50 km below the ammonia cloud tops.

[b] Although the plasma scale height cannot be determined with certainty in the Pioneer Saturn radio occultation data, the ingress data are more reliable. The data for $N_e < 3 \times 10^3$ cm^{-3} in these observations may be spurious.

IV. All measurements were made close to the terminator, i.e., at solar zenith angle ≈ 90°.

The Pioneer radio occultation measurements revealed an ionosphere extending up to 30,000 km from the planetary limb (Kliore et al. 1980a). The entry (ingress) data are more reliable than the exit data. Data for concentrations < 3000 electrons cm^{-3} may not be indicative of the local electron concentration on Saturn; they are more likely due to electron fluctuations of the interplanetary solar wind (Kliore et al. 1980b). Fig. 13 shows the entry and exit ionospheric data up to a radius of 70,000 km. Despite differences in details, the two profiles show the same general characteristics. A peak electron concentration of ~ 10^4 cm^{-3} at ~ 1800 km occurs in the entry profile, while a similar peak concentration is found nearly 1000 km higher in the exit data. The magnitude of the electron fluctuations of the interplanetary solar wind associated with uncertainties in the orbit, and the oscillator drift render the exit data only marginally useful, and then only for qualitative comparison with the entry data (Kliore et al. 1980a). Due to insufficient information about the topside, it is also not possible to deduce a unique plasma scale height from these data. It is, however, apparent that the plasma temperature in the 63,000 to 68,000 km range is at least 500 K, and perhaps as high as 1000 K.

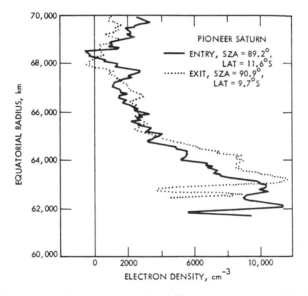

Fig. 13. Electron concentration in the ionosphere of Saturn between equatorial radii of 60,000 and 70,000 km. Solid curve is a profile produced from the closed-loop data taken during entry. The dashed curve is a profile obtained from exit data by using an artificial drift function and should be used only for comparison of features and not for magnitude of electron concentration (after Kliore et al. 1980b).

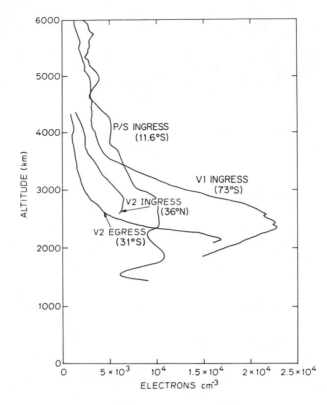

Fig. 14. Voyager 1 ingress, Voyager 2 ingress and egress, and Pioneer Saturn ingress ionospheric data plotted on a common altitude scale, the zero of which is at the 1-bar pressure level. The Voyager plots were prepared by L. Tyler and V. Eshleman on behalf of the Radio Science Team.

The Voyager 1 and 2 ionospheric measurements are shown in Fig. 14. For comparison, the Pioneer Saturn immersion data have also been plotted in this figure on the same scale as the Voyager data. The radio occultation experiment of Voyager 1 covered a latitude range from 73°S to 79°.5S over a 14° range of longitudes (Tyler et al. 1981). The immersion ionospheric measurements were carried out very near the beginning of this exercise; the latitude of the ionospheric region probed was 73°S. A local peak at 2.4×10^4 cm^{-3} in the electron concentration was measured around 2500 km; above and below the peak, the electron profile appears to drop off rapidly. The topside plasma scale height of 560 km appears to be much smaller than that deduced from the Pioneer Saturn data. In either situation, it is difficult to determine the true plasma temperature from the scale height since the identity of the topside ion is not known with certainty (Sec. II.D). One can make only a qualitative comparison between the Pioneer Saturn and the Voyager data because of the different observing frequencies and technique, and the fluctuations in the interplanetary solar wind electrons and the oscillator drift problems in the former. Furthermore, Voyager 1 data are particularly unsuitable for comparison as they represent polar conditions while the Pioneer data are for the equatorial region. The Voyager data analysis in the region below 2000 km is incomplete, and a considerably more complex structure is expected there (Tyler et al. 1982a) due, perhaps, to the presence of short-lived hydrocarbon ions and long-lived metallic ions as proposed for Jupiter (Atreya et al. 1974; Atreya and Donahue 1976). The metallic ions could be extraplanetary in origin such as from meteorites.

The Voyager 2 ionospheric measurements are for nearly midlatitude conditions, with immersion at $36°5N$ and emersion at $31°S$ (egress). Both measurements are made near the terminator, and both have peak electron concentrations somewhat lower than those measured on Voyager 1 in the polar region. The apparent peaks in the two Voyager 2 measurements are separated by nearly 700 km. The immersion data below 2500 km have not yet been analyzed, thus it is not entirely evident whether the observed peak is the main peak or simply a local maximum in the electron concentration profile. The topside scale height in the 2800–4000 km region of the ingress data is 1000 km, approximately twice the topside scale height for Voyager 1. The Voyager 2 egress ionospheric profile has approximately the same topside scale height (~ 1100 km) as the ingress one; the scale height just above the peak (2150 km) in the egress, however, is 260 km. Since the ionospheric data below ~ 2000 km have not been analyzed, it is suspected but not known whether the Saturn lower ionosphere would exhibit the type of multilayered structure seen on Jupiter.

C. Review of Theoretical Models

McElroy's (1973) review paper on the ionospheres of the major planets was the first theoretical attempt at modeling the Saturn ionosphere. He considered a neutral atmosphere composed of predominantly H_2, with He/H_2 and CH_4/H_2 ratios by volume of 0.3 and 10^{-3}, respectively. Earlier work by Gross and Rasool (1964) and Hunten (1969) had shown that in Jupiter's atmosphere every photon absorbed by H_2 leads to the production of two hydrogen atoms, either directly by

$$H_2 + h\nu \rightarrow H + H \tag{7}$$

or indirectly by

$$H_2 + h\nu \rightarrow H_2^+ + e^- \tag{8}$$

followed by

$$H_2^+ + H_2 \rightarrow H_3^+ + H \tag{9}$$

$$H_3^+ + e^- \rightarrow H_2 + H . \tag{10}$$

Since the atomic hydrogen produced by such processes could only be lost by 3-body recombination processes deep within the atmosphere, it was necessary to solve a diffusion equation to obtain the atomic hydrogen distribution. Knowledge of the H density distribution was required, since atomic hydrogen was the major source for the long-lived H^+ ions, the dominant ion species in past theoretical models of the Jovian ionosphere (Rishbeth 1959; Zabriskie 1960; Gross and Rasool 1964; Hunten 1969; Shimizu 1971). However, McElroy pointed out that all of the earlier studies had ignored the potential importance of dissociative ionization of H_2 as a source of H^+ ions. Following McElroy's suggestion, Atreya et al. (1974) included the above dissociative ionization process in their model of the Jovian ionosphere; it turned out to be the major source of topside ionization.

A comprehensive study of the ionospheres of other outer planets, i.e. Saturn, Uranus and Neptune, was carried out by Atreya and Donahue (1975a) using the same model atmosphere as McElroy, but including several important new chemical reactions, such as the 3-body recombination of H^+ and a new rate for H_3^+ electron recombination. This model again neglected ion diffusion, as had McElroy's model, but the changes in the chemical-reaction scheme resulted in some significant changes in the ionospheric profile. Atreya and Donahue were also the first to suggest the interaction of the ring particles with the ionosphere. In a follow-up study of the Saturn ionosphere carried out by Atreya and Donahue (1975b), their model considered the effect that reactions of H^+, H_3^+ and He^+ with CH_4 have on the structure of the ionosphere following another suggestion of McElroy (1973). Moderate values for the eddy diffusion coefficient ($K = 2 \times 10^6$ cm^2 s^{-1}) resulted in a pronounced hydrocarbon ion ledge below the ionospheric peak.

Further modeling of the Saturn ionosphere was carried out by Capone et al. (1977). They argued that due to the expected relatively weak magnetic fields of the outer planets and the decrease in solar extreme-ultraviolet radiation with increasing heliocentric distance, an ionospheric model of the outer planets is fundamentally incomplete without inclusion of galactic cosmic-ray ionization. However, the peak produced in their model by galactic cosmic rays occurred so deep in the atmosphere (at $\sim$ 0.5- to 1-bar level) that it had no influence on the main ionosphere but simply produced a low-lying ionospheric ledge.

From the Jovian ionospheric measurements it was apparent that high exospheric temperatures (Kliore et al. 1974; Fjeldbo et al. 1976; Atreya and Donahue 1976; Eshleman et al. 1979) and a wide range of values for the eddy diffusion coefficient (Cochran and Barker 1979) were also quite possible in the Saturn atmosphere. Waite et al. (1979) constructed a model to study the effect of variation of the eddy diffusion coefficient and exospheric temperature on ionospheric structure. This model used the ionospheric chemical scheme of the Jovian ionosphere (Atreya and Donahue 1976), but included the diffusion of the major ion H^+. The results of the Waite et al. model showed that for a cold isothermal thermosphere, values of the eddy diffusion coefficient from 10^4 to 10^6 cm^2 s^{-1} resulted in an ionosphere composed of H^+ ions with a peak electron density of $\sim 10^5$ cm^{-3}. Large values of the eddy diffusion coefficient ($\sim 10^9$ cm^2 s^{-1}) resulted in a large abundance of CH_4 being present at the level of maximum H_2 ionization, leading to a peak electron density of 10^3 cm^{-3}. High exospheric temperatures moderated this response through increased separation of the solar extreme-ultraviolet ionization region from the strong chemical loss of H^+ and H_2^+ in the methane layer.

D. Comparison Between Theoretical Models and Measurements

Since 1979 Pioneer Saturn and Voyager have measured the ionospheric structure of Saturn, as discussed in Sec. II.B. A major discrepancy exists between the measurements and the model prediction. All previous models predicted peak electron densities on the order of 10^5 cm^{-3}. The measurements indicate peak electron densities of $\sim 10^4$ cm^{-3}. Possible explanations include: decreased solar insolation due to shielding of the solar extreme-ultraviolet radiation by the rings (Waite 1981); the existence of a Saturn equatorial anomaly (Kliore et al. 1980b); photochemical loss due to reactions with CH_4 or OH (Shimizu 1980); and removal of topside ion H^+ by vibrationally excited H_2 (Atreya et al. 1979b; Atreya and Waite 1981; Waite 1981). A self-consistent ionospheric model which considers the above-mentioned effects has been developed by Waite et al. (1983a); its major points are discussed below.

Atreya and Waite (1981) attempted an interpretation of the Voyager radio science data by extending the ionosphere model developed by Waite (1981). They discussed the loss of H^+ via vibrationally excited H_2 mentioned above and the effect of vertical drifts on the magnitude and location of the peak in ionization. This calculation, however, assumed a preliminary atmospheric model. With the analysis of δ Sco stellar occultation data (Sec. II) it has been possible to construct a model of the ionosphere (Waite et al. 1983a) based on a neutral atmosphere measured simultaneously at nearly the same latitudes as the ionosphere. A calculation of the H_2 vibrational temperatures has also been included in this model to study the loss of H^+ via vibrationally excited H_2. Below, the model is discussed briefly, followed immediately by a summary of the calculated results.

The theoretical model used in the calculations of Waite et al. (1983a) consists of (1) a model neutral atmosphere including the major neutral species (H_2, He, H, CH_4, C_2H_6, C_2H_2, C_2H_4, and CH_3); (2) an ionospheric model based on the ion chemistry of earlier Jovian ionospheric models by Atreya and Donahue (1976) using the latest available rate constant as listed in Waite (1981) and Atreya and Waite (1981); (3) a 2-stream electron transport code including both photoelectrons and precipitating electrons; and (4) an ion and electron temperature model. All components of this theoretical model are coupled together to provide a self-consistent solution to the composition, structure, and temperature of the Saturn ionosphere. A detailed description of the numerical model can-be found in Waite (1981). For background on the conservation equations of concentration, momentum and energy governing the ionospheric physics, the reader is referred to Banks and Kockarts (1973).

The details of the thermal structure, eddy diffusion coefficient, and the hydrocarbon mixing ratios are discussed in Sec. II and Fig. 5. Both the Festou and Atreya (1982) and the Smith et al. (1983) model atmospheres were used in the ionospheric calculations. Solar ultraviolet fluxes appropriate for the Voyager observations were provided by H. E. Hinteregger (personal communication, 1981).

The important sources and sinks of ionization are listed in Table V, and significant chemical reactions schematized in Fig. 15. Following ionization of the major atmospheric species H_2, He and H by solar extreme-ultraviolet radiation or electron impact, the resulting ions charge exchange with the neutral species giving rise to the numerous intermediate ions such as H_2^+, CH^+, CH_2^+, etc. The eventual topside ion is H^+ or H_3^+; immediately below the peak it is H_3^+, and in the deep atmosphere several hydrocarbon ions such as CH_5^+, $C_2H_5^+$, $C_3H_8^+$, etc. are expected to be prevalent. The terminal ions in the various regions of the atmosphere are all lost by electron recombination. We discuss below the characteristics of the equatorial/midlatitude and auroral ionosphere calculations.

For the low and midlatitude regions, solar extreme-ultraviolet radiation provides the dominant source of ionization. Fig. 16 shows for the Festou and Atreya (1982) model atmosphere, a model of the ionosphere based on this ionization source (Waite et al. 1983a). There is considerable disagreement between the measurements and the calculations with regard to both the height and the magnitude of the ionospheric peak. The model calculations indicate a peak electron density of 2×10^5 cm^{-3} located 1200 km above the ammonia cloud tops whereas the measurements indicate peak densities of 1×10^4 cm^{-3} at altitudes between 2000 and 3000 km. A similar disagreement also resulted when the Smith et al. (1983) model atmosphere was used. In this case the peak electron density was 3.7×10^5 cm^{-3} and located at 1450 km. The scale height of the ionosphere was significantly less than that of the ionosphere model from the Festou and Atreya (1982) model atmosphere. For this model atmosphere, there is a rough agreement with the plasma scale heights from Voyager 2 measurements which indicate temperatures on the order of 600 to 1000 K, if we assume that $T_e = T_i$. This is a reasonable assumption since plasma temperatures remained in thermal equilibrium with the neutral temperature in the model calculations at altitudes below 4750 km due to strong H_2 vibrational cooling of the electrons.

Previous calculations have suggested that vertical drifts (Kliore et al. 1980b; Waite 1981; Atreya and Waite 1981; Atreya et al. 1982b; McConnell

TABLE V
Important Chemical Reactions in the Ionosphere of Saturn[a]

Reaction Number	Reaction	Rate Constant	References
Ion Production			
p1	$H_2 + h\nu \rightarrow H_2^+ + e$		
p2	$\rightarrow H^+ + H + e$		McElroy (1973).
p3	$H_2 + e \rightarrow H_2^+ + 2e$		
p4	$\rightarrow H^+ + H + 2e$		
p5	$H + h\nu \rightarrow H^+ + e$		
p6	$H + e \rightarrow H^+ + 2e$		
p7	$He + h\nu \rightarrow He^+ + e$		
p8	$He + e \rightarrow He^+ + 2e$		
p9	$CH_3 + h\nu \rightarrow CH_3^+ + e$		Atreya and Donahue (1975b).
Charge Exchange			
e1	$H_2^+ + H_2 \rightarrow H_3^+ + H$	2.0×10^{-9}	Theard and Huntress (1974).
e2	$H_2^+ + H \rightarrow H^+ + H_2$	6.4×10^{-10}	Karpus et al. (1979).
e3	$He^+ + H_2 \rightarrow H_2^+ + He$	$\leq 2.0 \times 10^{-14}$ $\left.\right\}$ sum	Johnsen and Biondi (1974).
e4	$\rightarrow HeH^+ + H$	$\leq 2.0 \times 10^{-14}$ $\quad 1 \times 10^{-13}$	
e5	$\rightarrow H^+ + H + He$	$< 8.0 \times 10^{-14}$	
e6	$He^+ + CH_4 \rightarrow CH^+ + H_2 + H + He$	2.4×10^{-10}	Huntress (1974).
e7	$\rightarrow CH_2^+ + H_2 + He$	9.3×10^{-10}	Huntress (1974).
e8	$\rightarrow CH_3^+ + H + He$	9.6×10^{-11}	Adams and Smith (1976).
e9	$\rightarrow CH_4^+ + He$	1.6×10^{-11}	Adams and Smith (1976).
e10	$H^+ + H_2 + H_2 \rightarrow H_3^+ + H_2$	3.2×10^{-29}	Miller et al. (1968).

TABLE V (continued)
Important Chemical Reactions in the Ionosphere of Saturn[a]

Reaction Number	Reaction	Rate Constant	References
e11	$H^+ + H_2(v' \geq 4) \rightarrow H_2^+ + H$	$k = fn(T_v)$	(see text)
e12	$H^+ + CH_4 \rightarrow CH_3^+ + H_2$	2.3×10^{-9}	Huntress (1974).
e13	$\rightarrow CH_4^+ + H$	1.5×10^{-9}	Huntress (1974).
e14	$HeH^+ + H_2 \rightarrow H_3^+ + He$	1.85×10^{-9}	Theard and Huntress (1974).
e15	$H_3^+ + CH_4 \rightarrow CH_5^+ + H_2$	2.4×10^{-9}	Huntress (1974).
e16	$CH^+ + H_2 \rightarrow CH_2^+ + H$	1.0×10^{-9}	Huntress (1974).
e17	$CH_2^+ + H_2 \rightarrow CH_3^+ + H$	1.6×10^{-9}	Smith and Adams (1977).
e18	$CH_3^+ + CH_4 \rightarrow C_2H_5^+ + H_2$	1.2×10^{-9}	Smith and Adams (1977).
e19	$CH_4^+ + CH_4 \rightarrow CH_5^+ + CH_3$	1.5×10^{-11}	Smith and Adams (1977).
e20	$CH_4^+ + H_2 \rightarrow CH_5^+ + H$	3.3×10^{-9}	Smith and Adams (1977).
Electron-Ion Recombination			
r1	$H_3^+ + e \rightarrow H_2 + H$	$2.8 \times 10^{-7} \left(\dfrac{200}{T_e}\right)^{0.7}$	Leu et al. (1973).
r2	$H_2^+ + e \rightarrow H + H$	$<1.0 \times 10^{-8}$	Auerbach et al. (1977).
r3	$HeH^+ + e \rightarrow He + H$	$\sim 1.0 \times 10^{-8}$	Hunten (1969).
r4	$H^+ + e \rightarrow H + h\nu$	$4.0 \times 10^{-12} \left(\dfrac{250}{T_e}\right)^{0.7}$	Bates and Dalgarno (1962).
r5	$He^+ + e \rightarrow He + h\nu$	$4.0 \times 10^{-12} \left(\dfrac{250}{T_e}\right)^{0.7}$	Bates and Dalgarno (1962).
r6	$CH_5^+ + e \rightarrow CH_4 + H$	3.9×10^{-6}	Maier and Fessenden (1975).
r7	$C_2H_5^+ + e \rightarrow C_2H_2 + H + H_2$	3.9×10^{-6}	Maier and Fessenden (1975).

[a]Table adapted from Atreya and Donahue 1976 and Atreya et al. 1979.

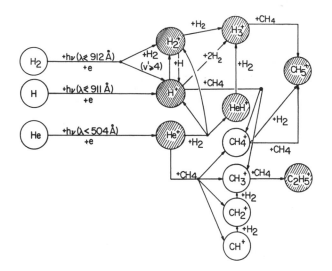

Fig. 15. Schematic of the Jovian and Saturnian ionospheric reactions (after Atreya and Donahue 1982).

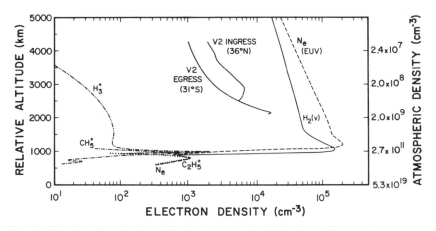

Fig. 16. Calculated ionospheric profiles for the nominal midlatitude model, curve N_e, and the Voyager data for the midlatitudes. The effect of vibrationally excited H_2 (by extreme ultraviolet) on the electron concentration is shown by the curve $H_2(v)$ (after Waite et al. 1983*a*).

et al. 1982*b*) and high H_2 vibrational temperatures (Atreya and Waite 1981) may be important in determining the height and magnitude of the ionospheric peak on Saturn. McConnell et al. (1982*a*) have also carried out calculations for Jupiter which show that reasonable theoretical fits to the ionospheric data can be obtained with appropriate choices for the vertical drift and H_2 vibrational temperature. The difficulty with previous calculations is the lack of justification for the somewhat arbitrary choice of the H_2 vibrational temperature profile and vertical drift rates needed to provide agreement between theory and measurement.

To quantify the uncertainty, Waite et al. (1983a) have carried out H_2 vibrational level calculations for the Saturn upper atmosphere for midlatitudes using in these calculations the Festou and Atreya (1982) model atmosphere. At an altitude of 1200 km, where the calculated electron density is at a maximum, the vibrational temperature T_{vib} for the fourth level is about a factor 3 larger than the neutral temperature. For a reaction rate for $H^+ + H_2$ (v′ $\geq$ 4) of 2×10^{-9} cm^3 s^{-1}, this degree of vibrational excitation is sufficient to reduce the calculated electron density from 2.3×10^5 cm^{-3} to 1.5×10^5 cm^{-3} (see the profile for H_2(v) in Fig. 16). At an altitude of 2500 km, where the observed electron density is $\sim 10^4$ cm^{-3}, T_{vib} (v′ = 4) is twice the neutral temperature and reaction of H^+ with H_2(vib) is the dominant sink of H^+. The calculated N_e is now 4×10^4 cm^{-3}, in better agreement with the radio occultation results (see Fig. 16).

Model calculations were also carried out for the auroral ionosphere (Waite et al. 1983a). The neutral atmosphere was taken from Festou and Atreya (1982), with the exception that the altitude scale is now somewhat compressed due to the larger gravity at higher latitudes. The effects of precipitating electrons were included using the 2-stream electron transport code (Waite 1981; Waite et al. 1983b). Monoenergetic electrons at 1 and 10 keV with a total energy flux of 0.67 ergs cm^{-2} s^{-1} were introduced at the top of the model atmosphere (6500 km). The total calculated Lyman and Werner band emission intensity was 7.1 kR in the 10 keV case and 6.4 kR for the 1 keV case, which is close to the observed average intensity of the auroral emission (Broadfoot et al. 1981; Sandel et al. 1982b). The peak emission altitude in the 10 keV case was only 1 to 2 scale heights above the level of unit optical depth of CH_4 ($\tau^\uparrow_{CH_4}$ = 1 level) but the emission for 1 keV was well above the $\tau^\uparrow_{CH_4}$ = 1 level. Precipitating auroral electrons most probably lie within this range of energies since Voyager 2 measurements by the ultraviolet spectrometer show little hydrocarbon extinction of the Lyman and Werner bands but some self-absorption (Sandel et al. 1982b). The resulting ionospheric profiles (Waite et al. 1983a) are shown in Fig. 17. The nominal 1 keV case (marked 0 K) indicates a peak electron density of 4×10^6 cm^{-3} at 1200 km, whereas the nominal 10 keV case indicates a peak density of only 3×10^5 cm^{-3}. The difference is due to the different atmospheric levels at which auroral energy is deposited. For the 1 keV electrons, the maximum electron impact ionization rate is at 1050 km (near the level of extreme-ultraviolet H^+ production peak at 950 km); for the 10 keV electrons it is at 750 km, which is well within the region of strong proton loss via reaction with methane.

At high latitudes where the auroral precipitation dominates, the 1 keV and 10 keV electron precipitation generates very large amounts of vibrationally excited H_2. However, there are two other factors which tend to mod-

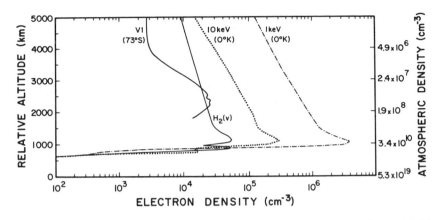

Fig. 17. Calculated ionospheric profiles in the auroral zone of Saturn for the 1-keV and 10-keV electron beams (dot-dashed and dotted lines, respectively). The Voyager 1 measurement for the high latitudes is marked V1 (73°S). The effect of the H_2 vibrational distribution on the ionospheric H^+ is also shown by the curve marked $H_2(v)$ for the 1 keV case (after Waite et al. 1983a).

erate the ionospheric effects of the large vibrational production rates. In the first place, the electron precipitation also produces copious amounts of H^+, so that any vibrationally excited H_2 must work harder to remove it. Second, there is also a large production rate of H resulting in H densities almost a factor of 100 larger than at midlatitudes—if this atomic hydrogen is not redistributed to lower latitudes by thermospheric winds. Large H densities tend to keep the lid on T_{vib} because vibrationally excited H_2 is readily quenched by H. The high-latitude electron density observed at 2250 km can be reproduced by the reaction of H^+ with H_2 ($v' \geq 4$) (see the $H_2(v)$ profile, the 1 keV auroral case in Fig. 17), if the H density at auroral latitudes is reduced near the midlatitude values by thermospheric redistribution (Waite et al. 1983a).

In conclusion, calculations indicate that vibrationally excited H_2 probably plays a major role in controlling the ionosphere of Saturn. It provides the most likely explanation, though not the only one, for the discrepancy between the observed electron densities and the ones calculated using the nominal ionospheric model with vibrationally excited H_2. However, several outstanding problems must be resolved in order to further quantify the effects of vibrationally excited H_2 on H^+ in the Saturn ionosphere. First and foremost, the reaction rate for the proposed $H^+ + H_2$ ($v' \geq 4$) reaction must be measured in the laboratory. Furthermore, a realistic evaluation of the effects of vertical drifts and vibrationally hot H_2 on the concentrations would require the knowledge of electron profiles at low altitudes.

Other potential possibilities for reconciling the observed ionospheric structure and the calculations include loss of the major ion H^+ on reaction with CH_4 or OH. Waite (1981) has found that for eddy diffusion coefficients less than a few times 10^9 cm^2 s^{-1}, the loss due to reaction with CH_4 is not important because CH_4 is not mixed to high enough altitudes unless the ionosphere undergoes strong diurnal vertical drifts that move the peak below the methane homopause. Such a scenario might help explain the apparent diurnal variations of the peak electron density from 100 cm^{-3} to 2×10^5 cm^{-3} as inferred from Saturn's electrostatic discharges (Kaiser et al. 1983; chapter by Kaiser et al.). Significant diurnal variations of electron density could then be explained in terms of the height of the ionospheric peak with respect to the methane layer. Shimizu (1980) has argued for a loss mechanism involving reaction of H$^+$ with the hydroxyl radical OH. Indeed, the source of OH may be in the rings of Saturn, as is evident from the laboratory experiments of W. L. Brown et al. (1982) on the energetic charged particle erosion of water ice. Although it is potentially an important loss mechanism for H$^+$, in order to explain the observed electron concentrations, an unreasonably large concentration of the OH radicals (on the order of 10^4 to 10^5 cm^{-3}) is required throughout the entire altitude range. Perhaps a way could be found to arrive at large OH fluxes into Saturn's atmosphere, without affecting the stability of the rings. One could conceive of other extraplanetary sources of water, such as meteorites or the icy satellites of Saturn.

E. Thermospheric Heat Sources

Several potential heat sources may play a role in determining the thermal structure in the upper atmosphere of Saturn. They include photoelectrons, gravity waves, energetic electron precipitation, and Joule heating; each source is discussed below.

1. Solar Extreme-Ultraviolet Heating. The calculations by Strobel and Smith (1973) estimated a small (10 K) rise in the temperature of the upper atmosphere of Jupiter as a result of solar extreme-ultraviolet heating. The major heating mechanism is chemical heating due to the formation of H_2^+ and subsequent reactions which result in the recombination of H_3^+. The overall process releases 10.95 eV of heat per H_2 ionization. The second most important source of extreme-ultraviolet heating is photoelectron impact dissociation of H_2 followed by a host of lesser sources such as indirect and direct vibrational excitation of H_2 (Cravens 1974).

The total column-integrated extreme-ultraviolet heat source in the upper atmosphere is only 3×10^{-3} erg cm^{-2} s^{-1} (Waite 1981), a factor of 100 smaller than that required to reproduce the exospheric temperature measured by Voyager 2 (Festou and Atreya 1982). The heating rate of column-integrated thermal electrons from photoelectrons is only 6×10^{-4} erg cm^{-2} s^{-1} and is insufficient to raise the electron temperature above the neutral temperature at any altitude in the ionosphere.

2. Inertia-Gravity Waves. The breaking of inertia-gravity waves as a source of heat in the upper atmosphere of the major planets was first suggested by the stellar occultation measurements of Jupiter (Veverka et al. 1974). This mechanism could produce a temperature profile similar to the inferred profile, if breaking of inertia-gravity waves 4 to 6 scale heights above the methane homopause was capable of depositing ~ 0.3 erg cm^{-2} s^{-1}. Although the dissipation of this much energy is consistent with the 3.4 erg cm^{-2} s^{-1} that French and Gierasch (1974) estimated to be available from the inertia-gravity waves propagating upward in the Jovian ionosphere, the magnitude of the inertia-gravity wave source for Saturn cannot be determined until a classical tidal calculation is carried out. Furthermore, it is not apparent if the waves would dissipate and deposit heat (and if so, at what height) or would simply be reflected.

3. Electron Precipitation. Electron precipitation is also a possible heat source, as discussed in the context of Jupiter by Hunten and Dessler (1977). Recent Voyager ultraviolet spectrometer measurements of Saturn's airglow estimate an average auroral electron energy influx of 2×10^{11} W between 78° and 81°5 latitude in both the northern and southern hemispheres (Sandel et al. 1982b). Model calculations of the ultraviolet spectrum from e + H_2 excitation by Shemansky and Ajello (1983) indicate that the energy of the electrons responsible for the auroral Lyman and Werner band emission is between 1 and 10 keV and that the Saturnian auroral spectrum can be produced solely by direct products of the e + H_2 process.

The altitude of peak heating is also a function of the precipitating electron energy. More energetic electrons deposit their energy at greater atmospheric densities. The altitude separation between the heat source and the infrared cooling level (i.e., the hydrocarbon level) determines the exospheric temperature for a thermosphere where vertical conduction is the dominant heat transfer process. Therefore, 1 keV electrons result in higher exospheric temperatures for a given column heating value than 10 keV electrons because the effects of increased heating efficiencies at higher electron energies are more than offset by the effect of the smaller altitude separation between the heat source and sink. The same is true for soft electrons below 100 eV which deposit their energy near the 10^8 cm^{-3} density level.

A one-dimensional heat conduction equation was solved for both a 10 keV and 1 keV electron aurora with a total energy flux of 0.67 erg cm^{-2} s^{-1}. Similar calculations were done for Jupiter by Hunten and Dessler (1977). The 1 keV electrons generate an exospheric temperature of ~ 1600 K at 1500 km, and the 10 keV electrons produce an exospheric temperature of ~ 650 K at 900 km on Saturn. Auroral deposition would appear to be an important source of high-latitude heating in Saturn's thermosphere. However, even if the entire auroral energy of 2×10^{11} W were distributed with a 100% efficiency over the entire planet, it would amount to < 0.01 erg cm^{-2} s^{-1}, which is inadequate for raising the low-latitude thermospheric temperature over the homopause value by any appreciable degree. It should be noted that in the case of Jupiter energetic oxygen and sulfur ions (10–30 MeV/nuc) can supply 10^{13} to 10^{14} W into the auroral region (Gehrels and Stone 1983). However, no such energetic heavy ions diffusing inward to Saturn have been identified.

Electron precipitation can also produce heating of the thermal electron population. However, electrons with energies between 1 and 10 keV with a total energy flux of up to 10 erg cm^{-2} s^{-1} produce no departure from thermal equilibrium at altitudes where the neutral density is $> 10^7$ cm^{-3} (Waite 1981). Soft electrons (< 100 eV) which deposit sufficient energy above the 10^8 cm^{-3} atmospheric density level can, however, cause substantial departures of the plasma temperature from the neutral temperature. Small electron precipitation energy fluxes of the order of 0.25 erg cm^{-2} s^{-1} can result in electron temperatures on the order of 5000 to 10,000 K in the topside ionosphere with accompanying order of magnitude increases in the electron density at altitudes from 2000 to 3000 km (Waite et al. 1983b). The eventual fate of this electron heating is to be conducted down to the 10^7–10^8 cm^{-3} atmospheric density level where it cools to the neutral atmosphere.

4. Joule Heating. Joule heating arises from the presence of electric currents in the ionosphere. In its most rudimentary form, the Joule heating rate for the neutral gas is given by

$$J = \sigma_p E^2 \tag{11}$$

where σ_p is Pedersen conductivity (Banks and Kockarts 1973) and E is electric field. Joule heating has been shown to be a major source of thermospheric heating on Earth. A study by Heaps (1976) of the Jovian atmosphere indicated that Joule heating was not expected to be an important heat source for that planet. However, Heaps underestimated the electric fields. Atmospheric electric field values extrapolated from the measured departure of the magnetospheric plasma from planetary corotation (McNutt et al. 1979) produce Joule heating rates of 5 erg cm^{-2} s^{-1} in the Jovian thermosphere at high latitudes (Waite 1981). In fact, detailed Joule heating calculations by Nishida and Watanabe (1981) indicate heating rates > 10 erg cm^{-2} s^{-1} in the high-latitude ionosphere of Jupiter.

McNutt (1983a) has suggested that there is a significant (10%) departure of the Saturn magnetosphere from planetary corotation in the outer magnetosphere. We can estimate the implied Joule heating rate at 80° latitude in the Saturn ionosphere using the method of Nishida and Watanabe (1981). The height-integrated Pedersen conductivity at these latitudes is 58 mhos according to a calculation by Waite et al. (1983a) (Fig. 18). Therefore, a 10% departure of the magnetosphere from the planetary rotation rate will result in a Joule heating rate of 0.14 erg cm^{-2} s^{-1}. This rate is comparable to the heating

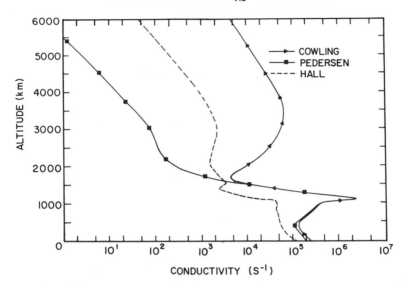

SATURN: I keV AURORA (T_{VIB} : 2500→4000 K)

Fig. 18. Calculated conductivity profiles for the 1-keV ionospheric calculation shown in Fig. 17. The height integrated conductivities are: 75.4 mhos (by Cowling); 58.0 mhos (by Pedersen); and 9.1 mhos (by Hall). By way of comparison, the height-integrated Pedersen conductivity in the auroral region on Jupiter is calculated by Strobel and Atreya (1983) to be ~ 10 mhos.

rates inferred for the auroral zone, ~ 0.3 erg cm^{-2} s^{-1}, from the measured H$_2$ Lyman and Werner band emissions. It should be emphasized, however, that the above-mentioned Joule heating rate is the upper limit since it is based on a model ionosphere which gives electron concentrations far too high in comparison with the measurements. Once the lower ionospheric data have been analyzed, it may turn out that the model electron concentrations are still higher than the data. Until all the data have been analyzed, and a good model fit obtained, Joule heating for Saturn remains of potential value. Possible interaction of the plasmasphere with the rings may be a significant additional source of Joule heating for the high- and low-latitude thermosphere (Waite 1981).

III. SUMMARY AND FUTURE PROSPECTS

A self-consistent picture of the thermal structure, composition, vertical mixing, ionospheric distribution, and energy budget of the upper atmosphere of Saturn has begun to emerge from the recent Voyager measurements of the neutral and plasma environments of Saturn. It is found that the average temperature in the middle atmosphere is approximately 140 K, thus not drastically different from that in the corresponding region on Jupiter. The exosphere of Saturn, at 600–800 K, however, is considerably cooler than the Jovian exosphere where temperatures on the order of 1100 K were recorded at about the same epoch and latitude. Unlike Jupiter, however, auroral particle precipitation does not directly supply sufficient energy for the observed thermospheric/exospheric heating on Saturn, and it also does not contribute significantly to the global distribution of the hydrogen atoms. Joule heating appears to be a promising candidate for accounting for the observed exospheric temperature on Saturn. The vertical mixing at the Saturn homopause is found to be about a factor of 100 greater than the value at the Jovian homopause, again around the same period. Greater eddy mixing on Saturn could result from a possible different stratospheric/mesospheric thermal structure, or the turbulence associated with proposed separation from hydrogen and rain out of helium toward the core. A discrepancy between the observed electron density profiles and the models can be resolved provided that large H_2 vibrational temperatures and the ion vertical drifts are present. Despite major new advances in the upper atmospheric physics and chemistry of Saturn made from the interpretation of Voyager observations, further breakthroughs will require understanding of the dynamics, seasonal effects, thermochemistry, and evolution of Saturn about which relatively little is now known, either theoretically or observationally. Following is a brief list of specific issues which need further investigation:

1. Additional analysis of the Voyager ultraviolet spectrometer data in the wavelength region 1100–1700 Å is required to determine the height profiles of C_2H_2 and C_2H_6. This is a particularly difficult problem since it involves absorption by a mixture in which the proportions of individual components are rapidly changing with height. The identity of the absorbers in the 1328–1412 Å range needs to be established. A possible candidate is H_2O (from an extraplanetary source of O, OH, or H_2O), but one must beware of its low vapor pressure at the relevant temperatures in Saturn's atmosphere. Once the height profiles of the hydrocarbons and other species have been obtained, photochemical models should be calculated for comparison, and for studying the validity of the chemical schemes.

2. Resolution of the discrepancy between the exospheric temperatures obtained from the solar occultation at 30° and the stellar occultation at 4° requires additional work. To fully understand the stellar occultation data, the calculations of Smith et al. (1983) should be extended to include the variation of temperature in the line of sight, absorption at $[H_2] > 10^{21}$ cm^{-2} (line-of-sight column), and additional sets of wavelength ranges in which H_2 absorption occurs. Some of these calculations necessitate the use of large computers.

3. Since the Voyager measurements generally refer to a given location on Saturn, the global distributions and seasonal variations can be only modeled from considerations of thermospheric and mesospheric dynamics.

Data for upper atmospheric dynamics are lacking; future spacecraft missions should ensure inclusion of instruments for dynamics measurements.

4. No satisfactory heat source has yet been proposed for Saturn's thermosphere. Both Joule heating and the inertia-gravity wave mechanisms need further investigation.

5. The modeling of ionospheric structure and the resulting conductivities require a thorough analysis of the radio occultation data below $\sim$ 2000 km.

6. Laboratory rate-constant measurements for the reaction between H^+ and the vibrationally excited H_2 must be carried out to understand its role in Saturn's topside ion chemistry.

Acknowledgment. We thank D. M. Hunten and an anonymous referee for valuable suggestions. We acknowledge individually or jointly useful conversations with V. Eshleman, M. Festou, J. Holberg, B. Sandel, D. Shemansky, G. Smith, D. Strobel, and L. Tyler. This work was supported by NASA Planetary Atmospheres Program, NASA Voyager Contract, and by the Natural Sciences and Engineering Research Council of Canada.

PART III
Magnetosphere

ENERGETIC PARTICLES IN THE INNER MAGNETOSPHERE OF SATURN

J. A. VAN ALLEN

The University of Iowa

Our present knowledge of energetic particles $E > 0.03$ *MeV in Saturn's inner magnetosphere, radial distances* $r \lesssim 10\ R_S$ *(Saturn radii), is based on observations made during the flybys of Pioneer 11 (September 1979), Voyager 1 (November 1980), and Voyager 2 (August 1981). In physical dimensions and gross characteristics of the trapped particle population, the magnetosphere of Saturn is intermediate between those of Earth and Jupiter but each of the three exhibits specific and unique features. Although many puzzles are evident, a number of salient features of the energetic particle population of the inner magnetosphere of Saturn can be listed: (1) The rings of particulate matter and the many large and small satellites inside* $r = 10\ R_S$ *reduce the population of particles* $E > 0.5$ *MeV to values of the order of* 10^3 *times less than would otherwise be present. On the other hand, sputtering and outgassing of their surfaces injects gas into the system; by some process not yet identified, particles of the resultant plasma are accelerated to energies of the order of tens of kilo-electron volts. (2) All trapped particles lie outside the magnetic shell through the outer edge of Ring A. (3) The radial distribution of very energetic protons* $E_p >$ *tens of MeV exhibits three major peaks at* $r = 3.37, 2.68,$ *and* $2.44\ R_S,$ *each of which is bounded by nearly complete voids associated with rings and satellites. The source of these protons is cosmic-ray-albedo neutron decay (CRAND), the relevant neutrons being among the products of the interactions of cosmic ray nuclei with material in the dense rings of the planet and, to a much lesser extent, with gas in the planet's atmosphere. (4) A second component of the proton population* $E_p \sim 1$ *MeV lies outside the orbit of Enceladus and is apparently diffused into the inner magnetosphere from the magnetopause. (5) A third component of the proton population, in the energy range* $E_p < 0.25$ *MeV, apparently has an internal origin associated with Dione, Tethys, Enceladus, Ring E, Mimas, and Ring G. This component is* intimately associated with the distribution of low-

*energy ions of atomic and molecular hydrogen and of helium, oxygen, and pos-
sibly other heavier species. (6) The distribution of electrons* $E_e > 0.040$ *MeV
extends throughout the magnetosphere with an internal boundary at the outer
edge of Ring A and exhibits spectral features that are indicative of relatively
unimpeded resonant diffusion across the orbits of satellites. The source of these
electrons is thermalized solar wind at the magnetopause. (7) A population of
lower-energy electrons is associated, as is the population of lower-energy protons, with satellite-emitted gas. (8) A variety of absorption microsignatures of
satellites have been observed. These provide a basis for estimating diffusion coefficients. A summary of provisional values of such coefficients is given.*

I. INTRODUCTION

The magnetic field and magnetosphere of Saturn were discovered by instruments on Pioneer 11 during the first encounter of a spacecraft with this
planet in late August—early September 1979. It had been established previously by direct observation that, as expected (Dessler 1967), the flow of the
solar wind continues out to and beyond the orbit of Saturn (Collard et al.
1982) but there was only meager (Brown 1975; Kaiser and Stone 1975) and
apparently mistaken (see chapter by Kaiser et al.) observational evidence on
the basic question of whether or not Saturn is magnetized. At an earlier date,
Luthey (1973) calculated the synchrotron emission from each of a family of
hypothetical radiation belts of relativistic electrons at Saturn and showed that
high intensities of such trapped electrons could exist (outside of the ring system) without violating the known radio emission spectrum of the Saturn
system.

The initial evidence for the existence of Saturn's magnetosphere was obtained during Pioneer 11's approach to the planet on 31 August 1979. Three
bow shock signatures were observed by the plasma instrument (Wolfe et al.
1980) and magnetometer (E. J. Smith et al. 1980a) near the noon meridian of
the planet at 24.0, 23.0, and 19.9 R_S. (1 R_S is taken to be 60,330 km.) The
first of these was also identified by a brief burst of electrons $E_e > 0.040$ MeV.
The magnetopause was observed by the same instruments as 17.2 R_S and was
confirmed as the effective boundary for trapping by the energetic particle
instruments.

A survey of the planet's external magnetic field and the distribution of
energetic particles therein was completed during the subsequent eight days of
Pioneer 11's encounter. Further observations of Saturn's magnetosphere, both
remotely and *in situ*, were made by instruments on Voyagers 1 and 2 (Saturn
encounters in November 1980 and August 1981, respectively). In the context
of the present chapter, the Voyager data provide noteworthy extensions of
knowledge of (1) the energy spectra and angular and spatial distributions of
electrons and protons in the sub-MeV range, (2) the energy spectra and spatial
distributions of identified ions having $Z > 1$, and (3) the absorption signatures
of satellites and values of diffusion coefficients inferred therefrom. Citations
of specific papers will be made at appropriate points in later sections.

Potential sources of energetic particles in a planet's magnetosphere are as follows:

(a) The solar wind;
(b) Solar energetic particles;
(c) Cosmic rays;
(d) Secondary particles from cosmic ray interactions in the planet's atmosphere, rings, and satellites;
(e) Ionized gas from the planet's ionosphere;
(f) Gas sputtered from rings and satellites by particle and photon bombardment;
(g) Gas emitted volcanically or outgassed from satellites.

It is likely that each of these sources contributes in some measure to the particle population of Saturn's magnetosphere, but only a preliminary understanding of the relative importance of the various sources is available at present. Injected particles are energized and diffused spatially by fluctuating magnetic and electric fields (including those in plasma waves) and convected by quasi-DC electric fields.

In considering the relationship of the radial diffusion coefficients of various species of energetic charged particles to the power spectra of fluctuating electric and magnetic fields (Schulz and Lanzerotti 1974), one must note that the situation at Saturn is different from that at either Earth or Jupiter. At Earth, quasi-thermal plasma corotates with the ionosphere out to the plasmapause at $\sim 4 \, R_E$, but the longitudinal drift periods of energetic particles are dominated by gradient and curvature drifts, are only slightly affected by the corotational electric field, and are, therefore, dependent on energy, species, and radial distance. At Jupiter, the corotational electric field is the dominant cause of longitudinal drift and therefore longitudinal drift periods are essentially the same for all species of particles at all relevant energies, though intermittent mass loading of the magnetic field by quasi-thermal plasma may cause the corotational angular velocity of plasma to be less than the planetary value (Hill 1980). At Saturn, the corotational angular velocity is comparable in magnitude to that attributable to gradient and curvature drifts in the inner magnetosphere and hence electrons may drift either eastward or westward and at widely varying rates depending on their energies, whereas protons of all energies drift eastward but also at widely varying rates (Thomsen and Van Allen 1980). The overall effect is that radial diffusion by violation of the third adiabatic invariant may be at quite different rates for particles of different species, for different energies, and for different radial distances because of response to different spectral ranges in the fluctuational power spectra of electric and magnetic fields. The theory of the radial dependence of diffusion coefficients has not yet been developed for Saturn.

In physical dimensions and gross characteristics of the trapped particle population, the magnetosphere of Saturn is intermediate between those of

Earth and Jupiter but each of the three exhibits many specific and unique features. The inner magnetosphere of Saturn is profoundly affected by the presence of rings of particulate matter and many large and small satellites.

. Bursts of nonthermal radio noise from Saturn exhibit a cyclic probability of occurrence corresponding to a sidereal period of 10 hr 39.4 min (Desch and Kaiser 1981a; Carr et al. 1981). This period falls at the upper end of the range 10 hr 14 min (equatorial) to 10 hr 40 min ($+57°$ latitude) derived for the prograde rotation of identifiable cloud features in the planet's upper atmosphere (Alexander 1962; Dollfus 1963) and is generally regarded as the internal rotational period appropriate to magnetospheric considerations. In the radial range 5 to 13 R_S, a less accurate but consistent mean rotational period of 10 hr 06 min ($+$ 1 hr 32 min, $-$ 1 hr 11 min) has been inferred from the Pioneer 11 data on the anisotropy of $0.61–3.41$ MeV protons if rigid corotation of the distribution of such particles is assumed, though there are significant departures from the mean value (Thomsen et al. 1980; Simpson et al. 1980b). The analysis of plasma distributions requires rigid corotation as a condition for internal consistency out to $\sim$ 10 R_S but "beyond 10 R_S, the bulk speed varies and appears to be $\sim$ $0.3–0.8$ of the expected rigid corotation speed . . . " (Frank et al. 1980).

At the orbit of Saturn (9.54 AU) the mean solar wind flow is characterized by a proton number density 0.06 cm^{-3}, a proton temperature of 2×10^5 K, and a bulk speed of 420 km s^{-1}. The total energy flow through a disk of radius 20 R_S, the presented area of Saturn's magnetosphere, is 5×10^{19} erg s^{-1}. This energy flow, which is essentially the same as the corresponding energy flow at Earth, is presumably the upper limit on that available to power magnetospheric processes directly by the solar wind.

To first order, the external magnetic field of Saturn is that of a dipole of moment $M_S = 4.6 \times 10^{28}$ Gauss cm^3, parallel to and in the same sense as the planet's angular momentum vector and located near the geometrical center of the planet (E. J. Smith et al. 1980b; Acuña et al. 1980,1981a; Ness et al. 1981,1982a). For radial distances $r \gtrsim 5$ R_S, the magnetic field is distended importantly by a system of magnetospheric currents, including an equatorial sheet current as shown by the directly measured magnetic vector field (Connerney et al. 1981), by the angular distributions of charged particles (Van Allen et al. 1980c), and by the radial positions of satellite absorption signatures as observed at off-equatorial points (Chenette and Davis 1982; Vogt et al. 1981,1982; Carbary et al. 1983a).

The encounter trajectories of Pioneer 11 and Voyagers 1 and 2 in the inner magnetosphere of Saturn were complementary. Pioneer 11 stayed quite close to the equator, Voyager 1 reached closest approach at latitude $-40°$, while Voyager 2 entered the magnetosphere at $\sim$ $+20°$ latitude, reached closest approach just above the equatorial plane, and exited the magnetosphere at $\sim$ $-20°$ latitude. The trajectories of Voyagers 1 and 2 in terms of magnetic shell coordinates (B,L) inside $L \sim 9$ are shown in Fig. 1, assuming the cen-

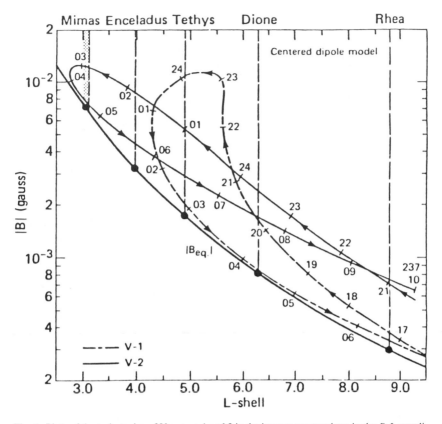

Fig. 1. Plots of the trajectories of Voyagers 1 and 2 in the inner magnetosphere in the B,L coordinate system, using a centered, axially-aligned dipolar model for the magnetic field (see, e.g.,
E. J. Smith et al. 1980a). In this coordinate system (ignoring magnetospheric currents) the
vertical lines show the positions of the L shells of Mimas, Enceladus, Tethys, Dione, and
Rhea. Note that the two Voyager trajectories cross each other at four points. The trajectory of
Pioneer 11 (not shown here) is nearly coincident with the magnetic equator with periapse at
1.34 R_S (Krimigis et al. 1983).

tered, aligned dipole model of the magnetic field (E. J. Smith et al. 1980a).
Pioneer 11 approached the planet to within a radial distance of 1.34 R_S, thus
crossing the L shells of all known (and several previously unknown) Saturnian
satellites and Rings A and B, reaching the radial distance of Ring C. Voyager
2 crossed the equatorial plane inside the orbit of Mimas, while Voyager 1
crossed through the L shell of Tethys at relatively high latitudes, but obtained
measurements close to the equatorial plane outbound. The local time coverage of the three hyperbolic flyby trajectories extended from $\sim$ noon (inbound)
through dusk to midnight and thence to $\sim$ 06 hours (outbound).

II. PROTONS OF ENERGY E_p > TENS OF MeV

The inner magnetosphere of Saturn is distinguished, as is that of Earth, by the presence of a stable population of trapped protons having energies E_p > tens of MeV. The spatial and angular distributions of the absolute intensities of such particles have been well determined, with concordant results, by both Pioneer 11 and Voyager 1 and 2 investigators. The inner portion of the spatial distribution near the equator is exemplified by the lower curve of Fig. 2 (Fillius et al. 1980).

The radial dependence of the phase-space density of protons having a representative value of the first adiabatic invariant $\mu = 8.3 \times 10^3$ MeV Gauss^{-1} is shown in Fig. 3. The nature of this curve corresponds to an internal source (Van Allen et al. 1980a). Angular distributions of high-energy protons have been well measured and an approximate spectral form has been determined (Fillius and McIlwain 1980; Vogt et al. 1982; Krimigis and Armstrong 1982). The omnidirectional intensity of protons E_p > 80 MeV has an absolute maximum value at the equator at $r = 2.66$ R$_S$ of 2.6×10^4 cm^{-2} s^{-1} (Van Allen et al. 1980a). Sketches of the spectrum by Krimigis and Armstrong are shown in the lower right-hand corner of Fig. 4.

The prevailing hypothesis is that the observed high-energy protons are the decay products of neutrons produced in nuclear interactions of primary cosmic rays, principally in the material of Rings A and B, as first advocated by Fillius et al. (1980) in analogy with the well-known Crand (cosmic-ray-albedo neutron decay) process in the Earth's inner radiation belt (White 1973). The presence of absorbing bodies in Saturn's inner magnetosphere provides a powerful diagnostic tool not available at Earth.

Analysis of the observational data yields a ratio of source strength $\mathcal{S}$ of protons E_p > 80 MeV to diffusion coefficient D of $\sim 6.9 \times 10^{-24}$ cm^{-5} at $r = 2.66$ R$_S$ (Van Allen 1983). Then using the theoretically estimated lower limit on $\mathcal{S}$ of 3.3×10^{-15} cm^{-3} s^{-1} (Blake et al. 1983), one finds a corresponding lower limit on D as 1.3×10^{-11} R$_S^2$ s^{-1} and an upper limit on the mean residence lifetime T against diffusion from the principal peak in the distribution as 15 yr. Other estimates of T in the principal peak are ~ 20 yr (Fillius and McIlwain 1980) and ~ 30 yr (Schardt and McDonald 1983). The latter estimate is based on the relatively very small intensity of energetic protons $48 < E_p < 63$ MeV and $63 < E_p < 160$ MeV in the Mimas gap, the calculated lifetime against absorption by this satellite, and the theoretical Crand source strength of Blake et al. In the gap in energetic proton intensity associated with the two small coorbital satellites 1979S1 (1979S2) (1980S3) and 1980S1 at $r \sim 2.5$ R$_S$, Van Allen (1982) finds an upper limit on the omnidirectional intensity of protons E_p > 80 MeV as 20 cm^{-2} s^{-1}, less by a factor of 10^3 than that at $r = 2.66$ R$_S$. This gives an upper limit on the source strength here of $\sim 1 \times 10^{-14}$ cm^{-3} s^{-1} inasmuch as the lifetime against absorption by each of the two satellites is of the order of 4×10^5 s (Rairden 1980).

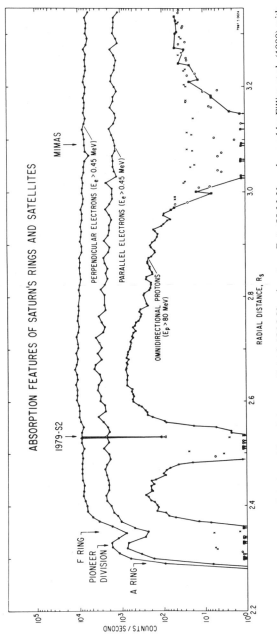

Fig. 2. Radial dependences of the intensities of electrons $E_e > 0.45$ MeV and protons $E_p > 80$ MeV as observed by Fillius et al. (1980) with Pioneer 11.

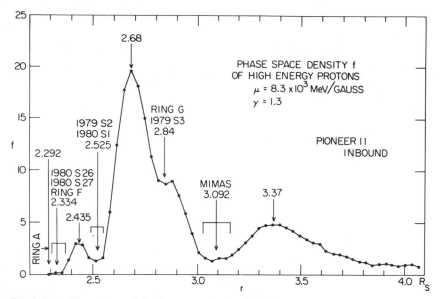

Fig. 3. Radial dependence of the phase-space density f (in arbitrary units) of high-energy protons at a constant, representative value of the first adiabatic invariant (figure adapted from Van Allen et al. 1980a).

The observed spectrum of the trapped energetic protons is consistent with that expected for a Crand source, though neither is well known. It is noted that nuclear interactions in Saturn's Rings A and B are estimated to be much more important sources of neutrons than those in its atmosphere (Fillius and McIlwain 1980; Van Allen et al. 1980a; Cooper and Simpson 1980; Blake et al. 1983) whereas at Earth those in the atmosphere are the sole source of outward flying neutrons (White 1973). In a recent paper, Cooper (1983) has made detailed numerical calculations of the source strength of Crand protons in Saturn's inner magnetosphere and has made substantial upward revisions of earlier estimates (cf. Cooper and Simpson 1980). Among other results he reports a residence time $T \sim 40$ yr, and by study of the relative intensities in the three inner zone peaks (cf. Fig. 3), he finds a diffusion coefficient $D \sim 10^{-15} L^9 R_S^2 \ \mathrm{s}^{-1}$ with L in units of R_S. At $L = 2.66$, his value of D is $7 \times 10^{-12} R_S^2 \ \mathrm{s}^{-1}$, which is similar to that quoted above.

III. PROTONS OF ENERGY $E_p \sim 1$ MeV

The radial distribution of protons having energies of the order of 1 MeV has been measured by a diversity of instruments on Pioneer 11, Voyager 1, and Voyager 2. Inbound and outbound curves of the spin-averaged absolute intensity of protons $0.61 < E_p < 3.41$ MeV are shown in Fig. 5. Angular

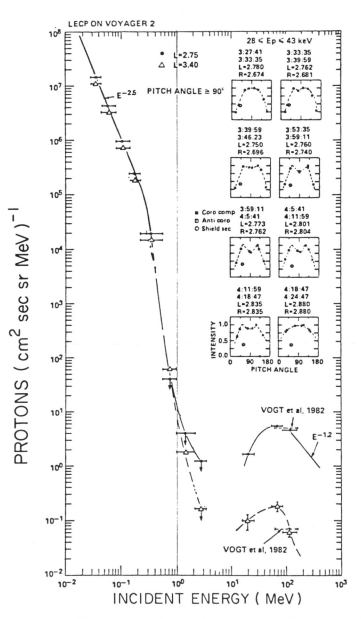

Fig. 4. Energy spectrum of ions (assumed to be protons) at $L = 3.40$ and 2.75 inside the orbits of Enceladus and Mimas, respectively. The Crand (cosmic-ray-albedo neutron decay) part of the spectrum at $E_p \gtrsim 16$ MeV is evident. Insets at right show development of pitch-angle distributions of protons $28 < E_p < 43$ keV as the spacecraft approached and crossed (~ 0418 UT) the ring plane. Dashed lines have been sketched in to guide the eye assuming symmetry about 90°. The open circles are the normalized counting rates when the detector is behind a 2 mm aluminum shield. These give a measure of the background (figure adapted from Krimigis and Armstrong 1982).

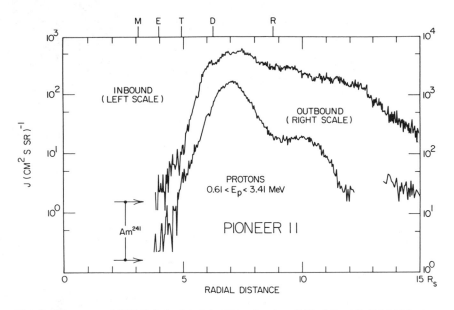

Fig. 5. Spin-averaged (115.5 s) absolute intensities of protons $0.61 < E_p < 3.41$ MeV for in-bound and outbound legs of Pioneer 11's encounter trajectory as a function of radial distance r (essentially equal to L). The horizontal arrows labeled Am^{241} show the equivalent proton intensity caused by the in-flight calibration source of α-particles. This contribution has not been subtracted in this presentation of the data (figure adapted from Van Allen et al. 1980a).

distributions have also been measured as have spectra in this general range of energies by several groups. The radial dependence of phase-space density f for protons having various values of the first adiabatic invariant μ as calculated by McDonald et al. (1980) is shown in Fig. 6. Families of $f(r)$ for various values of μ and the integral invariant K have been calculated by Rairden (1981) using University of Iowa intensity data $0.61 \le E_p \le 3.41$ MeV and Goddard Space Flight Center values of spectral indices. Sample results are shown in Fig. 7. Corresponding plots from Voyager 2 data are shown in Fig. 8.

In all three of Figs. 6, 7, and 8, it is seen that $\partial f/\partial r$ is essentially positive or zero for $r > 4$ R$_S$. It is clear that the radial distribution of protons $E_p \sim 1$ MeV is altogether different from that of protons of $E_p >$ few tens of MeV and hence that the two distributions are physically unrelated. Although it is conceivable that such a radial dependence of f for the lower-energy protons could be the result of internal sources, this is thought to be unlikely and the $f(r)$ curves provide evidence that the source of protons $E_p \sim 1$ MeV is at the magnetopause. There are two obvious possibilities for such a source: (a) the injection of quasi-thermalized solar wind protons, and (b) the capture of solar energetic particles from the interplanetary medium. Possibility (b) has

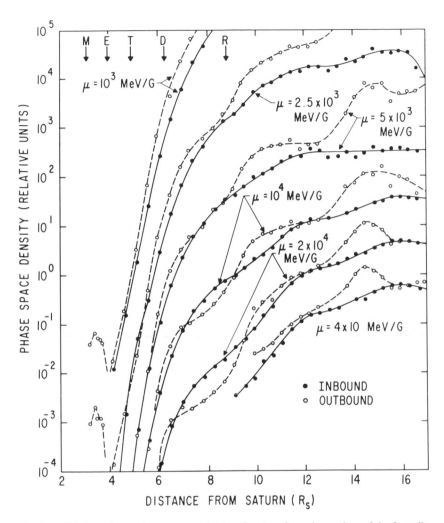

Fig. 6. Radial dependence of phase-space density of protons for various values of the first adiabatic invariant (McDonald et al. 1980).

a certain appeal because of the presence of enhanced intensities of solar energetic particles before, during, and after Pioneer 11's encounter. McDonald et al. (1980), Simpson et al. (1980b), and Vogt et al. (1981) present convincing evidence for this effect as being important in the outer magnetosphere. However, possibility (a) is the one characteristic of usual interplanetary conditions and, averaged over the inward diffusion time of such particles, is doubtless the more important for $L < 10$, as it is for the Earth's magnetosphere.

The other major feature of Figs. 6, 7, and 8 is the exhibition of strong

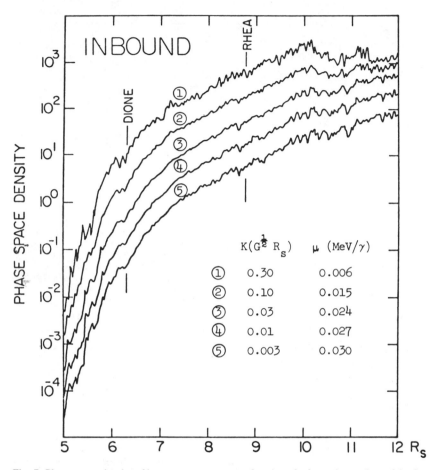

Fig. 7. Phase-space density of low-energy protons as a function of r for various values of the first adiabatic invariant μ and the integral invariant K (Pioneer 11 inbound) (figure from Rairden 1981).

losses of protons interior to $r \sim 10\ R_S$. Candidate causes of such losses are absorption by Rhea (8.739 R_S), Dione (6.258 R_S), Tethys (4.886 R_S), Enceladus (3.946 R_S), and the diffuse Ring E between 8 and 3 R_S (Baum et al. 1981), and pitch-angle scattering by wave-particle interactions in the ambient plasma. An analysis by Rairden (1981) reveals no clearly defined individual absorption signatures of Rhea, Dione, Tethys, or Enceladus but rather suggests that the losses are distributed with r. Also the angular distributions for $r < 10\ R_S$ show a greater relative depletion at equatorial pitch angles α_0 near 90° than at oblique pitch angles. The latter effect is indicative of partial absorption by satellites and dispersed ring material (Thomsen and Van Allen

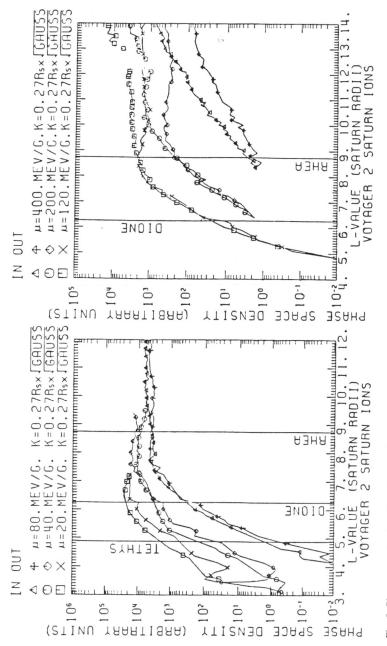
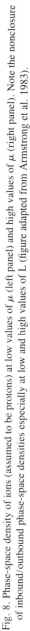

Fig. 8. Phase-space density of ions (assumed to be protons) at low values of μ (left panel) and high values of μ (right panel). Note the nonclosure of inbound/outbound phase-space densities especially at low and high values of L (figure adapted from Armstrong et al. 1983).

1979) and is opposite in nature to that caused by plasma wave-particle interactions (Lyons et al. 1972; Sentman and Goertz 1978). Hood (1981,1983) has
solved the lossy diffusion equation numerically for a composite model comprising partial absorption by Rhea, Dione, Tethys, Enceladus, and Ring E.
His treatment of the problem, as exemplified by sample results in Fig. 9, represents the current best state of the subject. A significant feature of the analysis is the "waterfall" effect of the overlapping absorptions of the successive
satellites. For protons having $\mu \approx 600$ MeV Gauss^{-1}, Hood (1983) reports
that $D(L) = 3$ to $7 \times 10^{-10} L^{3\pm1} R_S^2$ s^{-1} in the inner magnetosphere.

It now appears unlikely (Van Allen et al. 1980a; Krimigis and Armstrong 1982; Schardt and McDonald 1983) that there are significant intensities of protons $E_p > 0.6$ MeV interior to the orbit of Enceladus despite earlier reports to the contrary (Van Allen et al. 1980c; Simpson et al. 1980a;
McDonald et al. 1980).

IV. PROTONS OF ENERGY $0.028 < E_p < 0.5$ MeV

Krimigis and Armstrong's (1982) low-energy charged particle (LECP)
detectors on Voyagers 1 and 2 have yielded measurements of protons and
electrons in the energy range below that of detectors on Pioneer 11. A general
view of their Voyager 2 observations is given in Fig. 10. Absolute proton energy spectra from Voyager 1 data are given in Fig. 11 for $L = 4.26, 7.15$, and
7.83 and from Voyager 2 data, in Fig. 4 for $L = 2.75$ and 3.40. In the region
outside of the orbit of Enceladus the spectra of Fig. 11 yield values of spectral
indices γ in differential spectra of the form $j = K E^{-\gamma}$ of 5 to 8 as compared
to values of 3.8 to 4.8 from Pioneer 11 data (McDonald et al. 1980). Also the
integral unidirectional intensities J $(0.61 < E_p < 3.41$ MeV) inferred from
Fig. 11 differ from Pioneer 11 values shown in Fig. 5 by about an order of
magnitude. Such discrepancies appear to be irreconcilable except on the general presumption of temporal variation.

At energies $0.028 < E_p < 0.5$ MeV the LECP data are unique and no
independent confirmation is available. Such protons are found in abundance
interior to the orbits of both Enceladus and Mimas (Fig. 4). It is unlikely that
protons from an external source in this energy range have diffused across the
orbits of Enceladus and Mimas in the observed intensities. Hence, it appears
that these particles arise from internal sources and may be the high-energy tail
of the hot plasma observed in this region (Bridge et al. 1981a,1982; Sittler et
al. 1981). If so, they represent a third component of the energetic proton
population in the inner magnetosphere—the first being Crand protons $E_p >$
tens of MeV and the second being protons $E_p \sim 1$ MeV diffused inward from
the magnetopause as discussed in the previous section. No physical mechanism for the local acceleration of protons from quasi-thermal energy to ~ 0.1
MeV has been identified.

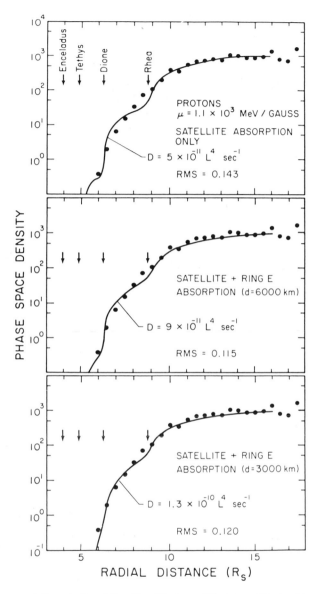

Fig. 9. A representative set of modeling fits of the lossy diffusion equation to observational values of the phase-space density of low-energy protons (Hood 1983).

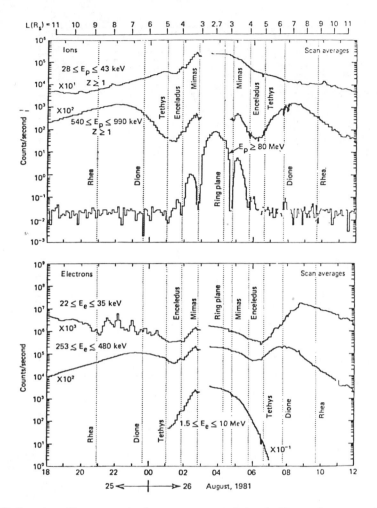

Fig. 10. Overview of inner magnetosphere measurements during the Voyager 2 encounter. The upper panel shows ion intensities in the indicated energy intervals, while the lower panel shows sample electron intensities. The times of crossings of satellite L shells, calculated on the basis of a centered, aligned dipole are shown by vertical dotted lines; R = Rhea, D = Dione, T = Tethys, E = Enceladus, M = Mimas. The data are scan averages, either 6 min 24 s or 48 s. The rates are corrected for background, an important consideration inside the orbits of Enceladus and Mimas. Data are not shown inside the orbit of Mimas for the 0.54 MeV protons (middle curve of the upper panel) because only an upper limit could be established due to the intense background, as indicated by the $\geq$ 80 MeV proton curve (bottom curve in the upper panel). The break between 0255–0326 UT is due to unavailability of appropriate background measurements for correction (figure courtesy of S. M. Krimigis).

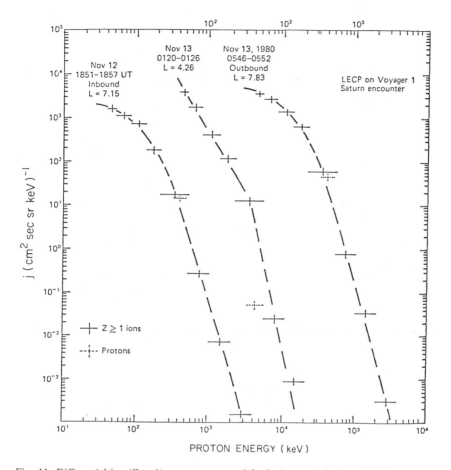

Fig. 11. Differential ion ($Z \geq 1$) spectra, corrected for background, for $L = 4.26$, 7.15, and 7.83. The ions are assumed to be protons. Points are plotted at the geometric mean of a particular passband (i.e., $E = \sqrt{E_1 E_2}$, where E_1 is the lower threshold and E_2 is the upper threshold of a particular channel). The horizontal bars indicate the channel width, while the vertical bars mark the location of the geometric mean and are not indicative of statistical errors. Statistics are typically better than 10% even for the highest energy channels. The dashed data points give the proton intensity (figure courtesy of S. M. Krimigis).

V. IONS HEAVIER THAN PROTONS

Wolfe et al. (1980) found quasi-thermal ions having a mass to charge ratio of 16 ± 4 (O^+ or OH^+, for example) in Saturn's inner magnetosphere at $L \sim 6.3$. A comprehensive analysis of the Pioneer 11 plasma data has been conducted by Frank et al. (1980). Outside of the orbit of Rhea (8.8 R_S), protons of typical number density 0.5 cm^{-3} and temperature 10^6 K dominate the distribution. In the radial range 4 to 7.5 R_S, a large torus of O^{++} and O^{+++} ions is found; maxima in the radial distribution are observed at the orbits of Dione (6.2 R_S) and Tethys (4.9 R_S) with number densities $\approx$ 50 cm^{-3}. Ion temperature increases from 2×10^5 K at 4 R_S to 5×10^6 K at 7.3 R_S. The authors of this work suggest that the heavy ions are dissociation products of water frost on the surfaces of satellites and Ring E particulates.

Data from the plasma instruments on Voyager 1 (Bridge et al. 1981a) and Voyager 2 (Bridge et al. 1982) confirm the existence of the Dione/Tethys torus of oxygen ions and add a wealth of data on the distribution of plasma in Saturn's magnetosphere.

In the specific energy range $\sim$ 1 MeV per nucleon, Simpson et al. (1980b) find the ratio of proton to helium ion abundances in equal energy per nucleon intervals to be $\sim$ 55 in the inner magnetosphere.

From an independent body of Pioneer 11 observations, McDonald et al. (1980) find that helium ions with energies $>$ 3 MeV per nucleon are present at distances $>$ 6 R_S. At greater radial distances the H/He ratio is 40 $\pm$ 11 in good agreement with the value observed in interplanetary space during the solar energetic particle event that was in progress during the Saturn encounter. These results confirm other evidence for the penetration of solar energetic particles into at least 10 R_S. These authors also find energetic ions having atomic number $Z > 2$; at about 6 R_S the proton to ion ($Z > 2$) ratio is 525 $\pm$ 90 for energies of 0.65 MeV per nucleon.

Krimigis et al. (1982a) find that at specific energies $\geq$ 0.2 MeV per nucleon, H_2 and H_3 (molecular ions), as well as atomic hydrogen, helium, carbon, and oxygen, are important constituents of the population of energetic particles in Saturn's magnetosphere. The presence of the molecular ions suggests an ionospheric source but the relative abundances of helium, carbon, and oxygen ions are consistent with a solar wind source. A sample energy spectrum of ions is given in Fig. 12.

It appears from the foregoing that energetic (specific energy $\geq$ 0.1 MeV per nucleon) ions of atomic and molecular hydrogen, helium ions, and $Z > 2$ atomic ions are indeed present in the inner magnetosphere in measurable intensities. Such ions are, of course, in addition to the much greater number densities of quasi-thermal ions. The interrelationships of all of the suggested sources (cf. Sec. I) and the various components of the particle population and the physical causes of acceleration are discussed cogently by Hamilton et al. (1983).

VI. ELECTRONS OF ENERGY $0.035 < E_e <$ A FEW MeV

The sharply defined trapping boundary at the sunward position of the magnetopause for electrons $E_e > 0.040$ MeV is illustrated in Fig. 13. The respective and quite different radial dependences of the intensities of electrons $E_e > 0.040$ and $E_e > 0.56$ MeV are shown for $r < 15$ R_S in Fig. 14. The ra-

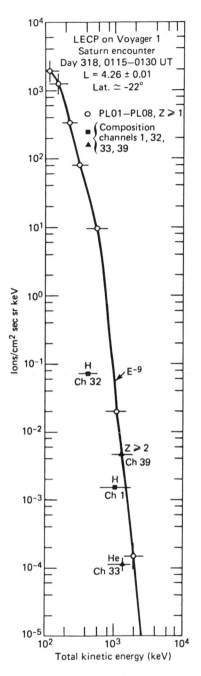

Fig. 12. Differential ion spectrum averaged over the indicated time interval from Voyager 1. Channels PL01–PL08 (open circles) are assumed to be responding to oxygen and are plotted at the appropriate total-energy passband. Channels 1 and 32 (squares) respond to protons only, while 33 (triangle) to He only; channel 39 ($Z \geq 2$) is plotted at the appropriate oxygen pass-band. All points have been corrected for background. Error bars on the two lowest energy channels reflect uncertainties in absolute oxygen efficiency rather than poor statistics (Krimigis et al. 1983).

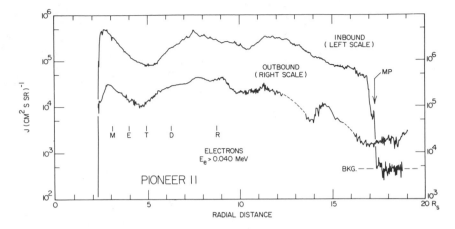

Fig. 13. Spin-averaged (115.5 s) absolute integral intensities of electrons $E_e > 0.040$ MeV for inbound and outbound legs of Pioneer 11's encounter trajectory as a function of radial distance r. The label MP means magnetopause and the labeled vertical ticks M, E, T, D, and R show radii of respective satellite orbits (figure adapted from Van Allen et al. 1980a).

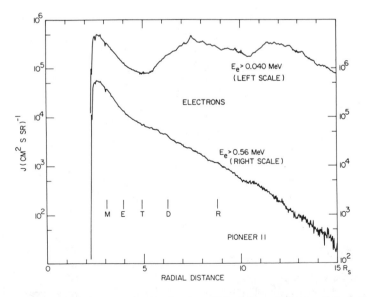

Fig. 14. Similar to Fig. 13 but showing absolute integral intensities of electrons $E_e > 0.040$ and $E_e > 0.56$ MeV (Pioneer 11 inbound) (figure adapted from Van Allen et al. 1980a).

dial dependence of the ratio of the two components is shown in Fig. 15. The ratio of the integral intensities

$$\frac{J\,(0.040 < E_e < 0.56 \text{ MeV})}{J\,(E_e > 0.040 \text{ MeV})} \tag{1}$$

approaches zero as r decreases inward from $\sim$ 10 R_S, and is zero, with an uncertainty of $\sim$ 10%, for $r <$ 5 R_S. A sample radial dependence of phase-space density f of electrons having a first adiabatic invariant μ = 525 MeV Gauss^{-1}, a spectrum of the form $j = k\,E^{-3.8}$ for $E > E^*$ and $j = 0$ for $E < E^*$, where E^* is the kinetic energy corresponding to a low cutoff $\mu^* = 525$ MeV Gauss^{-1}, is given in Fig. 16. Sample values of E^* are (a) in the solar wind, 0.0016 MeV; (b) at $r = 10$, 0.099 MeV; (c) at $r = 5$, 0.56 MeV; and (d) at $r = 3$, 1.59 MeV.

On the basis of Fig. 16 and of the above cited spectral evidence, Van Allen et al. (1980a) conclude that the source of electrons $0.040 < E_e <$ a few MeV is external, presumably thermalized solar wind in the magnetosheath; that diffusion into $r \sim 10$ R_S is essentially loss-free and source-free; and that at lesser radial distances there are strong losses. The latter are attributed to absorption by particulate matter in Ring E and by the large inner satellites Rhea, Dione, Tethys, Enceladus, and Mimas. The inner boundary condition on $f(r)$ is zero at the outer edge of Ring A, though $f(r)$ has already reached a low value at $r \sim 5$ R_S. No clear evidence for inner zone electrons from the decay of Crand neutrons (upper energy limit 0.78 MeV for a neutron decaying at rest) was found.

The inner satellites and rings of Saturn are central to the physics of its magnetosphere in at least three different ways: (a) as inert absorbers of energetic particles; (b) as sources of secondary particles; and (c) as sources of gas (neutral and/or partially ionized) and dust (Frank et al. 1980; Cheng and Lanzerotti 1978; Cheng et al. 1982; Lanzerotti et al. 1983a). Their role as absorbers of energetic particles provides a diagnostic capability in identifying the sources and sinks of such particles and in understanding their radial and pitch-angle diffusion. Examples have been given in preceding sections.

The motion of electrons in Saturn's radiation belts is such that absorption by solid bodies moving in Keplerian orbits is notably absent under so-called synchronous or resonant conditions. This is seen as follows. For any given radial distance from the center of the planet (i.e., for any given L shell), electrons having a particular energy and pitch angle have the same angular velocity of longitudinal drift as the Keplerian angular velocity of a satellite. Table I tabulates such resonant energies for a number of cases of interest (Thomsen and Van Allen 1980). Electrons having resonant energies diffuse across the orbit of the relevant satellite as though it were not there, whereas electrons of both lesser and greater energy make periodic encounters with the satellite (Fig. 17) and suffer a calculable probability of absorption at each en-

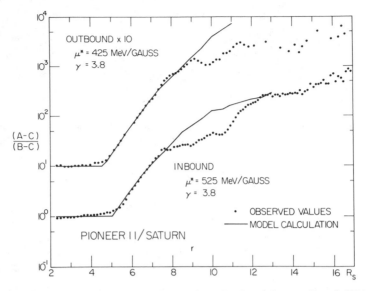

Fig. 15. Ratios of intensity of electrons $E_e > 0.040$ MeV to that of electrons $E_e > 0.56$ MeV as a function of radial distance (plotted points) and a model fit (solid line) (Van Allen et al. 1980a).

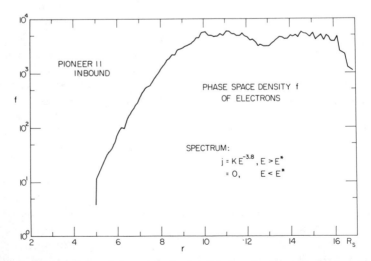

Fig. 16. Radial dependence of phase-space density of electrons having constant value of the first adiabatic invariant $\mu = 525$ MeV Gauss^{-1} and a low cutoff at the same value of μ (Van Allen et al. 1980a).

TABLE I
Resonant Energies E_R of Electrons [a]

Satellite	Semimajor axis of orbit (R_S)	Equatorial Pitch Angle (deg)	E_R (MeV)
1979S2	2.510	90	0.877
		60	0.924
		30	1.064
Mimas (S I)	3.075	90	1.087
		60	1.146
		30	1.318
Enceladus (S II)	3.946	90	1.082
		60	1.140
		30	1.312
Tethys (S III)	4.886	90	0.971
		60	1.024
		30	1.179
Dione (S IV)	6.258	90	0.804
		60	0.848
		30	0.977
Rhea (S V)	8.739	90	0.588
		60	0.620
		30	0.714

[a] Table adapted from Thomsen and Van Allen (1980) using the following values of basic constants: 1 R_S = 60,330 km; magnetic moment of Saturn = 4.61×10^{28} G cm³; angular velocity of Saturn = 1.638×10^{-4} rad s⁻¹.

counter (Van Allen et al. 1980*b*; Rairden 1980). Inward diffusion subjects electrons to a succession of band-pass energy filters. Between filters (i.e., between satellite orbits) the energies of the surviving particles increase in accordance with the conservation of μ. It is seen from Table I that the differential spectrum of electrons develops, by the orbit of Enceladus, a maximum value at $E_e \sim 1$ MeV by virtue of this sweeping process which causes a relative depletion of particles having either a lesser or a greater energy. Figure 18 shows the integral spectrum of electrons at $L = 3.1$ as assembled from the published literature by Chenette and Stone (1983). Differentiation of the solid curve in Fig. 18 yields a spectral maximum at $E_e = 1.5$ MeV and a half-width at $1/e$ of the maximum of ~ 0.7 MeV. The corresponding values inferred by Van Allen et al. (1980*b*) from a variety of considerations are 1.59 and 0.1 MeV, respectively. It is clear from the data of Fig. 18 that the apparent disagreement in width of the spectral peak is not significant. Differentiation of Chenette's dashed curve yields a spectral maximum at $E_e = 2.9$ MeV, a value that is decidedly incompatible with the inferences of Van Allen et al.

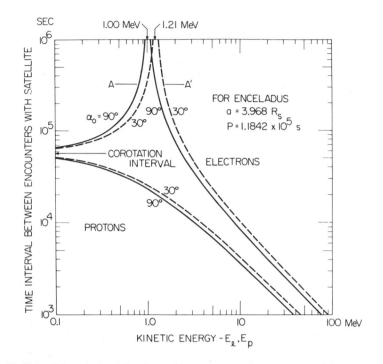

Fig. 17. Plots of the calculated time interval between successive encounters of electrons (upper two curves) and protons (lower two curves) with Enceladus as a function of kinetic energy of the particles. Of particular interest is the resonant, or synchronous, energy for electrons, $E_e =$ 1.00 MeV for $\alpha_0 = 90°$ and 1.21 MeV for $\alpha_0 = 30°$ (cf. Table I herein using a slightly different magnetic field model than that used by the original authors). Protons experience no such resonance at any energy. The horizontal line AA' illustrates a sample diffusion time past Enceladus so as to result in a narrow band-pass filtering effect on the spectrum of electrons diffusing inward across its orbit (1 R_S = 60,000 km in this figure) (figure from Van Allen et al. 1980b).

The unimpeded inward diffusion of electrons having energies at or near values of the resonant energy at successive satellites is apparently responsible for the survival of high intensities of energetic electrons ($E_e \sim 1$ MeV) inwards to the outer edge of Ring A whereas protons $E_p \sim 1$ MeV, which enjoy no such resonances, are essentially absent for $L < 4$. Nonetheless, absorption of electrons by the large inner satellites Rhea, Dione, Tethys, Enceladus, and Mimas and (to a much lesser degree) by 1979S2 and 1980S1 and Rings G, E, and F reduces the intensity of electrons by a factor of at least 10^3 below the value that would exist if these absorbers were absent. Similar statements are also applicable to both high-energy and low-energy protons (cf. Secs. II and III).

Inasmuch as the lifetimes of electrons against absorption by satellites are known, it should be possible to find absolute values of diffusion coefficients

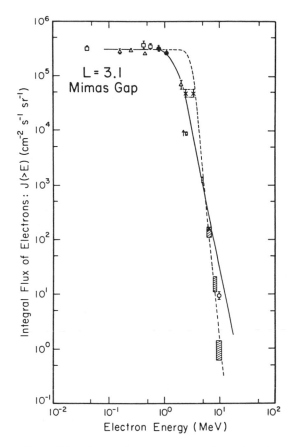

Fig. 18. An assemblage of data from various sources on the integral spectrum of electrons at
$L = 3.1$. The shaded boxes are data calculated from the cosmic ray instrument and the open
circle is from Krimigis et al. (1982a). Both of these data sets were obtained by instruments on
Voyager 2 in August 1981. Open squares are from Van Allen et al. (1980a); open triangles are
from Fillius and McIlwain (1980); solid circles are from McDonald et al. (1980); and crosses
are from Simpson et al. (1980b). These latter observations are from the Pioneer 11 spacecraft
in September 1979. Error bars assigned to points reflect the variability in the counting rates
during both inbound and outbound passes. The dashed and solid lines indicate model spectra
chosen as limiting cases of the true spectrum (Chenette and Stone 1983).

and their radial dependence in a manner analogous to that used by Hood
(1983) for protons. The analysis for electrons is more complex than that for
protons because of the strong energy dependence of the absorption lifetimes
for electrons; such an analysis has not yet been done.

The inner end of the radial dependence of the phase-space density of
nearly monoenergetic electrons having $\mu = 600$ MeV Gauss^{-1} is exhibited in
Fig. 19. The portion of the $f(r)$ curve interior to the orbit of Mimas is well fit

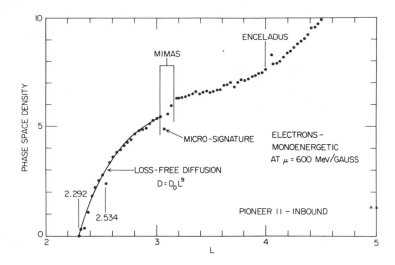

Fig. 19. Radial dependence of phase-space density of monoenergetic electrons having $\mu =$ 600 MeV Gauss^{-1} for $L < 5$ (Pioneer 11 data by Van Allen).

(with the exception of the microsignature of 1979S2) by a loss-free diffusion equation with $D = D_0 L^9$. A less satisfactory fit results if the exponent of L differs from 9 by $\pm$ 1. It is noted that this value of the exponent is markedly different from the values, 2 to 4, that are favored by Hood's analysis for protons (cf. Sec. I).

The low-energy charged particle (LECP) instruments on Voyagers 1 and 2 provide electron energy spectra having greater detail (Krimigis et al. 1981,1982*b*) than do those available from Pioneer 11 data. The radial dependences of the intensities of electrons in six different energy ranges are shown in Fig. 20. From these curves differential spectra are constructed for different L values. Examples of the latter at $L = 9.56, 6.8, 5.34, 4.4, 3.6,$ and 2.75 are shown in Fig. 21. At 9.56, the absolute differential energy spectrum is

$$j = 3.2 \times 10^3 \, E_e^{-2.96} \, (\text{cm}^2 \text{ s sr MeV})^{-1} \qquad (2)$$

with E_e in MeV. The corresponding integral intensities $J (E_e > 0.040 \text{ MeV})$ and $J (E_e > 0.56 \text{ MeV})$ differ by factors of the order of unity from the Pioneer 11 values. At progressively lesser values of L, the Voyager 2 spectra exhibit features at their high-energy ends that are qualitatively similar to that inferred from the Pioneer 11 data as supported by the resonant band-pass theory. However, the Voyager data exhibit an additional low-energy component having $E_e < 0.5 \text{ MeV}$. It appears that such lower-energy electrons somehow evade satellite absorption by virtue of a large diffusion coefficient *or*, more likely, that they have a local, internal source. Their presence is in mild conflict with

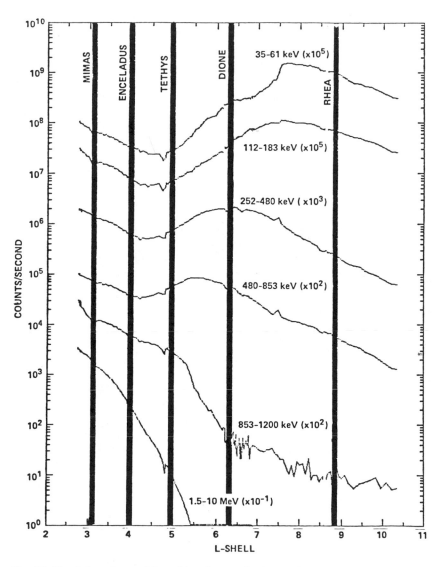

Fig. 20. The *L*-dependence of intensities of energetic electrons in several energy ranges, observed during the outbound trajectory of Voyager 2. Note the shift in peak intensity towards lower *L* values as the energy increases. The nominal *L* shells of major satellites are marked (Krimigis et al. 1982*a*).

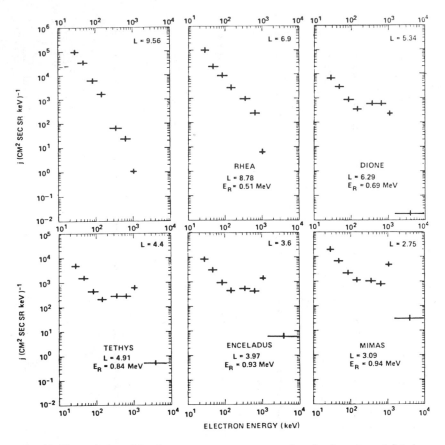

Fig. 21. The evolution of the electron energy spectrum at various L values, upper left to lower right as registered by LECP on Voyager 2. The L values are indicated in each panel, and the resonant electron energy E_R at each satellite is shown in successive panels (cf. Table I herein for recalculated values of E_R) (figure from Krimigis et al. 1982a).

the Pioneer 11 evidence, but the accuracy of the latter body of data is such that their absolute intensities can be tolerated in the low-energy end of the integral spectrum as shown in Fig. 18. No physical source of these lower-energy electrons has been yet identified or discussed in quantitative terms.

In view of the importance that synchrotron radiation has had in the history of our knowledge of Jupiter's radiation belt, it is of interest to use observed data on energetic electrons at Saturn to understand why no such radiation from Saturn has been observed.

Using a simplified model of the Pioneer 11 distribution of energetic electrons in Saturn's magnetosphere, Grosskreutz (1980) has calculated the synchrotron radiation thereof. The radiative lifetime of a 2 MeV electron at 2.64

R_S is $\sim 2.4 \times 10^4$ yr and the synchrotron spectrum has a maximum intensity at $\nu_{max} \sim 318$ kHz. The estimated flux density at ν_{max} in the equatorial plane of the planet is 2.3×10^{-28} W m^{-2} Hz^{-1} at a distance of 10 AU, far below any reasonable prospect of detection by an Earth-orbiting receiver. Even at a distance of 20 R_S from the planet, the estimated flux density is only 3.5×10^{-22} W m^{-2} Hz^{-1}, about an order of magnitude below the threshold detection level of the Voyager radio astronomy receiver. In fact, no such radiation was reported by Voyagers 1 or 2 (Warwick et al. 1981,1982*a*).

VII. ABSORPTION OF ENERGETIC PARTICLES BY INNER SATELLITES AND RINGS

This subject has been reviewed in recent papers by Van Allen (1982, 1983). The sketches in Fig. 22 illustrate the observable consequences of the presence of a satellite in a radiation belt of magnetically trapped charged particles. The time dependence of the diffusive fill-in of the particle shadow that leads or trails a satellite is illustrated in Fig. 23 as a function of the dimensionless parameter

$$\tau \equiv 4 \, Dt/b^2 \tag{3}$$

where D is the diffusion coefficient, t is the lapse of time from passage of the causative satellite through the distribution of particles, and b is the radius of the satellite, regarded as an inert absorbing body.

The radial traversal of a satellite's shadow yields an observed absorption signature. A time-dependent [$\partial(\text{intensity})/\partial t \neq 0$] signature that is localized in longitude (Fig. 22) is called a microsignature. The time-averaged or longitudinally averaged effect on the radial distribution of trapped particles is called the macrosignature of the causative satellite. There is no microsignature of a longitudinally uniform distribution of dispersed particulate material in the form of a ring. But there is, of course, a macrosignature. A longitudinally localized distribution of particulate matter masquerades as a satellite and is difficult or impossible to distinguish from a satellite by simple forms of the technique for observing particle absorption. (The above theorems from Van Allen [1982] are for the axially symmetric case, i.e., zero orbital eccentricity for a satellite or ring and axial symmetry for the magnetic field. Important modifications result for nonaxially symmetric situations. The author is developing the consequences of these modifications but does not yet have them in suitable form for the present review.)

The clearest case of the observation of a fresh microsignature of a satellite by Pioneer 11 is shown in Fig. 24. This satellite was initially designated 1979S2. Later it was shown that it is identical to 1979S1, found by the Pioneer imaging team (Gehrels et al. 1980) on the preceding day. In turn, it appears that this is one of the two (or more) satellites photographed by

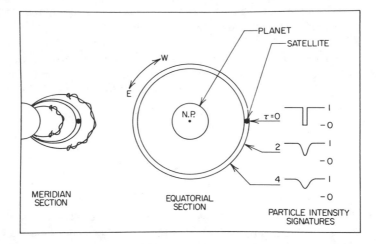

Fig. 22. Sketches illustrating the absorption by a planetary satellite of energetic charged particles trapped in the planet's magnetic field. The meridian section on the left shows the spiraling motion of charged particles between mirror points in opposite hemispheres and a tube of magnetic lines of force through the satellite (filled circle) that is depleted of particles. The equatorial section on the right shows the types of absorption signatures that may be observed as a spacecraft crosses the satellite's orbit at three different longitudes, i.e., as a function of the dimensionless parameter τ (Van Allen 1982).

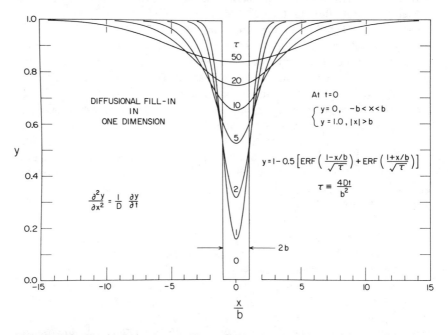

Fig. 23. A family of calculated curves illustrating the time dependence of the one-dimensional diffusive fill-in of the charged particle shadow of a satellite of radius b for particles whose radius of curvature is much less than b (Van Allen et al. 1980b).

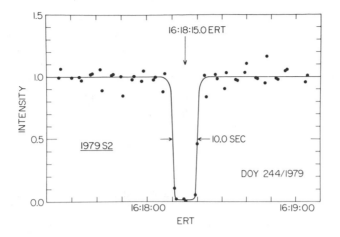

Fig. 24. The normalized absorption microsignature of satellite 1979S2 in electrons of energy E_e ~ 1.5 MeV (Pioneer 11, inbound). The radial component of the velocity of the spacecraft was 16.8 km s^{-1} (Van Allen 1982).

groundbased observers in 1966 (Dollfus 1967a; Aksnes and Franklin 1978; Fountain and Larson 1978; Larson et al. 1981b) and later observed, again by groundbased observers (Seidelmann et al. 1981; Dollfus and Brunier 1981), and comprehensively by the Voyager 1 imaging team (B. A. Smith et al. 1981; Synnott et al. 1981). The satellite provisionally designated by the latter authors as 1980S3 has been shown to be identical to 1979S2.

A charged particle absorption microsignature at the orbit of Mimas was discovered by Simpson et al. (1980a,b) and confirmed by Van Allen et al. (1980b). The first group of authors interpreted it as the absorption by a companion satellite of Mimas (or a longitudinally localized distribution of dust) trailing Mimas in a similar orbit by a longitude difference of $\sim 56°$ (near the trailing triangular Lagrangian point $L5$ of the Mimas-Saturn system). The second group of authors gave an altogether different interpretation, namely that the absorption feature was the microsignature of Mimas itself. A different microsignature at a radial distance near that of Mimas' orbit was observed by Vogt et al. (1982) in Voyager 2 data and attributed to a companion satellite of Mimas, but not one at the Lagrangian $L5$ point. The analysis of Chenette and Stone (1983) suggests that this microsignature could not be caused by Mimas itself but possibly by a longitudinally localized dust cloud coorbiting with Mimas. Carbary et al. (1983a) observed the same signature in the electron channels and found that the intensity decrease was also present in the low-energy ion channels. Chenette and Stone estimate that a lower limit on the diffusion coefficient for MeV electrons at $L = 3.1$ is $D > 10^{-8}\ \mathrm{R_S^2\ s^{-1}}$, based on their conclusion that there is no Mimas absorption signature in their data.

Absorption microsignatures have also been observed in apparent association with Rhea, Dione, Tethys, and Enceladus (Vogt et al. 1981,1982; Krimigis et al. 1982a; Carbary et al. 1983a). An example of the filling in of a Dione microsignature in electron intensity is shown in Fig. 25 (Carbary et al. 1983a). The fits (see Fig. 23) for the two lowest energy channels give values of the diffusion coefficient $D = 5 \times 10^{-8}$ R$_S^2$ s^{-1} for $\tau = 10$ and $D = 3 \times 10^{-7}$ R$_S^2$ s^{-1} for $\tau = 50$. But the value of D is clearly much greater for electrons $0.070 < E_e < 0.13$ MeV inasmuch as no absorption is perceptible at these slightly higher energies.

A microsignature of a rather puzzling nature was observed near the orbit of Enceladus (Krimigis et al. 1982a; Carbary et al. 1983a). The authors favor the interpretation that this absorption feature is attributable to clumps of finely divided material in Ring E—such as have been observed in Ring F—and not to Enceladus itself.

An absorption microsignature of Rhea was observed by Voyager 1 (outbound) at $L = 8.52$ (*vice* the expected value of 8.8) (Vogt et al. 1981). Also a clear absorption microsignature of Tethys (Fig. 26) was observed by Voy-

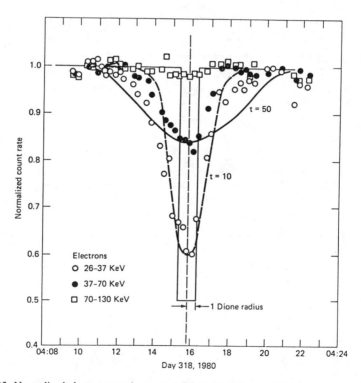

Fig. 25. Normalized electron counting rates at Dione. The Voyager 1 data were normalized by taking ratios of observed rates to those obtained by linearly extrapolating across the microsignature. The two theoretical fits to the data were taken from the equation shown in Fig. 23 (Carbary et al. 1983a).

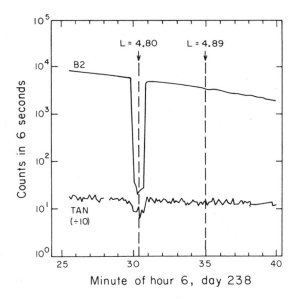

Fig. 26. Counting rates of electrons in nearly unshielded (B2) and shielded (TAN) detectors plotted against spacecraft event time near the orbit of Tethys as Voyager passed within 1° longitude of that satellite. The width of the absorption feature at 0630 corresponds to ~ 1100 km which is nearly equal to the 1050-km diameter of Tethys. The vertical dashed lines show the expected ($L = 4.89$) and observed ($L = 4.80$) coordinates of the absorption feature as calculated from a centered, untilted dipole model of Saturn's magnetic field. If the effect of the ring current suggested by Voyager 1 observations had been included in this calculation, the absorption feature would have been expected at 0632 (Vogt et al. 1982).

ager 2 as the spacecraft passed (outbound) within 1° of longitude of that satellite (Vogt et al. 1982), though again the distortion of the magnetic field by the equatorial ring current caused a radial displacement from the expected value of $L = 4.89$ to an apparent value of $L = 4.80$. Krimigis et al. (1982a) also observed a deep microsignature of Tethys.

A list of all microsignatures observed in association with Saturnian satellites is given in Table II. The implications of satellite microsignatures on models of the Saturnian magnetic field are discussed in the chapter by Connerney et al.

VIII. DIFFUSION COEFFICIENTS

Fundamental properties of a planetary magnetosphere in a known magnetic field are the sources and sinks of all charged particles therein (including, of course, the thermal and quasi-thermal plasma) and the mechanisms of diffusion by fluctuating electric and magnetic fields.

An idealized set of observations for determining the diffusion coefficient

TABLE II
Observed Microsignatures Caused by or Associated with the Orbits of Satellites as Listed

Satellite	Spacecraft	Relative Position of Spacecraft [a] (deg)	References
1979S2	P11/IN	−1.0 2.2 ± 2.3 W	Van Allen et al. 1980c; Van Allen 1982; Simpson et al. 1980a, 1980b; Fillius et al. 1980.
Mimas	P11/IN	+0.3 56.3 W	Simpson et al. 1980b; Van Allen et al. 1980b.
	V2/OUT	−6 147 E	Vogt et al. 1982; Carbary et al. 1983a; Chenette and Stone 1983.
Enceladus	V2/IN	+27 20 W	Carbary et al. 1983a; Vogt et al. 1982.
Tethys	V2/OUT	−19 ± 1	Vogt et al. 1982; Krimigis et al. 1982a; Carbary et al. 1983a.
Dione	V1/OUT	0 25 E	Carbary et al. 1983a.
	V2/IN	+27 93 E	Carbary et al. 1983a.
Rhea	V1/OUT	+8 3 E	Vogt et al. 1981; Carbary et al. 1983a.

[a] All satellites are assumed to be in the equatorial plane. The upper number in this column is the latitude (+N, −S) of the spacecraft at the time of the observed signature; the lower number is the longitude of the spacecraft either east (E) or west (W) of the satellite at that time.

of a particular species of particle of particular energy and at a particular radial distance would consist of a set of absorption microsignatures as a function of longitude from a particular satellite at that radial distance. The wealth of satellites (and rings) in Saturn's inner magnetosphere provides a splendid basis for such a program of diagnosis. The desired program of observations is within the capability of a spacecraft in a suitably eccentric, near-equatorial orbit about the planet. Then using (a) calculable values of the angular drift rate in longitude of the charged particles relative to the satellite and the absorption probability per encounter; (b) the radius of the satellite; and (c) observed values of the parameter τ, one can employ the analysis represented by Fig. 23 to determine D.

Meanwhile, the three hyperbolic encounter trajectories of Pioneer 11,

TABLE III
Some Provisional Values of Radial Diffusion Coefficients

Particle Species and Energy (or Magnetic Moment)	L	$\overset{D}{(R_S^2 \, s^{-1})}$	References
Protons $0.5 < E_p < 1.8$ MeV	3.1	$< 1 \times 10^{-8}$	Simpson et al. 1980a
Protons $E_p > 80$ MeV	3.1	$< 6 \times 10^{-7}$	Fillius et al. 1980
	2.5	$< 1 \times 10^{-7}$	
Protons $E_p > 80$ MeV	2.7	$1.3 \times 10^{-11} < D < 2 \times 10^{-10}$	Van Allen 1983
Electrons $E_e \sim 1$ MeV	4.0	$\sim 1 \times 10^{-10}$	Van Allen et al. 1980b
Protons $E_p \sim 1$ MeV	3.1	$\sim 4 \times 10^{-9}$ (episodic)	Simpson et al. 1980b
Electrons $E_e \sim 10$ MeV	3.1	$\sim 4 \times 10^{-8}$ (episodic)	Simpson et al. 1980b
Electrons $0.026 < E_e < 0.070$ MeV	6.3	$5 \times 10^{-8} < D < 3 \times 10^{-7}$	Carbary et al. 1983a
Electrons $\mu = 80$ MeV G^{-1}	6.3	$> 3 \times 10^{-7}$	Armstrong et al. 1983
Electrons $E_e \sim 0.050$ MeV	8.8	$\sim 4 \times 10^{-8}$	Krimigis et al. 1981
Electrons $E_e \sim 1.5$ MeV	2.5	1.5×10^{-10}	Van Allen 1982

TABLE III (continued)

Some Provisional Values of Radial Diffusion Coefficients

Particle Species and Energy (or Magnetic Moment)	L	D ($R_S^2\,s^{-1}$)	References
Protons $\mu \sim 600$ MeV G^{-1}	4 to 15	3 to $7 \times 10^{-10}\,L^{3\pm1}$	Hood 1983
Electrons $\mu = 600$ MeV G^{-1}	2.3 to 3.0	$D_0 L^{9\pm1}$	Van Allen, this chapter
Electrons $E_e \sim 1$ MeV	3.1	$> 1 \times 10^{-8}$	Chenette and Stone 1983
Protons $E_p > 30$ MeV	2.3 to 3.9	$10^{-15}\,L^9$	Cooper 1983

Voyager 1, and Voyager 2 provide a good beginning. Provisional, and disparate values of miscellaneous diffusion coefficients are collected in Table III. Some of the disparities may be attributable to true time variability because the time required for the diffusive fill-in of a satellite's microsignature is only of the order of an hour and the relevant state of fluctuating magnetic and electric fields may vary from one observed episode to another. For this reason diffusion coefficients derived from the general nature of time-averaged data may be more broadly significant. In either case, it is clear that a critical study of existing data is required before any coherent interpretation of particle diffusion in Saturn's magnetosphere can be developed.

Acknowledgments.   The author is indebted to S. M. Krimigis for discussion of observations with Voyagers 1 and 2 and to A. J. Dessler for critical reading of the manuscript.

MEASUREMENTS OF PLASMA, PLASMA WAVES, AND SUPRATHERMAL CHARGED PARTICLES IN SATURN'S INNER MAGNETOSPHERE

F. L. SCARF
TRW Space and Technology Group

L. A. FRANK, D. A. GURNETT
University of Iowa

L. J. LANZEROTTI
Bell Laboratories

A. LAZARUS
Massachusetts Institute of Technology

and

E. C. SITTLER, JR.
NASA Goddard Space Flight Center

The Pioneer 11 and Voyager 1, 2 traversals of Saturn's inner magnetosphere provided direct information on the complex and highly structured distributions of plasma and suprathermal charged particles present in this region. The Voyager wave instruments also yielded absolute electron density measurements in certain inner magnetosphere locations; the wave data were used to evaluate the magnitudes of several wave-particle interactions. We summarize the plasma and wave measurements for 24-hr periods centered around closest approach, evaluate pitch-angle scattering effects possibly associated with measured whistler mode turbulence, and speculate on the effects of wave-particle interactions associated with electrostatic waves.

Within a two-year period extending from September 1979 to August 1981 Saturn was visited by three spacecraft: Pioneer 11, Voyager 1, and Voyager 2. During short intervals associated with each encounter, the three spacecraft transmitted back to Earth local measurements that provide us with essentially all that is now known about the structure and dynamics of Saturn's magnetosphere. Some supplementary magnetospheric information of significance was obtained during the two Voyager approaches to the planet as the planetary radio astronomy investigations measured detailed characteristics of escaping Saturn radio emissions. However, these Saturn emissions are of such low intensity that remote sensing from Earth or from Earth orbit yields no magnetospheric information at all. Thus, the spacecraft *in situ* measurements from Pioneer and Voyager are of particular scientific importance.

While the Pioneer and Voyager encounters with Saturn supplied a wealth of definitive new knowledge, they have also introduced many fundamental questions that are still unanswered. On the positive side, we note that the Pioneer and Voyager instruments found a planetary magnetic field aligned with the planet's spin axis, an extended high-density plasma sheet, a hot ($\sim$ 35 keV) plasma in the outer magnetosphere, and a population of trapped energetic particles which interacts strongly with the inner satellites and the rings. It was determined that Titan has no intrinsic magnetic field and that this satellite is generally a significant source of plasma ions for the outer magnetosphere. It was also determined that the vast hydrogen torus extending from $\sim$ 8 to 25 R_S plays an important role as a source of plasma and as a medium interacting with ions which originate in other source regions. However, the relative importance of other plasma sources (such as the solar wind, the Saturn atmosphere, the ice-covered surfaces of the inner satellites and the rings) has not been conclusively established.

Since present understanding of the sources and sinks of plasma and suprathermal charged particles in Saturn's magnetosphere is highly incomplete, we concentrate here on measurement descriptions. Attention is focused on plasma physics problems in the inner magnetosphere, which we define as the region traversed during each 24-hr period centered around closest approach. In complementary reports, Schardt et al. discuss Saturn's outer magnetosphere ($L > 10$), Van Allen concentrates on the interactions of energetic charged particles with Saturn's rings and satellites, Connerney et al. describe Saturn magnetic-field models, and Kaiser et al. discuss Saturn radio emissions (see their respective chapters in this book). None of these topics are treated in detail here; the emphasis on questions of plasma physics at Saturn naturally leads to a strong focus on the Voyager observations since Pioneer 11 had no instruments to measure suprathermal charged particles or the characteristics of plasma waves.

I. PIONEER 11, VOYAGER 1 AND 2 TRAJECTORIES

All three spacecraft approached Saturn near the noon meridian. Both Voyager 2 and Pioneer 11 left the magnetosphere of Saturn near the dawn meridian, while Voyager 1 exited further down the tail at a local time of 0400.

Figure 1 gives information about the radial and latitudinal coverage of Saturn's magnetosphere by the three spacecraft. In each case the trajectory segment shown covers the 24-hr period centered about closest approach. The trajectories in Fig. 1 are given in a cylindrical Saturn-centered coordinate system, the Z axis is aligned along Saturn's spin axis, and R is the equatorial radial distance from Saturn. The lack of a tilt and offset for the planetary magnetic field makes the spin equator nearly congruent with the magnetic equator. While this makes modeling studies easier, it also has the unfortunate consequence that each spacecraft, during an encounter interval, does not make as many crossings of Saturn's plasma sheet as was the case during the Jupiter encounters (see Scarf et al. 1983, for further discussion of this point). The symmetry introduced by Saturn's internal magnetic field also makes the centrifugal and magnetic equatorial planes nearly coincident. Under this condition, the plasma, regardless of its thermal characteristics, will have mirror symmetry about the equatorial plane.

II. PLASMA OBSERVATIONS

Although the plasma probe on Pioneer 11 was designed for solar wind observations, this instrument was able to measure characteristics of magnetospheric ions within the energy-per-unit charge E/Q range of 100 eV to 8 keV on the inbound pass between $\sim 16\,R_S$ and $4\,R_S$ (Wolfe et al. 1980; Frank et al. 1980). In the inner magnetosphere ($\leq 10\,R_S$), the ions were found to be rigidly corotating with the planet, and at radial distances beyond the orbit of Rhea (8.8 R_S), these ions were identified as protons with typical average densities and temperatures near 0.5 cm^{-3} and 10^6 K, respectively. In some regions, the mass-per-unit charge M/Q of the dominant ions could also be determined by analysis of the angular distributions. Figure 2 shows Pioneer 11 profiles of the ion density and ion temperature adopted from the report by Frank et al. (1980); the ion composition identifications are also marked here.

Frank et al. concluded that a large torus of oxygen ions is located inside the orbit of Rhea, with densities $\gtrsim 10$ cm^{-3} over the radial distance range of $\sim 4\,R_S$ to 7.5 R_S. They noted that density maxima appear at the orbits of Dione and Tethys where oxygen ion densities are ~ 50 cm^{-3}. The dominant oxygen charge states were thought to be O^{2+} and O^{3+} in the radial distance ranges of $\sim 4\,R_S$ to 7 R_S and 7 R_S to 8 R_S, respectively. The observations suggest a decrease of ion energies to values less than the instrument energy threshold of $E/Q = 100$ eV at the apparent inward edge of the torus at 4 R_S. Measured ion temperatures increased rapidly from $\sim 2 \times 10^5$ K at 4 R_S to

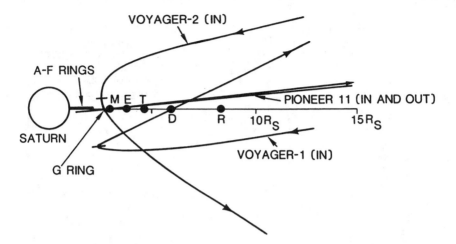

Fig. 1. Radius versus latitude trajectory plots for the Pioneer 11, Voyager 1, and Voyager 2 traversals of Saturn's inner magnetosphere. Each segment goes from 12 hr before closest approach to 12 hr beyond this point. The labels M, E, T, D, and R refer to the satellites Mimas, Enceladus, Tethys, Dione, and Rhea.

5×10^6 K at 7.3 R_S; it was conjectured that the most likely source of these plasmas is the photodissociation of water frost on the surface of the ring material along with ionization of the products and their subsequent outward radial diffusion.

The Voyager observations greatly extended our knowledge of plasma distributions in Saturn's magnetosphere. As shown in Fig. 1, the Voyager encounters yielded high-latitude information that was needed to estimate the thickness of the plasma sheet. In addition, the plasma probe on each spacecraft provided simultaneous high-resolution measurements of electrons and ions over an extended energy/charge range ($E/Q \approx 10$ V to 6 kV for both species), together with improved information on angular distributions. In a pair of initial reports (Bridge et al. 1981a,1982), the plasma probe measurements of Voyager were combined with Pioneer 11 observations to define the complex configuration of plasma in Saturn's magnetosphere. Bridge et al. (1981a,1982) discussed evidence for azimuthal flow of the plasma within $L = 15$, and they noted that the results were consistent with the magnetic-field rotation rate implied by Voyager radio observations (Kaiser et al. 1980). It was suggested that the ions were H$^+$ and O$^+$ (or N$^+$), although final identification was reserved for a later study. Three basic features in the plasma morphology were identified in these preliminary studies:

1. Highly variable regime outside $L = 15$, where abrupt changes in the plasma density by nearly an order of magnitude in < 96 s were observed;

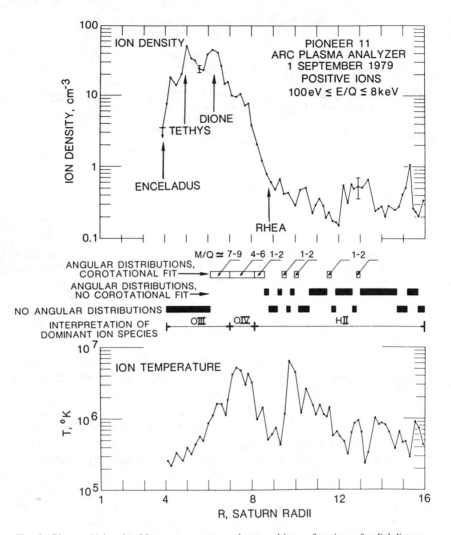

Fig. 2. Pioneer 11 ion densities, temperatures, and composition as functions of radial distance. The oxygen torus extends from ~ 4 R_S to 8 R_S. The rapid decline of densities at ~ 4 R_S may be due to ion energies decreasing with decreasing radial distance to values below the energy (E/Q) threshold of the plasma instrument at 100 eV-per-unit charge. Ion temperatures within the plasma torus at R_S to 7 R_S monotonically decrease with decreasing radial distance.

2. Extended plasma sheet between $L = 7$ and $L = 15$ with a half-thickness in the Z direction greater than 4 R_S;

3. Inner plasma torus inside $L = 7$ with a half-thickness in Z of 1 R_S at $L = 5$ and 0.2 R_S at $L = 2.88$.

A definite reduction in density scale height with decreasing L was observed; this was attributed to a reduction in plasma temperature with decreasing L and possibly a change in the pitch-angle distribution. Bridge et al. (1981a,1982) also estimated the total plasma content in the extended plasma sheet to exceed that in the inner plasma torus, which would rule out the inner plasma torus as the primary source for the extended plasma sheet. As noted in Sittler et al. (1983), it is premature at this time to make any conclusive comments about plasma sources until detailed scale height model calculations using both ion and electron measurements are performed.

Bridge et al. (1981a,1982) presented 15-min averages of the electron densities and used these data plots to discuss the configuration of Saturn's plasma sheet. Recently, Sittler et al. (1983) completed a more detailed analysis of the electron measurements. Figure 3 contains unaveraged plots of Voyager 1 and 2 electron density profiles, along with electron temperature profiles. In this figure all the principal features discussed by Bridge et al. can readily be discerned. For Voyager 1, the two broad regions of density enhancement (17 R_S to $\sim$ 7 R_S inbound; 3 R_S to 11 R_S outbound) suggest an equatorially confined plasma sheet surrounding Saturn. For the Voyager 2 pass, the corresponding density enhancements were found between 11 and 4 R_S inbound and between 4 and 16 R_S outbound.

Bridge et al. (1982) used these density profiles to construct empirical electron density models with simple functional forms such as $n_e L^4 \approx 5.5 \times 10^3$ cm^2 in the extended plasma sheet ($L \approx 8.5$ to 15) and $n_e L^3 \approx 2.4 \times 10^3$ cm^2 in the inner plasma sheet ($L \approx 4$ to 7.5). They showed that these models provided reasonable fits to most of the observations from Pioneer 11 (Fig. 2) and from both Voyagers (Fig. 3). This result suggests a reasonably stable spatial configuration for the Saturn plasma sheet, of the form sketched in Fig. 4.

An extremely interesting characteristic of Saturn's plasma is that the magnetospheric electrons have spectra with distinct hot (suprathermal) and cold (thermal) components. Figure 5 shows electron speed distributions measured near the Voyager 1 outbound ring-plane crossing. The dashed lines indicate Maxwellian fits to the cold component; the hot density n_H and hot temperature T_H were computed by setting $n_H = n_e - n_c$ and $T_H = (n_e T_e - n_c T_c)/n_H$. More examples are contained in the report by Sittler et al. (1983).

The Voyager electron measurements display large-scale radial gradients in electron temperature with T_e increasing from < 1 eV in the inner magnetosphere to as high as 800 eV in the outer magnetosphere. This increase in

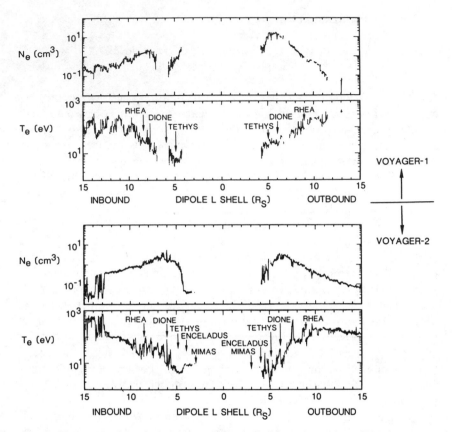

Fig. 3. Electron density and temperature profiles derived from observations of Voyagers 1 (top) and 2 (bottom) plasma probe (Sittler et al. 1983).

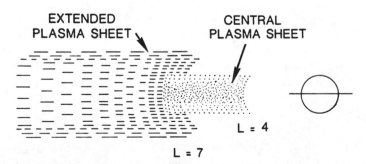

Fig. 4. Possible configuration of Saturn's plasma sheet.

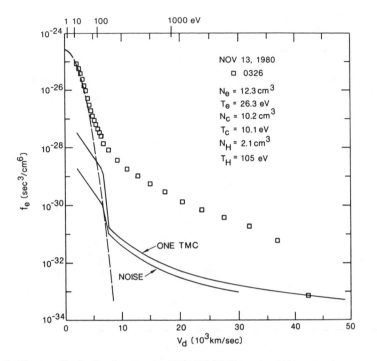

Fig. 5. Electron distribution function for 0326 UT, 13 November 1980, when Voyager 1 was outbound, approaching the ring plane near the Dione orbit. This shows the characteristic two-component spectrum, consisting of a suprathermal population and a cold distribution. Because the spacecraft was in ultraviolet ring occultation at this time, the spacecraft is negatively charged. Therefore, the computed electron density will be underestimated, which is probably down from the true value by a factor of 2 (see discussions in Sittler et al. 1983).

plasma temperature by nearly three orders of magnitude produces an observed increase in plasma sheet thickness of more than an order of magnitude with increasing radial distance from Saturn. Scale heights of cold heavy ions can be as low as 0.2 R_S in the inner magnetosphere and as high as 3 R_S in the outer magnetosphere. Many of the observed density variations can be attributed in part to changes in density scale height, with no need for a change in plasma flux-tube content.

In summary, Voyagers' plasma probe observations suggest that there are three fundamentally different plasma regimes: the hot outer magnetosphere, the extended plasma sheet, and the inner plasma torus. In the hot outer magnetosphere, densities range between ≲ 0.01 cm^{-3} and 0.1 cm^{-3} and electron temperatures range between 50 eV and 800 eV. Near noon local time, the plasma electrons of the outer magnetosphere are reported to have a highly time-dependent behavior (Sittler et al. 1983), and the observed density enhancements are generally regions with enhanced levels of cold plasma. On the average, 25% and 85% of the electron density and pressure, respectively, are partitioned to the suprathermals. Within the density enhancement where cold electrons are observed, cold high-Mach-number ions moving nearly in corotation directions are observed in the ion spectra. This highly time-dependent region, which resides outside the outer boundary of the extended plasma sheet, has been designated by Goertz (1983a) as a turbulence layer; the density enhancements have been attributed by Goertz to detached plasma "blobs" or "islands" which have broken off from the outer boundary of the plasma sheet because of the centrifugally driven flute instability. These density enhancements are subsequently (within a few rotation periods) lost down Saturn's magnetotail by way of a planetary wind mechanism, where Goertz has suggested that the low-density, hot-plasma regions measured by the low-energy charged particle (LECP) instrument (Maclennan et al. 1983) represent flux tubes recently emptied.

The extended plasma sheet is a region with enhanced levels of cold plasma relative to those detected in the hot outer magnetosphere. Densities range between 0.1 cm^{-3} and 2 cm^{-2} and electron temperatures range between 10 eV and 100 eV. Here, less than 15% of the density is partitioned to the suprathermals, although the suprathermals still contribute more than 80% of the electron pressure. Sittler et al. (1983) suggest that the extended plasma sheet is probably the primary contributor to the ring current modeled by the magnetometer team (Connerney et al. 1981), and estimate that a significant fraction of the ring current is probably produced by the low-energy plasma below 6 keV. Centrifugal stresses (dominated by cold heavy ions) and pressure gradients (dominated by hot ions) probably contribute equally to the ring current (see Lazarus and McNutt 1983).

The inner plasma torus is a region of reduced electron temperature T_e < 10 eV and scale height H_i < 1 R_S, with equatorial densities which can become as high as 100 cm^{-3}, and electron temperatures which can go as low as 1 eV. The inner plasma torus is also a region where the suprathermals are severely depleted relative to those in the extended plasma sheet; electron fluxes are generally confined below 50 eV compared to 6 keV in the extended plasma sheet. A strong attenuation of the suprathermals was observed inside $L \sim 5$ during the Voyager 2 encounter. Sittler et al. attributed this to the E Ring, which has an enhanced optical brightness or ring opacity inside L $\sim$ 5 (Baum et al. 1980). Localized reductions in electron temperature and enhanced depletion of the suprathermal electrons were also observed near the L shell of Tethys, Dione, and possibly Rhea. These effects signify an interaction between the electrons and localized concentrations of dust, neutral gas, plasma ions and/or plasma waves centered on the satellite positions. The energy dependence of the depletions supports an interaction with dust or plasma waves in some cases, while in other cases an interaction with neutral gas or plasma ions is favored. The observed depletions of suprathermal electrons seem to be associated with the energy-dependent signatures of electron fluxes

~ 20 keV measured by the LECP instrument (Krimigis et al. 1981,1982a).

Recently, Lazarus and McNutt (1983) presented a detailed description of the Voyager measurements of plasma ions in Saturn's magnetosphere. The full analysis of these ion observations is extremely complex. The authors had to deal with questions of changing plasma composition, varying scale heights and velocities for different species, and non-Maxwellian (hot and cold) distributions for any given species. In addition, there were concerns about the degree to which the plasma ions were strictly corotating with the planetary magnetosphere, and in some regions model fits suggested that the radial flow was also significant.

It is not appropriate to provide a summary of these extremely complex discussions in this chapter; we refer the reader to the report by Lazarus and McNutt (1983) for further details. Here, we simply extract from the full report a brief discussion comparing the Voyager 1 and 2 ion measurements near the ring-plane/magnetic-equator crossing. The bottom panel in Fig. 6 is similar to a figure published in the Voyager 2 report by Bridge et al. (1982), showing a sequence of full ion spectra taken from 0000 to 1000 UT on 26 August 1981, as the spacecraft swept down from 2.2 R_S above the ring plane to 2.2 R_S below it. The relative currents detected by one sensor are plotted against energy-per-unit charge; it can be seen that near the beginning of this interval the ions were observed only in the first few E/Q channels. Bridge et al. (1982) suggested that these spectra are associated with cold protons, and they noted that the sudden decrease in the currents after 0130 UT is related to an abrupt change in plasma properties, such as a density decrease.

Centered about the ring-plane crossing there is a narrow but intense enhancement in currents for the lowest channels; Bridge et al. (1982) stated that these observations can be explained by assuming an O^+ plasma with ion density of ~ 100 cm^{-3} and temperature near 10 eV. As indicated in Fig. 3, there were no measurable electrons at this time, suggesting $T_e < 1$ eV (Sittler et al. 1983; Lazarus and McNutt 1983).

The corresponding Voyager 1 plot is shown at the top of Fig. 6 (the side sensor is used because of a Voyager 1 roll manuever at 0044 UT; this is the interval with large currents in this figure). Simulations of some of these spectra with isotropic Maxwellian distributions convected with the rigid corotation velocity show that the Voyager 1 low-energy peaks in Fig. 6 are primarily H^+. Variations between the light and heavy ion peaks could be the result of noise or slight additions of very cold ions of high ionization states (e.g., O^{2+}, O^{3+}).

Lazarus and McNutt (1983) have shown that the proton number density reached its maximum at a distance of 1.15 R_S below the equatorial plane, with a thermal speed of ~ 40 km s^{-1} (~ 8 eV) and a density of 15 cm^{-3} (this number may be an overestimate because of spacecraft charging). The heavy-ion peak was still increasing at this time. The inferred heavy-ion number density exhibited a broad maximum between 0300 UT ($r = 5.0$ R_S, $z = -0.63$

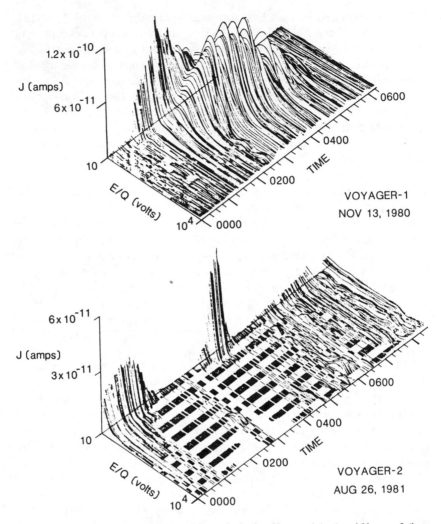

Fig. 6. Time-ordered sequences of ion spectra obtained as Voyager 1 (top) and Voyager 2 (bottom) crossed the ring plane/magnetic equator in the inner magnetosphere. In each scan, the relative current is plotted as a function of energy-per-unit charge.

R_S, $L = 5.1$) and 0400 UT ($r = 5.9$ R_S, $z = -0.14$ R_S, $L = 5.9$). This maximum occurs between Dione and Tethys and is consistent with these satellites being sources of plasma (Frank et al. 1980; Cheng et al. 1982; Lanzerotti et al. 1983a). The ion density is a function of radial distance from Saturn as well as of height above the ring plane. The ring plane was crossed at 0418 UT, just inside the orbit of Dione. There is no relative maximum in the density profile, which is consistent with the radial gradient in density being much greater than the gradient in z in this region.

The plasma ion results appear to be consistent with rigid corotation of the magnetosphere, and there is the possibility of more than one heavy ionic species. In a set of low-resolution spectra obtained at 0225 UT, Lazarus and McNutt found that a light-ion peak, a heavy-ion peak, and a possible warm-ion background were all visible.

The presence of a warm-ion component in the plasma science instrument (PLS) energy range is most prevalent in the outer magnetosphere (extended plasma sheet, hot outer magnetosphere). In the outer magnetosphere, for ion spectra that have been analyzed in detail, the warm-ion component dominates the ion pressure. The importance of the hot ions dominating the plasma pressure is reinforced by the significant ion pressures measured by the LECP instrument above 28 keV (Krimigis et al. 1983). As discussed in Lazarus and McNutt (1983) this warm component probably involves a heavy ion which is locally formed in the vicinity of the neutral hydrogen cloud. Sources of heavy neutrals in this neutral cloud are Titan and Saturn's main ring system (Eviatar et al. 1983). This hot-ion component probably makes a significant contribution to the ring current reported by Connerney et al. (1981), and the nearly coincident location of the ring current and the inner boundary of the neutral cloud at 8 R_S is probably not a coincidence (Sittler et al. 1983; Lazarus and McNutt 1983).

III. LOW-ENERGY RADIATION BELT PARTICLES

Pioneer 11 carried a number of instruments designed to study the characteristics of trapped radiation belts. Many details of these measurements are contained in the chapter by Van Allen in which he emphasizes particle interactions with the rings and the inner satellites. Figure 7 shows the buildup of electron flux and the characteristic hardening of the electron spectrum with decreasing distance from the planet. This also shows the electron cutoff at the ring boundary.

Krimigis et al. (1981,1982a) presented extensive preliminary discussions of the Voyager electron and ion measurements ranging down to low energies $\gtrsim$ 22 and 28 keV for electrons and ions, respectively. These Voyager measurements of low-energy particles greatly extend our knowledge of the physics of Saturn's magnetosphere. We refer the reader to initial reports for detailed discussions of the outer magnetosphere, the Titan encounter (see also Maclennan et al. 1982), and the traversals of the magnetopause and bow shock (Maclennan et al. 1983). In the remainder of this section we focus attention on the inner magnetosphere.

A comprehensive overview of the Voyager 1 electron and ion spectral intensities within approximately the orbit of Titan is presented in Color Plate 1. This plate shows differential ion and electron spectral intensities, respectively, as a function of time along the abscissa; the ordinates represent energy increasing downward, with the intensity color-coded in accordance

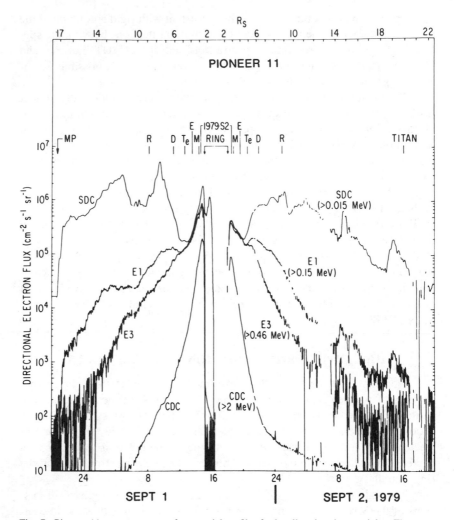

Fig. 7. Pioneer 11 measurements of equatorial profiles for locally mirroring particles. The magnetopause crossing is marked MP. As the spacecraft spins and the viewing orientation of the trapped radiation detector is perpendicular to the spin axis, the detectors become perpendicular to the magnetic field line twice during each spacecraft revolution, and the directional flux of particles with 90° pitch angle is shown here. Electron intensities above 4 thresholds are shown. The trace labeled SDC is a linear combinator of SPDC and SEDC outputs chosen to give the intensity of electrons only. As the detector is calibrated for energy flux (eV cm^{-2} s^{-1} sr^{-1}), an average energy of 15 keV was used to convert to number flux (cm^{-2} s^{-1} sr^{-1}). (See Fillius et al. [1980] for definitions of the channel names.)

with the color bars on the right-hand side. The ion data (15-min averages) have been smoothed using a logarithmic interpolation routine extending from ~ 40 keV to 1 MeV. The data have been corrected for background (Krimigis et al. 1983).

The decrease and recovery of the ion intensity at ~ 1015 UT (Day 317) have previously been interpreted as a feature of the boundary of the stably trapped region in the Saturn magnetosphere (Krimigis et al. 1981). Two other ion flux features particularly evident in the spectrogram are the sharp increase in intensity for energies ~ 150 keV beginning at the L shell of Rhea (peaking inside the orbit of Dione) and the relative depletion of 200 to 400 keV ions inside the orbit of Dione. The latter depletion was followed by a short recovery just before the crossing of the Tethys L shell (depletion of energetic ions $\gtrsim$ 80 keV). Between the orbits of Rhea and Tethys inbound the spectrum is relatively flat below ~ 100 keV, indicative of a hot plasma (Krimigis et al. 1983). The peak inside the orbit of Dione approximately coincides with the maximum in the low-energy electron density (Bridge et al. 1981a).

The electron spectrogram in Color Plate 1 (lower panel) has been constructed by applying a 5-point interpolation routine over the energy range 40 to 400 keV. The data have been corrected for background and averaged over 15-min intervals. The electron intensities varied considerably until Voyager entered into the durably trapped region at ~ 16 R_S at ~ 1015 UT on Day 317. The spectrum became progressively harder, especially inside the Rhea L shell. A sharp drop in low-energy electrons ($\lesssim$ 100 keV) occurred inside the orbit of Rhea. Depletion of fluxes was seen at the higher ($\gtrsim$ 150 keV) energies inside the Dione L shell and at the Tethys L shell. The electron intensities began to recover after Voyager crossed the Tethys L shell outbound, starting with the lower energies, progressing to higher energies, and reaching peak intensities at the orbit of Dione. In general the spectra become particularly hard at the higher ($>$ 100 keV) energies between the orbits of Rhea and Dione, both inbound and outbound. For the Voyager 2 observations, we refer the reader to the chapter by Van Allen which gives a corresponding summary discussion.

In the initial Voyager 2 report (Krimigis et al. 1982a), it was noted that the low-energy (22–35 keV) electrons exhibited considerable variability between the orbits of Rhea and Dione during the inbound pass. A detailed plot of the electron and ion intensities around Rhea and Dione during this time is shown in Fig. 8 at the highest available time resolution of the instrument (0.4 s per plotted point). The bottom set of points in each panel, occurring every 6.4 min, represents a measure of the background counting rate. The foreground/background ratio for all three channels is usually $>$ 10. The intensity fluctuations, largest at the lower energy, begin at the crossing of the Rhea L shell, continue through the Dione L shell, and terminate prior to crossing Tethys' L shell. The fluxes in the ion channel in the bottom panel do not show

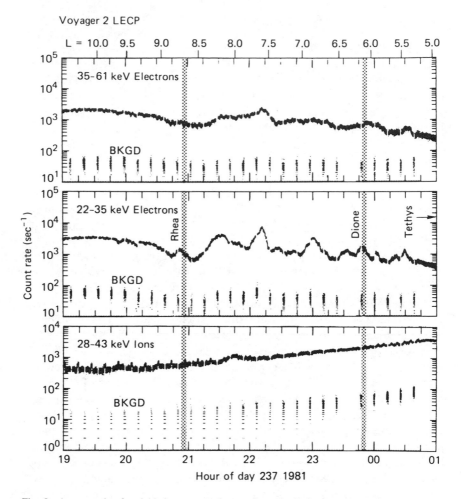

Fig. 8. An example of variable low-energy electron fluxes in the inner magnetosphere.

a variability similar to that of the electrons. Kurth et al. (1983) noted that these variations in electron rates are present in the general region where electrostatic waves are observed by the Voyager plasma-wave instrument (see Sec. IV).

A. Ion and Electron Spectra

The ion spectra observed by Voyagers 1 and 2 can be described well at lower energies by a Maxwellian distribution and at higher energies by a power law (Krimigis et al. 1981,1982*a*). A single expression used to fit the spectra is the k distribution (see e.g., Vasyliunas 1971)

$$f(\bar{v}) = \frac{n}{\pi^{3/2}\omega_0^2} \frac{\Gamma(k+1)}{k^{3/2}\Gamma(k-1/2)} \left| 1 + \frac{|\bar{v} - \bar{V}|^2}{k\omega_0^2} \right|^{-(k+1)} \tag{1}$$

where n is the particle density, $\bar{v}$ the particle velocity, $\bar{V}$ the bulk plasma velocity, and ω_0 (the most probable speed) is $\sqrt{2kT/M}$ where T is the plasma temperature and M is the particle mass. Using $j = p^2 f(\bar{v})$ where $j = dJ/dE$ is the differential ion intensity and p is the ion momentum for nonrelativistic particles

$$j(v, \phi) = \frac{nm^2 v^2 Q(k)}{\pi^{3/2} \left| \dfrac{2\,kT}{M} \right|^{3/2}} \left| 1 + \frac{v^2 - 2\,vV\cos\phi + V^2}{k(2kT/M)} \right|^{-(k+1)} \tag{2}$$

Here

$$Q(k) = \frac{\Gamma(k+1)}{k^{3/2}\Gamma(k-1/2)} \tag{3}$$

is approximately constant (0.90 to 0.97) for values of the exponent k ranging from ~4 to ~15. For $v \gg V$

$$j = j_0 \left| \frac{E}{kT} \right|^{-(k+1)} \tag{4}$$

a power law in energy. The Maxwellian distribution is a special case as $k \to \infty$.

Figure 9a shows hourly averaged ion spectra from Voyager 1 together with fits of Eq. (2) to the data points. The ions are assumed to be protons and are plotted at the appropriate proton energy. The values of L,B at the beginning and end of each hour interval are shown in the spectral plots.

The k function fits most of the observed distributions well. In the fits V was assumed to be equal to the corotation velocity ($V_c = 9.82\ R_S\ km\ s^{-1}$), although the fit is not very sensitive to the precise value. Typical values of n are in the range 10^{-2} to $10^{-3}\ cm^{-3}$ and, of course, include the contribution to the density from ions below the detector threshold.

Figure 9b plots ion spectra measured by Voyager 2. The spectra are substantially flatter at low energies than in the case of Voyager 1. The temperature values of kT are as high as 45 keV with corresponding values of k as high as 15, indicating that the spectrum can be characterized as mostly thermal at $L \sim 10$ (and at higher L as well; see Krimigis et al. 1983). The densities are in the same range as those observed by Voyager 1.

The functional forms of the electron energy spectra are different from those of the ions. Figure 10 shows a set of hourly-averaged electron spectra

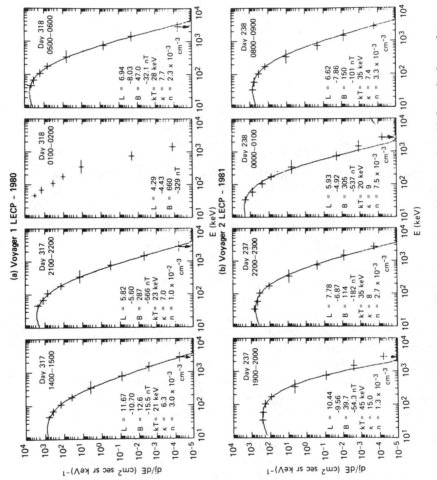

Fig. 9. Hourly averaged ion spectra from Voyagers 1 (a) and 2 (b), together with fits to the data points.

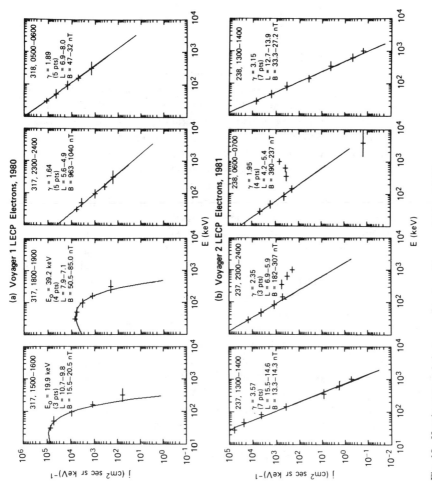

Fig. 10. Hourly averaged electron spectra from Voyagers 1 (a) and 2 (b).

for both spacecraft encounters at various L values. The data are fitted wherever possible to either a power law in energy

$$\frac{dJ}{dE} = CE^{-\gamma} \tag{5}$$

or an exponential of the form

$$\frac{dJ}{dE} = CE \exp(-E/E_0). \tag{6}$$

The Voyager 2 spectra (Fig. 10b) are best characterized by power laws, at least at the lower energies, whereas two of the Voyager 1 spectra are fit reasonably well by Eq. (5). In the Voyager 2 data (which had a broader electron energy coverage than did Voyager 1), at about the Dione L shell inbound, the spectra at low energies became substantially harder ($\gamma \sim 2.4$) and a secondary component began developing at energies $\gtrsim 200$ keV. This higher energy feature was present in the vicinity of the Tethys L shell outbound ($4.2 \leq L \leq 5.4$) as well, but exhibited a rather sharp drop above ~ 1 MeV. A similar spectral feature exists in the Pioneer data of McDonald et al. (1980).

The Voyager 1 and 2 trajectories in Saturn's inner magnetosphere crossed at four points in B,L space, inside the orbits of Dione and Tethys. Thus, the long-term (interval of ~ 9.5 months) variability of the fluxes of ions trapped deep within the magnetosphere of Saturn can be crudely assessed. Figure 11 shows a comparison of ion fluxes inbound at nearly identical points in B,L space. The Voyager 2 fluxes are lower almost everywhere than the intensities measured by Voyager 1, 286 days earlier. The decrease in intensity between the two encounters ranges from factors of 2 to 3 at the lower ($\lesssim 0.5$ MeV) energies to ~ 10 at the higher energies. Panels (a) and (b) include ions from a broad energy band (0.5–1.8 MeV) channel from the Pioneer encounter in September 1979 (Simpson et al. 1980b). Panel (c) contains Pioneer 11 measurements of McDonald et al. (1980). These Pioneer observations all show reasonable agreement with the Voyager results. Thus, there is a variability of ion fluxes over time deep within the magnetosphere, but an understanding of this will only come from detailed studies of the particle diffusion rates and sources. However, the existence of a variability does imply that the deep inner magnetosphere is not in a state of equilibrium.

B. Ion Energy Densities

The energy densities ε_i of the ions measured by the LECP instrument can be calculated as

$$\varepsilon_i = 4\pi \int_{E_t}^{E_{max}} dE \, \frac{j_i(\bar{x},E)}{v_i} \, E_i \cong 4\pi \sum_{k=1}^{8} \left[\Delta E_i \, \frac{j_i(E)}{v_i} \, \bar{E}_i \right]_k . \tag{7}$$

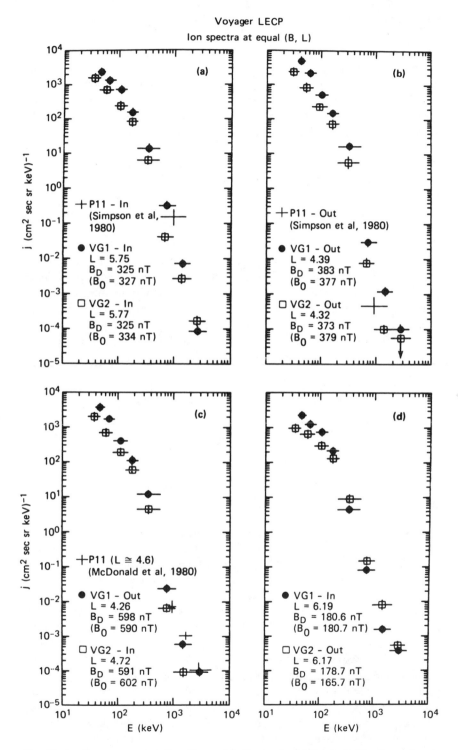

Fig. 11. Ion flux measurements from Pioneer 11, Voyager 1, and Voyager 2. The observations in each panel are associated with nearly identical values for B and L.

Discussions of this calculation as used for observations in the Jovian magnetosphere are given in Lanzerotti et al. (1980) and Krimigis et al. (1982*a*).
Here, $\bar{E}_i$ is the mean energy of a particular ion channel, v_i is the ion velocity, j_i
is the directional intensity, ΔE_i is the channel width, E_t and E_{max} represent the
lower and upper energy thresholds of the detector, and $k = 1$ to 8 represents
differential ion energy channels 1 through 8. Note that all quantities inside the
brackets are known or are observables. Here, all ions are assumed to be protons, and scan-averaged intensities are used because the first-order anisotropy
in Saturn's magnetosphere is quite small (Krimigis et al. 1981; Carbary et al.
1983*a*).

Figure 12 shows the proton energy densities for both encounters, to-
gether with the magnetic energy density $\varepsilon_M (B^2/8\pi)$ (Ness et al. 1981,1982*a*)
and the resulting value of $\beta = (\varepsilon_p/\varepsilon_M)$. The details of the energy balance
in the outer magnetosphere during both passes is given in Maclennan et al.
(1983) and Lanzerotti et al. (1983*a*). In the outer magnetosphere the β value
as determined from the ion measurements of the LECP instrument is quite
large and, during Voyager 2, was observed to be $\gg 1$ on one occasion (Maclennan et al. 1983). Inside ~ 10 R$_S$ the proton β is small in both passes, inbound and outbound, except for Voyager 1 outbound, where β did not decrease until the orbit of Rhea. It is in this inner magnetosphere region that
Frank et al. (1980) reported a high-β plasma detected by the plasma ion instrument on Pioneer 11. However, Krimigis et al. (1983) have noted that the
LECP ion fluxes in this inner magnetosphere region are likely to consist of a
large proportion of oxygen ions, in addition to hydrogen ions. For this situation, the total ion energy densities would be much larger than those estimated
assuming only proton contributions. The total β of the inner magnetosphere
would therefore have ion contributions from both the low-energy and high-
energy plasmas and would likely be close to one in magnitude.

IV. PLASMA WAVE OBSERVATIONS

Although the Pioneer 11 payload did not contain a dedicated plasma wave
investigation, measurements by the vector helium magnetometer showed that
large-amplitude magnetic field oscillations, which could be interpreted as
quasi-periodic electromagnetic waves, were present in a localized region near
the Dione L shell. Figure 13 shows the three magnetic field components along
with the field magnitude for one 12-min interval. Oscillations with a period of
~ 15 to 20 s are evident in the figure. These transverse waves were identified
as left-hand polarized ion cyclotron modes. Smith and Tsurutani (1983)
pointed out that these waves could be expected to interact strongly with very
low energy ions; for O^{++} and O^{+} the resonant ion energies would be $\simeq 300$
eV and $\simeq 5$ eV, respectively. Both Voyager spacecraft had electric field sen-
sors and comprehensive plasma wave instruments covering the frequency

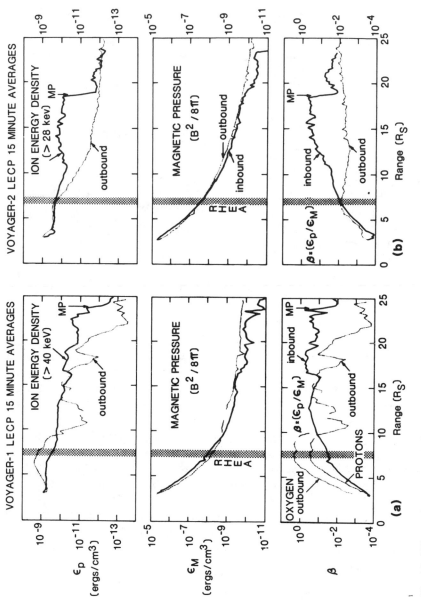

Fig. 12. Fifteen-minute averages of ion energy densities, magnetic pressure profiles, and energetic ion β variations for Voyagers 1 (left) and 2 (right).

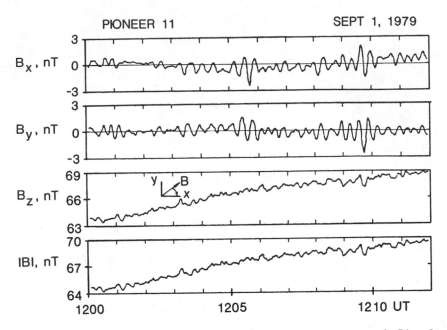

Fig. 13. Pioneer 11 observations of left-hand polarized ion cyclotron waves near the Dione L shell (figure from Smith and Tsurutani 1983).

range from 10 Hz to 56 kHz, plus radio astronomy instruments measuring wave activity up to a frequency of 40 MHz.

The initial reports on the Saturn plasma wave observations by Gurnett et al. (1981a) and Scarf et al. (1982) contained 24-hr plots of the Voyager 1 and 2 observations by the 16-channel spectrum analyzer centered about the times of closest approach. In each case, local values of the electron cyclotron frequency (f_c[Hz] = 28 B [gamma]) were derived from the magnetometer measurements (Ness et al. 1981,1982a), and f_c profiles were plotted along with the wave data. For Voyager 1, Gurnett et al. also provided a preliminary estimate of the electron plasma-frequency profile (f_p[kHz] = 9[N]$^{1/2}$, with N in cm^{-3}), derived largely from initial interpretations of the variations in wave intensity levels. A reanalysis of some high-frequency wave measurements by Pederson et al. (1981) suggested modifications in this density profile, and subsequent detailed studies by members of the Voyager plasma science team (Sittler et al. 1983) yielded greatly improved local electron density estimates. However, as noted by Sittler, the available results of the plasma probe still contain uncertainties caused by the presence of very cold electrons (which are not measured directly by the instrument), as well as by effects associated with spacecraft charging, deviations in ion flow direction, and variations in ion composition, all of which introduce problems in data analysis not easily resolved. Nevertheless, the electron measurements yield straightforward lower bounds for N, and these lower bounds, provided by the Voyager plasma team (see Fig. 3), were used to construct new f_p profiles shown in Figs. 14 and 15 for both encounters; Fig. 16 contains an unlabeled composite of the two wave data sets that provides a direct basis for comparison of the different inner magnetosphere traversals.

For the first 10-hr period on 13 November, the Voyager 1 f_p profile in Fig. 14 differs considerably from the preliminary one shown by Gurnett et al. (1981 a). Since plasma wave modes are identified by comparing the wave frequencies with local values for f_p and f_c, it is appropriate to review the Voyager 1 analysis using the new density curve. We note that at low frequencies ($f \leq 1.0$ kHz) the intense wave activity in Fig. 14 has $f < f_c$, $f \ll f_p$ and is peaked near the magnetic equator (ring-plane crossing) with a secondary maximum near closest approach. Gurnett et al. (1981 a) designated these emissions as "hiss and chorus," and Sittler et al. (1983) noted an association between these waves and fluxes of suprathermal electrons with energies be-

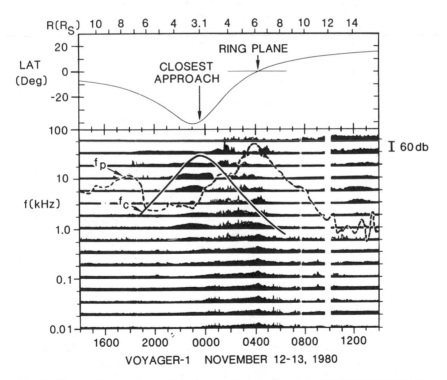

Fig. 14. Voyager 1 observations of the 16-channel spectrum analyzer for the 24-hr interval centered about closest approach. The f_c curve is derived from magnetometer data, and the f_p curve, derived from the plasma electron measurements, may represent a lower bound.

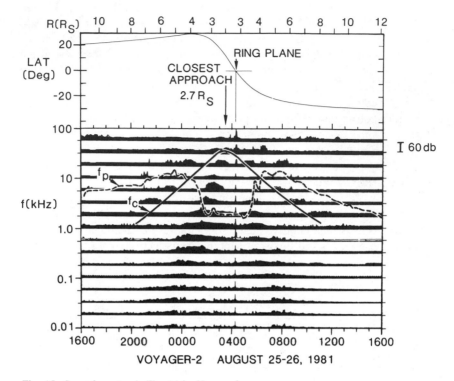

Fig. 15. Same format as in Fig. 14 for Voyager 2.

low 6 keV; this whistler mode identification is not affected by the changes in f_p. Similarly, the $(n + 1/2)f_c$ band designation for the 3 kHz enhancement near 0430 UT and the 5.6 kHz burst at 0300 UT, as well as the radio emission designation for the high-frequency waves (3.1 kHz $\leq f \leq$ 56 kHz) detected after 1000 UT, are unaffected by the change in the density curve. However, the new f_p profile is not in accord with some of the other identifications noted in the preliminary discussion by Gurnett et al. (1981 a), particularly for the intervals near 0100–0200 UT (13 November) and 1800–2230 UT (12 November).

The initial plasma wave report of Voyager 1 contained the following analysis for the 12 November encounter interval: (a) brief noise bursts were detected in the 10 kHz and 17.8 kHz channels between 1800 and 2000 UT inbound, and these impulsive events had characteristics of electron plasma oscillations; (b) after 2000 UT, a band of intense steady emissions was identified with a well-defined lower cutoff that first decreased smoothly from 10 kHz (at 2030 UT) down to 3 kHz (at 2300 UT), and then increased in frequency. It was conjectured that these waves were electromagnetic; it was

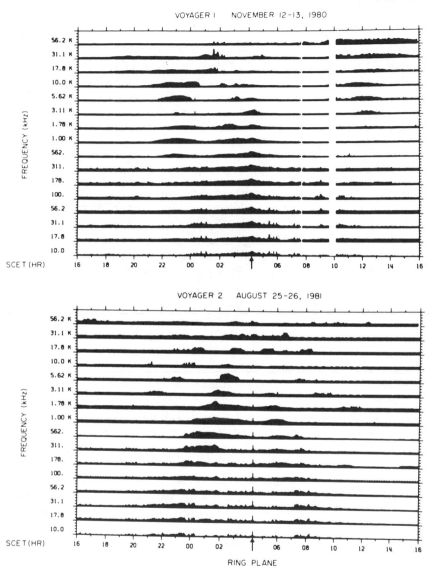

Fig. 16. Comparison of 16-channel measurements of Voyagers 1 (top) and 2 (bottom) in the vicinity of the ring plane/magnetic equator.

natural to assume that the smoothly varying lower cutoff represented a propagation cutoff at the local value for f_p.

A Voyager 1 wideband frame was recorded at 2252 UT on 12 November, showing that this radio emission had a very unusual spectral structure which contained a sequence of many narrowband emissions rather than the expected continuum. Gurnett et al. (1981b) used data from several additional wideband frames to show that the narrowband electromagnetic emissions were detected sporadically by Voyager 1 out to 59 R$_S$ during the outbound leg. It was suggested that the waves were generated by conversion of intense electrostatic waves to electromagnetic emissions in inner magnetospheric regions with strong plasma density gradients, although no definitive evidence for such a steep gradient was available. More recently, Gurnett et al. (1982a) analyzed Voyager 1 wideband data showing similar narrowband emissions at Jupiter. A generation process to produce such emissions appears to be a common one in both planetary magnetospheres.

The f_p profile shown in Fig. 14 does not explain the impulsive noise bursts measured between 1800 to 2000 UT on 12 November, or the low-frequency cutoff of the narrowband emissions observed between 2000 and 2300 UT. Moreover, on the basis of experience in the Earth's magnetosphere, Gurnett et al. (1981a) identified the very intense 31 kHz noise burst at 0130 UT (13 November) as an upper hybrid resonance emission. However, this identification is also quite inconsistent with the f_p profile in Fig. 14. It might be considered that (unmeasured) cold electrons were present in sufficient numbers to yield significantly higher total densities. However, between 1900 and 2300 UT it appears that the electrons were warm enough to be measured; the plasma probe data yield $T_c \lesssim 3$ eV, $n_e \simeq 0.5$ cm^{-3}, and $f_p \approx 6.3$ kHz. Thus, it can be concluded that the intense 31 kHz noise burst near 0130 is *not* an upper hybrid resonance emission. We have elaborated on this point to illustrate the present lack of understanding of several prominent wave emissions detected at Saturn.

It is known that very cold electrons must have been present during a part of the Voyager 2 encounter. As noted above (see Fig. 6), a high ion density was detected at the ring-plane crossing (Bridge et al. 1982), but the measured electron density was low. Thus, the f_p curve in Fig. 15 gives only a lower bound near the equator. However, for Voyager 2 there appear to be no serious problems associated with the new f_p profile beyond $L = 5$. For instance, Scarf et al. (1982) used upper hybrid resonance identifications to derive $f_p \approx 5.6$ kHz at 1640–1700 UT, $f_p \approx 10$ kHz at 2040–2120 UT, and $f_p \approx 18$ kHz at 2235–2315 UT. These values agree reasonably well with minimum values derived from the Voyager 2 plasma probe (note that the electron temperature was relatively high here). Moreover, the Voyager 2 detection of intense narrowband emissions at 5.6 kHz started just after 0202 UT, and this onset occurred fairly soon after a sharp dip in f_p. At 0200 UT, Lazarus et al. (1983) reported an H$^+$ density near 0.4 cm^{-3}, which gives $f_p \sim 5.6$ kHz; the density profile probably fell well below 0.4 cm^{-3} until the ring-plane crossing. Thus, one might relate the emission onset to the spacecraft motion into a propagation region where the local cutoff frequency was below the wave frequency. Figure 15 also shows Saturn kilometric radiation, $(n + 1/2)f_c$ bands, and whistler mode chorus and hiss. Detailed discussions of the electrostatic wave observations are contained in a report by Kurth et al. (1983).

Before proceeding with a discussion of wave-particle interactions, it is necessary to comment briefly on Voyagers' impulsive wave measurements near the ring-plane/magnetic-equator crossings. At Jupiter, it was possible to identify certain impulsive signals as Doppler-shifted ion acoustic waves that appeared to interact strongly with low-energy ions (Scarf et al. 1981*b*). We considered similar explanations for the Saturn ring-plane observations. However, Scarf et al. (1982) used high-resolution wideband measurements (one point per 35 microsec; see bottom panel in Color Plate 2) to show that noise impulses of this type represent dust impacts rather than plasma waves. In a related work, Gurnett et al. (1983) used the detailed observations by the plasma wave instrument to deduce significant characteristics of the ring particles impacting the spacecraft. Scarf et al. (1983) noted that similar impacts were detected on Voyager 1 near the ring plane and at other locations.

V. CHORUS AND ELECTRON SCATTERING

Figure 17 illustrates in detail how the 16-channel wave levels varied during the Voyager 1 ring-plane/magnetic-equator crossing. Here, the calibrated and unaveraged field amplitudes are plotted, and the new f_p curve (based on the Pedersen et al. [1981] analysis and electron plasma probe measurements [Sittler et al. 1983]), is again superimposed. The f_c, 3 $f_c/2$, and 5 $f_c/2$ curves generally serve to order the higher frequency noise bursts in terms of $(n + 1/2)f_c$ bands (f_c being the electron cyclotron frequency), but between 0100 and 0230 UT the variations in the 17.8, 31, and 56 kHz channels cannot be completely explained in terms of ordinary electron gyroharmonics.

At 0326 UT a wideband frame was recorded, and the color-coded spectrogram (Color Plate 3) shows that for this interval weak electron gyroharmonic bands and strong hiss and chorus were present. The 48-second-long spectrogram is actually made up of 800 successive spectral scans; the left side of Color Plate 3 contains a 0.6-second average of ten of these scans. This plot shows clearly that three fairly weak $(n + 1/2)f_c$ emissions were detected along with strong banded chorus and somewhat weaker hiss. The term chorus is used because the waves are banded with $f \lesssim f_c/2$ and have characteristic rising frequency versus time structures. Although the temporal variations may appear unusually slow, Burtis and Helliwell (1976) have described observations of similar emissions in the Earth's magnetosphere. The detailed structure of the Saturn chorus is displayed in the upper panel of Color Plate 2. Color Plates 2 and 3 show that the chorus bandwidth is several hundred Hertz and

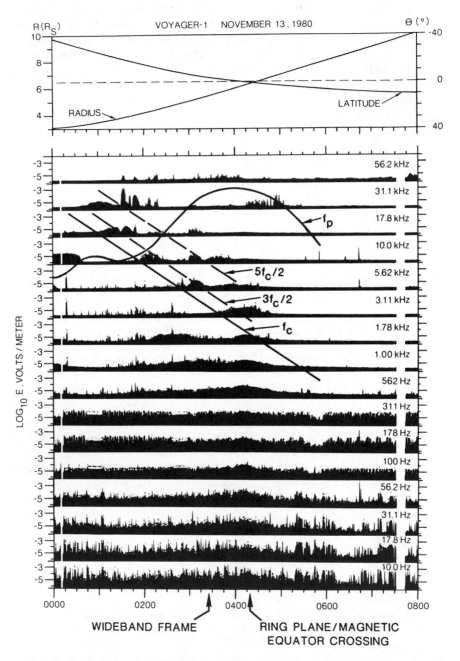

Fig. 17. Detailed plot of the Voyager 1 spectrum analyzer measurements near the outbound crossing of the ring plane and magnetic equator.

that the amplitude is orders of magnitude greater than for any of the $(n + 1/2)f_c$ modes.

Within the Io torus at Jupiter, Scarf et al. (1979), and Thorne and Tsurutani (1979) demonstrated that the whistler mode turbulence could be expected to produce significant loss of radiation-belt electrons. Subsequently, Coroniti et al. (1980) identified Jovian chorus in the Voyager data and showed that ~ 6 ergs cm^{-2} s^{-1} of electrons with energies on the order of a few keV would be precipitated into the atmosphere by whistler mode waves. For a quantitative analysis of the expected effects of Saturn chorus emissions, we consider the 1 kHz measurements at 0326 UT. Whistler mode waves that propagate parallel to the B field have an index of refraction n given by

$$n^2 = 1 + \frac{f_p^2}{f(f_c - f)} . \tag{8}$$

At 0326 UT, B was 120 γ while N was near 20 cm^{-3} (Sittler et al. 1983); thus $f_c = 3.36$ kHz, and $f_p = 40.25$ kHz so that $n(1$ kHz$) \approx 26$. The measured average E-field amplitude at 1 kHz was 1.76×10^{-6} V/m(Hz)$^{1/2}$; thus B' (the magnetic field amplitude of the wave) was nominally equal to 1.54×10^{-4} $\gamma/(\text{Hz})^{1/2}$.

The whistler mode waves resonate with the cyclotron motions of electrons having (see Coroniti et al. 1980)

$$\frac{v(\text{res})}{c} \cong \frac{f_c - f}{nf} \tag{9}$$

and the resultant pitch-angle diffusion coefficient $D_{\alpha\alpha}$ is

$$D_{\alpha\alpha} \cong \frac{1}{4}\left[\frac{e}{mc} 10^{-5}B'(f)\right]^2 \cdot \frac{f}{f_c + 2f} . \tag{10}$$

For $f = 1$ kHz, these expressions yield v(res) $= 2.55 \times 10^9$ cm s^{-1}, E(res) $= 1.85$ keV, and $D_{\alpha\alpha} = 3.5 \times 10^{-5}$ s^{-1}.

If we assume, as in the analysis for Jupiter, that the wave-particle interaction region extends over two Saturn radii along a flux tube centered at the magnetic equator, the bounce-average precipitation lifetime can be estimated to be $T_L \approx (\pi L/2)/D_{\alpha\alpha} \approx 2.5 \times 10^5$ s. The minimum precipitation lifetime is $T_M = 2L^4 R_S/v$ (Coroniti et al. 1980), which gives $T_M \approx 4000$ s; thus $T_M/T_L \approx 0.016$.

This very small value for T_M/T_L near the Saturn ring-plane crossing indicates that such wave-particle interactions would produce only weak diffusion and negligible precipitation. In order to assess this result, it is useful to recall that Coroniti et al. (1980) found $T_M/T_L \approx 0.3$ for Jovian chorus, while Scarf et al. (1979), and Thorne and Tsurutani (1979) found $T_M/T_L \approx 0.01$ to 0.2 for

Jovian whistler mode hiss. The weak diffusion result for Saturn appears to be quite consistent with electron measurements from the Voyager 1 plasma probe. Figure 5 yields $J(2.0 \text{ keV}) \approx 5 \times 10^5$ electrons cm^{-2} s^{-1} at 0326 UT, while the stable trapping limit is near 10^7 cm^{-2} s^{-1}. Thus, we should not expect strong diffusion or large precipitation fluxes.

VI. ELECTROSTATIC WAVES AND WAVE-PARTICLE INTERACTIONS

In the preceding sections, the existence of upper hybrid resonance emissions, electron plasma oscillations (Langmuir waves), and electron cyclotron harmonic waves in Saturn's inner magnetosphere was noted. Recently, Kurth et al. (1983) conducted a comprehensive study of these electrostatic waves at Saturn, and summarized all of the Voyager observations. It is of interest to evaluate the wave-particle interactions that could be associated with the electron cyclotron harmonic waves in particular, since at Earth these waves are thought to contribute to the diffusion of electrons into the loss cone to produce the diffuse aurora (Kennel et al. 1970; Lyons 1974).

Current theories for the generation of the electron cyclotron harmonic waves require electron distribution functions with two (or more) components; the analysis is based on the Harris dispersion relation (Harris 1959) for electrostatic waves propagating in a hot plasma with an imbedded magnetic field (Ashour-Abdalla and Kennel 1978; Hubbard and Birmingham 1978; Rönmark 1978). Young et al. (1973) demonstrated that relatively weak loss-cone distributions are unstable to the electron cyclotron waves, provided a cold electron component accompanied the hotter component possessing a free-energy source. The free-energy source apparently can take a variety of forms (Kurth et al. 1979; Kurth et al. 1980) but must have a positive slope with respect to $v_\perp$, i.e. $\partial f/\partial v_\perp > 0$.

At Earth, the free-energy source is thought to involve electrons in the range from about one to several keV (Rönmark 1978; Kurth et al. 1979; Kurth et al. 1980b); hence, the existence of $(n + 1/2)f_g$ emissions at Saturn could, in principle, imply substantial fluxes of 1 to 10 keV electrons with $\partial f/\partial v_\perp > 0$. Indeed, above 22 keV there is evidence for large electron fluxes in Saturn's inner magnetosphere. Figure 18 shows a portion of the LECP data discussed in Sec. III.A (Fig. 9), with plasma wave data added.

The variations observed in the count rate from the unshielded sectors probably represent temporal changes (Krimigis et al. 1982a,1983). Krimigis et al. (1982a) analyzed the electron data shown in Fig. 18 by comparing the rates in two mutually perpendicular directions and found similar intensities, suggesting a nearly isotropic pitch-angle distribution. They further argued that the temporal variations at electron energies $\lesssim 60$ keV may be evidence of electron acceleration. Kurth et al. (1983) speculated that a weak loss-cone feature could be present in the energetic electrons and argued that the 22 to 35

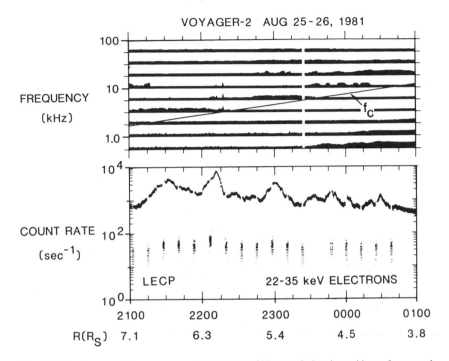

Fig. 18. Simultaneous plasma wave and LECP data of Voyager 2 showing evidence for several keV electron fluxes which may be interacting with the $(n + 1/2)f_c$ emissions seen in the upper panel.

keV electrons could be the high-energy end of the spectrum of electrons responsible for driving the electron cyclotron harmonic instability. Since all occurrences of the $(n + 1/2)f_g$ emissions are not accompanied by variable rates in the 22–35 keV channel, they conjectured that lower energy electrons, below the LECP instrument threshold and above the plasma science instrument (PLS) threshold, might drive the waves.

Barbosa and Kurth (1980) studied the quantitative relation between suprathermal electrons and electron cyclotron harmonic emissions in order to identify the electrons responsible for driving such waves. One important result of their analysis is that large convective growth rates in the lower portion of a harmonic band, as for the case of the Saturnian 3/2 band, can be the result of resonant electrons having a velocity V_r only a few times the thermal speed of the cool background. The lower panel of Fig. 19 contains a 4-second average wave spectrum from Voyager 2 outbound (0736, 26 August 1981). The frequency scale in Fig. 19 (upper panel) is the normalized frequency $\bar{\omega} = f/f_g$; it should be noted that the emission is confined to the very bottom of the first harmonic band.

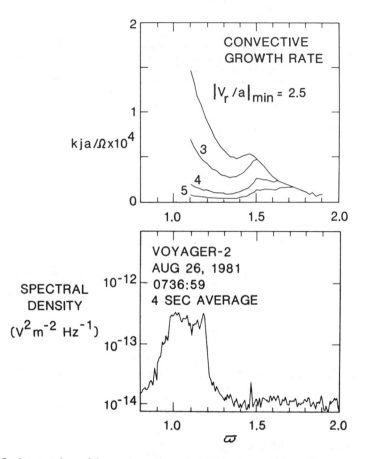

Fig. 19. A comparison of the spectrum of a typical 3/2 band observed at Saturn (lower panel) with convective growth rates (upper panel). The Saturn emission is confined to the lower portion of the band between f_c and $2f_c$, $\bar{\omega} \lesssim 1.3$. This implies that $|V_r/a|_{min}$ is small ($\lesssim 3$).

In the right-hand panel of Fig. 19 the convective growth rates are shown for $|V_r/a|_{min} = 2.5, 3, 4$, and 5 (here a is the thermal speed). Kurth et al. (1983) used $Tj_\perp = 1.1 \times 10^5\,\mathrm{cm}^{-2}\,\mathrm{s}^{-1}\,\mathrm{sr}^{-1}$, $n_c = 1.6\,\mathrm{cm}^{-3}$, and $T_c = 8\,\mathrm{eV}$. They also found that smaller values of $|V_r/a|_{min}$ yield growth rate curves nearly similar to the measured wave spectrum, although they caution the reader that spectral density cannot be directly related to growth rate. However, the comparison between theory and experiment did suggest $T_R = 9\,T_c$, where T_R and T_c refer to temperatures of resonant and cool electrons, respectively.

For the Voyager 2 inbound measurements near 2300 UT on 25 August, $T_c \approx 8\,\mathrm{eV}$ implying $T_R \sim 70\,\mathrm{eV}$. Kurth et al. noted that the actual value of T_R/T_c is not well specified, but concluded that T_R is on the order of 100 eV and not several keV as might be suggested by the enhanced fluxes of 22 to 35 keV electrons evident in Fig. 18. Thus, it seems that the electrons observed by LECP are not driving the electron cyclotron instability, although those electrons may very well be scattered by or perhaps even stochastically accelerated by the electrostatic waves. This conclusion is also supported by the theoretical analyses of Ashour-Abdalla and Kennel (1978), Hubbard and Birmingham (1978), and Hubbard et al. (1979) who calculated that values of $|V_r/a| \approx 3$ were suitable for optimal growth of 3/2 emissions at Earth.

Nevertheless, the plasma probe measurements did provide conclusive simultaneous evidence of a 100 to 200 eV suprathermal electron component with a density of ~ 0.16 cm^{-3}, so we must turn to theoretical arguments. For Jupiter, Barbosa and Kurth (1980) derived a quantity $Tj_{\perp}^{*}(T)$ defined as the critical flux of resonant electrons able to produce 10 e foldings in amplitude of electron cyclotron waves in the Jovian magnetosphere:

$$Tj_{\perp}^{*}(T) = 23 \left(\frac{0.1}{\delta \bar{\omega}} \right) pR^2 \bar{\omega} \ (\text{cm}^2 \ \text{s} \ \text{sr})^{-1}. \tag{11}$$

Here, T is the electron temperature in eV, $\bar{\omega}$ is the wave frequency normalized to f_g ($\bar{\omega} = f/f_g$), p is the pressure due to cool background electrons ($p = n_c T_c$) in units of eV cm^{-3}, and R is in planetary radii. Equation (11) can be modified for use at Saturn as follows:

$$Tj_{\perp}^{*}(T = T_R) = 23 \left(\frac{0.1}{\delta \bar{\omega}} \right) \left(\frac{4.2G}{B_s} \right) \left(\frac{7.13 \times 10^4 \ \text{km}}{R_S} \right) \left(\frac{T/T_c}{9} \right)^2 pR\bar{\omega}. \tag{12}$$

B_s is the surface field magnitude at Saturn, 0.21 G and R_S is 6×10^4 km. Following Barbosa and Kurth, Kurth et al. (1983) chose $\bar{\omega} = 1.3$ and $\delta \bar{\omega} = 0.25$ based on the measured frequencies and bandwidths of the Saturn emissions.

Kurth et al. (1983) evaluated Eq. (12) using wave and particle measurements for Voyager 2 inbound at 2324, 25 August ($R = 5 \ R_S$); they found $Tj_{\perp}^{*} = 1.1 \times 10^5$ (cm^2 s sr)$^{-1}$. The critical flux of resonant electrons from Eq. (12) was compared to the measured flux. Given $T_H = 100$ eV and $n_H = 0.16$ cm^{-3}, the measured flux of hot electrons is 7.6×10^6 (cm^2 s sr)$^{-1}$ or 76 times greater than the critical flux. This result seems satisfactory, since fluxes substantially larger than $Tj_{\perp}^{*}$ are probably required, because at this time Voyager was 28° above the equator, where conditions for wave growth would not be optimum. The preliminary conclusions reported by Kurth et al. (1983) suggest that at Saturn the suprathermal electrons detected by the plasma probe generate the electron cyclotron harmonic waves, and that these waves in turn may cause diffusion or even acceleration of electrons with energies in the range of tens of kilovolts.

VII. DISCUSSION

In order to evaluate the dynamics of Saturn's magnetosphere from the limited set of observations now available, certain ground rules must be adopted. In particular, when we use combined data from Pioneer 11, Voyager 1, and Voyager 2, we tacitly assume that the overall system is basically constant in time and that all important changes in measurements are associated with the different traversals of distinct spatial features. However, in terms of energetic particle measurements, it has been noted that Pioneer 11 encountered Saturn in an extremely disturbed state during one of the largest solar particle events of the current solar cycle (Simpson et al. 1980b). In contrast, both Voyager encounters in November 1980 and August 1981 occurred at times when the interplanetary medium in the vicinity of Saturn was relatively quiet for higher energy particles. Thus, the magnetosphere of Saturn was observed by Voyager in a more or less quiescent state, although significant temporal variations did occur between encounters and while the two encounters were in progress (Ness et al. 1981; Krimigis et al. 1982a; Maclennan et al. 1983).

In terms of phenomena of low-energy plasma physics, we note that the Saturn flyby of Voyager 2 was still unusual because at that time the solar wind pressure was low and variable, and the Saturn system was nearly lined up with Jupiter's extended magnetic tail. Scarf et al. (1983) presented evidence that Voyager was near the Jovian tail *after* the Saturn encounter. This suggested that Saturn may have been immersed in Jupiter's tail during a portion of the Voyager 2 encounter, as originally conjectured by Scarf (1979); Desch (1983) recently discussed changes in the radio emission pattern that strongly support this concept.

Despite these concerns about temporal variations, the measurements from the three spacecraft do appear to be reasonably consistent in terms of particle and field distributions in the inner magnetosphere. Although the Pioneer and Voyager trajectories and measurement capabilities did not overlap completely, there is no compelling reason to believe that the data sets are incompatible. Thus, one might expect that enough information was gathered during the three encounters to provide the foundation for constructing a dynamic model, based on synthesis of all the observations.

In fact, the data show that the Saturn system is extremely complex and that there are still a large number of fundamental uncertainties associated with interpretations of the spacecraft measurements. The rings and the many icy satellites in the inner magnetosphere are potential sources of plasma, and they also act as sinks. It has been conjectured that dust particles from satellites such as Enceladus cause plasma losses and even plasma cooling. The apparent modulation of radio emissions associated with the orbital phase of Dione (Gurnett et al. 1981a; Warwick et al. 1981; Kurth et al. 1981b) could even mean that this satellite was venting gases during the Voyager 1 encounter. Because the system is so complex, questions related to sources and sinks of plasma and suprathermal particles are largely unanswered; it is possible that charged particles observed in the inner magnetosphere originate in the solar wind, in Saturn's atmosphere, in the main ring system, the icy satellites, the atmosphere of Titan, or even in the tenuous E and G Rings. Since the spacecraft measurements of plasma composition are generally not precise enough to identify sources and sinks, it is likely that many significant questions involving magnetospheric dynamics will remain unanswered until a follow-up mission to the Saturn system is carried out.

MAGNETIC FIELD MODELS

J. E. P. CONNERNEY
Goddard Space Flight Center

L. DAVIS, JR.
California Institute of Technology

and

D. L. CHENETTE
The Aerospace Corporation

The planetary magnetic field of Saturn, as measured by the Pioneer 11, Voyager 1, and Voyager 2 spacecraft, is primarily that of a dipole of moment 0.21 Gauss–R_S^3 (where Saturn's radius R_S = 60,330 km). The field due to interior sources is defined in the Z_3 model by the zonal harmonic coefficients g_1^0 = 21535 nT, g_2^0 = 1642 nT, and g_3^0 = 2743 nT, with possible nonaxisymmetric terms below the level of observation. In contrast to the Earth and Jupiter, which have dipole tilts of 11°.5 and 9°.6, Saturn's magnetic axis lies well within 1° of its rotation axis. Thus Saturn is distinguished from Earth and Jupiter by the symmetry of its internal field about its axis of rotation (spin symmetry). And yet, the strong periodic modulation of Saturn's radio emission is puzzling and compelling evidence of some departure from axial symmetry.

This axial symmetry of Saturn's main field makes possible additional analyses of the dipole position and orientation based on the energetic charged particle absorption signatures observed by the Pioneer and Voyager spacecraft. These studies are consistent with the axisymmetric (zonal harmonic) models of Saturn's magnetic field and demonstrate that any equatorial displacement of the dipole is limited to < 0.01 R_S.

Saturn's magnetosphere appears to be intermediate in configuration to those of the Earth and Jupiter. An equatorial ring current of $\sim 10^7$ A, confined *to a 5-R_S-thick annulus with inner and outer radii of ~ 8 and 16 R_S, has a major effect on the geometry of Saturn's outer magnetosphere. Field lines are moderately but measurably stretched out in the equatorial plane. As at Earth and Jupiter, a well-developed magnetic tail results from the interaction of Saturn's magnetic field with the solar wind.*

I. INTRODUCTION

The first *in situ* observations of Saturn's magnetosphere were obtained by the Pioneer 11 spacecraft in September 1979. This spacecraft, launched in April 1973, was instrumented with two experiments to measure magnetic fields: a vector helium magnetometer (E. J. Smith et al. 1975) and a high-field fluxgate magnetometer (Acuña and Ness 1975). Subsequent observations of Saturn's magnetosphere were obtained by the Voyager 1 spacecraft in November, 1980 and by Voyager 2 in August, 1981. The magnetic field experiments on the Voyager 1 and 2 spacecraft are identical, consisting of dual low-field and high-field triaxial fluxgate magnetometer systems (Behannon et al. 1977). In addition to the magnetic field experiments, each spacecraft hosted a full complement of scientific investigations, many of which contributed greatly to our understanding of Saturn's magnetosphere. These are described elsewhere in this book and summarized by Dyer (1980) and Opp (1980) for Pioneer 11 and by Stone and Miner (1981,1982) for Voyagers 1 and 2, respectively.

The Pioneer 11 magnetometer investigations revealed a wealth of information regarding Saturn's magnetic field and magnetosphere (E. J. Smith et al. 1980a; Acuña and Ness 1980). These Pioneer 11 observations demonstrated for the first time the existence of an intrinsic magnetic field and associated magnetosphere of Saturn. Bow shock and magnetopause boundaries characteristic of the solar wind interaction with a planetary magnetic field were detected (E. J. Smith et al. 1980a) as well as an equatorial current sheet observed in the early morning sector of Saturn's magnetosphere. While the existence of an intrinsic magnetic field at Saturn was not unexpected, we should remember that the predictive abilities of dynamo theory are somewhat lacking (e.g., the surprise discovery [Ness 1979] of the intrinsic magnetic field of Mercury). Prior observational evidence of the Saturnian magnetosphere was limited to radio emissions detected by the Earth-orbiting satellite Imp 6 (Brown 1975) and associated with Saturn. However, that association was refuted by the Voyager planetary radio astronomy investigation (Kaiser et al. 1980) which first detected Saturn radio emission shortly after the Pioneer 11 encounter with Saturn (see chapter by Kaiser et al.).

Saturn's planetary magnetic field was found to be well approximated by a dipole of moment 0.20 G–R_S^3 (E. J. Smith et al. 1980a; Acuña and Ness 1980), rather less than expected on the basis of scaling laws (Busse 1979; Russell 1979). The polarity of Saturn's dipole, like Jupiter's, is opposite to that of the Earth. The most remarkable aspect of Saturn's planetary magnetic field proved to be the close alignment of Saturn's magnetic (dipole) and rotation axes (E. J. Smith et al. 1980a; Acuña and Ness 1980). The high degree of axisymmetry of Saturn's inner magnetosphere was also evident in the Pioneer 11 charged particle observations (Simpson et al. 1980a; Van Allen et al. 1980c; Fillius et al. 1980; Trainor et al. 1980). The simplicity of Saturn's magnetic field geometry and the presence of a multitude of satellites and rings in the inner magnetosphere contribute to make the charged particle observations a substantial supplement to the magnetometer data in studies of the magnetic field geometry.

The Voyager 1 magnetic field observations at Saturn (Ness et al. 1981) confirmed many of the general features of Saturn's magnetosphere established by Pioneer 11. Voyager studies (Ness et al. 1981; Acuña et al. 1981a) found the planetary magnetic field of Saturn to be well represented by a dipole of moment 0.21 G–R_S^3, tilted by $\lesssim$ 1° from the rotation axis, but it differed from the Pioneer studies (E. J. Smith et al. 1980b; Acuña et al. 1980) in the placement of the dipole along Saturn's rotation axis. A model of Saturn's inner magnetosphere, incorporating a $\sim$ 10^7 A equatorial ring current (Connerney et al. 1981,1983) was obtained from the Voyager 1 magnetometer observations. The presence of such a ring current was already implicit in the Pioneer 11 charged particle and magnetometer observations and corroborated by Voyager 1 plasma (Bridge et al. 1981a) and charged particle observations (Vogt et al. 1981).

Early analysis of the Voyager 2 observations (Ness et al. 1982a) confirmed the dipole moment (0.21 G–R_S^3) and orientation with respect to the rotation axis ($\lesssim$ 1°) and the continued presence of an equatorial ring current. A host of new satellite absorption features useful for studies of the magnetic field geometry were identified in the Voyager 2 charged particle observations (Krimigis et al. 1982a; Vogt et al. 1982). Analyses of the Voyager 2 magnetic field observations, in conjunction with the complementary observations obtained by Voyager 1, resulted in the Z_3 zonal harmonic model (Connerney et al. 1982b) of Saturn's planetary magnetic field. The relatively large axisymmetric octupole term of the Z_3 model distinguishes this model from the axially offset dipole models and successfully reconciles the magnetometer observations obtained by Pioneer 11, Voyager 1, and Voyager 2. This axisymmetric model magnetic field is also consistent with the many satellite absorption signatures evident in the Pioneer 11, Voyager 1, and Voyager 2 charged particle observations (Acuña et al. 1983). A model similar to Z_3 is among those recently obtained from the Pioneer 11 vector helium magnetometer observations (Davis and Smith, manuscript in preparation).

II. ENCOUNTER TRAJECTORIES

Saturn's lack of an appreciable dipole tilt results in a regular and limited sampling of magnetic latitudes by each spacecraft. Figure 1 shows the trajectories of Pioneer 11 and Voyagers 1 and 2 in a cylindrical (ρ, z) planet-

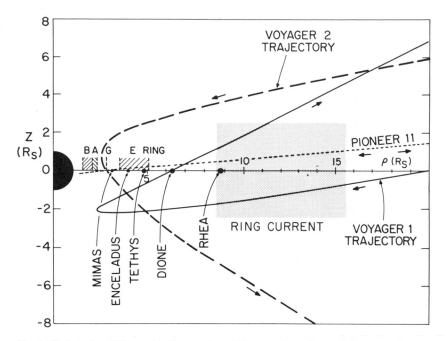

Fig. 1. Trajectories of Pioneer 11, Voyager 1, and Voyager 2 in a cylindrical planetocentric equatorial coordinate system. Positions of the major satellites are indicated as well as the region of (model) distributed ring currents in Saturn's magnetosphere. Satellite diameters are exaggerated by a factor of 10.

centered coordinate system, which in this case is practically identical to a magnetic equatorial coordinate system. Each spacecraft remains at essentially constant magnetic latitude for a large percentage of the encounter with only one (V2) or two (V1, P11) crossings of the magnetic equator. Pioneer 11 traced a path outbound from periapsis that was nearly identical in magnetic latitude and radial distance to that inbound. Figure 2 emphasizes the distribution of Saturnigraphic latitudes and longitudes sampled by each spacecraft. Pioneer 11, with a closest approach of 1.35 R_S, remained within $\sim$ 6° of the equator. The maximum field measured by Pioneer 11 was $\sim$ 8200 nT (E. J. Smith et al. 1980a) at $-6°$ latitude. Voyager 1 sampled relatively high ($-40°$) latitudes at close radial distances (3.07 R_S) but remained in the southern hemisphere during much of the near encounter period ($r < 8$ R_S). A maximum field of 1093 nT was measured by Voyager 1 at $-40°$ latitude and 184° SLS longitude, just before closest approach (Ness et al. 1981). Voyager 2's closest approach of 2.69 R_S occurred at 323° longitude, diametrically opposite from those of both Pioneer 11 and Voyager 1 (Fig. 3). Voyager 2 sampled relatively high latitudes in both hemispheres ($\approx$ 30°) and measured a maximum field of 1187 nT just prior to closest approach (Ness et al. 1982a).

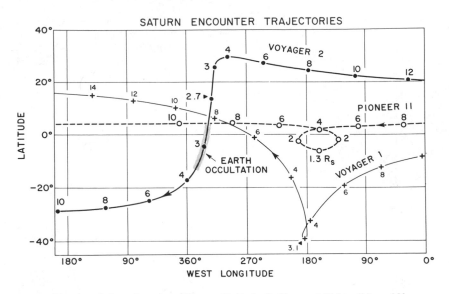

Fig. 2. The close flyby trajectories of Pioneer 11 (dashed), Voyager 1 (light solid), and Voyager 2 (bold) in Saturnographic coordinates. Shown are subspacecraft longitudes (SLS) and latitudes as a function of time for each encounter; the planetocentric radial distance of each spacecraft is indicated at ~ 2 R$_S$ intervals along the trajectory.

The availability of observations along three very different encounter trajectories is particularly important for models of Saturn's magnetic field. Each encounter trajectory by itself represents a rather singular trace through Saturn's magnetosphere. Within the context of a given physical model of the magnetic field, each tends to determine well a subset of parameters describing the field. But for any one flyby, certain parameters (or rather, combinations of parameters) are relatively poorly determined because of the spatial limitations of the observations (Connerney 1981). Thus, the availability of several trajectories and the ability to make use of the charged particle observations in field modeling (see, e.g., Chenette and Davis 1982), or as an independent test of field models (see, e.g., Acuña et al. 1983), is of great importance.

III. MODELS AND METHODS

The magnetic field observations themselves are limited spatially to the paths traversed by each spacecraft. In the absence of a model, our knowledge of Saturn's magnetic field is likewise limited to the regions of space explored by the spacecraft. With an adequate physical model, however, we can extrapolate with confidence from the available observations to unexplored regions of Saturn's magnetosphere. The ultimate test of any model is how well it predicts the field in regions far removed from the observations from which it was derived.

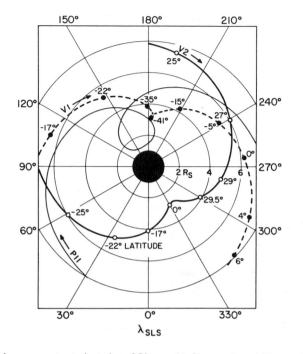

Fig. 3. The close encounter trajectories of Pioneer 11, Voyager 1, and Voyager 2 in a quasi-cylindrical Kronographic coordinate system. Radial distance is shown as a function of SLS longitude; latitudes of Voyager 1 and 2 are indicated at intervals along the trajectories. Pioneer 11 remained within ~ 6° of the equator throughout its encounter.

The observed magnetic field is the sum of contributions from various sources, foremost among them the main magnetic field presumably generated by a planetary dynamo in Saturn's interior. In decreasing order of importance, the main field is joined by contributions from Saturn's equatorial ring current (Connerney et al. 1981), magnetopause and magnetotail currents (E. J. Smith et al. 1980a; Behannon et al. 1981), and possibly more localized currents due to satellite and/or solar wind interaction with the magnetosphere.

Analyses of the magnetic field observations at Saturn utilize the traditional spherical harmonic expansion of a scalar potential V from which the magnetic field is obtained via $\bar{B} = -\nabla V$ (Chapman and Bartels 1940). The potential function V is the sum of two expansions in positive and negative powers of radial distance r and is given by

$$V = a \sum_{n=1}^{\infty} \left(\frac{r}{a} \right)^n T_n^e + \left(\frac{a}{r} \right)^{n+1} T_n^i \tag{1}$$

where a is Saturn's equatorial radius. The series T_n^e, representing contributions due to exterior sources, is

$$T_n^e = \sum_{m=0}^{n} P_n^m (\cos \theta) \{G_n^m \cos m\psi + H_n^m \sin m\psi\} \qquad (2)$$

and the contribution due to interior sources is represented by

$$T_n^i = \sum_{m=0}^{n} P_n^m (\cos \theta) \{g_n^m \cos m\phi + h_n^m \sin m\phi\}. \qquad (3)$$

The angles ϕ and θ are the conventional polar coordinates, θ measured from the axis of rotation and ϕ increasing in the direction of rotation. Longitudes ϕ for the internal source potential are referenced to the prime meridian defined by Desch and Kaiser (1981a) and assume the SLS rotation rate of $810°76$ day^{-1}. The external source potential may be expressed in the same coordinate system as the internal source potential or in an alternative coordinate system. In the former case, the longitudes ψ appearing in the external source potential are equal to those (ϕ) in the internal source potential. Another coordinate system employed for the external source potential (in this chapter, the models by Davis and Smith and E. J. Smith et al.) is defined by the Saturn-Sun vector and Saturn's rotation axis, i.e., a local time coordinate system. In this case the longitudes ψ are measured from the noon meridian, with positive ψ increasing towards dusk. The $P_n^m (\cos \theta)$ are the Schmidt-normalized associated Legendre functions and the g_n^m, h_n^m, G_n^m, H_n^m are the internal/external Schmidt coefficients. This representation is valid within a thick spherical shell bounded by the internal sources and the external sources and assumes that no currents are present ($\nabla \times B = 0$) within the shell. In situations where appreciable local currents exist (e.g., the Voyager 1 Jupiter encounter), other methods, utilizing explicit models of the magnetospheric currents in place of an external potential expansion, have been successfully employed (Connerney et al. 1982a). Such methods require detailed knowledge of the magnetospheric current systems. Thus far, in modeling Saturn's planetary field, fields of external origin have been approximated by some form of the traditional spherical harmonic expansion T_n^e.

The inversion methods by which the magnetic field models are derived from the observations vary and we will not attempt to describe them here. However, the field models obtained are influenced by several factors, including the selection of data to be inverted, the weight assigned to each observation, and the treatment of fields of external origin. These factors differ from model to model and are described in the appropriate references. A general discussion of the statistical basis of weighing observations and the nonuniqueness of magnetic field models derived from spacecraft observations is given by Connerney (1981).

Magnetometer Observations

A summary of spherical harmonic models of Saturn's magnetic field obtained to date from the Pioneer 11 magnetic field observations is given in Table I. Listed are the Schmidt-normalized spherical harmonic coefficients of three models obtained from the vector helium magnetometer observations (E. J. Smith et al. 1980*b*; Davis and Smith, manuscript in preparation) and one derived from the fluxgate observations (Acuña et al. 1980). The equivalent dipole offsets, computed from the spherical harmonic coefficients (Bartels 1936) are listed in Table I below the Schmidt coefficients (for an axially offset dipole, $\Delta Z = g_2^0/2g_1^0$ for small ΔZ). Five additional models based on the Pioneer 11 vector helium magnetometer observations have been presented (E. J. Smith et al. 1980*b*), but for reasons discussed by Smith et al. the "Nominal 79 dipole," designated N79D in Table I, and the "JGR 80 model," were preferred (but see the discussion of more recent work later in this section).

Smith et al. recommended the JGR 80 model for use in cases where accurate values of the magnetic field near the Pioneer trajectory are needed. For purposes not related to Pioneer Saturn data the N79D offset dipole model was recommended. The JGR 80 model was derived using the 10 hr 14 min rotation rate and longitude convention in existence prior to the definition of the SLS system. The large difference in the two rotation rates ($1°4$ hr^{-1}) precludes transformation of the JGR 80 model to the SLS system. The present uncertainty in the SLS rotation period is ± 2 s (M. D. Desch, personal communication), leading to a longitude uncertainty of $\sim 19°$ between the time of the Pioneer 11 encounter and the Voyager 1 encounter and $11°$ uncertainty between the two Voyager encounters. Uncertainty in Saturn's rotation rate at the time of the Pioneer 11 encounter was one factor leading Smith et al. (1980*b*) to recommend the N79D offset dipole model. Physically, the N79D model describes a dipole source of moment 0.20 G–R$_S^3$, located on and parallel to Saturn's rotation axis but displaced northward by ~ 0.05 R$_S$. The displacement of the dipole in the Pioneer 11 fluxgate model is slightly less, ~ 0.04 R$_S$.

Preliminary analyses of the Voyager 1 observations (Ness et al. 1981) did not confirm the dipole northward offset derived from the Pioneer 11 observations. Connerney et al. (1982*b*) found that the Voyager 1 and 2 observations could not be well represented by a second-order spherical harmonic model or equivalently an offset dipole. However, an axisymmetric, octupole model consisting only of zonal harmonics ($m = 0$) to order $n = 3$ provided a consistent representation of Saturn's planetary field (i.e., that due to internal sources). The three zonal harmonics (g_1^0, g_2^0, g_3^0) proved to be both necessary and sufficient to describe Saturn's planetary field as measured by the two Voyagers. Inclusion of tesseral ($n > m > 0$) or sectorial ($n = m$) harmonics resulted in no significant improvement in fitting the observations (Connerney et al. 1982*b*; Acuña et al. 1983). Indeed, these authors found no evidence in the Voyager magnetometer observations of any departure from axisymmetry of

TABLE I
Pioneer 11 Magnetic Field Models [a]

| | Vector Helium Magnetometer | | | Fluxgate [c,d] |
	JGR 80 [b]	N79D [b,c]	P11A [e]	
Coefficients				
g_1^0 (Gauss)	0.218	0.20	0.2118	0.20
g_1^1	0.002	—	—	—
h_1^1	0.002	—	—	—
g_2^0	0.022	0.020	0.0198	0.016
g_2^1	0.002	—	—	—
g_2^2	0.001	—	—	—
h_2^1	−0.001	—	—	—
h_2^2	0.012	—	—	—
g_3^0	0.028	—	0.0235	—
g_3^1	0.002			
g_3^2	−0.001			
g_3^3	−0.002			
h_3^1	−0.004			
h_3^2	0.002			
h_3^3	0.007			
G_1^0	−16.9	−15	−9.9	—
G_1^1 (nT)	−1.7	—	—	—
H_1^1	4.8	—	—	—
G_2^0	−0.5			
G_2^1	1.0			
G_2^2	−0.1			
H_2^1	−0.7			
H_2^2	−0.6			
Equivalent Offsets (R_S)				
ΔX	0.005	—	—	—
ΔY	0.003	—	—	—
ΔZ	0.050	0.050	0.047	0.040

[a] Models assume $R_S = 60,000$ km unless otherwise noted. The number of digits quoted for the model coefficients reflects their different sources and not the estimated uncertainties.
[b] E. J. Smith et al. 1980*b*.
[c] These models were defined as offset dipoles. The g_2^0 coefficient given in this table was computed from the model offset and g_1^0, e.g., $g_2^0 = 2 \cdot g_1^0 \cdot \Delta Z$.
[d] Acuña et al. 1980.
[e] Davis and Smith, manuscript in preparation. For this model, $R_S = 60,330$ km.

the magnetic field, at a level of 1 or 2 nT (0.1% and 0.2% of the total field at closest approach). Zonal harmonic models obtained from the Voyager 1 and 2 data sets independently and the combined Voyager 1 and 2 data set (Z_3 model) are listed in Table II. The first two columns of Table II demonstrate that two independent determinations (Voyager 1 and 2 encounters) of the parameters of the zonal harmonic model yield very similar results. The Voyager 1 and 2 coefficients differ by less than 150 nT; percentage differences are < 1% for the g_1^0 term and < 5% for the g_2^0 and g_3^0. Thus, the Z_3 model, fitted to the combined data sets, was proposed as a model of Saturn's planetary field (Connerney et al. 1982*b*).

The ratio $g_2^0 : 2g_1^0$ for the Z_3 model is 0.038, similar to that of the Pioneer 11 models. The relatively large g_3^0 terms, however, are inconsistent with a simple axially offset dipole representation. The ratio $g_3^0 : g_2^0$ is too large by a factor of ~ 30 to be attributed to a dipole offset; the model's large hemispherical asymmetry cannot be removed by a simple coordinate system translation. The magnetic field magnitude as a function of latitude at the surface for a dipole, offset dipole, and the Z_3 model field is shown in Fig. 4. The maximum polar surface field magnitudes given by the Z_3 model in the north and south

TABLE II
Voyager Magnetic Field Models [a]

Coefficients (nT)	V1	V2 Axisymmetric (n_{max} = 3)	Z_3 (V1 + V2)
g_1^0	21,576	21,433	21,535
g_1^1	—	—	—
h_1^1	—	—	—
g_2^0	1,715	1,643	1,642
g_2^1	—	—	—
g_2^2	—	—	—
h_2^1	—	—	—
h_2^2	—	—	—
g_3^0	2,686	2,583	2,743
G_1^0	−11	−8	−10
G_1^1	0	−1	−1
H_1^1	0	−2	0
rms (nT)	3.0	2.5	3.2

[a] Connerney et al. 1982*b*. These models use R_S = 60,330 km. Parameter uncertainties can best be estimated by comparison of the results of the independent Voyager 1 and 2 analyses.

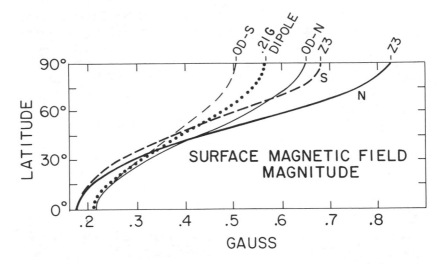

Fig. 4. Magnetic field intensity as a function of latitude on Saturn's oblate (flattening = 1/10.6) surface. Bold solid (dashed) line is the field intensity at north (south) latitudes corresponding to the Z_3 model. Light lines indicate the field intensity of the offset (0.04 R_S) dipole model and the dotted line corresponds to a simple centered dipole with 0.21 G–R_S^3 moment (figure after Connerney et al. 1982b).

are ~ 0.84 and ~ 0.65 Gauss, comparable with the Earth's ~ 0.7 Gauss polar intensities. The equatorial field magnitude is reduced to 0.18 Gauss by the relatively large g_3^0 coefficient. Saturn's zonal coefficients (normalized to the dipole term, g_1^0) are somewhat larger than those of the present-day Earth.

Several new models have been derived from the Pioneer 11 vector helium magnetometer observations (Davis and Smith, manuscript in preparation). One of these, the axially symmetric P11A model, uses the three zonal harmonic g_n^0 internal field terms and a constant axial external field G_1^0 to represent the field at radial distances $r < 8$ R_S. The coefficients obtained from this inversion, listed in Table I, are similar to those of the Z_3 model (Table II) and the axisymmetric terms of the JGR 80 model. This new model, P11A, fits the Pioneer 11 vector helium magnetometer observations with a residual of 1.8% compared to the 1.4% residual of the JGR 80 model listed by E. J. Smith et al. (1980b). Additionally, Davis and Smith (manuscript in preparation) have reconsidered a number of models including nonaxisymmetric field components and find that the Pioneer 11 data support the existence of a small nonaxisymmetric component of the internal field, but they cannot reliably determine the values of the corresponding spherical harmonic coefficients. In contrast, Connerney (1983) examined the same spacecraft data and found that the observations were consistent, within the accuracy of the Pioneer 11 vector helium instrument, with axisymmetry of the internal field. Whether or not a real departure from axisymmetry is evidenced in the magnetometer observations is currently under investigation.

Charged-Particle Observations

Additional information about the magnetic field geometry in Saturn's magnetosphere is available in the form of *in situ* energetic charged particle observations. Since charged particles closely follow magnetic field lines in the course of their latitudinal bounce motion, the signatures which result from satellite and ring-absorption effects can be used to infer the shape of field lines in latitude. Consideration of charged particle longitudinal drifts, together with the rotation of Saturn's magnetic field relative to the absorbers, can provide additional information about the variation with longitude of the magnetic field geometry.

An example of a satellite absorption signature is shown in Fig. 5. In this figure the 63- to 160-MeV proton flux measured by the cosmic ray system (CRS) on Voyager 2 (Vogt et al. 1982) shows the effects of absorption of these protons by Saturn's satellites Mimas and Enceladus. The absorption signatures visible in Fig. 5 are permanent, stable features of the energetic proton flux that exist at all longitudes. If Saturn's magnetic field were that of a dipole, these absorption signatures would appear at the same value of $L = r \cos^{-2}(\lambda)$ inbound and outbound, where r is the distance of the spacecraft from the dipole and λ is the spacecraft latitude measured from the dipole equator.

In Fig. 5a L was calculated assuming a magnetic dipole centered on Saturn and aligned with Saturn's rotation axis. For this assumed field geometry, absorption signatures of Mimas and Enceladus appear at greater values of L inbound when the spacecraft was in the northern hemisphere ($\lambda \sim 20°$), and at lesser values of L outbound in the southern hemisphere ($\lambda \sim -10°$). Similarly the Tethys absorption signature (Fig. 6) was observed almost 5 min earlier than a centered-dipole magnetic field model would have predicted.

Other magnetic field models which better organize the data of Fig. 5a can be found. An example of such a better model is shown in Fig. 5b, which illustrates the effect of incorporating a 0.05 R_S northward offset of the dipole center. Alternatively, tilting the dipole by 1°7 toward 180° SLS provides an equivalent improvement in organizing these data. This technique was used by Van Allen et al. (1974) to accurately determine the tilt angle (θ) and longitude (ϕ) of Jupiter's dipole field (see also E. J. Smith et al. 1974). A consideration of satellite (ring) absorption features observed by Pioneer 11 at Saturn enabled Simpson et al. (1980*a*) to set an upper limit of 0.01 R_S on the maximum equatorial offset of Saturn's magnetic dipole axis from its rotation axis.

Chenette and Davis (1982) developed a least-squares method to determine the coefficients of an axially symmetric spherical harmonic model of Saturn's magnetic field which best organize the charged particle signatures.

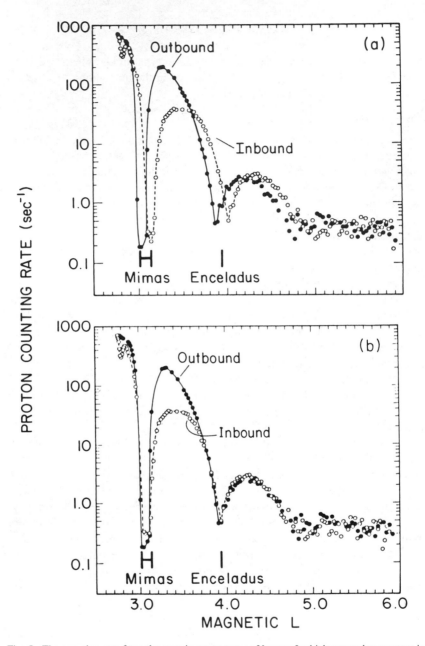

Fig. 5. The counting rate from the cosmic ray system on Voyager 2 which responds to protons in the energy range 63- to 160-MeV, plotted versus L, for two different dipole magnetic field models. In panel (a), the dipole was assumed to be centered on Saturn, while in panel (b) the dipole was assumed to be offset north by 0.05 Saturn radii. In both cases the dipole was assumed to be aligned with Saturn's axis of rotation. Alternately, an equally good fit is obtained by tilting a centered dipole $1°7$ towards SLS longitude 180°. Open symbols connected by dashed lines represent data obtained along the inbound pass. Filled symbols connected by solid lines represent data from the outbound pass. L values accessible to the satellites Mimas and Enceladus are indicated (figure from Chenette and Davis 1982).

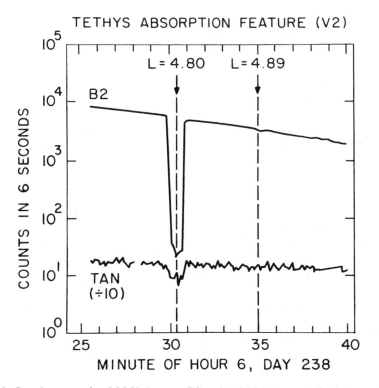

Fig. 6. Counting rates of ~ 2 MeV electrons (B2) and ~ 5 MeV electrons (TAN) versus time near the orbit of Tethys as the spacecraft passed within $1°$ longitude of that satellite. The width of the well-defined absorption feature at minute 30 corresponds to ~ 1100 km which is nearly equal to the 1050 km diameter of Tethys. The vertical dashed lines mark the time of the observed ($L = 4.80$) absorption feature and that expected ($L = 4.89$) for a centered, aligned dipole model of Saturn's magnetic field (figure from Vogt et al. 1982).

They adopted the usual spherical harmonic expansion representation, but limited it to include only the axially symmetric terms. They considered only the two lowest-order internal field terms, the dipole g_1^0 and quadrupole g_2^0 moments, and the lowest-order external field term, a uniform field with intensity determined by the coefficient G_1^0. Assuming axial symmetry, particle L shells are surfaces of revolution which enclose a constant magnetic flux. Each L shell is thus uniquely identified by the value of this function. For a spherical harmonic model of order 3, the flux linked by a particle drift shell is given by

$$\psi(r,\,\lambda) = \frac{\cos^2\lambda}{r} + \frac{(g_2^0/g_1^0)\,3\cos^2\lambda\,\sin\lambda}{2r^2} + \frac{(g_3^0/g_1^0)[5\sin^2\lambda - 1]\cos^2\lambda}{2r^3}$$
$$- \frac{(G_1^0/g_1^0)\,r^2\cos^2\lambda}{2} \qquad (4)$$

where r is the distance from Saturn in units of R_S and λ is the latitude relative to the magnetic dipole equator; the model parameters are enclosed in parentheses. From observations of charged particle absorption signatures made on the Pioneer 11, Voyager 1, and Voyager 2 spacecraft, Chenette and Davis constructed a data set consisting of pairs of points that should lie on the same L shell. The model parameters were determined in a least-squares sum method by minimizing the squared difference in ψ between the points of each pair, summed over all pairs. The details of the method and tabulations of the data sets are given by Chenette and Davis (1982).

Chenette and Davis considered various models and subsets of the available data and could not discriminate between the alternative models of a northward offset of the dipole and a small tilt of a centered dipole. Both models provided fits to these absorption features which were equally good. The magnitude of the tilt required, however, was inconsistent with the maximum tilt allowed by the Pioneer 11 and Voyager magnetometer data. Thus they concluded that models which incorporate a northward offset of the dipole are preferred over those in which the dipole is assumed to be centered on Saturn. The magnitudes of the derived dipole offsets were consistent with the values derived from the analysis of the Pioneer 11 magnetometer data (E. J. Smith et al. 1980b; Acuña et al. 1980).

A reanalysis of the charged particle observations, motivated by the introduction of the Z_3 model, was performed using the same techniques but allowing an octupole zonal harmonic g_3^0 in addition to the dipole g_1^0 and quadrupole g_2^0 considered previously. Inclusion of the octupole zonal harmonic substantially improves the fit of the model to the charged particle absorption signatures. The numerical values obtained for the model parameters (the ratios g_2^0/g_1^0 and g_3^0/g_1^0) in a least-squares fit to the charged particle observations depend to some extent on the data subsets considered.

The Voyager 2 charged-particle observations were obtained at high latitudes, far from the orbital plane of the absorbing satellites (see Fig. 1 and also Fig. 10 in Sec. III), and are thus ideally suited for studying these magnetic field models. The one signature currently available from the Voyager 1 encounter, that due to Rhea at $\sim 8.8\ R_S$, is too distant to be of value in studies of the planetary field. The magnetic field geometry at the orbit of Rhea is most sensitive to the (local) equatorial ring currents (Vogt et al. 1981; Connerney et al. 1981). The Pioneer 11 absorption signatures, obtained so close to the orbital plane of the absorbing satellites, are relatively less sensitive to differences among these axisymmetric magnetic field models. The Pioneer 11 signatures are, however, very sensitive to longitudinal variations in magnetic field geometry (see, e.g., Simpson et al. 1980a; Chenette and Davis 1982).

In Table III, we compare the zonal harmonic model (absorption) least squares fitted to the Voyager 2 absorption features with the Z_3 and P11A models and associated rms residuals. Estimated parameter uncertainties (1 standard deviation) are given in parentheses for the least-squares model. The Z_3

TABLE III
Comparison of Zonal Harmonic Models with Charged Particle Data

Model	rms Residual (arb. units)	g_2^0/g_1^0 (σ)	g_3^0/g_1^0 (σ)	G_1^0/g_1^0 (σ)
Absorption	1.1	0.076 ($\pm$ .011)	0.175 ($\pm$ .11)	-5.60 ($\pm$ 1.5) $\times 10^{-4}$
Z_3	1.2	0.076	0.127	-4.60 $\times 10^{-4}$
P11A	2.1	0.094	0.111	-4.40 $\times 10^{-4}$

model, which was derived from magnetic field observations only, fits the charged particle absorption observations almost as well as the least-squares solution. All three of the optimal model parameters are consistent with those of the Z_3 model. Two of the three are also within one standard deviation of the corresponding P11A model parameters. The exception is the relatively large quadrupole term of the P11A model. (Inclusion of the Pioneer 11 signatures in this analysis yields slightly different numerical values for the best-fitting zonal harmonic model parameters but does not alter any of the conclusions obtained above. Inclusion of the Pioneer 11 signatures results primarily in larger—by a factor of 2—estimated uncertainties for each of the model parameters.)

Acuña et al. (1983) have used the observed charged particle absorption signatures as an independent check of the model magnetosphere obtained from the Voyager magnetometer observations. The model magnetosphere consists of the Z_3 planetary field and the ring current (described in the next section). Each observed satellite absorption signature was found to be consistent with the predictions of the model magnetosphere. These studies of the charged particle absorption signatures of Saturn's satellites demonstrate (independent of the magnetic field data) that Saturn's planetary magnetic field is well represented by an axisymmetric, zonal harmonic expansion to order 3.

In recent work, Northrop and Hill (1983b) presented an analysis of charged particle motion quite unlike the particle absorption studies discussed thus far. They considered the stability of (high charge-to-mass ratio) charged dust particles in Saturn's ring plane. Northrop and Hill calculated a "marginal stability radius" inside of which these charged dust particles would become unstable in Saturn's ring plane, consequently depopulating the rings. They identified the optical inner edge of the B Ring at 1.524 R_S as a manifestation of this stability limit, which is very sensitive to the local magnetic field geometry. Northrop and Hill found that within the context of their theory, the zonal harmonic models obtained from the Voyager magnetometer observations predicted very accurately the radial position of the inner edge of the B Ring, whereas dipole and offset dipole models did not. Thus, Saturn's rings may provide yet another independent measure of the magnetic field geometry to complement the magnetometer observations.

IV. RING CURRENT

The next most important contribution to the magnetic field in Saturn's magnetosphere is due to an extensive equatorial ring current (Connerney et al. 1981,1983) of $\sim 10^7$ A flowing eastward at distances of ~ 8 to $16\,R_S$ from Saturn. The (model) ring current derived from the Voyager 1 observations is confined to a 5-R_S-thick ring in which the current density varies inversely with distance from Saturn (cf. Fig. 1). The near-axis ring-current field is essentially axial and antiparallel to Saturn's (equatorial) dipole field; at larger distances the two are parallel. Above the equatorial plane, the radial field is positive outward and below the equator, positive inward, increasing approximately linearly with distance Z above the equator for $Z < D$, the half thickness. The total field in this model is a sum of the field due to the ring current and that of Saturn's main field; smaller contributions due to magnetopause and tail currents have been neglected.

The Voyager 1 and 2 magnetic field observations at Saturn are shown in Figs. 7 and 8, in which the observed field magnitude is compared with that of a 0.21 G–R_S^3 centered dipole. Voyager 1 observed initially an enhanced magnetic field while inbound to periapsis near the outer edge of the model ring

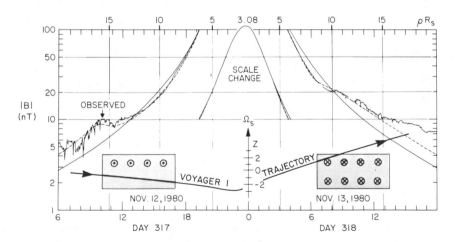

Fig. 7. Magnitude of the observed magnetic field within 20 R_S of Saturn for Voyager 1 encounter, compared with a centered dipole model (solid line) and the Voyager 1 ring-current model (dashed). The Voyager 1 trajectory is illustrated below, showing distance from the equator as a function of time. Note the different scale used for the vertical dimension of trajectory insert (figure from Connerney et al. 1981).

current as shown in Fig. 7. Because of Voyager 1's proximity to the equatorial plane, the field magnitude at lesser radial distances ($r < 10$ R$_S$) is diminished by the ring current. The localized field enhancement observed by Voyager 1 at $\sim$ 16 R$_S$ while inbound to periapsis is a consequence of the model and is understood as an edge effect. The lack of such a localized enhancement as Voyager 1 passed over the outer edge of the ring current while outbound from periapsis is, within the context of the model, a consequence of Voyager 1's higher latitude at this time. However, the development of an extensive magnetic tail (Behannon et al. 1981) would also be expected to remove any such edge effects. In the antisolar direction, the transition from eastward ring currents to dusk-dawn cross-tail currents eliminates the abrupt change in current that results in the edge effect as seen on the day side.

Similar behavior is evident in the Pioneer 11 observations illustrated in Fig. 9, in which the observed field magnitude is compared with a 0.2 G–R$_S^3$ centered dipole. Pioneer 11's trajectory (cf. Fig. 1) was more similar to the Voyager 1 inbound trajectory, as it remained close to the magnetic equator throughout encounter. Pioneer 11 observed initially an enhanced field magnitude while moving inbound toward periapsis, as did Voyager 1. This may be indicative of the enhancement of the field expected at the outer edge of the ring current, or alternately a contribution due to compression of the magnetopause. Inside of $\sim$ 10 R$_S$, both inbound and outbound, the observed field magnitude was less than the dipolar field, again very similar to the Voyager 1 observations and qualitatively consistent with the behavior expected of a ring

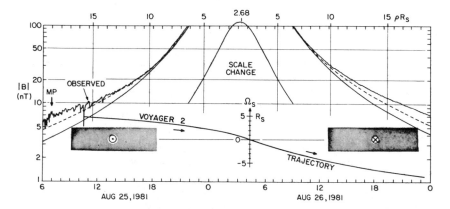

Fig. 8. Magnitude of the observed magnetic field within $\sim$ 20 R$_S$ of Saturn for Voyager 2 encounter, compared with a centered dipole (solid) and the Voyager 1 ring-current model (dashed). The Voyager 2 trajectory is illustrated below showing distance from the equator as a function of time. Note the different scale used for trajectory insert (figure from Ness et al. 1982a).

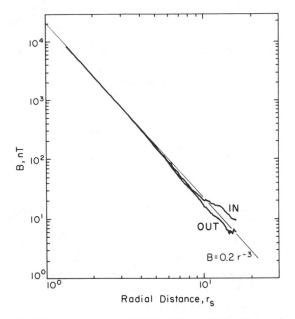

Fig. 9. Magnitude of the observed magnetic field within ~ 15 R$_J$ of Saturn for the Pioneer 11
    encounter, compared with (light line) that of a centered dipole. Logarithmic scales are used for
    both field magnitude and radial distance (figure from E. J. Smith et al. 1980a).

current. Outbound, along the dawn meridian, Pioneer 11 measured a reduced
field magnitude to ~ 15 R$_S$ and thereafter an enhanced field magnitude characteristic of an equatorial current sheet (E. J. Smith et al. 1980a).

The Voyager 1 ring-current model proved to be generally successful in
predicting the field magnitude observed by Voyager 2 (Fig. 8). The Voyager
1 ring-current model leads to an enhanced field magnitude along most of Voyager 2's trajectory, as a consequence of the relatively high latitude of Voyager
2. This is in contrast to the reduced field magnitudes observed by Pioneer 11
and Voyager 1 at lower latitudes. The agreement between the (predicted)
model field and the observed field is less good along the outbound trajectory
of Voyager 2 (near the dawn meridian). This lack of agreement may be due to
the neglect in this model of magnetotail and magnetopause currents associated
with the solar wind interaction (Ness et al. 1982a) or to solar wind induced
time variations of Saturn's magnetosphere (Ness et al. 1982a; Bridge et al.
1982; Vogt et al. 1982).

Magnetospheric Geometry

As a result of the ring current in Saturn's magnetosphere, the magnetic
field assumes a nondipolar geometry as is illustrated in Fig. 10. The nondipo-

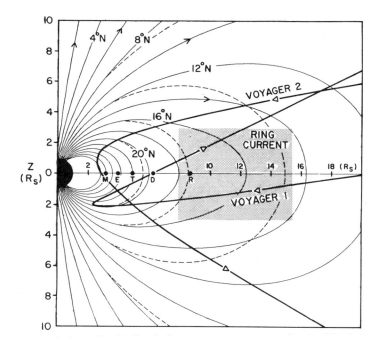

Fig. 10. Meridian plane projection of magnetosphere field lines for a dipolar field model (dashed) and a model containing a dipole and a distributed ring current (solid). Field lines are drawn for 2° increments in latitude. Positions of the major satellites Mimas, Enceladus, Tethys, Dione, and Rhea are as in Fig. 1 (figure from Connerney et al. 1983).

lar nature of Saturn's magnetosphere is evident beyond $\sim 5\ R_S$ but is particularly significant at the orbit of Rhea (8.78 R_S) and beyond (Connerney et al. 1981; Vogt et al. 1981). The stretching out of field lines in the equatorial plane is moderate when compared with Jupiter's magnetodisk geometry. However, even this modestly nondipolar field results in charged particle bounce periods and longitudinal drift rates that differ from the dipole values by as much as a factor of 2 at ring-current radial distances (Birmingham 1982).

The ring current in Saturn's magnetosphere reflects the dynamic balance between forces acting upon the entrapped, corotating magnetospheric plasma. The centripetal acceleration associated with the corotational motion of the plasma, the local pressure gradient, and the $\mathbf{J} \times \mathbf{B}$ force of the ring current must cancel if there is no acceleration of the plasma in the corotating frame. McNutt (1983b) examined the plasma force balance implied by the Voyager 1 plasma observations and the ring-current model of Saturn's magnetosphere and concluded that an approximate equilibrium is achieved at ring-current radial distances of ~ 9 to 16 R_S. Connerney et al. (1983) used the magnetohydrodynamic force balance equation to compute the plasma pressure and mass (ion) density implied by the azimuthal current density and magnetic field of the ring-current model. The computed ion density, assuming a heavy ion species with a mass-to-charge ratio of 16 (oxygen), is compared in Fig. 11 with the ion density obtained by Frank et al. (1980) from Pioneer 11 observations and the equatorial ion density inferred by Bridge et al. (1982) from Voyager 1 and 2 observations of electron density at higher latitudes. At ring-current radial distances there is a rather good agreement among all three estimates of ion density.

V. DISCUSSION

The most remarkable and enigmatic aspect of Saturn's planetary magnetic field is its high degree of axisymmetry with respect to the rotation axis. The Z_3 and P11A models possess axial symmetry, as does the offset dipole model; previous models suggested dipole tilts of $\lesssim 1°$. In contrast, the Earth and Jupiter have dipole tilts of $11°5$ and $9°6$. The tilt of Mercury's dipole field is less well known but at $14 \pm 5°$ (Ness 1979) comparable to that of Earth and Jupiter. This axisymmetry is remarkable because it is the only such dynamo known and it is therefore of central interest to dynamo theory. It is enigmatic because a number of phenomena have been reported which seemingly require a longitudinal asymmetry of the magnetic field.

Foremost among these is the periodic emission, or rotational modulation, of Saturn kilometric radiation (SKR) upon which the rotation rate of Saturn is based (Kaiser et al. 1980). SKR originates near local noon in a region of the northern hemisphere which maps to the surface along field lines into a narrow band at auroral latitudes (Kaiser and Desch 1982), possibly a dayside cusp region. These emissions are also highly correlated with solar wind variations (Desch 1982) suggesting a rather direct but as yet unknown relationship between SKR and the solar wind. The rotational modulation of SKR is taken to be indirect evidence of a small near-surface magnetic anomaly at $\sim 115°$ SLS longitude and $\sim 76°$ N latitude (Kaiser and Desch 1982; also see chapter by Kaiser et al.).

Additional reports of time-dependent phenomena include a periodic variation of optical spoke activity in the B Ring (Porco and Danielson 1982), polar auroral brightening (Sandel et al. 1982b), the periodic occurrence of Saturn electrostatic discharges (SED) (Warwick et al. 1981; Evans et al. 1981), and a quasi-periodic variation of low-energy charged particle fluxes (Carbary and Krimigis 1982). All of these with the exception of the SED periodicity have been ascribed to a possible magnetic anomaly. The SED periodicity is distinctly different from the SKR periodicity which defines the planetary rotation rate, but matches the periodicity of Saturn's superrotating equatorial atmosphere. A long-lived equatorial lightning storm has been proposed (Kaiser et al. 1983; chapter by Kaiser et al.) as the source of SED. The remaining periodicities are consistent with the SKR rate.

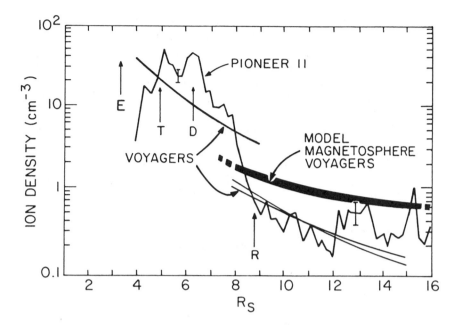

Fig. 11. Comparison of the ion density computed from the ring-current model magnetosphere with ion densities obtained by the Pioneer (Frank et al. 1980) and Voyager 1 and 2 (Bridge et al. 1982) plasma experiments. Radial positions of the satellites Enceladus, Tethys, Dione, and Rhea as indicated (figure from Connerney et al. [1983], adapted from Bridge et al. [1982]).

Thus far, none of the temporal variations described can be linked directly to a longitudinal asymmetry of the magnetic field. To do so requires an understanding of the underlying physical processes or mechanisms by which these phenomena occur. However, these observations are compelling evidence for some longitudinal asymmetry, or anomaly (magnetic or otherwise), which is not evident in the magnetic field observations.

The high degree of symmetry of the magnetic field about Saturn's rotation axis (spin symmetry) clearly distinguishes this dynamo from those of Mercury, Earth, and Jupiter. Whether Saturn's magnetic field is unique in this respect is difficult to judge from such a small sample. The surprising discovery of the spin symmetry of Saturn's magnetic field has already provoked speculation (Stevenson 1983a) that the magnetic fields of Uranus and Neptune may also be spin-symmetric. The Voyager 2 spacecraft, already en route to an encounter with Uranus in January 1986 and Neptune in 1989, will soon obtain observations sufficient to resolve this question.

The well-known Cowling anti-dynamo theorem (Cowling 1934), in its original form, excludes the possibility of sustaining a steady-state axisymmetric magnetic field by axisymmetric fluid motions. Thus, Hide (1981) suggested that an axisymmetric field is indicative of the ohmic decay phase (e.g., reversal) of a dynamo. The probability of actually observing an Earth-like dynamo in the decay phase is exceedingly small since the time scale for reversal is 10^2 or 10^3 yr, as compared with a frequency of reversal events of a few per 10^6 yr (determined from paleomagnetic studies). The corresponding probability of observing the decay phase of Saturn's magnetic field at the present time is unknown, due to a lack of knowledge of the time scales involved, so this possibility cannot be discarded. Paleomagnetic evidence suggests (Hoffman 1977) that the transition field characteristic of the reversal of the geodynamo is axisymmetric. Williams and Fuller (1981) modeled one such geomagnetic field reversal with an axisymmetric zonal harmonic model field in which higher order zonal harmonics grow at the expense of the dipole term. Interestingly, these transition field models are similar to the zonal harmonic models of Saturn's magnetic field.

An alternative explanation for the spin symmetry of Saturn's magnetic field has been proposed by Stevenson (1980). He suggests that Saturn's dynamo itself is not axisymmetric, but only outwardly appears as such. The nonaxisymmetric field components in his model are simply attenuated by the differential rotation of a metallic conducting shell above the active dynamo region. The formation of this shell is a consequence of the limited solubility of helium in hydrogen in Stevenson's model of Saturn's interior. The model also explains the depletion of helium in Saturn's atmosphere.

Whether Saturn's magnetic field is spin-symmetric, or only nearly so, it is certain to play an important role in contemporary dynamo theory and models of Saturn's interior.

Acknowledgments. The authors are grateful to their colleagues on the Pioneer 11 vector helium magnetometer team headed by E. J. Smith, the Voyager magnetometer team headed by N. F. Ness, and the Voyager cosmic ray experimenters led by R. E. Vogt for their contributions to the experiments and this review. We also thank F. Hunsaker and L. White for producing the illustrations. D. L. C. acknowledges support by the Aerospace Corporation's sponsored research project.

Note added in proof by L. Davis, Jr.: E. J. Smith and I (and independently G. R. Wilson, and D. E. Jones) have recognized that the field due to external sources may have changed significantly during the Pioneer 11 encounter and presumably at other times. This was to be expected because the solar wind and the interplanetary magnetic field changed substantially during the Pioneer 11 encounter. A simple model that crudely approximates the effects of these changes and that significantly improves the fit of the model to the observations uses one value of G_1^0 before periapsis and another value afterward. The parameters for this model in the SLS coordinate system are $g_1^0 = 0.2114$, $g_2^0 = 0.0196$, $g_3^0 = 0.0228$ Gauss with $G_{1\,in}^0 = -12.6$ and $G_{1\,out}^0 = -7.1$ nT. The weighted percent residual for data inside 8 R_S for this case is 1.63% as compared with 1.85% for the case with only one value of G_1^0.

J. E. P. Connerney, M. H. Acuña, and N. F. Ness (1984) have found that the Pioneer 11 data would fit their Z_S model very well if the data frame were rotated about Pioneer's spin axis by $-1°4$. Subsequently, Davis and Smith, using procedures similar to those used for their previous analyses, found that a suitable rotation did significantly improve the fit. Using the same weights as for the P11A model of Table I, they found that the best fit was obtained with a roll of $-0°8$, which gave a 1.15% residual with $g_1^0 = 0.2114$, $g_2^0 = 0.0175$, $g_3^0 = 0.0227$ Gauss, $G_{1in}^0 = -12.4$, $G_{1out}^0 = -7.1$ nT. A slightly better fit with a residual of 1.12% would be obtained with an inbound roll of $-1°1$ and a $-0°4$ roll outbound. The only significant change in the coefficients would be that g_2^0 became 0.0163 Gauss. The need for such a roll correction could stem from a flaw in the data processing, which E. J. Smth and L. Davis are convinced did not occur, or it could stem from the great difficulty of determining the roll attitude of the spacecraft at a time when Saturn, the Sun, and the Earth were nearly in line and the roll was supposed to be measured from the plane through the three bodies. If there were a flaw in the data processing, the roll correction should be constant throughout the encounter; if uncertainty in the roll orientation is the problem, the correction could well vary with time as the temperature of the spacecraft changed during occultation, and its moment of inertia and rotation rate would change.

If nonaxisymmetric internal source coefficients were added to these new models, the additional coefficients determined are considerably smaller than in the previous models and it now appears that any deviation from axial symmetry of the internal sources is below current levels of detection. It still seems probable that the changing external sources are not axially symmetric at the 10% level although it cannot be proved that the apparent deviations from axial symmetry are not actually due to changes with time.

In any attempt to model Saturn's magnetic field in the region inside 8 R_S, it must be kept in mind that the field due to internal sources appears to be constant on a time scale of at least a few years and to be accurately described by a small number of spherical harmonics. The field due to external sources—ring currents, magnetopause currents, and other external current systems—are complicated and change with time. With the available data only simple approximations to the actual external field sources can be used as models. There is always the danger that while a good fit may be obtained to the available data, the fit to the actual field in regions where it was not observed may be poor. The likelihood that this will occur increases rapidly as more parameters are introduced in the model. Changes with time may be improperly modeled as a complicated static model.

Note added in proof by L. R. Doose: The source of the roll-attitude corrections for Pioneer Saturn discussed in the second paragraph of the above note has been found. A close examination of data from the Imaging Photopolarimeter in conjunction with the best available spacecraft trajectory yields roll-attitude values shown in the figure below. Roll-attitude determinations for Pioneer Saturn were originally intended to be based on a sensor which provided a pulse each time the Sun passed through its field of view. At the time of encounter, however, the Sun was only 1° off the direction of the spin axis, and the system was subject to attitude errors as large as 10°. Experimenters were therefore forced to base their analyses on attitudes derived by comparison between predicted and observed limb fits to Saturn in the data of the Imaging Photopolarimeter. These data yielded attitudes shown by the solid curve. Improvements were made through careful limb fits using the final spacecraft trajectory. The new results, shown in the figure as triangles (low weight) and squares (high weight), supercede previously available values. The phase lag compared to the old attitude is about 0°8 six hours before closest approach. It decreases slowly until closest approach when it rapidly drops into close agreement with the older values.

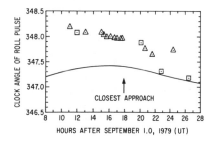

HOURS AFTER SEPTEMBER 1.0, 1979 (UT)

SATURN AS A RADIO SOURCE

M. L. KAISER, M. D. DESCH
NASA Goddard Space Flight Center

W. S. KURTH
University of Iowa

A. LECACHEUX, F. GENOVA, B. M. PEDERSEN
Observatoire de Paris

D. R. EVANS
Radiophysics, Inc.

Saturn has three nonthermal components to its radio spectrum: kilometric radiation, narrowband noise and related continuum, and electrostatic discharges. All three components were discovered by the Voyager mission. The Saturn kilometric radiation beams the isotropic equivalent power of 50 gigawatts at maximum intensity. The radiation takes place in the frequency range from as low as 3 kHz to ~ 1200 kHz, with peak intensity near 200 kHz in the kilometer wavelength band. The intense storms of km-wavelength radiation display periodicities on several time scales. The most fundamental of these periodicities is the tendency for storms to recur every 10 hr 39.4 min on the average; this is inferred to be the rotation period of Saturn's global magnetic field. Occasionally storms are also modulated at a period of ~ 66 hr, possibly due to the influence of the satellite Dione. Finally, a strong correlation exists between the solar wind and the intensity of Saturn's radio emission. This modulation induces a long-term periodicity in the radio output of ~ 25 days, consistent with the solar rotation period. Saturn's kilometric radio emission is strongly right-hand polarized (radio astronomical sense) when observed from above Saturn's northern hemisphere, and left-hand polarized when observed from above the southern hemisphere. These observations are interpreted in terms of radiation near the electron gyrofrequency escaping in the extraordinary mode from two source regions, one in the northern hemisphere and one in the southern hemisphere. The footprints of these sources are confined to small regions above Sat-

urn's auroral zones or polar cusps and near the noon meridian. In addition to this intense component of Saturn's magnetospheric radio spectrum, there are weak narrowband tones observable at very low frequencies (< 10 kHz) within Saturn's magnetosphere. These tones may be associated with plasma sheet gradients in the inner magnetosphere. Saturn is also the source of low-level continuum radiation at frequencies below the solar wind plasma frequency. This radiation is trapped in the low-density cavity formed by the outer magnetosphere in a manner similar to the trapped continuum radiation at Jupiter and the Earth. These emissions are presumed to be generated in the left-hand ordinary sense via a mode conversion process from electrostatic Bernstein waves near the upper hybrid resonance frequency. These two magnetospheric radio components are complemented by a third component, Saturn electrostatic discharges, which appeared as very brief, intense, broadband bursts in episodes separated by ~ 10 hr 10 min, distinctly faster than the magnetospheric emission's repetition rate. They are best explained as radio bursts from a 60°-wide lightning storm complex in Saturn's equatorial zone.

I. INTRODUCTION

The announcement by the Voyager planetary radio astronomy team (Kaiser et al. 1980) of the unequivocal detection of low-frequency nonthermal radio emission from Saturn ended two decades of frustration arising from the search for these emissions. Several researchers (Smith and Douglas 1957, 1959; Smith 1959; Smith and Carr 1959; Carr et al. 1961) reported bursts of possible Saturnian origin in the frequency range from 18 to 22 MHz, though all these authors pointed out that the association with Saturn was by no means certain. Observations in the wavelength range from 1 mm to about 70 cm failed to detect synchrotron emission similar to that observed in Jupiter's radiation belts (see Newburn and Gulkis 1973, and references therein). Subsequent *in situ* measurements of the magnetic field strength and the particle populations by Pioneer 11 and Voyagers 1 and 2 indicate that Saturn's radiation belts should not be a source of synchrotron emission. In a well-known paper, Brown (1975) reported the possible detection of Saturn radio bursts near 1 MHz using the radio astronomy instrument onboard the Earth-orbiting IMP 6 spacecraft. However, Kaiser (1977) was unable to confirm Brown's findings using data from a similar radio astronomy instrument onboard the RAE 2 spacecraft in lunar orbit. No radio astronomy investigations were carried aboard the Pioneer 11 spacecraft, and so Saturn's radio signature in the km-wavelength band was not revealed until early 1980, when the Voyager 1 spacecraft had approached to within 3 AU. Subsequent measurements of Saturn's varied radio spectrum by the planetary radio astronomy (PRA) and plasma wave science (PWS) instruments indicate that the radio bursts reported prior to 1980 were probably not associated with Saturn.

In this chapter, after a brief introduction (Sec. I) on the instrumentation and Voyager trajectories, we will describe the observable properties (Secs. II and III) and follow this with discussion of inferred source locations, emission theories and the future (Secs. IV, V and VI).

The PWS instrument overlaps the PRA frequency coverage at the low-frequency extreme ($<$ 60 kHz) and extends the Saturn observations down to $\sim$ 10 Hz. We will always be describing freely propagating nonthermal electromagnetic emission, i.e. emission that in principle could be detected from a fairly large distance from the planet (except for the case of the trapped continuum radiation, which is freely propagating only within the confines of the magnetospheric cavity). The *in situ* waves or plasma waves are described elsewhere in this book (see chapters by Scarf et al. and Neubauer et al.). Also, thermal radiation from Saturn's atmosphere will not be described here; refer to the review by Newburn and Gulkis (1973). After describing the observations we discuss the analysis of these observations completed as of May, 1983.

A. Instrumentation

Before reviewing the observations, we briefly describe the PRA and PWS instruments and discuss the trajectories of the two Voyagers as they approached and then receded from Saturn. These instruments have been fully described elsewhere (Warwick et al. 1977; Scarf and Gurnett 1977), so only those details of specific relevance to the Saturn observations will be discussed here.

Both instruments use the same pair of orthogonally mounted 10-m antennas; PRA uses them as separate monopoles, and PWS as a balanced dipole of 7-m effective length. With this antenna system the directivity at frequencies below a few MHz is very poor, so that direction-finding is purely inferential. The PRA instrument combines the signals from the monopoles in a 90° hybrid and measures incident radio wave power in the left-hand (LH) and right-hand (RH) circular or elliptical polarizations. The PWS instrument measures total received power in a given frequency band without reference to polarization.

The PRA instrument covers the frequency range from 40.5 MHz down to 1.2 kHz in two bands. The low-frequency band covers the range from 1326.0 to 1.2 kHz in 70 1-kHz-wide channels spaced 19.2 kHz apart. The high band covers the range from 40.5 to 1.5 MHz in 128 200-kHz wide channels spaced every 307.2 kHz. The sweep through the entire frequency range (both bands) is accomplished in 6 s with each individual channel requiring 25 msec of integration time plus 5 msec of settling time. The effective threshold of the instrument is limited by spacecraft-generated noise and corresponds to an ability to detect a signal in the low band at 60 kHz of 3×10^{-20} Wm^{-2}Hz^{-1} at 1 AU from the source. This detection capability improves by a factor of 3 above 150 kHz. However, the PRA high band suffers severely from spacecraft-generated noise so that the detection capability is typically only 10^{-18} Wm^{-2}Hz^{-1} at 10 MHz, which effectively limits observations to within a few hundred Saturn radii.

In the idealized case where the PRA monopoles may be considered as half dipoles, i.e. as monopoles erected above an infinite perfectly conducting plane, the responses of both antennas are elliptically polarized. The polarization ellipses are identical in shape, and have parallel major axes. The ellipticity varies with the source latitude above the plane of the monopoles. The polarization ellipses degenerate to a circle or to a straight line when the radio source direction is, respectively, normal to or constrained in the plane of the monopoles. The indicated circular polarization degree V_i is obtained simply as the ratio of the difference to the sum of the detected power flux in each polarization channel. This quantity V_i and the received total power summarize the information measured by the PRA instrument. V_i differs from the true circular polarization degree V (the fourth Stokes' parameter) by a proportionality coefficient that depends on the source latitude above the monopole plane and the 'cross-polarization' of the elliptical equivalent antennas. In addition, if the observed radiation contains a linearly polarized or unpolarized component, the antennas and wave polarization ellipses combine to give an over- or underestimated degree of circular polarization, depending on whether the ellipse orientations differ by more or less than 45°. Within this ideal approximation, and since the Saturn-spacecraft geometry is known, the polarization response of the PRA system could be determined unambiguously at each time. Unfortunately, the whole conducting structure of the Voyager spacecraft interacts electrically with the PRA monopoles and greatly complicates the interpretation of measurements of V_i. To date, only the sense of polarization, RH or LH, has been used in published analyses of Saturn's radio emissions.

The PWS instrument covers the frequency range from 10 Hz to 56.2 kHz in two different modes. In one mode, a 16 channel spectrum analyser is used. This spectrum analyser has four channels per decade, sampled two channels at a time every 0.5 s. The bandwidth is $\sim 15\%$ of the channel frequency. The PWS effective threshold at 50 kHz is $\sim 10^{-18}$ W m^{-2} Hz^{-1}. The other PWS mode consists of an automatic gain control amplifier sampled 28800 times per second with 4-bit resolution, which effectively provides a waveform analyser below 12 kHz.

B. The Voyager Trajectories

Gurnett (1974) showed that a good coordinate system for organizing the Earth's nonthermal radio emissions is the observer's magnetic latitude and solar hour angle (local time). This coordinate system is also useful for Saturn (Warwick et al. 1981a). For Saturn, the magnetic latitude is essentially the same as kronographic latitude because the Saturnian magnetic dipole moment is very nearly aligned with the rotation axis (see chapter by Connerney et al.).

Figure 1 shows the trajectories of both Voyager spacecraft projected onto the disk of Saturn. Here the solar hour angle is shown as local time with 12 hr corresponding to local noon, 24 hr to local midnight, and 06 hr and 18 hr to the dawn and dusk meridians, respectively. For the months prior to closest approach to Saturn, both Voyagers were approximately fixed just north of the noon meridian in this coordinate system. Likewise, for the several months

SUB-VOYAGER TRACKS ON SATURN

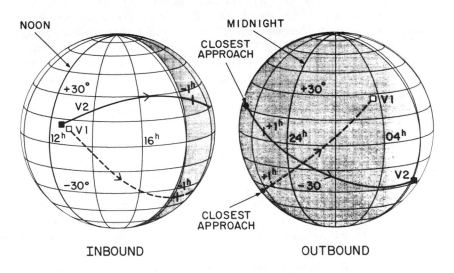

Fig. 1. The trajectories of Voyager 1 and Voyager 2 projected onto the disk of Saturn. The sub-spacecraft tracks remain nearly fixed in Saturn latitude and local time during the long inbound and outbound portions of the trajectories. Only during the few hours near the two closest approaches do the subspacecraft tracks show significant motion in this coordinate system.

after their respective encounters, the Voyagers stayed fixed with Voyager 1 at ~ 03.5 hr local time and +26° magnetic latitude, and Voyager 2 at 06 hr and −29°. Only during the few hours near closest approach did the two spacecraft move significantly in this coordinate system. For purposes of computing average properties of Saturn's radio emission such as intensity and polarization, the reader should realize that < 1/2% of the unit sphere has been sampled. Contrast this with the Earth (see e.g. Gallagher and Gurnett 1979) where spacecraft have observed the Earth's radio emissions repeatedly from essentially all combinations of local time and magnetic latitude.

II. MAGNETOSPHERIC RADIO EMISSIONS

Saturn's nonthermal radio spectrum has three components, two originating in the magnetospheric plasma surrounding the planet, and the third most likely originating from lightning storms in the atmosphere. Table I compares the overall properties of these three components, and they are described in detail below.

Studies of the temporal, spectral, and polarization behavior of Saturn's low-frequency (< 1200 kHz) radio emission have substantially increased our understanding of the magnetospheric environment with which these emissions

TABLE I
Comparison of Saturn Radio Components

Property	Kilometric Radiation	Electrostatic Discharge	Low-Frequency Emissions (escaping)	(trapped)
Frequency range	3 kHz–1200 kHz	~20 kHz–>40 MHz	3–100 kHz	300 Hz–3 kHz
Total power (isotropic)	10^8–10^{10} W	10^7–10^{10} W	~10^6 W	~10^7 W
Polarization	RH (northern) LH (southern)	usually unpolarized, sometimes mixed	ordinary mode	?
Dynamic spectral character	spectral arcs + other complex features	broadband	narrowband	amorphous
Recurrence periods	10 h 39.4 m 25 d	~ 10h 10m	~ 10h 40m	none
Origin	low- to mid-altitude dayside polar cusps	atmospheric lightning	near plasma sheet gradients	

interact. To date, researchers have examined the emissions on time scales from fractions of a minute to months and have found that intensity modulations occur at or near periods of 10.66 hr, 66 hr, and 25 day. We will discuss the nature and modulation source of each of these periodicities. Additionally, nonperiodic fluctuations have been observed and will be described.

Dynamic spectra of Saturn's kilometer wavelength emission clearly show several components. The first is a powerful wideband emission called Saturn kilometric radiation (SKR), first described by Kaiser et al. (1980). It is observed sometimes as low as 3 kHz (Gurnett et al. 1981*a*) and as high as 1200 kHz (Warwick et al. 1981*a*) with peak intensity at ~ 175 kHz. The other radio components overlap with SKR below 100 kHz, but have a different appearance, including narrowband tones and a low-level continuous noise. These low-frequency components are weak and sporadic, appearing only in the observations obtained near Saturn. They are described in Sec. II.C.

SKR was first observed in January, 1980 when the two spacecraft were at distances of > 2 and > 3 AU, respectively. Discrimination from Jovian emissions was made on the basis of polarization, spectral characteristics, and most importantly relative light travel time differences between the two spacecraft. Figure 2 shows the discovery event. SKR appears as a patch of emission centered near 150 kHz and persisting for ~ 1 hr. In this dynamic spectral display, the bottom panel of each pair shows total received power with increasing power indicated by increasing darkness. The top panel of each pair shows the sense of polarization of the received signal coded so that white corresponds to RH polarization, black to LH, and gray to unpolarized (or no emission). SKR appears as RH emission in this figure.

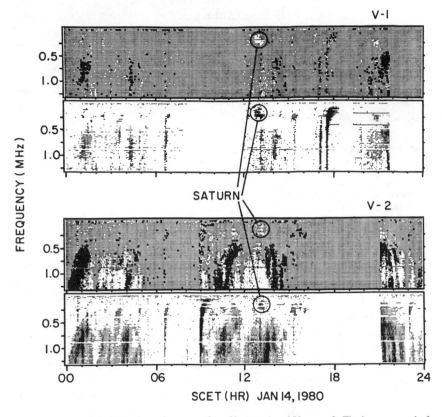

Fig. 2. Simultaneous 24-hr dynamic spectra from Voyager 1 and Voyager 2. The bottom panel of each set indicates total power, with increasing darkness proportional to increasing intensity. The top panel of each pair shows the sense of polarization with RH coded as white, LH coded as black. The indicated Saturn event occurs earlier and is more intense on the Voyager 1 spectra (from Kaiser et al. 1980). (SCET stands for spacecraft event time.)

From a distance of 100 R_S (R_S = Saturn radius = 60330 km). Fig. 3 shows a fairly typical 24 hr period of SKR, with two episodes of emission each lasting several hours. These intervals of emission are made up of a number of discrete features such as narrowband drifting structures and brief intensifications. The overall pattern of emission in the frequency-time plane is repetitive, but the individual features are not.

A. Saturn Kilometric Radiation (SKR) Modulations

Rotation. One of the first recognized properties of SKR was the approximate 10.66 hr interval between major emission episodes (Kaiser et al. 1980). This result was somewhat surprising, since the topology of Saturn's global magnetic field as modeled by the Pioneer 11 investigators (Smith et al.

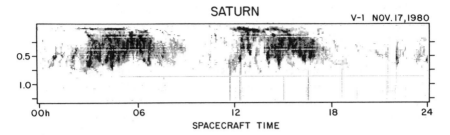

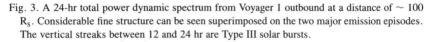

Fig. 3. A 24-hr total power dynamic spectrum from Voyager 1 outbound at a distance of ~ 100 R$_S$. Considerable fine structure can be seen superimposed on the two major emission episodes. The vertical streaks between 12 and 24 hr are Type III solar bursts.

1980a; Acuna and Ness 1980) was such that no modulation of SKR was expected. That is, because Saturn's dipole field lacks any significant tilt relative to the spin axis and also is axisymmetric, no modulation due to a nodding motion of the radio beam or to magnetic anomaly induced particle precipitation effects should occur. But observationally the spin modulation of SKR intensity (~ 20 dB) is immediately evident. Viewed from some distance from the planet, the emission intensity waxes and wanes, disappearing for several hours and then reappearing for several hours in episodes generally called radio storms. This can be seen in Fig. 3, where the centers of the two episodes are ~ 11 hr apart. The recurrence period of storms can vary anywhere from ~ 9 to 13 hr, placing Saturn between Earth and Jupiter in the degree to which it rotationally modulates its emission. For comparison, Jovian radio storms can be predicted, based on the central meridian longitude and the location of the satellite Io, to within ± several minutes at all frequencies, while the Earth's auroral kilometer-wavelength emission (AKR) is only weakly modulated at a 24 hr period.

In spite of the modest rotation control that Saturn exercises, the planet's magnetic rotation period can be determined to within several seconds if a long enough interval of time is analyzed. This calculation assumes that the radio emission is tied to the planetary magnetic field which rotates rigidly with the planetary core. Using a 9 month (600 rotation) time series and applying a method of power spectral analysis, Desch and Kaiser (1981a) measured a period 10 h 39 m 24 s ± 7 s, corresponding to a synodic rotation rate 810.76 deg day^{-1}. Carr et al. (1981) also measured Saturn's period using SKR data and obtained a value 2 s less than that derived by Desch and Kaiser but statistically not inconsistent with it. Since this was the first measurement of Saturn's true or magnetic rotation period, and presumably that of its deep interior, Desch and Kaiser also described a coordinate system within which investigators might consistently cast their results, and provided an equation to transform from the old system used by Pioneer investigators to the new system. The new Saturn longitude system, usually referred to as SLS (1980), has

its 0° prime meridian facing Saturn's vernal equinox on 1.0 Jan. 1980. Like
Jupiter's System III, it is a west longitude (left-hand) system; i.e., longitudes
at one point on the disk increase with time as viewed by a stationary observer.

The computation of SLS longitude proceeds thus: if t is the time of a
given observation, t_0 the system epoch (1.0 Jan. 1980), D the observer-Saturn
distance in km, and R the observer right ascension (angle in degrees in Saturn's equatorial plane from Saturn's vernal equinox to the observer), then the
longitude λ (SLS) of the meridian containing the observer is given by

$$\lambda \text{ (SLS)} = 810.76(t - t_0 - D \times 3.86 \times 10^{-11}) - R. \tag{1}$$

The longitude is that facing the observer at the time the light or radio signal
left the planet. Note that since the rotation rate constant 810.76 is in deg
day^{-1}, $t - t_0$ must be expressed in days and fractions thereof. The rotation
period of Desch and Kaiser and the coordinate system they proposed are now
widely used by Saturn researchers.

Figure 4 shows histograms of the occurrence of SKR as a function of
SLS. The top panel shows the SKR pattern observed before and after the Voyager 1 closest approach as a function of the spacecraft SLS. The two histograms are similar in shape, but offset by $> 120°$. Warwick et al. (1981a) and
Gurnett et al. (1981a) showed that this shift between the histograms was equal
to the angle through which Voyager 1 moved during the encounter period (see
Fig. 1). This was taken as direct evidence that the SKR radiation pattern is
fixed relative to the Sun and does not rotate with the planet. This is shown
more dramatically in the bottom panel of Fig. 4 where the data from both pre-
and post-encounter are displayed against subsolar SLS. The two histograms of
the top panel are merged into one with a strong peak in occurrence rate near
100° subsolar SLS.

Since the Saturn kilometric radiation is made up of discrete spectral
components, it is difficult to determine the average spectral behavior. Carr et
al. (1981) averaged long spans of pre-encounter data containing both Saturn
activity and inactive periods, to produce longitude versus flux profiles at each
PRA frequency channel and an average flux versus frequency spectrum. Kaiser et al. (1980) showed SKR spectra in terms of the probability of detecting
emission above a given threshold. Figure 5 shows the SKR spectra (normalized to an observer-Saturn distance of 1 AU) that are exceeded 50%, 10%,
and 1% of the time. These observations were made by the Voyager 1 PRA
instrument and are categorized into inbound and outbound spectra, corresponding to observations made from above the post-noon equator, and above
the post-midnight meridian, respectively (see Fig. 1). Saturn appears to be a
stronger radio source as observed from above its day side than above its night
side. The equivalent total isotropic radiated power is $\sim$ 200 MW, 3 GW, and
30 GW for the 50%, 10%, and 1% occurrence levels inbound, respectively.
The comparable values for the outbound observations are typically a factor of

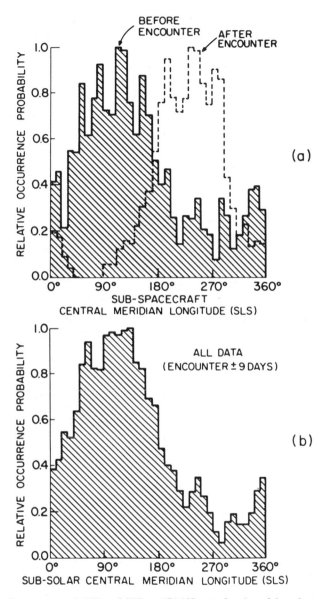

Fig. 4. (a) Occurrence probability of SKR at 174 kHz as a function of the subspacecraft SLS, showing both pre- and post-encounter distributions. (b) The same data as shown in panel a, but plotted as a function of subsolar SLS (from Warwick et al. 1981a).

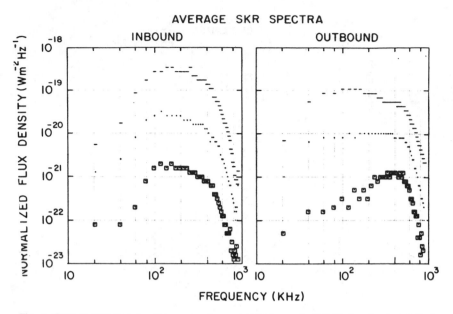

Fig. 5. Typical SKR flux density spectra observed 50% (square), 10% (dot) and only 1% (bar) of the time are shown for the inbound and outbound observations of Voyager 1 (from Kaiser et al. 1981).

3 to 4 less. Kaiser et al. (1981) pointed out that this difference between inbound and outbound intensity levels is greatest at low frequencies. The flux density values are comparable at frequencies > 500 kHz, but differ by a factor of ~ 5 at 100 kHz.

Satellite Control. Long-term (periods > 1 planetary rotation) modulations of the nonthermal radio emissions provide important evidence of phenomena occurring deep within the magnetosphere. Io modulation of the Jovian radio emission, for example, was the first indication of that satellite's special astrophysical importance. Saturn's magnetosphere, like Jupiter's, contains satellites within a corotating magnetic field, but Desch and Kaiser (1981*b*) reported no evidence of any persistent long-term modulation of SKR as might be revealed through power spectral analysis of long data spans. Near Voyager 1 Saturn encounter, however, Gurnett et al. (1981*a*) reported a short-lived modulation of SKR by Dione. SKR was observed to disappear every 66 hr, coincident with Dione's period of revolution. The satellite phase at the time of the disappearances was ~ 270° (subsolar SLS). Desch and Kaiser (1981*b*) showed further that the modulation was strongly frequency dependent, being most pronounced at the lowest frequencies, (below ~ 100 kHz) and not evident above ~ 250 kHz. This effect is shown in Fig. 6, where we compare the

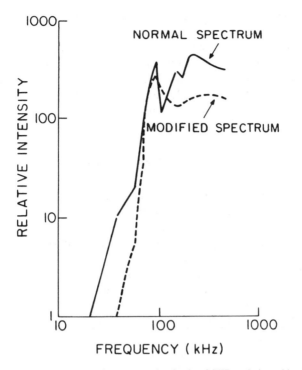

Fig. 6. Flux density spectra comparing a normal episode of SKR emission with a Dione modulated episode that occurred a short time earlier. The spectra have similar shapes except for the low-frequency portion which falls off much more rapidly in the Dione case.

power flux density spectrum of a normal, unmodulated SKR episode with an episode that displays apparent Dione control. Above ~ 80 kHz in this case the two spectra have similar shapes, peaking at ~ 100 kHz and remaining relatively flat at higher frequencies. Below 80 kHz, however, the Dione-modulated emission falls off much more rapidly than does the normal spectrum, consistent with a strong low-frequency propagation effect perhaps due to an intervening plasma of relatively high density. This idea was described by Kurth et al. (1981 a) who attributed the refraction of SKR to the presence of a Dione-related plasma torus. According to their model, the longitudinal asymmetry is introduced by photosputtering off of localized bright features on the surface of Dione. Preferential sputtering would occur when the trailing hemisphere of the satellite is illuminated, as is the case near 270° Dione phase.

Subsequent observations of putative Dione modulations have complicated this picture somewhat. Modulations near 66 hr period have also been observed from days 337–349, 1980 by Voyager 1 (Desch and Kaiser 1981b) and near Voyager 2 encounter on days 213–236, 1981 (Warwick et al. 1982a). Moreover, the Voyager 2 observations showed an increase in SKR activity every 66 hours rather than a decrease. Also, the phase of the later Voyager 1 and Voyager 2 modulations were 30° and 180°, respectively, not the 270° modulation phase of the earlier interval near Voyager 1 encounter. If these episodes were truly related to Dione, then the modulation is extremely complicated. All that can presently be said for certain is that the modulation at the Dione revolution period, which is strongly frequency dependent and phase variable, differs significantly from Io's modulation of the Jovian radio emission.

Solar Wind Control.   Saturn kilometric radiation is also observed to vary on a time scale of many days, as was noted by Warwick et al. (1982*a*) and Scarf et al. (1982). Sometimes this variation is quite dramatic; for example, just after Voyager 2 Saturn encounter SKR went completely undetected for 2 to 3 days after having appeared very strong just before and during encounter. At this time Voyager 2 was < 0.04 AU from Saturn, so the detection threshold was ∼ 4 orders of magnitude below nominal intensities at 200 kHz. These dropouts differ from the Dione-related modulations in that they generally last longer and occur over the entire ∼ 1 MHz natural radiation bandwidth. Similar pronounced fluctuations have appeared in the SKR data extending back at least several months before both Voyager encounters.

In attempting to explain these dramatic variations in the SKR emission level, we note that several independent lines of evidence have suggested a strong solar wind influence at Saturn. For instance, Bridge et al. (1982) tentatively invoked a plasma loss mechanism in Saturn's inner magnetosphere caused by a factor of two increase in solar wind pressure during the Voyager 2 flyby. Magnetic field observations (see chapters by Schardt et al. and Connerney et al.) led Behannon et al. (1981) and Ness et al. (1982*a*) to attribute magnetic tail and magnetosphere size fluctuations to variations in the solar wind flux at Saturn. Finally, Warwick et al. (1982*a*) and Scarf et al. (1982) hypothesized that the dramatic 2 to 3 day disappearance of SKR described above might be caused by the absence of solar wind flux at Saturn due to Saturn's presence in Jupiter's magnetic tail or tail filament (see e.g., Lepping et al. 1982; Kurth et al. 1982*b*). Kurth et al. (1982*a*) have shown evidence that Saturn may in fact have been immersed in Jupiter's tail at this time, based on the observation within Saturn's magnetosphere of Jovian-like low-frequency continuum radiation.

Evidence of direct solar wind control of Saturn's radio emission was provided by Desch (1982). A typical example adapted from his study of the response of the radio emission level to solar wind conditions is shown in Fig. 7. For example, a factor of 150 increase in solar wind pressure at Saturn and a 65% increase in bulk speed is accompanied by an order of magnitude increase in SKR emission level near 5 June 1981. Statistical analysis of two 160-day data intervals yielded a significant correlation with ram pressure and a slightly lower overall correlation with bulk speed (Fig. 8). In addition, just

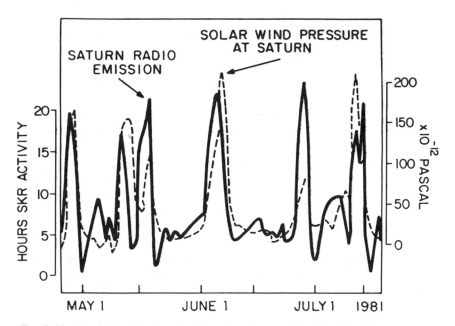

Fig. 7. Voyager 2 data showing close dependence between the solar wind pressure at Saturn (dashed) and the level of SKR (solid) for 90 days in 1981. Vertical scales are approximate (figure adapted from Desch 1982).

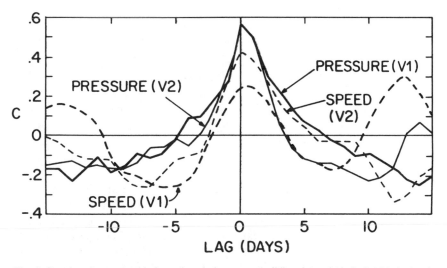

Fig. 8. Results of cross correlating solar wind pressure (solid) and speed (dashed) with the level of SKR for both Voyager 1 and Voyager 2 (figure from Desch 1982).

like the solar wind pressure, which exhibited a 25-day periodicity in magnitude over the extent of the analysis interval, SKR was also modulated in phase with the solar wind pressure at a period of 25 days. Since the SKR source location probably extends nearly to the cloud-top or ionosphere level on Saturn (see Sec. IV.A.), the solar wind influence must also reach deep into Saturn's magnetosphere.

Continuing this work, Desch and Rucker (1983) examined 13 solar wind quantities and their relationship with SKR and showed that the solar wind ram pressure is the primary driver of Saturn's radio energy output. They view their results as consistent with the continuous mass transfer of particles from the solar wind into Saturn's low-altitude cusp. This transport would proceed by means of eddy convection in the dayside polar cusp-magnetosheath region, as in a model developed for terrestrial magnetospheric processes (Haerendel et al. 1978).

Dynamic Spectra Variations.   Long-term modulations also appear in the shape of the dynamic spectrum itself, although apparently not with any well-defined periodicity. Figure 9 presents a compacted dynamic spectrum of Voyager 1 observations near encounter. The emission frequency range can be seen to vary considerably. In particular, the lowest frequency limit of the emission shifts from 20 kHz to more than 80 to 100 kHz. These fluctuations correspond to a distortion of the spectrum and not to a threshold effect due to temporary global intensity variations or change in distance. Genova et al. (1983) point out that this behavior is reminiscent of the Earth's radio emission during magnetically quiet times. They deduce that the changes in the SKR spectrum most likely reflect changes in the SKR source region, rather than, for example, variable propagation effects.

B. Polarization of Saturn Kilometric Radiation (SKR)

The interpretation of the recorded SKR polarization was simple and immediately suggested to Warwick et al. (1981a,1982a) the existence of two distinct radio sources, emitting in opposite senses of polarization and observed in various conditions of visibility by each spacecraft along its trajectory. This simple polarization behavior is quite different from that of the Jovian hectometer-wavelength emissions detected in the PRA low band, which showed frequent polarization reversals or events with mixed polarization (Alexander et al. 1981).

The SKR polarization behavior in terms of the parameter V_i (see Sec. I.A) is shown in Fig. 10 for several days around the planetary encounters. V_i indicates RH (LH) polarization sense when it is negative (positive). The polarization exhibits a nearly constant value except during the close encounter periods or the spacecraft rolls, and is nearly independent of frequency. Along the pre-encounter paths (see Fig. 1) the recorded dynamic spectra appear to be dominated by RH polarization, with only occasional periods of LH emis-

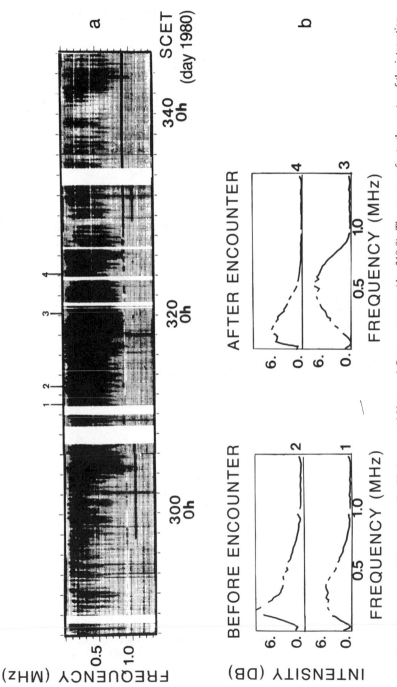

Fig. 9. (a) Compressed SKR spectrum for 58 days around Voyager 1 Saturn encounter (day 318.0). The arrows refer to the center of the integration intervals of the spectra shown in Fig. 9b. (b) Intensity spectra of SKR integrated over one rotation duration, thus eliminating intensity variations on shorter time scales.

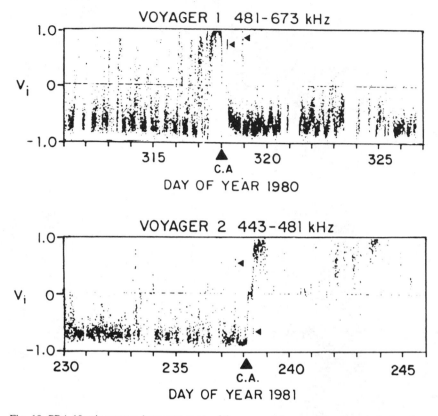

Fig. 10. PRA 10-min averaged measurements of the apparent degree of circular polarization during the 16 days centered on the Voyager 1 (upper panel) and Voyager 2 (lower panel) closest
approaches. Spacecraft rolls are indicated by dark arrows.

sion that are due to detections of the weaker LH polarized source. After the
planetary flybys, Voyager 1 observed exclusively RH polarization while Voyager 2 detected mostly LH emission. During the few hours around the Saturn
closest approaches the observations showed more complicated patterns, well
in agreement with visibility changes from one polarization source in one
hemisphere to the other, and with changes in the antenna-source geometry accompanying spacecraft rolls and motion along the flyby trajectory (Warwick
et al. 1981a). Because the direction to the radio source is generally not normal to the PRA antenna plane, V_i does not reach ± 1.0. However, note that V_i
reaches a value very close to unity during the second half of day 317 of 1980
(Voyager 1), when the source was received approximately in the direction normal to the PRA monopoles. This implies either pure circular SKR polarization or elliptical SKR polarization with a substantial linear component remaining for hours in the antenna plane.

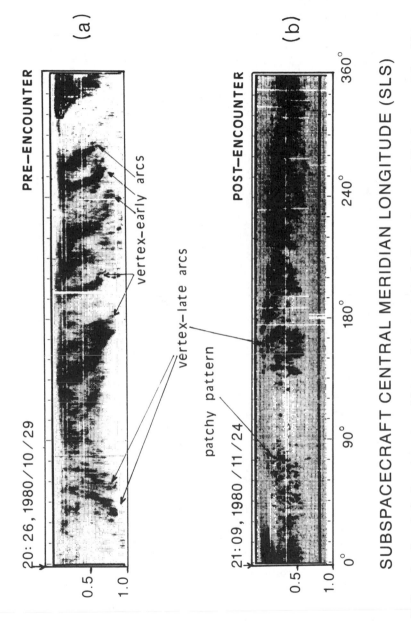

SUBSPACECRAFT CENTRAL MERIDIAN LONGITUDE (SLS)

Fig. 11. Examples of arcs in the dynamic spectrum of Saturn emission detected by the Voyager 1 spacecraft. Data from one rotation of Saturn before (a) and after (b) closest approach. The sub-Voyager 1 SLS longitude is indicated along with the beginning time of each rotation. VE and VL arcs together with an episode of patchy pattern are delineated.

C. Fine Structure

Saturn Kilometric Radiation Dynamic Spectral Arcs. In the SKR dynamic spectra, episodes of very bursty emission are observed with time and frequency scales < 48 s and $\sim 50-100$ kHz, producing patchy patterns in dynamic spectra, as seen for instance in Figs. 3 and 11b. These impulsive bursts have not been studied in any detail, so it is not known if they form a recognizable category in SKR phenomenology. However, at least one substructure in the SKR dynamic spectra has been studied, namely the arcs described by Boischot et al. (1981) and Thieman and Goldstein (1981).

Figure 11 shows examples of the two kinds of arcs: vertex early (VE) and vertex late (VL) arcs, i.e., arcs appearing respectively as opening and closing parentheses. The modulation amplitude of the arc emission is only a few decibels on average. The frequency extension of the arcs roughly follows the general behavior of SKR. A typical arc extends from 100 to 700 kHz and has a vertex near 400 kHz with symmetrical curvature around this vertex (see Fig. 2 in Thieman and Goldstein 1981). The total duration ranges from 10 to 40 min, with an average of ~ 30 min. There is no apparent difference in form between VE and VL arcs. SKR is not organized entirely into arc patterns, as Jupiter's emissions (Warwick et al. 1979) apparently are. Generally only a few arcs are observed within a rotation. Most often these arcs appear isolated, but sometimes nested arcs can also be observed, separated by 15 to 40 min.

Before the two encounters, in observations made from above the Saturnian day side, VE arcs were predominant. After the encounters, observations from above the night hemisphere show that VL arcs prevail (Boischot et al. 1981). Figure 12 illustrates this for Voyager 1. This figure also shows that the predominant species is preferentially observed in the SLS range where the overall emission probability is lowest (e.g., compare Fig. 12 with Fig. 4), whereas the less frequent type appears principally in the SLS range where this probability is highest. The latter type is generally observed at higher frequencies than the former. On Saturn no evident repetition of the arc pattern is observed from one rotation to the next in contrast to Jupiter.

Low Frequency Radio Emissions. In the low-frequency regime of a few kHz Saturn emits low-level electromagnetic nonthermal continuum radiation (using the terminology first applied to the terrestrial spectrum) in addition to SKR. At frequencies below ~ 2 or 3 kHz these emissions are trapped within the magnetospheric cavity, since the wave frequency is less than the electron plasma frequency f_p of the surrounding solar wind. Since electromagnetic waves cannot propagate below f_p, they are refracted away from the relatively high-density walls of the magnetopause and are trapped within the cavity. Waves generated at a frequency typically > 2 to 3 kHz will escape directly into the solar wind.

Figure 13 shows an example of Saturn's trapped radiation. The lower panel is a frequency-time spectrogram. The intense band near 300 Hz and the

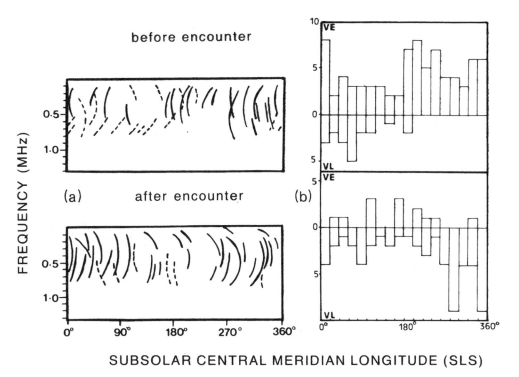

Fig. 12. Arcs observed by Voyager 1 before (upper panel) and after (lower panel) closest approach, as a function of subsolar SLS longitude. (a) Examples of arc shapes observed during several planetary rotations. (b) Histograms of the arc orientation during the same period. VE and VL arc histograms are plotted upwards and downwards, respectively (adapted from Boischot et al. 1981).

narrowband tones at 2.4 kHz and 700 Hz are spacecraft interference. The broad feature between 500 Hz and 4 kHz is the trapped radiation. The low-frequency cutoff of the continuum radiation corresponds very closely to the local f_p (inside the magnetosphere) as derived from the Voyager plasma instrument (Bridge et al. 1981a). The electron gyrofrequency f_g is $\sim$ 170 Hz, based on measurements by the Voyager magnetic field instrument (Ness et al. 1981). Hence, the emission lies well above f_g and has a cutoff very close to f_p thus is most likely an electromagnetic mode. As indicated by the dashed line in the upper panel, the radiation has a spectral index of $\sim$ -3; however, this has been observed to vary between -2.5 and -4.0. This trapped component is not as pervasive or intense at Saturn as at Jupiter (Kurth et al. 1982a), but its spectral density is similar to that at the Earth.

The escaping component of Saturn's low-frequency radiation was first described by Gurnett et al. (1981a) who used the term narrowband electromagnetic radiation. Figure 14 shows spectrograms taken from both Voyager

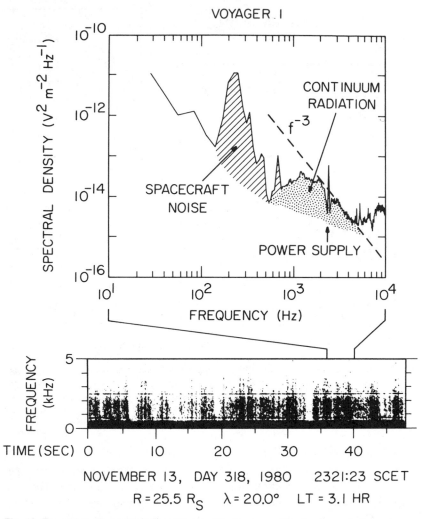

Fig. 13. Trapped continuum radiation at Saturn. The lower panel is a frequency-time spectrogram showing the diffuse continuum radiation spectrum between ∼ 500 Hz and a few kHz. The spectrum in the upper panel is 4 s average showing the low-frequency cutoff near f_p, and the spectral index of ∼ −3 at higher frequencies (from Kurth et al. 1982a).

1 and Voyager 2 at a distance of 3 R_S. Although the Voyager 1 spectrum is certainly more complex than that of Voyager 2, both show evidence of narrowband emissions with bandwidths in some cases as small as 1% of the center frequency. For the Voyager 1 spectrum f_p and f_g were 2.4 kHz and 25.3 kHz, respectively; hence, the bands above 2.4 kHz must be propagating in the free space electromagnetic ordinary (L–O) mode. The Voyager 2 spectrum

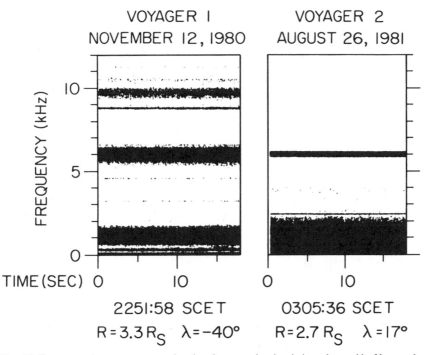

Fig. 14. Frequency-time spectrograms showing the narrowband emissions observed by Voyager 1 and Voyager 2. These emissions are thought to extend to several tens of kHz (figure adapted from Scarf et al. 1982).

was obtained in a region where f_g was 31 kHz (Ness et al. 1982a) and f_p was 2.2 kHz (Bridge et al. 1982), so the very intense band at 6 kHz is apparently propagating in the same mode.

   There is evidence that the narrowband emissions extend as high as 80 kHz. Warwick et al. (1981a) report narrowband emissions observed to be RH polarized throughout the Voyager 1 near-encounter period. Warwick et al. (1982a) report observations by Voyager 2 of a similar narrowband component near 40 kHz, although the observed polarization was LH during that encounter.

III. ELECTROSTATIC DISCHARGES

   For a few days around both Voyager encounters with Saturn, the planetary radio astronomy (PRA) instrument recorded intense, short-duration bursts of radio emission quite unlike the magnetospheric emissions described above. Figure 15a shows a dynamic spectrum recorded just at Voyager 1 closest approach. The SKR can be seen as the dark patch of emission centered at $\sim$ 500 kHz. The unusual emissions are the short, vertical streaks which are present throughout both the PRA low-frequency and high-frequency bands.

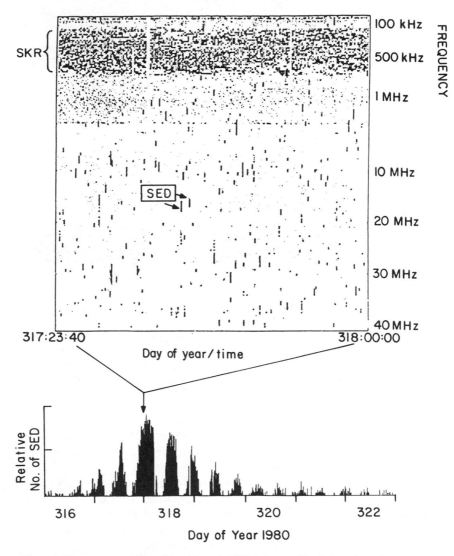

Fig. 15. The upper panel (from Warwick et al. 1981a) shows a 40-min-long dynamic spectrum at the time of Voyager 1 closest approach. Both the PRA low and high bands are shown. SED are the short, vertical streaks throughout the panel. SKR is the dark band near 500 kHz. The lower panel shows the overall occurrence of SED for the Voyager 1 encounter period as determined by a computerized detection scheme.

Warwick et al. (1981a) ruled out any sort of spacecraft interference or nearby discharging as the source of these bursts and concluded that the emissions were propagating to the spacecraft from the vicinity of Saturn. They coined the term "SED" for Saturn electrostatic discharges.

A. Burst Description

As can be seen in Fig. 15a, each individual SED appears on only a few consecutive PRA channels, yet SED are observed throughout the entire frequency range. Warwick et al. (1981*a*) concluded that the SED were short-duration (tens of milliseconds) bursts with a very large bandwidth (> 40 MHz). Since the PRA receiver steps from one frequency to the next every 30 msec, SED are detected only on the few channels being sampled during the 30 to 250 msec duration of a typical burst. Thus, the length of a streak in Fig. 15a reflects the duration of an SED burst (the entire vertical axis of the figure is swept through in 6 s). One series of very high time-resolution measurements at 10 MHz (a PRA alternate operating mode) showed that individual SED events like those of Fig. 15a are made up of many very short (< 1 msec) bursts superimposed on a strong continuum (Warwick et al. 1981*a*). Some structure at the time resolution limit of 140 microseconds (in the high-rate mode) led Warwick et al. (1981*a*) to estimate that the SED source was no larger than 40 km in size.

Figure 16 shows the distribution of SED durations during both the Voyager 1 and Voyager 2 encounter periods. Both sets of data are extremely well fit by an equation of the form

$$N = N_0 \exp{(-D/D_0)} \tag{2}$$

where N is the number of SED bursts, and D is their duration in milliseconds. Least squares fits give values of D_0 of 40 ± 1 and 37 ± 2 msec for the overall Voyager 1 and Voyager 2 encounters, respectively. The value of D_0 at closest approach was about 58 msec, because proximity to the source allows weak emission at the beginning and end of an SED burst to be detected, and this has the effect of lengthening the average SED burst. The maximum rate of SED detected by Voyager 1 at closest approach was ~ 0.2 per second, and the distribution of the number of SED per unit time was well matched by a Poisson distribution (Evans et al. 1983). Figure 16 also shows that the absolute occurrence of SED appears to be reduced by a factor of ~ 3 for the Voyager 2 encounter as compared to Voyager 1, in agreement with the findings reported by Warwick et al. (1982*a*). However, the PRA receiver on Voyager 2 was operating with 15 dB of attenuation during the 10 hr centered on closest approach which probably accounts for much of the difference between the total number of SED detected by the two spacecraft.

The total power dissipated in an SED burst has been estimated to be 10^7 to 10^8 W by Warwick et al. (1981*a*), and as high as 10^{10} W during some events (Evans et al. 1983; Zarka and Pedersen 1983). Total power is somewhat difficult to estimate because of the lack of knowledge of the instantaneous bandwidth and spectral shape of SED bursts. For the estimates appearing in the literature, the authors have used a bandwidth of 40 to 100 MHz and have assumed that the spectrum is flat. If SED are indeed as powerful as 10^{10}

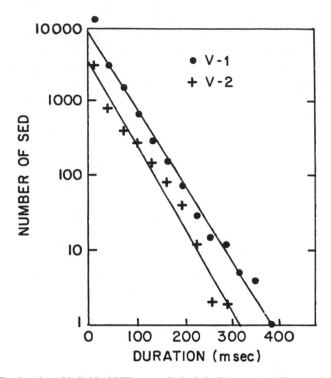

Fig. 16. The duration of individual SED events for both the Voyager 1 and Voyager 2 encounters. The events were cataloged by a computer algorithm which compared successive 6-s samples at the same observing frequency and recorded those samples which were larger by 25% or more from either the preceding or following sample. The SED events are exponentially distributed with an e-folding of ~ 40 msec for Voyager 1. The total number of SED detected by Voyager 2 is about a factor of three lower than those detected by Voyager 1, but the e-folding is nearly the same (figure adapted from Zarka and Pedersen 1983).

W, then groundbased radio telescopes suitably equipped should be able to monitor SED.

Knowledge about the polarization of SED is confused at best. Evans et al. (1981) reported that most SED bursts observed by Voyager 1 during the inbound trajectory were unpolarized, but, during the outbound leg, LH polarization was observed particularly at frequencies above 15 MHz. Zarka and Pedersen (1983) cautioned that the PRA antenna system does not measure polarization very reliably above 15 MHz, and suggested the possibility that SED may be unpolarized at all times.

Kaiser et al. (1983) analyzed the SED observations and found that SED were actually observed in the PRA low band only after the spacecraft had passed Saturn and were making observations from above the night hemisphere. Observations of SED during the inbound legs above the daylit hemisphere (see Fig. 1) never showed bursts below about 5 MHz (see Fig. 20a in Sec. IV.C).

B. Episode Periodicity

One of the most revealing aspects of the Saturn electrostatic discharges observations was the grouping of individual SED into episodes which recurred with a distinct periodicity of ~ 10 hr 10 min (Voyager 1), quite different from the 10 hr 39.4 min Saturn rotation period deduced from the Saturn kilometric radiation observations. These episodes appeared for a few days on either side of closest approach. The phase of the SED repetition period was found to be fixed relative to the line between planet and observer, implying that the source of SED rotates or revolves like a searchlight, and is not fixed relative to the Sun, as is the case for SKR (Warwick et al. 1982a). Zarka and Pedersen (1983) showed that the repetition of episodes for the Voyager 2 encounter were marginally faster (10 hr 00 min) than for the Voyager 1 encounter, but both periodicities have relatively large uncertainties. Kaiser et al. (1983) showed that the number of SED fell to zero for 3-hr periods in between episodes. This on-off behavior was quite different from the SKR which waxes and wanes with its 10 hr 39.4 min period but does not generally disappear. This pattern is depicted schematically for Voyager 1 in Fig. 20a in Sec. IV.C. The occurrence rate of SED during a given "on" period, however, is not uniform. An episode may contain several relative maxima and minima (e.g., see "double hump" episode described by Evans et al. 1981), although the general trend is for a buildup of the number of SED per unit time followed by a subsequent decline before total disappearance.

An additional puzzling aspect of the SED episodes was first reported by Evans et al. (1981), and concerned the onset of the episodes as a function of frequency. During all episodes but one, the onset of each episode was approximately independent of frequency. The lone exception was the episode which began just before the Voyager 1 closest approach where SED were first observed in the 30 to 40 MHz band and slowly, over a 4-hr interval, filled the PRA receiver down to the lowest frequency (Fig. 20a). Kaiser et al. (1983) counted the number of SED in the 30 to 40 MHz band as a function of time during this frequency-dependent onset and reported that the rate increased from one SED every 20–30 s to the maximum rate of one per 5 s over the same 4-hr interval required for the SED to spread throughout the entire PRA bandwidth.

IV. INFERRED SOURCE LOCATIONS

A. Saturn Kilometric Radiation (SKR)

The observation of predominantly RH polarized emission during the northern hemisphere inbound Voyager 1 trajectory led Kaiser et al. (1980) to conclude that the radio source was in the northern hemisphere and that the radiation was escaping in the right-hand extraordinary mode. This conclusion was founded on the assumption that the emission was being observed directly and not by way of a back lobe of the radiation pattern. It was also necessary to know, of course, that Saturn's main magnetic field, like Jupiter's, has its field lines emanating from the planet's northern hemisphere (Acuna et al. 1980; Smith et al. 1980a).

Post-encounter Voyager observations of the same RH polarized radio source showed that the subspacecraft SLS longitude of maximum detection probability had shifted by an amount equal to the local time change of the spacecraft. This important piece of data indicated that the source was locked in local time, i.e., emitting radiation into the same directions relative to the Sun whenever a particular magnetic longitude reached a fixed solar hour angle (Warwick et al. 1981a; Gurnett et al. 1981a). A searchlight type of radio source, one that rotates with the planet emitting all the time, was thus excluded by the observations. Only when the detection probability is a maximum at the same magnetic longitude regardless of observer local time is such a source indicated. Subsequent observation of a distinct LH polarized source at a time when Voyager 1 was south of Saturn's equatorial plane was taken as evidence of a southern-hemisphere source, also emitting in the extraordinary magnetoionic mode (Warwick et al. 1981a) and similarly locked into the dayside local time hemisphere.

These general conclusions regarding the hemisphere locations of the radio sources were later confirmed by Kaiser et al. (1981) and by Kaiser and Desch (1982) and Lecacheux and Genova (1983) who quantitatively modeled the precise source locations. In the first two of these papers, the observed ratios of pre-encounter to post-encounter radiation intensities were compared with the ratios that would be expected from beams located at many different places. The beam locations were not constrained in any way in the space around Saturn, but the beam axis orientation was required to be aligned with the magnetic field direction. The source altitude was set by assuming that emission took place at or near the local f_g. As had previously been concluded, the RH polarized source was found to be in the northern hemisphere and the LH polarized source in the southern. Further, Kaiser and Desch showed that for the northern hemisphere source, only radio sources on those magnetic field lines whose footprints were constrained to $\sim 70°$ to $80°$ latitude, $100°$ to $130°$ SLS longitude and 10 to 12 hr local time could satisfy the SKR observations. The southern hemisphere was less tightly constrained to $\sim -60°$ to $-85°$ latitude, $300°$ to $75°$ SLS longitude, and 7 to 16 hr local time. The footprints of the Kaiser and Desch source regions are shown in Fig. 17. Lecacheux and Genova (1983) found essentially the same source locations by studying the changes in the visibility of the two sources (RH and LH) as viewed from the spacecraft trajectories. It is important to note that these source locations are consistent with the surface locations of Saturn's ultraviolet aurorae (Sandel et al. 1982b), polar cap boundary (Ness et al. 1981),

SKR SOURCE LOCATION

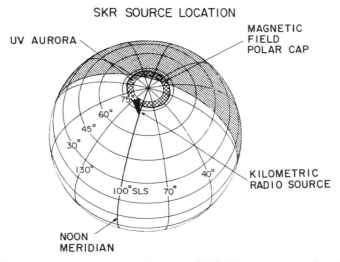

Fig. 17. The best estimates for the source footprints of SKR. RH emission comes from field lines mapping to a very small area in the northern hemisphere, and the LH emission is constrained to field lines mapping to a very narrow, long band at high southern latitudes (from Kaiser and Desch 1982).

and polar cusp (Behannon et al. 1981). Thus the source locations correspond to active regions in Saturn's magnetosphere, where the field lines tend to open up into the interplanetary medium and where intense particle precipitation and auroral stimulation occur. Localization of the source footprints in these high-latitude dayside regions of the magnetosphere is thus consistent with the strong SKR-solar wind correlation observed by Desch (1982).

Taken as a whole, there now seem several independent lines of evidence to suggest the existence of some irregularity or inhomogeneity in Saturn's high-latitude regions; in the northern hemisphere this inhomogeneity is probably near 115° SLS. The very existence of some source of rotation (10.66 hr) modulation of SKR, the localization of the radio sources to specific longitude ranges on the planet, and a similar confinement of the auroral activity all suggest a magnetic influence, possibly in the form of a magnetic anomaly in the near-surface field. However, no direct evidence of such an anomaly exists (see chapter by Connerney et al.). Saturn's global field is modeled to be azimuthally symmetric and lacks any substantial tilt away from the spin axis. This inability to account for magnetic longitude confinement of the aurorae and of the radio sources remains one of the major problems confronting investigators of Saturn.

B. Low-Frequency Emissions

The source location of the escaping narrowband emissions is not well known due to the small number of observations and also for the reason that there is no appreciable tilt in the magnetic dipole axis which might have permitted one to look for geometry-dependent propagation effects. Figure 18 shows the position of the Voyager spacecraft at times when narrowband emissions could be observed in the frequency range of 5 to 15 kHz. The two filled boxes represent the position of the spacecraft when the emissions were strongest and correspond to times represented by the two spectra shown in Fig. 14. The open boxes represent the detection of relatively weak narrowband emissions, primarily near 5 to 6 kHz. Gurnett et al. (1981*b*) reported a periodicity in the detection of this 5–6 kHz band which is close to that of Saturn's rotation period. The maximum intensity occurs when the spacecraft SLS longitude is near 290°. The observations depicted in Fig. 18 are consistent with a source near Saturn which illuminates high latitudes. Low latitudes are shielded by the high-density plasma sheet.

Gurnett et al. (1981*b*) also suggested another possible method for determining the source location of the escaping narrowband emissions. They no-

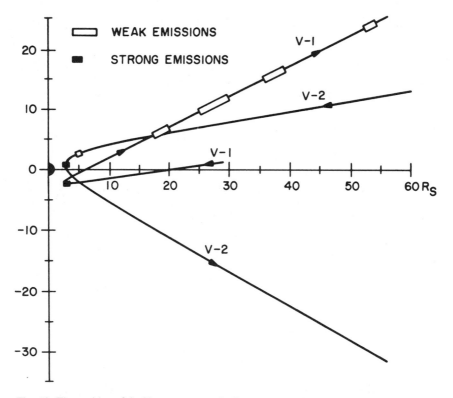

Fig. 18. The position of the Voyager spacecraft when narrowband emissions below 12 kHz were detected. These observations suggest a source located at small radial distances and illuminating high latitudes (figure adapted from Scarf et al. 1982).

ticed several different frequency spacings between the bands seen in the Voyager 1 spectrum shown in Fig. 14 and made the assumption that the spacing is related to f_g at the source. Three such sets of harmonically spaced lines were identified with the values of f_g at radial distances of 5.4, 7.3, and 8.8 R_S. These are near the orbits of Tethys, Dione, and Rhea. Gurnett et al. (1981*b*) conjectured the source of the escaping narrowband radiation was associated with the *L*-shells of these satellites and probably located above or below the equatorial plane on a density gradient of the plasma sheet. The bottom panel of Fig. 19 is a schematic representation showing one possible geometry implied by Gurnett et al. (1981*b*). Here, the Dione *L*-shell is shown as a preferred region and electrostatic waves near the upper hybrid resonance (UHR), $f_{UHR} = (f_p^2 + f_g^2)^{1/2}$ lying on the density gradient of the plasma sheet are

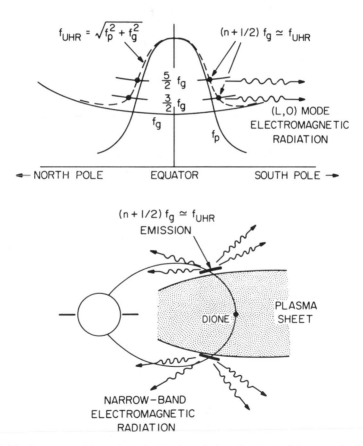

Fig. 19. The bottom panel is a schematic drawing of where the source of the narrowband emissions might be located. The upper panel shows the relationship between f_{UHR} and $(n + 1/2) f_g$ harmonics. It is thought that intense electrostatic waves at $f_{UHR} = (n + 1/2) f_g$ are the source of the narrowband emissions (figure adapted from Gurnett et al. 1981*b*).

thought to be the source of the electromagnetic emissions. This geometry also favors emission towards high latitudes.

The trapped component of the low-frequency spectrum was detected at relatively high latitudes ($>$ 20°) both above and below the equator in low-density regions of the pre-dawn magnetosphere (Kurth et al. 1982a). This geometry is consistent with the magnetotail lobes where a similar type of radiation is most prevalent at Earth and Jupiter. By analogy with the Earth and Jupiter, the radiation is probably emitted in a manner similar to the narrow-band emissions near f_{UHR} located on density gradients.

C. Saturn Electrostatic Discharge (SED)

Only two known locations in the Saturn system have rotation or revolution periods comparable to the 10 hr 10 min SED episode periodicity. One possible location, first suggested by Burns et al. (1983), is in the cloud tops of the equatorial atmosphere, where the Voyager imaging results (see chapter by Ingersoll et al.) showed wind velocities of 500 m s^{-1}, corresponding to a rotation period of the atmosphere at that latitude of $\sim$ 10 hr 10 min. The other location is at 1.8 R_S, in the middle of Saturn's B Ring, where the Keplerian revolution period is 10 hr 10 min. Warwick et al. (1981a) and Evans et al. (1981,1982,1983) concluded that the SED were generated by an object in the B Ring undergoing successive discharging. They reasoned that since SED were sometimes observed to frequencies well below 1 MHz (see Fig. 15a), the atmospheric location could be ruled out because the Saturnian ionosphere would prevent escape of frequencies that low (see chapter by Atreya et al.).

Burns et al. (1983) compared the characteristics of SED with the properties of lightning at Earth, Jupiter, and Venus and suggested that SED bursts originated from atmospheric lightning in the Saturnian equatorial region. They proposed that the emission escaped through a long-lived ionospheric hole created by the shadow of the ring system. Kaiser et al. (1983) noted that since SED bursts below 5 MHz were not observed when Voyager was above the day side, and low-frequency SED appeared only over the night side, the justification used by Warwick et al. (1981a) for ruling out the atmosphere was not as clear as previously assumed. They suggested that if the ionospheric plasma density near the noon meridian was $\sim$ 3 $\times$ 10^5 cm^{-3}, escape of atmospheric radio bursts below 5 MHz would be prevented. They further stated that if the nighttime ionosphere had low-density regions or holes, the observed low-frequency emissions could escape.

However, the argument of the day-night frequency extent for placing the source of SED in the equatorial atmosphere was secondary to the major argument presented by Kaiser et al. (1983). They pointed out that the on-off pattern of SED episodes was the signature to be expected from a source undergoing consecutive occultations by the planet. During the 3-hr intervals when SED bursts were absent, they reasoned that the source of SED was out of view of Voyager, behind the planet. The on-off pattern is shown schematically

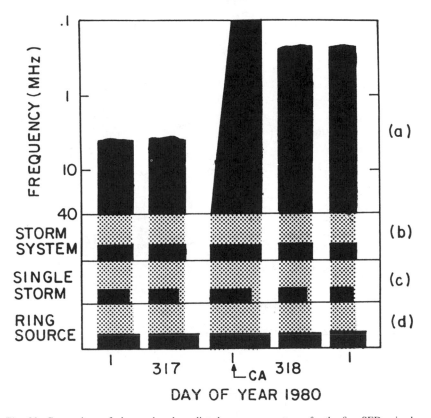

Fig. 20. Comparison of observed and predicted recurrence patterns for the five SED episodes centered on Voyager 1 closest approach. Panel (a) shows schematically the times and frequencies where SED were detected (black) and undetected (white). Before and almost up to closest approach (CA) no SED are seen below ~ 5 MHz. After (CA) SED are regularly observed down to and even below 500 kHz. In between the episodes, SED are virtually absent. Panels (b), (c), and (d) compare the predicted recurrence patterns for a 60°-wide atmospheric storm system, a single-point source atmospheric storm, and a single-point source in the rings at 1.8 R_s, respectively. The agreement between the observed and predicted start-stop times for the 60°-wide storm system (b) is clear. Note especially the coincidence between start and stop times for the episode centered on closest approach, which lasts ~ 3 hr longer than any other episode. The single-storm model (c) consistently predicts shorter episodes than those observed by ~ 2 hr, and the ring-source model (d) consistently predicts longer episodes by ~ 2 hr (after Kaiser et al. 1983).

in Fig. 20a. They were then able to calculate what the expected occultation pattern for a point source in the B Ring would be; this is shown in Fig. 20d. Clearly, the predicted SED "off" periods for a ring source are much shorter than those observed. Conversely, a single point source in the equatorial atmosphere should have much longer off times than observed (Fig. 20c). Figure 20b shows the occultation pattern predicted by an extended storm system

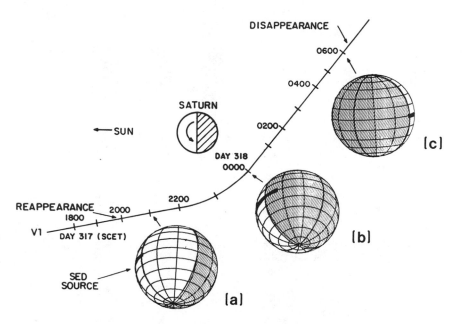

Fig. 21. Voyager 1 trajectory past Saturn shown projected into the equatorial plane. The shaded globes show the planetary aspect and SED source location as viewed by Voyager 1 at three times near closest approach. On day 317 at 1950 SCET (spacecraft event time), the source reappears on the west limb of the planet, marking the onset of the third episode in Fig. 20. At 2100 SCET (panel a), the leading edge of the source region is ~ 30° onto the visible face of Saturn; at 0000 SCET (b), the leading edge is ~ 50° from the limb and well into the night hemisphere, permitting escape of SED below 5 MHz; by 0600 SCET (c), the trailing edge of the source is near the eastern limb, close to disappearing beyond the spacecraft horizon. At 0625 SCET the episode ends as the source disappears from Voyager 1's view (after Kaiser et al. 1983).

aligned so that the onset of the first episode in the figure corresponds to the storm system coming into view over the limb. This panel shows that the only solution to the occultation pattern was for a storm in the equatorial atmosphere extended some 60° in longitude. Kaiser et al. (1983) were also able to show that the storm was confined to within a few degrees of the equator, because of the small spread (±5 min) in the repetition period.

The atmospheric storm source location of Kaiser et al. (1983) also explains the odd frequency-dependent episode onset at Voyager 1 closest approach. Figure 21 shows the trajectory of Voyager 1 past Saturn with 3 insets depicting the view of Saturn from the spacecraft at the indicated times. In inset (a), the equatorial storm system is just coming into view around the planet's morning terminator. At this time, however, the Voyager 1 spacecraft is moving so fast around Saturn that it nearly corotates with the planet. Thus, the storm takes a much longer time to rise, namely ~ 4 hr, than during any other episode. The radio emission from the lightning bursts necessarily travels through the dayside ionosphere before reaching the spacecraft, and experiences frequency-dependent refraction in the process. The low frequencies are either blocked completely, or are strongly refracted away from Voyager 1. The highest frequencies above 20 MHz, however, suffer very little refraction and can propagate to the spacecraft in more or less a straight line. Voyager 1 then observes first the high frequencies and slowly over the 4-hr interval successively lower and lower frequencies, until by the time of inset (b), the entire PRA frequency range can be viewed. Inset (b) also corresponds to the first occasion that Voyager 1 had to observe the storm on the night side of Saturn. By the end of the episode shown in inset (c), Voyager 1 is no longer corotating with Saturn, so the source quickly sets over the dawn meridian.

V. EMISSION THEORIES

A. Saturn Kilometric Radiation (SKR)

There are two phenomena that relate to SKR which have received attention from theorists. The first is the occurrence of the radio emission itself. The theoretical work in this area has not been done specifically for Saturn, but rather for the Earth's auroral kilometric radiation (AKR). The second area of interest for the theorists is the formation of discrete arcs in the dynamic spectra of both Saturn and Jupiter.

Grabbe (1981) presents a review of the theories for generation of AKR. All of these theories make use of the observed correlation between AKR and auroral electron precipitation. Intense beams of electrons in the few keV energy range are presumed to be the energy source for AKR. The theories are then concerned with just how this energy source gets converted into electromagnetic radiation. The AKR theories are divided into two categories, direct processes and conversion processes. The direct process theories convert the energy in the electron beam directly into an escaping electromagnetic mode without going through intermediate steps. The conversion process theories first convert the electron beam energy into an electrostatic mode which then subsequently gets converted into an escaping electromagnetic mode. Of the 15 or so theories proposed for AKR generation, many have been ruled out by more recent measurements, particularly those which have determined that AKR is emitted primarily in the extraordinary mode and not the ordinary mode (see Shawhan and Gurnett 1982). Only those AKR theories which are still viable as of this writing and which might have application at Saturn will be described. Since SKR is observed to be RH polarized above Saturn's northern hemisphere and LH above the southern hemisphere, Warwick et al. (1981a) have concluded that, like AKR, SKR is emitted in the extraordinary mode.

A problem for the direct process theories has been how to get the emission up to high enough frequency to escape. AKR is observed to be generated in regions where $f_p < 0.2 f_g$ (Calvert 1981), but the emission can only escape at frequencies above the extraordinary mode cutoff which is just above f_g. The direct process theories make use of an idea originally proposed by Ellis (1962), Ellis and McCulloch (1963), and Melrose (1973,1976). This idea uses the beam velocity to Doppler shift the emission, which is presumed to be generated at the local f_g value, to above the extraordinary mode cutoff.

Two recent theories show considerable promise in explaining much of the AKR phenomenology and could possibly be applied to Saturn's magnetospheric cusp or dayside auroral region. Wu and Lee (1979) and Lee et al. (1980) make use of a loss cone distribution (i.e., a distribution with low parallel velocity particles absent) which has been reported for the Earth's auroral electron beams. They determine that the free energy of this loss cone distribution can preferentially drive extraordinary mode radiation, and can produce the observed power levels. Grabbe et al. (1980), on the other hand, propose that the electromagnetic noise interacts with low-frequency coherent density fluctuations (which are created by electrostatic ion cyclotron waves). The result is a 3-wave process in which beat waves are produced that interact with the electron beam. When the wave frequency is just below the Doppler shifted beam f_g, the extraordinary mode emission undergoes a convective instability and wave growth occurs.

The major problem with the conversion process theories is that very high efficiency is required in each of the conversion steps so that the observed AKR power levels can be obtained from the energy available in the electron beam. One theory may be applicable to SKR. Roux and Pellat (1979) and James (1980) proposed a coherent 3-wave process in which two electrostatic waves combine to produce the observed electromagnetic wave. The best candidate waves for the resulting extraordinary mode emission are whistler waves at frequencies above the lower hybrid resonance, and so-called Z-mode waves in the range from f_g to f_{UHR}. Application of any of these theories to Saturn is strictly an arm-chair science, since no observations have been made *in situ* in Saturn's auroral zones or magnetospheric cusps. However, by analogy with the Earth, we can conjecture that conditions may well be similar.

The dynamic spectral arcs described in Sec. III.A show many similarities to arcs in the decameter wavelength range observed at Jupiter (Warwick et al. 1979). Two of the theories put forth to explain the Jovian arcs may be applicable to Saturn. The first theory (Pearce 1981; Goldstein and Thieman 1981) proposes that emission (Jovian) is beamed into a thin conical sheet. Rotation causes this beam to sweep past the observer so that emission is detected at a given frequency only when the line of sight is parallel to the sides of the conical sheet. Emission cones at successive frequencies have slightly differing orientations because they emanate from different points on a field line where the emission frequency is close to the local f_g. The combination of all these beams can produce the arc patterns observed. A major problem with this theory is the observation that the SKR source does not rotate with the planet, but stays fixed with respect to the Sun. However, Thieman and Goldstein (1981) believe that the SKR beam need only rotate through a relatively small angle for the mechanism to work.

The other theory (Lecacheux et al. 1981) suggests that arcs are caused by diffraction of radio waves as they pass through a phase-changing plasma structure. In the case of Jupiter, this plasma structure is the Io plasma torus which nods up and down (relative to the observer) with the rotation of the planet's tilted magnetic field. For Saturn, no nodding is possible because of the lack of appreciable tilt, but the torus associated with Dione and several other inner satellites (see chapter by Scarf et al.) could contain azimuthal variations in density which may be capable of generating SKR arcs.

B. Low-Frequency Emissions

For the case of Earth, Kurth et al. (1981a) have demonstrated that both the diffuse, trapped continuum radiation and the narrowband escaping radiation are generated by the same mechanism. Much progress has been made in understanding the source mechanism by studying the higher frequency escaping emissions, since the identity of the escaping emission is preserved by not having undergone multiple reflections within the magnetospheric cavity. A number of theories are currently proposed to explain the generation of continuum radiation. Most of these theories attempt to use intense electrostatic waves near f_{UHR} as the source. The theories generally fall into two categories: linear conversion to L–O mode radiation (Jones 1976; Okuda et al. 1982; Lembege and Jones 1982), and nonlinear 3-wave processes using low-frequency waves like the ion-cyclotron mode as the third wave (Melrose 1981). Barbosa (1982) provides a review of the status of the theories for the terrestrial and Jovian emissions.

Gurnett et al. (1981b) have applied the theoretical work done for the Earth and Jupiter to the Saturnian narrowband radiation. The schematic at the top of Fig. 19 shows the essential ideas of the emission mechanism. Kurth et al. (1979) established that intense electrostatic bands could often be detected at or near the terrestrial plasmapause when $f_{UHR} = (n + 1/2) f_g$. These are the upper hybrid waves commonly thought to be the source of the continuum radiation. Since evidence for the conversion from these electrostatic bursts to electromagnetic radiation is found primarily at a density gradient such as the plasmapause, and the conversion theories rely on a density gradient to enable the electromagnetic waves to escape, Gurnett et al. (1981b) proposed that the low-frequency Saturn emission is generated on the gradient at the edge of the plasma sheet. In this case, the gradient is approximately parallel to the magnetic field direction as can be seen in Fig. 19b. This is opposite to the geometry that exists at the Earth's plasmapause or the Io torus at Jupiter. This new geometry has not been analyzed theoretically to determine if the emission generation would proceed as in the more familiar perpendicular case. Alternatively, one could identify density gradients associated with the plasma sheet which are aligned nearly perpendicular to the magnetic field and which might be more suitable sites for the narrowband generation (see Fig. 6b of Gurnett et al. 1981b).

The ordinary mode inference by Gurnett et al. (1981b) is consistent with the linear conversion theory which predicts predominantly ordinary mode radiation. It is difficult to establish the consistency of the polarization observations of Warwick et al. (1981a) for RH and Warwick et al. (1982a) for LH since knowledge of the source location and field direction is required to relate these observations to a magnetoionic mode. Some mixture of modes, particularly in the trapped radiation, may be expected since both ordinary and extraordinary modes are commonly detected at the Earth. There it is thought the radiation is primarily in the ordinary mode and subsequent mode conversion is responsible for the extraordinary component.

C. Saturn Electrostatic Discharges (SED)

Virtually all researchers working with the SED data agree that the emission is produced by a discharge phenomenon. Warwick et al. (1981a,1982a) and Evans et al. (1981,1982,1983) proposed discharges by an unknown object within the Saturnian rings, although Kaiser et al. (1983) have shown that much of the work concerning the ring source (e.g. frequency extent and on-off episode structure) was based on incomplete analysis of the original data. Burns et al. (1983) and Kaiser et al. (1983) proposed that SED are the radio counterpart of atmospheric lightning flashes. Table II, adapted from Burns et al. (1983), compares some of the properties of SED with lightning observations at Jupiter, Venus and Earth. Many of the values in the table are poorly known. The quantities at both Jupiter and Venus have, in many cases, been deduced by using indirect observations and some assumptions. In the case of

TABLE II
Properties of Planetary Lightning and Saturn Electrostatic Discharge[a]

Property	Venus	Earth	Jupiter	Saturn (SED)
Flash rate $(km^{-2} yr^{-1})$	> 0.08 (R)[b] $0-45$ (O)[c]	$2-7$ (O)	$4-40$ (R) 4×10^{-3} (O)	0.02 (10^{-4})[d]
Event duration	250 msec	34 msec	< 35 s	$30-450$ msec
Stroke duration	?	~ 1 msec	?	~ 1 msec
Energy per flash (J)	7×10^7 (O)	4×10^8 (total)	2.5×10^9 (O)	10^7-10^{10} (R)

[a]Table adapted from Burns et al. (1983).
[b](R) = radio
[c](O) = optical
[d]Values for the assumed $60° \times 4°$ storm area and (in parentheses) for the entire planet.

the Earth, there is an overwhelming amount of data, but most of this data is concerned with cloud to ground lightning strokes, whereas, for comparison with SED, we are more interested in cloud to cloud strokes. Nevertheless, Table II suggests that SED bursts have many of the same properties as atmospheric lightning bursts. A good tutorial on discharge phenomena as applied to terrestrial lightning may be found in the text by Uman (1969).

The flash rates listed in Table II correspond to about 100 flashes per second for the Earth and 8 to 80000 flashes per second for Jupiter; thus the possibility of SED-like bursts from those planets should be investigated. Preliminary examination of the Voyager data obtained during the periods immediately following the launches in 1977 and the Jupiter encounters in 1979 has failed to reveal any signatures resembling SED. The reasons for this lack of detection are beyond the scope of this review, but should be studied as soon as possible.

VI. THE FUTURE

We have seen an explosion in knowledge of Saturn's magnetosphere since 1979, and much remains to be done with the Pioneer 11 and Voyager data sets. However, it seems as of this writing that there will be no new data recorded *in situ* from Saturn's magnetosphere during the remainder of this century. Thus, further progress in some areas of Saturn-related research will necessarily suffer. In the radio astronomy of Saturn, it is hoped that some additional insights can be gained by studying the radio emissions of the Earth and Jupiter. We have alluded to the similarities of some of the radio components from these three planets throughout this chapter, and there are likely other similarities to be found. Thus, one or more of the theories put forth to explain AKR may successfully be adapted to both Jupiter and Saturn. If that happens, perhaps we will then be in a position to "invert" the radio data and deduce just what conditions exist at the emission generation sites deep within the magnetospheres of these planets.

Acknowledgments. The authors thank J. K. Alexander for many valuable and constructive comments. The research at the University of Iowa and at Radiophysics, Inc. were supported by a contract from the National Aeronautics and Space Administration with the Jet Propulsion Laboratory. The research at the Observatoire de Paris in Meudon was supported by the Centre National d'Études Spatiales.

THE OUTER MAGNETOSPHERE

A. W. SCHARDT, K. W. BEHANNON, R. P. LEPPING
Goddard Space Flight Center

J. F. CARBARY
Mission Research Corporation

and

A. EVIATAR, G. L. SISCOE
University of California, Los Angeles

There are many similarities between the Saturnian and terrestrial magneto-spheres. Both have a bow shock and magnetopause with a standoff distance that scales with solar wind pressure as $p^{-1/6}$; however, the high-Mach number bow shock is similar to that of Jupiter and differs from the low-Mach shocks of the terrestrial planets. The magnetic field in the outer magnetosphere requires a planetary dipole field, a ring-current field, plus contributions from magneto-pause and tail currents. Saturn, like Earth, has a fully developed magnetic tail, 80 to 100 R_S in diameter. Pioneer 11 made multiple crossings of the tail current sheet during its near equatorial outbound trajectory in the dawn direction. One major difference between the two outer magnetospheres is the hydrogen and ni-trogen torus produced by Titan which maintains the plasma density between 10^{-2} and 5×10^{-1} ions cm^{-3} and a temperature of $\sim 10^6$ K. This plasma is, in general, convected in the corotation direction at nearly the rigid corotation speed. Energies of magnetospheric particles extend to above 500 keV. A large increase in the proton and electron population below $\sim$ 500 keV is found at the magnetopause. In contrast, interplanetary protons and ions above 2 MeV have free access to the outer magnetosphere to distances well below the Störmer cut-off. This access presumably occurs through the magnetotail. As a result, the

proton energy spectrum has two components: a low-energy power law compo-
nent $E^{-\gamma}$, with $\gamma \sim 7$, which becomes less steep below 100 keV; and a much
harder interplanetary high-energy component with $\gamma \sim 2$. In addition to the
H^+, H_2^+, and H_3^+ ions primarily of local origin, energetic He, C, N, and O ions
are found with solar composition. Their flux can be substantially enhanced over
that of interplanetary ions at energies of $0.2-0.4$ MeV nuc^{-1}. Proton pitch-
angle distributions are generally pancake (perpendicular to the field), while
electron pitch-angle distributions may be either dumbell (field-aligned) or pan-
cake. Field-aligned flow of protons and electrons was observed at various posi-
tions in the magnetosphere. Acceleration of particles into the several-hundred-
keV range must be taking place in the tail. Besides the field-aligned streaming,
Voyager 2 observed impulsive electron acceleration, and rapid changes in the
proton and electron population ($E \sim 1$ MeV) require, at times, the existence of
an intense source. No evidence has been found in the magnetotail for either ra-
dial outflow of plasma or the presence of a magnetic anomaly at Saturn, but the
spectrum of electrons and ions in the magnetotail appears to be modulated by
the Kronian period.

A planetary magnetosphere is formed by the interaction between the solar wind, its embedded magnetic field, and the magnetic field near the planet.[a]
The latter field can be due to ionospheric currents as at Venus or due to the
planetary dynamo field as at Earth, Jupiter, and Saturn. Various processes in
the magnetosphere require energy, such as heating the ambient plasma to a
superthermal temperature, accelerating electrons and ions to energies in the
MeV range, or creating an aurora. At Earth, the magnetospheric processes
derive their power primarily from the solar wind; at Jupiter, the magnetosphere probably extracts most of its energy from the rotational energy of
the planet. The Kronian magnetosphere is of special interest because its
parameters are intermediate, and both processes might make significant
contributions.

The outer magnetosphere and magnetotail are thought to be the primary
regions where the solar wind energy is transferred to the magnetosphere; they
are also regions where instabilities in the rapidly rotating plasma may be able
to extract energy from the planetary angular momentum. To date, we have
found good evidence for solar wind effects, but evidence for major rotational
energy sources at Saturn is still lacking. Because of the evolution of the solar
wind between 1 and 9.5 AU, the details of the interaction at Saturn may be
quite different from those at Earth, but the major features such as the existence of a bow shock and magnetopause are unchanged. The density of the
solar wind is only 1% of its density near Earth; the Alfvénic-Mach number
has increased substantially, and the density and velocity fluctuations have

---

[a]For clarity of presentation we have deleted all references from the introduction. References about the magnetosphere of Saturn are given later in the chapter. For more information on
the Jovian magnetosphere the reader should refer to Dessler (1983), and on the terrestrial magnetosphere to Hess (1968), or any number of other reviews. The changes in the properties of the
solar wind are discussed by Smith and Wolfe (1977).

damped out, while forward and reverse shocks have developed at the boundaries between solar wind streams with appreciable velocity differences.

The outer magnetosphere is generally defined as the region in which the field in the magnetosphere is no longer symmetric around the planet. The solar wind compresses the magnetosphere in the subsolar hemisphere and distends it on the downstream side where the outer magnetosphere merges gradually into the magnetotail. The energetic particles have soft spectra, and their fluxes show considerable temporal variability. Trapping lifetimes tend to be short, and near the magnetopause particles probably cannot complete a drift orbit around the planet. The boundary between the inner and outer magnetosphere is not sharp and covers the range from 6 to 10 R_S at Saturn (1 R_S = 60,330 km, Saturn's radius). Because the E Ring and Rhea absorption processes fall logically into the discussion of the inner magnetosphere, this chapter covers primarily the region from 10 R_S to the magnetopause and bow shock.

Saturn's outer magnetosphere is filled with thermal plasma with a density of 2×10^{-2} to 5×10^{-1} ions cm^{-3} near the equator. The density decreases at higher latitudes because the scale height is moderately small in the centrifugal potential. No such plasma exists in the outer terrestrial magnetosphere; there are two reasons for the difference. First, Titan is a plasma source in addition to the polar ionosphere. Photoionization of neutrals escaping from Titan's atmosphere are the dominant process, and direct interaction between the magnetosphere and Titan's ionosphere is another mechanism when Titan is inside the magnetopause. In addition, plasma is not convected out of the magnetosphere as rapidly as at Earth because the corotational electric field shields the plasma. Continuous readjustments of the outer magnetosphere to solar wind fluctuations are probably responsible for heating the plasma and producing a tail of superthermal particles. As a result, the energy density in the plasma is often comparable to that of the magnetic field, that is, $\beta \sim 1$. Such a plasma can be unstable and may significantly modify the energetic particle population.

The energetic particle population consists of electrons and protons plus some heavier ions. Their energies are mostly < 2 MeV at 10 R_S and < 1 MeV near the magnetopause, but higher energy cosmic rays have free access to the outer magnetosphere. On the average, the particle flux increases inward but is subject to large temporal changes. These fluxes and other properties of Saturn's magnetosphere resemble those of the terrestrial magnetosphere which will be used as a point of departure.

The reader should keep in mind that our knowledge about the Kronian magnetosphere is based on data from only three passes through that magnetosphere, those by Pioneer 11 and Voyagers 1 and 2. The relevant instruments on these spacecraft are listed in Table I. Trajectories together with the bow shock and magnetopause positions are shown in Fig. 1. The three spacecraft entered near the subsolar point at latitudes between 0° and 17°. The spacecraft

TABLE I

Experiments in Saturn's Outer Magnetosphere

Observations	Spacecraft	Designator	Instrument	Principal Investigator, Institution
Magnetic field	Pioneer	—	Vector helium magnetometer	E. J. Smith, JPL
	Pioneer	—	Flux gate magnetometer	M. H. Acuña, GSFC
	Voyager	MAG	Double triaxial flux gate magnetometer	N. F. Ness, GSFC
Low-energy plasma (0.1–8 keV)	Pioneer	—	Electrostatic analyzer	J. H. Wolfe, ARC
	Voyager	PLS	Plasma cup	H. S. Bridge, MIT
Low-energy charged particles (20–500 keV)	Voyager	LECP	Counter telescope and Si detector with magnetic particle selection	S. M. Krimigis, APL
Energetic charged particles (0.5–200 MeV)	Pioneer	—	Shielded GM counters and Si detector	J. A. Van Allen, U. Iowa
	Pioneer	—	Counter telescope, shielded Si detector and fission cell	J. A. Simpson, U. Chicago
	Pioneer	—	Cerenkov counter and solid state detectors	R. W. Fillius, UCD
	Pioneer	—	Counter telescopes	F. B. McDonald, GSFC
	Voyager	CRS	Counter telescopes	R. E. Vogt, CIT
	Voyager	LECP	Shielded detectors and counter telescope	S. M. Krimigis, APL
Plasma waves (10 Hz–56 Hz)	Voyager	PWS	Radiometer, electric dipole antenna	F. L. Scarf, TRW

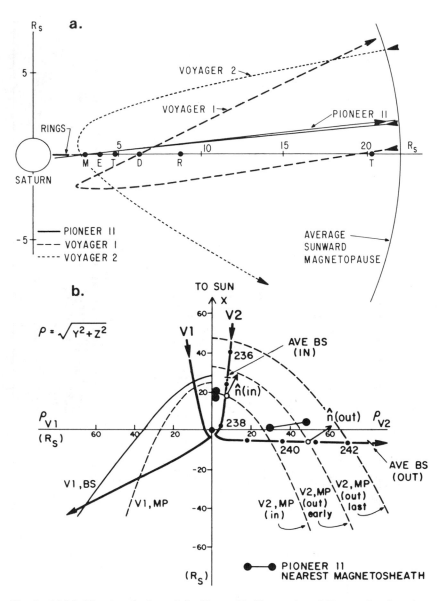

Fig. 1. (a) Meridional projections of the Pioneer 11, Voyager 1, and Voyager 2 trajectories.
(b) Saturn encounter trajectories in cyclindrical coordinates. This representation gives the
spacecraft position in the plane through the spacecraft and the Saturn-Sun line (X axis) with the
distance from that line plotted to the left for Voyager 1 and to the right for Voyager 2. For
Voyager 1, the model bow shock and magnetopause boundaries are given; Pioneer 11 followed
a very similar trajectory. For Voyager 2, observed average inbound and outbound shock location plus model magnetopause shapes are shown (Ness et al. 1982a). The outbound "early"
positions are based on the average of the first 5 outbound crossings; the outbound "last" curve
is based on the last crossing (Bridge et al. 1982) and preserves the earlier shape of the magnetopause (figure courtesy of Behannon et al. 1983).

left the magnetosphere towards dawn, Pioneer 11 and Voyager 2 about $-90°$ from the Saturn-Sun direction and Voyager 1 at about $-120°$. The latitude of the outbound trajectories was restricted to about $\pm20°$. We have no direct information about the polar region, the dusk side, or most of the magnetotail. Similarly, we lack data over a large range of interplanetary conditions. Clearly, our understanding of the magnetosphere based on such limited data is bound to contain uncertainties and incorrect interpretation of some of the observations.

The discussion in this chapter has been kept general, but some familiarity on the part of the reader with magnetospheric terminology and phenomena is assumed. The motion of trapped charged particles is described by adiabatic theory (see, e.g., Northrop 1963; Hess and Mead 1968) which breaks the particle motion down into three components: the motion of the particle around a field line in a gyrocircle; the motion of the center of this circle along the field line; and the slow drift around the planet. A constant of the motion, adiabatic invariant, is associated with each of the components. The McIlwain L parameter (McIlwain 1961) is a convenient way to characterize a drift shell and indicates approximately the distance in R_S at which the particle crosses the equator. In a rotating magnetosphere, a stationary observer sees also the bulk motion of the plasma which may be in rigid or partial corotation (Northrop and Birmingham 1982). Electric and magnetic field fluctuations permit particles to move radially across magnetic field lines, and because of the random nature of the fluctuations, this process can be described as a diffusion of the particle density (Schulz and Lanzerotti 1974; Birmingham and Northrop 1981).

I. SOLAR WIND–MAGNETOSPHERE INTERACTION

The interaction of the solar wind with an obstacle like Saturn's magnetosphere can be modeled by gas dynamic theory in which particle collisions in classical theory are replaced by coupling through electric and magnetic field fluctuations in the solar wind (Spreiter et al. 1966,1968). A detached bow shock forms if the flow is supersonic, and a transition region called the magnetosheath is created in which the shocked solar-wind plasma flows around the object. The magnetopause is the boundary between the solar wind plasma and the obstacle or magnetosphere. Because the magnetosphere is not rigid, the magnetopause occurs at a distance where the momentum flux of the impinging solar wind is balanced by internal pressure at the boundary of the magnetosphere. The subsolar distance where this occurs is proportional to the inverse 1/6 power of the solar wind ram pressure if the compressibility of the planetary magnetic dipole field controls the magnetopause position. This is the case at Earth and was also found to be true at Saturn (Bridge et al. 1981a,1982; Slavin et al. 1983). In contrast, the relation is $p^{-1/3}$ at Jupiter where the dynamic solar wind pressure is primarily balanced by the pressure and diamagnetic effect of the magnetospheric plasma rather than by the planetary field itself (Siscoe et al. 1980).

A variety of solar wind conditions existed during the Saturn encounters of Pioneer 11 and Voyagers 1 and 2. Table II lists the positions relative to Saturn at which the bow shock and magnetopause were encountered. These were identified in both the magnetic field and plasma signatures, except for some outbound crossings (especially for Voyager 2) during which the magnetic signatures were less distinct. A fast solar wind stream compressed the magnetosphere just prior to the time when Pioneer entered the magnetosphere at 17.3 R_S. The solar wind pressure relaxed while Pioneer 11 was in the magnetosphere and must have been quite variable to account for the many outbound crossings of the magnetopause and bow shock. The solar wind conditions were relatively stable during the Voyager 1 encounter and the bow shock and magnetopause were observed at "typical" distances, but the subsolar magnetosheath was significantly thinner than during the other crossings. However, even during this encounter, the dynamic solar wind pressure varied by at least a factor of two (Bridge et al. 1981a) as deduced from interplanetary Voyager 2 measurements made at the appropriate time to account for the propagation delay between the two spacecraft.

The Voyager 2 encounter with Saturn occurred during more disturbed interplanetary conditions than the others. An interplanetary shock wave had passed this region 3.5 day prior to encounter and an interplanetary current sheet 12 hr after the shock. The magnetic field jumped from < 0.7 nT to > 1.0 nT in the shock, and both the density and speed of the solar wind increased (E. C. Sittler, personal communication). Inbound, five bow shock crossings were observed between 31.5 and 23.6 R_S. Voyager 2 entered a compressed magnetosphere at 18.5 R_S, but the magnetosphere must have expanded while the spacecraft was in the magnetosphere because the last outbound magnetopause crossing did not occur until 70.4 R_S (Table II). Observations of the dynamic pressure of the solar wind during the preceding nine months gives a $\sim 3\%$ chance, based on the $p^{-1/6}$ scaling, that the magnetosphere could have expanded that much (Behannon et al. 1983; Bridge et al. 1982; Ness et al. 1982a). An alternate explanation for the large size is that Saturn passed through the extended Jovian magnetotail (Scarf et al. 1982; Warwick et al. 1982a). Prior to the Saturn encounter, recurring anomalous magnetic field, plasma, and plasma-wave features observed by Voyager 2 were interpreted as encounters of the Jovian tail at ~ 9000 R_J from Jupiter (Scarf 1979; Scarf et al. 1981c,1982; Lepping et al. 1982; Kurth et al. 1982b; Desch 1983), and the geometry for encountering the Jovian tail was still favorable while Voyager 2 was in Saturn's magnetosphere (Behannon et al. 1983).

These differences between encounters were also reflected in the particle population and other properties of the outer magnetosphere and will be discussed in the next sections. The observed large changes of the boundary location introduce uncertainties into the deduced shape of the bow shock and

TABLE II
Positions of Bow Shock and Magnetopause Crossings

Inbound	λ^a	δ^b	Distance in R_S (60,330 km) and direction of crossing [c]
(A) *Bow Shock*			
Pioneer 11	$-4°$	$5°$	4.1+, 23.1−, 20.0+
Voyager 1	$17°$	$2°$	26.1+
Voyager 2	$15°$	$15°$	31.5+, 29.0−, 27.9+, 26.6−, 23.6+
(B) *Magnetopause*			
Pioneer 11	$-1°$	$5°$	17.3+
Voyager 1	$18°$	$1°$	23.7+, 23.4−, 23.1+, 22.9−, 22.8+
Voyager 2	$22°$	$17°$	18.5+
Outbound			
(A) *Magnetopause*			
Pioneer 11	$-85°$	$5°$	30.3−, 33.2+, 34.7 [d], 35.9−, 39.0+, 39.8−
Voyager 1	$-140°$	$23°$	42.7−, 43.4+, 45.7−, 46.4+, 46.7−
Voyager 2	$-97°$	$-29°$	~48.1−, 48.4+, 51 [d], 51.8−, 51.9+, 52.0−, 52.7+, ~56.7−, 57.9+, 58.5−, 64.8+, 65.0−, ~66.2+, 70.4−
(B) *Bow Shock*			
Pioneer 11	$-81°$	$5°$	49.3−, 56.8+, 59.9−, ~63+, ~64.5−, 81.2+, 94.8−, ~95+, ~100−
Voyager 1	$-139°$	$24°$	77.4−
Voyager 2	$-94°$	$-29°$	77.5−, 77.7+, 77.8−, 79.5+, 83.4−, 86.2+, 87.0−

[a] Angle in the ecliptic plane relative to the Saturn–Sun line.
[b] Kronian latitude.
[c] A boundary moving outward past the spacecraft is indicated by + and moving inward by −.
[d] Probable multiple crossings.

magnetopause. The shapes appear to be similar to those in the terrestrial magnetosphere, but the bow shock and magnetopause at Saturn may be somewhat blunter in ecliptic plane projection (Slavin et al. 1983; Smith et al. 1980b). Observations were made only near the equatorial plane of Saturn, and no definitive information is available about the shape in a meridional plane.

The solar wind—magnetosphere interaction at Saturn occurs at sonic- and Alfvénic-Mach numbers of 10 to 20 and at pressures that are down by a factor of $\sim$ 100 from those at Earth. A major effect of the higher Mach number, as predicted by gas dynamic theory, is a larger jump in the plasma temperature across the shock, which should be proportional to the square of the Mach number (Spreiter et al. 1966). This increase was observed in the jump of electron temperatures T_2/T_1 (where T_1 and T_2 are electron temperatures in the solar wind and magnetosheath, respectively) which is $\sim$ 50 at Saturn, $\sim$ 15 at Jupiter, and only 3 to 4 at Earth (Scudder et al. 1981; Scudder, personal communications). Plasma wave turbulence was observed at the Saturnian bow shock, which suggests that the plasma instabilities associated with the higher Mach number shock differ from those of lower Mach numbers found at the terrestrial planets (Scarf et al. 1981a). The enhancement of the wave spectrum below the ion plasma frequency, characteristic of bow shocks at Earth and Venus, is not observed at Jupiter or Saturn.

In the subsolar hemisphere, the Saturnian bow shock is a strong, quasi-perpendicular shock, as is to be expected from the average spiral angle direction of the interplanetary field. Thus, the field magnitude changes at the shock but not its direction (Fig. 2). The jump in magnitude is significantly sharper when the magnetosphere contracts than when it expands (Lepping et al. 1981; Smith et al. 1980b). The shock is $\sim$ 2000 km thick corresponding to a few ion inertial lengths, much smaller than the ion cyclotron radius and larger than the electron cyclotron radius. Thus, cyclotron waves do not play a major role in the formation of the shock.

The level of plasma waves in the magnetosheath was found to be extremely low (Scarf et al. 1981a). At much lower frequencies, the magnetic field magnitude varied semiperiodically throughout the sheath (Fig. 2) during the Pioneer 11 and Voyager 1 inbound passes. These large field changes are anticorrelated with the plasma density in the sheath. Lepping et al. (1981) have interpreted these observations in terms of slow mode magnetosonic waves. An alternate explanation is based on the Zwan and Wolf (1976) mechanism. The large size and bluntness of the magnetopause may mean that flux tubes remain in the sheath for a relatively long time and the plasma could drain off by moving parallel to the field (Slavin et al. 1983; Smith et al. 1980b). In this explanation, an unspecified wave is required to account for the even spacing throughout the sheath. No such fluctuations in field magnitude were observed by Voyager 2 during its inbound magnetosheath passage.

The magnetopause of Saturn as well as that of the other planets may generally be characterized by a tangential discontinuity in the magnetic field. A

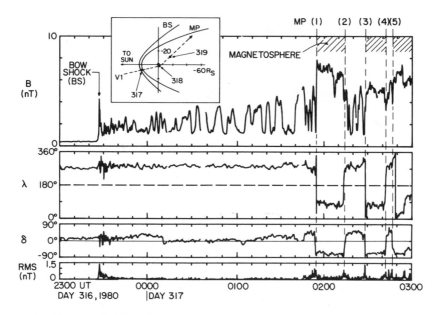

Fig. 2. The magnetic field (9.6 s averages) observed while Voyager 1 traversed the bow shock, magnetosheath and entered the magnetosphere of Saturn. Notice the absence of a systematic change in the field direction (λ and δ) at the bow shock and the major change in direction that occurred at the magnetopause. The semiperiodic changes in the field intensity in the magnetosheath are anticorrelated with the electron density. The angles λ and δ are expressed in a heliographic, spacecraft-centered coordinate system, such that $\lambda = \tan^{-1} B_T/B_R$ and $\delta = \sin^{-1} B_N/B$. The vector $\mathbf{R}$ is radially away from the Sun; $\mathbf{T}$ is parallel to the Sun's equatorial plane, normal to $\mathbf{R}$ and positive in the direction of Saturn's orbital motion; $\mathbf{N} = \mathbf{R} \times \mathbf{T}$ is within 2° of being normal to the ecliptic plane. The day numbers in the trajectory insert refer to spacecraft positions at the start of the referenced day (figure courtesy of Lepping et al. 1981).

minimum variance analysis applicable to tangential discontinuities (Siscoe et al. 1968) and the more general method of Sonnerup and Cahill (1967) give consistent results. These analyses give the direction of the normal to the magnetopause, and the directions agree with those predicted by the models of the magnetopause position (Lepping et al. 1981; Ness et al. 1982a; Smith et al. 1980b). The presence of surface waves on the magnetopause was deduced from the five regularly spaced crossings during the Voyager 1 inbound pass (Figs. 2 and 3). Such waves have also been seen at the Earth's magnetopause and were studied initially by Aubrey et al. (1971; see also Lepping and Burlaga 1979, and references therein). The waves at Saturn may be characterized by a period of 23 min, an amplitude of $\sim 0.5\ R_S$, and a wavelength of $\sim 5\ R_S$. It is believed that they propagate parallel to the equatorial plane of Saturn and tailward with a velocity of 180 ± 90 km s^{-1} (Lepping et al. 1981). These characteristics are consistent with a Kelvin-Helmholtz instability operating at

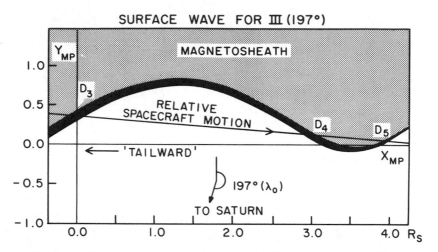

Fig. 3. Sketch of magnetopause surface wave observed during the inbound pass of Voyager 1. The X_{MP} axis is aligned with the unperturbed magnetopause, and the Voyager 1 trajectory is shown relative to the tailward moving wave. D_3, D_4, and D_5 refer to the 3rd, 4th, and 5th magnetopause crossing and the change in the width of the line indicates variability of the estimated magnetopause thickness (figure courtesy of Lepping et al. 1981).

the frontside of the magnetopause which is driven by the faster moving plasma in the magnetosphere. This is in contrast to the case at Earth, where the waves at the dawn and dusk magnetopause are thought to be driven by the more rapidly moving magnetosheath plasma (Miura and Pritchett 1982; Pu and Kivelson 1983).

The magnetopause constitutes a boundary to electrons, protons, and ions with energies < 0.5 MeV. An exception to this rule occurred during the Voyager 2 inbound pass when enhanced energetic particle fluxes were already observed 15 min prior to the magnetopause crossing as identified by magnetometer and plasma-cup data (Krimigis et al. 1982a; Vogt et al. 1982). This increase was accompanied by a change in the magnetic field of the magnetosheath from one with a vertical component opposite to the planetary field to one with a parallel vertical component. The first order anisotropy in the angular distribution of protons suggests that the magnetopause was moving inward at that time with a speed of 10 km s^{-1}. The enhanced density of hot plasma in the outer magnetosphere may have changed the magnetopause characteristics during the Voyager 2 observations (Maclennan et al. 1983).

II. COLD PLASMA IN THE OUTER MAGNETOSPHERE

The plasma density decreases at the magnetopause and its temperature increases (Fig. 4) as the spacecraft moves from the sheath into the hotter magnetospheric plasma (Bridge et al. 1981a). At Earth, in contrast, no thermal

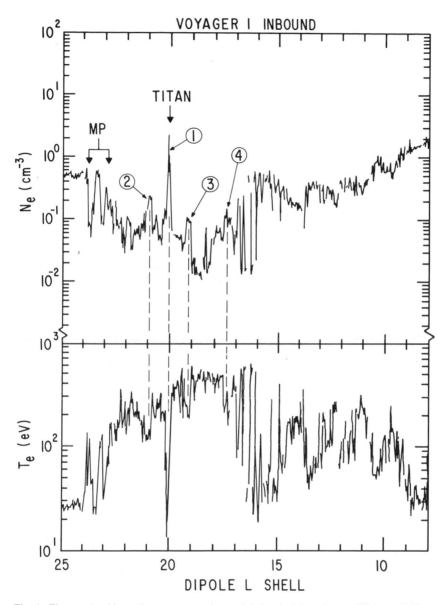

Fig. 4. Electron densities and temperatures observed during the inbound pass of Voyager 1. The high-density and low-temperature regions numbered 1 to 4 have been attributed to a plasma plume from Titan, with number 1 corresponding to the most recent interaction and number 4 to an interaction which occurred 3 Saturn periods earlier. The low-density region between $L = 18$ and 19 corresponded to an almost complete disappearance of the thermal plasma (figure adopted from Sittler et al. 1983).

plasma is found between ~ 6.6 R_E and the magnetopause. Before discussing the variations of the plasma parameters, let us first treat the average properties of the thermal plasma in the outer magnetosphere. A synthesis of the data from the three spacecraft shows a fairly thick sheet of plasma with a radial density profile proportional to $L^{-3.5 \pm 0.5}$ (Fig. 5). An ion spectrum taken ~ 0.5 R_S south of the equator at $L = 15$ (Fig. 6) shows a heavy ion component of either O^+ or N^+ which has three times the number density of protons (Bridge et al. 1981a). In the centrifugal potential of a rotating magnetosphere, the heavy ions should dominate at the center of the sheet because of ambipolarity. This effect is demonstrated by comparing ion spectra taken at different latitudes (Fig. 7). The Voyager 2 spectra taken at 18° contain almost no heavy ions. It may also be noted from Fig. 7 that the temperature of the hydrogen component was considerably higher during the Voyager 2 encounter (the distribution function peaked at a higher energy and was wider). In a two-component plasma, a higher temperature of the light component would be expected off the equator, but differences in the state of the magnetosphere may also have contributed to the difference between the two passes.

One major difference between the outer magnetospheres of Saturn and Earth is the presence of the giant satellite Titan and its dense atmosphere. At

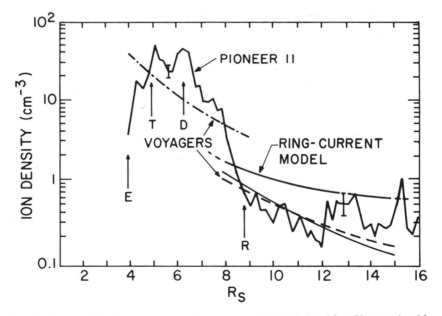

Fig. 5. Models of the electron density in Saturn's magnetosphere inferred from Voyagers 1 and 2 (Bridge et al. 1982) compared with the ion densities derived from Pioneer 11 plasma ion data (Frank et al. 1980). Also shown is the ion density computed from the ring-current model of Saturn's magnetic field (Connerney et al. 1983). The locations of the orbits of Enceladus, Tethys, Dione, and Rhea are indicated by arrows labeled E, T, D, and R (figure courtesy of Connerney et al. 1983).

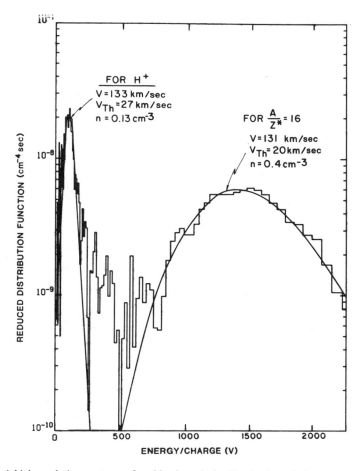

Fig. 6. A high-resolution spectrum of positive ions obtained by the thermal plasma cup on Voyager 1 during the inbound pass at $L \sim 15$. Although the heavy ion peak has been fitted with a curve corresponding to O^+, an equally acceptable fit is obtained for N^+ (figure courtesy of Bridge et al. 1981*a*).

any given time Titan can be in the magnetosphere, in the magnetosheath, or in the solar wind (Wolf and Neubauer 1982). The statistics of magnetopause locations given by Siscoe (1978) when extrapolated to the observed magnetic moment of Saturn lead to the prediction that the orbit of Titan will lie wholly within the magnetosphere $\sim 50\%$ of the time. In fact, of the three spacecraft that have flown past Saturn, only Voyager 1 found Titan's entire orbit enclosed within the magnetosphere. Titan itself, however, is generally inside the magnetosphere because it spends relatively little time near the subsolar part of its orbit.

Titan, its atmosphere, and the Titan magnetosphere interaction are dealt

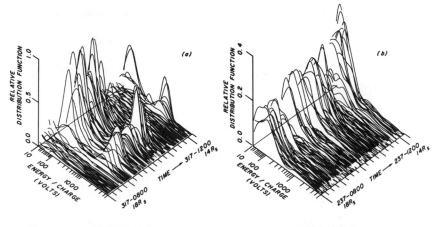

VOYAGER 1   VOYAGER 2

Fig. 7. Relative plasma distribution functions between 14 and 18 R_S observed during the in-bound passes of Voyagers 1 and 2 at Kronian latitudes of $-2°$ and $+18°$, respectively. The distribution functions were derived from low-resolution mode spectra of the side-looking plasma cup (D sensor). Peaks appearing at a low energy per charge (10–100 V) are attributed to H^+ and those at high values (> 800 V) are attributed to N^+. Note the high temperature of the H^+ ions at the latitude of Voyager 2 (18°) and the near absence of N^+ ions (figure courtesy of Bridge et al. 1982).

with elsewhere (see chapters by Hunten et al. and by Neubauer et al.). Our interest in Titan is restricted to its role as a source of plasma. Titan is the direct source of a stream of plasma extending in the local corotation direction. Estimates of the source strength range from 1.2×10^{24} ions s^{-1} (Gurnett et al. 1982b) to 6×10^{24} ions s^{-1} (Bridge et al. 1981a). This matter is apparently made up primarily of heavy ions of mass 14 (N^+) and possibly even molecular ions of mass 28 (N_2^+ or H_2CN^+) (Hartle et al. 1982) as would be expected from the high abundance of N_2 in Titan's upper atmosphere (Strobel and Shemansky 1982). This directly outflowing plasma or Titan plume tends to remain in the equatorial plane and to merge into the background, somewhat as a comet tail eventually merges into the interplanetary plasma.

In addition to the stream in the Titan wake (density peak 1 in Fig. 4), the plasma science (PLS) instrument on Voyager 1 also encountered density enhancements inward and outward of the orbit of Titan (density peaks 2, 3, and 4 in Fig. 4). These enhancements are also characterized by a high density and a low temperature and have been interpreted by Eviatar et al. (1982) to be multiple encounters with the extended Titan plume. They showed that the plume would be wrapped around Saturn by corotation and could last three Kronian periods before being reduced to the density of the background plasma. The plume would be dispersed by the combined effects of acceleration up to corotation speed, of radial expansion by the centrifugal interchange mode, and of heating which increases the vertical thickness of the plume. The radial displacements of the plume as a whole are related to magnetopause motions driven by variations in solar wind dynamic pressure. In addition to the density enhancements tentatively identified as the extended plume of Titan, other less pronounced enhancements occur between $L = 10$ and 17 (Fig. 4) which are also correlated with lower temperatures. At a higher latitude of $\sim 18°$, Voyager 2 observed similar density increases, but these correlate with higher rather than lower temperatures. It has been suggested that the flux tube content remains constant and that the decrease in scale height at lower temperatures concentrates the plasma near the equator (Sittler et al. 1983). This can account for the anticorrelation between temperature and density near the equator and correlation at a higher latitude.

An alternate explanation for the large variation of the plasma density in the outer magnetosphere has been proposed by Goertz (1983a). The plasma becomes potentially unstable against radial outflow when the centrifugal force on the plasma exceeds the magnetic field stress; this occurs when the corotation energy density of the plasma becomes greater than the magnetic field energy density (Hill et al. 1974; Kennel and Coroniti 1978). This condition is satisfied at some locations in the outer magnetosphere. For instance at $L = 15$ near the equator, Bridge et al. (1981a) find a heavy ion plasma rotating at 139 km s^{-1} with a density of ~ 0.4 cm^{-3}. Since the magnetic field is ~ 10 nT, the corotation energy density is about twice the field energy density for a nitrogen plasma. The solar wind pressure would inhibit radial outflow in the subsolar hemisphere, but plasma outflow into the tail will occur if the instability can grow sufficiently rapidly. In contrast to Eviatar et al. (1982), Goertz (1983a) concluded that the Titan plume and the high-density edge of the plasma sheet at $L \sim 15$ (Fig. 4) are sufficiently unstable to produce radial outflow into the magnetotail in the form of a planetary wind. In this picture, detached plasma blobs from the inner, stably-trapped plasma sheet account for most of the density enhancements observed by Voyager 1 and 2 spacecraft.

The plasma in the outer magnetosphere may also be moved around by means of a large-scale magnetospheric circulation, which is referred to as convection. In the case of Earth, the solar wind exerts a tangential drag on the magnetospheric plasma near the magnetopause. A convection pattern results which is fixed in the frame of reference in which the magnetopause is at rest. The depth of penetration of the convection into the magnetosphere is limited by a tendency of the resident plasma to corotate with the planet. The tendency to corotate increases toward the planet. In the case of the Earth, the convection pattern penetrates about half way from the magnetopause to the Earth. Locally generated plasma in this region is, in a manner of speaking, scoured out by the convection. Brice and Ioannides (1970) predicted, on the basis of a simple scaling argument, that at Jupiter rotation should dominate over convection all the way to the magnetopause. The same conclusion should also apply to Saturn (Scarf 1975); however, Kennel and Coroniti (1978) pointed out that the Brice and Ioannides calculation ignored the time dependence of the convection process, and this could occasionally expose the outer magnetosphere to convective transport. The possibility is more likely in the case of Saturn where rotational dominance is predicted to be smaller than at Jupiter. Since the refill time scale can be very long in the large volume of the Saturnian magnetosphere, the scars left by occasional scouring events would persist for a correspondingly long time (Siscoe 1979). It is possible that the discontinuity in the radial density profile between $L = 8$ and $L = 9$ seen in the Pioneer 11 data in Fig. 5 is such a scar. A more definitive test for the presence of the scouring process was possible with the Voyager data, which permitted a determination of the radial profile of the total number of electrons contained in a unit magnetic flux tube. In contrast to the number density profile, the content of the magnetic flux tube showed no appreciable discontinuity in the outer magnetosphere at least out to $L = 14$ (Bridge et al. 1982), but almost complete plasma dropouts were observed beyond $L = 14$ (Fig. 4). We may infer that rotation completely dominates solar-wind driven convection to this distance.

The above discussion has shown that several processes individually or in combination can account for the large fluctuations in plasma density observed outside of $L = 15$. On a phenomenological basis, this region has been described as the plasma mantle, and we find that magnetic field magnitudes and the fluxes of more energetic particles (Krimigis et al. 1982a) also show larger temporal variations in this region than inside $L \sim 15$.

In addition to the plume of Titan, there are several other candidate sources for the observed plasma. Ionization of the large neutral cloud of hydrogen reported by the ultraviolet spectrometer group (Broadfoot et al. 1981) will be a much stronger source of plasma than the Titan plume. Bridge et al. (1982) calculated the neutral hydrogen lifetime against ionization processes and found a maximum lifetime near Titan where photoionization is the dominant process. Closer to Saturn the plasma density increases, and the hydrogen is lost to both charge exchange and electron-impact ionization. In equilibrium, the rate at which plasma is generated from the neutral cloud must be equal to the rate at which the neutral hydrogen is added to the cloud by nonthermal escape from Titan, which occurs at the rate of $\sim 2 \times 10^{26}$ atoms s^{-1}. Also according to their calculation, nonthermal escape of atomic nitrogen occurs at the rate of $\sim 3 \times 10^{26}$ atoms s^{-1}. Under the assumption that all of the neutral matter entering the cloud is ionized, mass is added to the plasma from all of the Titan-related sources at a rate of the order of 7 kg s^{-1}. Protons could also originate from the atmosphere of Saturn, and both protons and oxygen ions would be produced by sputtering of the icy satellites, Tethys, Dione, and Rhea (Lanzerotti et al. 1983a) and of the A Ring. Thus, the composition at any one point is determined by the relative source strength, by possible loss mechanisms, and by transport that redistributes ions once they are formed.

In addition to the rapid transport mechanisms described above, a more uniform redistribution of the plasma can be achieved by means of diffusion. Diffusion can be driven by the centrifugal force (autotropic diffusion) (Siscoe and Summers 1981) or by the electric field generated in the dynamo region of the atmosphere and mapped into the magnetosphere along the highly conducting magnetic field lines (heterotropic diffusion) (Brice and Ioannidis 1970). In the Jovian magnetosphere, the Io torus provides a large mass on which the centrifugal force can act, thereby assuring the dominance of autotropic over heterotropic diffusion. At Saturn, on the other hand, the mass of locally generated plasma is much smaller, and it is not evident which mode of diffusion dominates. However, as Siscoe and Summers (1981) showed for Jupiter's magnetosphere, the radial density profile produced by the two types of diffusion are virtually indistinguishable. Both yield profiles of the form L^{-p} where p depends on precise details of the structure of the turbulence either in the atmosphere or in the magnetosphere. A value between 3 and 4 is consistent with predictions based on simple models of the driving turbulence in both cases.

For heterotropic diffusion, the applicable diffusion coefficient has the paradigmatic form

$$D = kL^3 \qquad (1)$$

where $k = 1 \times 10^{-10} \, R_S^2 \, s^{-1}$ has been suggested by extrapolation from Jupiter (Siscoe 1979). Krimigis et al. (1981) have derived a diffusion coefficient from electron observations near Rhea with $k \sim 6 \times 10^{-11} \, R_S^2 \, s^{-1}$. A steady-state model of the population of the region by neutral and ionized matter from Titan yields a similar value for the diffusion coefficient (Eviatar and Podolak 1983). These results are consistent with radial transport by means of simple heterotropic diffusion, but they do not constitute a definitive determination of the transport process. Based on the above diffusion coefficient, one can calculate the rate of outward transport of oxygen ions from the inner magnetosphere and balance this against the loss rate due to charge exchange in the neutral hydrogen torus from Titan. This is a very efficient loss mechanism for O^+ ions because of the near resonance of the reaction, and Eviatar et al. (1983) concluded that most of the O^+ ions could not diffuse far beyond the inner boundary (8 R_S) of the hydrogen torus.

We have not considered diffusion associated with solar wind fluctuations which can cause motion of the magnetopause and thereby violate the third adiabatic invariant. Such fluctuations are the prime source of radial diffusion of energetic particles in the Earth's magnetosphere (Fälthammar 1968) and produce magnetic impulses which generate a diffusion coefficient varying as L^{10} (Schulz and Lanzerotti 1974). If the diffusion reported by Krimigis et al. (1981) was attributed to such a process, we would find $k \sim 1.4 \times 10^{-17} \, R_S^2 \, s^{-1}$. Averaging this L dependence over the outer magnetosphere gives a mean radial transport rate three orders of magnitude greater than that calculated by Eviatar and Podolak (1983) and would require a cold plasma source strength vastly in excess of the fluxes observed during the Titan encounter (Hartle et al. 1982). It appears therefore that Saturn's outer magnetosphere resembles, in this respect, more the Jovian than terrestrial magnetosphere.

Another potential plasma source for the outer magnetosphere is the gas and ions surrounding the A Ring. A hydrogen cloud of 5×10^{33} atoms has been detected (Broadfoot et al. 1981). A charge exchange reaction between ions corotating with the planet and the neutral cloud would produce energetic neutrals with enough velocity to reach the outer magnetosphere but less than the local escape velocity. As Eviatar et al. (1983) have shown, these neutrals have a high probability of being reionized in the outer magnetosphere and may add significantly to plasma from Titan.

It is of interest to determine whether the plasma in Saturn's outer magnetosphere can corotate with Saturn. For the purpose of this calculation we will ignore the other plasma sources and use only the ~ 7 kg s^{-1} of plasma originating directly or indirectly from Titan. Corotation is retarded in Jupiter's magnetosphere by the outflow of plasma composed of matter originating on Io (McNutt et al. 1981). The braking results from the necessity to supply angular momentum to the outflowing matter by means of an electrical current system that links the magnetospheric plasma to the planet's ionosphere (Hill 1980). Beyond 20 R_S, the electrical circuit at Jupiter is no longer able to transfer the angular momentum required to maintain rigid corotation.

The neutral particle torus of Titan produces ions over a large radial range (6.5–23.5 R_S). Because we do not know the mass loading rate as a function of distance, we cannot derive rigorously the partial corotation velocity as a function of distance. However, we can estimate an average by treating the torus as a rigid unit. The resulting expression for the lag is (Eviatar et al. 1983)

$$\frac{V}{V_c} = \frac{\dot{M}c}{\dot{M}_c + \dot{M}} \tag{2}$$

in which

$$\dot{M}_c \equiv \frac{8\pi \ \Sigma \ B_s^2 \ R_s^2 \ \Delta L}{c^2 \ L^5} \tag{3}$$

(in Gaussian units) where V is the partial corotation speed; V_c, the rigid corotation speed, is evaluated at the center of the torus L; Σ is the ionospheric height integrated Pederson conductivity; B_s is the equatorial surface magnetic field strength; R_s is the planet's radius; ΔL is the radial thickness of the torus in units of R_S; and $\dot{M}$ is the mass loading rate. Eviatar et al. (1982) have calculated a conductance of $\Sigma = 1.7 \times 10^{10}$ cm s^{-1} (or 0.018 mho) on the basis

of electron and neutral particle densities in Saturn's ionosphere provided by
Voyager data. For $\Delta L = 17$ and $L = 15$, which are representative of the neutral hydrogen torus (Broadfoot et al. 1981), we obtain $\dot{M}_c \sim 14$ kg s^{-1}. Another recent estimate (Connerney et al. 1983) of $\Sigma \sim 9 \times 10^{10}$ cm s^{-1} gives
$\dot{M}_c \sim 70$ kg s^{-1}. Using the value of $\dot{M} \sim 7$ kg s^{-1} deduced earlier, we obtain
$V/V_c = 0.67$–0.91. Bridge et al. (1981a) found that corotation was main-
tained to within 10% out to 15 R$_S$, but probably began to lag between 15 and
18 R$_S$. Given the uncertainties in the determinations of Σ and $\dot{M}$, we can only
conclude that rigid corotation is marginal, and substantial departures may occur at times as seen by Frank et al. (1980).

III. MAGNETIC FIELD AND HOT PLASMA
IN THE OUTER MAGNETOSPHERE

The magnetic field configuration at Saturn resembles that found in the
outer terrestrial magnetosphere except that the direction of the field is reversed (Figs. 8 and 9). In the subsolar hemisphere the observed field is higher
than the model field because of compression by the solar wind, and in the
dawn direction we can observe the onset of the magnetotail (Fig. 8) which
will be discussed in Sec. V (Ness et al. 1981,1982a; Smith et al. 1980b). Figure 9 shows averaged vector fields for both Voyagers projected into X_{SM}–Z_{SM}
plane of the solar-magnetospheric coordinate system (X_{SM} toward the Sun,
Z_{SM} positive northward and oriented such that the planetary dipole axis lies
in the X_{SM}–Z_{SM} plane) (Ness 1965). The field magnitudes are scaled logarithmically, and the positions of a model magnetopause in the X_{SM}–Z_{SM} plane
are illustrated for both missions. The magnetic field observed by Voyager 1
outside 15 R$_S$ pointed almost perfectly southward, which is consistent with
the relatively quiet conditions of the solar wind during the early part of the
encounter.

Voyager 2 encountered a more compressed magnetosphere, and the
hourly averages shown in Fig. 9 were less steady. Initially, the field pointed
predominately southward ($B_Z = -6.6$ nT) but with substantial sunward (2.5
nT) and eastward (1.8 nT) components. Starting near 15 R$_S$, the field rotated
towards the radial direction with a decrease in the eastward component and
corresponding increase in the sunward component until B_X was comparable to
B_Z (at $\sim$ 10 R$_S$ in Fig. 9). These changes have been interpreted in terms of an
expansion of the magnetosphere in response to a major decrease of the external pressure (Ness et al. 1982a). Based on the position of the magnetopause
crossing observed during the outbound pass, such an expansion must have occurred while Voyager 2 was in the magnetosphere, and the field changes identify the time when the expansion occurred. It is quite possible that this expansion occurred because the distant Jovian magnetotail engulfed Saturn at that
time (Desch 1983). Significant changes in the energetic particle population
were observed concurrently with the field changes. Their flux became more

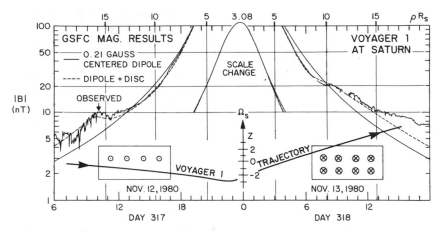

Fig. 8. Comparison of the magnetic field strength observed by Voyager 1 with pure dipole and dipole plus ring-current models. The Voyager 1 trajectory, illustrated below the observations, shows the distance from the equatorial plane. The ring current flows in the region denoted by the rectangles. Note the different scale used for the vertical spacecraft position (figure courtesy of Connerney et al. 1981).

variable and increased by an order of magnitude followed by a brief decrease by a factor of ~ 40 at 15.5 R_S (Vogt et al. 1982). The Voyager 2 observations are at least qualitatively similar to changes in the magnetic field magnitude and direction observed with Pioneer 11 which occurred most likely in response to a large decrease of the solar wind pressure (Smith et al. 1980b).

As can be seen in Fig. 8, the addition of a ring current field to the intrinsic planetary magnetic field improves substantially the fit between observations and the model field (Connerney et al. 1981,1983; Ness et al. 1982a). A good fit to both Voyager 1 and 2 data has been obtained with a ring current between $L = 8$ and 15.5 R_S, which falls off as L^{-1} and is confined to within $\pm$ 3 R_S of the equator. Under the assumption of a very cold plasma, Connerney et al. (1983) have derived the ion density in the outer magnetosphere (Fig. 5). Their values are in reasonable agreement with observations but consistently higher. The agreement between observed ion densities and densities derived from the model magnetosphere could be further improved by including the effect of the hot plasma.

The importance of the hot plasma is best expressed in terms of the ratio between the plasma and magnetic field pressures or energy densities. As Krimigis et al. (1983) have shown, this ratio in the outer magnetosphere near the equator is generally between 0.1 and 1.0 (the inbound pass in Fig. 10). When $\beta \sim 1$, the plasma is responsible for a major part of the pressure and can therefore significantly affect the field configuration. The plasma pressure is particularly high at the subsolar magnetopause with $\beta \sim 1$ and thus contributes half of the internal pressure required to balance the dynamic pressure of the solar wind (Smith et al. 1980b). A very similar situation exists at the mag-

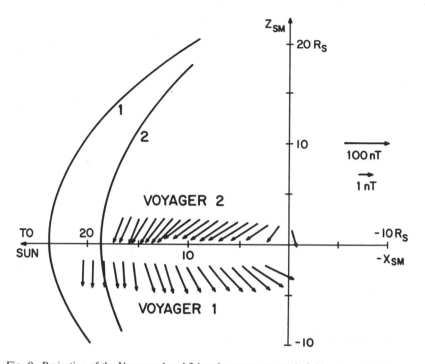

Fig. 9. Projection of the Voyagers 1 and 2 hourly average magnetic field vectors into the noon-midnight meridian plane (X-Z plane in solar-magnetospheric coordinates) (see Ness 1965). Intersection of the plane with a cylindrically symmetric magnetopause are also shown. During the Voyager 2 inbound pass, the magnetosphere was initially more compressed and then more dynamic when the spacecraft was between 10 and 15 R_S (figure courtesy of Behannon et al. 1983).

netopause of Jupiter (Lanzerotti et al. 1983b). Thus one might expect that the magnetopause and bow shock distances would also scale as at Jupiter which is $\propto p^{-1/3}$ rather than as at Earth, $\propto p^{-1/6}$. However, the significant quantity that controls the scaling is not the value of β but the origin of the magnetic field at the magnetopause. At Saturn this field is apparently due to the intrinsic planetary field and thus the magnetopause distance follows the scaling law of a dipole. At Jupiter when the subsolar magnetopause is beyond 50 R_J, the field in the outer magnetosphere is highly variable and apparently is primarily due to the magnetospheric plasma/current system. Because the plasma/currents will readjust themselves in response to changes of the solar wind pressure, a different scaling law applies.

IV. ENERGETIC PARTICLES IN THE OUTER MAGNETOSPHERE

A flux of energetic electrons, protons, and heavier ions with energies up to ~ 1 MeV is trapped by the magnetic field of the outer magnetosphere. Fig-

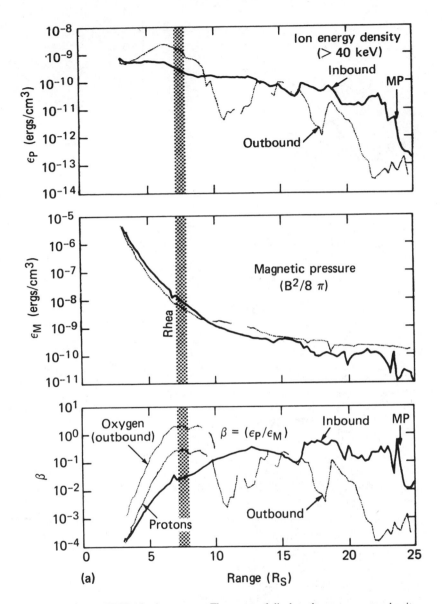

Fig. 10. Voyager 1 LECP 15-min averages. The top panel displays the proton energy density ε_p computed for protons with energies > 40 keV as a function of radial distance. Any heavy ion admixture would increase the energy density. The middle panel shows the magnetic field energy density ε_M based on *in situ* observations (Ness et al. 1981). The bottom panel shows β also under the assumption that the ions were O^+ rather than H^+. Data from the inbound pass are shown with a solid line and outbound data with a dotted line (figure courtesy of Krimigis et al. 1983).

ure 11 gives a three-dimensional overview of the ion (protons and heavier ions) and electron spectra observed with Voyager 2. A sharp increase in particle flux occurred at the magnetopause with the largest increase at lower energies. On the average, the flux increased further between the magnetopause and 10 R_S, although substantial spatial and/or temporal fluctuations were superimposed on the average increase. The most noticeable of these is the decrease in the low-energy ion flux inside 15 R_S which coincides with the magnetic field changes discussed previously and is presumably due to an expansion of the magnetosphere in response to a large decrease of the external pressure.

The observed fluxes are not symmetric between the inbound and outbound pass of Voyager 2. This asymmetry is partially due to the difference in latitude but also reflects to a large extent the asymmetry of the magnetosphere between the subsolar direction and the dawn direction (Fig. 1b). The onset of the magnetotail in the dawn direction is reflected in the more rapid decrease of particle flux with distance. These differences will be discussed in Sec. V on the magnetotail.

During the inbound pass of Voyager 1, the Kronian magnetosphere was less disturbed than during the Pioneer 11 and Voyager 2 encounters; therefore, our discussion is based primarily on Voyager 1 data with a mention of major differences relative to the other passes. Figure 12 shows the energetic electron intensities observed at different energies. Large variations are superimposed on the general decrease of the flux with distance from Saturn. These changes are most probably temporal and appear at the same time at all energies. They are thought to reflect the response of the magnetosphere to changing interplanetary conditions; however, other processes may have contributed. For instance, a flux minimum was found at the distance of Titan during both inbound and outbound passes. If the energetic particles constitute the non-Maxwellian tail of the thermal plasma, then the lower temperature in the Titan plume (Fig. 4) could be reflected in a lower flux of energetic particles. The two orders of magnitude flux dropout between 10 and 15 R_S observed outbound will be discussed in the next section. The energetic particle population was relatively stable during the Pioneer 11 pass (Fig. 12, lower panel), but large temporal changes were also observed by Voyager 2 especially during the period when the magnetosphere expanded.

The differential energy spectra of electrons in the outer magnetosphere follow a power law in energy, $E^{-\gamma}$, with $\gamma = 3.8$ to 4 during the Pioneer 11 pass (McDonald et al. 1980; Van Allen et al. 1980a); γ was in the range from 3.4 to 4 during the Voyager encounters (Krimigis et al. 1983). Such a spectrum is characteristic of magnetospheric electrons and is consistent with a non-Maxwellian tail of a thermal plasma produced by various acceleration processes.

During the Pioneer 11 encounter, the angular distributions of electrons above 0.4 MeV in the outer magnetosphere were field-aligned or "dumbbell"

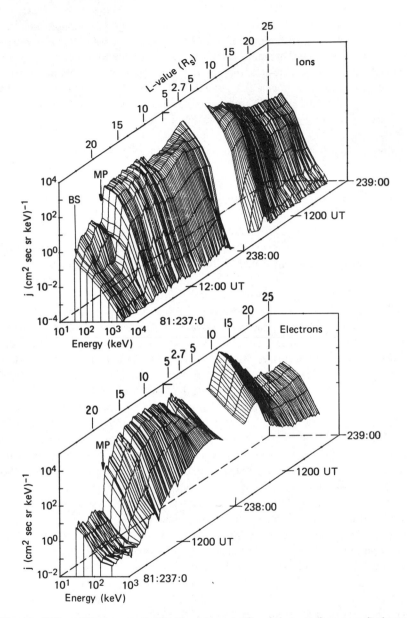

Fig. 11. Differential energy spectra for ions (upper panel) and electrons (lower panel) observed by Voyager 2. The last inbound bow shock and magnetopause crossings are indicated by BS and MP, respectively. Because of the high radiation background, spectra are not shown near closest approach (figure courtesy of Krimigis et al. 1982).

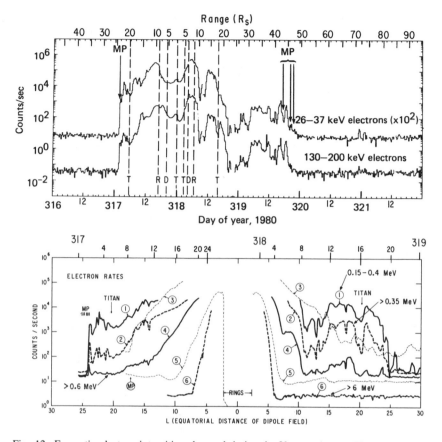

Fig. 12. Energetic electron intensities observed during the Voyager 1 pass. The upper panel shows 15-min averages observed with the LECP experiment (figure courtesy of Krimigis et al. 1983). The lower panel shows Voyager 1 data in heavy lines with curves 1, 2, 4, and 6 corresponding to electron energies of 0.15 to 0.4 MeV, > 0.35 MeV, > 0.60 MeV, and > 2.6 MeV, respectively. Curves 3 and 5 give Pioneer 11 rates for energies > 0.25 MeV and 2.0 MeV, respectively, normalized to the geometric factor of the Voyager instrument. The magnetopause crossings are indicated by MP with a circle around the MP for the Pioneer crossing (figure courtesy of Vogt et al. 1981).

(Fig. 13) near the magnetopause and slowly changed towards "pancake" (perpendicular to the field) distributions at 10 R_S (Bastian et al. 1980; Fillius et al. 1980; Krimigis et al. 1981,1982a; McDonald et al. 1980; Van Allen et al. 1980a). This change in the pitch-angle distribution is illustrated in Fig. 14, where the coefficient of the cos 2θ term in the Fourier expansion is plotted versus distances (θ is the angle between the look direction and the projection of the field into the scan plane). Because of the symmetry of particle motion around field lines, this term corresponds to a pitch-angle distribution that is

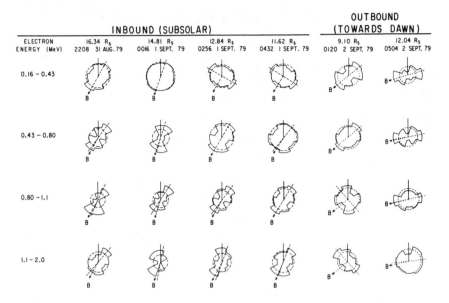

Fig. 13. Polar histograms of sectored electron counting rates (32-min average) observed with Pioneer 11. The dashed circle gives the spin average rate. The dashed arrow shows the projection of the magnetic field into the scan plane, and the dashed line gives the direction of the second-order anisotropy (figure courtesy of McDonald et al. 1980).

either field-aligned or perpendicular to the field. The distribution is proportional to $1 + b \sin^2\theta$ where $b = 2A_2/(1 - A_2)$ and A_2 is negative for dumbbell distributions. The behaviour of the 0.16 to 0.43 MeV channel (Fig. 14) is markedly different from that at higher energies, and the angular distribution changes back and forth between pancake and dumbbell distributions. At yet lower energies (37–70 keV) the pitch-angle distribution is pancake (Voyager 1) throughout the outer magnetosphere (Krimigis et al. 1983) and changes to dumbbell near Rhea where particles with pitch angles near 90° are preferentially absorbed.

The energy-dependent changes in the pitch-angle distribution are probably due to wave-particle interactions. Waves that could be responsible for these pitch-angle changes have not been observed directly, but bursts of electrostatic waves near the electron plasma frequency occur throughout the outer magnetosphere (Fig. 15) (Kurth et al. 1982b). These bursts resemble terrestrial emissions which are associated with chorus emissions, and the latter do couple to ~ 0.1 MeV electrons.

Butterfly pitch-angle distributions (maximum flux near $\theta = 45°$ and 135°) were observed at least at some energies near the inner boundary of the outer magnetosphere at ~ 10 R_S (Fig. 13), the 0.43 to 0.80 MeV channel at 11.6 R_S inbound and the 0.80 to 1.1 MeV channel at 9.1 R_S outbound. Such

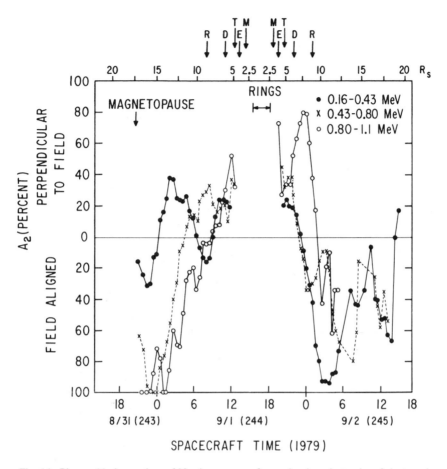

Fig. 14. Pioneer 11 observations of 32-min averages of second-order anisotropies of electrons in three energy intervals. The pitch-angle distribution represented by this term corresponds to $1 + b \sin^2\theta$, where $b = 2A_2/(1 - A_2)$. Pancake (perpendicular to the field) distributions result from $A_2 > 0$ and dumbbell (field-aligned) distributions from $A_2 < 0$. No corrections have been made for the inclination of the magnetic field relative to the scan plane. The positions of Saturn's satellites are indicated by arrows (figure courtesy of McDonald et al. 1980).

distributions can be produced by shell splitting (Roederer 1967) which occurs if the trapping field deviates from a dipole field and is asymmetric relative to the subsolar direction. In such a field the drift shells of particles with different mirror points separate as the particles drift around the planet. As a result, one can find regions in which the particle population is deficient at certain pitch angles. Based on the asymmetry of the observed magnetic field, shell splitting must occur, but it has not yet been demonstrated that the observed butterfly distributions are due to shell splitting rather than some other cause.

Proton fluxes are plotted in Fig. 16. Protons were identified above 2 MeV

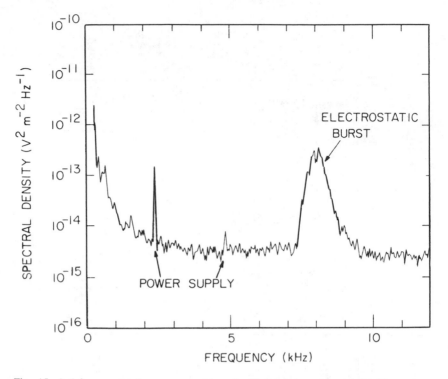

Fig. 15. A 1.8-s average frequency spectrum of wide-band data observed (12 Nov. 1981, 1001:31 SCET) by Voyager 1 inbound at L = 15.6 and a latitude of $-2°.3$. This emission is spread over a substantial frequency range and is therefore not due to the usual narrowband Langmuir waves (figure courtesy of Kurth et al. 1983).

by the $\Delta E-E$ technique (Stone et al. 1977), but the lower energy channels include contributions from heavier ions; however, composition measurements to be discussed later indicate that the heavy ion contribution is small. Outside the orbit of Rhea and below 1 MeV, the same general trend and temporal variations are present as in the electron flux. Above 2 MeV, the proton flux is unaffected by the presence of the magnetopause and remains at interplanetary values. This effect is clearly visible in the proton energy spectrum (Fig. 17) which consists of two components. The high-energy component is due to cosmic rays which gain access to the outer magnetosphere (Vogt et al. 1981). This component was much more intense during the Pioneer 11 encounter when a solar cosmic ray event was in progress and the observed proton-to-α-particle ratio was characteristic of cosmic rays (McDonald et al. 1980; Simpson et al. 1980b). Because the magnetic field near the magnetopause was undergoing time dependent changes, Simpson et al. (1980b) suggested direct penetration of the magnetopause by > 0.5 MeV ions and subsequent trap-

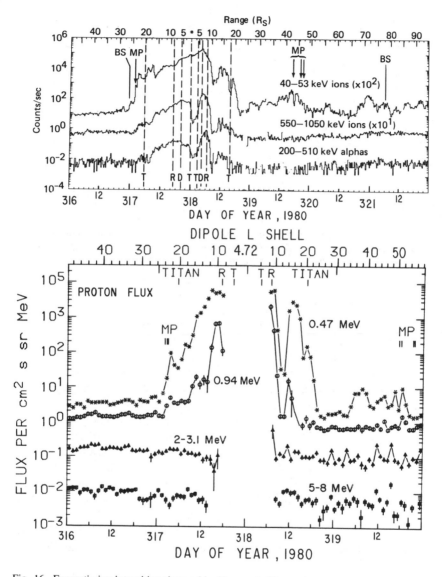

Fig. 16. Energetic ion intensities observed by Voyager 1. The upper panel shows 15-min averages observed with the low energy charged particle (LECP) experiment (figure courtesy of Krimigis et al. 1983) and the lower panel shows 1-hr average fluxes observed with the CRS experiment. The ions consist primarily of protons; the two highest energy rates in the lower panel responded only to protons.

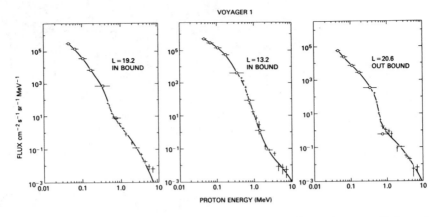

Fig. 17. Typical differential proton spectra in the outer magnetosphere and magnetotail observed by Voyager 1. The indicated dipole L values correspond to the position at the middle of the 1 hr averaging interval. The open circles represent LECP data and the crosses CRS data.

ping. McDonald et al. (1980) suggested that these particles gain access through the magnetotail and then drift into the subsolar hemisphere without being stably trapped; such access occurs in the terrestrial magnetosphere.

The low-energy part of the proton spectrum resembles a hot, convected Maxwellian distribution with a high-energy tail and can be fitted by a κ distribution (Krimigis et al. 1983). Temperatures in the outer magnetosphere fell into the range from 16 to 21 keV during the Voyager 1 inbound pass and 35 to 45 keV for Voyager 2; somewhat lower temperatures were observed during the outbound passes. The high-energy tail > 0.2 MeV follows a power law spectrum with $\gamma \sim 7$ (McDonald et al. 1980; Vogt et al. 1981). Near the equator, the diamagnetic pressure of this proton population constitutes a substantial fraction of the magnetic field pressure, and it appears that $\beta \sim 1$ when the pressure due to the low-energy thermal plasma is added.

The proton pitch-angle distributions were pancake, and during the Pioneer 11 inbound pass, the distributions became progressively flatter from the magnetopause to 11.5 R_S but then became almost isotropic at 10 R_S (Bastian et al. 1980; McDonald et al. 1980). Superimposed on this pitch-angle distribution is the Compton-Getting effect which is a first-order anisotropy due to the corotation of the plasma rest frame (Krimigis et al. 1981,1982a; Carbary et al. 1983b; Thomsen et al. 1980). The magnitude of the first-order anisotropy of the energetic particle population confirms that the plasma in the outer magnetosphere corotates nearly rigidly with Saturn out to $\sim 20\,R_S$. The changes in the pitch-angle distribution (second-order anisotropy) between the magnetopause and 11.5 R_S are qualitatively consistent with the inward diffusion and energization of protons starting at the magnetopause. The sudden disappearance of the anisotropy near 10 R_S would not be expected, and it is further surprising that this change in anisotropy was observed $\sim$ 2 R_S closer to Saturn on the outbound pass of Pioneer 11. Again wave-particle interactions are the most likely cause because absorption by a dust ring, which could cause a similar change of the angular distribution, should be symmetric between the noon and dawn directions.

A small flux of energetic ions heavier than protons are also found in the outer magnetosphere (Hamilton et al. 1983; Krimigis et al. 1981,1982a). The dominant species are H_2^+ and α particles (Fig. 18); Voyager 2 observed also a small flux of H_3^+. The presence of molecular ions indicates an ionospheric source which could be either the ionosphere of Titan or Saturn; Hamilton et al. (1983) consider the latter the more likely source. Molecular ions are photodissociated with a lifetime that depends on the vibrational state of the molecule. The longest lifetime for H_2^+ ions is $\sim$ 23 day, and this sets the time scale on which these ions have to be replenished. The spectrum of H_2^+ ions is very soft (Fig. 19), consistent with magnetospheric acceleration.

Ions from C, N, O through Fe are also found in the outer magnetosphere in the energy range from 0.2 to 0.4 MeV/nucleon. The composition of these ions and their relative abundances are consistent with a solar wind source (Fig. 18). From the composition of the magnetospheric plasma, ions accelerated from a local source should consist primarily of N^+ and O^+ with an almost complete absence of C. The abundance relative to He at equal energy per nucleon is also indicative of a solar wind source. Again, the soft spectra (Fig. 19) indicate magnetospheric acceleration. These observations are most easily explained if the acceleration mechanism is sensitive to a charge-to-mass ratio. The solar wind ions are more highly stripped and have a smaller charge-to-mass ratio than the ambient magnetospheric plasma (Hamilton et al. 1983).

Voyager 2 observed not only a much larger flux of heavy ions than Voyager 1 but also a proton-to-α-particle ratio that is consistent with a solar wind source for the protons. In contrast, a local source for protons below $\sim$ 1.5 MeV was observed by both Pioneer 11 and Voyager 1 (Hamilton et al. 1983; McDonald et al. 1980; Simpson et al. 1980b). At Earth, solar wind plasma populates the plasma sheet of the magnetotail; at Saturn the dense local plasma population could populate the plasma sheet in the near tail but probably not the far tail with the boundary between the two populations moving in and out in response to interplanetary conditions. If the particle acceleration occurs primarily in the near-tail region, one would normally expect to find a large excess of energetic protons and, occasionally under more disturbed conditions, a solar wind composition like that existing during the Voyager 2 encounter.

The phase space density of energetic electrons and protons at constant magnetic moment or first adiabatic invariant is approximately independent of the L value in the outer magnetosphere provided their energy falls below the cutoff energy (Armstrong et al. 1983; Krimigis et al. 1981; McDonald et al.

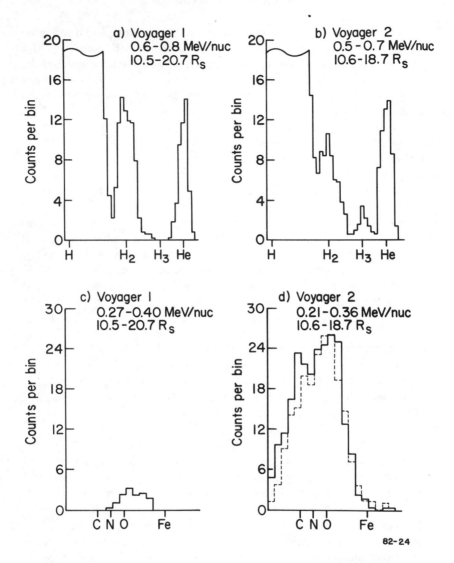

Fig. 18. Mass histograms of light and heavier ion species in the subsolar outer magnetosphere. The histograms were derived from two-dimensional pulse-height matrices accumulated over 10-hr periods by Voyagers 1 and 2. Note the nearly equal abundance of H_2 molecules and He ions. The flux of medium weight nuclei was much higher during the Voyager 2 flyby. A histogram of energetic solar ions is shown for comparison by a dashed line in panel (d). The relative lack of nitrogen suggests solar wind origin rather than the plasma torus in the outer magnetosphere (figure courtesy of Hamilton et al. 1983).

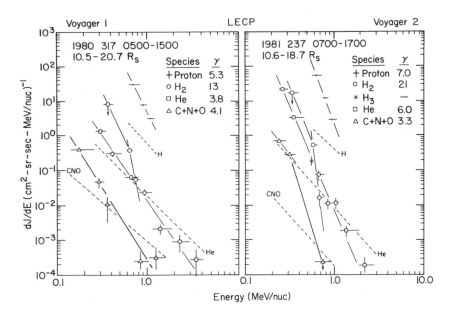

Fig. 19. Energy spectra of the most abundant ion species averaged over the same 10-hr period as for Fig. 18. A single power law in energy fits the H, He, and Voyager 1 C + N + O data. The H₂ spectrum (open circles) and Voyager 2 C + N + O spectrum (triangles) require two components with a cutoff above 0.4 MeV which falls at least as fast as E^{-8}. The H₃ ions were observed over such a narrow energy range that no spectrum could be derived. Interplanetary intensities and spectra of H, He, and C + N + O observed just prior to encounter are shown with dashed lines (figure courtesy of Hamilton et al. 1983).

1980; Van Allen et al. 1980a). This is consistent with loss-free diffusion from a source near the magnetopause or distributed sources throughout the outer magnetosphere.

V. THE MAGNETOTAIL

The outer magnetosphere in the predawn direction at −90° to −120° from the subsolar direction differs substantially from the subsolar outer magnetosphere; the magnitude and direction of the magnetic field (Fig. 8) and the energetic particle population (Figs. 12 and 16) are distinct. Minor differences are already observable at ∼ 6 R_S, and major differences start at ∼ 10 R_S. The asymmetry relative to solar aspect is due to the onset of a magnetotail resembling its terrestrial counterpart. The most direct evidence for the magnetotail comes from Voyager 1 which passed through the tail lobe. The magnetic field observations have been modeled by a cross-tail current (Fig. 20) which is driven by the convective electric field of the solar wind relative to the stationary magnetosphere (Behannon et al. 1981). This current passes through a plasma sheet located approximately in the magnetic equatorial plane near the planet and, farther from the planet, in the $X-Y$ plane of the solar-magnetospheric coordinate system. To simplify the model, the inclination of Saturn's magnetic axis relative to the ecliptic has been ignored in Fig. 20. The cross-tail current increases the radial component of the field which becomes almost parallel to the Saturn-Sun line. The tail presumably extends to a large distance from Saturn in the down-stream solar wind; for instance, the Jovian magnetotail was observed at $\sim$ 9 AU which is $\sim$ 4 AU behind Jupiter (Scarf et al. 1981c,1982; Lepping et al. 1982; Kurth et al. 1982b; Desch 1983).

The radial component of the field reverses across the plasma sheet (Fig. 20). That such a reversal does occur can be seen by comparing the Voyager 1 data north of the plasma sheet with Voyager 2 observations south of it (Fig. 21). Only Pioneer 11 was at a low enough latitude ($\sim$ 4°) to see direct evi-

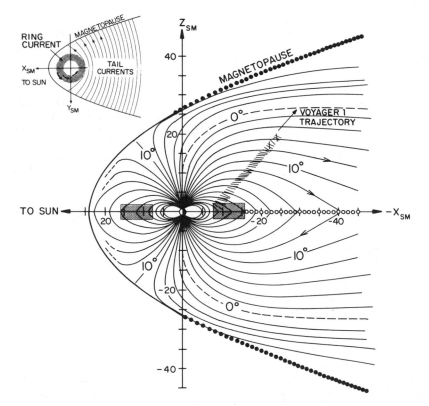

Fig. 20. Model of Saturn's magnetic field in the noon-midnight meridian plane (solar magnetospheric coordinates). The model is based on a centered dipole planetary field, an azimuthal ring current between 8 and 16 R_S (stippled) and a cross-tail current extending from 16 to 100 R_S which closes on the magnetopause boundary. Field lines are drawn every 2° of invariant latitude, and the projections of the observed magnetic field vectors (Voyager 1 outbound) are shown. The insert illustrates the cross-tail current in the solar magnetospheric X-Y plane (figure courtesy of Behannon et al. 1981).

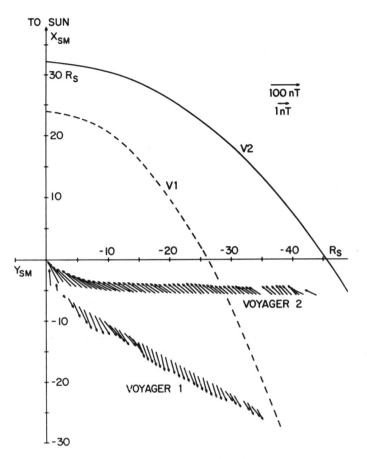

Fig. 21. Magnetic fields measured in Saturn's magnetotail and predawn magnetosphere by Voyagers 1 and 2 are shown in the plane through the spacecraft and the Saturn-Sun line (X axis). Hourly averaged field vectors have been rotated about the X axis into this plane and their length represents the field strength, scaled logarithmically. Greater temporal variations were observed during the Voyager 1 outbound pass than during the Voyager 2 pass. Because of the expanded magnetosphere during the Voyager 2 outbound pass, it observed significantly smaller field magnitudes than Voyager 1 even though the latter was at a greater distance down the tail (figure courtesy of Behannon et al. 1983).

dence for the existence of the plasma sheet. Because Saturn's magnetic axis is closely aligned with its spin axis, the position of the magnetic equator and plasma sheet does not wobble in latitude as is the case at Earth and Jupiter. Consequently, no plasma sheet crossings could be seen inside 25 R_S; however, near the magnetopause the solar wind can move the plasma sheet position to some extent and the sheet crossed Pioneer 11 several times as exemplified in Fig. 22.

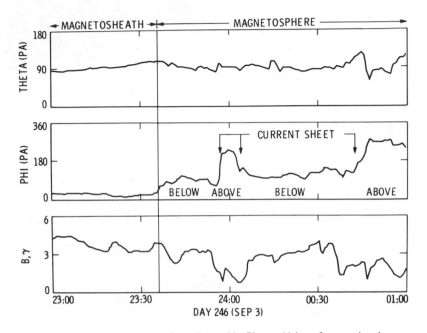

Fig. 22. Multiple current sheet crossings observed by Pioneer 11 just after entering the magnetosphere for the second time, when the magnetopause moved past the spacecraft near dawn. The data are presented in a principal axis (PA) system determined by minimum variance analysis (Sonnerup and Cahill 1967), which minimizes variation in the angle θ. Current sheet crossings were accompanied by a decrease in the field magnitude B (figure courtesy of Smith et al. 1980b).

Based on Voyager 1 data, the magnetotail diameter is $\sim 40\ R_S$ at 25 R_S behind Saturn with a typical field of 3 nT. By matching the total magnetic flux content in the tail with that in the polar cap, one finds the boundary of the polar cap in the ionosphere between 75° and 78°.5 (Ness et al. 1981). Based on similar arguments, Smith et al. (1980b) placed this boundary at $\sim 77°$. Optical observations place the southern auroral zone between 78° and 81°.5 (Sandel and Broadfoot 1981).

Voyager 2 observed a stable tail field (Fig. 21) during the entire outbound pass; the field varied in a smooth fashion in both magnitude and direction. The Voyager 1 data, in contrast, indicated that the tail configuration changed noticeably during the outbound pass of that spacecraft (Behannon et al. 1981). The data in Fig. 21 suggest that the magnetopause may first have moved outward when Voyager 1 was at $\sim 13\ R_S$ This caused a weakening of the tail field and a rotation towards the radial direction. Several hours later, in response to a compression, the field direction swung back to being nearly parallel to the model magnetopause surface (dashed curve in Fig. 21), and the field strength increased again. The greater average strength of the tail field is evident in the greater length of the Voyager 1 field vectors as compared to the Voyager 2 values (Fig. 21) even though Voyager 1 was at a greater distance down the tail.

There were no local simultaneous solar wind and interplanetary magnetic field observations; however, projections of the Voyager 2 interplanetary observations indicate that a temporary change in interplanetary field polarity and solar wind ram pressure may have occurred coincident with the changes observed while Voyager 1 was in the magnetotail (J. D. Sullivan and H. S. Bridge, personal communications; see Behannon et al. 1981, their Fig. 4). The interplanetary observations had to be projected over a distance of $\sim$ 1.7 AU, but the good correspondence between variations seen by both spacecraft pre- and post-encounter increases our confidence in the validity of the procedure. A two order of magnitude decrease in the energetic particle flux accompanied this perturbation as if the field lines through the spacecraft were suddenly connected to interplanetary field lines. (This is the dip at 10–15 R_S outbound, Figs. 12 and 16.) Simultaneously, the thermal electron density (eV energy range) dropped to very low values and the electron temperature increased markedly. These observations demonstrate the large influence of interplanetary conditions on the magnetotail and probably also on the subsolar outer magnetosphere through coupling with the magnetotail.

Adjacent to the tail magnetopause, Voyager 1 observed magnetic field and plasma behavior characteristic of "boundary layer" or "mantle" plasma flowing tailward (Behannon et al. 1983). A similarly directed flow is not evident in Voyager 2 outbound observations, possibly because of the very low pressure external to the magnetosphere when the observations were made.

A substantial population of 20 to 100 keV electrons and ions exist in the tail lobe at 20°–30° latitude (Figs. 12, upper panel and 16, upper panel); however, without data near the plasma sheet it is not known how their flux changes with latitude (Krimigis et al. 1981,1982a). The first-order anisotropy of the ions reflects full or partial corotation superimposed on the solar wind convective electric field which drives the cross-tail current (Fig. 23). The corotation velocity starts to fall below rigid corotation at $L \sim 27$ (Carbary et al. 1983b). The anisotropy also has a field-aligned component which results from a net flow of energetic electrons and ions from the equator towards the ionosphere. As in the subsolar magnetosphere, electron pitch-angle distributions are field aligned. Fluxes of more energetic electrons and protons up to $\sim$ 0.5 MeV are also found out to the magnetopause (Figs. 12 and 16, lower panel). These fluxes are subject to large temporal fluctuations. The electron flux disappears above 0.5 MeV. The ion flux above 0.5 MeV is due to cosmic rays entering the magnetosphere (McDonald et al. 1980; Simpson et al. 1980b; Vogt et al. 1981).

Just like the terrestrial and Jovian magnetotails, Saturn's tail is the site of particle acceleration. Below 100 keV, this results in a flow of particles towards the planet (Krimigis et al. 1981,1982a). At energies of $\sim$ 0.4 MeV,

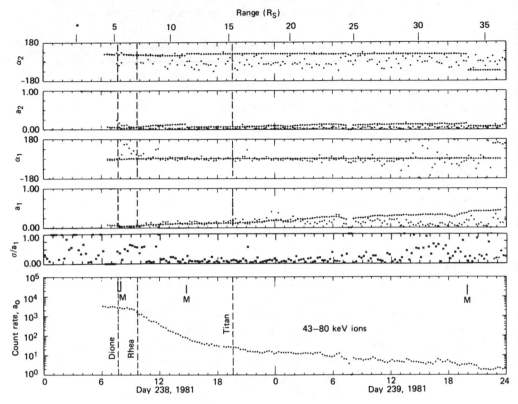

Fig. 23. Voyager 2 out-bound pass 15-min averages of fluxes and anisotropy parameters of 43 to 80 keV ions. The parameters are derived from a fit of the form $j = a_0[1 + a_1\cos(\alpha - \alpha_1) + a_2\cos2(\alpha - \alpha_2)]$. The angles α_1 and α_2 are measured relative to the projection of the corotation direction into the scan plane. The standard deviation of the fit to the data is given by σ and the ratio σ/a_1 is a measure of the significance of a_1. The results should be discounted when this ratio approaches 1. The plus symbols indicate the expected anisotropy parameters from rigid corotation for a convected κ distribution. M indicates spacecraft maneuvers. The dipole L-shell position is given at the top of the figure (figure courtesy of Carbary et al. 1983b).

ions were seen streaming away from Saturn by Voyager 1 between 35 and 45 R_S (Fig. 24, upper panel). Rather unusual bursts of electrons (Fig. 24, lower panel) accelerated to above 1 MeV were observed by Voyager 2 between 18 R_S and the magnetopause at $\sim 50\ R_S$ (Vogt et al. 1982). For each burst, the acceleration lasted ~ 5 min with all energies (0.15 to > 1 MeV) peaking simultaneously and an energy-dependent exponential decay with $\tau \sim 11$ min for 1–2 MeV electrons and ~ 19 min at ~ 0.4 MeV. Numerous acceleration mechanisms have been proposed to account for the observations in the Earth's

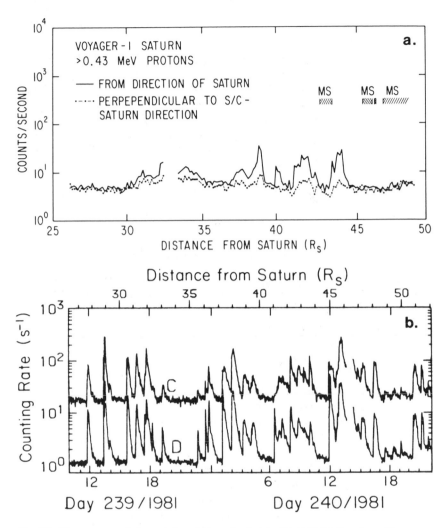

Fig. 24. The upper panel shows tailward-streaming bursts of > 0.43 MeV protons. These were observed by Voyager 1 primarily near the magnetopause between 35 and 45 R_S (figure courtesy of Vogt et al. 1981). The lower panel shows electron counting rates in the dawn side outer magnetosphere observed with Voyager 2 at a latitude of $-29°$. Curve C displays the rate of > 0.6 MeV electrons; and curve D, the rate of 1 to 2 MeV electrons ($\times$ 0.1). Typically, the electron fluxes increased by about an order of magnitude with a rise time of $\tau \sim 5$ min. The decay time was energy dependent with $\tau \sim 11$ min above 1 MeV and ~ 20 min at ~ 0.4 MeV (figure courtesy of Vogt et al. 1982).

tail, but the data at Saturn are still too fragmentary to identify specific mechanisms responsible for the observations.

The planetary field was found to be axisymmetric (see chapter by Connerney et al. on magnetic field models), but the particle data indicate that an asymmetry may yet exist in the polar region. This is based on the apparent modulation of the < 0.5 MeV electron and ion spectra with approximately Saturn's period (Carbary and Krimigis 1982). During the Voyager 2 outbound pass between 20 and 50 R_S about three modulation cycles were observed (Fig. 25) in the ratio of two electron channels [(22−35)/(183−500) keV], 10 hr 21 min ± 59 min period, and of two ion channels [(43−80)/(137−215) keV], 9 hr 49 min ± 59 min period. One cycle of modulation was observed by Voyager 1 in the same SLS (Saturn longitude system) longitude range of 0° to 90°. (For a definition of the SLS see Desch and Kaiser [1981a].)

VI. SUMMARY AND DISCUSSION

Saturn's outer magnetosphere looks superficially very much like the Earth's. Both have a bow shock, magnetopause, and magnetotail (Fig. 26), and substantial fluxes of energetic electrons and protons exist with soft spectra and large temporal variability. Acceleration of particles in the magnetotail is another common feature, and particle diffusion toward the planet with conservation of the first- and second-adiabatic invariants leads on the average to an increase in the particle flux above a fixed threshold energy. The size of both magnetospheres is controlled apparently by the pressure balance between the planetary magnetic field and dynamic ram pressure of the solar wind.

The most obvious differences between the two magnetospheres is the presence of thermal plasma out to the magnetopause at Saturn; in contrast, plasma is convected out of the terrestrial magnetosphere for $L > 6.5$. It appears that the β in the outer magnetosphere is ∼ 1; this implies the possibility of plasma instabilities which may affect many properties of the outer magnetosphere. Specific instabilities have not been identified, but many as yet unexplained phenomena may or may not have such a cause. Among these are the energy-dependent changes in the electron pitch-angle distribution (Fig. 14) and the almost regular changes in electron density and temperature observed between 10 and 15 R_S (Fig. 4). Plasma properties depend on solar aspect, which is consistent with the lack of symmetry of these phenomena between inbound and outbound passes.

Coincident minima in electron and ion densities were observed in the subsolar direction at the same dipole L values of 14 to 15 by Pioneer 11 and Voyagers 1 and 2, and at ∼ 19 by both Voyagers (Lazarus et al. 1983); coincident decreases occurred also in the energetic particle population. Because of the large ring current, the dipole L value is only an approximate measure of the field-line distance at the equator; still the coincidence is surprising. Lazarus et al. (1983) suggested the possibility of particulate or gaseous struc-

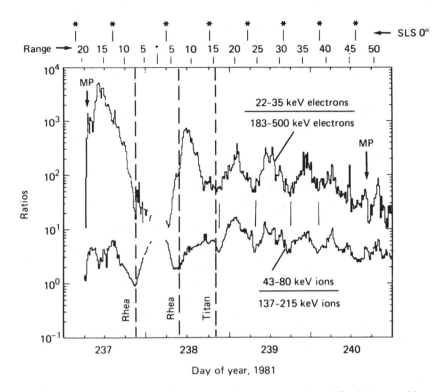

Fig. 25. Ratios of counting rates (15-min average) in two energy channels for electrons and ions. The dipole L shells of Rhea and Titan are shown as dotted lines, and the tick marks identify the minima used to determine the period. Times when the spacecraft was at an SLS longitude 0° are indicated at the top of the figure (figure courtesy of Carbary and Krimigis 1982).

ture to explain the observations, but the lack of symmetry between inbound and outbound passes presents problems for this proposal. Other explanations are based on the escape of plasma bubbles, on occasional penetration of solar wind convection into the magnetosphere, or on the existence of a magnetospheric anomaly at a fixed Kronian longitude.

The frequent and large changes in the energetic particle fluxes (Figs. 12 and 16) have not been explained in detail. One possibility is that the shock which develops at boundaries between slow and fast solar streams disrupts the magnetosphere substantially more than the smoother transition observed at Earth. This change in the character of the solar wind and how it affects a magnetosphere deserves further study.

It appears that the magnetospheres of Earth, Jupiter, and Saturn have the same ratio between the energy content E_B of the planetary magnetic field and the energy content E of the particles trapped in the field (Connerney et al. 1983). In each case the ratio between the quiet time ring-current field ΔB at

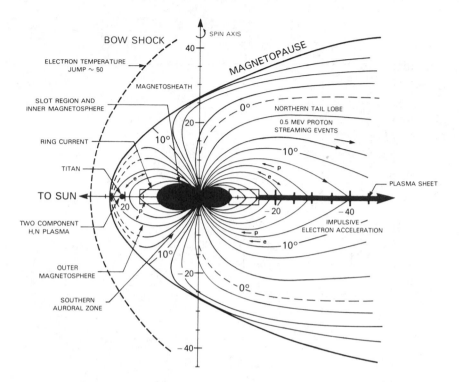

Fig. 26. Schematic of the global structure of Saturn's magnetosphere. Bow shock and magnetopause positions are "typical." A two-component H^+–N^+ plasma fills the subsolar outer magnetosphere, with the heavier ions concentrated near the equator. Energetic proton pitchangle distributions are pancake and electron distributions primarily dumbbell. An equatorial ring current extends into the outer magnetosphere and is continued by a plasma sheet in the magnetotail. Proton and electron acceleration occurs in the magnetotail. Streaming of low-energy electrons and protons towards Saturn was observed in the tail lobes as well as streaming away from Saturn at higher proton energies. Major solar wind-induced changes have been observed in the magnetic field properties, plasma densities, and energetic particle fluxes.

the equator and the equatorial planetary field B is $\Delta B/B \sim 1/2000$. According to the Dessler-Parker relation (Carovillano and Siscoe 1973, and references therein):

$$E/E_B \sim 1.5 \ \Delta B/B. \qquad (4)$$

The total energy contents of the three magnetospheres vary by orders of magnitude and the energy sources driving the magnetospheres are quite different. Still the energy in the plasma and energetic particles builds up to only $\sim 1/1000$ of the field energy in each of the magnetospheres. It would appear that one or another plasma instability is triggered at that point to prevent further buildup.

In summary, the three traversals of Saturn's magnetosphere have provided a good overall description, and as the analysis progresses, we will be able to fill in further details. However, our picture is bound to remain incomplete until long-term observations are performed with a Saturn orbiter. In contrast to Jupiter, the solar wind appears to be the dominant energy source and we need, therefore, observations of the magnetospheric response to various solar wind conditions.

Acknowledgments. The authors are greatly indebted to many scientists studying the magnetosphere of Saturn for helpful discussions, for prepublication copies of their work, and for permission to reproduce their illustrations. The help and support of the staff at the University of Arizona, especially M. S. Matthews, is gratefully acknowledged.

PART IV
Rings

SATURN'S RINGS: STRUCTURE, DYNAMICS, AND PARTICLE PROPERTIES

L. W. ESPOSITO
University of Colorado

J. N. CUZZI
Ames Research Center

J. B. HOLBERG
University of Southern California

E. A. MAROUF, G. L. TYLER
Stanford University

and

C. C. PORCO
California Institute of Technology

The rings of Saturn are not only beautiful, but also a laboratory for particle dynamics and a likely relic of the early solar system. Recent spacecraft results have brought a manifold increase in our knowledge and understanding of the rings. Spacecraft measurements have a resolution as good as 100 m on the rings, revealing unexpected microstructure down to the limit of resolution. Density waves and bending waves propagate through the rings. Analysis of these waves gives the mass, viscosity, and thickness of the rings. The mass of the rings is $\sim 5 \times 10^{-8}$ M_S, about the same as Mimas'. The rings are thinner than 100 m, but thicker than a single particle. The ring particles have a broad size distribution, ranging from 1 cm to 5 m, with some areas containing particles as *small as several microns. Moonlets (particles as large as 10 km) are quite rare. The ring particles are bright and strongly backscattering, consistent with rough, icy blocks. Recently observed features in the rings include dark, nearly radial lines (spokes) in the B Ring and noncircular ringlets and ring edges. Several regions in the rings contain fine dust, which allows a coupling between the magnetosphere and the observable rings. The important questions of the origin and evolution of Saturn's rings are still unresolved.*

For many of us, the first and most beautiful sight seen through a small telescope was Saturn with its rings. One of the most satisfying of the recent spacecraft results is that Saturn's rings are as beautiful seen close up. In addition, the rings provide a natural laboratory for study of many-particle dynamics in flat disks. They are likely a relic of the early days of the solar system, when the original small bodies had not yet coalesced into planets. Since rings have been discovered around Jupiter and Uranus as well, Saturn's rings provide the best-studied example of what is now seen as the common phenomenon of planetary rings.

A major advance in our understanding of Saturn's rings is due to the flyby encounters of Pioneer 11, Voyager 1, and Voyager 2. In the period of only a few years our best resolution improved by a factor of 10^4; we were able to view the rings from the side not illuminated by the Sun; we observed the rings as they reflected and transmitted light at angles never visible from Earth and at new wavelengths. These advances have made the limitations of prior groundbased observations quite clear. Nevertheless, the new results do not stand alone. A major effort which has just begun is to bring together and understand the diverse types of information, ground- and space-based, which we now possess. In this chapter we review our current state of knowledge, especially the burst arising from the spacecraft encounters, and attempt to point the direction for future analysis and observation.

In this chapter we emphasize the observed ring properties. We begin with a brief historical introduction (Sec. I); following this are discussions of the radial profile of the rings (Sec. II); ring dynamics (Sec. III); features without azimuthal symmetry (Sec. IV); the ring particle size distribution (Sec. V); and the physical properties of ring particles (Sec. VI). Finally, we close with a brief summary of our conclusions and our view of the most important open questions. A complementary review of Saturn's ring properties with greater emphasis on the active processes in the rings is given by Cuzzi et al. (1984) in the book *Planetary Rings* in this Space Science Series.

I. INTRODUCTION TO STUDIES OF SATURN'S RINGS

The following is a brief historical review of modern studies of Saturn's rings (the earlier history of ring studies may be found in the chapter by Van Helden). This section will also serve as an introduction to the following sections concerning detailed aspects of the rings. Earlier reviews have been made by Pollack (1975) and Alexander (1962).

The first "modern" quantitative observations of Saturn's rings were by Müller (1893). He observed that the Saturn system brightened strongly as the phase angle of observation (Sun-Saturn-Earth angle) decreased. Although he was not able to separate the light from Saturn and the rings, the increase in brightness is much larger than observed for Jupiter, and he correctly attributed this brightness increase near opposition as due to Saturn's rings.

Photometric measurement of this "opposition effect" has provided basic data for our understanding of the structure of Saturn's rings. Seeliger (1887) interpreted the effect as due to the shadows cast by the particles on one another. This implies a diffuse ring, where the average particle separation and total thickness are both much larger than the average particle size.

In a diffuse medium of scatterers that are large compared to the wavelength of observation, each particle casts a shadow which may fall on neighboring particles, if they are close enough. In general, when observing an ensemble of such particles, both the shadowed and illuminated portions on any particle are visible. However, exactly at opposition, when the phase angle is zero and the Sun is directly behind the observer, no shadows are visible; they lie exactly behind the illuminated particles which cast them. Thus, at opposition the rings appear considerably brighter.

The first quantitative calculation of the size of this effect is due to Seeliger (1887,1895). Later treatments essentially following his are due to Schoenberg (1929) and Irvine (1966), and several extensions to the theory are due to Bobrov (1970) and Esposito (1979). Franklin and Cook (1965), Bobrov (1970), Kawata and Irvine (1975), Esposito et al. (1979), and Lumme et al. (1983) have used this model to analyze Earth-based data. The major result of these analyses is to determine the fractional volume density D occupied by ring particles (total particle volume/total ring volume) in the brighter part of the B Ring. All of these treatments are in good agreement giving $D \sim 10^{-2}$; the rings are indeed diffuse. This model is the classical or many-particle-thick model. However, there is now some controversy as to whether this is the correct model (see Cuzzi et al. 1984).

Another photometric property of the rings is the "tilt effect." As Saturn moves in its orbit, the declination of the Earth and Sun relative to the ring plane are constantly changing; the apparent tilt of the rings to our line of sight B ranges from $0°$ to $26°.7$. For both the A and B Rings, the surface brightness has a minimum when $B = 0$ and increases as B increases. For ring A the surface brightness reaches a maximum and then decreases (Lumme and Irvine 1976a; Esposito and Lumme 1977). This behavior is easily explained by multiple scattering among the ring particles (expected in the classical diffuse model of the rings). Detailed calculations allow the optical properties of the individual particles to be inferred; the A and B Ring particles must be highly reflective and anisotropically scattering (Esposito and Lumme 1977; see Sec. VI).

Lyot was the first to detect the polarization of light scattered by Saturn's rings in 1923 (Lyot 1929). His work was later summarized by Dollfus (1961). Dollfus continued these measurements and published his results from the years 1958 to 1976 (Dollfus 1979a). He reported that the polarization from Saturn's B Ring varied with phase angle and position on the rings. Unexpectedly, he found a component of the polarization that did not lie in the scattering plane (plane containing the observer, scatterer, and Sun), but was related to the plane of the rings. This turning component T was significantly variable with time on a time scale of days (Dollfus 1979a,b). The polarization of the rings is comparable to that produced by coarse-textured objects (Dollfus 1982).

Kemp and Murphy (1973) measured the polarization of Saturn's rings from the near ultraviolet to the near infrared. This work was extended further into the infrared to 3.5 μm by Kemp et al. (1978) and to a range of phase angle in the visible by Johnson et al. (1980). No turning component T is found in their data; the polarization observed lies uniformly in the scattering plane. Johnson et al. (1980) use this fact to argue against the creation of any component related to the ring plane and put limits on the single particle polarization at 90° phase angle, $P(90°) < 3\%$. Esposito et al. (1980) pointed out that errors in their analysis allow this upper limit to be $P(90°) < 15\%$.

Spacecraft polarization measurements of the rings (Esposito et al. 1980; Esposito 1983) have the capability of making strong statements about the individual particle polarization at angles not observable from Earth. These data are still being analyzed.

In the decades of the 1960s and 1970s, the classical observations of the rings at visual wavelengths were supplemented by measurements in the infrared and at radio wavelengths; this work has been reviewed by Pollack (1975). Reflectivity spectra in the near infrared (1–3 μm) strongly indicate the presence of water ice (Pilcher et al. 1970; Kuiper et al. 1970). From this we can ascertain that the surface, at least, of the ring particles is dominantly water ice. Visible spectra show a sharp drop in reflectivity below 5000 Å; if this does not indicate the presence of another material on ring particle surfaces, it may be the result of radiation and particle damage to the crystal structure of ice (Lebofsky and Fegley 1976).

We have measurements of thermal radiation from Saturn's rings for wavelengths between 10 μm and 1 cm. At short wavelengths the measured temperatures slightly underestimate true ring temperatures; at longer wavelengths the rings appear much colder than their physical temperature. From this it was thought that all ring particles were far smaller than centimeter sized, and thus transparent to microwave radiations. Therefore, it was surprising to discover in 1973 the strong radar reduction from Saturn's rings (Goldstein and Morris 1973). Saturn's rings, in fact, have the largest radar cross section of any body in the solar system. An early explanation for the high return and the low microwave brightness was that the ring particles were made of metal (Goldstein and Morris 1973). This possibility is now excluded by Voyager data (see Sec. VI). Pettengill and Hagfors (1974) suggested that spherical particles could explain the backscattering brightness; however, in the likely event that the particles are even slightly nonspherical, the agreement is lost.

A third model was suggested by Pollack et al. (1973) which depends strongly on the assumption that the ring system is diffuse and many-particles-thick. In this situation, the individual return from a single ring particle is enhanced by multiple reflections of the scattered radiation from the other ring particles—this is the same reason a terrestrial water cloud is bright; this model is termed the bright-cloud model. The bright-cloud model depends on the particles being only weakly absorbing at microwave wavelengths and the total ring layer being substantially thick. Recent analysis shows this model to be entirely consistent with the data (Pollack 1975; Cuzzi and Pollack 1978; Cuzzi et al. 1980; Zebker 1983; see Sec. VI).

With the close encounter of Saturn by spacecraft has come an immense increase in our knowledge about Saturn's rings. Pioneer 11 flew by Saturn on 1 September 1979, Voyager 1 on 12 November 1980, and Voyager 2 on 26 August 1981. The following sections will discuss the new findings in detail. The spacecraft flybys have provided new information in three ways: resolution, aspect, and occultation studies.

The best resolution from groundbased telescopes is ~ 0.5 arcsec in the visible. At the mean distance of Saturn this translates into 3500 km; more typical observing conditions may give twice this value. The best Pioneer images (Gehrels et al. 1980) have a resolution of 400 km on the rings. The best Voyager images have a resolution of 1 to 2 km (B. A. Smith et al. 1981,1982); complete radial scans of both the bright (sunward) and dark ring faces are available with a resolution of ~ 10 km. This increase in resolution has not merely revealed more detail but entirely new phenomena.

Several new viewing geometries are available from spacecraft. Because of the distance to Saturn, the solar phase angle (Sun-Saturn-Earth) always lies between 0 and 6°3 for Earth-based observations. This means that only the backward-scattering properties of the ring particles are directly observable from the Earth. Spacecraft can view Saturn and its rings looking backward toward the Sun, allowing phase angles up to 180° which probe the rings' forward-scattering properties. From Earth, the maximum opening of the rings is 26°7. The declination of Earth relative to the ring plane ranges up to this value. Spacecraft approaching Saturn at higher latitudes can see a broader range of tilt angle and provide constraints on the vertical distribution of ring particles as well as information on the more opaque parts of the rings. Conversely, spacecraft can view the dark (i.e., away from Sun) side of the rings. Because the plane of Saturn's orbit is only inclined by 2°5 to the ecliptic, the Earth and Sun are rarely on opposite sides of the rings. Even when this does occur, the rings are nearly edge-on and in a poor position for observing. Because the opaque portions of the rings are dark when viewed in this geometry, the more transparent regions are most obvious. Pioneer 11 (Gehrels et al.

1980; Esposito et al. 1980) used this viewing aspect to study thinner parts of the rings.

When a spacecraft flies close to a planet, its apparent angular extent is quite large and thus blocks a larger portion of the celestial sphere. This provides many more opportunities for seeing a star occulted as it passes behind the rings. A stellar occultation observed by Voyager 2 allowed a resolution of 100 m along a single cut through the rings (Lane et al. 1982b). Similarly, the Earth was occulted by the rings as seen by the Voyager 1 spacecraft; this allowed a radio occultation of the spacecraft telemetry carrier (Tyler et al. 1981). The inversion of the radio occultation data gives the particle size distribution at several locations in the rings (see Sec. V).

In summary, Saturn's rings were one of the first discoveries of the new telescope. For nearly four hundred years the interplay between observation and explanation has led to increasing understanding. With the advent of spacecraft encounters, observation has leaped ahead, with our resolution improved by four orders of magnitude in three years. In the following sections we review our knowledge in several general areas, and suggest the directions in which theoretical explanations might follow.

II. THE RADIAL PROFILE OF THE RINGS

In 1675 J. D. Cassini of the Paris Observatory discovered the division in Saturn's ring system which now bears his name. Since that time, we must necessarily speak of rings in the plural. Before the 20th century, Saturn was known to have three classical rings: the outer A Ring (outside the Cassini Division), the bright B Ring, and the innermost C Ring (or crepe ring). These designations are due to O. Struve (Alexander 1962, p. 179). Surprisingly, no new rings or features in the rings were unequivocally recognized by the scientific community until after the first spacecraft encounters and the observations in 1979–1980 of the ring-plane crossing.

Groundbased visual observers had reported that the rings were not uniform; unfortunately photographic methods are not capable of substantiating these observations of ring features. Clearly, some of the reported features were spurious, as were the canals on Mars. Nevertheless, careful visual observers had achieved some credibility, notably Lyot and Dollfus (Dollfus 1970; Pollack 1975). Figure 1 is a tracing of the ring brightness observed by Dollfus (1970). We note that almost every feature in the brightness profile has a clear basis in the optical depth profile observed in the most recent studies; for comparison we show in Fig. 2 a plot of the vertical optical depth in the rings from the Voyager 2 photopolarimeter (PPS) stellar occultation (Esposito et al. 1983c).

Likewise, it was disputed whether or not gaps other than the Cassini Division existed. Dollfus and Lyot (see Alexander 1962) argued for the existence of a gap in the outer A Ring (the Encke Division) and one separating

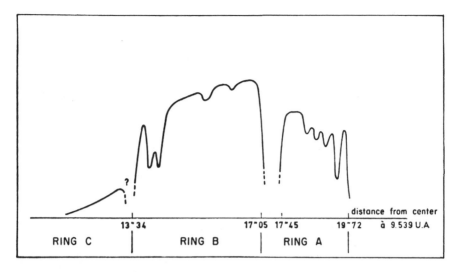

Fig. 1. Groundbased profile of ring brightness as a function of radial distance from the center of Saturn (Dollfus 1970).

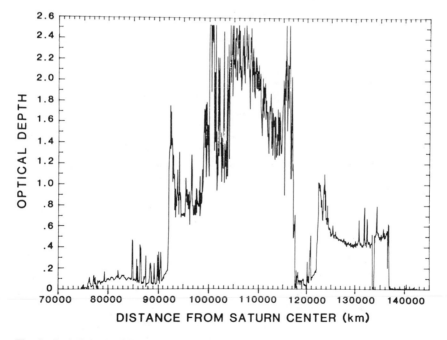

Fig. 2. Optical depth of Saturn's rings from the Voyager photopolarimeter spectrometer stellar occultation (Esposito et al. 1983c).

rings B and C. Kuiper in 1954 (reported in Alexander 1962) denied the existence of any such divisions. The existence of the Encke Division was finally established by observations of Iapetus as it was eclipsed by the rings (Reitsema 1978) and it was clearly seen in the detailed images from Pioneer 11 (Gehrels et al. 1980; Esposito et al. 1980; Burke and Ken Knight 1980). The division between rings B and C turned out to be a region of increased transparency containing a number of opaque ringlets, and thus not so similar to the other divisions. However, groundbased observers should not be held in fault; even the Pioneer imaging resolution was not sufficient to see this detailed structure (Gehrels et al. 1980). A high-resolution view of this region is given by Sandel et al. (1982b). Several smaller gaps, not visible from Earth, have now been officially recognized. These lie at 1.45 R_S ($R_S \equiv$ Saturnian equatorial radius $\equiv$ 60,330 km), the Maxwell Gap; at 1.95 R_S, at the inner edge of the Cassini Division, the Huygens Gap; and at 2.26 R_S, the Keeler Gap.

The existence of other rings was also hotly contested. Two new rings had been reported from groundbased observations. Ring D lay inside the classical rings (Guérin 1970a,b); ring E lay outside (Feibelman 1967). Smith (1978) reviewed the evidence for these rings prior to any spacecraft encounters. Taking ring D first, Smith found the evidence convincing, especially the observations and deconvolution of Larson (1979). Unfortunately, Pioneer 11 found no evidence for this ring at a level 10 times lower than predicted from the ground observations (Gehrels et al. 1980; Esposito et al. 1980). Fortunately, the more sensitive observations from the Voyager cameras found ring material inside the classical rings (B. A. Smith et al. 1981). Even though this amount of material would be invisible from Earth, the earlier proposed name still seems most appropriate. A Voyager photograph of this (new) ring D is shown in Fig. 3. The existence of ring E was also verified by Voyager (B. A. Smith et al. 1981) as well as by groundbased observers, who found it during the passage of Earth through Saturn's ring plane (see, e.g., Baum et al. 1981). The E Ring is shown in Fig. 4.

Two rings unsuspected from Earth were discovered by Pioneer 11. The imaging photopolarimeter on that spacecraft discovered ring F in an image specifically planned to seek new rings and satellites (Gehrels et al. 1980; Esposito et al. 1980). A satellite was also found. Ring G was discovered by its influence on charged particles. The Pioneer 11 investigators noted a drop in their counting rates at the location of this ring (Van Allen et al. 1980a; Simpson et al. 1980b). Ring G may also be seen in Fig. 4. Table I gives the locations and widths of all named ring features. Northrop (1983) has correlated some of these locations with stability boundaries for electromagnetic interactions with the ring particles; for more details see the chapter by Mendis et al.

In addition to the large-scale structure, it is now possible to discuss smaller scale features in the rings. It is somewhat surprising, given the extreme heterogeneity evident in the spacecraft images of the rings, that only a few years ago this was for the most part unexpected. There were several in-

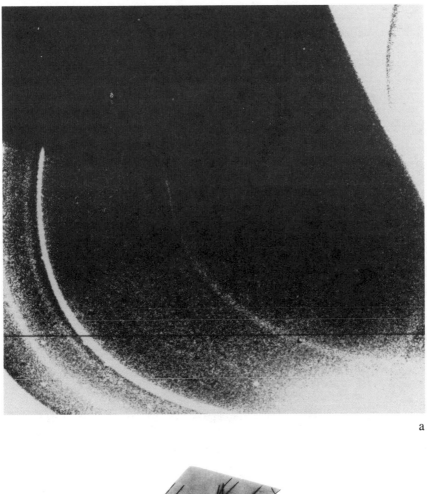

a

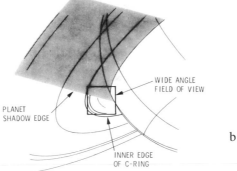

PLANET
SHADOW EDGE

WIDE ANGLE
FIELD OF VIEW

INNER EDGE
OF C-RING

b

Fig. 3. (a) The D Ring of Saturn recorded from a distance of 250,000 km by the Voyager imaging system. (b) Picture geometry (B. A. Smith et al. 1981).

a

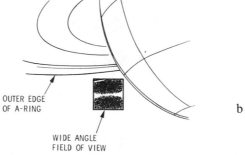

OUTER EDGE
OF A-RING

WIDE ANGLE
FIELD OF VIEW

b

Fig. 4. (a) Ring G as seen by the Voyager imaging cameras. (b) Picture geometry (B. A. Smith et al. 1981).

TABLE I
Dimensions of the Rings of Saturn

Feature	Radial Width (km)	Distance from Saturn Center $(R_S)^a$	Discoverer
D Ring inner edge		1.11	Voyager
C Ring inner edge		1.2348	Bond
Maxwell Gap	253	1.4500 (center)	
B Ring inner edge [b]		(1.5244–1.5259)	Galileo
B Ring outer edge [c]		1.9477	
Huygens Gap	430	1.9513 (center)	Voyager
Cassini Division	4540	1.9854 (center)	Cassini
A Ring inner edge [b]		(2.0230–2.0280)	
Encke Gap	328	2.2141 (center)	Encke
Keeler Gap	31	2.2628 (center)	Voyager
A Ring outer edge [c]		2.2670	
F Ring center [d]	50	2.3267	Pioneer
G Ring center		2.8	Pioneer
E Ring inner edge		3	Feibelman
E Ring outer edge		8	

[a] 1 R_S = 60,330 km.
[b] These edges are not sharp, but show a gradual transition in optical depth. The entire transition range is given.
[c] These edges are eccentric (see Sec. IV.A). The values given are for the distance measured by the Voyager 2 δ Scorpii occultation.
[d] Values from δ Scorpii occultation. Synnott et al. (1983) finds the F Ring to be noncircular with semimajor axis 2.3236 R_S and eccentricity 2.6×10^{-3}.

dications of more complicated structure. From edge-on ring measurements, Lumme and Irvine (1979b) inferred that ring A was nonuniform in opacity. Pioneer 11 data indicated that both ring A and ring B were horizontally inhomogeneous (Esposito et al. 1980). Hämeen-Anttila (1975) argued for horizontal inhomogeneities based on theoretical arguments. We know from the occultation experiments on Voyager that this fine-scale structure exists on all radial scales down to < 1 km; many thousands of significant features are seen (Esposito et al. 1983b). Perhaps the most valuable new information on planetary rings is the discovery of this structure. It is an important task to attempt both to define the observations and explain them. The Saturn ring system is the best available data set that we have on planetary rings; furthermore, we see many structures in this ring system analogous to those seen in the rings of Jupiter and Uranus. Thus, any conclusions are likely relevant to planetary rings in general.

Several possible models have been proposed to explain the multitude of structures seen in Saturn's rings. Density waves excited by the resonance of ring particles' motion with that of an external satellite were first proposed by Goldreich and Tremaine (1978); see Sec. III, below. Cuzzi et al. (1981) observed such a wave in the outer Cassini Division. The discovery of the F Ring shepherds 1980S27 and 1980S26 (B. A. Smith et al. 1981) showed that a narrow ring feature could be contained by satellites on both sides of it (Goldreich and Tremaine 1979). A straightforward generalization of this mechanism was proposed, postulating that especially large ring particles (moonlets) are imbedded in the ring system (Lissauer et al. 1981; Hénon 1981). Each moonlet resides in an empty gap which it has cleared, and between each pair of these sentinels we find a ringlet. This provides a static view of the ring system; as the rings spread due to viscous interactions, the relative position of ring features remains constant. A more dynamic possibility was suggested by Ward (1981) and Lin and Bodenheimer (1981). If the characteristics of interparticle collisions are suitable, the system of particles is unstable to clumping in the radial direction. The ring then segregates into regions of high-velocity dispersion with low optical depth, and between them, regions of low velocity and large optical depth.

Every physical mechanism proposed to explain the structure in Saturn's rings predicts certain characteristics for the optical depth distribution. Consider the moonlet explanation of the observed ring structure: embedded moonlets clear the regions between the observed ringlets. Moonlets large enough to create observable features must necessarily clear a swath in the rings at least as large as their diameter. The minimum diameter is ~ 2 km (Lissauer et al. 1981). Such empty regions would be apparent in the PPS occultation data even if the moonlets are too small and dark to be visible to the Voyager cameras. Figure 5 shows a histogram of number of occurrences of optical depth in a highly structured portion of the outer B Ring. Of 134,000 data points in this region, only 19 have optical depth $\tau_n \leq 0.01$. Thus, only 2 km of clear radius exists in this region. The measurements where the optical depth is not resolved (counting rate near zero counts) gives a large peak set nominally at 2.55, the largest measurable τ_n from the PPS occultation. This analysis shows no evidence of the kilometer-size gaps predicted by the moonlet theory (Esposito et al. 1983b).

Studying the power spectral density function of the ring optical depth is a general way of searching for preferred length scales and frequencies in the ring structure. The inferred optical depths from the Voyager 2 PPS stellar occultation represent a spatial series of 800,000 points. The power spectrum of this series contains information on the causes of the observed structure (Esposito et al. 1983b).

In order to study the entire ring, the PPS data were binned into 1-km bins, and 31 sets of 2048 1-km bins were Fourier transformed to yield a power spectrum. The sum of these 31 spectra is shown in Fig. 6 (Esposito et

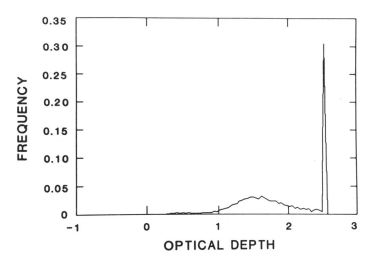

Fig. 5. Histogram of frequency of occurrences of optical depth in the outer B Ring 1.72–1.95 R_S. Data summed up in 1-km-long bins.

al. 1983b). The data were also binned to 8 km, which gave a single 8192-point series covering the ring system. The power spectrum of this series is shown in Fig. 7. Together these power spectra provide a search for periodic phenomena in the PPS occultation data spanning scales from 1 to 60,000 km.

The overall shape of the data power spectrum is approximately $1/f$, i.e., the power decreases inversely with frequency. There are several significant departures from this overall shape. First, for frequency greater than 0.06 cycles km^{-1} (wavelengths shorter than 15 km) the $1/f$ spectrum drops below the photon shot noise of our data. This flattening of the spectrum can be seen in Fig. 6. At these frequencies the spectrum only shows the noise in our data. Second, density waves are evident where their power exceeds the photon noise (see Sec. III). Third, a small but significant increase above $1/f$ is seen at frequency 0.04 to 0.05 cycles km^{-1} (wavelengths 20 to 25 km) in Fig. 7. We interpret this as due to unresolved density waves.

Of the 31 power spectra summed in Fig. 6, most are very similar to the average. They show a rapid decrease in power with frequency until the spectrum shows only photon noise. In the outer B Ring, the PPS counting rate is so low that the ring spectrum is not seen at all; the spectrum shows only white noise. The high optical depth means that the overall spectrum is dominated by photon noise. In the outer A Ring the PPS data show significant power above the noise out to frequency 0.4 cycles km^{-1} (wavelengths greater than 2.5 km). This is clearly due to the large number of density waves in this region. The other significant departure from the average shape is the Janus 2:1 density wave (see Sec. III, Table III) which propagates far enough to be a significant individual contributor to the power in a 2048-km bin.

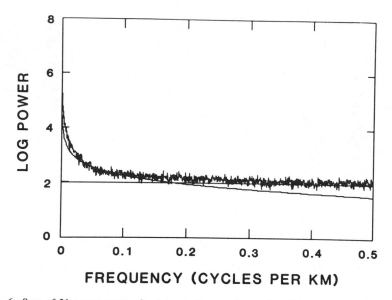

Fig. 6. Sum of 31 power spectra for Saturn's rings, each spanning 2048 km of ring distance. Vertical scale is the logarithm base ten of the power. Solid horizontal line: photon noise power. Solid decreasing line: $1/f$ power spectrum.

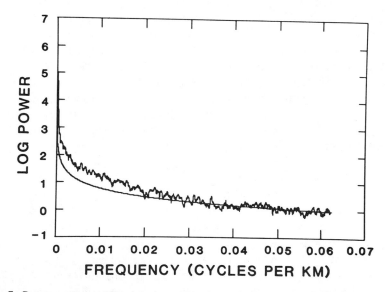

Fig. 7. Power spectrum of Saturn's rings. The raw photopolarimeter spectrometer data were summed in bins 8 km long, forming a series of length 8196. The vertical scale is the logarithm base ten of the power. Solid line: $1/f$ power spectrum.

Esposito et al. (1983b) propose a physical description consistent with these numerical findings. The large-scale structure of the rings as a basic organization or segregation into thicker and thinner rings (see Table II) explains most of the variance in optical depth. The power (or variance) in smaller structures and features steadily decreases with decreasing feature size until features as small as 15 km. At lengths below this, the power in the rings is less than the noise power of the data. This description from frequency space analysis is entirely consistent with frequency distributions of optical depth (Esposito et al. 1983c). Each of the classical ring segments shows a broad, unimodal distribution, but the mean τ_n differs from ring to ring.

This description does not rule out moonlets, waves, and other mechanisms for creating discrete structure in Saturn's rings. However, it does highlight several points.

1. The most important segregation of ring particles in the rings has occurred at the largest scales.
2. The mechanisms that create the majority of small structures in the rings are chaotic; they do not give rise to any preferred distance scales.

It seems that the suggested mechanisms for creating structure in Saturn's rings are together insufficient to explain the observed multitude of radial features. The general characteristics of the structure of the rings shown by its power spectrum and optical depth frequency distribution are not consistent with any currently published physical mechanism. It may be that the diffusive instability mechanism of Ward (1981) and Lin and Bodenheimer (1981) can be generalized to yield a broad, unimodal distribution for the optical depth and to produce structure at scales from 100 m to 15 km. This remains to be seen.

TABLE II
Average Optical Depth of Ring Regions

Region	Boundaries (R_S) [a]	Mean τ_n
Inner C	1.24 [b]–1.39	0.08
Outer C	1.39–1.52	0.15
Inner B	1.52–1.66	1.21
Middle B	1.66–1.72	1.76
Outer B	1.72–1.95	1.84
Cassini Division	1.95–2.02	0.12
Inner A	2.02–2.16	0.70
Outer A	2.16–2.27	0.57

[a] $1 R_S = 60{,}330$ km.
[b] First opacity observed by photopolarimeter spectrometer; actual inner edge at 1.23 R_S.

Another possibility which should be considered is that many mechanisms may be operative in creating structure. This could explain the chaotic nature of the power spectrum. We should also consider the possibility that much of the large-scale structure may be primordial, and thus represent initial conditions and not physical mechanisms. If the rings were created by the breakup of one or more large icy bodies, this could provide a natural explanation for segregation into distinct, classical rings (Shoemaker 1982). Another way that the current structure could be more representative of initial conditions is if effective barriers to particle diffusion are active. One example is guardian satellites: an embedded moonlet has the effect not only of creation of structure but maintenance of a primordial distribution of particles with radius against the diffusive spreading of the rings. The rings must spread due to interchange of angular momentum through interparticle interactions (including collisions, gravitational scattering, etc.).

This implies a need for further study of ring edges and gaps, the locations where ring particle diffusion is clearly impeded. Another obvious need is to extend diffusive instability models to include realistic particle distributions. The better knowledge we have on the dynamics of the ring system both in general and at specific locations, the better we can understand the source of ring structure.

In summary, Saturn's rings are subdivided into the classical rings (A,B,C) and recently discovered rings (D,E,F,G). The radial dimensions of the major ring features are given in Table I. An important discovery from spacecraft views of the rings is the amazing complexity of smaller scale structure within the major rings. This microstructure extends down to scales smaller than 1 km. At present we have no good explanation for this structure.

III. DYNAMIC PROCESSES EVIDENT IN THE RADIAL STRUCTURE OF THE RINGS

Voyager 1 and 2 observations of Saturn's rings now make it possible to compare directly the predictions of dynamical theories with the rings themselves. The observations most critical to ring dynamics are direct high-resolution imaging of the rings and ring occultations. A principal value of the occultation experiments is that they measure directly the opacity of the rings at high spatial resolution and with excellent geometrical fidelity. At radio wavelengths, the rings were probed at 3.6 and 13 cm by the signal from the Voyager 1 radio transmitter as the spacecraft passed behind the rings as viewed from the Earth. The Voyager 1 and 2 ultraviolet spectrometer (UVS) observed two stellar occultations extending out to the mid-C Ring while the Voyager 2 UVS and PPS jointly observed an occultation of the bright star δ Scorpii which traversed the entire ring system. Figure 8 shows an example of how the rings appear in various observations. The top panel shows an optical depth profile of the rings derived from a UVS stellar occultation and averaged to 15 km resolution. The bottom panel shows a corresponding sequence of Voyager 1 images of comparable resolution obtained in diffusely transmitted light.

A. Resonances

It has long been suspected that those regions of Saturn's rings subject to orbital resonances with the known satellites of Saturn would exhibit special properties. It is in these regions, where a commensurability between the orbital periods of the ring particles and the orbital period of an external satellite exists, that enhanced gravitational forcing of ring particle orbits can occur. Such periodic forcing tends to increase the frequency and velocity of collisions between ring particles so that clear gaps are expected to occur analogous to those observed in the Kirkwood gaps of the asteroid belt. This expectation is encouraged by the observation that the Cassini Division is located very near the strong 2:1 resonance with Mimas.

Kinematics of Resonances. It is standard to label resonances by a ratio of the form $\ell:(m - 1)$ where ℓ and m are integers. This ratio expresses the condition that,

$$\ell\Omega_s = (m - 1)\Omega, \tag{1}$$

where Ω_s and Ω are the orbital frequencies of the satellite and the ring particles, respectively. Although this notation serves to identify the resonances it also obscures the complex nature of the motions involved, especially for non-Keplerian orbits.

A more general and precise relationship between the resonant frequencies of ring particle motions and the satellite motion can be seen by looking in detail at orbits about an oblate planet. Consider a particle moving in a stable circular reference orbit of radius r about a planet. If the particle suffers a small perturbation from this reference orbit, so that it possesses a small inclination and eccentricity, then relative to its unperturbed position (or guiding center) it will exhibit periodic motion in three dimensions. In the plane of the reference orbit this relative motion can be resolved into a tangential component with the orbital frequency Ω, and a radial component with epicyclic frequency κ. Perpendicular to the orbit plane the motion has a vertical frequency μ. As will be seen, these frequencies can be regarded as the natural or characteristic frequencies of the orbit. For pure Keplerian motion these frequencies are all equal and the perturbed orbit a closed ellipse. For an oblate planet, however, all three frequencies differ slightly ($\mu > \Omega > \kappa$) and the perturbed orbit is not closed. In this case the motion can be described by an elliptic orbit having an apsidal precession rate $\Omega - \kappa$. The definition of these characteristic frequencies in terms of the planetary potential is given by Shu et al. (1983), whose general treatment we closely follow.

Consider the effect on a ring particle due to a small satellite orbiting out-

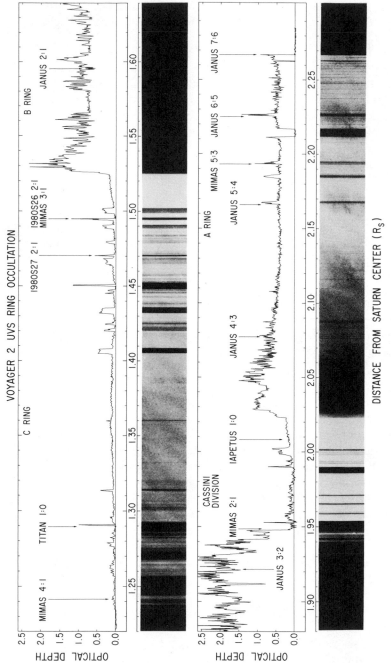

Fig. 8. A comparison of Voyager 1 ultraviolet spectrometer optical depths with a sequence of Voyager 1 images of the unilluminated side of the rings. The images were obtained beneath the ring plane, as seen from the Sun, and represent the rings as viewed in diffusely transmitted light. The brightest portions of the images correspond to intermediate optical depths while the dark portions represent either very large or low optical depths. The locations of a number of important satellite resonances are indicated by arrows. The distance scale is in units of Saturn radii ($R_S = 60,330$ km).

side the rings. Taking as a reference the plane of the rings, and adopting the convention of labeling satellite-associated quantities by a subscript s, the orbital inclination and eccentricities of the particle and the satellite are, respectively, i, e and i_s, e_s. To the planetary potential can now be added a small perturbing potential due to the satellite. This satellite potential, which is time dependent, can be decomposed in the plane of the rings into a Fourier series, the m-th component of which has argument of the form,

$$\omega_s t + m\theta \tag{2}$$

where ω_s is the gravitational forcing frequency due to the satellite and θ is the azimuthal angle between the particle and the satellite. The general form of the forcing frequency is a linear combination of the harmonics of the three characteristic frequencies (Ω_s, κ_s, and μ_s) of the satellite's orbit,

$$\omega_s = m\Omega_s + n\mu_s + p\kappa_s \tag{3}$$

where m, n, and p are positive integers. The lowest order, time-independent terms ($m = n = p = 0$), correspond to replacing the satellite with a time-averaged ring of material. These terms will contribute to the general precession of eccentric features within the rings but will not create nonaxisymmetric features. Since the m-th component of the potential has m-fold azimuthal symmetry the relative frequency with which a ring particle will experience gravitational forcing is $\omega_s - m\Omega$. When this relative frequency matches that of the characteristic epicyclic or vertical frequency of the ring particle orbit, we have the resonant conditions

$$\omega_s - m\Omega = -\kappa \tag{4a}$$

or

$$\omega_s - m\Omega = -\mu. \tag{4b}$$

Resonances of the type described by Eq. (4a) are called inner Lindblad resonances while those described by Eq. (4b) are called inner vertical resonances. Goldreich and Tremaine (1978) and Shu et al. (1983), respectively, show that it is the inner Lindblad and inner vertical resonances due to external satellites which are of primary importance to the rings of Saturn. The choice of a positive sign on the right-hand side of Eqs. (4a) and (4b) corresponds to outer Lindblad and outer vertical resonances which do not result in observable effects within the rings.

Combining Eq. (3) with Eqs. (4a) and (4b) yields a large number of potential resonances within the rings from any single satellite of Saturn. A tremendous reduction in this number is possible, however, through the realization that the magnitude of the gravitational forcing in a particular resonance is proportional to $(e_s)^p (\sin i_s)^n$ (Shu et al. 1983; see also Shu 1984 in *Planetary Rings*). Because the inner satellites of Saturn all have small eccentricities and inclinations, attention can be focused on those resonances where p and n are small. The highest order resonances in p and n definitely observed in the rings correspond to $p + n \leq 2$. Adopting this simplification, and setting $p = \ell - m = 0, 1, 2$ in order to conform to the notation of Eq. (1), the frequency conditions for inner Lindblad and inner vertical resonance are, respectively,

$$m\Omega - \kappa = m\Omega_s + (\ell - m - n)\kappa_s + n\mu_s \qquad (5a)$$

and

$$m\Omega - \mu = m\Omega_s + (\ell - m - n)\kappa_s + n\mu_s. \qquad (5b)$$

Equations (5a) and (5b) are sufficiently general to describe all currently recognized examples of resonances in Saturn's rings. It should be noted, however, that the only presently known example of a mixed mode ($p \neq 0$ and $n \neq 0$) resonance is the Mimas 8:5 ($p = 1$ and $n = 1$) vertical resonance (Shu et al. 1983). All known Lindblad resonances have $n = 0$ and all known vertical resonances correspond to $n = 1$. Equations (5a) and (5b) will reduce to Eq. (1) for pure Keplerian motion where $\mu = \Omega = \kappa$.

A special case of Eq. (5a) occurs when $m = \ell = 1$. This is called an apsidal resonance since the line of apsides of a ring particle orbit precesses at a rate equal to the mean motion of the satellite. Several important examples of such apsidal resonances occur within Saturn's rings (see Cuzzi et al. 1981).

The resonances are places within the rings where the characteristic frequencies of the ring particle orbits satisfy the resonant conditions of Eq. (5a) or (5b). In establishing the precise locations of resonances the satellite frequencies on the right-hand sides of Eqs. (5a) and (5b) are most accurately obtained from the observed mean orbital motion of the satellite. If the planetary potential is axisymmetric, Ω, κ, and μ depend only on the planetocentric distance r, the planetary mass, and the zonal harmonics of the planet's gravitational field. Lissauer and Cuzzi (1982) have calculated the locations and strengths of more than 250 of the stronger inner Lindblad resonances occurring within the rings produced by Saturn's known satellites.

In the vicinity of a resonance an external satellite exerts a net torque on the ring which removes angular momentum from the ring. One measure of the resonance strength is this satellite torque, which depends on a number of factors including satellite mass, distance, eccentricity, orbital inclination, and resonance order. What is the threshold strength for the production of observable effects within the rings of Saturn? In a search of the PPS occultation data, Esposito et al. (1983b) were able to identify about 50 wave structures, primarily in the A Ring, associated with the predicted locations of inner Lindblad resonances. In this search the smallest torque per unit of surface mass density identified was 4.5×10^{16} cm^4 s^{-2}, which implies the total number of observable resonances within the rings is not expected to exceed 75. On the other hand, Holberg et al. (1982b) and Holberg and Forrester (1982) have identified several locations in the C Ring where resonances of substantially lower strength are associated with singular, nonwave-like features, while Cuzzi et al. (1981) have identified a wave-like structure in the Cassini Division associated with the Iapetus apsidal resonance which is over three orders of magnitude weaker than the PPS limit. The expression of such weaker resonances is most likely a function of optical depth within the rings. The possibility of higher order resonances within the rings also has been considered by Wiesel (1982) from the point of view of classical celestial mechanics. Under the influence of the potential of an oblate planet higher order resonances form a band composed of nondegenerate discrete resonances. The association of such bands with ring features is not yet demonstrated.

The locations and strengths of vertical resonances within Saturn's rings have been discussed by Shu et al. (1983). The total number of potentially observable vertical resonances is rather limited due to the fact that only Mimas has sufficient mass and orbital inclination to produce strong vertical resonances. Only three vertical resonances, the Mimas 8:5, 5:3, and 4:2, have so far been identified within Saturn's rings (Shu et al. 1983).

Dynamics of Resonances. The density waves considered by Goldreich and Tremaine (1978) are slowly propagating, tightly wound spiral waves of mass density which lie in the plane of the rings and originate at inner Lindblad resonances. At the location of a vertical resonance an analogous phenomenon occurs in the vertical dimension resulting in bending waves (Shu et al. 1983). These waves are slowly propagating, tightly wound spiral waves of material vertically displaced from the mean ring plane.

Density waves propagate radially outward (positive direction) from an inner Lindblad resonance in the form of a long trailing spiral. Bending waves exhibit the opposite behavior; waves propagate inward away from the inner vertical resonance in the form of a long trailing bending wave. The magnitudes of the group velocities for density and bending waves are similar and, depending on the exciting resonance and the local surface mass density, are on the order of a few mm per second. For both density waves and bending waves the m-fold symmetry of the resonant potential leads to m spiral arms which have their origins at the resonance and maintain fixed phases with respect to the resonant satellite. The clearest illustration of the joint occurrence of density waves and bending waves is provided in Fig. 9 which shows a Voyager 2 image of the region of the rings containing the Mimas 5:3 inner Lindblad resonance and the inner vertical resonance. A somewhat different view of the Mimas 5:3 density wave is also given in Fig. (10b) which shows the optical depth profile of the density wave as seen in a UVS occultation.

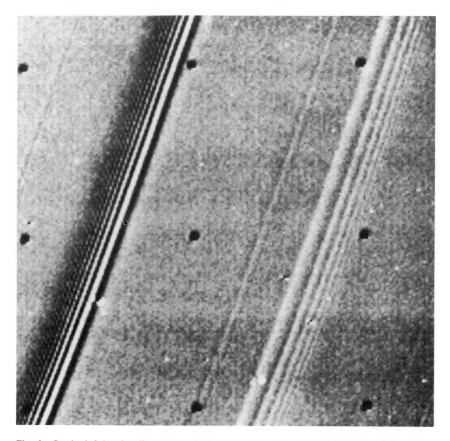

Fig. 9. On the left is a bending wave, which creates 1- to 2-km-high ripples in the ring, and on the right is a spiral density wave. Both are generated by the same 5:3 gravitational resonance with the satellite Mimas, but are separated by the gravitational effect of Saturn's oblateness. Date noise has spotted the image.

Resonance Locations Associated with Waves.   One unambiguous theoretical prediction concerning resonances is their location. Satellite orbital periods are known to better than six significant figures and the gravitational moments of Saturn are sufficiently precise that nonapsidal resonances ($m \neq 1$) have locations which can be predicted to within an accuracy of a few km. From observed density waves or bending waves it is possible to derive the location of the wave source. Analyses of this type have been performed by Holberg et al. (1982*b*), Esposito et al. (1983*b*) for more than one dozen density waves in the A and B Rings, and by Shu et al. (1982) for two bending waves in the A Ring. Four examples of such source determinations for density waves are shown in Fig. 10. In all instances where the uncertainty in the source location was less than 25 km, the predicted and observed locations differ by 12 km

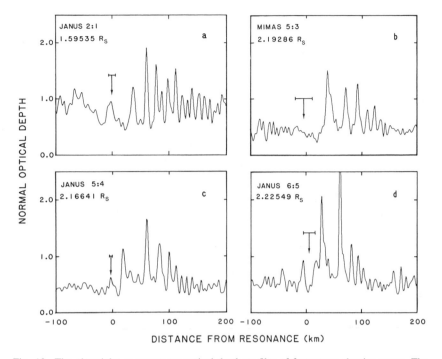

Fig. 10. The ultraviolet spectrometer optical depth profiles of four strong density waves. The origin of each horizontal axis is the expected location of the inner Lindblad resonance while the error bars and arrows represent the location of the wave source as deduced from the analysis of the wave train. These waves were analyzed by Holberg et al. (1982b). All show strong evidence of nonlinear or shocked behavior. Negative distances are toward Saturn.

or less. This is close to the precision of the absolute distance scale for the ring occultation observations of 10 km (Simpson et al. 1983).

In all, approximately 50 resonance-associated wave features due to at least five satellites have been identified in the Voyager data. Nearly all are density waves and occur within the A Ring and the majority of these are due to the shepherd satellites 1980S26 and 1980S27 (Holberg 1982; Esposito et al. 1983b). Three bending waves due to Mimas are currently recognized. Table III gives the predicted locations for 23 waves of both types which have been studied in detail.

Resonance Locations Not Associated with Waves. The Mimas 2:1 and the Janus 7:6 resonances are the two strongest resonances within the rings; they are located close to the outer edges of the B and A Rings, respectively. The Mimas 2:1 resonance has long been assumed to be associated with the formation of the Cassini Division. In the UVS occultation data (Holberg et al. 1982b) the outer edge of the B Ring was observed to be 48 km interior to the location of the Mimas 2 : 1 resonance. This distance is in good agreement with both the oval shape and orientation of the outer edge of the B Ring determined from Voyager images (see Sec. IV.A).

The explanation for the location of the outer edge of the A Ring has been less apparent. At the time of the Voyager encounter, it was frequently stated that the newly discovered Atlas (1980S28) orbiting just outside the A Ring, was responsible for confinement of the outer edge of the A Ring through the shepherding mechanism (see Sec. III.B). However, Atlas is probably not massive enough to confine the A Ring and a more likely explanation is the close proximity of the Janus 7:6 resonance. Holberg et al. (1982b) found that this resonance was located $\sim$ 7 km exterior to the outer edge of the A Ring and suggested that it is responsible for the A Ring outer edge, since the separation is well within the stated uncertainty of the distance scale. Of particular interest in this regard is the precise shape of the outer edge of the A Ring, which would then show a 7-lobed pattern having a fixed orientation with respect to Janus (see Sec. IV.A).

Although the 7:6 resonance with Janus may provide an answer for the location of the outer edge of the A Ring, it also serves to illustrate a major problem in our understanding of ring dynamics. The rings, through resonances, exert a positive torque on Janus causing its orbit to expand. The time scale for this outward evolution of Janus' orbit, and hence the A Ring, is distressingly short compared to the age of the solar system. Summing the torques contributed by the six known Janus resonances in the rings yields a characteristic expansion time for Janus' orbit of 6×10^8 yr. As is discussed by Goldreich and Tremaine (1982) such a short evolutionary time scale is not only a problem with respect to Janus but for all five of the satellites interior to Mimas. Unless they are locked, directly or indirectly, into orbital resonances with more massive external satellites, then the rings and the inner satellite orbits are presently in a state of rapid evolution. Lissauer et al. (1982b) have discussed one such possibility. An alternate possibility discussed by Goldreich and Tremaine (1982), that the large ring torques exerted on the inner satellites could be diminished by the clearing of gaps at inner Lindblad resonances, has now been shown by Voyager data to be inapplicable since no such gaps are seen. However, some decrease in torque is still possible (Borderies et al. 1983).

In spite of the fact that there exist several strong resonances in the C Ring, it is curious and perhaps significant that no density waves (or wavelike features) are observed there. What is observed is that resonances are associated with narrow well-defined features of enhanced optical depth. Holberg and Forrester (1982) and Holberg et al. (1982b) have shown that the five strongest resonances in the C Ring all occur associated with sharp-edged features, and in several cases with clear gaps in the C Ring. The location of these features is shown in Fig. 8 and corresponding $\sim$ 3 km resolution views of three of these features are given in Fig. 11. At even higher $\sim$ 0.1 km resolu-

TABLE III
Analyzed Wave Structures Within the Rings of Saturn

Resonance	Predicted [a] Location (R_S)	σ [b]	τ [c]	$\kappa \times 100$ [d]	References/Remarks [e]
Janus 2:1	1.59533	70(10),70	0.73	1.0(0.2),0.1	HFL, EOW
Iapetus 1:0	2.00867	16(3)	0.14	0.9	CLS
1980S26 5:4	2.02740	30	0.62	2.1	EOW
1980S27 6:5	2.04833	81	0.99	1.2	EOW
Janus 4:3	2.07635	139	0.53	0.4	EOW
1980S26 6:5	2.08259	43	0.55	1.3	EOW
1980S27 7:6	2.08668	53	0.70	1.3	EOW
1980S27 8:7	2.11523	57	0.53	0.9	EOW
1980S26 8:7	2.15064	27	0.46	1.7	EOW
1980S27 10:9	2.15490	49(13)	0.43	0.9(0.2)	H
Janus 5:4	2.16641	56(12),57	0.48	0.8(0.2),0.8	HFL, EOW
1980S26 9:8	2.17309	49	0.46	0.9	EOW
1980S27 12:11	2.18115	53(10)	0.43	0.8(0.2)	H
Mimas 5:3 [f]	2.1863	45(4)	0.8	1.8(0.2)	SCL(bending wave)
Mimas 5:3	2.19286	45(8),40	0.85	1.0(0.2),2.1	HFL, EOW
1980S26 11:10	2.20556	44(9),26	0.41	0.9(0.2),1.8	H, EOW
1980S27 15:14	2.20725	55(15)	0.42	0.8(0.2)	H
1980S27 17:16	2.21948	58(15)	0.48	0.8(0.2)	H
Janus 6:5	2.22549	40(6),96	0.59	1.0(0.2),0.6	HFL, EOW
1980S26 13:12	2.22791	42(9)	0.51	1.2(0.3)	H

TABLE III (Continued)

Resonance	Predicted [a] Location (R_S)	σ [b]	τ [c]	$\kappa \times 100$ [d]	References/Remarks [e]
1980S27 19:18	2.22912	52(14)	0.55	1.1(0.3)	H
1980S27 20:19	2.23321	76(17)	0.50	0.7(0.2)	H
Mimas 8:5 [f]	2.2483	42(13)	0.6	1.4(0.4)	SCL(bending wave)

[a] Inner Lindblad resonance locations (Lissauer and Cuzzi 1982); vertical resonance locations (Shu et al. 1983); $R_S = 60,330$ km.

[b] σ surface mass density (g cm^{-2}).

[c] τ optical depth.

[d] κ mass extinction coefficient $\times 100$ (cm^2 g^{-1}).

[e] CLS = Cuzzi, Lissauer and Shu (1981); HFL = Holberg, Forrester and Lissauer (1982b); H = Holberg (1982); EOW = Esposito, O'Callaghan and West (1983b); SCL = Shu, Cuzzi and Lissauer (1983).

[f] Vertical resonance (Shu et al. 1983).

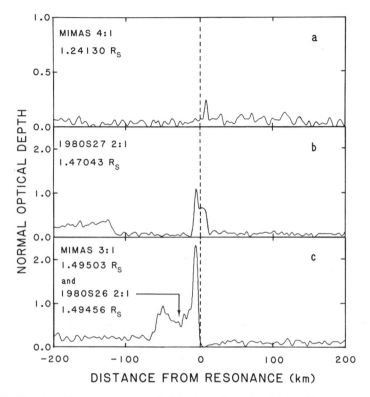

NORMAL OPTICAL DEPTH

DISTANCE FROM RESONANCE (km)

Fig. 11. The ultraviolet spectrometer optical depth profiles in the vicinity of four resonances in the C Ring. All resonance-associated features in the C Ring appear to be sharp, narrow, optically thick ringlets and not density waves. The Mimas 4:1 resonance (a) is the weakest resonance seen in the rings and in the photopolarimeter spectrometer high-resolution data it is resolved into four narrow, sharp, equally spaced features of higher optical depth (Lane et al. 1982b). The Mimas 3:1 resonance is the dominant resonance in the bottom panel (c), being about an order of magnitude stronger than the 1980S26 2:1 resonance (arrow). The origin of each horizontal scale corresponds to the expected location of the resonance with Saturn to the left.

tion (Lane et al. 1982b) these features reveal a very complex structure composed of sharp narrow regions of high and moderate optical depth. In one particularly striking example the narrow feature associated with the Mimas 4:1 resonance (Fig. 11) breaks up in the high resolution PPS data into four very narrow, equally spaced features of substantially higher optical depth. The explanation as to why the rings respond to the presence of resonances in two quite different fashions is not clear. It could be that gaps and narrow optically thick features form in the C Ring because of some property peculiar to that ring. A descriptive reason, suggested by Holberg et al. (1982b) is that local optical depth controls the response of the rings. Nearly all examples of density and bending waves occur in the A and B Rings where optical depths are in the range $0.4 < \tau_n < 1.0$ while in the C Ring, where waves are not seen, optical depths are much lower ($\tau_n < 0.2$). A possible exception to this statement could be the density waves associated with the Iapetus $1:0$ resonance in the Cassini Division where $\tau \sim 0.14$ (Cuzzi et al. 1981). In this instance, however, the resonance is exceedingly weak and the wavelengths exceptionally long.

Now consider the location of the Titan apsidal resonance. The uncertainties in the locations of apsidal resonances are relatively large since apsidal precession rates depend to first order on the J_2 gravitational moment of Saturn which is currently known only to an accuracy of $\sim 0.1\%$ (Null et al. 1981). This translates into a possible uncertainty of up to 50 or 60 km including uncertainties in Saturn's higher order gravitational moments. The Titan apsidal resonance lies nearly 80 km interior to a curious eccentric ringlet in the C Ring. This ringlet precesses with a rate indistinguishable from Titan's orbital period and maintains its major axis nearly aligned with Titan (Porco et al. 1984*b*; Holberg and Forrester 1982). Study of the location and dynamics of this and several other eccentric ringlets offers a unique opportunity to independently constrain the values of Saturn's higher gravitational moments.

Although resonances seem somehow related to the formation of as many as four sharp narrow features in the C Ring, they provide no explanation for the numerous broader and more diffuse opaque features seen in Fig. 8. Some dynamic mechanism must prevent the radial spreading of these features due to viscous forces. The viscous instability mechanism (see Sec. II) would appear to be ruled out on the grounds that optical depths in the C Ring are too low to support it. Likewise, embedded moonlets (see Sec. III.B) which could produce such features and are also expected to clear gaps, are not seen.

A further important aspect of the rings requiring explanation are the inner edges of the classical rings. The sharp outer edges of A and B Rings are apparently related to very strong resonances with *exterior* satellites. The corresponding inner edges have no ready explanation; no resonances of significance exist nearby. A resonant mechanism could involve a source lying interior to the edges and thus the transfer of *positive* angular momentum to the rings. Also, in contrast to outer edges, inner edges are not sharp, and are difficult to define since high resolution data show several stages of transition to the higher optical depths.

B. Embedded Moonlets

Following the Voyager 1 encounter, Lissauer et al. (1981) and Hénon (1981) proposed satellites (moonlets) within the rings. Such moonlets would be responsible for the clear gaps observed in the A and C Rings and particularly the Cassini Division. Most of these gaps occur near no known external satellite resonances. Some internal mechanism is needed to maintain the gap against viscous interactions which cause the rings to spread radially and fill the gaps in a few years. The moonlets were presumed to be somewhat smaller than the smallest observed satellite of Saturn (radius $\lesssim$ 20 km) but still massive enough to clear a gap in the rings in excess of their own size by the shepherding mechanism described by Goldreich and Tremaine (1979). The moonlet exerts gravitational torques on the surrounding ring material such that the material is forced away from the moonlet orbit creating a clear gap on each side. The size of a gap created by the gravity of a moonlet is determined by the balance of this torque and the viscous interaction. Lissauer et al. (1981) applied this mechanism to Saturn's rings and obtained a lower limit of $\sim$ 1 km in radius for moonlets capable of creating gaps larger than themselves.

Lissauer et al. (1981) used the moonlet hypothesis to explain the alternating succession of clear gaps and broad bands of low optical depth seen in Voyager 1 images of the inner Cassini Division. Their analysis yielded a specific prediction of the masses, radii, and radial locations for five moonlets. An attempt to observe the larger moonlets in these gaps with Voyager 2 imaging of the Cassini Division (B. A. Smith et al. 1982) was entirely negative down to a 6 km upper limit for possible moonlet radii. The existence of moonlets in the Cassini Division gaps is not completely ruled out by this negative result since it is still possible to accommodate them by appealing to extreme, but not impossible, combinations of moonlet density and albedo together with lower permissible limits to the dispersion velocity in the rings. It may also be possible to produce the gaps by a limited number of smaller moonlets rather than one large moonlet per gap.

The Encke and Keeler Gaps in the A Ring are also obvious candidates for explanation by the moonlet hypothesis. Voyager imaging coverage was insufficient to place any strong constraint on the sizes of possible moonlets within these gaps. The Encke Gap, however, does exhibit several features which favor such an explanation. The discontinuous and serpentine appearance of the kinky ringlets within the gap (B. A. Smith et al. 1982) is suggestive of the similar behavior seen in the F Ring, where such structure has been attributed to the shepherd satellites 1980S26 and 1980S27. Also, both inner and outer edges of the Encke Gap are observed to be noncircular, exhibiting a regular wave-like pattern of azimuthal undulations having a full amplitude of $\sim$ 4 km (Cuzzi and Scargle 1984). Such behavior is what would be expected from several small $\sim$ 10 km moonlets orbiting with the Encke Gap.

C. Analysis of Spiral Wave Structure

The analysis of density and bending waves observed within the rings leads to estimates of such basic ring parameters as surface mass density, mass extinction coefficient, kinematic viscosity, and particle dispersion velocity. These quantities have proved very difficult to estimate by other means. Two basic measurements are required to extract information from wave structures: the wavelengths and wave amplitudes as a function of the distance from the wave source. Since the discovery and analysis of the first density wave train in the Cassini Division by Cuzzi et al. (1981) there have been over twenty other density waves analyzed in the A and B Rings (see Table III). In the first analysis of density waves from occultation data, Lane et al. (1982b) determined the surface density and estimated the kinematic viscosity from the Janus 2:1 density waves in the B Ring. Using the UVS occultation data Holberg et al. (1982b) estimated wave source locations and surface densities of four strong density waves in the A and B Rings. In a related study Holberg (1982), using assumed source locations, estimated surface mass densities for eight density waves excited by the shepherd satellites 1980S26 and 1980S27. Voyager imaging data of two bending waves in the A Ring have also been discussed by Shu et al. (1983) and Lissauer (1982).

A general relation between the observed wavelength λ (km), distance from resonance d (km), and surface mass density σ (gm cm^{-2}) for both density waves and bending waves in Saturn's rings is,

$$\sigma = \frac{0.325\ (m - 1)\ \lambda d}{R^4} \qquad (m > 1) \qquad (6a)$$

where R is the resonance location in Saturn radii (Holberg et al. 1982b; Shu et al. 1983). For density waves at apsidal resonances, the analogous relation (Lissauer 1982) is,

$$\sigma = \frac{1.85 \times 10^{-2}\ \lambda d}{R^6} \qquad (m = 1) \qquad (6b)$$

Both relations are of the form,

$$\lambda_n = \frac{A}{d_n - d_0} \qquad (7)$$

where d_n and λ_n are the distance and wavelength of the n-th peak or trough of the wave. The constant d_0 has been introduced to allow an arbitrary origin, and the constant A is proportional to the local surface mass density. Fitting spacecraft measurements of λ_n and d_n to Eq. (7) determines the wave source location d_0 and the surface mass density.

In practice complications arise if the surface mass density σ changes significantly over the analyzed portion of the wave train. This was found for three strong A Ring resonances (Holberg et al. 1982). All three resonances show strong evidence of shock formation; wave amplitudes approached or exceeded the mean optical density of the A Ring. Under such circumstances it is reasonable to assume that the local surface mass density is also significantly perturbed.

One approach to the problem of variations in σ is to assume that the ratio of optical depth to surface mass density (τ/σ) remains constant throughout the wave (Holberg et al. 1982b). An example of the observed and expected dependence of wavelength on distance is shown in Fig. 12 for the Janus 5:4 density wave, where assumption of constant value of τ/σ has been made (Holberg and Forrester 1983).

It is in principle possible to measure the damping of these waves, obtaining a local estimate of the kinematic viscosity within the rings. In practice, difficulties arise with attempts to use this technique for density waves. The first use of density waves to determine kinetic viscosity in the rings was made by Cuzzi et al. (1981) who obtained a value of $\nu + 3/7\zeta \sim 180$ cm^2 s^{-1} (where ν is the shear viscosity and ζ the bulk viscosity). This result comes from the damping of the Iapetus 1:0 density waves in the Cassini Division. It was later realized (Lissauer et al. 1982b) that surface mass density was not constant throughout the region of the wave but was likely increasing in the direction of propagation. The effect of such a density gradient would be to overestimate the viscosity. The Janus 2:1 density waves (Lane et al. 1982b) give a kinematic viscosity of $\nu + 3/7\zeta = 20$ cm^2 s^{-1} in a $\tau \sim 0.7$ region of the B Ring. However, as these authors caution, the appearance of the waves is clearly nonlinear and the linear theory used to derive damping may not apply. Holberg et al. (1982b) found the same difficulty and did not attempt to measure damping. It may be possible to develop models to the point where such difficulties can be handled or to find density waves which satisfy the necessary conditions.

On the other hand, bending waves in Saturn's rings are damped by viscous forces before becoming nonlinear. Shu et al. (1983) have analyzed the Mimas 5:3 bending waves and Lissauer et al. (1982b) have reported a measure of the kinematic shear viscosity from this analysis. Although approximate amplitudes of the bending waves were obtained from the imaging data, most of the damping took place near or below the resolution of the images. It proved possible, however, to estimate the damping length from the radial extent of shadowing in the wave train caused by the low Sun angle. Using this effect to determine the distance at which maximum wave slopes decreased below 19°, Lissauer et al. (1982b) were able to obtain an estimate of $\nu = 260^{+250}_{-100}$ cm^2 s^{-1} for the region of the Mimas 5:3 bending wave. Using an optical depth of $\tau \sim 0.8$ obtained from the stellar occultation experiments, Lissauer et al. (1982b) obtained a value of $c = 0.4$ cm s^{-1} for the dispersion velocity. This observed dispersion velocity is well in excess of the vertical component of random velocity which would be expected if the rings were a monolayer (see Sec. VII).

As Lissauer et al. point out this value is in excess of the random velocities which are expected to result solely from viscous stress within the rings and therefore an additional source of energy for random motions is required. In the A Ring, where this measurement was made, such an additional source of random motion could be provided by the energy dissipated in the enhanced collisions of ring particles in damping the spiral density waves. Spi-

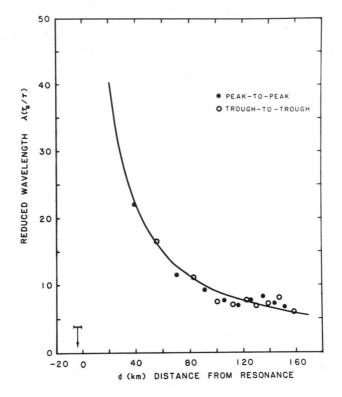

Fig. 12. The reduced wavelength (in km) as a function of distance from the expected location from the Janus 5:4 resonance. Reduced wavelength is adjusted by the observed optical depth τ/τ_0 where τ_0 is the local mean optical depth of this region of the A Ring ($\tau_0 = 0.41$). This is equivalent to the assumption of constant mass extinction coefficient.

ral density waves occur quite frequently in the A Ring, especially toward the outer edge. Lissauer et al. (1982*b*) estimate that the energy dissipated, averaged over the A Ring, would result in random velocities on the order of c ~ 0.5 cm s^{-1}, in good agreement with observation. This dispersion velocity would give a ring thickness of 35 m.

Esposito et al. (1983*b*) argue that the values for damping derived from density waves are valid despite the insufficiency of the linear model. Calculated from this idealized model, the ring thickness is less than 10 m in the B Ring and increases to ~ 30 m in the outer A Ring (see Fig. 13). This latter value is in good agreement with Lissauer et al. (1982*b*).

Zebker and Tyler (1984) conclude from microwave measurements of near forward scattering that ring A has a vertical thickness of 20–50 m. The good agreement among all these determinations can give us some confidence that (at least in ring A) the rings are very thin, but not as thin as a single particle (see also Sec. V).

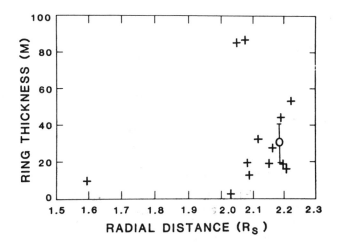

Fig. 13. Ring thickness implied by observed wave damping. Crosses: upper limits from density waves studied by Esposito et al. (1983b). Open circle: measured value from photopolarimeter spectrometer observations of Mimas 5:3 bending wave. Esposito et al. argue that the upper limits in the outer A Ring are reliable estimates of the actual value of the ring thickness.

D. Mass of the Rings

In spite of the substantial appearance of Saturn's rings, one fundamental ring property which has eluded successful measurement is the total mass of the rings. The total mass of the material contained within Saturn's rings is of interest both for understanding the internal dynamics of the rings and providing clues to, and constraints on, the origin of the rings. Since the end of the last century, estimates of the mass of the rings have relied primarily upon attempts to detect the gravitational perturbation of the rings on the motions of the external satellites. The most recent use of this method to determine the mass of the rings is that of McLaughlin and Talbot (1977) who found a value of $6.2 \pm 2.4 \times 10^{-6}$ M_S (M_S = the mass of Saturn = 5.7×10^{29} g). Unfortunately the gravitational effect of a disk of material orbiting within the equatorial plane is difficult to distinguish from the gravitational effects due to the oblateness of the planet itself, so that many early estimates of the mass of the rings were orders of magnitude too large. The successful Pioneer 11 flyby considerably improved our knowledge of Saturn's gravitational field and provided a unique opportunity to measure directly the mass of the rings. In spite of the fact that Pioneer 11 passed within 4000 km of the outer edge of the A Ring, spacecraft tracking data failed to reveal any measurable gravitational perturbations due to the rings. Null et al. (1981) were able to place an upper limit of 1.7×10^{-6} M_S on the mass of the rings by combining the Pioneer 11 tracking data with groundbased studies of satellite motions.

Voyagers 1 and 2 never passed close enough to the rings to improve the Pioneer 11 upper limit. It is nevertheless possible to estimate indirectly the mass of the rings through two novel and independent methods. By using density waves to yield local surface densities, it is possible to estimate the total mass of the ring system. The basic assumption is that the mass extinction coefficient (the ratio between optical depth and surface mass density) is constant throughout the rings. Although the mass extinction coefficient can be expected to vary somewhat with changes in the particle size distribution, observations have shown that a value $\kappa = 0.01$ cm^2 g^{-1} is characteristic of the A Ring as well as several locations in the Cassini Division and the B Ring (Table III). Given this fact, it is a simple matter to integrate the mass of the ring system from the observed optical depths. Holberg et al. (1982b), using the UVS optical depths for the entire ring, obtained an estimate of 6.4×10^{-8} M_S using this technique and a $\kappa = 0.01$ cm^2 g^{-1}. Esposito et al. (1983b) made a similar estimate of 5×10^{-8} M_S using the PPS optical depths and a slightly larger value of $\kappa = 0.013$ cm^2 g^{-1}. The observed masses and mass distributions are given in Table IV. In each case the two principal sources of error are the assumption of a uniform value for the mass extinction coefficient and the fact that the most opaque portions of the B Ring yield only a lower limit for the peak optical depth. The PPS and UVS observations were insensitive to optical depths above ~ 2.5. Therefore, if the optical depths in this portion ($\gtrsim 40\%$) of the B Ring substantially exceed this value, the mass of the B Ring would be correspondingly higher.

An independent estimate of the mass of the rings is made from the Voyager 1 radio occultation experiment. The mass in this case is calculated directly from the derived size distribution of the ring particles (see Sec. V). By assuming a specific gravity of 0.9 g cm^{-3} (for solid ice at ~ 70 K) and integrating the mass contained within the particle size distribution found to be characteristic of the rings, Tyler et al. (1982b) have estimated a total mass for the rings of $\sim 3 \times 10^{-8}$ M_S. This method suffers from the same two systematic errors mentioned in regard to the UVS and PPS estimates, namely the extrapolation of local measurements throughout the rings and the inability to penetrate the thickest portion of the B Ring. In addition, this method relies on an assumed value for the specific gravity of the ring particles. Taking these factors into account Tyler et al. estimate the mass is uncertain by a factor of 4 to 8. All three determinations give the mass of the rings as similar to that of Mimas.

We find the agreement among these three estimates encouraging. The knowledge of the ring mass is most certain for the A Ring, which is well sampled by both techniques, and less certain for the C Ring, where no density waves are seen and few mass extinction coefficients have been obtained. The B Ring, which contains the bulk of the mass of the rings, is most uncertain.

Summary. Voyager observations have provided a large and diverse body of evidence against which various theories of the dynamics of Saturn's rings can be judged and refined. The evidence for density waves and bending

TABLE IV
Ring Masses

Ring	Boundaries (R_S)	Mass − UVS $(\times\ 10^{-8}\ M_S)$	Mass − PPS $(\times\ 10^{-8}\ M_S)$
C	1.235–1.526	0.2	0.14
B	1.526–1.949	5.0	3.4
Cassini Division	1.949–2.023	0.1	0.06
A Ring	2.023–2.267	1.1	1.1
Total		6.4	5

waves associated with satellite resonances is clear and their appearance largely in accord with theoretical predictions. The status of embedded moonlets in the rings is less certain in light of the failure of Voyager 2 to observe them. There is, however, strong indirect evidence which points to the existence of such moonlets within the Encke Gap.

In the A Ring, density waves and bending waves can account for most of the major (and a considerable amount of the minor) features of the observed structure. Other than providing the expected explanation for its outer edge, satellite resonances are of little consequence in interpreting the B Ring. The C Ring seems most in need of understanding at present. Some features are clearly associated with the four or five strongest resonances in the C Ring but these features are narrow singular ringlets and not wave structures. Most striking of all are the optically thick ringlets which appear embedded within an underlying C Ring. They are not associated with any known resonances and their formation probably cannot be explained by known mechanisms.

IV. FEATURES WITHOUT RADIAL SYMMETRY

A. Eccentric Rings

Observations. The first evidence of eccentric rings came from Uranus occultation studies: 6 of the 9 known Uranian rings are eccentric and precessing in the Uranian gravitational field (see Elliot and Nicholson 1984). The Voyager cameras have observed several eccentric features in the Saturnian rings (B. A. Smith et al. 1982) which have been studied by Esposito et al. (1983a) and Porco (1983). The theoretical background for these studies is a series of papers by Goldreich and Tremaine, reviewed in Goldreich and Tremaine (1982).

Four Voyager experiments have provided information on these eccentric rings. Images were acquired with both the narrow- and wide-angle cameras, mostly through the clear filter (~ 5000 Å). The resolution ranges from 60 to 5 km/pixel. The occultation of δ Scorpii observed on Voyager 2 by the PPS and

UVS provides another data set, as well as several occultations observed by the UVS (Holberg and Forrester 1983). The resolution of these measurements was 0.1 to 7 km. Voyager 1 provided a radio occultation of the rings (see Sec. V). Initially diffraction effects limited the resolution to $\sim$ 10 km, but later processing produced occultation data at 1 km resolution in some areas (Tyler et al. 1982b). The two Voyager encounters were separated by 286 days. The diverse sets of data at different times and longitudes allow a determination of both the shape and motion of any eccentric features.

All evidence indicates that the eccentric features in Saturn's rings are moving as units; there is no differential precession between the inner and outer edges. All these features may be modeled as uniformly precessing, low eccentricity ellipses, each coplanar with the rings and with focus at Saturn's center. That is, if r and θ are the observed Saturnian distance and longitude of the feature, then

$$r = a\{1 - e \cos[\theta - \widetilde{\omega}_0 - \dot{\widetilde{\omega}}(t - t_0)]\} \tag{8}$$

where a is the semimajor axis, e is the eccentricity, $\widetilde{\omega}_0$ is the longitude of periapse at $t = t_0$, and $\dot{\widetilde{\omega}}$ is the rate of apsidal advance. [Longitudes are measured clockwise in Saturn's ring plane relative to an inertial system defined by the position of the subsolar point on the rings at epoch t_0 Voyager 1 closest approach—1980 November 12, 23 hr 46 min UTC/Spacecraft Event Time. This longitude system is referred to as the solar longitude system, or SOL.] This model has been fitted to eccentric features at 1.29, 1.45, and 1.95 R_S. The outer edge of the B Ring (near the 2:1 inner Lindblad resonance with Mimas; see Sec. III) was fitted by the 2-lobed figure:

$$r = a\{1 - e \cos 2[\theta - \widetilde{\omega}_0 - \dot{\widetilde{\omega}}(t - t_0)]\}. \tag{9}$$

The outer A Ring edge was fitted by a 7-lobed figure. Four parameters were allowed to vary: a, e, $\widetilde{\omega}_0$, $\dot{\widetilde{\omega}}$. The shape parameters, a and e, can be well constrained in all cases; but the precession rate and thus $\widetilde{\omega}_0$ are not unique. This is because a multiplicity of possible values are allowed for $\dot{\widetilde{\omega}}$, each corresponding to an additional complete revolution of the feature between the Voyager 1 and 2 encounters. An individual encounter is too brief to observe enough precession to constrain the precession rate to a single value. In work to date, this ambiguity was removed by requiring that the observed precession rate be near the expected value. The latest derived values for these parameters, along with 1 σ standard errors of the mean, are given in Table V. For the eccentric ringlets found in gaps, the gap dimensions are also given.

The eccentric ringlet at 1.29 R_S lies close to the (1:0) apsidal resonance with Titan (see Sec. III). We refer to this as the Titan Ringlet. For particles perturbed by Titan, the expected ring particle orbits are Keplerian ellipses closed in a reference frame rotating with the mean motion of Titan. Inside of the point of exact resonance the particles' periapses are aligned with Titan;

TABLE V

Noncircular Features in Saturn's Rings

Feature	Saturn Distance a (km) [e]	Eccentricity e (10^{-4})	Longitude of Periapse $\tilde{\omega}_0$ (deg)	Precession Rate $\dot{\tilde{\omega}}$ (deg/day)	Range of Semimajor Axis δa (km)	Range of Eccentricity δe (10^{-4})	Gap Dimensions a_{in} (km) [e]	a_{out} (km) [e]	Mass (g)	Surface Mass Density Σ (g cm^{-2})	Mass Extinction Coefficient κ (cm^2 g^{-1})
Titan Ringlet (1.29 R_S)	77871(7)	2.6(0.2)	129(5) [d]	22.577 [d]	25(3)	1.4(0.4)	77735(12)	77919(8)	2.1×10^{18}	17	7.5×10^{-2}
Maxwell Ringlet (1.45 R_S)	87491(8)	3.4(0.4)	342(9)	14.69(.03)	64(3)	3.4(0.6)	87322(8)	87590(8)	6.1×10^{18}	17	5.2×10^{-2}
Huygens Ringlet (1.95 R_S)	117828(11)	2.7(0.8)	213(26)	5.11(.11)	~50	—	117577 [f]	117936(5)	—	—	—
F Ring [a] (2.32 R_S)	140185(30)	26.6	230(15) [b]	—	—	—	—	—	—	—	—
A Ring edge	136773(8)	.49(.08)	197(10)	518.31(.04)	—	—	—	—	—	—	—
B Ring edge	117577(15)	6.3(0.8)	215(6) [c]	381.997 [c]	—	—	—	—	—	—	—

[a] Analysis by Synnott et al. (1983).

[b] Epoch: Voyager 2 encounter; all others, SOL system epoch.

[c] Mimas: $n = 381.997$ deg/day; $\lambda_{\text{SOL}} = 213°$.

[d] Titan: $n = 22.577$ deg/day; $180° + \lambda_{\text{SOL}} = 142°$.

[e] Excepting the F Ring, errors include the systematic uncertainty in absolute scale of ± 6 km.

[f] Inner edge of gap is edge of B Ring.

outside, the apoapses are aligned with Titan. The observed precession rate for a solution in which $\tilde{\omega}$ is free to vary is $22°57 \pm 0:06/\text{day}$, identical (within uncertainties) to Titan's mean motion, $22°577/\text{day}$. Figure 14 compares the observed shape and orientation of this ringlet with a model in which $\tilde{\omega}$ is held fixed at $22°577/\text{day}$. The apoapse of the ringlet is $13° \pm 5°$ away from the direction to Titan. Though the ringlet's behavior is apparently dominated by the influence of Titan, this small angular lag may be the manifestation of viscous effects in the ringlet or a small-amplitude libration around the Titan-Saturn line.

The eccentric ringlet at 1.45 R_S is located in the Maxwell Gap in the C Ring and is called the Maxwell Ringlet. A multi-instrument study of this region is given by Esposito et al. (1983a). In contrast to the Titan Ringlet, no known resonances are near this feature. Porco et al.'s (1984b) result in Table V of a precession rate $\tilde{\omega} = 14°69 \pm 0.03/\text{day}$ is consistent with a coplanar ringlet having a semimajor axis of $87{,}491$ km and precessing solely under the influence of Saturn's oblateness, in good agreement with the result of Esposito et al. ($14°66/\text{day}$).

The shape of the outer edge of ring B is predominantly influenced by Mimas; it straddles the $(2:1)$ inner Lindblad resonance with Mimas. The particles interior to this (just within the B Ring) are expected to have closed orbits in Mimas' reference frame, with the periapse of their orbits on the Saturn-Mimas line. The collective behavior of such orbits is described by Borderies et al. (1982). Voyager observations verify this description within their formal uncertainties. The derived precession rate $382°00 \pm 0.04/\text{day}$ is identical to Mimas' mean motion. Given the doubled-peaked ($m = 2$) behavior evident in Fig. 15, the connection to Mimas cannot be doubted.

Like the outer B Ring edge, the behavior of the outer A Ring edge is determined by its proximity to a strong low-order inner Lindblad resonance, the $7:6$ resonance with the coorbital satellite system. This resonance is complicated by the presence of many components of slightly different pattern speeds and positions which arise from the libration of the two satellites (Porco et al. 1984a). To within the uncertainties, the location of the A Ring edge coincides with the position of several of the strongest resonance components. The radius-azimuth data are consistent with a 7-lobed pattern rotating with the mass-weighted mean motion of the coorbital satellites, and having an amplitude of 6.7 ± 1.5 km.

Examination of Fig. 15 shows that the eccentric ringlet just outside the B Ring edge (in the Huygens Gap) is, conversely, not predominantly influenced by Mimas. Its shape is a simple Keplerian ellipse. The derived precession rate $5°11 \pm 0.11/\text{day}$ (Porco 1983) is within 1 σ of the precession rate of $5°02/\text{day}$ expected from Saturn's oblateness alone. We can conclude from study of all five eccentric features that each is precessing uniformly under the influences of either Saturn's oblateness or resonance with a Saturnian satellite.

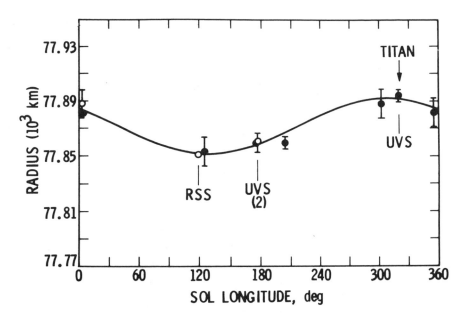

Fig. 14. The coplanar Keplerian ellipse, with apse precession rate fixed at Ω (Titan), which best fits the radius-longitude data for the Titan Ringlet at 1.29 R_S. Parameters are given in Table V. Data points have been precessed to the SOL system epoch. Open circles are Voyager 1 data: filled circles, Voyager 2. The position of Titan in the SOL system at epoch is indicated.

Fig. 15. Four images of the Cassini Division taken within 7 hr of each other. The middle two panels are taken from two images of the east ansa; the outer panels from the west ansa, 180° away. Resolution is ~ 8 km/pixel. The Huygens Ringlet at 1.95 R_S is clearly visible in a gap whose width varies due to variations in the B Ring edge. Note that the B Ring edge has azimuthal frequency of variation equal to twice that of the ringlet.

Uniform Precession and the Mass of Eccentric Ringlets.   The mass of each ringlet can be calculated from the difference in semimajor axis and eccentricity across the ringlet. The precession of an orbit is a function of distance from Saturn and therefore differential precession is expected between the inner and outer ringlet edges. Since the observed rings are uniformly precessing, the local gravitational potential must be perturbed; Goldreich and Tremaine (1978) suggest that the self-gravity of the ring particles themselves is sufficient to provide this modification. The tendency for positive differential precession is balanced by the tendency due to self-gravity for negative differential precession, the latter being dependent on the surface mass density and the eccentricity gradient across the ring. Thus, knowledge of the ring's dimensions and shape allows a determination of its mass.

One easily observed manifestation of a self-gravitating narrow ringlet is a positive eccentricity gradient. The ringlet will be wider at larger semimajor axes. Figure 16 shows this is true for the Maxwell Ringlet. The Titan Ringlet also shows this effect.

The masses in Table V are derived from this explanation for uniform precession, along with the assumption of constant mass density throughout a ringlet. The mean surface mass density σ is immediately derived. Comparison with optical depths measured by the stellar occultation (Lane et al. 1982*b*; Sandel et al. 1982*b*) gives the mass extinction coefficient $\kappa = \tau/\sigma$, which is inversely proportional to the particle size. It can be seen that the derived κ is larger than that inferred from density waves (Sec. III); this could imply smaller particles (Esposito et al. 1983*a*; Porco et al. 1984*b*).

Spreading of Ringlets and Confining Mechanisms.   Despite the good agreement noted above, a basic problem exists with narrow ringlets, eccentric and otherwise. Interactions between ring particles invariably cause ring features to spread rapidly. In the absence of confining mechanisms, any narrow feature will rapidly be smeared out in a time orders of magnitude less than the age of the solar system.

One important mechanism for ring spreading is due to collisions between the particles. Even in a monolayer ring this cannot be ignored. Collisions between ring particles dissipate energy (which is supplied by their orbital motion) but conserve angular momentum. This results in a broadening in the distribution of semimajor axis: the ring spreads. Using the surface mass density derived from the eccentric ringlets themselves, or from analysis of density waves (Sec. III) and the most favorable assumptions minimizing the collision rate, the time for a narrow feature to double in width is on the order of 10^6 yr (Porco 1983).

Radiation drag on the particles (Poynting Robertson effect) also spreads a narrow ring, though its effect is largest on the smaller particles. All particles smaller than 10 cm will be lost to a narrow ring in approximately 10^6 yr. Plasma drag may be important for the smallest particles (if they exist) if the

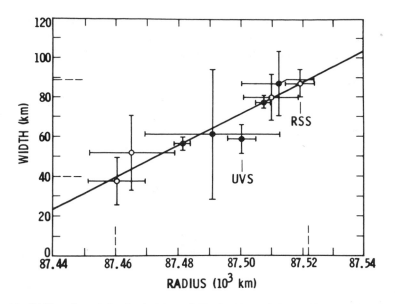

Fig. 16. Width-radius relation for the Maxwell Ringlet. Open circles are Voyager 1 data, filled circles are Voyager 2 data. The straight line is a weighted least-squares fit to these data. The maximum and minimum model widths and radii are indicated on the ordinate and abscissa, respectively.

number of charged particles in the region of the rings is significant. Most likely, this effect is smaller than collisional spreading.

If eccentric ringlets are not a recent phenomenon, some mechanism must confine them against the action of spreading effects. The most reasonable suggestion is that small moonlets on either side of an eccentric ringlet exert oppositely directed torques on the ring material between (Goldreich and Tremaine 1979). This material is collected into a narrow ringlet. An obvious example confirming this mechanism is the action of the F Ring shepherds 1980S26 and 1980S27. Unfortunately no such moonlets have been observed around the eccentric ringlets. A serious search for such moonlets in the Huygens Gap by Voyager 2 imaging cameras was entirely negative (B. A. Smith et al. 1982). Further, in areas where moonlets are found or suspected (F Ring and Encke Division) the confined ringlets are highly irregular (Cuzzi 1982). Another possible confining mechanism is diffusive instability (Ward 1981; Lin and Bodenheimer 1981), discussed in Sec. II. As noted, this idea has not been effectively worked out for a realistic model of Saturn's rings. Further, it does not seem particularly effective for small optical depths $\tau < 0.1$, common in the C Ring.

In summary, five eccentric features (three narrow ringlets and the outer edges of the B and A Rings) have been observed by Voyager. All the ringlets are uniformly precessing; mass estimates are derived from this fact. Two of the features are precessing at rates consistent with Saturn's gravitational field alone. The others are dominated by Titan, Mimas, and the coorbitals, and have their apses locked to the mean motion of those satellites.

B. Spokes

Thirty-seven days before its closest approach to Saturn, Voyager 1 discovered dark, nearly radial, wedge-shaped features in Saturn's B Ring; these finger-like markings are called spokes (B. A. Smith et al. 1981; see Figs. 17 and 18). The most important characteristics of spokes are (B. A. Smith et al. 1981):

1. Observations of spokes at large phase angles indicate they contain a large proportion of micron-sized particles.
2. Spokes are typically 10,000 to 20,000 km long and 2,000 to 5,000 km wide. They are found in the B Ring in a region which includes the part in the rings where the ring particle rotation rate equals that of the magnetic field.
3. Spokes move typically at the Keplerian rotation velocity and not that of the magnetic field (see Fig. 19).
4. Most spokes are tilted so that their evolution is consistent with their once having been radial. Backward mapping in time to radial position shows neither a preferred location or direction.
5. Spokes are most often seen on the morning (west) ansa, where ring particles have recently emerged from Saturn's shadow.

The small size of the particles in the spokes leads naturally to a possible connection with electromagnetic effects. It is worth noting that the Voyager planetary radio astronomy experiment (PRA) discovered broadband electrostatic discharges (SED) and a strong source of kilometric wavelength radiation (SKR). Power spectral analysis of spoke activity (Porco 1983) yields a significant periodicity at 640.6 ± 3.5 min, consistent with the 639.4 min period of rotation of Saturn's magnetic field (Kaiser et al. 1981). In addition, there is a strong correlation of maximum and minimum spoke activity with specific magnetic longitudes associated with the emission (or lack thereof) of the SKR (Porco and Danielson 1982; Porco 1983; Fig. 20). This further strengthens the idea that the appearance of spokes is related to Saturn's magnetic field. This observation joins a growing body of evidence, including observations of auroral brightenings (Sandel and Broadfoot 1982) and charged particle periodicities in the magnetosphere (Carbary and Krimigis 1982), which suggests a large-scale near-field anomaly in Saturn's magnetic field.

One of the most puzzling aspects of spokes is that they extend across the B Ring for many thousands of kilometers. Whether spoke particles actually move or are observed close to their points of initial elevation has not yet been demonstrated. B. A. Smith et al. (1982) observed a narrow radial spoke grow to a length of 6000 km in < 5 min. If the spokes are the manifestation of a dis-

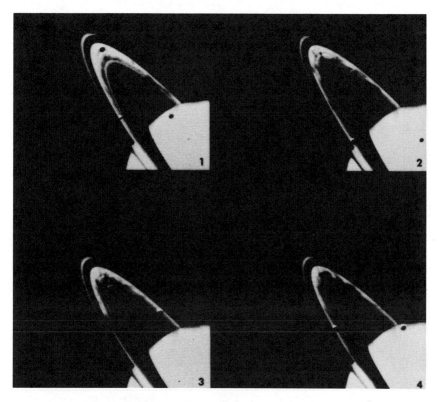

Fig. 17. Voyager narrow-angle frames taken through the clear filter (λ_{eff} = 4970 Å) exemplifying various degrees of spoke activity on the morning ansa. Spokes are seen here in backscattered light.

Fig. 18. High-resolution Voyager image taken in forward-scattering light showing wispy sheared spokes in the outermost B Ring, typical wedge-shaped structure, and the random array of bright and dark narrow ringlets across which these features lie.

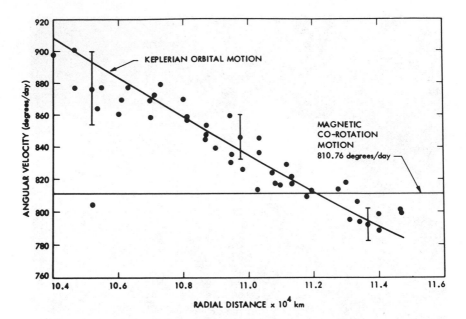

Fig. 19. Measurements taken from three Voyager 1 images (spanning 30 min) of the angular velocity of spokes plotted against radial distance. The angular velocity of the magnetic field and the variation of Keplerian motion with radius are indicated.

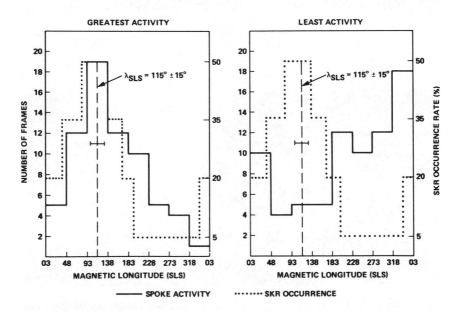

Fig. 20. Frequency distributions showing the number of frames in the categories of greatest and least spoke activity which fall into the magnetic longitude bins given on the abscissa. The center of the SKR source region (115° ± 15°) is indicated along with the 100-kHz SKR occurrence rates (dotted distribution) taken from Kaiser et al. (1981).

charge which proceeds along the length of a spoke, then this formation time implies a minimum disturbance speed of 2×10^6 cm s^{-1}. This is much greater than any expected mechanical propagation speed for the rings; e.g., velocity dispersion of the larger ring particles is $\lesssim 0.5$ cm s^{-1}.

Recent measurements have shed some light on this. By measuring the motions of two narrow-forming spokes (including the one reported in B. A. Smith et al. 1982), Grün et al. (1983) determined that inside the corotation point the trailing edges travel with the angular speed of the magnetic field, thereby remaining radial, while the leading edges move at Keplerian speeds. After formation, both edges tilt away from radial at the Keplerian rate. (Both edges of all other spokes measured in that study moved with Keplerian motion.) The angle of the wedge produced by the differential motion between edges is taken to be a measure of the time during which the spoke was active. Typical active times inferred from wedge-angle measurements of old spokes were 1 to 3 hr.

However, other recent measurements of spoke kinematics are apparently in conflict with these. Two different spokes were observed to form as diffuse features, grow in radius, darken, and move (both leading and trailing edges) with corotation motion for ~ 2 hr, after which their motions became Keplerian (Eplee and Smith 1982). All spokes not in the process of forming that were observed in this study moved with Keplerian speeds.

In total, only 4 spokes have been observed to form. While it appears from two analyses that the angular velocities of forming spokes can significantly differ from Keplerian, there is still some disagreement concerning the kinematics of formation. More measurements of this nature are necessary to clarify this important issue.

In summary, the spokes, which are most likely composed of micron-sized particles, have been observed only in the outer B Ring. Their distribution in orbital longitude at a fixed time generally peaks on the morning ansa; their appearance on the ring plane is modulated by the rotation of Saturn's magnetic field. Mature spoke features exhibit Keplerian motion; forming spokes apparently do not. The kinematics of spoke formation are still not understood.

C. Azimuthal Brightness Variations in Ring A

In addition to the dependence of ring brightness on phase and tilt (see Sec. I), photometry shows a dependence (in ring A, at least) of ring brightness on azimuth. This effect, first reported by Camichel (1958), is such that the first and third ring quadrants (measured prograde from superior geocentric conjunction) are dimmer than the second and fourth quadrants. A recent review has been made by Lumme and Irvine (1982).

Observations. These azimuthal variations in ring A have been observed from the ground by Ferrin (1978), Lumme and Irvine (1976a,b), Reitsema et al. (1976), Lumme et al. (1977,1979), and Thompson et al. (1981). These studies confirm Camichel's statement that the brightness difference between dark and bright quadrants was on the order of 10%. Pioneer 11 images of the rings' unlit face revealed similar azimuthal variations, showing a position angle of maximum brightness consistent with that seen from the Earth (Esposito et al. 1980). The solar phase angle of 8 to 10° was unfortunately not large enough to distinguish between the relative importance of viewing and illumination angles in determining the positions of maxima and minima. The Pioneer measurements were of diffusely *transmitted* light (not of the bright side of the rings visible from Earth). The fact that the brightness variation is in the same sense on both the illuminated and unilluminated faces of the ring provides a constraint not yet incorporated into models of the brightness asymmetry. Extensive observations of the rings by Voyager must still be analyzed. B. A. Smith et al. (1981) comment, however, that Voyager 1 data show a 15% effect, consistent with Earth-based observations.

International Planetary Patrol Survey.   The most extensive and complete study of the azimuthal variations was a five-year program using the International Planetary Patrol (IPP) Network operated by Lowell Observatory (Baum et al. 1970). Forty thousand photographic observations from Lowell, Mauna Kea, Perth, and Cerro Tololo provide the basic data set. About 1% of these were selected on the basis of quality for digitization, filtering, correction for atmospheric effects, and analysis. The results were summarized by Thompson et al. (1981):

1. The brightness was invariant in a 180° rotation about Saturn.
2. The maximum brightness difference is at intermediate tilt angles, with the declination of the Earth $6° < B < 16°$. The peak brightness is seen $\sim 20°$ ahead of each ansa. The exact location of the maxima and minima is uncertain from groundbased data because of large corrections for atmospheric smearing at azimuths away from the ansae.
3. Indications that the size of this effect depends on phase angle are not confirmed.
4. The variations are larger in red and green than at shorter wavelengths, although differential corrections for atmospheric effects allow some of this difference to be merely artifactual.

Explanations.   Two types of explanations have been proposed. The first involves intrinsic properties of the individual ring particles: elongated shapes, systematic leading-trailing albedo differences, synchronously rotating particles (Lumme 1969; Ferrin 1978; Reitsema et al. 1976). All these seem somewhat unlikely. The second class of explanations invokes a gravitational interaction: the mutual encounters among ring particles perturb their trajectories so that the particles preferentially align. Because of the differential Keplerian rotation, this alignment is in the form of a trailing wake. This phenomenon has been considered for both low-density (Franklin and Colombo 1978) and high-density regimes (Colombo et al. 1976). The exact manner in which this asymmetry in the particle distribution translates into a brightness asymmetry has not been shown. Lumme and Irvine (1979) suggest modeling the trailing wakes as blobs. Thompson (1982) gives a Monte Carlo calculation of the brightness from biaxial ellipsoids (blobs) embedded in a uniform (interblob) medium. Esposito (1978) proposed that the distribution of particles be modeled by a pair-correlation function; this qualitatively reproduces some of the observed characteristics of the brightness variations. Progress on the theoretical side has slowed down in expectation of the Voyager results, which could be definitive and are not yet fully reduced. A successful explanation of the azimuthal brightness variations would provide another input to our understanding of the collective dynamics of Saturn's ring particles.

V. RING PARTICLE SIZES

Our understanding of the size of the particles in Saturn's rings comes from a variety of sources: reflectance observations in the ultraviolet, visible, infrared, and microwave regions of the spectrum, infrared observations of thermal relaxation, microwave emission, and both stellar and microwave occultation. For a more complete review of the pre-Voyager state of knowledge of particle size, see Bobrov (1970), Pollack (1975), and Cuzzi (1978). A brief summary is given below.

Upper bounds on particle sizes were obtained theoretically by considerations of tidal stresses and collisional fragmentation (Jeffreys [1947] and Greenberg et al. [1977] emphasized that a distribution of sizes was expected) and observationally by the occultation of stars, which were reported to have been seen clearly through the ring (Bobrov 1970) so that the maximum particle size would be less than that of the apparent stellar disk at the distance of Saturn. Arguments along these lines give upper bounds of as large as several hundred kilometers, to as small as a few tens of kilometers. Such large particles are now known to be exceedingly rare in terms of the total particle population (B. A. Smith et al. 1982).

Radio and radar astronomy studies from the Earth in the last decade have provided opposing bounds on the effective size of the particles. On the one hand, the surprisingly low emission temperatures obtained by radio interferometry (see Sec. VI) placed an upper bound on the sizes, since increasing size increases the propagation distance within slightly lossy particles and rapidly raises their effective emissivity. Prior to 1973, the low temperatures were variously interpreted to limit the particle size to much less than a centimeter in radius.

The detection of radar echoes from the rings (Goldstein and Morris 1973) provided the first firm experimental evidence that there were a large number of particles larger than a few centimeters present in the rings. Comparisons with later observations of the rings at 3.5 and 12.6 cm wavelength showed no significant difference in scattering properties (Goldstein et al. 1977). From initial analysis of the radar results it was concluded that a large fraction of the particles must be comparable to, or larger than, the radar wavelengths employed (Goldstein and Morris 1973; Pollack et al. 1973; Pettengill and Hagfors 1974; Pollack 1975; Goldstein et al. 1977); however, all interpretations were in terms of a single effective size whose estimate varied from few centimeters to few meters depending on the model assumed.

Subsequent studies (Cuzzi and Pollack 1978; Epstein et al. 1980; Cuzzi et al. 1980) have shown that the high reflectivity and low emissivity results are also consistent with the existence of a distribution of sizes in which the number of particles in any size range varies approximately as an inverse cubic power law of the radius. This broad distribution makes the expected radar return nearly constant with frequency, providing a physical explanation for the observations. The upper size bound was somewhat uncertain because of difficulties with modeling the uncertainties in the intrinsic electromagnetic properties of the ring material (see Sec. VI).

The Pioneer 11 flyby in 1979 and the two Voyager encounters in 1980 and 1981 provided the first opportunities to extend the traditional observations to new geometries, and to implement experiments effectively impossible from Earth. These included high-resolution stellar occultation studies in the ultraviolet, photometric observations over a wide range of phase angles, including several in the forward-scattering hemisphere, and coherent occultation studies of the microwave extinction by the rings and of the principal forward-scattered diffraction lobe of the particle ensemble.

The new information regarding particle sizes is based on (a) observation of scattering in the visible over a substantial range of phase angles, (b) radio occultation, and (c) limited comparison of radio occultation results with observations in the visible and ultraviolet. Each of these data types contains information about a restricted size range. Observations of phase function in the visible sense micron-sized and larger material; for instance, Voyager images, collected at several phase angles (Cuzzi et al. 1984; see Sec. VI) show optically thin structures containing micron-size particles. Examples include the ring B spokes, thin ringlets within gaps, ring E, and ring F. Radio occultation is sensitive to both cm-sized material through differential extinction measurements, and to 1- to 20-m radius material from observations of the primary diffraction lobe of the particles. Comparison of radio and stellar occultations gives information on the total amount of material between $\sim 1~\mu m$ and 1 cm in size, but provides no size resolution within this range.

The Voyager 1 radio occultation experiment has greatly added to our knowledge of the particle size distribution in the rings, including its variation with radial distance. Although similar in principle to other occultation experiments which rely on a natural source (e.g., star, Iapetus, Saturn), it is distinguished by the use of a coherent, dual-wavelength, point source of radiation of precisely known properties. The wavelengths for the radio source onboard Voyager 1 were 3.6 and 13 cm, which provided a good match to the centimeter-to-several-meters-size range of the ring particles and permitted dispersive measurements over almost two octave separation in wavelength. During occultation, two distinct components may be identified in the spectrum of the received signal. The first is the direct (or coherent) signal, which is the coherent remnant of the incident signal after diminution by interaction with the particles. The microwave optical depth τ and phase shift ϕ are the two observables derived from this signal. The second is the scattered signal, which, uniquely for radio occultation, can be separated from the direct signal on the basis of the Doppler effects resulting from the relative motion between the particles and the spacecraft. The scattered signal is characterized in terms of variation of the differential radar cross section $\sigma_d(\theta)$ with scattering angle θ. The ability to resolve the individual ring features and to recover information about the size distribution of the resolved features is discussed below. For more details on the theory of radio occultation by rings, the reader is referred to Marouf et al. (1982). Results of the Voyager experiment are given by Tyler et al. (1982b) and Marouf et al. (1983).

The scattered signal observed during occultation is the incoherent superposition of contributions from all ring areas within the field of view of the spacecraft antenna. This can be tens of thousands of kilometers in extent. The contribution of an individual region is identified in the frequency spectrum of the received signal using Doppler mapping (Marouf et al. 1982). The mapping was possible because the spacecraft trajectory was designed so that contours of constant Doppler shift closely align with circles of constant distance from Saturn, thus mapping the signal scattered from an individual ring feature to a geometrically predetermined frequency window. For Voyager 1, the radial resolution of scattered signal measurements varied from a few thousand kilometers in ring C to several hundred kilometers in ring A.

The resolution of measurements of the (complex) extinction of the direct signal, on the other hand, is limited by the stability of the reference oscillator onboard the spacecraft and the signal-to-noise ratio of the measurement. After correction for diffraction effects (Marouf and Tyler 1982) the Voyager 1 resolution varied from subkilometer values for the thin parts of ring C and Cassini Division to several kilometers in the relatively thick part of ring A and in inner ring B. The thicker outer two-thirds of the B Ring was almost entirely opaque to both the direct and scattered signals.

Particles within different size ranges affect the observables ϕ, τ, and $\sigma_d(\phi)$ in different ways. For example, subcentimeter-sized particles in a size distribution $n(a)$, where a is the radius, are predominantly phase shifters (Marouf et al. 1983). Parts of the ring primarily composed of such small particles collectively behave like a refractive atmosphere and affect the radio signal much in the same way as a planetary atmosphere during occultation (Marouf et al. 1983). Phase measurements during Voyager 1 radio occultation by features in ring C have been reported (Fig. 8 in Tyler et al. 1982b) but analysis of such measurements is still underway.

On the other hand, contribution to τ is mainly due to particles larger in radius than approximately $\lambda/(m - 1)$, where m is the refractive index. For $m = 1.78$ (water ice), the radius is 1.5 and 5.3 cm for $\lambda = 3.6$ and 13 cm, respectively. Thus, measurements of the differential opacity $\Delta \equiv \tau(3.6$ cm$)$ $- \tau(13$ cm$)$ is strongly controlled by particles between these two values. A similar argument also applies to measurements of the differential, for example, visual and microwave opacities which could be controlled in this case by particles in the micron-to-centimeter size range.

Because at most two independent pieces of information are provided by measurements of $\tau(3.6$ cm$)$ and $\tau(13$ cm$)$, one approach to determining the size distribution $n(a)$ is to assume a distribution of some realistic, and hopefully physical, known functional form which is described in terms of unknown parameters, to be estimated from the measurements. For a large spread in both size and number density, the power law distribution

$$n(a) = \begin{array}{ll} n(a_0)\ (a_0/a)^q & a_{min} \leq a \leq a_{max} \\ 0 & \text{otherwise} \end{array} \qquad (10)$$

is commonly adopted (see, e.g., Bobrov 1970; Cuzzi and Pollack 1978; Greenberg et al. 1977; Epstein et al. 1980; Marouf et al. 1982; Marouf et al. 1983). The four parameters a_{min}, a_{max}, q, and a reference density $n(a_0)$ at some $a = a_0$ completely describe this model. Assuming $a_{min} < 1$ cm and a_{max} to be known from measurements of $\sigma_d(\phi)$ (see below), the observables $\tau(3.6$ cm$)$ and $\tau(13$ cm$)$ can be used to determine the power law index q and the scale factor $n(a_0)$ of the distribution (Marouf et al. 1982,1983).

Information about a suprameter-sized particle is emphasized, on the other hand, in measurements of $\sigma_d(\phi)$ because $\sigma_d(\phi)$ is an a^4 weighted average of the distribution while τ is an a^2 weighted average. Only particles that are larger than about a meter in size have forward diffraction lobes strong enough to be detectable in the background noise (Marouf et al. 1982). For the Voyager radio occultation, measurements of the differential scattering cross section over scattering angles between 0 and $0°7$ allow for explicit recovery of $n(a)$ over the size range $1 \lesssim a \lesssim 15$ m (Marouf et al. 1983; Zebker et al. 1983), thus eliminating the need for model fitting over this size range. Particles of radius $a \lesssim 1$ m scatter nearly uniformly over scattering angles less than $0°7$ and contribute little to the observed shape of $\sigma_d(\phi)$; a minimum sampling angle of $\sim 0°17$ ensures that the diffraction patterns of particles of radius as large as 15 m were properly sampled. The recovered distribution over $1 \lesssim a \lesssim 15$ m can be checked for consistency with the synthesized power law model for $a \gtrsim 0.01$ m. Together they specify a complete size distribution over the range $0.01 \lesssim a \lesssim 15$ m.

The above concepts can also be extended to comparison of the occultation results from the ultraviolet, the visible, the infrared, and the radio. To date only the most preliminary comparisons of this kind have been carried out. We find that some regions of the ring are apparently devoid of millimeter and smaller material. These regions include most of ring C and the Cassini Division. Other areas, including ring A and especially the inner boundaries of rings A and B, however, have substantially larger opacities in the ultraviolet and visible than in the microwave spectrum, thus indicating the presence of a substantial population of very small particles. But as the occultation data are somewhat incomplete, especially in the B Ring, these preliminary results should not be interpreted as bearing on the size distribution in all regions of the rings. More complete particle size distributions have been estimated for four ring features: two in ring C and one each in the Cassini Division and in ring A (Table I in Marouf et al. 1983). These are denoted C1.35, C1.51, CD2.01, and A2.12, where the numbers indicate the approximate radial distance from the center of Saturn in units of Saturn's radius. For all four features, positive detection of differential extinction of the direct signal and of a 3.6 cm-λ scattered signal (Marouf et al. 1983) strongly indicates the existence of a continuous size distribution that spans the centimeter (or less) to the several-meter size range. On the other hand, the inference of the meter size and large particle distribution from the data is somewhat model dependent. However, one conclusion that is nearly insensitive to the model is that, independent of the feature examined, there exists an upper size cutoff of the distribution at a particle radius of ~ 5 m beyond which both the number and mass density distributions fall rapidly. This conclusion does not preclude the existence of larger bodies, however, but such bodies would have to coexist in much smaller number densities relative to that of the particles in the main distribution.

Our current state of knowledge of the centi-to-decameter size distributions is summarized in Fig. 21, where we show the estimated range of possible distributions based on the radio data and the power law model for the size range 0.01 to 1.0 m. For a given feature and a given radius a, the upper and lower boundaries of the densely shaded regions determine upper and lower bounds on estimates of the *cumulative* number density of particles of radius $> a$, respectively. These bounds are chosen on the basis of the range in variation of the specific solutions from the radio data with specific models of the rings and the formal uncertainties in the data. All solutions obtained to date lie within the dense shading. The outer region for each location (light shading) provides a conservative estimate as to the variation which might be obtained by extremes in models of the distribution of material normal to the ring plane. The distributions inferred for the two models of a monolayer and slab vertical profiles fall within these bounds; thus, such bounds provide tight constraints on the size distributions of other intermediate vertical profiles. The existence of an upper size cutoff in the distributions is clearly apparent

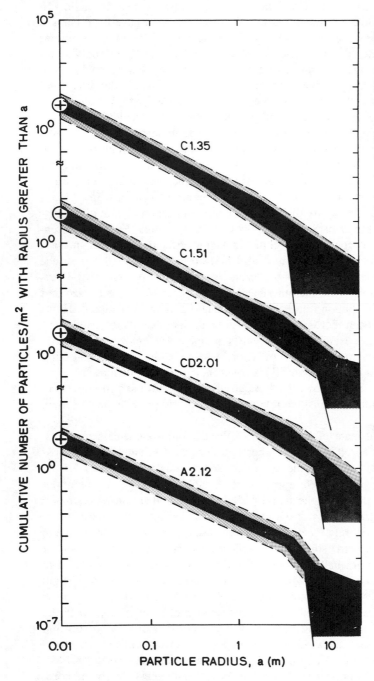

Fig. 21. Cumulative distribution of ring particles by size. Estimates of possible centi-to-decameter cumulative size distributions in 4 features of Saturn's rings, each labeled by its location in the ring system. Note the 5 orders of magnitude displacement of the vertical origin of each feature. The spread represented by the densely shaded regions reflects the variation of the specific solutions obtained by inversion of the radio data with specific ring models and the formal uncertainties of the data. The larger spread incorporating the lightly shaded regions provides a conservative estimate as to the variation which might be obtained by extremes in models of the distribution of material normal to the ring plane. The existence of an upper size cutoff of a few meters is apparent and a power-law index of roughly 3 characterizes all features.

and a power law index of roughly 3 characterizes all features, although the actual best estimates of slopes change from slightly larger values for C1.35 and C1.51 to slightly smaller values for CD2.01 and A2.12. We emphasize here, however, that model computations to date are based on radiative transfer theory (see, e.g., Hansen and Travis 1974) adapted specifically for the forward-scattering geometry of the occultation (Marouf et al. 1982; Marouf et al. 1983; Zebker et al. 1983). This theory neglects coherent interactions between the particles, hence constraining the volume fraction to be $\ll 1$. Further studies that examine the implications of coherent interactions on the inferred distributions when the volume fraction is not small would be quite valuable.

VI. INDIVIDUAL PARTICLE PROPERTIES

A. Properties from Reflectance

This section reviews our knowledge of the properties of the individual particles: brightness, color, composition, scattering phase function. In most cases, the properties of the individual particles must be inferred from remote observations which sample the ring particles acting collectively. This inference is less direct and more uncertain than for single particles like moons which are viewed individually. All these observations are of the remote sensing variety, and rely on radiation either reflected (this subsection) or emitted (Sec. VI.B) by the ring particles at wavelengths from ultraviolet to microwave. The analysis consists of inverting the radiative transfer equation, which gives the collective (observed) behavior (e.g., ring brightness) in terms of the individual (intrinsic) particle properties (albedo, phase function). The solution is not unique, and is naturally dependent on an assumed model for local ring structure. For instance, standard radiative transfer methods assume a locally plane-parallel layer within which particles are randomly distributed over a vertical extent many times their own size (see, e.g., Chandrasekhar 1960; Hansen and Travis 1974; Irvine 1975). For Saturn's rings, geometrical shadowing is an important complication (Irvine 1966; Esposito 1979). In this many-particle-thick model, photons suffer multiple reflections between various particles before either becoming absorbed or escaping to be observed, unless the overall particle spatial density is quite small. On the other hand, if the particles are all collapsed into a single monolayer, photons are much more likely to escape after only one or two scatterings, even if the particle spatial density is large. Furthermore, because of the different variation with zenith angle of particle cross section integrated over the line of sight, variation of reflectivity with illumination and viewing geometry will also be different. There are, in principle, advantages as well as uncertainties associated with these differences. It may be possible to decide between the two models of vertical structure on the basis of observed behavior, as well as to infer individual particle properties. There has been only limited progress to date in developing generally accepted, fully valid monolayer models of radiative transfer, although various workers have approached the problem (Hämeen-Anttila and Vaaraniemi 1975; Cuzzi and Pollack 1978; Froidevaux 1981). Therefore, there has been as yet no unambiguous discrimination between the physical models. We note that, because the likely vertical structure is on a scale smaller than observed by any spacecraft instrument to date, analysis of remotely sensed data may be our only way of obtaining constraints on the local vertical structure of the rings.

Several kinds of observations exist, each constraining the particle properties in different ways. The reflectivity has been measured from Earth at constant tilt angle as a function of solar phase angle (opposition effect), and at constant phase angle as a function of ring tilt angle (tilt effect), as well as over a large but sparsely sampled range of phase and elevation angles by the Pioneer and Voyager spacecraft. The ring reflectivity and polarization have also been measured as a function of wavelength with both broad and narrow spectral resolution. These observations are at both visible (ultraviolet to near infrared) and microwave (millimeter to decimeter) wavelengths. At intermediate infrared wavelengths, the particles are pure emitters; their thermal emission is discussed in Sec. VI.B.

Particle Phase Functions.   Particle scattering phase functions depend on particle size and refractive indices at the wavelength of observation, as well as surface structure if the surface structure has dimensions larger than the wavelength. Particles of almost any material which are much smaller than the wavelength are Rayleigh scatterers, fairly isotropic in total intensity but not in polarization (see, e.g., van de Hulst 1981). Particles which are comparable to, or several times larger than, the wavelength have phase functions dominated by a strong forward diffraction lobe, the width of which is an excellent diagnostic of the particle size (Kerker 1969). If such particles are fairly smooth and spherical, they also exhibit a strong, narrow backscattering peak (Hansen and Travis 1974). However, moderate irregularity destroys this peak if the particles are more than a few times the wavelength in size (Pollack and Cuzzi 1980). Particles much larger than the wavelength are typically backscatterers as a view of the unlit forward face reveals only shadow. The degree of sharpness of the backscattering depends largely on surface microstructure and associated microshadowing (Veverka 1977*a*). Such differences in single particle phase functions produce very different slab reflectivity variations. For convenience in radiative transfer modeling, the Henyey Greenstein phase function with asymmetry parameter g is commonly used (see, e.g., Irvine 1975):

$$p(\theta) = \frac{(1 - g^2)}{(1 + g^2 - 2g \cos \theta)^{3/2}} \qquad (11)$$

where θ is the scattering angle. Given a distribution of particles having a variety of different sizes and phase functions, the net phase function, albedo, and overall efficiency are area-weighted averages of those of the constituents. Thus, an ensemble containing both large, rough particles and wavelength-sized particles may have a phase function with both forward and backward peaks. This may be modeled by head-to-tail Henyey Greenstein functions (see, e.g., Lumme et al. 1983).

Most observations of particle scattering properties are from the ground at visible wavelengths and some inferences have been drawn as to the particle phase functions. However, due to the small range of phase angles covered, these inferences are necessarily of larger uncertainty that those from spacecraft observations, that cover a large range of phase angles. Spacecraft observations also offer the possibility of spatially resolving regions having different particle properties. Unfortunately, the spacecraft trajectories produce coupled variation of phase angle and ring tilt angle, and certain crucial angles are undersampled, making model analysis difficult. Also, photometric calibration of the Voyager imaging data is not as yet fully satisfactory. For this reason, only the broad outlines of the particle phase function are now known. In Fig. 22 are shown a set of observations of the rings by the Voyager 1 wide-angle camera (B. A. Smith et al. 1981) over a range of times and phase angles. The Voyager wide-angle clear filter has a passband similar to the B filter of classical photometry (Smith et al. 1977). It may be seen that the rings are 10 to 100 times brighter at low phase angles (backscatter) than at high phase angles (forward scatter). The smooth curves show model calculations for slabs of various optical depth τ composed of particles of various albedo A. The phase function assumed was that of Lambert-surface spheres, characteristic of particles much larger than the wavelength with rough surfaces without intricate surface microstructure (see Fig. 22).

This phase function, which is quite strongly backscattering, provides an acceptable first-order fit to the observations and is similar to that of "snowbanks" (Veverka 1977a). Such a phase function is also consistent with Pioneer and groundbased observations. However, neither Pioneer nor groundbased observations were able to demonstrate the overall lack of a significant forward-directed component of the phase function (Esposito and Lumme 1977; Esposito et al. 1980; Lumme et al. 1983). This is easy to understand when one realizes that conservative assumptions as to the optical depth of ring B ($\tau \lesssim 1$) produced insufficient multiple scattering from a purely backscattering phase function (see, e.g., Esposito and Lumme 1977; also see Table VII). The implication of these results is that, overall, only a small area fraction of the main ring (A, B, C, and Cassini Division) particles may be of visible wavelength size or smaller. This result is consistent with independent observations (see Sec. V). Ongoing analysis indicates that the actual particle phase function is even more sharply peaked in the backward direction than that of the Lambert-surface sphere, and more similar to those of the Galilean satellites. This indicates a rough surface.

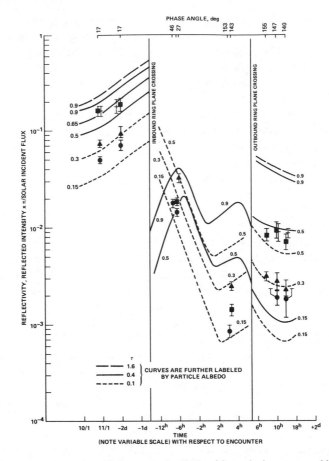

Fig. 22. Voyager 1 observations of the geometric albedo of Saturn's rings over a wide range of phase angles (top axis), showing the strong predominance of backward scattering in most ring areas. The points are plotted in time relative to encounter, with high phase angles occurring after encounter: (+) inner third of B Ring; (■) central half of A Ring; (●) central half of C Ring; and (▲) Cassini Division. Smooth curves are theoretical calculations for the optical depths indicated, assuming the phase functions of Lambert-surface spheres, and are each further labeled by the particle single-scattering albedo (from B. A. Smith et al. 1981).

The question of the particle phase function at very small phase angles is intimately tied to the implications of the opposition effect of the rings. The more of the opposition effect that can be explained by individual particle phase functions, the less need be explained by the collective effect of mutual shadowing. Recent analysis of the opposition brightness by Lumme et al. (1983) indicates that although a realistic microstructure improved the fit to observations for phase angles greater than a degree or two, a substantial fraction of the brightening is confined to too small a range of phase angles to be explained entirely by any known bright particle surface. Those calculations support the classical idea that it is primarily a volume density effect between individual well-separated particles that produces the intense brightening at phase angles less than $\sim 2°$. Johnson et al. (1980) have also observed the phase dependence of the polarization of the rings. Their results, which are in basic agreement with other observations of the rings at small phase angles (see, e.g., Dollfus 1979a,b), indicate that the rings are weakly polarized in the negative sense (polarization plane parallel to Sun-object-Earth plane). Also, Johnson et al. (1980) cite an opposition effect in the polarization with angular width comparable to the brightness opposition effect. Although Johnson et al. (1980) cite the work of Wolff (1975) as requiring the negative polarization arising from shadowed surface microstructure, we point out that Wolff's work implies only that shadowing of some kind produces the negative branch of polarization. Essentially, the argument notes that in-plane, double reflections produce positive polarization and orthogonal double reflections produce negative. Because any shadowing, whether of inter- or intra-particle origin, will lie preferentially in the in-plane direction, a slight preference for negative polarization will result. Thus, contrary to the conclusions reached by Johnson et al. (1980), we believe that the existence of a polarization-opposition effect is consistent with either a monolayer or many-particle-thick model; in balance, we believe that the various opposition effects of the rings are due to a combination of surface microstructure and the low volume density of a locally many-particle-thick ring. The particles themselves, however, are at most weakly polarizing at visible wavelengths; even at the large phase angle (70°) observed by Pioneer 11, the polarization is quite small ($< 10\%$; Esposito et al. 1980; Burke and KenKnight 1980). The presence of a (weak) negative branch and weak polarization at large phase angles is consistent with the scattering properties of bright objects with grainy, snowbank-like surfaces (Veverka 1970), especially if multiple scattering, which is depolarized, contributes significantly to the brightness at large phase angles.

However, the above results, which apply to the main rings in general, do not rule out small overall fractions ($< 10\%$), or higher localized concentrations, of micron-sized particles. Further, they do not apply to the outlying E, F, and G Rings or the clumpy, kinky ringlets in the Encke Division. The E, F, and G Rings are primarily of low optical depth with the exception of the opaque, km-wide core of the F Ring (Lane et al. 1982b), which contributes little to its visible reflectivity. Therefore, the reflectivity is dominated by single scattering and $p(\theta)$ can be determined directly. All three of these rings are much brighter when observed at high phase angles, indicating that a substantial fraction of their area is composed of small (i.e., visible-wavelength-sized) particles. However, the D Ring, of a similarly low optical depth, does not show this characteristic forward-scattering behavior. The E and F Rings are the most completely studied over a range of phase angles. Brightness variation at large phase angles ($\sim 140°$) is most unambiguously interpreted in terms of particle size, because phase functions in the forward lobe are essentially shape independent (see, e.g., Sec. V for a discussion of forward scattering and size determination). Pollack (1981b) has determined that the F Ring phase function implies a substantial fraction of very small ($\leq .01$ μm) sized particles. Because the highest phase angles observed did not extend into the forward lobe characterizing micron-sized particles (20°), no direct statement may be made as to the presence of particles of these sizes. However, the optical depth measured by the stellar occultation is significantly larger than that estimated from these scattering calculations; it is also much larger than measured at radio wavelengths (Tyler et al. 1982b). Therefore, it would seem that the F Ring contains substantial numbers of particles from ~ 0.1 μm to 10^3 μm and possibly consists entirely of such particles with the exception of its optically thick core. Pollack also finds evidence for a backscattering (large-particle) component. Whether the E and G Rings also contain a component of large, effectively backscattering particles remains to be seen. It may be that *in situ* measurements of absorption of magnetospheric particles, or impact events on the Pioneer and Voyager spacecrafts (Humes et al. 1980; Gurnett et al. 1983) will provide additional clues.

Inferences as to particle properties from backscatter phase functions are more risky, as backscatter peaks may be produced by either large, rough particles with surface microstructure or smooth, wavelength-sized particles. The lack of a backscatter peak, however, may still be consistent with wavelength-sized particles if the particles are fairly irregular (Pollack and Cuzzi 1980).

In the case of the E Ring, the observed backscattered brightness peak is too large to be explained by either surface microstructure or interparticle shadowing such as in the main rings (Fig. 23). In this case, fairly smooth, equidimensional wavelength-sized particles have been inferred (Pang et al. 1982). This inference is supported by wavelength variation of reflectivity in the near infrared (Terrile and Tokunaga 1980). Encouraged by the concentration of the E Ring material at the orbit of Enceladus, Pang et al. (1982) cite spherical particle shape as evidence for a liquid origin on Enceladus of the E Ring particles. We note, however, that the large scatter in the data obscures detailed information about the shape of the phase function. Also, even slightly roughened or irregular particles may exhibit phase variation not enormously different from spheres, if they are not too large compared to the wavelength. The optical size of a particle of radius $r \sim 1$ μm, is $x = 2\pi r/\lambda \sim 10$ at visible wavelengths; laboratory experiments (see, e.g., Zerull and Giese 1974) have shown that it is possible for roughened spheres of this size to exhibit backscatter peaks. The particles may also be smoothed by electromagnetic erosive effects (see, e.g., Burns et al. 1979). Consequently, sphericity and liquid origin are not necessarily required of the E Ring particles by these data alone.

Within the main rings themselves, and superposed on the general backscattering behavior, there is evidence for small, and regionally variable, area fractions ($\sim 10\%$) of particles which are less backward scattering than the

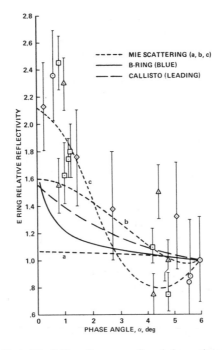

Fig. 23. Opposition effect of the E Ring, from groundbased observations by Dollfus and Brunier (1982): △□; and Larson (1982): ◇○. The long dashed and solid curves show, respectively, the opposition effect in the brightness of Callisto's trailing face and the B Ring, taken from Pang et al. (1982) and Franklin and Cook (1965); the short dashed curves show the phase functions near backscatter of wavelength-sized spheres of water ice, of optical size $x = 2\pi r/\lambda$, for: (a) $x = 5$; (b) $x = 10$; (c) $x = 20$.

general rule. That is, the brightness in certain areas *decreases less* from low to high phase angles than in other areas. Because a backscattering, rough-surface behavior at visible wavelengths would characterize particles larger than a few millimeters in size, the appearance of such preferentially forward scattering indicates a dust component in the rings. Areas in the main rings which show this preferential brightening relative to their surroundings at large phase angles include the entire outer A and B Rings in general, as well as certain areas associated with density waves. Spokes (Sec. IV.B) also exhibit this relative behavior. The fact that these areas are nonetheless darker in an absolute sense at high than at low phase angles indicates that the forward scattering dust contributes only a small fraction of the total optical depth. The detailed shape of the phase function of such a second component, and its precise fractional area, are strongly coupled to the shape of the overall phase function, which is poorly known. Thus, more analysis must be done before more precise statements can be made as to the size or optical depth of the dust material. However, the short lifetime of about one orbit period for fine dust to be swept up by larger ring particles suggests that areas with a significant area fraction of fine dust must produce it at an equally rapid rate, perhaps as a result of locally vigorous dynamics or large particle relative (random) velocities. This is in good general agreement with all of the associations of dusty areas in the A Ring, both with isolated, identified resonances and with the overall expected increase in resonant activity toward the outer A Ring.

Recall that a particle of a given physical size (e.g., 1 to 10 cm radius) may be a strong backscatterer at visible wavelengths ($x \sim 10^3 - 10^5$), but a strong forward scatterer at microwave wavelengths ($x \sim 1 - 100$). This situation describes the Saturn ring particles, as has been clearly demonstrated by the Voyager bistatic radar experiment (Marouf et al. 1983; see Sec. V). Forward scattering behavior of the ring as a whole has been modeled theoretically (Cuzzi and Van Blerkom 1974; Cuzzi et al. 1980; Muhleman and Berge 1982) and hinted at observationally (dePater and Dickel 1982). The effect is substantial in determining the brightness of the rings at centimeter wavelengths. The ring particle albedos are so high at centimeter wavelengths (see below) that the ring brightness is primarily determined by diffuse reflection of thermal emission from Saturn. If the ring particles are strong forward scatterers, the ring brightness in front of the disk is substantially larger than at the ring ansae or other off-axis azimuths. This must be kept in mind when deriving ring optical depths from microwave observations of the brightness of the rings where they occult the planet.

In summary, spacecraft and groundbased determinations of the particle phase function show the main ring particles to be primarily backscattering at visual wavelengths, implying rough-surfaced objects much larger than a visual wavelength in size. However, restricted local regions in the main rings contain a small but noticeable fraction of forward-scattering, possibly dust-sized, particles. The optically thin, outlying E, F, and G Rings are dominated in optical depth by such micron-sized, forward-scattering particles.

Particle Albedo

Broadband reflectivity measurements at visible and microwave wavelengths and high spectral resolution, near-infrared reflectance measurements directly constrain composition. The particle albedo is a function of the intrinsic particle composition and of the optical size $x = 2\pi r/\lambda$ of the particle (Pollack 1975). When $x \gg 1$, as for the ring particles at visible wavelengths, the particle single-scattering albedo is essentially independent of particle size or surface structure. When $x \sim 1$, as for the ring particles at microwave wavelengths, the single-scattering albedo may be a strong function of x as well as of the intrinsic composition.

In Table VI we show particle albedos A derived from broadband observations at visible wavelengths, along with their associated determinations or assumptions as to phase integral $q = 2\int_0^\pi p(\theta)\sin\theta \, d\theta$, where $p(\theta)$ is the phase function. Each of the results quoted is a model result, and it is not straightfor-

TABLE VI

Ring Particle Bond Albedos and Phase Integrals

Ring Region	Mean Radius (R_S)	λ	Albedo (A)	Phase Integrals (q)	Source
A	2.10	V	0.63	0.57 [a]	Cook et al. (1973)
		V	0.75–0.95	1.57 [b]	Esposito and Lumme (1977)
		V	0.5–1.0	1.0–1.5 [b]	Lumme and Irvine (1976a)
		≈B	0.6	1.5 [c]	B. A. Smith et al. (1981)
Cassini Division	1.97	≈B	0.3	1.5 [c]	B. A. Smith et al. (1981)
B	1.85	V	0.63	0.57 [a]	Cook et al. (1973)
		V	0.8–1.0	1.0–1.5 [b]	Lumme and Irvine (1976a)
		V	0.8–0.95	1.57 [b]	Esposito and Lumme (1977)
		V	0.75–0.9	1.4–2.0 [b]	Kawata and Irvine (1974)
		≈V	0.7–0.9	0.4–1.5 [b]	Lumme et al. (1983)
		B	0.49	0.57 [a]	Cook et al. (1973)
		B	0.45–0.75	1.0–2.0 [b]	Kawata and Irvine (1974)
		B	0.5–0.7	0.4–1.5 [b]	Lumme et al. (1983)
C	1.35	≈B	0.25	1.5 [c]	B. A. Smith et al. (1981)

[a] Assumed.
[b] Modeled using only low-phase-angle data (0–6°).
[c] Adopted as approximate fit to low- and high-phase-angle data (17–160°).

ward to adjust them all to the same value of A, q, or p. It is evident that the particle albedo is high; however, the more recent results based on current knowledge of the phase function (B. A. Smith et al. 1981) are slightly lower than previously believed. The A and B Ring particles have "blue" albedo $A \sim 0.6$, fairly bright as might be expected for macroscopic icy objects; however, the C and Cassini region particles are significantly darker ($A \sim 0.2$) at the same wavelength. In addition, preliminary analysis of Voyager data (Cuzzi 1982) suggests that substantial radial variations in particle albedo may be seen even within the B Ring. These facts have important implications for the thermal behavior of the particles, discussed subsequently.

The particle albedo is, in general, significantly lower at shorter B (of the *UBV* system) wavelengths. This is borne out by higher resolution spectrophotometry over the entire wavelength range from $\sim$ 2000 to 13,000 Å (see Figs. 24 and 25). These latter results have not yet been quantitatively corrected for multiple scattering effects, and the large overall increase of reflectivity from 0.2 to 1 μm surely includes an increasing multiply-scattered

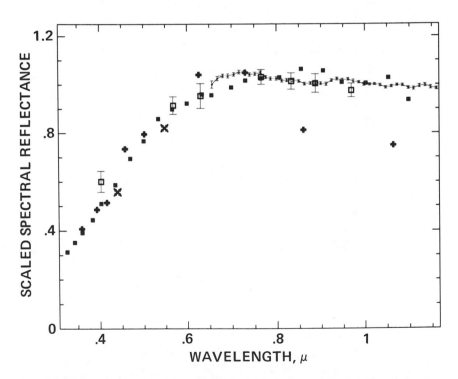

Fig. 24. Various observations of the spectral reflectivity of Saturn's rings at visible wavelengths, as summarized by Clark and McCord (1980); Singer (1977): □; Lebofsky and Fegley (1976): ■; Irvine and Lane (1973): +; Irvine and Lane (1973) B and V filters: ×. Of note is the decreasing reflectivity shortward of 0.65 μm wavelength.

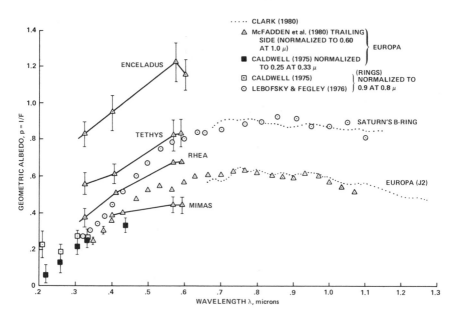

Fig. 25. Spectral reflectivity of Saturn's rings from 0.2 μm to 1.3 μm, compared with reflectivity of Europa and several Saturnian satellites. Figure adapted from Caldwell (1975), Lebofsky and Fegley (1976), Clark (1980), and McFadden et al. (1981).

component which serves to further steepen or accentuate the increase. For particles of albedo $A \sim 0.6$ to 0.8, and backscattering phase functions, the multiple scattering contribution can be from 10 to 30% of the total (see, e.g., Fig. 5 of Kawata and Irvine 1974). It is therefore at least possible that the overall color of the individual particles in the main rings is quite similar to the color of the various satellites at visible wavelengths (Fig. 25). However, this still must be confirmed by quantitative calculations. In the ultraviolet, a possibly significant difference is seen between the ring and satellite spectral reflectivities, with that of the rings seeming somewhat flatter (Caldwell 1975; see Fig. 25). In addition to the different brightness of their constituent particles, the colors of the A/B and C/Cassini Division (CD) regions are different with the C/CD regions being slightly more neutral in color, although still distinctly reddish (B. A. Smith et al. 1981,1982). Comparison of these effects in regions of different optical depth demonstrates that this albedo-color effect is related to intrinsic particle properties and not to multiple-scattering contributions in the optically thicker A and B Rings; for instance, color differences are observed in several instances between regions of the same optical depth (e.g., among the bright bands in the outer C Ring). At longer wavelengths, absorption features of water ice are seen (Fig. 26; see also below).

The color of the E Ring is distinctly blue, markedly different from that of the main rings, or any other icy object in the outer solar system (see Fig. 27;

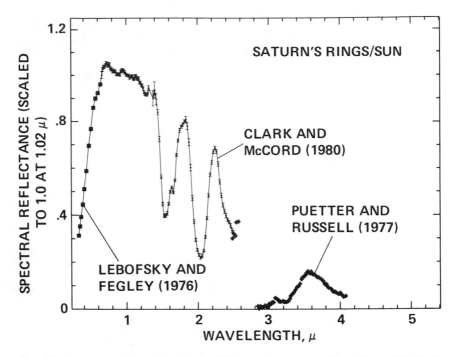

Fig. 26. A summary of ring reflectivity from 0.3 μm to 4 μm (adapted from Clark and McCord 1980).

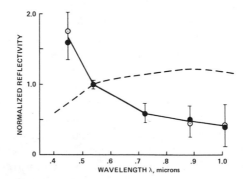

Fig. 27. Broadband photometry of Saturn's E Ring (solid line), showing its distinctly bluish color (from Larson 1982) as compared with the B Ring (dashed line) (Lebofsky and Fegley 1976). Open circles: calibrated on HD111631; filled circles: calibrated on Hyperion.

Larson 1982; Larson et al. 1981*a*). Taken in combination with the opposition brightening shown by the phase function of the E Ring discussed previously, this relative spectral variation suggests the effects of a generally small particle size rather than of intrinsic composition, i.e., it may be the "blue-sky" effect. Without better knowledge of the E Ring optical depth, we may not determine the absolute albedo of its constituent particles. Detailed calculations are needed to determine if the particle sizes indicated by the phase function and wavelength dependence are consistent.

The reflectivity of the rings has also been measured at microwave wavelengths $\gtrsim 1$ cm, most importantly by using groundbased radar techniques (Goldstein and Morris 1973; Goldstein et al. 1977; Ostro et al. 1980,1982). The reflectivity is quite high, with geometric albedo ~ 0.34, and is not a strong function of wavelength or ring tilt angle (see Table VII and Fig. 28). The albedo of an equally reflective Lambert surface parallel to the ring plane would be ~ 0.8 (Cuzzi and Pollack 1978). The degree of variation of ring radar reflectivity with ring tilt angle is somewhat underconstrained due to polarization effects. Radar transmissions, unlike sunlight, are initially 100% polarized, either linearly or circularly. The rings are extremely efficient depolarizers (see Table VII), and the energy reflected in both orthogonal polarizations must be measured in order to adequately measure the total reflectivity of the ring. A byproduct of the two measurements is the polarization ratio, which itself contains information on individual particle properties. The available observations are summarized in Table VII and Fig. 28. In several cases, reflectivity was measured in only a single polarization, and therefore the total reflectivity characterizing a particular observation must be parameterized by the ratio δ of reflectivity in the unobserved polarization orthogonal to the observed polarization. For most single solar system objects such as planets, this ratio is fairly small (Pettengill 1978). For the rings, observations at several wavelengths and tilt angles place it between 0.4 and 1.0 (Goldstein and Morris 1973; Goldstein et al. 1977; Ostro et al. 1980,1982). Note, however, that $\delta > 1$ for several Galilean satellites, an as yet unresolved mystery (Campbell et al. 1977; Ostro et al. 1980). Therefore, only under certain assumptions may the dependence of total ring reflectivity on tilt angle be addressed, as has been done by Ostro et al. (1980) and discussed further below.

Early reports of anomalously high reflected power at low radar Doppler shifts obtained at large ring opening angles (Goldstein and Morris 1973; Goldstein et al. 1977), which could arise either from regions with orbital velocities lower than within the rings or due to azimuthally asymmetric reflectivity, have not been confirmed in more recent observations at lower tilt angles (Ostro et al. 1980,1982). No obvious explanation has been found either for the effect or for its disappearance, nor are there any suggestions as to a contributing source of systematic error. We may have to wait until the ring tilt angle becomes large once again in order to resolve the question.

The radar reflectivity of the rings as a function of radius has been deter-

TABLE VII

Summary of Radar Results of Saturn's Rings [a]

B (deg)	λ (cm)	α	μ_C	σ_{oc}	σ_{sc}	μ_L	σ_{sL}	Reference
26.4	12.6			0.68±0.17				Goldstein and Morris (1973)
24.4	3.5	0.34±0.06	1.00±0.25	0.68±0.13	0.68±0.13			Goldstein et al. (1977)
	12.6					1.0±0.3		
21.4	3.5		0.75±0.08					Ostro et al. (1982)
	12.6						0.83±0.21	
18.2	12.6	0.24±0.06	0.57±0.12	0.61±0.15	0.35±0.09			Ostro et al. (1980)
11.7	12.6	0.27±0.07	0.40±0.05	0.76±0.19	0.30±0.08			
5.6	12.6	≤0.27		≤0.52	≤0.55			
6.3	3.5			0.54±0.15				Goldstein (1982)

[a] Values of geometric albedo α, normalized radar cross section σ, circular polarization ratio μ_C, and linear polarization ratio μ_L, measured at ring-plane tilt angle B, and wavelength λ, are listed in chronological order of the radar observations. Receiver polarization is designated as "sc" (same sense of circular polarization as transmitted), "oc" (sense of circular polarization orthogonal to that transmitted), or "sL" (same sense of linear polarization as transmitted).

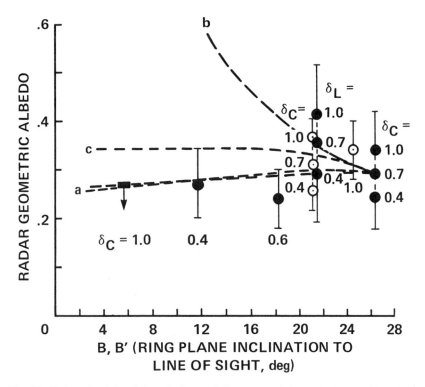

Fig. 28. Radar reflectivity of Saturn's rings at 3.5 cm (open circles) and 12.6 cm (filled circles) wavelengths, taken from Ostro et al. (1980,1982) and summarized in Table VII. Smooth curves indicate (a) the theoretically calculated reflectivity of a many-particle-thick ring of centimeter-to-meter-sized particles, (b) a monolayer of strong backscatterers (neglecting shadowing), and (c) the effect of shadowing on (b) using the results of Froidevaux (1981). Figure is adapted from Cuzzi and Pollack (1978), and Ostro et al. (1980). The depolarization ratio δ must be assumed when it is not measured in order to determine the total reflectivity.

mined by Ostro et al. (1982). The A and B Rings provide most of the reflectivity, and their reflectivities are comparable. The C Ring reflectivity is significantly lower. The relative reflectivities of the A and B Rings are consistent with similar particle properties and the observed visual optical depths; for the incidence angles of the observations, a doubling of ring optical depth produces less than a 40% increase in reflectivity. Whether or not the C Ring particles may have significantly different radar albedos, as perhaps implied by their different visible-wavelength albedos, is not yet determined.

In addition to the radar backscatter data, groundbased interferometric observations of the intrinsic microwave brightness temperature of the rings (Briggs 1974; Cuzzi and Dent 1975; Janssen and Olsen 1978; Schloerb et al. 1979a,b,1980; Cuzzi et al. 1980; de Pater and Dickel 1982) show it to be far lower than the particle physical temperature of $\sim$ 90 K (see below, Sec.

VI.B). More importantly, they and the Voyager radio occultation also showed that the optical depth of the rings at microwave wavelengths must be substantial, in fact, comparable to the value at visible wavelengths (see Sec. V).

The large radar geometric albedo of the rings, combined with their significant opacity and low intrinsic brightness temperature at the same wavelength, require the individual particles to have albedos unfamiliarly high ($A \gtrsim 0.95$) for rocky or icy objects of macroscopic sizes. Several imaginative ways of attaining the requisite high particle albedo and depolarizing ability have been advanced, including the possibility that the particles are made of metal, which is intrinsically a perfect reflector of microwave radiation (Goldstein and Morris 1973). However, the groundbased observations are also fairly easy to understand if the particles are composed of a material substantially, but not unrealistically, less lossy or absorptive than silicates. Given values of absorptivity reasonable for common icy materials (Whalley and Labbé 1969; Mishima et al. 1983), Pollack et al. (1973) and subsequent workers (Briggs 1974; Pollack 1975; Cook and Franklin 1977; Cuzzi and Pollack 1978; Ostro et al. 1980; Cuzzi et al. 1980) pointed out that such particles which are also comparable to the wavelength in size have extremely high single-scattering albedos. Detailed radiative transfer calculations (Cuzzi and Pollack 1978) show that multiple scattering among realistic particles with irregular shapes can contribute a large portion of the observed backscatter reflectivity at large ring opening angles. The large multiply scattered component, which is essentially depolarized, offers a natural explanation for the large depolarization of the total reflectivity (Cuzzi and Pollack 1978). The data indicate the depolarization may be wavelength dependent, being larger at 3.5 cm than at 12.6 cm wavelength (Ostro et al. 1980). This decrease with decreasing tilt angle might be expected due to the decreasing contribution of multiple scattering. A measurement of the depolarization at very small tilt angles would be extremely useful. Under a fairly broad range of assumptions for the depolarization of observations in which it was not measured, Ostro et al. (1980) conclude that a monolayer of large, Galilean-satellite type objects is inconsistent with the relatively constant or possibly decreasing reflectivity with tilt angle. The situation is somewhat complicated by the fact that the theoretical monolayer reflectivity used by Ostro et al. (from Cuzzi and Pollack 1978) did not account for interparticle shadowing (see, e.g., Froidevaux 1981). The reflectivity of a slab model of centimeter-to-meter-sized particles is, however, consistent with the observations (see Fig. 28). Furthermore, the rough wavelength independence of the ring reflectivity in itself implies a fairly broad size distribution (Cuzzi and Pollack 1978) very much like the results independently obtained by Voyager.

The Voyager radio occultation and bistatic scattering experiments have generally confirmed the broad size distribution and refined the size limits. These results are fully consistent with analyses of the groundbased observations. Furthermore, the high reflectivity and depolarization of the groundbased observations are even easier to understand now, in the light of the larger ring optical depths measured by Voyager (see, e.g., Holberg et al. 1982b; Esposito et al. 1983b,c; Tyler et al. 1982b) than assumed in the initial modeling analyses (Cuzzi and Pollack 1978).

Finally, the high microwave reflectivity of the rings calculated from the bright cloud model of Pollack et al. (1973) with an improved size distribution is adequate to explain essentially all of the intrinsic brightness of the rings at centimeter-and-longer wavelengths as diffuse reflection of thermal emission from the planet, as discussed in Sec. VI.B.

In summary, the ring-particle albedos are fairly high at visible wavelengths, but are a strong function of wavelength and location, ranging from 0.2 at blue wavelengths in the C Ring and Cassini Division to 0.65 at longer wavelengths (Voyager clear filter) in the A and B Rings. The color of the A and B Ring region particles is the most reddish, the C and Cassini region particles are more neutral, and the E Ring is distinctly bluish; the latter may, however, be produced by a particle size effect. The particle albedo at microwave wavelengths $\lambda > 1$ cm is quite close to unity, explaining the large radar reflectivity and low microwave emissivity of the rings.

Particle Composition. In many ways composition is the ultimately sought intrinsic aspect of ring particles. Water ice has long been known to be significant, at least on the surfaces of the particles, from identification of several ice absorption features in near-infrared reflectance spectra (Kuiper et al. 1970; Pilcher et al. 1970; Lebofsky et al. 1970; Pollack et al. 1973; Clark and McCord 1980; Clark 1980). Where water ice exists, various clathrate-hydrates of ammonia or methane (Miller 1961) could also exist. Near-infrared spectra do not distinguish ice from such clathrates (Smythe 1975), but spectra at longer wavelengths could (Bertie et al. 1973).

All of the high-quality spectral reflectance data are of the A and B Rings. The high particle albedos in these rings at visible wavelengths are in good agreement with water ice. However, some other constituents are also required to produce the decline in A and B Ring reflectivity shortward of 0.65 μm wavelength (see Fig. 24). Gradie et al. (1980) showed that various allotropes or combinations of sulfur could well provide the blue absorber needed not only for the rings but for practically all outer solar system objects as well. Lebofsky and Fegley (1976) point out that irradiated H_2S-H_2O frosts also demonstrate spectral properties reminiscent of ring spectra at visible wavelengths. However, such frosts become fairly dark near ~ 0.9 to 1 μm, unlike the ring spectra that are bright out to ~ 1.5 μm wavelength (Fig. 26).

Recent observations of ring and laboratory frost reflectivity spectra in the near infrared (1.5 to 4 μm) have been carefully analyzed by Clark (1980) and Clark and McCord (1980). The most obvious features are still the water absorption bands at 1.5 and 2.0 μm. The detailed shape of these features (Fig. 29) is well matched by laboratory spectra of medium-grained water frost

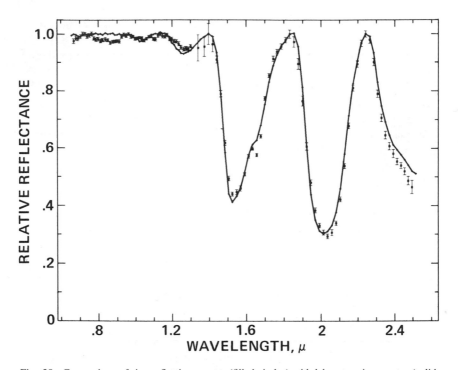

Fig. 29. Comparison of ring reflection spectra (filled circles) with laboratory ice spectra (solid curve). Slowly varying continua have been divided out. Note the excellent agreement of the 1.6 and 2.0 μm bands with the spectrum of pure water frost, and the additional presence in the ring spectrum of an absorption feature at ~ 0.85 μm (Clark 1980).

at ~ 100 K (Clark 1980; Pollack et al. 1973), as contrasted with even the spectra of frost overlying solid ice like that characterizing Europa (Clark 1980). Clark (1980) also points out that although the transparency of ice at short wavelengths allows impurity features to be very visible (the blue absorption shortward of 0.6 μm as well as a weak feature at 0.85 μm), the extremely low ring reflectivity at ~ 3 μm wavelength (Peutter and Russell 1977) where ice is opaque (see Fig. 26) implies that these impurities are well mixed with the ice, consistent with the conclusions of Lebofsky et al. (1970). The impurity feature at 0.85 μm (see Fig. 32 in Sec. VI.B), as well as the blue edge, are consistent with a variety of silicate materials. Silicate materials have an apparent advantage over sulfur-type impurities, in that their reflectivity is high and constant from 0.9 μm to 1.5 μm, like the rings. Further, silicates are more likely from cosmochemical abundances. Of course, all of the above observations refer only to the surface layers of the particles.

Radar and radio wavelength observations sample the bulk material of a particle to depth on the order of 10^2 wavelengths, and generally imply icy-material refractive indices. The possibly lower apparent microwave absorption coefficient for the ring particles than those measured for hexagonal-phase ice (Ih) cooled to 85 K (Whalley and Labbé 1969; Mishima et al. 1983; Epstein et al. 1984), and the extremely low infrared thermal inertia of the ring particles discussed in Sec. VI.B, could possibly be explained if the ice in the ring particles was in the amorphous phase state which is thermodynamically stable at low temperatures. Smoluchowski (1978b) has pointed out that amorphous-phase ice has substantially lower thermal conductivity than hexagonal-phase ice, and we note that thermal and electrical conductivities are related. The mass fraction of silicate impurities, if uniformly dispersed within the ring material, is limited to less than 10% based on ground-based radar and radio-wavelength observations (Cuzzi and Pollack 1978), given the absorption coefficient of Whalley and Labbé (1969). However, more recent lower values for ice absorption coefficient (Mishima et al. 1983) could allow this fraction to increase somewhat. Alternatively, the rocky material could be present in greater amounts if it were nonuniformly distributed, or segregated from icy material in cores of particles or separate chunks (Janssen and Olsen 1978; Epstein et al. 1984). Although local inhomogeneity is not seen by Voyager thermal infrared observations within the C Ring (see the previous section), the C Ring and Cassini Division particles in general are much darker and more neutral in color than the A and B Ring particles (B. A. Smith et al. 1981), and could be of different composition. Furthermore, local regions of darker particles are seen within the B Ring (Cuzzi 1982). The rejection of separated populations of ice and silicate particles by Lebofsky et al. (1970) would not hold if the ice fraction itself were colored by impurities. However, even the rocky fraction must have a substantial surface cover of ice, at least in the B Ring, in order to explain the very low reflectivity at 2.9 μm (Peutter and Russell 1977).

Metal, a possible candidate composition based on high microwave reflectivities, may now be rejected on particle density arguments. A combination of Voyager particle size distributions (Marouf et al. 1983), measurements of ring optical depth (Lane et al. 1982b; Holberg et al. 1982b), and determinations of ring surface mass density from observed density waves (Holberg et al. 1982b; Esposito et al. 1983b; Sec. III.D) point to a particle mass density of $\rho \lesssim 1$ g cm^{-3}.

In summary, the A and B Ring particles are probably primarily water ice, possibly containing clathrate hydrates, although minor amounts of colored material (possibly sulfur or iron-bearing silicate compounds) are also present. It is possible to imagine an overall ring composition not too different from that of the various icy satellites, consistent with the overall spectral similarities between the rings and satellites (Fig. 25b). Further spectral reflectance observations which resolve the C Ring would be very interesting, as would further range-resolved radar observations and millimeter-wavelength interferometric observations. Obviously, further laboratory measurements of the microwave absorption coefficients of various low-temperature ice phases (e.g., amorphous) would be of great value.

B. Thermal Radiation from the Ring Particles

Thermal radiation from the ring particles has been observed at wavelengths from $\sim 10~\mu$m to ~ 1 cm (see Table VIII). Observations of brightness temperatures at wavelengths shorter than $\sim 40~\mu$m are probably direct measurements of particle physical temperature, after allowance for filling factor $[1 - \exp(-\tau/\sin B)]$ where τ is normal optical depth and B is the viewing elevation angle (see, e.g., a review by Morrison 1977b). The physical temperatures of the A and B Ring particles are about equal, and the C Ring and Cassini Division particles have higher temperatures, as seen in Fig. 30 (Hanel et al. 1982; see also Reike 1975; Froidevaux and Ingersoll 1981). This behavior is qualitatively consistent with the lower visible albedos observed for the C and Cassini region particles (Cook et al. 1973; B. A. Smith et al. 1981). Closer inspection reveals interesting discrepancies, as discussed below. Interestingly, Voyager infrared spectroscopy spectra of the lit face of the C Ring are quite flat (Hanel et al. 1981a) indicating a remarkable regional degree of thermal uniformity that might not exist if the particles were a heterogeneous mixture of dark and bright ones, either intermixed or vertically segregated. Voyager IRIS spectra of the A and B Rings, where particle albedo inhomogeneity might be present, are not as yet available.

As shown in Fig. 31, the A and B Rings exhibit a significant decrease in $20~\mu$m brightness temperature with decreasing solar inclination, representing diminishing particle temperature due to diminishing solar input. The C Ring brightness temperature increases with decreasing solar inclination angle due to its low optical depth and an increasing filling factor. The details of the variation are not in good agreement with current thermal models of a homogeneous many-particle-thick ring (Kawata and Irvine 1975; see Fig. 31), and various alternatives have been proposed. Kawata (1979,1982) and Nolt et al. (1980) have suggested that vertically heterogeneous mixtures of dark and bright particles improve the fits to the data. Preliminary Voyager imaging results do indicate some variation of particle albedos in the B Ring (Cuzzi 1982). Another approach has been taken in which the fair agreement of the observations with the simple $(\sin B)^{1/4}$ dependence of a Lambert surface (Pollack 1975; Tokunaga et al. 1980b) is used to motivate a well-developed monolayer theory of ring emission including interparticle shadowing (Froidevaux 1981). The results for A and B Ring brightness variation with tilt angle may also be fairly well matched by such a model with proper choice of albedo, as shown in Fig. 31. However, problems arise in modeling the C Ring, as discussed below.

Variation of the particle temperature during eclipse, as studied theoretically by Aumann and Kieffer (1973) and in new groundbased eclipse exit

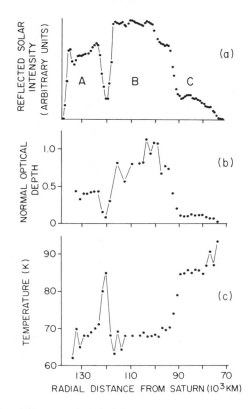

Fig. 30. Panels (a) and (b) present the 0.8–2.5 ring reflectivity and normal optical depth, respectively, as measured by the Voyager infrared spectrometer instrument. Voyager 2 observations of lit face brightness temperature are given in panel (c) (Hanel et al. 1982), and show that the C Ring and Cassini Division particles are warmer than those in the A and B Rings, as might be expected from their darker surfaces.

observations by Froidevaux et al. (1981) demonstrate that the particles must (a) be larger than ~ 1 mm, and (b) for both monolayer and many-particle-thick models, have a very low thermal inertia $= (K\rho c)^{1/2}$ where K is thermal conductivity, ρ is bulk density, and c is specific heat (see Morrison 1977b; Froidevaux and Ingersoll 1981). In conjunction with these results and the C Ring particle albedos (B. A. Smith et al. 1981), the implication of the monolayer thermal calculations of Froidevaux (1981) is that the particles' spin rate must be much larger than the Keplerian orbital rate. Because particle spins are probably in equilibrium with their random velocities, the larger spin rate implies that the particle random velocities are much larger than their minimum monolayer value in these optically thin regions. However, this leads to invalidation of the monolayer model assumption. Furthermore, published descriptions of Voyager IRIS eclipse entry observations note a larger drop in temperature at low phase angles (as in the groundbased case) than at high phase

TABLE VIII

Brightness Temperature Observations of Saturn's Rings as a Function of Wavelength

Wavelength	T_B (K)			B_0 (deg)	R (AU)	Reference
	A	B	C			
10–12 μm	86±3 [	]*	—	17	9.12	Allen and Murdock (1971)
		92±3	—	26.4	9.03	Morrison (1974b)
	90±2	94±2	—	26	9.01	Rieke (1975)
18–25 μm	56±1 [	71–76]	86	~0	9.42	Tokunaga et al. (1980b)[b]
	51–61[b]	75.5	87.7	4	9.51	Hanel et al. (1981a)[a]
			87	6.5	9.31	Nolt et al. (1980)[a]
	69	69	83.3	8	9.59	Hanel et al. (1982)[a]
	75.5	78.6	84	10	9.24	Froidevaux et al. (1981)[a]
		81	—	11.5	9.22	Sinton et al. (1980)[a]
		78.9	83	11.8	9.21	Nolt et al. (1980)[a]
		84	82.6	16.2	9.14	Sinton et al. (1980)[a]
	83.7	85.7	88±4[d]	16.3	9.14	Nolt et al. (1978)[a]
		91.0	—	24.8	9.07	Murphy et al. (1972a)[a]
	88±3	93±2	—	26	9.04	Murphy (1973)[a]
		95	—	26.4	9.03	Morrison (1974b)[a]
	87	90	80	26	9.01	Rieke (1975)[a]
30–45 μm	64±6	68±4	86±2	2.5	9.24	Froidevaux and Ingersoll (1980)
	~55 [	]	—	15.5	9.16	Courtin et al. (1979)
	88±7 [	]	—	21.2	9.07	Haas et al. (1982)
	80±3	85±3	—	24.5	9.03	Nolt et al. (1974,1977)[m]
	86±3	91±3	—	26	9.01	Rieke (1975)
45–100 μm	~62 [	]	—	4.3	9.51	Daniel et al. (1982)
	~55		—	10.5, 15.5	9.20, 9.16	Courtin et al. (1979)
	60–80 [	]	—	20.0	9.09	Melnick et al. (1983)[h]
	60–80 [	]	—	21.2	9.07	Haas et al. (1982)[i]

100–200 μm	[~55]	—	15.5	9.16	Courtin et al. (1979)
300–900 μm	[72±20 [f], 92±20 [g]]	—	21 [f], 27 [g]	9.07 [f], 9.02 [g]	Whitcomb et al. (1980)
	[78±4 [f], 84±14 [g]]	—	21 [f], 27 [g]	9.07 [f], 9.02 [g]	Cunningham et al. (1981)
1.0 mm	[32±7]	—	22.0	9.07	Werner et al. (1978) [e]
1.4 mm	[67±22]	—	26.6	9.02	Rather et al. (1974) [e]
1.4 mm	[48±11]	—	26.4	9.02	Courtin et al. (1977) [e]
1.7 mm	[45±8]	—	20.1	9.10	Rowan-Robinson et al. (1978) [e]
2.1 mm	[22±12]	—	26.4	9.03	Ulich (1974)
2.7 mm	[~20±10]	—	10	9.62, 9.68	Muhleman and Berge (1982)
3.3 mm	[17±3]	—	0–26	range	Epstein et al. (1980,1984)
3.5 mm	[8±7]	—	26.5		Ulich (1974) [e]
8.6 mm	[11±3]	—	21–25	9.04, 9.06	Janssen and Olsen (1978) [j]
1.3 cm	[7±1]	—	15.3	9.12	Schloerb et al. (1980)
1.3 cm	[8.5±3 [k], 24±9 [ℓ]]	—	12.5	9.63	de Pater and Dickel (1982)
2.0 cm	[6±3 [k], 21±6 [ℓ]]	—	12.5	9.63	de Pater and Dickel (1982)
3.7 cm	[5±1 8±2]	6±2	26.5	9.02	Schloerb et al. (1979a,b)
3.7 cm	[7±1 8.5±1]	8±1	20.8	9.09	Schloerb et al. (1979a,b)
	[11±2 10±2]	8±4	25.6	9.05	Briggs (1974) [j]
	[6±3 15±4]	8±4	26.2	9.03	Cuzzi and Dent (1975) [j]
6.0 cm	[7±3]	—	26.4	9.07	Jaffe (1977) [j]
	[5±2] [k]	—	5.9–12.5	9.59–9.63	de Pater and Dickel (1982)
	[18±4, 22±9] [ℓ]	—	12.5, 5.9	9.63, 9.59	de Pater and Dickel (1982)
11.1 cm	[<12.5]	—	25.6	9.05	Briggs (1974) [j]
21.0 cm	[6±3]	—	25.2	9.08	Briggs (1974) [j]

* Brackets indicate the extent of the observations.

[a] Normalized to 9.25 AU and T_d (Saturn) = 92 K; all are ± 2 K, except as noted.

[b] Unlit face.

[c] Actual observational result, not normalized.

[d] Highly model-dependent (Morrison 1974b).

[e] From Cunningham et al. (1981).

[f] Edge-on observations compared with Lowenstein et al. (1977).

[g] Edge-on observations compared with Hudson et al. (1974).

[h] Edge-on observations compared with Ward (1977).

[i] Edge-on observations compared with Erickson et al. (1978).

[j] Re-analysis by Cuzzi et al. (1980).

[k] At ansae.

[ℓ] In front of planet, possibly includes both direct transmission and forward scattering.

[m] From Courtin et al. (1979).

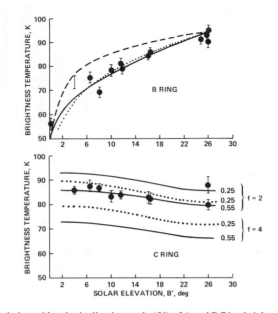

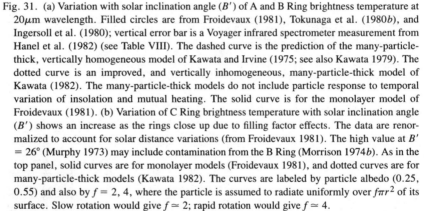

Fig. 31. (a) Variation with solar inclination angle (B') of A and B Ring brightness temperature at 20μm wavelength. Filled circles are from Froidevaux (1981), Tokunaga et al. (1980b), and Ingersoll et al. (1980); vertical error bar is a Voyager infrared spectrometer measurement from Hanel et al. (1982) (see Table VIII). The dashed curve is the prediction of the many-particle-thick, vertically homogeneous model of Kawata and Irvine (1975; see also Kawata 1979). The dotted curve is an improved, and vertically inhomogeneous, many-particle-thick model of Kawata (1982). The many-particle-thick models do not include particle response to temporal variation of insolation and mutual heating. The solid curve is for the monolayer model of Froidevaux (1981). (b) Variation of C Ring brightness temperature with solar inclination angle (B') shows an increase as the rings close up due to filling factor effects. The data are renormalized to account for solar distance variations (from Froidevaux 1981). The high value at B' = 26° (Murphy 1973) may include contamination from the B Ring (Morrison 1974b). As in the top panel, solid curves are for monolayer models (Froidevaux 1981), and dotted curves are for many-particle-thick models (Kawata 1982). The curves are labeled by particle albedo (0.25, 0.55) and also by f = 2, 4, where the particle is assumed to radiate uniformly over $f\pi r^2$ of its surface. Slow rotation would give $f \simeq 2$; rapid rotation would give $f \simeq 4$.

angles (Hanel et al. 1981a), which would support the hypothesis that the dark faces are colder than the lit faces, implying slow rotation. The response of a many-particle-thick ring is, however, systematically different. In this case, some particles are shadowed by other particles and the temperature decreases away from the illuminated face. Kawata (1982) has constructed new thermal models for vertically inhomogeneous many-particle-thick rings. Predictions of even vertically homogeneous cases of these models are in fair agreement with the groundbased observations (Fig. 31) for C Ring particle albedos and slow rotation. The difference between these and the monolayer models lies in the specifics of interparticle shadowing effects (Cuzzi et al. 1984). Furthermore, a great deal of physics still needs to be included in this type of model, most significantly time variation. Current slab models (Kawata and Irvine 1975; Kawata 1979,1982) are thermal equilibrium models and do not account for the temporal variation of the insolation and thermal environment of the particles. Preliminary results of work in progress (Cuzzi, unpublished) indicate that the observed form of the tilt angle variation may also be accounted for by the inclusion of such effects into slab ring models. Furthermore, the variation of temperature with viewing angle observed in the rings by Pioneer 11 (Froidevaux and Ingersoll 1981) is most naturally understood in terms of a many-particle-thick layer. It may thus be premature to rule out vertically homogeneous many-particle-thick models.

It is clear that the infrared data and thermal models are approaching a stage of maturity which will soon allow significant constraints to be placed on ring structure. At present, the many-particle-thick model (for the C Ring at least) is slightly favored. More detailed analysis of spacecraft and ground-based observations is certainly indicated.

Somewhere between 100 μm and 1 mm wavelength, the ring brightness temperature (Fig. 32) drops sharply below blackbody behavior, and long-wards of 1 cm wavelength the rings behave as a nearly perfect diffusing screen, reflecting the planetary emission and the cold of space (Cuzzi and Van Blerkom 1974; Schloerb et al. 1979a,b,1980; Cuzzi et al. 1980). The high particle reflectivity and low emissivity at centimeter wavelengths is, of course, precisely the behavior which produces the large reflectivity to terrestrial radar, and the brightness at $\lambda \gtrsim 1$ cm is relatively insensitive to material refractive indices in the range characterizing dielectric, icy material (Cuzzi et al. 1980). The intermediate spectral range from 100 μm to 1 cm (see Table VIII and Fig. 32) contains crucial information on material properties, as it is the most sensitive to refractive indices. It is also, unfortunately, the most difficult to treat theoretically. To a fair degree of accuracy, the thermal emission begins to appear at a wavelength of ~ 1 cm and increases as $\sim 1/\lambda$ (see Fig. 33) due to the nature of emission from an optically thick slab of high-albedo particles (Cuzzi et al. 1980). For such low-loss particles, the single-scattering albedo $A \sim 1 - K_\lambda c_p$ is about unity (Irvine and Pollack 1969) where $K_\lambda(\text{cm}^{-1})$, proportional to $1/\lambda^2$ is the absorption coefficient (Whalley and Labbé 1969). Because the emission from a slab of such scatterers is proportional to $(1 - A)^{1/2}$ (Horak 1950), the emission is proportional to $1/\lambda$. Also, note that a decrease in K_λ by some factor is equivalent to an increase in particle size by the same factor.

Several recent results for the thermal component of ring brightness are shown in Fig. 33. It is apparent that the permitted particle sizes inferred are substantially smaller than observed by Voyager (Sec. V). This may be due to several reasons. First, the absorption coefficient used for the calculations of Fig. 33 was $K_\lambda = 9.5 \times 10^{-4} \lambda^{-2}$ (cm) for solid ice at 100 K, from extrapolation of theoretical curves by Whalley and Labbé (1969). Recent data indicate

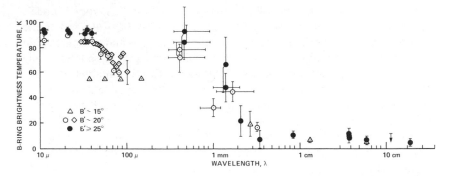

Fig. 32. Brightness temperature of the rings from 10 μm to 10 cm wavelength. (See also Table VIII.) Solar elevation > 25°: (●); solar elevation ≃ 20°: (○)(◇); solar elevation ≃ 15°: (△). The transition from essentially blackbody emission ($\lambda \lesssim 100$ μm) to essentially pure reflection ($\lambda \gtrsim 1$ cm) occurs because of a combination of particle composition (low-loss dielectric) and size distribution (primarily cm to m in radius).

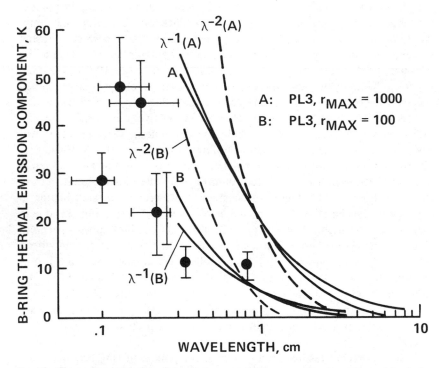

Fig. 33. Thermal emission from Saturn's rings at microwave wavelengths, with model calculations. Curves A and B are model calculations for ice particles distributed in power laws of the form $n(r) = n_0 r^{-3}$, with different upper-size cutoffs r_{max} in cm (adapted from Cuzzi et al. 1980). The data points are total ring brightness with a reflected component subtracted out. It appears that the data imply a λ^{-1} dependence rather than a λ^{-2} dependence, as explained in the text.

that K_λ for solid ice near 3 mm wavelengths at 85 K is $\sim 4.3 \times 10^{-4} \lambda^{-2}$ (Mishima et al. 1983). This effectively increases the size labels on the curves by about a factor of two (Epstein et al. 1984). If the particles are porous, i.e., more frosty than icy, the bulk absorption coefficient will decrease further (Campbell and Ulrichs 1969), and the representative sizes commensurately increase. Other phases of ice, such as amorphous ice, may be even less absorptive, as discussed above (see *Particle Composition* in Sec. VI.A). The range of uncertainty in these parameters has been cited by Epstein et al. (1980,1984) and Muhleman and Berge (1982) as capable of allowing the 3-mm observations to be consistent with ice particles following the Voyager size distribution.

Exactly where the spectrum turns over at shorter wavelengths in the far infrared is unknown. Broadband observations seem to imply that the spectrum remains flat out to ~ 400 μm, before particle-size-related effects decrease the emissivity. This would imply a low fractional area of particles smaller than ~ 1 mm, consistent with results from measurements of ring particle eclipse cooling (Froidevaux et al. 1981; see also Morrison 1977b). Also the small fraction of micron-sized particles implied by Voyager imaging observations at high phase angles would be insufficient to explain a significant emissivity variation at far infrared wavelengths. However, several narrowband studies of the rings near 100 μm (see Fig. 32) seem to show a significant decrease in brightness temperatures, which could be produced by a large fraction of particles smaller than 1 mm. This specific behavior may be due to a surface-grain-size effect rather than independent particles (see Simpson et al. 1981). For surface grain radii r_g typically ~ 30 μm (Pollack et al. 1973), strong scattering resonances could occur for $2\pi r_g \lambda^{-1} \sim 1$–10 at 100 μm wavelength, just as observed for the ring particles as separate entities at centimeter-to-meter wavelengths. Further observations, preferably spatially resolved, in the 100 μm to 1 cm region would be desirable.

In summary, the rings reveal their true physical temperatures at 10 to 30 μm wavelength. There is evidence from the tilt-angle variation of 20 μm brightness for a more complex situation than a simple, static, homogeneous slab of particles, but there is no single preferred explanation yet. Thermal emission decreases sharply at wavelengths larger than a few hundred microns, due to the increasing particle albedo and consequently decreasing emissivity. The rings do not become transparent (i.e., optically thin) at these longer wavelengths, but rather become an opaque diffusing mirror, reflecting the planetary emission and the cold of space.

VII. CONCLUSIONS AND OPEN QUESTIONS

A. New Findings

In the past several years we have made considerable progress in our knowledge of the physical nature of Saturn's rings. The following is a brief summary of these advances.

Ring Opacity Measurements.   The transmission of Saturn's rings has been measured in the visible by the Voyager imaging system, in the ultraviolet by the Voyager UVS and PPS, and in microwave wavelengths by the Voyager 1 radio science investigation. The resolution of these measurements ranges from 0.1 to 10 km: 100 to 10,000 times superior to that available from the ground. Entirely new and unexpected phenomena are visible in these data.

Ring Microstructure.   Saturn's main rings contain smaller features whose size ranges down to the limit of our resolution. Although many of these smaller regions have been called ringlets, this nomenclature is perhaps too restrictive, implying persistence for these features that is unproven. This fine structure is not yet explained, and may well be dynamic. Simple calculations of viscous interactions in the rings give lifetimes very much shorter than the age of Saturn for such features.

Waves.   Voyager observations give snapshots of waves propagating in the rings. Many density waves have been observed, along with several inclination (bending) waves (see Table III). These observations verify theoretical predictions and have inspired nonlinear study of ring dynamics. The processes of growth and damping are visible in the wave structure.

Physical Parameters of the Ring System.   Analysis of the observed waves has given estimates of the surface mass density and viscosity of the rings, along with verifying an absolute distance scale for the ring system. Comparison with the measured optical depths from stellar occultation gives a mass extinction coefficient $\kappa = \tau/\sigma \sim 0.01$ cm^2 g^{-1} for the measured regions. Estimates of the viscosity of the rings range from a minimum value near that of a perfect monolayer up to values consistent with the random motions of particles pumped up through the action of density waves in the outer A Ring.

Mass of the Rings.   Several independent methods have given an estimate of the total ring mass $M_R \sim 5 \times 10^{-8}$ M$_S$ (M$_S \equiv$ Saturn's mass). All these methods involve an extrapolation from local measurements, that is especially uncertain in the B Ring which contains most of the ring mass. The mass of the rings is close to the mass of the small satellite, Mimas.

New Rings.   Pioneer Saturn and Voyager, along with groundbased observations during the ring-plane crossing of the Earth in 1980 have identified the new rings D, E, F, and G to be added to the classical rings A, B, C (see Table I). The origin, nature, and future of these rings is less well known (see Sec. VI).

Size of Ring Particles.   The ring particle size distribution is quite broad, ranging from ~ 1 cm to ~ 5 m radius (Secs. V and VI). In some parts of the rings, perhaps 10% of the particle cross section is contributed by particles as small as several microns. Moonlets (particles as large as 10 km) are quite rare in gaps where they would be easily visible, although one or more might be found in the Encke Gap. The size distribution is consistent with a power law of index -3, as concluded by pre-spacecraft analyses.

Albedo of Ring Particles and Its Variation.   The photometric properties of the ring particles are now much better known because of (1) the greater range of angles observed by the spacecraft flybys, and (2) the more exact knowledge of ring optical depth from occultations. The A and B Ring particles are quite bright and strongly backscattering, consistent with cm-sized and larger chunks of ice. The albedo is much lower in the C Ring and Cassini Division and may vary substantially within the A and B Rings.

Spokes and Eccentric Ringlets.   These unexpected features are yet to be totally explained. Spokes are visible because of the micron-sized particles they contain and appear to have some connection with the planetary magnetic field. Eccentric ringlets may be contained by as yet undiscovered moonlets.

Dust.   Several regions in the rings show strong evidence of small particles: spokes, rings E, F, G, and albedo variations in ring B. This material has a short lifetime in the ring environment. The existence of such material allows a coupling between the magnetosphere and the observable rings. Some interesting interactions have been noted.

B. Open Questions

Vertical Thickness and Vertical Structure of the Rings.   How thin are the rings? The classical interpretation is that the rings' opposition effect is due to mutual shadowing among particles in a ring many particles thick. Direct observations of ring edges provide only an upper limit to the ring thickness of 100 to 200 m (Lane et al. 1982*b*; Marouf and Tyler 1982) which is comfortably larger than the size of a single particle. Dynamic expectations are that the loss of energy in collisions among the ring particles damps out vertical motions, leading to a monolayer. Indirect measurements of particle velocities inferred from density and bending waves generally indicate thickness larger than a monolayer.

A resolution to this conflict may lie in the fact that the particle size distribution is quite broad. Cuzzi et al. (1979) show that the larger particles in such a distribution have low velocity and are essentially a monolayer, while the smaller particles (with comparable random velocities and vertical excursions via their gravitational scatterings by the large particles) fill a layer much thicker than their own size. The result is that most of the mass of the rings is essentially in a monolayer while the majority of scattering cross section is distributed through a larger volume. Given the new constraints from spacecraft data, it is an open question whether this resolution is possible. Work is now needed both in the dynamics of a distribution of particle sizes and the light scattering from such a medium.

Dynamics.   Several important questions remain. What is the cause of the observed microstructure? How can we reconcile narrow ringlets and even the broad classical rings with the exceedingly short time scales for ring spreading due to interparticle collisions? Why have the transfers of angular momentum (observed in the waves in the rings) not pushed the smaller inner satellites further from Saturn?

Particle Composition.   Despite all the spacecraft advances, we know little more about the identity of the ring particles. They have water ice on their surfaces. Most of them cannot be predominantly silicate. The mass density is not far from 1, from the comparison of ring mass estimates from occultations (see Secs. III and V). Variations in albedo may imply composition differences. Since the Voyager photometry and polarimetry analyses are unfinished, new findings on the particle composition may come from future studies at millimeter wavelengths, in addition to space telescope observations, and future space missions.

C. Origin and Evolution of the Rings

This is one of the ultimate questions, but progress has slowed down while the new burst of information is being digested. The majority view is still that the rings are detritus not swept up or accreted into satellites because of the tidal influence from Saturn. Thus, the rings of Saturn provide a remnant of the early solar system and the ring particles have not suffered accretion and heating (see, e.g., Safronov 1984). The fact that tidal forces have stopped accretion means that the rings are like a snapshot of the solar system as it passed through an equivalent stage. Further, the dynamical processes now active in the rings were important at that stage in the early solar system. For both these reasons knowledge of the current rings is quite relevant to our understanding of the solar system's evolution. Unfortunately, how current data will advance our understanding of the ring's evolution is still murky.

Voyager data have also enhanced the appeal of a competing theory for ring origin: the impact theory (Pollack et al. 1973; Pollack 1975). In this theory, the ring material is all that remains of larger satellites shattered by meteor bombardment. Shoemaker (1982) extrapolates from the cratering history of the outer Saturnian satellites that the inner satellites must have been broken to pieces several times and reaccreted. Perhaps the rings themselves are just some of the debris which did not accrete afterwards. Support for this idea is the variability of particle optical properties, especially color within the rings

(Cuzzi 1982). This variability could be the result of different source bodies for different regions of the rings.

We note below several questions that, if answered, would improve our understanding of the evolution of Saturn's rings:

1. Are the rings young or old? Almost everyone believes the rings to be as old as the rest of the solar system, but then why are the rings so compact, and why have the inner Saturn satellites not moved out, conserving the angular momentum drawn from density waves?
2. Are the rings thick or thin? Perhaps they are a little of each; still, this is a basic input to every model of the rings.
3. What are the sources and sinks for ring particles, random kinetic energy, and angular momentum? Not only theorists but spacecraft observers should search their minds and data for the answers.

The rapid pace of observational advance in the past few years has sorely strained the system. Terrile (public communication at Voyager 2 Press Conf. 1981) has compared it to drinking from a fire hose. In the coming years, unencumbered by new data from space missions, we can hope to digest the new information, pull together the data from Earth and space, answer some of the current burning questions (and some others, too, that we are now too ignorant to deem important), and prepare the outlines of the next space missions. These new missions, as those before, will carry the questing human intellect outward to view at close range one particular aspect of the beauty of the universe, Saturn's rings.

Acknowledgments.   We have had helpful reviews from J. Pollack, R. Greenberg, and W. Irvine. The writing of this chapter was supported by the Voyager project and the NASA Planetary Atmospheres Program. J. B. H. and C. C. P. are now affiliated with the Lunar and Planetary Laboratory at the University of Arizona.

ELECTRODYNAMIC PROCESSES
IN THE RING SYSTEM OF SATURN

D. A. MENDIS, J. R. HILL
University of California, San Diego

W.-H. IP
Max Planck Institut für Aeronomie

C. K. GOERTZ
University of Iowa

and

E. GRÜN
Max Planck Institut für Kernphysik

A variety of novel magnetospheric processes are associated with the presence of an extended ring system of particulate matter within the Saturnian ring system. Being immersed in a radiative and plasma environment, the ring material is electrically charged to some potential, which may be positive or negative. While this electrical charging may have only physical effects on the larger bodies, it may also affect the dynamics of the smaller grains. While the rings can also act both as sources and sinks of the magnetospheric plasma, the relative motion between the charged dust and the surrounding plasma constitutes a new type of dust-ring current having interesting consequences for the upper ionosphere, magnetosphere, and the rings themselves. We discuss a number of recently observed ring phenomena including Voyager 1 and 2 observations of the rotating near-radial spokes in the B Ring; waves and braids of the F Ring; and discrete episodic bursts of broadband radio emission, claimed by some authors to originate in the ring. In addition, we discuss several other phenomena, including the origin and evolution of the diffuse E Ring and G Ring (which appear to be composed of fine dust), as well as the existence of a number of sharp discontinuities in the main ring system, within the context of gravito-electrodynamics of charged dust in the magnetosphere.

This is indeed a new and immature field of research. While the single-particle theory of dust particle motion is reasonably well understood, the role of collective dust plasma effects is still under discussion. If electrodynamic processes are responsible for certain phenomena observed in the ring system at the present time, it is possible that they may have also played a role in the past during its formation and early evolution. We conclude by discussing some prevailing ideas about the role of electrodynamic processes in the cosmogony of the ring system.

I. INTRODUCTION

The presence of an extended ring system of particulate matter within the inner Saturnian magnetosphere introduces a plethora of new magnetospheric processes. By virtue of being immersed in a radiative and plasma environment, the ring material (embedded moons, boulders, dust, etc.) will be electrically charged to some charge (positive or negative). While this charging has only physical effects on the larger bodies, it can also affect the dynamics of the smaller (micron- and submicron-sized) grains.

The relative motion between the charged dust and surrounding plasma constitutes a novel dust ring current, which under a variety of circumstances to be discussed below may close through the ionosphere via Birkeland currents, causing interesting effects in the upper ionosphere, the magnetosphere, and in the rings themselves.

The rings can also act both as sources and as sinks of the magnetospheric plasma. In fact, the narrow inner ring of Jupiter was first detected as an absorption feature in the measurements of energetic trapped particles by Pioneer 11 (Fillius et al. 1975) and interpreted as such by Acuña and Ness (1976). Similar absorptions of the energetic trapped particles in Saturn's magnetosphere were also observed by detectors on Pioneer 11, and Voyagers 1 and 2 (e.g., see Fig. 1).

In this chapter we will discuss a number of recently observed ring phenomena in the above context. These include the Voyager 1 and 2 observations of the rotating near-radial spokes in the B Ring; the waves and braids of the F Ring; and the discrete episodic bursts of broadband radio emission, termed Saturn electrostatic discharges (SED) which have been claimed by some authors to originate in the ring. We will also discuss several other phenomena, including the origin and evolution of the diffuse E and G Rings (which appear to be composed of fine dust) within the context of gravito-electrodynamics of charged dust within the magnetosphere.

This is a very new and immature field of research. Whereas the single-particle theory of dust particle motion is reasonably well understood, the

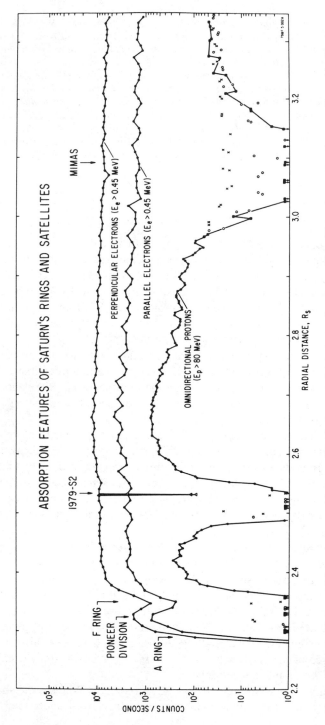

Fig. 1. Absorption features of Saturn's rings and inner satellites in the energetic electron and proton profiles of Saturn's magnetosphere, as observed by Pioneer 11 (figure from Fillius et al. 1980).

physics of a dusty plasma is not, and the role of collective dust plasma effects is still very much under discussion. It may thus seem premature to review the field. However, many papers have been published which propose electro-dynamic processes in the ring system of Saturn as an explanation for other-wise unexplained phenomena. Even though many of these explanations may need revisions at a later date, they are inherently interesting as a basis on which future research may be built.

If electrodynamic processes are responsible for certain phenomena ob-served in the ring system at the present time, it is likely that they may have also played a role in the past, during its formation and early evolution. We will therefore discuss some prevailing ideas about the role of electrodynamic pro-cesses in the cosmogony of the ring system. Grün et al. (1984) have made a general comparative review of the electrodynamic processes in planetary ring systems. That review is complementary to this chapter dealing only with Saturn.

The detailed structure and physical properties of Saturn's ring system, as well as its plasma environment, are discussed in detail elsewhere (see chap-ters by Esposito et al. and Scarf et al.). For the purposes of our discussion here, a schematic picture of the plasma and neutral gas environment of the ring system is given in Fig. 2. The inner plasma torus I ($L \lesssim 7$) is composed largely of O^+ with densities and energies of ~ 100 cm^{-3} and ~ 10 eV, re-spectively, as measured by the Voyager 2 plasma instrument near ring cross-ing at $L \approx 2.7$ (Bridge et al. 1982). Inward of $L \approx 4$ (near the orbit of En-celadus), this plasma torus appears to collapse into a thin sheet about the equatorial plane. Beyond $L \approx 7$ is a much hotter ion torus II ($kT_e \gtrsim 100$ eV; Bridge et al. 1982).

A crucial physical parameter for the discussion that follows is the size distribution of the ring particles. Groundbased radio, radar, and infrared ob-servations all indicate particle sizes in the range ~ 1 cm to 1 m in the main (A and B) ring system (see, e.g., Cuzzi and Pollack 1978; Epstein et al. 1980; see also Ip 1980 for a review). Radio observations by Voyager 1 (Tyler et al. 1981) confirm the existence of the larger (~ 1-m-sized) particles. At least the B Ring must also contain small (micron- and submicron-sized) grains, as shown by the scattering characteristics of the episodic spokes seen in rotation across the inner B Ring by Voyagers 1 and 2 (B. A. Smith et al. 1981,1982). Similar observations of the F Ring (B. A. Smith et al. 1981,1982) indicate that it as well is composed of fine dust.

Infrared observations of the diffuse extended E Ring ($3 \lesssim L < 8$; $10^{-8} \lesssim \tau \lesssim 10^{-6}$) indicate grains with diameters < 5 μm (Terrile and Tokanaga 1980). In addition, during the ring plane crossing of Voyager 2 at $L = 2.88$, slightly outside the G Ring, the plasma wave instrument detected intense im-pulsive noise which has been attributed to spacecraft impact by grains with radii in the range 0.3 μm to 3.0 μm (Gurnett et al. 1983). Considering the proximity of this observation to the G Ring ($L \approx 2.80$), it is plausible that the

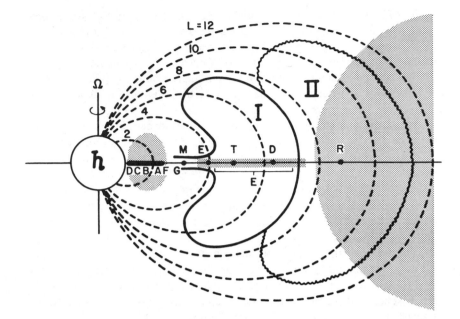

Fig. 2. A schematic meridional cross section of Saturn's magnetosphere in the vicinity of the rings, adapted from Stone and Miner (1982), with the addition of the atomic hydrogen cloud of the main ring system (A B C D F) and the tenuous E Ring (Ip 1982). The kidney-shaped regions I and II represent the inner plasma torus of 0^+ and the adjoining hot ion torus, respectively.

G Ring also contains such small grains. With regard to the innermost D Ring, there are no estimates of the grain size. However, due to its diffuse nature ($\tau \lesssim 10^{-2}$) and its proximity to Saturn ($1.11 \lesssim L \lesssim 1.23$), one suspects that it is largely composed of fine dust.

In the following sections we discuss the various relevant observed phenomena. As a prelude to that, however, we begin with a brief discussion of the general motion of charged dust in Saturn's magnetosphere.

II. THE DYNAMICS AND ORBITAL STABILITY
OF CHARGED DUST

While the motion of the plasma (both thermal and energetic) within a planetary magnetosphere is almost totally controlled by electromagnetic forces, with planetary gravitation playing only a very secondary role (e.g., causing slow azimuthal drifts), the motion of the larger bodies such as satellites is overwhelmingly controlled by planetary gravitation, with gravitational perturbations by neighboring satellites playing a secondary role. Even for cm- and mm-sized grains that populate the rings, the electromagnetic effects are negligible in relation to gravity for any acceptable values of the charge. It is only when we consider grains of micron size (0.1 μm to 5 μm) that these two forces become comparable for plausible values of the grain charge, as we will see presently.

We have already pointed out that observations indicate the presence of fine (micron- and submicron-sized) dust in the F, G, and E Rings, the spokes, and possibly also in the D Ring. Since these grains, by virtue of being within the planetary plasma sphere, are likely to be charged to some electrostatic potential, it is important to study both the charging processes and the dynamics of grains once they are charged.

The equation of motion of a charged dust grain in the planet-centered inertial frame is given, with the usual notation, by

$$\ddot{\mathbf{r}} = \frac{q(t)}{m}\left(\mathbf{E} + \frac{\dot{\mathbf{r}}}{c} \times \mathbf{B}\right) - \frac{GM_p}{r^3}\,\mathbf{r} + \mathbf{F} \tag{1}$$

where $\mathbf{F}$ represents forces associated with the collisions with photons, plasma, and other grains.

Within the rigidly corotating regions of the planetary magnetospheres, where the ring systems are observed,

$$\mathbf{E} = -\frac{1}{c}\,(\mathbf{\Omega}_p \times \mathbf{r}) \times \mathbf{B} \tag{2}$$

where $\mathbf{\Omega}_p$ is the angular velocity of the planet. This is, of course, strictly applicable only when the Debye spheres of neighboring particles do not intersect, otherwise the electric fields of neighboring particles will also have to be included in $\mathbf{E}$. In this case the charge that can be acquired by the individual grains will be decreased (see below) (Grün 1982).

The condition for the nonintersection of the Debye spheres may be written as $\eta = \lambda_D/d < 1$, where λ_D is the Debye length and d is the mean separation of the grains. It is easily shown (Grün et al. 1984) that

$$\eta \approx 2.3 \times 10^5 \sqrt{\frac{E_e(\text{eV})}{n_e}\left(\frac{\tau}{hR_g^2(\mu\text{m})}\right)^{1/3}} \tag{3}$$

where $E_e(\text{eV})$ and $n_e(\text{cm}^{-3})$ are, respectively, the electron energy and density, while τ and $h(\text{cm})$ are, respectively, the optical depth and the scale height of the grain distribution, and $R(\mu\text{m})$ is the radius of the grains.

Taking $E_e \approx 10$, $n_e \approx 10^2$, and $R_g \approx 1$ for the E Ring, the F Ring, and the spokes, we get

$$\eta \approx 7.3 \times 10^4\left(\frac{\tau}{h}\right)^{1/3}. \tag{4}$$

For the E Ring, $\tau \approx 10^{-6}$ and $h = 10^5$ km, giving $\eta \approx 0.3$; for the F Ring, $\tau = 0.1$ and $h \approx 1$ km giving $\eta \approx 700$; and for the spokes $\tau = 0.1$ and $h \approx 30$ km, giving $\eta \approx 200$ (Grün et al. 1984). Consequently, only in the case of the E Ring is the condition of the nonintersection of the Debye spheres met.

When several dust particles are present, one must consider the screening of the electric field due to the presence of other dust particles. Goertz (1983b) has argued that this causes a significant reduction of the charge on each dust particle. He considers an ensemble of dust grains (radius R_g, density $N = \tau/\pi\, R_g^2 h$) in a plasma (density n, temperature $T_e = T_i = T$). In static equilibrium the surface potential ϕ of each grain is given implicitly by the condition of zero total current

$$F_e(\phi) - F_i(\phi) - F_\nu(\phi) = 0 \qquad (5)$$

where $F_e(\phi)$ is the plasma electron flux, $F_i(\phi)$ the plasma ion flux, and $F_\nu(\phi)$ the photoelectron flux at the surface of the dust grain. For large plasma densities F_ν can be neglected and $\phi = (kT/e)\ell\mathrm{n}(m_e/m_i)$. Note that Eq. (5) determines the surface potential but *not* the charge on a dust grain. The potential ϕ is not due to the charge on the grain alone but due to the Debye screened potentials of all other grains as well

$$\phi = \frac{Q}{R} + \sum \frac{Q_i}{r_i}\, e^{-r_i/\lambda_D} = \frac{Q}{R}(1 + \gamma) \qquad (6)$$

where r_i is the distance to the ith grain and $\lambda_D = (kT/4\pi ne^2)^{1/2}$ is the plasma Debye length. We see that $Q = \phi R/(1 + \gamma)$, i.e., less than the free space value $Q_0 = \phi R$. To estimate the correction factor γ, we assume that the dust particles all have the same charge and size, and that they can be represented by a uniform density distribution. Then

$$\gamma = 4\pi R \int_d^\infty N\, e^{-r/\lambda_D}\, r\, dr$$

$$= 4\pi N R \lambda_D^2\, e^{-1/\eta}\left(1 + \frac{1}{\eta}\right) = \frac{4\tau\lambda_D^2}{hR}\, e^{-1/\eta}\left(1 + \frac{1}{\eta}\right). \qquad (7)$$

In the E Ring we have $\gamma \sim 10^{-7}$ and the single-particle gravito-electrodynamic theory is applicable. However, in the F Ring $\gamma = 1600$ and in the spoke region $\gamma = 50$. In these regions the single-particle gravito-electrodynamic theory based on the free-space charge Q_0 is not relevant. However, for extremely dense plasma conditions, like that described by, e.g., Goertz and Morfill (1983), the value of γ may be small. Also the value of the optical depth affects γ and for sufficiently optically thin regions the correction due to screening may be neglected. An important factor is the charge-to-mass ratio

Q/m of a grain. It is relatively easy to show that to obtain the same ratio one only needs to multiply the radius by a factor $\rho = 0.5\ \gamma[(1 + 4/\gamma^2)^{1/2} - 1]$ which reduces to $\rho = 1/\gamma$ for large values of γ.

Whipple et al. (1983) have disputed the expression for the grain potential ϕ (Eqs. 6 and 7) obtained by Goertz (1983). They too have considered a grain in a dusty plasma, subject to appropriate boundary conditions, and have concluded that the capacitance of the grain increases rather than decreases from its free-space value. According to these authors the grain potential ϕ is given, instead, by

$$\phi = \frac{Q}{R}\ F = \frac{Q}{R}\ (1 + \gamma) \tag{8}$$

where

$$(1 + \gamma) = F \tag{9}$$

$$= \frac{(R/\lambda + \ell/\lambda - 1)\exp(\ell/\lambda) + (R/\lambda + \ell/\lambda + 1)\exp(-\ell/\lambda) - 2R/\lambda}{(R/\lambda + \ell/\lambda - 1)(1 + R/\lambda)\exp(\ell/\lambda) + (R/\lambda + \ell/\lambda + 1)(1 - R/\lambda)\exp(-\ell/\lambda)}$$

where ϕ is approximately half the intergrain distance.

This disagreement between the results of Goertz on the one hand and Whipple et al. on the other underscores the point made in the introduction that the collective effects in a dusty plasma are still very much under discussion.

In the case of Saturn, where the magnetic moment and the spin vector are known to be parallel to within $0°.8$ (Ness et al. 1981,1982) and may even be strictly parallel within the observational uncertainties, assuming that they are indeed strictly parallel, it is easy to show when $\mathbf{F} = \mathbf{0}$ (Mendis et al. 1982; Northrop and Hill 1983a) that Eqs. (1) and (2) admit circular orbits in the equatorial plane moving with angular velocity Ω_G given by

$$\Omega_G = \frac{\omega_0}{2}\left[-1 \pm \sqrt{1 + 4\left(\frac{\Omega_p}{\omega_0} + \frac{\Omega_K^2}{\omega_0^2}\right)} \right] \tag{10}$$

where Ω_K is the local Kepler angular velocity, and ω_0 is given by

$$\omega_0 = -\frac{QB}{mc}. \tag{11}$$

Eq. (8) shows that two different motions are possible for a given grain. The plus sign in front of the radical corresponds to direct (or prograde) motion, while the negative sign corresponds to indirect (or retrograde) motion, for a negatively charged grain. For a positively charged grain, the minus sign in front of the radical gives a prograde motion while the plus sign gives a prograde or retrograde motion depending on the value of ω_0.

The values of χ (the ratio of the electric force to the gravitational force, $|F_e/F_G|$ in the nonrotating frame) for negatively charged grains in such prograde circular orbits, having different sizes and potentials at three different radial distances from Saturn, have been calculated by Mendis et al. (1983) neglecting, however, the correction factor $(1 + \gamma)$. Thus, the calculations of Mendis et al. (1983) are relevant only in the optically thin E and G Rings. These results are shown in Fig. 3. The case $L (= r/R_p) = 1.72$ corresponds to the inner boundary of the observed spokes, while $L = 2.33$ corresponds to the position of the F Ring and $L = 5.0$ corresponds to a position within the broad E Ring, which extends from about $L = 3$ to $L = 8$. If one wishes to include the screening effect, the size scale should be multiplied by the factor ρ.

It is seen that, other things being equal, χ is smallest when $L = 1.72$. This is due to its proximity to the synchronous orbit ($L = 1.86$) where $\chi = 0$ even when the grain is charged. Clearly, electromagnetic forces are most important for submicron- and micron-sized grains ($0.1~\mu$m $< R < 0.5~\mu$m) in the E Ring ($0.2 < \chi < 2.5$) for reasonable values of ϕ.

In the F Ring and the spoke region, the importance of electromagnetic effects is debatable. According to Goertz (1983), $\gamma >> 1$ in these cases, and consequently electromagnetic effects are unimportant even for submicron-sized grains, whereas according to Whipple et al. (1983), $F \approx 1$ (and $\gamma \approx 0$), making electromagnetic effects important for submicron-sized grains.

We next consider the stability of charged grain orbits. If a charged dust grain, moving in a circular orbit in the equatorial plane of the planet (whose magnetic moment and spin are assumed to be strictly parallel) is subject to a small perturbation in that plane (e.g., by the gravitational tug of a nearby satellite), Mendis et al. (1982) show that the grain will perform a motion described as an elliptical gyration about a guiding center which is moving uniformly in a circle with the angular velocity given by Ω_G (Eq. 8). The gyration frequency ω about the guiding center is given by

$$\omega^2 = \omega_0^2 + 4\omega_0\Omega_G + \Omega_G^2. \tag{12}$$

If a and b are the semimajor and semiminor axes of this ellipse, it is shown that

$$\frac{b}{a} = -\frac{\omega}{2\Omega_G + \omega_0} \tag{13}$$

with the minor axis aligned in the radial direction.

Not all the grain orbits are radially stable. Those which are, must satisfy the condition $\omega^2 > 0$. One can use Eqs. (8), (10), and (11), together with this condition, to obtain the stable orbits at any given distance from the planet. The classes of stable and unstable orbits within the rigidly corotating portion

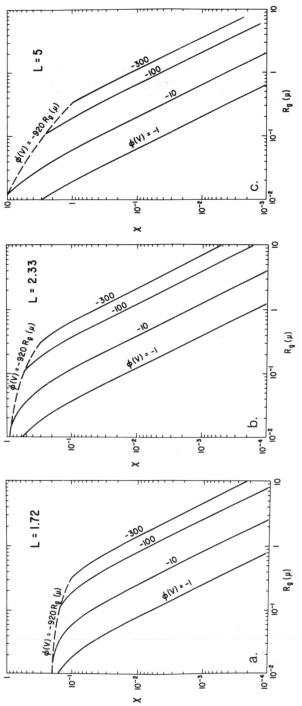

Fig. 3. The variation of χ (the ratio of the electric to the gravitational force) on grains of different sizes and different potentials at various positions within Saturn's magnetospheres, neglecting the screening effect of other dust particles ($\gamma = 0$). Panel (a) corresponds to L ($= r/R_g$) = 1.72, which is the inner boundary of the spokes; (b) corresponds to L = 2.33, which is the center of the F Ring; and (c) corresponds to L = 5.0 (a position within the broad E Ring (figure from Mendis et al. 1983).

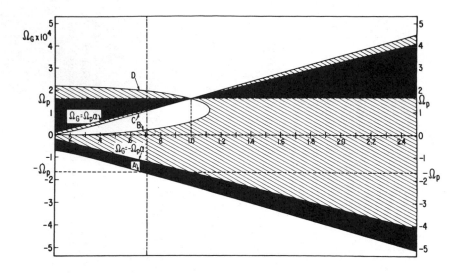

Fig. 4. The variation of Ω_G with α (= Ω_K/Ω_p). The curves marked A, B, C, D indicate the values of Ω_G when $\omega^2 = 0$, for various values of α. The shaded regions are where $\omega^2 > 0$, and the unshaded regions are where $\omega^2 < 0$. The dark shading corresponds to negative particles while the light shading corresponds to positive particles. The lines marked $\Omega_G = \pm\Omega_p\alpha$ represent Kepler particles, while the dashed line marked $\Omega_G = \Omega_p$ represents the corotating particles. The values of α = 0.1, 0.7, 1, and 2.4 correspond to the limit of rigid corotation in Saturn's magnetosphere, the F Ring, the synchronous orbit, and Saturn's surface, respectively (figure from Mendis et al. 1982).

of Saturn's magnetosphere are exhibited in a general fashion in Fig. 4. Here α(= Ω_K/Ω_p) is the independent variable and Ω_G is the dependent variable. It is seen, for instance, that at the distance of the F Ring, negative particles of all sizes, from the smallest (corotating) particles to the largest (essentially Keplerian) particles moving in the prograde sense, are stably trapped. Interestingly, it is seen that there are several other distributions of particles that can also be stably trapped there.

Figure 4 has been used by Hill and Mendis (1982a) to fix the polarity of the grains that compose the B Ring spokes. This point will be discussed in Sec. V.

Northrop and Hill (1982) have also studied the stability of negatively charged dust grains in the equatorial plane of Saturn subject to perturbation normal to the ring plane. They find that a critical radius exists which depends on the specific charge of the grain, such that grains interior to that are unstable to normal perturbations. They have used this fact to discuss the existence of a sharp boundary between the inner and outer parts of Saturn's B Ring. We discuss this point in some detail in Sec. VI.

III. MAGNETO-GRAVITATIONAL RESONANCES

From Fig. 4 it is clear that negatively charged prograde grains outside the synchronous orbit move with an orbital angular velocity Ω_G which is larger than the Kepler angular velocity at that distance. Since Ω_G depends on ω_0, and therefore on the grain size R_g (in μm) via the relation

$$\omega_0 = -4.40 \times 10^{-8} \, \phi(v)/\rho_g \, R_g^2 \, (1 + \gamma) \tag{14}$$

for a given potential, there would be grains of a certain size moving with the same angular speed Ω_s of a satellite interior to the grain orbit. This means an exact 1:1 orbit-orbit resonance with such a satellite. A similar situation clearly does not arise in the purely gravitational case. This magneto-gravitational resonance would, for instance, arise between a certain size grain in the F Ring and its nearby inner satellite 1980S27.

It must be stressed here that, unlike a pure gravitational resonance, which affects particles of all sizes equally, the magneto-gravitational resonance picks out a particular grain size R_{gc} for a given potential ϕ. Of course, grains with sizes close to R_{gc} (on either side of it) will also be strongly affected because their angular velocities will be close to that of the perturbing satellite, and will therefore remain in the vicinity of that satellite for long periods of time. If we consider such a particle with a gyration frequency ω about a guiding center moving with angular velocity Ω_G, it can be shown (Mendis et al. 1982) that the grain will move in an undulating orbit having a wavelength λ given by

$$\lambda = \frac{2\pi r}{\omega} \, (\Omega_G - \Omega_s) \tag{15}$$

where Ω_s is the angular velocity of the perturbing satellite and r is the planeto-centric distance of the guiding center of the grain. Furthermore, as each successive grain of the same size in the ring moves over the satellite, it will be subject to the same perturbation and will therefore follow the same path as its predecessor in the frame of the satellite. Consequently, all the grains will move in phase to form a wavy pattern with the wavelength λ in the frame of the perturbing satellite. It has been proposed (Mendis et al. 1982) that the waves observed in the F Ring (see Fig. 5) are formed in this way.

Mendis et al. (1982) also obtained a dispersion relation between λ and ω for these waves, from which λ can be expressed as a function of the grain radius R_g for a given value of ϕ. The case when $\phi = -38$ V, which is calculated by assuming $kT_e = 10$ eV and $n_e = 10^2$ cm^{-3}, is shown in Fig. 6. (For detailed discussions of the calculation of the electrostatic potentials of isolated grains in space see the papers by Wyatt [1969], Mendis [1981], and Whipple [1981].) A wavelength of $\lambda = 1500$ km corresponds to $R_g = 0.51 \, \rho$ or $0.55 \, \rho$, whereas $\lambda = 5000$ km corresponds to $R_g = 0.39 \, \rho$ or $0.65 \, \rho$. The value calcu-

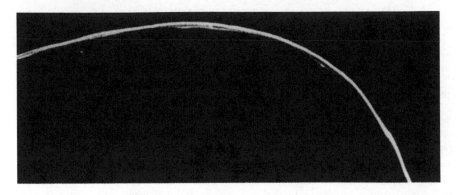

Fig. 5. The waves and braids in the F Ring, as observed by Voyager 1 (figure from B. A. Smith et al. 1981).

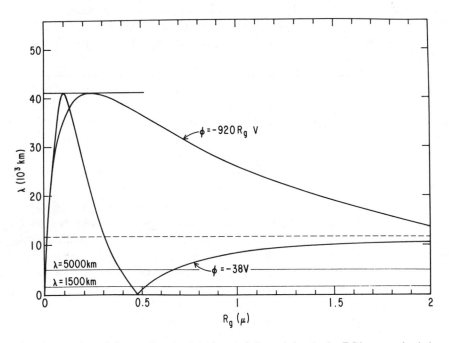

Fig. 6. Variation of the wavelength of the locus of charged dust in the F Ring near the 1:1 magneto-gravitational resonance with the inner shepherding satellite 1980S27. One curve represents the case when the grains are charged to the equilibrium potential of $\phi = -38$ V. The other curve represents the case when the grain potential is limited by field emission. The limit of the field emission itself is -920 V for a 1 μm grain and is directly proportional to the radius in the range of sizes considered. The broken line ($\lambda = 11,400$ km) represents the asymptote to which the curves tend for large values of the grain radius R_g. The two solid lines ($\lambda = 1500$ km, $\lambda = 5000$ km) correspond to two quoted values of the uncertain wavelength of the F Ring waves (figure from Mendis et al. 1982).

lated by Mendis et al. (1982) for Q in this case may be much too negative, due to the intersection of the grain Debye spheres (see, e.g., Grün 1982; Goertz 1983b; also see Whipple et al. 1983).

Attempts have also been made to explain the F Ring waves and braids in purely gravitational terms. Dermott (1981) proposed that the braids are caused by the larger of the two shepherding satellites, which has a single first-order resonance within the ring. As grains on opposite sides of this resonance oscillate radially 180° out of phase, a standing wave pattern of equally spaced loops in the frame of the perturbing satellite is formed. Dermott believes that the F Ring's braids are produced in this way. Following Goldreich and Tremaine (1979), Dermott also believes that the narrow F Ring is maintained in equilibrium by the balance of the tidal torques exerted by the two so-called shepherding satellites. This requires that the ratio of their masses be proportional to the ratio of the squares of their distances from the ring—a condition that does not appear to be satisfied. Further, Showalter and Burns (1982) have pointed out that since the F Ring is sufficiently wide and the shepherding satellites sufficiently close, several overlapping first-order resonances would be produced, resulting in the smearing of the two-braid structure obtained by Dermott.

Showalter and Burns (1982) have made a detailed numerical simulation of the evolution of a set of dust particles in the F Ring under the perturbations of the two shepherding satellites, allowing for the fact that the satellites, as well as the ring particles, have eccentric orbits. The results of the simulation are quite interesting; both wavy and clumpy features are produced after a few ring periods. However, they are eventually washed out after some 100 ring periods. Consequently this process does not lead to the formation of persistent waves. Furthermore, as the authors themselves admit, they do not observe any tendency towards braid formation.

In the magneto-gravitational picture with a radial segregation of grain size, one expects, in principle, several intertwined waves of different wavelengths. Two well-defined braids would merely reflect a strongly bimodal size distribution. On the other hand, as pointed out earlier, the application of gravito-electrodynamics, with its single-particle approach to the F Ring, is very questionable, and its quantitative predictions are therefore to be regarded with caution. Consequently, it is fair to say that the processes responsible for the waves and braids of the F Ring remain obscure at the present time.

IV. ISOLATED, ECCENTRIC RINGLETS
AND GYROORBITAL RESONANCES

There are several gaps in the Saturn ring system which contain narrow eccentric ringlets. The first (innermost) one is in the C Ring, at a radius of 87,100 km (1.444 R_S), located in a gap 270 km wide and eccentric by $\sim$ 50 km. The second ringlet, which is 20 km wide, is in a 450 km gap at the inner side of the Cassini Division 117,600 km (1.949 R_S) from Saturn. The third example is in the 340 km wide Encke Division. It is a kinky ringlet eccentric by $\sim$ 150 km and at 133,500 km (2.213 R_S) from Saturn. This ringlet is found near the center of the division in one Voyager 2 image and near the edge in another. Finally, the F Ring is also eccentric, with a 400 km variation in its 140,600 km radius orbit.

All these peculiar ringlets share one property, namely they are located within relatively broad regions of the ring system that are otherwise highly transparent. As a result of being located in such highly transparent regions, where presumably the magnetospheric plasma density is relatively high, we expect the grains that constitute these rings to be negatively charged. In addition, observations from forward-scattered and backscattered light have indicated that the F Ring is composed of micron- and submicron-sized particles, as already discussed. Similar observations by Voyager 1 indicated that the ringlet in the Encke Division is composed of such small grains (B. A. Smith et al. 1981). Presently there is no reliable evidence that the other eccentric rings also have a significant component of fine dust, but Hill and Mendis (1982a) have made such an assumption. In that case they will be affected by electric forces within the magnetosphere, provided that they are electrically charged.

Hill and Mendis (1982a) assumed that individual grains (i.e., they set γ = 0) are immersed in a thermal plasma of density $n_e \approx 100$ cm^{-3} at an energy $kT_e \approx 10$ eV. This gives $\phi \approx -37.6$ V on the sunlit side of the planet, as discussed earlier. However, as the grains move into the shadow, the potential changes by $\sim$ 0.5 V. Hill and Mendis (1982a) have shown that this change in the potential results in a distortion of the ring's otherwise circular shape. The ringlet in Encke's Division has been modeled by starting with a circle of 0.4 μm dust grains. Its shape after 20 hr and 30 hr is shown in Fig. 7. Note, of course, that these curves do not correspond to the real shapes of the rings, which are very close to circular; a scale has been used which greatly exaggerates the noncircularity. Each grain was initially in an equilibrium circular orbit (with v_g = 1.019 v_{Kepler} and a potential of -37.6 V). The orbital radius of these grains varies between 133,435 km and 133,565 km (Δr = 130 km). Due to the screening effect, Fig. 7 really applies to much smaller particles (R_g $\sim 3 \times 10^{-3}$ μm) and whether or not such small particles exist in the F Ring is questionable.

Hill and Mendis (1982a) have also argued that, due to sputtering by energetic particles, the physical lifetime of these small grains is small ($\lesssim 10^3$ yr). Consequently, they need to be continuously supplied into the gaps, and it seems that uneven charging of the grains in the main ring system could do just that. In the first instance, the continuous collisional grinding of the larger bodies there, as well as perhaps the sporadic electrostatic blow-off of dust from their surfaces, could provide a continuous supply of small (micron- and submicron-sized) grains within the main ring system. For grains lying close to one of the edges of a gap such as the Encke Division, a small

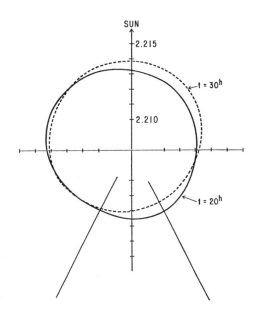

Fig. 7. The loci of grains in the Encke Division whose potential ϕ changes by $\Delta\phi \simeq 0.5$ V as they pass through the shadow of the planet. The solid line and the dotted line, respectively, represent these loci after 20 hr and 30 hr of orbital evolution. The wedge-shaped shadow is a result of the distortion of the radial scale, which is expanded beyond 2.210 R_S so as to magnify the small eccentricity of the grain orbit (figure from Hill and Mendis 1982a).

change in the surface potential, $\Delta\phi \simeq 0.5$ V, as the grains go into the shadow, can cause an eccentricity $\Delta r \geq 100$ km for grains with radii $\leq 0.4\, \rho$ [μm]. This will cause these grains to make an excursion into the gap, and cause them to move in an orbit that will alternately take them away and back towards the edge of the gap, as observed in the Encke Division (see Fig. 7).

Grains with $R_g \leq 0.2\, \rho$ to $0.3\, \rho$ are eventually removed from the ringlet (Hill and Mendis 1982a) because of a new type of "gyroorbital resonance" effect which occurs for grains of certain sizes in this range. For example, when the electron temperature $T_e \simeq 10$ eV, particles with $R_g = 0.160\, \rho$ [μm] have $N (= \Omega_g/\omega) = 2$, while particles with $R_g = 0.103\, \rho$ [μm] have $N = 3$. These resonances produce orbital eccentricities which grow until they are larger than the width of the gap and are thereby removed. Figure 8 shows this in the case of the Encke Division when grain orbits have evolved for two days. Since the grain sizes producing resonances depend on the grain potential, which in turn depends on the ambient electron temperature, an increase of this value to 20 eV will move the resonances up to $0.22\, \rho$ [μm] and $0.14\, \rho$ [μm], respectively. Hill and Mendis (1982a) point out that if such temperature changes, which are likely, do occur with a time scale that is larger than the rotation period of the grain around the planet, then the movement of these

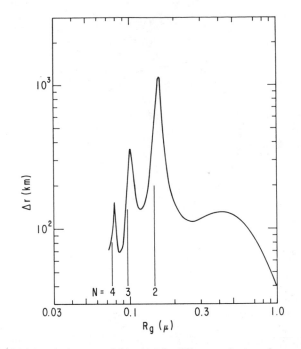

Fig. 8. Encke Division ringlet eccentricities (Δr) for different grain sizes after two days of orbital evolution. The peaks in this curve correspond to the gyroorbital resonance N ($= \Omega_G/\omega$) $= 2$, 3, and 4, and are shifted slightly from the exact values due to the continuous evolution of the orbits (figure from Hill and Mendis 1982a).

resonances across the grain size distribution will remove all the grains with R_g $\leq 0.25\ \rho\ [\mu\mathrm{m}]$.

Furthermore, grains of different sizes are moving with different velocities, and if they are fluffy their collisions would be highly inelastic. Consequently, there would be a tendency to equalize orbits. This so-called negative diffusion (Baxter and Thompson 1973) would result in all the grains being incorporated into orbits near those of the dominant size range. If grains break up during such collisions into fragments with $R_g \leq 0.25\ \rho\ [\mu\mathrm{m}]$, they are removed from the gap in less than two days.

Hill and Mendis (1982a) claim that this process not only exhibits the continuous supply of fine dust into the aforementioned gaps, but it also explains why the size distributions in these gaps are expected to be so sharply peaked. In fact, the narrower the gap, the more sharply peaked this distribution will be. The major drawback of their explanation may be, once again, the neglect of collective effects of the charged dust, as pointed out in Sec. III, and the resulting overestimate of the dust grain charge.

For values of γ obtained from Eq. (7) (Goertz 1983b), ρ turns out to be very small. Consequently, only very small grains would be affected by this resonance. On the other hand, use of Eq. (9) (Whipple et al. 1983) gives $F \approx 1$ and consequently $\rho \approx 1$, so that grains of the order of $R_g \lesssim 0.2$ would be affected, as originally suggested by Hill and Mendis (1982a).

V. FORMATION AND EVOLUTION OF THE SPOKES

As already mentioned in Sec. I, among the most interesting features of the Saturnian ring system, first discovered by Voyager 1 (B. A. Smith et al. 1981) and subsequently confirmed by Voyager 2 (B. A. Smith et al. 1982), are the near-radial spokes. Most of these spokes are confined to the dense central B Ring with inner edge at 1.52 R_S and outer edge at 1.95 R_S. The spokes themselves have an inner boundary at ~ 1.72 R_S and an outer boundary at approximately the outer edge of the B Ring. Against the background of the B Ring they appear dark in backscattered light and bright in forward-scattered light, indicating that they are composed of small (micron- and sub-micron-sized) grains.

A typical spoke pattern is seen in Fig. 9. The spokes exhibit a characteristic wedge shape (clearly apparent in the central spoke of the figure) with the vertex coinciding with the position of the synchronous orbit at ~ 1.86 R_S. Movies have been made showing the formation and dynamical evolution of these spokes (B. A. Smith et al. 1981). They are more easily seen, and seem to be more sharply defined on the morning ansa as they are rotated out of Saturn's shadow.

Based on a simple phenomenological model, which assumes that each spoke is initially formed along a radial line, and evolves as individual particles composing it move at their respective Kepler speeds, Grün et al. (1983) have made a histogram showing the local time at which these spokes are formed. While some spokes which have been observed at the morning ansa form in the shadow region, most of them seem to have formed in the region 3:00 to 7:00 local time, which is just outside the shadow of Saturn. Spokes were actually seen in the process of formation. One new spoke was clearly seen to form in the middle of an existing pattern on the sunlit side, with a time scale for formation as short as 5 min.

Grün et al. (1983) have classified the spokes as "extended" (which are broad with diffused edges), "narrow" (which are sharply defined, wedge-shaped, with the wedge apex at the synchronous orbit), and "filamentary." The motion of the well-defined narrow spokes has been measured by comparing images of the rings taken tens of minutes apart. The high-resolution images of Voyager 2 (B. A. Smith et al. 1982) show that the leading and trailing edges of a spoke have distinctly different angular velocities. Inside the synchronous orbit (where the spokes are mostly seen) both edges, which are tilted to the radial direction, have essentially a Keplerian rate. However, in some cases the trailing edge is more or less radial, and has approximately the

Fig. 9. Image of spokes in the B Ring. Note the distinct wedge shape of the spoke at the upper left portion of the picture. The vertex of the wedge is near the synchronous orbit. The rotation is anticlockwise with the leading edge being the slanted one (Voyager 2 photograph, from B. A. Smith et al. 1982).

corotational rate. Also, based on the observing geometry and the observed morphology of the spokes, B. A. Smith et al. (1982) have argued that the material constituting the spokes is elevated above the ring plane. Based on edge-on views of the ring system at the ring plane crossing of Voyager 2, Grün et al. (1983) have estimated that the spoke material may be elevated to a maximum height of less than 80 km above the ring plane. However, due to some uncertainty about the exact position of the spacecraft during this viewing, this inference must be regarded as tentative.

Porco and Danielson (1982) have made a detailed analysis of Voyager 1 (and, to a lesser extent, Voyager 2) data set spanning about 586 rotations, and concluded that there is a variation of the spoke activity with a periodicity of 621 ± 22 min, which, within the formal error, is compatible with the 639.4 min rotation period of Saturn determined by the periodicity of the Saturn kilometric radiation (Desch and Kaiser 1981a). They also find that the maximum in spoke activity is most likely to occur on the morning half of the rings when a particular magnetic field sector coincides with this area. This magnetic sector contains the region $\lambda_{SLS} = 155° \pm 15°$. (SLS refers to the Saturn longitude system defined by Desch and Kaiser [1981a] which is aligned with local noon at the time of maximum emission of SKR [Saturn kilometric radiation; see e.g., Warwick et al. 1981].) This is shown schematically in Fig. 10.

Several theories have already been proposed for the formation and evolution of these spokes (Gold 1980; Hill and Mendis 1981,1982b; Terrile et al. 1981; Bastin 1981; Thomsen et al. 1982; Carbary et al. 1982; Grün et al. 1983; Goertz and Morfill 1983; Morfill et al. 1983a). The one thing all these theories have in common is that they invoke electric forces in order to produce the spokes. In the theories of Gold (1980), Bastin (1981), and Carbary et al. (1982), local radial electric fields in the ring plane, produced one way or another, are used to electrically polarize and align elongated grains. They then show up as spokes owing to the change in light-scattering properties of aligned grains in relation to randomly oriented ones. On the other hand, the theories of Hill and Mendis (1981,1982b), Terrile et al. (1981), Thomsen et al. (1982), Grün et al. (1983), Goertz and Morfill (1983), and Morfill et al. (1983a) require small grains to be charged and to be electrostatically levitated off the surfaces of larger bodies in the ring plane.

Weinheimer and Few (1982) have critically evaluated these electrical processes in light of the fact that the spoke particles, like the rest of the ring particles, are most likely to be composed of water ice. The grain-alignment theories require that the grains be sufficiently conducting. However, Weinheimer and Few (1982) have concluded that, unless the ice becomes ferroelectric, the electrical torque used to align the particles is many orders of magnitude too small with the electric fields expected there. They also believe that it is very unlikely that the ice would become ferroelectric at the ring temperatures. Consequently, they argue that the grain alignment theories are untenable. As we have already pointed out, there is also some evidence, although not compelling, to believe that the spoke material is elevated above the ring plane, as proposed by the alternate theories.

The most detailed theory so far for the formation of the spokes is due to Goertz and Morfill (1983). First, they show that with the known average optical depth in the B Ring ($\langle \tau \rangle \approx 0.6$), the interparticle distance is expected to be smaller than the Debye shielding distance, unless the plasma density is large ($n_e \lesssim 300$ cm^{-3}, when $kT_e \approx 2$ eV). Under such circumstances the ring must be treated as a uniformly charged disk. Also, assuming that the electron density in the vicinity of the ring plane is provided largely by the photoelectrons from the rings themselves, they conclude that the ring potential is around $+5$

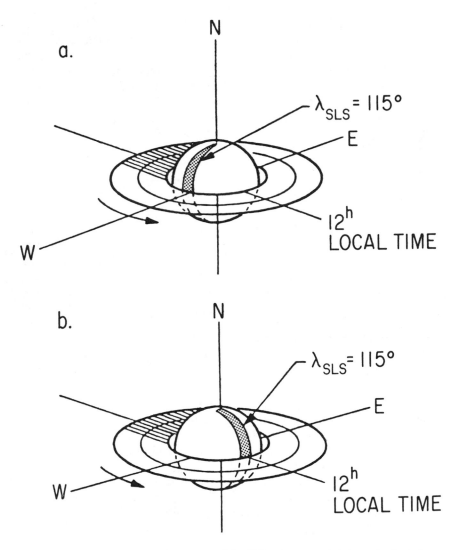

Fig. 10. (a) The configuration of the center of the SKR-active sector (relative to the Sun) showing the most likely location for the appearance of high-contrast spokes on the morning half of
the rings. (b) The configuration showing where the most intense SKR emission occurs (figure
from Porco and Danielson 1982).

to +10 V. Under these circumstances only a small fraction of micron-sized
grains lying on the surfaces of the larger embedded moons will have even one
electronic charge and would in any case lose it before emerging from the
Debye sheath. A much denser source of plasma in the vicinity of the rings is
needed to make the electrostatic levitation of grains a feasible process.

Noting also that spoke formation is a sporadic process, Goertz and Morfill propose that such dense localized plasma columns are produced by meteor impacts with the ring particles. Initially a very dense ($n_e \approx 10^{15}$ cm^{-3}) plasma cloud is formed, which expands rapidly while diamagnetically excluding the magnetic field until the diamagnetic factor $\beta \approx 1$. In addition, the neutral gas cloud produced by meteor impact will be photoionized. Losses through the loss cone of Saturn's atmosphere and via ring absorption will continue. The plasma column will initially move with the speed of the meteor target until $n_e \lesssim 10^2$ cm^{-3}, whereafter it will tend to corotate. During such conditions not only can small fine dust (typically $R_g \approx 0.1$ μm) be electrostatically levitated off the ring plane, but the plasma column itself will move radially. This is due to the following reason. The relative motion between the charged dust (which is close to Keplerian) and the surrounding plasma (which is corotating) constitutes a current. Since this current is confined azimuthally to the plasma column, it must close through the ionosphere via a Pederson current which is connected to the dust-ring current by a pair of field-aligned currents (see Fig. 11). The Pederson current sets up an electric field in the ionosphere, which then maps back onto the ring plane and causes a radial $\mathbf{E} \times \mathbf{B}$ drift of the plasma cloud. Knowledge of the ionospheric Pederson conductivity ($\Sigma_p \approx 0.1$ mho) allows one to calculate the radial velocity V_R of the plasma column which ranges from ~ 45 km s^{-1} at $L = 1.7$ (the inner edge of the spoke region) to zero at the synchronous orbit ($L = 1.86$). At $L = 1.9$, the outer edge of the B Ring, $V_R \approx 23$ km s^{-1}. As such a plasma column moves from its position of creation, it will leave behind a radial trail of dust much like a car racing across a dusty plain. The authors believe that the observed spokes are formed in this way.

This process requires that the plasma column survive long enough to travel large radial distances to form the long radially elongated spokes. The authors point out that this is more likely if the plasma is formed in the "free zone" on the same side of the ring as the magnetic equator. In that case, plasma with sufficiently large pitch angles would survive for some time. While it now seems possible that the magnetic dipole of Saturn is perfectly aligned with the spin vector (Acuña et al. 1982; Connerney et al. 1982b), it appears that the dipole is sufficiently offset to the north. In that case also there would be a free zone although one not confined to a particular longitude sector as in the tilted dipole case.

While this is an attractive model, two points must be stressed. In the first place, the entire process can take place only when there is relative motion between the dust and the plasma (i.e., when $n_e \lesssim 100$ cm^{-3}). Secondly, the field lines need not be equipotentials when parallel currents flow. If the parallel current density exceeds a certain critical value, part of the potential drop in the circuit may be localized as an electrostatic "double layer" somewhere on the current-carrying field line (e.g., see Alfvén 1981a). Consequently, the entire potential drop across the ionosphere may not be mapped back to the ring

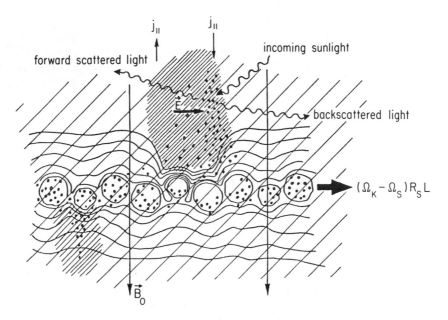

Fig. 11. A schematic view of the model for formation of spokes. Underneath a dense plasma
column (indicated by heavy shading) the equipotential contours are compressed and a large
surface electric field exists. Dust particles are lifted off the rings and drift through the cloud
which polarizes. A pair of field-aligned currents closes the current (figure from Goertz and
Morfill 1983).

plane; therefore, the calculated radial speeds V_R are upper limits. This model
also implies that the plasma clouds should not move across the synchronous
orbit because $V_R \rightarrow 0$ there, whichever way the plasma column moves (outward outside the synchronous orbit and inward inside for negatively charged
grains). Inspection of spoke photographs indicates that at least some of them
cross the synchronous orbit. This requires that some plasma clouds be formed
quite close to the synchronous orbit so that they overlap it and expand radially
away from it in both directions.

    An alternative mechanism for spoke formation has been proposed by Hill
and Mendis (1981,1982b). While they too propose that small grains constituting the spokes are electrostatically levitated and blown off the surfaces of
larger bodies when they are charged to high electrostatic potentials, they suggest that the sporadic field-aligned beaming of energetic electrons accelerated
at electrostatic double layers high up in Saturn's ionosphere may produce the
high surface potentials necessary for this process. They suppose that these
double layers are highly confined in longitude but extended in latitude, so as
to account for the radial extent of the thin spokes. While double layers are
known to occur in the Earth's magnetosphere, they do so on field lines con-

nected to the auroral region, where field-aligned currents flow from the magnetotail. The field lines threading the B Ring of Saturn are closed ones, and the question arises as to how such double layers could be generated there, because the flow of field-aligned currents constitutes a necessary condition for their formation on the field lines. The answer perhaps lies in the fact that the dust-ring current is diurnally modulated by Saturn's shadow, as discussed by Ip and Mendis (1983; see Fig. 18 in Sec. VII). The continuity of current then requires field-aligned currents to flow near the dawn and dusk terminators. If the grains are negatively charged the current system flows in the sense opposite to that shown in Fig. 19 (Sec. VII), which is for positive grains. Consequently, double layers formed high up in the ionosphere will accelerate electrons down onto the equatorial plane in the morning ansa, close to the terminator. This is consistent with the observations of high spoke activity in this region. The observations of Porco and Danielson (1982) concerning the correspondence of maximum spoke activity with a particular magnetic sector being in the morning hemisphere may simply be associated with a magnetic/ionospheric anomaly there. The fact that spokes cross the synchronous orbit poses a similar problem for this model as for that of Goertz and Morfill (1983). At the synchronous orbit, the azimuthal current is zero and no field-aligned currents flow into or out of the ionosphere; thus no double layers can be set up at synchronous orbit in this way.

Hill and Mendis (1982b) have shown that, on the basis of stable orbit populations discussed by Mendis et al. (1982), the observed wedge-shape morphology, with the leading edge (inside the synchronous orbit) slanted in the direction of the motion, requires that the grains be negatively charged. They have also discussed the detailed evolution of the charged fine dust grains constituting the spokes, once they are electrostatically blown off the surfaces of their parent bodies, and they have numerically calculated the orbits of different-sized grains as they move around the planet. The projections of these orbits on the ring plane are shown in Fig. 12. Note that these orbits are again calculated for single dust particles ($\gamma = 0$) and their relevance to the optically fairly thick spokes is thus open to question. The positions of the grains in their respective orbits at different times after the instant of projection are indicated, and the locus connecting grains of different sizes one hour after projection is shown by the broken line referred to as a "synchrone." Note that for a realistic spoke ($\tau = 0.1$), the radii labeling the curves should be reduced by a factor of ~ 50.

Four synchrones of grains ejected from four different positions on a radius vector (azimuth angle $\phi = 0°$ at $t = 0$), inside the synchronous orbit one hour after the ejection, are shown in Fig. 13a. The particles that are moving with the corotation speed (electrons and ions) will lie on the radius vector $\phi \approx 33°8$, since the rotation period of the planet is ~ 10 hr 39 min. The Kepler particles (i.e., those which are negligibly influenced by the electric field) move fastest within the synchronous orbit, and their positions are indicated by

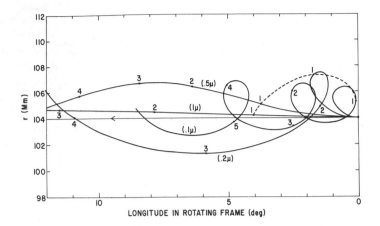

Fig. 12. Equatorial projection of the orbits in the corotating frame of different-sized charged grains which have been electrostatically blown off from the surface of an embedded moon at a given time. All these grains have escaped in the direction normal to the ring plane. The grain radius is indicated within parentheses, while the positions of the grains at various times after the ejection are shown by the tick marks. The numbers associated with these marks refer to the time lapse in hours. The dotted line is a synchrone, which is the locus of grains of different sizes one hour after their ejection (figure from Hill and Mendis 1982b).

the points at the largest azimuthal angle on each curve. The synchrones are essentially of the same shape, and are bounded by a series of envelopes that are essentially straight lines, all intersecting the radial line at the synchronous orbit. All these envelopes radiating out from the synchronous point S pass through grains of essentially the same size on each synchrone. Note that if a set of straight lines are drawn through S intersecting the synchrones, the maximum density of grains would correspond to the envelope lines. This is more clearly brought out in Fig. 13b, which is a schematic drawing showing a larger number of synchrones. In this diagram it is more apparent that the density of grains within the wedge is not uniform, and that the maxima in the grain density correspond to the envelope lines.

Hill and Mendis (1982b) therefore suggest that the so-called wedge is more like a lady's fan, with several ribs (each corresponding to a region of enhanced grain density) all intersecting at the synchronous point S. However, it should be noted that the average charge on the spoke dust grains may be considerably smaller for the same potential ϕ as assumed in the calculations of Hill and Mendis. The unfolding of the fan effect will only apply to very small dust particles, $R_g \approx 0.01$ μm, on the basis of Eq. (7) but would apply to larger $R_g \approx 0.5$ μm particles on the basis of Eq. (9).

A different scenario for the evolution of the wedge-shaped spokes has been discussed by Morfill et al. (1983a), who point out that once the charged dust particles of a radially aligned spoke emerge from the planet's shadow, the

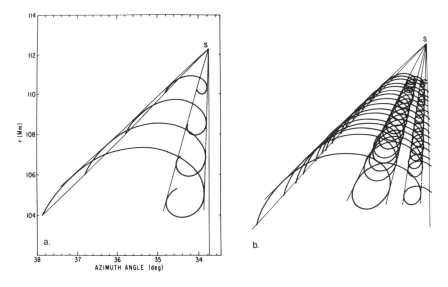

Fig. 13. (a) Four different synchrones of grains ejected from four different positions on a radius vector. The abscissa represents the azimuthal angle traversed from the time of projection one hour earlier. The points at the largest azimuthal distances represent the Kepler particles. The vertical line represents the trailing (corotational) edge, and S is its intersection with the Keplerian edge. The other lines represent envelopes to the synchrone curves. They all pass through S and are almost straight lines. Notice that the outermost envelope line is in fact ahead of the Keplerian line. (b) Schematic drawing illustrating the situation in (a) with a much larger number of synchrones. Notice the larger density of points along the envelope lines (figures from Hill and Mendis 1982b).

photoeffect will cause a rapid discharge of the electrons. Part of these electrons will be reabsorbed by the ring after one bounce and cause the electrostatic levitation of new dust particles. Since the electrons move at the corotation speed, whereas the dust particles move at nearly Keplerian speed, the spoke would grow azimuthally, with the growing edge being radially aligned, whereas the dust elevated previously lags or leads the radial depending on whether or not they are outside the synchronous orbit.

VI. THE B RING DISCONTINUITIES: EQUATORIAL CONFINEMENT OF CHARGED DUST AND PLASMA

Infrared observations of the B Ring by Pioneer 11 (Froidevaux and Ingersol 1980) and Voyager 2 (Hanel et al. 1982a) indicate a sharp boundary between the outer (brighter) and inner (fainter) portions of the B Ring at ~ 1.633 R$_S$. The same boundary also shows up in the imaging pictures of Voyager 1 (B. A. Smith et al. 1981) in forward scatter as well as in the Voyager 2 occultation of δ Scorpii by the rings in the ultraviolet (Sandel et al. 1982*b*). In the last case it shows up as a factor of 2 change in the optical depth starting very close to this radius.

Northrop and Hill (1982) have attempted to explain this discontinuity in terms of the stability limit of negatively charged grains (of large specific charge and zero magnetic moment) to perturbations normal to the ring plane. They show that if a negatively charged grain, moving in its equilibrium circular orbit in the ring plane, is subject to a perturbation normal to the ring plane, it will oscillate about the ring plane provided its equatorial radial distance $r > r_c$, where

$$r_c^3 = \frac{2}{3} \frac{GM_s}{\left(\Omega - \frac{GM_s}{3} \frac{mc}{Q\mu}\right)^2} \tag{16}$$

with Q and m being the charge and mass, respectively, of the grain, and μ the magnetic moment of the planet (which is assumed to be strictly parallel to the spin vector). If $r < r_c$, the perturbed grain leaves the plane altogether, and eventually enters the planetary atmosphere, with its guiding center moving along the pseudo-magnetic field line $\tilde{\mathbf{B}}$ given by (Northrop and Hill 1983*a*)

$$\tilde{\mathbf{B}} = \mathbf{B} + \left(\frac{2mc}{Q}\right)\Omega \tag{17}$$

where $\mathbf{B}$ is the dipole field of the planet and Ω is its angular velocity.

The orbits in the r,z plane for a grain having $Q/m = -200$ Coulombs kg^{-1} and projected normal to the plane with a velocity of 0.1 km s^{-1}, inside the marginal stability radius are shown in Fig. 14. The instability of these grains is apparent. The marginal stability radius in this case is ≈ 1.54 R$_S$.

In the limit of large negative charge (i.e., $Q/m \rightarrow -\infty$),

$$r_c \rightarrow r_{c\infty} = \left(\frac{2}{3} \frac{GM_s}{\Omega^2}\right)^{1/3} = 1.625 \text{ R}_S \tag{18}$$

which is very close to the position of the observed discontinuity in the B Ring ($R \approx 1.633$ R$_S$). Northrop and Hill (1982) suggest that the instability of negatively charged grains of very large specific charge inside this radius is the cause of this discontinuity. The limit considered by Northrop and Hill (1982) corresponds more to plasma than to charged grains, and the existence of this marginal stability limit for plasma (electrons as well as positive ions) $r_{c\infty}$ within rigidly corotating magnetospheres was first established by Alfvén (1954) and subsequently discussed also by Angerami and Thomas (1964), Melrose (1967), Ioannidis and Brice (1971), and Mendis and Axford (1974).

Ip (1983*b*) has used this fact to argue that the observed discontinuity in

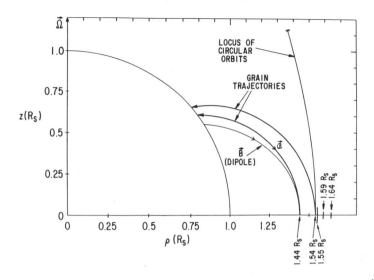

Fig. 14. Orbits in r,z plane for a grain having a specific charge of -200 Coulombs kg^{-1}. The initial velocity vector in each case has the local prograde circular velocity for its azimuthal component, 0.1 km s^{-1} velocity in the z direction, and no radial velocity. The theoretical marginal stability is 1.5404 R_S, and the instability of grains inside this radius is apparent (figure from Northrop and Hill 1982).

the B Ring is caused not by the loss of negatively charged grains of large specific charge inside $r_{c\infty}$, as suggested by Northrop and Hill (1982), but rather by the siphoning off of plasma inside this distance. Outside $r_{c\infty}$, plasma will be confined to the equatorial plane (see Fig. 15). Ip calculated the amount of residual ionization produced during hypervelocity impact by meteoroids with ring particles on the basis of laboratory studies (see, e.g., Hornung and Drapatz 1981; Grün and Reinhard 1981) and showed that with the expected flux at Saturn, $\sim 50\%$ of the ring material interior to 1.625 R_S, could be converted to plasma and siphoned away during the age of the solar system. The observed decrease in optical depth by a factor of 2 interior to 1.633 R_S is attributed to this case.

Finally, Ip (1983b) has shown that this channeling of impact-generated ring plasma into the mid-latitude ionosphere could cause the large reduction in the ionospheric electron content there, as observed by the Pioneer 11 and Voyager 2 radio science experiments. This is brought about by a chain of charge-exchange and dissociative recombination reactions involving the ring ions and heavy molecules present in the mid-latitude ionosphere.

More recently, Northrop and Hill (1983b) extended the stability analysis of charged dust grain orbits in Saturn's magnetosphere when the magnetic moment of the grains is nonzero. They considered several different models of the magnetic field of the planet, including a centered dipole, an offset dipole,

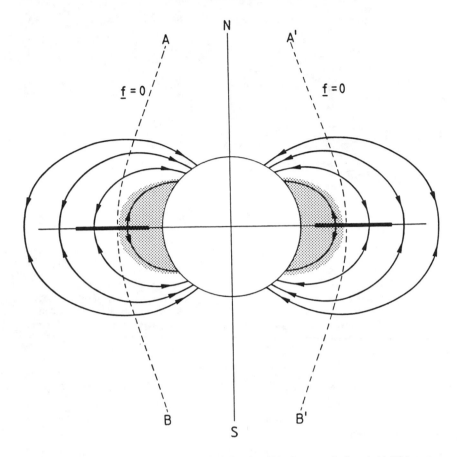

Fig. 15. Division of the rotating ionosphere by the dashed curves A–B and A′–B′ into two plasma regimes according to the consideration of the balance of centrifugal and gravitational forces: (1) the upward siphon flow region denoted by the shaded area with equatorial distance $r \leq 1.6252\ R_S$; (2) the equatorial confinement region with $r > 1.6252\ R_S$ (figure from Ip 1983b).

and a zonal harmonic model developed by Connerney et al. (1982b) from the Voyager 1 and 2 observations. They also used both a spherically symmetric gravitational field as well as one having zonal harmonics, as developed earlier from Pioneer data. They used this analysis to explain the inner edge of the B Ring, which has a very sharp transition in optical depth between 0.2 (in the C Ring) and ~ 1 (in the B Ring) at a planetocentric distance of 91,970 km ($= 1.524\ R_S$). They find that this corresponds to the stability limit of charged dust grains (of large specific charge) that are launched in the ring plane at local Kepler velocity. While the discrepancy between the observed and computed values is typically around 1000 km for most of the models, it is reduced to only 3 km when the zonal harmonics of both the magnetic and gravitational fields are used. While this agreement between observation and theory is most impressive, the theory suffers from one serious drawback because the zonal harmonic magnetic field lines are not perpendicular to the ring plane. Therefore, in addition to the magnetic mirror, gravitational, and centrifugal forces, another force (parallel to the field lines, towards the ring plane) must be postulated to produce equilibrium in the ring plane and make the stability analysis meaningful. The authors concede that at the present time the physical basis for this extra force remains unknown. While they have speculated on a possible attractive electrostatic force towards the ring plane, its existence has so far not been demonstrated.

In regard to meteor-impact-produced ring plasma, and a centered dipole magnetic field, Ip (1983a) has shown that the limiting radius of upward siphoning of plasma from the ring plane moves inwards from 1.625 R_S, for plasma with nonzero magnetic moment, and corresponds to the inner edge of the B Ring for particles launched at the local Kepler velocity. Plasma produced by meteor impact would indeed initially start off close to the Kepler speed (the speed of the parent ring bodies).

Finally, it should be pointed out that there is, in fact, another electrodynamic process proposed (Alfvén 1981b) for the formation of the inner B Ring boundary. This is the so-called cosmogonic shadowing process associated with the initial emplacement of the ring dust at cosmogonic times (Alfvén 1954). This process is discussed in more detail in Sec. X. While this process has been used to explain several other ring edges (these too will be discussed in Sec. X), and in addition, seems to qualitatively explain the radial brightness profiles of these edges, the agreement of the theoretical prediction with the inner edge of the B Ring is accurate only to within ~ 2000 km.

VII. DUST-RING CURRENTS AND THEIR CONSEQUENCES

As we have pointed out earlier, the charged dust in the rings moves with a speed different from the surrounding plasma, which is assumed to be in rigid corotation in the region occupied by the dust grains. In addition, for a given grain potential, size, and orbital position, the grain's velocity is known (Eq. 3). Consequently, if the dust distribution and extent of a ring are known, the total current carried by the ring can be estimated from

$$I = \int QN \, v_{rel} \, dh \, dr = \frac{33.3}{\pi} \frac{\phi(v)}{R_g(\mu m)} \frac{1}{1 + \gamma} \int_{L_1}^{L_2} v_{rel} \, \sigma dL \qquad (19)$$

where v_{rel} is the relative velocity between the grains, which are all assumed to be of radius $R_g(\mu m)$, and the plasma, and σ is the normal optical depth of the ring. This assumes that the plasma particles in the Debye sphere surrounding the charged dust do not move with the dust grains, but are constrained by the magnetic field to corotate.

The variation of optical depth across the F Ring is now accurately known from occultation data of the star δ Scorpii, obtained during the Voyager 2 mission (Lane et al. 1982b). These data indicate a number of thin strands (typically a few kilometers wide) of optical depths varying between ~ 0.1 and 0.4, with the maximum optical depth being near the outer edge of the ring. Hill and Mendis (1982c) have assumed that all the grains in this ring are typically of radius 0.5 μm, which gives the grain column density normal to the plane. Approximating the observed distribution to a sawtooth-shaped one with base length of 60 km, and assuming that all the grains are charged to -40 V and neglecting the correction factor $1 + \gamma$, these authors obtain a closed-form solution for the magnetic field produced by this current distribution. The addition of this field to the dipole field of the planet then gives the total field near the F Ring, which is shown in Fig. 16. Clearly the magnetospheric magnetic field would be noticeably changed by the dust-ring current, which amounts to $\sim 10^5$ A if $\gamma = 0$ (as estimated by Whipple et al. 1983). However, if $\gamma \approx$ 1600 in the F Ring (as estimated by Goertz [1983b]), the real distortion should be much smaller than that shown in Fig. 16. Goertz and Morfill (1983) estimate a much smaller total current for the B and F Rings because they use a numerically smaller potential ($+5$ V) and larger average particle size and take the screening effect into account. Their estimates indicate negligible effects on the magnetic field.

Ip and Mendis (1983) have pointed out that the diurnal modulation of the grain potential by Saturn's shadow will also cause a change in the dust-ring current near the terminators. Continuity of current there requires that this current be connected to the planetary ionosphere by field-aligned Birkeland currents and to close through the ionosphere via a Pederson current. This is illustrated schematically in Fig. 17 for grains having positive potentials. If the grains were negative, the sense of the current system is reversed. These authors have considered, in particular, the modulation of the dust-ring current associated with the faint innermost D Ring where γ is small ($1.11 \lesssim L \lesssim 1.23$) and have shown that a Pederson current on the order of 10^5 A could be generated even for small grain potentials $|\phi| \approx 3$ V. This Pederson current sets up ionospheric electric fields on the order of 0.1 V m^{-1}. Even if a significant fraction of the potential drop across the ionosphere is mapped by the field lines (which, of course, need not be equipotentials when parallel currents flow) onto the equatorial plane, the corresponding E field would drive strong magnetospheric convection near the D Ring. This convection pattern is shown schematically in Fig. 18. It is pointed out that the plasma reaching beyond the D Ring is likely to be absorbed and neutralized by the C Ring, which has a much higher optical depth than the D Ring.

It becomes clear that the density and temperature of the plasma in the vicinity of the rings, as well as the average particle size, are the most crucial parameters for estimating the magnitude of electrodynamic effects. Unfortunately, no direct measurements of these quantities are available as yet. We

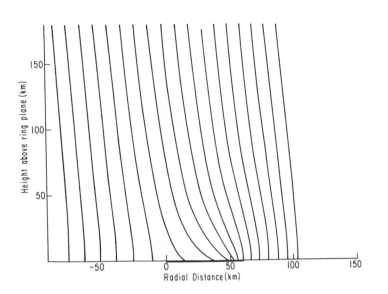

Fig. 16. The magnetic field morphology in the vicinity of Saturn's F Ring. The distortion of the magnetosphere dipole field is a result of the current caused by the relative motion of the negatively charged dust and corotating plasma. In this figure the dust current is located between 0 and 60 km and directed into the page, while the dipole field is directed downward parallel to the negative y direction (figure from Hill and Mendis 1982c). This figure does not include the screening effect discussed in the text.

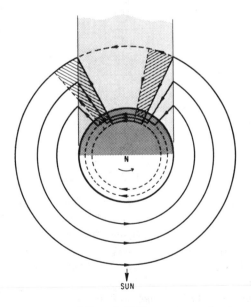

Fig. 17. Idealized current system of Saturn's D Ring, corresponding to grains having a small positive electrostatic surface potential on both the day and night sides. The cross-hatched areas show where the Birkeland currents largely flow. The current system would be reversed if the grains had a negative potential or if they were charged to a very large positive potential (which is unlikely) (figure from Ip and Mendis 1983).

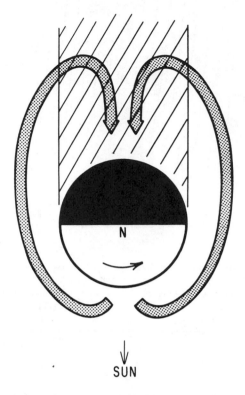

SUN

Fig. 18. The idealized convection pattern established in the equatorial upper ionosphere as a result of the D Ring Birkeland current system (figure from Ip and Mendis 1983).

have discussed possible electrodynamic effects for certain given plasma conditions and particle sizes. The observations of such effects may thus provide a means of estimating the plasma conditions and particle size distributions.

VIII. THE E AND G RINGS

Besides the easily observed and well-defined components of the main Saturnian ring system, there is also a widespread tenuous dust distribution exterior to it. The very tenuous E Ring, first detected from Earth in 1966 when the rings were seen edge-on, has been subsequently studied by both groundbased and spacecraft observations (Feibelman 1967; Smith 1978; Baum et al. 1981; Sittler et al. 1981). It extends from $L \approx 3$ to about $L = 8$ or 9, with an optical depth τ in the range 10^{-8} to 10^{-6} and a maximum at $L \approx 4$ near the orbit of Enceladus. Infrared observations indicate that E Ring grains have diameters $\lesssim 5 \ \mu$m (Terrile and Tokanaga 1980).

The thin diffuse G Ring, which is located at $L \approx 2.8$, also appears to be composed of fine dust. During the ring plane crossing of Voyager 2 at $L =$ 2.88, slightly outside the G Ring, the plasma wave instrument detected intense impulsive noise which has been attributed to spacecraft impact by grains with radii in the range 0.3 μm to 3.0 μm (Gurnett et al. 1983). Due to the proximity of this observation to the G Ring, it is suspected that the G Ring also contains such small grains. The interpretation of the energetic particle absorption signatures near the G Ring reported by Van Allen et al. (1980a) also indicates a mass density per unit area normal to the ring plane of $\gtrsim 7 \times 10^{-9}$ g cm^{-2}. For dust grains with an average radius of 1 μm and an average density of 1 g cm^{-3}, this corresponds to an optical depth $\gtrsim 5 \times 10^{-5}$, which is in good agreement with the value of $\tau \approx 10^{-4}$ to 10^{-5} reported by the Voyager imaging team (B. A. Smith et al. 1982). Ip (1982) has pointed out, however, that in order to produce the observed energetic charged particle signatures, a component of larger particles with a radius $\gtrsim 200$ μm, which is the penetration range of 1 MeV protons, may be required. Clearly the component of smaller (1-micron-sized) particles is also required, because in its absence the optical depth would be reduced to $\sim 2 \times 10^{-7}$.

Besides energetic charged particles, thermal plasma is also affected by the ring absorption effect. Both Voyager 1 and 2 observations of the electron distribution have indicated a bite-out (or depletion) of superthermal electrons ($kT_e > 700$ eV) at the E Ring (Sittler et al. 1981). Because of the much longer bounce period, the thermal electrons of a few eV energies are not as strongly absorbed. Sittler et al. (1981) have shown that the bite-outs extend to ~ 9 R$_S$ and are also not azimuthally symmetric about Saturn. Similar superthermal electron absorption signatures have been observed around $L \approx 13.5$, 15, and 17 (Sittler, personal communication; Lazerus et al. 1983). Although the size distribution of the dust causing the electron bite-outs is not known, it is possible that much of it is fine dust. Also, whether this dust distribution is contiguous with the E Ring, or whether it forms another distinct ring in the gap between the satellites and Titan, remains an open question.

Here we confine ourselves to the question of the origin of the E and G Rings. Morfill et al. (1983b) have considered the interaction of the E Ring grains with their plasma and radiative environment. They find that of the possible destructive processes, such as evaporation, photosputtering, and high- and low-energy charged particle sputtering, the last is the most efficient. Grains of 1 μm size are sputtered to extinction by the corotating thermal plasma ($E_{\rm rot} \approx 60$ to 500 eV), in 10^2 to 10^4 yr. Also, sputtering by protons above 50 keV leads to lifetimes of $\sim 10^4$ yr for 1-μm-sized grains (Cheng et al. 1982). This shows that there must be a continuing source for the replenishment of the E Ring grains. The coincidence of the maximum in the E Ring dust density with the orbit of Enceladus perhaps indicates that this satellite is that source (B. A. Smith et al. 1981,1982). Morfill et al. (1983b) suggest that the impact of interplanetary meteoroids entering Saturn's magnetosphere with this small satellite could result in the escape of a sufficient amount of fine dust fragments to explain the observed distribution. If this is true, one would expect a similar enhancement in the dust density near Mimas (a satellite which is even smaller than Enceladus). The failure to observe such a feature throws some doubt on the validity of Morfill et al.'s suggestion.

An alternative explanation for the escape of particulate matter from Enceladus has been forwarded by Pang et al. (1982). According to these authors, as the Voyager 2 spacecraft passed through the gap between the G and E Rings, data for face-on, oblique and edge-on views of the rings at high phase angle were obtained by the Canopus star tracker. Comparison of this measurement with the groundbased observations in the optical and the infrared wavelengths then suggests that the E Ring particles may be ice spheres with an effective diameter of 2 to 2.5 μm and a narrow size distribution with an effective variance of 0.1 to 0.15. The authors conclude that the spherical shape of the E Ring particles and their narrow size distribution imply that they had a molten origin followed by quick freezing. Consequently, volcanic eruptions on Enceladus may be a possible source of E Ring grains. Some support for this view comes from surface morphology of Enceladus and its very high albedo, which has led to the suggestion that the surface is relatively young and that melting may have taken place recently (B. A. Smith et al. 1981). In addition, the interior heating mechanism may be associated with the tidal interaction between Enceladus and Dione (see Yoder 1979).

Pang et al. (1982; see also Voge et al. 1982) have pointed out that the edge-on scans of the E Ring by the Voyager 2 star trackers have revealed at least 30 distinct layers. Pang and Ip (personal communication, 1983) suggest that this normal stratification may be related to episodic ejection of material from Enceladus. Morfill et al. (1983b) have studied in considerable detail the effects of plasma drag and stochastic variations in the grain charge due to fluctuations in the plasma density and/or mean energy. They find that while both these lead to the radial transport of the dust grains, plasma drag is the dominant effect with a time scale $\gtrsim 10^4$ yr. They suggest that the broad extent of the E Ring is mainly due to the effect of the plasma drag on the micron-sized collisionally-produced dust ejecta from Enceladus.

There is yet another process for the radial transport of charged dust grains in planetary magnetospheres. This has to do with the finite capacitance of the grain, which results in a phase lag of the grain potential with respect to its orbital position. This so-called radial gyro-phase drift (Northrop and Hill 1983a) has been invoked to explain the origin and structure of the bright dust ring and contiguous broad dust disk around Jupiter (Hill and Mendis 1979, 1980). This model has been used by Hill and Mendis (1982d) to study the radial transport of charged dust ejecta from Enceladus. Using the preliminary electron and ion densities and temperatures in Saturn's inner magnetosphere (Bridge et al. 1982), these authors approximate it using a multithermal electron distribution and a two-component ion distribution obtained by fitting the simple curves (Fig. 19, inset) to the data. The temperature of the cold plasma

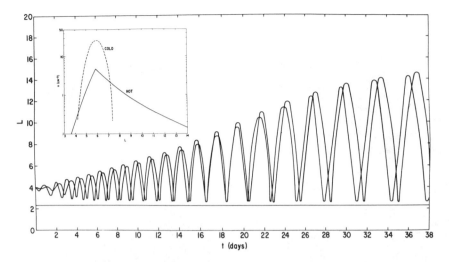

Fig. 19. The variation of L with time for two grains launched from Enceladus, each with a 1 km s^{-1} excess over the escape velocity from the satellite. One grain was launched normal to the satellite orbit, while the other was launched directly towards Saturn. The multithermal plasma distribution in Saturn's inner magnetosphere used in this calculation is shown in the inset (figure from Hill and Mendis 1982d).

component has been chosen to be 2 eV, while that of the hot component is taken as 10 eV inside $L = 4$, and 100 eV outside. The hot ions are assumed to be hydrogen and the cold ions are assumed to be oxygen.

The motion and charge variation of a single 1 μm diameter grain was followed for periods greater than $\sim$ 40 days after launch from Enceladus. The motion was in qualitative agreement with the predictions of the adiabatic theory of charged grain motion. The variation of L with time for two grain orbits is shown in Fig. 20. One was launched north (normal to the orbital plane) while the other was launched (in the orbital plane) towards Saturn, each with a 1 km s^{-1} excess over the escape velocity from the satellite. While the orbits show a phase shift resulting from the initial conditions, they are otherwise similar. The radial diffusion of the grain orbit is obvious. Both grains eventually approach a minimum L of $\sim$ 2.65 and a maximum L of nearly 15, in 30 to 40 days. One should note that this dispersion time scale is many orders of magnitude smaller than the one calculated by Morfill et al. (1983b); viz., $\sim 10^4$ yr. However, a more realistic calculation along these lines will have to include the modulation of the grain potential by the shadow of Saturn as well as the possible sinks provided by the satellites Tethys ($L = 4.88$), Dione ($L = 6.26$), and Rhea ($L = 8.74$).

Assuming an average optical depth of 10^{-6} and an average grain radius of 1 μm, we estimate that the total mass of the E Ring between 3 and 8 R_S is

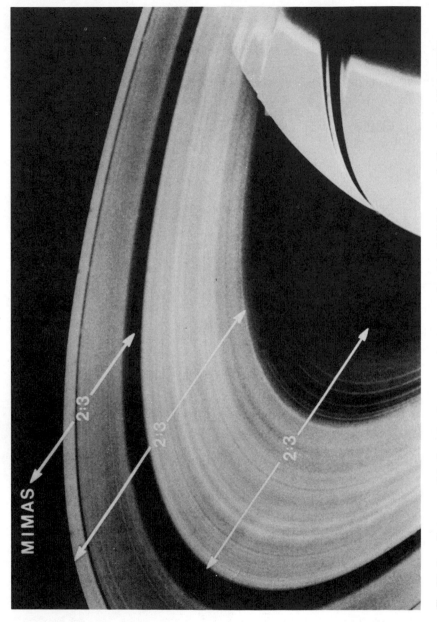

Fig. 20. Overall structure of the Saturnian ring system, showing examples of the 2/3 fall-down ratio (figure from Alfvén 1981*b*).

$\sim 2 \times 10^{10}$ g. Then, a dispersal time of 40 day corresponds to a mass loss rate from Enceladus $\dot{M}_E \approx 4 \times 10^4$ g s^{-1}, which is 2 to 3 orders of magnitude less than the mass loss rate from Io. However, since the mass of Enceladus M_E is only $\sim 7 \times 10^{22}$ g, this implies a characteristic time scale for the dispersion of Enceladus of 2×10^{10} yr. While this is uncomfortably close to the age of the solar system (4.5×10^9 yr), it should be pointed out that this derived dispersal time scale may be too short for several reasons. Both the electrostatic potential and the ejection velocity of the grains may be significantly smaller than the values used by Hill and Mendis (1982d). This would lengthen the dispersal time scale. Also one should note that the initial rapid dispersion of the charged dust in the model of Hill and Mendis (1982d) is a response to the large density gradients of the plasma in the inner region (see inset of Fig. 20). As the orbits of the grains migrate outward, the plasma density gradients encountered become smaller. Consequently, the grain orbits almost stop expanding at $L \approx 15$, and therefore the time taken for actual loss of dust from the magnetosphere according to this model is much longer than 40 day, the time taken to reach $L \approx 15$. As a result, the dust dispersal time scale would also be correspondingly longer. In summary, it is fair to state that while the E Ring essentially determines the dusty nature of the Saturnian magnetosphere, both its origin and overall structure remain open questions, despite the interesting theories that have been proposed.

Finally, regarding the origin of the thin G Ring at $L = 2.8$, it is possible that it too is composed of the fine dust ejecta from Enceladus and brought into place by one or the other of the radial transport processes described above. However, due to its narrowness it has been suggested that it is kept in place by shepherding satellites, as in the case of the F Ring (Morfill et al. 1983b). There is, however, no observational evidence for the existence of such satellites. On the other hand, if the dust-particle density at any given radial distance were to increase to a point that the interparticle collision time becomes shorter than any of the dispersive time scales considered, and if the collisions are sufficiently inelastic, "negative diffusion" (Baxter and Thompson 1973) could hold the ring together without the aid of any shepherding satellites. Unfortunately, with the observed optical depth $\tau \sim 5 \times 10^{-5}$, the collision time scale appears to be much too long for negative diffusion to be effective. Consequently, the G Ring remains an enigma at this time.

IX. SKR AND SED

During both Voyager encounters with Saturn, the planetary radio astronomy experiment (PRA) detected two distinct kinds of radio emission from Saturn (Warwick et al. 1981,1982a). The first is strongly polarized, bursty Saturn kilometric radiation (SKR), which appears at radio frequencies below ~ 1.2 MHz and is tightly correlated with Saturn's rotation period. This radiation, which appears to come from a strong dayside source at auroral latitudes in the northern hemisphere and a weaker source at complementary latitudes in the southern hemisphere, exhibits complex dynamic spectral features reminiscent of Jupiter's kilometric radiation detected earlier. The second kind of radio emission is unpolarized, extremely impulsive and very broad band, appearing throughout the observing range of the experiment (20.4 kHz to 40.2 MHz). These episodic radio bursts have been named Saturn electrostatic discharges (SED). They appear to come from the direction of the ring plane and they have no counterpart in the Jovian system. For a detailed treatment of this subject see the chapter by Kaiser et al.

The mean period of the SED is well defined, and has been determined as 10 hr 10 ± 5 min and 10 hr 11 ± 5 min for Voyagers 1 and 2, respectively (see Evans et al. 1982). If the ring is the source of the SED, the source region should be located at a distance range of 107,990 to 109,200 km from the planet, according to the measured periodicity. Interestingly, the Voyager 2 photopolarimeter star occultation data (at $\lambda = 2650$ Å) reveals a sharp, narrow ($\lesssim 200$ m) gap in the dense B Ring at 108,942 km. While the mean optical depth of the B Ring in the proximity of this gap is ~ 1.8, the optical depth in the gap itself is only ~ 0.15 (Evans et al. 1982). These authors suggested that the source of the SED is located in this gap. This gap, however, does not show up in the Voyager 2 ultraviolet spectrometer, which was bore-sighted with the photopolarimeter and observed the ring occultation of δ Scorpii in the spectral range 900 to 1700 Å. The above authors, however, believe that a real gap exists and the reason that it does not show up in the ultraviolet data is probably due to the existence of small ($r_g \lesssim 100$ Å) Rayleigh scatters. Warwick et al. (1982b) have proposed a mechanism which corresponds to sporadic discharges across capacitor plates, one plate being a highly charged satellite in the gap, the other being a small charged block of ice (a few meters in size). However, the mechanism for charging these bodies to the high potentials required remains obscure.

An alternative mechanism for the production of SED's has been proposed by Burns et al. (1983). They claim it is caused by lightning in Saturn's atmosphere (which has cloud-top wind velocities corresponding to 10 hr 10 min rotation period at equatorial latitudes) and that the low-frequency radio waves are permitted to escape due to the elimination of the equatorial ionosphere by the ring shadow, which was nearly equatorial during both Voyager encounters.

The fact that the radio power radiated from terrestrial lightning in the high-frequency region observed by the PRA instrument has a very different spectrum from that of SED, as well as the fact that the PRA experiment did not detect any SED-type discharges at Jupiter, even when independent evidence showed that Jupiter's atmosphere was creating super lightning bolts (Warwick et al. 1982b), throws some doubt on the atmosphere lightning source for SED's. However, Kaiser et al. (1983; see also the Kaiser et al. chapter in this book) have recently reanalyzed the SED episodes paying particular attention to their occurrence in both time and frequency as a function of the spacecraft-Saturn geometry. This analysis strongly supports an atmospheric source rather than a ring source. It shows that the wide-band emissions at low frequencies were largely observed only when the spacecraft was facing Saturn's night side. On the other hand, when the spacecraft was facing the day side of the planet there was indeed a low-frequency cutoff of the SED at ~ 5 MHz. They also showed that, while the periods of activity were centered 10 hr 10 min apart, they were also separated by periods of near silence for ~ 3 hr. They maintain that these were the times when the SED source was hidden from the spacecraft's view. They argue that a point source in the rings at 108,942 km would have been "occulted" for < 2 hr since the ring source would have been visible around a greater portion of Saturn's circumference. On the other hand, a localized lightning source in the Saturnian atmosphere would have been occulted for ~ 5 hr. Consequently, they conclude that the SED source was a large atmospheric thunderstorm system extending over 60° in longitude on the planet.

Both the SKR emission and, to a lesser extent, SED's seem to reach a maximum when a particular magnetic sector defined by ($\lambda_{SLS} = 155° \pm 15°$) is aligned with the local noon. It has also been shown (Porco and Danielson 1982) that the maximum spoke activity occurs when this same sector is in the morning portion of the rings. While these observations indicate an association among the three phenomena, with the planetary magnetic field clearly providing the common link, the basic nature of this electrodynamic coupling remains rather obscure at the present time.

The only attempt so far to explain all three phenomena under a single scheme is due to Goertz et al. (1981). The idea is to devise a mechanism which can produce electrodynamic coupling between the planetary ring system and magnetosphere in which the SKR, SED, and spoke activity are subject to longitudinal control. Goertz et al. (1981) point to the fact that the slight vertical displacement of the planetary dipole from the planet's center enables charged particles to mirror just above the ring plane without absorption by the rings; this would result in a thin plasma disk. The usual (centrifugal and pressure gradient) longitudinal drifts of the ions and electrons in this region would constitute a current in the azimuthal direction. A small ($\lesssim 0°.7$) tilt of the dipole axis with respect to the spin axis towards $\lambda_{SLS} = 331°$ would cause the magnetic and equatorial planes to intersect along a line. Consequently, the azimuthal ring current would be interrupted at certain longitudes, causing them to close via field-aligned Birkeland currents and Pederson currents in the ionosphere. The maximum parallel currents are shown to occur at $\lambda_{SLS} = 250°$ and $\lambda_{SLS} = 120°$, with the maximum flux from the ionosphere to the ring occurring at $\lambda_{SLS} = 250°$. Goertz et al. (1981) also argue that large parallel potential drops of a few kilovolts may be established in Saturn's shadow at $\lambda_{SLS} \approx 250°$, when the beaming of energetic electrons along the field line onto the plane may charge the ring bodies to high (kV) potentials. As soon as such ring particles move into the sunlight there will be a rapid electrostatic discharge through the emission of photoelectrons (accelerated to keV energies by the negative surface potential). According to this view the maximum intensity of SED should occur when the spot at $\lambda_{SLS} \approx 250°$ emerges from the shadow, at which time $\lambda_{SLS} = 110°$ faces the Sun. While the accelerated electrons could give rise to a maximum in SED, and perhaps SKR, the possible electrostatic levitation and blow-off of fine loose dust from the surfaces of the larger ring bodies on being charged to large electrostatic potentials could generate the spokes (Hill and Mendis 1981,1982b).

However, the modulation periods of SKR, SED, and spoke activity all differ by up to 20 min, which makes it implausible that the same mechanism is related to all three phenomena. In addition, as we pointed out earlier, there is some uncertainty at present about the magnetic field model of the planet, particularly with regard to the inclination of the spin and magnetic axes, which may even be exactly zero. Also, it is not clear that the parallel potential drops invoked by Goertz et al. (1981) will ever reach large enough values, because the average plasma density near the ring plane is probably significantly less than the 2000 cm^{-3} used by these authors. Nevertheless, the concept of Birkeland currents driven by azimuthal current gradients remains attractive, but needs further study. In this connection the Birkeland currents driven by the diurnal modulation of the dust-ring current discussed by Ip and Mendis (1983; see Sec. VII) may also be of interest.

X. THE POSSIBLE ROLE OF ELECTROMAGNETIC PROCESSES IN THE FORMATION OF THE RING SYSTEM

We have discussed the importance of plasma processes and the associated electromagnetic effects on the Saturnian ring system at the present time. As we have already shown, these effects are rather limited and are of importance only for fine micron- and submicron-sized grains located in certain regions of the ring system. There is, of course, the intriguing question of whether such processes played an important role in the past, both in the origin and in the early development of this ring system, as well as of others.

As we have already pointed out, there is considerable uncertainty regarding the plasma conditions in the ring plane of Saturn even at present, despite the findings of the Pioneer and Voyager spacecraft. Although purely speculative, it is, however, legitimate to question if the present structure of the ring system may point to the operation of electromagnetic processes in its formation and early evolution. This was first done by Alfvén (1954). He assumed that this system, like the rest of the planetary and satellite systems, started off as a dusty plasma, and evolved to its present state by a continuous process of agglomeration. However, unlike the larger bodies (planets and satellites), the Saturnian ring system, by virtue of being situated within the modified Roche limit of Saturn where tidal forces dominate, could not proceed all the way to form one or more larger satellites. Alfvén regards the Saturnian ring as a "cosmogonic time capsule" which has remained in a state of arrested development, and claims that its overall structure provides evidence for the operation of electromagnetic forces in its formation. The overall structure that Alfvén (1981b) considers is the following: the Cassini Division is at 2/3 the distance of Mimas; the inner edge of the B Ring is at 2/3 the distance of the outer edge of the A Ring; and the inner edge of the C Ring is at 2/3 the distance of the outer edge of the B Ring (see Fig. 20).

In order to understand this, consider the model for the formation of planetary and satellite systems proposed by Alfvén (1954) and further developed by Alfvén and Arrhenius (1976). The basic model consists of a spinning central body (Saturn or a precursor state of it in this case). It is assumed to have a strong dipolar magnetic field and to be surrounded by an ionized gas, which is a remnant of the infalling cloud out of which the central body formed. Under the action of the gravitational, centrifugal, and electromagnetic forces, a plasma element achieves an equilibrium in a circular orbit having a kinetic energy = 2/3 the kinetic energy of the Keplerian motion at that distance. If, consequently, condensation takes place and the resulting grain (with a low specific charge) essentially decouples from the magnetic field, it will fall through the central plane in an ellipse of eccentricity 1/3 and semimajor axis 3/4 the radial distance, r_0, at which the condensation took place. Furthermore, all these grains will pass the equatorial plane at a distance = 2/3 r_0, where the interparticle collisions are maximized. Inelastic collisions between these particles (which are regarded as fluffy snowballs rather than hard elastic spheres) will cause them to accumulate eventually in a circular equatorial disk.

The overall structure that this disk is expected to exhibit is shown in Fig. 21. The condensation is assumed to take place essentially from the neighborhood of the equatorial plane. The upper part of the figure shows the state where Mimas or its precursor jet stream is assumed to have already condensed and therefore to have swept up the plasma in that area, while the remaining plasma is the source regions for the A, B, and C Rings. (Incidentally, this is the first proposal of plasma sweep-up by planetary rings, which has been subsequently verified by the Pioneer and Voyager observations.) As the A, B, and C Rings form by plasma condensation on subsequent fall-down, it is clear that due to direct condensation (or absorption) of plasma onto the grains in the A and B Rings, there would be a depletion of plasma from the source region of the C Ring. In this way the boundary between the B and C Rings represents a "cosmogonic shadow" of the outer edge of the A Ring, and should be at exactly 2/3 that distance, as is indeed the case, as discussed above. The situation is further illustrated in Fig. 22, where the photometric curve of the ring system is shrunk by a factor of 2/3, inverted and superimposed in order to demonstrate the shadow it produces.

There is now another theory for the formation of the inner edge of the B Ring (Northrop and Hill 1983b; see also Ip 1983a). This is also an electro-

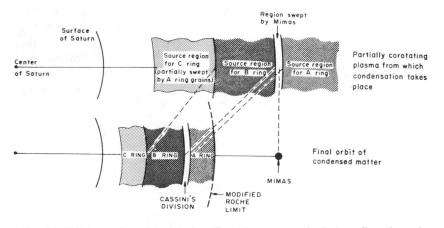

Fig. 21. Fall-down of condensed grains, illustrating cosmogonic shadow effect (figure from Alfvén and Arrhenius 1976).

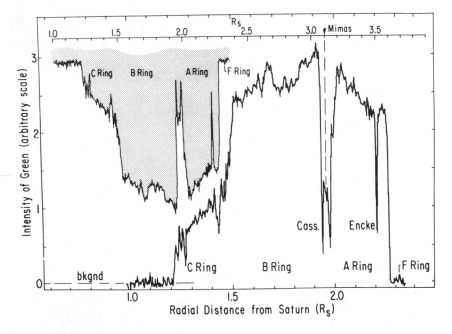

Fig. 22. Photometric curve of the ring system of Saturn according to Voyager 1 measurements, with a superimposition, at the upper left, of the same curve shrunk by a factor of 2/3 and inverted (figure from Alfvén 1981*b*).

dynamic theory, and has already been discussed in detail in Sec. VI. As pointed out there, the zonal harmonic model of Northrop and Hill (1983b) provides a much closer agreement with the observed position of the B Ring than Alfvén's cosmogonic shadow model. However (as pointed out in Sec. VI), it suffers from the drawback of having to postulate the existence of an extra force whose nature is unknown. Furthermore, the cosmogonic shadow model seems to apply to not one but several discontinuities or edges in the Saturnian ring system. Alfvén (personal communication, 1983) also points out the apparent similarity in the radial brightness profiles of all these edges, presumably indicating a similar cause for their formation.

Besides the overall structure of the Saturnian ring system, there is also the question of its newly discovered fine structure. An early idea that the narrow ringlets are produced and maintained by numerous small (shepherding) satellites has so far not been borne out by observations. A few of these ringlets are shown to be associated with spiral density waves that are similar to the waves believed to be responsible for the spiral structure of galaxies (Cuzzi et al. 1981). A few others are shown to be associated with spiral bending waves (Shu et al. 1983) of the type that are believed to be responsible for the warping of the disks of spiral galaxies. Both these types of waves are apparently excited by resonances with the various satellites of Saturn. Several other ringlets, including a few isolated ones in the C Ring, may also be associated with satellite resonances (Holberg et al. 1982a; Esposito et al. 1983). However, this accounts for only a very small fraction of the observed ringlet structure. Another suggestion is that the present overall ringlet structure is also a consequence of its origin as a dusty plasma disk (Houpis and Mendis 1983). Since the charged dust grains in this disk will be moving with a speed different from the plasma, the entire disk constitutes a dust-ring current sheet. The dusty plasma analog of the well-known finite resistivity "tearing" instability in such a current sheet is shown to tear up the disk into ringlets with widths and separation typically of the order of the thickness of the disk (Houpis and Mendis 1983). These authors have also shown that this instability grows at a much faster rate than another type of instability, like the one induced by viscous diffusion in a differentially rotating disk, proposed as an alternative intrinsic mechanism for ringlet formation (Lin and Bodenheimer 1981; Ward 1981a).

The combination of electromagnetic and gravitational forces may be responsible for a number of observational phenomena associated with the fine dust immersed in the Saturnian magnetosphere. However, our present understanding of this area is far from secure. This is partly due to the uncertainties associated with the immediate environment of the grains and the nature of the grains themselves. Even more it is due to the fact that the state of theoretical development of this new subject is as yet immature. Clearly more work is needed in the future particularly in relation to collective dusty plasma.

PART V
Satellites

ORBITAL RESONANCES AMONG SATURN'S SATELLITES

RICHARD GREENBERG
Planetary Science Institute

When orbital properties constituted most of our knowledge of Saturn's satellites, resonances provided important clues to physical properties and history. With the discoveries of the 1980s, including close-up direct observations of physical and geological properties, resonances are more important than ever. This chapter reviews the kinetics of the various classes of resonances represented among the classical and newly discovered satellites and describes how resonances have played a fundamental role in shaping the Saturn system as it is seen today.

Even before the discoveries of the past few years, the Saturn system was known to provide its share of examples of the resonant structure of the solar system. The satellite pairs Mimas and Tethys, Enceladus and Dione, and Titan and Hyperion showed how commensurable orbital periods (ratios of small whole numbers) could enhance mutual gravitational effects. These satellite resonances provided tantalizing clues about the origin and evolution of the system. And the location in Saturn's rings of the Cassini division at the 2:1 resonance with Mimas hinted strongly that such resonances might dominate the radial structure of the rings.

The recent flurry of new discoveries with spacecraft and with ground-based instruments provided striking evidence of the structural importance of resonances in the Saturn system. All of the eight new satellites recently discovered or confirmed (all in 1980) have peculiar dynamical behavior or effects related to resonance. The classical satellites, especially Enceladus, have geology apparently governed by unknown heat sources; tidal heating driven by orbital resonances is a leading candidate. The structure of Saturn's rings, now known to very high resolution, exhibits many features undoubtedly associated with resonances, although much of the structure is still not well explained.

In this chapter, I critically discuss the role of resonances in some interpretations of the new data, concentrating on the relations between resonances and the physical properties and dynamical histories of Saturn's satellites. I have discussed most of the fundamentals of orbital resonance mechanisms, including qualitative physical explanations, in other books in the Space Science Series (Greenberg 1977a; Greenberg and Scholl 1979; Greenberg 1982; see also Greenberg 1977b), and will not repeat them here. Interpretation of ring structure involves the subtle combination of resonant driving processes with still incompletely understood collective gravitational and collisional effects within the rings. Such interpretation will be the subject of the book *Planetary Rings* (Greenberg and Brahic 1984) and will not be included here. However, through the ring connection, a considerable amount of terminology from galactic dynamics involving resonances has come to be applied to the Saturn system. In order to elucidate the connection between strictly satellite resonances and resonances in the rings, I include a kinematic discussion (Sec. III) that relates the different terminology used in those areas.

I. OVERVIEW OF THE STRUCTURE OF THE SATURN SYSTEM

In order to introduce the extent to which resonances govern the Saturn system, I show in Fig. 1 the orbital positions of each of the satellites from the rings through Hyperion, in terms of the log of the sidereal mean motion n. Some ratios of mean motions are also shown. Each of these ratios is within a few percent of a ratio of small whole numbers. They suggest the commensurability scheme shown in the lower part of Fig. 1. Qualitatively, there appears to be a remarkable correlation between the actual positions (upper row of tick marks) and the idealized scheme (lower row). The greatest difference between actual and idealized position is for Mimas and Tethys, but even there the ratio of these two satellites' mean motions (2:1) is preserved in the idealized scheme. The statistical significance of such commensurability schemes has been rightly called into question (e.g., Hénon 1969). Any ratio is arbitrarily close to some ratio of whole numbers if one admits large whole numbers. The degree of impressiveness of the correlation in Fig. 1 depends on one's standards for how small the whole numbers should be and how close the actual position should be to commensurable positions.

An objective criterion for identifying significant commensurabilities is to require consequent observable resonant behavior. Such behavior can help to determine physical properties of the satellites and to constrain scenarios of their history. The three classical resonances, Mimas-Tethys, Enceladus-Dione, and Titan-Hyperion, have well-known effects that have allowed determination of masses and that act to maintain the commensurabilities. This quality of being locked into resonance is an important clue to origins and past orbital evolution. 1980S6, discovered by Laques and Lecacheux (IAU 1980a), orbits on average at the same rate as Dione. Due to this 1:1 reso-

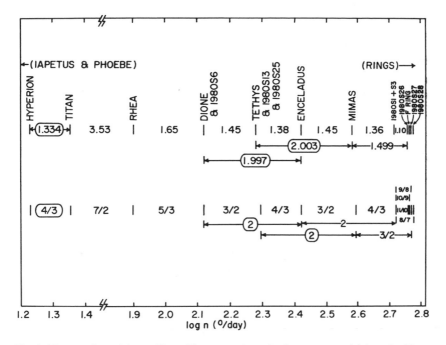

Fig. 1. Mean motions of the satellites of Saturn are shown by the upper row of tick marks. Numbers between these marks give the ratios of mean motions of adjacent orbits. Ratios between the pairs Mimas-Tethys, Enceladus-Dione, and 1980S26-Mimas are also indicated. Ovals around ratios identify the classical resonances of known dynamical significance. For comparison, the lower row of marks shows a scheme of precisely commensurable positions that closely match the actual satellite positions. No claim is made for dynamical or statistical significance of this correlation except as discussed in the text.

nance, the small satellite is locked near a position 60° ahead of Dione. Similarly, 1980S13 and 1980S25 are each in 1:1 resonance with Tethys (IAU 1980*b,c*), locked 60° ahead of and behind Tethys, respectively. Synnott reports that there may be additional satellites in similar 1:1 resonances with Dione and Tethys (IAU 1980*d,e*). 1980S1 (Janus) and 1980S3 (Epimetheus), the coorbital satellites, are in a 1:1 resonance with one another, with periodic effects that prevent mutual collisions. The coorbitals had been identified as a single satellite by Dollfus (1968); later studies of the data revealed that there is more than one satellite (Fountain and Larson 1978; Aksnes and Franklin 1978), but only after the 1980 ring plane crossing was their coorbital nature revealed (Smith et al. 1980).

Smith et al. (1982) suggest that the coorbital satellites are fragments of a collisionally disrupted parent body. Since they are very close to the 2:1 commensurability with Enceladus (Fig. 1), it is natural to speculate that a parent body might have been in a resonance involving Enceladus. Such a resonance might have promoted tidal heating in much the same way as resonance in the Jupiter system keeps Io hot, and may explain the relatively recent endogenic activity on Enceladus (Cowen 1982; Lissauer et al. 1982a; see also Sec. V of this chapter).

The coorbital satellites exert substantial confining torque on the outer edge of the A Ring, because ring particles there are in a 7:6 resonance with those satellites. Similarly, 1980S26, 1980S27, and 1980S28, known as the shepherd satellites, exert torques on the neighboring ring edges which tend to confine the rings. This shepherding process is made possible by the close clustering near the edge of a ring of commensurabilities of the form $j:(j + 1)$ with a nearby shepherd satellite. The combination of these resonance effects is generally accepted as the explanation for why the rings have not spread as far as collisional evolution might otherwise dictate. However, a problem remains: given the computed strengths of the confining torques, the inner satellites should have been pushed out by the reactive torques to distances from Saturn far beyond their actual positions. The original explanation (Synnott et al. 1981) was that 1980S26 must be locked in a resonance with Mimas, which holds the shepherd in even as the ring torque tries to push it out. Such a resonance seemed plausible because, as shown in Fig. 1, 1980S26 is very close to the exact 3:2 commensurability with Mimas, at least as close as any of the three classical resonances is to exact commensurability. However, the criterion for resonant behavior is somewhat more subtle than simply being close to having commensurable sidereal mean motions, and it turns out that 1980S26 is "not presently stabilized by any known resonance" (Goldreich and Tremaine 1982). The commensurability is apparently a coincidence according to Goldreich and Tremaine, although Lissauer and Cuzzi (1982) still call it a possible resonance (see Sec. IV for further discussion).

II. KINEMATICS

The precise criterion for resonance of an orbit is the same as for any dynamical system: the natural frequency of the system is matched by the frequency of a periodic force or a harmonic of that force. The relevant natural oscillations in a satellite's orbital motion might be (a) the epicyclic motion (corresponding to orbital eccentricity) relative to a perfectly circular reference orbit, or (b) the oscillations out of the reference plane (corresponding to orbital inclination). The forcing function, in this case the force of another satellite, has a period equal to the synodic period of the two satellites. In Keplerian motion, the natural (a) epicyclic and (b) vertical periods equal the orbital period, so one expects resonance where a satellite's sidereal mean motion equals a whole-number multiple (harmonic) of the synodic mean motion with another satellite:

$$n = j(n - n') \Rightarrow n/n' = j/(j - 1) \tag{1}$$

where j is a whole number and primes refer to the perturbing satellite. The epicyclic or vertical motion is not precisely sinusoidal, except to first order in eccentricity e or inclination i. Thus, more generally, including higher order effects in e and i, resonance may occur where any multiple of the natural frequency equals any multiple of the forcing frequency:

$$kn = j(n - n') \Rightarrow n/n' = j/(j - k). \tag{2}$$

It may appear from Eq. (2) that resonance does occur where sidereal mean motions are precisely commensurable. The flaw is that in the Saturn system a satellite's motion, even in the absence of a perturbing neighbor, is not Keplerian. Saturn's oblateness yields a force field that differs significantly from the idealized r^{-2} law. The natural frequencies of oscillation thus differ from the orbital mean motion; i.e., the orbits precess. The epicyclic frequency κ is related to the apsidal precession rate $\tilde{\omega}$ by $\kappa = n - \tilde{\omega}$. Consider a case where such precession is important. For simplicity, first assume that both orbits are equatorial and the perturber's is circular. The resonance criterion (Eq. 2) now takes the form

$$k(n - \tilde{\omega}) = j(n - n') \Rightarrow jn' - (j - k) n - k \tilde{\omega} = 0. \tag{3}$$

In Eq. (3) we have an example of how precession can shift a resonance away from a position of exactly commensurable sidereal mean motions. (Note, however, that even in the case of Eq. (3) the mean motions would be precisely commensurable if they were measured relative to the major axis of the perturbed satellite instead of being sidereal.) This type of shifting explains why 1980S26, despite its close sidereal commensurability, is not near any known resonance.

If we take eccentricity of the perturbing satellite into account, the forcing functions contain additional harmonics besides $j(n - n')$. The gravitational field of the perturbing satellite, Fourier decomposed longitudinally, takes the form

$$\sum_{m=0}^{\infty} \sum_{p=-\infty}^{\infty} f(r)\sin[\omega t - m\theta + \delta] \tag{4}$$

where $\omega \equiv mn' + p\kappa'$, δ is a phase constant, and (r, θ) is the position in polar coordinates. Thus, each component of the field behaves like a wave moving longitudinally at angular rate ω/m and with wavelength $360°/m$. A satellite moving with angular velocity n experiences oscillations in the field with frequency $m \mid (n - \omega/m) \mid$. Resonance occurs where the perturbed satellite's natural frequency κ matches that value, i.e. where

$$\kappa = m \mid (n - \omega/m) \mid . \tag{5}$$

Because of the absolute value signs, there are two values of n (and thus two orbital radii) at which resonance can occur: one for $n < \omega/m$, and one at smaller radius where $n > \omega/m$. These resonances are called outer and inner Lindblad resonances, respectively, in the terminology of galactic dynamics, which is increasingly applied to planetary systems due to the structural similarity between rings and galaxies. In terms of mean motions and apsidal precession rates, the Lindblad resonances from Eq. (5) take the forms:

$$(m + p)n' - (m - 1)n - \tilde{\omega} - p\tilde{\omega}' = 0 \text{ (inner Lindblad)} \qquad (6)$$

$$(m + p)n' - (m + 1)n + \tilde{\omega} - p\tilde{\omega}' = 0 \text{ (outer Lindblad)}. \qquad (7)$$

For a given harmonic of the perturbing potential (given m and p), the two Lindblad resonances may be nowhere near one another. For example, for $m = 3$ and $p = 2$ the inner Lindblad resonance is near the $5:2$ commensurability and the outer one is near $5:4$. The inner Lindblad resonance for $m = 5$, $p = 0$ is also near $5:4$, but at a slightly different position due to the different precession rates $\tilde{\omega}$ and $\tilde{\omega}'$. Note that the words inner and outer here do not refer to position relative to the perturbers; in the example above, both the inner and outer Lindblad resonances are inside the orbit of the perturber. Note also that for $p = 0$ (a harmonic independent of the perturber's eccentricity), Eqs. (6) and (7) reduce to Eq. (3) with $k = 1$ (first order in e).

Another kind of dynamical singularity, called corotation resonance, occurs where the mean motion of a perturbed satellite matches the angular velocity ω/m of a component of the field, i.e. where $n = \omega/m$ (cf. Eq. 5), or with the definition of ω

$$(m + p)n' - mn - p\tilde{\omega}' = 0. \qquad (8)$$

Perturbations at such a position are independent of the perturbed satellite's eccentricity, but do depend on the perturber's eccentricity to generate the appropriate harmonics.

The most general type of in-plane (eccentricity) resonances must include higher order terms in the eccentricity of the perturbed satellites. By analogy with Eq. (2), such resonances occur where multiples of the natural frequency match a component of the perturbing force:

$$k \kappa = m \mid (n - \omega/m)\mid \qquad (9)$$

which yields

$$(m + p)n' - (m - k)n - p\tilde{\omega}' - k\tilde{\omega} = 0 \qquad (10)$$

and

$$(m + p)n' - (m + k)n - p\tilde{\omega}' + k\tilde{\omega} = 0. \qquad (11)$$

The Lindblad resonances are the cases with $k = 1$, and the corotation resonances have $k = 0$. Thanks to precession of apsides, there is a cluster of separate resonant positions near each commensurability of sidereal mean motions.

An even more general resonance condition results if we consider motion not restricted to two dimensions. In this more general case, components of the gravitational field Eq. (4) rotate at rates

$$\omega = mn' + p\kappa + q\kappa_v \tag{12}$$

where q is an odd integer and κ_v is the rate of vertical oscillation of the perturber. (In the literature κ_v is sometimes called μ.) By definition, $\kappa_v \equiv n - \dot{\Omega}$ where $\dot{\Omega}$ is nodal precession rate. A resonance may occur wherever the forcing frequency $|mn - \omega|$ equals a multiple of either the vertical or horizontal natural frequency ($k\kappa_v$ or $k\kappa$). In the most general form, candidate positions for resonances are wherever

$$j_1 n' + j_2 n + j_3 \dot{\tilde{\omega}}' + j_4 \dot{\tilde{\omega}} + j_5 \dot{\Omega}' + j_6 \dot{\Omega} = 0 \tag{13}$$

where j's are integers.

In classical celestial mechanics, the gravitational potential of one satellite at the location of another is expressed in terms of orbital elements of both satellites, and then Fourier analyzed. The terms in the Fourier expansion are of the form

$$Gm'F(a,a') \; e'^{|j_3|} e^{|j_4|} i'^{|j_5|} i^{|j_6|} \cos\theta \tag{14}$$

where

$$\theta \equiv j_1 \lambda' + j_2 \lambda + j_3 \tilde{\omega}' + j_4 \tilde{\omega} + j_5 \Omega' + j_6 \Omega \tag{14a}$$

and λ's are mean longitudes, e's are eccentricities, and i's are inclinations. The θ's generally vary rapidly, on time scales comparable to orbital periods, because λ's vary at rates given by mean motions; the integrated perturbations cannot be very large. However, near a resonance of the general form of Eq. (13), one θ varies slowly; its rate of variation is given by the left side of Eq. (13). There, perturbations can be very large. Thus, the expanded disturbing function can serve as a catalog of possible resonances (see e.g., Brouwer and Clemence 1962). The coefficient of each term gives an indication of the strength of the corresponding resonance. We see from Eq. (14) that resonances involving small integer coefficients (j_3, j_4, j_5, j_6) for the precession terms are strongest, because they have the small e's and i's to the lowest power. According to d'Alembert's rules (see e.g., Danby 1962), the sum of the j's must be zero (e.g., Eq. 10 or 11). Thus, a resonance near the mean motion ratio 4:3 must be of only first order in eccentricity or inclination, but one near 5:3 must be second order and thus weaker. In this way, the expansion reflects the kinematically reasonable idea that resonances are only important near low order commensurabilities.

III. THE CLASSICAL RESONANCES

The first indication of a resonance among Saturn's satellites was the discovery (Hall 1884) of the annual 20° regression of Hyperion's apsides, rather than the expected advance due to Saturn's oblateness. Newcomb (1891) showed that the 4:3 resonance with Titan was responsible. The resonance condition in this case is

$$\theta \equiv 4\lambda_H - 3\lambda_T - \widetilde{\omega}_H = 180°. \tag{15}$$

In other words

$$\dot{\theta} \equiv 4n_H - 3n_T - \dot{\widetilde{\omega}}_H = 0 \tag{16}$$

and conjunction of the two satellites ($\lambda_T = \lambda_H$) always occurs at Hyperion's apocenter. In ring-galactic terminology, this is an outer Lindblad resonance (Eq. 7) with $p = 0$, $m = 3$. A dominant effect of Titan on Hyperion is thus an extra force at apocenter towards Saturn, which causes the apsidal regression. Newcomb also showed that Titan acted to stabilize the resonance by adjusting Hyperion's mean motion, such that θ librates about 180°, and Titan forces Hyperion's eccentricity to its large value of 0.1. Eichelberger (1891) reduced observational data to obtain the libration, and Woltjer (1928) refined those results to obtain a theory (and a mass value for Titan) that closely matches the observations (libration of 36° amplitude and 21 month period).

Similarly, the Enceladus-Dione resonance was revealed by Enceladus' apsidal precession being much slower than seemed consistent with the planet's oblateness. Woltjer (1922) explained this behavior by invoking the 2:1 resonance,

$$2\lambda_D - \lambda_E - \widetilde{\omega}_E = 0. \tag{17}$$

Here, too, resonance acts to maintain the relationship (Eq. 17), so conjunction librates about Enceladus' pericenter. The libration amplitude (deviation of Eq. 17 from zero) is $< 1°$, with period 12 yr. Here, unlike the Titan-Hyperion case, neither satellite is of negligible mass, and their mutual perturbations are important. From the point of view of perturbations on Enceladus, this is an inner Lindblad resonance with $p = 0$, $m = 2$; but at Dione this is a corotational resonance with $p = 1$, $m = 2$. The theory of the mutual interactions was refined by Jeffreys (1953) and Kozai (1957), who obtained masses of the two satellites from their mutual perturbations (see also Jefferys and Ries

1975). Without the resonance, mutual effects would have been too weak to yield values for the masses.

The Mimas-Tethys resonance was invoked by Struve (1890) to explain oscillations in Mimas' mean longitude (relative to uniform motion) of $\pm$ 44° with period 71 yr. This resonance is near the 2 : 1 commensurability, with conjunction of the two satellites librating about the mean of their ascending nodes on Saturn's equatorial plane. The stable configuration is thus

$$4\lambda_T - 2\lambda_M - \Omega_T - \Omega_M = 0. \tag{18}$$

By comparison with Sec. II, this resonance is a vertical analog of an inner Lindblad resonance at Mimas, and of an outer Lindblad resonance at Tethys. (In the terminology of Shu et al. [1983], these resonances would simply be called inner and outer vertical resonances.) In satellite resonance terminology, this is usually called an inclination-type resonance, and it is the only known example among satellites. (Apparently Mimas does drive vertical resonances in the rings, however [Shu et al. 1983].) The mechanism of this resonance is discussed by Allan (1969). Greenberg (1973a) gave a heuristic explanation of the significance of the midpoint between nodes as a stable orientation for satellite conjunction. The theory was updated by Jefferys and Ries (1979).

The Mimas-Tethys resonance maintains the relationship of Eq. (18) by periodically exchanging energy between the two satellites. The libration of the left side of Eq. (18) has amplitude 97°, most of which is due to the 44° variation in λ_M. From the period of the libration and the relative contributions of variations in λ_M and λ_T, Jeffreys (1953) and Kozai (1957) derived masses with standard deviation of only a few percent.

The Voyager 2 radio experiment has called those mass results into question; Tyler et al. (1982a) find from variations in the spacecraft's motion that Tethys' mass is 21% greater than the value derived from resonance theory. The standard deviation of the Voyager measurement is $\sim$ 15%, so it still seems likely that the resonance-based value is correct. Nevertheless, Tyler et al. believe that there may be some systematic error in the resonance theory. They cite as sources of such error nonspherical figures of the satellites themselves, effects of undiscovered satellites, and unnamed contributions to the libration, although such effects seem too weak or too *ad hoc* to compel belief that the resonance theory needs revision. Because Tyler et al. believe Tethys' mass needs to be revised upward, they compute a new increased mass for Mimas as well, in order to maintain the relative magnitude of oscillations in λ_T and λ_M. It is not clear, however, why they are willing to accept that part of the resonance theory, yet are also willing to adopt masses that are in conflict with the libration period.

Tyler et al. also suggest a revision of the resonance-determined mass for Dione, which is on even shakier ground than the Mimas and Tethys revisions. They find that the densities of the satellites of Saturn, especially with their revised masses for Mimas and Tethys, fit a particular cosmogonic scheme (Prentice 1978) fairly well, except that Dione's mass is too high. Hence, they propose that Dione's mass may be 10% to 15% less than the classical value (see e.g., Kozai 1957). Yet, the classical value is probably accurate to $\sim$ 2%. The classical theories are linear in the mass ratios of satellite to planet, but higher order effects are probably not important, given the small satellite masses involved. While it certainly could not hurt to check the classical determination of the masses, it is more conventional to suspect cosmogonical theory that does not fit physical measurements than vice versa.

IV. RESONANCES INVOLVING NEWLY DISCOVERED SATELLITES

The shepherd satellites 1980S26, 27, and 28, exert torques on nearby rings by forcing ring particles into epicyclic oscillations. To the extent that such oscillations are damped by collisions among ring particles, or by density-wave propagation, this response lags behind the forcing functions. The damping thus introduces an asymmetry, which allows a net torque to be exerted. For a given amount of damping, the torque is greatest on ring particles in resonance with the satellite, because the forced response has a large amplitude. Resonances dominate in shepherding because, in a ring near a satellite, commensurabilities of the form $j/(j + 1)$ are very densely packed in orbital radius (Goldreich and Tremaine 1980; Greenberg 1983). Averaging over many resonances, one finds that the torque is independent of the degree of damping; less damping means a decrease in phase lag, but this is balanced by increased amplitude at resonance.

While shepherding torques act to help confine rings, the reaction tends to drive satellites away. 1980S26, for example, would have been driven well beyond its present orbit in only $\sim 2 \times 10^8$ yr (Goldreich and Tremaine 1982), unless some unknown counteracting process is in effect. Similarly, a torque between the coorbital satellites and the outer edge of the A Ring, via the 7:6 resonance, seems to drive those satellites out at an unacceptable rate.

Synnott et al. (1981) wrote that 1980S26 is stabilized by the 3:2 resonance with Mimas. However, the closest resonance is the corotation resonance: $3n_M - 2n - \tilde{\omega}_M = 0$. At S26, the left-hand side has the value $-0.55°$ day^{-1}, not close enough for a significant resonance torque to be exerted (see Franklin et al. [1984] for a discussion of widths of resonances). The mechanism that stabilizes the shepherding satellites remains unknown. A somewhat improbable alternative is that these satellites have only recently come out of the rings, and their present orbits are quite temporary.

The coorbital satellites 1980S1 and 1980S3 are in a dynamically significant 1:1 resonance. In 1:1 resonances, the standard expansion (Eq. 14) of the disturbing function is usually invalid. If one satellite of the pair were of negligible mass, it might be possible to invoke some results from the restricted three-body problem. But, because these satellites are of comparable mass, and in order to get a more complete solution including perturbations from other satellites, the orbits were integrated numerically by Harrington and Seidelman (1981). Analytical theories have been constructed by Dermott and Murray (1981a,b) and Yoder et al. (1983). Each satellite follows a horseshoe-shaped orbit in a coordinate system that is centered on Saturn and that rotates about Saturn with the angular velocity of the other satellite.

This behavior can be understood in terms of the following physical description. When the satellites are far apart in longitude, they have negligible effect on one another. In fact, at such times slight deviations from Keplerian motion are dominated by Enceladus and Dione through the 2:1 and 4:1 resonances (see Fig. 1). Smith et al. (1980) had speculated that resonant perturbations by Enceladus and Dione were necessary to prevent collision by keeping the coorbitals 180° apart. However, those resonances can only enhance perturbations somewhat; they are not strong enough to lock the coorbital satellites into commensurability. The slight difference in periods between the two coorbitals, 0.005% of the 16.7 hr period, means that every few years these satellites must approach one another. As the faster one begins to approach the leading slower one, it pulls back on the slow one. The slower satellite loses orbital energy, and the faster one gains energy. An energy loss means a decreased semimajor axis and increased mean motion; similarly, an energy gain means a decreased mean motion. Thus, the fast trailing satellite slows down and the slow leading one speeds up. This effect gets stronger as the satellites get closer to one another. Just before they would come together, the leading satellite begins to move faster than the trailing one, and the satellites start to separate. This mechanism prevents collision. The satellites again have negligible mutual effect, until several years later when the now faster one begins to overtake the other, the energy exchange is reversed, and the cycle repeats. This energy exchange at close encounter has also been described analytically by Goldreich and Tremaine (1982).

The dynamics of the small satellites locked to Lagrange points of the much bigger satellites, Dione and Tethys, are more directly understood in terms of the restricted three-body problem. Reitsema (1981a,b) has fit observations of these satellites to Erdi's (1978) theory of the Trojan asteroids, which are similarly locked to the motion of Jupiter. The Dione Trojan (1980S6, or Dione B) is found to librate about the stable Lagrange point 60° ahead of Dione with an amplitude of + 17° and − 13° and with period 785 d. Reitsema notes two systematic effects, both related to resonances, that might affect these results. First, the libration period is sensitive to the mass of Dione, such that a 2% error in mass would correspond to a 1% change in period. This effect may eventually provide a sensitive measure of Dione's mass. Second, Enceladus is of course locked into a 2:1 commensurability with Dione B via the 2:1 resonance with Dione. Thus, resonance-enhanced perturbations by Enceladus might significantly modify Dione B's motion.

For the leading and trailing Trojans of Tethys (1980S13 and 1980S25), the former seems to fit a very small libration amplitude of ± 2° and period 693 d, but the latter does not seem to fit a unique orbit relative to the Lagrange point (Reitsema 1981*b*). Moreover, two data points do not seem to fit any librating trajectory. Reitsema suggests that some observations may be spurious, or may represent additional satellites. Alternatively, 1980S25 may be perturbed by Mimas through the 2:1 commensurability or it may have some other complex behavior, rather than straightforward libration about the trailing Lagrange point. Eventually, study of the motions of the Trojans of Tethys may allow its mass to be determined and resolve the controversy discussed in the previous section.

V. HISTORY OF RESONANCES

The existence of so many resonances must be a key to the Saturn system's past. The fact that resonances tend to stabilize commensurabilities is crucial. A process that acts to change orbital periods may bring satellites into commensurability, after which, under certain circumstances, resonances can preserve the commensurability despite the continuing presence of the period-changing mechanism. This process is called resonance capture. Goldreich (1965) proposed that tides raised on the planet yielded the required changes in orbital periods, and details of possible resonance captures were investigated by Allan (1969), Greenberg (1973*b*), Sinclair (1972), and Yoder (1975) for the classical cases in the Saturn system. That work is reviewed by Greenberg (1977*a*).

Capture into the Titan-Hyperion resonance by this mechanism required tidal dissipation in Saturn so great that inner satellites, such as Mimas, could not have remained as close to the planet as they are. This paradox may be resolved:

(i)   by invoking smaller values of the tidal dissipation parameter Q (i.e., dissipation of a greater fraction of the tidal strain) for tides raised by Titan than for the relatively high frequency tides raised by Mimas;
(ii) by invoking for Titan some other dissipative orbit-changing process such as early gas drag;
(iii) by assuming that Hyperion has always been in resonance with Titan.

To elaborate on the latter possibility, one might speculate that after formation of one large satellite, resonance might have promoted favorable accretion of another one at commensurability. Alternatively, Hyperion might be a product of natural selection, possibly one of many original small bodies on highly eccentric orbits near Titan's, but the only one prevented by resonance from moderately close approach to Titan. Recall that conjunction with Titan can only occur near Hyperion's apocenter.

Motivated by Voyager imaging of Hyperion, Farinella et al. (1983) have proposed an interesting version of the natural selection hypothesis. Hyperion's irregular shape suggests to them that the satellite is a collisional fragment of a larger proto-Hyperion. In addition, Smith et al. (1982) estimate that all the satellites from Dione inward have been collisionally disrupted by cometary impact. Yet in the inner satellite region most of the disrupted material has been reaccreted into nearly spherical bodies. Why did reaccretion not occur for Hyperion? Farinella et al. suggest that such reaccretion was impossible for the Hyperion debris because of the gravitational effects of Titan. They argue that, after the debris was ejected onto various orbits, only the largest fragment (the present Hyperion) was protected by resonance from encounter with Titan. The smaller pieces were then scattered by Titan out of the Saturn system or accreted by Titan, before they could have been reaccreted by Hyperion. Farinella et al. also suggest that some debris from Hyperion may have been scattered to contribute to the cratering flux on the inner satellites.

This model is intriguing, but raises questions that need to be addressed quantitatively (D. Davis, personal communication). Is a chunk of the size and shape of Hyperion really a plausible product of collisional disruption? A body as large as the proto-Hyperion would have been gravitationally bound, so a collision energetic enough to disrupt it would have had much more than enough energy to catastrophically fragment the material. Such super-catastrophic fragmentation would yield only very tiny pieces. This problem is also encountered in trying to explain the formation of Hirayama family asteroids from disruption of a parent asteroid (Davis et al. 1983). One solution might be that small fragments reaccreted to form Hyperion. But that solution raises questions about how Hyperion got its irregular shape and whether reaccretion could occur before the material was scattered or accreted by Titan. An alternative possible solution proposed by Davis et al. (1983) is that material of the large protobody was effectively strengthened by internal pressure, so that some fragments could be large.

Another aspect of the model by Farinella et al. that requires quantitative study is the likelihood that Hyperion fragments contributed significantly to the cratering of the inner satellites. Preliminary calculations by D. Davis show that such cratering occurs only if the fragments are ejected by the disruption into orbits with pericenter well inside Titan's orbit. Fragments with pericenter near Titan's orbit are either swept up by Titan or ejected from the Saturnian system.

Smith et al. (1982) suggest a natural selection origin for the Trojan satellites associated with Dione and Tethys, closely analogous to the scenario of Farinella et al. for the Hyperion-Titan case. Because the inner satellites have probably undergone a history of repeated collisional disruption and reaccretion, they suggest that each Trojan represents residual debris from the last disruption of the coorbiting larger body. In each case, most of the debris was reaccreted by the larger body, except for the smaller satellites protected by the 1:1 resonance. This mechanism for Trojan formation requires catastrophic disruption of the parent body. Furthermore, it requires that the larger body only reforms after some debris has spread out to the Lagrange points. Production of smaller satellites by cratering or chipping from a parent that essentially remains intact would probably not yield Trojans, unless some mechanism could be found to confine the satellites subsequently to the neighborhood of a Lagrange point. It is more likely that such satellites would simply be reaccreted by their parent.

Smith et al. (1982) address the question of how the Trojan satellites have remained nonspherical. The concern was that in the heavy bombardment environment required for their scenario of Trojan formation the Trojans themselves would undergo fragmentation and reassembly that would render them spherical. Smith et al. suggest that most of the debris from these events was ejected out of the protective 1:1 resonance with Dione or Tethys, such that the debris was reaccreted by the larger satellite rather than by the Trojans. This model leaves the Trojans small and irregular, just as the scenario of Farinella et al. may explain Hyperion's shape.

Even before the discovery of satellite Trojans, Colombo et al. (1974) discussed the possibility of Trojans locked to the L_4 and L_5 points of Titan. They noted that in the course of its resonant librations, Hyperion regularly swings into the Trojan zones. They concluded that encounters with Hyperion would eliminate such Trojans, because of either strong perturbations or actual collisions. Perhaps this was the source of the impact that gave Hyperion its irregular shape. Colombo et al. speculated that Hyperion itself may have been formed as a Trojan and later transferred to its present orbit, perhaps by collisions with other Trojans.

Smith et al. believe that the coorbital satellites (1980S1 and 1980S3), as well as the three shepherd satellites and possibly the rings themselves, "are probably products of a long, complex history of collisional disruption" and may be fragments of a common larger parent body that broke up early in the period of heavy bombardment.

Cowen (1982) and Lissauer et al. (1982a) have noted the proximity of the coorbitals to the 2:1 commensurability with Enceladus. They propose that the parent body may have been in resonance with Enceladus, or perhaps in a three-way resonance with

$$n_{\text{parent}} - 3n_E + 2n_D = 0, \tag{19}$$

analogous to the Laplace resonance among the Galilean satellites of Jupiter. Such a resonance, if it lasted until recently, would help solve two fundamental problems of the Saturn system. First, it would explain how the coorbital satellites could exert a confining torque on the rings without having been pushed much farther out from Saturn. The resonance lock with Enceladus and Dione could have anchored the parent body and thus the rings. Second, it would provide an additional heating mechanism to help explain the remarkable geology of Enceladus.

That Enceladus, like Io (Peale et al. 1979), might have peculiar geological properties due to tides driven by resonance had been predicted by Yoder (1979) before the Voyager encounters. Qualitatively, the analogy with Io seemed very close. Enceladus' free eccentricity is very small, indicating that significant tidal dissipation had occurred within the satellite; the eccentricity forced by resonance with Dione is much larger, providing a continuing cause for tidal working of the body. Voyager 2 did find evidence of long-term and possibly ongoing heating of Enceladus. Endogenic resurfacing of the satellite continued up to 10^9 yr ago, and possibly even later (Smith et al. 1982). Cook and Terrile suggested that the E Ring may be replenished by current volcanic eruptions from Enceladus (Smith et al. 1982), driven by tidal heating.

Yet, there are serious qualitative problems with the hypothesis of resonance-driven tidal dissipation (Yoder 1981). The eccentricity forced by Dione is simply too small to deliver the amount of heat required to drive continuing volcanism. The rate of heating is probably less than the radiogenic contribution from a modest rocky component. Yoder suggests that the resonance may drive heating in an episodic manner, similar to that proposed by Greenberg (1982) for Io. In such models, there is an oscillatory interaction between geophysical changes in the satellite, which affect the dissipation, and orbital changes; the orbital changes include variation in closeness to exact resonance, and hence variation in magnitude of forced eccentricity. Yoder also suggests an alternative resonance-driven model in which Tethys may have been in resonance with Enceladus (presumably at the 4:3 commensurability, as in the bottom of Fig. 1) up to ~ 10^9 yr ago, a resonance presumably disrupted by tidal evolution. Perhaps the disruption was related to capture of Tethys into the resonance with Mimas, as Mimas evolved outward from Saturn. Certainly, any model that includes an Enceladus-Tethys resonance must address the role of Mimas, which at some time would also have been commensurable with Enceladus.

Another issue that needs to be addressed in all these resonance-driven tidal heating models, including the one involving the parent body of the coorbitals, is how such mechanisms continued up to an aeon ago, or even up to the present if one wants to generate the E Ring in this way. Smith et al., for example, envisioned the coorbitals' parent body being destroyed much earlier, during the period of heavy bombardment.

Better constraints on the chronology of the satellites will be obtained after various scenarios for orbital evolution have been explored in detail. Orbital evolution models, with detailed studies of behavior as various commensurabilities are reached, will need to be considered. One example is the effect that Mimas might have had as it moved outward from Saturn through the 4:3 commensurability with the coorbital satellites. Preliminary numerical models by Franklin (personal communication) suggest that, if the time of passage through resonance were long compared with the period of the horseshoe libration, the coorbiting satellites could not survive as such (i.e., the horseshoe configuration would not remain). (In accord with the theory of Sinclair [1972] and Yoder [1975], the satellites were not captured into resonance with Mimas either.) These results, if correct, indicate that Mimas' tidal evolution may have been very slow (Saturn's tidal $Q \geqslant 10^6$) so that such passage through resonance never occurred. Another possibility is that passage through resonance occurred before the parent of the coorbital satellites was disrupted, perhaps while the parent was still in 2:1 resonance with Enceladus. Or perhaps Mimas played a role in setting up the 1:1 resonance between the coorbital satellites; could the horseshoe configuration have been induced by Mimas? These and other questions need to be examined in detail.

There can be no doubt that resonances have played a major role in shaping the dynamical structure of the Saturn system and the physical properties of the satellites themselves. Reconstruction of these relationships will involve exciting interplay between theoretical studies and the wealth of observational information now available. This process, as reviewed here, has only begun.

Acknowledgments. Referees F. Franklin and S. Tremaine each made critical suggestions and corrections regarding the content of this chapter. This work was supported by NASA's Planetary Astronomy Program. The Planetary Science Institute is a division of Science Applications, Inc.

SATELLITES OF SATURN: GEOLOGICAL PERSPECTIVE

DAVID MORRISON
University of Hawaii

TORRENCE V. JOHNSON
Jet Propulsion Laboratory

EUGENE M. SHOEMAKER, LAURENCE A. SODERBLOM
U.S. Geological Survey

PETER THOMAS, JOSEPH VEVERKA
Cornell University

and

BRADFORD A. SMITH
University of Arizona

The Voyager encounters have added the satellites of Saturn to those planetary bodies for which geological studies are possible. These satellites are surprisingly heterogeneous, in spite of their apparently common composition of more than half water ice. All are more or less heavily cratered, but most also show clear evidence of substantial endogenic modification and resurfacing during the first few hundred million years of their existence. In the case of Enceladus, this internal activity has been most dramatic and may have persisted into the past billion years of solar system history. Iapetus also remains a major mystery with its still unexplained hemispheric dichotomy, for which both external and internal causes have been suggested. In this chapter we review the current knowledge of the satellites from a geologic perspective, using primarily Voyager imaging data. Emphasis is placed on the six larger bodies (except Titan, which is treated elsewhere in this book): Rhea, Iapetus, Dione, Tethys, Enceladus, and Mimas.

I. HISTORY

The concept of a satellite system originated with Galileo's discovery in 1610 of the four satellites of Jupiter that bear his name. The first telescopic observers also searched for companions of Saturn, but it was not until 1655 that Christian Huygens in Holland discovered Titan, in the same year that he also was able to deduce the true nature of the rings. The discovery two decades later by Cassini of four additional satellites (Iapetus, Rhea, Tethys, and Dione) established the existence for Saturn of a regular satellite system comparable to that of Jupiter (see chapter by Van Helden).

By the start of the twentieth century, the nine "classical" satellites of Saturn had been discovered, and it was not until 1966 that a tenth object was tentatively identified. Observations in France by A. Dollfus and colleagues and at the University of Arizona primarily by G. Kuiper, J. Fountain, and S. Larsen yielded images of what would later be identified as two coorbital objects (Janus and Epimetheus) between the classical satellites and the rings. At the time considerable controversy surrounded the determination of their orbits, and most satellite listings between 1966 and 1980 included a single "tenth satellite" in an orbit not corresponding to either Janus or Epimetheus. In 1979 the Pioneer 11 spacecraft imaged the smaller of these two coorbitals (Epimetheus) and, one satellite orbit later, passed within a few thousand kilometers, intersecting its magnetic wake. However, the full interpretation of these observations, as well as the discovery of about a dozen additional Saturn satellites, did not occur until 1980, with another ring-plane crossing taking place in the same year as the first of the two Voyager Saturn encounters.

Physical studies of the satellites of Saturn began with dynamical determinations of mass (e.g., Duncombe et al. 1974). The determination of satellite sizes before the spacecraft era was necessarily restricted to indirect methods, except for the 1954 efforts of Kuiper to measure their subarcsecond sizes with a visual diskmeter (cf. Cruikshank 1979b). Murphy et al. (1972b) and Morrison (1974a) used the simultaneous measurement of reflected visual and emitted thermal radiation to estimate sizes for Iapetus, Rhea, and Dione, while the identification of water frost on the surfaces of several satellites (Morrison et al. 1976; Fink et al. 1976) also suggested high reflectances and correspondingly small sizes. The best pre-spacecraft diameters, however, were obtained from high-speed photometry of a 1974 occultation of Saturn by the Moon (Elliot et al. 1975; Veverka et al. 1978), supplemented for Hyperion by a value from the photometric/radiometric technique (Cruikshank 1979a). Finally, the sizes of Tethys and Enceladus, both highly reflective, were rather tightly constrained by the simultaneous requirement that their albedos be less, and their mean densities greater, than unity (Morrison 1977). A judicious combination of all of these pre-Voyager data yielded diameters generally accurate to ± 10% for the larger Saturn satellites (see e.g., Cruikshank 1979b; Morrison 1980), while the combination of diameter and mass values suggested densities < 2 g cm⁻³, consistent with the observed icy surfaces.

The photometry of the satellites and the spectral evidence concerning their surface compositions are discussed in the chapter by Cruikshank et al. It is sufficient to note here that by the end of the 1970s it had been clearly established that satellite albedos were high and that water ice was the dominant spectrally active component of the surfaces of Enceladus, Tethys, Dione, Rhea, and the bright hemisphere of Iapetus. Hyperion also showed the spectral signature of water ice, although with indications of a greater admixture of a darker contaminant (Cruikshank and Brown 1982). The special case of two-faced Iapetus had been analyzed by Morrison et al. (1975), who concluded that the bright-side material extended over the poles, with the dark component forming a large spot centered on the apex of orbital motion. Spectral evidence suggested that the dark material was reddish, perhaps carbonaceous in nature (Cruikshank et al. 1983). Marginal evidence suggested that Phoebe also might be dark (Degewij et al. 1977), while most workers guessed that Mimas and the coorbital satellites were icy, like their larger neighbors among the regular satellites, and the rings themselves.

By the time the two Voyager spacecraft were approaching Saturn after their successful encounters with the Jovian system, much less was known about the Saturn satellites than about the Galilean satellites of Jupiter at a comparable time. With only one object, Titan, in the size class of the Galilean satellites, the Saturn system was already at a disadvantage, compounded by greater distance, lower levels of solar illumination, lower temperatures that tended to shift thermal radiation outside the terrestrial atmospheric windows, and a high level of scattered light from the rings. With the exceptions of Titan and Iapetus, which were already recognized for their uniqueness, there seemed little other than size distinguishing the regular, icy satellites. As late as a week before the Voyager 1 encounter of 12 November 1980, they retained their anonymity, no more than tiny, featureless disks. All this changed with incredible rapidity; by the Voyager press conference of 13 Novemb r, Soderblom could describe five distinctive, geologically active worlds, with "only Mimas your basic unprocessed ice moon," and the following day Shoemaker was prepared to attempt a synthesis to yield a geologic history of the Saturn satellite system (cf. Morrison 1982a; Smith 1982).

The remainder of this chapter reviews the properties of these new worlds as revealed by the Voyager cameras during the encounters of November 1980 and August 1981. This is in the nature of a progress report, standing between the preliminary scientific interpretation in the imaging team 30-day reports (B. A. Smith et al. 1981,1982) and more detailed analysis resulting from IAU Colloquium No. 77 on the natural satellites, held in Ithaca in July 1983. The discussion here is limited primarily to the interpretation of imaging data, and

TABLE I
Satellites of Saturn

Satellite	$a(R_s)$	P(day)	e	i(°)	$V(1,0)$ [c]	p_v [c]	R(km) [d]	$M(10^{23}$ g) [e]	ρ(g cm^{-3})
S17 Atlas	2.276	0.602	0.002	0.3	9.5	0.4	20 × ? × 10	—	—
S16 1980S27	2.310	0.613	0.004	0.0	6.2	0.6	70 × 50 × 37	—	—
S15 1980S26	2.349	0.629	0.004	0.1	6.9	0.5	55 × 45 × 33	—	—
S10 Janus	2.51 [a]	0.69 [a]	— [b]	— [b]	4.9	0.6	110 × 95 × 80	—	—
S11 Epimetheus					6.1	0.5	70 × 58 × 50	—	—
S1 Mimas	3.08	0.94	0.020	1.5	3.3	0.6	197 ± 3	0.46 ± 0.05	1.4 ± 0.2
S2 Enceladus	3.95	1.37	0.004	0.0	2.2	1.0	251 ± 5	0.8 ± 0.3	1.2 ± 0.5
S3 Tethys	4.88	1.89	0.000	1.1	0.7	0.8	530 ± 10	7.6 ± 0.9	1.2 ± 0.1
S13 Telesto	4.88	1.89	—	—	9.1	~1.0	15 × 10 × 8	—	—
S14 Calypso	4.88	1.89	—	—	9.4	0.7	12 × 11 × 11	—	—
S4 Dione	6.26	2.74	0.002	0.0	0.8	0.6	560 ± 5	10.5 ± 0.3	1.4 ± 0.1
S12 1980S6	6.26	2.74	0.005	0.2	8.8	0.6	17 × 16 × 15	—	—
S5 Rhea	8.73	4.52	0.001	0.4	0.1	0.6	765 ± 5	24.9 ± 1.5	1.3 ± 0.1
S6 Titan	20.3	15.95	0.029	0.3	−1.2	0.2	2575 ± 2	1345.7 ± 0.3	1.88
S7 Hyperion	24.6	21.28	0.104	0.4	4.8	0.3	205 × 130 × 110	—	—
S8 Iapetus	59	79.3	0.028	14.7 [b]	0.7–2.5	0.4–0.08	730 ± 10	18.8 ± 1.2	1.2 ± 0.1
S9 Phoebe	215	550	0.163	150	6.8	0.06	110 ± 10	—	—

[a] No orbital determinations have been made for these coorbital satellites since their 1982 "orbital exchange."
[b] Variable.
[c] Thomas et al. (1983a); Morrison (1982b); B. A. Smith et al. (1982); and chapter by Cruikshank et al. (where uncertainties are given).
[d] B. A. Smith et al. (1982); Thomas et al. (1983a,b,c); Davies and Katayama (1983). Uncertainties for the tri-axial dimensions of the small satellites are generally ~ 10% (Thomas et al. 1983a,c).
[e] Tyler et al. (1982), except Enceladus from Kozai (1976).

should be considered in parallel with other groundbased and spacecraft data presented in Cruikshank et al.'s chapter. Titan is not included (see chapter by Hunten et al.), and for treatment of the smallest satellites we defer to the chapter by Cruikshank et al. Unless otherwise credited, most of the data are from B. A. Smith et al. (1981,1982). We first describe the individual objects, followed by discussion of the common evolutionary processes affecting the entire satellite system.

II. SURVEY OF THE SATELLITES

In this section, the basic geological data are presented for each satellite, with emphasis on the Voyager imaging results. Comparative studies of the satellites will be presented in the final part of this chapter. Table I summarizes the most important orbital and physical data on each satellite.

Iapetus

Perhaps the most mysterious of the Saturn satellites, Iapetus is unique in its range of surface albedos, from $\sim$ 0.5 (a value typical of icy objects) to < 0.05 in the interior parts of its leading hemisphere. Before the event of Voyager, groundbased observations (see e.g., Millis 1973,1977; Morrison et al. 1975; Veverka et al. 1978) had established the size and approximate albedo range for Iapetus, as well as the virtually perfect longitudinal symmetry of the dark material with respect to direction of orbital motion. When the first spacecraft images were obtained, they confirmed the predicted general appearance, including the bright polar areas suggested by previous photometry.

As a consequence of Iapetus' great distance from Saturn (59 R_S) and substantial orbital inclination (14°7), neither Voyager spacecraft was able to approach it closely. Voyager 1 came only within 2,470,000 km, providing images at a maximum resolution of 50 km/line-pair. At the time of the encounter, Iapetus was near orbital longitude 0°, showing the Saturn-facing hemisphere and the boundary between the leading (dark) and trailing (light) sides. At this resolution, little topographic detail could be seen, but a large equatorial dark ring $\sim$ 300 km in diameter centered near longitude 300° (i.e., extending into the light side) strikingly interrupted the symmetry of the albedo boundary. Voyager 2 passed closer (909,000 km) and imaged primarily the complementary (anti-Saturn) hemisphere. The highest-resolution images (17 to 20 km/line-pair) were of the region stretching from the north pole to the equator and from longitude 190° to 290°, and showed primarily the high albedo surface (Fig. 1). These images revealed large numbers of craters and helped to clarify the character of the boundary of the dark region.

The airbrush map of Iapetus (Fig. 2) obtained from the Voyager images shows that the dark material is strongly concentrated not just on the leading hemisphere, but distributed symmetrically about the apex of orbital motion. Photometric analyses (Squyres and Sagan 1982; Goguen et al. 1983*b*) indi-

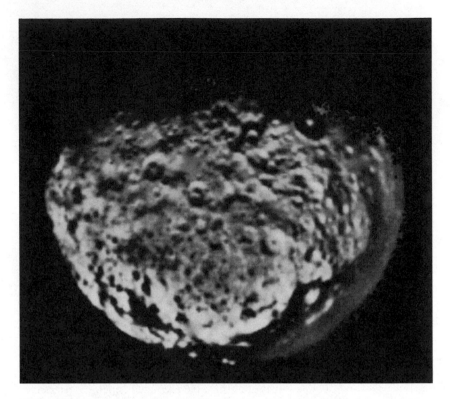

Fig. 1. Best Voyager 2 view of Iapetus. This image shows the northern hemisphere, primarily on the trailing (bright) side, at a resolution of 19 km/line-pair. Filtering has emphasized the heavily cratered topography of the bright side. Voyager/JPL image 260-1477.

cate that the reflectance of the surface is strikingly bimodal, with values in the range 0.4–0.5 for the bright side and 0.02–0.04 for the dark side. The albedo is lowest near the apex, but the brightness area is not at the antapex. In that part of the boundary imaged by Voyager 2, the width of the transition zone from bright to dark surface material is only a few hundred kilometers.

Preliminary Voyager 2 broadband visible spectrophotometry indicates that both the light and the dark materials are better reflectors of red light than blue, in agreement with groundbased results, but with substantial color variation from region to region. Typically green/violet reflectivity ratios are ~ 1.6 for the dark regions and ~ 1.2 for bright areas. Groundbased infrared spectra demonstrate the presence of water ice on the bright side, but the chemical nature of the dark material remains debatable (see detailed discussion in chapter by Cruikshank et al.).

The highest resolution Voyager images reveal that the bright side (and particularly the north polar region) is heavily cratered, with a surface density

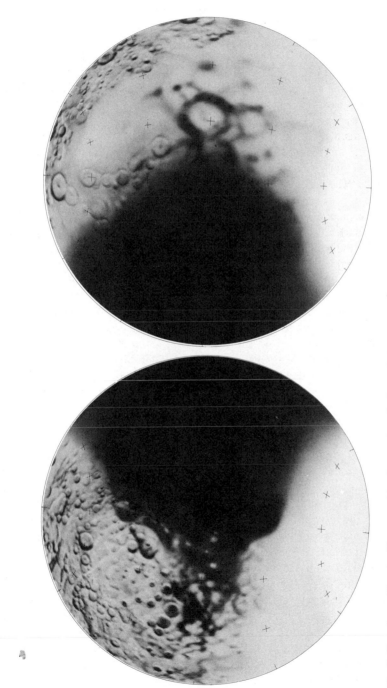

Fig. 2. Two views of Iapetus. These two airbrush maps show the Saturn-facing hemisphere (on the right) and opposite hemisphere (on the left). In both cases north is at the top. The dark region on the leading hemisphere faces in the direction of the satellite's motion about Saturn. These U.S. Geological Survey maps are presented in an equal area projection (maps prepared by J. L. Inge and E. M. Lee under the direction of R. M. Batson).

of 205 ± 16 craters ($D > 30$ km) per 10^6 km^2 (Plescia and Boyce 1983). Extrapolated to 10-km diameter, this density corresponds to > 2000 craters ($D > 10$ km) per 10^6 km^2, a value comparable to that on other densely cratered objects, such as Mercury, Callisto, or the lunar highlands. Unfortunately, due to low signal to noise in the images, no craters or other topographic detail can be seen unambiguously in the dark part of the surface, so we cannot establish whether this large crater density extends over both hemispheres of Iapetus.

A distinctive feature of the dark-light boundary on Iapetus is the presence of a number of dark-floored craters in the brighter material, but the absence of light floored or halo craters (or other white spots) within the dark material. The dark-floored craters and other dark spots below the resolution of the pictures extend near the equator to longitude 270°, the antapex point (see airbrush map, Fig. 2). There are some craters (for instance, near 200°) that exhibit dark areas in their floors and perhaps their walls suggestive of deposits that face the boundary. These oriented dark deposits and the appearance of the contact regions generally suggest that the dark material is superimposed on the bright terrain and is therefore younger. The absence of bright craters within the dark material indicates that the deposit is either thick (> 10 km) or that freshly exposed material from below is darkened or buried on a time scale short relative to the interval between major cratering events (perhaps $\sim 10^8$ yr).

An important additional clue to the relationship between dark and light materials is provided by the density of Iapetus, 1.16 ± 0.09 g cm^{-3}. This value is similar to that of the other icy satellites of Saturn, and consistent with models in which water ice is the primary bulk constituent. Thus the zebra is revealed as a white animal with dark stripes, not the reverse.

In view of the poor Voyager coverage of the leading hemisphere, the nature and origin of the dark material on Iapetus remains problematical. The strong symmetry with respect to the direction of orbital motion virtually demands some external control, if not an external origin, for the dark material. Yet the topographic relationship of dark deposits in the floors of bright-side craters suggests an internal origin for at least these deposits. It may be that two or more mechanisms are involved.

If the dark deposits are primarily externally controlled, a mechanism is required for their production. It has been suggested that the dark material is a coating of dust eroded from the surface of Phoebe (or some undiscovered other external satellite in a retrograde orbit) and drifting inward under Poynting-Robertson drag. Alternatively, impacts of material from Phoebe or elsewhere may modify the surface of the leading hemisphere with the result that low-albedo material becomes concentrated at the surface; in such a scenario the dark material is endogenic but the process that concentrates it is exogenic impacts. These ideas are discussed in more detail elsewhere (chapter by Cruikshank et al.; Squyres and Sagan 1982; Cruikshank et al. 1983).

Rhea

A near twin of Iapetus in size, but without its dark material, Rhea may represent a relatively uncomplicated archetype of the icy satellites of the outer solar system. Speculation before Voyager concerning this class of objects, apparently composed in substantial part of water ice and having diameters in the thousand kilometer range, were generally confirmed when Voyager 1 returned pictures of a bright, heavily cratered surface for Rhea with little obvious evidence of internal activity. Although closer analysis has revealed much of interest about this object, this basic initial impression still holds.

Rhea has a diameter of 1530 km (compared to 1460 km for Iapetus) and a density of 1.34 ± 0.09 g cm^{-3}. Its geometric albedo is 0.6, similar to that of the poles and trailing hemisphere of Iapetus. Although in no way mimicking the albedo differences of Iapetus, Rhea does exhibit its own type of longitudinal asymmetry, which is paralleled by its neighbor Dione.

On a hemispheric scale, the asymmetry of Rhea shows up as a brightness variation of 0.2 mag well aligned with orbital position, but having the opposite phase as Iapetus, with a broad maximum brightness on the leading side and a sharper minimum on the trailing (orbital longitude $270° \pm 5°$) (Noland et al. 1974; Cruikshank 1979). Even at low resolution (100–200 km/line-pair) Voyager pictures also showed the asymmetry: the leading side was fairly uniform in brightness, while the trailing side appeared mottled, with spots or patches of different albedo, leading to an overall lower value.

The best resolution of much of the trailing hemisphere was only ~ 50 km/line-pair, obtained by Voyager 1 several days before closest encounter. These images show that the highest albedo areas are wispy in form, giving the impression of cloud bands or broad irregular rays of ejecta superimposed on a darker, cratered terrain. Although roughly similar to some of the bright streaks on Ganymede, these Rhea features do not appear in radial patterns, nor is there any indication that they originate at impact centers. There are too few well-defined streaks for a statistical analysis of their orientations, but their breadth and irregular curvilinear shapes are not diagnostic of either external or internal processes. Similar wispy terrain was seen at higher resolution on Dione, where it seems more clearly endogenous, as will be discussed below.

The highest resolution images of any of the icy Saturn satellites were obtained of Rhea by Voyager 1. A special image motion compensation technique employed at a range of only $\sim 100{,}000$ km yielded a mosaic of images with resolution ~ 1 km/line-pair covering much of the north polar region (Fig. 3). The orbital position of the satellite was such that primarily the bright leading hemisphere was illuminated, but these images include the transition to darker terrains at about longitude 315°. Throughout both terrains the surface is dominated by well-formed impact craters and resembles the highland provinces of the Moon and Mercury. These craters are notably different from the flattened, relaxed forms common on Ganymede and Callisto, apparently be-

Fig. 3. Mosaic of the north polar region of Rhea. These images were acquired during the close approach of Voyager 1 to the satellite from a distance of roughly 59,000 km. Due to the high angular velocity of Rhea relative to the spacecraft, image motion had to be compensated for by slewing the scan platform during the exposure of each image, resulting in an effective resolution of $\sim$ 1 km/line-pair. The north pole is roughly centered along the terminator. Voyager/JPL image P-23177.

cause of the stiffer ice crust and lower surface gravity on Rhea. Also apparently absent are compact ejecta blankets, possibly another consequence of low surface gravity. The most unusual feature of these craters is the presence on some crater walls of curious bright patches. These may be exposures of relatively fresh ice, but the exact mechanism of emplacement is unclear.

In their original analysis of the distribution of craters on Rhea, Smith et al. (1981) pointed out a striking absence of craters with $D > 30$ km in the imaged polar area east of longitude $\sim 315°$, in contrast to the presence of craters with $D < 130$ km to the west of this boundary. Noting that the break in crater size distributions coincided approximately with the albedo transition from the bright leading to darker trailing sides seen in low-resolution images, they suggested that both the albedo and cratering boundaries were evidence of a resurfacing event west of $315°$ that must have occurred after the depletion of larger objects in the impacting flux. However, a more detailed examination (Plescia and Boyce 1982) suggests a more complicated story.

Plescia and Boyce (1982) have analysed the crater size-frequency distributions in three areas of Rhea: the two polar regions discussed above and an equatorial area near longitude $30°$ in the leading hemisphere. From their crater counts, they conclude that the demarcation between the polar terrain that has retained large craters and that in which they are absent is very near $0°$ (the sub-Saturn longitude) rather than near $315°$, associated with the boundary. (The albedo boundary is not easily seen in the high-resolution images but must be determined from observations obtained approximately one Rhea rotation earlier.)

All the areas of Rhea in which crater counts have been carried out are very heavily cratered for sizes below ~ 30 km. The cumulative size-frequency curves are, however, quite distinct. The absence of large craters in the polar region west of $0°$ is striking, and large craters are similarly lacking near the equator on the leading side. The part of the equatorial region studied seems subdued in topography, with an apparent loss of small ($D < 10$ km) craters. In both cases, a sequence of resurfacing events, or one or more episodes in which the crust softened and large-scale topography was destroyed by creep, seems indicated. Plescia and Boyce argue in particular for the presence of mantled regions near the equator in which the apparent thickness of the deposit is ~ 2 km, sufficient to bury most craters of $D < 10$ km.

The observed variations in cratering on Rhea also tell us something about the impacting flux. Whatever event was responsible for the loss of the large craters in the polar region east of $0°$, it appears that projectiles capable of producing craters of $D > 30$ km were largely absent after this time, even though the smaller craters were able to build back up to a saturated level. On the basis of such arguments, B. A. Smith et al. (1981,1982) argued for two populations of debris impacting the Saturn satellites, the first including large projectiles, the second lacking them. We will return to this problem later (Sec. III).

Dione

Although smaller (D = 1120 km) than Rhea, Dione exhibits most of the same general features, plus a few new wrinkles of its own. It has the highest well-determined density among the inner satellites: $1:43 \pm 0.04$ g cm^{-3}. Like Rhea, it has a bright icy surface (albedo ~ 0.5), is brighter on its leading hemisphere than its trailing, and has a mottled trailing hemisphere with brighter wispy terrain superimposed on a darker background. The lightcurve amplitude is 0.6 mag, the greatest of any large satellite except Iapetus. Dione also shows considerably more evidence of internal activity than does Rhea.

Although the Voyager imaging resolution (~ 2 km/line-pair) is not as great as on Rhea, the coverage is more complete, thanks to Dione's more rapid rotation (Fig. 4). Craters can be clearly identified in the trailing hemisphere, and while none is as large as the biggest on the leading side, there is no demonstrably significant difference in the cratering population. The most striking features of this side of Dione are, however, those that make up the

Fig. 4. Varied terrains on Dione. This Voyager 1 image at a resolution of ~ 4 km/line-pair shows a portion of the Saturn-facing hemisphere. Near the terminator in the leading hemisphere is a heavily cratered region including many large craters, while some of the enigmatic wispy terrain can be seen on the right. Voyager/JPL image P-23101.

wispy terrain. Imaged at a resolution of a few km, their contrast exceeds a factor of three, with reflectivities ranging from 0.2 for parts of the background to 0.7 in the bright streaks (Buratti 1983). One set of markings appears radial to a large ($D \sim 50$ km) crater, but generally they are curvilinear and complex and do not resemble crater rays.

Where they can be traced westward into higher-resolution images, the wispy bright markings appear more in the nature of relatively narrow bright lines, some of which extend into the leading hemisphere, where they lose contrast against the brighter background. A few of these narrow streaks are associated with topographic features—narrow linear troughs and ridges—cutting across the leading hemisphere. Some of these features, particularly in Palatine Chasma in the south polar region, strongly resemble grabens and horsts (Moore 1983). This association suggests that many if not all of the bright features may be controlled by a global tectonic system of fractures or faults. They may consist of surface deposits of ice outgassed along the fracture system, perhaps carried up from the interior by more volatile substances such as ammonia or methane.

Dione has a low overall crater population relative to Rhea. It also particularly lacks craters with $D > 30$ km, suggesting that the impacting flux recorded on its surface may be the same as that responsible for cratering the apparently younger terrains on Rhea. In addition, the crater densities are much more varied on Dione, indicating a more active geologic history. Terrain types in the leading hemisphere have been described as ranging from heavily cratered to smooth, lightly cratered plains.

Plescia and Boyce (1982) have analysed the crater size-frequency distributions for three areas: (1) a rough terrain with numerous large craters, centered at 50°S, 20° longitude; (2) a plains unit with intermediate crater density centered on the equator at 50° longitude; and (3) a smooth plains unit on the equator at 65° longitude. Even the heavily cratered area has a 20-km cumulative density of 270 ± 50, significantly less than the least cratered regions investigated on Rhea. The 20-km densities for the intermediate and lightly cratered plains are both 45 ± 25, but these curves rise steeply, and the three units may not be distinguishable for crater diameters below 10 km. The slopes of the curves are about -2 for the heavily cratered terrain and perhaps as high as -4 for the plains unit, suggesting to Plescia and Boyce that the two surfaces may reflect different populations of impacting bodies as well as different ages.

The presence of extensive resurfaced plains units on Dione as well as the existence of troughs or valleys hundreds of kilometers in length indicate a relatively high level of post-accretional alteration of the surface. In spite of its small size, Dione's level of endogenic activity exceeded in intensity and duration that of the larger Rhea. This lack of correlation between observed geologic history and expected thermal relaxation times represents one of the major geologic challenges of the Voyager data for the regular satellites of Saturn.

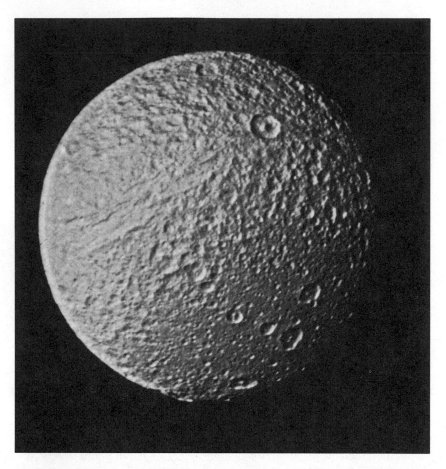

Fig. 5. Best global view of Tethys. This Voyager 2 image has a resolution of ~ 4 km/line-pair, similar to that of Fig. 4. The boundary between a heavily cratered unit (top right) and a more lightly cratered plain (bottom right) is easily distinguished. Ithaca Chasma stretches diagonally across the upper part of the image, with a large fresh-looking crater superimposed on it. Voyager/JPL image P-24065.

Tethys

With a diameter of 1060 km and a density of 1.21 ± 0.16 g cm^{-3}, Tethys is a near twin of Dione. Its albedo (0.8) is notably higher, however, than any satellite encountered so far in this review, and its trailing hemisphere lacks the darker regions and conspicuous wispy terrain of Rhea and Dione. All parts of the surface are densely cratered. Voyager 1 imaged primarily the Saturn-facing hemisphere with a resolution of ~ 15 km/line-pair, while Voyager 2 viewed principally the opposite hemisphere at resolutions ≤ 5 km/line-pair (Fig. 5). In addition, Voyager 2 obtained a single fractional frame at 2 km resolution, intended to be part of a high-resolution mosaic lost due to scan platform problems (Fig. 6).

Two outstanding topographic features dominate the images of Tethys: a giant crater and a trench or valley of global proportions. The crater, named Odysseus, is 400 km in diameter ($\sim$ 40% of the diameter of the satellite) and lies in the leading hemisphere centered at 30°N, 130° longitude. Odysseus is the largest crater in the Saturn system, with only the dark ring on the Saturn-facing side of Iapetus of even comparable dimensions. In fact, it is the largest crater with a well-developed central peak found anywhere in the solar system; the larger impact basins seen elsewhere generally lack distinctive central features.

Images of Odysseus on the limb of Tethys show that the floor has re-bounded by tens of kilometers, so that today it presents a convex face with about the same radius of curvature as the rest of the surface, rather than the concave shape associated with smaller craters. Such a rebound by creep or viscous flow in the icy lithosphere is not unexpected, with a thermal gradient of $\sim$ 0.1 K km^{-1} sufficient for ice to flow on a body of this size (Passey 1983).

Ithaca Chasma, the great trough or valley system of Tethys, stretches around at least 3/4 of the circumference along a great circle centered roughly on Odysseus and passing near the north pole. Its width is $\sim$ 100 km, and its total area amounts to nearly 10% of the surface of Tethys. Over much of its length, Ithaca Chasma is complex, with multiple subparallel walls and troughs. Rough estimates suggest that the walls are several kilometers high.

The best global views of Tethys show that it, like Dione and Rhea, ex-hibits terrains of different geologic ages. The well-imaged region north of the equator near longitude 60° is rough, hilly, and densely cratered (cumulative density for $D > 20$ km of 500 ± 92 per 10^6 km^2), with most of the larger craters highly degraded (Plescia and Boyce 1983). Part of the trailing hemi-sphere is less rugged and is best described as a lightly cratered plains unit (crater density 800 ± 66 per 10^6 km^2 for $D > 10$ km; Plescia and Boyce 1983). As with Dione, it appears that episodes of surface flooding persisted after the accretionary era.

Because of its large area, it has been possible for Plescia and Boyce (1983) to measure crater densities inside Ithaca Chasma. The values are simi-lar to those in the adjacent plains, with 20-km crater numbers of 120 ± 50. Thus, while Ithaca Chasma is the least cratered and possibly the youngest ma-jor unit identified on Tethys, its age is not dramatically less than that associ-ated with the plains deposits on either Tethys or Dione.

Enceladus

Mimas and Enceladus, the innermost of the classical satellites, are both small, with well-determined diameters of 394 and 502 km, respectively (Davies and Katayama 1983). However, they are not twins; in fact, the strik-

Fig. 6. Highest resolution view of Tethys. This is one of the last Voyager 2 satellite images; it is all that survived of what was intended to be a mosaic at 2 km/line-pair resolution. The region shown is very heavily cratered.

ing differences between them provide one of the major mysteries of the Saturn system.

Even before the Voyager missions it was apparent that Enceladus had a remarkably high albedo, and the spacecraft images quickly established that the surface was strongly backscattering, with normal reflectance > 1.1, substantially greater than that of any common natural surface, such as freshly fallen snow. The mean geometric albedo is ~ 0.9, and the Bond albedo is ~ 0.8, resulting in the lowest surface temperature among the Saturn satellites (cf. chapter by Cruikshank et al.).

Although it came no closer than 200,000 km, Voyager 1 established that the surface of Enceladus was deficient in large craters relative to its neighbors. Voyager 2 came within 90,000 km and imaged the northern half of the trailing hemisphere at a resolution of 2 km/line-pair. This picture (Fig. 7) revealed a surface quite unlike anything else seen in the Saturn system, with striking indications of large-scale endogenic activity.

Enceladus has a wide diversity of terrains, ranging from heavily cratered (cumulative density for $D > 10$ km of 800 ± 300 per 10^6 km^2, similar to the intermediate regions of Dione and Tethys) down to regions in which all impact craters have been erased (density for $D > 10$ km of < 100 per 10^6 km^2). In addition, the craters themselves display a range of forms indicating various degrees of crustal relaxation, and a system of peculiar curvilinear ridges up to 1 km in height is a prominent feature of the crater-free areas.

At least five distinct terrains have been identified on the basis of crater populations and other landforms: ct_1 with highly flattened craters 10 to 20 km in diameter; ct_2 with well-preserved craters in the same size range; cp with bowl-shaped craters 5 to 10 km in diameter but at a lower density; sp_1, a lightly cratered smooth plains unit with a rectilinear groove pattern; and finally the youngest units of crater-free smooth plains (sp_2) and ridged plains (rp). The locations of these units are shown in the geologic sketch map (Fig. 8). In unit rp, the spacing between ridges ranges from 7 to 15 km, while the ridge heights determined from photoclinometry (Passey 1983) range from a few hundred meters to 1.5 km. For comparison, the groove spacing on Ganymede is 3 to 10 km, and the relief is < 1 km.

Where the ridged terrain stretches in a corridor nearly to the north pole, its contacts with the older cratered plains units reveal a complex history. Evidently the sequence of events included replacement of the older units, formation and subsequent flattening of craters on the new surface, and later emplacement of the ridged terrain. Additional intermediate stages in the evolution of the crust are suggested by some of the details of the contact between the younger and older terrains. While absolute ages cannot be assigned, the crater-free terrains are almost certainly younger than 10^9 yr, and may be younger than 10^8 yr.

Passey (1983) has used photoclinometry to measure the profiles of large craters on Enceladus. On the basis of these profiles, he distinguishes different

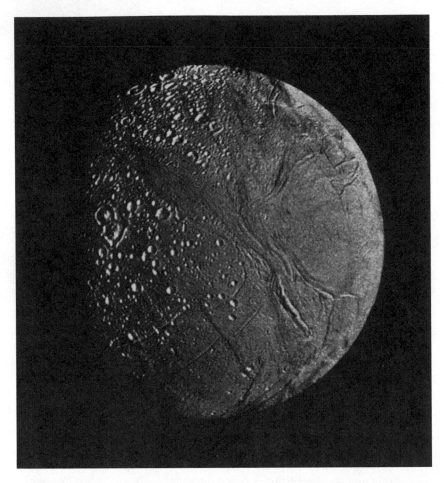

Fig. 7. High-resolution mosaic of Enceladus. These are the best Voyager 2 images of this satellite, with a resolution of 2 km/line-pair. The varied terrain types are delineated on a geologic
sketch map (see Fig. 8). Voyager/JPL image P-23956.

degradation histories even for regions of similar crater density. For example,
in one heavily cratered region all of the larger craters are extremely flattened
and have bowed-up floors, indicative of relaxation by lithospheric creep. In
another part of ct_1, $\sim$ 75% of the craters with $D > 8$ km are flattened, while
25% appear fresh and unrelaxed. Passey concludes that the viscosity of the
lithosphere in the heavily cratered regions has been between 10^{24} and 10^{25} P.
The exact time scale for viscous relaxation is unknown, but Passey argues that
it should be between 10^8 and 4×10^9 yr. The variations in degree of crater
relaxation from one region to another suggest that the heat flow has varied
from one terrain unit to the next.

In spite of the great range of surface ages indicated by the varied topography, the photometric properties of the surface of Enceladus are remarkably uniform. Buratti et al. (1983) have noted that this uniformity applies to
color and scattering properties as well as albedo. The total range of albedo
over the surface is $\sim 20\%$, with the trailing hemisphere lighter, but even these
small differences are not correlated with the ages of the major trailing-side
terrain units discussed above, which differ in albedo by ≤ 1 to 2%. As Buratti
et al. argue, both the unusual nature of the surface optical properties and the

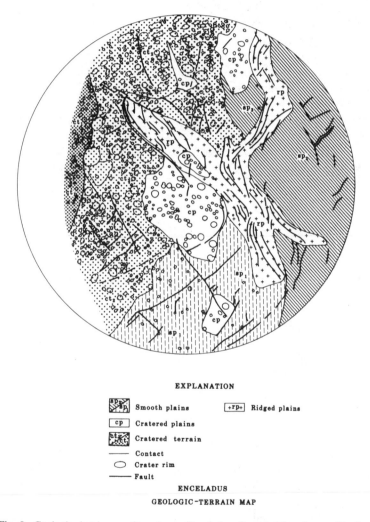

EXPLANATION

Smooth plains   Ridged plains

cp   Cratered plains

Cratered terrain

——— Contact

◯ Crater rim

——— Fault

ENCELADUS
GEOLOGIC-TERRAIN MAP

Fig. 8. Geologic sketch map of terrains on Enceladus. Compiled from images like that in Fig. 7,
the units are identified on the basis of the abundance of superposed impact craters or sinuous
ridges where impact craters are absent.

uniformity of their distribution suggest that the optical properties are determined by a ubiquitous surface layer of relatively recent age.

A further intriguing clue to the unique history of Enceladus may lie in the relationship to the tenuous E Ring of Saturn, which has a peak in its intensity near the orbit of the satellite (Baum et al. 1981). Photometric studies (Pang et al. 1982) indicate that the E Ring is composed of spherical micrometer-size particles, which should have lifetimes of only $\sim 10^4$ yr (Haff et al. 1983). It is therefore suggested that the E Ring is a recent phenomenon associated with Enceladus. Possibly the same event that produced the uniform, uncontaminated optical surface of the satellite also sprayed out ice particles to form the E Ring. Whatever the nature of this event, it probably occurred within the past few thousand years.

The extraordinary nature of Enceladus requires an explanation, presumably involving an unexpectedly large internal heat source. Since the mass of the satellite is not well determined from either groundbased or spacecraft data, its density remains relatively uncertain, but there is no reason to expect an excess of radionucleides, certainly not at the level required to maintain the satellite in a molten state to the present time. Tidal heating has been suggested (Yoder 1979), but this appears to provide an insufficient source also. We will return to these questions in Sec. III, after the completion of our survey of the satellites.

Mimas

Mimas, with a diameter of 394 km, is the smallest and innermost of the classical regular satellites of Saturn. It was imaged best by Voyager 1, which provided coverage primarily of the southern hemisphere at resolutions of ~ 2 km/line-pair. Its surface is exceedingly heavily cratered, with little indication of major endogenic modifications. The albedo is fairly uniform, and the light-curve amplitude is < 0.1 mag (Buratti 1983).

The most striking feature of Mimas is a very well preserved 130-km crater, Herschel, nearly centered on the leading hemisphere (Fig. 9). The crater walls are ~ 5 km high, and parts of the floor may be ~ 10 km deep. A large central peak has a base 20 to 30 km across and rises to ~ 6 km above the crater floor. Unlike the large crater Odysseus on Tethys, Herschel does not appear significantly modified by crustal creep or relaxation. This freshness, together with the size of the crater ($\sim 1/3$ the diameter of the satellite), make the leading hemisphere of Mimas one of the most memorable sights in the Saturn system.

The next largest crater on Mimas is less than half the size of Herschel, and most are < 50 km in diameter. The cratering, although everywhere heavy, is not uniform. On the leading hemisphere, west of Herschel, there are many craters with $D > 40$ km, while the south polar and adjacent areas lack craters with $D > 20$ km and are devoid of craters > 30 km. Plescia and Boyce (1982) measured a cumulative density ($D > 10$ km) of 1250 ± 490 near the south

Fig. 9. Leading hemisphere of Mimas. The 125 km diameter crater Herschel is prominent in a heavily cratered surface. Resolution is ~ 4 km/line-pair. Voyager/JPL image P-23210.

pole and suggested that the values in the leading hemisphere, near Herschel, might be several times higher yet. They note an inflection in the size-frequency distribution curves between $D = 15$ and $D = 40$ km which suggests a mixing of two populations of craters, perhaps the result of resurfacing. Evidence exists, therefore, for some modification of the surface during the period of heavy bombardment, perhaps similar to the processes seen more clearly on Rhea, Dione, and Tethys.

Small Satellites

The "small" satellites of Saturn are defined here to include two classical satellites, Hyperion and Phoebe, as well as eight objects in the inner part of the system, between the orbit of Dione and the outer edge of the A Ring. These include the ring shepherds, the coorbitals, and several Lagrangian satellites (see Table I). This list is not intended to represent a complete census of the small satellites of Saturn, but is limited to those objects sufficiently well observed to derive some physical characteristics.

Phoebe, the outermost satellite, orbits Saturn at a distance of 215 R_S in a retrograde sense. Voyager 2 observed Phoebe from a range of $\sim 2 \times 10^6$ km a week after its Saturn encounter, providing images over a 24-hr period at a resolution of $\sim$ 40 km/line-pair. Our discussion of these images is based on the analysis of Thomas et al. (1983c); additional material may be found in the chapter by Cruikshank et al.

Phoebe is a dark, approximately spherical object with a diameter of $\sim$ 220 km and a geometric albedo of 0.05–0.06. Its rotation is nonsynchronous, with a period of 9.4 ± 0.2 hr (prograde) determined from both the disk-integrated Voyager lightcurve and by tracking individual markings. The most prominent surface markings are scattered brighter patches with contrast up to 50% relative to the darker, bland areas. The Voyager color data agree with groundbased observations in showing a flatter spectrum than that of the dark side of Iapetus, excluding Phoebe-type dust as the sole darkening agent for Iapetus (see chapter by Cruikshank et al.; Cruikshank et al. 1983).

At the Voyager resolution no craters or other topography can be identified unambiguously. This observation does not mean that Phoebe is uncratered; with only 11 pixels across the satellite image and the low phase angles of the Voyager images, Thomas et al. argue that even the largest expected craters ($D \sim$ 100 km) would be only marginally visible. The maximum equatorial diameter is $\sim$ 20 km greater than the polar diameter, although this difference is no larger than the probable uncertainty of the measurements. Generally, its dark surface and retrograde orbit suggest that Phoebe is a captured object with primitive surface composition, but there is little direct information in the Voyager observations concerning either its origin or its geologic history.

Hyperion is the largest of the small satellites, occupying a moderately eccentric orbit between Titan and Iapetus. Voyager 2 images at resolutions $\sim$ 10 km/line-pair reveal it to be an irregularly shaped object $\sim$ 350 × 240 × 200 km, with angular features and facets. While nearly as large as Mimas or Enceladus, its rugged profile is in dramatic contrast to their smooth, spherical shapes. It is surely the fragmented remnant of a larger parent body. As discussed in the chapter by Cruikshank et al., Hyperion has a dirty ice surface, with a lower albedo and weaker ice bands than those of the inner, regular satellites. Hyperion's surface includes one crater 120 km in diameter with $\sim$ 10 km vertical relief. Other deep craters 40 to 50 km across can be recognized, and several major scarps (one nearly 300 km long) are also present. Cratering and spallation (see e.g., Thomas et al. 1981) appear to have been the dominant processes in its geologic history.

A major controversy for Hyperion concerns its rotation period, which is apparently not synchronous. The high-resolution Voyager images do not cover a sufficient time base to establish the period by tracking individual features, but analysis of the lightcurve from low-resolution images yields a

13-day periodicity (Thomas et al. 1981). A similar 13-day period has been derived from 1983 groundbased photometry by J. Goguen (personal communication, 1983). A recent dynamical analysis, however, suggests that this period may not be stable, and that the rotation of Hyperion should be chaotic (Wisdom et al. 1983). In such a situation, the satellite is expected to be tumbling irregularly, with a period that could change by tens of percent in a few weeks.

The inner, small satellites (Table I) all have dimensions smaller than Hyperion or Phoebe. The largest are the two coorbitals (Janus: $220 \times 190 \times 160$ km; Epimetheus: $140 \times 115 \times 100$ km). Next are the two F-Ring shepherds (~ 100 km), while the others do not exceed 40 km in major diameters. With imaging resolutions that generally amount to only a few pixels per diameter, the degree of geologic information that can be gleaned for these objects is clearly limited (Thomas et al. 1983b).

Photometrically, the analysis of Thomas et al. (1983b) shows that all of these inner satellites have bright, presumably icy surfaces, with geometric albedos ranging from ~ 0.4 up to perhaps as high as 1.0 for Telesto, the Tethys leading Lagrangian. The irregular shapes and cratered surfaces of these satellites suggest a violent past, and it seems likely that all are remnants of once larger bodies. Thomas et al. also discuss the possible roles of tidal deformation; it is interesting in this context to note that all of these satellites are more elongated than Roche ellipsoids of density 1.0 g cm^{-3}.

The two coorbitals are large enough to permit crater counts. Thomas et al. (1983b) have identified 12 craters on Janus and 16 on Epimetheus. Within the (rather large) statistical uncertainties, the crater densities for these two satellites agree with the values for heavily cratered surfaces (~ 1000 per 10^6 km for $D \geq 10$ km) on the other Saturn satellites.

All of the small inner satellites are in orbits that are remarkable in some way: as coorbitals, shepherds, or Lagrangians. Various scenarios have been suggested for their origin, generally involving the breakup of larger parent objects. We will return to some of these ideas in the final sections of this chapter.

III. GEOLOGIC HISTORIES

Although any attempt to trace a detailed geologic history of the Saturn satellite system is premature, it is instructive to compare these satellites on the basis of existing data and to sketch some of the processes that might have influenced their development. In this section, we discuss such processes under three headings: origin, exogenic processes, and endogenic processes.

Origins

The regular satellite system of Saturn presumably had a common origin in a circum-Saturnian nebula. The ubiquitous presence of water ice in the ring and satellite system indicates that temperatures throughout the nebula were cold enough to permit the condensation of water, unlike in the Jovian case. In addition, the presence of both methane and nitrogen in the atmosphere of Titan suggests that ammonia and methane were incorporated, perhaps as clathrates, at least in the outer part of the system. If these two volatiles are present on Titan, they were probably also included in the building blocks of Hyperion and Iapetus, and perhaps even in those of the inner satellites.

Some constraints on bulk composition of the Saturn satellites are provided by their densities, which are for the most part adequately determined after the Voyager missions (Table I; Fig. 10). The mass for Mimas has not been directly measured by Voyager, but the mass ratio Mimas/Tethys is known from resonance studies, and therefore Mimas can be scaled to the new Tethys value (Tyler et al. 1982). The mass of Enceladus, unfortunately, remains poorly determined, but the values for the other measured objects are all consistent to within the stated errors with a uniform density of 1.3 ± 0.1 g cm^{-3}. This density is consistent with a composition of 60 to 70% ice, if the remaining fraction is chondritic (Lupo and Lewis 1979). As shown in Fig. 10, this ice/rock proportion is also consistent with the compressed density for Titan of 1.90 g cm^{-3}. Although minor stochastic differences may exist among the compositions of the measured satellites of Saturn, the contrast with the regular progression in densities displayed by the Galilean satellites of Jupiter is striking. Either the Saturn protosatellite nebula was fairly uniform in com-

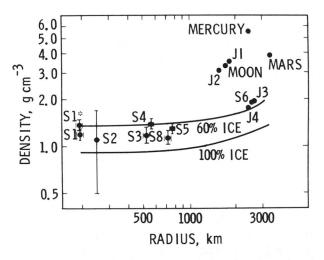

Fig. 10. Density versus radius trends for the icy Saturnian satellites. Taken from B. A. Smith et al. (1982), these data are compared with model curves for compression of rock and ice mixtures calculated by Lupo and Lewis (1979). Two values are shown for S1 (Mimas). The lower is the classical value based on mass determinations from Earth-based astronomical observations; the upper from Voyager radio science measurements (Tyler et al. 1982).

position, or else the process of accretion involved sufficient mixing to homogenize the satellites as they formed.

The satellites could have been heated and melted by accretional energy, tidal despinning, or radioactive energy from their presumed chondritic fraction. Because of the very low melting temperatures of either pure H_2O ice or various clathrates, it seems nearly certain that the larger bodies must have differentiated early in their history. However, smaller satellites may still retain their primitive composition and structure.

As will be discussed in the next section, the accretionary and post-accretionary bombardments of the inner satellites must have been intense and probably led to many cases of fragmentation and even disintegration of some of the smaller early satellites. Presumably the various coorbital, shepherd, and Lagrangian satellites assumed their present identities at that time. The rings themselves could have been formed as the consequence of similar bombardment-triggered disruptive events.

Finally, we would guess (primarily on dynamical grounds) that Phoebe is a captured object. Its moderately regular shape suggests that it was captured without catastrophic collisions, probably at a time when gas drag from the protosatellite nebula was still significant (see e.g., Pollack et al. 1979).

Exogenic Processes

The heavily cratered surfaces seen throughout the Saturn system testify to the role of impacts in satellite evolution. The observed crater densities, adapted from Plescia and Boyce (1982,1983) are summarized in Table II. However, in order to interpret the impacts, we must consider the effects of Saturn and the satellites themselves in determining the flux of bombarding particles. These effects were first emphasized for the Galilean satellites by Shoemaker and Wolfe (1982), and have been further developed for the Saturn system by B. A. Smith et al. (1981,1982).

The number of impacts and their energy for sources of projectiles outside the Saturn system are strongly influenced by the gravitational attraction of the planet, and will be significantly enhanced for satellites close to the planet. The present cratering rate on Rhea from such sources should be about twice that on Iapetus, and for Mimas the rate is 20 times that on Iapetus. The cratering rates are also affected by the masses of the satellites themselves, but for objects as small as the Saturn satellites, the dominant effect is that of the deep gravitational well of the planet. An additional effect for synchronous satellites is the enhancement of cratering rates on the leading hemisphere relative to the trailing. The variation from apex to antapex increases with orbital velocity: at Rhea the factor is 6, and at Mimas it is nearly 20 (B. A. Smith et al. 1981). However, we note that no dependence of crater density upon longitude has been seen for any Saturn satellite (Plescia and Boyce 1983), suggesting that they may not have been in synchronous rotation during periods of heavy bombardment.

TABLE II
Crater Density Cumulative Values per 10^6 km^2 [a]

| Satellite | Terrain | Crater Diameter | |
		10 km	20 km
Iapetus	Bright side (average)	[2000] [b]	740 ± 30
Rhea	North polar	1100 ± 20	300 ± 100
	Equatorial (unmantled)	700 ± 150	200 ± 100
	Equatorial (mantled)	160 ± 45	80 ± 40
Dione	Heavily cratered	[1000]	270 ± 50
	Intermediate	700 ± 100	45 ± 25
	Lightly cratered	260 ± 50	45 ± 25
Tethys	Heavily cratered	[2000]	500 ± 92
	Intermediate	690 ± 72	170 ± 26
	Plains	800 ± 66	140 ± 28
Enceladus	Heavily cratered	800 ± 300	[300]
	Fractured plains	200 ± 100	[50]
	Smooth plains	< 100	—
Mimas	South polar	1200 ± 500	< 100

[a] Adapted from Plescia and Boyce (1982,1983).
[b] Brackets in each case indicate extrapolation.

Present cratering rates due to long-period and Saturn-family comets have been estimated (B. A. Smith et al. 1982) to range from ~ 1 (cumulative crater count $D > 10$ km per 10^6 km^2 per 10^9 yr) for Iapetus to ~ 16 in the same units for Mimas. If these calculations correctly represent the cratering rates that have applied, they can be used to estimate surface ages. For instance, the rate for Enceladus is ~ 10, equal to the observed upper limit for the crater-free terrain, yielding an age for that terrain < 10^9 yr. Such calculations are, however, probably at best adequate to give an order-of-magnitude upper limit to surface ages. In fact, the Saturn system presumably experienced much higher impact fluxes in the past, as shown by the fact that at estimated current rates it would require ~ 10^{12} yr to produce the crater density observed on Iapetus.

A less assumption-dependent analysis of the significance of cratering statistics can be obtained by adopting Iapetus as a fiducial mark. The bright side of Iapetus has a cumulative crater density of ~ 2000 per 10^6 km^2 for $D > 10$ km. On the assumption that these craters were produced by projectiles from an external source, B. A. Smith et al. (1982) calculated the cumulative impacts for the other Saturn satellites over the same time period, and thus obtained a lower limit to the total crater density for each object. In all cases the calculated cratering exceeds the observed value, implying saturation of the surface or resurfacing subsequent to the stabilization of Iapetus' surface. Furthermore, the cratering rates extrapolated in this way from Iapetus are so high in the inner part of the system as to imply impact disruption sometime in the past for all satellites interior to Rhea. This exercise indicates, for instance, that Enceladus and Mimas should each have been disrupted and reaccreted four or five times during the same time period in which Iapetus was developing its current density of craters, if the Iapetus craters indeed represent an impact flux external to the Saturn system.

In a sense, these two exercises bracket the probable cratering history of the satellites. So long as we assume that all the satellites were impacted by the same population of externally derived projectiles that cratered the bright side of Iapetus, we are led (with B. A. Smith et al.) to the conclusion that disruptive impacts were common for the inner satellites. However, the implied cratering rates are orders of magnitude higher than the currently estimated rates for the known external source, i.e., the comets. An alternative supposition is that a major part of the satellite cratering, including that of Iapetus, might have been due to some other population of debris more closely associated with the Saturn system. In that case, catastrophic impacts might have been less common.

There is evidence in the cratering record for different populations of impacting bodies, as noted by B. A. Smith et al. (1981,1982) and Plescia and Boyce (1982,1983). Certainly, the observed crater size-frequency distributions are different on surfaces of different age, and it is reasonable to assign these differences to the different characteristic impacting populations. At the very least there seem to be two: population 1, representing the tail-off of a post-accretional heavy bombardment, and population 2, similar to the secondary populations on the terrestrial planets and possibly resulting from debris generated by collisions within the system. Both of these populations are, of course, distinct from the external, cometary population causing impacts today.

The Voyager observations indicate that population 1 craters are found in abundance on Rhea (the part of the polar terrain that includes the larger craters), on the older parts of Dione, and in the hilly, heavily cratered terrain of Tethys. Population 2 characterizes the younger terrains on all three of these satellites, plus most of the surface of Mimas. Even the oldest observed terrains on Enceladus are population 2, which is greatly depleted in projectiles that can generate craters with $D > 30$ km.

Although none of the satellites shows an apex-to-antapex crater gradient of the sort predicted for the flux on a synchronous satellite, this focusing of impacts by orbital motion may still have played an important role in the surface evolution of Rhea and Dione (as well as, possibly, Iapetus). Suppose that those satellites, and perhaps Tethys as well, developed a global wispy terrain as a result of an early episode of internal activity. It is then reasonable that subsequent impacts reworked the leading hemispheres of Rhea and Dione, and all of Tethys, to the point where the evidence of endogenic activity was erased, whereas the wispy terrain was preserved on the protected trailing hemispheres of Rhea and Dione. This scenario seems quite reasonable for the inner satellites, but in this simple form it does not appear to be consistent with the lack of longitudinal control of densities of craters of $D > 10$ km as counted in the Voyager images.

A very recent impact event in the Saturn satellite system has been suggested by McKinnon (1983) as the source of the E Ring and the remarkable photometric properties of Enceladus. McKinnon argues that the sphericity and narrow size distribution of the E-Ring particles (Pang et al. 1983) are indicative of condensation from a liquid, and that a water cloud would most readily be generated from the impact melt associated with a crater-forming event on Enceladus. He calculates a current mass for the E Ring of a few times 10^{10} g, equivalent to an ice sphere of ~ 20 m radius. This is not an unreasonable mass for even a small cometary impact, such as might have occurred during the past 10^3 to 10^4 yr. In this hypothesis, the optical properties of Enceladus result from reaccretion onto the satellite of these micrometer-size ice particles, and are unrelated to the unique geologic history of this satellite. Such "E Rings" would be produced periodically in the Saturn system, and it is coincidence that we find one now in association with Enceladus.

We should not leave the topic of exogenic processes without again raising the issue of the dark material on Iapetus. None of the Voyager data have altered the conclusion based on the symmetry of the lightcurve that the albedo distribution on this satellite is externally controlled (Millis 1973; Morrison et al. 1975). If anything, the spacecraft images have strengthened the case by demonstrating that the albedo contours approximately follow calculated contours of equal impacts for a synchronous satellite (Squyres and Sagan 1982; Goguen et al. 1983b). What remains uncertain is the mechanism for darkening the surface, as discussed in more detail in the chapter by Cruikshank et al.

Crater densities have not been measured within the dark regions. On the other satellites, the Voyager images have shown that albedo is not correlated with crater density, as suggested for the Galilean satellites by Hartmann (1980). A number of satellites do show leading/trailing asymmetries in average albedo, suggesting possible external influences, but the causes are more subtle than simply gardening by large-scale impact craters.

Endogenic Processes

The thermal evolution of small bodies, and particularly of small icy bodies, is a topic that owes its recent interest to the Voyager observations of the Jovian and Saturnian satellite systems. Early discussions by Lewis (1971) and Consolmagno and Lewis (1976,1977) dwelt primarily with the Galilean satellites, but are also applicable to the Saturn system, especially their suggestions of the possible role of small quantities of methane and ammonia in stimulating early melting and differentiation. These two volatiles may also be implicated in the outgassing of icy satellites as they evolve.

One of the clearest cases for thermal evolution from among the Saturn satellites is provided by Tethys, with its fracture system, Ithaca Chasma. If the satellite were once a ball of liquid water covered by a thin frozen crust, freezing of the interior would have produced expansion of the surface comparable to the area of the chasm. It is not clear, however, why the extension would be concentrated in a single narrow lane rather than distributed among multiple faults over the surface. The smaller-scale valleys or graben of Dione, similar in size and density to Tethys, might also be due to expansion of an icy object as it freezes. Moore (1983) has noted the concentration of these tectonic features on heavily cratered terrain on Dione, suggesting that they predate the resurfacing events widely seen on this satellite. There are suggestions of similar grooves on Mimas, but these have been nearly destroyed by subsequent impacts.

All of the inner satellites also show evidence of several stages of partial resurfacing, presumably by liquid escaping from the interior. In the absence of unexpected internal heat sources, it has been suggested that low temperature melts might have been provided by a water-ammonia eutectic, which melts at 170 K. Similar explanations may apply to the putative outgassing along fracture lines that may have produced the wispy terrain of Rhea and Dione. Clearly several different styles of internal activity are indicated, none of them very well understood. Studies of thermal evolution and of volcanism on icy bodies are still in their infancy.

Enceladus provides the strongest evidence for internal activity, with its multiple localized resurfacing events and the plastic deformation of the crust, probably extending to the current geologic epoch (e.g., $< 10^9$ yr). This dramatic and continuing evolution appears to require an internal heat source orders of magnitude larger than would be expected from known causes.

The most detailed analysis of lithospheric processes on Enceladus has been made by Passey (1983). His studies of the flattening of craters and the bowing up of their floors suggest a history of viscous relaxation of topography. The modified craters are located in distinct zones adjacent to other units, with similar crater densities, where craters have not been flattened. These modifications of craters appear to require viscosities one to two orders of magnitude less than for Ganymede or Callisto. Passey concludes that such a large difference in viscosity requires a difference in lithospheric material, as well as a substantial heat source on Enceladus. The presence of ammonia as well as water ice in the lithosphere of Enceladus would satisfy this requirement.

Even more dramatic than the alteration of craters is the formation on Enceladus of the plains terrains, which are nearly crater-free and may have ages $< 10^9$ yr. The contacts between cratered regions and the younger plains are often linear and abrupt, truncating preexisting craters. The curvilinear grooves that characterize part of these plains are morphologically similar to those on Ganymede. Squyres et al. (1983b) believe that the grooves resulted from open extension fractures in a relatively brittle crust. Widespread flooding of parts of the surface with water "magma" probably occurred at about the same time.

At some of the contacts between younger and older terrain, craters appear to have been distorted by a horizontal strain associated with extensional faulting. Kargel (1983) has also noted examples where there appears to have been strike-slip faulting with $\sim$ 20 km of offset, and he argues for past plate tectonic activity, including a possible analogy between the grooves near longitude $0°$ and terrestrial midocean ridges. Along spreading centers he expects flooding and even pyroclastic release of water in the form of large volcanic fountains. These ideas are intriguing, but perhaps go beyond what can firmly be supported on the basis of current data. Even the distortion of craters near geologic contacts is debatable, and Passey (1983) has suggested that a fortuitous superposition of craters seen near the limit of resolution provides an acceptable alternative interpretation.

Although the time scale for geologic activity on Enceladus is not known, it is clear that some terrains on this object are much younger than any seen elsewhere among the Saturn satellites, indicating a much more persistent activity. Such long-lived activity on so small on object requires a heat source considerably greater than that expected from radioactive decay in a presumed silicate core, which is only $\sim 10^8$ W, two orders of magnitude below that required for melting (Squyres et al. 1983b). Yoder (1979) suggested that tidal heating might be important for Enceladus as the result of increased eccentricity forced by a resonance with Dione, but calculations by Peale et al. (1980) indicated that this source was $< 10^8$ W, and therefore insufficient to produce the observed effects. The present forced eccentricity of 0.0044 would have to be increased by roughly a factor of 20 for any melting to occur in an H_2O lithosphere (Squyres et al. 1983a). Furthermore, Squyres et al. note that a similar analysis indicates that the heating rate for Mimas, which shows no evidence of persistent internal activity, should be twice as great as that for Enceladus. If tidal heating is the explanation of the unique activity level of Enceladus, it implies a past forced eccentricity much higher than can be maintained or episodically generated by the Dione resonance (Squyres et al. 1983a).

More generally, the thermal evolution of bodies of the size and composition of the Saturn satellites is a topic of great interest and no little complexity. The Voyager data indicating that substantial evolution has taken place will surely focus interest on this topic. Studies such as those of Peale et al. (1980), Passey (1983), and Squyres et al. (1983a) indicate that solid-state convection is probably the primary way to transport energy from the interior toward the crust. However, a critical unknown for the development of quantitative models is the exact chemistry of the satellite interiors.

Although the mean densities are consistent with comparable mixtures of water ice and chondritic rock, we do not know the actual density of the rocky component, and therefore we cannot establish the relative proportions of rock and ice. Further, we have very little data on the presence of non-H_2O ices, even a small admixture of which can greatly alter the rheology and phase changes of the mixture. Spectrometric data on surface composition cannot resolve this question in a quantitative way, since even at 90 K such species as ammonium clathrate and methane clathrate are less stable than H_2O and could be selectively depleted in the regolith. Thus, while it seems clear that the presence of these other ices may be implicated in the more active geologic histories of some of the Saturn satellites, this speculation cannot be demonstrated based on present data.

Acknowledgments.   We are grateful to our colleagues on the Voyager Imaging Science Team for many stimulating discussions, and to the staffs at Jet Propulsion Laboratory and the U.S. Geological Survey for photographic and cartographic products. This research was supported by the National Aeronautics and Space Administration and by Project Voyager.

SATELLITES OF SATURN: OPTICAL PROPERTIES

DALE P. CRUIKSHANK
University of Hawaii

JOSEPH VEVERKA
Cornell University

and

LARRY A. LEBOFSKY
University of Arizona

With a large body of groundbased observations of the satellites of Saturn and a wealth of new data from the Voyager 1 and 2 spacecraft, we have begun to examine the optical properties of this diverse family of outer solar system bodies in a systematic way. The optical properties include information on surface chemical composition and microstructure, plus basic data on sizes, shapes, rotation periods, and surface geologic structures. This review concentrates on composition and microstructures derived from photometric, spectrophotometric, and thermal observations from spacecraft and from groundbased telescopes. Photometric phase integrals of Mimas, Enceladus, Tethys, Dione, and Rhea were determined from Voyager photometry over a wide range in phase angles, and similar values were derived from Voyager infrared observations. General agreement is found for most objects with the exception of Rhea; the cause of this discrepancy is not resolved. A lunar-like law of surface optical scattering is found for some satellites, but others are somewhat more complex, involving a significant component of Lambert-like scattering. The high albedos of the small, innermost satellites suggest a water-ice composition, as has been derived for the larger bodies from groundbased spectrophotometric work. Low mean densities of the larger satellites (including Titan) indicate a bulk composi-

tion dominated by water, but the surface exposures of water ice are contaminated in varying degree by dust or other material fairly neutral or reddish in color. The leading hemisphere of Iapetus represents the extreme in contamination with dark material, the origin and nature of which are not yet known. Phoebe appears asteroidal in character, and is probably a captured body.

The optical properties of solar system bodies derived from Earth-based and spacecraft observations contain information on chemical composition and surface microstructure, as well as on shapes, sizes, rotation periods, and surface geologic structures. Pioneer 11 and Voyagers 1 and 2 have given us close-up views of several satellites of Saturn that previously had been accessible to study only from Earth-based telescopes over a relatively narrow range ($\pm$ 6°) of solar phase angle. The new body of data includes images with substantial spatial resolution of seven of the long-known satellites; images showing the shapes and some topography on nine others; thermal infrared observations of four larger Saturn satellites, including spatially resolved temperatures on Rhea; multicolor photometry of nearly all the satellites; and abundant polarimetric data on Titan (Pioneer 11). While the far-infrared spectrometric equipment aboard Voyager 1 and 2 spacecraft provided an entire new suite of compositional information on Titan's atmosphere, the spacecraft experiments did not include instrumentation suited to compositional studies of the solid surfaces known to characterize the other satellites of Saturn. For this information, we must still rely on groundbased observations.

Even during the epoch of *in situ* spacecraft observations of Saturn's satellites (1979–1981), groundbased observations continued on an intensive scale, and additional telescopic observations of certain satellites are still of great value. This review synthesizes the spacecraft and groundbased studies of the optical properties of Saturn's satellites and presents these bodies in the context of the other small objects in the outer solar system. A useful perspective on the growth of our knowledge of the satellites of Saturn over one decade can be gained from the review papers by Morrison and Cruikshank (1974), Morrison (1982a), and the chapter by Morrison et al. in this book.

I. PHOTOMETRY OF SATURN'S SATELLITES

Pre-Voyager knowledge of the photometric properties of Saturn's satellites has been reviewed by, among others, Veverka (1977a) and Cruikshank (1979b). Harris (1961) reported some of the earliest photometric results on the satellites, but only for Iapetus was it then clear that the rotation is synchronous. Blanco and Catalano (1971) established the synchronism of Rhea's rotation from its lightcurve. The study by Noland et al. (1974) established the synchronous lightcurves of Tethys and Dione, and provided new data for Rhea and Iapetus as well. Franz and Millis (1975) improved on these results and added new data for Enceladus. All these investigators contributed to our understanding of the phase effects in the satellite photometry. The work of Noland et al. (1974), Franklin and Cook (1974), and Franz and Millis (1975) shows the opposition effects on all the satellites. While spectrophotometric work, starting with McCord et al. (1971), progressed during the 1970s, the next significant advance in photometric studies was made by Voyager. A major contribution of the Voyager observations is that it is now possible to study individual areas on most satellites and determine variations in surface reflectance and scattering characteristics. In addition, the Voyager data provide measurements at phase angles much larger than are observable from Earth, and in the cases of small, faint satellites and those close to the rings, this is the only reliable photometric data available. Finally, the Voyager observations produced accurate diameters and geometric albedos for most of the satellites.

A. Definitions

In this review we use standard photometric terminology in which p is the geometric albedo of a satellite, q its phase integral, and A_B the Bond albedo (Harris 1961). The symbol r_n is used to denote the normal reflectance of a surface element, that is its bidirectional reflectance for $i = \varepsilon = 0°$ (where i and ε are the incidence and emission angles, respectively). At opposition, the phase angle $\alpha = 0°$, and $p = r_n$ for a satellite, if (as is approximately true for the Moon) the disk is uniformly bright (no limb darkening). If limb darkening is present at $\alpha = 0°$, as is the case for Europa and for some of the brighter satellites of Saturn, p will be less than r_n. A simple relationship exists among p, r_n, and the Minnaert limb darkening parameter k measured near $\alpha = 0°$, namely: $r_n = (0.5 + k)p$. Thus for Europa, for which $k \simeq 0.65$ (Buratti and Veverka 1983a), $r_n = 1.15\ p$.

The dependence on phase angle of the light scattering properties of a satellite's surface can be expressed in at least two equivalent ways. One may plot the disk-integrated flux of scattered light (properly normalized to a common distance) against phase angle to yield the object's disk-integrated phase curve. The slope of this curve (usually close to opposition) is commonly expressed in units of stellar magnitudes per degree and is called the phase coefficient of the satellite. One can construct an analogous phase curve for any surface element of the satellite, and define an *intrinsic phase coefficient*. This quantity is always numerically lower than the phase coefficient of the satellite as a whole, since the latter involves an integration of the surface scattering properties over the satellite's visible disk (see Veverka 1977a, for details).

B. Mimas, Enceladus, Tethys, Dione, and Rhea

Since Voyager coverage in phase angle extends to 75° in some cases, reasonably reliable determinations of the phase integrals are possible (Table I). Generally, these values agree with estimates derived from other sources (groundbased radiometry and Voyager IRIS measurements). The apparent exception is Rhea, for which groundbased and Voyager IRIS radiometry suggest

TABLE I

Photometric and Radiometric Properties of Saturn's Satellites

Satellite	Voyager Phase Angle Coverage	Voyager Clear Filter: $\lambda \sim 0.47~\mu m$ [a]			Voyager Infrared (IRIS) Results [b]	
		Geometric Albedo p (± 0.1)	Phase Integral q (± 0.1)	Bond Albedo $p \cdot q$ (± 0.1)	Subsolar Point Temperature (K)	Bolometric Bond Albedo A_{bol}
Rhea	2–68°	0.65	0.70	0.45	100 ± 2	0.65 ± 0.03
Dione	7–74°	0.55	0.80	0.45	—	—
Tethys	10–42°	0.80	0.75	0.60	93 ± 4	0.73 ± 0.05
Enceladus	12–42°	1.00	0.85	0.9	75 ± 3	0.89 ± 0.02
Mimas	12–75°	0.75	0.80	0.60	—	—

[a] After Buratti et al. (1982) and Buratti (1983).
[b] After Hanel et al. (1981a,1982).

a q of close to 1, whereas the photometry of Voyager images implies a value near 0.7. This discrepancy remains to be resolved. Most of the observed photometric properties of Rhea seem consistent with a lunar-like scattering law (and low q), for small phase angles showing negligible limb darkening (Buratti 1983). It should also be noted that in terms of albedo and general photometric properties, Rhea shows many similarities to Ganymede, which is known to have a relatively small value of $q = 0.8$ (Squyres and Veverka 1981).

Rotational brightness variations derived from Voyager data for Rhea, Dione, and Tethys by Buratti et al. (1982) agree well with the results obtained from telescopic observations; for all these satellites, the lightcurves are crudely sinusoidal with the leading hemisphere being brighter. For Enceladus, the Voyager data show that the trailing side is brighter (lightcurve amplitude $\simeq 0.2$ mag) confirming an earlier suggestion by Franz and Millis (1975). For Mimas, no rotational brightness variation is established by the Voyager data. The lightcurve amplitude is definitely < 0.1 mag, and if nonzero, then the trailing side is probably slightly brighter. Thomas et al. (1983c) also found that the trailing face of the smaller coorbital may be brighter than the leading side.

Whatever patterns may exist in the lightcurve systematics, it is evident from the Voyager images that these cannot be attributed directly to crater densities on the surfaces. There is definitely no correlation in the Saturn satellite system between albedo and crater density of the type proposed by Hartmann (1980) for the satellites of Jupiter; nor is there any correlation of crater density with distance from the apex of motion (Plescia and Boyce 1983).

While the photometric properties of some satellites such as Dione and Rhea can be described adequately by a lunar-like scattering law, those of Tethys, Mimas, and especially Enceladus, are more complicated and involve a significant Lambert-like component. Buratti et al. (1982) found that the brightness distribution on the satellites can be fitted adequately by a simplified photometric function of the form

$$\frac{I}{F} = a\, f(\alpha)\frac{\mu_0}{\mu + \mu_0} + (1 - a)\,\mu_0 \tag{1}$$

where I is the reflected intensity, πF the incident solar flux at $i = 0°$, $f(\alpha)$ is a function of the phase angle α, and μ_0 and μ are the cosines of the angles of incidence and emission, respectively. The parameter a is a measure of the importance of the Lambert-like component. For a lunar-like photometric function $a = 1$, while $a = 0$ for Lambert's law.

Buratti et al. (1982) found that the importance of the Lambert term is a definite function of albedo. Voyager data indicate that for Rhea and Dione $a \simeq 1$, for Tethys and Mimas $a \simeq 0.7$, whereas for Enceladus $a \simeq 0.4$. The above equation implies that near $\alpha = 0°$, there will be no limb darkening if $a = 1$, and that the importance of limb darkening will increase as a decreases.

In fact, Voyager data show that close to opposition limb darkening is negligible for Rhea and Dione, noticeable on Mimas and Tethys, and substantial in the case of Enceladus (Buratti 1983). The importance of the $(1 - a)$ term can be related directly to the importance of multiple scattering within the surface layer (Buratti and Veverka 1983b). Note that unless $a = 1$, $p \neq r_n$. For $a < 1$, p is less than r_n, reaching a value of 2/3 r_n for $a = 0$.

C. Hyperion

The photometric properties of Hyperion have not been fully elaborated, in part because of a continuing uncertainty about its spin rate. Data from Voyager showed brightness variations that were inconsistent with a synchronous spin, a result also found by groundbased observers. The best surface-resolved images of Hyperion were obtained by Voyager 2 over a time span of less than two days during which interval most of the apparent motion of surface features resulted from the motion of the spacecraft rather than the spin of the satellite. Thomas et al. (1983d) have, however, used a large number of other Voyager images, obtained at a large distance from the satellite, but from which accurate photometric measurements could be derived, to establish a lightcurve for Hyperion. The Voyager 2 data are best fitted by a 13-day period, and Thomas et al. (1983d) suggest that the rotational axis of Hyperion lies close to the orbital plane. Groundbased photometric work in 1983 (Goguen et al. 1983a) also yields a period of 13 days, though the residuals in a period analysis of the telescopic data are not inconsistent with the possibility of an unstable period. Wisdom et al. (1984) have deduced from a study of Hyperion's orbit and spin vector that the satellite may exhibit large and random variations in its spin rate and in the orientation of the rotational axis. The present observational evidence suggests that at the time of the Voyager 2 encounter (August 1981), and in the interval May–August 1983, the rotation period of Hyperion was 13 days. Observational confirmation of the hypothesis by Wisdom et al. (1984) that Hyperion is tumbling chaotically will be difficult.

B. A. Smith et al. (1982) noted that Hyperion is a very irregularly shaped body; according to P. Thomas (personal communication) and Morrison et al. (see their chapter), the principal diameters are 350 ± 30, 240 ± 20, and 200 ± 20 km. Voyager phase coverage extends from 20° to 70°. The intrinsic phase coefficient is ~ 0.013 mag deg^{-1}. Extrapolation of the Voyager data to $\alpha = 0°$ suggests that the geometric albedo is ~ 0.25 (> 0.20 and probably < 0.32). This value is derived from Voyager surface photometry and involves no assumptions about the dimensions of the satellite. Several determinations of the geometric albedo available from Earth-based observations agree with the Voyager estimate. Based on radiometry, Cruikshank and Brown (1982) found a value of 0.3. Using updated values of the mean opposition magnitude ($V_0 = 14.4$), and dimensions given by B. A. Smith et al. (1982), Tholen and Zellner estimate a range of 0.19 to 0.25 (depending on the aspect of Hyperion during the observations). Using the dimensions derived by Thomas (personal communication) quoted above, the Tholen/Zellner estimate becomes 0.20 to 0.36.

Albedo variations of some 10–20% caused by surface variegation or mottling are visible in the Voyager images (B. A. Smith et al. 1982). There is no evidence of a hemispheric albedo asymmetry, but until the rotation state is determined it is difficult to judge what fraction of Hyperion's surface has been imaged. Hyperion's irregular shape should lead to periodic brightness fluctuations of ~ 0.4 mag, which ought to be readily detectable by telescopic observers for this 14th magnitude object; however, the problem is made more difficult by the relatively long rotation period of at least several days. Figure 1 shows Voyager images of three different views of Hyperion.

The limited Voyager color data show that between 0.35 and 0.59 μm the color of Hyperion is more similar to that of the dark side of Iapetus than to that of Phoebe. This conclusion is consistent with more extensive measurements made by telescopic observers. Tholen and Zellner (1982) noted that between 0.35 and 1.1 μm Hyperion has a reddish spectrum somewhat similar to that of the dark side of Iapetus and to those of D-asteroids (but has a higher albedo than either). They also stressed that the spectrum is unlike that of Phoebe, and proposed that the surface of Hyperion consists of a mixture of water frost and reddish Iapetus-like dark material. A similar conclusion was reached by Cruikshank and Brown (1982) and by Brown (1983). Cruikshank (1981) detected absorption bands due to water frost in 1.5–2.5 μm spectra of Hyperion. We return to this in Sec. II.A.

D. Iapetus

Cruikshank et al. (1983) quote $V_0 = 12.1 \pm 0.1$ for the dark hemisphere of Iapetus; using the radius determined by Voyager (720 $\pm$ 20 km) this gives a geometric albedo for the dark hemisphere of 0.081 $\pm$ 0.008. Since the light-curve amplitude is $\sim 1.7–1.8$ mag, V_0 for the bright face is 10.3–10.4, and the geometric albedo, 0.41 $\pm$ 0.04.

Voyager images confirm that the albedo distribution has a dominantly hemispheric asymmetry (leading hemisphere darker) as had been inferred from several analyses of the observed lightcurve (e.g., Morrison et al. 1975). The Voyager observations now make it possible to discuss the actual distribution of reflectances on the two hemispheres, rather than merely the hemispherically averaged values. The phase coverage of the Voyager data is from 23° to 89°. The coverage in longitude is extensive but incomplete with inadequate photometric data for regions near the antapex of motion (270°, 0°). Furthermore, the photometric quality of the data for the darkest areas of the leading hemisphere is mediocre due to low signal-to-noise ratios (the exposures were selected to assure that features on the bright hemisphere would not be overexposed). It must also be realized that the spatial resolution of the Voyager images of Iapetus is low, being ≥ 17 km per line pair. In spite of these

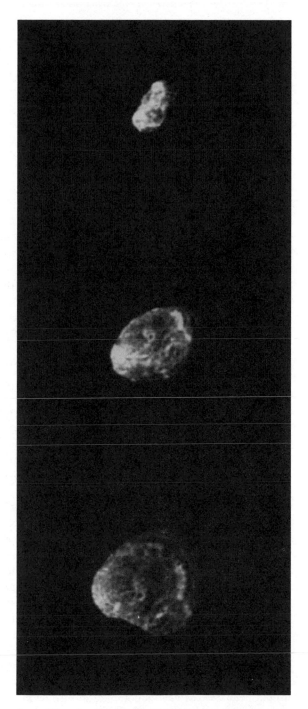

Fig. 1. Three views of Hyperion from Voyager 2. Top image was taken from a range of 1,200,000 km, the middle from 700,000 km, and the bottom from 500,000 km. (NASA photograph courtesy of JPL: P-23932.)

limitations, the Voyager data provide important new information on the distribution of reflectances on Iapetus, on the absolute reflectances of the darkest and brightest areas, and on the nature of the "boundary" between the two hemispheres. These data are being analyzed by Squyres et al. (1983*a*) and Goguen et al. (1983*b*) on whose work the following preliminary summary is based.

The albedo pattern on the dark hemisphere is simpler than the one on the bright face. On the dark (leading) hemisphere albedo contours are concentric about the apex of motion. They are approximately elliptical, being flattened at the poles and elongated along the equator. The reflectance increases gradually from the apex of motion where the lowest albedo occurs (the lowest reflectance is certainly no higher than 0.04, and may be lower; derivation of a precise value is being hindered by low signal-to-noise ratios). No bright spots occur within the dark hemisphere. The reflectance increases gradually from the apex: from ~ 0.04 to 0.15 along the equator, and from ~ 0.04 to 0.20 toward higher latitudes (Squyres et al. 1983*a*; Goguen et al. 1983*b*).

The albedo pattern on the bright (trailing) hemisphere is more complex. Reflectances increase gradually away from the interface with the dark hemisphere; values between 0.3 and 0.4 are common. Highest values probably reach 0.5, indicating that the total range in reflectances on Iapetus is at least a factor of ten. While there is no direct imaging of the antapex of motion, available contours show a definite trend suggesting that Iapetus is brightest near the poles, rather than near the antapex of motion.

It is noteworthy that the transition in reflectance contours between the two hemispheres is gradual, not abrupt. Similarly, histograms of reflectances in images centered on the boundary are not bimodal. While the gradation of albedo is gradual, the change in color seems more abrupt. Voyager data show that the dark material is distinctly redder than the bright material, and that within the accuracy of the Voyager color measurements no color gradient is discernible within either hemisphere.

The colors and spectrophotometry of Iapetus as they pertain to the surface composition are discussed in Sec. II.A. It is well established from ground-based observations that the dark side of Iapetus has a very steep phase curve (see, e.g., Veverka 1977*a*). Out to $\alpha = 6°$, measured phase coefficients range from 0.05 to 0.06 mag deg^{-1} (roughly twice the lunar value), indicating that the dark material has a very pronounced opposition effect. Although the slope of the phase curve decreases away from opposition, the relatively steep nature of the phase curve is maintained out to large phase angles (Squyres et al. 1983*a* and suggests that the phase integral of the dark hemisphere may be lower than the commonly quoted value of 0.6 (Morrison 1977*b*). Earth-based observations show that even the bright hemisphere has a relatively steep phase curve considering its albedo (typical phase coefficient = 0.03 mag deg^{-1}) suggesting that both hemispheres have a complex surface texture. These results are consistent with Voyager measurements at higher phase angles. The

Voyager data indicate a phase integral of ~ 0.9 for the bright hemisphere.

Morrison et al. (see their chapter) discuss the hemisphere albedo and compositional asymmetries in terms of the geologic structures imaged by Voyager 1 and 2 spacecraft.

E. Phoebe

Voyager 2 data provide reliable measurements of the radius (110 ± 10 km), and spin period (9.4 ± 0.2 hr, prograde) of Phoebe, as well as a good estimate of the geometric albedo (Thomas et al. 1983b). The surface shows albedo markings, the normal reflectances of which center around 0.046 and 0.060. Over the range of phase angles covered by Voyager (8° to 34°) the phase coefficients are 0.036 and 0.033 mag deg^{-1} for the darker and brighter areas, respectively. These equivalent disk-integrated values are lower than the average value for Phoebe reported by Degewij et al. (1980a) from Earth-based measurements between $\alpha = 1°$ and $\alpha = 6°$, suggesting a pronounced opposition effect. The mean opposition magnitude quoted by Cruikshank (1979a), $V_0 = +16.4$, combined with a radius of 110 km gives a geometric albedo of 0.069 in the V passband of the UBV system, comparable to the Voyager value (~ 0.06) and again consistent with the presence of an opposition effect. Phoebe's Bond albedo is very low, ~ 0.02 (Thomas et al. 1983c).

F. Other Small Satellites

Voyager observations of these small bodies (shown in Figs. 2 and 3) are reviewed by B. A. Smith et al. (1982) and by Thomas et al. (1983b). Many of these objects are irregular in shape and probably heavily cratered; available spacecraft data are consistent with synchronous spin states in which the largest axis points toward the planet. Normal reflectances (and presumably the geometric albedos) range from about 0.4 to at least 0.8. The reflectances and colors span the range observed for the larger ice-covered satellites of Saturn. Thomas et al. (1983b) discuss the possibility that all are icy bodies with surfaces (at least) contaminated by small amounts of a dark, opaque material.

Voyager coverage in phase angle is scattered and is large only in a few cases (from 24° to 139°). The lack of spacecraft data below $\alpha = 24°$ makes it difficult to determine precise geometric albedos; the values quoted are based on extrapolating the phase behavior observed at larger phase angles to opposition (see Thomas et al. 1983b, for details). Earth-based photometric observations of these bodies are still rather crude and do not help much in constraining the geometric albedos.

The two satellites in this group for which phase coefficients can be determined well are the larger coorbital (Janus) and the outer F-Ring shepherd (Fig. 4). Their intrinsic phase coefficients are 0.02 and 0.092 mag deg^{-1}, respectively. For comparison, the intrinsic phase coefficients of Mimas, Enceladus, Tethys, Dione, and Rhea range from 0.007 to 0.018 mag deg^{-1} (Buratti et al. 1982; Buratti 1983); among these bodies there is no clear depen-

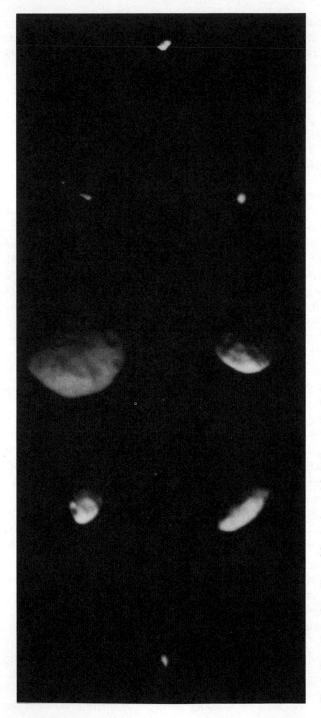

Fig. 2. Several small satellites of Saturn imaged from Voyager 2, all shown at a uniform scale. Starting at top center and proceeding clockwise these satellites may be identified as Janus, Telesto, 1980S6 (S12), Calypso, Epimetheus, 1980S27 (S15), Atlas, and 1980S26 (S16). (NASA photograph courtesy of JPL: P-24061.)

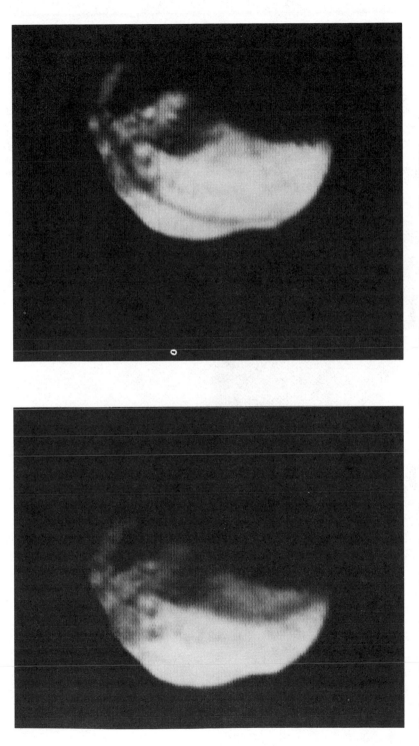

Fig. 3. Two Voyager 1 images of the smaller (trailing) coorbital satellite, Epimetheus. (NASA photograph courtesy of JPL: P-23104.)

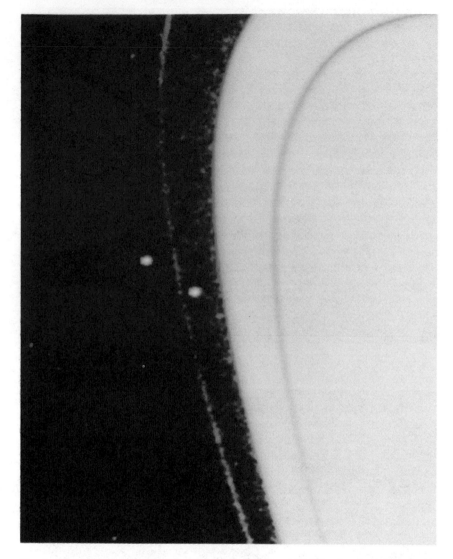

Fig. 4. The two small satellites shepherding the F Ring (Janus and Epimetheus) are shown in this Voyager 2 image. The two satellites are less than 1800 km apart in this image. (NASA photograph courtesy of JPL: P-23911.)

dence of the value of the phase coefficient on surface reflectance. Table II summarizes the Voyager data for the sizes, albedos, and opposition magnitudes for the small satellites derived by Thomas et al. (1983b).

The mean opposition magnitudes derived from Voyager data are based on the photometry of spacecraft images using Mimas as the primary reference

TABLE II
Properties of Small, Inner Satellites of Saturn [a]

No.	Name	Type	Orbital Radius ($\times 10^3$ km)	Cross Sectional Radius at Elongation A × C (km)	Geometric Albedo	$+\Delta m$ (Relative to Mimas)	Mean Opposition Magnitude V_0
S17	Atlas	A-Ring Shepherd	138	20 × 10	0.4	—	—
S16	1980S27	Inner F-Ring Shepherd	139	70 × 37	[a] 0.49–0.68	[b] 3.1–2.7	[b] 15.6–15.2
S15	1980S26	Outer F-Ring Shepherd	142	55 × 33	0.52	3.6	16.1
S10	Janus	Larger Coorbital	151	110 × 80	0.63	1.6	14.1
S11	Epimetheus (1980S3)	Smaller Coorbital	151	70 × 50	0.42–0.60	3.0–2.6	15.5–15.1
S1	Mimas	Regular	186	196 × 196	0.60 ± 0.1	0	12.5
S14	Calypso	Tethys Leading Lagrangian	295	12 × 11	0.53–0.75	6.3–5.9	18.8–18.4
S13	Telesto	Tethys Trailing Lagrangian	295	15 × 8	0.81–1.15	6.0–5.6	18.5–18.1
S12	1980S6	Dione Leading Lagrangian	377	17 × 15	0.46–0.66	5.7–5.3	18.2–17.8

[a] After Thomas et al. (1983b) in part.

[b] Where a range is given, the two estimates correspond to two different assumptions about the phase coefficient below $\alpha = 30°$: (a) intrinsic phase coefficient of 0.0092 mag deg^{-1} (derived for Outer F Shepherd); and (b) intrinsic phase coefficient of 0.022 mag deg^{-1} (derived for Larger Coorbital). In this table, we have assumed that the geometric albedo equals the normal reflectance (i.e., no limb darkening at $\alpha = 0°$, or $k = 0.5$). If $k = 0.6$ were more appropriate, the geometric albedos in the table should be decreased by ~ 10% and the values of V_0 increase by 0.1 mag.

source. Recent analyses of Voyager observations (Buratti and Veverka 1983b) indicate that the mean opposition magnitude of Mimas is +12.5, and not +12.9 as has been previously quoted. The values of V_0 for the satellites observed by Voyager and given in Table II are thus referenced to the new value of V_0 for Mimas. The Voyager data are in good agreement for the value of V_0 = 18.5 ± 0.5 found from groundbased observation of the Tethys leading Lagrangian satellite by Larson et al. (1981a).

II. SPECTROPHOTOMETRY OF SATURN'S SATELLITES

Ices and rock-forming minerals common in the solar system have characteristic absorption bands in the visual and near-infrared (0.3–5 μm) spectrum, and because these bands tend to be broad, they can be discerned in low resolution spectrophotometry, and in some cases, in broadband, multicolor photometry of very low spectral resolution.

The historical development of spectrophotometric studies of Saturn's satellites has been reviewed elsewhere (Morrison and Cruikshank 1974; Johnson and Pilcher 1977; Cruikshank 1979b) and rather than repeat what has already been published, we concentrate here on the most recent information on the nature of these bodies derived from studies of their spectral reflectances. Saturn's family of satellites can be categorized as icy, rocky, or gas-enshrouded (Titan only). Of the 17 known satellites, there is direct spectral evidence relating to surface composition for only six (plus Titan), as given in Table III. Information on the remaining bodies, particularly albedos, gives clues to their surface compositions. For example, the high albedo plus low mean density of Mimas assure us that its surface is icy, and this is surely true of the small, inner bodies of high albedo listed in Table II as well.

A. The Icy Satellites

Near-infrared photometry (Johnson et al. 1975; Morrison et al. 1976) and especially the spectroscopy by Fink et al. (1976) showed the characteristic water-ice absorption spectrum of Rhea, Dione, Tethys, and the trailing hemisphere of Iapetus. At the time these observations were made, the presence of water ice on two of the Galilean satellites and the rings of Saturn was well established from comparisons of the telescopic data with laboratory samples. Spectrophotometry with high photometric precision in the range 0.65– 2.5 μm has been published for the leading and trailing hemispheres of Tethys, Dione, Iapetus, and Enceladus (leading only) by Clark et al. (1984; see also Clark and Owensby 1982). Except for the leading hemisphere of Iapetus, which we discuss separately below, all the data show the strong characteristic water-ice features. The data for some of the satellites have signal precision sufficient to reveal the 1.04-μm band that is especially diagnostic of mean optical path within the surface layer of ice. The depth of this band on Rhea corresponds to a 1 mm mean optical path length in the ice, suggesting a sur-

TABLE III
Properties of the Principal Satellites of Saturn

Satellite No.	Name	Orbital Radius ($\times 10^3$ km)	Radius (km)	Mean Density (g cm^{-3})	Geometric Albedo	Mean Opposition Magnitude V_0	Surface Composition
S1	Mimas	186	196	1.4 ± 0.3	0.8	12.5	Ice + neutral
S2	Enceladus	238	255	1.2 ± 0.3	1.0	11.8	Ice
S3	Tethys	295	530	1.21 ± 0.16	0.8	10.3	Ice + neutral
S4	Dione	377	560	1.43	0.6	10.4	Ice + neutral
S5	Rhea	527	765	1.34 ± 0.09	0.6	9.7	Ice + neutral
S6	Titan	1222	2575	1.90	0.2	8.4	Atmosphere of N$_2$ + CH$_4$
S7	Hyperion	1481	$205 \times 130 \times 110$	—	0.3	14.4	Ice + reddish dust
S8	Iapetus	3561	730	1.16 ± 0.09	0.4–0.08	10.3–12.1	Ice + reddish material
S9	Phoebe	12954	110 ± 10	—	0.06	16.4	Carbonaceous?

face relatively free of opaque mineral contaminants. Clark et al. (1984) reach the same conclusion about ice purity on Tethys, Dione, and the trailing hemisphere of Iapetus. The absorptions of water ice are stronger on the leading hemisphere than on the trailing. In general, there is considerable variability in the strengths of the ice bands among the Saturn satellites, and Clark et al. (1984) find that the band depths and shapes are somewhat different from those seen at the same resolution on the Galilean satellites. From this they infer that the Saturn satellites have surfaces that are more frosty than Jupiter's satellites. The main criteria upon which are based inferences as to frost versus ice and the nature of contamination by foreign matter are the slope of the continuum, the depth of the strong absorption bands, and the appearance of the 1.04-μm ice band, all in the spectral range 0.6–2.5 μm (Clark 1980,1981a,b, 1982; Clark and Rousch 1982).

In their study of the leading-trailing hemispheric differences in the ice-band strengths, Clark et al. (1984) suggest that magnetospheric interaction with the satellites influences the band appearances (see Figs. 5 and 6). Similar, but more pronounced effects are found in the case of the Galilean satellites of Jupiter (Clark et al. 1983). Nelson (1983) has made a strong case for the ultraviolet darkening of the trailing hemisphere of Europa through magnetospheric interaction. While the mechanisms are not yet clear, ion implantation (Lane et al. 1981) and other possibilities (Wolff and Mendis 1983) have been suggested. While the magnetosphere of Saturn at the positions of the satellites that Clark et al. studied is weaker than the Jovian magnetosphere at the Galilean satellites, the leading-trailing asymmetry is observed and requires explanation. Until the exact mechanism of ice-band modification is elaborated, it would be wise to keep open the possibility that some other effects are operating in both the Jovian and Saturnian systems.

The spectrum of Hyperion shows water-ice features (Cruikshank and Brown 1982) comparable in strength to those seen at the same resolution on Ganymede and Rhea, though there are distinct differences in the band shapes. The differences are probably related to the fraction of the surface that has exposed water frost and the degree of foreign matter contamination of the frost. In another study, Brown and Cruikshank (1983) point out the similarity in the spectra of Hyperion and Uranus's satellite, Ariel. Though it may be fortuitous, the independently derived albedos of these two bodies are very similar; for Ariel $p_V = 0.30 \pm 0.06$, and for Hyperion $p_V = 0.28$. [The V subscript to the geometric albedo p indicates measurements in the V photometric band at 0.56 μm wavelength.] Brown (1983) showed that the spectra of Hyperion and Ariel from 0.8 to 2.5 μm are very similar, and that both can be matched by an intimate mixture of fine-grained water frost and carbon-containing dark material in proportions of $\sim 65 \pm 15\%$ of dark material by weight. Such results leave open the possibility that both the leading hemisphere of Iapetus and the surface of Hyperion are contaminated by external dark material, which, however, almost certainly does not come from Phoebe (see Cruikshank et al.

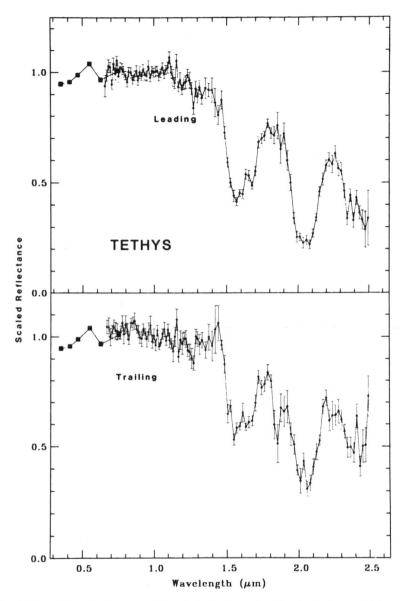

Fig. 5. Near-infrared reflectance spectra of the leading and trailing hemispheres of Tethys are shown at 1.5% spectral resolution. The six data points in the photovisual spectral region are from Noland et al. (1974). The narrow features in the trailing side spectrum at 1.8 to 2.0 μm are probably systematic errors in the telluric water extinction correction and are not real features in the satellite spectrum. (From Clark et al. 1984.)

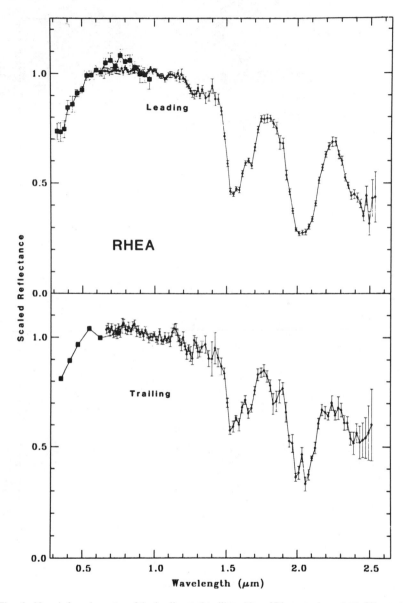

Fig. 6. Near-infrared spectra of the leading and trailing sides of Rhea are shown at 1.5% spectral resolution, along with data in the photovisual spectral region from Bell et al. (1981; leading side) and Noland et al. (1974). The Noland et al. data are only plotted with the trailing side spectrum, but they did not distinguish between leading and trailing sides. (From Clark et al. 1984.)

1983). Since Hyperion is apparently not in a synchronous spin state, a Iapetus-like hemispheric coating would not result. The implication of such a model is that, like Iapetus, by bulk Hyperion is an icy satellite.

The high-albedo icy surfaces of the satellites so far discussed as derived from infrared data are consistent with the photovisual region (0.4–1.1 μm) spectrophotometry published by McCord et al. (1971). Though there are variations from object to object, the spectral reflectances of Tethys, Dione, Rhea, and Iapetus (trailing hemisphere) are relatively neutral in this region. Specifically, they lack the steep downward slope toward the violet that is so characteristic of many other solar system bodies, even some having relatively ice-rich surfaces (Europa and Ganymede, and the rings of Saturn).

The albedo range of the eight newly discovered satellites of 0.4 to 0.8 indicates, by analogy with the other inner Saturn satellites, predominantly ice surface compositions. Further analysis of Voyager data may reveal the photovisual spectral reflectances of these satellites for comparison with similar data for the larger bodies, but there is presently no specific information on the nature of the ice and its possible contaminating mineral components.

Iapetus is predominantly an icy satellite as judged from its low mean density of 1.16 (B. A. Smith et al. 1982; chapter by Morrison et al.) and the spectral signature of its trailing hemisphere. Clark et al. (1984) show, in addition to the usual strong H_2O bands, the weaker 1.04-μm H_2O band in the spectrum of the icy trailing hemisphere. They suggest that the spectrum indicates essentially pure ice; any low albedo contaminants must occur in concentrations of < 1% by weight. The spectral reflectance and albedo of the leading hemisphere, however, are those of a surface dominated in composition by rocky or dusty material. The disk-averaged geometric albedo of the leading hemisphere, as noted above, is $p_V = 0.081 \pm 0.008$. Figure 7 shows the spectral reflectance of the dark hemisphere from 0.3 to 3.8 μm. Notable features of the spectrum include the steep rise from the ultraviolet toward 0.9 μm, with a possible inflection at 0.6 μm. A steep red slope is seen on some outer Jovian satellites (Cruikshank et al. 1982a; Tholen and Zellner 1984), and outer-belt and Trojan asteroids (Degewij and van Houten 1979), but is not common among asteroids and regular planetary satellites. It was also seen in earlier photometry of Iapetus by Millis (1973,1977). Absorptions in the Iapetus spectrum at 1.5, 2.0, and 2.4 μm are attributed to water frost, probably from the polar regions which intrude on the dark hemisphere (see the photometric map, Fig. 2 in the chapter by Morrison et al. and the Voyager map in B. A. Smith et al. [1982]). The very strong absorption in the 3-μm region may be in part due to polar cap ice, but Lebofsky et al. (1982) and Cruikshank et al. (1983) have given arguments for bound water in the surface minerals as the main cause of this feature.

Clark et al. (1984) have established that the dark material on the leading hemisphere of Iapetus can contain very large amounts of water ice if a sufficient amount of dark contaminating material is intermixed. The strong ice

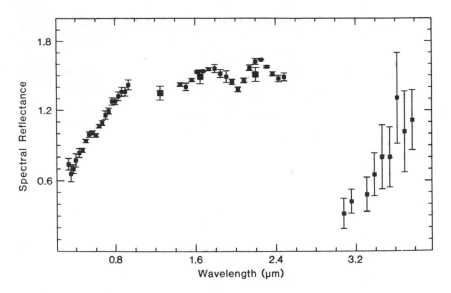

Fig. 7. The spectral reflectance of the dark hemisphere of Iapetus from 0.3 to 3.8 μm, showing the steep red slope in the short wavelengths, and strong absorption at 2.8 μm attributed to water of hydration in the surface minerals. (From Cruikshank et al. 1983.)

bands at 1.5 and 2.0 μm could remain invisible even if the surface consists of as much as 70% by weight ice with large (10–100 μm) grains of dark contaminant, while the quantity of ice could reach some 95% if the dark contaminant occurs in submicrometer grains (Clark and Lucey 1984).

When the anomalous water-ice features in the dark hemisphere spectrum (0.3–2.5 μm) are accounted for, the resulting curve is reasonably well matched by the spectrum of a carbon-rich residue from the Murchison C2 carbonaceous chondritic meteorite. This conclusion is consistent with that of Tholen and Zellner (1982) that the dark material on Iapetus is similar to that of D (or RD) asteroids (cf. Gradie and Veverka 1980). Further, all spectral and color data, including those from Voyager, agree in showing that the dark material on Iapetus is much redder than, and spectrally distinct from, the material on the surface of Phoebe. Details of the spectral comparison and implications for models of the genesis of the dark hemispheric coating on Iapetus are discussed by Cruikshank et al. (1983), who also review other models of dark hemisphere emplacement. Images of Iapetus from Voyager 2 with spatial resolution up to ~ 20 km per line pair do not readily settle the questions of the origin of the dark material, though they help clarify the nature of the bright/dark boundary. Recent debates on the exogenic or endogenic origin of the dark material are found in B. A. Smith et al. (1982), Squyres and Sagan (1983), Soderblom and Johnson (1982), Cruikshank et al. (1983), and in the chapter by Morrison et al.

B. The Rocky Satellites

Phoebe is the only known satellite of Saturn that has a globally uniform low albedo. The geometric albedo at 0.48 μm wavelength derived from Voyager images varies with longitude from 0.046 to 0.060, with relatively bright patches at high latitudes.

There are three independent sets of multicolor photometric data for Phoebe, all mutually inconsistent, at least in the violet spectral region. As shown in Fig. 8, the two groundbased sets diverge by the greatest amount. The eight-color data of Tholen and Zellner (1982) are relatively flat except for the strong absorption at 0.33 μm, while the Degewij et al. (1980a) data for UBV show a distinctly different slope. The spectrophotometric curve constructed from Voyager 2 photometry (Thomas et al. 1983a) shows approximately the same slope as the Degewij et al. data. The Tholen and Zellner data also indicate a downturn toward longer wavelengths, suggestive of the presence of a solid state absorption seen in other mineral surfaces and attributed to the Fe^{++} ion; the other filter points at 0.81 and 1.25 μm (Degewij et al. 1980b) are too widespread in wavelength to resolve the question. Because of the inconsistencies in the three data sets, it is difficult to characterize the surface of Phoebe from its reflectance. The two curves most nearly in agreement, taken together with the low albedo, suggest that Phoebe's surface is comparable to C-type asteroids (see Cruikshank et al. 1983).

Inasmuch as Phoebe has been considered a possible source of the dark material lying on the leading hemisphere on Iapetus, comparisons of the colors of Phoebe and Iapetus have been made, but the reflectance of Iapetus in the photovisual region is distinctly steeper toward the red than Phoebe. Whether or not the interior of Phoebe is icy cannot be determined with existing data, but we must allow the possibility that it is essentially an icy object with a substantial carbonaceous or other dark mineral contaminant in the surface layer. On the other hand, various investigators (see, e.g., Pollack et al. 1979) have discussed the probability that Phoebe is an object captured by the Saturn system from some other source, in which case its composition could be representative of an entirely different class of objects. As Cruikshank et al. (1983) have noted, Phoebe may be only one of several captured satellites of Saturn; the most thorough search for additional exterior satellites extended to magnitude 19 (T. Gehrels, personal communication) with the Palomar 1.2-m Schmidt telescope. At magnitude 19, a body of albedo $p_V = 0.04$ at Saturn's distance would have a radius of 40 km. Phoebe and any other captured objects in Saturn's system are probably comets, but may also represent asteroids from the inner solar system that have been expelled from the main belt or its vicinity.

C. Titan

Titan is the only satellite of Saturn known to have or suspected of having an atmosphere (see chapter by Hunten et al.). Spectrophotometry of Titan reveals strong bands of methane that were first found photographically by Kui-

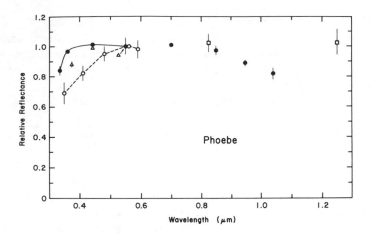

Fig. 8. The reflectance of Phoebe. Solid dots are from Tholen and Zellner (1982), open circles from Voyager photometry (Thomas et al. 1983*b*), triangles from Degewij et al. (1980*b*), and squares from Degewij et al. (1980*b*). The data are all normalized to a reflectance of 1.0 at 0.55 μm. The data points from Degewij et al. (1980*b*) were corrected for the solar flux as determined from the eight-color survey (D. J. Tholen, personal communication).

per (1944). The aerosol haze comprising various layers in the upper atmosphere affects the shape and strength of the methane absorption bands, but the study of this important topic is more the subject of spectroscopy at relatively high resolution rather than spectrophotometry.

The atmosphere of Titan has been found to be variable on a time scale of about one day in the *JHK* (1.25–2.2 μm) filter bands (Cruikshank and Morgan 1980), from which one might expect a spectrophotometric variation also to appear (repeated low-resolution scans by Cruikshank throughout this spectral region have so far shown no changes).

Spectral data for Titan expressed in geometric albedo versus wavelength have been assembled by Podolak and Giver (1979) using data from Younkin (1974), Caldwell (1975), and Barker and Trafton (1973) for various spectral regions from 0.25 to 1.1 μm. Similarly, Fink and Larson (1979) have published a compilation of the albedo spectrum from 0.5 to 2.5 μm using their own spectra and the spectrophotometry by Wamsteker (1975).

III. RADIOMETRY

A. Groundbased Results

Measurements of the thermal emission from airless satellites and asteroids, when combined with measurements of the sunlight reflected from them, yield important information on the albedos, dimensions, and optical properties of the bodies. The photometric/radiometric method used at groundbased telescopes has become the prime means of determining the dimensions of asteroids (see, e.g., Morrison and Lebofsky 1979) and satellites prior to close proximity spacecraft measurements (Murphy et al. 1972b; Morrison 1974a). Most recently, the diameters of four satellites of Uranus were determined by this method (R. H. Brown et al. 1982a), and the thermal flux of Triton has also been detected (Lebofsky et al. 1982). The theory and applicability, as well as inherent errors and assumptions in the method, have been described at length (see, e.g., Morrison 1977a; Morrison and Lebofsky 1979), and the most recent calibration of the technique has been made by R. H. Brown et al. (1982b).

In the case of Titan, broadband radiometry at 10 and 20 μm at first suggested a greenhouse thermal structure in the atmosphere (Morrison et al. 1972). The combination of this information with narrowband observations in the same wavelength regions (Gillett et al. 1973) showed the importance of emission from specific molecules and from the aerosol layer, resulting in the correct identification of a temperature inversion in the atmosphere (Danielson et al. 1973). Further details about Titan are found in the chapter by Hunten et al.

The dimensions and albedos of Rhea and Iapetus were determined from the first thermal flux measurements at 20 μm (Murphy et al. 1972b); these have subsequently been revised (Morrison 1974a; Morrison et al. 1975) and superceded by Voyager imaging results, but the groundbased values are correct to within the error estimates of the determinations. Radiometry of Iapetus (Murphy et al. 1972b) showed conclusively that the brightness variations can be attributed to a bimodal distribution of albedo rather than an irregular shape. Likewise, 20-μm measurements yielded the effective diameter and albedo of Hyperion (Cruikshank 1979a; Cruikshank and Brown 1982), but the values have been superceded by results based on Voyager images.

An eclipse of Iapetus in 1978 by Saturn's rings was observed at photovisual and thermal wavelengths at various observatories. Analysis of the combined data in terms of a two-component model of the satellite's surface representing portions of the high-albedo and low-albedo hemispheres is in progress (G. S. Aldering et al., in preparation).

B. Spacecraft Results

Except for some radiometric observations of Titan made with the Pioneer 11 spacecraft in 1979, the principal proximate spacecraft data on the Saturn satellites come from Voyager 1 (1980) and Voyager 2 (1981). The data were acquired with a radiometrically calibrated Michelson interferometer (IRIS) (Hanel et al. 1977). We do not consider the IRIS observations of Titan here; the essential results are discussed in the chapter by Hunten et al.

Because of the relatively large field of view (0°.25) of the Voyager infrared spectrometer (IRIS), and the small apparent sizes of the satellites during the flybys, only low spatial resolution and full-disk thermal data were acquired. Rhea was resolved to approximately one-fifth of the disk, and full-disk measurements were obtained for Enceladus, Tethys, and Iapetus. Observations of the remaining satellites were not of use because of low signal-to-noise ratios (Hanel et al. 1981a,1982).

Near full phase (phase angle $\alpha = 15°$) the central regions of the disk of Rhea give brightness temperatures of 100 $\pm$ 2 K at 9.516 AU, with a gradual decrease to 93 $\pm$ 2 K near the limbs; at $\alpha = 77°$ the temperatures at the same geographic positions are 90 $\pm$ 4 K and 83 $\pm$ 4 K, respectively. This anisotropy in the thermal emission is quantitatively similar to that of the Moon (Saari and Shorthill 1972) suggesting that both the large- and small-scale roughness of Rhea are also similar. Judging by Voyager images (see Fig. 3 in the chapter by Morrison), this appears true on a topographic scale of a few kilometers (B. A. Smith et al. 1981).

Additional information on surface conditions on Rhea is obtained from disk observations of an eclipse; relative to an observation made before eclipse entry, the disk flux dropped by $\sim$ 75%. The brightness temperature spectra of the observation are shown in Fig. 9. The nearly wave-number-independent pre-eclipse brightness temperature of 96 $\pm$ 2 K is consistent with the spatially resolved measurements discussed above. The significant slope in brightness temperature measured in eclipse indicates lateral inhomogeneities in the surface thermal properties of the satellite. Decomposition of the spectrum (assuming a mixture of two surface temperatures) indicates that roughly half the surface cools to $\sim$ 75 K, while the remainder cools to $<$ 55 K. This indicates the presence of components having high and low thermal inertia. The high-inertia component is probably solid material with a block size comparable to or greater than the thermal skin depth $(\tau K/\pi \rho C)^{1/2}$, where τ is the period of the insolation change and K, ρ, and C are the thermal conductivity, density, and heat capacity of the material. For the eclipse observation, assuming the material is water ice, the thermal skin depth $\simeq$ 10 cm. The low-inertia component is most likely a thin frosty or pulverized layer, like that present on Europa and Ganymede.

The bolometric albedo of a satellite can be estimated from thermal spectra in cases where the disk is spatially resolved. The subsolar point spectrum of Rhea is very nearly that of a blackbody with a temperature $T_{ss} = 100 \pm 2$ K. Assuming unit surface emissivity, the bolometric albedo is

$$A = 1 - R^2 T_{ss}^4/T_0^4 = 0.65 \pm 0.03 \qquad (2)$$

where $R = 9.156$ AU is the heliocentric distance of Saturn, and $T_0 = 401 \pm 6$ K is the subsolar point temperature at 1 AU, allowing for the peaking of infrared flux at small phase angles (see Hanel et al. 1981a). This value of the bolometric albedo is significantly higher than the Bond albedo derived from analysis of Voyager images (Table I); this difference has not yet been reconciled.

To determine albedos from full-disk observations, the phase integral or a

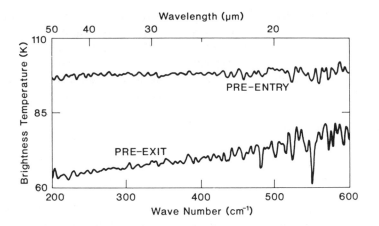

Fig. 9. The thermal spectrum of Rhea in and out of eclipse. The brightness temperatures are shown before entry into Saturn's shadow and before exit. The slope of the pre-exit curve indicates a range of surface temperatures caused by a range of thermal inertia values of the material on the satellite's surface. (From Hanel et al. 1981a.)

surface temperature model is required. A simple temperature model for an airless spherical object is

$$T(\theta) = T_{ss} \cos^{1/n} \theta \tag{3}$$

where θ is the zenith angle of the Sun. For a nonrotating Lambert sphere $n = 4$; for the Moon $n \approx 6$. A fit to the Voyager data for Rhea gives $n = 7 \pm 1$. With this temperature distribution integrated over the disk and then corrected for the 19° phase angle, the pre-eclipse disk observation of Rhea yields a subsolar point temperature of 102 ± 3 K, in good agreement with the spatially resolved measurement. Application of this model to Voyager 2 observations of Tethys and Enceladus ($R = 9.601$ AU) yields the subsolar point temperatures and bolometric albedos shown in Table I; both albedos are consistent with Bond albedos derived from Voyager images.

Disk observations of Iapetus which included much of the dark (leading) hemisphere (see Fig. 2 in chapter by Morrison) were also obtained. A subsolar point temperature in excess of 100 K is indicated, consistent with the very low albedo of this material.

IV. OPTICAL PROPERTIES FROM POLARIMETRY

Polarimetric observations of solid bodies in the solar system can yield information on the geometric albedo and the microstructure of their surfaces if data can be acquired over a sufficiently large range of phase angles. For the satellites of Saturn the range observable from Earth is not sufficient for a satisfactory solution to the albedo, and gives only imperfect information on surface structure. Bowell and Zellner (1974) and Veverka (1977b) have reviewed the subject of polarimetry as it applies to solid solar system bodies, and there has been no substantial progress either with the observations or the mathematical theory of their interpretations since that time. Zellner (1974) has added some new observations of Iapetus, noting that the polarization follows the expected law of reciprocity between polarization and albedo. From this he infers that the surface microstructure is similar on the bright and dark hemispheres of the satellite. Observations of Dione, Rhea, and the two hemispheres of Iapetus show that there is a pronounced negative branch in the polarization versus phase angle curve, but this parameter is insufficient to give a value of the geometric albedo. The presence of a deep negative branch in the polarization curve appears to indicate only that the surface from which the sunlight is reflected has a complex microstructure at the wavelength of the light. This microstructure can take the form of powder in a regolith and/or a complex roughness at the same scale on large fragments or chips of rock (Veverka 1977b). Multiple scattering in complex surfaces of high albedo, such as in frost, tends to quench the negative branch as well as the total surface polarization. The strengths of the negative branches in the three satellites of Saturn that have been observed are consistent with known albedos, but give little else in the way of useful information.

V. COMPARISONS WITH OTHER PLANETARY SATELLITES

From the point of view of the optical properties, particularly the surface compositions and albedos, the satellites of Saturn can be compared with other satellite systems. With the exception of Phoebe, all the Saturnian satellites have high albedos indicative of a significant water-ice component on their surfaces. The range in albedo, from ~ 0.45 to 1.0, apparently reflects a variation in the degree of contamination of the ice or frost by matter that is either neutral or reddish in color and low in albedo. Enceladus, with a geometric albedo of 1.0, and the dark hemisphere of Iapetus ($p_V = 0.07$) apparently represent the extremes of purity and contamination for the icy satellites. A parallel for purity in the Jovian system is Europa, whose surface may be rejuvenated frequently, while the icy surfaces of Ganymede and Callisto are contaminated to varying degrees with neutral material (Clark 1980,1981a,b,1982).

The other major system of icy satellites, that of Uranus, has just begun to be explored. The several papers by Brown and colleagues cited earlier point out that these icy bodies reveal varying degrees of surface contamination, though the range in geometric albedos is relatively narrow, from ~ 0.18 (Oberon) to 0.30 (Ariel). The innermost known Uranian satellite, Miranda, may have an albedo as high as 0.5, according to recent spectrophotometry by Brown and Clark (1984) in which water ice was detected.

The special characteristics of Phoebe invite comparison with dark-surfaced asteroids. If Phoebe is indeed a captured body, its bulk composition may be more characteristic of an asteroid or a giant unprocessed cometary nucleus than of a planetary satellite. In size, it ranks with the largest 30 or 40 asteroids, many of which have dark, neutral spectral reflectances.

Acknowledgments.   We thank J. C. Pearl for important contributions to this chapter.

PART VI
Titan

TITAN

D. M. HUNTEN and M. G. TOMASKO
University of Arizona

F. M. FLASAR and R. E. SAMUELSON
Goddard Space Flight Center

D. F. STROBEL
Naval Research Laboratory

and

D. J. STEVENSON
California Institute of Technology

Titan is the only satellite with a substantial atmosphere, and this atmosphere is denser than that of any terrestrial planet except Venus. Also, Titan resembles Venus in having a global haze and even a small greenhouse warming of the surface. Methane clouds are probably present, but this is not definitely established. Though methane and other hydrocarbons dominate the infrared spectrum, the major gas is N_2 which shows up in strong ultraviolet emissions. The upper atmosphere is remarkably close to isothermal all the way from the stratosphere to the exosphere, and surface temperatures are globally uniform within a few degrees. The dark-orange or brown stratospheric aerosol particles are slightly > 0.1 μm in radius, and there may be larger particles at greater depths. The aerosols are thought to be the end product of methane photochemistry and to accumulate on the surface (or dissolve in liquid methane or ethane). Reasonable photochemical paths can be set out for production of the observed hydrocarbons and organic molecules. A puzzling aspect of the upper atmosphere is the ultraviolet emissions confined to the day side but too bright to be excited by

the available solar energy. Hydrogen escapes rapidly to populate the observed
torus, accompanied by some nitrogen ejected by electron impact dissociation of
N_2. A global wind system can be constructed on the basis of the observed temperature field. The interior is made up of roughly equal parts of rocky and
icy material. The methane supply must have been adequate to replenish the
atmospheric amount many times over.

Each section was originated by one author: I. Overview (Hunten); II. Atmospheric Composition and Thermal Structure (Samuelson); III. Upper Atmospheric Composition and Temperature (Strobel); IV. Properties of Aerosols
(Tomasko); V. Meteorology and Atmospheric Dynamics (Flasar); VI. Interior
and Surface (Stevenson).

I. OVERVIEW

Titan is the second largest satellite in the solar system (Ganymede's
radius is 70 km greater) and is considerably larger than Mercury. Its dense
atmosphere, dominantly nitrogen but with a substantial component of methane, makes it unique. Its surface and possible clouds are hidden by a dense,
uniform dark-orange haze, though in view of Titan's low albedo the color may
be more accurately called brown. Basic properties of Titan are summarized in
Table I; many of these data were obtained by the close pass of Voyager 1 on
12 November 1980. The considerable history of pre-Voyager observations
has been discussed in two NASA workshop reports (Hunten 1974a; Hunten
and Morrison 1978) and in reviews by Hunten (1977), Caldwell (1977c), and
Owen (1982b). Here we will only mention that limb darkening implying the
presence of an atmosphere was reported visually by Comas Sola (1908); near-
infrared methane bands were discovered by Kuiper (1944); and then a burst of
observational activity in the early 1970s created a broad interest in Titan studies and led to a view of its nature that was confirmed, though enormously
extended, by the Voyager 1 encounter. The key methods were near-infrared
spectroscopy, thermal-infrared and radio radiometry, polarimetry, and ultraviolet spectroscopy from Earth orbit. Lewis (1971) made the influential realization that the low mean density implies an interior composition much richer
in ices than that of the terrestrial planets, despite the many parallels between
Titan and these bodies. Lewis' paper also suggested that the atmosphere
might be rich in nitrogen from photolysis of ammonia. Models based on this
idea were presented by Hunten (1978) and Atreya et al. (1978). Ambiguous
spectral evidence was interpreted by Lutz et al. (1976) and Hunten (1978) as
implying a large amount of transparent gas (nitrogen or argon) in addition to
the observed methane, but there were alternative interpretations (Podolak and
Giver 1979).

The atmospheric temperature (Fig. 1) falls from its surface value of 94 K
at a pressure of 1500 mbar to a minimum of 71 K at a height of 42 km and
pressure of 128 mbar; this level may be called the tropopause. It then rises

TABLE I
Properties of Titan

Surface radius, R_T	2575 km
Mass, M	1.346×10^{26} g $= 0.022 \times$ Earth
GM	8.976×10^{18} cm^3 s^{-2}
Surface gravity, g_s	135 cm s^{-2}
Mean density	1.881 g cm^{-3}
Rock/ice ratio (by mass)	$\sim 52{:}48$
Distance from Saturn	1.226×10^6 km $= 20\ R_S$
from Sun	9.546 A U
Orbit period	15.95 day
around Sun	30 yr
Obliquity	$26°.7$ (assumed)
Temperature, K[a]	
Surface	94
Effective	86
Tropopause (42 km)	71.4
Stratopause (200 km)	170
Exobase (1600 km)	186 ± 20
Bond albedo	0.20
Solar flux	1.1% $\times$ Earth
Surface pressure	1496 ± 20 mbar

[a] Atmospheric temperatures below 200 km assume mean molecular weight 28 and scale with this value (Lindal et al. 1983).

fairly rapidly to 160–170 K, apparently becoming constant from 200 km (a little less than 1 mbar) to the exosphere at 1600 km. Over this range the acceleration of gravity varies by a factor of > 2. The temperatures in the troposphere are no closer than 5 K to the condensation temperature of N_2, coming closest at 30 km. It is therefore unlikely that nitrogen clouds can form. Considerable attention has been devoted to the formation of clouds by condensation of CH_4, which would occur if the mixing ratio below the clouds exceeded 1.5% by volume. If pure liquid methane were present on the surface, this mixing ratio would be 12% and continuous cloud up to 30 km would be expected. Detailed examination of the temperature profile in the bottom few km suggests that liquid methane is present at most locally, and does not form a global ocean (Eshleman et al. 1983; Flasar 1983). However, an ethane ocean ~ 1 km deep has been suggested by Lunine et al. (1983).

At high altitudes there are small variations of temperature with latitude, as observed by Voyager 1. But the troposphere is almost certainly highly uni-

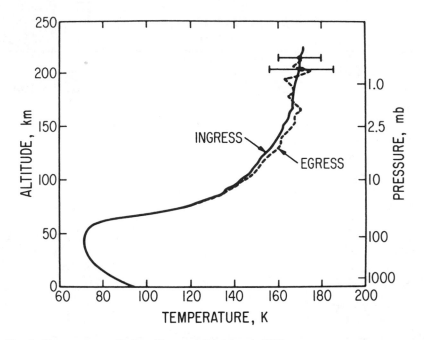

Fig. 1. Temperature profile from Voyager 1 (Lindal et al. 1983).

form with latitude, time of day (Lindal et al. 1983), and season, as assured by the large density of the atmosphere which, along with the low temperature, provides a long thermal time constant. Relatively sluggish motions are all that is required to give the uniform conditions.

The amount of methane, obtained from its near-infrared absorption, implies that dissociation is occurring, driven by solar radiation with wavelengths < 1400 Å. The products are hydrogen and free radicals such as CH_2. The hydrogen mixing ratio stabilizes at 0.2%, as shown in Table II, and reaches a steady state with escape. The radicals initiate the buildup of heavier hydrocarbons, many of which eventually condense to form particles. This chain of processes plausibly explains many of the observed molecules and the dark haze. Production of nitrogen compounds such as HCN requires dissociation of N_2, which is caused primarily by electron impact or other ionospheric processes like recombination. Electrons may come from the solar wind, the magnetosphere, or photoionization with excess energy.

The compounds produced in greatest abundances are the simplest ones, ethane C_2H_6 and acetylene C_2H_2; ethylene C_2H_4 is destroyed rapidly by photolysis. Acetylene is an important precursor to the heavier compounds (see Sec. III). Lunine et al. (1983) point out that ethane is likely to dominate the

TABLE II
Atmospheric Composition of Titan below 0.1 mbar

Gas	Band	Position (cm^{-1})	Mole Fraction Inferred Indirectly	Mole Fraction Measured Directly
Nitrogen	N$_2$	—	—	0.65–0.98	
Argon(?)	Ar	—	—	0–0.25	
Methane	CH$_4$	ν_4	1304	0.02–0.10	
Hydrogen	H$_2$	S$_0$	360		2×10^{-3} [a]
		S$_1$	600		
Carbon monoxide	CO	(3–0) X$^1\Sigma^+$	6350		$6 \times 10^{-5} - 1.5 \times 10^{-4}$ [b]
Ethane	C$_2$H$_6$	ν_9	821		2×10^{-5} [c]
Propane	C$_3$H$_8$	ν_{21}	748		4×10^{-6} [d]
Acetylene	C$_2$H$_2$	ν_5	729		2×10^{-6} [c]
Ethylene	C$_2$H$_4$	ν_7	950		4×10^{-7} [c]
Hydrogen cyanide	HCN	ν_2	712		2×10^{-7} [c]
Methyl acetylene	C$_3$H$_4$	ν_9	633		3×10^{-8} [c]
		ν_{10}	328		
Diacetylene	C$_4$H$_2$	ν_8	628		$\sim 10^{-8} - 10^{-7}$ [c]
		ν_9	220		
Cyano-acetylene	HC$_3$N	ν_5	663		$\sim 10^{-8} - 10^{-7}$ [c]
		ν_6	499		
Cyanogen	C$_2$N$_2$	ν_5	233		$\sim 10^{-8} - 10^{-7}$ [c]
Carbon dioxide	CO$_2$	ν_2	667		1.5×10^{-9} [e]

[a] Samuelson et al. 1981; [b] Lutz et al. 1983; [c] Kunde et al. 1981; [d] W. C. Maguire, personal communication; [e] Samuelson et al. 1983b.

material falling to the surface. At Titan's surface temperature, condensed ethane is liquid and, with CH$_4$, N$_2$, and other heavier compounds produced in the atmosphere, it can form a global ocean $\sim$ 1 km deep. Lunine et al. estimate that 100–200 m of solid acetylene lies beneath such an ocean. Their proposed ocean is composed of 70% C$_2$H$_6$, 25% CH$_4$, and 5% N$_2$, although, as discussed in Sec. V, there remains considerable latitude in estimates of the relative C$_2$H$_6$–CH$_4$ composition. Similarly, the depth estimated in Sec. III is $\gtrsim$ 500 m.

The methane in the atmosphere must be in a steady state between photolytic destruction and some source from the surface or interior. If the ethane-rich ocean is global, methane from the interior must be passing through it.

Eshleman et al. (1983) and Flasar (1983) rejected the possibility of a global ocean of pure methane because the vapor pressure would be too high, implying a lapse rate inconsistent with the observations shown in Fig. 2 (cf. Sec. V).

The outer atmosphere was observed in considerable detail by the ultraviolet spectrometer (UVS) on Voyager 1. This instrument gives the only direct evidence that N_2 is the major constituent, although the mean mass of 28.0 amu strongly implies the same thing. The temperature of 186 K assures that hydrogen and helium will escape freely at a rate limited by diffusion from below. This rate is readily calculated, and for H_2 is compatible with the observed mixing ratio and the proposed source by photolysis of methane (Sec. III). The escaping gas is probably a mixture of H and H_2; it goes into orbit about Saturn, forming a torus. The Lyman-α radiation scattered by the atoms was observed by both Voyagers; it extends from 8 to 25 R_S (Saturn radii), enveloping Titan's orbit at 20 R_S. Its vertical thickness is 14 R_S. The lifetime of atoms in the torus is estimated at 3 yr, so the required source strength is $\sim 3 \times 10^{27}$ atoms s^{-1}. The supply from Titan is 8×10^{27} atoms s^{-1} (Sec. III), but both estimates are rather uncertain and the disagreement is tolerable.

Saturn's magnetosphere, on the average, extends to or slightly beyond Titan's orbit in the solar direction and fully envelops it in most other directions. The solar wind probably reaches Titan often. During the closest approach of Voyager 1, Titan was well inside the magnetopause, and magnetospheric plasma was being swept past by Saturn's rotation. No Titanian magnetic field was detected, but there was a distinct tail in the downstream direction. The currents induced in Titan's ionosphere probably deflected the plasma in much the same way that Venus affects the solar wind. This may be the best evidence for the existence of an ionosphere, which could not be detected by radio occultation. The radio occultation necessarily sampled at a very large solar zenith angle, and the null result is reasonable if the recombination coefficient is as large as expected (Sec. III).

The radio-occultation experiment has given detailed profiles of the ratio of temperature to mean molecular weight (Lindal et al. 1983). The results shown in Figs. 1 and 2 assume a mean mass of exactly 28. As discussed in Sec. II, comparison of radio and infrared data permits a value of 28.6, although slightly larger values are favored. The only gas that might raise it above 28 is ^{36}Ar, accreted as clathrate (Owen 1982a,b). Opinions differ as to its plausibility. Over the history of planetary studies, large amounts of noble gases have been advocated for Mars and Venus, but were ruled out when the measurements were made.

Figures 1 and 2 are supplemented by a numerical tabulation in Lindal et al. (1983). Above the tropopause is a well-defined stratosphere, with the temperature increasing to an asymptotic value ~ 170 K at ~ 200 km and pressure ~ 1 mbar; this level is the stratopause. The presence and general structure of this region were inferred by Danielson et al. (1973) from its strong infrared

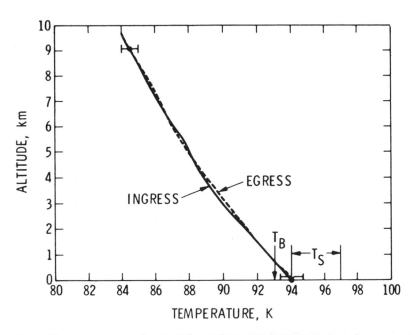

Fig. 2. Temperatures in the lowest 10 km (Lindal et al. 1983). T_S is the surface temperature range derived from infrared data (Samuelson et al. 1981), and T_B is a preliminary value for the brightness temperature in the 500–600 cm^{-1} band, where the atmosphere is most transparent.

emission. They pointed out the likely heat source in absorption of solar radiation by the brown high-altitude haze. The haze particles are too small to be efficient thermal radiators and therefore transfer the heat to the gas (see Sec. II).

At much higher altitudes (1265 km), the density and temperature were measured in solar occultation by the Voyager UVS (G. R. Smith et al. 1982). These results do not depend on the mean mass. With this density point and the one at the stratopause, the hydrostatic equation can give a mean temperature for the entire intervening region, the mesosphere and thermosphere. This temperature is 165 K, and the density profile is shown in Fig. 3 taken from G. R. Smith et al. (1982). Their measured local temperature was 186 K, at the region of the exobase. The curvature in Fig. 3 arises from the variation of gravity with height. Formulae to handle this situation are given by Smith et al. (1983). An alternative is to replace the geometric height z by geopotential (or Titanopotential) height:

$$z_1 = r_s z/(r_s + z) \tag{1}$$

where r_s is the radius of a reference level where the gravity g has value g_s.

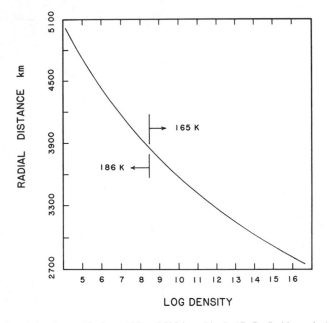

Fig. 3. Inferred density profile from 125 to 2525 km altitude (G. R. Smith et al. 1982). The 186 K temperature and the densities in that neighborhood were measured in solar occultation. The right-hand part of the curve is an interpolation.

The hydrostatic equation then takes its normal exponential form with $g = g_s$ (U.S. Standard Atmosphere Supplements 1966).

Although this isothermal interpolation seems simple, it is implausible because many physical processes are involved in the region from 1 mbar to 6 nanobar. At the higher pressures, most of the heating is via the smog particles and the radiation by the observed molecular bands, particularly of methane, ethane, and acetylene. A constant temperature at lower pressures requires a constant ratio of heating to cooling, in the face of effects such as reduction of the smog content, vibrational relaxation of the radiating molecules, and additional heating by ionizing wavelengths < 1000 Å as well as molecular absorptions at longer wavelengths. Some structure in the temperature profile seems almost inevitable, but a detailed study is needed before any useful predictions can be made (cf. Friedson and Yung 1984).

The structure of the upper atmosphere is greatly affected by diffusive separation. On Titan this process enriches methane upwards and is almost neutral for C_2 hydrocarbons with masses 26–30. The level at which separation begins is the homopause or turbopause, and its height can be derived from the methane mixing ratio observed in the solar occultations (G. R. Smith et al. 1982; see also Sec. III); the result is a little below 1000 km.

Global temperature structure at three levels has been assessed from the

Voyager infrared interferometer spectrometer (IRIS) data by Flasar et al. (1981b); see also Secs. II and V. They found no sign of diurnal changes and very little meridional structure, as illustrated by Fig. 4. The three curves, top to bottom, refer to heights of 130, 40, and 0 km, or pressures of 1, 100, and 1500 mbar. The thermal time constants are longer than the 30-yr seasonal cycle up to the 50-mbar level. Further, the circulation of the troposphere is easily able to keep the temperature of the polar regions within a degree or two of that of the equator, as observed at least to 70° latitude. A similar conclusion was reached by Leovy and Pollack (1973), though there are substantial differences of detail. This strong tendency to global symmetry makes it appropriate to treat the heat balance in terms of global means, in Sec. II.D. The consequences of the remaining structure for the global circulation are discussed in Sec. V. Most of the topics of Sec. II (composition, thermal structure, and radiative equilibrium) have already been introduced. Section III discusses aer-

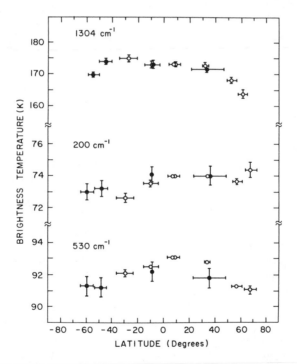

Fig. 4. Latitude distributions of brightness temperature at 1304, 200, and 530 cm^{-1}. The 1304 cm^{-1} data are normalized to emission angle 47°5, while the other two are normalized to 52°7. ○: daytime data; ●: nighttime data. Vertical bars include uncertainties due to noise, calibration, and emission angle. Horizontal bars indicate latitude range over which data comprising each point extend (figure after Flasar et al. 1981b).

onomy over a wide range of height, including possible sources of the smog particles. Sources of atoms and molecules for the torus are also considered. Section IV describes the wealth of observational data on high-altitude aerosols and the interpretation in terms of particle properties and distribution. Regrettably, almost nothing is known about possible methane condensates, although their potential thermal opacity is discussed in Sec. II.

Although Sec. III is largely theoretical, observational constraints are involved. Sections V and VI are even more theoretical. Titan's meteorology maintains a high degree of global uniformity, but that is almost all we can observe. The interior structure is constrained only by the mean density; attempts to deduce it must therefore employ hypotheses of origin, evolution, and analogy, along with copious laboratory data on possibly relevant substances.

II. ATMOSPHERIC COMPOSITION AND THERMAL STRUCTURE

A. Chemical Identifications and Abundances

Methane has been known to be present on Titan since its discovery by Kuiper (1944), who subsequently estimated an abundance of $\sim$ 200 m-amagat from comparison with laboratory data (Kuiper 1952). In a reanalysis of Kuiper's work Trafton (1972b) suggested that the 6190 Å CH_4 band used by Kuiper fell on the square-root portion of the curve of growth, and that consequently only the product of the abundance and the mean broadening pressure could be inferred. For a pure CH_4 atmosphere Trafton suggested a surface pressure and abundance of 16 mbar and 1.6 km-amagat respectively, but adequate laboratory data were not available at that time. Later, upon analyzing weak pressure-independent bands in the visible region of the spectrum, Lutz et al. (1976) derived an abundance of 80 m-amagat and a surface pressure $\sim$ 200 mbar, implying that methane is only a minor constituent. Subsequent analyses (Danehy et al. 1978; Podolak and Giver 1979; Fink and Larson 1979) have placed the abundance at $\sim$ 1–3 km-amagat. All determinations suffer from model dependence, because the vertical distribution of scatterers providing the spectral continuum is not well defined.

In a major breakthrough, Gillett et al. (1973) and Low and Rieke (1974) obtained narrow-band radiometric observations, from 8 to 13 μm and from 1.6 to 34 μm, respectively. Positive identifications of the 821 cm^{-1} ν_9 band of C_2H_6 and the 1304 cm^{-1} ν_4 band of CH_4 were established. Shortly thereafter Gillett (1975) filled in the 8–13 μm spectral region with high-precision spectrophotometric observations, clearly identifying the 1150 cm^{-1} ν_4 band of CH_3D (monodeuterated methane) and the 950 cm^{-1} ν_7 band of C_2H_4. There was also a hint of structure beyond 13 μm, suggesting that the 729 cm^{-1} ν_5 band of C_2H_2, or possibly the 748 cm^{-1} ν_{21} band of C_3H_8 (propane) might be contributing. The identification of C_2H_2 was later confirmed by Tokunaga (1980).

Here the observations stood until the November 1980 near encounter of Voyager 1, when a wealth of new information became available from the UVS and IRIS experiments on board (Broadfoot et al. 1981; Hanel et al. 1981*a*). The UVS observed atomic hydrogen in Lyman-α at 1216 Å, atomic nitrogen between 1085 and 1243 Å, and molecular nitrogen between 800 and 1020 Å, including emission in the Rydberg bands at 960 and 980 Å (Fig. 9 in Sec. III.B). We discuss the UVS results in Sec. III.

Whereas much of the information obtained by UVS pertains to altitudes between 500 and 1500 km, IRIS is sensitive to conditions between 300 km and the surface. A remarkable array of features in the IRIS spectra, associated mainly with organic compounds including several nitriles and higher hydrocarbons, can be ascribed to emission from a warm stratosphere (Hanel et al. 1981*a*; Kunde et al. 1981; Maguire et al. 1981). Molecular hydrogen is the only compound seen in absorption, placing its effective emission level in the troposphere (Samuelson et al. 1981).

An overview of Titan's spectral appearance between 200 and 1400 cm^{-1} is shown in Fig. 5. The disk average consists of 346 spectra acquired when the instrument field of view was $< 2/3$ the apparent angular diameter of Titan. Most of the spectra correspond to low latitudes although some data from latitudes as high as $\pm 60°$ are included. The north polar region spectrum is an average of 30 spectra taken with the center of the field of view above $+70°$ latitude. Finally, the limb spectrum is an average of three taken at grazing incidence at the edge of the north polar hood. The inclusion of space in the field of view depresses the continuum relative to the sharp-line features in this average.

Relative abundances of those gases either observed directly or inferred to be in the atmosphere below ~ 300 km are listed in Table II. A few (N$_2$, Ar, CO, H$_2$) are expected to be uniformly mixed throughout the lower atmosphere without undergoing phase changes. N$_2$ was observed directly by UVS (Broadfoot et al. 1981), but its amount in the lower atmosphere can only be inferred by comparing radio occultation and IRIS data (Tyler et al. 1981; Samuelson et al. 1981; Lindal et al. 1983; see also Sec. II.B below). Owen (1982*b*) and Strobel (1982) have argued that if argon is present then the observed N$_2$ is probably primordial, and that CO should also be present, having originated in the same clathrate structure as argon and nitrogen. Once CO$_2$ was identified (Samuelson et al. 1983*b*), the argument for CO was greatly strengthened, though for a different reason. CO was later discovered by Lutz et al. (1983) from groundbased observations in the near infrared. Preliminary analysis has suggested the mole fraction range listed in Table III, which is comparable to the value $\sim 1.1 \times 10^{-4}$ predicted by Y. Yung (in Samuelson et al. 1983) on the basis of photochemical modeling (cf. also Sec. III).

The presence of argon can only be inferred. A gas heavier than N$_2$ may be required to make the radio occultation and IRIS results internally consistent (Samuelson et al. 1981), although Lindal et al. (1983) and Flasar (1983)

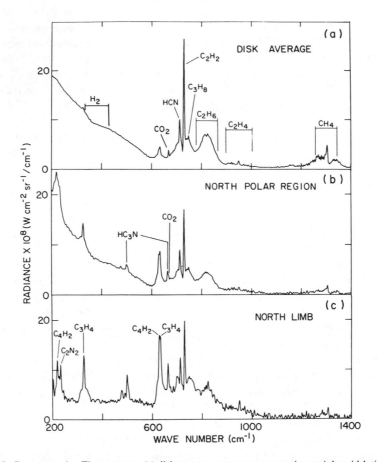

Fig. 5. Representative Titan spectra: (a) disk average spectrum representing mainly mid-latitude regions on Titan; (b) north polar spectrum associated with a region above +70° latitude; (c) north limb spectrum from just off the edge of the north polar hood. Space contamination depresses the continuum relative to the sharp emission features (figure after Samuelson et al. 1983*b*).

have suggested that a pure nitrogen atmosphere (possibly augmented by a small amount of methane) may also be acceptable. In the nominal model of Samuelson et al., a tropospheric mean molecular weight of $m \sim 28.6$ amu (with a large error margin) appears to be required, and the other two gases of large abundance there (N_2 and CH_4) have molecular weights of 28 and 16 amu, respectively. This model requires the temperatures for the surface and the atmosphere immediately above it to be equal. In Sec. II.B we show that these temperatures may in fact differ by up to 0.5 K, implying that the requirement for argon may be less severe than the nominal model indicates.

On the other hand, if the mean molecular weight m does exceed 28 amu (or slightly less if a significant fraction of CH_4 is present), argon would seem to be the only heavy gas available with a sufficiently high vapor pressure to be present in the amounts required. Owen (1982a,b) has pointed out that the Ar/N_2 mixing ratio of ~ 0.15 inferred by Samuelson et al. is consistent with the primordial ratio expected for the proto-Saturn nebula, allowing for subsequent loss of atmospheric N_2 (to space and through photolysis) over geologic time. According to Owen, almost no ^{40}Ar should be available; only primordial argon in the isotopic ratio $^{36}Ar : ^{38}Ar = 4$ would be present. However, Lunine et al. (1983) argue that ethane oceans are capable of retaining ~ 3 times as much argon as the atmosphere can retain. Thus, if their ethane ocean model for Titan is correct (cf. also Sec. V), only a modest amount of atmospheric argon should remain. The last uniformly mixed gas that does not undergo phase changes is H_2, and its mole fraction of 0.002 was measured directly.

Methane should also be uniformly mixed globally, but since it may undergo phase changes in the troposphere it may not be vertically uniform. As discussed in Sec. II.B, CH_4 clouds in the troposphere seem necessary to account for the observed opacity between 200 and 400 cm^{-1}. Thus, even though a CH_4 mole fraction of only ~ 0.02 can be supported at the tropopause, its value can increase to nearly 0.10 at the surface if the troposphere is 70% saturated there (cf. Sec. V).

After methane, the next four most abundant hydrocarbons are C_2H_6, C_3H_8, C_2H_2, and C_2H_4 in that order (for detailed discussions see Kunde et al. 1981 and Maguire et al. 1981). Their abundances appear approximately constant with latitude, although no systematic quantitative study has yet been made, and real variations are possible. Each must condense at some level in the stratosphere and precipitate out, effectively reducing its gas phase abundance to smaller amounts below this level. The mole fractions listed in Table II assume condensation at 50 mbar and uniform mixing above.

The five least abundant organics all show strong latitudinal variations, being most abundant in the north polar region and least abundant at high southern latitudes. Methylacetylene C_3H_4, diacetylene C_4H_2, and hydrogen cyanide HCN show this steady southerly decline down to $-60°$ latitude, beyond which reliable data are not available. Cyanoacetylene HC_3N and cyanogen C_2N_2 cannot be identified below $\sim +60°$. The reason for this variation is not understood, though the effect is probably seasonal, with considerable phase lag implied because the data were taken near equinox where the illumination is symmetrical.

Finally, even though the mole fraction for CO_2 is lower than that for any other identified gas, it appears to be uniformly distributed in latitude. This anomalous behavior is a characteristic that distinguishes it from the trace hydrocarbons and nitriles, and thus led to its identification (Samuelson et al. 1983b). As with other trace constituents, it will condense out in the stratosphere and precipitate to the surface.

Thus, Titan's atmosphere is continuously evolving. Methane is released into the atmosphere from the surface, which may be partly ocean composed of a mixture of liquid methane, ethane, and propane (cf. Lunine et al. 1983, and Sec. V). Some of the atmospheric methane is cycled back to the surface as precipitation, and some passes through the cold trap at the tropopause and diffuses upward into the stratosphere. There it is decomposed by photolysis and particle bombardment. The ions and radicals thus produced recombine chemically, forming higher hydrocarbons. Cosmic ray and particle bombardment together initiate additional chemistry, producing nitriles from N_2 and CO_2 from CO and H_2O (water, though not yet detected, presumably originates in meteoritic debris). A $\sim$ 10 yr time for the atmosphere to overturn either by eddy diffusion or planetary scale motions cycles these low vapor pressure products down to levels where they condense. From these levels, both they and the aerosols produced higher precipitate to the surface, forming a mixture of tars and snow. Exceptions are propane and ethane, which remain liquid at the surface. Vapor pressures are too low to permit recycling back into the atmosphere, except for methane, for which the net molecular flux through the $\sim$ 72 K cold trap at the tropopause is sustained by an equal net upward flux from the surface. Molecular hydrogen, formed as a photochemical byproduct, diffuses upward and escapes to space. Eventually the atmosphere will become almost totally depleted in methane, leaving behind on the surface $\sim$ 1 km of frozen or liquid residue.

B. Vertical Temperature Structure

The radio occultation experiment (RSS) is central to determining the vertical thermal structure at low latitudes. According to a detailed analysis by Lindal et al. (1983), the surface radius and pressure are 2575.0 $\pm$ 0.5 km and 1496 $\pm$ 20 mbar, respectively. They assume a pure nitrogen atmosphere and compute the temperature profile shown in Fig. 1; the corresponding surface temperature is 94.0 $\pm$ 0.7 K. The inferred lapse rate of 1.38 $\pm$ 0.10 K km^{-1} near the surface is consistent with the adiabatic value expected for 100% N_2. This lapse rate changes abruptly to 0.9 $\pm$ 0.1 K km^{-1} at an altitude $\sim$ 3.5 km, above which it decreases slowly with height to a value of zero near the tropopause.

However, because only the ratio of temperature to mean molecular weight T/m can be inferred directly from the data, an independent knowledge of either T or m is required at some point to put the temperature profile on an absolute scale (Lindal et al. assumed $m = 28$ amu for their examples). A variation of m with altitude increases the complexity of the problem, as do deviations of the real atmosphere from the perfect gas law. The data are reliable from the surface ($\sim$ 1500 mbar) to about the 10 mbar pressure level. At higher altitudes the inferred thermal structure becomes increasingly sensitive to assumed initial conditions.

In principle, IRIS can provide both the reference required to put the radio occultation profile on an absolute scale, and independent thermal profiles. In practice this requires a known distribution of opacity sources. Methane is the only useful gas, and, while we can assume that it is distributed uniformly in the stratosphere, its abundance is not directly measured but must be inferred from physical principles. Another difficulty with methane is that only the 1304 cm^{-1} ν_4 band is available, and it is useful only in the 0.1–3 mbar pressure range on Titan. Thus, the temperature structure inferred from inversion of the transfer equation using methane as the opacity source fails to overlap the profile inferred from radio occultation where it is reliable, and normalization of one to the other is extremely uncertain.

A normalization procedure that does not depend critically on the exact nature of the opacity sources was developed by Samuelson et al. (1981). In contrast with other spectral regions, the low latitude IRIS data at 200 and 540 cm^{-1} show very little variation with emission angle. This indicates that stratospheric limb brightening compensates tropospheric darkening almost exactly, or that both tend to zero. The latter condition requires that the troposphere is either almost completely opaque or very transparent at these wavenumbers. Radiative transfer analyses demonstrate that the troposphere is in fact extremely opaque at 200 cm^{-1}, with a brightness temperature $\lesssim 74$ K. As this temperature cannot be less than the minimum atmospheric temperature, it becomes an upper limit to the temperature at the tropopause. Extrapolation along the shape of the RSS thermal profile then yields an upper limit at the surface of $\sim$ 96.3 K. Conversely, the troposphere is found to be most transparent at 540 cm^{-1}, where the brightness temperature is $\sim$ 94 K. Since this value cannot exceed the surface value, it gives a lower limit there.

A few corrections to the upper and lower bounds of the surface temperature are required. In connection with the lower limit, there is a finite tropospheric opacity at 540 cm^{-1} due to the 0.002 mole fraction of H_2 contained in the atmosphere (cf. Table II). Also, as shown in Fig. 3 of Samuelson et al. (1981), an additional stratospheric opacity is needed at 540 cm^{-1} in order to make the corresponding wavenumber dependence of opacity appear more plausible, but the amount is uncertain. A conservative lower bound to the surface temperature would appear to be 94.5 ± 0.4 K, though the error is only an estimate based on signal-to-noise considerations in addition to the opacity uncertainties.

This surface temperature, however, may be slightly higher than the air temperature immediately above. The radiative equilibrium model discussed below in Sec. II.D indicates a temperature difference of $\sim$ 1 to 3 K. In Sec. V we show that turbulence will reduce this difference to a factor of $>$ 5 below the purely radiative solution. Thus the lower limit to the air temperature immediately above the ground is $\sim$ 94.0 ± 0.4 K, and this is the temperature to which RSS is sensitive.

The upper limit to the surface temperature is more uncertain because it requires using the shape of the radio occultation profile to extrapolate from 74 K at the tropopause (an upper limit) to the surface. If a constant mean molecular weight is assumed, the upper limit to the surface temperature is 96.3 K with estimated uncertainty $\pm$ 0.7 K, independent of any uncertainty in the radio occultation profile. A Van der Waals correction to the equation of state of $\sim$ 2% would increase the upper limit to $\sim$ 98 K (Lindal et al. 1983).

Variable amounts of CH_4 in the lower troposphere will modify the radio occultation analysis. According to the discussion in Sec. V, a CH_4 mole fraction as high as $q \sim$ 0.1 at the surface can be made consistent with a near surface temperature in the range 94–98 K. Higher values of q force the thermal gradient to become superadiabatic and hence nonphysical. Thus a fraction of saturation $f \sim$ 0.7 near the surface can be tolerated by the data, and would inevitably lead to saturation and attendant condensation higher in the troposphere. This value of f is compatible with global oceans composed of comparable amounts of methane and ethane.

If, on the other hand, the surface is composed mainly of a methane clathrate, the equilibrium vapor pressure is lower by a factor $\sim$ 10^{-2} (Hunten 1978) and large amounts of CH_4 in the atmosphere are impossible. In this case clouds would fail to form, and a large fraction of saturation at the surface would be unnecessary. Calculations of pressure-induced absorption by various $CH_4–N_2$ combinations show that these gases by themselves fail to provide sufficient opacity between 200 and 400 cm^{-1} to match the observations of Titan (Courtin 1982b). Hunt et al. (1983) make a similar finding, although they speculate that sufficient pressure-induced opacity may yet be found. Data on the opacity of liquid methane (Jones 1970; Weiss et al. 1969; Savoie and Fournier 1970) and what little data is available on the solid (Savoie and Fournier 1970) strongly suggest that methane clouds would account for the missing opacity. The observational evidence thus appears to favor their presence in the troposphere. In either case the surface temperature is $\sim$ 95 K or so, the value required to normalize the radio occultation profile.

The thermal structure between 0.1 and 3 mbar has been estimated in two ways. Both methods require a smooth extrapolation to the radio occultation profile as normalized by the method described above. The first method involves assuming various temperature profiles and computing the shape of the 1304 cm^{-1} ν_4 band of CH_4. The profile that leads to the best overall fit is adopted as correct. The second method relates the brightness temperatures measured at 1304 and 1260 cm^{-1} to their effective emission levels, assumed to be where the respective weighting functions are maximum. Agreement between the two methods is good. More rigorous thermal inversion techniques are beginning to be applied, although no published information is yet available. All methods must assume a methane mole fraction for the stratosphere, since it cannot be measured directly. Because methane clouds appear to be required near the tropopause, saturation is also required there. If the tropopause temperature is 74 K, the CH_4 mole fraction in the stratosphere is 0.027. If the tropopause temperature is lower, the CH_4 mole fraction becomes somewhat less. Putting all this together, we obtain the low latitude temperature profile covering the pressure range 0.01–1500 mbar shown in Fig. 1. The surface temperature is $\sim$ 94 K and the stratospheric temperature approaches a limiting value $\sim$ 174 K as P $\rightarrow$ 0, although the data are not adequate to insure the isothermal character of the profile near the top. The higher altitudes have been discussed in Sec. I, and are shown in Fig. 3.

C. Horizontal Temperature and Opacity Structure

Once the vertical thermal structure at low latitudes has been established, a determination of the global horizontal structure becomes feasible using the bulk of the high spatial resolution IRIS data. A preliminary reconnaissance has been carried out by Flasar et al. (1981b), taking advantage of the fact that the thermal fields associated with the upper stratosphere, tropopause, and surface could be monitored at wavenumbers 1304, 200, and 530 cm^{-1}, respectively. (Although the atmosphere is slightly more transparent at 540 cm^{-1} than 530 cm^{-1}, they chose the latter because of superior signal-to-noise ratio.) The results are shown in Fig. 4.

The overall trends are fairly obvious. The 530 cm^{-1} lower troposphere channel implies a thermal structure basically symmetric about the equator, with the poles $\sim$ 3 K cooler. The 200 cm^{-1} tropopause channel implies a temperature essentially independent of latitude, although there is some suggestion of a slight increase in brightness temperature from south to north. The 1304 cm^{-1} channel implies (though the extrapolation is rather large) an equator-to-pole variation of $\sim$ 20 K, and a $\sim$ 4 K latitudinal asymmetry at mid-latitudes, with the southern hemisphere warmer.

On a finer scale some thermal and/or opacity anomalies appear. There is no apparent diurnal variation except for a 1.2 K day-night difference at 35° north latitude at 530 cm^{-1}. The difference is outside the error range and may be real. The 530 cm^{-1} daytime brightness temperatures appear slightly higher in the north between 0° and 40° latitudes than their southern counterparts. Similarly, the 200 cm^{-1} brightness temperatures appear to be $\sim$ 1° cooler in the southern hemisphere than in the north, with the transition at $\sim$ −15° latitude. The radio experiment, however, finds almost no difference between dawn and dusk, at latitudes of +6°.2 and −8°.5.

Sromovsky et al. (1981) find a comparable asymmetry in the violet, blue, and green albedos from Voyager 1 images, with the southern mid-latitudes brighter. It is tempting to suggest a physical relation between these effects, associating the higher southern hemisphere albedos with a higher thermal opacity leading to slightly lower brightness temperatures compared to those in the northern low and mid-latitudes. Such an opacity variation might be ascribed to different thicknesses of tropospheric methane clouds in the two hemispheres.

Conversely, Sromovsky et al. (1981) have suggested that the albedo asymmetry originates in the stratosphere. They argue that the observed wavelength dependence of albedo requires that the visible absorption coefficient of the responsible aerosol increase with decreasing wavelength. The contrast of an underlying cloud should therefore be reduced more at the shorter wavelengths, as the path length is longer. However, because the violet end of the spectrum shows greater contrast, an underlying cloud cannot be responsible for the observed wavelength variation of contrast, and consequently the variation must arise from the stratospheric aerosol. Of course this does not negate the suggestion that there is a hemispheric asymmetry of thermal opacity in the troposphere, but it does imply that the connection between the thermal and visible asymmetries is indirect.

On the other hand, there should be a direct relationship among the stratospheric thermal, optical, and abundance asymmetries. According to Flasar et al. (1981b), the gaseous radiative damping time is too short to account for the latitudinal asymmetry of temperature observed there (Fig. 4). However, if stratospheric heating and cooling rates are governed principally by aerosols (cf. Sec. II.D), there is a strong connection between temperature and aerosol particle size; the latter in turn depends on particle growth rates and hence upon the abundances of gaseous hydrocarbons and nitriles. Physical correlations among the various asymmetries are thus assured, although details of the interactions have yet to be worked out.

D. Radiative Equilibrium

The variation of temperature with latitude is extremely small throughout the troposphere and moderate in the stratosphere. On the other hand, the atmosphere appears to be stable against convection throughout most of the troposphere, at least with respect to the dry adiabat, and of course is extremely stable in the stratosphere. Thus the global vertical thermal structure of the atmosphere may well be governed by radiative equilibrium, even though the horizontal structure is largely controlled by large-scale circulation (Leovy and Pollack 1973; Flasar et al. 1981b; see also Sec. V).

Two basic types of model have been considered. In the greenhouse model, solar radiation is transmitted through a transparent atmosphere and partially absorbed by the surface. There the energy is converted to thermal radiation which is repeatedly absorbed and reradiated in an atmosphere that is optically opaque at thermal wavelengths, thereby heating it at depth to form a troposphere. On the other hand, the inversion model requires deposition of solar radiation and attendant heating high in the atmosphere with little if any transmission to lower regions, leading to the establishment of a temperature reversal.

Both models have been investigated for Titan. Various authors have considered the greenhouse model, including Allen and Murdock (1971), Morrison et al. (1972), Hunten (1972), Sagan (1973), and especially Pollack (1973), who constructed nongrey radiative-convective models which included pressure-induced absorption by methane and hydrogen and absorption due to rotational transitions in ammonia. The models containing half methane and half hydrogen compared favorably (though fortuitously) with the broadband data available at the time. Once narrow-band data in the 8–14 μm spectral window became available (Gillett et al. 1973), it was evident that a temperature inversion existed in Titan's atmosphere. Danielson et al. (1973) first deduced the conditions that would explain the thermal emission features of CH_4 and C_2H_6 found in the narrowband spectra. Their basic conclusion was that a dust (presumably photochemical in origin) absorbs solar radiation high in the atmosphere. Because this aerosol consists of very fine particles, it is an extremely inefficient radiator in the thermal infrared, requiring a particle temperature of ~ 160 K before sufficient thermal energy can be radiated to balance the solar input. As Danielson et al. point out, the gas would be in local thermodynamic equilibrium with the aerosol, which is a quite general characteristic required for gas-aerosol mixtures in all planetary atmospheres (Samuelson 1970). Thus, in the case of Titan the gaseous atmosphere quickly evolves to a steady-state thermal structure controlled by the internal temperature field of the particles, and the optically active gases are seen in emission against the ~ 74 K background.

The vertical thermal structure deduced from the Voyager 1 results (Figs. 1 and 6) demonstrates that both mechanisms operate in Titan's atmosphere. Because the atmosphere also appears to be statically stable, it is natural to pose the questions: (1) to what extent can the globally averaged vertical thermal structure be explained solely as a result of radiative processes? and (2) what basic optical and physical parameters of the atmosphere can consequently be derived?

In a recent study, Samuelson (1983) attempts to answer these questions from as general a viewpoint as possible, by limiting the number of physical parameters to the minimum required to reproduce the observed results. The values of these parameters are therefore in some sense fundamental to Titan, and serve as general constraints in the development of more sophisticated models. In the simplest acceptable model the solar radiation field is divided into two channels, of which one is both absorbed and scattered (and leads to heating within the atmosphere), while the other is conservatively scattered (no absorption), and is diffusely transmitted to the surface and partially absorbed there. The first channel is associated with the violet end of the spectrum where the albedo is low, while the second is associated with the red and near infrared continuum between methane absorption lines where the albedo is maximum. Heating by the violet channel causes the thermal inversion, whereas heating at the surface by the red channel gives rise to a greenhouse effect in the lower atmosphere; the thermal infrared field is maintained as a single separate channel which serves to redistribute the thermal energy and radiatively cool to space. The scattering properties of the atmosphere are assumed to be invariant with optical depth, although cloud layering with altitude is permitted.

In spite of its simplicity the model gives close agreement with the actual atmosphere in several respects. The tropopause temperature of ~ 69 K com-

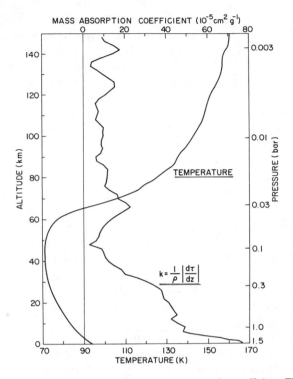

Fig. 6. Vertical profiles of temperature and thermal mass absorption coefficient. The temperature structure is a globally averaged composite derived from Voyager 1 RSS and IRIS data. The profile for the mass absorption coefficient was obtained by fitting the theoretical temperature profile to the observed one (figure after Samuelson 1983).

pares favorably with the measured value of ~ 72 K, especially since atmospheric motions will tend to modify the purely radiative solution somewhat. The mean thermal infrared optical thickness of the stratosphere $\tau_m^{\text{IR}} \sim 0.09$ is fairly close to the range $(0.05 \leq \tau_m^{\text{IR}} \leq 0.08)$ inferred from the data for homogeneously mixed conditions (Samuelson et al. 1981), and quite far from values computed for concentrated layers high in the stratosphere. Distinguishing between such classes of model is important in assessing aerosol production and heating rates, which are essential to understanding the chemistry and dynamical state of the stratosphere.

Several general conclusions about the state of Titan's atmosphere can be inferred from the model. The ratio of violet-to-thermal infrared absorption coefficients in the stratosphere is

$$\frac{\chi_A^{\text{V}}}{\chi_A^{\text{IR}}} = \beta \, \frac{1 - \widetilde{\omega}_0^{\text{V}}}{1 - \widetilde{\omega}_0^{\text{IR}}} \sim 45 \qquad (2)$$

where $\beta = \chi_E^V/\chi_E^{IR}$ is the corresponding extinction coefficient ratio and $\widetilde{\omega}_0^V$ and $\widetilde{\omega}_0^{IR}$ are the respective single-scattering albedos. Equation (2) confirms the idea of significant absorption in the stratosphere by very small particles (Danielson et al. 1973; Caldwell 1977c). By extending an analysis by van de Hulst (1957), the limiting value of the absorption cross section for small particles is found to be

$$\chi_A \sim \frac{2\pi^2 r^3}{\lambda} \frac{24 n_r n_i}{(n_r^2 + 2)^2} \tag{3}$$

where λ is the wavelength, r is the particle radius, and n_r and n_i are the real and imaginary parts of the refractive index (assuming $n_i \ll n_r$). According to several lines of evidence (Tomasko and Smith 1982; Sec. IV) the dominant particle size is of order 0.1 μm, making Eq. (3) acceptable at all wavelengths. Thus, if we assume that n_r is roughly independent of wavelength, and that optically active gases do not dominate the absorption properties of the stratosphere, we obtain from Eqs. (2) and (3)

$$\frac{\lambda^{IR}}{\lambda^V} \frac{n_i^V}{n_i^{IR}} \sim 45. \tag{4}$$

The ratio λ^{IR}/λ^V is ~ 100, demonstrating that $n_i^{IR} \sim 2n_i^V$; i.e., the bulk absorption properties of the aerosol material are comparable in the blue-violet and thermal infrared.

The violet stratospheric optical thickness τ_m^V is found to be quite large. In the limit, as $\tau_m^V \to \infty$, the relation between $\widetilde{\omega}_0^V$ and the spherical albedo a_v is given by

$$\left(\frac{1 - \widetilde{\omega}_0^V}{1 - \widetilde{\omega}_0^V \langle \cos\theta \rangle} \right)^{1/2} = \frac{1 - a_v}{1 + a_v} \tag{5}$$

where $\langle \cos\theta \rangle$ is the asymmetry factor for single scattering. Mie scattering calculations for particle radii ~ 0.1 μm and reasonable refractive indices suggest $\langle \cos\theta \rangle \sim 0.6$, yielding $\widetilde{\omega}_0^V \sim 0.6$ for $a_v = 0.12$. Substituting back into Eq. (2) yields $\beta \sim 100$, from which it follows that $\tau_m^V = \beta \tau_m^{IR} \sim 9$.

On the other hand, the bulk material must be considerably more transparent at red and near infrared wavelengths. The assumption of conservative scattering gives rise to an expression for the total red atmospheric optical thickness τ_1^r,

$$(1 - \langle \cos\theta \rangle) \sqrt{3} \, \tau_1^r = \frac{2(a_r - A)}{(1 - a_r)(1 - A)} \tag{6}$$

where A is the surface albedo and a_r is the albedo at $\tau_c = 0$, the top of the conservatively scattering layer. Substituting $a_r = 0.42$ and $A = 0.1$ yields $\tau_1^r \sim 2$.

If a pure absorbing layer is added on top, so that $\tau_c > 0$, a_r and $\langle \cos \theta \rangle$ can be increased to ~ 0.65 and ~ 0.8, respectively, yielding $\tau_1^r \sim 18$. This latter value probably cannot be increased substantially, however, as a fairly deep penetration of red radiation is required to explain the $1-3$ km-amagat of CH_4 found by Podolak and Giver (1979) and Danehy et al. (1978). In addition, the troposphere is a poor insulator between 400 and 600 cm^{-1}, leading to a rather inefficient greenhouse on Titan. The only gas capable of closing the window is H_2, but its measured abundance is only ~ 200 m-amagat, far too little to have an appreciable effect. As a result, the fraction of solar radiation required to reach and heat the surface is fairly substantial. The net solar flux at the surface is approximately

$$\phi_r \sim - \frac{1}{4} \, \phi_s \, e^{-\tau_c \sqrt{3}} \, (1 - q)(1 - a_r) \qquad (7)$$

where $(1 - q)$ is the fraction of the solar constant ϕ_s in the red channel (the violet channel cannot contribute). To maintain radiative equilibrium, ϕ_r must equal the negative of the thermal net flux at the surface; the analysis yields a range $0.13 < (1 - q)(1 - a_r) < 0.33$. However, according to Eq. (6) $a_r \to 1$ as $\tau_1^r \to \infty$ independently of A. As a result τ_1^r cannot become very large before most of the conservative solar field is backscattered and the surface heating rate is reduced below acceptable limits. This is to be contrasted with conditions on Venus, where only a trickle of solar radiation need reach the surface to maintain the very efficient greenhouse established by H_2O and pressure induced CO_2 absorption (Tomasko et al. 1980; Pollack et al. 1980).

Finally, a comparison of the theoretical and observed vertical temperature profiles gives the variation of thermal optical depth τ^{IR} with altitude z. Thus we can calculate the vertical variation of the infrared mass absorption coefficient k through the expression

$$d\tau^{IR} = -k \, \rho \, dz \qquad (8)$$

where the atmospheric gas density ρ is also known as a function of z. The results are shown in Fig. 6. The analysis indicates that $k \sim 9 \times 10^{-5}$ cm^2 g^{-1} throughout the stratosphere, but increases approximately linearly with decreasing z in the troposphere, becoming $\sim 7.5 \times 10^{-4}$ cm^2 g^{-1} near the surface. So, from Eq. (8) we see that the opacity is approximately proportional to ρ in the stratosphere and to ρ^2 in the troposphere. The former condition is consistent with a stratospheric opacity provided by a uniformly mixed aerosol, while the latter implies pressure-induced absorption in the troposphere due to a gaseous $N_2-CH_4-H_2$ mixture. In addition, there appear to be two regions of augmented opacity covering the pressure ranges $25-40$ mbar and $200-800$ mbar (the variations above 110 km cannot be trusted, as they depend critically on very uncertain temperature fluctuations at these levels). The former region corresponds to the levels in the lower stratosphere at which acetylene, ethane, and propane are expected to condense, while the latter range can be associated with tropospheric methane clouds, perhaps slightly enriched with ethane. Although the magnitudes of the opacity enhancements are sensitive to uncertainties in the thermal structure, their corresponding locations are suggestive enough to render the interpretations mentioned above plausible.

The 200–800 mbar region also corresponds closely to an altitude range over which unusually strong scintillations of the radio occultation signal were detected. Hinson and Tyler (1983a,b) independently have suggested that refractivity variations resulting from layered clouds could be one (though not the only) possible explanation.

III. UPPER ATMOSPHERIC COMPOSITION AND TEMPERATURE

A. Observations

During the Voyager 1 encounter with Titan, the UVS performed a solar occultation experiment which measured the differential absorption of sunlight as a function of altitude and wavelength in Titan's upper atmosphere (Broadfoot et al. 1981; G. R. Smith et al. 1982). The differential absorption is displayed in Fig. 7 at six wavelengths as a function of radial distance, in terms of the optical depth $\tau = -\ell n(I/I_0)$ where I is the transmitted intensity and I_0 is the unattenuated solar intensity. The optical depth is equal to the product of the absorption cross section and column density summed over all absorbers at that wavelength along the path of sunlight through the atmosphere. The absorbers are most easily identified from relative I/I_0 spectra at selected altitudes, as shown in Fig. 8. The sharp onset of absorption at ~ 800 Å is characteristic of an N_2 or Ar atmosphere. From the airglow data discussed below, the absorber must be N_2. At longer wavelengths the relative I/I_0 spectra are most consistent with CH_4 (the absorption edge at 1350 Å) and C_2H_2 (the absorption peaks at 1300 and 1470 Å), although in the latter case a contribution from polyacetylenes may also be present (Yung et al. 1983). From the optical depth data in Fig. 7, the column densities of absorbers as a function of altitude can be obtained. Differentiation of these column densities with respect to altitude yields the local number densities and scale heights, which give temperature. G. R. Smith et al. (1982) find from a detailed analysis of the UVS solar occultation data that the average temperature obtained from the N_2 scale heights at 3840 km is 186 ± 20 K at the terminators, ~ 20 K warmer than the thermal inversion region. Although their data suggest that the morning terminator is slightly warmer than the evening terminator, the difference was not statistically significant. Friedson and Yung (1984) have calculated a diurnal amplitude of the exospheric temperature of ~ 30 K in agreement with the solar occultation data.

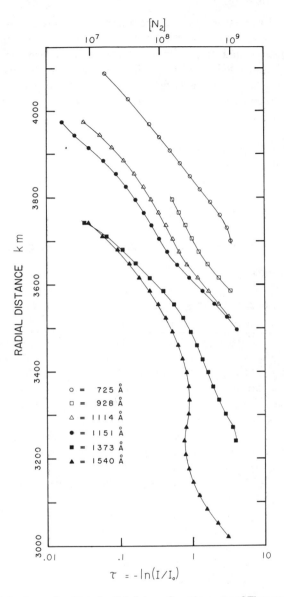

Fig. 7. Optical depth as a function of radial distance from the center of Titan at indicated wavelengths. The lines represent a smooth curve through the data points. The N_2 density scale refers only to the 725 Å curve (figure after G. R. Smith et al. 1982).

CHANNEL

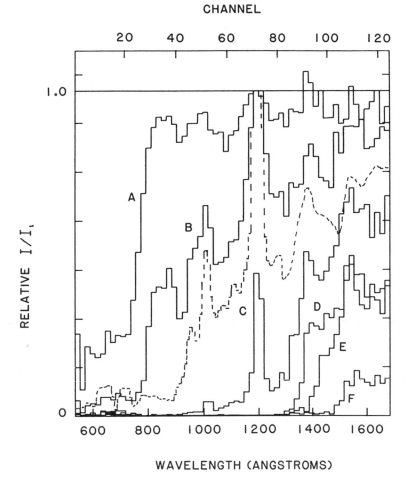

Fig. 8. Relative I/I_1 spectra at various altitudes in Titan's atmosphere where I_1 is solar intensity at 3800 km. A = 3730, B = 3620, C = 3500, D = 3310, E = 3220, F = 3050 km. The dashed curve corresponds to a model atmosphere with measured CH_4 and 1% C_2H_2 at 3670 km (figure after G. R. Smith et al. 1982).

At 3840 km where the optical depth at 725 Å is unity (cf. Fig. 7) the N_2 density is $2.7 \pm 0.2 \times 10^8$ cm^{-3}. A CH_4 density of $1.2 \pm 0.4 \times 10^8$ cm^{-3} is obtained at 3700 km, which yields a $[CH_4]/[N_2]$ mixing ratio of 0.08 ± 0.03 at this altitude provided the atmosphere is isothermal between 3700 and 3840 km. Detailed modeling was necessary to recover the $[C_2H_2]/[N_2]$ mixing ratio of 1–2% above 3400 km and 0.1–0.3% below 3250 km (G. R. Smith et al. 1982). This strong increase in the C_2H_2 mixing ratio with altitude, as illustrated by the optical depth at 1540 Å in Fig. 7, is consistent with conversion of C_2H_2 to C_4H_2 as discussed by Allen et al. (1980).

In the solar occultation data there are a number of absorption layers that must be molecular in nature, because particles would fall out too fast. For example, at 3660 km there is a sharp change in slope or scale height in all data at $\lambda > 800$ Å (cf. Fig. 7). The variation occurs within a small fraction of an N_2 scale height, which suggests that it is an edge of an absorbing layer rather than a result of gravitational diffusive separation (G. R. Smith et al. 1982). These layers make occultation data analysis very difficult because it is not possible to accurately follow a constituent over many scale heights.

Determination of the homopause location and eddy diffusion profile requires a simple hydrocarbon photochemical model and appropriate upper and lower boundary conditions from the ultraviolet solar occultation data and infrared thermal inversion data, respectively. G. R. Smith et al. (1982), who adopted a CH_4 mixing ratio in the lower stratosphere of 0.026 on the basis of Titan's cold trap temperature at the tropopause (Hanel et al. 1981a), concluded that the homopause is at ~ 3500 km, with an eddy diffusion coefficient $\sim 10^8$ $cm^2 \ s^{-1}$. Below the homopause the eddy coefficient decreases to $\sim 10^3 \ cm^2$ s^{-1} in the lower stratosphere (G. R. Smith et al. 1982). Titan's thermal inversion region is thus as stable and stagnant as the Earth's lower stratosphere.

Since the solar occultation data indicate that CH_4 and C_2H_2 are the dominant absorbers of solar radiation below 2000 Å, a large upward flux of CH_4, $\phi_1(r_0) \sim 5 \times 10^9 \ cm^{-2} \ s^{-1}$ at the homopause (discussed below) is required to balance the CH_4 lost by photolysis. To maintain the observed CH_4 mixing ratio of ~ 0.08 at 3700 km, the tropopause mixing ratio must exceed 0.02. This minimum CH_4 mixing ratio, f_{min}, is given by the last term in Eq. (13) of G. R. Smith et al. (1982)

$$f_{min} \sim \frac{H_{a0}\phi_1(r_0)}{[N_2]_0 K(1 - H_a/H_1)} = 6.9 \times 10^{-19} H_{a0}\phi_1(r_0) \sim 0.023 \qquad (9)$$

where H_{a0} is the N_2 scale height at the homopause ($= 67$ km).

B. Extreme Ultraviolet Airglow

The sunlit atmosphere of Titan exhibits a spectrum characteristic of electron-excited N_2, shown in Fig. 9; the prominent atomic hydrogen Lyman-α line at 1216 Å has been subtracted (Broadfoot et al. 1981; Strobel and Shemansky 1982). On the dark hemisphere only the H Lyman-α line is detectable. During the Voyager 1 encounter the sunlit hemisphere was oriented away from Saturn, and the quadrant facing into the corotating magnetospheric plasma had the brightest emission. There were no opportunities during this encounter to observe Titan when the sunlit hemisphere faced Saturn. The UVS detected the $c_4' \ ^1\Sigma_u^+ - X^1\Sigma_g^+$ (0–0) and (0–1) Rydberg bands of $N_2(N_2(RYD))$, the $NI(^4S^0-^4P)$ multiplet at 1134 Å, and the NII ($^3D^0-^3P$) multiplet at 1085

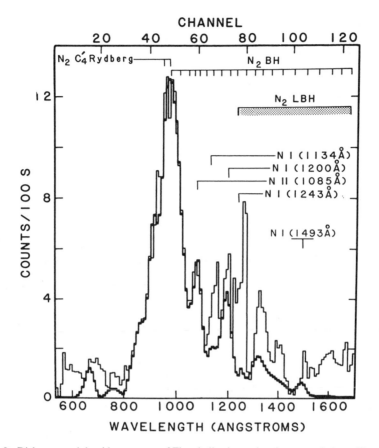

Fig. 9. Disk averaged dayside spectrum of Titan indicating major nitrogen emissions. The spectrum has been reduced by removal of internal instrumental scattering and a 1216 Å synthetic H Lyman-α line, to show the presence of blended features. The heavy overplotted spectrum is a model calculation composed of N_2, NI, and NII emissions excited by electron impact (figure after Strobel and Shemansky 1982).

Å with the average dayside intensities given in Table III for the Voyager 1 encounter (Strobel and Shemansky 1982). During the Voyager 2 encounter these intensities were typically a factor of 2 higher (Sandel et al. 1982). Although Titan is occasionally outside the Saturnian magnetosphere, there is no definite evidence that the emission intensity and characteristics depend on Titan's location with respect to the magnetosphere.

The nominal resolution of the UV spectrometer is $\sim$ 30 Å, and detection of individual bands and closely spaced multiplets is not possible. Emitters and emission features must be inferred partly by generating synthetic spectra based on known emitters and electron impact cross section, as discussed by Strobel and Shemansky (1982). Their model calculations include the signifi-

TABLE III
Titan EUV Emission Intensities[a]

λ(Å)	Observed Intensity (R)	$N = 2 \times 10^{15}$ cm^{-2} [b]	$N = 9 \times 10^{16}$ cm^{-2} [b]
N_2 c$_4'$(0,0) 958 Å	12	61	—
N_2 c$_4'$(0,1) 981 Å	8.6	10	—
N_2 c$_4'$(0,2) 1003 Å	7.0	1	—
N_2 BH(1,v'')	25	9.7	—
N_2 LBH	96	100	100
NI (1493 Å)	15	17	5.4
NI (1200 Å)	30	42	12
NI (1134 Å)	7.9	7.2	2
NII (1085 Å)	12	16	4

[a] Disk averaged.
[b] Calculated volume emission intensity with 1 keV primary electrons at N_2 column densities.

cant multiplets of NI and NII and the most important band systems of N_2. In spite of these numerous bands and multiplets, a number of emission features are unaccounted for in the spectrum of Fig. 9: 657 Å, 1160 Å, 1250 Å, 1400 Å, $\sim$ 1510 Å, and the 1550–1670 Å region, to which possible contributors are CI, NI(^{2}D), and CII (Strobel and Shemansky 1982). The fit of their model to the spectrum depends rather critically on the strong presence of the N_2 BH bands in the 980–1050 Å region, with intensity relative to the N_2 LBH bands a factor of $\sim$ 3 in excess of laboratory measured cross sections (cf. Table III). This problem can be partially resolved by attributing some of this emission to Lyman β at 1026 Å excited by electron impact on atomic hydrogen, which is close to the strongest BH bands: (1,1) at 1009 Å and (1,2) at 1033 Å. This e* + H process could also account for part of Titan's prominent emission at Lyman α.

As computed from the known solar spectrum, photoelectrons in Titan's upper atmosphere could account for no more than 1 R of the 12 R observed for NII (1085 Å) (Table III). Since the threshold for NII (1085 Å) emission from N_2 is 36 eV, energy in excess of the magnetospheric plasma corotation energy is required. The most viable candidates are energetic magnetospheric electrons and accelerated photoelectrons (Bridge et al. 1981a). Because the intensity of the NI (1134 Å) multiplet is not comparable to NI (1200 Å) and is much weaker than NII (1085 Å), the dominant excitation source for NI and NII emission features must be electron impact on N_2 rather than on N(^{4}S) or N(^{2}D).

Perhaps the most remarkable feature of Titan's spectrum is the strong emission in the N_2 RYD bands. In the Earth's atmosphere the (0–0) band at

958 Å has never been seen and the (0–1) band at 981 Å is at most a marginal detection (Feldman and Gentieu 1982). But in laboratory spectra under optically thin conditions (N_2 column densities $\lesssim 10^{13}$ cm^{-2}) the (0–0) band is the brightest emission of N_2; it gradually disappears as optically thick conditions are approached (N_2 column densities $\gtrsim 10^{16}$ cm^{-2}; Zipf et al. 1982). The N_2 RYD bands have a small but significant probability of predissociation (0.15; Zipf and McLaughlin 1978). The branching ratio of the (0–0) band to the (0–1) band is 1.0:0.17 (Zipf et al. 1982). Thus, the observation of the 0–0 band at 958 Å places an upper limit on the permissible optical depth between the excitation source and the Voyager spacecraft. On the basis of its oscillator strength and laboratory data (E. C. Zipf, personal communication, 1980) the N_2 RYD (0–0) band emission in the disk-average spectrum of Fig. 9 must originate in the upper 10^{15} cm^{-2} of N_2 in Titan's atmosphere, i.e., essentially the exosphere. The oscillator strength of 0.22 (Zipf and McLaughlin 1978) gives an average cross section over the (0–0) band of $\simeq 10^{-15}$ cm^2. Individual rotational lines have cross sections at line center which exceed 10^{-12} cm^2. These constraints eliminate solar photons and photoelectrons as primary excitation sources for the observed emission in the N_2 Rydberg bands.

From the UVS the upper limits on [Ne]/[N_2] and [Ar]/[N_2] are 0.01 and 0.06 at 3900 km, respectively (Broadfoot et al. 1981). When coupled with the G. R. Smith et al. (1982) eddy diffusion profile the mixing ratios at the surface are ≤ 0.002 for Ne, ≤ 0.6 for ^{40}Ar, and ≤ 0.3 for ^{36}Ar. Most argon on Titan would be ^{36}Ar if it were present with an [Ar]/[N_2] mixing ratio of > 0.1; otherwise Titan's interior would have to be excessively enriched in ^{40}K (Owen 1982b).

C. Magnetospheric Interaction

As pointed out in the previous section, solar photons cannot be the principal source of the ultraviolet glow because they do not release their energy at a high enough altitude. In addition, the following discussion shows that the energy required to excite the glow is about 5 times greater than that available from the Sun. The required power dissipation can be estimated from the cross sections for excitation of an observed spectral feature (σ_X), and ionization (σ_I) and dissociation (σ_D):

$$P \simeq I(x) \frac{\sigma_I + \sigma_D + \sigma_X}{\sigma_X} \Delta E \qquad (10)$$

where I is the intensity of the spectral feature and ΔE is the photon energy. From the N_2 RYD bands and the NII (1085 Å) multiplet, the inferred power dissipation is $\sim 0.025-0.03$ ergs cm^{-2} s^{-1} or (2.5–3) $\times 10^9$ W. For the LBH bands secondary electrons are involved and each emitted photon corresponds to an electron energy loss of 35 eV. Thus 100 R of N_2 LBH band emission requires a power dissipation rate of ~ 0.04 ergs cm^{-2} s^{-1} or 4×10^9 W. Other calculations suggest that slightly more power must be dissipated (5×10^9 W) to produce 100 R (Strobel and Shemansky 1982). Solar EUV energy below 1000 Å, the minimum energy for excitation of N_2 RYD bands, is only 0.006 ergs cm^{-2} s^{-1}. By comparison with the Earth's airglow, photoelectrons could produce ~ 25 R of LBH on Titan. We conclude that the EUV emission indicates a power dissipation of $\sim 3 \times 10^9$ W in Titan's atmosphere which is shown below to be consistent with the energy content of Titan's L shell. However, there is some concern whether the magnetosphere can replenish plasma with enough energy at a sufficient rate to maintain this power dissipation rate (Eviatar et al. 1982).

Voyager plasma science measurements taken in the wake of Titan indicated that only the energetic component of magnetospheric electrons (> 700 eV) was removed via interaction with Titan's atmosphere (Bridge et al. 1981a; see also chapter by Neubauer et al.). This preferential removal of energetic electrons has been explained by invoking inward centrifugal drift of electrons through the ionopause to the denser regions of the exosphere (Hartle et al. 1982). The centrifugal drift velocity and radius increase with increasing energy, and it can be shown that electrons below 500 eV are confined to within 1 scale height of the ionopause. Mass loading by ionospheric plasma slows down the magnetospheric plasma sweeping by Titan. Magnetospheric electrons have more time to traverse Titan's exosphere as they bounce between magnetic mirror points, and the N_2 column density sampled by magnetospheric electrons is significantly enhanced. If an electron is confined to the ionopause its probability of an inelastic collision is < 0.5, whereas if it drifts to the exobase it will invariably lose most of its energy. An essential requirement is multiple bounces through the exosphere as a result of flux tube mass loading by ionospheric plasma; this retards the rate at which plasma sweeps by Titan. On the night side where the ionospheric plasma density is low and mass loading negligible, this interaction is unimportant, and the magnetospheric plasma sweeps by rapidly.

A conservative estimate of the suprathermal tail removed from the magnetospheric electron distribution by interaction with Titan is $n_e \sim 10^{-2}$ cm^{-3} at $T_e \sim 1$ keV. This corresponds to a local energy flux of $n_e v_e T_e \sim 0.03$ ergs cm^{-2} s^{-1} or 3×10^9 W integrated over Titan's sunlit hemisphere. The total energy content in Titan's L shell bounded by the ionopause, with some allowance for centrifugal drift, is

$$\int n_e \, T_e \, \mathrm{d}V \simeq 80\pi \, R_S^2 \, R_I \, L \, n_e \, T_e \tag{11}$$

where $\mathrm{d}V$ is the volume element of the flux tubes, R_I is the ionopause radius and the flux tube is assumed to be full (Bridge et al. 1982). The magnetospheric plasma sweeps by Titan at ~ 125 km s^{-1} (Bridge et al. 1981a) which corresponds to a rotation rate relative to Titan of

$$\frac{V_{plasma}}{2\pi \ 20 \ R_S} \simeq 1.6 \times 10^{-5} \ s^{-1}. \qquad (12)$$

Thus the power dissipation in Titan's atmosphere P could be as large as

$$P = \frac{V_{plasma}}{40\pi \ R_S} \int n_e \ T_e \ dV = 2 \ R_S \ R_I \ L \ n_e \ T_e \ V_{plasma} \qquad (13)$$

which comes to $\sim 2 \times 10^9$ W if all the energy in Titan's L shell were dissipated, which is a highly probable result of centrifugal drift and mass loading. This upper limit is significantly larger than that obtained by Eviatar et al. (1982), who ignored the substantial retardation of flux tube speed by mass loading and used the bounce period for the transit time. There seems to be enough power available to generate the optical emission.

In addition, a plausible case can be made that the excitation is confined to very high altitudes and low optical depths. Calculations by Strickland et al. (1976) for terrestrial aurora excited by incoming electrons of 1 keV energy are also applicable to Titan. Intensities from N_2 in the optically thin limit were computed for N_2 column densities of 2×10^{15} and 9×10^{16} cm^{-2}. At larger column densities the primary (keV) electron flux is negligible in comparison to the secondary electron flux. These calculations are compared with the observed intensities in Table III; only at N_2 column density $\sim 2 \times 10^{15}$ cm^{-2} are the volume intensities consistent with observations. An important contribution to excitation is required of NI and NII multiplets from the primary electrons. This is caused by the energetic electrons bouncing between magnetic mirror points with multiple traversals of Titan's exosphere. Both the primary and degraded secondary electrons occupy the same volume element because of magnetic flux tube constraints; the resulting intensities exhibit this property.

Calculations by Conway (1983) suggest that excitation of the N_2 RYD bands must occur above 3800 km to obtain the limb brightening observed by the UVS experiment (Strobel and Shemansky 1982). The maximum apparent emission rate was initially reported at ~ 3660 km, which corresponds to a slant N_2 column density $\sim 10^{18}$ cm^{-2} and a very optically thick atmosphere for the N_2 $c_4'(0-0)$ band. Improved pointing information now indicates the peak emission occurs at ~ 3900 km, where it is consistent with the theory outlined above if a significant part of the emission is from the $c_4'(0-1)$ band. In addition, the limb brightening must be entirely due to $c_4'(0-1)$ band emission.

D. Thermal Escape of H and H_2

The extended atmosphere of Titan, with its weak gravitational acceleration and warm thermosphere, ensures rapid escape of H and H_2. In many instances, as Hunten (1973) pointed out, escape is limited by the maximum upward diffusion flux permitted through the background atmosphere. In a

spherical atmosphere the limiting flux for a minor constituent at the homopause is (Hunten 1973)

$$\phi = \frac{bf}{H_a} \left(1 - \frac{m}{m_a} \right) \tag{14}$$

where b is the binary collision coefficient in N_2, H_a is the N_2 scale height, m_a
is the N_2 mass, f is the volume mixing ratio, and m is the mass. From G. R.
Smith et al. (1982) the homopause is at ~ 3500 km, where $H_a \sim 67$ km and
$T \sim 165$ K. From Marrero and Mason (1972) and Hunten (1973) $b = 1.88$
$\times 10^{17} T^{0.82}$, for H_2 diffusing through N_2. Since the H_2 mixing ratio is $f =$
0.002 ± 0.001 (Samuelson et al. 1981), the H_2 escape flux in cm^{-2} s^{-1} at the
homopause is

$$\phi_{H_2} = (3.4 \pm 1.7) \times 10^9. \tag{15}$$

This escape rate is comparable to the rate of integrated dissociation of CH_4 by
direct photolysis, 1.2×10^9 cm^{-2} s^{-1} and by catalytic photolysis (Allen et al.
1980; Yung et al. 1983), $\sim 4 \times 10^9$ cm^{-2} s^{-1} based on Strobel (1982), with a
combined C_2H and C_2 quantum yield of 0.2 (Okabe 1981,1983). This equiv-
alence of photolysis rate and escape rate is established by hydrocarbon photochemical models (Strobel 1974; Allen et al. 1980; Yung et al. 1983) as noted
by Hunten (1974b), and corresponds to the buildup rate of heavier hydrocarbons, primarily C_2H_6. A comparison of the Jeans escape velocity with the
maximum upward H_2 diffusion velocity at the homopause indicates that the
actual H_2 escape flux is close to the limiting flux.

Analysis of the Titan Lyman-α data (Broadfoot et al. 1981) has not been
completed to determine precise H atom escape rates. From a theoretical
model of Titan's hydrocarbon photochemistry, Allen et al. (1980) estimate the
H atom escape flux to be comparable to the H_2 escape flux. A combined
escape rate of H atoms in all forms of $\sim 1 \times 10^{10}$ cm^{-2} s^{-1} (referred to
3500 km) is obtained, and implies that $\sim 1 \times 10^{27}$ H atoms cm^{-2} have escaped over geologic time. Alternatively, the assumption that C_2H_6 is the principal product of CH_4 photolysis combined with the integrated CH_4 dissociation rate leads to an escape rate of $\sim 5 \times 10^9$ cm^{-2} s^{-1}.

The escape of H and H_2 is sufficient to account for the extensive atomic
hydrogen torus detected by Voyager UVS at Lyman α (Broadfoot et al. 1981),
and generates the chemical environment required to irreversibly convert CH_4
to less hydrogen-rich hydrocarbons, e.g. C_2H_6 and C_3H_8. The atomic hydrogen torus is radially confined between 8 and 25 R_S by the charged particle
distribution, with a lifetime against ionization by charge exchange and electron impact of $\sim (3-4) \times 10^7$ s within the torus and $\sim 10^6$ s at the radial
boundaries (Broadfoot et al. 1981; Shemansky and Smith 1982). Either ballistic velocities of up to 1.9 km s^{-1} or a temperature in the collision limit of

260 K is required to explain the large observed vertical torus thickness of ~ 14 R_S (Sandel et al. 1982b). From the peak Lyman-α intensity 170 R, an H density of ~ 20 cm^{-3} has been inferred in the optically thin scattering limit (Sandel et al. 1982b). Thus, collisions are probably important in understanding the H distribution in torus since its lifetime is comparable to the mean time between collisions with H atoms. The possibility of an even larger H_2 density would ensure the importance of collisions. The required supply rate is ~ 3 $\times$ 10^{27} H s^{-1}. Based on the escape rates discussed above, it would appear that Titan supplies H + 1/2 H_2 to the torus at a rate of $\gtrsim 8 \times 10^{27}$ s^{-1}, in excess of the required amount. The discrepancy is, however, probably not outside the combined error of the two estimates. The H_2 in the torus can be removed by charge exchange which produces fast neutrals (denoted by an asterisk) which escape from the Saturn system.

$$e^* + H_2 \rightarrow H_2^+ + 2e$$

$$H^+ + H_2 \rightarrow H_2^+ + H^*$$

$$H_2^+ + H_2 \rightarrow H_2^+ + H_2^* \tag{16}$$

E. Nonthermal Escape Processes

At the exobase only 0.33 eV is needed for N atoms to escape Titan. Electron impact dissociation of N_2, the dominant dissociation process in Titan's exosphere, occurs preferentially through electronic singlet states 12.1–19 eV above the ground state and yields N atoms with substantial translational energy. It is possible to estimate the fraction of N atoms produced at energies > 0.33 eV with the dipole-allowed oscillator strengths of Wight et al. (1976) (Strobel and Shemansky 1982). Because N_2 predissociation leads to N(^{4}S) + N(^{2}D) products (cf. Oran et al. 1975) only N_2 singlet states excited to 9.76 + 2.38 + 2(0.33) = 12.8 eV and above will yield escaping N atoms. The fraction of N atoms originating from states in the 12.8–19 eV region is ~ 0.85 (Strobel and Shemansky 1982). With the usual assumption that one half are directed into the upward hemisphere, the conclusion is that $\simeq 40\%$ of the N atoms produced in the exosphere escape. From the disk averaged intensities for the N_2 RYD bands (given by Strobel and Shemansky 1982), which must originate from the exosphere, a magnetospheric dissipation rate in the exosphere of $\simeq 0.02$ ergs cm^{-2} s^{-1} was inferred. Thus 7×10^8 N atoms cm^{-2} s^{-1} are produced in the exosphere, and $\sim 2\pi R_c^2 \times 7 \times 10^8 \times 0.4 \simeq 3 \times 10^{26}$ N atoms s^{-1} (R_c = radius of the exobase) are supplied to the magnetosphere and/or solar wind from the hemisphere where energetic electron interaction with Titan preferentially occurs. This is equivalent to the loss of 0.1 of the present N_2 atmosphere over the age of the solar system, but this estimate of long term loss is subject to the considerable uncertainty in the inferred magnetospheric and solar wind power dissipation rates.

In addition to significant N_2 dissociation approximately 3.5×10^8 N_2^+ cm^{-2} s^{-1} are produced in the exosphere by electron impact. These ions can react with CH_4 and H_2 as well as dissociatively recombine:

$$N_2^+ + e \rightarrow N(^2D) + N(^4S), \qquad \alpha \sim 2 \times 10^{-7} \ cm^3 \ s^{-1} \qquad (17)$$

$$\begin{aligned} N_2^+ + CH_4 &\rightarrow CH_3^+ + N_2 + H, \\ &\rightarrow CH_2^+ + N_2 + H_2, \end{aligned} \qquad k = 1 \times 10^{-9} \ cm^3 \ s^{-1} \qquad (18)$$

$$N_2^+ + H_2 \rightarrow N_2H^+ + H, \qquad k = 1.7 \times 10^{-9} \ cm^3 \ s^{-1}. \qquad (19)$$

The recombination coefficient is an estimate based on the assumption of elevated electron temperatures; the other rate constants are taken from Huntress et al. (1980). At the exobase $[CH_4] \sim 2 \times 10^6$ cm^{-3} and $[H_2] \sim 7 \times 10^5$ cm^{-3} if the Jeans escape flux equals the limiting flux (G. R. Smith et al. 1982). Strobel and Shemansky (1982) estimate an electron density of 5×10^3 cm^{-3} at this level if the recombination coefficients are $\sim 1 \times 10^{-6}$ cm^3 s^{-1}. The respective rates for Reactions (17–19) are $\sim 1 \times 10^{-3}, 2 \times 10^{-3}$, and 1×10^{-3} s^{-1}. Thus, unless the H_2 densities are substantially larger, N_2^+ reacts preferentially with CH_4 to dissociate it into H and H_2 and form hydrocarbon ions. The CH_4 dissociation rate by ion chemistry may be comparable to the direct photolysis rate.

If all N_2^+ reacted preferentially with H_2 because of an extended corona, then subsequent recombination of N_2H^+ would yield translationally hot H atoms and a nonthermal H atom escape rate of $\sim 2 \times 10^{26}$ s^{-1} (average flux 1×10^8 cm^{-2} s^{-1} at the homopause), which would leave the Saturnian system with the H atoms' large energy. This may be compared with the thermal H atom flux $\sim 2 \times 10^9$ cm^{-2} s^{-1}.

F. Photochemistry

Hydrocarbons. Strobel (1974) first studied Titan's hydrocarbon photochemistry and concluded that C_2H_6 and C_2H_2 were the principal products of CH_4 photolysis. Allen et al. (1980) constructed an updated model which included catalytic dissociation of CH_4 by C_2H_2 and the formation of polyacetylenes as the precursor molecules for the haze layers on Titan. Neither of these studies considered N_2 as the dominant background gas. One important effect of N_2 on hydrocarbon chemistry is the rapid quenching of $^1CH_2$, produced in direct CH_4 photolysis, to $^3CH_2$ with rate constant 7.9×10^{-12} cm^3 s^{-1} (Ashford et al. 1980). Thus $^3CH_2$ is the principal product of CH_4 photolysis in an N_2 atmosphere, whereas $^1CH_2$ dominates in CH_4 and H_2 atmospheres and yields C_2H_6 by subsequent reactions. At high altitudes $^3CH_2$ reacts rapidly with itself to yield C_2H_2, and with CH_3 to form C_2H_4 which rapidly dissociates to C_2H_2. In addition, $^3CH_2$ recombines with C_2H_2 to form allene CH_2CCH_2 and methylacetylene CH_3C_2H.

The hydrocarbon photochemistry is schematically presented in Fig. 10;

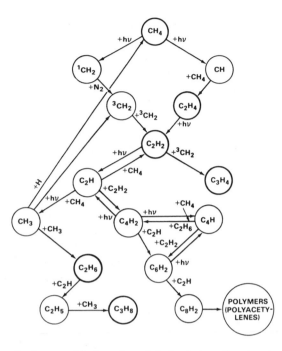

Fig. 10. Schematic diagram of hydrocarbon photochemical reactions based on Strobel (1974) and Allen et al. (1980) (figure after Strobel 1982).

for details see the original studies of Strobel (1974) and Allen et al. (1980) as well as the recent detailed photochemical model of Yung et al. (1983). From the Voyager solar occultation data the relative densities of CH_4 and C_2H_2 above the homopause indicate that C_2H produced from C_2H_2 photolysis reacts preferentially with C_2H_2 to form C_4H_2 (G. R. Smith et al. 1982). This process leads to the destruction of C_2H_2 until a density ratio of C_2H_2/CH_4 is reached where C_2H reacts preferentially with CH_4 to catalytically dissociate CH_4 as shown by Fig. 10; this is called secondary photolysis by Allen et al. (1980).

According to the work of Allen et al. (1980) and Yung et al. (1983) photolysis of polyacetylenes is particularly efficient in the dissociation of CH_4

$$C_{2n}H_2 + h\nu \rightarrow C_{2n}H + H$$
$$C_{2n}H + CH_4 \rightarrow C_{2n}H_2 + CH_3 \qquad n = 1, 2, 3 \qquad (20)$$
$$\text{net} \qquad CH_4 \rightarrow CH_3 + H$$

since polyacetylenes absorb solar radiation out to > 3000 Å, in contrast to absorption by CH_4 at $\lambda < 1450$ Å. Formation of C_2H_6 occurs primarily by CH_3 radical recombination after catalytic CH_4 dissociation (Reaction 20) in N_2 atmospheres. Methane is efficiently recycled only by the reaction $H + CH_3 + M \rightarrow CH_4 + M$.

Substitution of C_2H_6 for CH_4 in Reaction (20) leads to net dissociation of C_2H_6 into $C_2H_5 + H$. Reactions of $CH_3 + C_2H_5$ and $C_2H_5 + C_2H_5$ lead to the formation of higher alkanes, C_3H_8 and C_4H_{10}, respectively. A more complete and extensive discussion of Titan photochemistry may be found in Yung et al. (1983).

In Sec. III.D we estimated the CH_4 photolysis to be $\sim 5 \times 10^9 \, \mathrm{cm^{-2} \, s^{-1}}$ at the homopause. Based on the present-day source strength for photochemistry, $\sim 30 \, \mathrm{kg \, cm^{-2}}$ of C_2 or heavier hydrocarbons (principally C_2H_6 and polymers) would have accumulated on Titan's surface over geologic time. According to Lunine et al. (1983) deposition occurs primarily as an ethane ocean containing dissolved CH_4 and N_2 with a solid acetylene layer on the ocean floor. The minimum ocean depth would be 0.5 km on the basis of present-day photochemistry. The missing hydrogen from these less saturated hydrocarbons initially escaped to the torus and subsequently escaped the Saturnian system by charge exchange and/or recombination as an ion.

N_2. The discovery of an N_2 atmosphere on Titan by the combined Voyager infrared, ultraviolet, and radio occultation experiments suggests that N_2 photochemistry plays a central role in Titan's aeronomy (Broadfoot et al. 1981; Hanel et al. 1981a; Tyler et al. 1981). Since the N_2 bond is extremely hard to break, the three principal energy sources able to break it are energetic magnetospheric electrons interacting with Titan's upper atmosphere, solar radiation below 1000 Å, and cosmic-ray-produced secondary electrons in the lower stratosphere (Strobel and Shemansky 1982; Capone et al. 1980). Based on the analysis in Sec. III.C, magnetospheric electrons dissipate ~ 5 times as much energy as solar EUV radiation and 3 times that of cosmic rays.

The ion chemistry of N_2^+ preserves the N_2 bond in reactions with CH_4 and H_2 (Reactions 17–19; see Huntress et al. 1980). N^+ reacts rapidly with CH_4 to yield

$$N^+ + CH_4 \rightarrow CH_3^+ + N + H \qquad 53\%$$
$$\rightarrow CH_4^+ + N \qquad 4\%$$
$$\rightarrow H_2CN^+ + H + H \qquad 32\%$$
$$\rightarrow HCN^+ + H_2 + H \qquad 10\% \qquad (21)$$

or with H_2 to form

$$N^+ + H_2 \rightarrow NH^+ + H$$
$$NH^+ + N_2 \rightarrow N_2H^+ + N \qquad (22)$$

followed by

$$HCN^+ + CH_4 \rightarrow H_2CN^+ + CH_3 \qquad 84\%$$

$$\rightarrow C_2H_3^+ + NH_2 \qquad 16\% \qquad (23)$$

and then recombination

$$H_2CN^+ + e \rightarrow HCN + H$$

$$N_2H^+ + e \rightarrow N_2 + H \qquad (24)$$

(Huntress et al. 1980; McEwan et al. 1981). Thus N^+ primarily forms HCN and N atoms at high altitudes.

For either terminal ion in Reaction (23) the rapid recombination at a rate of $\sim 1 \times 10^{-6}$ cm^3 s^{-1} ensures the quick decay of the ionosphere at sunset and a negligible nighttime ionosphere (Mul and McGowan 1979,1980). On the day side, if we assume that ionization by magnetospheric interaction is spread over 500 km in altitude, then with the above recombination rate an average dayside electron density $\sim 4 \times 10^3$ cm^{-3} would be maintained. The actual electron density could be less, due to mass loading of magnetospheric flux tubes. This is consistent with the upper limits obtained by the Voyager radio occultation experiment: 5×10^3 cm^{-3} on the morning terminator and 3×10^3 cm^{-3} on the evening terminator (Lindal et al. 1983). On the evening terminator the EUV emissions were significantly less intense than on the morning terminator.

In addition to ionization, dissociation of N_2 by impact of magnetospheric electrons yields substantial quantities of $N(^2D)$ and $N(^4S)$, the ground state (Oran et al. 1975). The N_2 photochemistry is schematically outlined in Fig. 11. Note that the fate of $N(^2D)$ is crucial in determining the net HCN production. It is uncertain whether the $N(^2D)$ reaction with CH_4 yields HCN, as Capone et al. (1980) assumed, leads to deactivation to the ground $N(^4S)$ level, or yields NH; the latter would seem most probable. The most likely fate of $N(^4S)$ may be reactions with radicals CH_3 and $^3CH_2$ to form HCN. Solar excitation of $N_2(A^3\Sigma)$ (which may be important on the Earth [Zipf 1980]), and subsequent quenching by hydrocarbons through complex, unknown mechanisms to yield HCN, may also be important (see Fig. 11).

Dissociation of HCN yields CN, which is isoelectronic with C_2H and can then react with a number of radicals as well as with C_2H_2, C_2H_4, HCN, and CH_4 as shown in Fig. 11. Only the latter reaction recycles HCN. Based on the C_2H_2/CH_4 density ratio inferred from the Voyager solar occultation experiment (G. R. Smith et al. 1982) and the rate coefficients measured by Schacke et al. (1977), there should be significant production of HC_3N formed by CN $+ C_2H_2$, in agreement with the Voyager infrared data. Also detected was C_2N_2, formed by CN + HCN (Kunde et al. 1981). Photolysis of HCN can lead to the catalytic dissociation of CH_4 in a manner similar to C_2H_2 photolysis, but at a much slower rate according to Yung et al. (1983).

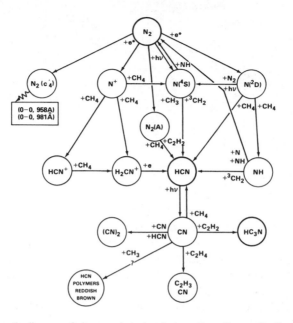

Fig. 11. Schematic diagram of nitrogen photochemical reactions (figure after Strobel 1982).

In their study of cosmic-ray induced formation of organic molecules, Capone et al. (1980) calculated an HCN production rate of $(2-40) \times 10^6$ cm^{-2} s^{-1}. The maximum production rate applies when all $N(^2D)$ reacts with CH_4 to yield HCN directly, which we consider improbable, and should be compared with a globally averaged production rate of $N(^2D)$ by magneto-spheric electrons of 5×10^8 cm^{-2} s^{-1}. The main cosmic-ray deposition region is in the lower stratosphere where the temperature gradient is steep and condensation of complex molecules is highly probable. The distinct advantage of cosmic ray production is that the molecules are formed deep in the atmosphere where dissociating solar ultraviolet radiation has been attenuated (Capone et al. 1980).

An absorbing haze layer in Titan's upper atmosphere was first proposed by Danielson et al. (1973) and long suspected to be a photochemical smog. Note that laboratory produced HCN polymers (Mizutani et al. 1975) have optical properties similar to those of aerosols on Titan (Rages and Pollack 1980). Mizutani et al. (1975) have observed that CH_3 radicals are necessary to initiate formation of HCN polymers. Addition of CH_3 radicals to carbon chains will generally lower their vapor pressure and enhance the possibility of condensation. The addition of CN radicals to polyacetylenes, which may be formed on Titan (Allen et al. 1980), could also enhance their absorption properties in the near ultraviolet and visible wavelength regions. The solar heating of the aerosols creates an inversion layer in Titan's stratosphere which is extremely stable ($K \leq 10^3$ cm^2 s^{-1}). The accumulation and subsequent sublimation of complex molecules maintains the photochemical smog with positive feedback (Yung et al. 1983). The rapid escape of H and H_2 ensures a photochemically produced density ratio of unsaturated (C_2H_2, C_2H_4) to saturated (CH_4) hydrocarbons of $\geq 10^{-3}$ above 3300 km. Such environmental conditions are favorable to building up complex molecules, for example

$$\begin{matrix} CN \\ C_2H \end{matrix} + C_2H_2 \rightarrow \begin{matrix} HC_3N \\ C_4H_2 \end{matrix} + H \qquad (25)$$

rather than

$$\begin{matrix} CN \\ C_2H \end{matrix} + CH_4 \rightarrow \begin{matrix} HCN \\ C_2H_2 \end{matrix} + CH_3. \qquad (26)$$

The formation of polyacetylenes in the model of Allen et al. (1980), of which Reactions (25) are the first step, resulted from this favorable density ratio. In addition, Reactions (26) have a substantial activation energy ~ 2 kcal mole^{-1}, whereas Reactions (25) do not. Thus, lower temperatures actually favor Reactions (25) over (26). This can explain why the colder north polar hood region on Titan contains larger abundances of complex hydrocarbons than the equatorial regions (Hanel et al. 1981a). In fact, the products of Reactions (25) are only evident in infrared spectra of the north polar cap (Kunde et al. 1981; see also Sec. II).

CO and CO$_2$. The recent discoveries of CO_2 and CO in Titan's atmosphere may indicate either formation of the atmosphere with significant CO, constant input of oxygen-bearing meteoritic material, or sputtering of oxygen off Saturn's icy satellites (Samuelson et al. 1983b; Lutz et al. 1983). The existence of oxygen in Titan's reducing atmosphere leads to interesting photochemistry (Samuelson et al. 1983b; Yung et al. 1983). CO_2 is converted readily to CO by reactions such as

$$\begin{aligned} CO_2 + h\nu &\rightarrow CO + O(^1D, {}^3P) \\ {}^3CH_2 + CO_2 &\rightarrow H_2CO + CO \\ H_2CO + h\nu &\rightarrow H_2 + CO \\ &\rightarrow H + HCO \\ HCO + h\nu &\rightarrow H + CO \\ HCO + H &\rightarrow H_2 + CO. \end{aligned} \qquad (27)$$

Likewise, CO can be restored to CO_2 by the reaction

$$CO + OH \rightarrow CO_2 + H \tag{28}$$

where the OH is formed by

$$H_2O + h\nu \quad \rightarrow H + OH$$
$$\rightarrow H_2 + O(^1D \text{ or } ^1S) \tag{29}$$
$$O(^1D) + CH_4 \rightarrow OH + CH_3$$

and the $O(^1D)$ may also be derived from

$$e^* + CO \rightarrow C + O(^1D) \tag{30}$$

high in the atmosphere. The present abundance of CO on Titan is a balance between H_2O input and sublimation loss of CO_2 and H_2CO to the surface and escape of O from the exosphere (Samuelson et al. 1983b; Yung et al. 1983).

If Titan's atmosphere initially contained appreciable CO, then precipitation of magnetospheric electrons e^* could initiate the chemical destruction of CO by

$$e^* + CO \rightarrow C + O(^1D) + e$$
$$\tag{31}$$
$$e^* + N_2 \rightarrow N_2(A^3\Sigma) + e$$

The $N_2(A^3\Sigma)$ electronic energy can be transferred almost resonantly to CO;

$$N_2(A^3\Sigma) + CO \leftrightharpoons CO(a^3\Pi) + N_2 \tag{32}$$

followed by

$$CO(a^3\Pi) + CO \rightarrow CO_2 + C. \tag{33}$$

Samuelson et al. (1983b) estimate that an initial CO/N_2 mixing ratio of 0.1 results in the deposit of ~ 1 m of dry ice on the surface over geologic time.

IV. PROPERTIES OF AEROSOLS

A. Observations

Titan's aerosols play an important role in determining the temperature structure of the atmosphere. We review the various observations that constrain the nature and distribution of these aerosols, and summarize our present understanding of their properties.

An important source of information concerning the aerosols is the variation of Titan's geometric albedo with wavelength. Figure 12 (Rages and Pol-

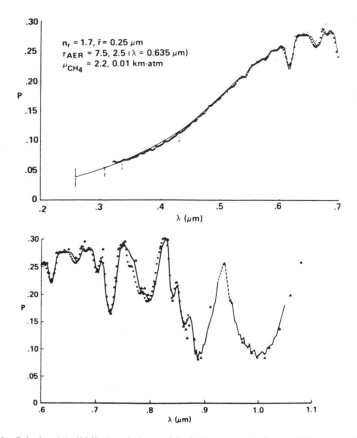

Fig. 12. Calculated (solid line) and observed (points) geometric albedo of Titan from 0.259 to 1.08 μm (from Rages and Pollack 1980). A radius 2700 km was used. Tomasko and Smith (1982) indicate that a radius of $\sim$ 2800 km is more appropriate in the red, and $\sim$ 50 km larger in the blue. For some purposes, the variation with time (which can amount to 10%) also needs to be included. The data points in the figure came from Caldwell (1975), $\bigcirc$; Nelson and Hapke (1978), $\bullet$; and Younkin (1974), $\triangle$.

lack 1980) summarizes the available data from several sources. As pointed out by Podolak and Danielson (1977), the $\sim$ 2 km-amagats of methane required to produce the absorptions seen in the spectrum longward of 0.6 μm would produce a geometric albedo of 0.22 at 0.26 μm in the ultraviolet due to Rayleigh scattering even if the gas were above a purely absorbing ground. The presence of a much larger quantity of N_2 only strengthens this argument. In contrast, Fig. 12 shows an observed geometric albedo of only < 0.05 at this wavelength (see also Caldwell et al. 1981). Thus, a significant amount of material that absorbs in the ultraviolet must be present high in Titan's atmosphere. At red wavelengths, however, this material must permit sunlight to sample the underlying methane. Podolak and Danielson (1977) concluded that the aerosols must be reasonably small (with radii of the order of magnitude 0.1 μm) for the backscattering cross section of the particles to decrease from blue to red wavelengths as rapidly as the observations require. Several other authors have successfully modeled these data with particles having radii of a few 0.1 μm or less, as discussed below.

In addition to the variation of brightness with wavelength, groundbased observations (Noland et al. 1974) and observations by Pioneer 11 (Tomasko and Smith 1982) and Voyager (West et al. 1983b) have measured the variation of brightness of the integrated disk of Titan at several wavelengths as functions of phase angle. The groundbased data are limited to phase angles $\lesssim 6°$, but cover a wide range of wavelengths. The Pioneer 11 data were obtained in red (0.64 μm) and blue (0.44 μm) channels and range from $\sim 22°$ to 96° phase (see Fig. 13). These data give information on the shape of the single-scattering phase function for the range of scattering angles observed.

On both the Pioneer 11 and Voyager 2 missions linear polarization of the integrated disk of Titan was observed. The Voyager polarization measurements, while less precise than those of Pioneer, were obtained at 2640 Å and 7500 Å and at phase angles from 2°.7 to 154° and thus extend both the phase angle and wavelength coverage provided by Pioneer. The observed polarization of the integrated disk is strikingly large near 90° phase in all the data sets, about 60% at 2640 Å, 56% at 4400 Å, 47% at 6400 Å, and 42% at 7500 Å. When modeled by scattering from spherical particles, the data imply particle radii as small as 0.05 μm at the top of the atmosphere and increasing particle size with depth into the atmosphere. However, these size estimates depend on the assumption of spherical particle shape, and would be different (and hard to determine) for irregular particles.

The Voyager imaging system vidicon recorded images of Titan at phase angles up to 160° at high spatial resolution from which limb scans and limb darkening information as well as the brightness of the integrated disk can be obtained. The data at the highest phase angles from Voyager 1 show a large increase in brightness of the brightest portions of the disk between 129° and 160° phase, implying that the particles at the top of the haze must be large enough (radii $\gtrsim 0.2$ μm) to be strongly forward scattering (Rages et al. 1983). These data are especially important because the lower limit on particle size that they give depends on diffraction and is relatively insensitive to the detailed shape of the particles. The particle size limit implied by these data remains to be reconciled with the large polarizations measured for Titan.

The limb scans from the Voyager images show a thin detached haze encircling the planet. The extinction profile as a function of altitude can be determined from these data for vertical optical depths $\lesssim 0.1$ before multiple-scattering effects in a spherical atmosphere greatly complicate the analysis (Rages and Pollack 1983). More data on the vertical variation of aerosol properties at deeper levels are contained in the variation of brightness all the way

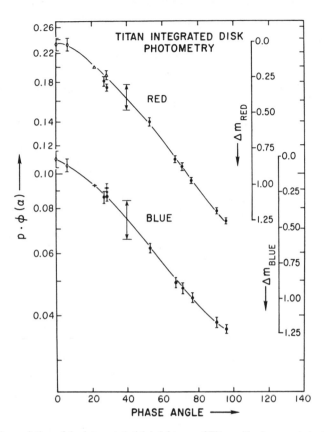

Fig. 13. The variation of the integrated disk brightness of Titan with phase angle in the red and
blue Pioneer 11 channels (from Tomasko and Smith 1982). The left-hand scale indicates a geometric albedo p at zero phase with a phase law $\phi(\alpha)$ to represent the decrease in brightness at
increasing phase angle α. A relative magnitude scale is included for reference. The error bar at
40° phase represents a one sigma uncertainty in the absolute calibration of all the Pioneer observations. The individual bars are the relative measurement errors.

across Titan's disk in both the Pioneer images (Tomasko and Smith 1982) and
the higher resolution Voyager images.

Finally, the Voyager images (B. A. Smith et al. 1981, 1982) show a considerable variation in the brightness of Titan as a function of latitude. Both the
Pioneer and Voyager data show a $\sim 25\%$ decrease in brightness at northern
relative to southern mid-latitudes at blue wavelengths. Further in the red (at
6400 Å), the Pioneer data show the effect to be much smaller, $\lesssim 5\%$. The
location and movement with time of the transition between the relatively
bright and dark regions has important implications for the cause of Titan's
secular brightness variation (Smith et al. 1982; Sromovsky et al. 1981).

B. Deduced Aerosol Properties

Single-Scattering. The single-scattering process can be described by multiplying the Stokes vector describing the incident radiation (I, Q, U, V) by a phase matrix $P_{ij}(\theta)$ to give a Stokes vector (I', Q', U', V') of the radiation scattered through an angle θ. When circular polarization (involving Stokes parameter V) is neglected, it has been shown (Hansen 1971) that reducing the equations to 3×3 matrices yields a considerable savings in computing time while preserving reasonable accuracy in the scattered intensity and linear polarization. Fairly general symmetry considerations, such as assuming equal numbers of randomly oriented particles of arbitrary shape and their mirror images, imply that many of the 3×3 matrix elements are zero. For example, only the elements along the diagonal and $P_{12}(\theta)$ and $P_{21}(\theta)$ (which are equal) are nonzero for the above case. The ratio $-P_{12}(\theta)/P_{11}(\theta)$ is the degree of polarization introduced in a single scattering event for incident unpolarized light. The function $P_{11}(\theta)$ (single-scattering phase function) gives the angular distribution of scattered light when the incident beam is unpolarized. When this function is normalized so that $\int P_{11}(\theta) \, d\Omega = 4\pi$, the product $F\widetilde{\omega}P_{11}(\theta)/4$ gives the intensity scattered by the particle into a direction at angle θ to the direction of an unpolarized narrow illuminating beam of flux πF. Here $\widetilde{\omega}$ represents the single-scattering albedo of the particle.

In addition to functions $P_{11}(\theta)$ and $P_{12}(\theta)$, the functions $P_{22}(\theta)$ and $P_{33}(\theta)$ remain to be specified. For spherical particles $P_{22}(\theta) = P_{11}(\theta)$, and $P_{33}(\theta)$ is similar in magnitude; it is often assumed that $P_{11}(\theta) = P_{22}(\theta) = P_{33}(\theta)$ when synthetic phase matrices are considered for irregular particles as well.

If the atmosphere is assumed to consist of a thick homogeneous cloud of particles, one can use observations of the intensity and linear polarization of the multiply scattered light observed at phase angle α to constrain the phase matrix elements $P_{11}(\theta)$ and $-P_{12}(\theta)/P_{11}(\theta)$ at the corresponding scattering angle $\theta = 180 - \alpha$, without assuming that the particles are spherical. Trial values of these single-scattering functions can be used to compute the multiply scattered intensity and polarization for comparison with observations. After some trial and error, the elements of the single-scattering phase matrix can be found which reproduce the multiple-scattering observations. The derived single-scattering phase function and polarization can then be compared with the single-scattering properties of candidate aerosol particles to constrain the physical properties of the aerosols without assumptions regarding particle shape.

Tomasko and Smith (1982) followed this approach in analyzing the Pioneer data shown in Fig. 13. They described the function $P_{11}(\theta)$ by a three-parameter double Henyey-Greenstein function $P_{11}(\theta) = f P(g_1, \theta) + (1 - f) P(g_2, \theta)$ where $P(g, \theta) = (1 - g^2)/(1 + g^2 - 2g \cos \theta)^{3/2}$. Here g_1 controls the steepness of the forward-scattering peak, g_2 (taken to be negative) controls the steepness of a back-scattering peak, and f controls the relative contribution of each. The single-scattering albedo $\widetilde{\omega}$ is determined by requiring the model to reproduce the absolute brightness as well as the shape of the curve shown in Fig. 13.

Because the Pioneer data include no observations at phase angles $> 96°$ (corresponding to scattering angles $< 84°$) the forward peak in the phase function is not well constrained by these data alone—any value of $g_1 \gtrsim 0.4$ in the blue and $\gtrsim 0.6$ in the red is acceptable. However, at scattering angles from 80° to 180° the phase function is well constrained. (See Table IV and Fig. 14.)

The ratio of the maximum brightness in two violet (0.42 μm) Voyager 1 images obtained at phase angles of 160° to 129° has been measured to be 5.2 $\pm$ 0.7 by Rages et al. (1983). They show that this requires $P_{11}(20°)/P_{11}(51°)$ $= 7.2 \pm 1$. Thus the phase function must be fairly strongly forward scattering in the blue (g_1 must be $\gtrsim 0.65$). When Tomasko and Smith used $g_1 = 0.70$ for the blue Pioneer channel, they derived $g_2 = -0.35$, $f = 0.933$, and $\widetilde{\omega} = 0.791$. In the red g_1 is indirectly constrained by the polarimetry to be $\lesssim 0.60$, and this value gives $g_2 = -0.60$, $f = 0.980$, and $\widetilde{\omega} = 0.915$ (see Table IV).

The phase function derived depends significantly on the single-scattering polarizing properties of the particles used in the calculation, when the particle polarizations are as large as required for Titan. In analyzing the Pioneer data, the single-scattering polarizing properties of the particles were shown to be well described by a constant fraction times the single-scattering polarization produced by Rayleigh scattering, $\sin^2 \theta/(1 + \cos^2 \theta)$, and these polarizing properties were used in determining g_1, g_2, and f.

Because multiple scattering tends to scramble the polarization state of scattered light, the polarization produced in a single-scattering event by individual aerosol particles must be larger than the polarization values measured directly (in the multiply-scattered light) on the Pioneer and Voyager missions. In addition, since the single-scattering albedo is larger at red than at blue wavelengths, the relative contribution of multiple to single scattering in the observed scattered light, and hence the dilution of the single scattering, will be largest at red wavelengths. Table IV shows the polarization produced in single scattering at a scattering angle of 90° which is required to match the observed polarization in the multiply scattered light. Note that while the observed polarizations are greatest at short wavelength, the required single-scattering polarizations are least there, although they are very high ($\gtrsim 75\%$) at all these wavelengths. In fact, unless $g_1 \lesssim 0.6$ at the red Pioneer wavelength, the single-scattering albedo required to match the observed geometric albedo will be so large and the dilution by multiple scattering so great that even particles that are 100% polarizing at 90° phase will be unable to produce as much polarization as is observed.

Size and Shape. Historically, several types of observations have been used to estimate the size of the aerosols in Titan's atmosphere. Danielson et al. (1973) realized that Titan's low albedo in the ultraviolet implied that the

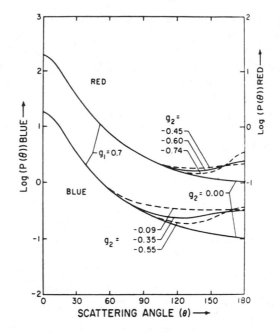

Fig. 14. The single scattering phase functions, $P_{11}(\theta)$, derived for Titan's aerosols in the Pioneer red and blue channels. The red curves are displaced a factor of 10 for clarity and use the scale on the right (figure from Tomasko and Smith 1982). The dashed curves indicate the range of values permitted by the data of Fig. 13 for scattering angles $> 80°$. At smaller scattering angles other arguments are used to estimate the values of g_1 shown in Table IV (see text).

TABLE IV
Single Scattering Properties of Titan's Aerosols at Four Wavelengths[a]

(μm)	g_1	g_2	f	$\widetilde{\omega}$	P_{90} SS [c]	P_{90} MS [d]
0.264	0.70	-0.35	0.933	0.52	0.75	0.60
0.264 [b]	(0.5 μm Mie + H.G.)			0.85	0.84	0.60
0.44	0.70	-0.35	0.933	0.791	0.93	0.56
0.64	0.60	-0.60	0.980 [e]	0.915	1.00	0.47
0.75	0.60	-0.60	0.90	0.85	0.78	0.42
0.75 [b]	(0.5 μm Mie + H.G.)			0.91	0.93	0.42

[a] The phase function $P_{11}(\theta) = f P(g_1, \theta) + (1 - f)P(g_2, \theta)$ where $P(g, \theta) = (1 - g^2)/(1 + g^2 - 2g \cos \theta)^{2/3}$ and $\widetilde{\omega}$ is the single-scattering albedo.
[b] At 0.264 μm and 0.75 μm a second phase function, equal to that for Mie scattering by spheres of effective radius 0.5 μm at $\theta < 80°$ and the indicated Henyey-Greenstein function at $\theta > 80°$ is also shown (from West et al. 1983b).
[c] The single scattering polarization, $-P_{12}(\theta)/P_{11}(\theta)$ is given by P_{90} SS $\sin^2 \theta/(1 + \cos^2 \theta)$.
[d] P_{90} MS is the observed polarization at 90° phase in the multiply scattered light.
[e] This value was erroneously given as 0.90 in Tomasko and P. H. Smith (1982).

aerosols had to be high in the atmosphere and therefore reasonably small. If smaller than thermal wavelengths, they would not emit efficiently in the infrared and so could be responsible for a warm stratosphere, explaining measurements of high brightness temperatures at several thermal wavelengths. Podolak and Danielson (1977) developed these ideas to interpret the visible spectrum of Titan in detail. They were reluctant to require the imaginary index of refraction of the aerosol material to vary more rapidly than λ^{-3} from red to blue wavelengths, based on laboratory measurements by Khare and Sagan (1973). By selecting a maximum particle radius of ~ 0.1 μm, they were able to take advantage of the decreasing backscattering lobe of the particle phase function toward shorter wavelengths to produce some of the required decrease in the geometric albedo toward the blue while still having an imaginary index which varied no faster than $\lambda^{-2.5}$. For particles with radii ≤ 0.1 μm, the total optical depth of the cloud also decreased rapidly from blue to red wavelengths, permitting the formation of the observed methane bands longward of 0.6 μm. Podolak and Giver (1979) adjusted the vertical distribution of aerosols of this size range to produce an excellent fit to Titan's visible spectrum.

Nevertheless, particle radii of 0.1 μm are not uniquely required by the geometric albedo observations, as pointed out by Rages and Pollack (1980). These authors used the observations by Noland et al. (1974) of the wavelength dependence of Titan's phase coefficient between 0 and 6°.4 observed from the Earth to find a relation between the size and real index of refraction of spherical particles that would reproduce these data. They concluded that the effective radius of the aerosols was between ~ 0.20 and 0.35 μm for refractive indices between 2.0 and 1.5. For a given refractive index and size, they then simply determined the necessary imaginary index at each wavelength (which was no longer a simple power law) to reproduce the shape of Titan's albedo from red to blue wavelengths. For particles of this size range, the optical thickness of the cloud varied more slowly from blue to red wavelengths than in the models with smaller particles, but by distributing most of the aerosol above most of the methane the shape of the methane bands could still be well fitted. While strikingly successful in reproducing Titan's spectrum, a potential shortcoming in the size determination was the assumption of spherical particle shape in the calculations for the wavelength dependence of the particle phase function between 174° and 180° scattering angles. The presence of surface waves for spherical particles is known to produce glory features at backscattering geometries which would be different for other particle shapes.

Nevertheless, simply because it is difficult to compute the single-scattering properties of general irregular particles, new observations are often compared with models based on scattering by spherical particles. For spherical particles, the large single-scattering polarizations determined from the Pioneer 11 and Voyager observations at 90° phase place strong constraints on particle size. If the particles turn out to be small enough compared to the wavelengths of observation (in or near the Rayleigh scattering domain) the polarization may not be too dependent on the detailed particle shape.

The single-scattering polarizations extracted from the Pioneer 11 and Voyager data shown in Table V have been compared with calculations for spherical particles by Tomasko and Smith (1982) and West et al. (1983b). These authors find that the polarization data cannot be reproduced by a thick, homogeneous cloud of spherical particles of any refractive index. Tomasko and Smith (1982) point out that a thick cloud of spherical particles can reproduce these data (as well as the photometry observations of Fig. 13) if there is an increase of particle size with increasing depth into the cloud. Cloud physics models by Toon et al. (1980) indicate that particle size may be expected to behave roughly as a power law of optical depth near the cloud top. Assuming this form, Tomasko and Smith found that a model with an area-weighted effective radius $a = (0.117 \ \mu m) \ (\tau_{red}/0.5)^{0.217}$ would reproduce the Pioneer polarimetry, photometry and limb darkening data reasonably well. Calculations by West et al. (1983b) indicate that such a model also fits the Voyager polarimetry well. The particles in this model are sufficiently small for their cross sections to decrease rapidly from blue to red, and so could probably be made to match the shape of Titan's spectrum as well by adjusting their vertical distribution.

However, the Voyager images at highest phase angles show a large increase in brightness with increasing phase, implying much stronger forward scattering than this model produces. One powerful advantage of these observations is that the size limit they impose is relatively independent of particle shape, as can be seen from Fig. 15 taken from Rages et al. (1983). The observed ratio in the single-scattering phase function between scattering angles of 20° and 51° (phase angles 160° and 129°) is due to diffraction by the particles and requires a particle radius $\gtrsim 0.20 \ \mu m$ regardless of whether the particle is treated as a flat disk, sphere, or prolate spheroid. Thus there seems little doubt that the aerosol particles near the top of Titan's haze layer have radii $\gtrsim 0.2 \ \mu m$.

Since these high phase-angle observations refer to light which has not penetrated very deeply into Titan's atmosphere, Tomasko and Smith (1982) suggested that the high phase-angle data might be the result of a detached layer of optical thickness ~ 0.05 of particles with radius $\sim 0.25 \ \mu m$, 150 km above the main haze of particles which would have the size distribution inferred from the polarization data. However, Rages and Pollack (1983) and Rages et al. (1983) have derived the optical thickness of the detached haze as only 0.01 at several locations around Titan's limb, too low to give the high brightness at 160° phase if the forward-scattering particles are confined to the detached haze layer. Apparently the strongly forward-scattering particles are present in the main haze region.

It seems unlikely that larger particles would be found above smaller par-

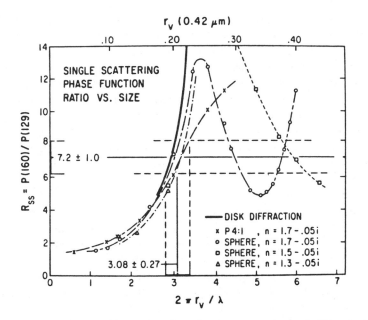

Fig. 15. The ratio of the single scattering phase function at scattering angle 20° to that at 51° (corresponding to phase angles 160° and 129°) for various spheres, a prolate spheroid with axial ratio 4:1, and a flat disk, as a function of size parameter $2\pi r/\lambda$. For the prolate spheroid r is taken as the semiminor axis of the ellipse. The single-scattering ratio needed to match the Voyager 1 data at high phase is marked at the left, and the scale at the top gives the radius corresponding to the size parameter along the bottom scale for the effective wavelength (0.42 μm) of the violet Voyager images. Aerosol radii $\gtrsim 0.20$ μm seem required regardless of shape.

ticles within the same haze region, so we must examine other ways of reconciling the sizes derived from large phase-angle photometry with the high single-scattering polarizations observed at 90° scattering angles. Spherical particles with $r > 0.10$ μm are simply incapable of producing the required polarization. Broad or bimodal size distributions have been investigated only briefly (Smith 1982) but also do not seem capable of satisfying both the large phase photometry and the polarimetry constraints over a wide wavelength range.

It is easy to suggest that nonspherical particles are responsible for the photometric and polarimetric properties of Titan's aerosols, but quite another matter to find measurements or calculations for a particular particle type that has the required properties. West et al. (1983b) summarize several results in the literature for scattering by nonspherical particles. A variety of calculations and measurements for long cyclinders or hexagonal prisms or for thin hexagonal plates generally do not show sufficient polarization except when such particles are extremely absorbing (have imaginary indices of refraction

$\gtrsim 0.2$). Microwave scattering experiments by Zerull and Giese (1974) on slightly absorbing cubes and octahedrons also showed too little polarization. On the other hand, for large irregular particles that are compact and fairly dark, the microwave measurements of Giese et al. (1978) show that the light scattered outside the narrow forward-scattering diffraction peak is dominated by Fresnel reflection from the irregular surface of the particle, and the polarization is very large with a peak value of $\gtrsim 85\%$. In fact, in their attempts to fit the lower peak polarization ($\sim 30\%$) observed for interplanetary dust, Giese et al. were led to consider fluffy particles in which the light entering the particle had a reasonable chance of reemerging before being absorbed, and so could dilute the very large polarization produced by external Fresnel reflection.

At first glance, such large, compact, irregular dark particles seem to have, many of the properties required of Titan's aerosols, especially at ultraviolet and blue wavelengths. An important question is whether the high polarization will drop below the required values as the absorption in the particle is decreased to the fairly low values ($1 - \widetilde{\omega} \approx 0.1$) required at red wavelengths. For example, the most favorable type of particles of the few types given by Zerull and Giese is described as "compact" with $n_r = 1.45$, $n_i = 0.05$, and the same volume as a sphere with size parameter $2\pi r/\lambda = 17.3$. The optical depth for absorption across such a sphere is ~ 3.4 and the single-scattering albedo of the sphere of equal volume is barely > 0.5. The single-scattering albedo of the irregular particle is not given, but it would seem that the product of n_i and the size parameter would have to be nearly an order of magnitude smaller to reproduce Titan's geometric albedo in the red. The polarization of the other cases given by these authors, which have n_i a factor of 10 smaller, are all much too low.

On the other hand, some calculations for oblate spheroids (Asano and Sato 1980) with ratios of axes of 5:1 can give high polarization for conservative scattering if the size parameter of the short dimension is not too large (~ 1). The radius of the long axis would have a size parameter of ~ 5, giving a radius of 0.33 μm at wavelength 0.42 μm. Such a flattened oblate spheroid would behave much like a disk with regard to diffraction, and would give the required brightness at large phase while producing high polarization at this wavelength. It remains to be seen whether such particles, or other nonspherical particles, could continue to give high single-scattering polarizations over the range from 0.264 μm to 0.75 μm, and be large enough to give high brightness at high phase in the blue and not too absorbing to reproduce Titan's albedo vs. wavelength curve.

The feeling emerges that while particles with dimensions $\gtrsim 0.4$ μm are necessary to give the high brightness at large phase in the blue, some dimension of the aerosol particles, either due to surface irregularities or elongation or flattening, must be $\lesssim 0.1$ μm to form the observed large single-scattering polarizations without excessive absorption at red wavelengths. At shorter wavelengths, the increased particle absorption may permit Fresnel reflection to contribute to the necessary large polarization even though the size parameter becomes larger. While such suggestions seem reasonable, additional calculations coupled with laboratory scattering measurements are still needed before a particle size and shape is demonstrated to be able to reproduce all the Titan observations.

Vertical Distribution. Two types of observations have been used to derive information on the vertical distribution of Titan's aerosols. From the highest resolution Voyager images, Rages and Pollack (1983) inverted the profile of brightness vs. height to give optical depth or extinction coefficient vs. height for a reasonable phase function and single-scattering albedo. Such inversions are possible down to vertical optical depth ~ 0.1 before the multiple-scattering effects in a spherical atmosphere greatly complicate the technique. Rages and Pollack found a local peak in the extinction coefficient at vertical optical depth ~ 0.01 and altitude 340–360 km (above a surface at radius 2570 km). The region around optical depth 0.1 was reached at altitudes of 200 to 230 km depending on latitude (see Fig. 16). In this region the aerosol scale height was ~ 30 km, a little less than that of the gaseous atmosphere.

The vertical distribution at deeper levels is less certain. The best information comes from fitting the strengths of Titan's red methane bands, which allows the optical thickness of the aerosols to be tied to the methane abundance above a given level. Calculations have been made for particles of radii ≤ 0.1 μm by Podolak and Giver (1979) and for particles of radius 0.25 μm by Rages and Pollack (1980). In both models most of the aerosols must be distributed above most of the methane to match strong and weak absorption bands simultaneously. Podolak and Giver have an aerosol optical thickness of 5 (at 0.5 μm) mixed uniformly with 50 m-amagat of CH_4 above a clear region containing 1.95 km-amagat of CH_4, above a "ground" of reflectivity ~ 0.5. In Rages and Pollack's model, an aerosol optical depth of 2.5 is uniformly mixed with 10 m-amagat of CH_4 above a region in which the particle optical depth is 7,5 (distributed uniformly with height) and the CH_4 abundance is 2.2 km-amagat (exponentially distributed with height).

Figure 16 shows the pressure level of the boundary between the upper and lower layers of these two models, assuming that the methane to nitrogen mixing ratio in the stratosphere is constant at 1/4% or 2%. The action of a cold trap at a temperature as low as 75 K at a pressure level as high as 360 mbar may limit the stratospheric methane mixing ratio to values $\leq 2\%$. On the other hand, if the CH_4/N_2 ratio is $<<1/4\%$, the model of Rages and Pollack which has particles large enough to reproduce the high phase angle brightness would have aerosol optical depths in the ultraviolet not much greater than the Rayleigh scattering optical depth of the nitrogen atmosphere when the total optical depth is ≥ 1, and so would be unable to reproduce Titan's low ultraviolet geometric albedo. In addition, for a CH_4/N_2 mixing ratio $<< 1/4\%$

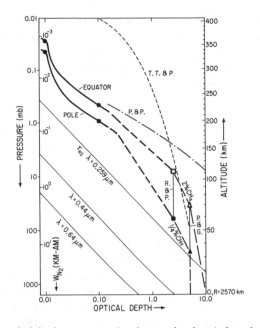

Fig. 16. Aerosol optical depth vs. pressure (or nitrogen abundance) along the left scales, or altitude along the right scale. The dark points at $\tau = 0.01$ and 0.1 (and the approximate solid
curves) give the aerosol extinction optical depth at 0.5 μm derived by Rages and Pollack
(1983) from Voyager 2 images of Titan's limb. The square points labeled R.&P. are from
Rages and Pollack's (1980) fits to Titan's geometric albedo for either 2% or 1/4% CH_4 in Titan's stratosphere. The triangles labeled P.&G. are the corresponding points from Podolak and
Giver's (1979) similar models. The dashed curves are meant to indicate reasonable bounds for
the vertical distribution of Titan's aerosols. The short dashed line labeled T.T.&P is one model
from Toon et al. (1980), while the dot-dashed curve is the model by Podolak and Podolak
(1980) with sticking coefficient 10^{-4} (both at wavelengths 0.635 μm). The other curves give
the optical depth of the nitrogen atmosphere for Rayleigh scattering at the indicated
wavelength.

in the stratosphere, the optical depth of aerosol above 100 km altitude becomes
significantly below unity, and the ability of the aerosols to absorb enough sunlight to produce the high temperatures at these levels becomes questionable.

If there were no CH_4 cloud, the required 2 km-amagat of CH_4 in the
lower layer of the models that match Titan's spectrum could be distributed
down to the surface. The resulting CH_4/N_2 ratio in the troposphere would be
$\sim 2.5\%$, and at least this much seems required. If there were a CH_4 cloud it
would prevent photons from penetrating so deeply, and this would have to be
compensated for by a higher CH_4/N_2 mixing ratio in the lower atmosphere.
The extent to which some of the absorption features could be formed in solid
methane cloud particles is unknown.

Cloud physics models predict that the aerosol particles should be larger at increasing depth. If they are composed of the same material at all heights, then the larger particles at lower altitudes will have lower single-scattering albedos. Such effects have yet to be included in models that aim to reproduce Titan's spectrum. In principle, if cloud physics considerations were used to constrain the vertical variation in single-scattering properties of the aerosols (whose "average" properties are given in Table IV), models of Titan's spectrum might provide more precise constraints on the CH_4/N_2 mixing ratio in the atmosphere.

The total optical thickness of Titan's photochemical aerosol haze is difficult to determine by using optical measurements from outside the atmosphere. Nevertheless, in view of their low single-scattering albedo the aerosols cannot have optical thickness much greater than the values of 5–10 used in the above models if enough solar flux is to reach the surface to provide even the modest observed rise of surface temperature above Titan's effective temperature. This seems especially true if the aerosols at low altitudes are larger and darker than those at higher levels that primarily determine Titan's geometric albedo. If a significant CH_4 cloud is required to provide the thermal opacity inferred from Voyager IRIS data, the constraints on the aerosol layer's thickness may be even more severe.

C. Comparison of Deduced Aerosol Properties with Physical Cloud Models

A few studies of the physical processes that determine the production rate, size, and vertical distribution of aerosols in Titan's atmosphere have been carried out and can be compared with the available information concerning the aerosols. Several authors have studied the photolysis of CH_4 on Titan (Strobel 1974; Bar-Nun and Podolak 1979; Podolak et al. 1979; Allen et al. 1980) which leads to the production of polymers that could constitute the visible aerosols. In addition, laboratory studies attempting to simulate the production of these aerosols have been done using energetic particles (Scattergood and Owen 1977) or ultraviolet light on mixtures of CH_4, NH_3 and H_2S (Khare and Sagan 1973) or on C_2H_4, C_2H_2 or HCN in CH_4 (Podolak and Bar-Nun 1979). Podolak and Bar-Nun suggested an upper limit to the aerosol mass production rate of 3.5×10^{-14} g cm^{-2} s^{-1} based on the quantum yields of C_2H_2 and C_2H_6 that they measured in the photolysis of CH_4 using an ultraviolet lamp. However, from calculations of Titan photochemistry, Strobel (1974) and Allen et al. (1980) suggest that the actual aerosol mass production rate on Titan might be somewhat higher.

Podolak and Podolak (1980) and Toon et al. (1980) have derived steady-state aerosol size distributions at various altitudes from a parameterized aerosol mass production function. Following Toon et al. (1980) the continuity equation governing the number density of aerosol of volume v, $n(v)$, can be written as

$$\partial n(v)/\partial t = -\frac{\partial}{\partial z}(\phi_S + \phi_D) + \frac{1}{2}\int_0^V K(v', v - v')n(v')n(v - v')dv'$$
$$- n(v)\int_0^\infty K(v, v')n(v')dv' + q(v). \tag{34}$$

Here the time rate of change of the concentration of aerosols of volume v at each altitude z is the sum of: the gradients in sedimentation and diffusion fluxes ϕ_S and ϕ_D; the creation rate of particles of volume v due to coagulation of smaller particles; the loss rate of particles of volume v due to coagulation with all other particles; and the production rate of particles of volume v by photochemical processes. Except in the lower atmosphere, the ratio of molecular mean free path to the aerosol particle radius (the Knudsen number K) is large, and Toon et al. (1980) take the sedimentation flux as

$$\phi_S(v) = -0.74 \, X_s \, g \, r \frac{\rho}{\rho_g}\left(\frac{8 \, k \, T}{\pi \, m_g}\right)^{-1/2} \tag{35}$$

Here X_s is a factor that allows for particles of different shape and is unity for spheres. The factors ρ_g and m_g are the local gas density and mass of a gas molecule, while ρ is the aerosol density (usually near unity unless fluffy or open irregular particle shapes are considered), r is the aerosol radius and the other symbols have their usual meanings.

Toon et al. (1980) assume that the small particles diffuse as a gas and take

$$\phi_D(v) = K_E \, n_g \frac{\partial}{\partial z}\left(\frac{n}{n_g}\right) \tag{36}$$

where n_g is the concentration of gas molecules and K_E is an eddy diffusion coefficient to be specified. Podolak and Podolak (1980) show that coagulation due to Brownian motion dominates over coalescence (particle growth in collisions due to different sedimentation velocities for particles of different size) practically everywhere in the atmosphere. For large Knudsen number, the coagulation kernel for Brownian motion is given as

$$K(v_1, v_2) = X_k \, \alpha \, 8 \, \pi \, r^2\left(\frac{k \, T}{\pi \, m}\right)^{1/2} \tag{37}$$

by Toon et al. (1980). Here α is a sticking coefficient near unity except for charged aerosols. X_k, like X_s, is a factor that accounts for particle shape.

Podolak and Podolak assumed the particle production rate was proportional to the number of 2000 Å photons that survived to some atmospheric level, and to the atmospheric pressure at that level raised to a power. Toon et al. characterized the production rate as a Gaussian function in pressure centered at some low pressure with $\sim$ 30% spread in pressure. In constructing

their model, Podolak and Podolak set the time derivative in Eq. (34) to zero
and used numerical techniques to obtain the aerosol size distribution as a
function of height. Toon et al. began a time dependent calculation with no
particles in the atmosphere, and continued stepping in time until a steady state
was achieved.

Several interesting points emerged from these studies. Podolak and
Podolak were reluctant to use mass production rates much greater than the
value of 3.5×10^{-14} g cm^{-2} s^{-1} derived from laboratory experiments. With
this production rate, they concluded that a thick cloud of particles with radii
on the order of 0.1 μm could not be sustained at pressures $\lesssim$ 2 mbar, and
therefore that Titan's stratosphere must consist of $\lesssim$ 5% methane. Further,
achieving reasonable aerosol sizes and optical depths with this production
rate required a sticking coefficient $\alpha \sim 10^{-4}$ implying electrically charged
aerosols.

Toon et al. (1980) made calculations using aerosol production rates that
were factors of 10 to 20 higher than Podolak and Bar-Nun's (1979) upper
limit, and showed that aerosols of the required sizes could be supported at
quite low pressures even for sticking coefficients of unity with these higher
mass production rates.

Since both these aerosol models were constructed before the Voyager
mission established the predominance of nitrogen in Titan's atmosphere, they
should not be expected to correspond too closely to the vertical aerosol profile
near the top of Titan's atmosphere, without further adjustments. Nevertheless, we show both models (taken from Toon et al. 1980) in Fig. 16. The
model of Podolak and Podolak (1980) was constructed assuming 5% CH$_4$, an
isothermal stratosphere at 160 K, an aerosol mass generation rate 3.5×10^{-14}
g cm^{-2} s^{-1}, a sticking coefficient $\alpha = 10^{-4}$ throughout the atmosphere, and
an eddy diffusion coefficient of zero. The agreement with the observed pressure level at an aerosol optical depth 0.1 and with the rate of increase in optical depth with pressure in this region is remarkable. However, this model
leads to large aerosol optical depths (> 10) at pressures > 10 mbar due to the
slow sedimentation rates of the small aerosols found at these levels if the sticking and diffusion coefficients remain small at depth.

The model by Toon et al. (1980) also uses a diffusion coefficient of zero,
but uses a sticking coefficient of unity throughout the atmosphere and an aerosol production rate 10 times higher than that of the model by Podolak and
Podolak. In this case, the aerosol optical depths are too large at low pressure,
but fall within a reasonable range at pressures > 10 mbar. Toon et al. suggest
that the sticking coefficient could be < 1 at low pressure due to charging of
aerosols by ultraviolet photons via the photoelectric effect, but this might well
not continue to deep levels. Figure 16 suggests that this could be one way to
obtain a vertical aerosol profile closer to that required by the observations.

Despite their uncertainties, such models are quite interesting and could
also offer additional leverage for constraining the methane mixing ratio. The

visible spectrum allows the aerosol optical depth to be tied to the CH_4 abundance. If the physical models could help tie the aerosol optical depth to height (effectively, to total pressure) perhaps the CH_4/N_2 ratio could be constrained more closely.

Figure 17 shows aerosol size as a function of optical depth for the two aerosol models discussed above, and also shows the two points determined from the red and blue Pioneer 11 polarimetry assuming the aerosols polarize as spheres. For optical depth < 1 the models and polarimetry points agree to a factor of ~ 2. The upper model curves with high mass input also can produce the high phase angle brightness observed by Voyager. Thus, aside from the remaining photometry–polarimetery problem, which seems likely to be resolved only by a better understanding of scattering from nonspherical particles, the particle sizes computed in the physical models are in reasonable agreement with those required by the observations.

Future work is needed in several areas to improve our understanding of Titan's aerosols. Microwave studies are needed to determine the sizes and shapes of particles that could satisfy the single-scattering polarization and phase function constraints determined from the optical observations. Improved cloud physics calculations are needed, using the temperature-pressure structure of Titan's atmosphere determined by Voyager with appropriate aerosol mass generation rates (which still need to be refined). Combining physical

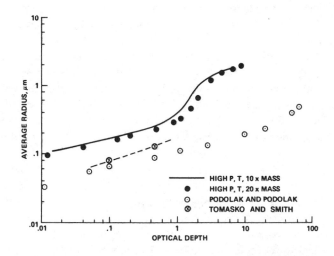

Fig. 17. The average (cross-section weighted) particle radius vs. extinction optical depth at 0.635 μm for the models of Toon et al. (1980) using an isothermal stratosphere at 160 K and 10 and 20 times the mass production rate of Podolak and Podolak (1980), as marked. Also shown is the model by Podolak and Podolak (1980). The dependence of aerosol radius on optical depth derived by Tomasko and Smith (1982) from the Pioneer polarimetry is marked. The aerosol size estimates agree to within a factor of ~ 2 at optical depths less than unity.

models that tie aerosol optical depth to total pressure with calculations of Titan's spectrum in the visible CH_4 bands could better constrain the CH_4/N_2 mixing ratio. Finally, the combination of thermal models with calculations at solar wavelengths which include reasonable variations of particle size with depth might give additional constraints on the total thickness of aerosol haze required for thermal balance in Titan's lower atmosphere.

V. METEOROLOGY AND ATMOSPHERIC DYNAMICS

A. Introduction

In dynamical studies of Titan's atmosphere from a limited set of observations, one must rely on the extensive body of work on the meteorologies of Earth and the other planets. In many respects Titan's meteorology lies between those of Earth and Venus. The atmospheres of both Titan and Venus are characterized by large thermal inertias and small horizontal temperature contrasts in the lowest few scale heights, although on Titan the former are attributable more to low temperatures than to high densities. Like Venus, Titan rotates slowly on its axis and appears to have a global cyclostrophic wind system. Unlike Venus, Titan has a large obliquity similar to Earth's, and its stratosphere should exhibit large seasonal changes. The contribution of planetary-scale waves and eddies to heat transport is apparently small in both the troposphere and stratosphere of Titan, in marked contrast to Earth's and perhaps Venus' atmospheres. Condensation processes may be important in Titan's atmosphere, with methane rather than water as the condensable. Titan therefore is an additional natural laboratory for comparative study of the dynamics of planetary atmospheres.

We first discuss the global aspects of the dynamics of Titan's atmosphere: the role of the radiative time constant, the meridional advection of energy, the zonal cyclostrophic wind system and the dynamical transports required to maintain it. Then we discuss smaller-scale dynamical processes near the surface: the turbulent fluxes of heat in the atmosphere and the possible occurrence of condensation near the surface. Finally, Sec. VII lists some future observations needed to clarify several elusive aspects of Titan's atmospheric dynamics.

Before proceeding, we define the notation used in this section and give the fundamental dynamical equations.

λ, Λ longitude, latitude
z a measure of height $\equiv -H \ln (p/p_0)$
H pressure scale height $\equiv RT_0/g$, assumed constant
D vertical scale of dynamical circulation (usually assumed $\simeq H$)
T_0 a constant "mean" temperature
g gravitational acceleration
p pressure

p_0 a constant reference pressure
u, v   eastward and northward velocities
w vertical velocity $\equiv dz/dt$
T local temperature
Φ local geopotential $\equiv gz$
R gas constant for Titan's atmosphere (universal gas constant/mean atmosphere mass)
C_p specific heat per gram at constant pressure
θ potential temperature $\equiv T(p_0/p)^{R/C_p}$
θ_e potential temperature corresponding to radiative equilibrium
Ω rotational frequency of Titan $\simeq 2\pi/16$ days
Ω_s "annual" frequency of Titan $= 2\pi/$Saturn's orbital period, $\simeq 2\pi/29.5$ yr
R_T radius of Titan
τ_r radiative time constant
M angular momentum per unit mass $\equiv R_T \cos\Lambda(\Omega \, R_T \cos\Lambda + u)$
f_λ, f_Λ force terms in the momentum equations representing the effect of friction and eddies
f_h heating or cooling (in K s^{-1}) by eddies
ν eddy viscosity
ϕ_s incident solar flux
A Titan's albedo for solar radiation ~ 0.2
ρ mass density
H_s turbulent flux of sensible heat into the base of the atmosphere
C bulk transfer coefficient

The zonally averaged dynamical equations are (Holton 1975):

$$\frac{d\bar{M}}{dt} = \bar{f}_\lambda \, R_T \cos\Lambda \tag{38a}$$

$$\frac{d\bar{v}}{dt} + \left(2\Omega + \frac{\bar{u}}{R_T \cos\Lambda}\right) \bar{u} \sin\Lambda + \frac{1}{R_T} \frac{\partial \bar{\phi}}{\partial \Lambda} = \bar{f}_\Lambda \tag{38b}$$

$$\frac{\partial \bar{\phi}}{\partial z} = \frac{R\bar{T}}{H} \tag{38c}$$

$$\frac{1}{R_T \cos\Lambda} \frac{\partial}{\partial \Lambda} (\bar{v} \cos\Lambda) + \frac{\partial \bar{w}}{\partial z} - \frac{\bar{w}}{H} = 0 \tag{38d}$$

$$\frac{d\bar{\theta}}{dt} = \frac{\bar{\theta}_e - \bar{\theta}}{\tau_r} + \bar{f}_h \tag{38e}$$

$$\frac{d}{dt} \equiv \partial_t + \frac{\bar{v}}{R_T} \partial_\Lambda + \bar{w} \partial_z \tag{38f}$$

where the overbar denotes an average over longitude. Radiative heating and cooling have been parameterized in terms of a linear relaxation to an equilibrium solution, Eq. (38e). The absence of zonal (east–west) thermal structure inferred from IRIS data suggests that planetary-scale eddies do not contribute to $\overline{f_h}$. In discussing global-scale flows we will usually neglect vertical transport of heat by small-scale eddies and assume $\overline{f_h} = 0$. In the remainder of this section the overbars will be omitted for simplicity, but zonal averages are to be assumed.

B. Radiative Time Constants

Consider first the purely radiative solution of Eq. (38e), θ_r, which holds when the dynamical advective terms are absent ($w = v = 0$):

$$\partial_t \theta_r = \frac{\theta_e - \theta_r}{\tau_r} . \tag{39}$$

Since Eq. (39) is linear in θ_r, the response to a time varying thermal forcing $\theta_e(t)$, can be analyzed via a Fourier expansion:

$$\theta_r(\mathbf{x}, t) = \sum_{n=0}^{\infty} \theta_r^{(n)}(\mathbf{x}) e^{in\omega t}, \quad \theta_e(\mathbf{x}, t) = \sum_{n=0}^{\infty} \theta_e^{(n)}(\mathbf{x}) e^{in\omega t} \tag{40}$$

where ω is the temporal frequency of interest (e.g., diurnal or seasonal). The solution to Eq. (39) is

$$\theta_r^{(n)}(\mathbf{x}) = \frac{\theta_e^{(n)}(\mathbf{x}) e^{-i\phi_n}}{[1 + (n\omega\tau_r)^2]^{1/2}} \quad ; \tan\phi_n = n\omega\tau_r. \tag{41}$$

The quantity $\omega\tau_r$ gives a measure of the thermal inertia of the atmosphere. Its magnitude determines both the amplitude and phase lag of the time varying components ($n > 1$) of $\theta_r(x, \tau)$. The steady ($n = 0$) component of θ_r is independent of $\omega\tau_r$ and is equal to the instantaneous radiative equilibrium solution (θ_e) averaged over the period $2\pi/\omega$.

In the lowest scale height of Titan's atmosphere the radiative time constant is quite large. Following Gierasch et al. (1970), B. A. Smith et al. (1981) estimated it as the ratio of the heat content of the atmosphere to the globally averaged rate of absorption of solar radiation

$$\tau_r \simeq \frac{(P_0/g) \, C_p T_0}{(\phi_{s0}/4)(1 - A)} \tag{42}$$

where ϕ_{s0} is the solar constant at 9.55 AU, 1.51×10^4 ergs cm^{-2} s^{-1}. Equation (42) gives $\tau_r \sim 4 \times 10^9$ s, which is quite long compared to a season:

$\Omega_s \tau_r \sim 26$. With a value this large, the lowest ($n = 0$) term in the expansion of θ_r, Eq. (40) dominates, and the annual term ($n = 1$) is reduced by a factor $(\Omega_s \tau_r)^{-1}$. Smith et al. noted that such a large value of $\Omega_s \tau_r$ implies a phase lag in the annual term close to one full season. They argued that the north-south asymmetry in Titan's albedo observed by the Voyager spacecraft near Titan's northern spring equinox could be explained in terms of this seasonal lag in temperature. At this time the temperatures in the lower atmosphere should be warmer in the south than in the north, which could induce a cross-equatorial thermally direct circulation with rising motion in the south and subsidence in the north. By analogy with terrestrial circulations, the upwelling in the south could lead to enhanced condensation and to the higher brightness observed. The difficulty with this interpretation is that the radiative time constant is so large that it is questionable whether a north-south thermal asymmetry of sufficient magnitude would develop. The observed brightness temperatures at 550 cm^{-1} suggest an asymmetry of ≤ 1 K (cf. Sec. II).

Estimating τ_r for the lower troposphere (Eq. 42) requires no exact knowledge of the radiative properties of the atmosphere, since the bulk of the atmosphere mass is contained in the lowest scale height and the heat content of this region derives ultimately from solar heating. However, at higher altitudes a detailed knowledge of the infrared opacity is necessary to estimate τ_r. Fortunately, in much of the stratosphere a simplification is possible because the local radiative balance is between absorption of solar radiation and direct infrared cooling to space. Table V presents the cooling rates N (in s^{-1}) of the principal gaseous emitters at several levels in Titan's stratosphere (Flasar et al. 1981b). Mole fractions of 1×10^{-2} for CH$_4$, 2×10^{-5} for C$_2$H$_6$, and 3 $\times 10^{-6}$ for C$_2$H$_2$ were assumed, but the cooling rates scale weakly with abundance, approximately as the square root (Cess and Caldwell 1979). Radiative response times attributable to gaseous cooling were obtained by summing the cooling rates for these three species. Aerosols also contribute to stratospheric cooling. Samuelson et al. (1981) concluded that some of the observed emission at 200–600 cm^{-1} is from this region, probably from aerosols in the lower stratosphere. Table V lists the radiative time constants associated with aerosol emission at two assumed pressure levels in the lower stratosphere. The estimates assume that the aerosol emission is confined to a slab one scale height thick, centered about the specified pressure level.

Radiative time constants within the stratosphere range from values that are small compared to the length of a season at high altitudes (~ 0.3–1 mbar) to values near the tropopause for which $\Omega_s \tau_r \sim 1$ or even $\gg 1$. The uncertainty arises because the vertical distribution of infrared aerosol emitters in the lower stratosphere and the tropopause region is not known. At all levels τ_r is large compared to the presumed 16-day diurnal cycle on Titan; this is consistent with the absence of detectable structure in longitude in the IRIS data (Flasar et al. 1981b; see below).

Since $\Omega_s \tau_r \ll 1$ in the upper stratosphere, Eq. (41) predicts $\theta_r^{(n)} = \theta_e^{(n)}$

TABLE V

Radiative Response Times in Titan's Stratosphere[a]

| P(mbar) | T(K) | Gaseous Radiative Cooling Rate N (s^{-1}) | | | τ_r (s) | | $\Omega_s \tau_r$ | |
		CH_4	C_2H_6	C_2H_2	Gas	Aerosol	Gas	Aerosol
0.3	173	2×10^{-8}	6×10^{-9}	2×10^{-8}	2×10^7	—	0.1	—
1	170	1×10^{-8}	7×10^{-9}	1×10^{-8}	3×10^7	—	0.2	—
10	149	1×10^{-9}	4×10^{-9}	4×10^{-9}	1×10^8	4×10^7	0.7	0.2
30	104	3×10^{-12}	2×10^{-10}	5×10^{-10}	1×10^9	7×10^7	9.	0.5

[a] Table after Flasar et al. 1981b.

and $\phi_n \simeq 0$. The purely radiative response is in phase and is in radiative equilibrium with the solar forcing. If the opacities for solar and thermal radiation are uniformly distributed in latitude, meridional temperatures should be symmetric about the equator, since at equinox the solar insolation is symmetrically distributed. Although gross symmetry is evident in the 1304 cm^{-1} temperatures in Fig. 4, temperatures are 3 to 4 K warmer in the southern hemisphere at mid-latitudes than in the north. The cause of this difference is unknown; a hemispheric asymmetry in the opacity may exist.

C. Energy Transport by Large-Scale Motions

The thermal structure of Titan's atmosphere cannot be controlled by radiative processes alone, because the observed meridional contrasts are too small. In the lower troposphere, the inferred contrast between equator and pole is $\sim$ 2 to 3 K. A crude estimate of the thermal contrast associated with the steady radiative solution can be made from

$$\frac{\theta_e(\Lambda)}{\theta_e(0)} \sim \left(\frac{\langle \phi_s(\Lambda) \rangle}{\langle \phi_s(0) \rangle} \right)^{1/4} \tag{43}$$

where $\langle \phi_s(\Lambda) \rangle$ denotes the annual mean solar flux incident on the atmosphere at latitude Λ. For a rotating body with equatorial plane inclined 27° to the ecliptic, $\phi_s(\geq 60°) \sim 1/2 \, S(0°)$ (Brinkman and McGregor 1979). Equation (43) predicts that temperatures at latitudes $\geq 60°$ should be $\sim$ 15 K cooler than those near the equator. The small meridional contrast observed in brightness temperatures at 200 cm^{-1} suggests that the meridional variation in atmospheric temperature near the tropopause is also small compared to that implied by a purely radiative solution, although the magnitude of the latter cannot be estimated without a detailed model. In the upper stratosphere the meridional structure of the radiative solution can be roughly estimated by assuming a uniform distribution of opacity with latitude. Solar heating then varies with latitude at equinox as $\cos\Lambda$. An assumed local balance between absorption of solar radiation and direct cooling to space by CH$_4$, C$_2$H$_6$, and C$_2$H$_2$ predicts a thermal contrast between 0° and 60° latitude $\simeq$ 7 K greater than that actually observed.

Evidently dynamical processes redistribute heat and potential energy, and reduce the meridional contrast in temperature from that expected if only radiative processes were acting. On Earth, much of this redistribution is effected by planetary-scale eddies known as baroclinic eddies, the temperature of which varies with longitude. The limited global coverage of Titan by the IRIS experiment failed to produce any convincing evidence of such structure down to horizontal scales $\sim$ 30° in longitude. Flasar et al. (1981b) assigned upper limits of 1 K to longitudinal temperature variations near the surface and at the tropopause, and 3 K in the upper stratosphere. In addition, the temperature profiles derived from the RSS ingress and egress soundings, which were separated by only 15° in latitude but nearly 180° in longitude, are remarkably similar (Lindal et al. 1983).

Leovy and Pollack (1973) had previously concluded on theoretical grounds that baroclinic eddies are ineffective in transporting heat in Titan's troposphere. The preferred horizontal scale of baroclinic eddies is given by

$$L \sim \left(\frac{g}{T_0} \frac{\partial \theta}{\partial z} \right)^{1/2} \frac{D}{\Omega} . \tag{44}$$

Over the lowest $D \sim 1\text{--}1/2$ scale heights (28 km) θ increases by 22 K, and Eq. (44) implies $L/R_T \sim 8$. Titan is thus too small for baroclinic eddies to fit in its troposphere. A similar computation can be made for the stratosphere, if the rotation rate of the atmosphere below is used in Eq. (44) instead of the rotation rate of Titan's surface. As a worst case, consider the stratosphere near 1 mbar and assume $\Omega_{\text{eff}} R_T \sim 75$ m s^{-1}, comparable to the zonal wind speed inferred from the temperature data (see below); Eq. (44) still implies $L/R_T \sim 2$. Thus Leovy and Pollack and later Flasar et al. (1981b) were led to consider transport of energy by zonally symmetric, thermodynamically direct meridional circulations.

Some evidence of zonally symmetric flows has been found in the Voyager images. In addition to the hemispheric asymmetry marked by a sharp change in albedo near the equator, discussed earlier, a broad-banded structure in the northern hemisphere appears in contrast-enhanced images (B. A. Smith et al. 1981, 1982; Sromovsky et al. 1981). Voyager 2 images have also shown a dark ring at 70° surrounding the north pole. Such structure may indicate differential vertical motion in Titan's atmosphere, but the processes responsible for the albedo differences, as well as the altitudes at which these processes operate, are not known.

Consider an idealized meridional cell confined to altitudes z_0 and $z_0 + D$, and latitudes 0 and $\pi/2$, divided into upper and lower parts such that each "box" contains equal mass. Let θ_U and θ_L denote the average of θ with respect to mass in the upper and lower boxes, respectively, and define the mass weighted vertical contrast $\delta_m \theta \equiv \theta_U - \theta_L$. Similarly divide the region into equatorial and polar boxes with a boundary at 30° latitude. Let θ_E and θ_P denote the averages with respect to mass of θ in these two boxes, and define the mass-weighted horizontal contrast $\Delta_m \theta \equiv \theta_E - \theta_P$. Performing these averages on the heat equation (Eq. 38e) and differencing yields (Stone 1974):

$$\partial_t (\Delta_m \theta) + \frac{V}{R_T} \delta_m \theta \sim \frac{\Delta_m \theta_e - \Delta_m \theta}{\tau_r} \tag{45a}$$

$$\partial_t (\delta_m \theta) - \frac{W}{D} \Delta_m \theta \sim \frac{\delta_m \theta_e - \delta_m \theta}{\tau_r} . \tag{45b}$$

Factors of order unity have been neglected. V is a mass-weighted meridional velocity whose magnitude is obtained by averaging over the upper or lower leg of the cell. Likewise, W is a characteristic vertical velocity. In a thermodynamically direct cell, when equatorial latitudes are warmer than polar ($\Delta_m\theta$ > 0), there is upwelling at low latitudes and poleward motion in the upper leg of the cell; in this situation both V and W are positive. According to Eq. (45a), when the atmosphere is stably stratified ($\delta_m\theta$ > 0) the atmosphere aloft moving poleward has more heat and potential energy than the mass nearer the surface moving toward the equator, the net transport of energy is poleward, and the circulation acts to reduce $\Delta_m\theta$ ($\partial_t\Delta_m\theta$ < 0). Similarly, since $\Delta_m\theta$ > 0, the upwelling mass at low latitudes is warmer than the sinking mass at high latitudes, resulting in a net transport of heat upward which acts to increase the static stability ($\partial_t\delta_m\theta$ > 0) (cf. Eq. 45b).

Flasar et al. (1981*b*) used Eq. (45a) to calculate meridional velocities in the lower troposphere and upper stratosphere, the two regions for which they had sufficient information on τ_r, $\Delta_m\theta_e$, $\Delta_m\theta$, and $\delta_m\theta$. In both regions the term $\partial_t(\Delta_m\theta)$ could be neglected. In the lower troposphere conditions are nearly steady ($\partial_t \equiv 0$). In the upper stratosphere where seasonal changes are significant, $\partial_t \sim \Omega_s << 1/\tau_r$; the upper stratosphere is in radiative-dynamical balance with the current value of the solar forcing. Estimates of the meridional velocities V and turnover times, R_T/V, are given in Table VI, along with the values of D, τ_r, $\Delta_m\theta_e$, $\delta_m\theta$, and $\Delta_m\theta$ used in the estimates. The meridional velocities inferred for the troposphere and stratosphere are quite sluggish by terrestrial standards (Oort and Rasmusson 1971; Crane et al. 1980), and result from the much larger radiative time constants on Titan.

Information on the vertical structure of Titan's atmosphere can also be inferred from Eqs. (45a,b). Using continuity to set

$$\frac{W}{D} \sim \frac{V}{R_T} \tag{46}$$

and setting $\partial_t = 0$, the equations can be combined to provide an estimate of the atmosphere's departure from radiative equilibrium:

$$\delta_m\theta - \delta_m\theta_e \sim \frac{\Delta_m\theta}{\delta_m\theta}(\Delta_m\theta_e - \Delta_m\theta). \tag{47}$$

Estimates are listed in Table VI. The departure from radiative control of the static stability of the atmosphere effected by global circulations is relatively small. This discussion has implicitly assumed a dry atmosphere. The estimated turnover time in the troposphere and the interpretation of Eq. (47) could change if the transport of latent heat, e.g. by convection in CH_4 cumulus clouds, is significant. As yet there is no compelling evidence of such activity on Titan (see Sec. E below).

Figure 18 is a schematic diagram of an idealized meridional circulation in Titan's atmosphere at northern spring equinox. Uniform distribution of opacity with latitude is assumed. At low levels, the symmetry about the equator follows from the long radiative time constant, ensuring steady conditions. The cross-equatorial circulation in the lower stratosphere and tropopause re-

TABLE VI
Radiative and Dynamical Characteristics of Titan's Atmosphere

	Lower Troposphere ($p \sim 1000$ mbar)	Upper Stratosphere ($p \sim 1$ mbar)
Vertical scale: D (km)	28	42
Observed vertical temperature contrast: $\delta_m\theta$ (K) [a]	11	30
Horizontal temperature contrast (K)		
Observed: $\Delta_m\theta$ [a]	1	6
Radiative equilibrium: $\Delta_m\theta_e$ [a]	8	13
Radiative time constant: τ_r(s)	4×10^9	3×10^7
Meridional velocity: V (cm s^{-1})	0.04	2
Turnover time: R_T/V (s)	7×10^9	1×10^8
Deviation of static stability from radiative equilibrium: $\delta_m\theta - \delta_m\theta_e$ (K) [a]	0.6	1.4
Zonal velocity: u (m s^{-1})	25	75
Vertical viscosity: ν (cm^2 s^{-1})	$\lesssim 10^3$	—

[a] Potential temperatures at the 1000 mbar and 1 mbar levels are defined with $p_0 = 1500$ mbar and 1 mbar, respectively.

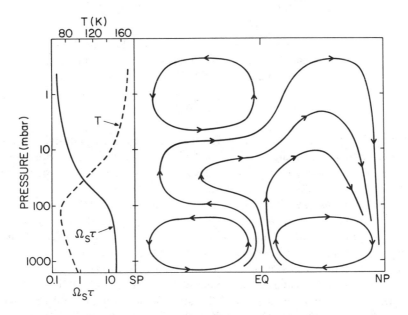

Fig. 18. Schematic diagram of the possible meridional flow in Titan's atmosphere at northern spring equinox. Mean vertical profiles of temperature and radiative relaxation time are indicated at left (figure after Flasar et al. 1981b).

gion reflects the seasonal lag in temperature expected when $\Omega_s\tau_r$ is on the order of unity. (When $\Omega_s\tau_r = 3$ the phase lag of the annual component [$n = 1$] is already close to its asymptotic value of one full season.) In the upper stratosphere the circulation reflects the relatively small time constant ($\Omega_s\tau_r \ll 1$) and the symmetry of the solar heating about the equator. The topology indicated is quite speculative. Thermodynamically indirect circulations may well exist at certain altitudes.

D. Zonal Flow and Angular Momentum Transports

The meridional contrasts in temperature observed in the troposphere and stratosphere imply, from the hydrostatic Eq. (38c), a north-south gradient in geopotential which must be balanced. Experience with other planetary atmospheres suggests that this gradient is balanced by the second term on the left-hand side of Eq. (38b), representing the Coriolis and centrifugal forces acting on the zonal wind u:

$$\frac{\tan\Lambda}{R_T} \, u^2 + 2\Omega \, \sin\Lambda \, u = - \frac{1}{R_T} \frac{\partial\phi}{\partial\Lambda} \, . \tag{48}$$

This can be combined with Eq. (38c) to yield the thermal wind relation

$$\frac{\partial}{\partial z} \left(\frac{\tan\Lambda}{R_T} \, u^2 + 2\Omega \, \sin\Lambda \, u \right) = - \frac{1}{R_T} \frac{R}{H} \frac{\partial T}{\partial\Lambda} \, . \tag{49}$$

A characteristic zonal velocity can be estimated from Eq. (49) by integrating it over one scale height. Because Titan rotates so slowly, the first term on the left-hand side dominates, and an estimate of u follows from

$$u \sim \sqrt{R\Delta T} \tag{50}$$

where $\Delta T = T(90°) - T(0°)$. In the lower troposphere the equatorial regions are warmer than high latitudes, $\Delta T \sim 2$ K and $u \sim 25$ m s^{-1}. In comparison, the equatorial rotation speed at the surface is $\Omega R_T = 11$ m s^{-1}. In the upper stratosphere $\Delta T \sim 20$ K and $u \sim 75$ m s^{-1}.

Observations at visible wavelengths have not confirmed the existence of a cyclostrophic zonal wind system ($u/\Omega R_T \gg 1$). Visually Titan's atmosphere is quite hazy, and no localized features usable as tracers for zonal wind measurements have been identified (B. A. Smith et al. 1981,1982; Sromovsky et al. 1981). The question naturally arises whether the meridional gradient in geopotential could be balanced in another way than in Eq. (48). Two possibilities can be suggested. The first is that the advective terms in Eq. (38b) balance the gradient in geopotential. Gierasch et al. (1970) and Stone (1974) considered this balance for a nonrotating planet and applied it to the lower atmosphere of Venus. Characteristic meridional velocities are given by v $\sim \sqrt{R\Delta T}$, analogous to Eq. (50). The difficulty with this model is that it cannot reconcile the heat equation with the observed meridional contrasts in temperature. The meridional flows transport heat, and velocities $\sim \sqrt{R\Delta T}$ would quickly reduce ΔT to orders of magnitude below the $\sim 1-10$ K observed in Titan's atmosphere. The other possible model is a balance between the geopotential gradient and friction. Approximating f_Λ in Eq. (38b) as a Navier-Stokes term:

$$f_\Lambda = \nu \, \partial_{zz} v \qquad (51)$$

the assumed balance becomes:

$$\frac{\nu v}{H^2} \sim \frac{1}{R_T} R\Delta T. \qquad (52)$$

Using values of the meridional velocity from Table VI, which are consistent with the heat equation, vertical viscosities $\nu \sim 10^{12}$ cm^2 s^{-1} are estimated for both the lower troposphere and upper stratosphere; they are quite large compared to those in comparable regions of other planetary atmospheres including the Earth's. A balance with horizontal diffusion would imply viscosities a factor $(R_T/H)^2 \sim 10^4$ larger. Perhaps the most compelling argument for global cyclostrophic winds on Titan is that they have been directly observed in Venus' atmosphere. Leovy (1973) has suggested that such flows will occur in the atmosphere of any slowly rotating planet in which equatorial regions are heated more than the poles, and which cannot develop baroclinic instabilities to efficiently redistribute energy; we have noted that this situation probably exists on Titan. However, Titan differs from Venus in that it has a large effective obliquity. As a result of the short radiative time constant at high altitudes, its stratospheric zonal winds will vary seasonally.

According to Eq. (48), in the cyclostrophic limit the zonal flow can be either prograde (eastward flowing) or retrograde. Surface friction does ensure that winds near the surface are small compared to those aloft (Flasar et al. 1981b), and integration of Eq. (49) from near the surface upward through the geostrophic regime ($u/\Omega R_T \ll 1$) indicates prograde winds at high altitudes, provided $\Delta T > 0$. This integration cannot be so readily extended to the upper stratosphere because the meridional temperature contrasts in the lower stratosphere are not well known. However, Leovy (1973) has argued on general grounds, by considering the spin up of an atmosphere from an initial state of no relative motion to a final cyclostrophic state, that the angular momentum of the atmosphere will have the same sense as that of the solid body. Thus, it seems reasonable that the strong zonal winds in the upper stratosphere are also prograde.

Although the mechanisms producing cyclostrophic flows are not well understood, we can make a few general observations concerning the transports that maintain them. The thermally direct, zonally symmetric circulations discussed earlier transport angular momentum as well as energy. An idealized situation relevant for the circulations in the lower troposphere and, at the equinoxes, in the upper stratosphere is depicted in Fig. 19 (cf. also Fig. 18). From the thermal wind equation, the angular momentum per unit mass increases with height and the symmetric circulations effect a net poleward transport of angular momentum. Since $\partial_z u > 0$, turbulent eddies will also transport momentum downward (Leovy 1973). For the equatorial angular momentum to be maintained, these fluxes must be balanced. (For reasons similar to those given for the heat equation, the term $\partial_t M$ in Eq. (38a) is not expected to be significant.) One possibility is a compensating flux of angular momentum toward the equator by eddies, for instance by lateral mixing of vorticity or angular velocity (Gierasch 1975). This mixing probably must be quasi-barotropic; in other words, the eddies should redistribute momentum but not heat. Otherwise they tend to laterally mix heat so efficiently as to destroy the meridional temperature gradients in Eq. (49) implied by the zonal winds (Kalnay de Rivas 1975; Rossow and Williams 1979).

Lateral mixing by quasi-barotropic eddies in Titan's atmosphere is consistent with the absence of detectable zonal thermal structure on planetary scales, but it does require small meridional contrasts in temperature so that the north-south eddy motions do not give rise to significant zonal perturbations in temperature. The meridional contrast in temperature is small in the lower troposphere and at latitudes $< 40°$ in the upper stratosphere (Fig. 4). At latitudes $> 40°$ in the stratosphere meridional gradients in temperature may

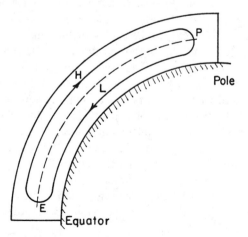

Fig. 19. Schematic of meridional cell. If the angular momentum per unit mass is greater at point E than at P, the meridional cell produces an upward flux of angular momentum. If it is greater at point H than at L, there is a poleward transport (figure after Gierasch 1975).

be too large for quasi-barotropic eddies to exist. An alternative mechanism for balancing the equatorial losses of angular momentum, particularly in the stratosphere, is an upward flux of zonal momentum at low latitudes associated with vertically propagating waves. This was suggested for Venus by Leovy (1973) and Fels and Lindzen (1974). However, the absence of detectable thermal structure with longitude suggests that the horizontal scale of such waves in Titan's atmosphere may be smaller than the spatial resolution of the IRIS observations ($\simeq 30°$ in longitude, $\simeq 1300$ km horizontal scale). Hinson and Tyler (1983a,b) have interpreted weak intensity scintillations, observed in the RSS ingress and egress soundings between 25 and 90 km altitude, as indicating vertically propagating internal gravity waves. They derive vertical and horizontal wavelengths of 1 and 4 km, respectively, and estimate that the momentum flux associated with the waves could build up the inferred zonal winds in the upper stratosphere in as little as 10^7 s, comparable to τ_r and to the mean flow advective time scale (Table VI). Whether such waves can propagate to the upper stratosphere ($z \geq 150$ km), and are not absorbed at lower altitudes, depends on the detailed vertical profile of u, which is not currently available. Another possibility is that there are vertically propagating waves with large horizontal wavelengths, but that the amplitudes required to maintain the angular momentum balance are below the IRIS detection limit (~ 3 K in the upper stratosphere). This remains to be investigated.

Consideration of the globally averaged vertical transports of angular momentum enabled Flasar et al. (1981b) to place an upper limit on the vertical eddy viscosity of the troposphere. Assuming that the vertical mixing of angular momentum was controlled by downgradient diffusion of M, and neglecting the flux of momentum into the stratosphere, they averaged Eq. (38a) over latitude and differenced the result in the vertical. For steady conditions ($\partial_t = 0$) they obtained

$$W \, \Delta_m M \sim 2\nu \frac{\delta_m M}{D} \qquad (53)$$

where $\delta_m M$ and $\Delta_m M$ denote mass weighted contrasts as before. Eq. (53) states that vertical diffusion of angular momentum must be balanced by vertical transport of M by the symmetric circulation. In Titan's lower troposphere $\delta_m M > 0$, and the diffusive flux of M is downward. The symmetric cell can balance this only if $\Delta_m M > 0$, i.e. if there is an excess of angular momentum at low latitudes relative to the polar regions (Fig. 19). A zonally symmetric circulation such as that in Fig. 19 tends to conserve angular momentum in its upper, poleward-moving branch, giving $\Delta_m M \sim 0$; vigorous lateral mixing of angular velocity tends to produce $\Delta_m M \sim \langle M \rangle$, the tropospheric average. The thermal wind relation discussed earlier implies that $\delta_m M \sim \langle M \rangle \sim \Omega R_T^2$. Flasar et al. conclude that $\Delta_m M \leq \delta_m M$ in Eq. (53) and estimate the rather small upper limit of 10^3 cm^2 s^{-1} for the vertical viscosity.

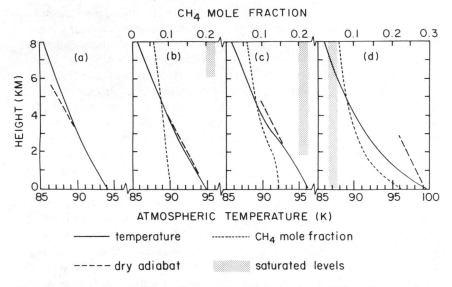

Fig. 20. Temperature and CH$_4$ profiles of model Titan atmospheres: (a) Pure N$_2$ atmosphere (Lindal et al. 1983); (b)–(d) N$_2$-CH$_4$ atmospheres for various relative humidities at the top of the surface layer: (b) 75%, (c) 90%, (d) 98.4%. The profiles were derived from the RSS ingress sounding, but those corresponding to egress are nearly identical.

E. Meteorology Near the Surface

The existence of a troposphere, with temperature decreasing monotonically with increasing altitude up to 40 km, implies that sunlight is absorbed at Titan's surface. The radiative equilibrium calculations by Samuelson (1983), discussed in Sec. II, predict a temperature discontinuity of 1–3 K at the surface. This temperature discontinuity ensures that the net upward flux of thermal radiation balances the net downward flux of solar radiation; the discontinuity is unstable in the sense that the surface is warmer than the atmosphere just above it. In reality, turbulent fluxes of sensible and possibly latent heat into the base of the atmosphere will supply much of the required upward energy flux and reduce (by a factor > 5) the discontinuity in temperature predicted by the radiative models. The reduction can be estimated from the bulk transfer relation that expresses the turbulent flux into the atmosphere of sensible heat H in terms of the drop in temperature across the "surface boundary layer":

$$H = \rho_a \, C_p \, C|\mathbf{v}_a| \, (T_s - T_a). \tag{54}$$

The quantities ρ_a, $\mathbf{v}_a$, T_s, and T_a denote density, wind velocity relative to the surface, surface temperature, and temperature at an "anemometer" level typically taken to be a few meters above the surface for terrestrial measurements; the exact height is not too important provided that the vertical gradients in temperature and wind speed at this level are small relative to those at the surface. For a positive flux of sensible heat into the atmosphere $T_s > T_a$. Much of the physics in Eq. (54) is involved in the value of C, which depends on several factors including the vertical profiles of mass density and wind speed near the surface.

A plausible upper limit to the global average of H is the average solar radiation absorbed at the surface, $r\phi_{s0}/4$, where $\phi_{s0} = 15.1$ W m^{-2} and r is the fraction of solar irradiation absorbed at the surface. Samuelson's (1983) calculations impose the limits $0.13 \leq r \leq 0.33$. Under unstable conditions ($T_s > T_a$) the minimum value of $C\,v_a$ for fixed $T_s - T_a$ occurs in the free convection limit where $v_a \rightarrow 0$ (Leovy 1969; Leovy and Mintz 1969):

$$C\,|v_a| \rightarrow 0.204 \left(\frac{\kappa\,g(T_s - T_a)}{T_s} \right)^{1/3} \text{cm s}^{-1} \qquad (55)$$

where $\kappa \simeq 0.016$ cm^2 s^{-1} is the molecular thermal diffusivity (Forsythe and Powell 1957). The lower limit on $C|v_a|$ and the upper limit on H impose an upper limit on $T_s - T_a$ in Eqs. (54) and (55). For the range $r = 0.13-0.33$, $T_s - T_a \leq 0.24-0.48$ K; the radiative discontinuity is reduced by a factor $\gtrsim 5$.

The turbulent flux of heat into the atmosphere given by Eq. (54) will establish a layer of buoyancy-driven turbulence above the surface layer. Vertical mixing will ensure that the lapse rate in this layer is close to adiabatic; because of the long radiative time constant, diurnal changes in temperature are probably small. The thermal structure above this well-mixed turbulent layer is controlled by a combination of radiative and larger-scale dynamical processes. In the Earth's atmosphere, lapse rates in this region are characteristically subadiabatic. Lindal et al. (1983) have derived temperatures for a pure N$_2$ atmosphere from the RSS ingress and egress soundings. These indicate lapse rates close to adiabatic up to an altitude of 3–4 km and markedly subadiabatic above (Fig. 20a). It seems reasonable to conclude that the adiabatic region defines the vertical extent of the well-mixed turbulent layer.

These considerations have assumed that only dry processes are operating. Alternatively, moist convection could occur and the change in lapse rate observed at altitudes 3–4 km could mark the base of convective clouds. Methane would presumably be the condensate, and CH$_4$ oceans would probably be required to maintain the moist convection on a global scale. Eshleman et al. (1983) have argued that such activity is unlikely, because the temperature in the subadiabatic region decreases with altitude faster than the moist pseudoadiabat (i.e., condensation products fall out) up to altitudes $\simeq 15$ km, well above the cloud base hypothesized to occur at 3–4 km. The potential difficulty with this argument is that temperature lapse rates in regions of cumulus activity on Earth typically lie between the dry adiabat and the moist pseudoadiabat (Augstein 1978; Augstein et al. 1980; Ludlam 1980, pp. 146–147), because of the heterogeneity of cumulus convection. In regions of cumulus activity, active clouds typically cover only a few percent of the surface area and convective transports within these are effected by means of moist updrafts and drier downdrafts.

Flasar (1983) has derived temperature profiles from the RSS soundings for several model atmospheres composed of N_2 and CH_4. In these models the mole fraction q of CH_4 is specified just above the surface layer and decreases with altitude to the small value $q = 0.017$ near the tropopause, corresponding to saturation at the cold trap. In the lowest part of the atmosphere where the methane is unsaturated, its mole fraction is specified to decrease slowly with altitude. At higher altitudes saturation typically occurs, and the mole fraction follows the saturation curve. Models with high relative humidities near the surface (Figs. 20c and d) exhibit lapse rates in temperature which exceed the dry adiabat. These models must be excluded since they are statically unstable to convective overturning. In these models methane saturates in the lowest few km of the atmosphere and is constrained by the saturation curve to decrease rapidly with altitude. This in turn produces a rapid variation in the refractive properties of the N_2–CH_4 mixture in such a way as to yield the steep gradients in the derived temperatures. Models with surface relative humidities $\lesssim 70\%$, which remain unsaturated in the lowest 3–4 km, produce profiles that do not exceed the adiabatic lapse rate and are consistent with the upper bound on the tropopause temperature and the lower bound on surface temperatures inferred from IRIS data (Samuelson et al. 1981; see also Sec. II).

Flasar (1983) also examined the possibility of methane oceans on Titan. By considering the turbulent flux of latent heat into the atmosphere in the free convection limit, using arguments analogous to those presented in the discussion of Eqs. (54) and (55), he concluded that if global methane oceans exist on Titan, the atmosphere above them must exhibit relative humidities $> 98\%$. This is much closer to saturation than occurs in the atmosphere above the Earth's tropical oceans. Several factors account for this difference, but one of the most crucial is the 100-fold reduction in the solar constant at Titan relative to Earth. Because the radio occultation soundings are only consistent with surface relative humidities $\lesssim 70\%$, Flasar concluded that the soundings could not have occurred over a methane ocean and suggested instead that they occurred over land. Another possibility is an ocean largely composed of heavier hydrocarbons. For instance propane and ethane are quite soluble in methane at Titan's surface temperature. C_2H_6 is of particular interest, because it is the main hydrocarbon product of CH_4 dissociation in the upper atmosphere (see Sec. III). Dissolving large amounts of C_2H_6 in a CH_4 ocean acts to depress the saturation vapor pressure of CH_4 at the surface, but the vapor pressure of C_2H_6 itself is so low that it is virtually absent from the gas. In effect the CH_4 saturation vapor pressure curve at the surface is depressed relative to that in the atmosphere. (The limited solubility of N_2 in CH_4 has been ignored here, but it is not crucial to these arguments.) Recent studies in progress by J. Pearl and F. M. Flasar indicate that CH_4 relative humidities $\leq 70\%$ near the surface can be maintained by a CH_4–C_2H_6 ocean with a C_2H_6 mole fraction $\gtrsim 35\%$. As discussed in Sec. III, it is estimated that ~ 500 m of C_2H_6 have precipitated over geologic time. This estimate is subject to several uncertainties, the most serious of which is the long-term history of the solar ultraviolet flux. Taking the estimate at face value, the above arguments suggest an upper limit of 1200 m to the depth of a hydrocarbon ocean, with the apparently reasonable assumption that it is well mixed over geologic time. A lower limit of 400 m follows from the argument that the past tidal dissipation must be small enough for consistency with Titan's present orbital eccentricity (Sagan and Dermott 1982).

Recently, Lunine et al. (1983) have proposed an ethane-dominated ocean in which the mole fraction of CH_4 is only 25%. They compute the solubility of N_2 in such an ocean and derive a nitrogen mole fraction of 5%, with a factor of 2 uncertainty.

VI. INTERIOR AND SURFACE

A. Introduction

Few statements can be made with certainty about Titan's interior. The average density 1.88 g cm^{-3} and radius 2575 km (Lindal et al. 1983) are neatly bracketed by the corresponding values for Jupiter's satellites Ganymede and Callisto (Table VII), bodies that are believed to consist of rock and water ice (25–50% by mass) but probably no other ices (Consolmagno and Lewis 1976; Cassen et al. 1980). The term 'rock' refers to silicates and iron (as sulfide, oxide, and/or elemental metal). The large uncertainty in ice content arises because it is not known whether the silicates are hydrated. The similarities between the three satellites listed in Table VII might lead one to suspect that they have similar interior properties. Table VIII shows why this would be a premature conclusion: other candidate ices, most notably $NH_3 \cdot H_2O$ and $CH_4 \cdot nH_2O$, have very similar densities to water ice. If Titan formed at lower temperatures than Ganymede or Callisto then it may have incorporated one or more of these more volatile ices without greatly affecting its average density. Quantification of this possibility cannot be carried out yet with confidence because the densities of these additional ices at relevant internal pressures (up to ~ 40 kbar), and the processes determining the ratio of silicate to ice in large satellites, are poorly known. All three satellites in Table VII are silicate-enriched relative to cosmic abundances, which predict 60% water ice and 40% anhydrous chondritic rock by mass. The enrichment is even more striking if other ices are included (see Anders and Ebihara 1982, for the most recent compilation of cosmic abundances).

The most interesting issue for Titan concerns the relationship of the atmosphere, the surface, and the interior. In terrestrial planets, the relationship

TABLE VII
Comparison of Callisto, Titan, and Ganymede

	Callisto	Titan	Ganymede
Mass (10^{26} g)	1.075	1.345	1.482
Radius (km)	2410 ($\pm$10)	2575 ($\pm$0.5)	2638 ($\pm$10)
Average Density (g cm^{-3})	1.83 ($\pm$0.02)	1.881 ($\pm$0.005)	1.93 ($\pm$0.02)
Ice Mass Fraction [a] (Hydrated Silicates)	0.34	0.30	0.27
Ice Mass Fraction [a] (Anhydrous Silicates)	0.475	0.45	0.42

[a] Assumes $\bar{\rho}_{ice} = 1.2$ g cm^{-3}, $\bar{\rho}$ (hydrated silicates) = 2.5 g cm^{-3}, $\bar{\rho}$ (anhydrous silicates + iron) = 3.5 g cm^{-3}. These are nominal values only.

TABLE VIII
Ice Densities in g cm^{-3} at Low Pressure

H_2O (Ice I$_h$)	0.93	(200 K)
$NH_3 \cdot H_2O$ [a]	0.95	(200 K)
$CH_4 \cdot 5\text{-}3/4\ H_2O$ [b]	0.925	(270 K)
$(CO, N_2) \cdot 5\text{-}3/4\ H_2O$ [b]	1.03	(200 K)
NH_3 [c]	0.86	(200 K)
CH_4 [d]	0.49	(90 K)
N_2 [d]	0.95	(60 K)
CO [c]	0.96	(60 K)

[a] Crystallographic determination. See Landolt-Bornstein 1978.
[b] Based on clathrate systematics; Byk and Fomina 1968. See also the subsection of VI.B on *Clathrates*.
[c] Mellor 1964.
[d] Cheng et al. 1975.

between atmosphere and interior is tenuous and complex because the atmosphere consists of very minor constituents of the planet. In the giant planets, the atmosphere is intimately related to the deep interior. Titan is intermediate, yet possibly closer to the giant planets in the sense that the atmospheric constituents may also be important internal constituents (but not necessarily in the same chemical form). It is also conceivable, however, that Titan's atmosphere is derived from a volatile-rich veneer added at the end of accretion.

B. Constituents

The bulk composition of Titan is probably determined by the range of condensates that formed in a dense, gaseous disk around proto-Saturn (Pollack et al. 1976; Stevenson 1982*a*). Precise identification of these condensates is not possible because of the likelihood of some thermodynamic disequilibrium (sequestration and kinetic inhibition). An example of sequestration would be water ice snowballs immersed in a gas phase containing NH_3 under conditions where $NH_3 \cdot H_2O$ is the preferred condensate. The snowball would develop a thin layer of the compound but might not equilibrate internally. An example of kinetic inhibition would be N_2 and CO formed at high temperatures and quenched into a hydrogen-rich environment in which NH_3 and CH_4 are thermodynamically preferred but kinetically prevented (Lewis and Prinn 1980; Prinn and Fegley 1981). We consider below the behavior of each potentially important constituent or class of constituents. We make no assumption of thermodynamic equilibrium in the following analysis.

Rock. "Rock" is the dominant constituent by mass in any model of Titan, but its detailed chemical make-up is unimportant. However, three questions must be addressed: how much radiogenic heat is produced? are the rock and ice components intimately mixed? are the silicates hydrated?

The rock is probably close to carbonaceous chondritic material (K/U $= 2 \times 10^4$ g/g) although potassium enrichment as in ordinary chondrites (K/U $= 8 \times 10^4$ g/g) cannot be excluded. If these are limiting cases, then the maximum primordial radiogenic heat flow of Titan (assuming that very short-lived isotopes had already decayed) would have been $15-45$ erg cm^{-2} s^{-1}. The present-day heat flow is $4-8$ erg cm^{-2} s^{-1}.

If the ice and silicates are intimately mixed and the silicate inclusions are small, then this heat production may be transported by convection in solid ice (see below). The silicates would then probably still be hydrated. If melting of the ice takes place, or if ice separates from silicates before or during satellite formation, then a rock core forms. Convection in this core requires much higher temperatures and the heat production is more than adequate to dehydrate and partially melt the silicates (as noted for Ganymede by McKinnon [1981]). The issues of mixing and hydration are thus intimately related.

Water Ice. The behavior of satellites containing only rock and water ice has been extensively studied (Huaux 1951; Consolmagno and Lewis 1976; Reynolds and Cassen 1979; Parmentier and Head 1979; Cassen et al. 1980; Thurber et al. 1980; Squyres 1980; Schubert et al. 1981) and the relevant thermodynamic properties of the water-ice phase diagram are well known. The rheological aspects are less well understood, yet crucial for understanding the structure and evolution of Titan. Can solid state convection in water ice eliminate the radiogenic heat? As Poirier (1982) discussed, the effective viscosity of water ice just below the melting point is likely to be low at all pressures. A plausible parameterization for the kinematic viscosity is

$$\nu = \nu_0 \exp \left[A(T_m/T - 1) \right] \qquad (56)$$

with $\nu_0 \sim 10^{14}$ cm^2 s^{-1} for clean ice I at the stresses of interest (the rheology is likely to be nonlinear), $A \sim 25$ (but could be as low as 18 or as high as 30); T_m and T are the melting temperature and actual temperature. According to parameterized recipes for subsolidus convection (reviewed by Schubert 1979), the heat flux transported by convection through a layer of thickness d in the presence of a superadiabatic temperature drop ΔT is given by

$$\phi \cong \frac{k\Delta T}{d} \cdot \left(\frac{\text{Ra}}{\text{Ra}_c}\right)^{1/3}$$

$$\text{Ra} \equiv g\alpha\Delta T d^3/\kappa\nu \tag{57}$$

where k is the thermal conductivity, Ra $>$ Ra$_c$ is the Rayleigh number, Ra$_c$ $\sim 10^3$ is the critical value of Ra for convective onset, α is the volumetric coefficient of thermal expansion, and κ is the thermal diffusivity. Note that ϕ is independent of d. The satellite is envisaged as having an outer conductive shell underlaid by a convective core, the outer part of which is a thermal boundary layer of thickness $\sim d(\text{Ra}_c/\text{Ra})^{1/3}$. The temperature drop ΔT is across this boundary layer and the deeper convecting region is essentially adiabatic. The viscosity ν appearing in Eq. (57) should be evaluated within the boundary layer rather than in the adiabatic region. If T_c is the temperature at the top of the adiabatic zone, then a reasonable choice (Reynolds and Cassen 1979; Friedson and Stevenson 1983) is $T = T_c - \Delta T/2$ in Eq. (56). It is then possible to find the best choice for ΔT by maximizing ϕ at constant T_c. The result for the surface heat flow of Titan in erg cm^{-2} s^{-1}, assuming $A = 25$, is

$$\phi \cong 30\ x^{4/3}\ \exp\left[-8.6\left(\frac{1}{x} - 1\right)\right]\left(\frac{10^{14}}{\nu_0}\right)^{1/3}$$

$$x \equiv T_c/T_m. \tag{58}$$

Since $x \leq 1$ for solid state convection, the heat flux that must be exceeded to initiate melting of pure ice within Titan is ~ 30 erg cm^{-2} s^{-1}. Since the present heat flux is certainly much less than this, an extensive pure liquid water mantle is not possible. The difficulty of sustaining a liquid layer in a water ice satellite was first noticed by Reynolds and Cassen (1979). It is also unlikely that the primordial radiogenic heating rate was sufficient by itself to cause melting, especially if the satellite began cold, since most of the early heating would then go into increasing the temperature rather than providing a surface heat flow. The situation is marginal, however, and only a modest increase in viscosity may suffice to initiate melting. Friedson and Stevenson (1983) show that the effects of suspended silicate particles and dispersion hardening may increase ν_0 or A enough for partial melting to occur by radiogenic heating alone. Complete melting of the water ice can follow, because the gravitational energy available from the downward displacement of silicate-rich material is rapidly released and is more than sufficient to provide the latent heat of melting for all the ice. Despite the uncertainties concerning degree of differentiation, the present-day mantle of Titan is predicted to have a temperature ~ 220 K in its uppermost (Ice I) region, if the rheology is dominated by water ice. This is important for understanding the state of other volatiles in Titan.

Ammonia.   After allowance for the oxygen incorporated in rock, the mole fraction of ammonia in a cosmic NH_3–H_2O mixture is $(18 \pm 6)\%$, assuming that all nitrogen is in the form of NH_3. This cosmic composition is more water-rich than the eutectic composition of $\sim 32\%$ (Rollet and Vuillard 1956). A solid cosmic mixture would, upon heating within a satellite, partially melt into a eutectic fluid coexisting with pure water ice when the temperature reached 173 K. If cooling is rapid enough, as in a volcanic eruption, then a eutectic glass forms (van Kasteren 1973).

We have not found any published measurement of the density of the eutectic liquid but it is estimated at $0.91-0.92$ g cm^{-3} from the known density of 0.89 g cm^{-3} for the $50-50$ mixture at 205 K (Hildenbrand and Giauque 1953) together with the approximate volume additivity and (temperature dependent) coefficient of thermal expansion indicated by the International Critical Tables (Washburn 1928). The eutectic liquid is almost certainly buoyant relative to water ice and would rise diapirically (i.e. as large blobs, viscously deforming the neighboring ice) or along cracks to provide volcanic activity in any icy satellite containing ammonia (Lewis 1971; Stevenson 1982b). Along the liquid-ice I coexistence curve towards pure H_2O, a temperature must be reached at which liquid and ice densities are equal. This is estimated to be $(210 \pm 20$ K), probably lower than the temperature at which a cosmic mixture would be completely fluid. It may therefore be possible to float an ice I layer on a water-ammonia ocean, provided the ocean is not too warm or too cold. It is interesting that Titan's present-day convectively self-regulated internal temperature, calculated above for an ice I rheology is ~ 220 K, which lies in the range of temperatures for which ice and liquid densities become equal. A liquid H_2O–NH_3 mantle in the present Titan is conceivable; unfortunately, the data are not accurate enough to reach a stronger conclusion. The assumption of a water-ice dominated rheology is reasonable since most of the ammonia should be in the coexisting liquid, and small amounts of NH_3 in water ice have little effect on rheology (Nakamura and Jones 1973).

Figure 21 shows the pressure dependence for the congruent melting curve of $NH_3 \cdot H_2O$ and the water-rich eutectic ($\sim 32\%$ NH_3). The superimposed temperature profile for the present Titan is based on Eq. (58).

Clathrates.   A clathrate is a crystal lattice containing cages (or voids) that can accommodate guest molecules. In the clathrate hydrates, the host lattice is provided by water ice and the guest molecules can be chosen from ~ 50

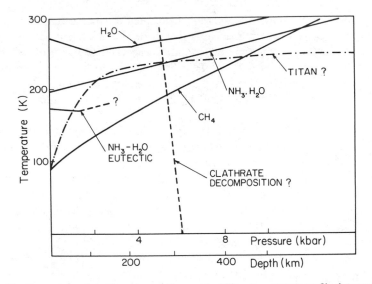

Fig. 21. Comparison of melting curves with a nominal Titan temperature profile that assumes a
water ice-dominated rheology and neglects the effects of phase transitions. All the ices may be
solid deep enough within Titan, except possibly the H_2O-NH_3 mixture. The almost vertical
dashed line is a lower bound to the pressure above which clathrate decomposes. The data for
NH_3-H_2O are from Nicol and Schwake (unpublished). The CH_4 curve is from Cheng et al.
(1975).

possibilities (Byk and Fomina 1968; Davidson 1973; Davidson and Ripmeester 1978) including many cosmically abundant molecules (CH_4, N_2, CO,
CO_2, noble gases, H_2S, PH_3). The possible importance of clathrate hydrates
in the solar system was first emphasized by Miller (1961,1973). They are non-
stoichiometric compounds (but with a maximum ratio of guest molecules to
water molecules of $1:5\frac{3}{4}$) with almost ideally substitutional cage sites (i.e.
more than one species of guest molecule will be accommodated if the coexisting phase is mixed). Their occurrence on the Earth is increasingly well
documented for polar and deep-sea sediment environments (Makogon 1974;
Kvenvolden and McMenamin 1980; Claypool and Kvenvolden 1983; Shoji
and Langway 1982). They form easily in the laboratory in the presence of an
aqueous phase, but with difficulty (kinetically inhibited) if only water ice and
gas are present (Barrer and Ruzicka 1962). Nevertheless, an accretion environment involving collision and grinding of snowballs over a long period of
time can be expected to enable clathrate formation. (Grinding of ice in the
presence of methane gas is the procedure Miller used to make methane clathrate at low temperature.) Notwithstanding the lack of direct evidence for
clathrates elsewhere in the solar system, the frequently expressed doubts
about their formation in nonaqueous environments appear ill-founded. It is
equally important that aqueous environments may have existed on the primor-

dial Titan for which there would be no way of preventing clathrate formation.

Owen (1982*b*) has stressed the possible importance of clathrates in determining Titan's atmosphere. Three aspects of this have been quantified (Lunine and Stevenson 1982*a*, and in preparation) using well-established statistical mechanical formulations for the stability fields and compositions of clathrates (van der Waals and Platteeuw 1959; Parrish and Prausnitz 1972): clathrate formation in a proto-Saturnian nebula, clathrate formation in a primordial NH_3–H_2O ocean coexisting with a N_2–CH_4 atmosphere, and clathrate decomposition in the deep interior of Titan.

Consider first a nebula of cosmic composition and pressure $\sim$ 1 bar, with most of the carbon assumed to be in the form of CH_4 but most of the nitrogen in the form of N_2 (perhaps a substantial overestimate; see Prinn and Fegley 1981). Clathrate forms at $T \lesssim 100$ K and although most of the guest molecules will be CH_4, significant amounts of N_2 and noble gases are retained. If all the water ice is transformed into clathrate, then upper bounds to the total masses of clathrate guest molecules incorporated into Titan are obtained: 1×10^{25} g of CH_4; 7×10^{21} g of N_2; 8×10^{20} g of Ar; 8×10^{19} g of Kr; and $\sim 10^{20}$ g of Xe (the latter being limited only by the finite reservoir rather than by clathrate thermodynamics). These amounts are marginally sufficient, in principle, to supply the present atmospheric mass of Titan ($\sim 8 \times 10^{21}$ g). However, the high relative abundance of CH_4 in this simple scenario is a problem since any oceanic or internal reservoir of methane would also tend to sequester the other molecules. (Nitrogen is very soluble in liquid methane.) If carbon is primarily in the form of CO rather than CH_4, then Titan can retain much larger amounts of N_2 (because CO is a poor clathrate former relative to CH_4) but the CO:N_2 ratio in the present atmosphere should then be of order unity, and it is not. (As Vegard [1930] first noticed, CO and N_2 have very similar physical chemistries.)

Consider, instead, a scenario where the present atmospheric composition and pressure are determined by buffering with clathrates at Titan's surface. It is found that the present atmosphere would be in equilibrium at ~ 200 K with a clathrate that contains roughly comparable amounts of N_2 and CH_4. This corresponds to about the temperature at which water ice can just float on a water-ammonia ocean. Suppose Titan started out hot, because of accretional heating, with a water-ammonia ocean overlaid by a massive and hot CH_4-rich atmosphere in which modest amounts of N_2 have been supplied by photolysis of NH_3 (Atreya et al. 1978). As the ocean and atmosphere cool, clathrate forms in the upper level of the ocean and sinks, scavenging mainly CH_4 from the atmosphere. Substantial amounts of N_2 would not be incorporated into the clathrate until the CH_4 mole fraction drops to ~ 5–10%, thus perhaps explaining the N_2-rich atmosphere. At ~ 200 K, with $P(N_2) \sim 1$ bar and $P(CH_4) \sim 0.1$ bar (these partial pressures are predicted by clathrate thermodynamics), a water-ice crust forms and further massive clathrate formation may be kinetically inhibited. This may explain the present atmosphere.

However, the ongoing photolysis of CH_4 in Titan's atmosphere (Strobel 1974,1982) prompts us to consider continuing sources of CH_4 and the possibility of clathrate decomposition at depth. A lower limit to the clathrate decomposition pressure can be obtained by assuming that the host lattice does not contract with pressure, but allowing for the high density phases of coexisting pure water ice. This lower bound is found to be $\sim$ 6 kbar, but a more precise estimate is $\sim$ 10 kbar (Lunine and Stevenson, in preparation). High pressure data, which exist only for the tetrahydrofuran clathrate hydrate (Ross and Andersson 1982), suggest a high pressure modification of the clathrate structure which may greatly extend the range of pressures in which they are stable. Nevertheless, it seems likely that clathrates decompose at the higher pressures ($\sim$ 30–40 kbar) encountered deep within Titan, because the densest water ices (ices VI–VIII) are "self-clathrates" (Kamb 1973). The decomposition will be aided by radiogenic heating; the fate of the released volatiles is discussed below.

More Volatile Ices and Liquids.   Even if clathrates form, there is insufficient water ice to incorporate the cosmic abundance of all clathrate-forming guest molecules. At lower temperatures ($\sim$ 40 K for $P \sim$ 1 bar, a plausible pressure for a proto-Saturnian nebula), ices such as CH_4 and perhaps even N_2 or CO can form. These molecules readily form substitutional (nonstoichiometric) ice compounds. For example, it is not possible to have pure methane ice in the presence of any nitrogen vapor (Omar et al. 1962; see also Hiza et al. 1975 for a bibliography of phase diagrams). They also readily form liquid mixtures.

In the interior of Titan, a volatile ice can remain indefinitely in fine, disseminated solid form if the actual temperature (determined by subsolidus convection of the water ice-dominated matrix) is low enough. Figure 21 shows that this may be possible for CH_4 at $P \gtrsim$ 10 kbar because of the steepness of the methane melting curve. If N_2 is also present, the situation may be complicated because CH_4–N_2 may form a low melting point eutectic system. If macroscopic amounts of CH_4 and/or N_2 are present as either solid or liquid, then they can rise diapirically from the interior. The ascent velocity is given by Stokes' formula:

$$v \simeq \frac{2}{9} g\Delta\rho \frac{R_d^2}{\mu} \qquad (59)$$

where $g \sim$ 150 cm s^{-2} is the gravitational acceleration, $\Delta\rho \sim$ 0.1–0.5 g cm^{-3} is the density difference between diapir and surrounding ice, R_d is the diapir radius and μ is the self-regulated viscosity of the surrounding ice. For $\mu \sim 10^{19}$ P, appropriate to the present-day interior (Eqs. 56, 58), the rise time to travel 10^3 km is $\sim 3 \times 10^7$ (1 km/R)^2 yr. Near the surface of Titan, the viscosity must increase enormously (perhaps to $\sim 10^{26}$ P or even more) and the diapir may be unable to reach the surface directly. However, cracks can nucleate (Stevenson 1982b) and the CH_4 and/or N_2, now necessarily fluid, can migrate to the surface in volcanic events.

If a methane-rich ocean accumulates at Titan's surface then the phase diagram of CH_4–N_2 (Omar et al. 1962) dictates that it must contain a substantial amount ($\sim 20\%$) of N_2. No solid phase (e.g. solid CH_4) is possible on Titan because the presence of N_2 depresses the freezing point. The closeness of the present surface conditions of Titan to the triple point of methane must be a coincidence because nitrogen is not a passive bystander. If the total mass of the ocean is much greater than that of the atmosphere (i.e. depth $\gtrsim 100$ m) then it is an excellent thermostat, barostat, and regulator of atmospheric composition. This assertion depends on the reasonable assumption that the ocean is well mixed on a geologic time scale. Note, however, that this requires a large oceanic reservoir of N_2, which is much harder to supply from any primordial reservoir than a large reservoir of CH_4. For this reason alone, a deep methane ocean would be difficult to explain. The recent proposal (Lunine et al. 1983) of an ethane-rich ocean avoids this difficulty, the large amount of ethane being supplied by photolysis of CH_4. Consideration of tidal dissipation (Sagan and Dermott 1982) indicates that a global ocean would have to be $\gtrsim 400$ m in depth. Any ocean would also be an important reservoir of many other molecular species, most notably nonpolar species (noble gases, hydrocarbons) and weakly polar species (e.g. CO). This may have testable implications for the photochemical processes in the atmosphere.

C. Formation and Primordial State

There are three possible origin scenarios: cold accretion (meaning the temperature rise during formation is negligible), hot accretion in the absence of a dense gas phase, and hot accretion in the presence of a dense gas phase. Each can be further subdivided into heterogeneous and homogeneous accretion, making six possibilities altogether. A dense gas phase is an accretional environment in which the energy of accretion must be eliminated by convection through an optically thick, gravitationally bound primordial atmosphere. Heterogeneous accretion is defined here to mean the formation of a rock core before the accumulation of ice.

In any formation process, a large gravitational energy is involved. The total gravitational energy of formation is $\delta GM^2/R$ where the body has mass M and radius R, and where $0.4 \lesssim \delta \lesssim 0.6$ for plausible interior models of Titan. Most of this energy is converted into heat, and for Titan it corresponds to ~ 1600 K temperature rise if there were no losses. If this energy were lost by radiation in a time τ, then the surface temperature during accretion, in the absence of insolation, would have been $\sim (700 \text{ K})(10^4 \text{ yr}/\tau)^{1/4}$. If the energy were lost by convection, the primordial atmosphere would have to share the same specific entropy as the nebula (Lunine and Stevenson 1982b). Aside from latent heat effects, the amount of the temperature increase between the nebula and the primordial satellite is then

$$\Delta T \equiv \int_R^\infty - \left. \frac{dT}{dr} \right|_{ad} dr = \int_R^\infty \frac{g \, dr}{C_{p,g}} = \frac{GM}{RC_{p,g}} \tag{60}$$

where $dT/dr|_{ad}$ is the adiabatic temperature gradient and $C_{p,g}$ is the specific heat of the gas. For the present mass and radius of Titan, $\Delta T \sim 300$ K.

These considerations argue against cold accretion. First, a cold accretion must be devoid of gas because of the large warming effect implied by Eq. (60). Second, cold accretion must be slow so that the radiative equilibrium temperature is low. This is difficult to achieve because accretion of satellites is expected to be rapid (Safronov and Ruskol 1977; Harris 1978), much faster than for planets. Third, analysis of impact processes strongly suggests that a significant retention of energy occurs because planetesimal sizes are likely to be large compared to thermal skin depths (Safronov 1972; Kaula 1980).

Schubert et al. (1981) considered gas-free accretion of the Galilean satellites and parameterized the accretional temperature rise by a numerical constant h, the fraction of accretional energy retained as internal heat. Since Titan is so like Ganymede and Callisto in bulk characteristics (Table VII), similar results apply. Even for $h \sim 0.15$ and an ambient temperature ~ 100 K, a large fraction of Titan's water ice mass would have undergone accretional melting. The post-accretion state, after settling of a silicate layer, is a rock core overlaid by a rock-ice mixture overlaid in turn by an outer mantle of almost pure ice. It is unlikely that h is small enough for Titan to accrete in undifferentiated form, especially if it contains low melting point ices.

A similar conclusion applies for accretion in a gaseous environment (Lunine and Stevenson 1982a,b). However, gas-free and gas-rich scenarios have important differences concerning the surface environment. In gas-rich accretions, a massive atmosphere (up to hundreds of bars surface pressure) always occurs and can last a long time (up to $\sim 10^7$ yr), during which the environment is warm enough for large amounts of NH_3 to be converted to N_2. This time scale can be roughly estimated by assuming a water-ammonia ocean at ~ 200 K overlaid by an atmosphere that radiates at, e.g., ~ 100 K, maintained at this temperature by heat from below. The time to cool an ocean (of mass $\sim 5 \times 10^{25}$ g) by 20 K is then $\sim 10^7$ yr. In gas-free models, no immediate atmospheric formation is likely unless very volatile ices (CH_4, N_2, CO) are included in the accreting material. The accretion time is probably short enough ($\lesssim 10^6$ yr) to ensure significant radiative surface temperatures (~ 200 K) and volatilization of these ices. (However, the resulting gases might be blown off during accretion.)

The issue of heterogeneous vs. homogeneous accretion is probably unimportant since it seems likely that differentiation can proceed during accretion anyway. The presence of low melting point ices argues for at least partial differentiation. The complete differentiation of rock from water ice is less certain but probable because of the enhanced viscosity of rock-ice mixtures (Friedson and Stevenson 1983).

The accretion discussed here concerns relatively low velocity impacts (comparable to Titan's escape velocity) associated with Saturn-orbiting planetesimals. Sun-orbiting bodies (comets) may have bombarded Titan over an extended period of time (10^8–10^9 yr) with much higher impact velocity. It can be shown (von Herzen and Stevenson 1982, unpublished) that the net effect of high velocity (v $\gtrsim$ 10 km s^{-1}) impacts on Titan is loss of mass (especially atmosphere). Comets are not a good atmospheric source for a body the size of Titan, except perhaps for small ones (radius $\lesssim$ 1/2 km) that can be "stopped" by a preexisting atmosphere.

D. Models, Predictions and Problems

Table IX is a matrix of possible Titan models, categorized according to formation process and the list of constituents. Although qualitative, the assessments given in the matrix are firmly based on the preceding discussion. Clearly, scenarios involving accretion of all major condensables, including clathrates or methane ice, are desirable, but it might also be possible to explain Titan with just a volatile-rich veneer overlying a water-ammonia ice + rock interior. Figure 22 is a plausible schematic pie diagram for Titan.

Table X in the next section is a list of crucial observations needed for real progress in our understanding of Titan as a whole. Most of them require *in situ* sampling. These questions are further discussed below.

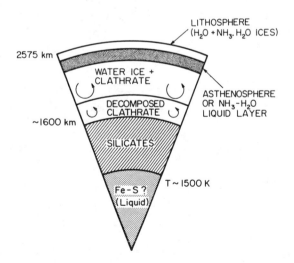

Fig. 22. Schematic diagram of a possible Titan interior. A hot, dehydrated rock core is likely, as is partial melting of NH$_3$-H$_2$O, but the detailed arrangement of ice layers is speculative at present. Convection prevails everywhere except in the outermost shell.

TABLE IX
Matrix of Titan Models

	Rock + H_2O	Rock, H_2O, NH_3	Rock, H_2O, NH_3 + volatile rich veneer [a]	Rock, H_2O, NH_3 + clathrates or methane [b]
Cold accretion	Final state like Callisto. Atmosphere unexplained.	NH_3–H_2O volcanism after a few × 10^8 yr, but atmosphere unexplained.	Atmosphere possible but CH_4/N_2 ratio has no obvious explanation.	Prompt outgassing of CH_4 but N_2/CH_4 is likely to be too small.
Hot, gas-free accretion	Final state like Ganymede. Atmosphere unexplained.	Prompt NH_3–H_2O volcanism and ocean but N_2 buildup is difficult. Atmosphere unexplained.	Atmosphere possible but N_2 production from NH_3 vapor may be difficult. CH_4/N_2 ratio unconstrained?	May be difficult to supply enough N_2. CH_4/N_2 ratio unexplained.
Hot, gas-rich accretion	Final state like Ganymede. Atmosphere unexplained.	Primordial atmosphere overlying NH_3–H_2O ocean. N_2 production but no source of CH_4.	N_2 production possible and CH_4 supply adequate. N_2/CH_4 ratio explained by clathrate production in NH_3–H_2O ocean?	Massive primordial atmosphere. Adequate N_2 made. N_2/CH_4 ratio explained by clathrate production in NH_3–H_2O ocean?

[a] Containing methane.
[b] In large amounts, as dictated by cosmic abundances.

VII. QUESTIONS FOR THE FUTURE

A close flyby of Voyager 1, supplemented by the previous decade of studies from Earth, has provided a remarkably complete picture of Titan and its atmosphere. Nevertheless, many questions remain and many more can now be asked for the first time. For some, additional theoretical work is required; for many others more observations and measurements are needed.

Although nitrogen is the most abundant gas in the atmosphere, the amounts of the next two, methane and argon, are not well known. Argon can probably be measured only by instruments on an atmospheric probe. Optical absorptions due to methane are prominent, but quantitative interpretation is frustrated by the absorbing and scattering haze. If methane clouds form in the troposphere, the abundance below these clouds must be greater.

The absorbing layers shown in Fig. 7 are most likely molecular, but the composition is unknown, as well as the process that forms layers instead of uniform distributions.

The globally-averaged heat balance of the surface and troposphere is well described by the simple model given in Sec. II, but the far-infrared opacity source is not fully identified. Methane clouds seem to be the best choice, but this is far from certain. The meteorological processes that maintain the global uniformity are discussed below. Above 200 km the energy balance is not well understood and global uniformity is not assured; although this region can be modeled as nearly isothermal at $\sim$ 170 K, there is probably considerable structure about this mean.

We seem to have a good working description of methane photochemistry and the production of more complex compounds and smog particles, although many details remain to be understood. Here again, we are faced with the uncertainty in the methane abundance, but probably more significant is the rate at which methane is transported upwards to replace that lost by photolysis. The faster this transport (modeled as an eddy diffusion process), the less is the methane depleted at high altitudes. At even higher altitudes we have the puzzle of the ultraviolet glow, confined to the day side (but too bright to be produced by solar radiation alone) and located at optical depth much less than unity. In Sec. III it is argued that magnetospheric charged particles are impinging, but are held at high altitudes by mirroring in the magnetic field. This field is of magnetospheric origin but modified by currents induced in Titan's ionosphere. This type of process can be easily studied by instruments on an orbiter, preferably of Titan but possibly of Saturn as long as it encounters Titan frequently. The principal measurements would be of Titan's ionosphere, the magnetospheric plasma, and the magnetic field. There are known effects on the magnetosphere as well (Sec. III; chapter by Scarf et al.) which would be elucidated by the same data.

The observed H in the torus is undoubtedly accompanied by H_2 and N, and perhaps other atoms and molecules. The radial and vertical distributions in the torus have been modeled only in two unrealistic limits, collisionless and fully collisional. Treatments of heating have been equally crude. While there is little prospect for more data in the near future, there is a need for a realistic treatment of sources (including Saturn, rings, and other satellites), collisions, and sinks.

Despite the large amount of information collected on recent missions regarding the single-scattering properties of Titan's aerosols, many important questions concerning their nature and spatial distributions remain. We know that the particles at the top of the atmosphere are large enough to produce strong forward scattering, but we still do not have calculations or laboratory measurements for particles with the proper size or shape to give the very large polarizations which Titan's aerosols produce along with strong forward scattering. We suspect that the size of the aerosols must increase with increasing depth into the atmosphere, but until we can relate the single-scattering phase matrix of the appropriate nonspherical aerosols to their size and shape, it will continue to be difficult to convert the optical observations of Titan to constraints on the variation of aerosol size and optical depth (to say nothing of composition) with pressure. Laboratory measurements of nonspherical and porous particles might reconcile the seemingly incompatible properties inferred. We know that the optical depth of the aerosols permits greater penetration of sunlight at red than at blue wavelengths, but without estimates of the aerosol size at various pressures, it is difficult to evaluate the penetration of solar and thermal radiation through the atmosphere across a wide spectral range. The availability of accurate temperature profiles from the Voyager radio occultations and the IRIS infrared spectra offer a way to constrain the aerosol and cloud particle optical depths below the cloud tops, but such work is still at a relatively early stage and must be pursued further.

Physical models of the photochemical formation, growth, and fallout of the aerosols also need further exploration. Especially useful would be better estimates for the mass production rate of the aerosols, and some estimate for the charge likely to be carried on the particles which strongly determines their ability to stick together upon collision. The detached haze layers and the dark polar regions seen in the Voyager images suggest several additional questions. How are the aerosols in the polar regions related to the aerosols seen at lower latitudes and are these seasonal effects? Are the low latitude aerosols produced by ultraviolet photolysis and transported to high latitudes, or are the aerosols primarily produced by enhanced energetic particle fluxes near the poles before being carried to lower latitudes? Are the aerosols in either region of the same composition as the small ultraviolet absorbing aerosols seen in the stratospheres of Saturn and Jupiter? In the context of evolution and stability of Titan's stratosphere, what are the feedback processes among inversion temperature, aerosol formation rate, condensation rate, solar heating, and vertical mixing?

Questions concerning the optical depth, vertical extent, and size distribution of particles in the methane cloud also remain to be answered. The IRIS data suggest that a methane cloud is. required to provide the needed infrared opacity, but the cloud cannot have too great an optical depth if it is to permit penetration of sufficient solar flux to the surface to support the weak but significant greenhouse effect. In addition to the effect on radiation fluxes, methane condensation can release latent heat and affect Titan's meteorology. Measurements *in situ* may be needed, but limb scans in the $200-600$ cm^{-1} region would also be useful.

The existing treatment of meteorology is based entirely on observed temperatures; wind measurements are completely lacking. Although almost any further measurements would be valuable, the following list represents the more important ones:

1. Determination of the meridional thermal structure, particularly in the lower stratosphere and upper troposphere. This might be best accomplished by repeated radio occultations from a Titan orbiter.
2. Seasonal coverage of the stratosphere. This includes observing not only temperature, but also chemical constituents such as C_2H_6 and C_2H_2, in order to better understand the radiative processes operating. Seasonal monitoring of zonal winds is also important (see no. 3).
3. Direct confirmation of a cyclostrophic zonal wind system. A descent probe would unambiguously determine the vertical profile of u. However, longer term seasonal coverage of the stratospheric winds is also desirable. This might be accomplished through the use of a heterodyne spectrometer with sufficient spatial resolution to determine accurately the Doppler shifts of infrared emission lines in the stratosphere (e.g., those of C_2H_6 near 830 cm^{-1}).
4. Search for eddies or waves with a thermal signature. The limited coverage by Voyager IRIS permitted only crude upper limits to be assigned to the variation of temperature with longitude. Such eddies are potentially important for the meridional transport of both heat and momentum. The presence or absence of such eddies or waves in Titan's atmosphere would have important implications for the maintenance of global cyclostrophic wind systems, including those on Venus.
5. Determination of the vertical structure and composition of tropospheric clouds and the nature of the surface. These are important for assessing the role that moist processes involving CH_4 play in Titan's meteorology. Cloud measurements have been discussed above. The surface can be examined by radar observations (see e.g. Sagan and Dermott 1982; Eshleman et al. 1983). Even if a near-global ocean is present, there could be islands or even continents.

Concerning the relationships of interior, surface, and atmosphere, the most important unresolved questions are:

1. Why is the atmosphere nitrogen-rich?
2. What dictates the abundances of other atmospheric constituents (especially methane and argon)?
3. Why is the atmospheric pressure $\sim$1.5 bar?
4. How is CH_4 continuously supplied to offset the ongoing photolysis in the atmosphere?

Here are some possible answers and comments to these questions.

1. The most probable source of N_2 is by photochemical production from NH_3 in a hot primordial atmosphere. Direct supply of N_2 from a clathrate is difficult because it tends to provide an overabundance of CH_4. This CH_4 cannot be hidden in an ocean because any massive CH_4 ocean necessarily incorporates most of the N_2.
2. The relative abundances of CH_4 and N_2 might be dictated by the formation of clathrates near the surface of a primordial NH_3–H_2O ocean. These clathrates preferentially scavenge CH_4 until N_2:$CH_4 \sim 10$:1. This process would also fractionate ^{36}Ar in the direction of increasing the atmospheric abundance. The magnitude of this fractionation has not yet been predicted.
3. The total pressure of $\sim$1.5 bar may reflect the vapor pressure coexisting with a CH_4–N_2 clathrate at the time of closure at $T \sim 200$ K when a primordial ammonia-water ocean froze over.
4. The continued resupply of CH_4 is least well understood. It could come from an ocean (but this also implies a large reservoir of N_2) or from continued outgassing.

TABLE X
Diagnostic Observations

Possible Observation	Implication
Substantial ^{36}Ar in atmosphere and relative overabundance of heavier noble gases (Kr, Xe).	Evidence for primordial clathrates and efficient fractionation relative to CH_4.
Presence of an ocean.	Large reservoir of CH_4. Reservoir of N_2 depends on CH_4:C_2H_6 ratio. Efficient atmospheric buffer and supplier of CH_4.
$^3He/^4He$ and $^{36}Ar/^{40}Ar$.	Measures of outgassing efficiency and role of primordial vs. outgassed volatiles.
Improved deuterium abundance.	May indicate primordial NH_3–H_2O ocean.
Surface topography, including impact structures (if no ocean).	Indicator of substrate rheology (water ice vs. H_2O–NH_3) and tectonics.

In the final analysis, these issues can only be understood by further ob-servations, especially *in situ* sampling. Table X lists crucial observations and the ways they could be interpreted.

Acknowledgments. The work of Hunten, Strobel, Samuelson, and Flasar was supported by NASA's Planetary Atmospheres Program. Tomasko thanks P. H. Smith and R. West for many useful discussions regarding results of the imaging science and photopolarimeter experiments on Voyager. Ste-venson acknowledges support by NASA's Geophysics and Geochemistry Pro-grams and thanks J. I. Lunine for useful discussions.

TITAN'S MAGNETOSPHERIC INTERACTION

FRITZ M. NEUBAUER
Universität zu Köln

DONALD A. GURNETT
University of Iowa

JACK D. SCUDDER and RICHARD E. HARTLE
Goddard Space Flight Center

Voyager 1 encountered Titan on 12 Nov. 1980 around 0540:20 spacecraft event time when Titan was located in Saturn's outer magnetosphere. In this chapter we review the plasma, magnetic field and plasma wave observations and analyses from the plasma dynamist's point of view. In particular we discuss the following: (1) The incident Saturn magnetoplasma was characterized by Alfvénic and sonic Mach numbers 1.9 and 0.57, respectively, and a fast magnetohydrodynamic Mach number of 0.55; (2) The incident plasma was found to interact with the atmosphere of Titan rather than an internal magnetic field leading to the formation of an induced magnetosphere by mass pickup and associated field-line draping in a similar way to that of Venus; (3) In the tail of the induced magnetosphere a central region characterized by a strong reduction of ≥ 700 eV electrons and substantial mass loading due to the addition of N^+ or N_2^+/H_2CN^+ associated with strong field-line draping could be distinguished from the surrounding regions of H^+ mass loading and relatively weak magnetic field-line draping; (4) In the central, strong draping region a four-lobe magnetotail has been identified where the lobes can either be north or south of the neutral sheet of the tail and be due to draping over either the dayside or nightside hemisphere of Titan; (5) An upper limit of 7×10^{20} Gauss cm^3 has been derived for an internal rotationally aligned magnetic dipole moment; (6) A loss rate of 10^{24} N^+ or N_2^+/H_2CN^+ ions per second has been estimated from observations of the tail.

Titan, as every planet or satellite in the solar system, will be found to be ultimately enclosed by a streaming magnetized plasma if an observer were to move outward from its center through its interior and subsequently through its extensive atmosphere (chapter by Hunten et al.). In this chapter the interaction of Titan with the incident plasma flow is discussed. It is especially interesting for various reasons. In the first place, this interaction determines an important part of the mass balance of the atmosphere. In addition to the losses and gains of atmospheric species through the surface (e.g. by condensation and evaporation etc.) and the gain of mass by incoming particulate matter, the mass balance of the atmosphere is determined by losses (and sometimes perhaps gains). Losses due to the electrodynamic interaction of the satellite with its plasma environment, as well as classical Jeans escape of neutrals. The losses at the same time represent a gain for the surrounding plasma. If Titan is located in Saturn's magnetosphere, it will contribute to the plasma population of the magnetosphere (see chapter by Schardt et al.).

Another interesting consideration is Titan's magnetospheric interaction as a special case of a fundamental problem in space plasma physics, i.e. the interaction between a magnetized plasma flow and a planet or satellite. The detailed physics of this interaction depends on the properties of the incident flow and the characteristics of the body. Several cases have been studied in the past: the case of a planet without an appreciable atmosphere but with a substantial global magnetic field in the case of Mercury; the case of a planet with an atmosphere but without a substantial magnetic field like that of Venus; the case of a planet with an atmosphere and a magnetic field like Earth's; and the case of neither an atmosphere nor a global magnetic field like that of the Earth's moon, to mention only a few.

It has been known for almost four decades that Titan possesses a substantial atmosphere (Kuiper 1944). The Voyager mission has now given us a detailed picture of this atmosphere (see chapter by Hunten et al.; *Science* 10 Apr. 1981; Smith et al. 1982; Strobel and Shemansky 1982; Lindal et al. 1983). The question is then whether the incident plasma interacts directly with Titan's atmosphere or whether the atmosphere is, so to speak, protected by a substantial internal magnetic field diverting the plasma flow, before it reaches the atmosphere. For favorable conditions in Titan's interior constrained by general cosmochemical considerations and the observed bulk density and size of the satellite, one can derive a global magnetic field of 100 nT at Titan's equator (Neubauer 1978). One of the objectives of the close encounter of Voyager 1 with Titan was to investigate for the first time the detailed nature of Titan's interaction with the surrounding plasma flow and to answer the question of the existence of an internal field. Our present observational knowledge of Titan's interaction is based almost exclusively on the data from the single encounter of Voyager 1. However, Titan's orbital radius of 20.2 Saturn radii (R_S) is such that the satellite may be located in the solar wind, in the magnetosheath of Saturn, or in Saturn's magnetosphere depending on the dynamic state of the magnetosphere and Titan's orbital phase (Acuna and Ness 1980). It therefore experiences widely varying plasma conditions in the incident flow (Wolf and Neubauer 1982). Nevertheless, many significant results of general validity have been derived from the Voyager 1 encounter.

Our review of the electrodynamic interaction between Titan and Saturn's magnetosphere based on Voyager 1 encounter data and general physical principles will begin with a brief description of Voyager 1 experiment characteristics necessary for the synthesis of the observations (Sec. I). This is followed by a summary of the encounter characteristics (Secs. II and III). We shall then consider the possibility of an internal magnetic field (Sec. IV). The extensive discussion of Titan's induced magnetosphere (Sec. V) will constitute the main body of the chapter.

I. EXPERIMENT DESCRIPTIONS

Several experiments contributed to the observations defining the plasma flow associated with Titan. *In situ* measurements were performed by the magnetometer, plasma science, plasma wave, planetary radio astronomy and low-energy charged particles experiment onboard Voyager 1. The ultraviolet spectrometer experiment also contributed to the overall picture. Results have been published in special journal issues of *Science* (10 Apr. 1981) and *Journal of Geophysical Research* (1 Mar. 1982). We shall emphasize here the use of the magnetic field, plasma and plasma wave data. For detailed descriptions of the experiments we refer the reader to a special issue of *Space Science Reviews* (Nov. Dec. 1977).

During Titan encounter the magnetometer experiment was operating at its highest sampling rate, one vector measurement every 0.060 s and its highest sensitivity, 8.8 nT full scale, $\pm$ 0.0022 nT quantization step size. The accuracy is estimated to be $\pm$ 0.05 nT.

The Voyager plasma instrument consisted of four potential modulated Faraday cups: three (A, B, C) arranged as a triad equally disposed about an axis pointing in the earthward (nearly solar) direction, supplemented by a single cup (D) whose cone of sensitivity had its axis almost perpendicular to that of the triad. The D-cup axis pointed towards the direction of cororational flow in Saturn's magnetosphere for the inbound part of the Saturn encounter trajectory. All four cups made both high- and low-energy resolution measurements of the ion velocity distribution functions; the D-cup measured ion and electron distribution functions alternately. The energy per charge range for both electrons and ions extended from 10 eV to 5950 eV with contiguous energy converage. It was broken into two intervals from 10 to 140 eV (E_1-mode) and 140 to 5950 eV (E_2-mode) for electrons. A complete measurement cycle with electron and low energy resolution ion measurements (L-mode) was carried out every 96 s, and for high energy resolution ions every (M-mode) 2 $\times$ 96 s. The E_1-, E_2- or L-mode in order E_1, L, E_2 were completed in 3.84 s each with 20.2 s between two consecutive measuring intervals. Note the substantial gaps in time coverage particularly for the electrons. The M-mode was not used for the Titan work.

The plasma wave experiment used the two orthogonal 10-m electric antennas also used by the planetary radio astronomy experiment. A 16-channel spectrum analyzer with four channels per decade from 10 Hz to 56.2 Hz provided the necessary frequency resolution. One scan through all channels was performed every 4 s.

II. ENCOUNTER GEOMETRY

The Titan encounter occurred on 12 November 1980 at a minimum flyby distance of 6969 km from Titan's center at Saturn local time 1330. Closest approach occurred at 0540:20 spacecraft event time (SCET) in UT and the crossing of the orbital plane at Titan at 0543:07 SCET. The fact that the orbital plane crossing of Titan occurred very near closest approach turned out to be very useful for our investigations.

The flyby trajectory relative to Titan is shown projected on the XY-plane in Fig. 1. The axes of our coordinate system X, Y, Z with its origin at the center of Titan point towards the direction of ideal corotational flow, away from Saturn and perpendicular to the orbital plane, respectively. The flyby velocity was 17.3 km s^{-1}. Because of the 8°7 tilt of the trajectory with respect to the orbital plane, the velocity components in the XY-plane and along the $(-Z)$ direction were 17.1 and 2.6 km s^{-1}, respectively. The flyby data therefore cover only a few minutes. Figure 1 also shows two ideal wake regions under the assumption of a flyby in the magnetosphere with ideal corotational flow and flyby in an ideal solar wind with flow radially from the Sun. The terminator is also shown for Titan. Since the encounter occurred a few months after Saturn's equinox, the northern pole of Titan was sunlit.

III. PROPERTIES OF INCIDENT PLASMA FLOW DURING TITAN ENCOUNTER

Because the encounter occurred close to local noon an interesting question was where Titan would be found: in the magnetosphere, in the magnetosheath or even in the solar wind. It turned out, that Titan was well within the outer magnetosphere of Saturn with Saturn's magnetopause crossings observed 5 times between 22.8 and 23.7 R_S (Ness et al. 1981; Bridge et al. 1981a; Gurnett et al. 1981a). The magnetic field data displayed in Fig. 2 indicate a magnetic field of magnitude $B = 5$ nT and a direction perpendicular to the XY-plane in the region surrounding Titan's plasma wake. Table I gives a summary of magnetic field and plasma properties considered to be most representative of the incident flow conditions. They have been derived from ob-

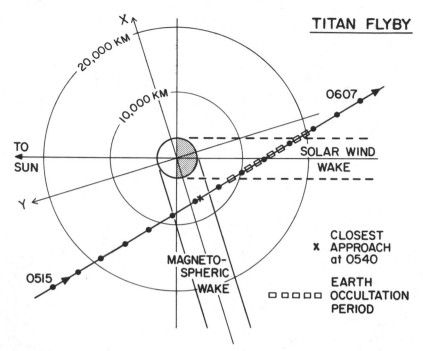

Fig. 1. Voyager trajectory past Titan projected into Saturn's equatorial plane with hypothetical magnetospheric and solar wind wake regions also indicated (from Ness et al. 1982b). The coordinate axes X and Y are also shown.

servations just outside the interaction region. For example, the ion results constitute a synthesis of the data points marked by one and eight shown in Fig. 6 in Sec. V.A. The velocities should be compared with the velocity of ideal corotational flow at Titan's orbit, i.e. 200 km s^{-1}. The ion and electron measurements (Hartle et al. 1982) yield densities which are consistent with the plasma wave electron densities (Gurnett et al. 1982b). The variability is considered small enough for our following discussions.

Although the ion fractions and temperatures may still change to some extent after further analysis, the general picture which Table I provides is not expected to change. We note that the pressure is dominated by the plasma pressure in the incident flow. The two characteristic speeds in plasma flow problems, i.e. the Alfvén speed V_A and the speed of sound V_s are such that the respective Mach numbers are 1.9 and 0.57. Choosing the lower-limit electron density $n_e = 0.1$ cm^{-3} and no N$^+$-ions $M_A \approx 0.5$. The expression for the sonic speed corresponds to the result of collisionless theory for perpendicular propagation. It is interesting to note that the combination of Mach numbers derived above has not been encountered before in any study of the interaction between a planet or satellite and its magnetoplasma environment.

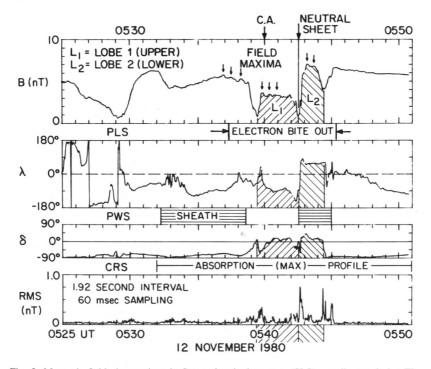

Fig. 2. Magnetic field observations in Saturn longitude system (SLS) coordinates during Titan close approach (C.A.). B, λ and δ are magnitude, longitude and elevation, respectively. For definition of SLS coordinates see Desch and Kaiser (1981a). The values plotted are 1.92 s averages. Also shown are the Pythagorean mean rms values. Significant intervals in the plasma science (PLS), plasma wave (PWS) and cosmic ray experiment (CRS) observations are also indicated (from Ness et al. 1982b).

Whether a fast magnetohydrodynamic (MHD) bow shock is formed depends on the fast Mach number which turned out to be 0.55. Hence no bow shock is expected. An independent test of this result will be presented in Sec. IV. The flow can nominally be characterized by superalfvénic, subsonic and sub-fast. Since the Mach numbers are relatively close to one, it is useful to speak of transalfvénic and transsonic flow. The physics of such cases is particularly complex since none of the three terms in the plasma momentum balance equation, i.e. the acceleration term, the plasma pressure gradient or the magnetic force, can be neglected. The problem is further complicated, because the ion gyroradii are comparable to the radius of Titan R_T. Some of the consequences of these results will be discussed in Sec. V. The deviation of plasma flow leading to an inward flow component may be interpreted in terms of a temporary inward motion of Saturn's magnetopause during the Titan encounter.

TABLE Ia
Incident Magnetoplasma Properties

Magnetic field magnitude B	5 nT
Direction	approximately perpendicular to orbital plane
Plasma flow speed	80–150 km s^{-1}
Plasma speed adopted V	120 km s^{-1}
Plasma flow direction	20° from corotational direction inward
Proton number density n_p	0.1 cm^{-3}
N$^+$ number density n_{N^+}	0.2 cm^{-3}
Electron temperature $k\,T_e$	200 eV
Proton temperature $k\,T_p$	210 eV
N$^+$ temperature $k\,T_{N^+}$	2.9 keV
Electron number density n_e	0.3 cm^{-3}

TABLE Ib
Derived Quantities*

Mass density ρ	2.9 amu cm^{-3}
Total plasma pressure p	1.1×10^{-9} dynes cm^{-2}
$\beta = p/(B^2/2\mu_0)$	11.1
Alfvén speed V_A	64 km s^{-1}
Sound speed $V_s = 2p/\rho$	210 km s^{-1}
Fast magnetoacoustic speed $(V_A^2 + V_s^2)^{1/2}$	220 km s^{-1}
Alfvén Mach number $M_A = V/V_A$	1.9
Sonic Mach number $M_s = V/V_s$	0.57
Fast Mach number $M_f = V/(V_A^2 + V_s^2)^{1/2}$	0.55
Gyroradius of newly created proton	248 km
Gyroradius of thermal proton	413 km
Gyroradius of newly created N$^+$ − ion	3470 km = 1.35 R$_T$
Gyroradius of thermal N$^+$ ion	5790 km = 2.25 R$_T$
Gyroradius of thermal electron	9.5 km
Motional electric field $B \cdot V$	0.6 volt km^{-1}

*Titan radius used R$_T$ = 2570 km.

IV. QUESTION OF AN INTERNAL MAGNETIC FIELD OF TITAN

Since the flyby lasted only a few minutes, we use the magnetic field data for a first overview to ensure high spatial and temporal resolution. These data are shown in Fig. 2 for a 30-min interval including the time of closest approach 0540 UT. Close inspection of the highest time resolution data (16 2/3 vectors per second) shows that no information is lost by our choice of 1.92 s averages. The most conspicuous features of Fig. 2 are the various dips in magnetic field magnitude. The deep minimum at 0529 UT and the minor one at 0533 UT are distinct from the sharp minima around closest approach at 0539:29, 0542:29 and 0544:21 UT. At least the deep minimum at 0529 UT has probably nothing to do with the actual interaction. First, the magnetic field direction does not differ appreciably from the direction of Saturn's dipole field at $\sim \delta \approx -90°$ whereas the sharp minima around closest approach are characterized by directions $\delta \approx 0°$, i.e. in Saturn's equatorial plane. Second, the former minimum is not uncommon in the outer magnetosphere of Saturn after the last magnetopause crossing. In fact, the magnetic field behavior typical of the outer magnetosphere of Saturn observed during the Voyager 1 encounter continues to at least 0532 UT and resumes at about 0546 UT. We note that the rapid variations of λ in the earlier part of the interval shown in Fig. 2 are due to magnetic field vectors being close to the south pole of the magnetic vector coordinate system.

Independent evidence to define the specific interaction region can be obtained from the electron data (Bridge et al. 1981a; Hartle et al. 1982). Figure 3 shows log f_e as a function of electron velocity, where f_e is the velocity space electron distribution function for times from 0521 UT to 0554 UT. We note that the measurements in electron mode E_2 begin 44.2 s after the end of E_1 the time which is used in the plot. Two features in Fig. 3 are significant. At low energies we observe a maximum in log f_e whereas at high energies log f_e displays a minimum with increasing steepness of the flanks of the minimum towards lower energies. The high-energy behavior is somewhat delayed with respect to the low-energy behavior. As a function of energy or velocity it abruptly stops below 16,000 km s^{-1} or ~ 700 eV. The core of this "bite-out" region lasts from ~ 0538 to 0543 UT, where the times have been corrected for the shift between modes E_1 and E_2. The figure confirms our assertion, that at least the B minimum at 0529 UT is not connected with the actual interaction region. The figure allows an easy deduction of the electron temperature from the width of the distribution.

Our next step is to look for a possible fast bow shock, which would show up as a steep jump in B towards the center of the wake. Careful inspection of Fig. 2 shows no evidence for a fast MHD shock either inbound or outbound. This independently confirms the result shown in Table I, that the fast Mach number M_f is < 1, i.e.

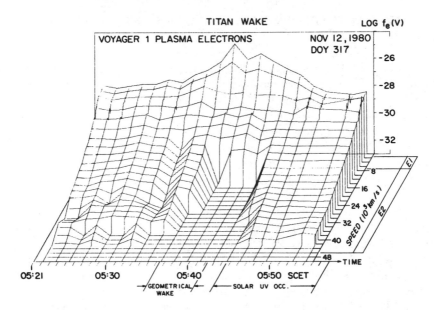

Fig. 3. Electron distribution function f_e as a function of velocity for the Titan encounter period. Note the electron bite-out region above ~ 16,000 km s^{-1} (from Hartle et al. 1982).

$$M_f = \frac{V}{(V_A^2 + V_s^2)^{1/2}} \lesssim 1. \tag{1}$$

Inserting the expressions for V_A and V_s we obtain a useful alternate interpretation of M_f, i.e.

$$M_f = \left(\frac{\frac{1}{2} \rho \, v^2}{p_t} \right)^{1/2} \tag{2}$$

where $p_t = p + B^2/2\mu_0$ is the total pressure.

The question of the identification of an internal magnetic field of Titan depends to some extent on our knowledge of what a magnetosphere in sub-fast ($M_f < 1$) flow would look like. Nothing is known about a magnetosphere under these conditions. Note, that the variation of the magnetic field due to a simple dipole along the trajectory will be appreciably modified by the field of plasma currents. The following discussion has therefore to be taken with a grain of salt.

Let us assume that a magnetosphere in sub-fast flow also develops a magnetospheric tail opposite to the direction of incident flow, with a corresponding cusp region known as the midnight cusp in the case of the Earth. If the magnetic moment of Titan were very strong, the Voyager 1 encounter trajectory would have passed inside the cusp. In this case two magnetopause crossings should have been observed with an approximately dipolar field region in between. Figure 2 shows three dips in magnetic field magnitude and other deviations from the expected behavior, and therefore excludes this possibility.

For a smaller dipole moment the cusp would be located inside the encounter trajectory, that would then pass through the hypothetical magnetospheric tail due to the global magnetic field of internal origin as postulated. If the dipole moment were approximately parallel or antiparallel to the rotational axis of Titan, we would expect a neutral sheet in the equatorial plane of Titan which coincides with its orbital plane. According to the trajectory characteristics described above, we could interpret the first dip in magnitude B as the inbound magnetopause, the second one as due to crossing the neutral sheet of the tail and the third one as the outbound magnetopause. The regions L_1 and L_2 in Fig. 2 are then the northern tail lobe and southern tail lobe, respectively. The magnetic field data can be displayed more clearly by projecting the vectors into the XY-plane and the YZ-plane. Figure 4 shows every fifth 1.92 s average magnetic field vector in this representation. (No significant detail is lost by the decimation.) We see that the vectors in tail lobe L_1 point toward Titan whereas those in L_2 point away from the satellite. The hypothetical di-

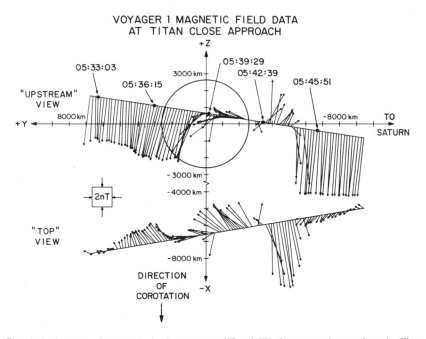

Fig. 4. Projections of magnetic field vectors on YZ and XY planes superimposed on the Titan flyby trajectory. The 1.92 s observations have been decimated by a factor of five (from Ness et al. 1982b).

pole moment of Titan would therefore point in a direction antiparallel to the magnetic moment of Saturn. According to present knowledge in this field, reconnection should essentially go on continuously in this case. However, no evidence for hot plasma-sheet electrons is found in the electron data of Fig. 3. On the contrary, a strong reduction of electron temperatures is indicated by the electron observations. In addition, the substantial differences in field magnitude between the two tail lobes are difficult to understand, if they are due to a magnetic field of internal origin parallel or antiparallel to the Z-axis. We can give a very conservative upper limit for an internal field of Titan by attributing the magnetic flux through a tail with a somewhat arbitrary extension of $\pm$ 5 R_T in the $\pm Z$ direction and the observed field along the trajectory to the total flux through one hemisphere. The result is $M \lesssim 7 \times 10^{20}$ Gauss cm^3 corresponding to a surface equatorial magnetic field of $\lesssim 4.1$ nT. The magnetic field is therefore much less than the 100 nT field derived by one of the authors (Neubauer 1978) for the most favorable case for dynamo action. Using Eq. (1) of Neubauer (1978), the upper limit to the radius of an inner metallic core would be 410 km. This upper limit is only significant under the condition that there is enough energy to drive a dynamo, a condition which need not necessarily be fulfilled. The small fraction of Titan's radius, which this upper limit implies, i.e. 0.16, also leads to the expectation, that if a dynamo field exists, it would be dipolar to a good approximation.

   In conclusion, there is no evidence for the interaction between Titan and Saturn's sub-fast magnetoplasma being due to an internal magnetic field. Since an extensive atmosphere is known to exist, the interaction must therefore be atmospheric as in the case of Venus with the formation of an induced magnetosphere of the tail which has been traversed by Voyager 1. If an internal field close to our upper limit should exist, there would be some limited influence on the flow system.

V. INDUCED MAGNETOSPHERE OF TITAN

A. Mass Loading and Field-Line Draping

   In this section we shall review the interpretation of the encounter data of Voyager 1 in terms of an induced magnetosphere, which relies on the physical concepts of mass loading by mass pickup and magnetic field-line draping. Mass loading and draping are intimately connected when the frozen-in field concept is applicable. Before we can apply these concepts, we must be sure that the frozen-in field concept applies in the case under consideration. The electron gyroradius in Table I indicates that at least the electrons will be frozen to the magnetic field lines down to the level where the collision frequencies approach the electron cyclotron frequency. The large ion gyroradii would tend to allow deviations from the frozen-in field conditions. However, estimates of the current densities based on a simple application of Ampère's law to the magnetic field data combined with typical densities, indicate rela-

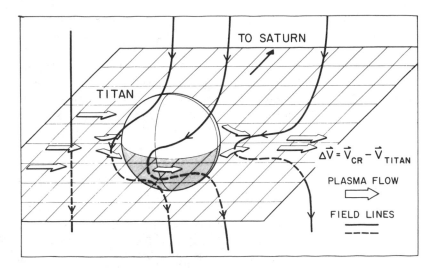

Fig. 5. Sketch illustrating magnetic field-line draping at Titan (from Ness et al. 1981).

tive speeds between ions and electrons which are small enough to constrain the motion of ions to follow the frozen-in field concept in a smoothed-out sense.

In analogy with a comet and the case of Venus, the formation of an induced magnetosphere at Titan can then be described as follows (see Fig. 5). Consider an initially straight flux tube of small diameter along the $(-Z)$ direction filled with the incident plasma, as characterized in Table I, as it moves towards the satellite. Because of the sub-fast character of the flow, the signal in the fast MHD-mode caused by the partial shortcircuiting of the motional electric field by Titan's ionosphere can propagate in all directions and prepare the incident plasma to slow down and deviate around Titan, before collisions with the neutral atmosphere become important. As an initially straight flux tube enters the hydrogen corona of Titan or even the nitrogen dominated atmosphere close to the satellite, mass loading due to the production of electrons and ions in the flux tube will slow down the flux tube near $Z = 0$. In the mass loading or mass pickup process, a newly generated ion is accelerated by the electric field from the small or zero initial velocity of the parent neutral in Titan's frame to a new average velocity corresponding to the $\mathbf{E} \times \mathbf{B}$ drift speed. The necessary momentum is extracted from the flow leading to a decelerating effect. Here the ion production is due to photoionization and collisional ionization by electrons in the high-energy tail, above 700 eV of the magnetospheric electron distribution function having a bulk temperature of 200 eV. Whereas the part of the flux tube near $Z = 0$ is slowed down by the mass loading, it proceeds in an unimpeded way at large values of $|Z|$. In addition to mass loading, elastic collisions between plasma particles and neutral gas will impede the flow around $Z = 0$ particularly closer to Titan, where elastic collisions become relatively more important. As the mass loading continues, the tip of the flux tube lags more and more behind the unperturbed field line at large $|Z|$. Figure 5 illustrates this process referred to as magnetic field-line draping, first described by Alfvén (1957) in an application to comets. More precisely, the mass loading near $Z = 0$ generates an Alfvén wave signal leading to a bending of the flux tube, as the Alfvén wave propagates up and down the undisturbed flux tube along $+Z$ and $-Z$. The Alfvén wave signals affect a region limited by the Alfvén Mach angles $\tan^{-1} M_A^{-1} = 28°$.

Because of the increased mass density due to the mass loading, the Alfvén speed is relatively small near $Z = 0$ leading to a particularly pronounced stretching of the draped flux tube. As the draped field line enters the wake region, the draping will become stronger and stronger until the tension in the field line pulls it away from the ionosphere. Figures 5 and 6 show the flow in the XY-plane. This tension will eventually lead to an acceleration of the plasma on the field line near $Z = 0$ until the initial velocity is recovered. The tensional force can also be described in terms of a $\mathbf{j} \times \mathbf{B}$ force with $\mathbf{j}$ being directed along $+Y$. Since the overall current system must be source-free, there must also be a system of Alfvénic currents (Drell et al. 1965; Neubauer 1980; Goertz 1980; Eviatar et al. 1982) towards the plane $Z = 0$ at negative Y and away from $Z = 0$ at positive Y. This has been displayed schematically in Fig. 9 in Sec. V.D.

If we compare our case with Venus, we expect the flux tubes penetrating deepest into Titan's atmosphere to be finally guided around the satellite along an ionopause boundary layer. At this boundary there is total pressure equilibrium between the outer mass-loaded, stretched-out magnetic field and the inner ionosphere of Titan, which is due to ionization by solar extreme-ultraviolet (EUV) photons and collisional ionization by high-energy charged particles with gyroradii big enough to penetrate the boundary. The distance of this boundary from Titan has been estimated at 4400 km by Hartle et al. (1982) for the day side. We also mention here, that the formation of an ionopause layer is not always possible. If the ionization inside any assumed ionopause is insufficient, e.g. due to very low EUV-photon fluxes to produce enough pressure to balance the mass loaded plasma, we expect a gradual decrease in velocity with field lines even penetrating to the inside of the ionosphere.

Taking the electron density to be 3×10^3 cm^{-3} in the ionosphere of Titan (Strobel and Shemansky 1982), a temperature sum $T_e + T_i = 4600$ K is necessary to balance the estimated ram pressure using the parameters in Table I. An ionopause should therefore exist at least around part of Titan's atmosphere-ionosphere system. Upper bounds of 3×10^3 cm^{-3} and 5×10^3 cm^{-3} on the electron density have been derived by Lindal et al. (1983) near the evening and morning terminators, respectively.

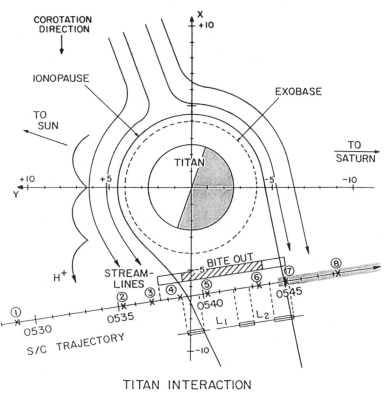

TITAN INTERACTION
NOV 12, 1980

Fig. 6. Idealized plasma flow around Titan. L_1 and L_2 refer to the northern and southern magnetic tail lobes. The shaded bars refer to the minima corresponding to magnetopause and neutral sheet crossings. The trajectory of a proton is approximately to scale in the observed magnetic field (after Hartle et al. 1982).

The draping picture described above suggests, that the magnetic field projected into the XY-plane has a component towards the flow north of the equatorial plane and a component parallel to the flow south of it, i.e. at $Z <$ 0. This expectation is born out by information in Fig. 4, since the spacecraft was first north and after 0543:07 UT south of the equatorial plane. Starting at 0533 UT the magnetic field begins to develop a projection towards the direction of incident flow in the XY-plane while the total vector is predominantly in the southern direction. Then the XY-projection starts to rotate in a counterclockwise sense, until at about 0538:30 UT the total vector suddenly begins to rotate completely into the XY-plane. In the XY-plane after further clockwise rotation, it is first directed essentially antiparallel to the incident flow direction, as expected for draping north of the symmetry plane in Fig. 5. After 0542:39 UT the field is directed roughly parallel to the direction of incident

TABLE II
Results of Analysis of Ion Spectra

Numbers in Fig. 6	Ion Species Time Interval 05, SCET	H^+			N^+			N_2^+/H_2CN^+		
		V, km s^{-1}	n, cm^{-3}	kT, eV	V, km s^{-1}	n, cm^{-3}	kT, eV	V, km s^{-1}	n, cm^{-3}	kT, eV
4	38:35–38:39	60	2.5	2.1	10	15	2.1	5	30	2.1
5	40:11–40:15	25	1.9	3	5	7	1.8	3.5	10	0.9
6	43:23–43:27	50	1	3	10	3	2.1	4	10	2.4
7	44:59–45:03	40	1	33	20	22	29	—	—	—

flow, comparable to the draping south of the symmetry plane in Fig. 5. The switch in the magnetic field direction in the XY-plane from antiparallel to parallel to the direction of incident flow is almost exactly coincident with the crossing of Titan's equator. At about 0544:30 UT the strong southerly component appears again with a relatively small component parallel to the incident flow direction. We can therefore distinguish a region of weak draping with predominantly southerly fields surrounding a region of strong draping near Titan with fields in the XY-plane.

Let us now discuss the consequences of the mass pickup process by use of the plasma science observations on the encounter trajectory. As we approach Titan, the weak draping region can be attributed to the pickup of protons in Titan's hydrogen corona. The subsequent transition from weak to strong draping is consistent with the onset of heavy ions (N^+ or N_2^+/H_2CN^+) of ionospheric origin. The observed energy spectra of these ions are increasingly consistent with the pickup energy distribution expected for newly ionized neutral gas (cf. Hartle et al. 1982, their Fig. 4; Hartle et al. 1973). This provides a direct experimental association of the observed plasma in the wake with processed portions of Titan's neutral atmosphere, and indirectly implies copious momentum exchange, as the mass loading of newborn ions slows the incident magnetized plasma. The results of fitting convected Maxwellians to the ion data in and near the region of strong draping are given in Table II. The locations on the encounter trajectory are also indicated by the numbers four to seven in Fig. 6. As will be shown in Sec. V.B, where electron densities consistent with all data are derived, the heavy ion interpretations must be considered the realistic ones. The fits then clearly define a strong shear layer between at least 0537 and 0539 UT with the plasma moving only 5 to 10 km s^{-1} just before the dip in the magnetic field preceding the region L_1 in Fig. 2, indicating the inbound magnetopause. This should be contrasted with the 80 to 150 km s^{-1} flow observed in the undisturbed magnetosphere of Saturn. The analysis of observed low-energy charged particles by Maclennan et al. (1982) suggests an even earlier onset of the low-velocity region indicating the shear layer.

Throughout the inbound velocity shear layer the electrons collected by the plasma science instrument are observed to cool as a result of the addition of cold electrons which are the by-products of the ionization process. This has already been noted in connection with Fig. 3. The electron velocity distributions f_e through this region begin to show evidence of inelastic interactions (e.g., electron impact ionization) of the hot magnetospheric plasma with neutral gas, where the strong bite-out region lasts for the entire duration of the strong draping region as defined by the magnetic field.

The ion analysis is consistent with a neutral exosphere model, which has N_2 as the dominant molecule above the exobase at 4000 km with a transition to a hydrogen-dominated exosphere at 5000 km. The exobase temperature has been taken to be 160 K (Hartle et al. 1982).

B. Electron Density Variation in Titan's Wake Along the Voyager 1 Encounter Trajectory

The important variation of total electron density can only be obtained by a synthesis of the plasma science electron and ion data and the wave data obtained by the plasma wave and planetary radio astronomy experiments. We shall use Fig. 7 to develop the solutions, which will also indicate the complexity of distribution functions resulting from magnetoplasma-atmosphere interactions.

First, the electron densities obtained by the plasma experiment in modes E_1 and E_2 are shown by octagons. They are based on direct integration of the electron distribution function f_e observed above 10 eV and approximated by a Maxwellian fit to the lower energy channels for energies < 10 eV. We may use the term hot electrons for these electrons in contrast to a distinct cold electron population at energies well below 10 eV which coexists with hot electrons, as we will argue later in this section. The corresponding densities and temperatures can be called n_H, n_c and T_H, T_c, respectively. Figure 7 indicates a maximum in n_H in the shear layer which is combined with a minimum in the temperature T_H, also suggested by Fig. 3. The temperature T_H is cooler than the magnetospheric value throughout the entire region of draping. The strong bite-out region is also indicated in the upper panel of Fig. 7. The lighter stippling reflects the temporal uncertainties of these boundaries.

The plasma science ion data, part of which are given in detail in Table II, give a first hint that electrons are hidden outside the energy range of 10 to 6950 eV. The electron densities required to balance the ion densities under the three assumptions H^+, N^+, N_2^+/H_2CN^+ are also shown in Fig. 7 and turn out to be greater than the densities n_H.

The additional use of the electron density information contained in the observation of upper hybrid resonance (UHR) emissions allows the reconstruction of the total electron density. UHR emissions occur, when a cold electron component of density n_c exists together with a hot, loss-cone type electron distribution of density n_H under suitable conditions of the temperatures T_c and T_H and other parameters. For a thorough discussion we refer the reader to Hubbard and Birmingham (1978) and Birmingham et al. (1981). In the important special case $n_c \gg n_H$ the UHR emissions occur at $f_{UHR} = (f_{pc}^2 + f_{ce}^2)^{1/2}$ where f_{pc} is the cold plasma frequency given in Hertz by $f_{pc} = 9 \times 10^3 \cdot (n_c)^{1/2}$ with n_c in electrons per cm^3 and f_{ce} the electron cyclotron frequency given by $f_{ce} = 28$ Hz $\cdot B$ where B is given in nanoteslas. Hence f_{UHR} yields n_c or because $n_H \ll n_c$ approximately, the total electron density $n_e = n_c + n_H \approx n_c$. For $B \lesssim 7$ nT and $f_{UHR} \gtrsim 5.62$ kHz we always have $f_{ce} \lesssim 196$ Hz and therefore $f_{UHR} = (f_{pc}^2 + f_{ce}^2)^{1/2} \approx f_{pc}$, i.e., the frequency scale f_{UHR} can immediately be turned into a density scale for n_c. If $n_c \gg n_H$ is not fulfilled, the relationship between f_{UHR} and the densities n_c or n_H is more complicated (Hubbard and Birmingham 1978), although n_c generally is the directly determined quantity.

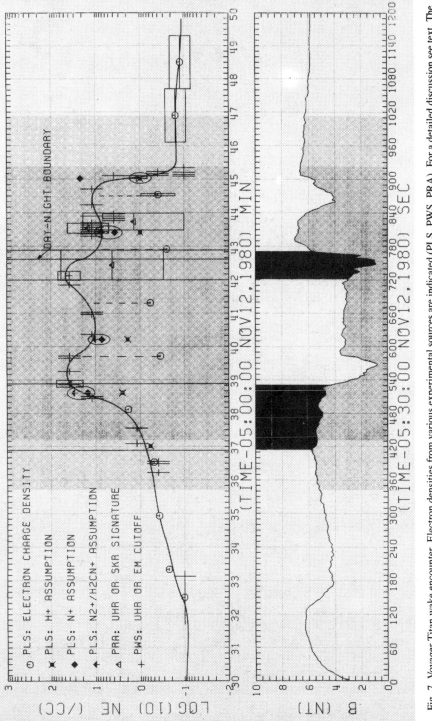

Fig. 7. Voyager Titan wake encounter. Electron densities from various experimental sources are indicated (PLS, PWS, PRA). For a detailed discussion see text. The solid curve represents total electron density consistent with the data and physical concepts developed in this chapter. The magnetic field magnitude and PRA drop-out regions are also shown for guidance.

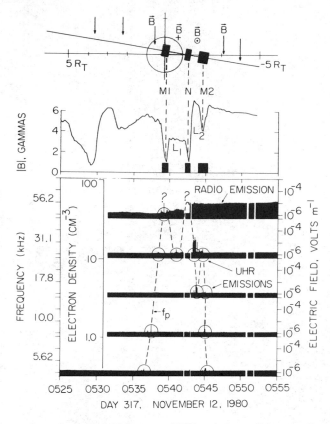

Fig. 8. Observations of upper-hybrid resonance (UHR) emissions and Saturn radio emissions in the upper frequency range of the Voyager plasma wave experiment (after Gurnett et al. 1982b).

Let us now briefly discuss the plasma wave observations relevant to this problem. They are shown in Fig. 8 for the frequency channels from 5.62 kHz to 56.2 kHz. Gurnett et al. (1982b) point out that the UHR emissions are excited as a narrowband emission near the local f_{UHR}. In addition to the upper hybrid resonance emissions we also observe nonlocal radio emissions from Saturn, the observation of which requires both the activity of the source and an unobstructed path of propagation. The possibility, that these emissions may be generated in Titan's vicinity (Gurnett et al. 1981a), has been ruled out by Daigne et al. (1982). Since the identification of the UHR emissions depends to some extent on their burst-like character, it is generally not difficult to distinguish local UHR emissions and nonlocal Saturn radio emissions, except for cases when the latter are very intense or the propagation path connecting the source with the spacecraft varies rapidly. The observed candidates for UHR emissions are encircled in Fig. 8. The question marks indicate possible extensions to even higher frequencies or densities. The first density maximum at 56.2 kHz is derived by one UHR spike only indicating the maximum f_{UHR} to be contained in the channel corresponding to a maximum density of ~ 40 cm^{-3}. The second density maximum at about 0542:30 UT could be as high as ~ 60 cm^{-3} according to combined plasma wave and planetary radio astronomy data. We shall come back to this point later (Sec. V.C). A third maximum of 12 cm^{-3} occurs in the electron density determined from the UHR emissions at about 0544:30 UT.

Let us now return to Fig. 7. The electron densities derived from the upper hybrid resonance (UHR) frequencies observed by the plasma wave (PWS) experiment are shown by vertical-horizontal crosses. The planetary radio astronomy (PRA) data are shown as open triangles. There are appreciable differences between the plasma science (PLS) electron densities and the PWS-PRA set of inferred values. However, the times of the measurements agree in only one case, when the PLS electron data were obtained in the low-energy mode E_1 lasting from 0539:42 to 0539:43.9 UT. However, the reliability of the relevant UHR data point could be questioned since the corresponding spike occurred at a time when Saturn radio emissions start to appear and UHR emissions start to move from 56.2 kHz to 31.1 kHz. It may therefore be due to intensity scintillations of 56.2 kHz Saturn radio waves, as the source rises above some plasma horizon as seen from the spacecraft. Note that no corresponding spikes are seen at the frequency of 59 kHz by the PRA experiment. Apart from this one possible point, we could in principle explain the differences between the PLS electron densities and the UHR densities as being due to temporal or, better still, to spatial variations in density. To be consistent with the physical models for the generation of UHR emissions (Hubbard and Birmingham 1978), this electron population would then have to be called the cold population whereas a hot destabilizing population would have to be postulated at temperatures above 6.95 keV. In other words, in this interpretation the PLS measurements have fortuitously missed the times of high electron densities. Although we cannot completely exclude this possibility, we prefer an alternate explanation because of general considerations of the physics involved. As suggested by Gurnett et al. (1982b) and independently by physical arguments, in this interpretation the electron population, resulting from a relatively strong interaction between the plasma and Titan's atmosphere (see chapter by Hunten et al.) and delineated by the core of the electron bite-out region, contains a cold ionospheric part at energies well below 10 eV, the lower energy limit of the PLS experiment. The electrons observed by the PLS-sensor in this strong interaction region, approximately coinciding with the region of strong draping, must be considered as a mixture of cooled magnetospheric electrons, secondary electrons resulting from collisional ionization and photoelectrons. The PLS electrons are then the hot electrons responsible for the UHR emissions. They can be characterized by the density n_H and temperature T_H. Since the UHR emissions require loss-cone type distributions, the values of n_H in Fig. 7 contain a small systematic error because of the isotropy assumption in deriving them. The PWS-PRA densities then correspond to the cold electrons characterized by n_c and T_c in the region of strong draping. Outside this region the condition $n_H \ll n_c$ is less well fulfilled and the assignment of n_c to the PWS densities is less accurate. Gurnett et al. (1982b) have estimated an ionospheric temperature of 8600 K by assuming total pressure equilibrium in the tail of the induced magnetosphere, which coincides with the strong draping region. Analysis of the response of the PLS instrument to a cold electron population of 8600 K shows that the densities required by the PWS observations are completely consistent with the PLS data.

We have also drawn a solid line in Fig. 7 which expresses our preferred total electron density variation. The vertical dashed lines represent the cold ionospheric densities not seen by PLS. The ion data shown by solid symbols in Fig. 7 nicely fit into this picture, if we assume the heavy ions N^+ or N_2^+/H_2CN^+ to be dominant in the strong interaction region, except for the time 0545 UT (Hartle et al. 1982). We have therefore emphasized these data points. Unfortunately, the three possibilities regarding the composition cannot be distinguished.

Let us finally turn to the electron densities beyond the dip in B terminating the magnetic lobe L_2 in Fig. 2, which in our picture of the induced magnetosphere of Titan will be termed the outbound magnetopause crossing. The electron densities above the night side of Titan after about 0545:30 UT do not show a gradual falloff but drop immediately to magnetospheric densities. This is also consistent with Fig. 3 and the reduced plasma sheath noise in Fig. 10 that will be discussed in Sec. V.E. The smaller extent of mass loading on the night side may be due to a less pronounced hydrogen corona above the night side due to much cooler atmospheric temperatures. The less pronounced draping in the magnetic field of Fig. 4 after 0545:30 UT may be due to this diminished mass loading but could also be explained by a closer distance to the symmetry plane of the draped field lines, that may have moved in the $(-Z)$ direction. Also we mention, that the electric field direction is such that newly formed ions are accelerated into the dense atmosphere on the night side and away from it on the day side (Bridge et al. 1981a).

C. Magnetopause and Neutral Sheet Boundaries of the Induced Magnetosphere and Associated Density Structure

After having reviewed the observations along the Voyager 1 encounter trajectory and their interpretation in terms of mass loading and draping, we shall now describe the resulting structure in three dimensions. The induced magnetosphere is bounded by a plasma layer, which may be referred to as the magnetopause. The magnetopause is produced by draped field lines that get their mass load in the transition layer from the H^+ loading region to the N^+ or N_2^+/H_2CN^+ loading region; in our model this region corresponds essentially to the transition from the outer weak draping region to the inner strong draping regime. Here high-plasma densities develop in the mass loading process, which are dragged away from Titan by the draped field lines. As the plasma moves outward, it adjusts to the smaller surrounding total pressure. It accomplishes this by an expansion leading to approximate pressure equilibrium with the surrounding plasma and a reduction of the frozen-in magnetic field strength; this results in boundary layers of minimum magnetic field downstream of the dayside and nightside atmosphere of the satellite. The Voyager 1 spacecraft crossed the inbound magnetopause around the central time 0539:29 UT and the outbound magnetopause about 0544:21 UT. We have noted the associated dips in magnetic field magnitude in Fig. 2 and the maxima in plasma electron density in Fig. 7. The important orientation of these magnetopause boundaries almost perpendicular to the XY-plane has been shown by detailed analysis of the transition data within the boundaries (Ness et al. 1982b). The duration of the magnetopause crossings was 59 s and 79 s for the inbound and outbound magnetopause, respectively; this leads to thicknesses of 1000 km and 1350 km, respectively, if we use the projection of the flyby speed on the XY-plane. Following arguments proposed by Daigne et al. (1982), we can even infer the density structure of the inbound magnetopause away from the trajectory. These authors have shown from the drop-out of Saturn radio waves, that the electron density of 40 cm^{-3} extends to a level of 150 km above the spacecraft trajectory corresponding to $Z \approx 0.25$ R$_T$. A decrease in electron density away from $Z \approx 0$ is expected, since in a draped magnetic flux tube the plasma accumulates near the tip of the tube at $Z = 0$.

The field lines, that have penetrated into the atmosphere most deeply, are observed well inside the induced magnetosphere. Here draping is essentially complete and the field lines are well aligned with the XY-plane. The neutral sheet, which is approximately parallel to the XY-plane, then divides field lines with a direction towards the flow in the north and approximately aligned with it in the south. The neutral sheet has been observed around 0542:29 UT with an associated maximum in electron density given by the UHR emission observed by the PWS instrument. Since the grazing incidence hypothesis suggested (see Fig. 6 in Daigne et al. 1982) requires very special conditions, (i.e. a smooth density isocontour without any corrugations and a very special tilt angle), we prefer a blockage interpretation that leads to higher densities in agreement with the PWS data. Our discussion has also provided information on the electron density above and below the Voyager 1 trajectory, i.e., in two dimensions. In this picture maximum electron densities are associated with the magnetopause and neutral sheet boundaries. The minima in density between the boundary crossings in Fig. 7 are then due to a maximum distance from any of the boundaries inside the induced magnetotail.

The approximate picture of the electron density distribution, which for singly ionized ions corresponds to the ion number density distribution, allows us to estimate the mass loss from Titan through its induced tail. From Figs. 4 and 6 a width of 2.5 R$_T$ can be deduced for the induced tail inside the velocity shear layers mentioned above. The discussion above suggests an extent of $\pm$ 0.25 R_T of the major region of mass loss perpendicular to the XY-plane. From Table II and Fig. 7 we have estimated an average flux density of 5×10^6 cm^{-2} s^{-1}. This leads to a total loss of $\sim 10^{24}$ ions per second from Titan's atmosphere; the ions would be N$^+$ or N$_2^+$/H$_2$CN$^+$. The uncertainties are such as to allow even somewhat higher loss rates. This should be compared with other atmospheric loss rates (see e.g. chapter by Hunten et al.). Strobel and Shemansky (1982) also have given a N atom escape rate due to electron impact dissociation of N$_2$ molecules of 3×10^{26} s^{-1} and an H atom escape rate of 2×10^{26} s^{-1}.

D. Flow Asymmetry and the Picture of Four Tail Lobes

Some additional features of the encounter observations warrant explanations. First, the center of the disturbance region is somewhat offset from the plane $Y = 0$ which would be the symmetry plane for ideal corotational flow and an ionosphere symmetric with respect to the plane $Y = 0$. This may be partly explained by the 20° deviation of the flow and in addition, by the asymmetry of the two lobes of the induced tail.

Since the solar direction, which controls the photochemistry of the ionosphere, is at a large angle to the general direction of incident flow, there will be a strong asymmetry between the parts of the tail lobes resulting from field-line draping around the dayside hemisphere and the nightside hemisphere. The direction of incident flow combined with the 1330 local time yields $90 - 20 - 22.5 = 47°5$ for the angle between the incident flow direction and the solar direction. This also provides an explanation for the maximum of the ultraviolet nitrogen emissions (Strobel and Shemansky 1982) in the quadrant of the sunlit hemisphere facing the ideal corotational flow. The ionization rate will be obviously greater on the day side. As already mentioned, newly created ions will move out of the atmosphere on the day side and run into the deeper atmosphere on the night side because of the sense of gyration determined by the southward magnetic field (Bridge et al. 1981a).

As illustrated in Fig. 9 we expect to find a north-south trending boundary or vertical boundary (with respect to the XY-plane) in the tail between field lines due to dayside draping and nightside draping. This is in addition to the east-west trending or horizontal boundary (with respect to the XY-plane) between northern and southern field polarities produced by the draping and referred to as the neutral sheet. As a consequence we have four tail lobes, which must be carefully distinguished. Since the two dayside lobes (northern and southern) are expected to be loaded with much more plasma than the nightside lobes, some expansion of its magnetoplasma will occur on the day side to enforce total pressure equilibrium across the tail. This will lead to a reduction in magnetic field magnitude in lobe L_1 as observed in Fig. 2. A casual inspection of Figs. 1 and 4 might suggest that the spacecraft crossed the intersection between the day-night boundary and the north-south boundary simultane-

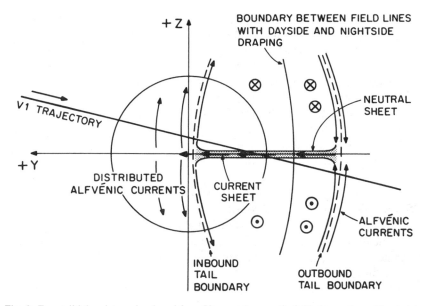

Fig. 9. Four tail lobe picture developed from Voyager 1 magnetic field observations. The sketch also shows the current distribution consistent with the magnetic field topology. The view is towards the direction of idealized corotational plasma flow (from Ness et al. 1982b).

ously. This would seem very improbable. However, a close inspection of Figs. 2 and 4 shows that the low magnetic field of lobe L_1 characteristic of the dayside lobe extends somewhat into the southern lobe after the neutral sheet crossing. Voyager 1 has therefore traversed the following tail lobes: northern dayside, southern dayside and southern nightside, where the latter two correspond to L_2 in Fig. 2. A north-south asymmetry could also be expected due to seasonal variations in Titan's atmosphere-ionosphere system (B. A. Smith et al. 1981; Hanel et al. 1981; chapter by Hunten et al.). Five months before the encounter the Sun had passed above the equatorial plane. The above considerations show that both the offset and the asymmetry between L_1 and L_2 can be explained by the atmospheric interaction model. Although the various electron density data cannot resolve the difference between the neutral sheet and day-night boundary, we have schematically indicated in Fig. 7, what one might expect.

E. Dynamical Phenomena

The physical model developed organizes so many data, that it can be expected to be basically correct. However, much more work will be required to understand the interesting dynamical phenomena which are briefly reviewed here.

An interesting feature of the magnetic field data begins at about 0537:00 UT and lasts until 0539:30 UT. If we consider the sequence of magnetic field vectors as due to some stationary structure, they could be extrapolated to correspond to a clockwise rotation of magnetic field vectors on circles around a fixed axis. Both the assumption of stationary structure and the continuity of vectors around a circle are not much more than plausible working hypotheses. The feature could be explained as due to a bulging of stream lines due to severe deviations of the flow from symmetry (see Fig. 11 in Ness et al. 1982*b*), local vortices in the flow or Titan's counterpart of auroral curls which are characterized by parallel magnetic field and current density in the polar magnetosphere of the Earth (e.g. Hallinan 1976).

Another interesting dynamic feature occurs in the same general region. An oscillation with a fixed period of 33 s occurs in the magnetic field magnitude as indicated by small vertical arrows in Fig. 2 between 0536:50 and 0541 UT. It is interesting to consider the possibility of these oscillations as due to the shedding of vortices in a hypothetical vortex strait following Titan.

The plasma wave experiment showed a number of interesting wave phenomena. Figure 10 shows the encounter observations from 10 Hz to 1000 Hz for the time interval from 0525 to 0555 UT. Two types of broadband wave emissions have been detected. The so-called sheath noise occurs in the inbound region of hydrogen mass pickup where we also observed weak draping of magnetic field lines. As illustrated by Fig. 9, the pickup current in the Y direction decreases with increasing Y, which requires a discharge of the current by Alfvénic currents in the $+Z$ direction. This is confirmed by a straightforward application of Ampère's law to the field projections into the XY plane. However, in addition to the proposal by Bridge et al. (1981*a*), who postulate waves due to the unstable ion distributions produced by the mass pickup process (i.e., essentially rings in velocity space), we propose that these waves may also be due to the approximately field-aligned currents, which generate electrostatic emissions as the auroral electrostatic turbulence reported by Gurnett and Frank (1977).

The so-called tail noise can be considered a similar type to that observed in the magnetospheric tail of the Earth (Gurnett et al. 1976). The mechanism of this type of emissions is not well understood and therefore the diagnostic value here is somewhat limited. There seems to be a sharp change in intensity in the region of the day-night boundary of draped field lines.

F. Relevant Future Investigations

It would be interesting to learn whether effects of the interaction between Titan and its plasma environment have been observed by other spacecraft. One might also look for other examples in the solar system, where at least some of the physical aspects of the interaction problem have been studied in some detail.

Jones et al. (1980) have reported a possible magnetic wake in connection with a bow shock of Titan during the Pioneer 11 encounter of Saturn. Since the fast Mach number in our case was $M_f = 0.55$, a corresponding variation in

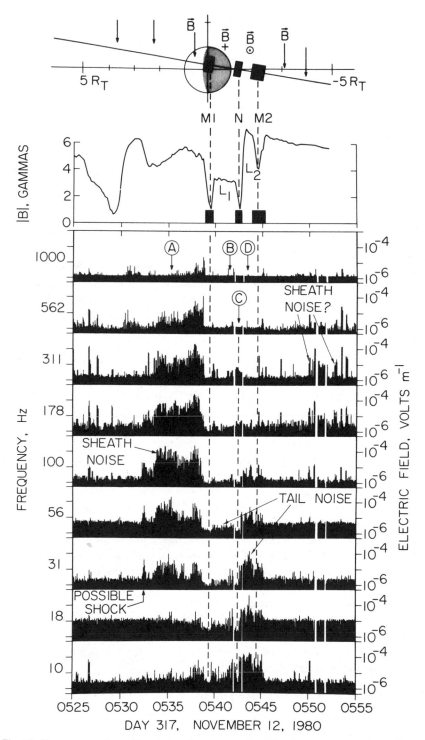

Fig. 10. Plasma wave observations in low-frequency range of the plasma wave experiment (from Gurnett et al. 1982b).

plasma and magnetic field parameters to explain $M_f > 1$ is certainly possible. Although the distinction between a Titan wake and some magnetospheric disturbance of different origin is difficult to make, the possibility remains interesting.

Another way to learn more about the interaction is to study some aspects of the physics in similar cases. Atmospheric interactions at Io and Venus have also been studied; we have already used the analogy with Venus in several instances. At Io the Mach numbers turned out to be $M_A = 0.15$ (Acuna et al. 1981b) and $M_s = 3.0$ (Bagenal and Sullivan 1981); the case is therefore sub-fast as at Titan. The ordering of the bulk kinetic, magnetic and thermal energy densities is different, however. In Io's case the magnetic field energy density dominates whereas in the case of Titan the plasma thermal energy is dominating although not very strongly. At Io the analysis showed the existence of an Alfvén wing or Alfvénic current tube (Drell et al. 1965; Neubauer 1980; Goertz 1980) carrying a current of 2.8×10^6 amperes driven by a voltage of 500,000 volts across Io (Acuna et al. 1981b). In the case of Titan we also expect the existence of Alfvén wings but carrying a much smaller current due to the modest driving voltage of 4800 volts across the diameter of the exobase. At small Mach numbers $M_A < 1$ the Alfvénic currents can be clearly distinguished at large $|Z|$, whereas in our case of $M_A = 1.9$ their disturbance field is expected to merge with disturbances due to the slow magnetoacoustic mode. In the case of Venus, Mach numbers of $M_A \approx 8$ and $M_s \approx 8$ are typical leading to strong bow shocks (Russell et al. 1981) in contrast to the sub-fast case of Titan. However, the effect of the fast Mach number should not be very large above the hemisphere of plasma incidence, since the shock decelerates the plasma to sub-fast speeds $M_f < 1$. The difference is expected to be substantial in the tail region in the case of Venus because of the acceleration of the flow to fast speeds along the flanks of the tail.

At Venus the process called draping has first been studied in detail by *in situ* observations and found to yield a good description of the magnetic field configuration in the wake in terms of an induced tail with a tail neutral sheet perpendicular to the magnetic field frozen into the incident plasma (see e.g. Luhmann et al. 1981). For Venus the atmosphere seems to be much harder than at Titan in the sense of a small ratio scale height/radius at Venus compared with Titan; Titan turned out to be a much softer target than Venus at least as far as dayside conditions are concerned. In this sense the case of Titan's day side is intermediate between the case of Venus and that of a cometary interaction (Biermann et al. 1967; Wallis 1973). The Venus solar wind interaction has recently been compared with Titan's magnetospheric interaction by Kivelson and Russel (1983).

VI. SUMMARY AND CONCLUSIONS

The Voyager 1 encounter of Titan has yielded a unique set of magnetic field, plasma and plasma wave data which allows a first attempt to develop a comprehensive picture of the interaction between the magnetospheric plasma of Saturn and the atmosphere of Titan. This is complicated by the large angle between the solar direction and the direction of incident flow, the large ion gyroradii and the transsonic and transalfvénic character of the incident flow. A 3-dimensional sketch of the induced magnetotail of Titan as observed during the Voyager 1 encounter and reviewed in this chapter is shown in Color Plate 4. The exosphere of Titan is depicted in blue. Ionization of the upper atmosphere of Titan and the associated mass-loading of magnetic field lines and field-line draping define two regions of enhanced plasma density due to dayside and nightside draping. The boundaries of these regions are shown in red and green, respectively. Note the much more extended dayside draping region due to hydrogen ionization. The figure also gives views of the four tail lobes mentioned above. Whereas the structure near the equatorial plane separating northern and southern lobes has been obtained by data analysis, the north-south boundaries of the lobes have been inferred by more theoretical arguments. Cuts through the enhanced plasma regions are shown crosshatched. The boundary between the dayside and nightside tail lobes is shown in gray. Several thin magnetic field filaments have also been drawn to illustrate the draping process.

Ideally a physical problem is considered solved, when a quantitative model explaining in detail the observations has been developed. As shown in this review, Titan's magnetospheric interaction is certainly one of the most complex problems among the many possibilities for the interaction between a planetary body and the surrounding plasma environment. For example, an unanswered question is the contribution of Titan's magnetospheric interaction to the energy balance of the atmosphere in view of the power dissipation rate of 5×10^9 W by magnetospheric electrons (Strobel and Shemanski 1982). Although first estimates concerning the processes involved have been published (Eviatar et al. 1982), a comprehensive model may not be available for some time.

Acknowledgments. The results reviewed in this chapter have essentially been derived by a collaborative effort between the Voyager magnetometer, plasma science, plasma wave and radio science teams. We acknowledge the support and significant contributions of the principal investigators N. F. Ness, H. S. Bridge and F. L. Scarf. In addition we acknowledge the particular contributions of our colleagues M. H. Acuña, S. K. Atreya, K. W. Behannon, J. W. Belcher, G. Daigne, M. D. Desch, A. Eviatar, M. L. Kaiser, W. S. Kurth, A. J. Lazarus, K. W. Ogilvie, B. M. Pedersen, G. L. Siscoe and E. C. Sittler. The research was supported by grants from NASA Headquarters, JPL, the Office of Naval Research and the Bundesministerium, Forschung, Technologie, F. R. Germany.

ORGANIC MATTER IN THE SATURN SYSTEM

CARL SAGAN, B. N. KHARE
Cornell University

and

JOHN S. LEWIS
University of Arizona

Organic matter is probably widespread in the Saturnian system. Low-temperature condensates contain small amounts of organic matter produced by spontaneous reactions in solar composition gases; radioactive decay in fine-grained ice-rock mixtures produces organic matter via gamma and beta radiolysis; thermal evolution of ice-plus-rock bodies to temperatures near 1000 K can cause Fischer-Tropsch reactions to occur; and outgassed reducing gases can, as on Titan, be processed by ultraviolet radiation and charged particles to form complex organic species. The albedo pattern of Iapetus is suggestive of organic matter produced by photolysis of ices and distributed by collisional volatilization of ices and ballistic diffusion.

On Titan, the Voyager IRIS experiment has identified nine organic species heavier than methane. Their detection confirms earlier deductions that synthesis of complex organic molecules is occurring in the atmosphere of Titan. The observed atmospheric composition is roughly consistent with expectations from kinetic calculations and with laboratory experiments on irradiated mixtures of methane and nitrogen. A complex organic heteropolymer is produced by passing an electrical discharge through such mixtures; it exhibits a reflectivity at visible wavelengths matching the observed reflectivity of Titan at wavelengths uncontaminated by methane absorption bands, and is consistent with available observations of Titan in the infrared and ultraviolet. The pyrolyzates of this material include many organic molecules central to terrestrial biology.

Contemporary production rates on Titan imply, over the age of the solar system, the deposition of a surface layer of condensed organic matter at least 100 m thick.

TITAN

The Voyager discovery (Hanel et al. 1981a; Maguire et al. 1981; Kunde et al. 1981) of at least nine organic molecules heavier than methane in the gas phase in the Titanian atmosphere has brought an important speculative subject to a firm experimental stage, and has effected a convergence between work in prebiological organic chemistry on Earth and solar system astronomy. The molecules range from ethane at 20 ppm down to butadiyne, cyanoacetylene, and cyanogen at 0.1 to 0.01 ppm. They include the alkanes, ethane and propane; the alkene, ethylene; the alkynes, acetylene, butadiyne (diacetylene) and methylacetylene; and the nitriles, hydrogen cyanide, cyanoacetylene and cyanogen. The heavy atom numbers range up to $n = 4$. These molecules are minor constituents in an atmosphere composed mainly of N_2 and CH_4 (in a roughly 20:1 ratio), with about 0.2% H_2, about 0.01% CO (Lutz et al. 1983), and 0.01 ppm CO_2 (Samuelson et al. 1983a).

Since one-third of the organic molecules discovered are nitrogenous, these molecules and more complex ones yet to be identified in the Titanian atmosphere cannot in general be described as hydrocarbons, although of course they all contain carbon. The word organic has been in general use for a century and a half for such molecules and is adopted here. The word carries no implication of biological origin. While such molecules are almost certainly involved in the events that led to the origin of life on Earth (the pentamer of HCN, for example, is adenine, an essential nucleic acid building block, and cyanoacetylene is an important condensation agent for organic polymerization), it is conceivably confusing to describe them as prebiological or prebiotic in a Titanian context, and we will avoid such usage here. Nevertheless, it is a reasonable speculation that the organic chemistry on Titan holds clues to the processes which on Earth some 4.0×10^9 years ago led to the origin of life.

The modern experimental study of prebiological organic chemistry began with the celebrated paper by Miller (1955), in which a mixture of CH_4, NH_3, and H_2O, circulated through boiling water, was subjected to an electrical discharge, and the results examined mainly by paper chromatography. A number of simple alpha-amino and alpha-hydroxy acids were found in the aqueous phase. In addition, Miller found that a sticky brownish residue accumulated on the tungsten electrodes; not much attention was devoted to it at the time. In a related experiment, Sagan and Miller (1960) subjected a roughly equimolar mixture of CH_4, NH_3, and H_2O, with a 10:1 H_2 excess to an electrical discharge in an attempt at rough simulation of the environments of the Jovian planets. Among the gas phase molecules detected by mass spectroscopy were ethane, ethylene, acetylene, hydrogen cyanide, and acetonitrile,

as well as formaldehyde and acetaldehyde. Thus, laboratory work on prebiological organic synthesis on the primitive Earth and on organic synthesis in
reducing planetary atmospheres both date back to the 1950s, and even then
were recognized as complementary investigations. Subsequently, a large literature has accumulated, especially in the context of terrestrial prebiological
organic chemistry (see, e.g., Miller and Orgel 1974).

Kuiper's (1944) finding of methane, the simplest organic molecule, as a
significant constituent of the atmosphere of Titan was the first critical discovery on the road to indicating a rich organic chemistry on Titan. (The short-
wavelength ultraviolet irradiation even of pure methane has long been known
to produce higher-molecular-weight hydrocarbons; see, e.g., Calvert and Pitts
1966). Kuiper's identifications were based principally on the 0.619 and 0.725
μm bands. Longward of 0.6 μm and into the near infrared, the methane bands
dominate the Titan reflection spectrum. Broadband photometry of Titan (Harris 1961) indicated a very red, ultraviolet-absorbing material at the level to
which we can see in Titan's atmosphere. The absence of a negative branch at
low phase angles in the polarization curve of Titan led Veverka (1970,1973),
and, independently, Zellner (1973) to deduce an opaque aerosol layer or cloud
deck, overlain by an optically thin Rayleigh atmosphere. By this time it was
well known (Woeller and Ponnamperuma 1969; Khare and Sagan 1973) that
ultraviolet irradiation or electrical discharge in CH_4-rich mixtures of cosmically abundant reducing gases produces a reddish brown organic heteropolymer, related to the residue on Miller's electrodes; it was no major step to
propose that the red cloud or aerosol in the methane-rich Titanian atmosphere
was composed, at least in part, of complex organic molecules (Sagan 1971,
1973,1974; Khare and Sagan 1973).

It was also suggested (Danielson et al. 1974) that this aerosol might be
similar to that inferred in the upper atmospheres of Jupiter and Saturn from
the absence of a dramatic near-ultraviolet increase in reflectivity due to
Rayleigh scattering (Axel 1972). There are two circumstances which make
the Titanian atmosphere a more clement environment for organic molecules
than the atmospheres of Jupiter and Saturn: (1) the 1000:1 hydrogen dilution
in the atmospheres of the major planets tends to drive chemical reactions towards the fully saturated hydrides, although this effect is far slower than linear with hydrogen dilution (Raulin et al. 1979; Scattergood 1982) and may not
be adequate to reduce organic production rates by several orders of magnitude
below those on Titan. Different organic products, of course, have different
sensitivities to hydrogen dilution. (2) The surface of Titan, at a temperature
around 94 K, provides a deep-freeze environment for the preservation of synthesized organic molecules that sediment out of the atmosphere. On Jupiter
and Saturn, on the other hand, convection carries any organic molecules produced in the upper atmosphere down to pyrolytic depths in about one month
(Sagan and Salpeter 1976). By comparison, the sedimentation time for Titanian aerosol particles to reach the cold surface is less than about one year

(Toon et al. 1980). Thus, other things being equal, organic molecules in the atmospheres of the Jovian planets should be considerably less abundant than they are on Titan.

There are at least three possible approaches to predicting the species and abundances of complex organic matter in a given planetary environment. The first, thermochemical equilibrium, always reflects the intrinsic thermodynamic instability of organic matter. The equilibrium mole fractions of biochemically interesting species under normal temperature and pressure conditions are always negligible; however, appreciable traces of HCN and acetylene may be present at high temperatures (Sagan et al. 1967; Lippincott et al. 1967; Lewis 1969,1980). These high-temperature shock processes may make products that, when quenched to normal temperatures, are thermodynamically unstable but also kinetically unreactive. Thus chemical equilibrium theory applied to thunder- or meteorite-shocked gases may make useful predictions about the departure of cooled post-shock gas from equilibrium. For applications of these ideas to Jupiter, see Sagan et al. (1967), Bar-Nun (1975), and Lewis (1980). For Titan, thermodynamic equilibrium processes are of dubious applicability, in part because ultraviolet photochemistry—known to be a major mechanism for the production of Titanian organics (see below)—is a distinctly disequilibrium process, while thunder and shock waves from impacting objects are probably secondary processes energetically. No calculations have been made which purport to show that the observed relative abundances of gas phase organics on Titan are predicted by thermodynamic equilibrium.

The second approach is that of absolute reaction rate kinetic theory. Here, a network of reaction pathways connecting parent and product molecules is laid out. To the degree that the reaction rate constants are known or can be estimated, the set of differential rate equations can be solved numerically. This technique has been applied to Titan by Allen et al. (1980), Strobel (1978,1982), Strobel and Shemansky (1982), and Yung et al. (1983), and is especially fruitful when applied to molecules containing few carbon atoms, for which the laboratory kinetic data are more nearly complete. For more complex products, the dearth of rate constant data and the computational difficulty presented by the proliferation of reaction paths combine to render this approach ineffective.

This leads us to the third approach, the technique of last resort: experimental simulation. Ideally, a gas mixture identical in composition with that on Titan, held at a temperature and pressure appropriate to the selected level in the atmosphere of Titan, is subjected to appropriate disequilibrating energy sources capable of breaking chemical bonds. Experiments intended to simulate the effects of ultraviolet light, electrical discharges, proton and electron bombardment, and alpha particle and gamma ray irradiation have been carried out. But simulation of the most important of these processes on Titan, ultraviolet photolysis, is particularly daunting. Realistic pressures for absorption of short-wavelength ultraviolet photons cannot be simulated. At any pressure, but especially at the low pressures at which methane photolysis takes place, the reaction vessel introduces wall effects, the surface of the vessel catalyzing or inhibiting chemical reactions. (The surface of Titan itself will provide a wall effect, although one probably different from that provided by silicate glass.) In closed-cycle experiments, reaction products (e.g., H_2) will build up, changing the chemistry of the principal reactants late in the experiment. Prudent experimenters will monitor the reactant gases and terminate the experiment before their relative proportions change significantly, or will perform open-cycle experiments. Simulation of the actual temperature on Titan (which may strongly influence the product distribution) is very difficult and almost never attempted. The problems faced by attempts to simulate conditions in reducing atmospheres in the outer solar system have been critically surveyed by Prinn and Owen (1976).

For Titan, the first of these techniques is of undemonstrated relevance; while the second and third provide partially overlapping and partially complementary insights into parts of the problem. For complex organics, the experimental approach, for all its shortcomings, becomes the method of choice. It is clear that an accurate simulation, if possible, would be of major importance. We thus are led to ask what energy sources are present on Titan, how much energy they dissipate, and where in the atmosphere the influence of each such process is most strongly felt.

In Table I is a compilation of energy sources available (some, such as interplanetary electrons in the solar wind, only intermittently) for organic synthetic processes known to operate at radial distances < 3900 km in the Titanian atmosphere. The energy sources are listed in rank order of decreasing altitude. The calculations are discussed more fully in Sagan and Thompson (1984). The penultimate entry in the table covers the case that ultraviolet photons at wavelengths as long as 2200 Å can be gainfully employed in higher organic synthesis with a quantum yield ~ 0.1. Even neglecting this possibility, we see that ultraviolet photolysis is by far the most effective mechanism for organic synthesis in the atmosphere of Titan, with charged particles from the Saturnian magnetosphere running a distant second. Photodissociation occurs mainly in the 10^{-7} to 10^{-5} mbar pressure level, pressures so low that a reaction vessel several thousand km in diameter would be required to achieve photodissociation optical depth unity. This would be an unwieldy as well as an expensive experimental apparatus. Thus no experiments have been performed which adequately simulate irradiation by the CH_4 photodissociation continuum. Even a reaction vessel designed to simulate the putative photodissociation at $\simeq 2000$ Å would have to be some tens of meters in diameter. Consequently, laboratory simulations of ultraviolet photolysis on Titan (and indeed of the interaction of any other principal energy source with the Titanian atmosphere) are forced to use inappropriate pressures, at which 3-body reactions occur at rapid rates, unlike the actual circumstances in the

TABLE I

Rates of Organic Synthesis on Titan ($r < 3900$ km) [a]

Energy Source	Altitude of Peak Dissociation (from Center of Titan) (km)	Pressure Level of Peak Dissociation (mbar)	Dissociation Rate ($cm^{-2} s^{-1}$)	Synthesis Rate of Organic Molecules, Integrated Over the History of Titan ($g \, cm^{-2}$)
N_2 photolysis, $\lambda \leq 800$ Å	3700	10^{-7}	1.2×10^9	3×10^3 [b]
Interplanetary electrons	3550	10^{-6}	6×10^7	20
CH_4 photolysis, $\lambda \leq 1550$ Å (mainly H Lyman-α)	3500	10^{-6}	4×10^9	10^4
Saturn magnetosphere, energetic electrons	3300	3×10^{-5}	5×10^8	10^2
Saturn magnetosphere, energetic protons	3150	5×10^{-3}	4×10^7	10
Photolysis, higher organics, $\lambda > 1550$ Å, $\phi \sim 0.1$	2850	10^{-1}	2×10^{11}	5×10^5
Cosmic rays	2650	20	2×10^8	50

[a] Table after Sagan and Thompson 1984.
[b] Production rate of organics from N_2 photolysis depends on uncertain efficiencies (Strobel 1982).

upper atmosphere of Titan. The ratio of the rate of 3-body reactions to the rate of competing 2-body reactions is proportional to the pressure, and thus the relative importance of these types of reactions is changed by a factor of 10^5 by moving from the 10^{-8} bar typical of the methane photolysis region on Titan to the 10^{-3} bar which is the lowest in any Titan laboratory simulation yet attempted. It should also be mentioned that certain classes of important reactions, such as radical-molecule abstraction reactions, have large activation energies and therefore rate constants that are strongly temperature-dependent.

So far as we know, the first published results on organic synthesis in N_2/CH_4 mixtures are by Dodonova (1966), who employed a hydrogen discharge lamp emitting a continuum between 1250 and 1700 Å and reactants at both a few mbar and low hundreds of mbar. In the former, but not the latter, pressure range she reported synthesis of HCN. This result has not been confirmed in later ultraviolet irradiation of CH_4/N_2 mixtures (see, e.g., Chang et al. 1979), but those subsequent experiments have not simultaneously duplicated Dodonova's wavelength and pressure ranges.

Bombardment by low-energy charged particles provides $\sim 10\%$ of the disequilibrating energy flux on Titan (see Table I). Although there is at present no evidence for lightning in the atmosphere of Titan, the effects of other charged-particle processes may perhaps be simulated to some degree with electrical discharges. The first discharge experiment on CH_4/N_2 atmospheres was performed by Sanchez et al. (1966) at ~ 1 bar. Principal products were ethane, butadiyne, HCN, and cyanoacetylene, all of which have since been detected on Titan by the Voyager 1 IRIS experiment. Balestic (1974) performed a 17 MeV electron irradiation of an N_2/CH_4 mixture at a pressure of 950 mbar and produced, in the gas phase, ethylene, ethane, propane, hydrogen cyanide, acetonitrile, and propionitrile. Toupance et al. (1975) employed a silent electrical discharge for a few seconds in N_2/CH_4 mixtures at a pressure of about 28 mbar, producing ethane, ethylene, acetylene, and other hydrocarbons up to C_4, cyanogen, hydrogen cyanide, cyanoacetylene, acetonitrile, propionitrile, and acrylonitrile. In this experiment, essentially all of the organics later found by the Voyager IRIS experiment were produced. Scattergood et al. (1974,1975), in a proton irradiation of a CH_4/N_2 mixture at pressures approaching 1 bar, reported hydrocarbons up to C_3, butadiyne, and acetonitrile.

In a post-Voyager series of studies (Gupta et al. 1981), N_2/CH_4 mixtures with CH_4 mixing ratios ranging from 1 to 4% and total pressures ranging between a few mbar and a few hundred mbar were subjected, in separate experiments, to ultraviolet irradiation, electrical discharge, gamma irradiation, and proton and electron beams. To first order, all Voyager molecules were produced with charged-particle irradiation as the energy source, but only a third to a half of these products were produced by electromagnetic radiation. Relative abundances of organic products in these experiments are qualitatively similar to those found by the Voyager IRIS experiment, but no detailed quantitative agreement has yet been reported. A pre-Voyager review of organic chemistry on Titan (Chang et al. 1979) is now somewhat obsolete, but still useful.

We note that energy deposition from charged particles in the Saturnian magnetosphere and in the solar wind tends to be concentrated at high altitudes and very low pressures (Table I). It is not apparent that laboratory experiments at ~ 1 bar are an adequate simulation of events that occur on Titan at $\sim 10^{-8}$ bar. The similarity of products produced in the two cases, however, may indicate a much weaker dependence of results on pressure and temperature than is obvious. Also, the cosmic ray energy deposition (Table I) is nearly the most productive charged-particle energy source and does occur in the lowest part of the atmosphere of Titan.

Sagan (1971) and Khare and Sagan (1973) have proposed that a material similar to the complex organic solid first noted by Miller (1955) may be the principal reddish chromophore observed in the atmospheres of the outer solar system. This suggestion has been disputed for the Jovian planets by Lewis and Prinn (1970) and by Prinn and Lewis (1975), who have argued that the chromophores on the Jovian planets are inorganic polymers of sulfur and phosphorus, derived from the photolysis of H_2S and PH_3. However, for Titan, which contains no detectable trace of either of these gases, the organic hypothesis achieves its greatest cogency. It is on Titan, if anywhere, that the presence of complex organic materials should be expected.

For convenience, these materials need a name. They are chemically different from the kerogens of biological origin in terrestrial sediments and from the organic matter in carbonaceous chondrites. They are not, strictly speaking, polymers, since they do not seem to be a simple repetition of a small number of common monomers. The generic term tholins,* derived from the Greek for "muddy," has been proposed by Sagan and Khare (1979); it is independent of any detailed assumptions on the chemistry of the materials so described, and seems preferable both to terms with specific chemical referents (asphalt, tar, etc.) and to terms of excessive generality (goo, gunk, etc.). Explicitly, tholins are defined as organic solids produced by the irradiation of mixtures of cosmically abundant reducing gases.

In Table II are listed the properties of the three categories of tholins to which the greatest attention has been devoted. Some of the optical properties of spark tholin were used for a proposed upper atmospheric aerosol in Danielson et al.'s (1974) early model of the ultraviolet reflection spectrum of Titan. The Titan tholin was originally produced from a simulated Titanian atmosphere (at a total pressure > 10 mbar), and is the solid organic material most relevant to the properties of the Titanian clouds (Sagan and Khare 1981,1982). Subsequent production has been accomplished at pressures near 0.2 mbar

*This term invented by Sagan and Khare is not universally accepted in the literature at this time.

TABLE II
Summary of Some Tholin Properties [a]

Tholin	Ultraviolet	Spark	Titan
Precursors	CH_4, NH_3, H_2O, C_2H_6, H_2S (roughly equimolar)	CH_4, NH_3, H_2O ($\sim$ 2.5%)	CH_4, N_2 (1:10 ratio)
Energy source	2537 Å Hg discharge	high-frequency Tesla coils	high-frequency Tesla coils
TGA, 50% thermal dissociation temperature	$\cong$ 300°C	$\cong$ 900°C	$\cong$ 550°C
Atomic composition, apart from S:	H, 3.8%; C, 37%; N, 11%; O, 48% [b]	H, 5.0%; C, 47%; N, 36%; O, 12% [b]	H, 6.4%; C, 48%; N, 29%; O, 17% [b]
Principal GC/MS pyrolyzates	alkanes, alkenes, aromatic hydrocarbons, alkylthiophenes, alkylmercaptans, alkyldisulfides, isothiocyanates, and some nitriles	alkanes, alkenes, abundant nitriles, aromatic hydrocarbons, alkylbenzenes, indenes, indanes, pyrroles, pyridine, and pyrazines	alkanes, alkenes, abundant nitriles, aromatic hydrocarbons, alkylbenzenes, indenes, pyrroles, pyridines, pyrazines, amines, purines, and abundant pyrimidines

[a] See also Khare et al. 1978, 1981, 1982.
[b] Tholin particles exposed to air are known to acquire an oxidation and/or hydration film, which shows on sequential GC/MS pyrolysis as a large low-temperature CO_2 peak.

(Khare et al. 1983). A material similar to Titan tholin has also been produced by Balestic (1974) and by Scattergood and Owen (1977). Balestic noted a chromatographic peak corresponding to mass number 970, and the presence of CN, CH_2NH, and $CH = N$ functional groups, although the material was not further characterized. Scattergood and Owen irradiated a CH_4/N_2 mixture with 3-MeV protons at a pressure of 600 mbar, and produced a "reddish-brown oil," which on the basis of its infrared spectrum was characterized as a mixture of complex hydrocarbons with amino groups attached. The spectrum showed no sign of the 4.6-μm nitrile functional group, and the authors state that no evidence of nitriles was found. No further physical or chemical characterization of the oil was attempted. Our Titan tholin is a solid, not a liquid; exhibits the 4.6-μm nitrile absorption feature; and differs in several other respects from the product of Scattergood and Owen. The reasons for these discrepancies, conceivably due to the difference between electrical discharge and proton irradiation, remain unknown to us, and to Scattergood (personal communication, 1981).

In a typical experiment, indicated schematically in Fig. 1, a mixture of 9% CH_4 and 91% N_2 was exposed for $\sim 10^7$ s at a total pressure of 73 mbar to high-frequency electrical discharge (corresponding approximately to 50-keV electrons). A thin, red tholin film gradually builds up in the reaction vessel. Scanning electron microscopy reveals a particle size distribution function peaking between 0.1 and 0.5 μm, with occasional irregular large particles present to tens of μm. A preliminary laboratory polarimetric determination of the wavelength-dependent real part of the refractive index of this tholin is $\cong 1.6 \pm 0.1$. These particle sizes and refractive indices are consistent with those deduced for the Titan aerosols from Earth-based observations (Rages and Pollack 1980); from Pioneer 11 observations (Tomasko and Smith 1982); and from Voyager observations (Rages and Pollack 1983; Rages et al. 1983). Subsequent work shows the real and imaginary values of the complex refractive indices of these materials to be in excellent agreement with those derived for the Titan aerosols (Khare et al. 1983).

Since, for whatever reason, the particle size distribution of Titan tholin appears to match that deduced for the Titanian clouds, it is appropriate to compare the reflectivity of the laboratory sample with that measured by groundbased spectrophotometry for Titan. A comparison of the spectral reflectivity of both Titan and spark tholins with that of Titan between 0.38 and 1.1 μm is shown in Fig. 2. The individual points of measurement have been displayed to indicate probable errors of measurement. The reflectivities are normalized to unity at 0.56 μm. Between 0.38 and 0.60 μm, where the CH_4 bands become prominent, the reflectivity of the tholin follows, within the probable errors, the measured reflectance of Titan. At longer wavelengths the tholin, not masked by CH_4 absorption, continues to increase in reflectivity. The comparison is exhibited in expanded scale for $\lambda < 0.60$ μm in Fig. 3, while in Fig. 4 we see a representative and entirely unsuccessful comparison

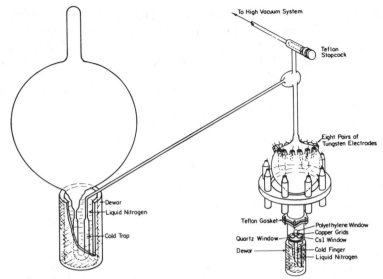

Fig. 1. Experimental apparatus for Titan chemistry simulation, Cornell University.

of two quite different materials, an allotrope of elemental sulfur and one of red phosphorus. Our preliminary conclusion is that the Titan tholin provides a natural and plausible candidate for the principal chromophore in the Titanian clouds.

We do not claim that what is exhibited in Figs. 2 and 3 is a unique spectral agreement. For example, Podolak et al. (1979) have achieved a fair agreement between the observed Titan reflection spectrum and the spectrum of ultraviolet-irradiated polyethylene. However, while ethylene is a constituent of the Titanian atmosphere, polyethylene, a pure hydrocarbon, is not expected to be an abundant solid-phase product produced from an atmosphere which is composed mainly of molecular nitrogen, and the Titan tholins are, of course, not mainly polyethylene (cf. Table II). Our principal contention is, we think, fairly plausible: the chromophores in the Titanian clouds are those produced on irradiating a simulated Titanian atmosphere. This is certainly the simplest and most straightforward hypothesis.

The ultraviolet spectrum of Titan tholin between 0.2 and 0.3 μm is essentially of low albedo and featureless, in rough accord with the IUE measurements of Caldwell et al. (1981). A determination of the complex refractive indices of Titan tholin from ultraviolet to microwave frequencies has recently been completed (Khare et al. 1983); ultimately, this will permit a simulation of a detailed radiative transfer model of Titan's atmosphere and clouds, with methane absorption added, to test further the hypothesis that something very similar to Titan tholin is responsible for the coloration of the clouds of Titan (cf. Sagan et al. 1983).

Infrared reflectance spectroscopy of the three tholins described in Table

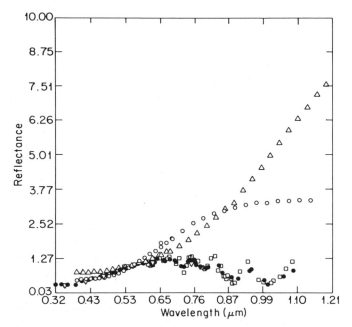

Fig. 2. A comparison of the reflectivity of Titan tholin (circles) and spark tholin (triangles), as measured on the Cornell goniometer, with spectrophotometric observations of Titan by McCord et al. (1971) (filled circles); Noland et al. (1974) (dels); and Younkin (1974) (squares). The agreement, until onset of CH_4 absorption near 0.6 μm, is satisfactory.

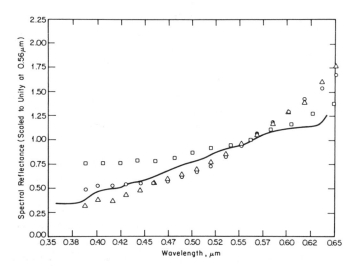

Fig. 3. Comparison of the measured groundbased spectral reflectivity of Titan (McCord et al. 1971: solid line) with three tholins performed on the Cornell goniometer. The three are Titan tholin: ○; spark tholin: □; ultraviolet tholin: △. Data are normalized to 0.56 μm.

II has also been performed. Longward of 6 μm, where Voyager IRIS data become available, the Titan tholin shows a low albedo of almost featureless aspect. Short of 5 μm, the spectrum of all three tholins is as displayed in Fig. 5, to which suggested functional groups responsible for absorption features have been added. All three tholins show the 4.6-μm nitrile feature; it is present in the Titan tholin very clearly. The very large absorption feature near 3 μm is due to C–H and N–H stretching modes in the Titan tholin (and possibly to O–H from adsorbed atmospheric water vapor in the laboratory). Groundbased or airborne spectroscopy of Titan to confirm the presence of the 3-μm and 4.6-μm features is feasible and important to carry out. In the longer term, a near-infrared mission spectrometer (NIMS) of the sort to be carried on the Galileo mission to Jupiter would be a promising addition to a Titan flyby or orbiter spacecraft.

The analysis of such tholins, once described as intractable polymers, is extremely difficult due to their complexity and involatility. Preliminary sequential and nonsequential pyrolytic gas chromatography/mass spectrometry (GC/MS) of the Titan tholin reveals a very rich array of organic pyrolyzates, as summarized in Table II. The GC/MS protocols extend only up to 700°C, while we know from thermogravimetric analysis (TGA) that one-third of the tholin mass remains involatile at this temperature. The molecules listed in Table II therefore are derived from only the most volatile fraction. Of course, organic molecules at temperatures near 1000 K may rearrange and react extensively, and some of the pyrolyzates may be synthesized by pyrolysis, not by the process that formed the tholin. On the other hand, cosmic ray reprocessing (cf. Table I) of tholins sedimented on the surface (or beneath the putative hydrocarbon oceans) of Titan may yield products similar to those produced by pyrolysis. In any event, it is very likely that the Titan tholin is a richly complex organic heteropolymer, including as components a large number of organic molecules that are central to terrestrial biology. The determination of which of the observed species are present in the unheated tholin is an exciting area for future research. This question is doubly important because of the likelihood that the theory of absolute reaction rate kinetics will be unable, at least in the near future, to make predictions about the production of such complex species.

In the experiments of Scattergood and Owen (1977) the 3-MeV irradiation of CH_4/H_2 mixtures produced a colorless liquid, from which the authors deduced that "species containing elements other than carbon and hydrogen are necessary for the production of color." Among other possibilities, they proposed nitrogen (as NH_3 or N_2) and sulfur (as H_2S) as candidate additional materials for the Titanian atmosphere in order to produce cloud chromophores. Since the discovery of N_2 in Titan's atmosphere by Voyager 1, this argument has been referred to as a prediction of the presence of N_2 on Titan.*

*The other principal early argument for nitrogen followed from Trafton's (1972a,b) an-

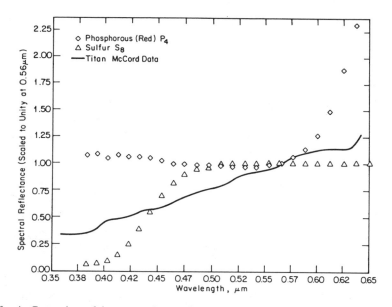

Fig. 4. Comparison of the measured groundbased spectral reflectivity of Titan (solid line) with S_8 and P_4, performed on the Cornell goniometer. Data are normalized to 0.56 μm.

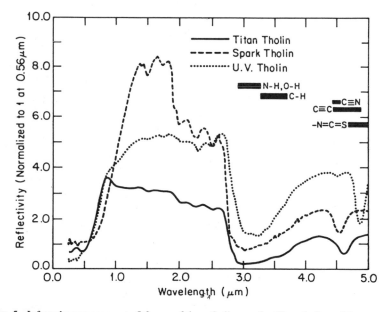

Fig. 5. Infrared spectroscopy, to 5.0 μm, of three tholin samples (Titan tholin: solid curve; spark tholin: dashed curve; ultraviolet tholin: dotted curve), in specular reflection with proposed organic functional group assignments. The 1.5 and 2.0 μm features are probably due to adsorbed water.

It is therefore of interest to ask what the primary sources are of visible light opacity, and particularly blue absorption, in common terrestrial organic materials:

1. Chlorophyll, hemoglobin, and their congenors are metal-chelating tetra-pyrroles, the pyrroles being five-membered nitrogenous rings.
2. Indole derivatives and products of which perhaps the best known is melanin, the principal coloring agent of human skin. Indoles are nine-membered nitrogenous heterocyclic compounds.
3. Conjugated polyenes. These are aliphatic hydrocarbons with alternating single and double bonds which begin to absorb visible light at about carbon number six (Sondheimer et al. 1961). They are pure hydrocarbons.
4. Polycyclic aromatic hydrocarbons. Here the visible light absorption is caused by π electrons in the benzene rings. The most common examples are asphalts and tars, which are among the darkest materials known.

These examples suffice to demonstrate that only if we are restricted to very low carbon numbers does it follow that hydrocarbons are uncolored in the visible. In fact, (unchelated) pyrroles, indoles, conjugated polyenes, and polycyclic aromatics have all been detected as pyrolyzates in CH_4/NH_3 or CH_4/N_2 tholins (Table II), although their presence in unpyrolyzed tholin is of course not yet fully demonstrated. It is therefore still unclear whether the principal chromophores of the Titan aerosols are nitrogenous.

From the organic molecular production rates estimated in Table I, we see that at least a hundred-meter thickness of tholins should have accumulated on Titan's surface over geological time. Because of the low surface temperatures and the low radiation fluxes at the surface, relatively little reworking of this material should have occurred. The key component present on the primitive Earth and likely to be absent on Titan is liquid water as an interaction medium for these molecules (although it is not out of the question that episodic endogenous or exogenous melting events on Titan have led to the temporary availability of liquid water). At the very least, the surface of Titan seems to hold a rich storehouse of information on the earliest stages of the complex organic chemistry that on Earth some 4×10^9 yr ago led to the origin of life. It is an extraordinarily promising objective for an entry probe and ultimately a lander spacecraft mission equipped for sophisticated analytic organic chemistry.

---

nouncement of large quantities of molecular hydrogen in Titan's atmosphere. This would have led to rapid blow-off of the entire atmosphere (Sagan 1973; Hunten 1973) contrary to observation; one of several proposed resolutions of this problem was to increase the mean molecular weight of the atmosphere significantly with a major admixture of N_2 (Hunten 1978). However, the evidence for large quantities of H_2 on Titan has evaporated even more rapidly than the hydrogen-rich atmosphere would have (Munch et al. 1977; Hunten 1978).

IAPETUS

Voyager images (B. A. Smith et al. 1981,1982) reveal Iapetus to have the largest known albedo variance in the solar system, from 0.03 near the center of the leading hemisphere to more than an order of magnitude brighter towards the antapex of motion and, especially, near the poles. Photometric contours of albedo variation are very similar to the calculated variation with position of the predicted cometary impact flux; the darkest areas correspond to regions of highest flux (Squyres and Sagan 1983).

The dark regions are also found to be red on Voyager images (Squyres and Sagan 1983) consistent with groundbased measurements (Cruikshank et al. 1983). The optical frequency spectral reflectance of the trailing hemisphere of Iapetus is less red, consistent with a mixture of leading hemisphere dark material and ice; infrared spectra clearly show the presence of water ice (Cruikshank et al. 1983). Spectral properties of the dark-side material are inconsistent with basalts and with other dark silicates, and also with C-type asteroids and carbonaceous chondrites; but they are consistent with organic residue from the Murchison C-2 carbonaceous chondrite (Cruikshank et al. 1983). More recently, Iapetus' dark side has been matched by a mixture of hydrated silicates and "10% simulated meteoritic organic polymers" resembling the Murchison residue (Bell et al. 1983). The low albedo and red spectral properties of this material are typical of many varieties of complex organic compounds. However, the argument for abundant organic matter especially on the leading hemisphere is still inferential. A search on Iapetus for the 2900 cm^{-1} C–H bond—known, not surprisingly, to be present in water-soluble or organic extracts of carbonaceous chondrites (Briggs 1961; Meinschein et al. 1963)—would be very useful.

The origin of the presumptive organic material on Iapetus and the reasons for its curious areal distribution are matters of some debate. Proposals involving material ejected from Phoebe and dragged towards Saturn by Poynting-Robertson drag (Soter 1974) run afoul of the large color differences between Phoebe and Iapetus and the presence of dark-floored craters on the trailing hemisphere of the latter (Squyres and Sagan 1983). Proposals involving the extrusion of organics preferentially to the surface on the leading hemisphere run afoul of the close agreement between photometric contours and the meteorite flux isopleths (Squyres and Sagan 1983).

Squyres and Sagan suggest two models consistent with the observed distribution of albedo. In the first, a thick layer of organics is generated—e.g., by hypervelocity impacts, or possibly (see below) by gamma radiolysis—from CH_4-rich ice, and is then covered by condensed volatiles which are then preferentially eroded from the leading hemisphere. In the second model, which they tentatively prefer, organic chromophores are produced *in situ* by solar ultraviolet irradiation of CH_4-rich ice. They calculate that the albedo pattern can be understood by impact volatilization of exposed methane hydrate and ballistic diffusion of surface material in response to the impact flux gradient. Similar processes may be occurring on Hyperion and, especially (because CH_4-rich ice is even more likely there) on the Uranian satellites (Squyres and Sagan 1983). NIMS-type spectroscopy of Iapetus and other satellites with a presumptive partly organic surface would be a very useful component of future spacecraft missions.

ORGANIC MATTER ON THE SMALLER SATELLITES

More generally, we have two valuable lines of evidence regarding the presence of organic matter in low-temperature solar system materials. First, we have the extensively studied carbonaceous chondrites. The most volatile-rich and most highly oxidized members of this class, the C1(I) chondrites, contain up to 6% by weight of organic matter in intimate association with hydrated salts, clay minerals, sulfates, phosphates, carbonates, and magnetite. It has long been supposed that the carbonaceous chondrites were produced by low-temperature equilibration in the solar nebula, since several of the major minerals in them are in fact thermodynamically stable in the nebula at temperatures below about 360 K (Larimer and Anders 1967; Lewis 1974). It was then reasonable to suppose that the organic matter, which is intimately associated with precisely those meteorites whose mineralogy suggests the lowest temperature of origin, was also produced in the nebula. The production of this material in Fischer-Tropsch (F-T) type reactions, in which CO and hydrogen react catalytically on mineral grain surfaces, was suggested as a nebular process by Anders and coworkers (see, e.g., Studier et al. 1968). However, several recent studies of the petrology of the C1 chondrites have suggested strongly that the characteristic minerals in them were formed by secondary oxidation and hydration reactions within the meteorite parent body, at temperatures near the melting point of ice and pressures not much less than (and perhaps much greater than) one atmosphere (Kerridge and Bunch 1979; McSween 1979; Bunch and Chang 1980). This growing consensus, combined with the severe difficulty of making organic matter via F-T reactions in the nebula with ambient CO pressures near 10^{-9} bar, suggests that the origin of the organic matter in these meteorites be reassessed as well.

One recent suggestion (Hartman et al. 1983) attributes the water-soluble organic matter to gamma radiolysis of a fine-grained mixture of rock dust and ices, the extractable hydrocarbons to F-T reactions within the meteorite parent body, and the intractable polymer to presolar (possibly interstellar) processes. This is in accord with the recent discovery that these three classes of organic matter differ substantially in isotopic composition, and cannot be made from each other by any straightforward physical or chemical fractionation mechanism (Robert and Epstein 1982; Becker and Epstein 1982). In this hypothesis the hydrocarbon (F-T) products are actually synthesized under conditions closely similar to those used in the laboratory experiments of Studier et al. If this suggestion is correct, then the biochemically interesting species in the carbonaceous chondrites are of secondary origin. They were made by the most rudimentary expression of the natural radioactivity of solar system solids, and these processes must be at work wherever a condensed mixture of ices and rock powder is present. This mixture should have been present in asteroids in the outer part of the belt (see the optical evidence on asteroid compositions as a function of heliocentric distance presented by Gradie and Tedesco [1982]). It should also have been present in ice-rich planetary satellites and in comets.

It is the spectroscopic study of cometary volatiles that provides the second line of evidence. Instead of exhibiting a large content of methane and ammonia, which would be expected if cometary ices were formed in chemical equilibrium in a cold solar-composition gas (Lewis 1972b), comets display substantial contents of CO, N_2, HCN, possibly CO_2, and only a little ammonia (Delsemme 1982). This is easily understood as a consequence of the kinetic inaccessibility of the equilibrium state of the system at low nebular temperatures. A theoretical study of the kinetics of homogeneous and heterogeneously catalyzed reactions in the solar nebula predicts that the condensation of water ice should be accompanied by condensation of the ionic salt ammonium bicarbonate and the organic salt ammonium carbamate in amounts up to a few percent of the mass of ice (Lewis and Prinn 1980). The latter salt contains a covalent C–N bond. Volatilization of this mixture provides HCN, carbon dioxide, and ammonia in addition to abundant water vapor. Because of the high densities expected in circumplanetary subnebulae, the formation of methane and ammonia will be much more rapid, and large amounts of these substances will be present along with traces of molecular nitrogen and the oxides of carbon (Prinn and Fegley 1982). Thus, condensates formed at similar temperatures in the solar nebula and in circumplanetary nebulae may differ significantly in composition. In both cases, however, gamma irradiation by decaying potassium-40 (or aluminum-26, if present) will drive a substantial amount of organic synthesis.

Sufficiently small ice-plus-rock bodies, if heated only by long-lived radionuclides like potassium-40, will never warm up enough to melt and differentiate, and will never develop core temperatures high enough for Fischer-Tropsch reactions to occur. Only heat sources capable of replenishing heat faster than it is lost to space (by conduction and solid-state convection followed by radiation) can force melting of small, initially cold bodies. These short-lived heat sources include aluminum-26 decay, heating by the T-Tauri phase solar wind, heat released by the accretion of a nearby planet, and tidal dissipation of rotational or orbital energy. It appears that bodies less than a few hundred kilometers in diameter could melt and differentiate only if one or more of these short-lived heat sources were available. Even if no melting occurs, however, primordial organic matter and gamma radiolysis products should be present.

The organic content of undifferentiated icy bodies may therefore be quite significant. If we then consider the possible formation of additional organic matter by energetic processes operating on the exposed surfaces of these bodies, it would seem that organic matter from several different sources must be widespread in the outer solar system. Solar ultraviolet photolysis of the simple hydrocarbon component of ices such as clathrate hydrates, and shock heating by small impact events (both discussed in the previous section) are likewise possible sources.

IMPLICATIONS

We have emphasized the importance of faithful simulation of planetary environmental conditions, such as pressure, temperature, initial composition, and energy sources in laboratory studies. For this reason, we feel obliged to exercise caution in the application of experience with Titan to other planetary bodies. The atmospheres of Jupiter and Saturn approximate the elemental composition of the Sun, and deviate very strikingly from the nitrogen-plus-methane environment on Titan. We have cited in passing a number of simultion experiments that have been carried out under conditions closer to those on the Jovian planets than to those on Titan. Even there, absolute temperatures twice those appropriate for the planets are usually used for the sake of convenience, and hydrogen is usually very severely depleted relative to its cosmic abundance. Energy sources many orders of magnitude larger than those available on the planets are almost always used, wall effects may be important, and energy sources are often used on gas mixtures and at pressures that in reality would never be exposed to such sources. If we were in possession of a detailed and reliable theory of the effects of, for example, ultraviolet irradiation, then we could adjust the product yields from simulation experiments to take into account the differences in environmental conditions between the simulation and the planet of interest. We are as yet unable to do this well, even for Jupiter. We are in a much less secure position with respect to Titan. To translate from Titan's environment to Jupiter's or vice versa is far beyond our skill. To translate either of these discussions carelessly into the warm oceans of the early Earth would be unjustified. Nevertheless, the emerging picture that Titan presents to us strengthens the contention that large-scale synthesis of complex organic molecules occurred on the early Earth, and was antecedent to the origin of life.

With these disclaimers in mind, let us try to identify the conclusions from this survey that may have more general significance. First, Titan has brought to our attention a new environment in which abiological chemical evolution can begin. It broadens our appreciation of the ability of energy fluxes to create complex molecules out of the tedious chaos of a near-equilibrium gas. We are led to wonder what materials actually are present on the surface of Titan; whether complex compounds of C, H, N, and other atoms are dominant;

whether a chemically complex ocean of hydrocarbons (Dermott and Sagan 1982; Lunine et al. 1983) and lipophilic organic compounds is present; and to what degree the aerosol that dominates the reflection spectrum of Titan is in fact chemically similar to that synthesized in laboratory experiments on nitrogen-methane mixtures.

Second, the remainder of the Saturnian system reminds us in a less direct way that primordial organic matter was almost certainly very widespread in the outer solar system at the time of planetary formation, probably accounting, as in carbonaceous chondrites, for more than 1% by weight of the solid materials available for the formation of satellites, rings, and comets. We are led to look more closely at the chemistry of fine-grained mixtures of ices and rock, in which the radiolysis of the ices by natural radionuclides in the rock dust provides a driving energy flux capable of breaking several percent of the chemical bonds in the original solids. Other energy sources may also be implicated. Experiments on such systems are easy to carry out, and should be undertaken without delay.

Third, our attention is drawn to the analogies between Titan, Triton, and Pluto. The logical consequences of the emerging chemistry of Titan must have their parallels on any body with a methane atmosphere or surface. The crucial difference is that the latter bodies may have never been warm enough for evaporation and subsequent photolysis of ammonia. They may be dependent on primordial clathrate hydrates of nitrogen for their present content of free nitrogen. We thus should expect no more than parallels in the chemistry, certainly not an identity.

Our principal conclusion, however, is that the unfolding view of Titan's chemistry has strong ramifications for future solar system exploration. Voyager has opened, not closed, a host of fundamental and exciting issues. Perhaps the most tantalizing of these is the clear realization that there is indeed a complex and diverse array of nonbiological organic syntheses at work in the outer solar system. Proper understanding of these synthetic processes requires further spacecraft study of Titan's lower atmosphere and surface, the surfaces of the smaller satellites and ring particles of the outer planets, and Triton and Pluto as well. The only part of this problem currently scheduled for spacecraft exploration, the clouds of Jupiter, is in many ways a less attractive arena than any of these for the detection and identification of organic matter—although the spectroscopic identification of simple gas phase hydrocarbons and HCN suggests that Jupiter may also be of some interest for the emerging field of organic chemistry in the outer solar system.

Acknowledgments. This research was supported in part by grants from the National Aeronautics and Space Administration. We are indebted to J. Gradie for assistance in the laboratory goniometer measurements and to W. R. Thompson for helpful discussions.

PART VII
Origin

ORIGIN AND EVOLUTION OF THE SATURN SYSTEM

JAMES B. POLLACK
Ames Research Center

and

GUY CONSOLMAGNO
Massachusetts Institute of Technology

As was the case for Jupiter, Saturn formed either as a result of a gas instability within the solar nebula or the accretion of a solid core that induced an instability within the surrounding solar nebula. In either case, the protoplanet's history can be divided into three major stages: early, quasi-hydrostatic evolution (stage 1); very rapid contraction (stage 2); and late, quasi-hydrostatic contraction (stage 3). During stage 1, Saturn had a radius of several hundred times that of its present radius R_S, while stage 3 began when Saturn had a radius of $\sim 3.4\ R_S$. Stages 1 and 2 lasted $\sim 10^6$–10^7 yr and ~ 1 yr, respectively, while stage 3 is continuing through the present epoch.

Saturn's current excess luminosity is due, in part, to the loss of thermal energy built up by a faster contraction that marked the earliest phases of stage 3. But, in contrast to the situation for Jupiter, this internal energy source fails by a factor of several in producing the observed excess luminosity. The remainder is most likely due to the gravitational separation of helium from hydrogen due to its partial immiscibility in the outer region of the metallic hydrogen zone.

The irregular satellite Phoebe was most likely captured by gas drag experienced in its passage through a bloated Saturn, just prior to the onset of stage 2. During stage 2, a nebular disk formed from the outermost portions of Saturn, due to a progressive increase in their rotational velocity as the planet contracted. This increase may have been enhanced significantly by a transfer of angular momentum from the inner to the outer regions of the planet. The nebu-

lar disk served as the birthplace of Saturn's regular satellites and probably the ring material. Viscous dissipation within the nebula caused an inward transfer of mass, and thus may have determined the nebula's lifetime, and an outward transfer of angular momentum. It is not clear what the relative roles of Saturn's luminosity and viscous dissipation were in determining the nebula's radial temperature structure and its evolution with time.

As Saturn's excess luminosity declined or less viscous dissipation in the nebula occurred during the early portion of stage 3, water was able to condense at progressively closer distances to the center of the system and water clathrates and hydrates were able to form throughout much of the nebula, especially in its outer regions. It is the likely presence of ices other than pure water ice in at least some of the regular satellites of Saturn that makes them chemically distinct from the large icy satellites of Jupiter. If Saturn's nebula had a high enough pressure (greater than several tens of bars) in its inner region, a liquid solution of water and ammonia, rather than water ice, would have been the first "icy" condensate to form.

Despite the comparatively small size (hundreds to about a thousand kilometers) of the inner satellites of Saturn, a number (especially Dione and Rhea) may have experienced significant expansion and melting during the first $\sim 1 \times 10^9$ yr due to the presence of substantial quantities of ammonia monohydrate (~ 10–20% by weight). The occurrence of the youngest known surfaces in the Saturn system on the comparatively small-sized Enceladus is most readily attributed to strong tidal heating created by its forced orbital eccentricity. But a significantly larger eccentricity is required at some time in its past for tidal heating to be quantitatively capable of initiating melting, with the current eccentricity being perhaps large enough to maintain a molten interior.

During the early history of the Saturn system, giant impact events may have catastrophically disrupted most of the original satellites of Saturn. Such disruption, followed by reaccretion, may be responsible, in part, for the occurrence of Trojans and coorbital moons in the Saturn system, the apparent presence of a stochastic component in the trend of satellite density with radial distance, and the present population of ring particles.

Titan's atmosphere formed from the hydrates and clathrates—especially ammonia monohydrate and/or nitrogen clathrate and methane clathrate—that constituted the satellite. Over the age of the solar system, a nontrivial amount of atmospheric nitrogen (about several tens of percent of the current atmospheric inventory) and much more methane than is presently in the atmosphere have been lost, through a combination of N and H escape to space and the irreversible formation of organic compounds. These considerations imply quasi-real time buffering of atmospheric methane by a near-surface methane reservoir and the existence of a layer of 0.1 to 1 km thickness of organic compounds close to or on Titan's surface.

The Saturn system resembles the Jupiter system in a number of ways. The planet Saturn is massive (~ 95 Earth masses) by the standards of the inner solar system and is composed chiefly of hydrogen and helium. However, like Jupiter, it has a central core of "rock" (silicate and iron compounds) and perhaps "ice." * The planet has an internal heat source that causes it to radiate

---

*By "ice," we mean compounds derived from low-temperature condensates, such as water ice, ammonia hydrate, and methane clathrate. Naturally, at the conditions of Jupiter's interior, an "ice" component would be a supercritical fluid.

to space about twice as much energy as the amount of sunlight it absorbs. Surrounding Saturn is a miniature solar system composed of at least 16 regular satellites, one captured satellite (Phoebe), and, of course, its magnificent set of rings.

In all the above regards, the Saturn system differs markedly from the planets of the inner solar system. The terrestrial planets are much less massive; they are made almost entirely of "rock"; their internal heat source is orders of magnitude smaller than the amount of sunlight they absorb; and there are only three satellites, with perhaps all of them (or at least Phobos and Deimos) being captured objects.

Upon closer examination, however, the Saturn system differs in a number of important ways from the Jupiter system. First, Saturn's central core represents a larger fraction of the planet's mass, although the two core masses are quite similar in absolute value. Second (as discussed below) precipitation of helium may provide the chief energy source for Saturn's internal heat flux, whereas thermal cooling may act as the chief energy source for Jupiter. Water ice appears to be a major constituent of all of Saturn's regular satellites, whereas it is at best a minor component of Jupiter's innermost satellites. The Jupiter system has two families of irregular satellites, but the Saturn system has only one known irregular satellite. Finally, Saturn's rings are much more prominent and much more massive than Jupiter's.

In this chapter, we review our current understanding of the origin and evolution of the Saturn system and its individual components. We will seek not only to view Saturn in isolation, but also to compare theories of its history with those for Jupiter, the other giant planets, and the terrestrial planets. In so doing, we will attempt to understand the similarities and differences among these objects, as noted above, and to test the internal self-consistency of alternative theories.

I. CRITICAL CONSTRAINTS

Certain key observational data on the Saturn system are relevant to understanding its history. Some of these constraints were derived from Earth-based observations. However, measurements conducted from the Pioneer 11 and Voyager 1 and 2 spacecraft have greatly supplemented these results.

Models of Saturn's interior, constrained to match the planet's mass, radius, and J_2 and J_4 gravitational moments, suggest that it consists of two major, compositionally distinct zones: a central core with a mass of 20 ± 5 $M_{\oplus}$ (Earth masses) made of an unknown mixture of "rock" and "ice"; and an outer fluid envelope with a mass of 75 ± 5 $M_{\oplus}$ made of an *approximately* solar mixture of elements (Slattery 1977; Podolak 1978; Hubbard and Mac-Farlane 1980b; Grossman et al. 1980; Stevenson 1982a). This value for the core mass lies within a factor of two of the core masses of Jupiter, Uranus, and Neptune, despite a factor of 20 variance in their total masses.

The hydrogen-helium dominated envelope consists of two major regions that are distinguished by the phase of hydrogen in them. At pressures less than a few megabars, hydrogen is present as molecular hydrogen H_2, a good electrical insulator, while at higher pressures it occurs as metallic hydrogen, a good conductor. Saturn's metallic H zone constitutes about 20% of its envelope's mass, in contrast to a value of $\sim$ 70% for Jupiter.

There are several tentative deviations of the composition of Saturn's observable atmosphere from that expected for a solar mixture of elements. First, the helium mass fraction Y in Saturn's atmosphere appears to be significantly smaller than the corresponding value for Jupiter, 0.11 ± 0.03 versus 0.19 ± 0.05 (Hanel et al. 1981a). The latter value is close to the solar value. If true, this difference could be due to a preferential segregation of He toward the bottom of Saturn's envelope, driven by the partial immiscibility of He in metallic H at low temperatures (Salpeter 1973; Stevenson and Salpeter 1977). Note, however, that the calculation of the conditions under which immiscibility commences is very difficult and one recent calculation would rule out such an effect for the temperature and pressure conditions characteristic of Saturn's metallic H zone (MacFarlane and Hubbard 1983). Other deviations from solar elemental ratios may include an approximately twofold enhancement of the C/H ratio in Saturn's atmosphere (Gautier and Owen 1983b).

Saturn radiates to space about 1.79 ± 0.10 times as much energy as the amount of sunlight it absorbs (Hanel et al. 1983). The implied excess luminosity, due to an internal heat source, is $\sim$ 4 times smaller than Jupiter's excess luminosity, is $\sim$ 1.5 orders of magnitude greater than Neptune's, and is at least 1.5 orders of magnitude greater than Uranus' as yet undetected excess (Stevenson 1982a).

The rings of Saturn are composed of a myriad of particles in independent orbits about the planet. The principal rings, A, B, and C, are located between 1.21 and 2.26 R_S from the center of Saturn, where R_S is the planet's equatorial radius. Thus, these rings lie within the classical Roche tidal radius, as do the rings of Jupiter and Uranus. However, fainter rings of the Saturn system are located both closer to the planet and farther away, including distances outside the Roche limit (Stone and Miner 1981, see their Table 3).

Most of the particles in the main rings have radii that lie between 0.1 cm and a few meters (Pollack et al. 1973; Cuzzi and Pollack 1978; Cuzzi et al. 1980; Tyler et al. 1981). Much smaller, micron-sized particles are present in some portions of the rings, such as the F Ring and the spokes of the B Ring (Pollack 1981b; B. A. Smith et al. 1981). The latter particles have lifetimes that are probably much less than the age of the solar system, due to catastrophic impacts with micrometeoroids, and are thus most likely the products of the continued erosion of the larger ring particles by micrometeoroid impact (Pollack 1981b). Water ice is the dominant material making up both the surfaces and interiors of the centimeter- to meter-sized ring particles (Pilcher et al. 1970; Pollack et al. 1973; Cuzzi et al. 1980). Although rock may be present as a minor coloring agent (Lebofsky et al. 1970), an upper limit on its bulk abundance appears to lie far below that expected from solar abundance considerations (Pollack et al. 1973; Cuzzi et al. 1980). Finally, the rings have a mass of $\sim 6.4 \times 10^{-8}$ Saturn masses, a value comparable to that of the satellite Mimas (Holberg et al. 1982a).

The Saturn system contains 17 known satellites, whose locations range from just outside the A Ring (1980S28) at 2.28 R_S to 215 R_S (Phoebe) and whose sizes range from a diameter of 30 km (1980S28) to 5150 km (Titan) (see Table I). All the satellites but Phoebe travel in prograde orbits having low eccentricities and low inclinations to Saturn's equatorial plane. (Iapetus' inclination of 15° may be due to the Sun's tidal torque on Saturn's protosatellite nebula [Ward 1981b]. Such a torque significantly warped the plane of the nebula toward the Laplacian plane at great distances from the planet.) Thus all the regular satellites were presumably formed coevally and from the same cloud that gave birth to Saturn, while Phoebe is most likely a captured object.

As summarized in Table I and Fig. 1, all the regular satellites whose mean densities have been measured have densities below 2 g cm^{-3}. The corresponding uncompressed mean densities fall well below those of the inner Galilean satellites of Jupiter and well below those characterizing unhydrated (> 3 g cm^{-3}) and hydrated (≥ 2.3 g cm^{-3}) rock. Thus, ice (mostly water) represents a major component of their interiors, in addition to rock. On the average, the uncompressed mean density of the Saturnian satellites is comparable to that expected for a solar abundance of ice and rock (about equal masses) and similar to those of the outer Galilean satellites. There is no obvious trend in the mean densities as a function of distance from Saturn, in contrast to the situation for the Galilean satellites. Furthermore, there may be significant, but stochastic, departures of individual values from their mean value.

Water ice is also a major constituent of the surfaces of the regular satellites. This conclusion is based on the presence of water ice absorption features in the near-infrared reflectivity spectra of the larger satellites (Cruikshank 1979b; Fink et al. 1976) and the high albedo in the visible of all the regular satellites but Titan (whose surface is masked by an optically thick smog layer) and the leading side of Iapetus (see Table I). In the case of the smaller and presumably undifferentiated satellites, a water-rich surface implies a water-rich interior.

The very low albedo of Phoebe and the shape of its visible and near-infrared reflectivity spectrum imply that it is made of carbonaceous chondritic-like material (Degewij et al. 1980b). Its reflectivity spectrum closely matches those of Jupiter's irregular satellites, all studied Trojan asteroids, and C-type objects in the main asteroid belt.

The geologic histories of all the Saturnian satellites have been strongly influenced by meteoroid impact events, while tectonism and resurfacing events have also played important roles for the larger satellites (B. A. Smith

TABLE I
Observed and Inferred Properties of Saturn's Satellites [a]

Name	Orbit (km)	Orbit (R_S)	Radius (km)	Mass (10^{23} g)	Density (g cm^{-3})	I/F at 0°	Central pressure (kbar)	Percent ice by mass
1980S28	137,670	2.282	10 × 20			0.4		
1980S27	139,350	2.310	70 × 50 × 40			0.6		
1980S26	141,700	2.349	55 × 45 × 35			0.6		
1980S3	151,422	2.510	70 × 60 × 50			0.4		
1980S1	151,472	2.511	110 × 100 × 80			0.4		
Mimas	185,540	3.075	196 ± 3	0.375 ± 0.008 [b] (0.455 ± 0.054) [b]	1.19 ± 0.05 (1.44 ± 0.18)	0.7	0.1	45–60
Enceladus	238,040	3.946	250 ± 10	0.74 ± 0.30	1.2 ± 0.4	1.0	0.1	50–100
Tethys	294,670	4.884	530 ± 10	7.55 ± 0.90	1.21 ± 0.16	0.8	0.6	60–90
1980S13	294,670	4.884	17 × 14 × 13			0.6		
1980S25	294,670	4.884	17 × 11 × 11			0.8		
Dione	377,420	6.256	560 ± 5	10.5 ± 0.3	1.43 ± 0.06	0.7	0.9	50–60
1980S6	378,060	6.267	18 × 16 × 15			0.5		
Rhea	527,100	8.737	765 ± 5	24.9 ± 1.5	1.33 ± 0.09	0.6	1.4	55–70
Titan	1,221,860	20.253	2575 ± 2	1345.7 ± 0.3	1.88 ± 0.01	0.2	32.7	~40
Hyperion	1,481,000	24.55	205 × 130 × 110			0.2		
Iapetus	3,560,800	59.022	730 ± 10	18.8 ± 1.2	1.16 ± 0.09	0.5 0.05	1.0	70–90
Phoebe	12,954,000	214.7	110 ± 10			0.06		

[a] From B. A. Smith et al. (1982) and Consolmagno (1984b).
[b] These masses are based on analyses of groundbased measurements of the positions of the Saturn satellites and on using the ratio of Mimas' and Tethys' masses derived from these data and the Voyager value of Tethys' mass, respectively (see B. A. Smith et al. 1982).

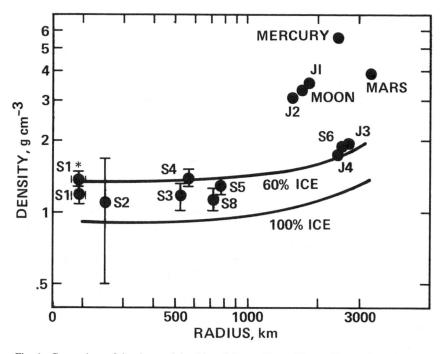

Fig. 1. Comparison of the sizes and densities of the satellites of Saturn (S), satellites of Jupiter (J), and other solar system objects with theoretical curves for objects made of 100% water ice (lower solid curve) and a solar elemental mixture of rock and ice (from B. A. Smith et al. 1982). S1 and S1* represent two alternative choices of the density. See Table I.

et al. 1981,1982). The high crater density on parts or all of the surfaces of almost all the Saturnian satellites has commonly been interpreted, by lunar analogy, as reflecting an early period (first $\sim 1 \times 10^9$ yr) of heavy meteoroid bombardment. Furthermore, there is substantial evidence implying that some of the original satellites were catastrophically disrupted during this period: Several objects have almost identical orbits, including the small coorbital satellites 1980S1 and 1980S3, Tethys and its Trojans (1980S13 and 1980S25), and Dione and its Trojan (1980S6); objects as big as Hyperion have distinctly nonspherical shapes; extrapolation of the crater density on Iapetus to the inner parts of the Saturn system, with an allowance for gravitational focusing, results in a near-unit probability of catastrophic disruption of the smaller, innermost satellites.

Several of the larger satellites, including Tethys, Dione, and Rhea, have global scale fractures of their surfaces, indicative of extensional tectonism. The relatively low crater densities on this portion of their surfaces imply that tectonism extended to the end of or beyond the epoch of heavy bombardment. There are also surface morphologies on these bodies indicative of resurfacing events in the late or post heavy bombardment epoch. Perhaps most surprisingly of all, the comparatively small satellite Enceladus has a very low crater density over much of its surface, exhibits grooves somewhat reminiscent of those on Ganymede, and has experienced extensive resurfacing. On the basis of crater densities, these extensional tectonic and resurfacing events have extended up until relatively recent times (within the last $\sim 1 \times 10^9$ yr). Finally, there is the classical puzzle of at least an order of magnitude variation in the brightness of Iapetus from the dark leading hemisphere to the bright trailing hemisphere.

Titan is the only satellite known to have a substantial atmosphere. Measurements made from the Voyager 1 spacecraft yielded a surface pressure of 1.6 bar (Tyler et al. 1981), with molecular nitrogen being the dominant gas (Tyler et al. 1981; Samuelson et al. 1981). The volume mixing ratio of N_2 lies between 80% and 95%. Other major constituents include methane (few percent) and molecular hydrogen (0.2%). There is also indirect evidence for the possible presence of substantial quantities of Ar ($\sim 10\%$) based on the value of the mean molecular weight, derived from comparing radio occultation and infrared interferometer spectrometer temperature profiles (IRIS, see Samuelson et al. 1981). Trace gases that are present at the ppm level include low-order hydrocarbons, such as C_2H_2 and C_2H_6, and nitrogen-containing organics such as HCN (Hanel et al. 1981a). One very recent surprise has been the detection of oxidized gases at the ppb (CO_2) and ppm (CO) levels (Samuelson et al. 1983; Lutz et al. 1983). Finally, a pervasive smog layer is present throughout the lowest several hundred kilometers of the atmosphere (B. A. Smith et al. 1981,1982). The smog particles are probably made of complex organic polymers that are end products of the photochemistry occurring in Titan's atmosphere.

II. FORMATION AND EVOLUTION OF SATURN

A. Origin

As discussed in Sec. I, Saturn consists of a 20 $M_\oplus$ core made of rock and ice and a 75 $M_\oplus$ envelope containing approximately a solar mixture of elements. Such a structure suggests two alternative theories for the origin of the Saturn system:

1. Gas Instability Theory: a gaseous condensation first formed within the solar nebula and it later acquired core material.
2. Core Instability Theory: solid body accretion occurred first, with the core mass eventually becoming large enough to effectively concentrate an even greater mass of solar nebula gas about itself.

During the last 10 yr, both theories have been studied in some detail.

According to the gas instability theory, the solar nebula was sufficiently massive ($\gtrsim 0.1$ $M_\oplus$) that it was unstable to global, azimuthal perturbations (Cameron 1978) (see Fig. 2). As a result of this gravitational instability, rings

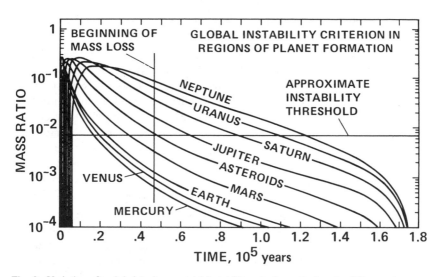

Fig. 2. Variation of a global (axisymmetric) instability criterion with time for different planetary formation regions. The vertical scale is the mass ratio of the protoplanet to that of the forming Sun. For values of this mass ratio above the horizontal line labeled "approximate instability threshold," the solar nebula is unstable to ring formation (after Cameron 1978).

of elevated density formed and grew progressively narrower. Eventually, the gas density within the ring became high enough for local gravitational instabilities to occur. The resultant giant gas balls within a given ring may have either merged with one another or gravitationally deflected one another to different orbits. The resultant giant gaseous protoplanets subsequently gravitationally contracted to form the outer planets.

The above scenario, according to which giant gaseous protoplanets form from a series of instabilities in a massive solar nebula, is far from proven. For example, spiral density waves rather than ring instabilities may be the fastest growing mode in an unstable massive nebula (P. Cassen, personal communication).

There are two ways in which the giant gaseous protoplanets may have acquired their cores. First, as they contracted and grew hotter, pressure and temperature conditions within their deep interior may have permitted solid grains to become liquid grains, and subsequently coalesce efficiently into large-sized particles. Such particles would have large terminal velocities and tend to precipitate to the protoplanet's center to form a core (Slattery et al. 1980). This mechanism is important for Saturn's core only if there is a large exchange of material between the solar nebula and the protoplanet. Saturn's core represents about 20% of the planet's mass, whereas heavy elements constitute only about 1% of the mass in a solar mixture of elements. Furthermore, standard interior models require an excess of heavy elements in the core rather than a mere redistribution of elements. Thus, one must postulate a continual replenishment of refractory materials from the nearby solar nebula so that a 20 $M_\oplus$ core can be constructed.

Only the more refractory materials become liquid inside protoplanets and so contribute to the cores, according to the calculations of Slattery et al. (1980). But solids can also coagulate, although somewhat less efficiently than liquids, so that it is not clear that the grains need to go through a liquid phase to sink to the protoplanet's center. In any case, it seems unlikely that cores containing ices could be formed through this mechanism in view of the high temperatures in the deep interior of the protoplanets during almost all of their lifetime.

A second mechanism by which giant gaseous protoplanets could acquire cores is through gas drag capture of solid planetesimals that formed nearby in the solar nebula (Pollack et al. 1977). In the next section, it will be argued that this mechanism is the most likely means by which Phoebe was captured. However, an irregular satellite rather than core material is the end product of gas drag capture only for a very short time interval ($\sim$ 10 yr) in the history of the protoplanet and only for bodies lying within a restricted mass range. Thus, if Phoebe was captured by gas drag, many orders of magnitude more mass would have been contributed to the core by other captured planetesimals. Whether as much as a 20 $M_\oplus$ core could have been acquired in this manner is open to question. To the degree that Phoebe is representative of the planetesimals present near Saturn during its very early history, we would expect that core material resulting from gas drag capture would be lacking significant amounts of nitrogen compounds and would have an excess of rock over water and carbon compounds. Thus, neither core formation mechanism is expected to lead to significant quantities of nitrogen compounds in the core and only limited amounts of water and carbon compounds at best.

According to the alternative "core instability" model (Perri and Cameron 1974; Mizuno 1980), accretional processes involving solid planetesimals led to the growth of planet-sized objects in the outer solar system as well as in the inner solar system. As the solid cores grew larger, they concentrated solar nebula gas more and more effectively within their sphere of influence. Eventually, the envelopes became sufficiently massive that they underwent a rapid contraction or collapse onto the central core. Such a collapse insures the ultimate survival of the envelope against tidal disruption by the forming Sun; leads to a compact object; and can result in a further significant increase in the mass of the envelope as additional material from the solar nebula enters within the object's gravitational sphere of influence.

Estimates of the core mass at which instability first occurs have been obtained by constructing static equilibrium models of core/envelope configurations and determining the largest core mass for which such a model can be constructed. The first models of this type were constructed by Perri and Cam-

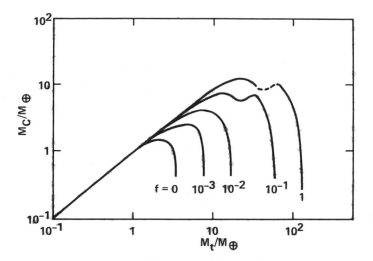

Fig. 3. Core mass as a function of total mass (core plus envelope) for conditions of hydrostatic equilibrium. The parameter f denotes the ratio of assumed grain opacity to that expected in a cold region of the solar nebula. Note that no equilibrium solutions exist for core masses exceeding certain critical values (after Mizuno 1980).

eron (1974), who assumed that the temperature structure was adiabatic throughout the envelope. They obtained critical core masses of $\sim 70\ M_\oplus$ for Saturn, with the value of the critical core mass being somewhat sensitive to the boundary conditions and thus position within the solar nebula. Both the large value of the critical mass and its sensitivity to position resulted in a poor match with the inferred core masses of Saturn and the other giant planets.

Both problems have been overcome in more recent models, in which the temperature structure has been calculated rather than assumed. The occurrence of a zone of radiative equilibrium in the outer portion of the envelope results in both a lower critical core mass and an insensitivity of this mass to boundary conditions (Mizuno 1980). Throughout all but the hot inner parts of the envelope, the radiative opacity is dominated by grains, whose properties therefore determine the temperature structure of the envelope. Figure 3 illustrates the relationship between core mass and total mass, where the parameter f denotes the ratio of the actual grain opacity to that expected from grains in a cool solar nebula. For each choice of f, there is a maximum value of core mass ("critical value"). It is not possible to find a static equilibrium configuration for core masses exceeding this critical value. As can be seen, this value does not depend sensitively on f. Critical core masses comparable to that of Saturn are obtained for models with $f \sim 1$.

Unfortunately, these models of Mizuno and more generally the earlier ones as well suffer from several potentially serious problems. First, the inability to construct a hydrostatic model for large core masses is not equivalent to such models undergoing a hydrodynamic or otherwise rapid collapse. These models could, in principle, undergo a slow contraction or expansion and indeed such behavior would be expected in general for accretionary objects radiating to space. In a similar vein, no fundamental physical change in the interior of these models is cited to suggest that hydrodynamic collapse should occur for the large core models, as claimed by Mizuno (1980) and others. Second, it is not clear that the tidal radius is the appropriate outer boundary of these primordial planets, as assumed in all the above calculations. An alternative choice is the accretionary radius, where the thermal energy of the gas balances in absolute magnitude its gravitational energy. Certainly for small enough core masses, the accretionary radius is smaller than the tidal radius and is, therefore, the more appropriate choice. Consequently, there may be more sensitivity to boundary conditions than when only the tidal radius is considered.

Bodenheimer and Pollack (1984) have carried out a series of calculations of the accretionary growth of core models that overcome the above problems. They find that when the core mass grows sufficiently large, a very rapid contraction, but not a hydrodynamical collapse occurs. Critical core masses of ~ 10 $M_\oplus$ characterize their models at the point of rapid contraction, but with this mass varying by factors of several depending on the assumed nebular boundary conditions and core accretion rate.

Once rapid contraction occurs, gases may be added much more rapidly to the protoplanet than planetesimals to the core. Thus, the envelope mass may grow much more rapidly after this point than does the core mass. Preliminary studies of this and the earlier phases have been conducted by Safronov and Ruskol (1982). They distinguish six stages of formation and growth. Stage 1 is equivalent to the accretional growth of the core and ends with it reaching the critical core mass. Since the core as well as the envelope can grow somewhat during the subsequent stages, the critical core mass is as-assumed to be a few $M_\oplus$ rather than 20 $M_\oplus$. During stage 2, accretion of gas from the solar nebula is limited by the time required to radiate away part of the gravitational energy of accretion so that thermal pressure does not prevent the added mass from lying within the protoplanet's sphere of influence. During stage 3, an unrestricted rapid growth occurs. During stage 4, accretion is limited by the size of the sphere of influence. During stages 5 and 6 accretion is further limited by the need to resupply gas from distant parts of the solar nebula to parts lying close to the protoplanet's orbit and by the dissipation of the solar nebula, respectively. Estimates of the duration and amount of mass added to Jupiter and Saturn during each stage are summarized in Table II. According to these estimates, several times 10^8 yr are required before Saturn's growth is completed. A slightly smaller time scale is found for Jupiter.

Both of the above models—core and gas instability—have their strong and weak points. If only the outer planets formed as a result of gas instabilities in the solar nebula, it seems likely that Jupiter might have formed before

TABLE II
Duration of Different Stages of Accretion for Jupiter and Saturn [a]

	Duration (yr)		Change in Mass, (units of Earth Masses)	
Stage [b]	Jupiter	Saturn	Jupiter	Saturn
1	3×10^7–1×10^8	2×10^8	1–3	2–3
2	10^8	10^6	10	8
3	10^5–10^6	—[c]	40	—[c]
4	2×10^2	6×10^4	85	20
5	10^4	10^6–10^7	120	35
6	10^6–10^7	3×10^8	60	30

[a] According to Safronov and Ruskol (1982).
[b] Stages are characterized by (1) the accretional growth of a solid core; (2) the occurrence of an instability in the surrounding envelope, followed by a slow accretion from the surrounding solar nebula; (3) an epoch of rapid accretion; (4) continued accretion, but from a restricted space inside the Hill sphere (Bodenheimer et al. 1980); (5) a still slower accretion limited by diffusion of the solar nebula into the now depleted feeding zone for the giant planets; and (6) the final accretion during the dissipation of the solar nebula.
[c] Saturn does not pass through stage 3.

much accretional growth of solid bodies took place in the inner solar system. In this event, Jupiter could have interfered with the accretional growth of solid bodies within the asteroid belt and in the vicinity of Mars' orbit, by gravitationally perturbing orbits of nearby planetesimals into ones that crossed those of the more distant asteroidal and Martian planetesimals at high relative velocities. In this way, the absence of a single planet-sized body in the asteroid belt and the relatively small size of Mars could be accounted for. However, if gaseous protoplanets formed throughout the solar nebula, as has been proposed by Cameron (1978), then the ability of this theory to explain the size of Mars and the multiplicity of asteroids is less clear. If the terrestrial planets initially had massive gaseous envelopes, their envelopes could have been eliminated by the tidal action of the forming Sun (Cameron 1978) or thermal evaporation as the solar nebula heated up (Cameron et al. 1982). Because of their greater distance from the Sun, the outer planets were less susceptible to tidal stripping and thermal evaporation.

The gas instability model has a number of serious problems. First, it does not provide an obvious explanation for the similarity of the core masses of the four giant planets. Second, if the inner planets also formed as a result of gas instabilities and if they acquired their cores by precipitation of liquids or solids in their envelopes, then it is not clear why the uncompressed densities of the terrestrial planets tend to decrease from Mercury to Mars or why the abundance of primordial rare gases systematically increases by orders of magnitude from Mars to Venus. Finally, it remains to be demonstrated that the large enrichment of heavy elements in the giant planets can be achieved through a combination of precipitation and gas drag capture. This problem may be especially serious for Uranus and Neptune, whose cores constitute 80–90% of the planets' total mass.

We next assess the viability of the core instability model. As explained above, the critical core mass required to induce rapid accretion from the surrounding solar nebula is somewhat insensitive to the boundary conditions at the tidal radius and hence to pressure and temperature conditions within the solar nebula. In this sense, it provides a natural explanation for the similarity of the core masses of the giant planets as well as the actual value of the core masses. Second, a much larger variance is expected in the masses of the envelopes, as in fact is observed, due to the subsequent preferential accretionary growth of the envelope after the rapid onset of contraction. Third, it seems likely that the smaller bodies of the solar system—asteroids, comets, and satellites—formed by accretional growth. On aesthetic grounds, it might seem preferable for this same process to account for the mean density (Lewis 1972a) and for the rare gas content of the terrestrial planets (Wetherill 1981; Pollack and Black 1979, 1982) as well as the formation of the giant planets.

However, the core instability theory also has problems. In particular, there are problems connected with time scales. Because Jupiter's core probably formed in a region of the solar nebula of lower density than did Mars or the asteroid belt, and because its core mass is much greater than even the mass of the Earth, Jupiter should have formed after Mars and the asteroid parent bodies did. In this event, Jupiter could not have interfered with their growth. Also, the time scales for core growth progressively increase with distance from the center of the solar nebula due to a combination of the lower volume densities and longer orbital periods of planetesimals at greater distances. This time scale approaches the age of the solar system for Uranus and Neptune and therefore apparently exceeds the lifetime of the solar nebula for these planets (Safronov and Ruskol 1982). However, this estimate of the time scale for core growth is based on a theory of planetesimal assembly in a gas-free environment, surely an invalid assumption for the giant planet models under discussion. Gas drag might significantly expedite core growth.

Finally, to some extent, the two origin theories can be distinguished on the basis of the mass of the solar nebula they imply. A nebular mass in excess of $\sim 0.1\ M_\odot$ is required for gas instabilities to occur with a 1 solar mass Sun at the nebula's center, whereas nebular masses ranging from $0.01\ M_\odot$ to $0.1\ M_\odot$ are typically invoked in models having solid accretional growth. The expected mass of the solar nebula is related to the angular momentum of the cloud from which the solar system formed (Cameron 1978). While Cameron (1978) advances astrophysical arguments in favor of a large angular momentum and hence a large nebula mass, almost all of the angular momentum must be lost to match the current angular momentum of the solar system. It seems difficult to discriminate between these two theories on the basis of the implied nebular mass or cloud angular momentum in view of the poor independent constraints on either of these quantities.

B. Evolution

According to both theories of its formation, Saturn underwent three major phases in its evolution during and following formation: early quasi-hydrostatic contraction or expansion (stage 1), very rapid contraction (stage 2), and late, quasi-hydrostatic contraction (stage 3). During stage 1, the envelope was in hydrostatic equilibrium to a very good first approximation. In the case of the gas instability model, the envelope slowly contracted from about 1600 R_S to about 60 R_S on a time scale of 4.6×10^6 yr (Bodenheimer et al. 1980). The initial radius is set by Saturn's mass and the minimum density at which a local instability occurs for a solar nebula in which most of the Sun's mass was not yet concentrated near the center of the nebula. The initial radius would be a factor of 8 smaller if the Sun had already fully formed prior to the gas instability.

The time scale for this stage is set by the rate at which the protoplanet radiates to space part of the gravitational energy released by contraction, i.e., it is a Kelvin-Helmholtz time scale. This time scale can be altered by a factor of several by the outer boundary conditions imposed by the solar nebula and by the growth of the Sun at the center of the nebula (Cameron et al. 1982). On the one hand, a nonzero pressure at the boundary results in a faster contraction, while a nonzero temperature has the opposite effect. Since the time scale for stage 1 is comparable to the time scale over which the solar nebula evolves significantly, there is even the possibility that the protoplanet may be tidally disrupted by the forming Sun before it contracts to a small enough radius.

In the case of the core instability model, both the core mass and the envelope mass increase with time during stage 1 (see Fig. 3). The outer boundary of the envelope is initially an accretion radius at which thermal and gravitational energies of a gas element are equal and a tidal disruption limit, determined by the combined gravitational effects of the solar nebula and the Sun. As the mass of Saturn M increases during stage 1, the outer boundary expands as $M^{1/3}$ for the latter boundary. The duration of stage 1 is set by the accretional time scale required for the core mass to reach its critical value. According to Safronov and Ruskol (1982), stage 1 lasts $\sim 10^8$ yr for Saturn.

At the end of stage 1 for the gas instability model it is no longer possible for pressure gradient forces within the envelope to balance gravitational forces and a hydrodynamical collapse is initiated marking the start of stage 2. The collapse is initiated when the temperatures near the base of the envelope become high enough (~ 2500 K) for molecular hydrogen to dissociate (Bodenheimer et al. 1980). Stage 2 lasts for only about 0.1 year for Saturn, with this duration being comparable to a free-fall time. Collapse first ceases near the center of Saturn due to an increasing stiffness (incompressibility) of the equation of state. Soon, the infalling, outer lying material attains supersonic velocities at the boundary with the static central material, and a shock wave develops (Bodenheimer et al. 1980). Infalling continues until hydrostatic equilibrium has been attained again throughout the entire planet. According to Bodenheimer et al. (1980), Saturn's radius was $\sim 3.4\,R_S$ at the end of stage 2, or almost 20 times smaller than the value at the start of this stage, and the density at the base of envelope increased by almost four orders of magnitude to a value of $\sim 1/3$ g cm^{-3}.

Figure 4 summarizes the time history of Saturn during stages 1 and 2, according to Bodenheimer et al. (1980). These results pertain to a gas instability origin. Also, zero pressure at the exterior boundary and spherical symmetry are assumed. A somewhat different equatorial radius at the start of stage 2 characterizes models in which the rotation and angular momentum of the protoplanet are taken into account (Bodenheimer 1977).

In the case of the core instability model, the core reaches its critical mass at the end of stage 1 and the early parts of stage 2 are therefore characterized by a very rapid contraction of constant mass surfaces within the envelope and rapid accretion of gas from the surrounding solar nebula. It is not yet clear whether or not a hydrodynamical collapse occurs at some point during this stage. However, rapid accretion eventually slows down and then ceases and eventually, a much more compact structure is achieved, with the planet having essentially the same dimensions as it did at the end of stage 2 for the gas instability model. Subsequent evolution is identical for the two models.

A nebula disk within which the regular satellites and perhaps the rings formed may have come into existence during stage 2, on the basis of the current dimensions of the ring/satellite system. This disk remained attached to the planet, but continued to extend into the region later occupied by satellites due to its initially large specific angular momentum and the outward transfer of angular momentum by viscous dissipation. Because Titan's specific angular momentum is about two orders of magnitude larger than that of Saturn, enough angular momentum redistribution within the protoplanet had to occur during stage 1 and during stage 3 to concentrate the angular momentum in the outermost parts of the system. (Note that Titan dominates the mass of the Saturn ring-satellite system.) Such a redistribution could have occurred by viscous processes, since much of the protoplanet was in convective equilibrium during stage 1, which, in turn, was caused by the high infrared opacity of grains mixed with the gas of the envelope (Bodenheimer et al. 1980). Essentially no angular momentum *transport* occurred during the very short duration of stage 2, but creation of the nebula disk during this stage was fostered by the decreasing radius of the protoplanet and hence the increasing angular velocity demanded by conservation of angular momentum. Further outward transfer of angular momentum is expected during stage 3 as long as the nebula persisted, as discussed in more detail in Sec. IV.

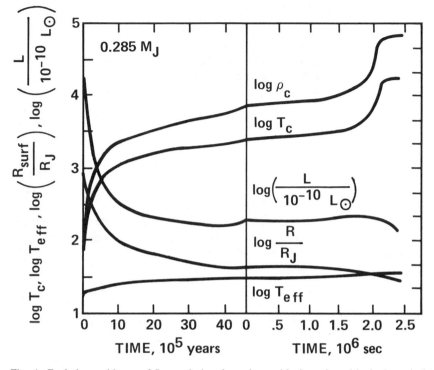

Fig. 4. Evolutionary history of Saturn during the early quasi-hydrostatic and hydrodynamical collapse phases for a gas instability model. T_c, T_{eff}, R_{surf}, L, and ρ_c refer to the central temperature, effective radiating temperature, surface radius, internal luminosity, and central density, respectively. Note the scale change on the time axis between the right and left sides of the figure that correspond to the two different phases of evolution (after Bodenheimer et al. 1980).

After the attainment of complete hydrostatic equilibrium that marks the beginning of stage 3, Saturn continued to contract, but at a very slow rate. This contraction occurred because the planet was radiating to space and because its interior was not totally incompressible. During the earliest portion of stage 3 ($\sim 10^6$ yr), contraction took place sufficiently rapidly that the released gravitational energy caused a continued warming of Saturn's interior, despite the planet radiating to space at an intrinsic luminosity of 10^{-5} $L_\odot$, where $L_\odot$ is the Sun's present luminosity. However, the interior became increasingly incompressible so that throughout almost all of stage 3 ($\sim 4.6 \times 10^9$ yr), the interior underwent a progressive cooling (Pollack et al. 1977; Bodenheimer et al. 1980; Grossman et al. 1980). At the end of the interior warming phase, the central temperature was $\sim 20,000$ K, whereas at present it is $\sim 10,000$ K. During the entirety of stage 3, the radius and intrinsic luminosity steadily declined, achieving values that are 3.4 times and 4.5 orders of magnitude smaller at the present epoch than their corresponding values at the commencement of stage 3.

C. Present Internal Heat Source

As noted in Sec. I, Saturn radiates to space at present ~ 1.8 times as much energy as the amount of sunlight it absorbs. Here, we examine our current understanding of the source of the excess luminosity. Even when allowance is made for the presence of a heavy element core, the intrinsic luminosity that can be realized from the decay of long-lived radioactive isotopes of K, U, and Th falls short of the observed value by about two orders of magnitude (Flasar 1973). Only gravitational energy release is capable of supplying the required intrinsic luminosity (ibid). The major question concerns the nature of the gravitational energy release. In particular, what are the relative roles of present gravitational contraction, present cooling of the interior that was warmed during earlier, more rapid contraction phases, and gravitational segregation, especially of He from H?

Detailed evolutionary calculations of Saturn during stage 3 have provided a partial answer to the above question. Figure 5 compares the observed intrinsic luminosities of Jupiter and Saturn with predicted values at a time of 4.5×10^9 yr after the start of stage 3 (Pollack et al. 1977). These results are

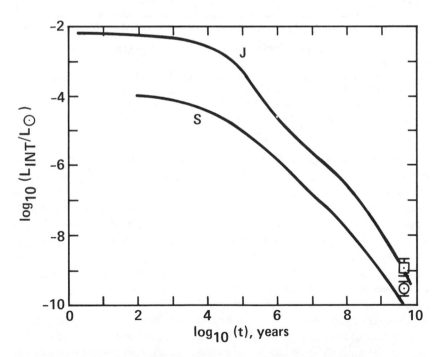

Fig. 5. Excess luminosity of Jupiter (J) and Saturn (S), in units of solar luminosity, as a function of time during the late quasi-hydrostatic stage of evolution. Observed values at the 4.5×10^9 yr time point are indicated by the square and circle for Jupiter and Saturn, respectively (adopted from Pollack 1978).

insensitive to the choice of initial conditions. They were obtained for homogeneous compositional models that lacked central heavy element cores. However, the predicted excess does not change significantly when a central core is included (Grossman et al. 1980). According to this figure, the observed intrinsic luminosity of Jupiter can be closely replicated, but the predicted value for Saturn falls short by about a factor of 3. About 1/3 of the predicted luminosity for both planets resulted from their present contraction rates, while the remaining 2/3 was derived from cooling of their interiors (Pollack et al. 1977).

In view of the success of the homogeneous model for Jupiter, it appears that its lack of success for Saturn is most likely due to gravitational segregation. Only He is sufficiently abundant for its separation from H to supply the needed deficit. As early as 1967, the importance of gravitational layering for generating internal energy in the outer planets was realized (Kieffer 1967). Exploration of the physical processes by which this layering could occur were begun in 1973 by Salpeter and studied further by Stevenson and Salpeter (1977). According to Salpeter (1973), He could be expected to first become partially immiscible in H in the metallic H zone. While diffusion of He atoms occurs too slowly, even over the age of the solar system, to release significant amounts of gravitational energy, rapid segregation could be realized through the nucleation of He droplets and their continued growth to droplet sizes characterized by large terminal velocities.

An explicit calculation of the temperature and pressure conditions required for He immiscibility to occur were made by Stevenson and Salpeter (1977), who found that it would first begin in the outermost part of the metallic H zone when the temperature fell to $\sim 10,000$ K and that it would spread to progressively deeper regions of the metallic H zone with time. A major theoretical uncertainty at that time, and even today, is the nature of the phase transition from molecular to metallic H. If it is a first-order transition, a density discontinuity occurs at the boundary between the two phases. As a result, He might not be exchanged effectively between the molecular and metallic zones, or some of the He removed from the upper part of the metallic H zone might be transferred to the molecular H region. An even more basic problem is the considerable uncertainty in the conditions under which He becomes immiscible in metallic H. According to MacFarlane and Hubbard (1983), the critical temperature, below which He becomes partially immiscible in metallic hydrogen, is a factor of about ten smaller than that found by Stevenson and Salpeter (1977). If so, no He separation would be expected in Saturn's interior and, thus, the source of its excess energy would not be understood.

The large discrepancy between the two estimates of the critical temperature for helium immiscibility stems from the need to know the Gibbs free energy of hydrogen-helium mixtures to an accuracy of much better than a percent. Unfortunately, such accuracy is very difficult to achieve. Stevenson and Salpeter (1977) used quantum mechanical perturbation theory to determine the Gibbs free energy, while MacFarlane and Hubbard relied on the Thomas-Fermi-Dirac (TFD) theory of high-pressure materials. Because atoms are more strongly screened in the TFD theory than in the theory used by Stevenson and Salpeter, the critical temperature occurs at a lower value in the TFD theory. More accurate calculations are required to resolve this crucial discrepancy.

Salpeter (1973) proposed that the gravitational energy released by He immiscibility was the source of Jupiter's excess luminosity. Pollack et al. (1977) first pointed out that gravitational segregation was probably not operative for Jupiter, but was probably operative for Saturn. They based this conclusion on two points. First, their homogeneous contraction models were capable of reproducing Jupiter's observed excess luminosity, but not Saturn's, as pointed out above. Second, the temperatures in Jupiter's metallic H zone at present lie a factor of ~ 2 above the temperature at which immiscibility is first reached, according to Stevenson and Salpeter (1977), while this point is reached in their Saturn models after only $\sim 1 \times 10^9$ yr of evolution (cf. Fig. 6). They also pointed out that planet-wide segregation of He was re-

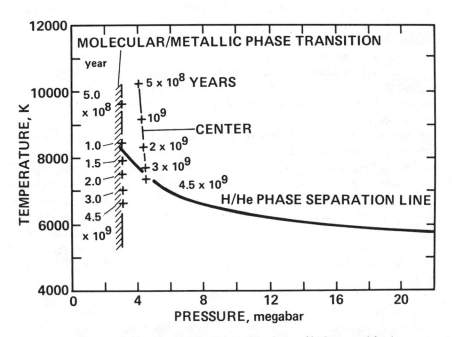

Fig. 6. Phase boundaries for the molecular and metallic phases of hydrogen and for the separation of helium from hydrogen. For points lying below the separation line, helium becomes partially immiscible in hydrogen, according to the calculations of Stevenson and Salpeter (1977). Also shown is the evolutionary track of the center and molecular/metallic boundary of a homogeneous, solar mix Saturn model of Pollack et al. (1977). Numbers next to the crosses indicate time from the start of the late, quasi-hydrostatic stage.

quired to produce the observed intrinsic luminosity. Hence, He had to be efficiently exchanged and uniformly mixed between the molecular H zone and the top of the metallic zone. They predicted that He is depleted in the observable atmosphere of Saturn by several tens of percent. Somewhat more refined estimates of this depletion factor of ~ 40% were obtained by Stevenson (1980) and Hanel et al. (1981a). These values compare favorably with the apparent deficiency of He in Saturn's atmosphere deduced from the IRIS observations of Hanel et al. (1981a), although the He mixing ratio in Saturn's atmosphere differs from that in Jupiter's atmosphere at only the 1+ standard deviation level.

There is a strongly coupled servo loop connecting He segregation, internal temperatures, and excess luminosity. The amount of He segregation occurring over any given time interval is determined by the drop in temperature that takes place in the metallic H zone. But this decrease is determined by the amount of excess luminosity radiated to space during this interval, which in turn is determined chiefly by the amount of gravitational energy released by He segregation. Thus, only as much He segregation occurs as is needed to self-consistently match the luminosity needs.

In summary, homogeneous contraction and thermal cooling appear to be inadequate sources of Saturn's excess luminosity, although they can fully account for Jupiter's excess luminosity. A plausible additional energy source is the gravitational energy released by the partial immiscibility of He in metallic H at low temperatures. The apparently observed depletion of He in Saturn's observable atmosphere, in comparison to solar elemental ratios and the He abundance in Jupiter's atmosphere, offers some evidence in support of this hypothesis. But the theoretical foundation for He immiscibility has been called into question by the thermodynamic calculations of MacFarlane and Hubbard (1983).

III. ORIGIN OF PHOEBE

There are several characteristics of Phoebe that suggest that this outermost satellite of Saturn is a captured object, rather than one that formed within the Saturn system during its early history. First, it has a retrograde orbit of high inclination and eccentricity, in contrast to all the other satellites. Second, as discussed in Sec. I, its low visible albedo and spectral reflectivity properties stand in marked contrast to the corresponding properties of the other satellites of Saturn, but are similar to those of carbonaceous chondrites, the irregular satellites of Jupiter, Jupiter's Trojan asteroids, and C-type objects in the asteroid belt. Indeed, if Phoebe formed within the Saturn system, we would expect it to be at least as rich in ices as the other satellites, given its greater distance, rather than essentially ice free.

There are three major classes of capture theories: Lagrange point capture, collision between a stray body and a natural satellite, and gas drag capture. According to the first of these, a body initially in orbit about the Sun can transfer to an orbit about a planet by passing at very low velocity relative to the planet through its interior Lagrange point (Heppenheimer 1975). However, as can be seen from the symmetry of the equation of motion with respect to the sign of time, such capture is only temporary, with the body returning to a solar orbit through the Lagrange point after only a few or at most about a hundred orbits about the planet (Heppenheimer and Porco 1977). Permanent capture could occur if Saturn's mass increased by several tens of percent or if the Sun's mass decreased by a comparable percentage during the time of temporary capture (1–100 yr) (ibid). However, no such rapid change in either body's mass is predicted by current theories of the formation of the solar system. For example, the calculations of Safronov and Ruskol (1982) for the core instability model have time scales of $\gtrsim 6 \times 10^4$ yr during which Saturn's mass increases by several tens of percent (cf. Table II). Until plausible mechanisms for effecting permanent capture are proposed, the Lagrange point mechanism cannot be viewed as an entirely satisfactory theory for the capture of Phoebe.

Colombo and Franklin (1971) suggested that the two families of irregular satellites of Jupiter originated from the collision of a stray body with a regular satellite of this planet, with the prograde irregular satellites being fragments of the original satellite and the retrograde satellites being fragments of the stray body. One potentially serious problem with applying this theory to Phoebe is the absence of a larger *irregular* satellite of Saturn having a prograde orbit (recall that Phoebe has a retrograde orbit). Also, no other satellite of Saturn is made of the same material as Phoebe.

Early theories for gas drag capture of satellites were advanced by See (1910) and Kuiper (1951). Pollack et al. (1979) presented a modern version of this theory that took into consideration the evolutionary phases of the outer planets discussed in Sec. II. According to this latter model, Phoebe formed independently in the outer solar system by accretional processes and was captured at the very end of stage 1 (early hydrostatic) when it passed through the outer portion of proto-Saturn and had its relative velocity reduced by gas drag. Under most circumstances, gas drag would continue to act on the captured body and it would soon spiral into the center of the protoplanet to be incorporated into its core. This may, in fact, have been the fate of many other stray bodies that were captured at less opportune times. However, the onset of collapse shortly after Phoebe's putative capture quickly removed gas from the body's orbit, thus allowing it to remain a satellite and permitting its orbit to retain a significant eccentricity and inclination. Capture within ~ 10 yr prior to the start of stage 2 is required to achieve this end state.

Several predictions of the gas drag theory can be compared with the observed properties of Phoebe. First, its semimajor axis should be comparable to or somewhat less than the radius of proto-Saturn at the onset of the collapse. For the gas instability model, stage 2 begins at a radius of ~ 60 R_S for a spherically symmetric protoplanet (Bodenheimer et al. 1980). A somewhat different value may characterize a protoplanet having a nonzero angular velocity. For the core instability model, the corresponding radius is equal to the tidal radius or $\gtrsim$ 125 R_S, where the equality holds if the Sun had already formed. The above estimates of the outer boundary of proto-Saturn at the commencement of stage 2 do not appear to be inconsistent with Phoebe's semi-major axis, 215 R_S. In the case of the core instability model, it may be necessary for the Sun not to have fully formed at the time of stage 2 in order for Phoebe to achieve, ultimately, its present orbital distance. Also, all or almost all of the growth of the envelope would have had to occur during the rapid contraction phase prior to the onset of collapse, so that no or little gas was added after the onset of collapse.

Phoebe's radius of 110 km (B. A. Smith et al. 1982) may be compared with the range of sizes for which gas drag capture would be effective. A stray body has to pass through an amount of nebular mass that is at least $\sim$ 10% of its own mass in order to be significantly slowed down (Pollack et al. 1979). Also, the captured object cannot be too small or else it would be carried along by gas drag during the collapse phase. Using Bodenheimer's (1977) model of proto-Jupiter near the end of stage 1 and scaling this to proto-Saturn, Pollack et al. (1979) predicted that bodies having radii between 0.1 and several hundred kilometers could be captured by proto-Saturn and survive as satellites. The observed size of Phoebe is consistent with this rather broad range of sizes.

The above discussion indicates that Phoebe was most likely captured by gas drag, although this theory is far from proven. Let us for the moment accept this theory as being correct and consider its consequences. First, asteroid-sized bodies were present in the outer solar system close ($\sim$ 10^6– 5×10^6 yr) to the times at which Jupiter and Saturn originated. Presumably, these bodies formed through accretional processes. This deduction lends credence to, but does not prove, the core instability model for the origin of the giant planets. Second, the number density of planetesimals near Jupiter's and Saturn's orbits at these early times may have been similar, since each planet captured one, several hundred kilometer-sized, stray body (in Jupiter's case, this was the parent body of the prograde satellites, which fragmented into these satellites, due to dynamic gas pressure [cf. Pollack et al. 1979]). Third, temperatures within the solar nebula at these times may have favored the formation of carbonaceous chondrite-like material near Jupiter's and Saturn's locations. While as much as several meters of water ice could have been ablated off the leading side of Phoebe during the capture process, ice located at greater depths would have been excavated by cratering events at subsequent times and brought to the surface to significantly brighten it. This expectation is inconsistent with Voyager 2 pictures of Phoebe, which show its surface to be very dark (B. A. Smith et al. 1982). Therefore, the presently observed surface composition of Phoebe was not significantly altered by the capture process.

IV. ORIGIN OF THE SATELLITE SYSTEM

As discussed in Sec. II, gravitational energy released by contraction during the first 10^6 yr of the late, quasi-hydrostatic stage provided Saturn with a luminosity of 10^{-5} $L_\odot$. This planetary luminosity together with viscous dissipation in the nebula preferentially heated the inner portions of the satellite nebula, which formed during the short-lived ($\sim$ 1 yr) hydrodynamic stage. One can construct models of the evolution of this nebula; by seeing which classes of models produce satellites consistent with their observed properties (e.g., composition, size), one can derive constraints on the nature of the nebula (e.g., temperature structure, lifetime).

Such modeling must examine three phases in the history of the satellites: the equilibration of solids with the gas, to determine the chemical composition of the material from which the satellites were formed; the accretion of these solids into satellites; and the thermal and geological evolution of these satellites after their accretion. This last phase, which is treated in Sec. V, is not only interesting in its own right, but also necessary if we wish to use characteristics of the satellites as observed today to constrain the conditions in the regions where they were formed 4.6 billion years earlier.

Lewis (1972a) examined in detail the chemical species which would exist in local thermodynamic equilibrium in a gas of solar elemental composition at low temperatures and a variety of pressures. In contrast to the higher temperature conditions prevailing in the inner part of the solar nebula, ices as well as rocks could have condensed in the outer portion of the solar nebula and in portions of the nebulae of the outer planets. An examination of how these phases could be assembled into icy bodies was presented by Consolmagno and Lewis (1977). Figure 7 shows the results of Lewis' chemical equilibrium model, outlining the regions of stability of various ices in pressure-temperature space. In general, water ice is the highest temperature ice condensate and is the most abundant ice species, since oxygen is more abundant than carbon or nitrogen in a gas of solar elemental composition. At sufficiently high pressures or low temperatures, water occurs instead in solution with ammonia monohydrate, either as a liquid (high pressure) or solid solution (low temperature). In the presence of condensed water and at temperatures significantly below those at which water first condenses, gaseous methane forms a clathrate, effectively exhausting the available water; the remaining methane freezes out at very low temperatures.

The above results are significantly altered if nonequilibrium species, formed at higher temperatures or conceivably in regions of nonsolar chemical abundances, survive at these low temperatures. Lewis and Prinn (1980) showed that CO and N_2, the stable carbon- and nitrogen-bearing species in the warmer parts of the solar nebula, could be kinetically stable for times comparable to the age of the solar nebula against reactions that convert them to methane and ammonia in the outer solar nebula. If there were large-scale

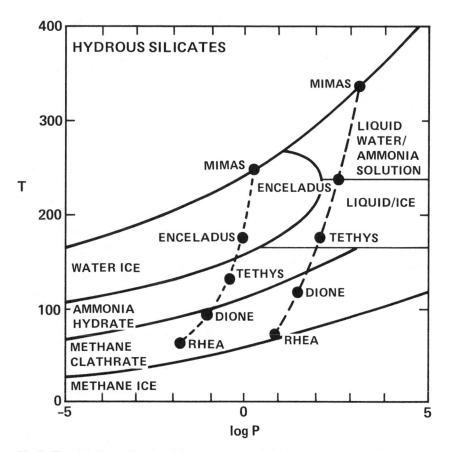

Fig. 7. The adiabatic profiles of a minimum mass (short dashes) and maximum mass (long dashes) nebula are shown against the regions of stability of various ices in pressure-temperature space, based on the work of Lewis (1972a). The minimum mass nebula has just sufficient material to make the inner regular satellites; the maximum mass nebula has one Saturn mass of material in the region of satellite formation. Note the possibility of liquid condensates in the latter case. Beyond Rhea, the nebula is likely to be isothermal since temperatures cannot drop below the ambient temperature at Saturn's distance from the Sun.

convection throughout the nebula, gas carried into the warm regions and converted into CO and N_2 would not have time to change back into CH_4 and NH_3 when it was carried back into the colder, outer regions. Thus, in the outer solar system, out to the regions where comets reside today, one would expect, for example, CO ice rather than methane ice.

However, in circumplanetary nebulae, the situation may be different. If the gas pressure is sufficiently high, methane and ammonia may be stable at temperatures sufficiently high to allow reactions forming these species from CO and N_2 to go to completion on time scales much less than the lifetime of the nebula (10^6–10^7 yr), especially in the presence of dust grains which could act as catalysts. Prinn and Fegley (1981) examined this question with regard to the circum-Jovian nebula, and concluded that methane and ammonia gas could well be formed there; pressures were sufficiently high that kinetic inhibition would not have been a problem, although N_2 and CO may have also been present there in nontrivial amounts.

An important question to be addressed, then, is the pressure and temperature structure of the Saturnian nebula. This will tell us what chemical constituents were present within the nebula, and whether or not kinetic inhibition of certain species might have been important.

The satellites themselves put limits on the mass of the nebula. An obvious minimum mass is the mass of the satellites, plus sufficient gas to bring the condensed material into cosmic elemental abundance, with hydrogen and helium being the most abundant gas species. To obtain an upper limit we will assume that the nebula was not more massive than Saturn itself. Given cosmic abundances (see, e.g., Cameron 1973), the mass of condensibles in a low-temperature ($\sim$ 50–250 K), solar-composition nebula is $\sim$ 1% of the total mass; thus, multiplying the mass of the satellites by 100, we find the minimum nebula mass is of the order of 10^{28} g, while the maximum mass is of the order of 10^{30} g. Spread uniformly in a disk as thick as Saturn and extending to the orbit of Iapetus, the density of the nebular gas lies between 10^{-3} and 10^{-5} g cm^{-3}. These densities imply pressures of the order of 0.1 to 10 bar, for a temperature of 200 K. (Considering a smaller minimum mass nebula extending only out to Rhea, using the masses of the inner five regular satellites and neglecting Titan, one finds a similar nebular density.) Even the minimum mass nebula would have been much more dense than the solar nebula in the region of the outer planets.

Of course, one would expect density, pressure, and temperature to be higher near Saturn, and lower farther out in the circum-Saturnian nebula. Exactly how these quantities vary with position is model dependent, however.

The dependence of nebular pressure P, density ρ, and temperature T on radial distance r, vertical distance above the midplane z, and time t is determined by force balances and energy transport in the r and z directions. To a very good approximation, hydrostatic equilibrium determines the variation of ρ with z, with the vertical component of gravity balancing the vertical component of the pressure gradient. Note, however, that the latter might include a nonnegligible component due to turbulence having a characteristic velocity not much smaller than the speed of sound. Thus, the pressure scale height H is proportional to $r^{3/4}T^{1/2}$ (Lin 1981; Lunine and Stevenson 1982b). For typical values of T, $H \simeq 0.1\ r$. In the radial direction, the key forces are gravity (mostly Saturn's), centrifugal force, the radial component of the pressure gradient, and Reynolds stresses due to viscosity. In many cases, the first two forces are the dominant ones, although, as indicated below, viscous dissipation may play a key role in the evolution of the nebula.

There are two major sources of heating for the nebula: the heat emitted by Saturn and that produced by viscous dissipation. When Saturn's luminosity controls the temperature structure of its nebula, the dependence of T on r and z is determined by the opacity of the nebula. If the opacity is small, $T \sim r^{-1/2}$ and is approximately independent of z (Pollack et al. 1977). In this case, individual particles are in radiative balance with Saturn's luminosity and help to heat the surrounding gas. When the nebular opacity is large in both the r and z directions, the radiative equilibrium temperature gradient is likely to be convectively unstable, leading to $T \sim r^{-1}$ and similarly T decreases steeply with increasing z. Naturally, at large enough values of r and z, the opacity becomes small and hence subadiabatic temperature gradients are achieved, with the temperature asymptotically approaching that of the surrounding solar nebula. When small grains of ice and silicate are present in the Saturnian nebula, as they would be during most of its lifetime, they serve as the dominant source of opacity and insure that the inner portions of the nebula near its midplane are optically thick in both the r and z directions. Even in their absence, the nebula is optically thick in the radial direction and perhaps even in the z direction due to the pressure-induced transitions of hydrogen that are effective at the high pressures of the nebula (Lunine and Stevenson 1982b).

During the last several years, it has become fashionable to apply the concepts of viscous accretion disks around stars (Lynden-Bell and Pringle 1974) to models of the solar nebula (see, e.g., Cameron 1978; Lin and Papaloizou 1980; Lin 1981). Conceivably, these concepts are also relevant to nebulae around the outer planets, although to date only Lunine and Stevenson (1982b) have used them for the Jovian nebula. In such models, viscous dissipation within the nebula plays a central role in heating the nebula and in causing it to evolve with time. Unfortunately, the source of the viscous dissipation and the relationship between the kinematic coefficient of viscosity and other state variables have been key problems for these models. But Lin and Papaloizou (1980) and Lin (1981) have suggested that thermal convection in the vertical direction is the dominant source of the turbulence that leads to viscous dissipation and have used mixing length theory to explicitly determine it. According to these calculations, $T \sim r^{-3/2}$ in the inner, opaque portions of viscous accretion disks. Since the implied temperature gradient in the radial direction is superadiabatic for the solar nebula and nebulae of the outer planets, it is possible that thermal convection in the radial direction may result, in which case a $T \sim r^{-1}$ relationship would be established. But the rotation of the accretion disk may hinder the occurrence of thermal convection in the radial direction, in which case the $T \sim r^{-3/2}$ relationship may be maintained.

Due to viscous dissipation, the mass distribution of the nebula varies secularly with time. In particular, there is an inward directed mass flux in the inner portions of the nebula and an outward flow at large distances, and an outward direction flux of angular momentum at all locations. Thus, the mass of the nebula decreases with time as the central object (e.g., the Sun or Saturn) accretes inflowing material. Also, the outer parts of the nebulae expand due to the outflux of angular momentum. A typical lifetime for a viscous accretion disk model of the solar nebula is $\sim 10^5$ yr (Lin 1981). Analogous lifetimes for nebulae of the outer planets could well be much shorter, on the basis of a simple scaling of formulae given by Lin (1981). If so, it is not clear that satellites could be assembled in so short a time scale.

In summary, viscous accretion disks differ in their properties in the following ways from optically thick disks, whose temperature structure is controlled by the central object's luminosity: a superadiabatic temperature gradient (vs. an adiabatic one) may be present in the inner parts of the nebula. The temperature of the disk may be semidiscontinuous (vs. continuous) near the central object's photosphere; the total mass and mass distribution in the nebula continuously vary with time (vs. are nearly constant); and the lifetime of the nebula may be very short, $\ll 10^5$ yr (vs. 10^6 yr).

In all published papers to date concerning the Saturnian nebula, only models for which the planet's luminosity is the dominant heat source have been investigated. Thus, of necessity, the detailed discussion given below reviews nebula models based on this type of heat source.

The first explicit calculation of the temperature structure of a Saturnian nebula and the history of condensation products in it was made by Pollack et al. (1977). They considered two limiting cases in evaluating the variation of temperature with radial distance in the nebula: optically thick and thin nebulae. As discussed above, $T \sim r^{-1}$ in the former case and $T \sim r^{-1/2}$ in the latter case.

Figure 8 illustrates how temperature varies with time and distance in the high-opacity case (the more likely situation during the early phases of the nebula, before grains underwent successive aggregation). Because Saturn's luminosity decreased with time during the late, quasi-hydrostatic stage, a given ice species first condensed at progressively later times at closer distances to Saturn. Thus, satellite and ring formation need not be coeval throughout the nebula. For the sake of simplicity, we will assume below that condensation of satellite-forming material occurred at some discrete time, which can be characterized by a single radial temperature profile. This time may be the last, coldest epoch of condensation, just prior to the dissipation of the nebula. But the possibility that satellites formed at different times should be kept in mind.

A lower limit for the lifetime of a nebula heated by Saturn's luminosity can be derived from Fig. 8 by finding the time at which ices first condensed at a given location. Water ice first condensed in the region of the B Ring at 5×10^6 yr from the end of the hydrodynamic collapse phase and methane clathrate first formed at Titan's distance at 10^6 yr.

Prentice (1980,1982) developed an elaborate theory of the physics of the Saturnian nebula, from which temperatures and pressures in satellite-forming gas rings could be determined. Of necessity, however, his theory was based on many assumptions concerning initial conditions and the behavior of the gas-dust mixture as the nebula evolved. Some parts of his formalism, most nota-

SATURN—HIGH OPACITY

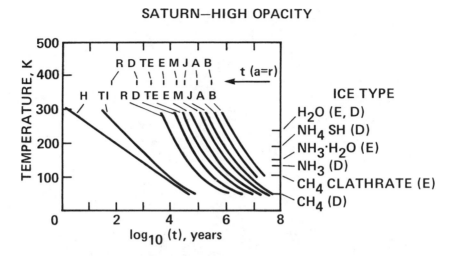

Fig. 8. Temperature of a condensing ice grain as a function of time from the start of the late, quasi-hydrostatic stage. Each curve refers to a fixed distance from the center of the planet and is labeled by the first letter of the name of the satellite or ring segment, which is currently at that distance. The right-hand, vertical scale denotes the temperature at which various ice species condense under equilibrium (E) or disequilibrium (D) conditions between the gas and solid phases. The Saturnian nebula has been assumed to be highly opaque due to grain opacity (from Pollack et al. 1976).

bly supersonic turbulence within the nebula, are quite controversial and, indeed, appear to be implausible.

Weidenschilling (1982) developed a model for the temperature and pressure structure of the nebula by assuming that the mass distribution in the nebula was in some way reflected by the distribution of mass in the satellites today. Given a distribution of nebular density, one can then find a pressure profile either by solving for a hydrostatic, rotating gas disk or by assuming the nebula was convecting and adiabatic.

Weidenschilling (1982) used the formalism of Safronov (1972) for evolving nebulae for which rotational and gravitational energies were in balance, and applied the resulting relationship between temperature and distance from the nebula center to a nebula massive enough to produce the Saturnian satellites. The Safronov relationships neglect disk self-gravity, and lead to a temperature gradient in the nebula which is subadiabatic, and hence stable against convection. Such a shallow temperature gradient seems unlikely in view of the large grain and gas opacities in the radial direction, as discussed above.

Alternatively, one can follow the work of Prinn and Fegley (1981) for Jupiter, and Pollack et al. (1976) for Saturn and assume that the nebula was convecting and adiabatic. The density of the nebula can be assumed, or taken

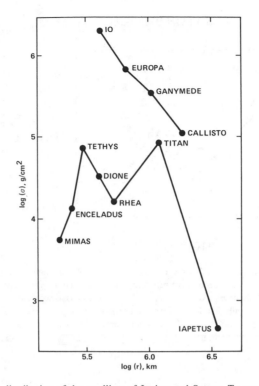

Fig. 9. The mass distribution of the satellites of Jupiter and Saturn. To create this figure, the mass of each satellite (supplemented with sufficient gas and ice to match cosmic abundances) was spread into an annulus centered on that satellite's orbital radius and extending to the midpoints between the satellite orbits. Thus satellite mass, divided by the area of that annulus, represents the minimum surface density of material in the protoplanetary nebula. In this loglog plot, the slope of the Jovian satellites' line is -1.9 and that of the Tethys-Dione-Rhea line is -2.6, indicating that the mass distribution is controlled by a simple power law.

from the surface density, whose variation in space can be defined by the $1/r^{2.5}$ trend seen for Tethys, Dione, and Rhea in Fig. 9. If one assumes that the position of Mimas coincided with the condensation of water ice, the temperature and pressure at that position can be determined, and an adiabatic gradient followed out to the position of Rhea (cf. Table III). Beyond Rhea, such an adiabat would predict temperatures lower than the ambient temperature at Saturn's distance from the Sun; the nebula thus might be presumed to be isothermal beyond that point.

In general, one can postulate a wide variety of Saturnian nebula models. There still does not exist any generally accepted theory describing the evolution of such disks of gas, nor is there even agreement as to the proper set of starting conditions. But in nearly all of the above models of the nebula, a similar condensation sequence results. Rocky material plus water ice are assumed to be stable at the positions of Mimas and Enceladus; ammonia monohydrate is also present at the distances of Tethys and Dione, with methane clathrate possible at Rhea's location and likely at Titan's and Iapetus'. Such a pattern is illustrated in Fig. 7, where the left-hand curve marks an adiabat of a minimum-mass nebula as described above. The right-hand curve signifies a maximum-mass adiabat, which represents a significantly different condensation scheme. There, ammonia monohydrate is present even at Mimas' and Enceladus' distance.

Once solid material has condensed in a circum-Saturnian nebula, it settles into the midplane of the nebula, collects itself into rings, and eventually accretes into planetesimals, from which the satellites, in turn, are accreted. Analytical and numerical models of this process have been developed, most

TABLE IIIa
Mass Distribution in a Minimum Mass Nebula [a]

Satellite	σ obs (g cm^{-2})	σ calc (g cm^{-2})	ρ [c] (g cm^{-3})	T (K)	log P
Mimas	5,500	240,000	2.8×10^{-4}	250	0.46
Enceladus	13,000	123,000	1.2×10^{-4}	175	−0.08
Tethys	69,000	70,000	5.3×10^{-5}	125	−0.55
Dione	36,500	37,000	2.2×10^{-5}	90	−1.1
Rhea	15,000	15,000	6.6×10^{-6}	60	−1.8
Titan	86,000 [b]	4,600			
Hyperion	20	2,800			
Iapetus	400	290			

[a] σ, ρ, T, and P are the nebular column mass density, volume mass density, temperature, and pressure (in bars) at the central plane, respectively.
[b] Assumes Titan's feeding zone extends to Iapetus.
[c] ρ is derived from σ: $\rho = \sigma/H$, where $H = 0.12\ r$ is the scale height and r is radial distance from the center of the nebula (Prinn and Fegley 1981).

TABLE IIIb
Maximum Mass Nebula [a]

Satellite	ρ (g cm^{-3})	T (K)	log P
Mimas	8.6×10^{-2}	340	3.1
Enceladus	3.5×10^{-2}	240	2.6
Tethys	1.6×10^{-2}	175	2.1
Dione	6.7×10^{-3}	120	1.6
Rhea	2.0×10^{-3}	75	0.8

[a] Assumes $\sigma = 10^{35}\ r^{-2.6}$ g cm^{-2}.

recently by Weidenschilling (1981) and Coradini et al. (1981), looking specifically at the case of accretion in nebulae of the outer planets.

As we saw above, the gas densities in such nebulae are much higher than in the solar nebula; thus, gas-dust dynamics are likely to be quite important. Gas drag, which causes an inward spiraling of particles, leads to very rapid evolution times for accretion of dust to planetesimals. Consequently, one expects rapid and efficient accumulation of satellites in the nebula. In this situation, satellite formation is bounded by the time scale over which Saturn's luminosity decreased significantly or the viscous dissipation lifetime of the nebula. Satellites could have formed with a zoned structure if the accretion time scale is much less than the latter time scales, with the lower temperature condensates being located closer to their surfaces (cf. Consolmagno and Lewis 1977).

In such a straightforward nebula, satellite masses might be expected to reflect the mass of material available in the nebula, and the bulk chemical properties of each satellite to be fixed by the local pressure and temperature conditions. The distribution of mass within the Saturnian satellite system is quite irregular, however, especially in contrast to the Galilean satellites of Jupiter, as noted by Weidenschilling (1982). He took the masses of each satellite, augmented them by sufficient material to include the mass of uncondensed gas, spread these masses into annuli centered on the orbits of each satellite, and thus determined nebular surface density as a function of distance from Saturn. This is illustrated in Fig. 9.

Figure 9 also includes the results of the exercise for the Galilean satellites. In that case, a monotonic and basically linear decrease in density is seen. Such a decrease is also seen for Tethys, Dione, and Rhea, but Mimas and Enceladus are much too low in mass, and Titan much too massive, for this simple gradation of densities.

Weidenschilling (1982) suggested that aerodynamic effects separating the first-condensed materials in the dense proto-Saturnian nebula may be responsible for the reduced masses of the inner two satellites, similar to his explanation (Weidenschilling 1978) for the small mass and high density of Mercury compared with the other terrestrial planets. Alternatively, Pollack et al. (1976) pointed out that the commencement of satellite formation in the innermost region of the nebula may have been delayed, due to the high initial temperatures there and the large initial size ($3-5$ R_S) of Saturn. Satellites that began accreting early (e.g., Tethys, Dione, and Rhea) may have accreted more efficiently due to their larger sizes than ones that began later (e.g., Mimas and Enceladus). Finally, these satellites may simply be fragments of initially larger bodies, broken up by cataclysmic collisions (cf. B. A. Smith et al. 1981).

The case of Titan presents the opposite problem. Prentice (1982) suggested that Titan was a planetesimal from the solar nebula which was captured by Saturn. No mechanism has yet been worked out, however, which could bring such a massive body from a solar orbit into a regular, circular, non-inclined, prograde orbit about Saturn (cf. the discussion in Sec. III concerning the possible capture of Phoebe).

On the other hand, one might speculate that Titan is the only survivor of an accreting satellite system which underwent intense bombardment and disruption of accreting planetesimals. A 10-km radius comet impacting at a relative velocity of 10 km s^{-1} (a typical orbital speed for inner Saturnian satellites) would have sufficient energy to shatter a 2500-km radius icy satellite (cf. Greenberg et al. 1977). Such impacts would have to be much more common at Saturn's distance from the Sun than at Jupiter's during the period of satellite formation to account for the survival of the four Galilean satellites. However, the Valhalla basin of Callisto shows that large impacts did occur in that system, and such impacts would be less likely to shatter Ganymede if they occurred when that satellite was significantly melted. Another difficulty with the above catastrophic disruption hypothesis is that the satellites other than Titan represent collectively only $\sim$ 10% of the mass of Titan. If they are survivors of a family of Titan-class objects, a significant amount of their original mass must have been broken into very small fragments that were lost from the system via plasma drag, gas drag, or sublimation, perhaps to be captured eventually by Saturn or by Titan itself.

Given the chemical trends predicted by the work described above, one would expect to see a regular and predictable trend in satellite bulk densities, just as occurs for the Galilean satellites of Jupiter. But there is no pattern to the densities of the Saturnian satellites (cf. Fig. 1). In part, this difference between the two satellite systems may reflect the more numerous types of ices possible in the Saturnian system than near Jupiter and the occurrence of a major ice component throughout the Saturnian system. There is much less of a density contrast among various ice species, as may occur for the Saturnian system, than for varying mixtures of rock and water ice, as occurs for the Jovian system. In addition, the large uncertainties in the densities of Mimas and Enceladus make it difficult to draw hard and fast conclusions about trends in density. But it is clear that Tethys is less dense, and Dione more dense, than a theory of simple condensation and accretion would predict.

This lack of regularity has led to suggestions that the satellites we see today may be reaccreted fragments of protosatellites which were broken up during an early period of heavy bombardment. B. A. Smith et al. (1981) noted that the cratering flux implied by the densely cratered plains on Iapetus and Rhea should have led to at least some impact events, deeper in Saturn's gravity well (i.e., in the inner satellite system) violent enough to completely disrupt these satellites; and, indeed, both Mimas and Tethys have an enormous crater on their surface whose impact energy must have been very nearly sufficient to accomplish such a disruption. Note, however, that this argument depends on an extrapolation to larger sizes of the population of impacting bodies responsible for the craters on Iapetus and Rhea.

What if the first satellites formed were disrupted into a few large chunks? Random reaccretion of such chunks could account for the deviations of satellite density from a regular pattern. It can be shown that treating the densities of Tethys and Dione as one-sigma deviations from a mean density would imply that they would have been assembled from only 30 or so planetesimals of pure rock or ice. Indeed, fragmentation experiments on ice (though clearly not of planetary size) show that the largest fragments are on the order of 1/30 of the original mass. Thus, superimposed on a radial variation of the proportion of low-temperature ices, there may be a random component of varying amount of rock and ice.

V. EVOLUTION OF THE REGULAR SATELLITES

A. Interior Evolution

Given the composition of the satellites, as implied by the models of the previous section, one can construct evolutionary models of them and test their behavior against the observed surface features of the satellites. Such models attempt to predict the temporal and radial variation of temperature, composition, and size of the satellites' interiors, which are determined by the strength of heat sources, bulk composition, size, and initial conditions. Key heat sources for the Saturnian satellites, as for the Jovian ones, include gravitational energy released during satellite formation (accretional heat), radioactive decay of long-lived radionuclides (U, K, and Th) contained in the rock or silicate component, and tidal friction. The buildup or loss of heat from the satellites' interiors is controlled by liquid- and solid-state convection as well as ordinary solid-state conduction. Thus, the smaller satellites evolve more quickly than the larger ones, in cases where tidal heating is unimportant, to thick crusts and cool interiors and hence "geologically dead" surfaces. Tidal heating can be important for satellites that are situated close to their primaries and have large, forced eccentricities. In such cases, even small satellites can remain geologically active over much of their lifetimes.

Constraints on thermal history models and hence on properties of the satellites' interiors are placed by the observed morphology of their surfaces. Tectonic features provide markers of epochs of increases or decreases in the size of satellites due to phase changes and/or compositional segregation occurring in their interiors, while resurfacing events denote times when near-surface liquids have been able to reach the surface. Crude time markers for these features are provided by the density of craters on them. All of the large satellites of Saturn, whose surfaces have been photographed at good spatial resolution (i.e., Mimas, Enceladus, Tethys, Dione, and Rhea), have experienced both *extensional* tectonism and resurfacing, although the timing and extent of these processes have varied greatly among these inner five satellites. The occurrence of a thick atmosphere, which obscures its surface, and its large size imply a geologically rich history for Titan. Iapetus' surface has not been photographed at an adequate resolution to define its surface geology, although its comparatively large size would, by analogy to Dione and Rhea, suggest that its surface has undergone some tectonism and resurfacing.

The first evolutionary models for the Saturnian satellites were developed by Consolmagno (1975), who computed the time-dependent flow of heat from small icy satellites with a cosmic composition of rock, water, and ammonia. Heating was due to the decay of radioactive nuclides in the rocky material; heat transport was by conduction within the ice, and convection within melted ammonia-water regions. This work was further developed by Consolmagno and Lewis (1978), who noted that the small icy satellites of Saturn would probably not be stable against solid-state convection.

The first thermal model to explicitly include solid-state convection in outlining the evolution of Saturn's satellites was that of Ellsworth and Schubert (1981). Ellsworth and Schubert (1983) constructed models for bodies of water ice and silicates but with no ammonia or methane and hence no eutectic melting. They proposed that the presence or absence of solid-state convection could be the deciding factor in determining whether tectonic features appear on the surfaces of these satellites. According to their models, only Dione and Rhea would have experienced substantial convection; these are the two satellites with the most widespread extensional features (neglecting Enceladus, which presumably was subjected to tidal heating and is not, therefore, included in these models). But it is not clear why solid-state convection would produce only extensional, but not compressional, tectonic features nor whether the occurrence of a limited amount of extensional tectonic features on Tethys and Mimas is consistent with the behavior of these models.

One major issue suggested by their work is the need for a quantitative estimate of the surface stresses which would result from the predicted internal convection. The first step, an estimation of the velocities in convecting icy satellites, has been taken by Golitsyn (1978) but connecting these velocities with surface features remains to be carried out.

A different approach to the problem of the origin of surface tectonic features is to examine the stresses produced as these satellites expand and contract upon heating and cooling. Such stresses would be greatest where phase changes occur inside these bodies.

As a first step in such modeling, Consolmagno and Huang (1982) constructed thermal models of five of the large regular Saturnian satellites; they used Voyager 2 masses to define the satellites' mean densities and explicitly examined the density changes of the various components upon heating and cooling. Enceladus, which clearly has been resurfaced in the recent past, presumably has a special heating source, such as tidal heating (see below), and so it is excluded from this thermal model; Titan, which is massive enough to contain mostly high-pressure phases, represents a separate problem as well. In the first case considered, the satellites were assumed to be made of water and rocky material only (no ammonia), whose relative abundances were determined from the observed mean densities. Two sets of models were investigated: one in which these bodies formed slowly enough that they were initially a cold ($\sim$ 80 K), homogeneous mixture of water ice and rock; and one in which they accreted quickly enough to start their histories fully differentiated, with liquid water mantles overlying rocky cores.

These two sets of models fail to provide expansion after the first 100 million years, and fail to provide interior temperatures warm enough to melt ice or to keep water liquid for very long (cf. Table IV). But Dione and Rhea are observed to have lightly cratered extensional features, which are, presumably, at least a billion years younger than the oldest surfaces of these satellites; some evidence of resurfacing by liquid material also exists in this time frame. Thus, these models are inconsistent with the satellites' surface morphology.

The most severe shortcoming of these models, as well as that of Ellsworth and Schubert (1983), is that they neglect the presence of ammonia monohydrate, which is expected in many or all of these satellites (depending on conditions in the nebula, as described in Sec. IV). A more complex model has been constructed by Consolmagno (1984a), based on the model of Consolmagno (1975) which did include ammonia eutectic melting, but which was limited at the time by a lack of information on the masses and sizes of the satellites in question. This model was expanded to include changes in radius due to changes in density as ice melted and as the composition of the melt changed (with the addition or freezing out of water). The mean density of each satellite was matched with a composition including rocky material of density 3.0 g cm^{-3}, ammonia monohydrate, and water ice, with the latter two having jointly cosmic proportions of N to O. The initial conditions for this model corresponded to those appropriate for slow accretion: a temperature everywhere of 80 K and a uniform distribution of rocky material.

The above model succeeded in matching some of the surface morphology inferred for these bodies from the Voyager images (cf. Table IV and Fig. 10): The model of Tethys showed a small degree of tectonism, while the models of Dione and Rhea showed evidence for extensional stresses and the possibility of flows on their surfaces. The presence of a low-density, eutectic melt having a low freezing point ($\sim$ 170 K) and the passages opened by large-scale expansional stresses combine to favor the formation of surface flows. Surface extensional stresses of 25 bar for Dione and 60 bar for Rhea were obtained; the critical stress for extensional failure of ice at 80 K is not well known, but is probably on the order of 20 bar.

Another satisfying match of this model to observation is the time scale of such activity: well past the period of early heavy bombardment, but early enough to allow cratering by a later population of impactors. In this sense the evolution is comparable in time scale to that of the emplacement of mare material on the Earth's Moon. However, this model does not explain all the observed features on these satellites, including the large Ithaca Chasma on Tethys (which may be of impact origin [McKinnon 1982; Moore and Ahern 1982]), and areas that appear to be resurfaced on Tethys and Mimas (Plescia

TABLE IV
Thermal Models

Water Ice and Rock	Water Ice, Ammonium Monohydrate, Rock
MIMAS (present surface: old, heavily cratered, unaltered)	
Never melts, never convects; if starts molten, then refreezes by 0.1 byr.	Never melts, never convects.
ENCELADUS (present surface: very young, resurfaced, relaxed craters, grooves)	
With tidal heating 5–7 times current rate, molten interior possible; crust 10–30 km thick? (20 × current tidal heat needed to melt from ice originally.)	With tidal heating, thinner crust more likely than in pure water case; needs 7-fold enhancement of current tidal heating to melt initially. The present tidal heating may suffice to keep its interior molten thereafter.
TETHYS (present surface: old, cratered; one large extensional groove)	
No interior melting, little interior convection, no significant expansion. If started molten, refreezes within 0.1 byr.	33% rock, 20% NH_3H_2O. At 0.6 byr, small melting at core, 1/3 km expansion; maximum melt to 130 km radius; 30 km rocky core; refreezing by 1.2 byr.
DIONE (present surface: complex older cratered sections, and younger resurfaced areas; evidence for considerable expansion)	
May start with ice II core; heating changes to ice I with expansion within 0.1 byr; if starts molten, refreezes by then to ice I; convection occurs, but no further melting. Later cooling may produce ice II core and contraction.	50% rock, 15% NH_3H_2O. At 0.3 byr melts, contracts 0.5 km; maximum melting at 1 byr to within 200 km of surface, with 150 km rocky core; expansion by 0.4 km until refrozen at 3 byr; slow contraction to present.
RHEA (present surface: complex regions of differing crater histories; resurfaced plains, evidence of considerable expansion)	
Large ice II core early (unless starts molten) changing to ice I by 0.1 byr; at 4.0 byr, change back to ice II leads to 15 km contraction. Convection but no internal melting.	Expands 1 km until melt at 0.3 byr; contracts 1.5 km until maximum melt at 1.7 byr; slow 1 km expansion to present day. If ice II forms, get recent significant contraction.
TITAN (present surface: obscured by atmosphere)	
Evolution similar to Ganymede; early convection and melting possible, but likely refrozen by present.	Melts within 0.15 byr; substantially differentiated by 0.5 byr. Thin crust, molten convecting interior to present.
IAPETUS (present surface: light and dark hemispheres, heavily cratered)	
Smaller ice II core than Rhea; less rock so less heating and convection. No melting.	Expands by 0.8 km, melting at 0.5 byr; 0.5 km contraction until refrozen at 1.2 byr, with 300 km crust and 100 km rocky core. Slow half-kilometer contraction to present.

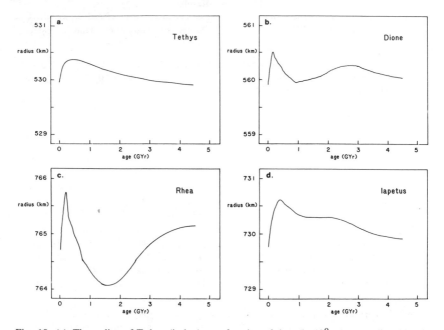

Fig. 10. (a) The radius of Tethys (in km) as a function of time (in 10^9 yr) as predicted by the thermal model of Consolmagno (1984a). The initial expansion is due to the warming of the ice; after a small degree of melting and refreezing, the planet slowly cools off and contracts. (b) The radius of Dione (in km) as a function of time. Because Dione is denser, it presumably has more radionuclide-bearing silicates and so heats up faster than Tethys. Once it starts to melt (0.5 Gyr) it contracts rapidly, then expands slowly upon refreezing until the entire body is refrozen, at 3 Gyr. After that time, it slowly cools down and contracts. (c) The radius of Rhea as a function of time. Rhea is the largest of the satellites modeled by Consolmagno (1984b) and melts the most. As with Dione, the period of melting is one when the satellite contracts, and refreezing makes the satellite expand again. Since refreezing may not be complete yet, it is possible that Rhea is still expanding at the present time. (d) The radius of Iapetus as a function of time. Iapetus is large but not very dense, hence poor in radionuclides, and so it does not melt as much as Rhea. The contraction and expansion upon melting and cooling of interior sections is masked by the general cooling and contraction of the outer portions of the satellite, which contain the bulk of the volume of the body.

and Boyce 1982). Also, the model predicts a greater resurfacing for Rhea than for Dione; in fact, the opposite is observed.

Variations on this model do not yield markedly better results. If the initial temperatures inside the bodies are raised, melting may occur somewhat earlier, but neither the degree nor the duration of the melting is substantially different from the models that start at 80 K everywhere. This is not surprising, since these satellites are small enough to reach thermal equilibrium very rapidly. There is not a significantly long time lag between the production of heat in their interiors and its transport to the surface, and, hence, any thermal spike introduced into these bodies is soon conducted away.

Satellites that were initially differentiated into rocky cores and icy mantles could result from rapid accretion, due either to their being completely melted by accretional heat or being assembled in an "onion layer" fashion (Consolmagno and Lewis 1977). Models run with such starting conditions predict roughly twice the contraction and expansion of Rhea as occurs for the model with homogeneous starting conditions; the times when these stresses appear remain the same. A model of an initially differentiated Dione shows a surface evolution very similar to the homogeneous case. However, an initially differentiated Tethys model is virtually the same as the corresponding Dione model. Given that such a starting condition would imply a Rhea that expands much more than Dione, and a Tethys indistinguishable from Dione, both in contradiction with observations, we conclude that such models seem less satisfactory than models starting with a homogeneous mixture of rock and ice.

Conceivably, models that lie between the extremes of being initially fully differentiated and fully undifferentiated might yield a better match to the observations. For example, such models might display sufficient expansion and near-surface melting for Tethys to be compatible with its surface morphology, while at the same time implying a more active geology for Dione during its first $\sim 10^9$ yr than for Tethys.

Bianchi et al. (1982) also modeled the time-dependent melting of these satellites. They included the effects of ammonium monohydrate in producing a eutectic melt, as did Consolmagno (1984a), but in addition incorporated the heating implied by their accretion models. As of this writing, this work remains in progress.

Models of Enceladus and Titan need separate, special treatment. As discussed in Sec. I, portions of the surface of Enceladus have been subjected to extensional tectonism and resurfacing over much of its history, with its youngest surfaces being less than $\sim 1 \times 10^9$ yr old. Squyres et al. (1983b) proposed that the tectonism was the result of the freezing of liquid water (recall ice I is less dense than liquid water), and that the resurfacing was due to the eruption of a liquid NH_3/H_2O mixture onto the surface from a hot interior. Their suggestion that the resurfacing agent was a mixture of NH_3 and H_2O rather than pure water was motivated by the lower melting point of the mixture (as much as 100°C) and by the density of a eutectic mixture being less than that of solid water ice, thus facilitating the ability of the eutectic mixture to reach the surface. In contrast, liquid water is denser than solid water ice.

It is quite possible that resurfacing events on Enceladus are occurring even at the present epoch: The peak density of the E Ring is located very close to the orbit of Enceladus (Baum et al. 1981), implying that Enceladus is the source of the particles in the E Ring. In addition, the lifetime of the micron-sized particles of this ring is quite short ($\sim 10^3$ yr) (Cheng et al. 1982). Hence this material must be constantly replenished, unless the E Ring is a transitory feature.

The above discussion implies that the interior of Enceladus has contained liquid zones in its recent past, including perhaps the present, as well as at other times in its lifetime. But significant fluctuations in the size of these zones over its lifetime are implied by the presence of extensional tectonism, if solidification of liquid water is responsible for the satellite's expansion. As a result of Enceladus' small size and high-bulk ice content, radiogenic heating fails by at least two orders of magnitude to provide the heating necessary to keep ices molten over much of the satellite's lifetime (Squyres et al. 1983b). Indeed, if radiogenic heating was capable of meeting this requirement, it would even more easily cause extensive interior melting of the interiors of the larger satellites Tethys, Dione, and Rhea, contrary to observation.

The most likely source of the required heating of Enceladus is tidal dissipation engendered by the satellite's forced eccentricity (Yoder 1979), just as tidal heating is suspected to be responsible for the active volcanism on Io. But there are quantitative problems with this mechanism for the current epoch. At present, Enceladus has a forced eccentricity of only 0.0044 due to a gravitational orbital resonance with Dione. Using this eccentricity and as low a tidal Q as seems likely ($Q = 20$), Squyres et al. (1983b) obtained a tidal heating rate for a homogeneous, solid Enceladus that was comparable to the radiogenic heating rate and thus incapable of melting the satellite. Furthermore, for the same material properties, Mimas was found to have a tidal heating rate 30 times that of Enceladus, contrary to the strong differences in the ages of the surfaces on the two satellites.

Squyres et al. (1983b) overcame the above difficulties faced by the tidal heating mechanism by proposing that Enceladus was initially melted by tidal friction at a time when the satellite's forced eccentricity was a factor of 7 to 20 higher than its current value and hence tidal heating was a factor of 50 to 400 times larger than at present. Furthermore, they point out that a much smaller tidal heating rate is needed to maintain a molten interior once it is established, since tidal dissipation for a thin, rigid-shelled object is generally much larger than for a totally solid object. Minimum eccentricities ranging from the current value to seven times this value are adequate for maintaining a molten zone, depending on the composition of the melt (H_2O vs. NH_3/H_2O) and on the thermal conductivity of the outer rigid shell. If the thermal conductivity of pure solid water ice is used, minimum eccentricities of five to seven times the present value are required. But if the thermal conductivity is about an order of magnitude smaller, due to the presence of hydrates and clathrates, and/or a frost insulating layer on the surface, minimum eccentricities close to the present value may be adequate. The failure of tidal heating to keep the interior of Mimas molten over most of its history can be attributed to a combination of different material properties (e.g., a higher thermal conductivity) and the absence of an earlier, higher forced eccentricity.

An intriguing mechanism for generating a higher forced eccentricity in the recent past for Enceladus has been suggested by Lissauer et al. (1983). They point out that Janus is quite close to a $2:1$ orbital resonance with Enceladus. Furthermore, Janus' orbit should be evolving rapidly outward due to angular momentum it is exchanging with the main rings by means of spiral density waves. Hence, only 10^7 to 10^8 yr ago Janus may have been temporally locked in a $2:1$ resonance with Enceladus. In this case, the exchange of angular momentum between the rings and Janus would have resulted in a pumping up of Enceladus' eccentricity to a high enough value to have caused a melting of its interior. This resonance lock may have been broken either by a catastrophic collision or by the establishment of the resonance between Enceladus and Dione. In summary, tidal heating appears to be responsible for the occurrence of molten zones inside Enceladus over much of its lifetime, including perhaps the present epoch. A larger forced eccentricity than the present value is required to initiate melting. This may have occurred due to the transfer of angular momentum from the rings to Janus to Enceladus. Smaller eccentricities, perhaps even encompassing the present value, suffice to maintain a molten interior.

Few detailed models of Titan's internal structure have been attempted. The major stumbling blocks involve the atmosphere, whose origin and age are unclear but which must clearly have affected its thermal evolution, and the unknown behavior of ammonia monohydrate and methane clathrate at high pressures. The presence of a nitrogen-methane atmosphere implies that both kinds of ice could be present, now or in the past.

The occurrence of polymorphs of water ice (especially ices VII and VIII, which are in effect self-clathrate forms of water ice) would seem to preclude the existence of other clathrates. One might speculate that the growth of internal lithospheric pressures as Titan accreted led to the destruction of clathrates and the formation of an atmosphere coincident with accretion. But preliminary work by D. Stevenson (personal communication) indicates that ammonia monohydrate may inhibit the formation of the high-pressure polymorphs of ice.

A thermal model of a Titan-sized body of ammonia hydrate, water ice, and rock in cosmic proportions was attempted by Consolmagno (1975), who concluded that large-scale internal melting and differentiation would occur within 100 million years after a cold formation, and that a thin (< 100 km) crust could be maintained to the present day. In the absence of more experimental work on the high-pressure behavior of the appropriate ices, and lacking any knowledge of the geology of Titan's surface, such models must remain quite speculative.

B. Surface Evolution

The surfaces of the Saturnian satellites are altered not only by the internal forces discussed above, but also by forces acting from outside these bodies. The most prominent of these forces is, of course, impact cratering; this topic has been examined extensively by several authors in this book and will not be discussed here. Other effects include the sublimation of ices and their destruction or alteration by solar ultraviolet radiation and high-energy particles of Saturn's magnetosphere and the solar wind.

These latter effects on satellite surfaces were examined by Consolmagno and Lewis (1978). Briefly, any methane-containing ice is unstable against sublimation, at Saturn's distance from the Sun; water ice, on the other hand, is clearly stable. The situation for ammonia ice is not clear cut, given the uncertainty in the vapor pressure data found by Lebofsky (1975); but, in any case, solar ultraviolet radiation is efficient enough to break N–H bonds (with the subsequent loss of the hot H atoms so produced) to eliminate ammonia from the visible top 100 μm of the surface in the space of 100 yr. Thus, the observed absence of these ices in the spectra of the Saturnian satellites (Cruikshank 1979b) is not surprising.

The effect of high-energy particles on the satellite surfaces has not been directly addressed, although such work is currently being undertaken for the satellites of Jupiter (cf. Sieveka and Johnson 1982). Given the lower magnetospheric fluxes around Saturn's satellites, however, it is clear that they will be less important here than near Jupiter.

An interesting comparison can be made between the reflectivity of the surfaces of the Jovian and Saturnian icy satellites. Jupiter's ice-rich satellites have old surfaces which are consistently darker than the younger surfaces. Almost all the icy satellites of Saturn have bright surfaces, irrespective of their ages. The source of the darkening component on Ganymede and Callisto has been unclear; it may be external dust which has accumulated on their surfaces, or it may be indigenous material which was left on the surfaces while water ice was preferentially removed by sputtering and sublimation. The Saturnian and Jovian satellites have been subjected to similar fluxes of darkening material from outside these systems, but the inner satellites of Saturn have been largely isolated from the flux of impact ejecta emanating from Phoebe due to an effective sweeping up of this material by Iapetus and Titan, while the outer Galilean satellites of Jupiter have experienced the full flux of impact ejecta from the irregular satellites of Jupiter. Furthermore, the ejecta originating from the prograde, irregular satellites of Jupiter impact the surfaces of the larger satellites of Jupiter at a much smaller velocity (approximately the escape velocity of the satellites) than the ejecta from retrograde Phoebe impact the surfaces of the Saturnian satellites. Hence, the regolith of the former satellites can be expected to contain a larger fraction of exogenous ejecta. We conclude that the difference in the brightness of the surfaces of the icy satellites of Jupiter and Saturn cannot be attributed to a flux of darkening material from outside these systems, but could be due to a flux of darkening material from the irregular satellites.

On the other hand, one could attribute this difference to a preferential removal of water ice from the older surfaces of the icy Jovian satellites due to enhanced rates of sputtering and/or sublimation. Certainly, the Jovian satellites are exposed to a higher flux of magnetospheric high-energy particles and are hotter (because of their closer distance to the Sun) than are the satellites of Saturn. But the older surfaces on Ganymede are brighter than similarly aged surfaces of Callisto, although Ganymede experiences a larger flux of magnetospheric particles. If sublimation is the controlling factor, it is not clear why the old surfaces of Callisto are much darker than the old surfaces of Ganymede.

In speaking of the reflectivity of the surfaces of the Saturnian satellites, some comment must, of course, be made about the famous two-faced satellite, Iapetus.

1. While a number of satellites exhibit modest (approximately several tens of percent) differences in the brightness of their leading and trailing hemispheres, Iapetus is unique in having a trailing hemisphere that is about an order of magnitude brighter than its leading hemisphere (B. A. Smith et al. 1982) (fact 1);

2. Additional constraints on the nature and source of this extreme brightness asymmetry include: At least one dark floored crater is present on the bright trailing hemisphere but no bright floored craters are obvious on the dark leading hemisphere (B. A. Smith et al. 1982) (fact 2);

3. Albedo contours in both hemispheres closely match impact flux contours (Squyres and Sagan 1983) (fact 3);

4. Visible and near-infrared spectral reflectivity of the dark hemisphere are significantly different from that of Phoebe (Cruikshank et al. 1983) (fact 4);

5. Mean density of Iapetus is 1.16 ± 0.09 g cm^{-3}, indicative of an ice-rich object (B. A. Smith et al. 1982). Furthermore, the discussion in Sec. IV implied that methane, ammonia, and perhaps nitrogen and carbon monoxide may be present in these ices, besides the dominant water component (fact 5).

Below, we compare the predictions of a number of theories for the brightness asymmetry with the above facts.

Even prior to the establishment of fact 3, a number of theories were advanced in which impacting meteoroids played a prominent role in generating the brightness asymmetry. They differed in the composition of the meteoroids (icy vs. carbonaceous) and in whether the key effect of the postulated impacts was erosion of native surface material or deposition of meteoroid material.

Peterson (1975) suggested that impacting icy meteoroids stuck to the trailing hemisphere of an initially dark Iapetus, but were vaporized on the leading side since their kinetic energy of impact was much greater on the leading side. This hypothesis requires an initially dark surface on Iapetus and is thus in apparent contradiction with fact 5 and the related fact that the surfaces of the inner regular satellites are bright. It is also inconsistent with fact 2.

Cook and Franklin (1970) suggested that the preferential erosion of a superficial layer of ice on the leading hemisphere exposed an underlying dark surface. This hypothesis is also inconsistent with fact 5 and, in a subtle sense, with fact 2. Most of the biggest craters on the bright side have bright interiors.

Soter (1974) proposed that dark material from Phoebe was ejected into circum-Saturnian orbits by impact events, spiralled inward under Poynting-Robertson drag, and preferentially coated the leading hemisphere of Iapetus, the first satellite between Phoebe and Saturn. But in this form, Soter's hypothesis is inconsistent with fact 4.

Other hypotheses have involved purely internal origins for the dark material. For example, B. A. Smith et al. (1982) suggested that dark carbonaceous material may have been preferentially extruded on the leading hemisphere. While motivated by fact 2, such hypotheses, in their pure form, are inconsistent with fact 3.

The above discussion suggests that some type of hybrid hypothesis (involving both impacting bodies and a "native" source of the dark material) may best explain all the facts. In particular, such a hybrid hypothesis might explain both facts 2 and 3. The probable occurrence of carbon-containing ices (methane and/or carbon monoxide) and the role of carbon in making dark carbonaceous chondrites and the highly absorbing aerosols in Titan's atmosphere imply that these ices are a key source for the dark material on Iapetus' surface. Squyres and Sagan (1983) proposed that this dark material contains organic chromophores produced *in situ* by ultraviolet irradiation of methane-containing ices. However, laboratory simulations need to be performed to determine whether such organic synthesis can occur under environmental conditions relevant for Iapetus and, if so, the efficiency of the production of compounds that strongly absorb visible radiation.

An alternative mechanism for generating dark organic material may be provided by impacting events. Matson and Gaffney (1979) pointed out that impacts into icy satellites could provide pressures sufficient to create high-density polymorphs of ice. Such impacts into a methane-clathrate surface might thus liberate methane, and possibly subject that methane to pressures and temperatures sufficient to form more complex organic materials. Such synthesis is known to occur in mixtures of gaseous methane, ammonia, and water vapor that are subjected to shock waves (Bar-Nun et al. 1970), but the nature and efficiency of the production of dark organic matter under conditions of high-velocity impact into $CH_4/NH_3/H_2O$ ice mixtures are unknown.

Squyres and Sagan (1983) proposed that the albedo contrasts across Iapetus' surface arise from variations in impact flux with distance from the apex of the leading hemisphere. Variations in this flux from apex to antapex range from a factor of two for meteoroids from outside the Saturn system to almost infinity for ejecta from Phoebe. The flux variation due to a combination of both sources may be as little as a factor of two or as large as a factor of 100. According to Squyres and Sagan (1983), the net transport of material from regions of higher impact fluxes leads to a net erosion of these surfaces and hence a continual exposure of fresh methane-containing ices and an easier ability of subsurface methane to diffuse to the surface. Ultraviolet photolysis of the methane ice-containing surfaces results in the production of visible light absorbing organics. Conversely, surfaces of lower impact flux experience a net deposition of ejecta. Squyres and Sagan (1983) postulate that this ejecta deposit is devoid of methane ice because methane ice is readily vaporized and could thus be lost to space during repeated impact events involved in the net ballistic transport down the flux gradient. If so, regions of lower impact flux would be bright due to a nonmethane ice ejecta cover. The occurrence of a dark floored crater on the bright side could be attributed to the impact event being of recent origin. Thus, it exposed fresh methane ice and there has not been enough time yet for it to be buried by ballistic ejecta from elsewhere.

While the above hypothesis is certainly a step in the right direction, its assumptions need to be carefully scrutinized, tested, and perhaps altered. For example, might the brightness gradient be equally well produced by a combination of the generation of light-absorbing chromophores by impacts rather than by ultraviolet photolysis and the net ballistic transport down the flux gradient? Do either of these mechanisms produce light-absorbing chromophores in a sufficient yield? Finally, will repeated impacts lead to the volatilization and loss of ices other than methane and, if so, will they perhaps cause a concentration of light-absorbing chromophores in the deposited ejecta?

The central role that carbon-containing ices appear to play in creating dark places on Iapetus has some interesting implications. First, the intermediate value of Hyperion's albedo might be due to a combination of processes analogous to those postulated for Iapetus and Hyperion's nonsynchronous rotation (Squyres and Sagan 1983). Second, the high albedo of the surfaces of the satellites interior to Titan may imply that they lack carbon-containing ices.

VI. ORIGIN AND EVOLUTION OF TITAN'S ATMOSPHERE

There are three potential sources of Titan's atmosphere: direct retention of the gases of Saturn's primordial nebula disk by Titan's gravity ("primitive atmosphere" hypothesis), impact with volatile-rich stray bodies, especially comets ("stray-body" hypothesis), and outgassing of volatiles contained in the solid material that formed the satellite ("secondary atmosphere" hypothesis). The first two of these hypotheses, with a minor exception or two, can be readily dismissed. According to solar elemental abundances (Cameron 1973), $Ne/N \sim 1$ in the Saturn nebula. Furthermore, much of the N may have been in condensed form at the temperatures likely at Titan's distance (Pollack et al. 1976) and much of the N may have occurred as NH_3 rather than N_2 in Saturn's nebula (Prinn and Fegley 1981). Therefore, the Ne/N_2 ratio in nebula gas captured by Titan would probably have been much greater than unity and, in any event, not less than 2. Since Ne is too heavy to readily escape from Titan's atmosphere, the mean molecular weight of its present atmosphere would be expected to be significantly less than 28, in contrast to an observed value of

28.6 (Samuelson et al. 1981). Hence, nebular gases are not a major source of Titan's atmosphere (Owen 1982).

Because comets probably formed in very cold regions of the solar nebula, they may contain ice hydrates and clathrates. Upon impact with Titan, compounds contained in these ices could be released into the atmosphere. Because of kinetic inhibitions for the low pressures of the solar nebula, CO and N_2 are expected to be the dominant C- and N-containing species throughout the solar nebula (Lewis and Prinn 1980). Thus, cometary impact with Titan would be expected to produce an atmosphere containing substantial quantities of CO and N_2. While this prediction is consistent with the occurrence of substantial amounts of N_2 in Titan's present atmosphere, it is inconsistent with CH_4 and not CO being the observed dominant C-containing gas species. However, cometary impacts may be the source of the small amount of CO ($\sim$ 50 ppm) recently detected in Titan's atmosphere. In this case, the atmospheric CO is derived directly from CO-containing ices and from chemical reactions between cometary H_2O and atmospheric CH_4 and its derivatives.

As discussed in Sec. IV, the temperature may have been low enough in Saturn's primordial nebula disk at Titan's distance to permit the condensation of ices less volatile than water ice. Such volatiles were the major sources of Titan's atmosphere, being released when these ices were decomposed, either upon impact during accretion or in hot portions of the interior, followed by migration to the surface. Because the pressures in Saturn's nebula probably were much higher than in the solar nebula, for a fixed temperature, NH_3 and CH_4 were probably the dominant N- and C-containing gases in Saturn's nebula (Prinn and Fegley 1981). For plausible nebula temperatures at Titan's distance—$\sim$ 60 K (Pollack et al. 1976)—the following ices, hydrates, and clathrates may have occurred: H_2O, NH_3, $CH_4 \cdot 7\ H_2O$, $NH_3 \cdot 1\ H_2O$, $N_2 \cdot 7\ H_2O$, $CO \cdot 7\ H_2O$, $Ar \cdot 7\ H_2O$ (Lewis 1972a; Pollack et al. 1976; Owen 1982b; Strobel 1982). As detailed below, these compounds may have served as the primary sources of atmospheric Ar, N_2, CH_4, and perhaps CO.

Ar. As discussed in Sec. I, Ar may be a major constituent of the atmosphere, although the evidence for its presence is indirect and subject to some uncertainty. If this inference is correct, very little could be due to ^{40}Ar resulting from the decay of ^{40}K in the interior. Using a K abundance characteristic of carbonaceous chondrites ($\sim$ 400 ppm) for the rock component of Titan and assuming complete degassing of this radiogenic Ar, we obtain a mixing ratio of only 3×10^{-4}, in contrast to a value of $\sim 10^{-1}$ obtained by Samuelson et al. (1981) for putative atmospheric Ar. Almost all of the inferred Ar would have had to have been derived from the decomposition of Ar clathrate (Owen 1982b). In this case, ^{36}Ar and ^{38}Ar would be the dominant Ar isotopes, in an approximate ratio of 5/1.

N_2. Atmospheric N_2 may have been derived, alternatively, from the decomposition of N_2 clathrate (Pollack 1981a; Owen 1982; Strobel 1982) or

from the decomposition and volatilization of NH_3-containing ices, followed by photolysis of NH_3 to N_2 (Atreya et al. 1978). Both hypotheses face potential quantitative difficulties. The amount of atmospheric N_2 derived from N_2 clathrates is limited by the following factors: NH_3 and not N_2 was probably the dominant N-containing gas in Saturn's nebula; only a fraction f of the available nebula N_2 may have been incorporated as N_2 clathrate due to the limited amount of H_2O ice exposed to nebular gases and the formation of higher temperature clathrates from some of the exposed H_2O, and incomplete outgassing. In order for N_2 clathrates to be the dominant source of Titan's atmosphere, $f \cdot g \cdot h \cong 6 \times 10^{-3}$, where f is defined above, g is the ratio of $N_2/(NH_3 + N_2)$ in the nebula, and h is the fractional degree of outgassing. According to Prinn and Fegley (1981), $g \sim 5 \times 10^{-4}$ to 10^{-1}. Thus, N_2 clathrates are the chief source of Titan's atmospheric N_2 only if g is close to its upper bound and f and h are not much smaller than unity.

An alternative source of Titan's atmospheric N_2 is the photolysis of NH_3 vapor, resulting from the volatilization of NH_3-containing ices (Atreya et al. 1978). Ammonia is irreversibly converted to N_2 by a series of photochemical reactions initiated by ultraviolet radiation at wavelengths shortward of 2300 Å. Even at Titan's distance from the Sun, there is a large enough flux of ultraviolet photons in this wavelength domain to generate several tens of bars of N_2 over the age of the solar system, *provided that enough NH_3 vapor exists in Titan's atmosphere* (ibid). However, temperature conditions in Titan's atmosphere may drastically diminish the production of N_2 from NH_3, perhaps to a level of insignificance. First, a negligible conversion takes place in the present-day stratosphere, since the temperature minimum at the tropopause limits the NH_3 mixing ratio at higher altitudes to values $\lesssim 10^{-14}$. Second, insignificant production occurs in the troposphere due to a large attenuation of ultraviolet radiation by the highly absorbing photochemical smog layer and to the condensation of N_2H_4, an intermediate product in the photolysis cycle, at the temperatures of the troposphere.

While present-day conditions in Titan's atmosphere appear to preclude significant production of N_2 from NH_3, this may not have always been the case. Of particular interest is the possibility that the atmosphere may have been much hotter during the period of Titan's formation, due to accretional heating of the surface (Lunine and Stevenson 1982*b*). Furthermore, the solar ultraviolet output may have been much higher at these early times (Canuto et al. 1982). But if the Saturn nebula and/or the solar nebula were present at these times, as assumed by Lunine and Stevenson (1982*b*), grain opacity in these nebulae may have totally attenuated the solar ultraviolet radiation. In the post-accretional epoch, it may have been extremely difficult to significantly elevate the temperature at the tropopause above its present value and thus permit significant quantities of NH_3 to enter the stratosphere; the tropopause temperature is essentially a "skin" temperature, determined by the amount of sunlight Titan absorbs. Titan already has a very low albedo (~ 0.2) (B. A.

Smith et al. 1982) and the wavelength-integrated solar luminosity was probably somewhat smaller in the past (Pollack and Yung 1980).

Nonthermal escape processes may have led to the loss of a significant fraction of Titan's N_2 to space over the age of the solar system (Strobel 1982). Near the level of the exobase, N_2 is dissociated to N atoms at a substantial rate through impacts with energetic magnetospheric electrons. About 40% of these atoms have enough energy and are traveling in the right direction to escape from Titan's low gravitational field. Using Voyager UVS data to estimate the flux of magnetospheric electrons, Strobel (1982) estimated that Titan lost an amount of N equal to $\sim$ 20% of the present atmospheric content due to the operation of this process over the age of the solar system. Additional loss of N to space, perhaps as much as the electron dissociation source, could have resulted from the ionization of N_2 near the exobase by magnetospheric electrons, followed by the dissociative recombination of N_2^+. Some further loss of atmospheric N_2 occurred due to the irreversible conversion of N_2 to HCN-containing polymers, which sedimented out of the atmosphere (ibid).

CH_4. Atmospheric methane was probably derived from the decomposition of methane clathrates. In this case, nebular temperatures below $\sim$ 100 K at Titan's distance are required (Pollack et al. 1976). Absorption of solar ultraviolet radiation by methane initiates a sequence of photochemical reactions that results in a significant rate of destruction of methane. This occurs because one of the photolysis products H_2 (and its derivative H) escapes readily to space and because unsaturated hydrocarbons formed in this low H_2 mixing ratio atmosphere ultimately produce aerosols, which sediment out of the atmosphere. Photochemical calculations suggest that $\sim$ 6 to 40 kg cm^{-2} of atmospheric CH_4 has been so lost over the age of the solar system (Strobel 1982); but the present atmosphere contains only $\sim$ 0.6 kg cm^{-2} of CH_4 (Samuelson et al. 1981). A comparison of these two sets of numbers implies that atmospheric CH_4 is being replenished on a quasi-continual basis, either as a result of outgassing or buffering by near-surface liquid methane/ethane.

CO. As discussed earlier, plausible sources of atmospheric CO (and its derivative CO_2) include cometary CO and H_2O and volatilized native CO clathrate. Because such an oxidized gas would be quickly eliminated in the reduced atmosphere of Titan, it must be quasi-continually resupplied from either or both of these sources.

The occurrence of a substantial atmosphere on Titan, as contrasted to the situation for all the other satellites of the solar system, is due to the concurrent operation of the following factors (Pollack 1981*a*):

1. A low enough nebular temperature to permit the occurrence of C- and N-containing ices (in contrast to the comparable-sized Galilean satellites of Jupiter);

2. A close enough distance to the Sun to avoid the freezing out of some major volatiles (in contrast to Triton);

3. A large enough mass to permit the gravitational retention of all but the lightest gases and a substantial devolatilization of ices (in contrast to the other satellites of Saturn).

VII. ORIGIN OF THE RINGS

Three principal theories for the origin of the rings involve formation of the present ring particle population through, alternatively, (1) tidal disruption of a large body, (2) condensation of material in Saturn's primordial nebula, and (3) collisional disruption of one or several large bodies by high-velocity stray bodies. The tidal disruption theory was first advanced by Roche (1847), who showed that no stable equilibrium configuration existed for a liquid satellite (i.e., one having zero tensile strength) at distances less than a critical value from its primary. This critical distance, the Roche limit, depends weakly on the ratio of the densities of the primary and satellite. If this ratio is not very different from unity, the Roche limit lies close to the outer boundary of the A Ring. Roche proposed that a satellite evolved inward and was tidally disrupted when it crossed this limit, thereby generating Saturn's ring system. Alternatively, a stray body from outside the Saturn system could have been disrupted within the Roche limit when it passed close to Saturn.

There are, however, some very serious problems connected with a tidal origin for the rings (Pollack and Cuzzi 1981). First, suppose that the ring material formed from a satellite of a low-mass Saturn nebula or from an external object. If so, the parent body would have been a solid body, not a liquid one. Aggarwal and Oberbeck (1974) showed that the tidal disruption limit of a solid body lies much closer to the primary than that for a liquid body as a result of the finite strength of the solid body. In particular, it is not possible to tidally disrupt a solid body smaller than ~ 100 km at any positive altitude above Saturn's atmosphere. In the case of a larger body, the tidal limit moves out only as far as $1.4\ R_S$ from the center of Saturn. It seems likely that if the parent body was a satellite of Saturn, it would initially have had a nearly circular orbit and its orbit would have continued to have had a low eccentricity as it evolved inward. Thus, tidal disruption of a large solid satellite would be expected to produce a ring system that was situated well inside the A and B Rings, the principal rings of the Saturn system. The inner boundary of the B Ring is located at $1.53\ R_S$.

If the parent body was a large stray body, its orbit and that of its tidally produced fragments would cross the region of the main rings. Thus, its tidal disruption could occur at closer distances without causing the problem encountered with a satellite as the parent body. However, tidal disruption can be expected to induce only very small relative velocities among the fragments. Thus, the fragments, like their parent body, would have kinetic energies in excess of the absolute values of their gravitational energies and they would escape from the Saturn system. Collisions among the fragments would cause their velocities to approach that of their parent and so not obviate the above problem.

While the above geometrical arguments offer the most severe challenge to the tidal disruption theory for solid parent bodies, it may also experience some difficulties in accounting for the size distribution of the ring particles. Harris (1975) pointed out that solid fragments of a tidally disrupted body would experience further breakup due to collisions among themselves. The kinetic energy of the colliding fragments is supplied chiefly by the gravitational potential of their partners. Greenberg et al. (1977) made several significant improvements to the calculations of Harris (1975) and deduced that collisional evolution would result in a broad spectrum of particle sizes ranging from 1 mm to ~ 100 km. They suggested that the size distribution would be roughly a power law of −3.3 covering this entire range, but with most of the cross section in the centimeter-scale particles and an excess of the larger size particles, relative to the power law. Although there is evidence for the occurrence of a few "moonlets" of ~ 10 km size within the rings, the measured size distribution of the particles shows an abrupt change in slope from −3 for particles below ~ 10 m to a much steeper slope at larger sizes (Esposito et al. 1983b). Such a size distribution does not appear to be in accord with the predictions of Greenberg et al. (1977).

Next, suppose that the ring material originated from condensed material in a high-mass Saturn nebula. In this case, the condensates may have been composed of a liquid solution of ammonia and water, rather than solid water (cf. Fig. 7). Hence, the classical Roche theory is directly applicable: droplets forming inside the Roche limit would have been unable to *gravitationally* aggregate and a liquid satellite that formed outside the Roche limit and evolved through it would have been tidally disrupted. However, the total mass of the present-day rings is comparable to that of Mimas (Holberg et al. 1982a). Thus, the invocation of a high-mass nebula has the potential problem of requiring the elimination of almost all the condensates that formed near the region of the rings. An even more fundamental objection to this hypothesis involves the mechanism by which the satellite evolved inward to be tidally disrupted. The most obvious mechanism for radial migration is secular acceleration due to tidal interactions between the postulated satellite and Saturn. But such tidal forces cause an inward migration of the satellite only if it is situated closer to the planet than the synchronous distance at which an object's orbital period equals the planet's rotational period. At present, the synchronous distance is located *within* the Roche limit (in the outer part of the B Ring). Thus, unless Saturn rotated less rapidly in its early history, any hypothetical liquid satellite forming outside the classical Roche limit would evolve *outward* due to tidal forces and so would not be tidally disrupted.

According to the "condensation" theory (see, e.g., Pollack 1975; Pollack et al. 1977), the ring particles formed within Saturn's nebula by the same initial condensation and aggregation processes that eventually led to the formation of the regular satellites. However, only incomplete accretion occurred within the region of the rings due to tidal disruption and, hence, many small-sized particles resulted rather than a single large body. The formation of water ice particles within the region of the current rings did not begin until the latter part of the satellite formation period for two reasons. First, the planet's radius extended beyond the outer boundary of the main rings at the start of stage 3. It took about 5×10^5 yr of further evolution for the planet's radius to reach the inner edge of the B Ring (Pollack et al. 1977). Second, the planet's luminosity had to decrease enough that temperatures within the region of the rings fell below the condensation temperature of water vapor (~ 240 K). According to Fig. 8, such a temperature was not reached until $\sim 5 \times 10^6$ yr after the start of stage 3. If the condensation model for the rings is correct and if the nebular temperature in the region of the rings was determined primarily by Saturn's luminosity and not viscous dissipation, the above times represent a lower bound on the lifetime of Saturn's nebula. Whether the nebula could have lasted this long, in light of viscous dissipation and the accompanying inward transfer of mass, is problematical.

Once temperatures fell below the condensation point, water ice particles began to form throughout the rather extensive thickness ($\sim 10\%$ of the radial dimension) of the nebula. Due to a combination of the vertical component of the gravitational force and gas drag, they gradually sank toward its central plane. During the settling, they continued to grow by vapor phase deposition as the nebula cooled still further and by coagulation and coalescence. Very crude calculations suggest that they could have achieved a size on the order of centimeters by the time they reached the central plane of the nebula (Pollack 1975), in approximate accord with their observed values of 0.1 to 10^3 cm. A limited amount of further growth may have been possible as ice particles gently collided with one another and remained attached to one another by chemical sticking forces. However, it may have been more difficult for them to remain attached by their mutual gravity because of tidal disruption. If two particles have the same dimension, then the tidal disrupting force exceeds their mutual gravitational attraction at precisely the classical Roche limit (Pollack 1975; Smoluchowski 1978a). But if they differ greatly in their dimensions, the tidal disruption limit lies at a much closer distance to the planet, near the inner edge of the C Ring (Smoluchowski 1978a). Thus, it is not exactly clear why the ring particles have undergone as limited an amount of growth as they evidently have, although disruption by micrometeoroid impacts and other erosional agents might act to counter accretional growth.

Although moonlets having a size of ~ 10 km do not appear to be abundant in the rings, there is good, indirect evidence for the occurrence of at least a few moonlets of this dimension within Saturn's rings, e.g., in Encke's Division in the A Ring (Esposito et al. 1983b). At first glance the occurrence of any such large-sized objects would appear to be inconsistent with the "condensation" theory. However, it is conceivable that aggregation of centimeter-to meter-sized ring particles to much larger-sized objects, although inhibited in most places, can occur in a few select places within the rings. For example, higher random velocities of ring particles are expected near resonance positions, such as Encke's Gap. Higher velocities may promote the efficient growth of "spongy snowballs."

Because Saturn initially extended beyond the region of the rings, silicate-containing grains and/or planetesimals may have been underabundant in this region by the time it became part of the nebula; silicate-containing grains initially present there, but within Saturn's envelope, would have tended to remain with the planet. Also, silicate grains initially located at somewhat greater distances in the innermost parts of the nebula may have aggregated into planetesimals that were large enough not to be carried inward with the nebula gas as Saturn contracted. This expectation of a depletion of rocky material in the rings is consistent with analyses of microwave data discussed in Sec. I.

The condensation theory places emphasis on constructional events in the formation of the ring particles. However, they are also susceptible to disaggregation processes, particularly collisions with high-velocity, stray bodies. The high crater density on the surface of Iapetus and an enhancement of the stray-body flux at closer distances to Saturn due to gravitational "focusing" by Saturn imply that small satellites in the inner part of Saturn's system have a high probability of being completely fragmented over the age of the solar system (B. A. Smith et al. 1981). As discussed in Sec. I, evidence for such catastrophic disruptions is provided by the occurrence of coorbital satellites and the shapes of some of the satellites.

The significance of such destructive collisions for the rings is not entirely clear. Shoemaker (see his chapter) suggests that one or several large satellites were originally present in the region of the rings and that they and their fragments suffered catastrophic collisions, resulting eventually in the present, very much smaller ring particles. This hypothesis neglects reaccretion of the fragmented material in the long intervals between destructive collisions. If tidal forces somehow prevent reaccretion, it is not clear how the original parent bodies formed in the first place. Nevertheless, collisional processes, in perhaps a much less dramatic sense than envisioned in Shoemaker's scenario, may have had an important influence on the present size distribution of ring particles.

VIII. CONCLUSIONS

Here, we summarize our opinions concerning the major issues discussed in this chapter and cite principal areas of uncertainty.

A. Formation and Evolution of Saturn

It is not yet possible to make a definitive distinction between the gas and core instability models for the origin of the giant planets in general and Saturn in particular. However, we tend to favor the core instability model for several reasons: first, it provides a natural explanation for the similar values of the core masses of all four giant planets; second, the occurrence of irregular satellites for Jupiter and Saturn implies that accretion of asteroid-sized bodies was occurring in the outer solar system close to the time of formation of the outer planets. On aesthetic grounds, we prefer the idea of one major formation mechanism for all the bodies of the solar system, rather than two major mechanisms. But, as detailed above, the core instability model has potentially very serious problems; for example, the formation times for the cores of Uranus and Neptune appear to exceed the lifetime of the solar nebula.

Numerical models, fashioned after stellar evolution models, have provided a good definition of the evolution of proto-Saturn subsequent to its formation. Major steps of evolution include an early, quasi-hydrostatic stage (stage 1), a very rapid contraction stage (stage 2), and a later, quasi-hydrostatic contraction stage (stage 3). A major problem, that has hardly been addressed so far, has to do with the transfer of angular momentum between different regions of proto-Saturn from near the end of stage 1 to the beginning of stage 3. This transfer played a crucial role in the formation and nature of the Saturnian nebula within which the regular satellites formed.

Thermal cooling and contraction of a chemically homogeneous Saturn at the present epoch produce a factor of 3 less internal luminosity than is observed. The reality of this difference is strengthened by the approximate agreement of theory and observation for Jupiter. The most likely additional source of internal luminosity for Saturn is the gravitational energy released by helium sinking toward the center of Saturn in the outer part of the metallic hydrogen zone due to its partial immiscibility there. However, the thermodynamic basis for this immiscibility has been called into question. The origin of the planet is also discussed in Sec. VIII.C of the chapter by Hubbard and Stevenson.

B. Origin of Phoebe

Phoebe was most likely captured by gas drag when it passed through proto-Saturn just prior to the onset of stage 2. However, more refined models of the Saturn nebula at this epoch need to be developed to provide more definitive predictions of the orbital distance and object mass for which such a capture would operate. Conceivably, such models might provide a test of the two formation mechanisms, as our cursory study indicates that they may yield significantly different predictions, especially regarding the orbital distance of capture.

C. Origin and Evolution of the Regular Satellite System

As mentioned above, the regular satellites formed within a nebular disk that developed from proto-Saturn during the time of stage 2. In contrast to the situation for Jupiter's nebula and regular satellites, temperatures within Saturn's nebula became cool enough that water was able to condense in all regions of satellite formation and ices containing ammonia and methane were able to form at progressively greater distances from the center of the nebula. An intriguing possibility, which is discussed at length below, is the formation of a liquid ammonia/water solution as the highest temperature condensate, rather than water ice. Such condensation may occur for the highest pressure (approximately or greater than tens of bars) and most massive (about Saturn's mass) nebular models that appear to be plausible.

The temperature of Saturn's nebula may have been controlled by Saturn's internal luminosity, which was $\sim$ 5 orders of magnitude larger at the start of stage 3 than its present value. The steady decline of Saturn's internal luminosity with time led to progressively cooler nebular temperatures. Alternatively, the nebula's temperature structure may have been determined largely by internal viscous dissipation; in this case, temperatures at a given location would have also declined with time, although on a much shorter time scale. If accretion was sufficiently rapid, as perhaps suggested by recent calculations that include nebular drag, chemically zoned satellites could have resulted, especially in the outer parts of the nebula.

Unfortunately, it is much more difficult to obtain estimates of Saturn's nebular density from the masses of its regular satellites than for Jupiter's nebula. A much less ordered pattern emerges (cf. Fig. 9). This difficulty might reflect the role that catastrophic collisions played in the early history of the Saturn system. All the present regular satellites, except Titan and possibly Iapetus, may be the reaccretion products of fragments of earlier, possibly larger-sized satellites. Clearly, a central problem of today is to further assess the role of catastrophic bombardment in shaping the Saturn system.

A minimum mass nebula is constructed by using the "systematic" nebular densities derived from the masses of Tethys, Dione, and Rhea. A maximum mass nebula is generated by assuming its mass is the same as Saturn's. Nebulae having masses toward the high-mass limit have several attractive features. They allow for enough mass to produce Titan along with the other regular satellites. At the high pressures of the inner portions of these nebulae, liquid droplets of ammonia/water may condense. Satellites made in part of these droplets would start out already differentiated into rocky cores and rock-free mantles. If such satellites were broken up, the fragments would form the two populations of rocky and icy planetesimals whose random reaccretion could lead to the observed, apparently random variation in satellite densities.

One problem with massive nebulae is getting rid of the large gas mass and the extra condensed mass not currently residing in the satellites of today. In principle, most of the gas can be expected to be added to Saturn due to viscous processes within the nebula, while the remainder could be lost to space, due to angular momentum transfer. Getting rid of an amount of condensed mass in excess of that currently within the Saturn system is a more formidable problem.

In discussing the mass of Saturn's nebula, as above, it should be borne in mind that all the nebula's properties, including its mass, continually changed with time over its entire lifetime due to viscous dissipation. Thus, the above estimates are merely meant to be some crude, time-averaged values that are useful for very approximate nebular models.

Despite their comparatively small sizes, satellites such as Tethys, Dione, and Rhea experienced major tectonism (principally expansion) and resurfacing events during their first $\sim 1 \times 10^9$ yr of history due to the presence of substantial quantities of ammonia in their interior makeup. Enceladus has undergone tectonism and resurfacing within the last $\sim 1 \times 10^9$ yr, despite its still smaller size, probably as a result of tidal heating. If so, a significantly larger, forced orbital eccentricity than its present value appears to be required at some time in its past to *initiate* melting. A smaller, forced eccentricity, perhaps including the present value, suffices to *maintain* a molten interior. Origins are also discussed in Sec. III of Morrison et al.'s chapter.

D. Origin of Titan's Atmosphere

Titan's atmosphere was derived from C-, N-, and possibly Ar-containing clathrates and hydrates that helped to form the satellite. While atmospheric methane can readily be ascribed to the dissociation of methane clathrates that occurred during accretion and in the high-temperature and/or high-pressure environment of the satellite's interior, the source of atmospheric nitrogen is less clear. The latter gas was derived either from the direct thermal and pressure dissociation of nitrogen clathrate and/or the dissociation of ammonia hydrate, followed by ultraviolet photolysis of ammonia vapor. The first of these sources has the potential problem of being quantitatively inadequate because of the low N_2/NH_3 ratio in the Saturnian nebula, the only partial incorporation of nebular N_2 into clathrates, and the incomplete outgassing of the satellite. The second source is totally inadequate under current atmospheric conditions due to a very effective tropopause cold trap and the opacity of the aerosol layer, but it could be important for past, especially very early, atmospheres. The occurrence of a significant quantity of atmospheric Ar has not been firmly established, but, if so, this Ar was most likely derived from the dissociation of Ar clathrate. Finally, the source of the surprisingly large amount of atmospheric CO and its derivative, CO_2, remains a puzzle, although cometary and/or clathrate sources are possible.

Over the age of the solar system, large amounts of atmospheric N_2 (approximately a few tens of percent of the current abundance) and CH_4 (many times the current abundance) have been permanently eliminated from the atmosphere, through a combination of escape of selected atoms (H, N) to space and the irreversible conversion of these gases to aerosol-forming, complex organic molecules. Despite its heavy loss, atmospheric methane is maintained by the vapor pressure buffering of a near-surface reservoir of methane. While this reservoir need not be in continual contact with the atmosphere, it cannot be isolated for long periods of time. Contact may occur as a result of the presence of surface oceans of ethane/methane, vapor diffusion through a regolith, or intermittent outgassing. The huge deposit of complex organic molecules residing near the surface (equivalent depth of ~ 0.1 to 1 km) represents a resource of enormous potential value to future generations. A discussion of the origin of Titan may be found in Sec. VI.C of the chapter by Hunten et al.

E. Origin of the Rings

It seems very likely that the ring material formed within the Saturn nebula toward the end of the satellite formation period, when temperatures permitted the condensation of water in the region near the rings. An intriguing possibility, suggested by Fig. 7, is that the condensate was not water ice, but a liquid water/ammonia solution. The latter forms in the high-mass nebulae discussed above. If so, the classical Roche theory would be strictly applicable. In this case, the generation of a multitude of small ring particles could be attributed to the inability of liquid particles to accrete within the Roche limit or the tidal disruption of a liquid satellite, which formed outside the limit and migrated inside it. However, such a hypothesis still does not explain why satellite accretion did not occur after the particles solidified. Another open question concerns the role of early catastrophic bombardment in generating the present population of ring particles. On this subject see also Sec. VII.C of Esposito et al.'s chapter.

Acknowledgments. We are grateful to A. G. W. Cameron, P. Bodenheimer, S. Squyres, and R. Reynolds for their careful reading of the manuscript and helpful comments. This work was supported by the Planetary Geology Program.

Color Section

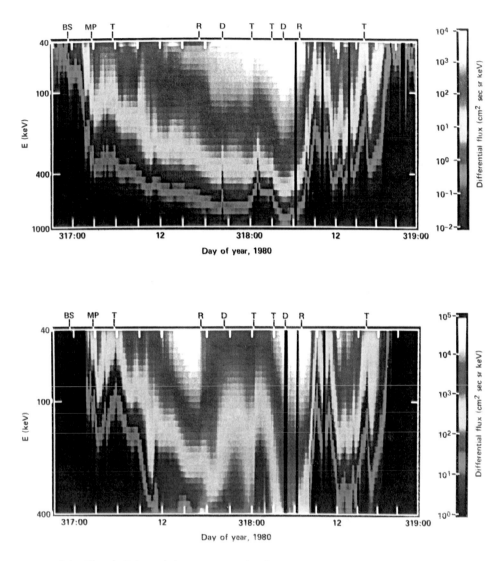

Color Plate 1. Color-coded energy versus time ion (upper) and electron (lower) spectrograms for the Voyager 1 encounter with Saturn. See Sec. III in the chapter by Scarf et al. for further discussion.

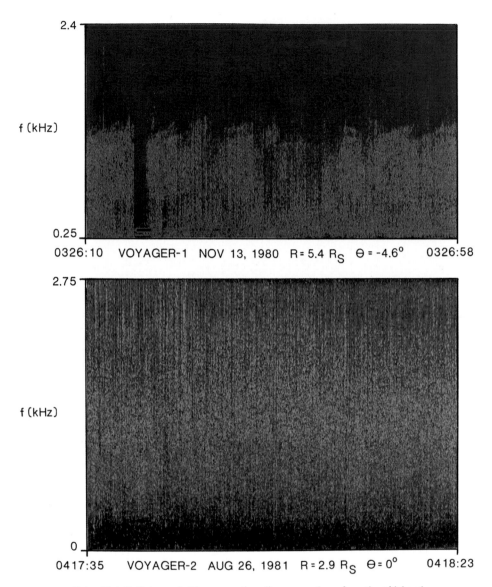

Color Plate 2. Color-coded frequency-time diagram made up from the wideband
measurements obtained near the magnetic-equator crossing (Voyager 1) and at
the ring-plane crossing (Voyager 2). See p. 345 in the chapter by Scarf et al.
for discussion.

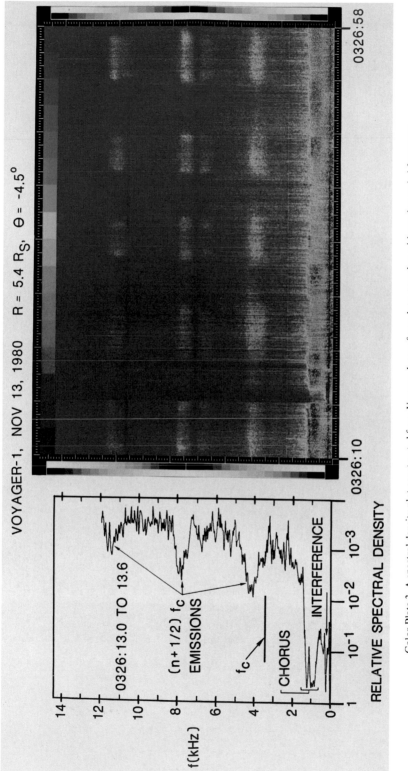

VOYAGER-1, NOV 13, 1980   R = 5.4 R_S, Θ = -4.5°

Color Plate 3. A spectral density plot constructed from Voyager 1 waveform data, together with a color-coded frequency diagram. See p. 345 in the chapter by Scarf et al. for discussion.

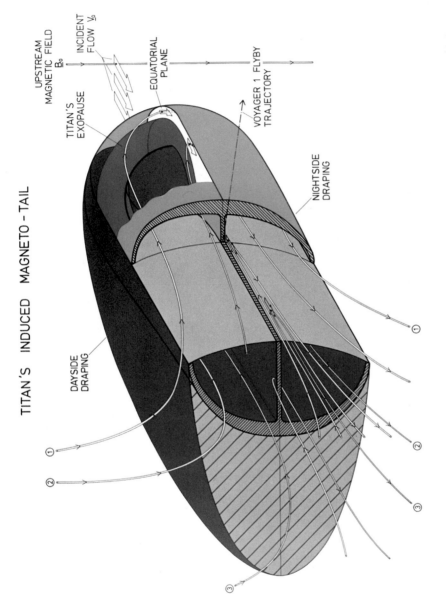

TITAN'S INDUCED MAGNETO-TAIL

UPSTREAM
MAGNETIC FIELD
$\underline{B_o}$

INCIDENT
FLOW $\underline{V_o}$

TITAN'S
EXOPAUSE

EQUATORIAL
PLANE

VOYAGER 1 FLYBY
TRAJECTORY

DAYSIDE
DRAPING

NIGHTSIDE
DRAPING

Color Plate 4. A three-dimensional sketch of Titan's induced magnetotail as observed during the Voyager encounter with the satellite. See p. 787 in the chapter by Neubauer et al. for a discussion of this figure.

Bibliography

BIBLIOGRAPHY
(Compiled by Terry S. Mullin)

Acuña, M. H., and Ness, N. F. 1975. The Pioneer 11 high field fluxgate magnetometer. *Space Sci. Inst.* 1:177–188.

Acuña, M. H., and Ness, N. F. 1976. The main magnetic field of Jupiter. *J. Geophys. Res.* 81: 2917–2922.

Acuña, M. H., and Ness, N. F. 1980. The magnetic field of Saturn: Pioneer 11 observations. *Science* 207:444–446.

Acuña, M. H., Ness, N. F., and Connerney, J. E. P. 1980. The magnetic field of Saturn: Further studies of the Pioneer 11 observations. *J. Geophys. Res.* 85:5675–5678.

Acuña, M. H., Connerney, J. E. P, and Ness, N. F. 1981a. Topology of Saturn's main magnetic field. *Nature* 292:721–724.

Acuña, M. H., Neubauer, F. M., and Ness, N. F. 1981b. Standing Alfvén wave current system at Io: Voyager 1 observations. *J. Geophys. Res.* 87:8513–8521.

Acuña, M. H., Ness, N. F., and Connerney, J. E. P. 1982. The magnetic field of Saturn. Paper read at "24th COSPAR Meeting," Ottawa, Canada, May 1982.

Acuña, M. H., Connerney, J. E. P., and Ness, N. F. 1983. The Z_3 zonal harmonic model of Saturn's magnetic field: Analysis and implications. *J. Geophys. Res.* 88:8771–8778.

Adams, N. G., and Smith, D. 1976. Product-ion distributions for some ion-molecule reactions. *J. Phys. B.* 9:1439–1451.

Aggarwal, H. R., and Oberbeck, V. R. 1974. Roche limit of a solid body. *Astrophys. J.* 191: 577–588.

Aksenova, A. N., Atai, A. A., and Ibragimov, N. B. 1978. A study of hydrogen quadrupole lines in the spectra of Jupiter and Saturn. *Astron. Vestn.* 12:101–106 (Russian).

Aksnes, K., and Franklin, F. A. 1978. The evidence for faint satellites of Saturn reexamined. *Icarus* 36:107–118.

Alexander, A. F. O'D. 1962. *The Planet Saturn: A History of Observation, Theory, and Discovery* (London: Faber and Faber).

Alexander, J. K., Carr, T. D., Thieman, J. R., Schauble, J. J., and Riddle, A. C. 1981. Synoptic observations of Jupiter's radio emissions: Average statistical properties observed by Voyager. *J. Geophys. Res.* 86:8529–8545.

Alfvén, H. 1954. *On the Origin of the Solar System*, eds. N. F. Mott and E. C. Mullard (Oxford: Clarendon Press).

Alfvén, H. 1957. On the theory of comet tails. *Tellus* 9:92–96.

Alfvén, H. 1981a. *Cosmic Plasma* (Dordrecht: D. Reidel Publ.).

Alfvén, H. 1981b. The Voyager 1/Saturn encounter and the cosmogonic shadow effect. *Astrophys. Space Sci.* 79:491–505.

Alfvén, H., and Arrhenius, G. 1976. *Evolution of the Solar System* (Washington, D.C.: NASA SP-345).

Allan, R. R. 1969. Evolution of Mimas-Tethys commensurability. *Astron. J.* 74:497–506.

Allen, D. A., and Murdock, T. L. 1971. Infrared photometry of Saturn, Titan, and the rings. *Icarus* 14:1–2.

Allen, M., Pinto, J. P., and Yung, Y. L. 1980. Titan: Aerosol photochemistry and variations related to the sunspot cycle. *Astrophys. J.* 242:L125–L128.

Allison, M., and Stone, P. H. 1983. Saturn meteorology: A diagnostic assessment of thin layer configurations for the zonal flow. *Icarus* 54:296–308.

Anders, E., and Hayatsu, R. E. 1981. Organic compounds in meteorites and their origins: Cosmo- and geochemistry. In *Topics in Current Chemistry*, ed. F. L. Boschki (New York: Springer-Verlag), pp. 1–37.

875

Anders, E., and Ebihara, M. 1982. Solar-system abundances of the elements. *Geochim. Cosmochim. Acta* 46:2363–2380.

Angerami, J. J., and Thomas, J. O. 1964. Studies of planetary atmospheres. 1. The distribution of electrons and ions in the Earth's exosphere. *J. Geophys. Res.* 69:4537–4560.

Appleby, J. F. 1981. Atmospheric Structures of the Giant Planets from Radiative-convective Equilibrium Models. Unpublished Ph.D. thesis, State Univ. of New York, Stony Brook, NY.

Apt, J., and Singer, R. B. 1982. Cloud height differences on Saturn. *Icarus* 52:503–508.

Armstrong, T. P., Paonessa, M. T., Bell, E. V. II, and Krimigis, S. M. 1983. Voyager observations of Saturnian ion and electron phase space densities. *J. Geophys. Res.* 88:8893–8904.

Asano, S., and Sato, M. 1980. Light scattering by randomly oriented spheroidal particles. *Appl. Optics* 19:962–974.

Ashford, M. N. R., Hancock, G., Ketly, G. W., and Minshull-Beach, J. P. 1980. Direct measurements of a^1A, CH_2 removal rates. *J. Photo. Chem.* 12:75.

Ashour-Abdalla, M., and Kennel, C. F. 1978. Nonconvective and convective electron cyclotron harmonic instabilities. *J. Geophys. Res.* 83:1531–1543.

Atreya, S. K. 1981. Measurement of minor species (H_2O, Cl, O_3, NO) in the Earth's atmosphere by stellar occultation technique. *Adv. Space Res.* 1:127–141. Also in *Planetary Aeronomy and Astronomy*, eds. S. K. Atreya and J. J. Caldwell (Oxford: Pergamon Press).

Atreya, S. K. 1982. Eddy mixing coefficient on Saturn. *Planet. Space Sci.* 30:849–854.

Atreya, S. K., and Donahue, T. M. 1975a. Ionospheric models of Saturn, Uranus, and Neptune. *Icarus* 24:358–362.

Atreya, S. K., and Donahue, T. M. 1975b. The role of hydrocarbons in the ionospheres of the outer planets. *Icarus* 25:335–338.

Atreya, S. K., and Donahue, T. M. 1976. Model ionospheres of Jupiter. In *Jupiter*, ed. T. Gehrels (Tucson: Univ. of Arizona Press), pp. 304–318.

Atreya, S. K., and Waite, J. H. Jr. 1981. Saturn ionosphere: Theoretical interpretation. *Nature* 292:682–683.

Atreya, S. K., and Donahue, T. M. 1982. The atmosphere and ionosphere of Jupiter. *Vistas in Astron.* 25:315–335.

Atreya, S. K., Donahue, T. M., and McElroy, M. B. 1974. Jupiter's ionosphere: Prospects for Pioneer 10. *Science* 184:154–156.

Atreya, S. K., Donahue, T. M., and Kuhn, W. R. 1978. Evolution of a nitrogen atmosphere on Titan. *Science* 201:611–613.

Atreya, S. K., Donahue, T. M., Sandel, B. R., Broadfoot, A. L., and Smith, G. R., 1979a. Jovian upper atmospheric temperature measurement by the Voyager 1 UV spectrometer. *Geophys. Res. Letters* 6:795–798.

Atreya, S. K., Donahue, T. M., and Waite, J. H. Jr. 1979b. An interpretation of the Voyager measurement of Jovian electron density profiles. *Nature* 280:795–796.

Atreya, S. K., Kuhn, W. R., and Donahue, T. M. 1980. Saturn: Tropospheric ammonia and nitrogen. *Geophys. Res. Letters* 7:474–476.

Atreya, S. K., Donahue, T. M., and Festou, M. C. 1981. Jupiter: Structure and composition of the upper atmosphere. *Astrophys. J.* 247:L43–L47.

Atreya, S. K., Festou, M. C., Donahue, T. M., Kerr, R. B., Barker, E. S., Cochran, W D., Bertaux, J. L., and Upson, W. L. II 1982a. Copernicus measurement of the Jovian Lyman-alpha emission and its aeronomical significance. *Astrophys. J.* 262:377–387.

Atreya, S. K., Waite, J. H. Jr., Donahue, T. M., and Nagy, A. F. 1982b. Upper atmosphere and ionosphere of Saturn. Program "Saturn Conference," Tucson, AZ, May 1982 (abstract).

Aubry, M. P., Kivelson, M. G., and Russell, C. T. 1971. Motion and structure of the magnetosphere. *J. Geophys. Res.* 76:1673–1696.

Auerbach, D., Cacak, R., Candano, R., Gaily, T. D., Keyser, C. J., McGowan, J. W., Mitchell, J. B. A., and Wilk, S. F. J. 1977. Merged electron-ion beam experiments. I. Methods and measurements of $(e-H_2^+)$ and $(e-H_3^+)$ dissociative recombination cross sections. *J. Phys. B.* 10:3797–3820.

Augstein, E. 1978. The atmospheric boundary layer over the tropical oceans. In *Meteorology Over the Tropical Oceans*, ed. D. B. Shaw (London: Roy. Met. Soc.), pp. 73–103.

Augstein, E., Garstand, M., and Emmitt, G. D. 1980. Vertical mass and energy transport by cumulus clouds in the tropics. In *GATE: Oceanography and Surface Layer Meteorology in the B/C Scale*, Vol. 1, eds. G. Siedler and J. D. Woods (New York: Pergamon), pp. 9–21.

Aumann, H. H., and Kieffer, H. H. 1973. Determination of particle sizes in Saturn's rings from their eclipse cooling and heating curves. *Astrophys. J.* 186:305–311.

Avramchuk, V. V. 1967. Spectrophotometry of the methane absorption band 6190 Å on Saturn's disk. *Astron. Tsirkular USSR Acad. Sci.* 452:1–3 (Russian).

Avramchuk, V. V. 1968. Distribution of methane absorption in the band 6190 Å along the disk of Saturn in 1966. *Astrometry Astrophys.* 1:161–164 (Russian).

Avramchuk, V. V., and Krugov, V. D. 1972. The results of the photometric observations of Saturn. *Astron. Tsirkular USSR Acad. Sci.* 689:1–4 (Russian).

Axel, L. 1972. Inhomogeneous models of the atmosphere of Jupiter. *Astrophys. J.* 173:451–468.

Bagenal, F., and Sullivan, J. D. 1981. Direct plasma measurements in the Io torus and inner magnetosphere of Jupiter. *J. Geophys. Res.* 86:8447–8466.

Balestic, F. S. 1974. Synthèse Abiotique d'acides Amines par voie Radiochimique. Doctoral thesis, Univ. of Paris, France.

Banks, P. M., and Kockarts, G. 1973. *Aeronomy* (New York: Academic Press).

Barbosa, D. D. 1982. Low level VLF and LF radio emissions observed at Earth and Jupiter. *Rev. Geophys. Space Sci.* 2:316–334.

Barbosa, D. D., and Kurth, W. S. 1980. Superthermal electrons and Bernstein waves in Jupiter's inner magnetosphere. *J. Geophys. Res.* 85:6729–6742.

Barcilon, A., and Gierasch, P. J. 1970. A moist, Hadley cell model for Jupiter's cloud bands. *J. Atmos. Sci.* 27:550–560.

Barker, E. S., and Trafton, L. M. 1973. Ultraviolet reflectivity and geometrical albedo of Titan. *Icarus* 20:444–454.

Barker, E. S., Cazes, S., Emerich, C., Vidal-Madjar, A., and Owen, T. 1980. Lyman-alpha observations in the vicinity of Saturn Copernicus. *Astrophys. J.* 242:383–394.

Barnard, E. E. 1895. Micrometrical measures of the ball and ring system of the planet Saturn, and measures of the diameter of his satellite Titan. *Mon. Not. Roy. Astron. Soc.* 55:367–382.

Bar-Nun, A. 1975. Thunderstorms on Jupiter. *Icarus* 24:86–94.

Bar-Nun, A., and Podolak, M. 1979. The photochemistry of hydrocarbons in Titan's atmosphere. *Icarus* 38:115–122.

Bar-Nun, A., Bar-Nun, N., Bauer, S. H., and Sagan, C. 1970. Shock synthesis of amino acids in simulated primitive environments. *Science* 168:470–473.

Barrer, R. M., and Ruzicka, D. J. 1962. Non-stoichiometric clathrate compounds of water. 4. Kinetics of formation of clathrate phases. *Trans. Faraday Soc.* 56:2262–2271.

Barshay, S. S., and Lewis, J. S. 1978. Chemical structure of the deep atmosphere of Jupiter. *Icarus* 33:593–611.

Bartels, J. 1936. Eccentric dipole approximating the Earth's magnetic field. *Terr. Magn.* 41:225–250.

Bastian, T. S., Chenette, D. L., and Simpson, J. A. 1980. Charged particle anisotropies in Saturn's magnetosphere. *J. Geophys. Res.* 85:5763–5771.

Bastin, J. A. 1981. Note on the rings of Saturn. *Moon Planets* 24:467.

Batchelor, G. 1967. *An Introduction to Fluid Dynamics* (Cambridge: Cambridge Univ. Press).

Bates, D. R., and Dalgarno, A. 1962. Electronic recombination. In *Atomic and Molecular Processes*, ed. D. R. Bates (New York: Academic Press), pp. 245–271.

Baum, W. A., Mills, R. L., Jones, S. E., and Martin, L. J. 1970. The international planetary patrol program. *Icarus* 12:435–439.

Baum, W. A., Kreidl, T., Westphal, J. A., Danielson, G. E., Seidelmann, P. K., and Pascu, D. 1981. Saturn's E-ring. I. CCD observations of March 1980. *Icarus* 47:84–96.

Baxter, D., and Thompson, W. E. 1973. Elastic and inelastic scattering in orbital clustering. *Astrophys. J.* 183:323–335.

Becker, R. H., and Epstein, S. 1982. Carbon, hydrogen and nitrogen isotopes in solvent-extractable organic matter from carbonaceous chondrites. *Geochim. Cosmochim. Acta* 46:97–103.

Beebe, R. F., Ingersoll, A. P., Hunt, G. E., Mitchell, J. L., and Muller, J.-P. 1980. Measurements of wind vectors, eddy momentum transports, and energy conversions in Jupiter's atmosphere from Voyager 1 images. *Geophys. Res. Letters* 7:1–4.

Beer, R., and Taylor, F. W. 1973. The abundance of CH_3D and the D/H ratio in Jupiter. *Astrophys. J.* 179:309–327.

Beer, R., and Taylor, F. W. 1978. The D/H and C/H ratios in Jupiter from the CH_3D phase. *Astrophys. J.* 219:763–767.

Behannon, K. W., Acuña, M. H., Burlaga, L. F., Lepping, R. P., Ness, N. F., and Neubauer, F. M. 1977. Magnetic field experiment for Voyagers 1 and 2. *Space Sci. Rev.* 21:235–257.

Behannon, K. W., Connerney, J. E. P., and Ness, N. F. 1981. Saturn's magnetic tail: Structure and dynamics. *Nature* 292:753–755.

Behannon, K. W., Lepping, R. P., and Ness, N. F. 1983. Structure and dynamics of Saturn's outer magnetosphere and boundary regions. *J. Geophys. Res.* 88:8791–8800.

Bell, J. F., Clark, R. N., McCord, T. B., and Cruikshank, D. P. 1981. Reflection spectra of Pluto and three distant satellites. *Bull. Amer. Astron. Soc.* 11:570 (abstract).

Bell, J. F., Cruikshank, D. P., and Gaffey, M. J. 1983. The nature of the Iapetus dark material. *Bull. Amer. Astron. Soc.* 15:856 (abstract).

Bergstrahl, J. T. 1973. Methane absorption in the atmosphere of Saturn: Rotational temperature and abundance from the $3v_3$ band. *Icarus* 18:605–611.

Bergstrahl, J. T., Orton, G. S., Diner, D. J., Baines, K. H., Neff, J. S., and Allen, M. A. 1981. Spatially resolved absolute spectrophotometry of Saturn: 3390 to 8080 Å. *Icarus* 46:27–39.

Berlin, P. 1981. METEOSAT tracks Karman vortex streets in the atmosphere. *European Space Agency Bull.* 25:16–19.

Bertie, J. E., Othen, D.A., and Solinas, M. 1973. The infrared spectra of ethylene oxide hydrate and hexamethylenetetramine at 100°. In *Physics and Chemistry of Ice*, eds. E. Whalley, S. J. Jones, and L. W. Gold (Ottawa: Royal Society of Canada), pp. 61–65.

Bessel, F. W. 1835. Bestimmung der Länge und Grösse des Saturnringes, und der Figur und Grösse des Saturn. *Astr. Nach.* 12:153–170.

Bianchi, R., Coradini, A., Federico, C., Lanciano, P., and Magni, G. 1982. On the formation and thermal evolution of regular satellites. Program "Saturn Conference," Tucson, AZ, May 1982 (abstract).

Biermann, L., Brosowski, B., and Schmidt, H. U. 1967. The interaction of the solar wind with a comet. *Sol. Phys.* 1:254–284.

Binder, A. B., and McCarthy, D. W. 1973. IR spectrophotometry of Jupiter and Saturn. *Astron. J.* 78:939–950.

Birmingham, T. J. 1982. Charged particle motions in the distended magnetospheres of Jupiter and Saturn. *J. Geophys. Res.* 87:7421–7430.

Birmingham, T. J., and Northrop, T. G. 1981. Diffusion of cold magnetospheric ions. *J. Geophys. Res.* 86:8971–8976.

Birmingham, T. J., Alexander, J. K., Desch, M. D., Hubbard, R. F., and Pedersen, B. M. 1981. Observations of electron gyroharmonic waves and the structure of the Io torus. *J. Geophys. Res.* 86:8497–8507.

Birnbaum, G. 1978. Far-infrared absorption in H_2 and H_2-He mixtures. *J. Quant. Spectrosc. Radiat. Trans.* 19:51–62.

Birnbaum, G., and Cohen, E. 1976. Theory of line shape in pressure-induced absorption. *Can. J. Phys.* 54:594–602.

Bjoraker, G. L., Larson, H. P., Fink, U., and Smith, H. A. 1981. A study of ethane on Saturn in the 3 μm region. *Astrophys. J.* 248:856–862.

Bjoraker, G. L., Larson, H. P., Fink, U., and Kunde, V. G. 1982. The relative contribution of reflected solar and thermal emission to the 5 μm spectrum of Jupiter and Saturn. *Bull. Amer. Astron. Soc.* 14:730 (abstract).

Black, D. C. 1973. Deuterium in the early solar system. *Icarus* 19:154–159.

Blake, J. B., Hilton, H. H., and Margolies, S. M. 1983. On the injection of cosmic ray secondaries into the inner Saturnian magnetosphere. I. Protons from the CRAND process. *J. Geophys. Res.* 88:803–807.

Blanco, C., and Catalano, S. 1971. Photoelectric observations of Saturn satellites Rhea and Titan. *Astron. Astrophys.* 14:43–47.

Bless, R. C., Code, A. D., and Taylor, D. J. 1968. Filter photometry of Saturn in the spectral region 2950–2450. *Astrophys. J.* 154:1151–1153.

Bobrov, M. S. 1970. Physical properties of Saturn's rings. In *Surfaces and Interiors of Planets and Satellites*, ed. A. Dollfus (New York: Academic Press), pp. 376–461.

Bodenheimer, P. 1977. Calculations of the effects of angular momentum on the early evolution of Jupiter. *Icarus* 31:356–368.

Bodenheimer, P., and Pollack, J. B. 1984. Evolution of core/envelope models of the outer planets. Submitted to Icarus.

Bodenheimer, P., Grossman, A. S., DeCampli, W. M., Marcy, G., and Pollack, J. B. 1980. Calculations of the evolution of the giant planets. *Icarus* 41:293–308.

Boischot, A., Leblanc, Y., Lecacheux, A., Pedersen, B. M., and Kaiser, M. L. 1981. Arc structure in Saturn's radio dynamic spectra. *Nature* 292:727–728.

Bond, G. P. 1851. On the rings of Saturn. *Astron. J.* 2:5–11; also reprinted in *Am. J. Sci. and Arts* 12:97–105, 2nd series.

Bond, W. C. 1848. Discovery of a new satellite of Saturn. *Mon. Not. Roy. Astron. Soc.* 8:1–2.

Bond, W. C. 1851. On the new ring of Saturn. *Astron. J.* 2:5; also reprinted in *Am. J. Sci. and Arts* 12:133–134, 2nd series.

Bond, W. C. 1857. Observations of the planet Saturn. *An. Har. Col. Obs.* 2:48.

Borderies, N., Goldreich, P., and Tremaine, S. 1982. Sharp edges of planetary rings. *Nature* 299:209–211.

Borderies, N., Goldreich, P., and Tremaine, S. 1983. Dynamics of elliptical rings. Preprint.

Borucki, W. J., Bar-Nun, A., Scarf, F. L., Cook, A. F. II, and Hunt, G. E. 1982. Lightning activity on Jupiter. *Icarus* 52:492–502.

Bowell, E., and Zellner, B. 1974. Polarizations of asteroids and satellites. In *Planets, Stars, and Nebulae Studied with Photopolarimetry*, ed. T. Gehrels (Tucson: Univ. of Arizona Press), pp. 381–404.

Braginskii, S. I. 1965. Theory of the hydromagnetic dynamo. *Sov. Phys. JETP* 10:1462–1471.

Brault, J. W., and Smith, W. H. 1980. Determination of the H_2 4-O S(1) quadrupole line strength and pressure shift. *Astrophys. J.* 235:L177–L178.

Bregman, J. D., Lester, D. F., and Rank, D. M. 1975. Observation of the ν_2 band of PH_3 in the atmosphere of Saturn. *Astrophys. J.* 202:L55–L56.

Brice, N. M., and Ioannides, G. A. 1970. The magnetosphere of Jupiter and Earth. *Icarus* 13:173–183.

Bridge, H. S., Belcher, J. W., Lazarus, A. J., Olbert, S., Sullivan, J. D., Bagenal, F., Gazis, P. R., Hartle, R. E., Ogilvie, K. W., Scudder, J. D., Sittler, E. C., Eviatar, A., Siscoe, G. L., Goertz, C. K., and Vasyliunas, V. M. 1981a. Plasma observations near Saturn: Initial results from Voyager 1. *Science* 212:217–224.

Bridge, H. S., Scudder, J. D., and Sittler, E. C. Jr. 1981b. Distribution of neutral gas and dust near Saturn. *Nature* 292:711–714.

Bridge, H. S., Bagenal, F., Belcher, J. W., Lazarus, A. J., McNutt, R. L., Sullivan, J. D., Gazis, P. R., Hartle, R. E., Ogilvie, K. W., Scudder, J. D., Sittler, E. C., Eviatar, A., Siscoe, G. L., Goertz, C. K., and Vasyliunas, V. M. 1982. Plasma observations near Saturn: Initial results from Voyager 2. *Science* 215:563–570.

Briggs, F. H. 1974. The microwave properties of Saturn's rings. *Astrophys. J.* 189:367–377.

Briggs, M. H. 1961. Organic constituents of meteorites. *Nature* 191:1137–1140.

Brinkman, A. W., and McGregor, J. 1979. The effect of the ring system on solar radiation reaching the top of Saturn's atmosphere. *Icarus* 38:479–482.

Broadfoot, A. L., Belton, M. J. S., Takacs, P. Z., Sandel, B. R., Shemansky, D. E., Holberg, J. B., Ajello, J. M., Atreya, S. K., Donahue, T. M., Moos, H. W., Bertaux, J. L., Blamont, J. E., Strobel, D. F., McConnell, J. C., Dalgarno, A., Goody, R., and McElroy, M. B. 1979. Extreme ultraviolet observations from Voyager 1 encounter with Jupiter. *Science* 204:979–983.

Broadfoot, A. L., Sandel, B. R., Shemansky, D. E., Holberg, J. B., Smith, G. R., Strobel, D. F., McConnell, J. C., Kumar, S., Hunten, D. M., Atreya, S. K., Donahue, T. M., Moos, H. W., Bertaux, J. L., Blamont, J. E., Pomphrey, R. B., and Linick, S. 1981. Extreme ultraviolet observations from Voyager 1 encounter with Saturn. *Science* 212:206–211.

Brouwer, D., and Clemence, G. M. 1962. *Methods of Celestial Mechanics* (New York: Academic), pp. 490–494.

Brown, L. W. 1975. Saturn radio emission near 1 MHz. *Astrophys. J.* 198:L89–L92.

Brown, L. W. 1982. Experimental line strengths for the ν_4 and ν_2 bands of methane in the 7 μm range. *Bull. Amer. Astron. Soc.* 14:733–734 (abstract).

Brown, R. H. 1983. The Uranian satellites and Hyperion: New spectrophotometry and compositional implications. *Icarus* 56:414–425.

Brown, R. H., and Cruikshank, D. P. 1983. The Uranian satellites: Surface compositions and opposition surges. *Icarus* 55:83–92.

Brown, R. H., and Clark, R. N. 1984. Surface of Miranda: Indentification of water ice. Submitted to Icarus.

Brown, R. H., Cruikshank, D. P., and Morrison, D. 1982a. The diameters and albedos of the satellites of Uranus. *Nature* 300:423–425.

Brown, R. H., Morrison, D., Telesco, C. M., and Brunk, W. E. 1982b. Calibration of the radiometric asteroid scale using occultation diameters. *Icarus* 52:188–195.

Brown, W. L., Lanzerotti, L. J., and Johnson, R. E. 1982. Fast ion bombardment of ices: Laboratory measurements and astrophysical implications. *Science* 218:525–531.

Bugaenko, L. A. 1972. Monochromatic brightness coefficients of the major planets. *Astron. Vestn.* 6:19–21 (Russian).

Bugaenko, O. I., and Galkin, L. S. 1972. Photometry of major planets. II. Phase dependence of polarization for selected regions of the disk of Saturn. *Astron. Zh.* 49:837–843 (Russian).

Bugaenko, O. I., and Galkin, L. S. 1973. *Soviet Astron.* 16:681–686 (English trans.).

Bugaenko, O. I., and Morozhenko, A. V. 1981a. Physical characteristics of the upper layers of Saturn's atmosphere. *Adv. Space Res.* 1:183–186.

Bugaenko, O. I., and Morozhenko, A. V. 1981b. Oriented particles in the high atmosphere of Saturn. In *Physics of Planetary Atmospheres*, (Kiev: Naukova Dumka Press), pp. 108–112 (Russian).

Bugaenko, O. I., Galkin, L. S., and Morozhenko, A. V. 1971. The polarimetric observations of the Jovian planets. I. The distribution of the polarization on the disk of Saturn. *Astron. Zh.* 48:373–379 (Russian).

Bugaenko, O. I., Galkin, L. S., and Morozhenko, A. V. 1972. A study of molecular absorption in the atmosphere of major planets. *Astron. Vestn.* 6:223–227 (Russian).

Bugaenko, O. I., Dlugach, J. M., Morozhenko, A. V., and Yanovitskij, E. G. 1975. On optical properties of the Saturn cloud layer in the visible part of the spectrum. *Astron. Vestn.* 9:13–21 (Russian).

Bunch, T. E., and Chang, S. 1980. Carbonaceous chondrites. II. Carbonaceous chondrite phyllosilicates and light element geochemistry as indicators of parent body processes and surface conditions. *Geochim. Cosmochim. Acta* 44:1534–1577.

Buratti, B. 1983. Photometric Properties of Europa and the Icy Satellites of Saturn. Ph.D. thesis, Cornell Univ., Ithaca, New York.

Buratti, B., and Veverka, J. 1983a. Voyager photometry of Europa. *Icarus* 55:93–110.

Buratti, B., and Veverka, J. 1983b. Voyager photometry of Rhea, Dione, Tethys, Enceladus, and Mimas. *Icarus*. In press.

Buratti, B., Veverka, J., and Thomas, P. 1982. Voyager photometry of Saturn's satellites. In *Rept. Planetary Geology Program*, ed. H. E. Holt (NASA TM-84127), pp. 41–43.

Buriez, J. C., and de Bergh, C. 1981. A study of the atmosphere of Saturn based on methane line profile near 1.1 μm. *Astron. Astrophys.* 94:382–390.

Burke, J. J., and KenKnight, C. E. 1980. An extraordinary view of Saturn's rings. *J. Geophys. Res.* 85:5925–5928.

Burke, J. J., Gehrels, T., and Strickland, R. N. 1980. Cloud forms on Saturn. *J. Geophys. Res.* 85:5883–5890.

Burns, J. A., Lamy, P. L., and Soter, S. 1979. Radiative forces on small particles in the solar system. *Icarus* 40:1–48.

Burns, J. A., Showalter, M. R., Cuzzi, J. N., and Durisen, R. H. 1983. Saturn's electrostatic discharges: Could lightning be the cause? *Icarus* 54:280–295.

Burtis, W. J., and Helliwell, R. A. 1976. Occurrence patterns and normalized frequency. *Planet. Space Sci.* 24:1007–1024.

Busse, F. H. 1970. Differential rotation in stellar convection zones. *Astrophys. J.* 159:629–639.

Busse, F. H. 1976. A simple model of convection in the Jovian atmosphere. *Icarus* 29:255–260.

Busse, F. H. 1979. Theory of planetary dynamos. In *Magnetospheric Particles and Fields*, Vol. 2, eds. C. F. Kennel, L. J. Lanzerotti, and E. N. Parker (Amsterdam: North-Holland).

Byke, Sh. Sh., and Fomina, V. I. 1968. Gas hydrates. *Russ. Chem. Rev.* 37:469–491.

Cai, Q., and Liou, K. N. 1982. Polarized light scattering by hexagonal ice crystals: Theory. *Appl. Optics* 21:3569–3580.

Caldwell, J. 1975. Ultraviolet observations of small bodies in the solar system by OAO-2. *Icarus* 25:384–396.

Caldwell, J. 1977*a*. The atmosphere of Saturn: An infrared perspective. *Icarus* 30:493–510.

Caldwell, J. 1977*b*. Ultraviolet observations of Mars and Saturn by the TD1A and OAO-2 satellites. *Icarus* 32:190–209.

Caldwell, J. 1977*c*. Thermal radiation from Titan's atmosphere. In *Planetary Satellites*, ed. J. A. Burns (Tucson: Univ. of Arizona Press), pp. 438–450.

Caldwell, J., Gillett, F. C., Nolt, I. G., and Tokunaga, A. 1978. Spatially resolved infrared observations of Saturn. I. Infrared limb scans at 20 microns. *Icarus* 35:308–312.

Caldwell, J., Owen, T., Rivolo, A. R., Moore, V., Hunt, G. E., and Butterworth, P. S. 1981. Observations of Uranus, Neptune, and Titan by the International Ultraviolet Explorer. *Astron. J.* 86:298–305.

Caldwell, J., Tokunaga, A., and Orton, G. 1983. Further observations of 8 μm polar brightening of Jupiter. *Icarus* 53:133–140.

Calvert, J. G., and Pitts, J. N. 1966. *Photochemistry* (New York: John Wiley and Sons).

Calvert, W. 1981. The auroral plasma cavity. *Geophys. Res. Letters* 8:919–921.

Cameron, A. G. W. 1973. Abundances of the elements in the solar system. *Space Sci. Rev.* 15:121–146.

Cameron, A. G. W. 1978. Physics of the primitive solar accretion disk. *Moon Planets* 18:5–40.

Cameron, A. G. W. 1982. Elementary and nuclidio abundances in the solar system. In *Essays in Nuclear Astrophysics*, eds. C. A. Barnes, D. D. Clayton, and D. N. Schramm (Cambridge: Cambridge Univ. Press), pp. 23–43.

Cameron, A. G. W., DeCampli, W. M., and Bodenheimer, P. 1982. Evolution of giant gaseous protoplanets embedded in the primitive solar nebula. *Icarus* 49:298–321.

Camichel, H. 1958. Mesures photométriques de Saturne et de son anneau. *Ann. Astrophys.* 21:231–242.

Campani, G. 1664. *Ragguaglio di due Nuove Osservazioni* (Rome).

Campbell, D. B., Chandler, J. F., Pettengill, G. H., and Shapiro, I.I. 1977. The Galilean satellites: 12.6 cm radar observations. *Science* 196:650–653.

Campbell, M. H., and Ulrichs, J. 1969. The electrical properties of rocks and their significance for lunar radar observations. *J. Geophys. Res.* 74:5867–5881.

Campbell, W W. 1895. A spectrographic determination of velocities in the system of Saturn. *Astron. J.* 2:127–135.

Canuto, V. M., Levine, J. S., Augustsson, T. R., and Imhoff, C. L. 1982. UV radiation from the young sun and oxygen and ozone levels in the prebiological paleoatmosphere. *Nature* 296:816–820.

Capone, L. A., Whitten, R. C., Prasad, S. S., and Dubach, J. 1977. The ionospheres of Saturn, Uranus, and Neptune. *Astrophys. J.* 215:977–983.

Capone, L. A., Dubach, J., Whitten, R. C., Prasad, S. S., and Santhanam, K. 1980. Cosmic ray synthesis of organic molecules in Titan's atmosphere. *Icarus* 44:72–84.

Carbary, J. F., and Krimigis, S. M. 1982. Charged particle periodicity in the Saturnian magnetosphere. *Geophys. Res. Letters* 9:1073–1076.

Carbary, J. F., Bythrow, P. F., and Mitchell, D. G. 1982. The spokes in Saturn's rings: A new approach. *Geophys. Res. Letters* 9:420–422.

Carbary, J. F., Krimigis, S. M., and Ip, W.-H. 1983*a*. Energetic particle microsignatures of Saturn's satellites. *J. Geophys. Res.* 88:8947–8958.

Carbary, J. F., Mauk, B. H., and Krimigis, S. M. 1983*b*. Corotation anisotropies in Saturn's magnetosphere. *J. Geophys. Res.* 88:8937–8946.

Carlos, E. S. 1880. *The Sidereal Messenger of Galileo Galilei and a Part of the Preface of Kepler's Dioptrics* (London); also reprinted 1960 (London: Dawsons).

Carlson, B., Caldwell, J., and Cess, R. 1980. A model of Saturn's seasonal stratosphere at the time of the Voyager encounters. *J. Atmos. Sci.* 37:1883–1885.

Carovillano, R. L., and Siscoe, G. L. 1973. Energy and momentum theorems in magnetospheric processes. *Rev. Geophys. Space Phys.* 11:289–353.

Carr, T. D., and Desch, M. D. 1976. Recent decametric and hectometric observations of Jupiter. In *Jupiter*, ed. T. Gehrels (Tucson: Univ. of Arizona Press), pp. 693–737.

Carr, T. D., Smith, A. G., Bollhagen, H., Six, N. F., and Chatterton, N. E. 1961. Recent decameter wavelength observations of Jupiter, Saturn, and Venus. *Astrophys. J.* 134:105–125.

Carr, T. D., Schauble, J. J., and Schauble, C. C. 1981. Pre-encounter distributions of Saturn's low frequency radio emission. *Nature* 292:745–747.

Cassen, P., Peale, S. J., and Reynolds, R. G. 1980. On the comparative evolution of Ganymede and Callisto. *Icarus* 41:232–239.

Cassini, J. D. 1665*a*. *Quattro Lettere al Signor Abbate Falconieri sopra la Varietà delle Macchie osservate in Giove, e loro diurne Revoluzioni* (Rome).

Cassini, J. D. 1665*b*. *Tabulae Quotidianae Revolutionis Macularum Jovis* (Rome).

Cassini, J. D. 1666*a*. *Martis circa Axem Proprium Revolubilis Observationes* (Bologna).

Cassini, J. D. 1666*b*. *De Periodo Quotidianae Revolutionis Martis* (Bologna).

Cassini, J. D. 1671*a*. *Suite des Observations des Taches du Soleil Faites à l'Académie Royale, Avec Quelques Autres Observations Concernant Saturne* (Paris).

Cassini, J. D. 1671*b*. Observations concerning Saturn. *Phil. Trans.* 6:3024–3027.

Cassini, J. D. 1673*a*. *Découverte de deux Nouvelles Planètes autour de Saturne* (Paris).

Cassini, J. D. 1673*b*. Discovery of two new planets about Saturn, made in the Royal Parisian Observatory by Signor Cassini. *Phil. Trans.* 8:5178–5185.

Cassini, J. D. 1676. An extract of Signor Cassini's letter concerning a spot lately seen in the Sun: Together with a remarkable observation of Saturn by the same. *Phil. Trans.* 11:689–690, fig. opp. p. 710.

Cassini, J. D. 1677. Observations nouvelles de M. Cassini, touchant le globe et l'anneau de Saturne. *J. des Scavans* 1 March, pp. 32–33.

Cassini, J. D. 1686*a*. Nouvelle découverte des deux satellites de Saturne les plus proches. *J. des Scavans* 22 April, p. 97.

Cassini, J. D. 1686*b*. An extract of the *Journal des Scavans*. of April 22 . . . 1686. Giving an account of two new satellites of Saturn, discovered lately by Mr. Cassini at the Royal Observatory at Paris. *Phil. Trans.* 16:79–85.

Cess, R. D., and Caldwell, J. 1979. A Saturnian stratospheric seasonal climate model. *Icarus* 38:349–357.

Chakravarty, S., Rose, J. H., Wood, D., and Ashcroft, N. W. 1981. Theory of dense hydrogen. *Phys. Rev.* 24B:1624–1635.

Chamberlain, J. W. 1963. Planetary coronae and atmosphere evaporation. *Planet. Space Sci.* 19:675–684.

Chambers, G. F. 1867. *Descriptive Astronomy* (London).

Chandrasekhar, S. 1960. *Radiative Transfer* (New York: Dover Publ. Co.).

Chang, S., Scattergood, T., Aronowitz, S., and Flores, J. 1979. Organic chemistry on Titan. *Rev. Geophys. Space Phys.* 17:1923–1933.

Chapman, S. J., and Bartels, J. 1940. *Geomagnetism* (Oxford: Oxford Univ. Press).

Chedin, A., Husson, N., Scott, N. A., and Gautier, D. 1978. The V_4 band of methane ($^{12}CH_4$ and $^{13}CH_4$) line parameters and evaluation of Jovian atmospheric transmissions at 7.7 μm. *J. Molec. Spectrosc.* 71:343–368.

Chenette, D. L., and Davis, L. Jr. 1982. An analysis of the structure of Saturn's magnetic field using charged particle absorption signatures. *J. Geophys. Res.* 87:5267–5274.

Chenette, D. L., and Stone, E. C. 1983. The Mimas ghost revisited: An analysis of the electron flux and electron microsignatures observed in the vicinity of Mimas at Saturn. *J. Geophys. Res.* 88:8755–8764.

Cheng, A. F., and Lanzerotti, L. J. 1978. Ice sputtering by radiation belt protons and the rings of Saturn and Uranus. *J. Geophys. Res.* 83:2597–2602.

Cheng, A. F., Lanzerotti, L. J., and Pirronello, V. 1982. Charged particle sputtering of ice surfaces in Saturn's magnetosphere. *J. Geophys. Res.* 87:4567–4570.

Cheng, V. M., Daniels, W. B., and Crawford, R. K. 1975. Melting parameters of methane and nitrogen from 0 to 10 kbar. *Phys. Rev.* 11B:3972–3975.

Chopra, K. P., and Hubert, L. R. 1965. Mesoscale eddies in the wake of islands. *J. Atmos. Sci.* 22:652–657.

Clark, R. N. 1980. Ganymede, Europa, Callisto, and Saturn's rings: Compositional analysis from reflectance spectroscopy. *Icarus* 44:388–409.

Clark, R. N. 1981*a*. The spectral reflectance of water-mineral mixtures at low temperatures. *J. Geophys. Res.* 86:3074–3086.

Clark, R. N. 1981*b*. Water frost and ice: The near-infrared spectral reflectance 0.65–2.5 micrometers. *J. Geophys. Res.* 86:3087–3096.

Clark, R. N. 1982. Implications of using broadband photometry for compositional remote sensing of icy objects. *Icarus* 49:244–257.

Clark, R. N., and McCord, T. B. 1979. Jupiter and Saturn: Near-infrared spectral albedos. *Icarus* 40:180–188.

Clark, R. N., and McCord, T. B. 1980. The rings of Saturn—new infrared reflectance measurements and a 0.326–4.08 micron summary. *Icarus* 43:161–168.

Clark, R. N., and Owensby, P. D. 1982. The infrared spectrum of Rhea. *Icarus* 46:354–360.

Clark, R. N., and Rousch, T. L. 1982. Reflectance spectroscopy: Quantitative analysis techniques for remote sensing applications. *J. Geophys. Res.* In press.

Clark, R. N., and Lucey, P. G. 1984. Spectral properties of ice particulate mixtures: Implications for remote sensing. I. Intimate mixtures. Submitted to *J. Geophys. Res.*

Clark, R. N., Fanale, F. P., and Zent, Z. A. 1983. Frost grain side metamorphism: Implications for remote sensing of planetary surfaces. *Icarus* 56:233–245.

Clark, R. N., Brown, R. H., Owensby, P. D., and Steele, A. 1984. Saturn's satellites: Near-infrared spectrophotometry (0.65–2.5 μm) of the leading and trailing sides and compositional implications. *Icarus*. In press.

Clarke, J. T., Moos, H. W., Atreya, S. K., and Lane, A. L. 1981. IUE detection of bursts of H Lyman-alpha from Saturn. *Nature* 290:226–227.

Clarke, J. T., Moos, H. W., and Feldman, P. D. 1982. The far ultraviolet spectra and geometric albedos of Jupiter and Saturn. *Astrophys. J.* 255:806–818.

Claypool, G. E., and Kvenvolden, K. A. 1983. Methane and other hydrocarbon gases in marine sediment. *Ann. Rev. Earth Planet. Sci.* 11:299–327.

Clerke, A. 1893. *A Popular History of Astronomy during the Nineteenth Century* (London), 3rd edition.

Cochran, A. L., and Cochran, W. D. 1981. Longitudinal variability of methane and ammonia bands on Saturn. *Icarus* 48:488–495.

Cochran, W. D. 1982. Spatially resolved reflectivities of Saturn: 3000–6000 Å. *Astron. J.* 87:718–723.

Cochran, W. D., and Barker, E. S. 1979. Variability of Lyman-alpha emission from Jupiter. *Astrophys. J.* 234:L151–L154.

Colegrove, F. D., Hanson, W. B., and Johnson, F. S. 1965. Eddy diffusion and oxygen transport in the lower thermosphere. *J. Geophys. Res.* 70:4931–4941.

Collard, H. R., Mihalov, J. D., and Wolfe, J. H. 1982. Radial variation of the solar wind speed between 1 and 15 AU. *J. Geophys. Res.* 87:2203–2214.

Colombo, G., and Franklin, F. A. 1971. On the formation of the outer satellite groups of Jupiter. *Icarus* 15:186–191.

Colombo, G., Franklin, F. A., and Shapiro, I.I. 1974. On the formation of the orbit-orbit resonance of Titan and Hyperion. *Astron. J.* 79:61–72

Colombo, G., Goldreich, P., and Harris, A. W. 1976. Spiral structure as an explanation for the asymmetric brightness of Saturn's A ring. *Nature* 264:344–345.

Comas Sola, J. 1908. Observations des satellites principeaux de Jupiter et de Titan. *Astron. Nach.* 179:289.

Combes, M., de Bergh, C., Lecacheux, J., and Maillard, J. P. 1975. Indentification of $^{13}CH_4$ in the atmosphere of Saturn. *Astron. Astrophys.* 40:81–84.

Combes, M., Maillard, J. P., and de Bergh, C. 1977. Evidence for a telluric value of $^{12}C/^{13}C$ ratio in the atmospheres of Jupiter and Saturn. *Astron. Astrophys.* 61:531–537.

Combes, M. et al. 1981. Vertical distribution of NH_3 in the Jovian atmosphere from IUE observations. *Adv. Space Res.* 1:109–175.

Connerney, J. E. P. 1981. The magnetic field of Jupiter: A generalized inverse approach. *J. Geophys. Res.* 86:7679–7693.

Connerney, J. E. P. 1983. Saturn's zonal harmonic magnetic field. Paper presented at M.I.T. Jupiter-Saturn Conference. To be submitted to *J. Geophys. Res.*

Connerney, J. E. P., Acuña, M. H., and Ness, N. F. 1981. Saturn's ring current and inner magnetosphere. *Nature* 292:724–726.

Connerney, J. E. P., Acuña, M. H., and Ness, N. F. 1982*a*. Voyager 1 assessment of Jupiter's planetary magnetic field. *J. Geophys. Res.* 86:3623–3627.

Connerney, J. E. P., Ness, N. F., and Acuña, M. H. 1982b. Zonal harmonic model of Saturn's magnetic field from Voyager 1 and 2 observations. *Nature* 298:44–46.

Connerney, J. E. P., Acuña, M. H., and Ness, N. F. 1983. Currents in Saturn's magnetosphere. *J. Geophys. Res.* 88:8779–8789.

Connerney, J. E. P., Acuña, M. H., and Ness, N. F. 1984. The Z_3 model of Saturn's magnetic field and the Pioneer 11 vector helium magnetometer observations. Submitted to *J. Geophys. Res.*

Connes, J., Connes, P., and Maillard, J. P. 1969. Near infrared spectra of Venus, Mars, Jupiter, and Saturn. Colition du Centre Nationale de Recherche Scientifique, Paris.

Conrath, B. J., and Gautier, D. 1980. Thermal structure of Jupiter's atmosphere obtained by inversion of Voyager 1 infrared measurements. In *Remote Sensing of Atmospheres and Oceans*, ed. A. Deepak (New York: Academic Press), pp. 611–630.

Conrath, B. J., and Pirraglia, J. A. 1983. Thermal structure of Saturn from Voyager infrared measurements: Implications for atmospheric dynamics. *Icarus* 53:286–292.

Conrath, B. J., Flasar, F. M., Pirraglia, J. A., Gierasch, P. J., and Hunt, G. E. 1981a. Thermal structure and dynamics of the Jovian atmosphere. II. Visible cloud features. *J. Geophys. Res.* 86:8769–8775.

Conrath, B. J., Gierasch, P. J., and Nath, N. 1981b. Stability of zonal flows on Jupiter. *Icarus* 48:256–282.

Conrath, B. J., Gautier, D., Hornstein, J., and Hanel, R. 1983. The helium abundance of Saturn from Voyager. Submitted to *Astrophys. J.*

Consolmagno, G. J. 1975. Thermal History Models of Icy Satellites. M.S. thesis, Massachusetts Institute of Technology, Cambridge, MA.

Consolmagno, G. J. 1984a. Resurfacing Saturn's satellites: Models of partial differentiation and expansion. Submitted to *Icarus*.

Consolmagno, G. J. 1984b. Ice-rich planets and the physical properties of ice. *J. Chem. Phys.* In press.

Consolmagno, G. J., and Lewis, J. S. 1976. Structural and thermal models of icy satellites. In *Jupiter*, ed. T. Gehrels (Tucson: Univ. of Arizona Press), pp. 1035–1051.

Consolmagno, G. J., and Lewis, J. S. 1977. Preliminary thermal history models of icy satellites. In *Planetary Satellites*, ed. J. A. Burns (Tucson: Univ. of Arizona Press), pp. 492–500.

Consolmagno, G. J., and Lewis, J. S. 1978. The evolution of icy satellite interiors and surfaces. *Icarus* 34:280–293.

Consolmagno, G. J., and Huang, P. Y. 1982. Evolution of the icy moons of Saturn. Program "Saturn Conference," Tucson, AZ, May 1982 (abstract).

Conway, R. R. 1983. Multiple fluorrescent scattering of N_2 ultraviolet emissions in the atmospheres of the Earth and Titan. *J. Geophys. Res.* 88:4784.

Cook, A. F., and Franklin, F. A. 1970. An explanation of the lightcurve of Iapetus. *Icarus* 13:282–291.

Cook, A. F., and Franklin, F. A. 1977. Saturn's rings: A new survey. In *Planetary Satellites*, ed. J. A. Burns (Tucson: Univ. of Arizona Press), pp. 412–419.

Cook, A. F., Franklin, F. A., and Palluconi, F. D. 1973. Saturn's rings: A survey. *Icarus* 18:317–337.

Cooper, J. F. 1983. Nuclear cascades in Saturn's rings: Cosmic ray albedo neutron decay and origins of trapped protons in the inner magnetosphere. *J. Geophys. Res.* 88:3945–3954.

Cooper, J. F., and Simpson, J. A. 1980. Sources of high energy protons in Saturn's magnetosphere. *J. Geophys. Res.* 85:5793–5802.

Coradini, A., Frederico, C., and Magni, G. 1981. Gravitational instabilities in satellite disks and formation of regular satellites. *Astron. Astrophys.* 99:255–261.

Coroniti, F. V., Scarf, F. L., and Kennel, C. F. 1980. Detection of Jovian whistler mode chorus: Implications for the Io torus aurora. *Geophys. Res. Letters* 7:45–48.

Courtin, R. 1982a. La Structure Thermique et la Composition des Atmosphères des Planètes Géantes à Partir de leur Spectre Infra-rouge. Thesis, Univ. de Paris, France.

Courtin, R. 1982b. The spectrum of Titan in the far-infrared and microwave regions. *Icarus* 51:466–475.

Courtin, R., Coron, N., Encrenaz, T., Gispert, R., Bruston, P., LeBlanc, J., Dambier, G., and Vidal-Madgar, A. 1977. Observations of giant planets at 1.4 mm and consequences on the effective temperatures. *Astron. Astrophys.* 60:115–123.

Courtin, R., Lena, P., De Muizon, M., Rouan, D., Nicollier, C., and Wijnbergen, J. 1979. Far-infrared photometry of planets—Saturn and Venus. *Icarus* 38:411–419.

Courtin, R., Balutean, J. P., Gautier, D., Marten, A., and Maguire, W. 1982a. The minor constituents of Saturn's atmosphere from the Voyager 1 IRIS experiment. Program "Saturn Conference," Tucson, AZ, May 1982 (abstract).

Courtin, R., Gautier, D., Marten, A., and Maguire, W. 1982b. The acetylene and ethane abundances and the phosphine distribution in Saturn's atmosphere from the Voyager 1—IRIS experiment. *Bull. Amer. Astron. Soc.* 13:722 (abstract).

Courtin, R. Gautier, D., Marten, A., and Kunde, V. 1983a. The $^{12}C/^{13}C$ ratio in Jupiter from the Voyager infrared investigation. *Icarus* 53:121–132.

Courtin, R., Gautier, D., Marten, A., and Kunde, V. 1983b. The atmospheric composition of Saturn from the infrared Voyager experiment. I. Determination of C_2H_2, C_2H_6 and PH_3 vertical distributions. In preparation.

Courtin R., Gautier, D., Marten, A., and Maguire, W. 1983c. The atmospheric composition of Saturn from the infrared Voyager experiment. II. Determination of the C/H and D/H ratios. In preparation.

Cowen, R. 1982. A former inner satellite of Saturn? Preprint.

Cowling, T. G. 1934. The magnetic field of sunspots. *Mon. Not. Roy. Astron. Soc.* 94:39–48.

Crane, A. J., Haigh, J. D., Pyle, J. A., and Rogers, C. F. 1980. Mean meridional motions of the stratosphere and mesosphere. *Pure App. Geophys.* 118:307–328.

Cravens, T. E. 1974. Astrophysical Applications of Electron Energy Deposition in Molecular Hydrogen. Ph.D. thesis, Harvard University, Cambridge, MA.

Cruikshank, D. P. 1979a. The radius and albedo of Hyperion. *Icarus* 37:307–309.

Cruikshank, D. P. 1979b. The surfaces and interiors of Saturn's satellites. *Rev. Geophys. Space Phys.* 17:165–176.

Cruikshank, D. P. 1981. Near-infrared studies of the satellites of Saturn and Uranus. *Icarus* 41:246–258.

Cruikshank, D. P., and Morgan, J. S. 1980. Titan: Suspected near-infrared variability. *Astrophys. J.* 235:L53–L54.

Cruikshank, D. P., and Brown, R. H., 1982. Surface composition and radius of Hyperion. *Icarus* 50:82–87.

Cruikshank, D. P., Degewij, J., and Zellner, B. H. 1982a. The outer satellites of Jupiter. In *Satellites of Jupiter*, ed. D. Morrison (Tucson: Univ. of Arizona Press), pp. 129–146.

Cruikshank, D. P., Lebofsky, L. A., Pearl, T. C., and Veverka, J. 1982b. Satellites of Saturn: Optical properties. Program "Saturn Conference," Tucson, AZ, May 1982 (abstract).

Cruikshank, D. P., Bell, J. F., Gaffey, M. J., Brown, R. H., Howell, R., Beerman, C., and Rognstad, M. 1983. The dark side of Iapetus. *Icarus* 53:90–104.

Cunningham, C. T., Ade, P. A. R., Robson, E. I., Nolt, I. G., and Radostitz, J. V. 1981. The submillimeter spectra of the planets: Narrow-band photometry. *Icarus* 48:127–139.

Cuzzi, J. N. 1978. The rings of Saturn: State of current knowledge and some suggestions for future studies. In *The Saturn System*, ed. D. Morrison (Washington: NASA Publ. 2068), pp. 73–104.

Cuzzi, J. N. 1982. Physical properties of Saturn's rings. Proceedings "Planetary Rings/Anneaux des Planètes Conference," Toulouse, France, August 1982 (abstract).

Cuzzi, J. N., and van Blerkom, D. 1974. Microwave brightness of Saturn's rings. *Icarus* 22:149–158.

Cuzzi, J. N., and Dent, W. A. 1975. Saturn's rings: The determination of their brightness, temperature, and opacity at centimeter wavelengths. *Astrophys. J.* 198:223–227.

Cuzzi, J. N., and Pollack, J. B. 1978. Saturn's rings: Particle composition and size distribution as constrained by microwave observations. I. Radar observations. *Icarus* 33:233–262.

Cuzzi, J. N., and Scargle, J. 1984. The wavy edges of the Encke Gap. In preparation.

Cuzzi, J. N., Durisen, R. H., Burns, J. A., and Hamill, P. M. 1979. The vertical structure and thickness of Saturn's rings. *Icarus* 38:54–68.

Cuzzi, J. N., Pollack, J. B., and Summers, A. L. 1980. Saturn's rings: Particle composition and size distribution as constrained by observations at microwave wavelengths. II. Radio interferometric observations. *Icarus* 44:683–705.

Cuzzi, J. N., Lissauer, J. J., and Shu, F. H. 1981. Density waves in Saturn's rings. *Nature* 292:703–707.

Cuzzi, J. N., Lissauer, J. J., Esposito, L. W., Holberg, J. B., Marouf, E. A., Tyler, G. L., and Boischot, A. 1984. Saturn's rings: Properties and processes. In *Planetary Rings*, eds. R. Greenberg and A. Brahic (Tucson: Univ. of Arizona Press), pp. 72–198.

Daigne, G., Pedersen, B. M., Kaiser, M. L., and Desch, M. D. 1982. Planetary radio astronomy observations during the Voyager 1 Titan fly-by. *J. Geophys. Res.* 87:1405–1409.

Dalgarno, A., and Williams, D. A. 1962. Rayleigh scattering by molecular hydrogen. *Astrophys. J.* 136:690–692.

Danby, J. M. A. 1962. *Fundamentals of Celestial Mechanics* (New York: MacMillan).

Danehy, R. G., Owen, T., Lutz, B. L., and Woodman, J. H. 1978. Detection of the Kuiper bands in the spectrum of Titan. *Icarus* 35:247–251.

Daniel, R. R., Ghosh, S. K., Iyengar, K. V. K., Rengarijan, T. N., Tandon, S. N., and Varma, R. P. 1982. Far-infrared brightness temperature of Saturn's disk and rings. *Icarus* 49: 205–212.

Danielson, R. E., Caldwell, J. J., and Larach, D. R. 1973. An inversion in the atmosphere of Titan. *Icarus* 20:437–443.

Danielson, R. E., Caldwell, J. J., and Larach, D. R. 1974. An inversion in the atmosphere of Titan. In *The Atmosphere of Titan*, ed. D. M. Hunten (Washington: NASA SP-340), pp. 92–109. Also *Icarus* 20:437–443, 1973.

Davidson, D. W. 1973. Clathrates hydrates. In *Water: A Comprehensive Treatise* Vol. 2, ed. F. Franks (New York: Plenum), pp. 115–234.

Davidson, D. W., and Ripmeester, J. A. 1978. Clathrate ices: Recent results. *J. Glaciology* 21:33–49.

Davies, M. E., and Katayama, F. Y. 1983. The control networks of Mimas and Enceladus. *Icarus* 53:332–340.

Davis, D. R., Chapman, C. R., Greenberg, R., and Weidenschilling, S. J. 1983. Asteroid collisions: Effective body strength and efficiency of catastrophic disruption. *Lunar. Planet. Sci.* XIV:146–147.

Dawes, W. R. 1850. On the ring of Saturn. *Mon. Not. Roy. Astron. Soc.* 11:22–27, 49–53, 230.

de Bergh, C., Vion, M., Combes, M., Lecacheux, J., and Maillard, J. P. 1973. New infrared spectra of the Jovian planets from 1200 to 4000 cm^{-1} by Fourier transform spectroscopy. II. Study of Saturn in the $3\nu_3$ CH$_4$ band. *Astron. Astrophys.* 28:457–466.

de Bergh, C., Lecacheux, J., Combes, M., and Maillard, J. P. 1974. New infrared spectra of the Jovian planets from 1200 to 4000 cm^{-1} by Fourier transform spectroscopy. III. First-overtone pressure-induced H$_2$ absorption in the atmospheres of Jupiter and Saturn. *Astron. Astrophys.* 35:333–337.

de Bergh, C., Lecacheux, J., and Maillard, J. P. 1977. The 2-O quadrupole spectrum H$_2$ in the atmosphere of Jupiter and Saturn. *Astron. Astrophys.* 56:227–233.

Degewij, J:, and Van Houten, C. J. 1979. Distant asteroids and outer Jovian satellites. In *Asteroids*, ed. T. Gehrels (Tucson: Univ. of Arizona Press), pp. 417–435.

Degewij, J., Andersson, L. E., and Zellner, B. 1980a. Photometric properties of outer planetary satellites. *Icarus* 44:520–540.

Degewij, J., Cruikshank, D. P., and Hartmann, W. K. 1980b. Near-infrared colorimetry of J6 Himalia and S9 Phoebe: A summary of 0.3 to 2.2 μm reflectances. *Icarus* 44:541–547.

De la Rue, W. 1856. *Mon. Not. Roy. Astron. Soc.* 17:19, 221.

Delsemme, A. H. 1982. Chemical composition of cometary nuclei. In *Comets*, ed. L. L. Wilkening (Tucson: Univ. of Arizona Press), pp. 85–130.

DeMarcus, W. C. 1958. The constitution of Jupiter and Saturn. *Astron. J.* 63:2–28.

de Pater, I., and Dickel, J. R. 1982. New information on Saturn and its rings from VLA multifrequency data. Proceedings "Planetary Rings/Anneaux des Planètes Conference," Toulouse, France, August 1982 (abstract).

Dermott, S. F. 1979. Tidal dissipation in the solid cores of the major planets. *Icarus* 37: 310–321.

Dermott, S. F. 1981. The "braided" F ring of Saturn. *Nature* 290:454–457.

Dermott, S. F., and Murray, C. D. 1981a. The dynamics of tadpole and horseshoe orbits. I. Theory. *Icarus* 48:1–11.

Dermott, S. F., and Murray, C. D. 1981b. The dynamics of tadpole and horseshoe orbits. II. The coorbital satellites of Saturn. *Icarus* 48:12–22.

Dermott, S. F., and Sagan, C. 1982. The tide in the seas of Titan. *Nature* 300:731–733.

Desch, M. D. 1982. Evidence for solar wind control of Saturn radio emission. *J. Geophys. Res.* 87:4549–4554.

Desch, M. D. 1983. Radio emission signature of Saturn immersions in Jupiter's magnetic tail. *J. Geophys. Res.* 88:6904–6910.

Desch, M. D., and Kaiser, M. L. 1981*a*. Voyager measurement of the rotation period of Saturn's magnetic field. *Geophys. Res. Letters* 8:253–256.

Desch, M. D., and Kaiser, M. L. 1981*b*. Saturn's kilometric radiation: Satellite modulation. *Nature* 292:739–741.

Desch, M. D., and Rucker, H. O. 1983. The relationship between Saturn kilometric radiation and the solar wind. *J. Geophys. Res.* 88:8999–9006.

Deslandres, H. A. 1895. Recherches spectrales sur les anneaux de Saturne. *Comptes Rendus* 120:1155–1158.

Dessler, A. J. 1967. Solar wind and interplanetary magnetic field. *Rev. Geophys.* 5:1–41.

Dessler, A. J. ed. 1983. *Physics of the Jovian Magnetosphere* (Cambridge: Cambridge Univ. Press).

De Vico, 1838. *Memoria Intorno ad Alcune Osservazioni Fatte Alla Specola del Collegio Romano nel Corrente Anno 1838* (Rome).

DeWitt, H. E., and Hubbard, W. B. 1976. Statistical mechanics of light elements at high pressure. IV. A model free energy for the metallic phase. *Astrophys. J.* 205:295–301.

Dlugach, J. M., Morozhenko, A. V., Vidmichenko, A. P., and Yanovitsky, E. G. 1983. Investigations the optical properties of Saturn's atmosphere carried out at the Main Astronomical Observatory of the Ukrainian Academy of Science. *Icarus* 54:319–336.

Dodonova, N. Ya. 1966. Activation of nitrogen by vacuum ultraviolet radiation. *Russ. J. Phys. Chem.* 40:523–524.

Dollfus, A. 1961*a*. Visual and photographic studies of planets at Pic du Midi. In *Planets and Satellites*, eds. G. P. Kuiper and B. M. Middlehurst (Chicago: Univ. of Chicago Press), pp. 534–571.

Dollfus, A. 1961*b*. Polarization studies of planets. In *Planets and Satellites*, eds. G. P. Kuiper and B. M. Middlehurst (Chicago: Univ. of Chicago Press), pp. 343–399.

Dollfus, A. 1963. Mouvements dans l'atmosphère de Saturne en 1960. Observations coordonnées par l'Union Astronomique Internationale. *Icarus* 2:109–114.

Dollfus, A. 1967*a*. The discovery of Janus, Saturn's tenth satellite. *Sky Teles.* 34:136–137.

Dollfus, A. 1967*b*. Un nouveau satellite de Saturne. *Comptes Rendus Acad. Sci. Paris* 264(B): 822–824.

Dollfus, A. 1968. La Decouverte du 10e satellite de Saturne. *L'Astronomie* 82:253–262.

Dollfus, A. ed. 1970. Diamètres des planètes et satellites. In *Surfaces and Interiors of Planets and Satellites* (London: Academic Press), pp. 45–139.

Dollfus, A. 1979*a*. Optical reflectance polarimetry of Saturn's globe and rings. I. Measurements on B ring. *Icarus* 37:404–419.

Dollfus, A. 1979*b*. Optical reflectance polarimetry of Saturn's globe and rings. II. Interpretations for B ring. *Icarus* 40:171–179.

Dollfus, A. 1982. Optical reflectance polarimetry of Saturn's rings. Proceedings "Planetary Rings/Anneaux des Planètes Conference," Toulouse, France, August 1982 (abstract).

Dollfus, A., and Brunier, S. 1981. Observations of Saturn's inner satellites and the orbit of Janus in 1980. *Icarus* 48:29–38.

Dollfus, A., and Brunier, S. 1982. Observation and photometry of an outer ring of Saturn. *Icarus* 49:194–204.

Drake, S. ed. 1960. *Galileo. Discourse on Bodies in Water*, (Urbana: Univ. of Illinois Press), trans. T. Salusbury.

Drake, S. 1978. *Galileo at Work: His Scientific Biography* (Chicago: Univ. of Chicago Press).

Drell, S. D., Foley, H. M., and Ruderman, M. A. 1965. Drag and propulsion of large satellites. An Alfvén propulsion engine in space. *J. Geophys. Res.* 70:3131–3145.

Duncombe, R. L., Klepczynski, W. J., and Seidelman, P. K. 1974. The masses of the planets, satellites, and asteroids. *Fund. Cosmic Phys.* 1:119–165.

Dyer, J. W. 1980. Pioneer Saturn. *Science* 207:400–401.

Eichelberger, W. S. 1891. The orbit of Hyperion. *Astron. J.* 11:145–157.

Elliot, J. L., and Nicholson, P. D. 1984. The rings of Uranus. In *Planetary Rings*, eds. R. Greenberg and A. Brahic (Tucson: Univ. of Arizona Press), pp. 25–71.

Elliot, J. L., Veverka, J., and Goguen, J. 1975. Lunar occultation of Saturn. I. The diameters of Tethys, Dione, Rhea, Titan, and Iapetus. *Icarus* 26:387–407.

Ellis, G. R. A. 1962. Cyclotron radiation from Jupiter. *Aust. J. Phys.* 15:344–353.

Ellis, G. R. A., and McCulloch, P. M. 1963. The decametric radio emissions of Jupiter. *Aust. J. Phys.* 16:380–397.

Ellsworth, K., and Schubert, G. 1981. Thermal history models of Saturn's icy satellites. *Bull. Amer. Astron. Soc.* 13:704 (abstract).

Ellsworth, K., and Schubert, G. 1983. Saturn's icy satellites: Thermal and structural models. *Icarus* 54:490–510.

Elphic, R. C., and Russell, C. T. 1978. On the apparent source depth of planetary magnetic fields. *Geophys. Res. Letters* 5:211–214.

Encke, J. F. 1838. Über den Ring des Saturn. In *Mathematische Abhandlungen der Königlichen Akademie der Wissenschaften zu Berlin*, pp. 1–18.

Encrenaz, T., and Owen, T. 1973. New observations of the hydrogen quadrupole lines on Saturn and Uranus. *Astron. Astrophys.* 28:119–124.

Encrenaz, T., and Combes, M. 1982. On the C/H and D/H ratios in the atmospheres of Jupiter and Saturn. *Icarus* 52:54–61.

Encrenaz, T., Owen, T., and Woodman, J. H. 1974. The abundance of ammonia on Jupiter, Saturn, and Titan. *Astron. Astrophys.* 37:49–55.

Encrenaz, T. Combes, M., Zeau, Y., Vapillon, L., and Berezne, J. 1975. A tentative identification of C_2H_4 in the spectrum of Saturn. *Astron. Astrophys.* 43:355–356.

Eplee, R. E. Jr., and Smith, B. A. 1982. Dynamics of spokes in Saturn's B ring. Proceedings "Planetary Rings/Anneaux des Planètes Conference," Toulouse, France, August 1982 (abstract).

Epstein, E. E., Janssen, M. A., Cuzzi, J. N., Fogarty, W. G., and Mottman, J. 1980. Saturn's rings: 3-mm observations and derived results. *Icarus* 41:103–108.

Epstein, E. E., Janssen, M. A., and Cuzzi, J. N. 1983. Saturn's rings: 3 mm low-inclination observations and derived properties. *Icarus*. In press.

Epstein, P. S., and Carhat, R. R. 1953. The absorption of sound in suspensions and emulsions. I. Water fog in air. *J. Acoust. Am.* 25:553–565.

Erdi, B. 1978. The three-dimensional motion of Trojan asteroids. *Cel. Mech.* 18:141–161.

Erickson, E. F., Goorvitch, D., Simpson, J. P., and Strecker, D. W. 1978. Far infrared spectrophotometry of Jupiter and Saturn. *Icarus* 35:61–73.

Eshleman, V. R. 1973. The radio occultation method for the study of planetary atmospheres. *Planet. Space Sci.* 21:1521–1531.

Eshleman, V. R., Tyler, G. L., Anderson, J. D., Fjeldbo, G., Levy, G. S., Wood, G. E., and Croft, T. A. 1977. Radio science investigations with Voyager. *Space Sci. Res.* 21:207–232.

Eshleman, V. R., Tyler, G. L., Wood, G. E., Lindal, G. F., Anderson, J. D., Levy, G. S., and Croft, T. A. 1979. Radio science with Voyager 1 at Jupiter: Preliminary profiles of the atmosphere and ionosphere. *Science* 204:976–980.

Eshleman, V. R., Lindal, G. F., and Tyler, G. L. 1983. Is Titan wet or dry? *Science* 221:53–55.

Esposito, L. W. 1978. A multiple scattering explanation for the azimuthal variations in Saturn's rings. *Bull. Amer. Astron. Soc.* 10:584 (abstract).

Esposito, L. W. 1979. Extensions to the classical calculation of the effect of mutual shadowing in diffuse reflection: Applied to Saturn's rings. *Icarus* 39:69–80.

Esposito, L. W. 1983. Photometry and polarimetry of Saturn's rings. Proceedings "Natural Satellites," Ithaca, NY, July 1983 (abstract).

Esposito, L. W., and Lumme, K. 1977. The tilt effect for Saturn's rings. *Icarus* 31:157–168.

Esposito, L. W., Lumme, K., Benton, W. D., Martin, L. J., Ferguson, H. M., Thompson, D. T., and Jones, S. E. 1979. International planetary patrol observations of Saturn's rings. II. Four color phase curves and their analysis. *Astron. J.* 84:1408–1415.

Esposito, L. W., Dilley, J. P., and Fountain, J. W. 1980. Photometry and polarimetry of Saturn's rings from Pioneer 11. *J. Geophys. Res.* 85:5948–5956.

Esposito, L. W., Borderies, N., Cuzzi, J. N., Goldreich, P., Holberg, J. B., Lane, A. L., Lissauer, J. J., Marouf, E. A., Pomphrey, R. B., Terrile, R. J., and Tyler, G. L. 1983a. Voyager observations of an eccentric ringlet in Saturn's C ring. *Science* 222:57–59.

Esposito, L. W., O'Callaghan, M., and West, R. A. 1983b. The structure of Saturn's rings: Implications from the Voyager stellar occultation. *Icarus* 56:439–452.

Esposito, L. W., O'Callaghan, M., West, R. A., Hord, C. W., Simmons, K. E., Lane, A. L., Pomphrey, R. B., Coffeen, D. L., and Sato, M. 1983c. Voyager PPS stellar occultation of Saturn's ring. *J. Geophys. Res.* 88:8643–8649.

Evans, D. R., Warwick, J. W., Pearce, J. B., Carr, T. D., and Schauble, J. J. 1981. Impulsive radio discharges near Saturn. *Nature* 292:716–718.

Evans, D. R., Romig, J. H., Hord, C. W., Simmons, K. E., Warwick, J. W., and Lane, A. L. 1982. The source of Saturn electrostatic discharges. *Nature* 299:236–237.

Evans, D. R., Romig, J. H., and Warwick, J. W. 1983. Saturn electrostatic discharges: Properties and theoretical considerations. *Icarus* 54:267–279.

Evans, D. S., Deeming, T. J., Evans, B. H., and Goldfarb, S. 1969. *Herschel at the Cape: Diaries and Correspondence of Sir John Herschel, 1834–1838* (Austin: Univ. of Texas Press).

Everett, E. 1849. Report on the discovery of an eighth satellite of Saturn. *Mem. Amer. Ac. Arts. Sci.* 2:275–290 (new series).

Eviatar, A., and Podolak, M. 1983. Titan's gas and plasma torus. *J. Geophys. Res.* 88:833–840.

Eviatar, A., Siscoe, G. L., Scudder, J. D., Sittler, E. C. Jr., and Sullivan, J. D. 1982. The plumes of Titan. *J. Geophys. Res.* 87:8091–8103.

Eviatar, A., McNutt, R. L. Jr., Siscoe, G. L., and Sullivan, J. D. 1983. Heavy ions in the outer Kronian magnetosphere. *J. Geophys. Res.* 88:823–831.

Fälthammar, C.-G. 1968. Radial diffusion by violation of the third adiabatic invariant. In *Earth's Particles and Fields*, ed. B. M. McCormac (New York: Reinhold), pp. 157–169.

Farinella, P., Miloni, A., Nobili, A. M., Paolicchi, P., and Zappalà, V. 1983. Hyperion: Collisional disruption of a resonant satellite. *Icarus* 54:353–360.

Fegley, B., and Lewis, J. S. 1979. Thermodynamics of selected trace elements in the Jovian atmosphere. *Icarus* 38:166–179.

Fegley, B., and Prinn, R. G. 1984. Equilibrium and non-equilibrium chemistry of Saturn's deep atmosphere: Implications for the observability of PH_3, N_2, CO, and GeH_4. In preparation.

Feibelman, W. A. 1967. Concerning the "D" ring of Saturn. *Nature* 214:793–794.

Feldman, P. D., and Gentieu, E. P. 1982. The ultraviolet spectrum of an aurora 530–1520 Å. *J. Geophys. Res.* 87:2453–2458.

Fels, S. B., and Lindzen, R. S. 1974. The interaction of thermally excited gravity waves with mean flows. *Geophys. Fluid Dyn.* 6:149–191.

Ferrin, I. 1978. Azimuthal brightness variations of Saturn's ring A and size of particles. *Nature* 271:528–529.

Festou, M. C., and Atreya, S. K. 1982. Voyager ultraviolet stellar occultation measurements of the composition and thermal profiles of the Saturnian upper atmosphere. *Geophys. Res. Letters* 9:1147–1150.

Festou, M. C., Atreya, S. K., Donahue, T. M., Sandel, B. R., Shemansky, S. E., and Broadfoot, A. L. 1981. Composition and thermal profiles of the Jovian upper atmosphere determined by the Voyager ultraviolet stellar occultation experiment. *J. Geophys. Res.* 86:5715–5725.

Fillius, R. W., Mogro-Campero, A., and McIlwain, C. E. 1975. Radiation belts of Jupiter: A second look. *Science* 188:465–467.

Fillius, W., and McIlwain, C. E. 1980. Very energetic protons in Saturn's radiation belt. *J. Geophys. Res.* 85:5803–5811.

Fillius, W., Ip, W.-H., and McIlwain, C. E. 1980. Trapped radiation belts of Saturn: First look. *Science* 207:425–431.

Fink, U., and Larson, H. P. 1977. The 5 μm spectrum of Saturn. *Bull. Amer. Astron. Soc.* 9:535 (abstract).

Fink, U., and Larson, H. P. 1978. Deuterated methane observed on Saturn. *Science* 201:343–345.

Fink, U., and Larson, H. P. 1979. The infrared spectra of Uranus, Neptune, and Titan from 0.8–2.5 microns. *Astrophys. J.* 233:1021–1040.

Fink, U., Larson, H. P., Gautier, T. N. III, and Treffers, R. R. 1976. Infrared spectra of the satellites of Saturn: Identification of water ice on Iapetus, Rhea, Dione, and Tethys. *Astrophys. J.* 207:L63–L68.

Fink, U., Larson, H. P., and Treffers, R. R. 1978. Germane in the atmosphere of Jupiter. *Icarus* 34:344–354.

Fink, U., Larson, H. P., Bjoraker, G. L., and Johnson, J. R. 1983. The NH_3 spectrum in Saturn's 5 μm window. *Astrophys. J.* 268:880–888.

Firey, B., and Ashcroft, N. W. 1977. Thermodynamics of Thomas-Fermi screened Coulomb systems. *Phys. Rev.* 15A:2072–2075.

Fjeldbo, G. A. 1973. Radio occultation experiments planned for Pioneer and Mariner missions to the outer planets. *Planet. Space Sci.* 21:1533–1547.

Fjeldbo, G. A., Seidel, B., Sweetnam, D., and Woiceshyn, P. 1976. The Pioneer 11 radio occultation measurement of the Jovian ionosphere. In *Jupiter*, ed. T. Gehrels (Tucson: Univ. of Arizona Press), pp. 238–245.

Flammarion, C. 1901. Note on the rotation period of Saturn in 1896 and 1897. *Mon. Not. Roy. Astron. Soc.* 61:131–132.

Flasar, F. M. 1973. Gravitational energy sources in Jupiter. *Astrophys. J.* 186:1097–1106.

Flasar, F. M. 1983. Oceans on Titan. *Science* 221:55–57.

Flasar, F. M., Conrath, B. J., Pirraglia, J. A., Clark, P. A., French, R. G., and Gierasch, P. J. 1981*a*. Thermal structure and dynamics of the Jovian atmosphere. 1. The Great Red Spot. *J. Geophys. Res.* 86:8759–8768.

Flasar, F. M., Samuelson, R. E., and Conrath, B. J. 1981*b*. Titan's atmosphere: Temperature and dynamics. *Nature* 292:693–698.

Flierl, G. R., Larichev, V. D., McWilliams, J. C., and Reznik, G. N. 1980. The dynamics of baroclinic and barotropic solitary eddies. *Dyn. Atmos. Oceans* 5:1–41.

Flierl, G. R., Stern, M. E., and Whitehead, J. A. Jr. 1983. The physical significance of modons: Laboratory experiments and general integral constraints. *Dyn. Atmos. Oceans* 7:233–263.

Fontana, F. 1638. Florence, Biblioteca Nazionale, MSS. "Galileiana" 95, 81r.

Fontana, F. 1646. *Novae Coelestium Terrestriumque Rerum Observationes* (Naples).

Forsythe, W. E., and Powell, R. L. 1975. Thermal conductivity. In *American Institute of Physics Handbook*, ed. D. E. Gray (New York: McGraw-Hill), pp. 65–79.

Fountain, J. W., and Larson, S. M. 1977. A new satellite of Saturn? *Science* 197:915–917.

Fountain, J. W., and Larson, S. M. 1978. Saturn's ring and nearby faint satellites. *Icarus* 36:92–106.

Frank, L. A., Burek, B. G., Ackerson, K. L., Wolfe, J. H., and Mihalov, J. D. 1980. Plasmas in Saturn's magnetosphere. *J. Geophys. Res.* 85:5695–5708.

Franklin, F. A., and Cook, A. F. 1965. Optical properties of Saturn's rings. II. Two-color phase curves of the two bright rings. *Astron. J.* 70:704–720.

Franklin, F. A., and Cook, A. F. 1974. Photometry of Saturn's satellites: The opposition effect of Iapetus at maximum light and the variability of Titan. *Icarus* 23:355–362.

Franklin, F. A., and Colombo, G. 1978. On the azimuthal brightness variations of Saturn's rings. *Icarus* 33:279–287.

Franklin, F. A., Lecar, M., and Wiesel, W. 1984. Particle dynamics in resonances. In *Planetary Rings*, eds. R. Greenberg and A. Brahic (Tucson: Univ. of Arizona Press), pp. 562–588.

Franz, O. G., and Millis, R. L. 1975. Photometry of Dione, Tethys, and Enceladus on the UBV system. *Icarus* 24:433–442.

Franz, O. G., and Price, M. J. 1979. Saturn: UBV photoelectric pinhole scans of the disk. *Icarus* 37:272–281.

French, R. G., and Gierasch, P. J. 1974. Waves in the Jovian upper atmosphere. *J. Atmos. Sci.* 31:1707–1712.

Friedli, C., and Ashcroft, N. W. 1977. The band structure of molecular hydrogen at high pressure. *Phys. Rev.* 16B:662–674.

Friedson, A. J., and Stevenson, D. J. 1983. Viscosity of rock-ice mixtures and applications to the evolution of icy satellites. *Icarus* 56:1–14.

Friedson, A. J., and Yung, Y. L. 1984. The thermosphere of Titan. *J. Geophys. Res.* 89:85–90.

Froidevaux, L. 1981. Saturn rings: Infrared brightness variation with solar elevation. *Icarus* 46:4–17.

Froidevaux, L., and Ingersoll, A. P. 1980. Temperatures and optical depths of Saturn's rings and a brightness temperature of Titan. *J. Geophys. Res.* 85:5929–5936.

Froidevaux, L., Matthews, K., and Neugebauer, G. 1981. Thermal response of Saturn's ring particles during and after eclipse. *Icarus* 46:18–26.

Gaffney, E. S., and Matson, D. L. 1979. High pressure phases of ice in the outer solar system. *Bull. Amer. Astron. Soc.* 11:602 (abstract).

Gajsin, S. M. 1978. Brightness coefficients of the Saturn disk center. *Astron. Tsirkular USSR Acad. Sci.* 997:2–4 (Russian).

Gajsin, S. M. 1979. Brightness coefficients of Saturn. *Astron. Tsirkular USSR Acad. Sci.* 1048:7–8.

Galilei, G. 1890–1909. *Le Opere di Galileo Galilei*, Edizione Nazionale, ed. A. Favaro, 20 Vols. (Florence); reprinted 1929–1939, 1964–1966.

Gallagher, D. L., and Gurnett, D. A. 1979. Auroral kilometric radiation. Time-averaged source location. *J. Geophys. Res.* 84:6501–6509.

Gassendi, P. 1658. *Opera Omnia* 6 Vols. (Lyons), Vol. 4.

Gautier, D., and Grossman, K. 1972. A new method for the determination of the mixing ratio hydrogen to helium in the giant planets. *J. Atmos. Sci.* 29:788–792.

Gautier, D., and Owen, T. 1983a. Helium and deuterium abundances on Jupiter and Saturn: Cosmological implications. *Nature* 302:215–218.

Gautier, D., and Owen, T. 1983b. Cosmogonical implications of elemental and isotopic abundances in atmospheres of the giant planets. *Nature* 304:691–694.

Gautier, D., Lacombe, A., and Revah, I. 1977a. Saturn: Its thermal profile from infrared measurements. *Astron. Astrophys.* 61:149–153.

Gautier, D., Lacombe, A., and Revah, I. 1977b. The thermal structure of Jupiter from infrared spectral measurements by means of a filtered iterative inversion method. *J. Atmos. Sci.* 34:1130–1137.

Gautier, D., Conrath, B. J., Flasar, M., Hanel, R., Kunde, V., Chedin, A., and Scott, N. 1981. The helium abundance of Jupiter from Voyager. *J. Geophys. Res.* 86:8713–8720.

Gautier, D., Bezard, B., Marten, A., Baluteau, J. P., Scott, N., Chedin, A., Kunde, V., and Hanel, R. 1982. The C/H ratio in Jupiter from the Voyager infrared investigation. *Astrophys. J.* 257:901–912.

Gehrels, N., and Stone, E. C. 1983. Energetic oxygen and sulfur ions in the Jovian magnetosphere and their contribution to the auroral excitation. *J. Geophys. Res.* 88:5537–5550.

Gehrels, T. ed. 1976. *Jupiter* (Tucson: Univ. of Arizona Press).

Gehrels, T., Baker, L. R., Beshore, E., Blenman, C., Burke, J. J., Castillo, N. D., Da Costa, B., Degewij, J., Doose, L. R., Fountain, J. W., Gotobed, J., KenKnight, C. E., Kingston, R., McLaughlin, G., McMillan, R., Murphy, R., Smith, P. H., Stoll, C. P., Strickland, R. N., Tomasko, M. G., Wijesinghe, M. P., Coffeen, D. L., and Esposito, L. W. 1980. Imaging photopolarimeter on Pioneer Saturn. *Science* 207:434–439.

Genova, F., Pedersen, B. M., and Lecacheux, A. 1983. Dynamic spectra of Saturn kilometric radiation. *J. Geophys. Res.* 88:8985–8992.

Gierasch, P. J. 1975. Meridional circulation and the maintenance of the Venus atmospheric rotation. *J. Atmos. Sci.* 32:1038–1044.

Gierasch, P. J. 1976. Jovian meteorology: Large-scale moist convection. *Icarus* 29:445–454.

Gierasch, P. J. 1983. Dynamical consequences of orthohydrogen-parahydrogen disequilibrium on Jupiter and Saturn. *Science* 219:847–849.

Gierasch, P. J., and Goody, R. M. 1969. Radiative time constants in the atmosphere of Jupiter. *J. Atmos. Sci.* 26:979–980.

Gierasch, P. J., Goody, R. M., and Stone, P. 1970. The energy balance of planetary atmospheres. *Geophys. Fluid Dyn.* 1:1–18.

Gierasch, P. J., Ingersoll, A. P., and Williams, R. T. 1973. Radiative instability of a cloudy planetary atmosphere. *Icarus* 19:473–481.

Gierasch, P. J., Ingersoll, A. P., and Pollard, D. 1979. Baroclinic instabilities in Jupiter's zonal flow. *Icarus* 40:205–212.

Giese, R. H., Weiss, K., Zerull, R. H., and Ono, T. 1978. Large fluffy particles: A possible explanation of the optical properties of interplanetary dust. *Astron. Astrophys.* 65:265–272.

Gillett, F. C. 1975. Further observations of the 8–13 micron spectrum of Titan. *Astrophys. J.* 201:L41–L43.

Gillett, F. C., and Forrest, W. J. 1974. The 7.5 to 13.5 micron spectrum of Saturn. *Astrophys. J.* 187:L37–L38.

Gillett, F. C., Forrest, W. J., and Merrill, K. M. 1973. 8–13 micron observations of Titan. *Astrophys. J.* 184:L93–L95.

Gilman, P. A. 1977. Nonlinear dynamics of Boussinesq convection in a deep rotation spherical shell. *Geophys. Astrophys. Fluid Dyn.* 8:93–135.

Gilman, P. A. 1979. Model calculations concerning rotation at high solar latitudes and the depth of the solar convection zone. *Astrophys. J.* 231:284–292.

Giver, L. P., and Spinrad, H. 1966. Molecular hydrogen features in the spectra of Saturn and Uranus. *Icarus* 5:586–589.

Gladstone, G. R. 1982. Radiative Transfer and Photochemistry in the Upper Atmosphere of Jupiter. Ph.D. thesis, California Institute of Technology, Pasadena, CA.

Godfrey, D., Hunt, G. E., and Suomi, V. E. 1983. Some dynamical properties of vortex streets in Saturn's atmosphere from analysis of Voyager images. *Geophys. Res. Letters* 10:865–868.

Goertz, C. K. 1980. Io's interaction with the plasma torus. *J. Geophys. Res.* 85:2949–2956.

Goertz, C. K. 1983a. Detached plasma in Saturn's front side magnetosphere. *Geophys. Res. Letters* 10:455–458.

Goertz, C. K. 1983b. Limitation of electrostatic charging of dust particles in a plasma. Unpublished.

Goertz, C. K., and Morfill, G. E. 1983. A model for the formation of spokes in Saturn's ring. *Icarus* 53:219–229.

Goertz, C. K., Thomsen, M. F., and Ip, W.-H. 1981. Saturn's radio emission: Rotational modulation. *Nature* 292:737–739.

Goguen, J., Hammel, H., Cruikshank, D. P., and Hartmann, W. K. 1983a. The rotational light-curve of Hyperion during 1983. *Bull. Amer. Astron. Soc.* 15:854 (abstract).

Goguen, J., Morrison, D., and Tripicco, M. 1983b. Variation of the photometric function over the surface of Iapetus. Proceedings "Planetary Satellites Conference," Ithaca, NY, July 1983 (abstract).

Gold, T. 1980. Electric origin of spokes seen in Saturn's rings. Paper given at "Meeting on Planetary Exploration, R. Soc. London", London, England, Nov. 1980.

Goldreich, P. 1965. An explanation of the frequent occurrence of commensurable mean motion in the solar system. *Mon. Not. Roy. Astron. Soc.* 130:159–181.

Goldreich, P., and Soter, S. 1966. Q in the solar system. *Icarus* 5:375–389.

Goldreich, P., and Nicholson, P. D. 1977. Turbulent viscosity and Jupiter's tidal Q. *Icarus* 30:301–304.

Goldreich, P., and Tremaine, S. 1978. The formation of the Cassini division in Saturn's rings. *Icarus* 34:240–253.

Goldreich, P., and Tremaine, S. 1979. Towards a theory for the Uranian rings. *Nature* 277:97–99.

Goldreich, P., and Tremaine, S. 1980. Disk-satellite interactions. *Astrophys. J.* 241:425–441.

Goldreich, P., and Tremaine, S. 1982. The dynamics of planetary rings. *Ann. Rev. Astron. Astrophys.* 20:249–283.

Goldstein, M. L., and Thieman, J. R. 1981. The formation of arcs in the dynamic spectra of Jovian decameter bursts. *J. Geophys. Res.* 86:8569–8578.

Goldstein, R. M. 1982. Radar observations of the rings of Saturn. Program "Saturn Conference," Tucson, AZ, May 1982 (abstract).

Goldstein, R. M., and Morris, G. A. 1973. Radar observations of the rings of Saturn. *Icarus* 20:249–283.

Goldstein, R. M., Green, R. R., Pettengill, G. H., and Campbell, D. B. 1977. The rings of Saturn: Two-frequency radar observations. *Icarus* 30:104–110.

Golitsyn, G. S. 1978. Convection in the ice crusts of satellites and its effect on surface relief. *Geophys. Res. Letters* 6:466–468.

Govi, G. 1887. Della invenzione del micrometro per gli strumenti astronomici. *Bullettino di Bibliografia e di Storia delle Scienze Matematiche e Fisiche* 20:610–622.

Grabbe, C. L. 1981. Auroral kilometric radiation: A theoretical review. *Rev. Geophys. Space Sci.* 19:627–634.

Grabbe, C. L., Palmadesso, P., and Papadopoulos, K. 1980. A coherent nonlinear theory of auroral kilometric radiation. 1. Steady state model. *J. Geophys. Res.* 85:3337–3346.

Graboske, H. C. Jr., Pollack, J. B., Grossman, A. S., and Olness, R. J. 1975. The structure and evolution of Jupiter: The fluid contraction stage. *Astrophys. J.* 199:265–281.

Gradie, J., and Veverka, J. 1980. The composition of Trojan asteroids. *Nature* 283:840–842.

Gradie, J., and Tedesco, E. F. 1982. Compositional structure of the asteroid belt. *Science* 216:1405–1407.

Gradie, J., Thomas, P., and Veverka, J. 1980. The surface composition of Amalthea. *Icarus* 44:373–387.

Greenberg, R. 1973a. The inclination-types resonance of Mimas and Tethys. *Mon. Not. Roy. Astron. Soc.* 165:305–311.

Greenberg, R. 1973b. Evolution of satellite resonances by tidal dissipation. *Astron. J.* 78: 338–346.

Greenberg, R. 1977a. Orbit-orbit resonances among natural satellites. In *Planetary Satellites*, ed. J. A. Burns (Tucson: Univ. of Arizona Press), pp. 157–168.

Greenberg, R. 1977b. Orbit-orbit resonances in the solar system: Varieties and similarities. *Vistas in Astron.* 21:209–240.

Greenberg, R. 1982. Orbital evolution of the Galilean satellites. In *Satellites of Jupiter*, ed. D. Morrison (Tucson: Univ. of Arizona Press), pp. 65–92.

Greenberg, R. 1983. The role of dissipation in shepherding of ring particles. *Icarus* 53: 207–218.

Greenberg, R., and Scholl, H. 1979. Resonances in the asteroid belt. In *Asteroids*, ed. T. Gehrels (Tucson: Univ. of Arizona Press), pp. 310–333.

Greenberg, R., and Brahic, A. eds. 1984. Planetary rings: A dynamic evolving subject. In *Planetary Rings* (Tucson: Univ. of Arizona Press), pp. 3–11.

Greenberg, R., Davis, D. R., Hartmann, W. K., and Chapman, C. R. 1977. Size distribution of particles in planetary rings. *Icarus* 30:769–779.

Gregory, D. 1702. Royal Society of London, MSS. David Gregory, f. 63v.

Grigorjeva, Z. N., and Tejfel, V. G. 1979. Latitudinal variations of reflectivity and the limb darkening on Saturn's disk. *Trudy Astrophys. Inst. Acad. Sci. KazSSR* 35:98–103 (Russian).

Gross, S. H., and Rasool, S. I. 1964. The upper atmosphere of Jupiter. *Icarus* 3:311–312.

Grosskreutz, C. L. 1980. An estimate of the synchrotron radiation from Saturn. Univ. of Iowa Res. Rep. 80–37.

Grossman, A. S., Pollack, J. B., Reynolds, R. T., Summers, A. L., and Graboske, H. C. Jr. 1980. The effect of dense cores on the structure and evolution of Jupiter and Saturn. *Icarus* 42: 358–379.

Grün, E. 1982. Dynamics of charged particles. Proceedings "Planetary Rings/Anneaux des Planètes Conference," Toulouse, France, August 1982 (abstract).

Grün, E., and Reinhard, R. 1981. Ion-electron pair production by the impacts of cometary neutrals and dust particles on Halley probe shield. In *Cometary Halley Probe Environment Study* (ESA SP-155), pp. 7–21.

Grün, E., Morfill, G. E., Terrile, R. J., Johnson, T. V., and Schweln, G. J. 1983. The evolution of spokes in Saturn's B-Ring. *Icarus* 54:227–252.

Grün, E., Morfill, G. E., and Mendis, D. A. 1984. Dust-magnetosphere interactions. In *Planetary Rings*, eds. R. Greenberg and A. Brahic (Tucson: Univ. of Arizona Press), pp. 275–332.

Guèrin, P. 1970a. Sur la mise en évidence d'un quatrième anneau et d'une nouvelle division obscure dans le système des anneaux de Saturne. *Comptes Rend. Serie B. Sci. Phys.* 270:125–128.

Guèrin, P. 1979b. The new ring of Saturn. *Sky Teles.* 40:88.

Gulkis, S., and Poynter, R. 1972. Thermal radio emission from Jupiter and Saturn. *Phys. Earth Planet. Interiors* 6:36–43.

Gulkis, S., McDonough, T., and Craft, H. 1969. The microwave spectrum of Saturn. *Icarus* 10:421–427.

Gupta, S., Ochiai, E., and Ponnamperuma, C. 1981. Organic synthesis in the atmosphere of Titan. *Nature* 293:725–727.

Gurnett, D. A. 1974. The Earth as a radio source: Terrestrial kilometric radiation. *J. Geophys. Res.* 79:4227–4238.

Gurnett, D. A., and Frank, L. A. 1977. A region of intense plasma wave turbulence on auroral field lines. *J. Geophys. Res.* 82:1031–1050.

Gurnett, D. A., Frank, L. A., and Lepping, R. P. 1976. Plasma waves in the distant magnetotail. *J. Geophys. Res.* 81:6059–6071.

Gurnett, D. A., Kurth, W. S., and Scarf, F. L. 1981a. Plasma waves near Saturn: Initial results from Voyager 1. *Science* 212:235–239.

Gurnett, D. A., Kurth, W. S., and Scarf, F. L. 1981b. Narrowband electromagnetic emissions from Saturn's magnetosphere. *Nature* 292:733–737.

Gurnett, D. A., Kurth, W. S., and Scarf, F. L. 1982a. Narrowband electromagnetic emissions from Jupiter's magnetosphere. Submitted to *Nature*.

Gurnett, D. A., Scarf, F. L., and Kurth, W. S. 1982b. The structure of Titan's wake from plasma wave observations. *J. Geophys. Res.* 87:1395–1403.

Gurnett, D. A., Grün, E., Gallagher, D., Kurth, W. S., and Scarf, F. L. 1983. Micron-size particles detected near Saturn by the Voyager plasma wave instrument. *Icarus* 53:236–254.

Haas, M. R., Erickson, E. F., McKibbin, D. D., Goorvitch, D., and Caroff, L. J. 1982. Far-infrared spectroscopy of Saturn and its rings. *Icarus* 51:476–490.

Haerendel, G., Paschmann, G., Sckopke, N., Rosenbauer, H., and Hedgecock, P. C. 1978. The frontside boundary layer of the magnetosphere and the problem of reconnection. *J. Geophys. Res.* 83:3195–3216.

Haff, P. K., Eviatar, A., and Siscoe, G. L. 1983. Ring and plasma: The enigmae of Enceladus. *Icarus.* In press.

Hall, A. 1878. The period of Saturn's rotation. *Mon. Not. Roy. Astron. Soc.* 38:209–210.

Hall, A. 1884. The motion of Hyperion. *Mon. Not. Roy. Astron. Soc.* 44:361–365.

Hall, J. S., and Riley, L. A. 1969. Polarization measures of Jupiter and Saturn. *J. Atmos. Sci.* 26:920–923.

Hall, J. S., and Riley, L. A. 1974. A photometry study of Saturn and its rings. *Icarus* 23:144–156.

Hallinan, T. J. 1976. Auroral spirals. 2. Theory. *J. Geophys. Res.* 81:3959–3965.

Hamann, S. D., and Linton, M. 1966. Electrical conductivity of water in shock compression. *Trans. Faraday Soc. (GB)* 62:2234–2241.

Hämeen-Anttila, K. A. 1975. Statistical mechanics of Keplerian orbits. *Astrophys. Space Sci.* 37:309–333.

Hämeen-Anttila, K. A., and Vaaraniemi, P. 1975. A theoretical photometric function of Saturn's rings. *Icarus* 25:470–478.

Hamilton, D. C., Brown, D. C., Gloeckler, G., and Axford, W. I. 1983. Energetic atomic and molecular ions in Saturn's magnetosphere. *J. Geophys. Res.* 88:8905–8922.

Hanel, R. A., Conrath, B. J., Gautier, D., Gierasch, P., Kumar, S., Kunde, V. G., Lowman, P., Maguire, W., Pearl, J. C., Pirraglia, J. A., Ponnamperuma, C., and Samuelson, R. 1977. The Voyager infrared spectroscopy and radiometry investigation. *Space Sci. Rev.* 21:129–157.

Hanel, R. A., Conrath, B. J., Flasar, F. M., Kunde, V. G., Lowman, P., Maguire, W., Pearl, J. C., Pirraglia, J. A., Samuelson, R., Gautier, D., Gierasch, P., Kumar, S., and Ponnamperuma, C. 1979. Infrared observations of the Jovian system from Voyager 1. *Science* 204:972–976.

Hanel, R. A., Crosby, D., Herath, L., Vanous, D., Collins, D., Creswick, H., Harris, C., and Rhodes, M. 1980. Infrared spectrometer for Voyager. *Appl. Opt.* 19:1391–1400.

Hanel, R. A., Conrath, B. J., Flasar, F. M., Kunde, V. G., Maguire, W., Pearl, J. C., Pirraglia, J. A., Samuelson, R., Hearth, L., Allison, M., Cruikshank, D. P., Gautier, D., Gierasch, P., Horn, L., Koppany, R., and Ponnamperuma, C. 1981a. Infrared observations of the Saturnian system from Voyager 1. *Science* 212:192–200.

Hanel, R. A., Conrath, B. J., Herath, L., Kunde, V. G., and Pirraglia, J. A. 1981b. Albedo, internal heat and energy balance of Jupiter: Preliminary results of the Voyager infrared investigation. *J. Geophys. Res.* 86:8705–8712.

Hanel, R. A., Conrath, B. J., Flasar, F. M., Kunde, V. G., Maguire, W., Pearl, J. C., Pirraglia, J. A., Samuelson, R., Cruikshank, D. P., Gautier, D., Gierasch, P., Horn, L., and Ponnamperuma, C. 1982. Infrared observations of the Saturnian system from Voyager 2. *Science* 215:544–548.

Hanel, R. A., Conrath, B. J., Kunde, V. G., Pearl, J. C., and Pirraglia, J. A. 1983. Albedo, internal heat flux, and energy balance of Saturn. *Icarus* 53:262–285.

Hansen, J. E. 1971. Multiple scattering of polarized light in planetary atmospheres. Part II. Sunlight reflected by terrestrial water clouds. *J. Atmos. Sci.* 28:1400–1426.

Hansen, J. E., and Travis, L. D. 1974. Light scattering in planetary atmospheres. *Space Sci. Rev.* 16:527–610.

Hansen, J.-P., and Vieillefosse, P. 1976. Equation of state of the classical two-component plasma. *Phys. Rev. Letters* 37:391–393.

Harrington, R. S., and Seidelman, P. K. 1981. Dynamics of the Saturnian satellite 1980S1 and 1980S3. *Icarus* 47:97–99.

Harris, A. W. 1975. Collisional breakup of particles in a planetary ring. *Icarus* 24:190–192.

Harris, A. W. 1978. Satellite formation. II. *Icarus* 34:128–145.

Harris, D. L. 1961. Photometry and colorimetry of planets and satellites. In *Planets and Satellites*, eds. G. P. Kuiper and B. M. Middlehurst (Chicago: Univ. of Chicago Press), pp. 272–342.

Harris, E. G. 1959. Unstable plasma oscillations in a magnetic field. *Phys. Rev. Letters* 2:34.

Hartle, R. E., Ogilvie, K. W., and Wu, C. S. 1973. Neutral and ion-exospheres in the solar wind with applications to Mercury. *Planet. Space Sci.* 21:2181–2191.

Hartle, R. E., Sittler, E. C. Jr., Ogilvie, K. W., Scudder, J. D., Lazarus, A. J., and Atreya, S. K. 1982. Titan's ion exosphere observed from Voyager 1. *J. Geophys. Res.* 87:1383–1394.

Hartman, H., Lewis, J. S., Prinn, R. G., and Fegley, M. B. Jr. 1983. Origin of the carbonaceous chondrites. Preprint.

Hartmann, W. K. 1980. Surface evolution of two-component stone/ice bodies in the Jupiter region. *Icarus* 44:441–453.

Hathaway, D. H., Gilman, P. A., and Toomre, J. 1979. Convective instability when the temperature gradient and rotation vector are oblique to gravity. I. Fluids without diffusion. *Geophys. Astrophys. Fluid Dyn.* 13:289–316.

Hatzes, A., Wenkert, D. D., Ingersoll, A. P., and Danielson, G. E. 1981. Oscillations and velocity structure of a long-lived cyclonic spot. *J. Geophys. Res.* 86:8745–8749.

Hays, P. B., and Roble, R. G. 1973a. Observations of mesospheric ozone at low latitudes. *Planet. Space Sci.* 21:273–279.

Hays, P. B., and Roble, R. G. 1973b. Stellar occultation measurements of molecular oxygen in the lower thermosphere. *Planet. Space Sci.* 21:339–348.

Heaps, M. G. 1976. The role of particle precipitation and joule heating in the energy balance of the Jovian thermosphere. *Icarus* 29:273–281.

Hénon, M. 1969. A comment on "The Resonant Structure of the Solar System" by A. M. Molchanov. *Icarus* 11:93–94.

Hénon, M. 1981. A simple model of Saturn's rings. *Nature* 293:33–35.

Henry, R. J. W., and McElroy, M. B. 1969. The absorption of extreme ultraviolet solar radiation by Jupiter's upper atmosphere. *J. Atmos. Sci.* 26:912–917.

Heppenheimer, T. A. 1975. On the presumed capture origin of Jupiter's outer satellites. *Icarus* 24:172–180.

Heppenheimer, T. A., and Porco, C. 1977. New contributions to the problem of capture. *Icarus* 30:385–401.

Herschel, J. 1847. *Results of Astronomical Observations made during the Years 1834, 5, 6, 7, 8 at the Cape of Good Hope* (London).

Herschel, J. 1859. *Outlines of Astronomy* (London), 6th ed.

Herschel, W. 1790a. An account of the discovery of a sixth and seventh satellite of the planet Saturn; With remarks on the construction of its ring, its atmosphere, its rotation on an axis, and its spheroidal figure. *Phil. Trans.* 80:1–20.

Herschel, W. 1790b. On the satellites of the planet Saturn, and the rotation of its ring on an axis. *Phil. Trans.* 80:427–495.

Herschel, W. 1792. On the ring of Saturn, and the rotation of the fifth satellite upon its axis. *Phil. Trans.* 82:1–22.

Herschel, W. 1794. On the rotation of the planet Saturn on its axis. *Phil. Trans.* 84:48–66.

Herschel, W. 1805. Observations of the singular figure of the planet Saturn. *Phil. Trans.* 95:272–280.

Hess, S. L. 1953. Variation in atmospheric absorption over the disks of Jupiter and Saturn. *Astrophys. J.* 118:151–160.

Hess, S. L., and Panofsky, H. A. 1951. The atmospheres of the other planets. In *Compendium of Meteorology*, ed. T. F. Malone (Boston: Amer. Meteorological Soc.), pp. 391–398.

Hess, W. N. 1968. *The Radiation Belt and Magnetosphere* (Waltham, MA: Blaisdell Publ. Co.).

Hess, W. N., and Mead, G. D. eds. 1968. *Introduction to Space Science* (New York: Gordon & Breach, Science Publ.).

Hevelius, J. 1647. *Selenographia* (Gdansk).

Hevelius, J. 1656. *Dissertatio de Nativa Saturni Facie* (Gdansk).

Hevelius, J. 1659. Letter of Hevelius to Ismael Boulliau, 9 December 1659. Paris, Bibliothèque Nationale, MSS. FF. 13043, 90r.

Hide, R. 1967. On the dynamics of Jupiter's interior and the origin of its magnetic field. In *Magnetism and the Cosmos*, eds. W. Hindmarsh, F. Lomes, P. Roberts, and S. Runcorn (New York: American Elsevier Press).

Hide, R. 1978. How to locate the electrically-conducting fluid core of a planet from external magnetic observations. *Nature* 271:640–641.

Hide, R. 1981. Self-exciting dynamos and geomagnetic polarity changes. *Nature* 293:728–729.

Hide, R., and Mason, P. J. 1970. Baroclinic waves in a rotating fluid subject to internal heating. *Phil. Trans. Roy. Soc. London* A268:201–232.

Hildenbrand, D. L., and Giauque, W. F. 1953. Ammonium oxide and ammonium hydroxide: Heat capacities and thermodynamic properties from 15 to 300 K. *J. Phys. Chem.* 75:2811–2818.

Hill, J. R., and Mendis, D. A. 1979. Charged dust in outer planetary magnetospheres. 1. Physical processes. *Moon Planets* 21:3–15.

Hill, J. R., and Mendis, D. A. 1980. Charged dust in outer planetary magnetospheres. II. Trajectories and spatial distribution. *Moon Planets* 23:53–71.

Hill, J. R., and Mendis, D. A. 1981. On the braids and spokes in Saturn's ring system. *Moon Planets* 24:431–436.

Hill, J. R., and Mendis, D. A. 1982a. The isolated non-circular ringlets of Saturn. *Moon Planets* 26:217–226.

Hill, J. R., and Mendis, D. A. 1982b. The dynamical evolution of the Saturnian ring spokes. *J. Geophys. Res.* 87:7413–7420.

Hill, J. R., and Mendis, D. A. 1982c. On the dust ring current of Saturn's F ring. *Geophys. Res. Letters* 9:1069–1071.

Hill, J. R., and Mendis, D. A. 1982d. The origin of the E-ring of Saturn. Paper read at "AGU Fall Meeting", San Francisco, CA, Dec. 1982, *EOS* 63:1019 (abstract).

Hill, T. W. 1980. Corotation lag in Jupiter's magnetosphere: A comparison of observation and theory. *Science* 207:301–302.

Hill, T. W., Dessler, A. J., and Michel, F. C. 1974. Configuration of the Jovian magnetosphere. *Geophys. Res. Letters* 1:3–6.

Hind, J. R. 1852. *The Solar System* (New York: Putnam).

Hinson, D. P., and Tyler, G. L. 1983a. Spatial irregularities in Titan's atmosphere observed by Voyager radio occultation. Submitted to *Icarus*.

Hinson, D. P., and Tyler, G. L. 1983b. Internal gravity waves in Titan's atmosphere observed by Voyager radio occultation. *Icarus* 54:337–352.

Hiza, M. J., Kidney, A. J., and Miller, R. C. 1975. *Equilibrium Properties of Fluid Mixtures*, NSRDS Bibliographic Series (New York: Plenum).

Hoffman, K. A. 1977. Polarity transition records and the geomagnetic dynamo. *Science* 196:1329–1332.

Holberg, J. B. 1982. Identifications of 1980S27 and 1980S26 resonances in Saturn's A ring. *Astron. J.* 87:1416–1422.

Holberg, J. B., and Forrester, W. T. 1982. Voyager UVS observations of stellar occultations by the rings of Saturn. Proceedings "Planetary Rings/Anneaux des Planètes Conference," Toulouse, France, August 1982 (abstract).

Holberg, J. B., Forrester, W. T., and Lissauer, J. J. 1982a. Identification of resonance features within the rings of Saturn. *Nature* 297:115–120.

Holberg, J. B., Forrester, W. J., Shemansky, D. E., and Barry, D. 1982b. Voyager absolute far-ultraviolet spectrophotometry of hot stars. *Astrophys. J.* 257:656–671.

Holmes, A. W. 1981. Light Scattering from Ammonia and Water Crystals. Unpublished Ph.D. thesis, Univ. of Arizona, Tucson.

Holmes, A. W., Paxman, R., Stahl, P. H., and Tomasko, M. G. 1980. Light scattering by crystals of NH_3 and H_2O. *Bull. Amer. Astron. Soc.* 12:705 (abstract).

Holton, J. R. 1975. *The Dynamic Meteorology of the Stratosphere and Mesosphere* (Boston: Amer. Met. Soc.).

Holton, J. R. 1979. *An Introduction to Dynamic Meteorology* 2nd Edition (New York: Academic Press).

Hood, L. L. 1981. A comparison of characteristic times for satellite absorption of energetic protons trapped in the Jovian and Saturnian magnetic fields. *Geophys. Res. Letters* 8:976–979.

Hood, L. L. 1983. Radial diffusion in Saturn's radiation belts: A modeling analysis assuming satellite and Ring E absorption. *J. Geophys. Res.* 88:808–818.

Hooke, R. 1666. An observation about Saturn, June 29, 1665. *Phil. Trans.* 1:246–247, and fig. R.

Horak, H. G. 1950. Diffuse reflection by planetary atmospheres. *Astrophys. J.* 112:445–463.

Hornung, K., and Drapatz, S. 1981. Residual ionization after impact of large dust particles. In *Cometary Halley Probe Plasma Environment Study* (ESA SP-155), pp. 23–37.

Houpis, H. L. F., and Mendis, D. A. 1983. On the fine structure of the Saturnian ring system. *Moon Planets* 29:39–46.

Howland, G. R., Harteck, P., and Reeves, R. R. 1979. The role of phosphorus in the upper atmosphere of Jupiter. *Icarus* 37:301–306.

Hoyt, W. G. 1976. *Lowell and Mars* (Tucscon: Univ. of Arizona Press).

Huaux, A. 1951. Sur un modèle de satellite en glace. *Bull. Acad. Roy. Sci. Belg.* 37:534–539.

Hubbard, R. F., and Birmingham, T. J. 1978. Electrostatic emissions between electron gyroharmonics in the outer magnetosphere. *J. Geophys. Res.* 83:4837–4850.

Hubbard, R. F., Birmingham, T. J., and Hones, E. W. Jr. 1979. Magnetospheric electrostatic emission and cold plasma densities. *J. Geophys. Res.* 84:5828–5838.

Hubbard, W. B. 1974. Tides in the giant planets. *Icarus* 23:42–50.

Hubbard, W. B. 1977. The Jovian surface condition and cooling rate. *Icarus* 30:305–310.

Hubbard, W. B. 1981. Constraints on the origin and interior structure of the major planets. *Phil. Trans. Roy. Soc. Lond.* A303:315–326.

Hubbard, W. B. 1982. Effects of differential rotation on the gravitational figures of Jupiter and Saturn. *Icarus* 52:509–515.

Hubbard, W. B., and MacFarlane, J. J. 1980a. Theoretical predictions of deuterium abundances in the Jovian planets. *Icarus* 44:676–682.

Hubbard, W. B., and MacFarlane, J. J. 1980b. Structure and evolution of Uranus and Neptune. *J. Geophys. Res.* 85:225–234.

Hubbard, W. B., and Horedt, G. P. 1983. Computation of Jupiter interior models from gravitational inversion theory. *Icarus* 54:456–465.

Hubbard, W. B., Trubitsyn, V. P., and Zharkov, V. N. 1974. Significance of gravitational moments for interior structure of Jupiter and Saturn. *Icarus* 21:147–151.

Hubbard, W. B., MacFarlane, J. J., Anderson, J. D., Null, G. W., and Biller, E. D. 1980. Interior structure of Saturn inferred from Pioneer 11 gravity data. *J. Geophys. Res.* 85:5909–5916.

Hudson, H. S., Lindsey, C. A., and Soifer, B. T. 1974. Submillimeter observations of planets. *Icarus* 23:374–379.

Huffman, D. R., and Bohren, C. F. 1980. Infrared absorption spectra of non-spherical particles treated in the Rayleigh-ellipsoid approximation. In *Light Scattering by Irregularly Shaped Particles*, ed. D. Schuerman (New York: Plenum Press), pp. 103–112.

Humes, D. H., O'Neil, R. L., Kinard, W. H., and Alvarez, J. M. 1980. Impact of Saturn ring particles on Pioneer 11. *Science* 207:443–444.

Hunt, G. E. 1975. Spectroscopy of Jupiter and Saturn. In *Atmospheres of the Earth and Planets* (Dordrecht: Reidel), pp. 425–432.

Hunt, G. E., Godfrey, D., Muller, J.-P., and Barrey, R. F. T. 1982a. Dynamical features in the northern hemisphere of Saturn from Voyager 1 images. *Nature* 297:132–134.

Hunt, G. E., Muller, J.-P., and Gee, P. 1982b. Measurements of convective growth rates and equatorial features in the Jovian atmosphere using Voyager images. *Nature* 295:491–494.

Hunt, J. L., Poll, J. D., Goorvitch, D., and Tipping, R. H. 1983. Collision-induced absorption in the far-infrared spectrum of Titan. *Icarus* 55:63–72.

Hunten, D. M. 1969. The upper atmosphere of Jupiter. *J. Atmos. Sci.* 26:826–834.

Hunten, D. M. 1972. The atmosphere of Titan. *Comm. Astrophys. Space Phys.* 4:149–154.

Hunten, D. M. 1973. The escape of H_2 from Titan. *J. Atmos. Sci.* 30:726–732.

Hunten, D. M. ed. 1974a. *The Atmosphere of Titan* (Washington: NASA SP-340).

Hunten, D. M. 1974b. Titan's atmosphere and surface. In *Chemical Evolution of the Giant Planets*, ed. C. Ponnamperuma (New York: Academic Press), pp. 27–47.

Hunten, D. M. 1975. Vertical transport in atmospheres. In *Atmospheres of Earth and Planets*, ed. B. M. McCormac (Dordrecht: D. Reidel), pp. 59–72.

Hunten, D. M. 1977. Titan's atmosphere and surface. In *Planetary Satellites*, ed. J. A. Burns (Tucson: Univ. of Arizona Press), pp. 420–437.

Hunten, D. M. 1978. A Titan atmosphere with a surface temperature of 200 K. In *The Saturn System*, eds. D. M. Hunten and D. Morrison (Washington: NASA Conf. Publ. 2068), pp. 127–140.

Hunten, D. M., and Veverka, J. 1976. Stellar and spacecraft occultations by Jupiter. 1. In *Jupiter*, ed. T. Gehrels (Tucson: Univ. of Arizona Press), pp. 247–283.

Hunten, D. M., and Dessler, A. J. 1977. Soft electrons as a possible heat source for Jupiter's thermosphere. *Planet. Space Sci.* 25:817–821.

Hunten, D. M., and Morrison, D. eds. 1978. *The Saturn System* (Washington: NASA Conf. Publ. 2068).

Huntress, W. T. Jr. 1974. A review of Jovian ionospheric chemistry. In *Advances in Atomic and Molecular Physics*, eds. D. R. Bates and B. Beaderson (New York: Academic Press), pp. 295–340.

Huntress, W. T., McEwan, M. J., Karpas, Z., and Aniaich, V. G. 1980. Laboratory studies of some of the major ion-molecule reactions occurring in cometary comae. *Astrophys. J. Suppl.* 44:481–488.

Huygens, C. 1888–1950. *Oeuvres Complètes de Christiaan Huygens*, 22 Vols. (The Hague: Martinus Nijhoff).

Ibragimov, N. B. 1974. Investigation of Jupiter and Saturn with the 2-m reflector bands on the Saturn disk. *Astron. Zh.* 51:178–186 (Russian).

Ibragimov, N. B., and Avramchuk, V. V. 1971. Investigation of methane absorption bands on the Saturn disk. *Tsirkular. Shemahka Astrophys. Obs.* 12:10–15 (Russian).

Ingersoll, A. P. 1973. Jupiter's great red spot: A free atmospheric vortex? *Science* 182: 1346–1348.

Ingersoll, A. P. 1976. Pioneer 10 and 11 observations and the dynamics of Jupiter's atmosphere. *Icarus* 29:245–253.

Ingersoll, A. P. 1981. Jupiter and Saturn. In *The New Solar System*, eds. J. K. Beatty, B. O'Leary, and A. Chaikin (Cambridge: Cambridge Univ. Press), pp. 117–128.

Ingersoll, A. P., and Cuzzi, J. N. 1969. Dynamics of Jupiter's cloud bands. *J. Atmos. Sci.* 26:981–985.

Ingersoll, A. P., and Porco, C. C. 1978. Solar heating and internal heat flow on Jupiter. *Icarus* 35:27–43.

Ingersoll, A. P., and Cuong, P. G. 1981. Numerical model of long-lived Jovian vortices. *J. Atmos. Sci.* 38:2067–2076.

Ingersoll, A. P., and Pollard, D. 1982. Motion in the interiors and atmospheres of Jupiter and Saturn: Scale analysis, anelastic equations, barotropic stability criterion. *Icarus* 52:62–80.

Ingersoll, A. P., Orton, G. S., Münch, G., Neugebauer, G., and Chase, S. C. 1980. Pioneer Saturn infrared radiometer: Preliminary results. *Science* 207:439–443.

Ingersoll, A. P., Beebe, R. F., Mitchell, J. L., Garneau, G. W., Yagi, G. M., and Muller, J.-P. 1981. Interactions of eddies and mean zonal flow on Jupiter as inferred from Voyager 1 and 2 images. *J. Geophys. Res.* 86:8733–8743.

International Astronomical Union 1980*a*. IAU Circular No. 3457.

International Astronomical Union 1980*b*. IAU Circular No. 3466.

International Astronomical Union 1980*c*. IAU Circular No. 3496.

International Astronomical Union 1980*d*. IAU Circular No. 3651.

International Astronomical Union 1980*e*. IAU Circular No. 3656.

Ioannidis, G. A., and Brice, N. M. 1971. Plasma densities in the Jovian magnetosphere: Plasma slingshot or Maxwell demon? *Icarus* 14:360–373.

Ip, W.-H. 1980. Physical studies of planetary rings. *Space Sci. Rev.* 26:39–96.

Ip, W.-H. 1982. On planetary rings as sources and sinks of magnetospheric plasma. Proceedings "Planetary Rings/Anneaux des Planètes Conference," Toulouse, France, August 1982 (abstract).

Ip, W.-H. 1983*a*. On the equatorial confinement of thermal plasma generated in the vicinity of the rings of Saturn. Submitted to *J. Geophys. Res.*

Ip, W.-H. 1983*b*. On plasma transport in the vicinity of the rings of Saturn: A siphon flow mechanism. *J. Geophys. Res.* 88:819–822.

Ip, W.-H., and Mendis, D. A. 1983. On the equatorial transport of Saturn's ionosphere as driven by a dust-ring current system. *Geophys. Res. Letters* 10:207–209.

Irvine, W. M. 1966. The shadowing effect in diffuse reflection. *J. Geophys. Res.* 71:2931–2937.

Irvine, W. M. 1975. Multiple scattering in planetary atmospheres. *Icarus* 25:175–204.

Irvine, W. M., and Pollack, J. B. 1969. Infrared optical properties of water and ice spheres. *Icarus* 8:324–360.

Irvine, W. M., and Lane, A. L. 1971. Monochromatic albedos for the disk of Saturn. *Icarus* 15:18–26.

Jacob, W. S. 1853. Remarks on Saturn as seen with the Madras Equatorial. *Mon. Not. Roy. Astron. Soc.* 13:240–241.

Jaffe, W. 1977. 6 cm radio observation of Saturn. Preprint.

Jaki, S. L. 1976. The five forms of Laplace's cosmogony. *Amer. J. Phys.* 44:4–11.

James, H. G. 1980. Direction-of-arrival measurements of auroral kilometric radiation and associated ELF data from Isis I. *J. Geophys. Res.* 85:3367–3375.

Janssen, M. A., and Olsen, E. T. 1978. A measurement of the brightness temperature of Saturn's rings at 8-mm wavelength. *Icarus* 33:263–278.

Jefferys, W. H., and Ries, L. M. 1975. Theory of Enceladus and Dione. *Astron. J.* 80:876–884.

Jefferys, W. H., and Ries, L. M. 1979. Theory of Mimas and Tethys. *Astron. J.* 84:1778–1782.

Jeffreys, H. 1947. The effects of collisions on Saturn's rings. *Mon. Not. Roy. Astron. Soc.* 107:263–267.

Jeffreys, H. 1953. On the masses of Saturn's satellites. *Mon. Not. Roy. Astron. Soc.* 113:81–96.

Johnsen, R., and Biondi, M. A. 1974. Measurements of positive ion conversion and removal reactions relating to the Jovian ionosphere. *Icarus* 23:139–142.

Johnson, H. L. 1970. The infrared spectra of Jupiter and Saturn at 1.2–4.2 microns. *Astrophys. J.* 159:L1–L5.

Johnson, P. E., Kemp, J. C., King, R., Parker, T. E., and Barbour, M. S. 1980. New results from optical polarimetry of Saturn's rings. *Nature* 283:146–149.

Johnson, R. 1979. Ph.D. thesis, Texas A & M Univ., College Station, TX.

Johnson, T. V., and Pilcher, C. B. 1977. Satellite spectrophotometry and surface compositions. In *Planetary Satellites*, ed. J. A. Burns (Tucson: Univ. of Arizona Press), pp. 232–268.

Johnson, T. V., Veeder, G. J., and Matson, D. L. 1975. Evidence for frost on Rhea's surface. *Icarus* 24:428–432.

Jones, D. E. 1976. Source of terrestrial nonthermal radiation. *Nature* 260:686–688.

Jones, D. E., Tsututani, B. T., Smith, E. J., Walker, R. J., and Sonett, C. P. 1980. A possible magnetic wake of Titan: Pioneer 11 observations. *J. Geophys. Res.* 85:5835–5840.

Jones, M. C. 1970. Far infrared absorption in liquefied gases. *Nat. Bur. Stand. Tech. Note* 390.

Jortner, J., Kestner, N. R., Rice, S. A., and Cohen, M. H. 1965. Study of the properties of an excess electron in liquid helium. *J. Chem. Phys.* 43:2614–2625.

Journal of Geophysical Research 87, 1982. Special Issue, 1 March.

Judge, D. L., Wu, F. M., and Carlson, R. W. 1980. Ultraviolet photometer observations of the Saturnian system. *Science* 207:431–434.

Kaiser, M. L. 1977. A low-frequency radio survey of the planets with RAE 2. *J. Geophys. Res.* 82:1256–1260.

Kaiser, M. L., and Stone, R. G. 1975. Earth as an intense planetary radio source: Similarities to Jupiter and Saturn. *Science* 189:285–287.

Kaiser, M. L., and Desch, M. D. 1982. Saturnian kilometric radiation: Source locations. *J. Geophys. Res.* 87:4555–4559.

Kaiser, M. L., Desch, M. D., Warwick, J. W., and Pearce, J. B. 1980. Voyager detection of nonthermal radio emission from Saturn. *Science* 209:1238–1240.

Kaiser, M. L., Desch, M. D., and Lecacheux, A. 1981. Saturnian kilometric radiation: Statistical properties and beam geometry. *Nature* 292:731–733.

Kaiser, M. L., Connerney, J. E. P., and Desch, M. D. 1983. Atmospheric storm explanation of Saturn's electrostatic discharges. *Nature* 303:50–53.

Kalnay de Rivas, E. 1975. Further numerical calculations of the atmosphere of Venus. *J. Atmos. Sci.* 32:1017–1024.

Kamb, B. 1973. Crystallography of ice. In *Physics and Chemistry of Ice*, eds. E. Whalley, S. J. Jones, and L. W. Gold (Ottawa: Univ. of Toronto Press).

Kargel, J. S. 1983. Enceladus: An analog of terrestrial plate tectonism? *Lunar Planet. Sci.* 14:363–364 (abstract).

Karpus, A., Anicich, V. G., and Huntress, W. T. Jr. 1979. An ion cyclotron resonance study of reactions of ions with hydrogen atoms. *J. Chem. Phys.* 70:2877–2881.

Kater, H. 1830. On the division in the exterior ring of Saturn. *Mon. Not. Roy. Astron. Soc.* 1:177–179.

Kater, H. 1831. On the appearance of divisions in the exterior ring of Saturn. *Memoirs Roy. Astron. Soc.* 4:383–390.

Kaula, W. M. 1980. The beginning of the Earth's thermal evolution. In *The Continental Crust and Its Mineral Deposits*, ed. D. W. Strangeway (Geol. Assoc. of Canada Spec.), pp. 25–34.

Kawata, Y. 1978. Circular polarization of sunlight reflected by planetary atmospheres. *Icarus* 33:217–232.

Kawata, Y. 1979. The infrared brightness temperature of Saturn's ring based on the multi-particle layer assumption. In *Proceedings of the 12th Lunar and Planetary Symposium* (Tokyo: Univ. of Tokyo, Inst. of Space & Aeronautical Science).

Kawata, Y 1982. Thermal energy balance of Saturn's rings. Proceedings "Fifteenth ISAS Lunar & Planetary Symposium," Tokyo, Japan, July 1982 (abstract).

Kawata, Y., and Irvine, W. M. 1974. Models of Saturn's rings which satisfy the optical observations. In *Exploration of the Planetary System*, eds. A. Woszczyk and C. Iwaniszewska (Dordrecht: D. Reidel), pp. 441–464.

Kawata, Y., and Irvine, W. M. 1975. Thermal emission from a multiply scattering model of Saturn's rings. *Icarus* 24:472–482.

Keeler, J. E. 1888. First observations with the 36-inch Equatorial of the Lick Observatory. *Sidereus Nuncius* 7:79–83.

Keeler, J. E. 1889. The outer ring of Saturn. *Astron. J.* 8:175.

Keeler, J. E. 1895*a*. A spectroscopic proof on the meteoric constitution of Saturn's rings. *Astron. J.* 1:416–427.

Keeler, J. E. 1895*b*. Note on the spectroscopic proof of the meteoric constitution of Saturn's ring. *Astron. J.* 2:163–164.

Kemp, J. C., and Murphy, R. E. 1973. The linear polarization and transparency of Saturn's rings. *Astrophys. J.* 189:679–686.

Kemp, J. C., Rudy, R. J., Lebofsky, M. J., and Reike, G. H. 1978. Near-infrared polarization studies of Saturn and Jupiter. *Icarus* 35:263–271.

Kennel, C. F., and Coroniti, F. V. 1978. Jupiter's magnetosphere and radiation belts. In *Solar System Plasma Physics—A Twentieth Anniversary Review*, eds. C. F. Kennel, L. J. Lanzerotti, and E. N. Parker (Amsterdam: North-Holland Publ. Co.), pp. 105–181.

Kennel, C. F., Scarf, F. L., Fredricks, R. W., McGehee, J. H., and Coroniti, F. V. 1970. VLF electric field observations in the magnetosphere. *J. Geophys. Res.* 75:6136–6152.

Kerker, M. 1969. *The Scattering of Light and other Electromagnetic Radiation* (New York: Academic Press).

Kerridge, J. F., and Bunch, T. E. 1979. Aqueous activity on asteroids. In *Asteroids*, ed. T. Gehrels (Tucson: Univ. of Arizona Press), pp. 745–764.

Khare, B. N., and Sagan, C. 1973. Red clouds in reducing atmospheres. *Icarus* 20:311–321.

Khare, B. N., Sagan, C., Bandurski, E. L., and Nagy, B. 1978. Ultraviolet-photoproduced organic solids synthesized under simulated Jovian conditions: Molecular analysis. *Science* 199:1199–1201.

Khare, B. N., Sagan, C., Zumberge, J. E., Sklarew, D. S., and Nagy, B. 1981. Organic solids produced by electrical discharges in reducing atmospheres: Tholin molecular analysis. *Icarus* 48:290–297.

Khare, B. N., Sagan, C., Shrader, S., and Arakawa, E.T. 1982. Molecular analysis of tholins produced under simulated Titan conditions. *Bull. Amer. Astron. Soc.* 14:714 (abstract).

Khare, B. N., Sagan, C., Arakawa, E. T., Suite, F., Callicott, T. A., and Williams, M. W. 1983. Optical constants of Titan tholin aerosols. *Bull. Amer. Astron. Soc.* 15:842–843 (abstract).

Kharitonova, G. A. 1976. On longitudinal and temporal variations of the methane absorption in the atmosphere of Saturn. In *The Study of the Major Planets Atmospheres* (Alma-Ata), pp. 75–87 (Russian).

Kharitonova, G. A., and Tejfel, V. G. 1973. The spectral reflectivity of Saturn's rings and disk at 0.4–0.8 μm. *Astron. Tsirkular USSR Acad. Sci.* 747:1–3 (Russian).

Kieffer, H. H. 1967. Calculated physical properties of planets in relations to composition and gravitational layering. *J. Geophys. Res.* 72:3179–3197.

Kirk, R. L., and Stevenson, D. J. 1983. Hydromagnetic implications of the zonal winds of the giant planets. In preparation.

Kirkwood, D. 1866. On the theory of meteors. *Proc. A.A.A.S.* 15:8–14.

Kirkwood, D. 1872. On the formation and primitive structure of the solar system. *Proc. Amer. Phil. Soc.* 12:163–167.

Kivelson, M. G., and Russel, C. T. 1983. The interaction of flowing plasmas with planetary ionospheres: A Titan-Venus comparison. *J. Geophys. Res.* 88:49–57.

Klein, M. J., Janssen, M. A., Gulkis, S., and Olsen, E. T. 1978. Saturn's microwave spectrum: Implications for the atmosphere and the rings. In *The Saturn System*, eds. D. M. Hunten and D. Morrison (Washington: NASA Conf. Publ. 2068), pp. 195–216.

Kliore, A. J., Cain, D. L., Fjeldbo, G., and Seidel, B. L. 1974. Preliminary results on the atmospheres of Io and Jupiter from the Pioneer 10 S-band occultation experiment. *Science* 183:323–324.

Kliore, A. J., Lindal, G. F., Patel, I. R., Sweetnam, D. N., and Hotz, H. B. 1980a. Vertical structure of the ionosphere and upper neutral atmosphere of Saturn from the Pioneer radio occultation. *Science* 207:446–449.

Kliore, A. J., Patel, I. R., Lindal, G. F., Sweetnam, D. N., Hotz, H. B., Waite, J. H. Jr., and McDonough, T. R. 1980b. Structure of the ionosphere and atmosphere of Saturn from Pioneer 11, Saturn radio occultation. *J. Geophys. Res.* 85:5857–5870.

Kozai, Y. 1957. On the astronomical constants of Saturnian satellite system. *Ann. Tokyo Astron. Obs.* 5:73–106.

Kozai, Y. 1976. Masses of satellites and oblateness parameters of Saturn. *Publ. Astron. Soc. Japan* 28:675–691.

Krimigis, S. M., and Armstrong, T. P. 1982. Two-component proton spectra in the inner Saturnian magnetosphere. *Geophys. Res. Letters* 9:1143–1146.

Krimigis, S. M., Armstrong, T. P., Axford, W. I., Bostrom, C. O., Gloeckler, G., Keath, E. P., Lanzerotti, L. J., Carbary, J. F., Hamilton, D. C., and Roelof, E. C. 1981. Low energy charged particles in Saturn's magnetosphere: Results from Voyager 1. *Science* 212:225–231.

Krimigis, S. M., Armstrong, T. P., Axford, W. I., Bostrom, C. O., Gloeckler, G., Keath, E. P., Lanzerotti, L. J., Carbary, J. F., Hamilton, D. C., and Roelof, E. C. 1982a. Low energy hot plasma and particles in Saturn's magnetosphere. *Science* 215:571–577.

Krimigis, S. M., Carbary, J. F., Keath, E. P., and Armstrong, T. P. 1982b. Energetic electron spectra in Saturn's magnetosphere. *EOS* 63:1068 (abstract).

Krimigis, S. M., Carbary, J. F., Keath, E. P., Armstrong, T. P., Lanzerotti, L. J., and Gloeckler, G. 1983. General characteristics of hot plasma and energetic particles in the Saturnian magnetosphere: Results from the Voyager spacecraft. *J. Geophys. Res.* 88:8871–8892.

Krugov, V. D. 1972. Photometric characteristics of Jupiter and Saturn in the region of 0.48–0.33 μm. *Astron. Vestn.* 6:85–90 (Russian).

Kuiper, G. P. 1944. Titan: A satellite with an atmosphere. *Astrophys. J.* 100:378–383.

Kuiper, G. P. 1947. Infrared spectra of planets. *Astrophys. J.* 106:251–254.

Kuiper, G. P. 1951. On the origin of the irregular satellites. *Proc. Nat. Acad. Sci.* 37:717–721.

Kuiper, G. P. ed. 1952. Planetary atmospheres and their origin. In *The Atmospheres of the Earth and Planets* (Chicago: Univ. of Chicago Press), pp. 306–405.

Kuiper, G. P. 1955. Commission pour les observations physiques des planètes et des satellites. *Trans. IAU* 9:250–260.

Kuiper, G. P., Cruikshank, D. P., and Fink, U. 1970. The composition of Saturn's rings. *Sky Teles.* 39:14.

Kunde, V. G., Aikin, A. C., Hanel, R. A., Jennings, D. E., Maguire, W. C., and Samuelson, R. E. 1981. C_4H_2, HC_3N and C_2N_2 in Titan's atmosphere. *Nature* 292:686–688.

Kunde, V. G., Hanel, R. A., Maguire, W. C., Gautier, D., Baluteau, J. P., Marten, A., Chedin, A., Husson, N., and Scott, N. 1982. The tropospheric gas composition of Jupiter's North Equatorial Belt (NH_3, PH_3, CH_3D, GeH_4, H_2O) and the Jovian D/H isotopic ratio. *Astrophys. J.* 263:443–467.

Kuo, H. L. 1963. Perturbations of plane Couette flow in stratified fluid and origin of cloud streets. *Phys. Fluids* 6:195–211.

Kuratov, K. S. 1979. Preliminary results of photoelectric measurements of Saturn at 0.6–1.1 μm. *Trudy Astrophys. Inst. Acad. Sci. KazSSR* 35:93–97 (Russian).

Kurth, W. S., Craven, J. D., Frank, L. A., and Gurnett, D. A. 1979. Intense electrostatic waves near the upper hybrid resonance frequency. *J. Geophys. Res.* 84:4145–4154.

Kurth, W. S., Frank, L. A., Ashour-Abdalla, M., Gurnett, D. A., and Burek, B. G. 1980. Observations of a free-energy source for intense electrostatic waves. *Geophys Res. Letters* 7:293–296.

Kurth, W. S., Gurnett, D. A., and Anderson, R. R. 1981a. Escaping nonthermal continuum radiation. *J. Geophys. Res.* 86:5519–5531.

Kurth, W. S., Gurnett, D. A., and Scarf, F. L. 1981b. The control of Saturn's kilometric radio emission by Dione. *Nature* 292:742–745.

Kurth, W. S., Scarf, F. L., Sullivan, J. D., and Gurnett, D. A. 1982a. Detection of nonthermal continuum radiation in Saturn's magnetosphere. *Geophys. Res. Letters* 9:889–892.

Kurth, W. S., Sullivan, J. D., Gurnett, D. A., Scarf, F. L., Bridge, H. S., and Sittler, E. C. 1982b. Observations of Jupiter's distant magnetotail and wake. *J. Geophys. Res.* 87:10373–10383.

Kurth, W. S., Scarf, F. L., Gurnett, D. A., and Barbosa, D. D. 1983. A survey of electrostatic waves in Saturn's magnetosphere. *J. Geophys. Res.* 88:8959–8970.

Kvenvolden, K. A., and McManamin, M. A. 1980. Hydrates of natural gas: A review of their geologic occurrence. *U.S. Geol. Survey Circ.* 825.

Lam, C. S., and Varshni, Y. P. 1983. Ionization energy of the helium atom in a plasma. *Phys. Rev.* 27A:418–421.

Lambert, D. 1978. The abundances of the elements in the solar photosphere. VIII. Revised abundances of carbon, nitrogen, and oxygen. *Mon. Not. Roy. Astron. Soc.* 82:249–272.

·Landau, L., and Lifshitz, E. M. 1969. *Statistical Physics* (Menlo Park, CA: Addison-Wesley).

Landolt-Bernstein 1978. *Numerical Data and Functional Relations in Science and Technology*, Group III, Vol. 7c1 (Berlin: Springer Verlag).

Lane, A. L., Nelson, R. N., and Matson, D. L. 1981. Evidence for sulphur implantation in Europa's UV absorption band. *Nature* 292:38–39.

Lane, A. L., Graps, A. L., and Simmons, K. E. 1982a. The C ring of Saturn: A high resolution view of some of its structure. Proceedings "Planetary Rings/Anneaux des Planètes Conference," Toulouse, France, August 1982 (abstract).

Lane, A. L., Hord, C. W., West, R. A., Esposito, L. W., Coffeen, D. L., Sato, M., Simmons, K. E., Pomphrey, R. B., and Morris, R. B. 1982b. Photopolarimetry from Voyager 2: Preliminary results on Saturn, Titan, and the rings. *Science* 215:537–543.

Lanzerotti, L. J., Maclennan, C. G., Krimigis, S. M., Armstrong, T. P., Behannon, K., and Ness, N. F. 1980. Statics of the nightside Jovian plasma sheet. *Geophys. Res. Letters* 7:817–820.

Lanzerotti, L. J., Maclennan, C. G., Brown, W. L., Johnson, R. E., Barton, L. A., Reimann, C. T., Garrett, J. W., and Boring, J. W. 1983a. Implications of Voyager data for energetic ion erosion of icy satellites of Saturn. *J. Geophys. Res.* 88:8765–8770.

Lanzerotti, L. J., Maclennan, C. G., Lepping, R. P., and Krimigis, S. M. 1983b. On the plasma condition at the dayside magnetopause of Saturn. *Geophys. Res. Letters* 10:1200–1202.

Laplace, P. S. de 1878–1912. *Oeuvres Complètes de Laplace* (Paris), 14 Vols.

Larimer, J. W., and Anders, E. 1967. Chemical fractionation in meteorites. I. Condensation of the elements. *Geochim. Cosmochim. Acta* 31:1215–1238.

Larson, H. P. 1980. Infrared spectroscopic observations of the outer planets, their satellites, and the asteroids. *Ann. Rev. Astron. Astrophys.* 18:43–75.

Larson, H. P., Fink, U., Smith, H. A., and Davis, D. S. 1980. The middle-infrared spectrum of Saturn: Evidence for phosphine and upper limits to other trace atmospheric constituents. *Astrophys. J.* 240:327–337.

Larson, S. J. 1982. Observations of Saturn's E ring. Proceedings "Planetary Rings/Anneaux des Planètes Conference," Toulouse, France, August 1982 (abstract).

Larson, S. M. 1979. Observations of the Saturn D ring. *Icarus* 37:399–403.

Larson, S. M., Fountain, J. W., Smith, B. A., and Reitsema, H. J. 1981a. Observations of the Saturn E ring and a new satellite. *Icarus* 47:288–290.

Larson, S. M., Smith, B. A., Fountain, J. W., and Reitsema, H. J. 1981b. The 1966 observations of the coorbiting satellites of Saturn S10 and S11. *Icarus* 46:175–180.

Lassell, W. 1848a. Observations of Mimas, the closest and most interior satellite of Saturn. *Mon. Not. Roy. Astron. Soc.* 8:42–43.

Lassell, W. 1848b. Discovery of a new satellite of Saturn. *Mon. Not. Roy. Astron. Soc.* 8:195–197.

Lassell, W. 1852. Extract of a letter from Mr. Lassell. *Mon. Not. Roy. Astron. Soc.* 13:11–13.

Lazarus, A. J., and McNutt, R. L. Jr. 1983. Low energy plasma ion observations in Saturn's magnetosphere. *J. Geophys. Res.* 88:8831–8846.

Lazarus, A. J., Hasegawa, T., and Bagenal, F. 1983. Long-lived particulate or gaseous structure in Saturn's outer magnetosphere? *Nature* 302:230–232.

Lebofsky, L. A. 1975. Stability of frosts in the solar system. *Icarus* 25:205–217.

Lebofsky, L. A., and Fegley, M. B., Jr. 1976. Chemical composition of icy satellites and Saturn's rings. *Icarus* 28:379–388.

Lebofsky, L. A., Johnson, T. V., and McCord, T. B. 1970. Saturn's rings: Spectral reflectivity and compositional implications. *Icarus* 13:226–230.

Lebofsky, L. A., Feierberg, M. A., and Tokunaga, T. 1982. Infrared observations of the dark side of Iapetus. *Icarus* 49:382–386.

Lecacheux, A., and Genova, F. 1983. Source localization of Saturn kilometric radio emission. *J. Geophys. Res.* 88:8993–8998.

Lecacheux, A., de Bergh, C., Combes, M., and Millard, J. 1976. The C/H and $^{12}CH_4/^{13}CH_4$ ratios in the atmospheres of Jupiter and Saturn from 0.1 cm^{-1} resolution near-infrared spectra. *Astron. Astrophys.* 53:29–33.

Lecacheux, A., Meyer-Vernet, N., and Daigne, G. 1981. Jupiter's decametric radio emission: A nice problem of optics. *Astron. Astrophys.* 94:L9–L12.

Lee, L. C., Kan, J. R., and Wu, C. S. 1980. Generation of AKR and the structure of auroral acceleration region. *Planet. Space Sci.* 28:703–711.

Lembege, B., and Jones, D. 1982. Propagation of electrostatic upper hybrid emission and Z mode waves at the geomagnetic equatorial plasmapause. *J. Geophys. Res.* 87:6187–6201.

Le Monnier, P. C. 1741. *Histoire Céleste* (Paris).

Leovy, C. B. 1969. Bulk transfer coefficient for heat transfer. *J. Geophys. Res.* 74:3313–3321.

Leovy, C. B. 1973. Rotation in the upper atmosphere of Venus. *J. Atmos. Sci.* 30:1218–1220.

Leovy, C. B., and Mintz, Y. 1969. Numerical simulation of the atmospheric circulation and climate of Mars. *J. Atmos. Sci.* 26:1167–1190.

Leovy, C. B., and Pollack, J. B. 1973. A first look at atmospheric dynamics and temperature variations on Titan. *Icarus* 19:195–201.

Lepping, R. P., and Burlaga, L. F. 1979. Geomagnetopause surface fluctuations observed by Voyager 1. *J. Geophys. Res.* 84:7099–7106.

Lepping, R. P., Burlaga, L. F., and Klein, L. W. 1981. Surface waves on Saturn's magnetopause. *Nature* 292:750–753.

Lepping, R. P., Burlaga, L. F., Desch, M. D., and Klein, L. W. 1982. Evidence for a distant (>8700 R_J) Jovian magnetotail: Voyager 2 observations. *Geophys. Res. Letters* 9:885–888.

Leu, M. T., Biondi, M. A., and Johnsen, R. 1973. Dissociative recombination of electrons with H_3^+ and H_5^+ ions. *Phys. Rev.* 8:413–419.

Lewis, J. S. 1969. Observability of spectroscopically active compounds in the atmosphere of Jupiter. *Icarus* 10:393–409.

Lewis, J. S. 1971. Satellites of the outer planets: Their physical and chemical nature. *Icarus* 15:174–185.

Lewis, J. S. 1972a. The chemistry of solar system material. *Earth Plant. Sci. Letters* 15:286–290.

Lewis, J. S. 1972b. Low temperature condensation from the solar nebula. *Icarus* 16:241–252.

Lewis, J. S. 1974. The temperature gradient in the solar nebula. *Science* 186:440–442.

Lewis, J. S. 1980. Electrical discharge synthesis of organic compounds on Jupiter. *Icarus* 43:85–95.

Lewis, J. S., and Prinn, R. G. 1970. Jupiter's clouds: Structure and composition. *Science* 169:472–474.

Lewis, J. S., and Prinn, R. G. 1980. Kinetic inhibition of CO and N_2 reduction in the solar nebula. *Astrophys. J.* 238:357–364.

Limaye, S. S., Revercomb, H. E., Sromovsky, L. A., Krauss, R. J., Santek, D. A., and Suomi, V. E. 1982. Jovian winds from Voyager 2. Part I. Zonal mean circulation. *J. Atmos. Sci.* 39:1413–1432.

Lin, C. C., and Shu, F. H. 1968. Density wave theory of spiral structure. In *Astrophysics and General Relativity*, Vol. 2, eds. M. Chretian, S. Deser, and J. Goldstein (New York: Gordon & Breach), pp. 236–329.

Lin, D. N. C. 1981. Convective accretion disk model for the primitive solar nebula. *Astrophys. J.* 246:972–984.

Lin, D. N. C., and Papaloizou, J. 1980. On the structure and evolution of the primordial solar nebula. *Mon. Not. Roy. Astron. Soc.* 191:37–48.

Lin, D. N. C., and Bodenheimer, P. 1981. On the stability of Saturn's rings. *Astrophys. J.* 248:L83–L86.

Lindal, G. F., Wood, G. E., Levy, G. S., Anderson, J. D., Sweetnam, D. N., Hotz, H. B., Buckles, B. J., Homes, D. P., and Doms, P. E. 1981. The atmosphere of Jupiter: An analysis of the Voyager radio occultation measurements. *J. Geophys. Res.* 86:8721–8727.

Lindal, G. F., Wood, G. E., Hotz, H. B., and Sweetnam, D. N. 1983. The atmosphere of Titan: An analysis of the Voyager 1 radio occultation measurements. *Icarus* 53:348–363.

Lindzen, R. S. 1971. Tides and gravity waves in the upper atmosphere. In *Mesospheric Models and Related Experiments*, ed. G. Fiocco (Boston: D. Reidel), pp. 122–130.

Lippincott, E. R., Eck, R. V., Dayhoff, M. O., and Sagan, C. 1967. Thermodynamic equilibria in planetary atmosphere. *Astrophys. J.* 165:753–764.

Lipps, F. B. 1971. Two-dimensional numerical experiments in thermal convection with vertical shear. *J. Atmos. Sci.* 28:3–19.

Lissauer, J. J. 1982. Dynamics of Saturn's Rings. Ph.D. thesis, University of California, Berkeley.

Lissauer, J. J., and Cuzzi, J. N. 1982. Resonances in Saturn's rings. *Astron. J.* 87:1051–1058.

Lissauer, J. J., Shu, F. H., and Cuzzi, J. N. 1981. Moonlets in Saturn's rings? *Nature* 292:707–711.

Lissauer, J. J., Peale, S. J., Cuzzi, J. N., and Squyres, S. W. 1982*a*. Torque on the co-orbital satellites due to Saturn's rings and the melting of Enceladus. *EOS* 63:1022 (abstract).

Lissauer, J. J., Shu, F. H., and Cuzzi, J. N. 1982*b*. Viscosity in Saturn's rings. Proceedings "Planetary Rings/Anneaux des Planètes Conference," Toulouse, France, August 1982 (abstract).

Lissauer, J. J., Peale, S. J., and Cuzzi, J. N. 1984. Ring torque on Janus and the melting of Enceladus. *Icarus*. In press.

Liu, L.-G. 1982. Compression of ice VII to 500 kbar. *Earth Planet. Sci. Letters* 61:359–364.

Loewenstein, R. F., Harper, D. A., Moseley, S. H., Telesco, C. M., Thronson, H. A., Jr., Hildebrand, R. H., Whitcomb, S. E., Winston, R., and Stiening, R. F. 1977. Far-infrared and submillimeter observations of the planets. *Icarus* 31:315–324.

Low, F. J., and Rieke, G. H. 1974. Infrared photometry of Titan. *Astrophys. J.* 190:L143–L145.

Ludlum, F. H. 1980. *Clouds and Storms* (University Park: Penn State Univ.).

Luhmann, J. G., Elphic, R. C., Russell, C. T., Slavin, J. A., and Mihalov, J. D. 1981. Observations of large scale steady magnetic fields in the Venus nightside ionosphere and near wake. *Geophys. Res. Letters* 8:517–520.

Lumme, K. A. 1969. *Ann. Acad. Sci. Fenn. AVI* 324.

Lumme, K. A., and Irvine, W. M. 1976*a*. Photometry of Saturn's rings. *Astron. J.* 81:863–893.

Lumme, K. A., and Irvine, W. M. 1976*b*. Azimuthal brightness variations of Saturn's rings. *Astrophys. J.* 204:L55–L57.

Lumme, K. A., and Reitsema, H. J. 1978. Five-color photometry of Saturn and its rings. *Icarus* 33:288–300.

Lumme, K. A., and Irvine, W. M. 1979. A model for the azimuthal brightness variations in Saturn's rings. *Nature* 282:695–696.

Lumme, K. A., and Irvine, W. M. 1982. Azimuthal variations in Saturn's A ring. Proceedings "Planetary Rings/Anneaux des Planètes Conference," Toulouse, France, August 1982 (abstract).

Lumme, K. A., Esposito, L. W., Irvine, W. M., and Baum, W. A. 1977. Azimuthal brightness variations of Saturn's rings. II. Observations at an intermediate tilt angle. *Astrophys. J.* 216:L123–L126.

Lumme, K. A., Irvine, W. M., Martin, L. J., and Baum, W. A. 1979. Azimuthal brightness variations of Saturns rings: Observations at B ~ 11°.5. *Astrophys. J.* 229:L109–L111.

Lumme, K. A., Irvine, W. M., and Esposito, L. W. 1983. Theoretical interpretation of the ground-based photometry of Saturn's B ring. *Icarus* 53:174–184.

Lunine, J. I., and Stevenson, D. J. 1982*a*. Post-accretional evolution of Titan's surface and atmosphere. *Astrophys. J.* 238:357–364.

Lunine, J. I., and Stevenson, D. J. 1982*b*. Formation of the Galilean satellites in a gaseous nebula. *Icarus* 52:14–39.

Lunine, J. I., Stevenson, D. J., and Yung, Y. L. 1983. Ethane ocean on Titan. Submitted to *Science*.

Lupo, M. J., and Lewis, J. S. 1979. Mass-radius relationships in icy satellites. *Icarus* 40:157–170.

Luthey, J. L. 1973. Possibility of Saturnian synchrotron radiation. *Icarus* 20:125–135.

Lutz, B. L., Owen, T., and Cess, R. D. 1976. Laboratory band strengths of methane and their application to the atmospheres of Jupiter, Saturn, Uranus, Neptune, and Titan. *Astrophys. J.* 203:541–551.

Lutz, B. L., Owen, T., and Cess, R. D. 1982. Laboratory band strengths of methane and their application to the atmospheres of Jupiter, Saturn, Uranus, Neptune, and Titan. *Astrophys. J.* 258:886–898.

Lutz, B. L., de Bergh, C., and Owen, T. 1983. Titan: Discovery of carbon monoxide in its atmosphere. *Science* 220:1374–1375.

Lynden-Bell, D., and Pringle, J. E. 1974. The evolution of viscous disks and the origin of the nebular variables. *Mon. Not. Roy. Astron. Soc.* 168:603–637.

Lyons, L. R. 1974. Electron diffusion driven by magnetospheric electrostatic waves. *J. Geophys. Res.* 79:575–580.

Lyons, L. R., Thorne, R. M., and Kennel, C. F. 1972. Pitch angle diffusion of radiation belt electrons within the plasmasphere. *J. Geophys. Res.* 77:3455–3474.

Lyot, B. 1929. Recherches sur la Polarisation de la Lumière des Planètes et de quelques Substances Terrestres. Thèse, Ann. Obs. Meudon; English transl. NASA TTF-187.

Lyot, B. 1944. Planetary and solar observations on the Pic du Midi in 1941, 1942, and 1943. *Astrophys. J.* 101:255–259.

Lyzenga, G. A., Ahrens, T. J., Nellis, W. J., and Mitchell, A. C. 1982. The temperature of shock-compressed water. *J. Chem. Phys.* 76:6282–6286.

MacFarlane, J. J., and Hubbard, W. B. 1983. Statistical mechanics of light elements at high pressure. V. Three-dimensional Thomas-Fermi-Dirac theory. *Astrophys. J.* 272:301–310.

Maclennan, C. G., Lanzerotti, L. J., Krimigis, S. M., Lepping, R. P., and Ness, N. F. 1982. Effects of Titan on trapped particles in Saturn's magnetosphere. *J. Geophys. Res.* 87:1411–1418.

Maclennan, C. G., Lanzerotti, L. J., Krimigis, S. M., and Lepping, R. P. 1983. Low energy particles at the bow shock, magnetopause, and outer magnetosphere of Saturn. *J. Geophys. Res.* 88:8817–8830.

Macy, W. 1976. An analysis of Saturn's methane $3\nu_3$ band profiles in terms of an inhomogeneous atmosphere. *Icarus* 29:49–56.

Macy, W. 1977. Inhomogeneous models of the atmosphere of Saturn. *Icarus* 32:328–347.

Macy, W. Jr., and Smith, W. H. 1978. Detection of HD on Saturn and Uranus and the D/H ratio. *Astrophys. J.* 222:L137–L140.

Maguire, W. C., Hanel, R. A., Jennings, D. E., Kunde, V. G., and Samuelson, R. E. 1981. Propane and methyl acetylene in Titan's atmosphere. *Nature* 292:683–686.

Maier, H. N., and Fessenden, R. W. 1975. Electron-ion recombination rate constants for some compounds of moderate complexity. *J. Chem. Phys.* 162:4790–4794.

Main, R. 1853. Note on the form of the planet Saturn. *Mon. Not. Roy. Astron. Soc.* 13:152–156.

Makogon, Yu. F. 1974. *Hydrates of Natural Gas* (Moscow: Nedra). In Russian.

Mao, H. K., and Bell, P. M. 1979. Observations of hydrogen at room temperature (25°C) and high pressure (to 500 kbars). *Science* 203:1004–1006.

Marouf, E. A., and Tyler, G. L. 1982. Microwave edge diffraction in Saturn's rings: Observations with Voyager 1. *Science* 217:243–245.

Marouf, E. A., Tyler, G. L., and Eshleman, V. R. 1982. Theory of radio occultation by Saturn's rings. *Icarus* 49:161–193.

Marouf, E. A., Tyler, G. L., Zebker, H. A., Simpson, R. A., and Eshleman, V. R. 1983. Particle size distributions in Saturn's rings from Voyager 1 radio occultation. *Icarus* 54:189–211.

Marrero, T. R., and Mason, E. A. 1972. Gaseous diffusion coefficients. *J. Phys. Chem. Ref. Data* 1:3.

Marten, A., Courtin, R., Gautier, D., and Lacombe, A. 1980. Ammonia vertical density profiles in Jupiter and Saturn from their radioelectric and infrared emissivities. *Icarus* 41:410–422.

Marten, A., Rouan, D., Baluteau, J.-P., Gautier, D., Conrath, B. J., Hanel, R. A., Kunde, V. G., Samuelson, R., Chedin, A., and Scott, N. 1981. Study of the ammonia ice cloud layer in the equatorial region of Jupiter from the infrared interferometric experiment on Voyager. *Icarus* 46:233–248.

Martin, M. 1968. Photométrie photographique de Saturne. *J. Obs.* 51:179–191.

Martin, T. Z. 1975. Saturn and Jupiter: A Study of Atmospheric Constituents. Ph.D. thesis, Univ. of Hawaii.

Martin, T. Z., Cruikshank, D. P., Pilcher, C. B., and Sinton, W. M. 1976. Pressure-induced absorption by H_2 in the atmospheres of Jupiter and Saturn. *Icarus* 27:391–406.

Massie, S. T., and Hunten, D. M. 1982. Conversion of para and ortho hydrogen in the giant planets. *Icarus* 49:213–226.

Maxwell, J. C. 1859. *On the Stability of Saturn's Rings* (London); reprinted in *The Scientific Papers of James Clerke Maxwell* 2 Vols., 1890 (Cambridge: Univ. Cambridge Press), pp. 288–374.

Maxworthy, T., and Redekopp, L. G. 1976. A solitary wave theory of the Great Red Spot and other observed features in the Jovian atmosphere. *Icarus* 29:261–271.

Maxworthy, T., Redekopp, L. G., and Weidman, P. D. 1978. On the production and interaction of planetary solitary waves: Application to the Jovian atmosphere. *Icarus* 33:388–409.

McConnell, J. C., Sandel, B. R., and Broadfoot, A. L. 1981. Voyager UV spectrometer observations of the He 584Å dayglow at Jupiter. *Planet. Space Sci.* 29:283–292.

McConnell, J. C., Holberg, J. B., Smith, G. R., Sandel, B. R., Shemansky, D. E., and Broadfoot, A. L. 1982a. A new look at the ionosphere of Jupiter in light of the UVS occultation results. *Planet. Space Sci.* 30:151–167.

McConnell, J. C., Holberg, J. B., Smith, G. R., Sandel, B. R., Shemansky, D. E., and Broadfoot, A. L. 1982b. A new look at the ionosphere of Saturn in light of the UVS occultation results. Program "Saturn Conference," Tucson, AZ, May 1982 (abstract).

McCord, T. B., Johnson, T. V., and Elias, J. H. 1971. Saturn and its satellites: Narrowband spectrophotometry (0.3–1.1 μm). *Astrophys. J.* 165:413–424.

McDonald, F. B., Schardt, A. W., and Trainor, J. H. 1980. If you've seen one magnetosphere you haven't seen them all: Energetic particle observations in the Saturn magnetoshere. *J. Geophys. Res.* 85:5813–5830.

McElroy, M. B. 1973. The ionospheres of the major planets. *Space Sci. Rev.* 14:460–473.

McEwan, M. J., Anicich, V. G., and Huntress, W. T. 1981. An ICR investigation of ion-molecule reactions of HCN. *Intl. J. M. Spect. Ion Phys.* 37:273.

McFadden, L. A., Bell, J., and McCord, T. B. 1981. Visible spectral reflectance measurements 0.3–1.1 μm of the Galilean satellite at many orbital phase angles 1977–1978. *Icarus* 44:410–430.

McIlwain, C. E. 1961. Coordinates for mapping the distribution of magnetically trapped particles. *J. Geophys. Res.* 66:3681–3691.

McKinnon, W. B. 1981. Tectonic deformation of Galileo Regio and limits to the planetary expansion of Ganymede. *Proc. XII Lunar Planet. Sci. Conf.* 12:1585–1597.

McKinnon, W. B. 1982. Cratering the icy satellites of Saturn: Effects myriad and sundry. Program "Saturn Conference," Tucson, AZ, May 1982 (abstract).

McKinnon, W. B. 1983. Origin of the E Ring: Condensation of impact vapor or boiling of impact melt? *Lunar Planet. Sci.* 14:478–488 (abstract).

McLaughlin, W. I., and Talbot, T. D. 1977. On the mass of Saturn's rings. *Mon. Not. Roy. Astron. Soc.* 179:619–633.

McNutt, R. L. Jr. 1983a. Vector velocities in Saturn's magnetosphere. Proceedings "Fifth Conf. on the Physics of the Jovian and Saturnian Magnetospheres," Cambridge, MA (abstract).

McNutt, R. L. Jr. 1983b. Force balance in the magnetospheres of Jupiter and Saturn. *Adv. Space Res.* 3:55.

McNutt, R. L. Jr., Belcher, J. W., Sullivan, J. D., Bagenal, F., and Bridge, H. S. 1979. Departure from rigid co-rotation of plasma in Jupiter's dayside magnetosphere. *Nature* 280:803.

McNutt, R. L. Jr., Belcher, J. W., and Bridge, H. S. 1981. Positive ion observations in the middle magnetosphere of Jupiter. *J. Geophys. Res.* 86:8319–8342.

McSween, H. Y. 1979. Are carbonaceous chondrites primitive or processed? A review. *Rev. Geophys. Space Phys.* 17:1059–1078.

McWilliams, J. C., and Zabusky, N. J. 1982. Interactions of isolated vortices. *Geophys. Astrophys. Fluid Dyn.* 19:207–277.

Meinschein, W. G., Nagy, B., and Hennessy, J. D. 1963. Evidence in meteorites of former life: The organic compounds in carbonaceous chondrites are similar to those found in marine sediments. *Ann. N. Y. Acad. Sci.* 108:559–579.

Mellor, J. W. 1983. *Comprehensive Treatise on Inorganic and Theoretical Chemistry* (New York: J. Wiley & Sons).

Melnick, G., Russell, R. W., Gosnell, T. R., and Harwit, M. 1983. Spectrophotometry of Saturn and its rings from 60 to 180 microns. *Icarus* 53:310–318.

Melrose, D. B. 1967. Rotational effects on the distribution of thermal plasma in the magnetosphere of Jupiter. *Planet. Space Sci.* 15:381–393.

Melrose, D. B. 1973. Coherent gyromagnetic emission as a radiation mechanism. *Aust. J. Phys.* 26:229–247.

Melrose, D. B. 1976. An interpretation of Jupiter's decametric radiation and the terrestrial kilometric radiation as direct amplified gyroemission. *Astrophys. J.* 207:651–662.

Melrose, D. B. 1981. A theory for the nonthermal radio continua in the terrestrial and Jovian magnetosphere. *J. Geophys. Res.* 86:30–36.

Mendis, D. A. 1981. The role of electrostatic charging of small and intermediate sized bodies in the solar system. In *Investigating the Universe*, ed. F. D. Kahn (Dordrecht: D. Reidel), pp. 353–384.

Mendis, D. A., and Axford, W. I. 1974. Satellites and magnetospheres of the outer planets. *Ann. Rev. Earth Planet. Sci.* 2:419–474.

Mendis, D. A., Hill, J. R., and Houpis, H. L. F. 1982. The gravito-electrodynamics of charged dust in planetary magnetospheres. *J. Geophys. Res.* 87:3449–3455.

Mendis, D. A., Hill, J. R., and Houpis, H. L. F. 1983. Charged dust in Saturn's magnetosphere. *J. Geophys. Res.* 88:A929–A942. Special Saturn Issue.

Miller, S. L. 1955. A production of some organic compounds under possible primitive Earth conditions. *J. Amer. Chem. Soc.* 77:2351–2361.

Miller, S. L. 1961. The occurrence of gas hydrates in the solar system. *Proc. Nat. Acad. Sci. U.S.* 47:1798–1808.

Miller, S. L. 1973. The clathrate hydrates: Their nature and occurrence. In *Physics and Chemistry of Ice*, eds. E. Whalley, S. J. Jones, and L. W. Gold (Ottawa: Roy. Soc. Canada), pp. 42–50.

Miller, S. L., and Orgel, L. 1974. *The Origins of Life on the Earth* (Englewood Cliffs, NJ: Prentice Hall).

Miller, T. M., Moseley, J. T., Martin, D. W., and McDaniel, E. W. 1968. Reactions of H^+ in H_2 and D^+ in D_2: Mobilities of hydrogen and alkali ions in H_2 and D_2 gases. *Phys. Rev.* 173:115–123.

Millis, R. L. 1973. UBV photometry of Iapetus. *Icarus* 18:247–252.

Millis, R. L. 1977. UBV photometry of Iapetus: Results from five apparitions. *Icarus* 31:81–88.

Mishima, O., King, D. D., and Whalley, E. 1983. The far infrared spectrum of ice in the range $8-25$ cm^{-1}. Sound waves and difference bands, with applications to Saturn's rings. *J. Chem. Phys.* 78:6394–6404.

Mitchell, A. K., and Nellis, W. J. 1982. Equation of state and electrical conductivity of water and ammonia shocked to the 100 GPa (1 Mbar) pressure range. *J. Chem. Phys.* 76:6273–6281.

Mitchell, J. L., Beebe, R. F., Ingersoll, A. P., and Garneau, G. W. 1981. Flow fields within Jupiter's Great Red Spot and Whilt Oval BC. *J. Geophys. Res.* 86:8751–8757.

Miura, A., and Pritchett, P. L. 1982. Nonlocal stability analysis of the MHD Kelvin-Helmholtz instability in a compressible plasma. *J. Geophys. Res.* 87:7431–7444.

Mizuno, H. 1980. Formation of the giant planets. *Progr. Theor. Phys.* 64:544–557.

Mizutani, H., Mikuni, H., Takashi, M., and Noda, H. 1975. Study on the photochemical reaction of HCN and its polymer products relating the primary chemical evolution. *Origins of Life* 6:513.

Moffatt, H. K. 1978. *Magnetic Fields Generation in Electrically Conducting Fluids* (Cambridge: Cambridge Univ. Press).

Möller, J. C. 1819. *Beschreibung des Saturnringes und Anschauliche Darstellung der Ursachen Seiner Veränderlichen Lichtgestalt* (Altona).

Moore, J. M. 1983. The plains and lineaments of Dione. *Lunar Planet. Sci.* 14:511–512 (abstract).

Moore, J. M., and Ahern, J. L. 1982. The geology of Tethys. *J. Geophys. Res.* 88:A577–A584. Special Saturn Issue.

Moos, H. W., and Clarke, J. T. 1979. Detection of acetylene in the Saturnian atmosphere using the IUE satellite. *Astrophys. J.* 229:L107–L110.

Morfill, G. E., Grün, E., Goertz, C. K., and Johnson, T. V. 1983a. On the evolution of Saturn's spokes: Theory. *Icarus* 53:230–235.

Morfill, G. E., Grün, E., and Johnson, T. V. 1983b. Saturn's E, G, and F-rings: Modulated by the plasma sheet. *J. Geophys. Res.* 88:5573–5579.

Moroz, V. I. 1966. The spectra of Jupiter and Saturn in the 1.0–2.5 μm region. *Sov. Astron. J.* 10:457–468.

Morozhenko, A. V. 1977. Physical properties of upper layers of Saturn's atmosphere. *Astron. Astrophys.* 33:78–86.

Morrison, D. 1974a. Albedos and densities of the inner satellites of Saturn. *Icarus* 22:51–56.

Morrison, D. 1974b. Infrared radiometry of the rings of Saturn. *Icarus* 22:57–65.

Morrison, D. 1977a. Asteroid sizes and albedos. *Icarus* 31:185–220.

Morrison, D. 1977b. Radiometry of satellites and the rings of Saturn. In *Planetary Satellites*, ed. J. Burns (Tucson: Univ. of Arizona Press), pp. 269–301.

Morrison, D. 1980. Sizes and densities of the Saturn satellites: A pre-Voyager analysis. *Bull. Amer. Astron. Soc.* 12:727 (abstract).

Morrison, D. 1982a. The satellites of Jupiter and Saturn. *Ann. Rev. Astron. Astrophys.* 20:469–495.

Morrison, D. 1982b. *Voyages to Saturn* (Washington: NASA SP-451).

Morrison, D., and Cruikshank, D. P. 1974. Physical properties of the natural satellite. *Space Sci. Rev.* 15:641–739.

Morrison, D., and Lebofsky, L. A. 1979. Radiometry of asteroids. In *Asteroids*, ed. T. Gehrels (Tucson: Univ. of Arizona Press), pp. 184–205.

Morrison, D., Cruikshank, D. P., and Murphy, R. E. 1972. Temperatures of Titan and the Galilean satellites at 20 μm. *Astrophys. J.* 173:L143–L146.

Morrison, D., Jones, T. J., Cruikshank, D. P., and Murphy, R. E. 1975. The two faces of Iapetus. *Icarus* 24:157–171.

Morrison, D., Cruikshank, D. P., Pilcher, C. B., and Rieke, G. H. 1976. Surface compositions of the satellites of Saturn from infrared photometry. *Astrophys. J.* 207:L213–L216.

Mount, G. H., and Rottman, G. J. 1981. The solar spectral irradiants 1200–3184 Å near solar maximum: 15 July 1980. *J. Geophys. Res.* 86:9193–9198.

Mühleman, D., and Berge, G. 1982. Microwave emission from Saturn's rings. Proceedings "Planetary Rings/Anneaux des Planètes Conference," Toulouse, France, August 1982 (abstract).

Mul, P. M., and McGowan, J. W. 1979. Dissociative recombination of N_2H^+ and N_2D^+. *Astrophys. J.* 227:L157–L159.

Mul, P. M., and McGowan, J. W. 1980. Dissociative recombination of C_2^+, C_2H^+, $C_2H_2^+$, and $C_2H_3^+$. *Astrophys. J.* 237:749–751.

Müller, G. 1893. Helligkeitsbestimmungen der grossen Planeten und einiger Asteroiden. *Publ. Obs. Potsdam* 8:193.

Münch, G., and Spinrad, H. 1963. On the spectrum of Saturn. *Mém. Soc. Roy. Liège* 7:541–542.

Münch, G., Trauger, J. T., and Roesler, F. L. 1977. A search for the H_2 (3,0) S1 line in the spectrum of Titan. *Astrophys. J.* 216:963–966.

Murphy, R. E. 1973. Temperatures of Saturn's rings. *Astrophys. J.* 181:L87–L90.

Murphy, R. E., Cruikshank, D. P., and Morrison, D. 1972a. Limb darkening of Saturn and thermal properties of the rings from 10 and 20 micron photometry. *Bull. Amer. Astron. Soc.* 4:358 (abstract).

Murphy, R. E., Cruikshank, D. P., and Morrison, D. 1972b. Radii, albedos, and 20-micron brightness temperatures of Iapetus and Rhea. *Astrophys. J.* 177:L93–L96.

Nagy, A. F., Chameides, W. L., Chen, R. H., and Atreya, S. K. 1976. Electron temperatures in the Jovian ionosphere. *J. Geophys. Res.* 81:5567–5569.

Nakamura T., and Jones, S. J. 1973. Mechanical properties of impure ice crystals. In *Physics and Chemistry of Ice*, eds. E. Whalley, S. J. Jones, and L. W. Gold (Ottawa: Roy. Soc. Canada), pp. 365–369.

Nellis, W. J., Ree, F. H., Van Thiel, M., and Mitchell, A. C. 1981. Shock compression of liquid CO and CH$_4$ to 90 GPA. *J. Chem. Phys.* 75:3055–3063.

Nellis, W. J., Ross, M., Mitchell, A. C., Van Thiel, M., Young, D. A., Ree, F. H., and Trainor, R. J. 1983. Equation of state of molecular hydrogen and deuterium from shock wave experiments to 750 kbar. *Phys. Rev.* 27A:608–611.

Nelson, M. L. 1983. Spectral evidence for magnetospheric interaction with surfaces of the icy Galilean satellites. Proceedings "Planetary Satellites Conference," Ithaca, NY, July 1983 (abstract).

Nelson, R. M., and Hapke, B. W. 1978. Spectral reflectivities of the Galilean satellites and Titan, 0.32–0.86 micrometers. *Icarus* 36:304–329.

Ness, N. F. 1965. The Earth's magnetotail. *J. Geophys. Res.* 70:2989–3005.

Ness, N. F. 1979. The magnetic fields of Mercury, Mars, and Moon. *Ann. Rev. Earth Planet. Sci.* 7:249–288.

Ness, N. F., Acuña, M. H., Lepping, R. P., Connerney, J. E. P., Behannon, K. W., Burlaga, L. F., and Neubauer, F. M. 1981. Magnetic field studies by Voyager 1: Preliminary results at Saturn. *Science* 212:211–217.

Ness, N. F., Acuña, M. H., Behannon, K. W., Burlaga, L. F., Connerney, J. E. P., Lepping, R. P., and Neubauer, F. M. 1982*a*. Magnetic field studies by Voyager 2: Preliminary results at Saturn. *Science* 215:558–563.

Ness, N. F., Acuña, M. H., Behannon, K. W., and Neubauer, F. M. 1982*b*. The induced magnetosphere of Titan. *J. Geophys. Res.* 87:1369–1381.

Neubauer, F. M. 1978. Possible strengths of dynamo magnetic fields of the Galilean satellites and of Titan. *Geophys. Res. Letters* 5:905–908.

Neubauer, F. M. 1980. Nonlinear standing Alfvén wave current system at Io: Theory. *J. Geophys. Res.* 85:1171–1178.

Newburn, R. L. Jr., and Gulkis, S. 1973. A survey of the outer planets Jupiter, Saturn, Uranus, Neptune, Pluto and their satellites. *Space Sci. Rev.* 14:179–271.

Newcomb, S. 1891. On the motion of Hyperion. A new case in celestial mechanics. *Astron. Paper Amer. Ephem.* III:345–371.

Newcomb, S. 1911. Saturn. In *Encyclopedia Britannica* 11th Ed., Vol. 24, pp. 232–233.

Nier, A. O., and McElroy, M. B. 1977. Composition and structure of Mars upper atmosphere: Results from the neutral mass spectrometers on Viking 1 and 2. *J. Geophys. Res.* 82:4341–4349.

Nishida, A., and Watanabe, Y. 1981. Joule heating of the Jovian ionosphere by corotation enforcement currents. *J. Geophys. Res.* 86:9945–9952.

Noland, M., Veverka, J., Morrison, D., Cruikshank, D. P., Lazarewicz, A. R., Morrison, N. D., Elliot, J. L., Goguen, J., and Burns, J. A. 1974. Six-color photometry of Iapetus, Titan, Rhea, Dione, and Tethys. *Icarus* 23:334–354.

Nolt, I. G., Radostitz, J. V., Donnelly, R. J., Murphy, R. E., and Ford, H. C. 1974. Thermal emission from Saturn's rings and disk at 34 μm. *Nature* 248:659–660.

Nolt, I. G., Sinton, W. M., Caroff, L. J., Erickson, E. F., Strecker, D. W., and Radostitz, J. V. 1977. The brightness temperature of Saturn and its rings at 39 microns. *Icarus* 30:747–759.

Nolt, I. G., Gillett, F. C., Caldwell, J., and Tokunaga, A. 1978. The 22.7 micron brightness of Saturn's rings versus declination of the Sun. *Astrophys. J.* 219:L63–L66.

Nolt, I. G., Caldwell, J., Radostitz, J. V., Tokunaga, A. T., Barrett, E. W., Gillett, F. C., and Murphy, R. E. 1980. IR brightness and eclipse cooling of Saturn's rings. *Nature* 283:824–843.

Norrish, R. G., and Oldershaw, G. A. 1961. The flash photolysis of phosphine. *Proc. Roy. Soc. London* 262:1–9.

North, J. D. 1974. Thomas Harriot and the first telescopic observations of sunspots. In *Thomas Harriot: Renaissance Scientist*, ed. J. W. Shirley (Oxford: Clarendon Press), pp. 129–165.

Northrop, T. G. 1963. *The Adiabatic Motion of Charged Particles*, ed. R. E. Marshak (New York: Interscience Publishers).

Northrop, T. G. 1983. The inner edge of Saturn's B ring. Submitted to *Nature*.

Northrop, T. G., and Birmingham, T. J. 1982. Adiabatic charged particle motion in rapidly rotating magnetospheres. *J. Geophys. Res.* 87:661–670.

Northrop, T. G., and Hill, J. R. 1982. Stability of negatively charged dust grains in Saturn's ring plane. *J. Geophys. Res.* 87:6045–6051.

Northrop, T. G., and Hill, J. R. 1983a. The adiabatic motion of charged dust grains in rotating magnetospheres. *J. Geophys. Res.* 88:1–11.

Northrop, T. G., and Hill, J. R. 1983b. The inner edge of Saturn's B-Ring. *J. Geophys. Res.* 88:6102–6108.

Noy, N., Podolak, M., and Bar-Nun, A. 1981. Photochemistry of phosphine and Jupiter's Great Red Spot. *J. Geophys. Res.* 86:11985–11988.

Null, G. W., Lau, E. L., Biller, E. D., and Anderson, J. D. 1981. Saturn gravity results obtained from Pioneer 11 tracking data and earth-based Saturn satellite data. *Astron. J.* 86:456–468.

Odierna, G. B. 1657. *Protei Caelestis Vertigines seu Saturni Systema* (Palermo).

Ohring, G. 1975. The temperature profile in the upper atmosphere of Saturn from inversion of thermal emission observations. *Astrophys. J.* 195:223–225.

Ohring, G., and Lacser, A. 1976. The ammonia profile in the atmosphere of Saturn from inversion of its microwave emission spectrum. *Astrophys. J.* 206:622–626.

Okabe, H. 1981. Photochemistry of acetylene at 1470 Å. *J. Chem. Phys.* 75:2772–2778.

Okabe, H. 1983. Photochemistry of acetylene at 1849 Å. *J. Chem. Phys.* 78:1312–1317.

Okuda, H., Ashour-Adballa, M., Chance, M. S., and Kurth, W. S. 1982. Generation of nonthermal continuum radiation in the magnetosphere. *J. Geophys. Res.* 87:10457–10462.

Omar, M. H., Dokoupil, Z., and Schroten, H. G. M. 1962. Determination of the solid-liquid equilibrium diagram for the nitrogen methane system. *Physics* 28:309–329.

Oort, A. H., and Rasmusson, E. M. 1971. *Atmospheric Circulation Statistics*, NOAA Prof. Paper No. 5 (Rockville: U.S. Dept. Commerce).

Opp, A. G. 1980. Scientific results from the Pioneer Saturn encounter: Summary. *Science* 207:401–403.

Oran, E. S., Julienne, P. S., and Strobel, D. F. 1975. The aeronomy of odd nitrogen in the thermosphere. *J. Geophys. Res.* 80:3068–3076.

Orton, G. S. 1983. Thermal infrared constraints on ammonia ice particles as candidates for clouds in the atmosphere of Saturn. *Icarus* 53:293–300.

Orton, G. S., and Ingersoll, A. P. 1980. Saturn's atmospheric temperature structure and heat budget. *J. Geophys. Res.* 85:5871–5881.

Orton, G. S., Ingersoll, A. P., Terrile, R. J., and Walton, S. J. 1981. Images of Jupiter from the Pioneer 10 and 11 infrared radiometers: A comparison with visible and 5-μm images. *Icarus* 47:145–158.

Orton, G. S., Appleby, J. F., and Martonchik, J. V. 1982. The effect of ammonia ice on the outgoing thermal radiance from the atmosphere of Jupiter. *Icarus* 52:94–116.

Osterbrock, D. E., and Cruikshank, D. P. 1983. J. E. Keeler's discovery of a gap in the outer part of the A ring. *Icarus* 53:165–173.

Ostro, S. J., Pettengill, G. H., and Campbell, D. P. 1980. Radar observations of Saturn's rings at intermediate tilt angles. *Icarus* 41:381–388.

Ostro, S. J., Pettengill, G. H., Campbell, D. B., and Goldstein, R. M. 1982. Delay-Doppler radar observations of Saturn's rings. *Icarus* 49:367–381.

Owen, T. C. 1969. The spectra of Jupiter and Saturn in the photographic infrared. *Icarus* 10:355–364.

Owen, T. C. 1982a. Titan. *Sci. Amer.* 246:98–109.

Owen, T. C. 1982b. The composition and origin of Titan's atmosphere. *Planet. Space Sci.* 30:833–838.

Owen, T. C., McKellar, A. R. W., Encrenaz, T., Lecacheux, J., de Bergh, C., and Maillard, J. P. 1977. A study of the 1.56 μm NH_3 band on Jupiter and Saturn. *Astron. Astrophys.* 54:291–295.

Pang, K. D., Voge, C. C., and Rhoads, J. W. 1982. Macrostructure and microphysics of Saturn's E ring. Proceedings "Planetary Rings/Anneaux des Planètes Conference," Toulouse, France, August 1982 (abstract).

Pang, K. D., Voge, C. C., Rhoads, J. W., and Ajello, J. M. 1983. Saturn's E ring and satellite Enceladus. *Lunar Planet. Sci.* 14:592–593 (abstract).

Parmentier, E. M., and Head, J. W. 1979. Internal processes affecting surfaces of low-density satellites: Ganymede and Callisto. *J. Geophys. Res.* 84:6263–6276.

Parrish, W. R., and Prausnitz, J. M. 1972. Dissociation pressures of gas hydrates formed by gas mixtures. *Ing. Eng. Chem. Proc. Des. Develop.* 11:26–35.

Passey, Q. R. 1983. Viscosity of the lithosphere of Enceladus. *Icarus* 53:105–120.

Peale, S. J., Cassen, P., and Reynolds, R. T. 1979. Melting of Io by tidal dissipation. *Science* 203:892–894.

Peale, S. J., Cassen, P., and Reynolds, R. T. 1980. Tidal dissipation, orbital evolution, and the nature of Saturn's inner satellites. *Icarus* 43:65–72.

Pearce, J. B. 1981. A heuristic model for Jovian decametric arcs. *J. Geophys. Res.* 86: 8579–8581.

Pederson, B. M., Aubier, M. G., and Alexander, J. K. 1981. Low-frequency plasma waves near Saturn. *Nature* 292:714–716.

Peebles, P. J. 1964. Structure and composition of Jupiter and Saturn. *Astrophys. J.* 140:328–348.

Peirce, B. 1851. On the constitution of Saturn's ring. *Astron. J.* 2:17–19; reprinted *Amer. J. Sci. Arts.* 12:106–108, 2nd series.

Peirce, B. 1856. On the Adams prize-problem. *Astron. J.* 4:110–112.

Peirce, B. 1866. The Saturnian system. *Mem. Nat. Acad. Sci.* 1:263–286.

Perri, F., and Cameron, A. G. W. 1974. Hydrodynamic instability of the solar nebula in the presence of a planetary core. *Icarus* 22:416–425.

Peterson, C. 1975. An explanation for Iapetus' asymmetric reflectance. *Icarus* 24:499–503.

Pettengill, G. H. 1978. Physical properties of the planets and satellites from radar observations. *Ann. Rev. Astron. Astrophys.* 16:265–292.

Pettengill, G. H., and Hagfors, T. 1974. Comment on radar scattering from Saturn's rings. *Icarus* 21:188–190.

Peutter, R. C., and Russell, R. W. 1977. The 2–4 micron spectrum of Saturn's rings. *Icarus* 32:37–40.

Pickering, E. C. 1899a. A new satellite of Saturn. *Astron. J.* 9:173.

Pickering, E. C. 1899b. A new satellite of Saturn. *Astron. J.* 9:274–276.

Pickering, E. C. 1904. The ninth satellite of Saturn. *Astron. J.* 20:357–358.

Pickering, W. H. 1905a. The ninth satellite of Saturn. *Ann. Harv. Col. Obs.* 53:45–73.

Pickering, W. H. 1905b. *The Observatory* 28:227, 433.

Pilcher, C. B., Chapman, C., Lebofsky, L. A., and Kieffer, H. H. 1970. Saturn's rings: Identification of water frost. *Science* 167:1372–1373.

Pipes, J. G., Browell, E. V., and Anderson, R. D. 1974. Reflectance of amorphous-cubic NH_3 frosts and amorphous-hexagonal H_2O frosts at 77°K from 3000 Å to 1400 Å. *Icarus* 21: 283–291.

Pirraglia, J. A., Conrath, B. J., Allison, M. D. P., and Gierasch, P. 1981. Thermal structure and dynamics of Saturn and Jupiter. *Nature* 292:677–679.

Plescia, J. B., and Boyce, J. M. 1982. Crater densities and geological histories of Rhea, Dione, Mimas, and Tethys. *Nature* 295:285–290.

Plescia, J. B., and Boyce, J. M. 1983. Crater numbers and geological histories of Iapetus, Enceladus, Tethys, and Hyperion. *Nature* 301:666–670.

Podolak, M. 1978. Models of Saturn's interior: Evidence for phase separation. *Icarus* 33: 342–348.

Podolak, M., and Cameron, A. G. W. 1974. Models of the giant planets. *Icarus* 22:123–148.

Podolak, M., and Danielson, R. E. 1977. Axel dust on Saturn and Titan. *Icarus* 30:479–492.

Podolak, M., and Bar-Nun, A. 1979. A constraint on the distribution of Titan's atmospheric aerosol. *Icarus* 39:272–276.

Podolak, M., and Giver, L. P. 1979. On inhomogeneous scattering models of Titan's atmosphere. *Icarus* 37:361–376.

Podolak, M., and Podolak, E. 1980. A numerical study of aerosol growth in Titan's atmosphere. *Icarus* 43:73–84.

Podolak, M., Noy, N., and Bar-Nun, A. 1979. Photochemical aerosols in Titan's atmosphere. *Icarus* 40:193–198.

Poirier, J. P. 1982. Rheology of ices: A key to the tectonics of the ice moons of Jupiter and Saturn. *Nature* 299:683–688.

Pollack, J. B. 1973. Greenhouse models of the atmosphere of Titan. *Icarus* 19:43–53.

Pollack, J. B. 1975. The rings of Saturn. *Space Sci. Rev.* 18:3–93.

Pollack, J. B. 1978. Origin and evolution of the Saturn system: Observational consequences. In *The Saturn System*, eds. D. M. Hunten and D. Morrison (Washington: NASA Conf. Pub. 2068), pp. 9–30.

Pollack, J. B. 1981*a*. Titan In *The New Solar System*, eds. J. K. Beatty, B. O'Leary, and A. Chaikin (Cambridge: Sky Publ. Corp.), pp. 161–166.

Pollack, J. B. 1981*b*. Phase curve and particle properties of Saturn's F ring. *Bull. Amer. Astron. Soc.* 13:727 (abstract).

Pollack, J. B., and Black, D. C. 1979. Implications of the gas compositional measurements of Pioneer Venus for the origin of planetary atmospheres. *Science* 205:56–59.

Pollack, J. B., and Cuzzi, J. N. 1980. Scattering by nonspherical particles of size comparable to a wavelength: A new semi-empirical theory and its applications to tropospheric aerosols. *J. Atmos. Sci.* 37:868–881.

Pollack, J. B., and Yung, Y. L. 1980. Origin and evolution of planetary atmospheres. *Ann. Rev. Earth Planet. Sci.* 8:425–487.

Pollack, J. B., and Cuzzi, J. N. 1981. Rings in the solar system. *Sci. Amer.* 245:105–129.

Pollack, J. B., and Black, D. C. 1982. Noble gases in planetary atmospheres: Implications for the origin and evolution of planetary atmospheres. *Icarus* 51:169–198.

Pollack, J. B., Summers, A., and Baldwin, B. 1973. Estimates of the size of the particles in the rings of Saturn and their cosmogenic implications. *Icarus* 20:263–278.

Pollack, J. B., Grossman, A. S., Moore, R., and Graboske, H. C. Jr. 1976. The formation of Saturn's satellites and rings as influenced by Saturn's contraction history. *Icarus* 29:35–48.

Pollack, J. B., Grossman, A. S., Moore, R., and Graboske, H. C. 1977. A calculation of Saturn's gravitational contraction history. *Icarus* 30:111–128.

Pollack, J. B., Burns, J. A., and Tauber, M. E. 1979. Gas drag in primordial circumplanetary envelopes: A mechanism for satellite capture. *Icarus* 37:587–611.

Pollack, J. B., Toon, O. B., and Boese, R. 1980. Greenhouse models of Venus' high surface temperature as constrained by Pioneer Venus measurements. *J. Geophys. Res.* 85:8223–8231.

Pollack, J. B., Burns, J. A., and Moore, J. T. Jr. 1983. Properties of Saturn's F ring. In preparation.

Pollock, E. J., and Alder, B. J. 1977. Phase separation for a dense fluid mixture of nuclei. *Phys. Rev.* 15A:1263–1268.

Porco, C. C. 1983. Voyager Observations of Saturn's Rings. I. The Eccentric Rings at 1.29, 1.45, 1.95, and 2.27 R$_S$. 2. The Periodic Variation of Spokes. Ph.D. thesis, California Institute of Technology, Pasadena.

Porco, C. C., and Danielson, G. E. 1982. The periodic variation of spokes in Saturn's rings. *Astron. J.* 87:826–833.

Porco, C. C., Danielson, G. E., Goldreich, P., Holberg, J. B., and Lane, A. L. 1984*a*. Saturn's non-axisymmetric ring edges at 1.95 R$_S$ and 2.27 R$_S$. Submitted to *Icarus*.

Porco, C. C., Nicholson, P. D., Borderies, N., Goldreich, G. E., Holberg, J. B., and Lane, A. L. 1984*b*. The eccentric Saturnian ringlets at 1.29 R$_S$ and 1.45 R$_S$. Submitted to *Icarus*.

Pound, J. 1718. A rectification of the motions of the five satellites of Saturn. *Phil. Trans.* 30:768–774.

Prather, M. J., Logan, J. A., and McElroy, M. B. 1978. Carbon monoxide in Jupiter's upper atmosphere. *Astrophys. J.* 223:1072–1081.

Prentice, A. J. R. 1978. Towards a modern Laplacian theory for the formation of the solar system. In *The Origin of the Solar System*, ed. S. F. Dermott (London: Wiley), pp. 111–162.

Prentice, A. J. R. 1980. Saturn: Origin and composition of its inner moons and rings. *JPL Publ.* 80–80.

Prentice, A. J. R. 1982. Formation of the Saturnian system: The modern Laplacian theory. Program "Saturn Conference," Tucson, AZ, May 1982 (abstract).

Price, M. J., and Franz, O. G. 1980. Saturn: UBV photoelectric pinhole scans of the disk. II. *Icarus* 44:657–666.

Prickard, O. A. 1916. The 'Mundus Jovialis' of Simon Marius. *The Observatory* 39:367–381, 403–412, 443–452, 498–503.

Prinn, R. G., and Lewis, J. S. 1973. Uranus atmosphere: Structure and composition. *Astrophys. J.* 179:333–342.

Prinn, R. G., and Lewis, J. S. 1975. Phosphine on Jupiter and implications for the Great Red Spot. *Science* 190:274–276.

Prinn, R. G., and Owen, T. 1976. Chemistry and spectroscopy of the Jovian atmosphere. In *Jupiter*, ed. T. Gehrels (Tucson: Univ. of Arizona Press), pp. 319–371.

Prinn, R. G., and Barshay, S. S. 1977. Carbon monoxide on Jupiter and implications for atmospheric convection. *Science* 198:1031–1034.

Prinn, R. G., and Fegley, B. Jr. 1981. Kinetic inhibition of CO and N_2 reduction in circumplanetary nebulae: Implications for satellite composition. *Astrophys. J.* 249:308–317.

Prinn, R. G., and Olaguer, O. P. 1981. Nitrogen on Jupiter: A deep atmospheric source. *J. Geophys. Res.* 86:9895–9899.

Proctor, R. A. 1865. *Saturn and its System* (London).

Proctor, R. A. 1882. *Saturn and its System* (London), 2nd Ed.

Pu, Z. Y., and Kivelson, M. G. 1983. Kelvin-Helmholtz instability at the magnetopause: Solution for compressible plasmas. *J. Geophys. Res.* 88:841–852.

Purcell, E. M., and Pennypacker, C. R. 1973. Scattering and absorption of light by nonspherical dielectric grains. *Astrophys. J.* 186:705–714.

Rages, K., and Pollack, J. B. 1980. Titan aerosols: Optical properties and vertical distribution. *Icarus* 41:119–130.

Rages, K., and Pollack, J. B. 1983. Vertical distribution of scattering hazes in Titan's upper atmosphere. *Icarus* 55:50–62.

Rages, K., Pollack, J. B., and Smith, P. H. 1983. Size estimates of Titan's aerosols based on Voyager 1 high-phase-angle images. *Icarus*. In press.

Rairden, R. L. 1980. Satellite sweeping of electrons and protons in Saturn's inner magnetosphere. Univ. of Iowa Res. Rep. 80-29.

Rairden, R. L. 1981. Diffusion and Losses of ~1 MeV Protons in Saturn's Magnetosphere. M.S. thesis, Univ. of Iowa, Iowa City.

Rather, J. D. G., Ulich, B. L., and Ade, P. A. R. 1974. Planetary brightness temperature measurements at 1.4 mm wavelength. *Icarus* 23:448–453.

Raulin, F., Bossard, A., Toupance, G., and Ponnamperuma, C. 1979. Abundance of organic compounds photochemically produced in the atmospheres of outer planets. *Icarus* 38:358–366.

Read, P. L., and Hide, R. 1983. Long-lived eddies in the laboratory and in the atmospheres of Jupiter and Saturn. *Nature* 302:126–129.

Ree, F. H. 1979. Systematics of high pressure and high temperature behavior of hydrocarbons. *J. Chem. Phys.* 70:974–983.

Ree, F. H. 1982. Molecular interaction of dense water at high temperature. *J. Chem. Phys.* 76:6287–6302.

Ree, F. H., and Bender, C. F. 1979. Repulsive intermolecular potential between two H_2 molecules. *J. Chem. Phys.* 71:5362–5375.

Ree, F. H., and Bender, C. F. 1980. Erratum: Repulsive intermolecular potential between two H_2 molecules. *J. Chem. Phys.* 73:4712.

Reese, E. J. 1970. Jupiter's Red Spot in 1968–1969. *Icarus* 12:249–257.

Reese, E. J. 1971. Recent photographic measurements of Saturn. *Icarus* 15:466–479.

Reitsema, H. J. 1978. Photometric confirmation of the Encke division in Saturn's ring A. *Nature* 272:601–602.

Reitsema, H. J. 1981a. Libration of the Saturn satellite Dione B. *Icarus* 48:23–28.

Reitsema, H. J. 1981b. Orbits of the Tethys Lagrangian bodies. *Icarus* 48:140–142.

Reitsema, H. J., Beebe, R. F., and Smith, B. A. 1976. Azimuthal brightness variations in Saturn's rings. *Astron. J.* 81:209–215.

Reynolds, R. T., and Cassen, P. M. 1979. On the internal structure of the major satellites of the outer planets. *Geophys. Res. Letters* 6:121–124.

Rhines, P. B. 1975. Waves and turbulence on a beta plane. *J. Fluid Mech.* 69:417–443.

Ridgway, S. T., Larson, H. P., and Fink, U. 1976. The infrared spectrum of Jupiter. In *Jupiter*, ed. T. Gehrels (Tucson: Univ. of Arizona Press), pp. 384–417.

Rieke, G. H. 1975. The thermal radiation of Saturn and its rings. *Icarus* 26:37–44.

Rishbeth, H. 1959. The ionosphere of Jupiter. *Australian Phys. J.* 12:466–468.

Robert, F., and Epstein, S. 1982. The concentration and isotopic composition of hydrogen, carbon and nitrogen in carbonaceous chondrites. *Geochim. Cosmochim. Acta* 46:81–95.

Robnik, M., and Kundt, W. 1983. Hydrogen at high pressures and temperatures. *Astron. Astrophys.* 120:227–233.

Roche, E. 1984. *Acad. Sci. Letters Montpellier* 243.

Roche, E. 1849. Mémoire sur la figure d'une masse fluide soumise à l'attraction d'un point éloigné. In *Mémoires de l'Académie des Sciences et Lettres de Montpellier Section des Sciences* 1:243–262, 333–348.

Roederer, J. G. 1967. On the adiabatic motion of energetic particles in a model magnetosphere. *J. Geophys. Res.* 72:981–992.

Rollet, A.-P., and Vuillard, G. 1956. Sur un nouvel hydrate de l'ammoniac. *C. R. Acad. Sci.* 243:383–386.

Rönnmark, K., Borg, H., Christiansen, P. J., Gough, M. P., and Jones, D. 1978. Banded electron cyclotron harmonic instability: A first comparison of theory and experiment. *Space Sci. Rev.* 22:401–417.

Ross, M. 1974. A theoretical analysis of the shock compression experiments of the liquid hydrogen isotropes and a prediction of their metallic transition. *J. Chem. Phys.* 60:3634–3644.

Ross, M. 1981. The ice layer in Uranus and Neptune—diamonds in the sky? *Nature* 292:435–436.

Ross, M., and McMahan, A. K. 1976. Comparison of theoretical models for metallic hydrogen. *Phys. Rev.* 13B:5154–5157.

Ross, R. G., and Andersson, P. 1982. Clathrate and other solid phases in the tetrahydrofuran-water system: Thermal conductivity and heat capacity under pressure. *Can. J. Chem.* 60:881–892.

Rossow, W. B. 1978. Cloud microphysics: Analysis of the clouds of Earth, Venus, Mars, and Jupiter. *Icarus* 36:1–50.

Rossow, W. B., and Williams, G. P. 1979. Large-scale motions in the Venus stratosphere. *J. Atmos. Sci.* 36:377–389.

Rowan-Robinson, M., Ade, P. A. R., Robson, E. I., and Clegg, P. E. 1978. Millimetre observations of planets, galactic and extra-galactic sources. *Astron. Astrophys.* 62:249–254.

Roux, A., and Pellat, R. 1979. Coherent generation of the terrestrial kilometric radiation by nonlinear beating of electrostatic waves. *J. Geophys. Res.* 84:5189–5198.

Ruiz, V. H. G., and Rowland, F. S. 1978. Possible scavenging of C_2H_2 and C_2H_4 by phosphorus-containing radicals in the Jovian atmosphere. *Geophys. Res. Letters* 5:407–410.

Russell, C. T. 1979. Scaling law test and two predictions of planetary magnetic moments. *Nature* 281:552–553.

Russell, C. T., Luhmann, J. G., Elphic, R. C., and Scarf, F. L. 1981. The distant bow shock and magnetotail of Venus: Magnetic field and plasma wave observations. *Geophys. Res. Letters* 8:843–846.

Russell, R. W., and Soifer, B. T. 1977. Observations of Jupiter and Saturn at 5–8 μm. *Icarus* 30:282–285.

Saari, J., and Shorthill, R. 1972. The sunlit lunar surface. I. Albedo studies and full moon temperature distribution. *Moon* 5:161–178.

Safronov, V. S. 1972. Evolution of the protoplanetary cloud and formation of the Earth and planets. *NASA TT* F-677.

Safronov, V. S. 1982. Formation of the solar system. Proceedings "Planetary Rings/Anneaux des Planètes Conference," Toulouse, France, August 1982 (abstract).

Safronov, V. S., and Ruskol, E. L. 1977. The accumulation of satellites. In *Planetary Satellites*, ed. J. A. Burns (Tucson: Univ. of Arizona Press), pp. 505–512.

Safronov, V. S., and Ruskol, E. L. 1982. On the origin and initial temperature of Jupiter and Saturn. *Icarus* 49:284–296.

Sagan, C. 1971. The solar system beyond Mars: An exobiological survey. *Space Sci. Rev.* 11:73–112.

Sagan, C. 1973. The greenhouse of Titan. *Icarus* 18:649–656.

Sagan, C. 1974. Organic chemistry in the atmosphere. In *The Atmosphere of Titan*, ed. D. M. Hunten (Washington: NASA SP-340), pp. 134–142.

Sagan, C., and Miller, S. L. 1960. Molecular synthesis in simulated reducing planetary atmospheres. *Astron. J.* 65:499.

Sagan, C., and Salpeter, E. E. 1976. Particles, environments and hypothetical ecologies in the Jovian atmosphere. *Astrophys. J. Suppl.* 32:737–755.

Sagan, C., and Khare, B. N. 1979. Tholins: Organic chemistry of interstellar grains and gas. *Nature* 277:102–107.

Sagan, C., and Khare, B. N. 1981. The organic clouds of Titan. *Bull. Amer. Astron. Soc.* 13:701 (abstract).

Sagan, C., and Dermott, S. F. 1982. The tide in the seas of Titan. *Nature* 300:731–733.

Sagan, C., and Khare, B. N. 1982. The organic clouds of Titan. *Origins of Life* 12:280.

Sagan, C., and Thompson, W. R. 1984. Production and condensation of organic gases in the atmosphere of Titan. *Icarus*. In press.

Sagan, C., Lippincott, E. R., Dayhoff, M. O., and Eck, R. V. 1967. Organic molecules and the coloration of Jupiter. *Nature* 213:273–274.

Sagan, C., Thompson, W. R., and Khare, B. N. 1983. Reflection spectra of model Titan atmospheres and aerosols. *Bull. Amer. Astron. Soc.* 15:843 (abstract).

Salpeter, E. E. 1973. On convection and gravitational layering in Jupiter and in stars of low mass. *Astrophys. J.* 181:L83–L86.

Samuelson, R. E. 1970. Non-local thermodynamic equilibrium in cloudy planetary atmospheres. *J. Atmos. Sci.* 27:711–720.

Samuelson, R. E. 1983. Radiative equilibrium model of Titan's atmosphere. *Icarus* 53:364–387.

Samuleson, R. E., Hanel, R. A., Kunde, V. G., and Maguire, W. C. 1981. Mean molecular weight and hydrogen abundance of Titan's atmosphere. *Nature* 292:688–693.

Samuelson, R. E., Maguire, W. C., Hanel, R. A., Kunde, V. G., Jennings, D. E., Yung, Y. L., and Aikin, A. C. 1983. CO_2 on Titan. *J. Geophys. Res.* 88:8709–8715.

Sanchez, R. A., Ferris, J. P., and Orgel, L. E. 1966. Cyanoacetylene in prebiotic synthesis. *Science* 154:784–785.

Sandel, B. R., and Broadfoot, A. L. 1981. Morphology of Saturn's aurora. *Nature* 292:679–682.

Sandel, B. R., Shemansky, D. E., Broadfoot, A. L., Bertaux, J. L., Blamont, J. E., Belton, M. J. S., Ajello, J. M., Holberg, J. B., Atreya, S. K., Donahue, T. M., Moos, H. W., Strobel, D. F., McConnell, J. C., Dalgarno, A., Goody, R., McElroy, M. B., and Takacs, P. Z. 1979. Extreme ultraviolet observations from Voyager 2 encounter with Jupiter. *Science* 206:962–966.

Sandel, B. R., McConnell, J. C., and Strobel, D. F. 1982*a*. Eddy diffusion at Saturn's homopause. *Geophys. Res. Letters* 9:1077–1080.

Sandel, B. R., Shemansky, D. E., Broadfoot, A. L., Holberg, J. B., Smith, G. R., McConnell, J. C., Strobel, D. F., Atreya, S. K., Donahue, T. M., Moos, H. W., Hunten, D. M., Pomphrey, R. B., and Linick, S. 1982*b*. Extreme ultraviolet observations from the Voyager 2 encounter with Saturn. *Science* 215:548–553.

Santer, R., and Dollfus, A. 1981. Optical reflectance polarimetry of Saturn's globe and rings. IV. Aerosols in the upper atmosphere of Saturn. *Icarus* 48:496–518.

Savoie, R., and Fournier, R. P. 1970. Far-infrared spectra of condensed methane and methane d_4. *Chem. Phys. Letters* 7:1–3.

Scarf, F. L. 1975. The magnetospheres of Jupiter and Saturn. In *The Magnetospheres of Earth and Jupiter*, ed. V. Formisano (Dordrecht: D. Reidel Publ.) pp. 443–449.

Scarf, F. L. 1979. Possible traversals of Jupiter's distant magnetic tail by Voyager and by Saturn. *J. Geophys. Res.* 84:4422–4424.

Scarf, F. L., and Gurnett, D. A. 1977. A plasma wave investigation for the Voyager mission. *Space Sci. Rev.* 21:289–308.

Scarf, F. L., Coroniti, F. V., Gurnett, D. A., and Kurth, W. S. 1979. Pitch-angle diffusion by whistler mode waves near the Io plasma torus. *Geophys. Res. Letters* 6:653–656.

Scarf, F. L., Gurnett, D. A., and Kurth, W. S. 1981*a*. Plasma wave turbulence at planetary bow shocks. *Nature* 292:747–750.

Scarf, F. L., Gurnett, D. A., and Kurth, W. S. 1981*b*. Measurements of plasma wave spectra in Jupiter's magnetosphere. *J. Geophys. Res.* 86:8181–8198.

Scarf, F. L., Kurth, W. S., Gurnett, D. A., Bridge, H. S., and Sullivan, J. D. 1981*c*. Jupiter tail phenomena upstream from Saturn. *Nature* 292:585–586.

Scarf, F. L., Gurnett, D. A., Kurth, W. S., and Poynter, R. L. 1982. Voyager 2 plasma wave observations at Saturn. *Science* 215:587–594.

Scarf, F. L., Gurnett, D. A., Kurth, W. S., and Poynter, R. L. 1983. Voyager plasma wave measurements at Saturn. *J. Geophys. Res.* 88:8971–8984.

Scattergood, T. 1982. Paper presented at the "Sixth College Park Colloquium on the Origins of Life."

Scattergood, T., and Owen, T. 1977. On the sources of ultraviolet absorption in spectra of Titan and the outer planets. *Icarus* 30:780–788.

Scattergood, T., Lesser, P., and Owen, T. 1974. Production of organic molecules by proton irradiation. *Nature* 274:100–101.

Scattergood, T., Lesser, P., and Owen, T. 1975. Production of organic molecules in the outer solar system by proton irradiation: Laboratory simulation. *Icarus* 24:465–471.

Schacke, H., Wagner, H. G., and Wolfrum, J. 1977. Reaktionen von Molekulen in differenzierten Schwingungszustäden. IV. Reaktionen Schwingungsangeregter Cyanradikale mit Wasserstoff und einfachen Kohlenwasserstoffen. *Ber. Bunsenges Physik* 81:670.

Schardt, A. W., and McDonald, F. B. 1983. The flux and source of energetic protons in Saturn's inner magnetosphere. *J. Geophys. Res.* 88:8923–8935.

Schloerb, F. P., Mühleman, D. O., and Berge, G. L. 1979a. Interferometric observations of Saturn and its rings at a wavelength of 3.71 cm. *Icarus* 39:214–231.

Schloerb, F. P., Mühleman, D. O., and Berge, G. L. 1979b. An aperture synthesis study of Saturn and its rings at 3.71 cm wavelength. *Icarus* 39:232–250.

Schloerb, F. P., Mühleman, D. O., and Berge, G. L. 1980. Interferometry of Saturn and its rings at 1.30 cm wavelength. *Icarus* 42:125–135.

Schoenberg, E. 1929. Theoretische photometrie. In *Handbuch der Astrophysik*, eds. G. Eberhard, A. Kohlschütter, and I. Ludendorf (Berlin: Springer-Verlag), pp. 130–144.

Schröter, J. H. 1808. *Kronographische Fragmente zur Genauern Kentnisse des Planeten Saturn, Seines Ringes und Seiner Trabanten* (Göttingen).

Schubert, G. 1979. Subsolidus convection in the mantles of the terrestrial planets. *Ann. Rev. Earth Planet. Sci.* 7:289–342.

Schubert, G., Stevenson, D. J., and Ellsworth, K. 1981. Internal structures of the Galilean satellites. *Icarus* 47:46–59.

Schulz, M., and Lanzerotti, L. J. eds. 1974. *Particle Diffusion in the Radiation Belts* (New York: Springer-Verlag).

Science 212, 1981. Voyager 1 Special Issue, 10 April.

Scorer, R. S. 1978. *Environmental Aerodynamics* (New York: Halsted Press).

Scudder, J. D., Sittler, E. C. Jr., and Bridge, H. S. 1981. A survey of the plasma electron environment of Jupiter: A view from Voyager. *J. Geophys. Res.* 86:8157–8179.

See, T. J. J. 1910. *Researches on the Evolution of the Stellar Systems*, Vol. II. The Capture Theory of Cosmical Evolution (Lynn, MA: R. P. Nichols and Sons), Chapters 10 and 11.

Seeliger, H. 1887. Theorie der Beleuchtung staubförmiger Kosmischen Massen Insbesondere des Saturnringes. *Abh. Bayer. Akad. Wiss. Math. Naturwiss. Kl. II* 18:1–72.

Seeliger, H. 1895. Zur Theorie der Beleuchtung der Grossen Planeten, Insbesondere des Saturn. *Abh. Bayer. Akad. Will. Math. Naturwiss. Kl. II* 16:405–516.

Seidelmann, P. K., Harrington, R. S., Pascu, D., Baum, W. A., Currie, D. G., Westphal, J. A., and Danielson, G. E., 1981. Saturn satellite observations and orbits from the 1980 ring plane crossing. *Icarus* 47:282–287.

Sentman, D. D., and Goertz, C. K. 1978. Whistler mode noise in Jupiter's inner magnetosphere. *J. Geophys. Res.* 83:3151–3165.

Sharma, S. K., Mao, H. K., and Bell, P. M. 1980. Raman measurements of H_2 in the pressure range 0.2–630 kbar at room temperature. *Phys. Rev. Letters* 44:886–888.

Shawhan, S. D., and Gurnett, D. A. 1982. Polarization measurements of auroral kilometric radiation by Dynamics Explorer 1. *Geophys. Res. Letters* 9:913–916.

Shemansky, D. E., and Smith, G. R. 1982. Whence comes the "Titan" hydrogen torus. *EOS* 63:1019 (abstract).

Shemansky, D. E., and Ajello, J. M. 1983. The Saturn spectrum in the EUV–Electron excited hydrogen. *J. Geophys. Res.* 88:459–464.

Shimizu, H., Brody, E. M., Mao, H. K., and Bell, P. M. 1981. Brillouin measurements of solid $-H_2$, and $-D_2$ to 200 kbar at room temperature. *Phys. Rev. Letters* 47:128–131.

Shimizu, M. 1971. The upper atmosphere of Jupiter. *Icarus* 14:273–281.

Shimizu, M. 1980. Strong interaction between the ring system and the ionosphere of Saturn. Proceedings "13th Lunar and Planetary Symp.," Tokyo, July 1982, p. 709 (abstract).

Shmulovich, K. I., Mazur, V. A., Kalinicher, A. G., and Khodorerskaya, L. I. 1980. P-V-T and component-activity-concentration relations for systems of H_2O-nonpolar gas type. *Geochem. International* 17:18–30; Transl. *Geokhimiya* 11:1625.

Shoemaker, E. M. 1982. Collisional history of the Saturn system. Program "Saturn Conference," Tucson, AZ, May 1982 (abstract).

Shoemaker, E. M., and Wolfe, R. F. 1982. Cratering time scales for the Galilean satellites. In *Satellites of Jupiter*, ed. D. Morrison (Tucson: Univ. of Arizona Press), pp. 277–339.

Shoji, H., and Langway, C. C. Jr. 1982. Air hydrate inclusions in fresh ice cores. *Nature* 298:548–550.

Showalter, M. R., and Burns, J. A. 1982. A numerical study of Saturn's F-ring. *Icarus* 52:526–544.

Shu, F. H. 1984. Waves in planetary rings. In *Planetary Rings*, eds. R. Greenberg and A. Brahic (Tucson: Univ. of Arizona Press), pp. 513–561.

Shu, F. H., Cuzzi, J. N., and Lissauer, J. J. 1983. Bending waves in Saturn's rings. *Icarus* 53:185–206.

Sieveka, E. M., and Johnson, R. E. 1982. Thermal and plasma induced molecular redistribution on the icy satellites. *Icarus* 51:528–548.

Sigua, L. A. 1978. Electropolarimetric study of Saturn and its rings. *Astron. Tsirkular USSR Acad. Sci.* 985:5–7 (Russian).

Sill, G., Fink, U., and Ferraro, J. R. 1980. Absorption coefficients of solid NH_3 from 50 to 7000 cm^{-1}. *J. Opt. Soc. Amer.* 70:724–739.

Sill, G., Fink, U., and Ferraro, J. R. 1981. The infrared spectrum of ammonia hydrate: Explanation for a reported ammonia phase. *J. Chem. Phys.* 74:997–998.

Simpson, J. A., Bastian, T. S., Chenette, D. L., Lentz, G. A., McKibben, R. B., Pyle, K. R., and Tuzzolino, A. J. 1980a. Saturnian trapped radiation and its absorption by satellites and rings: The first results from Pioneer 11. *Science* 207:411–415.

Simpson, J. A., Bastian, T. S., Chenette, D. L., McKibben, R. B., and Pyle, K. R. 1980b. The trapped radiations of Saturn and their absorption by satellites and rings. *J. Geophys. Res.* 85:5731–5762.

Simpson, J. P. Cuzzi, J. N., Erickson, E. F. Strecker, D. W., and Tokunaga, A. T. 1981. Mars: Far infrared spectra and thermal emission models. *Icarus* 48:230–245.

Simpson, R. A., Tyler, G. L., and Holberg, J. B. 1983. Saturn's pole: Geometric correction based on Voyager UVS and radio occultations. *Astron. J.* 88:1521–1536.

Sinclair, A. T. 1972. On the commensurabilities amongst the satellites of Saturn. *Mon. Not. Roy. Astron. Soc.* 160:169–187. See also Part II 1974. *Mon. Not. Roy. Astron. Soc.* 166:165–179.

Sinton, W. M., Macy, W. W., Good, J., and Orton, G. S. 1980. Infrared scans of Saturn. *Icarus* 42:251–256.

Siscoe, G. L. 1978. Magnetosphere of Saturn. In *The Saturn System*, eds. D. M. Hunten and D. Morrison (Washington: NASA Conf. Publ. 2068), pp. 265–284.

Siscoe, G. L. 1979. Towards a comparative theory of magnetospheres. In *Solar System Plasma Physics*, Vol. III, eds. C. F. Kennel, L. J. Lanzerotti, and E. N. Parker (Amsterdam: North-Holland Publ.), pp. 319–402.

Siscoe, G. L., and Summers, D. 1981. Centrifugally driven diffusion of iogenic plasma. *J. Geophys. Res.* 86:8471–8479.

Siscoe, G. L., Davis, L. Jr., Coleman, P. J. Jr., Smith, E. J., and Jones, D. E. 1968. Power spectra and discontinuities of the interplanetary magnetic field: Mariner 4. *J. Geophys. Res.* 73:61–82.

Siscoe, G. L., Crooker, N. U., and Belcher, J. W. 1980. Sunward flow in Jupiter's magnetosheath. *Geophys. Res. Letters* 7:25–28.

Sittler, E. C. Jr., Scudder, J. D., and Bridge, H. S. 1981. The distribution of neutral gas and dust near Saturn. *Nature* 292:711–714.

Sittler, E. C. Jr., Ogilvie, K. W., and Scudder, J. D. 1983. Survey of low energy plasma electrons in Saturn's magnetosphere: Voyager 1 and 2. *J. Geophys. Res.* 88:8847–8870.

Slattery, W. L. 1977. The structure of the planets Jupiter and Saturn. *Icarus* 32:58–72.

Slattery, W. L., and Hubbard, W. B. 1976. Thermodynamics of a solar mixture of molecular H_2 and He at high pressure. *Icarus* 29:187–192.

Slattery, W. L., DeCampli, W. M., and Cameron, A. G. W. 1980. Protoplanetary core formation by rain-out of minerals. *Moon Planets* 23:381–390.

Slattery, W. L., Doolen, G. D., and DeWitt, H. E. 1982. N dependence in the classical one-component plasma Monte Carlo calculations. *Phys. Rev.* 26A:2255–2258.

Slavin, J. A., Smith, E. J., Gazis, P. R., and Mihalov, J. D. 1983. A Pioneer-Voyager study of the solar wind interaction with Saturn. *Geophys. Res. Letters* 10:9–12.

Slobodkin, L. S., Buyakov, I. F., Cess, R. D., and Caldwell, J. 1978. Near infrared reflection spectra of ammonia frost: Interpretation of the upper clouds of Saturn. *J. Quant. Spectrosc. Radiat. Transfer* 20:481–490.

Smith, A. G., and Carr, T. D. 1959. Radio frequency observations of the planets in 1957–1959. *Astrophys. J.* 130:641–647.

Smith, B. A. 1978. The D and E-rings of Saturn. In *The Saturn System*, eds. D. M. Hunten and D. Morrison (Washington: NASA Conf. Publ. 2068), pp. 105–111.

Smith, B. A. 1982. The Voyager encounters. In *The New Solar System*, eds. J. K. Beatty, B. O'Leary, and A. Chaikin (Cambridge, MA: Sky Publ. Corp.), pp. 105–116.

Smith, B. A., and Hunt, G. E. 1976. Motions and morphology of clouds in the atmosphere of Jupiter. In *Jupiter*, ed. T. Gehrels (Tucson: Univ. of Arizona Press), pp. 564–585.

Smith, B. A., Briggs, G. A., Danielson, G. E., Cook, A. F. II, Davies, M. E., Hunt, G. E., Masursky, H., Soderblom, L. A., Owen, T. C., Sagan, C., and Suomi, V. E. 1977. Voyager imaging experiment. *Space Sci. Rev.* 21:103–127.

Smith, B. A., Soderblom, L. A., Beebe, R. F., Boyce, J., Briggs, G. A., Carr, M. H., Collins, S. A., Cook, A. F. II, Danielson, G. E., Davies, M. E., Hunt, G. E., Ingersoll, A. P., Johnson, T. V., Masursky, H., McCauley, J. F., Morrison, D., Owen, T., Sagan, C., Shoemaker, E. M., Strom, R. G., Suomi, V. E., and Veverka, J. 1979a. The Galilean satellites and Jupiter: Voyager 2 imaging science results. *Science* 206:927–950.

Smith, B. A., Soderblom, L. A., Johnson, T. V., Ingersoll, A. P., Collins, S. A., Shoemaker, E. M., Hunt, G. E., Masursky, H., Carr, M. H., Davies, M. E., Cook, A. F. II, Boyce, J., Danielson, G. E., Owen, T., Sagan, C., Beebe, R. F., Veverka, J., Strom, R. G., McCauley, J. F., Morrison, D., Briggs, G. A., and Suomi, V. E. 1979b. The Jupiter system through the eyes of Voyager 1. *Science* 204:951–971.

Smith, B. A., Reitsema, H. J., Fountain, J. W., and Larson, S. M. 1980. Saturn's syzgistic co-orbital satellites. *Bull. Amer. Astron. Soc.* 12:727 (abstract).

Smith, B. A., Soderblom, L. A., Beebe, R. F., Boyce, J., Briggs, G. A., Bunker, A., Collins, S. A., Hansen, C. J., Johnson, T. V., Mitchell, J. L., Terrile, R. J., Carr, M. H., Cook, A. F. II, Cuzzi, J. N., Pollack, J. B., Danielson, G. E., Ingersoll, A. P., Davies, M. E., Hunt, G. E., Masursky, H., Shoemaker, E. M., Morrison, D., Owen, T., Sagan, C., Veverka, J., Strom, R. G., and Suomi, V. E. 1981. Encounter with Saturn: Voyager 1 imaging science results. *Science* 212:163–191.

Smith, B. A., Soderblom, L. A., Batson, R., Bridges, P., Inge, J., Masursky, H., Shoemaker, E. M., Beebe, R. F., Boyce, J., Briggs, G. A., Bunker, A., Collins, S. A., Hansen, C. J., Johnson, T. V., Mitchell, J. L., Terrile, R. J., Cook, A. F., II, Cuzzi, J. N., Pollack, J. B., Danielson, G. E., Ingersoll, A. P., Davies, M. E., Hunt, G. E., Morrison, D., Owen, T., Sagan, C., Veverka, J., Strom, R. G., and Suomi, V. E. 1982. A new look at the Saturn system: The Voyager 2 images. *Science* 215:504–537.

Smith, D., and Adams, N. G. 1977. Reactions of simple hydrocarbon ions with molecules at thermal energies. *Int. J. Mass Spectrum Ion Phys.* 23:123–135.

Smith, D. W., Shorthill, R. W., Johnson, P. E., Budding, E., and Asaad, A. S. 1981. Measurement of stratospheric aerosols on Saturn using an eclipse of Titan. *Icarus* 46:424–428.

Smith, E. J., and Wolfe, J. H. 1977. Pioneer 10, 11 observations of evolving solar wind streams and shocks beyond 1 AU. In *Study of Traveling Interplanetary Phenomena*, eds. M. A. Shea, D. F. Smart, and S. T. Wu (Hingham, MA: Reidel), pp. 227–257.

Smith, E. J., and Tsurutani, B. T. 1983. Saturn's magnetosphere: Observations of ion cyclotron waves near the Dione L shell. *J. Geophys. Res.* 88:7831–7836.

Smith, E. J., Davis, L. Jr., Jones, D. E., Coleman, P. J. Jr., Colburn, D. S., Dyal, P., Sonett, C. P., and Frandsen, A. M. A. 1974. The planetary magnetic field and magnetosphere of Jupiter: Pioneer 10. *J. Geophys. Res.* 79:3501–3513.

Smith, E. J., Conner, B. V., and Foster, G. T. Jr. 1975. Measuring the magnetic fields of Jupiter and the outer solar system. *IEEE Trans. on Magnetics* MAG-11:962–980.

Smith, E. J., Davis, L. Jr., Jones, D. E., Coleman, P. J. Jr., Colburn, D. S., Dyal, P., and Sonett, C. P. 1980a. Saturn's magnetic field and magnetosphere. *Science* 207:407–410.

Smith, E. J., David, L. Jr., Jones, D. E., Coleman, P. J. Jr., Colburn, D. S., Dyal, P. and Sonett, C. P. 1980b. Saturn's magnetosphere and its interaction with the solar wind. *J. Geophys. Res.* 85:5655–5674.

Smith, G. R., Strobel, D. F., Broadfoot, A. L., Sandel, B. R., Shemansky, D. E., and Holberg, J. B. 1982. Titan's upper atmosphere: Composition and temperature from the EUV solar occultation results. *J. Geophys. Res.* 87:1351–1359.

Smith, G. R., McConnell, J. C., Shemansky, D. E., Holberg, J. B., Broadfoot, A. L., and Sandel, B. R. 1983. Saturn's upper atmosphere from the Voyager 2 EUV solar and stellar occultation. *J. Geophys. Res.* 88:8667–8678.

Smith, H. J. 1959. Nonthermal solar system sources other than Jupiter. *Astron. J.* 64:41–43.

Smith, H. J., and Douglas, J. N. 1957. First results of a planetary radio astronomy program of the Yale Observatory. *Astron. J.* 62:247.

Smith, H. J., and Douglas, J. N. 1959. Observations of planetary nonthermal radiation. In *Paris Symposium in Radio Astronomy*, ed. R. N. Bracewell (Stanford: Stanford Univ. Press), pp. 53–55.

Smith, P. H. 1982. The case for a bimodal size distribution in Titan's upper haze layer. In *The Comparative Study of the Planets*, eds. A. Coradini and M. Fulchignoni (Dordrecht: D. Reidel), pp. 253–260.

Smith, P. H., and Tomasko, M. G. 1984. Photometry and polarimetry of Jupiter at large phase angles. II. Polarimetry of the South Tropical Zone, South Equatorial Belt, and the Polar Regions. *Icarus* vol. 58.

Smith, R. J., and Wolstencroft, R. D. 1983. High precision spectropolarimetry of stars and planets. II. Spectropolarimetry of Jupiter and Saturn. *Mon. Not. Roy. Astron. Soc.* 205:39–55.

Smith, W. H., and Macy, W. 1977. Observation of HD on Saturn and Uranus. *Bull. Amer. Astron. Soc.* 9:516 (abstract).

Smith, W. H., Macy, W., and Cochran, W. 1980. Ammonia in the atmospheres of Saturn and Titan. *Icarus* 42:93–101.

Smith, W. H., McCord, T. B., and Macy, W. 1981. High-spectral-resolution imagery of Saturn. *Icarus* 46:256–262.

Smoluchowski, R. 1967. Internal structure and energy emission of Jupiter. *Nature* 215:691–695.

Smoluchowski, R. 1975. Jupiter's molecular hydrogen layer and the magnetic field. *Astrophys. J.* 200:L119–L121.

Smoluchowski, R. 1978a. Width of a planetary ring system and the C-ring of Saturn. *Nature* 274:669–670.

Smoluchowski, R. 1978b. Amorphous ice on Saturnian rings and on icy satellites. *Science* 201:809–811.

Smythe, W. D. 1975. Spectra of hydrate frosts: Their application to the outer solar system. *Icarus* 24:421–427.

Soderblom, L. A., and Johnson, T. V. 1982. The moons of Saturn. *Sci. Amer.* 246:100–116.

Sondheimer, F., Ben-Efrain, D. A., and Wolovsky, R. 1961. Unsaturated macrocyclic compounds XVII. The phototropic rearrangement of linear 1.5 enynes to conjugated polyenes. The synthesis of a series of vinylogs of butadiene. *J. Amer. Chem. Soc.* 83:1675–1681.

Sonnerup, B. U. O., and Cahill, L. J. 1967. Magnetopause structure and attitude from Explorer 12 observations. *J. Geophys. Res.* 72:171–183.

Soter, S. 1974. Proceedings "Planetary Satellites Conference," Ithaca, NY.

Space Science Reviews 21, 1977. Voyager Special Issue, Nov.-Dec.

Spreiter, J. R., Summers, A. L., and Alksne, A. Y. 1966. Hydromagnetic flow around the magnetosphere. *Planet. Space Sci.* 14:223–253.

Spreiter, J. R., Alksne, A. Y., and Summers, A. L. 1968. External aerodynamics of the magnetosphere. In *Physics of the Magnetosphere*, eds. R. L. Carovillano, J. F. McClay, and H. R. Radoski (Dordrecht: D. Reidel), pp. 301–375.

Squyres, S. W. 1980. Volume changes in Ganymede and Callisto and the origin of grooved terrain. *Geophys. Res. Letters* 7:593–596.

Squyres, S. W., and Veverka, J. 1981. Voyager photometry of Ganymede and Callisto. *Icarus* 46:137–155.

Squyres, S. W., and Sagan, C. 1982. The abledo asymmetry of Iapetus. *Bull. Amer. Astron. Soc.* 14:739 (abstract).

Squyres, S. W., and Sagan, C. 1983. Albedo asymmetry of Iapetus. *Nature* 303:782–785.

Squyres, S. W., Buratti, B., Veverka, J., and Sagan, C. 1983a. A photometric map of Iapetus from Voyager data. Proceedings "Planetary Satellites Conference," Ithaca, NY, July 1983 (abstract).

Squyres, S. W., Reynolds, R. T., Cassen, P. M., and Peale, S. J. 1983b. The evolution of Enceladus. *Icarus* 53:319–331.

Squyres, S. W. et al. 1983c. Photometric properties of Iapetus from Voyager images. Submitted to *Icarus*.

Sromovsky, L. A., Suomi, V. E., Pollack, J. B., Krauss, R. J., Limaye, S. S., Owen, T., Revercomb, H. E., and Sagan, C. 1981. Titan brightness contrasts: Implications of Titan's north-south brightness asymmetry. *Nature* 292:698–702.

Sromovsky, L. A., Revercomb, H. E., Suomi, V. E., Limaye, S. S., and Krauss, R. J. 1982. Jovian winds from Voyager 2. Part II. Analysis of eddy transports. *J. Atmos. Sci.* 39:1433–1445.

Sromovsky, L. A., Revercomb, H. E., Krauss, R. J., and Suomi, V. E. 1983. Voyager 2 observations of Saturn's northern mid latitude cloud features: Morphology, motions and evolution. *J. Geophys. Res.* 88:8650–8666.

Stahl, H. P., Tomasko, M. G., Wolfe, W. L., Castillo, N. D., and Stahl, K. A. 1983. Measurements of the light scattering properties of water ice crystals. *IEEE Trans.* In press.

Starr, V. P. 1968. *Physics of Negative Viscosity Phenomena* (New York: McGraw-Hill).

Steklov, A. F., Vidmachenko, A. P., and Minyajlo, N. F. 1983. Seasonal changes in Saturn's atmosphere. *Astron Zh. Letters.* In press (Russian).

Stern, M. E. 1975. Minimal properties of planetary eddies. *J. Marine Res.* 33:1–13.

Stevenson, D. J. 1975. Thermodynamics and phase separation of dense, fully ionized hydrogen-helium fluid mixtures. *Phys. Rev.* 12B:3999–4007.

Stevenson, D. J. 1977. Hydrogen in the Earth's core. *Nature* 268:130–131.

Stevenson, D. J. 1979. Solubility of helium in metallic hydrogen. *J. Phys. F. Metal Phys.* 9:791–800.

Stevenson, D. J. 1980. Saturn's luminosity and magnetism. *Science* 208:746–748.

Stevenson, D. J. 1982a. Formation of the giant planets. *Planet. Space Sci.* 30:755–764.

Stevenson, D. J. 1982b. Interiors of the giant planets. *Ann. Rev. Earth Planet. Sci.* 10:257–295.

Stevenson, D. J. 1982c. Reducing the nonaxisymmetry of a planetary dynamo and an application to Saturn. *Geophys. Astrophys. Fluid Dyn.* 21:113–127.

Stevenson, D. J. 1983a. Planetary magnetic fields. *Rep. Prog. Phys.* 46:555–620.

Stevenson, D. J. 1983b. Anomalous bulk viscosity of two-phase fluids and implications for planetary interiors. *J. Geophys. Res.* 88:2445–2455.

Stevenson, D. J., and Ashcroft, N. W. 1974. Conduction in fully-ionized liquid metals. *Phys. Rev.* 9A:782–789.

Stevenson, D. J., and Salpeter, E. E. 1976. Interior models of Jupiter. In *Jupiter*, ed. T. Gehrels (Tucson: Univ. of Arizona Press), pp. 85–112.

Stevenson, D. J., and Salpeter, E. E. 1977a. The phase diagram and transport properties of hydrogen-helium fluid planets. *Astrophys. J. Suppl.* 35:221–237.

Stevenson, D. J., and Salpeter, E. E. 1977b. The dynamics and helium distribution properties for hydrogen-helium fluid planets. *Astrophys. J. Suppl.* 35:239–261.

Stevenson, D. J., and Fishbein, E. 1981. The behavior of water in the giant planets. *Lunar Planet. Sci.* 12:1040–1042 (abstract).

Stone, E. C., and Miner, E. D. 1981. Voyager 1 encounter with the Saturnian system. *Science* 212:159–162.

Stone, E. C., and Miner, E. D. 1982. Voyager 2 encounter with the Saturnian system. *Science* 215:499–504.

Stone, E. C., Vogt, R. E., McDonald, F. B., Teegarden, B. J., Trainor, J. H., Jokipii, J. R., and Webber, W. R. 1977. Cosmic ray investigation for the Voyager mission: Energetic particle studies in the outer heliosphere and beyond. *Space Sci. Rev.* 21:355–376.

Stone, P. H. 1972. A simplified radiative-dynamical model for the static stability of rotating atmospheres. *J. Atmos. Sci.* 29:405–418.

Stone, P. H. 1973. The dynamics of the atmospheres of the major planets. *Space Sci. Rev.* 14:444–459.

Stone, P. H. 1974. The structure and circulation of the deep Venus atmosphere. *J. Atmos. Sci.* 31:1681–1690.

Stone, P. H. 1976. The meteorology of the Jovian atmosphere. In *Jupiter*, ed. T. Gehrels (Tucson: Univ. of Arizona Press), pp. 586–618.

Straus, D. M., Ashcroft, N. W., and Beck, H. 1977. Phase separation of metállic hydrogen-helium alloys. *Phys. Rev.* 15B:1914–1921.

Streett, W. B. 1974. Phase equilibria in molecular H_2-He mixtures at high pressures. *Astrophys. J.* 186:1107–1125.

Strickland, D. J., Book, D. L., Coffey, T. P., and Fedder, J. A. 1976. Transport equation techniques for deposition of auroral electrons. *J. Geophys. Res.* 81:2755–2764.

Strobel, D. F. 1974. The photochemistry of hydrocarbons in the atmosphere of Titan. *Icarus* 21:466–470.

Strobel, D. F. 1975. Aeronomy of the major planets: Photochemistry of ammonia and hydrocarbons. *Rev. Geophys. Space Phys.* 13:372–382.

Strobel, D. F. 1977. Ammonia and phosphine photochemistry in the Jovian atmosphere. *Astrophys. J.* 214:L97–L99.

Strobel, D. F. 1978. Aeronomy of Saturn and Titan. In *The Saturn System*, eds. D. Hunten and D. Morrison (Washington: NASA Conf. Publ. 2068), pp. 185–194.

Strobel, D. F. 1982. Chemistry and evolution of Titan's atmosphere. *Planet. Space Sci.* 30: 839–848.

Strobel, D. F., and Smith, G. R. 1973. On the temperature of the Jovian atmosphere. *J. Atmos. Sci.* 30:718–725.

Strobel, D. F., and Yung, Y. 1979. The Galilean satellites as a source of CO in the Jovian atmosphere. *Icarus* 37:256–263.

Strobel, D. F., and Atreya, S. K. 1983. Ionosphere. In *Physics of the Jovian Magnetosphere*, ed. A. J. Dessler (Cambridge: Cambridge Univ. Press), pp. 51–67.

Strobel, D. F., and Shemansky, D. E. 1982. EUV emission from Titan's upper atmosphere: Voyager 1 encounter. *J. Geophys. Res.* 87:1361–1368.

Struve, F. G. W. 1826. Micrometrical observations of the planet Saturn, made with Fraunhofer's large refractor, at Dorpat. *Memoirs Roy. Astron. Soc.* 2:513–517.

Struve, H. 1890. Vorläufige Resultate aus den Beobachtungen der Saturnstrabanden am 30 zollingen Refractor. *Astron. Nachr.* 125:97–118.

Studier, M. H., Hayatsu, R., and Anders, E. 1968. Origin of organic matter in the early solar system. I. Hydrocarbons. *Geochim. Cosmochim. Acta* 18:151–173.

Suggs, R. 1982. Temporal albedo variations in Saturn's atmosphere. Program "Saturn Conference," Tucson, AZ, May 1982 (abstract).

Swedlund, J. B., Kemp, J. C., and Wolstencroft, R. D. 1972. Circular polarization of Saturn. *Astrophys. J.* 178:257–265.

Synnott, S. P., Peters, C. F., Smith, B. A., and Morabito, L. A. 1981. Orbits of the small satellites of Saturn. *Science* 212:191–192.

Synnott, S. P., Terrile, R. J., and Smith, B. A. 1983. Orbits of Saturn's F-ring and its shepherding satellites. *Icarus* 53:156–158.

Taylor, F. W. 1973. Preliminary data on the optical properties of solid ammonia and scattering properties in the infrared. *J. Atmos. Sci.* 30:677–683.

Tejfel, V. G. 1967. The CH_4 absorption in the equatorial zone of Saturn on photoelectric measurements. *Astron. Tsirkular USSR Acad. Sci.* 421:1–3 (Russian).

Tejfel, V. G. 1974a. The atmosphere of Saturn. In *Exploration of Planetary Systems* (IAU Symp. No. 65), pp. 415–440.

Tejfel, V. G. 1974b. Saturn's atmosphere from measurements in the near ultraviolet. *Astron. Vestn.* 8:3–12 (Russian).

Tejfel, V. G. 1975. On the limb darkening on Saturn's disk. *Astron. Zh.* 52:823–831 (Russian).

Tejfel, V. G. 1976. On the methane and ammonia abundance in the atmosphere of Saturn. *Astron. Zh. Letters* 2:584–588 (Russian).

Tejfel, V. G. 1977a. Latitudinal differences in the structure of Saturn's cloud cover. *Astron. Vestn.* 11:1–24 (Russian). English translation *Solar System Res.* 11:10–18.

Tejfel, V. G. 1977b. On the determination of the two-layer planetary atmosphere model parameters from the absorption bands observations. *Astron. Zh.* 54:178–189 (Russian).

Tejfel, V. G. 1980. Optical properties and structure of Saturn's atmosphere. *Astron. Vestn.* 14:3–19 (Russian).

Tejfel, V. G., and Kharitonova, G. A. 1970. The study of the absorption distribution on Saturn's disk in the CH_4 6190 Å band. *Astron Tsirkular USSR Acad. Sci.* 549:5–8 (Russian).

Tejfel, V. G., and Kharitonova, G. A. 1972a. The cloud cover altitude and equatorial acceleration in Saturn's atmosphere. *Astron. Tsirkular USSR Acad. Sci.* 735:4–6 (Russian).

Tejfel, V. G., and Kharitonova, G. A. 1972b. Some properties of the cloud cover on Saturn. *Astron. Tsirkular USSR Acad. Sci.* 683:4–7 (Russian).

Tejfel, V. G., and Kharitonova, G. A. 1974. Optical properties and the structure of Saturn's atmosphere. III. The vertical structure of the aerosol layer derived from data of the photographic and photoelectric spectrophotometry. *Astron. Zh.* 51:167–177 (Russian).

Tejfel, V. G., and Kharitonova, G. A. 1981. Geometric albedo of Saturn in 1980. *Astron. Tsirkular USSR Acad. Sci.* 1178:7–8 (Russian).

Tejfel, V. G., and Kharitonova, G. A. 1982. Zonal differences of the spectrophotometric gradients on Saturn's disk from the observations in 1979. *Astron. Tsirkular USSR Acad. Sci.* In press (Russian).

Tejfel, V. G., Usoltseva, L. A., and Kharitonova, G. A. 1971a. Optical properties and the structure of Saturn's atmosphere. I. Preliminary results on the study of the CH_4 absorption bands on the disk of the planet. *Astron. Zh.* 48:380–389 (Russian).

Tejfel, V. G., Usoltseva, L. A., and Kharitonova, G. A. 1971b. The spectral characteristics and probable structure of the cloud layer of Saturn. In *Planetary Atmospheres* (IAU Symp. No. 40), pp. 375–383.

Tejfel, V. G., Usoltseva, L. A., and Kharitonova, G. A. 1973. Optical properties and the structure of Saturn's atmosphere. II. The latitudinal variations of the absorption in the CH_4 band 0.62 μm and the peculiarities of the planet in the near ultraviolet. *Astron. Zh.* 50:167–171 (Russian).

Terrile, R. J., and Beebe, R. F. 1979. Summary of historical data: Interpretations of the Pioneer and Voyager cloud configurations in a time-dependent framework *Science* 204:948–951.

Terrile, R. J., and Tokunaga, A. 1980. Infrared photometry of Saturn's E-ring. *Bull. Amer. Astron. Soc.* 12:701 (abstract).

Terrile, R. J., Yagi, G., and Cook, A. F. 1981. A morphological model for spoke formation in Saturn's rings. *Bull. Amer. Astron. Soc.* 13:728 (abstract).

Theard, L. P., and Huntress, W. T. Jr. 1974. Ion-molecule reactions and vibrational deactivation of H_2^+ ions in mixtures of hydrogen and helium. *J. Chem. Phys.* 60:2840–2848.

Thieman, J. R., and Goldstein, M. L. 1981. Arcs in Saturn's radio spectra. *Nature* 292:728–731.

Tholen, D. J., and Zellner, B. 1982. Eight color photometry of Hyperion, Iapetus and Phoebe. *Icarus* 53:341–347.

Tholen, D. J., and Zellner, B. H. 1984. Multi-color photometry of outer Jovian satellites. Submitted to *Icarus*.

Thomas, P., Veverka, J., Morrison, D., and Davies, M. 1981. Photometry and topography of Saturn's small satellites from Voyager data. *Bull. Amer. Astron. Soc.* 13:720–721 (abstract).

Thomas, P., Veverka, J., and Davies, M. 1983a. Voyager observations of Hyperion. In preparation.

Thomas, P., Veverka, J., Morrison, D., Davies, M., and Johnson, T. V. 1983b. Saturn's small satellites: Voyager imaging results. *J. Geophys. Res.* 88:8743–8754.

Thomas, P., Veverka, J., Morrison, D., Davies, M., and Johnson, T. V. 1983c. Phoebe: Voyager 2 observations. *J. Geophys. Res.* 88:8736–8742.

Thomas, P., Veverka, J., Wenkert, D., Danielson, G. E., and Davies, M. 1983d. Voyager photometry of Hyperion: Rotation rate. Proceedings "Planetary Satellites Conference," Ithaca, NY, July 1983 (abstract).

Thomsen, M. F., and Van Allen, J. A. 1979. On the inference of properties of Saturn's ring E from energetic charged particle observations. *Geophys. Res. Letters* 6:893–896.

Thomsen, M. F., and Van Allen, J. A. 1980. Motion of trapped electrons and protons in Saturn's inner magnetosphere. *J. Geophys. Res.* 85:5831–5834.

Thomsen, M. F., Northrop, T. G., Schardt, A. W., and Van Allen, J. A. 1980. Corotation of Saturn's magnetosphere: Evidence from energetic proton anisotropies. *J. Geophys. Res.* 85: 5725–5730.

Thomsen, M. F., Goertz, C. K., Northrop, T. G., and Hill, J. R. 1982. On the nature of particles in Saturn's spokes. *Geophys. Res. Letters* 9:423–426.

Thomson, W. T. 1982. The 'Swarm' Model for the Azimuthal Brightness Variations in Saturn's Ring A. Ph.D. thesis, Univ. of Massachusetts, Amherst.

Thomson, W. T., Irvine, W. M., Baum, W. A., Lumme, K., and Esposito, L. W. 1981. Saturn's rings—azimuthal variations, phase curves and radial profiles in four colors. *Icarus* 46: 187–200.

Thorne, R. T., and Tsurutani, B. T. 1979. Diffuse Jovian aurora influenced by plasma injection from Io. *Geophys. Res. Letters* 6:649–652.

Thurber, C. H., Hsui, A. T., and Toksoz, M. N. 1980. Thermal evolution of Ganymede and Callisto: Effects of solid-state convection and constraints from Voyager imagery. *Proc. Lunar Planet. Sci. Conf.* 11:1957–1977.

Todoeschuck, J. P., Crossley, D. J., and Rochester, M. G. 1981. Saturn's magnetic field and dynamo theory. *Geophys. Res. Letters* 8:505–508.

Tokunaga, A. 1978. The thermal structure of Saturn: Inferences from ground-based and airborne infrared observations. In *The Saturn System*, eds. D. Hunten and D. Morrison (Washington: NASA Conf. Publ. 2068), pp. 53–72.

Tokunaga, A. 1980. The detection of C_2H_2 on Saturn and Titan. *Bull. Amer. Astron. Soc.* 12:669 (abstract).

Tokunaga, A., and Cess, R. D. 1977. A model for the temperature inversion within the atmosphere of Saturn. *Icarus* 32:321–327.

Tokunaga, A., Knacke, R. F., and Owen, T. 1975. The detection of ethane on Saturn. *Astrophys. J.* 197:L77–L78.

Tokunaga, A., Knacke, R. F., and Owen, T. 1977. 17–25 micrometer spectra of Jupiter and Saturn. *Astrophys. J.* 213:569–574.

Tokunaga, A., Knacke, R. F., Ridgway, S. T., and Wallace, L. 1979. High resolution spectra of Jupiter in the 744–980 cm^{-1} spectral range. *Astrophys. J.* 232:603–615.

Tokunaga, A., Beck, S. C., Geballe, T. R., and Lacy, J. H. 1980a. The detection of C_2H_2 on Saturn and Titan. *Bull. Amer. Astron. Soc.* 12:669 (abstract).

Tokunaga, A. T., Caldwell, J., and Nolt, I. G. 1980b. The 20-micron brightness temperature of the unilluminated side of Saturn's rings. *Nature* 287:212–214.

Tokunaga, A., Dinerstein, H. L., Lester, D. F., and Rank, D. M. 1980c. The phosphine abundance on Saturn derived from new 10 micron spectra. *Icarus* 42:79–85.

Tokunaga, A., Beck, S. C., Geballe, T. R., Lacy, J. H., and Serabyn, E. 1981a. The detection of HCN on Jupiter. *Icarus* 48:283–289.

Tokunaga, A., Dinerstein, H. L., Lester, D. F., and Rank, D. M. 1981b. Erraturm to: The phosphine abundance on Saturn derived from new 10 micron spectra. *Icarus* 48:540–541.

Tomasko, M. G., and Smith, P. H. 1982. Photometry and polarimetry of Titan: Pioneer 11 observations and their implications for aerosol properties. *Icarus* 51:65–95.

Tomasko, M. G., and Stahl, H. P. 1982. Measurements of the single-scattering properties of water and ammonia ice crystals. Proceedings "Fourth Annual Meeting of Planetary Atmospheres Principal Investigators," Ann Arbor, MI, April 1982 (abstract).

Tomasko, M. G., and Doose, L. R. 1984. Polarimetry and photometry of Saturn from Pioneer 11: Observations and constraints on the distribution and properties of cloud aerosol particles. *Icarus* vol. 57.

Tomasko, M. G., West, R. A., and Castillo, N. D. 1978. Photometry and polarimetry of Jupiter at large phase angles. I. Analysis of imaging data of a prominent belt and a zone from Pioneer 10. *Icarus* 33:558–592.

Tomasko, M. G., Doose, L. R., Smith, P. H., and O'Dell, A. P. 1980a. Measurements of the flux of sunlight in the atmosphere of Venus. *J. Geophys. Res.* 85:8167–8186.

Tomasko, M. G., McMillan, R. S., Doose, L. R., Castillo, N. D., and Dilley, J. P. 1980b. Photometry of Saturn at large phase angles. *J. Geophys. Res.* 85:5891–5903.

Tomasko, M. G., Doose, L. R., and West, R. A. 1984. The vertical distribution of aerosols on Saturn. To be submitted to *Icarus*.

Toon, O. B., Turco, R. P., and Pollack, J. B. 1980. A physical model of Titan's clouds. *Icarus* 43:260–282.

Toupance, G., Raulin, F., and Buvet, R. 1975. Formation of prebiochemical compound in models of the primitive Earth's atmosphere I: CH_4-NH_3 and CH_4-N_2 atmospheres. *Origins of Life* 6:83–90.

Trafton, L. M. 1972a. On the possible detection of H_2 in Titan's atmosphere. *Astrophys. J.* 175:285–293.

Trafton, L. M. 1972b. The bulk composition of Titan's atmosphere. *Astrophys. J.* 175:295–306.

Trafton, L. M. 1973. Saturn: A study of the $3\nu_3$ methane band. *Astrophys. J.* 182:615–636.

Trafton, L. M. 1977. Saturn: Long-term variation of H_2 and CH_4 absorptions. *Icarus* 31: 369–384.

Trafton, L. M. 1978. Saturn's atmosphere: Results of recent investigations. In *The Saturn System*, eds. D. M. Hunten and D. Morrison (Washington: NASA Conf. Publ. 2068), pp. 31–51.

Trafton, L. M. 1981. The atmospheres of the outer planets and satellites. *Rev. Geophys. Space Phys.* 19:43–89.

Trafton, L. M., and Macy, W. 1975. Saturn's $3\nu_3$ methane band: An analysis in terms of a scattering atmosphere. *Astrophys. J.* 196:867–876.

Trafton, L. M., Cochran, W., and Macy, W. 1979. On seasonal phenomena in Saturn's atmosphere: New observations of Saturn's $3\nu_3$ methane band. *Publ. Astron. Soc. Pacific* 91: 702–706.

Trainor, J. H., McDonald, F. B., and Schardt, A. W. 1980. Observations of energetic ions and electrons in Saturn's magnetosphere. *Science* 207:421–425.

Trauger, J., Roesler, F., and Michelson, M. 1977. The D/H ratios on Jupiter, Saturn, and Uranus based on new HD and H_2 data. *Bull. Amer. Astron. Soc.* 9:516 (abstract).

Tritton, D. J. 1977. *Physical Fluid Dynamics* (New York: Van Nostrand Reinhold).

Trubitsyn, V. P., Vasil'ev, P. P., and Efimov, A. B. 1976. Gravitational fields and figures of differentially rotating planets. *Astron. Zh.* 53:1278–1287.

Turner, J. S. 1973. *Bouyancy Effects in Fluids* (Cambridge: Cambridge Univ. Press).

Tyler, G. L., Eshleman, V. R., Anderson, J. D., Levy, G. S., Lindal, G. F., Wood, G. E., and Croft, T. A. 1981. Radio science investigations of the Saturn system with Voyager 1: Preliminary results. *Science* 212:201–206.

Tyler, G. L., Eshleman, V. R., Anderson, J. D., Levy, G. S., Lindal, G. F., Wood, G. E., and Croft, T. A. 1982a. Radio science with Voyager 2 at Saturn: Atmosphere and ionosphere and the masses of Mimas, Tethys, and Iapetus. *Science* 215:553–558.

Tyler, G. L., Marouf, E. A., Simpson, R. A., Zebker, H. A., and Eshleman, V. R. 1982b. The microwave opacity of Saturn's rings at wavelengths of 3.6 and 13 cm from Voyager radio occultation. *Icarus* 54:189–211.

Ulrich, B. L. 1974. Absolute brightness temperature measurements at 2.1 mm wavelength. *Icarus* 21:254–261.

Uman, M. A. 1969. *Lightning* (New York: McGraw-Hill).

U.S. Standard Atmosphere Supplements, 1966. Nat. Oceanic Atmos. Admin. S/T (Washington: U.S. Gov't Printing Office).

Van Allen, J. A. 1982. Findings on rings and inner satellites of Saturn by Pioneer 11. *Icarus* 51:509–527.

Van Allen, J. A. 1983. Absorption of energetic protons by Saturn's ring G. *J. Geophys. Res.* 88:6911–6918.

Van Allen, J. A., Baker, D. N., Randall, B. A., and Sentman, D. D. 1974. The magnetosphere of Jupiter as observed with Pioneer 10. 1. Instrument and principal findings. *J. Geophys. Res.* 79:3559–3577.

Van Allen, J. A., Randall, B. A., and Thomsen, M. F. 1980a. Sources and sinks of energetic electrons and protons in Saturn's magnetosphere. *J. Geophys. Res.* 85:5679–5694.

Van Allen, J. A., Thomsen, M. F., and Randall, B. A. 1980b. The energetic charged particle absorption signature of Mimas. *J. Geophys. Res.* 85:5709–5718.

Van Allen, J. A., Thomsen, M. F., Randall, B. A., Rairden, R. L., and Grosskreutz, C. L. 1980c. Saturn's magnetosphere, rings, and inner satellites. *Science* 207:415–421.

van de Hulst, H. C. 1957. *Light Scattering by Small Particles*, (New York: John Wiley & Sons). Revised 1981 (New York: Dover Publ.).

van der Waals, J. H., and Platteeuw, J. C. 1959. Clathrate solutions. *Adv. Chem. Phys.* 2:1–57.

Van Helden, A. 1968. Christopher Wren's *De Corpore Saturni*. *Notes and Records Roy. Soc. London* 23:213–229.

Van Helden, A. 1973. The Accademia del Cimento and Saturn's ring. *Physis* 15:237–259.

Van Helden, A. 1974a. Saturn and his Anses. *J. Hist. Astron.* 5:105–121.

Van Helden, A. 1974b. 'Annulo Cingitur': The solution to the problem of Saturn. *J. Hist. Astron.* 5:155–174.

Van Helden, A. 1977. The invention of the telescope. *Trans. Amer. Phil. Soc.* 67, no. 4.

van Kaskteren, P. H. G. 1973. The crystallization behavior and caloric properties of water/ammonia mixtures between 70 K and 300 K. *Bull. Inst. Int. Froid. Annexe* 4:81–87.

van Stratten, J., Wijngaarden, R. J., and Silvera, I. F. 1982. Low temperature equation of state of molecular hydrogen and deuterium to 0.37 Mbar: Implications for metallic H. *Phys. Rev. Letters* 48:97–100.

Vasil'ev, P. P., Efimov, A. B., and Trubitsyn, V. P. 1978. The effect of differential rotation on figures and fields of planets. *Astron. Zh.* 55:148–155.

Vasyliunas, V. T. 1971. Deep space plasma measurements. In *Methods in Experimental Physics*, Vol. 9, eds. R. Lovberg, and H. Griem (New York: Academic Press).

Vegard, L. 1930. Structur und leuchtfahigkeit von festem kohlenoxyd. *Z. Phys.* 61:185–190.

Veverka, J. 1970. Photometric and Polarimetric Studies of Minor Planets and Satellites. Ph.D. thesis, Harvard Univ., Cambridge, Massachusetts.

Veverka, J. 1973. Titan: Polarimetric evidence for an optically thick atmosphere. *Icarus* 18:657–660.

Veverka, J. 1977a. Photometry of satellite surfaces. In *Planetary Satellites*, ed. J. Burns (Tucson: Univ. of Arizona Press), pp. 171–209.

Veverka, J. 1977b. Polarimetry of satellite surfaces. In *Planetary Satellites*, ed. J. Burns (Tucson: Univ. of Arizona Press), pp. 210–231.

Veverka, J., Elliot, J., Wasserman, L., and Sagan, C. 1974. The upper atmosphere of Jupiter. *Astron. J.* 179:73–84.

Veverka, J., Burt, J., Elliot, J. L., and Goguen, J. 1978. Lunar occultation of Saturn. III. How big is Iapetus? *Icarus* 33:301–310.

Vdovichenko, V. D., Kuratov, K. S., and Kharitonova, G. A. 1979. Distribution of methane absorption on Saturn's disk in the spectral region 0.6–1.1 μm. *Trudy Astrophys. Inst. Acad. Sci. KazSSR* 335:86–92 (Russian).

Voge, C. C., Pang, K. D., and Rhodes, J. K. 1982. Voyager 2 star tracker observations of Saturn's E-ring. *EOS* 63:1019 (abstract).

Vogt, R. E., Chenette, D. L., Cummings, A. C., Garrard, T. L., Stone, E. C., Schardt, A. W., Trainor, J. H., Lal, N., and McDonald, F. B. 1981. Energetic charged particles in Saturn's magnetosphere: Voyager 1 results. *Science* 212:231–234.

Vogt, R. E., Chenette, D. L., Cummings, A. C., Garrard, T. L., Stone, E. C., Schardt, A. W., Trainor, J. H., Lal, N., and McDonald, F. B. 1982. Energetic charged particles in Saturn's magnetosphere: Voyager 2 results. *Science* 215:577–582.

von Zahn, U., Fricke, K. H., Hunten, D. M., Kraukowsky, D., Mauersberger, K., and Nier, A. O. 1980. The upper atmosphere of Venus during morning conditions. *J. Geophys. Res.* 85:7829–7840.

Waite, J. H. Jr. 1981. The Ionosphere of Saturn. Ph.D. thesis, Univ. of Michigan.

Waite, J. H. Jr., Atreya, S. K., and Nagy, A. F. 1979. The ionosphere of Saturn: Predictions for Pioneer 11. *Geophys. Res. Letters* 6:723–726.

Waite, J. H. Jr., Atreya, S. K., and Cravens, T. E. 1983a. A fact of atomic hydrogen and vibrationally excited molecular hydrogen on the ionosphere of Saturn. Submitted to *J. Geophys. Res.*

Waite, J. H. Jr., Cravens, T. E., Kozyra, J., Nagy, A. F., Atreya, S. K., and Chen, R. H. 1983b. Electron precipitation and related aeronomy of the Jovian thermosphere and ionosphere. *J. Geophys. Res.* 88:6143–6163.

Wallace, L. 1975. On the thermal structure of Uranus. *Icarus* 25:538–544.

Wallace, L., and Hunten, D. M. 1973. The Lyman-alpha albedo of Jupiter. *Astrophys. J.* 182:1013–1031.

Wallace, L., Caldwell, J., and Savage, B. D. 1972. Ultraviolet photometry from the Orbiting Astronomical Observatory. III. Observations of Venus, Mars, Jupiter, and Saturn longward of 2000 Å. *Astrophys. J.* 172:755–769.

Wallis, M. K. 1973. Weakly-shocked flows of the solar wind plasma through atmospheres of comets and planets. *Planet. Space Sci.* 21:1647–1660.

Wamsteker, W. 1975. A Spectrophotometric Study of the Major Planets and their Large Satellites. Ph.D. thesis, Leiden Univ.

Ward, D. B. 1977. Far-infrared spectral observations of Saturn and its rings. *Icarus* 32:437–442.

Ward, W. R. 1981*a*. On the radial structure of Saturn's rings. *Geophys. Res. Letters* 8:641–643.

Ward, W. R. 1981*b*. Orbital inclination of Iapetus and the rotation of the Laplacian plane. *Icarus* 46:97–107.

Warwick, J. W., Pearce, J. B., Peltzer, R. G., and Riddle, A. C. 1977. Planetary radio astronomy instrument for the Voyager missions. *Space Sci. Rev.* 21:309–319.

Warwick, J. W., Pearce, J. B., Riddle, A. C., Alexander, J. K., Desch, M. D., Kaiser, M. L., Thieman, J. R., Carr, T. D., Gulkis, S., Boischot, A., Harvey, C. C., and Pedersen, B. M. 1979. Voyager 1 planetary radio astronomy observations near Jupiter. *Science* 204:995–998.

Warwick, J. W., Pearce, J. B., Evans, D. R., Carr, T. D., Schauble, J. J., Alexander, J. K., Kaiser, M. L., Desch, M. D., Pedersen, B. M., Lecacheux, A., Daigne, G., Boischot, A., and Barrow, C. H. 1981. Planetary radio astronomy observations from Voyager 1 near Saturn. *Science* 212:239–243.

Warwick, J. W., Evans, D. R., Romig, J. H., Alexander, J. K., Desch, M. D., Kaiser, M. L., Aubier, M., Leblanc, Y., Lecacheux, A., and Pedersen, B. M. 1982*a*. Planetary radio astronomy observations from Voyager 2 near Saturn. *Science* 215:582–587.

Warwick, J. W., Romig, J. H., and Evans, D. R. 1982*b*. Saturn electrostatic discharges. Proceedings "Planetary Rings/Anneaux des Planètes Conference," Toulouse, France, August 1982 (abstract).

Washburn, E. W. ed. 1928. *International Critical Tables* (National Research Council: McGraw-Hill).

Weidenschilling, S. J. 1978. Iron/silicate fractionation and the origin of Mercury. *Icarus* 35:99–111.

Weidenschilling, S. J. 1981. Aspects of accretion in a circumplanetary nebula. *Lunar Planet. Sci.* 12:1170–1172.

Weidenschilling, S. J. 1982. Origin of regular satellites. In *Comparative Study of the Planets*, eds. A. Coradini, and M. Fulchignoni (Dordrecht: D. Reidel), pp. 49–59.

Weidenschilling, S. J., and Lewis, J. S. 1973. Atmospheric and cloud structure of the Jovian planets. *Icarus* 20:465–476.

Weinheimer, A. J., and Few, A. A. Jr. 1982. The spokes in Saturn's rings: A critical evaluation of possible electrical processes. Rice Univ., Dept. of Space Physics & Astronomy, Houston, preprint.

Weiser, H., Vitz, R., and Moos, H. 1977. Detection of Lyman alpha emission from the Saturnian disk and from the ring system. *Science* 197:755–757.

Weiss, S., Leroi, G. E., and Cole, R. H. 1969. Pressure-induced infrared spectrum of methane. *J. Chem. Phys.* 50:2267.

Werner, M. W., Neugebauer, G., Houck, J. R., and Hauser, M. G. 1978. One millimeter observations of the planets. *Icarus* 35:289–296.

West, R. A. 1983. Spatially resolved methane band photometry of Saturn. II. Cloud structure models at four latitudes. *Icarus* 53:301–309.

West, R. A., and Tomasko, M. G. 1980. Spatially resolved methane band photometry of Jupiter. III. Cloud vertical structures for several axisymmetric bands and the Great Red Spot. *Icarus* 41:278–292.

West, R. A., Hord, C. W., Simmons, K. E., Coffeen, D. L., Sato, M., and Lane, A. L. 1981. Near-ultraviolet scattering properties of Jupiter. *J. Geophys. Res.* 86:8783–8792.

West, R. A., Tomasko, M. G., Wijesinghe, M. P., Doose, L. R., Reitsema, H. J., and Larson, S. M. 1982. Spatially resolved methane band photometry of Saturn. I. Absolute reflectivity and center-to-limb variations in the 6190-, 7250-, and 8900-Å bands. *Icarus* 51:51–64.

West, R. A., Hord, C. W., Simmons, K. E., Hart, H., Esposito, L. W., Lane, A. L., Pomphrey, R. B., Morris, R. B., Sato, M., and Coffeen, D. L. 1983*a*. Voyager photopolarimeter observations of Saturn and Titan. *Adv. Space Res.* 3:45–48.

West, R. A., Lane, A. L., Hart, H., Simmons, K. E., Hord, C. W., Coffeen, D. L., Esposito, L. W., Sato, M., and Pomphrey, R. B. 1983*b*. Voyager 2 photopolarimeter observations of Titan. *J. Geophys. Res.* 88:8699–8708.

West, R. A., Sato, M. Hart, H., Lane, L. A., Hord, C. W., Simmons, K. E., Esposito, L. W.,

Coffeen, D. L., and Pomphrey, R. B. 1983c. Photometry and polarimetry of Saturn at 2640 and 7500 Å. *J. Geophys. Res.* In press.

Wetherill, G. W. 1981. Solar wind origin of [36]Ar on Venus. *Icarus* 46:70–80.

Whalley, E., and Labbe', H. J. 1969. Optical spectra of orientally disordered crystals. III. Infrared spectra of the sound waves. *J. Chem. Phys.* 51:3120–3127.

Whipple, E. C. 1981. Potentials of surfaces in space. *Reports Progress Phys.* 44:1197.

Whipple, E. C., Mendis, D. A., Ip, W.-H., and Hill, J. R. 1983. The capacitance of a charged grain in a dusty plasma (unpublished).

Whitcomb, S. E., Hildebrand, R. H., and Keene, J. 1980. Brightness temperatures at Saturn's disk and rings at 400 and 700 micrometers. *Science* 210:788–789.

White, R. S. 1973. High-energy proton radiation belt. *Rev. Geophys. Space Phys.* 11:595–632.

Wiesel, W. 1982. Saturn's rings: Resonances about an oblate planet. *Icarus* 51:149–154.

Wight, G. R., Vander Weil, M. J., and Brion, C. E. 1976. Dipole excitation, ionization, and fragmentation of N_2 and CO in the 10–60 eV region. *J. Phys. B.: Atom. Molec. Phys.* 9:675.

Wijngaarden, R. J., Lagendijk, A., and Silvera, I. F. 1982. Pressure dependence of the vibron in solid hydrogen and deuterium up to 600 Kbar. *Phys. Rev.* 26B:4957–4962.

Williams, A. S. 1894. On the rotation of Saturn. *Mon. Not. Roy. Astron. Soc.* 54:297–314.

Williams, A. S. 1895. On the rotation of Saturn in 1894. *Mon. Not. Roy. Astron. Soc.* 55:354–367.

Williams, G. P. 1975. Jupiter's atmospheric circulation. *Nature* 257:778.

Williams, G. P. 1978. Planetary circulations. 1. Barotropic representations of Jovian and terrestrial turbulence. *J. Atmos. Sci.* 35:1399–1426.

Williams, G. P. 1979. Planetary circulations. 2. The Jovian quasi-geostrophic regime. *J. Atmos. Sci.* 36:932–968.

Williams, G. P., and Robinson, J. B. 1973. Dynamics of a convectively unstable atmosphere: Jupiter? *J. Atmos. Sci.* 30:684–717.

Williams, I., and Fuller, M. 1981. Zonal harmonic models of reversal transition fields. *J. Geophys. Res.* 11:657–665.

Winkelstein, P., Caldwell, J., Owen, T., Combes, M., Encrenaz, T., Hunt, G., and Moore, V. 1983. A determination of the composition of the Saturnian stratosphere using the IUE. *Icarus* 54:309–318.

Wisdon, J., Peale, S. J., and Mignard, F. 1984. The chaotic rotation of Hyperion. Submitted to *Icarus*.

Witteborn, F. C., Pollack, J. B., Bergman, J. D., Goebel, H. J., Soifer, B. T., Puetter, R. C., Rudy, R. J., and Willner, S. P. 1981. Observations of Saturn in the 5 to 8 μm spectral region. *Icarus* 45:653–660.

Woeller, F., and Ponnamperuma, C. 1969. Organic synthesis in a simulated Jovian atmosphere. *Icarus* 10:386–392.

Wolf, D. A., and Neubauer, F. M. 1982. Titan's highly variable plasma environment. *J. Geophys. Res.* 87:881–885.

Wolfe, J. H., Mihalov, J. D., Collard, H. R., McKibben, D. D., Frank, L. A., and Intriligator, D. S. 1980. Preliminary results on the plasma environment of Saturn from the Pioneer 11 plasma analyzer experiment. *Science* 207:403–407.

Wolff, M. 1975. Polarization of light reflected from rough planetary surfaces. *Appl. Optics* 14:1395–1405.

Wolff, R. S., and Mendis, D. A. 1983. On the nature of interaction of the Jovian magnetosphere with the icy Galilean satellites. *J. Geophys. Res.* 88:4749–4769.

Woltjer, J. 1922. Is there libration in the system Enceladus-Dione? *BAN* 1:23 and 36.

Woltjer, J. 1928. The motion of Hyperion. *Ann. Sterrewacht Leiden* XVI, part 3.

Woodman, J. H., Trafton, L. M., and Owen, T. 1977. The abundances of ammonia in the atmospheres of Jupiter, Saturn, and Titan. *Icarus* 32:314–320.

Wu, C. S., and Lee, L. C. 1979. A theory of terrestrial kilometric radiation. *Astrophys. J.* 230:621–626.

Wyatt, S. P. 1969. The electrostatic charge of interplanetary grains. *Planet. Space Sci.* 17:155–171.

Yagi, G. M., Lorre, J. J., and Jepson, P. L. 1978. Dynamic feature analysis for Voyager at the image processing laboratory. In *Proceedings of a Conference on Atmospheric Environment of Aerospace Systems and Applied Meteorology* (Boston: Amer. Meteorological Soc.), pp. 110–117.

Yoder, C. F. 1975. Establishment and evolution of satellite-satellite resonance. *Cel. Mech.* 12:97.

Yoder, C. F. 1979. How tidal heating on Io drives the Galilean orbital resonance locks. *Nature* 279:767–770.

Yoder, C. F. 1981. Tidal friction and Enceladus' anomalous surface. *EOS Trans. Amer. Geophys. Union* 62:939 (abstract).

Yoder, C. F., Colombo, G., Synnott, S. P., and Yoder, K. A. 1983. Theory of motion of Saturn's coorbiting satellites. *Icarus* 53:431–443.

Young, D. A., and Ross, M. 1981. Theoretical calculation of thermodynamic properties and melting curves for hydrogen and deuterium. *J. Chem. Phys.* 74:6950–6955.

Young, D. A., McMahan, A. K., and Ross, M. 1981. Equation of state and melting curve of helium to very high pressure. *Phys. Rev.* 24B:5119–5127.

Young, T. S. T., Callen, J. D., and McCune, J. E. 1973. High-frequency electrostatic waves in the magnetosphere. *J. Geophys. Res.* 78:1082–1099.

Younkin, R. L. 1974. The albedo of Titan. *Icarus* 21:219–229.

Younkin, R. L., and Münch, G. 1964. Wavelength dependence of the band structures of Jupiter and of Saturn. *Astron. J.* 69:565.

Yung, Y. L., Allen, M., and Pinto, J. P. 1983. Photochemistry of methane, nitrogen, and carbon monoxide on Titan—Abiotic synthesis of organic compounds. Preprint.

Zabriskie, F. R. 1960. Studies on the Atmosphere of Jupiter. Ph.D. thesis, Princeton Univ., Princeton, NJ.

Zarka, P., and Pedersen, B. M. 1983. Statistical study of Saturn electrostatic discharges. *J. Geophys. Res.* 88:9007–9018.

Zasova, L. V. 1974. Observations of Saturn at 0.8–1.6 m. *Astron. Tsirkular USSR Acad. Sci.* 345:1–2 (Russian).

Zebker, H. A., and Tyler, G. L. 1984. Thickness of Saturn's rings from Voyager 1 observations of microwave scatter. Submitted to *Science*.

Zebker, H. A., Tyler, G. L., and Marouf, E. A. 1983. On obtaining the forward phase functions of Saturn ring features from radio occultation observations. *Icarus* 56:209–228.

Zellner, B. 1973. The polarization of Titan. *Icarus* 18:661–664.

Zellner, B. 1974. Comments on Veverka's paper. In *Planetary Satellites*, ed. J. A. Burns (Tucson: Univ. of Arizona Press), pp. 230–231.

Zerull, R. H., and Giese, R. H. 1974. Microwave analogue studies. In *Planets, Stars, and Nebulae, Studied with Photopolarimetry*, ed. T. Gehrels (Tucson: Univ. of Arizona Press), pp. 901–914.

Zerull, R. H., Giese, R. H., and Weiss, K. 1977. Scattering measurements of irregular particles vs. Mie theory. *Opt. Polar.* 112:191–199.

Zharkov, V. N., and Trubitsyn, V. P. 1978. *Physics of Planetary Interiors* (Tucson: Pachart Publ.).

Zharkov, V. N., Trubitsyn, V. P., and Makalkin, A. B. 1972. The gravitational moments of Jupiter and Saturn. *Astrophys. J.* 10:L159–L162.

Zharkov, V. N., Makalkin, A. B., and Trubitsyn, V. P. 1974*a*. Models of Jupiter and Saturn. I. *Sov. Astron.—AJ* 18:492–498 (Russian).

Zharkov, V. N., Makalkin, A. B., and Trubitsyn, V. P. 1974*b*. Models of Jupiter and Saturn. II. *Sov. Astron.—AJ* 18:768–773 (Russian).

Zipf, E. C. 1980. Concerning a stratosphere source of nitrous oxide. *EOS* 61:1053 (abstract).

Zipf, E. C., and McLaughlin, R. W. 1978. On the dissociation of nitrogen by electron impact and by EUV photoabsorption. *Planet. Space Sci.* 26:449–462.

Zipf, E. C., McLaughlin, R. W., and Gorman, M. R. 1982. Electron impact excitation of the sinlet states of N_2. II. The C'_4 $^1\Sigma_u$-X $^1\Sigma g^+$ system. Submitted to *J. Chem. Phys.*

Zwan, B. J., and Wolf, R. A. 1976. Depletion of plasma near a planetary boundary. *J. Geophys. Res.* 81:1636–1648.

Glossary

GLOSSARY*

Compiled by Terry S. Mullin

adiabatic lapse rate	the rate of temperature decrease in an atmosphere with altitude for a parcel that does not exchange heat with its surroundings.
AKR	auroral kilometer radiation of the Earth.
albedo	Bond albedo: fraction of the total incident light reflected by a spherical body. It is equal to the phase integral, multiplied by the ratio of its brightness at zero phase angle to the brightness it would have if it were a perfectly diffusing disk. Geometric albedo: ratio of observed planetary brightness at a given phase angle to the brightness of a perfectly diffusing disk with the same position and angular size.
Alfvén speed	the speed at which Alfvén waves are propagated along the magnetic field: $V_A = B/(r\pi\rho)^{1/2}$ where B is magnetic field and ρ is plasma density.
amagat	a unit of molecular volume at 0°C and a pressure of 1 atmosphere. This unit varies slightly from one gas to another, but in general it corresponds to 2.24×10^4 cm^3. Also, a unit of density equal to 0.0446 gram mole per liter at 1 atm pressure (about 2.687×10^{19} molecules cm^{-3} under standard conditions).
AMOS system	atmospheric feature tracking system of the Image Processing Laboratory at the Jet Propulsion Laboratory.
ansa	Galileo's term for what are now known as the rings of

*We have used various definitions from *Glossary of Astronomy and Astrophysics* by J. Hopkins (by permission of the University of Chicago Press, copyright 1980 by the University of Chicago) and from *Astrophysical Quantities* by C. W. Allen (London: Athlone Press, 1973).

Saturn when he first viewed them nearly edge on. It is now used for opposing extensions or knots of a celestial object, such as a lenticular galaxy or a planetary nebula.

ARC   NASA Ames Research Center.

asteroids, C-type   the most numerous physical class of asteroids, characterized by low albedo and neutral color, thought to be primitive.

atomic time   time based on the atomic second. International Atomic Time (TAI) was officially adopted on 1 January 1972. From 1 January 1972 to 1 January 1974, leap seconds had to be introduced to keep atomic time within 0.7 s of Universal Time. For the past several years, only 1 leap second per year has been inserted. *See* UT.

Avogadro's number   (6.02×10^{23}) the number of atoms in 12 g of ^{12}C; by extension, the number of atoms in a gram-atom (or the number of molecules in a mole) of any substance.

barotropic stability   condition for stability of a zonal velocity profile in a planetary atmosphere (Holton 1979).

Bernstein waves   electrostatic waves which occur at ($n + 1/2$) multiples of the ion- or electron-cyclotron frequency.

beta plane approximation   tangent-plane geometry used for simplifying the equations of atmospheric motion (Holton 1979).

Birkeland currents   electric currents that flow parallel to a magnetic field.

blackbody   an idealized body which absorbs all radiation of all wavelengths incident on it. The radiation emitted by a blackbody is a function of temperature only. Because it is a perfect absorber, it is also a perfect emitter.

Boltzmann's constant   the constant in the ideal gas law, $k = 1.38 \times 10^{-3}$ Joule K^{-1}. It also appears in the related expression for the Maxwell-Boltzmann distribution of momenta in an equilibrium-fluid swarm of particles, derived from the Boltzmann equation.

Born
approximation

an approach to collision problems by use of perturbation methods. In collisional excitation, the Born approximation becomes valid when the incident energy is some 50 times larger than the excitation energy.

Brownian motion

random jiggling motion of microscopic particles suspended in a gas due to collisions with the gas molecules.

Brunt-Vaisala
frequency

the frequency of oscillation of buoyant parcels displaced vertically in a planetary atmosphere; it is proportional to the vertical gradient of potential temperature.

Cassini Division

a gap about $1.95-2.02$ R_S wide between the A and B Rings. It was discovered by Cassini in 1675. It is 1.9854 R_S (center) away from Saturn.

Clausius-
Clapeyron
equation

the thermodynamics equation relating changes in vapor pressure of a gas over its condensed state to changes in temperature.

conductivity,
Cowling

the tensor components of the electrical conductivity of a collisional, magnetized plasma; the Cowling conductivity $\sigma_C = \sigma_P + \sigma^2_H / \sigma_P$. In first approximation, σ_P is along the electric field and σ_H is transverse to it.

coordinated
Universal Time
(UTC)

Universal Time coordinated with ephemeris time; i.e., the rate is defined relative to atomic clock rate, but the epoch is defined relative to Universal Time. UTC is defined in such a manner that it differs from International Atomic Time (TAI) by an exact whole number of seconds. The difference UTC minus TAI was set equal to -10 s starting 1 January 1972; this difference can be modified by 1 s, preferably on January 1 and in case of need on July 1, to keep UTC in agreement with the time defined by the rotation of the Earth with an approximation better than 0.7 s. *See also* atomic time. *See* UT.

Coriolis forces

a pseudo force that appears in the dynamical equations when set up for a rotating coordinate system. In the Earth's northern hemisphere, it tends to turn all velocity vectors to the right.

Coulomb plasma

a gas of charged particles, interacting via a potential which scales as (separation)$^{-1}$.

Danielson dust — the name given to small (0.1 μm) absorbing particles found high in the atmospheres of Jupiter, Saturn, and Titan whose presence was first suggested by R. E. Danielson and his students (see Axel 1972). These particles are thought to be the product of photochemical reactions rather than meteoric debris.

Debye length — a characteristic distance in a plasma beyond which the electric field of a charged particle is shielded by particles with charges of the opposite sign.

Dione — *see under* satellite data.

Doppler broadening — the broadening of spectral lines caused by the thermal, turbulent, or mass motions of atoms along the line of sight. Small displacements toward longer and shorter wavelengths of radiation absorbed or emitted by these atoms result in broadening of the lines, when summed over all line-of-sight velocities of the ensemble.

Enceladus — *see under* satellite data.

Encke Division — a region of decreased brightness, 876 km wide, in the A Ring.

exosphere — the outermost region of an atmosphere, where atoms and molecules are on ballistic or escaping trajectories and where collisions can usually be neglected.

Faraday cup — a plasma sensor which measures a current proportional to incident plasma flux. The Voyager sensor is comprised of three potential modulated Faraday cups described in detail by Bridge et al. (1977). By AC modulating the incident plasma flux and using synchronic current collection, the flux measurements can be made at well-defined particle energy per unit charge as determined by the particles' speed along the sensor normal. The AC synchronic current measurements make the measurements essentially insensitive to DC currents caused by penetrating radiation.

Fermi gas — a gas of free electrons, whose momentum distribution is governed by the Fermi exclusion principle.

Fickian diffusion	the diffusion down a gradient at a rate proportional to the magnitude of the gradient.
Fourier transform	a function of the form $g(k) = (2\pi)^{-1/2}\int_{-\infty}^{\infty} dx\ e^{ikx} f(x)$, where $g(k)$ is the Fourier transform of $f(x)$.
Fresnel diffraction	the diffraction of light produced when a light source and/or an observing screen are at a finite distance from an aperture or obstacle.
FTS	Fourier transform spectrometer.
Gaussian distribution	a statistical distribution defined by the equation $p = c\ \exp(-k^2 x^2)$, in which x is the statistical variable. It yields a bell-shaped curve. Accidental errors of measurement and similar phenomena follow this law.
geostrophic balance	the balance between horizontal pressure gradient and Coriolis forces.
Gibbs energy	the function $G = E - TS + PV$, where E is the internal energy, T is the temperature, S is the entropy, P is the pressure, and V is the volume.
GM counter	Geiger-Müller counter. A detector of energetic particles which responds equally to electrons, protons and other heavy ions.
graben	an elongate crustal depression bounded by normal faults on its long sides.
Grüneisen parameter	a logarithmic derivative giving a measure of the rate of increase of lattice vibrational frequencies with increase in density of a solid.
GSFC	NASA Goddard Space Flight Center.
GW	1 gigawatt $= 10^9$ watt.
Henkel transform	a special case of a double Fourier transform with application to axially symmetric problems.
Helmholtz free energy	the function $F = E - TS$, where E is the internal energy, T is the temperature, and S is the entropy.

Henyey- Greenstein function	a parameterized phase function for light scattering by particles. For scattering through angle r, $p(r) = (1 - g^2)/(1 + g^2 - 2g \cos r)^{3/2}$, where g is the asymmetry factor.
Hill sphere	a region where the planet's gravity dominates the motion of solid objects.
horst	an elongate, relatively uplifted crustal unit or block bounded by normal faults on its long sides.
Huygens Gap	see Table I in the chapter by Esposito et al.
hydrostatic balance	the balance between vertical pressure gradient and gravity.
Hyperion	*see under* satellite data.
Iapetus	*see under* satellite data.
ice I, II, etc.	different phases of water ice; I is the familiar form, and II is the first of several high-pressure forms.
IMP-6 spacecraft	Interplanetary Monitoring Platform, one of several launched Earth-orbiting spacecraft.
international system of units	a practical system of units of measurements adopted in 1960 by the 11th International General Conference of Weights and Measures. The seven base units are the meter, kilogram, second, ampere, kelvin, mole, and candela.
inviscid flow	fluid motion in which the effects of viscosity are negligible.
IPP	imaging photopolarimeter on Pioneers 10 and 11.
IRIS	infrared interferometer spectrometer on Voyager spacecraft.
IRR	infrared radiometer experiment on Voyager spacecraft.
IUE	International Ultraviolet Explorer, an Earth-orbiting observatory.

Janus   *see under* satellite data.

Jeans escape   the escape of gas from a planetary atmosphere caused by
                         the thermal motion of the gas molecules.

Joule heating   heating by dissipation of electric currents.

JSFC   NASA Johnson Space Flight Center.

KAO   G. P. Kuiper Airborne Observatory, an airplane operated
                         by the NASA Ames Research Center.

Kelvin-Hemholtz   the tendency of waves to grow on a shear boundary
  instability   between two regions in relative motion parallel to the
                         boundary.

Lagrange points   equilibrium points in the restricted three-body problem.
                         Two of the stable Lagrange points each form an equi-
                         lateral triangle with the two massive bodies (Sun and
                         Saturn, or Saturn and satellite).

Lambert reflecting   a surface which appears uniformly bright when viewed
  layer   from any direction. The brightness of the surface is pro-
                         portional to the cosine of the angle between the incident
                         illuminating beam and the normal to the surface.

Langmuir waves   an electrostatic plasma oscillation. It is produced when
                         electrons oscillate under the restoring force produced by
                         charge separation in a plasma. They have little spread in
                         frequency and occur at or just above the local plasma
                         frequency.

Laplace   the coefficients appearing in the Fourier expanded form
  coefficients   of the disturbing function, the gravitational potential due
                         to one orbiting perturbing body on another orbiting per-
                         turbed body. They are functions of the ratio of the semi-
                         major axes of the two orbiters.

lapse rate   the rate at which temperature in an atmosphere de-
                         creases with respect to altitude. *See* adiabatic lapse rate.

LBH, BH, RYD   ultraviolet band systems of N_2, named by the initials of
  (or Ryd) bands   their discoverers (Lyman, Birge, Hopfield, Rydberg).

LECP   low energy charged particle experiment on Voyager spacecraft.

Legendre   a complete set of polynomials in x, which are given by
polynomials $P_n(x) = (2^n n!)^{-1} [d^n (x^2 - 1)^n / dx^n]$, where n is an integer from 0 to infinity.

Lindblad   a resonance of first order in the eccentricity of the per-
resonance   turbed particle. This terminology comes from galactic dynamics; in the 1920s Lindblad invoked such reso-nances to explain spiral arms.

Lindemann melt-   an empirical rule which states that a lattice melts when
ing criterion   the root-mean-square amplitude of lattice particle vibra-tion exceeds a universal fraction of the lattice spacing.

lithosphere   approximately the equivalent of "crust."

Loschmidt's   the number of molecules of an ideal gas per unit volume
number $(2.687 \times 10^{19}$ molecules cm$^{-3})$.

L-shells   McIlwain's invariant shell parameter, whose units are expressed in planetary radii at the magnetic equator.

Lyman-α a line in the far-ultraviolet part of the spectrum at 1215.67 Å, strongly absorbed and easily emitted by hy-drogen atoms and associated with the ground state. It is a part of the Lyman spectral series.

Mach number   the ratio of the speed of a flow to the local speed of sound.

Maxwell Gap   see Table I in the chapter by Esposito et al.

Maxwell   an expression for the statistical distribution of velocities
distribution   among the molecules of a gas at a given temperature.

meridional   the northward component of the wind direction.
velocity

mesosphere   the atmospheric region lying above the stratosphere, with a negative temperature gradient. It extends typi-cally from 10^{-3} to 10^{-6} bar for planetary atmospheres. It is not known exactly for Saturn.

MHD   magnetohydrodynamics, the study of the collective mo-
                        tions of charged particles in a magnetic field.

mho   a unit of electrical conductivity, the reciprocal of the re-
                        sistance in ohms.

Mie scattering   the scattering of light by particles with size comparable
                        to the wavelength of the light. Exact albedo and phase
                        functions are highly oscillatory with direction and wave-
                        length; however, slight irregularities and distributions of
                        size smooth these features out, leaving predominantly
                        forward-scattering behavior.

Mimas   *see under* satellite data.

Minnaert limb   the exponent k for a surface whose scattered intensity I
darkening   is proportional to $\mu_0{}^k \mu^{k-1}$ where μ_0 and μ are the co-
parameter   sines of the zenith angles to the Sun and the observer,
                        respectively.

modon   a stable nonlinear vortex structure.

Monte Carlo   a numerical technique which calculates average values
simulation   within a statistical ensemble by randomly choosing
                        members of the ensemble according to a prescribed
                        procedure.

MSFC   NASA Marshall Space Flight Center.

MW   1 megawatt $= 10^6$ watt.

Navier-Stokes   hydrodynamic equations governing viscous flow.
equations

NEB   North Equatorial Belt.

OAO   Orbiting Astronomical Observatory.

optical depth   a measure of the integrated absorbing power of a plane-
                        tary atmosphere as a function of altitude. The intensity
                        ratio $I/I_0 = e^{-\tau}$, where τ is $N\sigma l$ (N is column density,
                        σ is cross section, and l is path length). Optical depth 1
                        corresponds to the "visible" surface and occurs when the
                        intensity is reduced by a factor e.

orthohydrogen	a molecule of hydrogen (H_2) in which the proton spins are parallel.
parahydrogen	a molecule of hydrogen (H_2) in which the proton spins are opposite.
Phoebe	*see under* satellite data.
pixel	a shortened term for picture element. It is a resolution element in a vidicon-type detector, photographic plate, or other two-dimensional detector.
Planck function	the energy distribution of blackbody radiation under conditions of thermal equilibrium at a temperature T: $B_\nu = (2h\nu^3/c^2)\,[\exp(h\nu/kT) - 1]^{-1}$, where h is Planck's constant and ν is the frequency.
PLS	plasma science instrument on Voyager spacecraft.
Poisson distribution	an approximation to the binomial distribution used when the probability of success in a single trial is small and the number of trials is very large.
potential temperature	the temperature of a fluid parcel at a reference pressure, e.g., 1 bar.
Poynting-Robertson drag (PR drag)	a loss of orbital angular momentum by orbiting particles associated with their absorption and reemission of solar radiation.
PPS	photopolarimeter stellar occultation, Voyager 2.
pythagorean variance	$\sigma = (\sigma_x^2 + \sigma_y^2 + \sigma_z^2)^{1/2}$ (with σ_i the standard variance of the ith magnetic field component) an indicator of angular and magnitude changes in the field vector during the collection time interval that it characterizes. These variances may reflect intrinsic ($\partial/\partial t$) or advective changes ($V_{sc}\nabla$) of the ambient magnetic field.
radiative equilibrium	the state of thermal equilibrium in which the radiative heat sources balance.
Rayleigh number, Ra	a dimensionless parameter involving the temperature gradient and the coefficients of thermal conductivity and

kinematic viscosity, which determines when a fluid, under specified geometrical conditions, will become convectively unstable.

Rayleigh scattering	the scattering at shorter wavelengths of light by molecules or small particles suspended in a planetary atmosphere. The scattering is inversely proportional to the fourth power of the wavelength. *Compare* Mie scattering.
RAE-2 spacecraft	Radio Astronomy Explorer, a lunar orbiting spacecraft.
regolith	the layer of fragmental incoherent rocky debris that nearly everywhere forms the surface terrain produced by meteoritic impact on the surface of a planet, satellite, or asteroid.
Reynolds number	a dimensionless number, $R = Lv/\nu$, where L is a typical dimension of the system, v is a measure of the velocities that prevail, and ν is the kinematic viscosity that governs the conditions for the occurrence of turbulence.
Rhea	*see under* satellite data.
Richardson number	a measure of the stability of a horizontal shear flow, specifically the square of the ratio Brunt-Vaisala frequency to the wind shear.
rms	root mean square, the square root of the mean square value of a set of numbers.
Roche limit	the minimum distance at which a satellite influenced by its own gravitation and that of a central mass, about which it describes a Kepler orbit, can be in equilibrium. For a satellite of negligible mass, zero tensile strength, and the same mean density as its primary, in a circular orbit around its primary, the critical distance at which the satellite will break up is 2.44 times the radius of the primary.
Rossby wave	a large-scale wave in which planetary rotation is the principal restoring force.
R_S	Saturn radius, 60330 km.

RSS   radio subsystem occultation experiment on Voyager
                                   spacecraft.

Satellite data   see tables on pp. 612, 634, 643, 653, 655, 673 (Titan),
                                   744.

Saturn data

Mean distance to Sun	9.555 AU
Present orbital eccentricity	0.056
Sidereal period	$29°.46$ yr
Inclination of equator to orbit	$26°.73$
Mean orbital velocity	9.64 km s^{-1}
Saturn equatorial radius, R$_S$, at 1 bar pressure	60,330 km
Saturn polar radius	54,180 km
Dynamic oblateness, at 1 bar	0.09
Mass	5.688×10^{29} g
Mean density	0.69 g cm^{-3}
Titan mass/Saturn mass	2.37×10^{-4}
Mean surface gravity	1050 cm s^{-2}
Gravitational acceleration: equatorial	905 cm s^{-2}
polar	1180 cm s^{-2}
Escape velocity	$\geqslant 36$ km s^{-1}
J$_2$	$16,479 \pm 18 \times 10^{-6}$
J$_4$	$-937 \pm 38 \times 10^{-6}$
J$_6$ (assumed)	84×10^{-6}
Mass of core, in Earth masses	~ 10
Rotation period: IAU System, equatorial atmosphere	$10^h 10^m$
kilometric	$10^h 39^m 24^s (\pm 7^s)$
Angular velocity, ω	1.638×10^{-4} rad s^{-1}
$q = \omega^2 R_S{}^3/GM$	0.1527
Bolometric Bond albedo	0.342 ± 0.030
Temperature, at 1 bar	135 ± 5 K
Effective temperature	95.0 ± 0.4 K
Thermal emission/insolation	1.78 ± 0.09
Internal heat flux	$2.01 \pm 0.14 \times 10^{-4}$ W cm^{-2}
Magnetic dipole	0.21 Gauss R$_S{}^3$
Magnetic dipole tilt	$<1°$
Quadrupole/dipole $(g_2°/g_1°)$	0.076
Octupole/dipole $(g_3°/g_1°)$	0.127
North pole	0.84 Gauss
South pole	0.65 Gauss
Offset (nearly equatorial)	<0.01 R$_S$

Saturnian coordinates	a body-fixed coordinate system aligned with Saturn's spin axis at 0° longitude (SLS) as defined by Desch and Kaiser (1981).
Saturnian ring data	see tables on pp. 473, 477, 487, 497, 499, 523, 528, 533.
S-band	a radio-frequency band at a wavelength of 13.6 cm (about 2.3 GHz).
scarp	a line of cliffs produced by faulting or erosion, or a cliff-like face or slope of considerable linear extent.
SCET	spacecraft event time, the UT of the event as observed by the spacecraft.
Schmidt coefficient	the coefficient of a spherical harmonic expansion, in which the Schmidt normalization factors (see, e.g., Chapman and Bartels 1940) commonly used in geomagnetism are employed.
SEB	South Equatorial Belt.
SED	Saturn electrostatic discharges. See chapter by Kaiser et al.
shear, wind	the cross-stream gradient of the wind speed.
SI unit	*see* international system of units.
Si detector	a solid state detector (silicon material) of energetic particles. The signal from this type of detector is proportional to the total ionization produced by a particle passing through it. Several detectors, arranged in a telescope configuration, can be used to measure the total energy and identify different atomic ions and electrons.
SKR	Saturn kilometric radiation. See chapter by Kaiser et al.
SLS	Saturn longitude system. See chapter by Kaiser et al. pp. 385–386.
SOL	solar longitude system.

solitary wave

a propagating disturbance with only one wave crest.

spalling

the chipping, fracturing, or fragmentation and the upward and outward heaving of rock caused by interaction of a shock (compressional) wave with a free surface.

Stokes law

the law stating that the force which retards a sphere moving through a viscous medium is directly proportional to the velocity and radius of the sphere, and to the viscosity of the medium.

Störmer cutoff

the minimum energy that an electron or ion needs to reach an observer inside an ideal magnetic dipole field. The cut-off energy depends on the incident direction of the particle as well as on the location of the observer in the dipole field.

stratosphere

the atmospheric region lying above the troposphere, with a positive temperature gradient. It extends typically from 200 to 1 millibar in planetary atmospheres.

Taylor-Proudman theorem

a theorem in fluid mechanics which states that all steady slow motions in a rotating fluid that has zero viscosity are necessarily two-dimensional, with no variations parallel to the rotation axis.

Tethys

see under satellite data.

thermal wind condition

the relation between vertical gradient of zonal velocity and latitudinal gradient of temperature.

thermopause

the upper boundary of the thermosphere.

thermosphere

the region of temperature rise due to ionospheric heating.

Thomas-Fermi model

a model of a compressed atom in which the electrons are assumed to be a Fermi gas with the local density determined by the local electrostatic potential. This model becomes quantitatively correct at very high pressure.

Thomas-Fermi-Dirac (quantum statistical) model

a Thomas-Fermi model with electron exchange interactions included.

Titan	*see under* satellite data.
Trojans	objects occurring in two Lagrange points. *See* Lagrange points.
troposphere	the convective region of an atmosphere and the region of all weather. Extends typically down from 200 millibars on Saturn.
Type III solar burst	a fast drifting solar radio burst caused by relativistic electrons streaming away from the Sun on open magnetic field lines.
UT	universal time. The measure of time defined as the Greenwich hour angle of a smooth mean motion of the Sun plus 12 hr. *See also* coordinated universal time.
UVS	ultraviolet spectrometer on Voyager spacecraft.
Van der Waals forces	the relatively weak attractive forces operative between neutral atoms and molecules.
Voight profile	the profile of a spectral line allowing for the effects of Doppler broadening combined with a Lorentz (damping) profile.
whistler waves	a plasma wave thought to be responsible for heating ionospheres. Also, the mode in which electromagnetic waves from lightning can penetrate the nightside ionosphere.
X-band	a radio band at a wavelength of 3.5 cm (8600 MHz).
Z-mode waves	also known as slow X-mode waves. A magnetic wave mode which exists between the upper hybrid resonance frequency and the $L=0$ cutoff. This wave is electrostatic near the upper hybrid frequency and is electromagnetic elsewhere.
zonal velocity	an eastward component of the wind vector.

LIST OF CONTRIBUTORS WITH
ACKNOWLEDGMENTS TO FUNDING AGENCIES

The following people helped to make this book possible, in organizing, writing, refereeing, or otherwise.

J. H. Abraham, Indianapolis, IN.

M. H. Acuña, NASA Goddard Space Flight Center, Greenbelt, MD.

A. Aikin, NASA Goddard Space Flight Center, Greenbelt, MD.

M. Allen, Jet Propulsion Laboratory, Pasadena, CA.

M. Allison, NASA Goddard Space Flight Center, Greenbelt, MD.

J. D. Anderson, Jet Propulsion Laboratory, Pasadena, CA.

T. P. Armstrong, Jet Propulsion Laboratory, Pasadena, CA.

S. K. Atreya, Atmospheric and Oceanic Science Department, University of Michigan, Ann Arbor, MI.

W. I. Axford, Max-Planck-Institut für Aeronomie, Heidelberg, West Germany.

F. Bagenal, Department of Physics, Massachusetts Institute of Technology, Cambridge, MA.

J. P. Baluteau, Observatoire de Paris, Meudon, France.

H. B. Barber, School of Physics and Astronomy, University of Minnesota, Minneapolis, MN.

A. Bar-Nun, Dept. of Physics and Planetary Sciences, Tel Aviv University, Ramat Aviv, Israel.

J. Basile, Lunar and Planetary Laboratory, University of Arizona, Tucson, AZ.

W. A. Baum, Lowell Observatory, Flagstaff, AZ.

D. B. Beard, Department of Physics and Astronomy, University of Kansas, Lawrence, KS.

K. Beatty, Sky and Telescope, Cambridge, MA.

R. F. Beebe, Department of Astronomy, New Mexico State University, Las Cruces, NM.

K. Behannon, NASA Goddard Space Flight Center, Greenbelt, MD.

L. Berens, Lunar and Planetary Laboratory, University of Arizona, Tucson, AZ.

G. L. Berge, Owens Valley Radio Observatory, California Institute of Technology, Pasadena, CA.

J. Bergstrahl, Jet Propulsion Laboratory, Pasadena, CA.

J. L. Bertaux, Service d'Astronomie, CNRS, Verrières-le-Buisson, France.

R. Bianchi, Instituto di Astrofisica Spaziale, CNR, Roma, Italy.

G. Bjoraker, Lunar and Planetary Laboratory, University of Arizona, Tucson, AZ.

J. B. Blake, Space Science Laboratory, Aerospace Corporation, Los Angeles, CA.

P. H. Bodenheimer, Lick Observatory, Santa Cruz, CA.

N. J. Borderies, CNES/CRGS, Toulouse, France.

M. A. Borucci, NASA Ames Research Center, Moffett Field, CA.

J. M. Boyce, Planetary Geology Program, NASA Headquarters, Washington, DC.

A. Brahic, Observatoire de Paris, Meudon, France.

H. S. Bridge, Department of Physics, Massachusetts Institute of Technology, Cambridge, MA.

A. L. Broadfoot, Lunar and Planetary Laboratory, University of Arizona, Tucson, AZ.

D. C. Brown, Physics and Astronomy, University of Maryland, College Park, MD.

M. W. Buie, Lunar and Planetary Laboratory, University of Arizona, Tucson, AZ.

A. S. Bunker, Jet Propulsion Laboratory, Pasadena, CA.

B. Buratti, Jet Propulsion Laboratory, Pasadena, CA.

L. F. Burlaga, NASA Goddard Space Flight Center, Greenbelt, MD.

J. A. Burns, Center for Radiophysics and Space Research, Cornell University, Ithaca, NY.

F. H. Busse, Institute of Geophysics and Planetary Physics, University of California, Los Angeles, CA.

J. Caldwell, Department of Earth and Space Sciences, State University of New York, Stony Brook, NY.

A. G. W. Cameron, Harvard College Observatory, Cambridge, MA.
D. B. Campbell, Arecibo Observatory, Arecibo, Puerto Rico.
J. K. Campbell, Jet Propulsion Laboratory, Pasadena, CA.
L. A. Capone, NASA Ames Research Center, Moffett Field, CA.
J. F. Carbary, Applied Physics Laboratory, Johns Hopkins University, Laurel, MD.
E. D. Carlson, Adler Planetarium, Chicago, IL.
T. D. Carr, Department of Physics, University of Florida, Gainesville, FL.
A. Carusi, Laboratorio di Astrofisica Spaziale, CNR, Frascati, Italy.
P. M. Cassen, NASA Ames Research Center, Moffett Field, CA.
C. R. Chapman, Planetary Science Institute, Tucson, AZ.
D. L. Chenette, Space Sciences Laboratory, Aerospace Corporation, Los Angeles, CA.
S. A. Collins, Jet Propulsion Laboratory, Pasadena, CA.
N. Conarro, Lunar and Planetary Laboratory, University of Arizona, Tucson, AZ.
J. E. P. Connerney, NASA Goddard Space Flight Center, Greenbelt, MD.
B. J. Conrath, NASA Goddard Space Flight Center, Greenbelt, MD.
G. J. Consolmagno, Dept. of Earth, Atmospheric, and Planetary Sciences, Massachusetts
   Institute of Technology, Cambridge, MA.
A. F. Cook II, National Research Council, Herzberg Institute of Astrophysics, Ottawa, Canada.
C. Cook, University of Arizona Press, Tucson, AZ.
B. H. Cooper, Division of Physics, Mathematics, and Astronomy, California Institute of
   Technology, Pasadena, CA.
J. F. Cooper, Enrico Fermi Institute for Nuclear Studies, University of Chicago, Chicago, IL.
A. Coradini, Instituto di Astrofisica Spaziale, CNR, Roma, Italy.
R. Courtin, Laboratoire de Physique Stellaire et Planétaire, Verrières-le-Buisson, France.
S. Cox, University of Arizona Press, Tucson, AZ.
S. K. Croft, Lunar and Planetary Institute, Houston, TX.
D. P. Cruikshank, Institute for Astronomy, Honolulu, HI.
C. C. Cunningham, Lunar and Planetary Laboratory, University of Arizona, Tucson, AZ.
J. N. Cuzzi, NASA Ames Research Center, Moffett Field, CA.
G. E. Danielson, Division of Geological and Planetary Sciences, California Institute of
   Technology, Pasadena, CA.
D. S. Davis, Stewart Observatory, University of Arizona, Tucson, AZ.
L. Davis, Jr., Division of Physics, Mathematics and Astronomy, California Institute of
   Technology, Pasadena, CA.
C. De Bergh, Observatoire de Paris, Meudon, France.
K. Denomy, Lunar and Planetary Laboratory, University of Arizona, Tucson, AZ.
I. de Pater, Lunar and Planetary Laboratory, University of Arizona, Tucson, AZ.
S. F. Dermott, Center for Radiophysics and Space Research, Cornell University, Ithaca, NY.
M. D. Desch, NASA Goddard Space Flight Center, Greenbelt, MD.
A. J. Dessler, NASA Marshall Space Flight Center, Huntsville, AL.
J. R. Dickel, Astronomy Department, University of Illinois, Urbana, IL.
D. J. Diner, Jet Propulsion Laboratory, Pasadena, CA.
J. M. Dlugach, Main Astronomical Observatory of the Ukrainian Academy of Sciences, Kiev,
   U.S.S.R.
A. Dollfus, Observatoire de Paris, Meudon, France.
T. M. Donahue, Department of Atmospheric and Oceanic Sciences, University of Michigan,
   Ann Arbor, MI.
L. R. Doose, Lunar and Planetary Laboratory, University of Arizona, Tucson, AZ.
R. J. Drean, Hughes Aircraft Company, Los Angeles, CA.
J. Dubach, University of Maryland, College Park, MD.
R. H. Durisen, Astronomy Department, Indiana University, Bloomington, IN.
T. C. Duxbury, Jet Propulsion Laboratory, Pasadena, CA.
J. Eberhart, Science News, Washington, DC.
R. E. Eplee, Lunar and Planetary Laboratory, University of Arizona, Tucson, AZ.
J. H. Eraker, Enrico Fermi Institute for Nuclear Studies, University of Chicago, Chicago, IL.
V. R. Eshleman, Center for Radar Astronomy, Stanford University, Stanford, CA.
L. W. Esposito, Laboratory for Atmospheric and Space Physics, University of Colorado,
   Boulder, CO.

D. R. Evans, Radiophysics Inc., Boulder, CO.
N. Evans, Department of Astronomy, University of Texas, Austin, TX.
A. Eviatar, Department of Atmospheric Sciences, University of California, Los Angeles, CA.
P. Farinella, Osservatorio Astronomico, Milano, Italy.
C. Federico, Instituto di Astrofisica Spaziale, CNR, Roma, Italy.
G. Fifer, Riverside, CA.
R. W. Fillius, University of California at San Diego, La Jolla, CA.
R. O. Fimmel, NASA Ames Research Center, Moffet Field, CA.
J. H. Fink, Geology Department, Arizona State University, Tempe, AZ.
U. Fink, Lunar and Planetary Laboratory, University of Arizona, Tucson, AZ.
F. M. Flasar, NASA Goddard Space Flight Center, Greenbelt, MD.
L. A. Frank, Department of Astronomy and Astrophysics, University of Iowa, Iowa City, IA.
F. A. Franklin, Planetary Science Division, Center for Astrophysics, Cambridge, MA.
K. L. Franklin, American Museum-Hayden Planetarium, New York, NY.
O. G. Franz, Lowell Observatory, Flagstaff, AZ.
M. Fulchignoni, Laboratorio di Astrofisica Spaziale, Frascati, Italy.
W. O. Garden, NASA Ames Research Center, Moffett Field, CA.
C. Garriott, G & S Typesetters, Inc., Austin, TX.
D. E. Gault, Murphy's Center of Planetology, Murphy, CA.
D. Gautier, Observatoire de Paris, Meudon, France.
P. R. Gazis, Department of Physics, Massachusetts Institute of Technology, Cambridge, MA.
J. Geake, Physics Department, University of Manchester, Manchester, Great Britain.
M. W. Gehrels, Lunar and Planetary Laboratory, University of Arizona, Tucson, AZ.
T. Gehrels, Lunar and Planetary Laboratory, University of Arizona, Tucson, AZ.
F. Genova, Observatoire de Paris, Meudon, France.
C. Gever, G & S Typesetters, Inc., Austin, TX.
P. J. Gierash, Astronomy Department, Cornell University, Ithaca, NY.
O. Gingerick, Center for Astrophysics, Cambridge, MA.
L. Giver, NASA Ames Research Center, Moffett Field, CA.
G. Gloeckler, Department of Physics and Astronomy, University of Maryland, College Park, MD.
D. A. Godfrey, Laboratory for Planetary Atmospheres, Department of Physics and Astronomy, University College London, London, Great Britain.
C. K. Goertz, Department of Physics and Astronomy, University of Iowa, Iowa City, IA.
P. Goldreich, Division of Geological and Planetary Science, California Institute of Technology, Pasadena, CA.
S. J. Goldstein Jr., Astronomy Department, University of Virginia, Charlottesville, VA.
R. M. Goldstein, Jet Propulsion Laboratory, Pasadena, CA.
T. R. Gosnell, Cornell University, Ithaca, NY.
R. Greeley, Geology Department, Arizona State University, Tempe, AZ.
R. Greenberg, Planetary Science Institute, Tucson, AZ.
E. Grün, Max-Planck-Institut für Kernphysik, Heidelberg, West Germany.
S. Gulkis, Jet Propulsion Laboratory, Pasadena, CA.
D. A. Gurnett, Department of Physics and Astronomy, University of Iowa, Iowa City, IA.
P. K. Haff, W. K. Kellogg Radiation Laboratory, California Institute of Technology, Pasadena, CA.
D. C. Hamilton, Department of Physics and Astronomy, University of Maryland, College Park, MD.
R. A. Hanel, NASA Goddard Space Flight Center, Greenbelt, MD.
B. Hapke, Department of Geology and Planetary Sciences, University of Pittsburgh, Pittsburgh, PA.
A. W. Harris, Jet Propulsion Laboratory, Pasadena, CA.
R. E. Hartle, NASA Goddard Space Flight Center, Greenbelt, MD.
W. K. Hartmann, Planetary Sciences Institute, Tucson, AZ.
M. O. Harwitt, Astronomy Department, Cornell University, Ithaca, NY.
T. Hasegawa, Nobeyama Radio Observatory, Nobeyama, Japan.
J. W. Head, Department of Geological Sciences, Brown University, Providence, RI.
F. L. Herbert, Lunar and Planetary Laboratory, University of Arizona, Tucson, AZ.

R. Hide, Geophysical Fluid Dynamics Laboratory, Bracknell, Berkshire, Great Britain.
J. R. Hill, Department of Electrical Engineering and Computer Sciences, University of California at San Diego, La Jolla, CA.
D. Hilton, Lunar and Planetary Laboratory, University of Arizona, Tucson, AZ.
H. H. Hilton, Space Sciences Laboratory, Aerospace Corporation, Los Angeles, CA.
D. P. Hinson, Center for Radar Astronomy, Stanford University, Stanford, CA.
D. Hirschi, Department of Physics and Astronomy, University of Kansas, Lawrence, KS.
J. B. Holberg, Lunar and Planetary Laboratory, University of Arizona, Tucson, AZ.
L. L. Hood, Lunar and Planetary Laboratory, University of Arizona, Tucson, AZ.
C. W. Hord, Laboratory for Atmospheric and Space Physics, University of Colorado, Boulder, CO.
J. S. Hornstein, Computer Sciences Corporation, Silver Spring, MD.
J. W. Hovenier, Department of Physics and Astronomy, Vrije Universiteit, Amsterdam, The Netherlands.
K. C. Hsieh, Physics Department, University of Arizona, Tucson, AZ.
P. Y. Huang, Massachusetts Institute of Technology, Cambridge, MA.
W. B. Hubbard, Lunar and Planetary Laboratory, University of Arizona, Tucson, AZ.
G. E. Hunt, Atmospheric Physics Group, Imperial College, London, Great Britain.
D. M. Hunten, Department of Planetary Sciences, University of Arizona, Tucson, AZ.
W. Huntress, Jet Propulsion Laboratory, Pasadena, CA.
R. Hutchins, Orion Publications, Leigh-on-Sea, Essex, Great Britain.
A. P. Ingersoll, Division of Geological and Planetary Sciences, California Institute of Technology, Pasadena, CA.
D. S. Intriligator, Carmel Research Center, Santa Monica, CA.
W.-H. Ip, Max-Planck-Institut für Aeronomie, Katlenburg, West Germany.
W. M. Irvine, Department of Astronomy, University of Massachusetts, Amherst, MA.
A. J. Jeffries, Twickenham, Middlesex, Great Britain.
D. E. Jennings, NASA Goddard Space Flight Center, Greenbelt, MD.
D. C. Jewitt, Division of Geological and Planetary Science, California Institute of Technology, Pasadena, CA.
W. Joffe, Space Telescope Science Institute, Baltimore, MD.
T. V. Johnson, Jet Propulsion Laboratory, Pasadena, CA.
M. L. Kaiser, NASA Goddard Space Flight Center, Greenbelt, MD.
C. R. Kennel, Physics Department, University of California, Los Angeles, CA.
R. A. Kerr, Science Magazine, Washington, DC.
B. Khare, Center for Radiophysics and Space Research, Cornell University, Ithaca, NY.
R. M. Killen, Department of Space Physics and Astronomy, Rice University, Houston, TX.
T. J. Kreidl, Lowell Observatory, Flagstaff, AZ.
L. Kresák, Astronomical Institute, Bratislava, Czechoslovakia.
S. M. Krimigis, Applied Physics Laboratory, Johns Hopkins University, Laurel, MD.
V. G. Kunde, NASA Goddard Space Flight Center, Greenbelt, MD.
W. S. Kurth, Department of Physics and Astronomy, University of Iowa, Iowa City, IA.
P. Lanciano, Instituto di Astrofisica Spaziale, CNR, Roma, Italy.
A. L. Lane, Jet Propulsion Laboratory, Pasadena, CA.
L. J. Lanzerotti, Bell Laboratories, Murray Hill, NJ.
H. P. Larson, Lunar and Planetary Laboratory, University of Arizona, Tucson, AZ.
S. M. Larson, Lunar and Planetary Laboratory, University of Arizona, Tucson, AZ.
A. J. Lazarus, Department of Physics, Massachusetts Institute of Technology, Cambridge, MA.
M. A. Leake, Department of Physics, Valdosta State College, Valdosta, GA.
Y. Leblanc, Observatoire de Paris, Meudon, France.
L. A. Lebofsky, Lunar and Planetary Laboratory, University of Arizona, Tucson, AZ.
A. Lecacheux, Observatoire de Paris, Meudon, France.
R. P. Lepping, NASA Goddard Space Flight Center, Greenbelt, MD.
J. S. Lewis, Lunar and Planetary Laboratory, University of Arizona, Tucson, AZ.
D. N. C. Lin, Lick Observatory, Santa Cruz, CA.
G. F. Lindal, Jet Propulsion Laboratory, Pasadena, CA.
J. L. Lissauer, NASA Ames Research Center, Moffett Field, CA.

K. Lumme, Observatory/Astrophysics Laboratory, University of Helsinki, Helsinki, Finland.
J. I. Lunine, Division of Geological and Planetary Sciences, California Institute of Technology, Pasadena, CA.
B. L. Lutz, Lowell Observatory, Flagstaff, AZ.
J. J. MacFarlane, Lunar and Planetary Laboratory, University of Arizona, Tucson, AZ.
G. Magni, Instituto di Astrofisica Spaziale, CNR, Roma, Italy.
W. Maguire, NASA Goddard Space Flight Center, Greenbelt, MD.
S. H. Margolis, McDonnell Center for the Space Sciences, Washington University, St. Louis, MO.
E. A. Marouf, Center for Radar Astronomy, Stanford University, Stanford, CA.
A. Marten, Laboratoire d'Astronomie Infrarouge, Observatoire de Paris, Meudon, France.
H. Masursky, U.S. Geological Survey, Flagstaff, AZ.
D. L. Matson, Earth and Space Sciences Division, Jet Propulsion Laboratory, Pasadena, CA.
M. A. Matthews, Lunar and Planetary Laboratory, University of Arizona, Tucson, AZ.
M. S. Matthews, Lunar and Planetary Laboratory, University of Arizona, Tucson, AZ.
J. C. McConnell, Physics Department, York University, Downsview, Canada.
F. B. McDonald, NASA Goddard Space Flight Center, Greenbelt, MD.
D. D. McKibben, NASA Ames Research Center, Moffett Field, CA.
R. B. McKibben, Enrico Fermi Institute, University of Chicago, Chicago, IL.
W. B. McKinnon, Department of Earth and Planetary Sciences, Washington University, St. Louis, MO.
G. McLaughlin, Lunar and Planetary Laboratory, University of Arizona, Tucson, AZ.
R. L. McNutt, Jr., Center for Space Research, Massachusetts Institute of Technology, Cambridge, MA.
G. Melnick, Harvard-Smithsonian Center for Astrophysics, Cambridge, MA.
D. A. Mendis, Electrical Engineering and Computer Sciences, University of California at San Diego, La Jolla, CA.
F. C. Michel, Space Physics and Astronomy Department, Rice University, Houston, TX.
A. Milani, Instituto di Matematica, Università di Pisa, Pisa, Italy.
R. L. Millis, Lowell Observatory, Flagstaff, AZ.
E. D. Miner, Jet Propulsion Laboratory, Pasadena, CA.
D. Mohlman, Institut für Kosmosforschung, Berlin-Adlershof, East Germany.
G. Morfill, Max-Planck-Institut für Extraterrestrische Physik, München, West Germany.
A. V. Morozhenko, Main Astronomical Observatory of the Ukrainian Academy of Sciences, Kiev, USSR.
D. Morrison, Institute for Astronomy, Honolulu, HI.
D. O. Muhleman, Division of Geological and Planetary Sciences, California Institute of Technology, Pasadena, CA.
T. L. Mullikin, Rockwell International, Pittsburgh, PA.
T. S. Mullin, Lunar and Planetary Laboratory, University of Arizona, Tucson, AZ.
A. F. Nagy, NASA Marshall Space Flight Center, Huntsville, AL.
N. F. Ness, NASA Goddard Space Flight Center, Greenbelt, MD.
F. M. Neubauer, Institut für Geophysik und Meteorologie, Universität zu Köln, Köln, West Germany.
K. C. Newland, Superior, AZ.
H. Newsom, Max-Planck-Institut, Mainz, West Germany.
A. M. Nobili, Instituto di Matematica, Università di Pisa, Pisa, Italy.
T. G. Northrop, NASA Goddard Space Flight Center, Greenbelt, MD.
S. K. Nozette, Department of Earth and Planetary Sciences, Massachusetts Institute of Technology, Cambridge, MA.
M. O'Callaghan, Laboratory for Atmospheric and Space Physics, University of Colorado, Boulder, CO.
K. W. Ogilvie, NASA Goddard Space Flight Center, Greenbelt, MD.
M. E. Olszewski, Lunar and Planetary Laboratory, University of Arizona, Tucson, AZ.
S. J. O'Meara, Sky Publishing Corporation, Cambridge, MA.
G. S. Orton, Jet Propulsion Laboratory, Pasadena, CA.
D. E. Osterbrock, Lick Observatory, Santa Cruz, CA.
S. J. Ostro, Department of Astronomy, Cornell University, Ithaca, NY.

T. C. Owen, Department of Earth and Space Science, State University of New York, Stony Brook, NY.

K. D. Pang, Jet Propulsion Laboratory, Pasadena, CA.

P. Paolicchi, Osservatorio Astronomico, Merate, Italy.

M. T. Paonessa, Department of Physics and Astronomy, University of Kansas, Lawrence, KS.

Q. R. Passey, Division of Geological and Planetary Sciences, California Institute of Technology, Pasadena, CA.

S. J. Peale, Physics Department, University of California, Santa Barbara, CA.

J. C. Pearl, NASA Goddard Space Flight Center, Greenbelt, MD.

B. M. Pedersen, Observatoire de Paris, Meudon, France.

G. H. Pettengill, Department of Earth and Planetary Sciences, Massachusetts Institute of Technology, Cambridge, MA.

J. A. Pirraglia, NASA Goddard Space Flight Center, Greenbelt, MD.

J. B. Plescia, Jet Propulsion Laboratory, Pasadena, CA.

M. Podolak, Department of Physics and Planetary Sciences, Tel-Aviv University, Romat-Aviv, Israel.

J. B. Pollack, NASA Ames Research Center, Moffett Field, CA.

C. Ponnamperuna, Department of Chemistry, University of Maryland, College Park, MD.

R. B. Pomphrey, Jet Propulsion Laboratory, Pasadena, CA.

C. C. Porco, Lunar and Planetary Laboratory, University of Arizona, Tucson, AZ.

S. S. Prasad, Jet Propulsion Laboratory, Pasadena, CA.

A. J. R. Prentice, Department of Mathematics, Monash University, Clayton, Victoria, Australia.

R. G. Prinn, Department of Meteorology, Massachusetts Institute of Technology, Cambridge, MA.

K. R. Pyle, Enrico Fermi Institute, University of Chicago, Chicago, IL.

K. Rages, NASA Ames Research Center, Moffett Field, CA.

B. A. Randall, Department of Physics and Astronomy, University of Iowa, Iowa City, IA.

A. S. P. Rao, Department of Geology, Osmania University, Hyderabad, India.

R. T. Reynolds, NASA Ames Research Center, Moffett Field, CA.

J. W. Rhoads, Jet Propulsion Laboratory, Pasadena, CA.

A. E. Roy, Department of Astronomy, Glasgow University, Glasgow, Great Britain.

R. W. Russell, Aerospace Corporation, Los Angeles, CA.

V. S. Safronov, Institute of Physics of the Earth, Moscow, USSR.

C. Sagan, Center for Radiophysics and Space Research, Cornell University, Ithaca, NY.

E. E. Salpeter, Newman Laboratory, Cornell University, Ithaca, NY.

R. E. Samuelson, NASA Goddard Space Flight Center, Greenbelt, MD.

J. Samz, Science World, New York, NY.

B. R. Sandel, Lunar and Planetary Laboratory, University of Arizona, Tucson, AZ.

R. Santer, Université de Lille, Lille, France.

F. L. Scarf, Space Sciences Department, TRW Defense and Space Systems Group, Redondo Beach, CA.

T. W. Scattergood, NASA Ames Research Center, Moffett Field, CA.

G. G. Schaber, U.S. Geological Survey, Flagstaff, AZ.

A. W. Schardt, NASA Goddard Space Flight Center, Greenbelt, MD.

N. Schneider, Lunar and Planetary Laboratory, University of Arizona, Tucson, AZ.

J. D. Scudder, NASA Goddard Space Flight Center, Greenbelt, MD.

H. Shaffer, University of Arizona Press, Tucson, AZ.

D. E. Shemansky, Lunar and Planetary Laboratory, University of Arizona, Tucson, AZ.

E. M. Shoemaker, U.S. Geological Survey, Flagstaff, AZ.

M. R. Showalter, Center for Radiophysics and Space Research, Cornell University, Ithaca, NY.

F. H. Shu, Astronomy Department, University of California, Berkeley, CA.

G. Sill, Lunar and Planetary Laboratory, University of Arizona, Tucson, AZ.

K. E. Simmons, Laboratory for Atmospheric and Space Physics, University of Colorado, Boulder, CO.

J. A. Simpson, Enrico Fermi Institute, University of Chicago, Chicago, IL.

R. A. Simpson, Center for Radar Astronomy, Stanford University, Stanford, CA.

G. L. Siscoe, Department of Atmospheric Sciences, University of California, Los Angeles, CA.

E. C. Sittler, Jr., NASA Goddard Space Flight Center, Greenbelt, MD.

B. A. Smith, Department of Planetary Sciences, University of Arizona, Tucson, AZ.

E. J. Smith, Jet Propulsion Laboratory, Pasadena, CA.

G. R. Smith, Lunar and Planetary Laboratory, University of Arizona, Tucson, AZ.

P. H. Smith, Lunar and Planetary Laboratory, University of Arizona, Tucson, AZ.

R. Smoluchowski, Department of Astronomy and Physics, University of Texas, Austin, TX.

L. A. Soderblom, Division of Geological and Planetary Science, California Institute of Technology, Pasadena, CA.

G. X. Song, Shanghai Observatory, Shanghai, China.

J. R. Spencer, Lunar and Planetary Laboratory, University of Arizona, Tucson, AZ.

S. W. Squyres, NASA Ames Research Center, Moffett Field, CA.

S. Staley, Indianapolis, IN.

D. J. Stevenson, Division of Geological and Planetary Science, California Institute of Technology, Pasadena, CA.

E. C. Stone, Jr., Division of Physics, Mathematics and Astronomy, California Institute of Technology, Pasadena, CA.

P. H. Stone, Center for Meteorology and Planetary Oceanography, Massachusetts Institute of Technology, Cambridge, MA.

D. F. Strobel, Naval Research Laboratory, Washington, DC.

R. G. Strom, Lunar and Planetary Laboratory, University of Arizona, Tucson, AZ.

R. Suggs, Astronomy Department, New Mexico State University, Las Cruces, NM.

J. D. Sullivan, Center for Space Research, Massachusetts Institute of Technology, Cambridge, MA.

D. N. Sweetnam, Jet Propulsion Laboratory, Pasadena, CA.

M. Sykes, Lunar and Planetary Laboratory, University of Arizona, Tucson, AZ.

S. P. Synnott, Jet Propulsion Laboratory, Pasadena, CA.

E. F. Tedesco, Jet Propulsion Laboratory, Pasadena, CA.

V. G. Tejfel, Astrophysical Institute, Academy of Sciences, Alma-Ata, USSR.

A. Thirunagari, Lunar and Planetary Laboratory, University of Arizona, Tucson, AZ.

D. J. Tholen, Institute for Astronomy, Honolulu, HI.

P. Thomas, Cornell University, Ithaca, NY.

M. F. Thomsen, Department of Physics and Astronomy, University of Iowa, Iowa City, IA.

R. M. Thorne, Department of Atmospheric Sciences, University of California, Los Angeles, CA.

D. Timmons, G&S Typesetters, Inc., Austin, TX.

M. G. Tomasko, Lunar and Planetary Laboratory, University of Arizona, Tucson, AZ.

T. A. Tombrello, W. K. Kellogg Radiation Laboratory, California Institute of Technology, Pasadena, CA.

M. Torbett, Department of Terrestrial Magnetism, Carnegie Institution of Washington, Washington, DC.

L. M. Trafton, Astronomy Department, University of Texas, Austin, TX.

S. D. Tremaine, Department of Physics, Massachusetts Institute of Technology, Cambridge, MA.

G. L. Tyler, Center for Radar Astronomy, Stanford University, Stanford, CA.

G. B. Valsecchi, Instituto di Astrofisica Spaziale, Roma, Italy.

J. A. Van Allen, Physics and Astronomy Department, University of Iowa, Iowa City, IA.

A. Van Helden, Department of History, Rice University, Houston, TX.

D. Venkatesan, Physics Department, University of Calgary, Alberta, Canada.

J. Veverka, Astronomy Department, Cornell University, Ithaca, NY.

A. P. Vid'machenko, Main Astronomical Observatory of the Ukrainian Academy of Sciences, Kiev, USSR.

C. C. Voge, Jet Propulsion Laboratory, Pasadena, CA.

R. Wagener, Earth and Space Sciences Department, State University of New York, Stony Brook, NY.

J. H. Waite, Jr., NASA Marshall Space Flight Center, Huntsville, AL.

R. A. Wallace, Jet Propulsion Laboratory, Pasadena, CA.
W. R. Ward, Jet Propulsion Laboratory, Pasadena, CA.
J. W. Warwick, Astrogeophysics Department, University of Colorado, Boulder, CO.
S. Weidenschilling, Planetary Sciences Institute, Tucson, AZ.
P. R. Weissman, Jet Propulsion Laboratory, Pasadena, CA.
R. A. West, Laboratory for Atmospheric and Space Physics, University of Colorado, Boulder, CO.
E. Whitaker, Lunar and Planetary Laboratory, University of Arizona, Tucson, AZ.
R. C. Whitten, NASA Ames Research Center, Moffett Field, CA.
W. E. Wiesel, Air Force Institute of Technology, Dayton, OH.
P. Winkelstein, Earth and Space Sciences Department, State University of New York, Stony Brook, NY.
A. Woronow, Geosciences Department, University of Houston, Houston, TX.
E. G. Yanovitskij, Main Astronomical Observatory of the Ukrainian Academy of Sciences, Kiev, USSR.
Y. L. Yung, Division of Physics, Mathematics and Astronomy, California Institute of Technology, Pasadena, CA.
V. Zappalà, Osservatorio Astronomico, Torino, Italy.
H. A. Zebker, Center for Radar Astronomy, Stanford University, Stanford, CA.
B. H. Zellner, Lunar and Planetary Laboratory, University of Arizona, Tucson, AZ.
V. N. Zharkov, Institute of Earth Physics, Moscow, USSR.

The editors acknowledge the support of NASA Grant NASW-3655, NASA Contracts NAS7-918, NAS2-6992, NAS2-6265, and NAS2-11647 for the preparation of the book. The following authors wish to acknowledge specific funds involved in supporting the preparation of their chapters.

Atreya, S. K.; NASA Grant NSG-7404 and NASA Contract 7-100
Beebe, R. F.; NASA Grant NAGW-167
Caldwell, J.; NASA Grants NSG-7320 and NAG-5275
Consolmagno, G.; NASA Grant NAGW-191
Davis, L.; NASA Grant NGR-05-002-160
Esposito, L. W.; NASA Grant NAGW-389
Evans, D. R.; NASA JPL Contract NAS7-100
Eviatar, A.; NASA Grants NAGW-333 and NAGW-87
Greenberg, R.; NASA Grant NSG-7045
Holberg, J. B.; NASA Grant NAGW-62
Hubbard, W. B.; NASA Grants NSG-7045 and NAGW-192
Ingersoll, A.; NASA Grant NAGW-58
Kaiser, M. L.; NASA JPL Contract NAS7-100
Kurth, W. S.; JPL Contract 954013
Lebofsky, L. A.; NASA Grant NSG-7114
Lewis, J. S.; NASA Grant NAGW-340
McConnell, J. C.; NASA Contract 7-100
Mendis, D. A.; NASA Grants NSG-7102, NSG-7623, and NAGW-399
Nagy, A. F.; NASA Grants NAGW-15 and NGR-23-005-015
Orton, G. S.; NASA-JPL Contract NAS7-100
Owen, T.; JPL Contract 953614 and NASA Grant NGR-33-015-141
Pollack, J. B.; NASA Grants 154-10-80-18 and 153-03-60-01
Porco, C. C.; NASA Grant NGL-05-002-003 and NSF Grant AST-80-20005
Prinn, R. G.; NSF Grant ATM-81-12341
Sagan, C.; NASA Grants NGR-33-010-101, NGR-33-010-082, and NGR-33-010-220
Scarf, F. L.; Voyager JPL Contract 954012
Siscoe, G.; NASA Grants NAGW-333 and NAGW-87
Stevenson, D. J.; NASA Grant NAGW-185
Stone, E. C.; JPL Contract NAS7-918 and NASA Grant NAGW-200

Van Allen, J. A.; Ames Research Center NAS2-11125 and Office of Naval Research
    N00014-76-C-0016
Veverka, J.; NASA Grants NAGW-193 and NSG-7156
Waite, J. H.; NASA Grant NAGW-312
West, R. A.; NASA Grant NAGW-197

INDEX

(Compiled by Martha A. Matthews)